T5-AFH-874

GEOPHYSICAL MONOGRAPH SERIES

Geophysical Monograph Volumes

1 **Antarctica in the International Geophysical Year** *A. P. Crary, L. M. Gould, E. O. Hulburt, Hugh Odishaw, and Waldo E. Smith (Eds.)*
2 **Geophysics and the IGY** *Hugh Odishaw and Stanley Ruttenberg (Eds.)*
3 **Atmospheric Chemistry of Chlorine and Sulfur Compounds** *James P. Lodge, Jr. (Ed.)*
4 **Contemporary Geodesy** *Charles A. Whitten and Kenneth H. Drummond (Eds.)*
5 **Physics of Precipitation** *Helmut Weickmann (Ed.)*
6 **The Crust of the Pacific Basin** *Gordon A. Macdonald and Hisashi Kuno (Eds.)*
7 **Antarctic Research: The Matthew Fontaine Maury Memorial Symposium** *H. Wexler, M. J. Rubin, and J. E. Caskey, Jr. (Eds.)*
8 **Terrestrial Heat Flow** *William H. K. Lee (Ed.)*
9 **Gravity Anomalies: Unsurveyed Areas** *Hyman Orlin (Ed.)*
10 **The Earth Beneath the Continents: A Volume of Geophysical Studies in Honor of Merle A. Tuve** *John S. Steinhart and T. Jefferson Smith (Eds.)*
11 **Isotope Techniques in the Hydrologic Cycle** *Glenn E. Stout (Ed.)*
12 **The Crust and Upper Mantle of the Pacific Area** *Leon Knopoff, Charles L. Drake, and Pembroke J. Hart (Eds.)*
3 **The Earth's Crust and Upper Mantle** *Pembroke J. Hart (Ed.)*
he Structure and Physical Properties of the rth's Crust *John G. Heacock (Ed.)*
Use of Artificial Satellites for Geodesy *W. Henricksen, Armando Mancini, and d H. Chovitz (Eds.)*
d Fracture of Rocks *H. C. Heard, , N. L. Carter, and C. B. Raleigh (Eds.)*
e Lakes: Their Problems and ntal Effects *William C. Ackermann, ite, and E. B. Worthington (Eds.)*
mosphere in Motion: A Selection Annotation *C. O. Hines and*
f the Pacific Ocean Basin and me in Honor of George P. *Sutton, Murli H. Manghnani, .)*
Nature and Physical *ock (Ed.)*
Magnetospheric *.)*

22 **Derivation, Meaning, and Use of Geomagnetic Indices** *P. N. Mayaud*
23 **The Tectonic and Geologic Evolution of Southeast Asian Seas and Islands** *Dennis E. Hayes (Ed.)*
24 **Mechanical Behavior of Crustal Rocks: The Handin Volume** *N. L. Carter, M. Friedman, J. M. Logan, and D. W. Stearns (Eds.)*
25 **Physics of Auroral Arc Formation** *S.-I. Akasofu and J. R. Kan (Eds.)*
26 **Heterogeneous Atmospheric Chemistry** *David R. Schryer (Ed.)*
27 **The Tectonic and Geologic Evolution of Southeast Asian Seas and Islands: Part 2** *Dennis E. Hayes (Ed.)*
28 **Magnetospheric Currents** *Thomas A. Potemra (Ed.)*
29 **Climate Processes and Climate Sensitivity (Maurice Ewing Volume 5)** *James E. Hansen and Taro Takahashi (Eds.)*
30 **Magnetic Reconnection in Space and Laboratory Plasmas** *Edward W. Hones, Jr. (Ed.)*
31 **Point Defects in Minerals (Mineral Physics Volume 1)** *Robert N. Schock (Ed.)*
32 **The Carbon Cycle and Atmospheric CO_2: Natural Variations Archean to Present** *E. T. Sundquist and W. S. Broecker (Eds.)*
33 **Greenland Ice Core: Geophysics, Geochemistry, and the Environment** *C. C. Langway, Jr., H. Oeschger, and W. Dansgaard (Eds.)*
34 **Collisionless Shocks in the Heliosphere: A Tutorial Review** *Robert G. Stone and Bruce T. Tsurutani (Eds.)*
35 **Collisionless Shocks in the Heliosphere: Reviews of Current Research** *Bruce T. Tsurutani and Robert G. Stone (Eds.)*
36 **Mineral and Rock Deformation: Laboratory Studies —The Paterson Volume** *B. E. Hobbs and H. C. Heard (Eds.)*
37 **Earthquake Source Mechanics (Maurice Ewing Volume 6)** *Shamita Das, John Boatwright, and Christopher H. Scholz (Eds.)*
38 **Ion Acceleration in the Magnetosphere and Ionosphere** *Tom Chang (Ed.)*
39 **High Pressure Research in Mineral Physics (Mineral Physics Volume 2)** *Murli H. Manghnani and Yasuhiko Syono (Eds.)*
40 **Gondwana Six: Structure, Tectonics, and Geophysics** *Gary D. McKenzie (Ed.)*
41 **Gondwana Six: Stratigraphy, Sedimentology, and Paleontology** *Garry D. McKenzie (Ed.)*
42 **Flow and Transport Through Unsaturated Fractured Rock** *Daniel D. Evans and Thomas J. Nicholson (Eds.)*
43 **Seamounts, Islands, and Atolls** *Barbara H. Keating, Patricia Fryer, Rodey Batiza, and George W. Boehlert (Eds.)*

Geophysical Monograph Series

Including

IUGG Volumes

Maurice Ewing Volumes

Mineral Physics Volumes

44 Modeling Magnetospheric Plasma *T. E. Moore and J. H. Waite, Jr. (Eds.)*

45 Perovskite: A Structure of Great Interest to Geophysics and Materials Science *Alexandra Navrotsky and Donald J. Weidner (Eds.)*

46 Structure and Dynamics of Earth's Deep Interior (IUGG Volume 1) *D. E. Smylie and Raymond Hide (Eds.)*

47 Hydrological Regimes and Their Subsurface Thermal Effects (IUGG Volume 2) *Alan E. Beck, Grant Garven, and Lajos Stegena (Eds.)*

48 Origin and Evolution of Sedimentary Basins and Their Energy and Mineral Resources (IUGG Volume 3) *Raymond A. Price (Ed.)*

49 Slow Deformation and Transmission of Stress in the Earth (IUGG Volume 4) *Steven C. Cohen and Petr Vaníček (Eds.)*

50 Deep Structure and Past Kinematics of Accreted Terranes (IUGG Volume 5) *John W. Hillhouse (Ed.)*

51 Properties and Processes of Earth's Lower Crust (IUGG Volume 6) *Robert F. Mereu, Stephan Mueller, and David M. Fountain (Eds.)*

52 Understanding Climate Change (IUGG Volume 7) *Andre L. Berger, Robert E. Dickinson, and J. Kidson (Eds.)*

53 Plasma Waves and Instabilities at Comets and in Magnetospheres *Bruce T. Tsurutani and Hiroshi Oya (Eds.)*

54 Solar System Plasma Physics *J. H. Waite, Jr., J. L. Burch, and R. L. Moore (Eds.)*

55 Aspects of Climate Variability in the Pacific and Western Americas *David H. Peterson (Ed.)*

56 The Brittle-Ductile Transition in Rocks *A. G. Duba, W. B. Durham, J. W. Handin, and H. F. Wang (Eds.)*

57 Evolution of Mid Ocean Ridges (IUGG Volume 8) *John M. Sinton (Ed.)*

58 Physics of Magnetic Flux Ropes *C. T. Russell, E. R. Priest, and L. C. Lee (Eds.)*

59 Variations in Earth Rotation (IUGG Volume 9) *Dennis D. McCarthy and Williams E. Carter (Eds.)*

60 Quo Vadimus Geophysics for the Next Generation (IUGG Volume 10) *George D. Garland and John R. Apel (Eds.)*

61 Cometary Plasma Processes *Alan D. Johnstone (Ed.)*

62 Modeling Magnetospheric Plasma Processes *Gordon R. Wilson (Ed.)*

63 Marine Particles: Analysis and Characterization *David C. Hurd and Derek W. Spencer (Eds.)*

64 Magnetospheric Substorms *Joseph R. Kan, Thomas A. Potemra, Susumu Kokubun, and Takesi Iijima (Eds.)*

65 Explosion Source Phenomenology *Steven R. Taylor, Howard J. Patton, and Paul G. Richards (Eds.)*

66 Venus and Mars: Atmospheres, Ionospheres, and Solar Wind Interactions *Janet G. Luhmann, Mariella Tatrallyay, and Robert O. Pepin (Eds.)*

67 High-Pressure Research: Application to Earth and Planetary Sciences (Mineral Physics Volume 3) *Yasuhiko Syono and Murli H. Manghnani (Eds.)*

68 Microwave Remote Sensing of Sea Ice *Frank Carsey, Roger Barry, Josefino Comiso, D. Andrew Rothrock, Robert Shuchman, W. Terry Tucker, Wilford Weeks, and Dale Winebrenner*

69 Sea Level Changes: Determination and Effects (IUGG Volume 11) *P. L. Woodworth, D. T. Pugh, J. G. DeRonde, R. G. Warrick, and J. Hannah*

70 Synthesis of Results from Scientific Drilling in the Indian Ocean *Robert A. Duncan, David K. Rea, Robert B. Kidd, Ulrich von Rad, and Jeffrey K. Weissel (Eds.)*

71 Mantle Flow and Melt Generation at Mid-Ocean Ridges *Jason Phipps Morgan, Donna K. Blackman, and John M. Sinton (Eds.)*

72 Dynamics of Earth's Deep Interior and Earth Rotation (IUGG Volume 12) *Jean-Louis Le Mouël, D.E. Smylie, and Thomas Herring (Eds.)*

73 Environmental Effects on Spacecraft Positioning and Trajectories (IUGG Volume 13) *A. Vallance Jones (Ed.)*

74 Evolution of the Earth and Planets (IUGG Volume 14) *E. Takahashi, Raymond Jeanloz, and David Rubie (Eds.)*

75 Interactions Between Global Climate Subsystems: The Legacy of Hann (IUGG Volume 15) *G. A. McBean and M. Hantel (Eds.)*

76 Relating Geophysical Structures and Processes: The Jeffreys Volume (IUGG Volume 16) *K. Aki and R. Dmowska (Eds.)*

77 The Mesozoic Pacific: Geology, Tectonics, and Volcanism—A Volume in Memory of Sy Schlanger *Malcolm S. Pringle, William W. Sager, William V. Sliter, and Seth Stein (Eds.)*

78 Climate Change in Continental Isotopic Records *P. K. Swart, K. C. Lohmann, J. McKenzie, and S. Savin (Eds.)*

79 The Tornado: Its Structure, Dynamics, Prediction, and Hazards *C. Church, D. Burgess, C. Doswell, R. Davies-Jones (Eds.)*

80 Auroral Plasma Dynamics *R. L. Lysak (Ed.)*

81 Solar Wind Sources of Magnetospheric Ultra-Low Frequency Waves *M. J. Engebretson, K. Takahashi, and M. Scholer (Eds.)*

82 **Gravimetry and Space Techniques Applied to Geodynamics and Ocean Dynamics** *Bob E. Schutz, Allen Anderson, Claude Froidevaux, and Michael Parke (Eds.)*
83 **Nonlinear Dynamics and Predictability of Geophysical Phenomena** *William I. Newman, Andrei Gabrielov, and Donald L. Turcotte (Eds.)*
84 **Solar System Plasmas in Space and Time** *J. Burch, J. H. Waite, Jr. (Eds.)*

Maurice Ewing Volumes

1 **Island Arcs, Deep Sea Trenches, and Back-Arc Basins** *Manik Talwani and Walter C. Pitman III (Eds.)*
2 **Deep Drilling Results in the Atlantic Ocean: Ocean Crust** *Manik Talwani, Christopher G. Harrison, and Dennis E. Hayes (Eds.)*
3 **Deep Drilling Results in the Atlantic Ocean: Continental Margins and Paleoenvironment** *Manik Talwani, William Hay, and William B. F. Ryan (Eds.)*
4 **Earthquake Prediction—An International Review** *David W. Simpson and Paul G. Richards (Eds.)*
5 **Climate Processes and Climate Sensitivity** *James E. Hansen and Taro Takahashi (Eds.)*
6 **Earthquake Source Mechanics** *Shamita Das, John Boatwright, and Christopher H. Scholz (Eds.)*

IUGG Volumes

1 **Structure and Dynamics of Earth's Deep Interior** *D. E. Smylie and Raymond Hide (Eds.)*
2 **Hydrological Regimes and Their Subsurface Thermal Effects** *Alan E. Beck, Grant Garven, and Lajos Stegena (Eds.)*
3 **Origin and Evolution of Sedimentary Basins and Their Energy and Mineral Resources** *Raymond A. Price (Ed.)*
4 **Slow Deformation and Transmission of Stress in the Earth** *Steven C. Cohen and Petr Vaníček (Eds.)*
5 **Deep Structure and Past Kinematics of Accreted Terrances** *John W. Hillhouse (Ed.)*
6 **Properties and Processes of Earth's Lower Crust** *Robert F. Mereu, Stephan Mueller, and David M. Fountain (Eds.)*
7 **Understanding Climate Change** *Andre L. Berger, Robert E. Dickinson, and J. Kidson (Eds.)*
8 **Evolution of Mid Ocean Ridges** *John M. Sinton (Ed.)*
9 **Variations in Earth Rotation** *Dennis D. McCarthy and William E. Carter (Eds.)*
10 **Quo Vadimus Geophysics for the Next Generation** *George D. Garland and John R. Apel (Eds.)*
11 **Sea Level Changes: Determinations and Effects** *Philip L. Woodworth, David T. Pugh, John G. DeRonde, Richard G. Warrick, and John Hannah (Eds.)*
12 **Dynamics of Earth's Deep Interior and Earth Rotation** *Jean-Louis Le Mouël, D.E. Smylie, and Thomas Herring (Eds.)*
13 **Environmental Effects on Spacecraft Positioning and Trajectories** *A. Vallance Jones (Ed.)*
14 **Evolution of the Earth and Planets** *E. Takahashi, Raymond Jeanloz, and David Rubie (Eds.)*
15 **Interactions Between Global Climate Subsystems: The Legacy of Hann** *G. A. McBean and M. Hantel (Eds.)*
16 **Relating Geophysical Structures and Processes: The Jeffreys Volume** *K. Aki and R. Dmowska (Eds.)*
17 **Gravimetry and Space Techniques Applied to Geodynamics and Ocean Dynamics** *Bob E. Schutz, Allen Anderson, Claude Froidevaux, and Michael Parke (Eds.)*
18 **Nonlinear Dynamics and Predictability of Geophysical Phenomena** *William I. Newman, Andrei Gabrielov, and Donald L. Turcotte (Eds.)*

Mineral Physics Volumes

1 **Point Defects in Minerals** *Robert N. Schock (Ed.)*
2 **High Pressure Research in Mineral Physics** *Murli H. Manghnani and Yasuhiko Syona (Eds.)*
3 **High Pressure Research: Application to Earth and Planetary Sciences** *Yasuhiko Syono and Murli H. Manghnani (Eds.)*

Geophysical Monograph 85

The Polar Oceans and Their Role in Shaping the Global Environment

The Nansen Centennial Volume

O. M. Johannessen
R. D. Muench
J. E. Overland
Editors

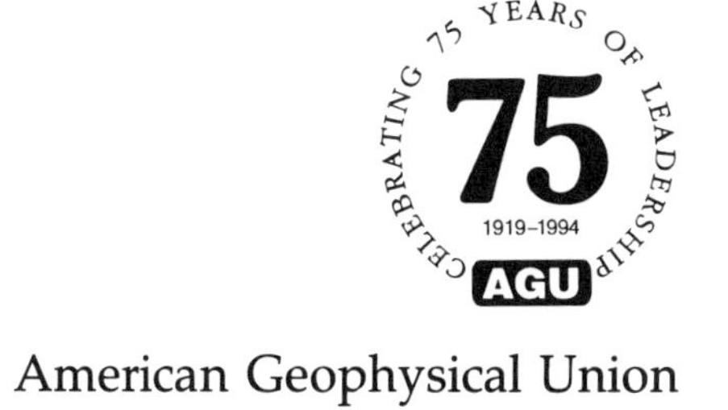

American Geophysical Union

Published under the aegis of the AGU Books Board.

Cover photograph courtesy of Robert W. Fett.

Library of Congress Cataloging-in-Publication Data

The polar oceans and their role in shaping the global environment :
the Nansen centennial volume / O. M. Johannessen,
R. D. Muench, J. E. Overland, editors.
p. cm. — Geophysical monograph ; 85)
Includes bibliographical references.
ISBN 0-87590-042-9
1. Arctic Ocean—Congresses. 2. Antarctic Ocean—Congresses.
3. Climatic changes—Congresses. I. Nansen, Fridtjof, 1861–1930.
II. Johannessen, Ola M. III. Muench, Robin D. IV. Overland, James
E. V. Nansen Centennial Symposium (1993 : Solstrand, Norway)
VI. Series.
GC401.P637 1994
551.46′8—dc20 94-35047
CIP

ISSN 0065-8448

ISBN 0-87590-042-9

This book is printed on acid-free paper. 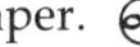

Printed in the United States of America.

CONTENTS

CONTENTS

CONTENTS

The 'Fram'

PREFACE
A Celebration of 100 Years of Polar Research

On June 24, 1993, one hundred years had passed since Fridtjof Nansen and his companions set out on one of the most daring and exciting research expeditions the world had ever seen. They allowed their vessel, the *Fram*, to be frozen into the ice close to the New Siberian Islands, in the Arctic Ocean. Three years were to elapse before the ice released its hold on the *Fram* and allowed her to return to Norway via the strait between Greenland and Spitsbergen, which later came to be known as Fram Strait. The research carried out during *Fram*'s drift in the ice altered forever our concept of the Arctic Basin.

Nansen discovered that the Arctic Ocean was deep, and he confirmed the existence of the Transpolar Current. During the drift, it was also observed that *Fram* and the ice pack drifted to the right of the wind direction. Nansen interpreted this drift as the effect of the Earth's rotation and formulated the concept of the Ekman Spiral, the foundation for the modern wind-driven circulation model. The *Fram* drift set the stage for research which, over the past century, has established that the polar oceans interact vigorously with global climate. Moreover, these bodies of water, covered with seasonally varying, dynamically and thermodynamically active ice covers, condition the deep bottom waters of the global ocean, a topic Nansen pioneered later in his career during investigations of the deep water formation and circulation in the Greenland-Norwegian Sea. Again, Nansen laid the foundation for currently important research in this region, where today's focus is on understanding the role of deep water formation in the conveyor circulation and global climate.

This volume provides a comprehensive treatment of the major advances made in the past decade in our understanding of the interactions between the polar oceans and the global atmosphere and ocean system. Topics covered are polar ocean circulation and dynamics, convective processes in the polar regions, polar ocean carbon cycle chemistry and biology, paleoceanography and paleoclimate, polar ocean and global climate interaction, and detection strategies for climate change in the polar regions.

O. M. Johannessen
Nansen Center
Bergen, Norway

R. D. Muench
SAIC
Bellevue, Washington

J. E. Overland
NOAA
Seattle, Washington

Editors

ACKNOWLEDGMENTS

The work presented here has been developed from contributions to the Nansen Centennial Symposium, held June 21-25, 1993, near Bergen, Norway, to honor Fridtjof Nansen and the pioneering drift of the FRAM a century earlier. The Symposium was organized by the Nansen Environmental and Remote Sensing Center, Bergen, and the University of Bergen. The editors thank the International Organizing Committee (O. M. Johannessen (Co-Chairman), Norway; R. Muench (Co-Chairman), USA; G. V. Alekseev, Russia; L. Anderson, Sweden; J. C. Gascard, France; A. L. Gordon, USA; K. Hasselmann, Germany; K. Ya. Kondratyev, Russia; J. E. Overland, USA; T. Platt, Canada; R. A. Shuchman, USA; T. Vorren, Norway; and J. D. Woods, UK) and acknowledge Bente Johannessen for her spirit in handling the organizational details. The editors are indebted to the reviewers for their diligence and to Ann Thomason for her tenacity as editor's assistant for the volume.

The Symposium was sponsored by the International Association for the Physical Sciences of the Ocean, the Norwegian Research Council, the Norwegian Space Center, the Norwegian Ministry of Foreign Affairs, Rieber Shipping A/S, the Office of Naval Research, the Commission of the European Communities, the European Science Foundation, the European Space Agency, and the Intergovernmental Oceanographic Commission of UNESCO.

This volume is dedicated to William (Bill) J. Campbell (1930-1992), a dear friend and inspiring colleague.

O. M. Johannessen

Fridtjof Nansen.

„Fram" 15/2·95.

Fridtjof Nansen - Scientist, Diplomat and Humanist

TORE GJELSVIK

The Fram Committee, Oslo, Norway

In my opinion, no Norwegian has made a deeper impression upon his contemporaries than Fridtjof Nansen. Even today, more than 60 years after he passed away, young people who want to make an extraordinary physical effort talk about walking in his footsteps - or skiing in his ski tracks. References to Nansen's humanitarian efforts after the first world war can be read almost daily in Norwegian newspapers in articles discussing aid to refugees, starving populations or ethnic minorities threatened with extermination. More than 200,000 people, many of them foreigners, visit Nansen's famous polar ship FRAM, housed on the peninsula Bygdøy, on the outskirt of Oslo.

Already on his first polar expedition, the crossing of the Greenland inland ice in 1888 when he was 27 years old and immediately after defending his pioneering doctoral thesis in zoology on the histology of the central nervous system of the hagfish, he demonstrated three aspects of his character: careful compilation and evaluation of all available information, detailed study of all practical and logistical aspects of the expedition, and a forceful execution of the expedition plan. On the practical side he improved the eskimo sledge to a model named after him, the Nansen sledge, which, with small modifications has been used up to this time. He developed cooking gear that used all excess heat to melt snow for drinking water.

Nansen adopted an expedition philosophy in direct opposition to the conventional wisdom of his time: he started from the unpopulated, hostile east coast of Greenland where retreat was impossible, rather than from the milder, inhabited west coast as earlier explorers had attempted. Nansen purposely burned the bridges behind him and left no alternative for him and his men but to move forward. They had to do or die. No wonder that he named his specially constructed polar vessel FRAM - which is the Norwegian phrase for "forward." When - in 1892 - he outlined his plan for a North Pole expedition at the Royal Geographic Society of London, he was met with a thunderstorm of criticism and opposition by people who were considered the most experienced arctic explorers of the time. One of them argued that Nansen violated the safety code for expeditions by not securing a safe retreat line. Nansen could not disagree more, and did not yield an inch.

The most important results of Nansen's FRAM expedition were: The discovery of a deep, Arctic Ocean and the confirmation of the existence of the Transpolar Current. Furthermore he observed that FRAM and the ice pack drifted approximately 30°to the right of the wind direction. This fact he interpreted as the effect of the earth's rotation, which laid the concept for the Ekman Spiral and the foundation for the modern wind-driven ocean circulation. In connection with his polar expeditions, he involved himself in the study of oceanography, meteorology, climatology, geomagnetism, glaciology, geology and polar history. His papers on most of this topics were of high quality and became widely quoted. As a geologist, I have been struck with his skill in studies of the morphology of the coastal shelves, and of other coastal and glaciomorphological problems. Some of his papers on this topic have also become classics, such as "The strandflat and isostasy," which was published in 1892.

As a scientist, Nansen was much aware of the need for precise and exact measurements, and during his later oceanographic investigations he found that some of the oceanographic measurements on the FRAM drift were not of sufficient precision. The invention of the Nansen bottle for sampling ocean water at various depths is a well-known example of his improved methods and instruments.

Nansen was a strong supporter of international cooperation in oceanography and was one of the founding fathers of the International Council for the Exploration of the Sea (ICES) in 1902.

The successful outcome of Nansen's North Pole expedition and his report on the daring attempt to reach the North Pole by skis and sledges made Nansen a national hero and gave him a world-wide reputation. Shortly before he returned from the expedition there had been a deep crisis in the Swedish Norwegian Union, and the Norwegians had to withdraw their demands. The success of the FRAM expedition

The Polar Oceans and Their Role in Shaping the Global Environment
Geophysical Monograph 85

made the Norwegians raise their heads again, and Nansen involved himself in the political debate of the Swedish Norwegian relationship. He played an important role in 1905 when the union was dissolved and Norway declared full independence. He was also instrumental in persuading the Danish Prince Carl to accept being king of Norway (the later King Haakon VII). Nansen wrote articles in leading newspapers of the world, explaining and defending the position of Norway. Because of his fame and reputation, the world listened. He was appointed Norway's first ambassador in London and succeeded in getting Great Britain and other great powers to recognize the new Norwegian state.

After two years in London he felt his job was done and returned to scientific work for some years. He never liked politicians and diplomats and did not want to be one of them. He had some good years studying the oceanography of the Norwegian Sea. One of the scientific questions he raised and tried to answer was the mode of formation of bottom water in the Greenland Sea, published jointly with Professor B. Helland-Hansen in the classic book "The Norwegian Sea."

The first world war very much changed the course of Nansen's life. He detested war, but came to use most of his later life cleaning up the mess which the war had left. He was very critical of the great powers and he wanted the international anarchy to be replaced by a system founded on the principles of lawful order and human rights. Disputes between states should be solved by negotiation and arbitration instead of by military force. In 1920 he was appointed Norwegian delegate to the League of Nations in Geneva, a forerunner of the United Nations. In central Russia and Siberia, a large but unknown number of war prisoners were living under appalling conditions - clothed in rags, freezing and starving, ravaged by disease and unable to reach their homelands; some of them lived as slaves. Nansen was asked by the League to be High Commissioner for repatriation of the war prisoners, and he undertook the task. A main problem was the bitter hostility and the lack of diplomatic contacts between the revolutionary government in Russia and the western governments. Although Nansen was strongly against the communist ideology, he realized the need for political and technical agreements with the Soviet-Russian government. Obtaining the confidence of the Soviet foreign minister, he succeeded in acquiring the necessary agreements. In less than two years all the prisoners, numbering nearly a half million, were back in their homelands.

During his travels in Soviet Russia, Nansen discovered that the Russian people were suffering from a severe famine because of lack of grain, and that without aid millions of people would inevitably die. Nansen appealed to the western governments to grant a loan to Soviet Russia for buying grain, but in vain. The International Red Cross asked Nansen to lead a relief action. In cooperation with private and religious charity organizations Nansen organized the "Nansen Aid" program to collect money and buy food for the starving Russian people. He was able to rescue about one million lives. Still several millions died during the famine in the early twenties. At the same time, 1.5 million Russians had fled after the October revolution and were living in desperate conditions in the cities of European countries. The International Red Cross asked the League of Nations to appoint Nansen as the first High Commissioner for refugees. Beside the material problems of the refugees, many were stateless, living in foreign states without national identity papers. As High Commissioner for refugees, Nansen issued identification papers in his own name, they were called Nansen passports and were recognized by more than 50 states. This gave the stateless refugees a new start in life. Again, Nansen had demonstrated his skill in solving difficult problems through unorthodox methods and means.

In 1922 Nansen was awarded the Nobel Peace Prize for his humanitarian work. The award became no sleeping pillow for Nansen - a new and terrible crisis had developed in Asia Minor in the wake of the Greek-Turkish war. Kemal Ataturk's armies swept westwards, crushing the Greek army and forcing 1.5 million Greeks out of their homes in Anatolia, where they had lived for several hundred years. They had to escape in a hurry, leaving everything behind, and taking little or no food and very little clothing with them. Many returned to their homeland, which accepted them; however, Greece was economically ruined and had no means of self help. Nansen quickly intervened with a bold and unconventional plan: The exchange of the Greek refugees with 400,000 Turks living in Greece. He also laid forth a plan describing how this costly operation could be organized and financed. At first the plan was met by opposition from both sides, but after a short time it was accepted and realized. Perhaps Nansen's scheme should be studied by those trying to appease the fighting ethnic groups in the former Yugoslavia today.

Nansen's last humanitarian action was an attempt to help the Armenian refugees achieve a national homeland by irrigating the slopes of Mount Ararat. Because he was refused the necessary funding, the plan had to be abandoned.

During his humanitarian activity Nansen applied the same philosophy as on his expeditions. Once he had studied all sides of a problem, made a plan and decided on his ways and means, he did not compromise. His only course was forward. His unconventional and uncompromising methods often met opposition from politicians and diplomats. Many considered him "The enfant terrible" of the League of Nations. His answer to them was: *Charity is practical politics.* The men and women who worked together with him or for him had the highest regard for him and found him to be a most effective leader.

The sufferings and tragedies of refugees and other war victims and the ceaseless charity work gradually wore Nansen down. Perhaps his worst frustrations were the heavy responsibilities and hard work that the governments loaded on him without providing the funding necessary for successfully carrying out the tasks.

When he returned to science during the last years of his life, Nansen took a special interest in the possibility of using airships as scientific platforms in the Arctic. He chaired an international group to study the project. Because they were not able to raise the necessary funding, the plan never materialized.

In 1930, Fridtjof Nansen died, 60 years old, of *heart failure.* A Norwegian journalist once said that this was not true: Nansen died because his heart *never failed.*

T. Gjelsvik, Framkomitéen, Postboks 5072, Maj., 0301 Oslo.

The Arctic Ocean and Climate: A Perspective

K. Aagaard

Polar Science Center, Applied Physics Laboratory, College of Ocean and Fishery Sciences, University of Washington, Seattle, Washington

E. C. Carmack

Institute of Ocean Sciences, Sidney, British Columbia, Canada

The most likely effects of the Arctic Ocean on global climate are through the surface heat balance and the thermohaline circulation. The former is intimately related to the stratification of the Arctic Ocean, while the latter may be significantly controlled by outflow from the Arctic Ocean into the major convective regions to the south. Evaluating these issues adequately requires detailed knowledge of the density structure and circulation of the Arctic Ocean and of their variability. New long time series of temperature and salinity (T/S) from the Canadian Basin show a grainy T/S structure, probably on a horizontal scale of a few tens of kilometers. The temperature field is particularly inhomogeneous, since for cold water it is not greatly constrained by buoyancy forces. The simultaneous velocity time series show that the grainy T/S structure results from a complex eddy field, often with vertically or horizontally paired counter-rotating eddies drifting with a slow larger-scale flow. The ocean is therefore not well mixed on these scales. Finally, we note that the ventilation of the interior Arctic Ocean from the adjacent shelves appears to be highly variable on an interannual basis, and indeed may not be robust on longer time scales. In particular we note the absence, or near-absence, of deep ventilation of the Canadian Basin during the last 500 years. Based on the ^{14}C model of Macdonald et al. [1993], however, we hypothesize that these same waters were ventilated prior to that time and that the deep convective shutdown about 500 years ago coincided with the end of the whale-hunting Thule culture. We further suggest that the two events had a common cause, viz., the increase of sea ice over the continental shelves during summer.

1. BACKGROUND

"It is evident that the oceanographical conditions of the North Polar Basin have much influence upon the climate, and it is equally evident that changes in its conditions of circulation would greatly change the climatic conditions." Thus *Nansen* [1902] ended his discussion of the oceanographic results of the *Fram* voyage.

A century later, with climate issues having become scientifically respectable and human society concerned about the consequences of global change, we are now seriously beginning to explore whether Nansen's claims have merit: Is the Arctic Ocean in fact an important cog in the global climate machinery? Can realistic changes in the density structure and circulation of the Arctic Ocean effect significant changes on a much larger spatial scale? While it is not yet clear that the answer to either question is an unequivocal and rigorously founded "Yes," we are at a stage where we can both frame the issues in sufficient detail to make them scientifically interesting, and probably also foresee meaningful answers within a reasonable time, say a decade or two.

At this point it appears that there are two ways in which the Arctic Ocean might play an important role in global climate. The first is through its effect on the surface heat

The Polar Oceans and Their Role in Shaping the Global Environment
Geophysical Monograph 85

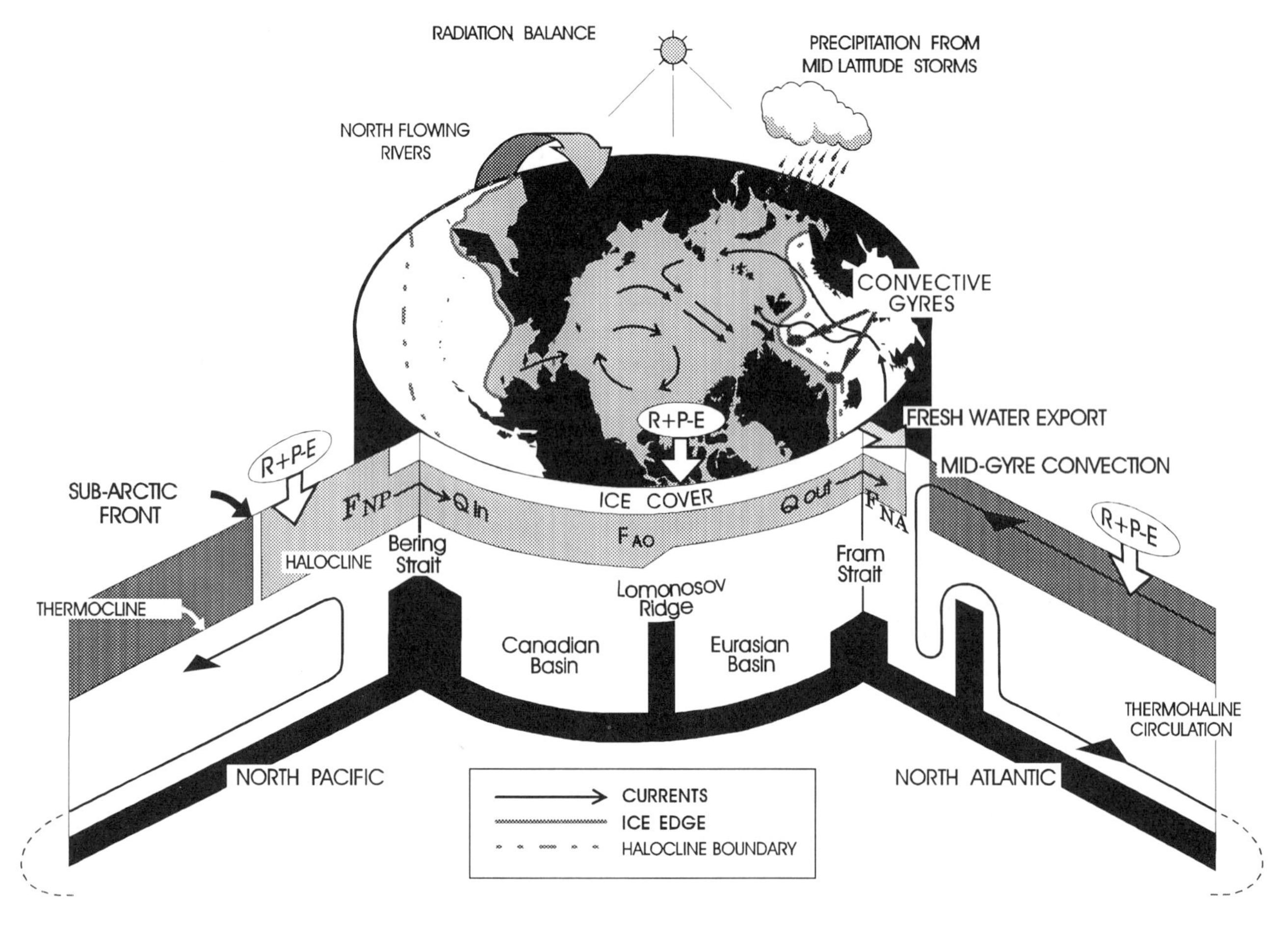

Fig. 1. Schematic Arctic Ocean climate connections. In the horizontal plane the extent of sea ice in winter is shown by the shaded region and the mean surface circulation by arrows. Sections from the North Pacific and the North Atlantic extend through Bering and Fram straits, respectively. The subarctic front separates the salt-stratified upper waters of the arctic and subarctic oceans from the temperature-stratified upper waters of the subtropical oceans. The components of the freshwater balance include runoff (R), precipitation (P), evaporation (E), storage in the upper North Pacific (F_{NP}), the Arctic Ocean (F_{AO}), and the North Atlantic (F_{NA}), and the horizontal freshwater fluxes (Q_{in} and Q_{out}). Shaded ovals indicate the present sites of convection in the Greenland and Iceland seas.

balance, i.e., through the role of the ocean in the local radiation balance and in disposing of the heat surplus acquired at lower latitudes. The second is through its effect on the large-scale ocean overturning, i.e., through the role of buoyancy (fresh water) exported from the Arctic Ocean in regulating the intensity of the global thermohaline circulation. Likely, the surface heat balance and the ocean overturning are coupled. Figure 1 schematically shows these various connections.

1.1 Surface Heat Balance

Consider first the ocean's effect on the surface heat balance. The Arctic Ocean transfers heat to the atmosphere primarily along the periphery, where leads, polynyas, and seasonally open seas allow large fluxes. In contrast, the annual net surface heat flux in the interior is very small, because at present the Arctic Ocean is markedly stratified between about 50-150 m. This results in very low effective vertical diffusion rates in the upper water column and insulation of the underlying warm Atlantic layer from the surface. The stratification also prevents winter convection and surface mixed layer deepening greater than about 50 m. This is fundamentally why the Arctic Ocean is more or less permanently ice covered. The upper ocean heat content, including what is absorbed through short-wave radiation in summer, is simply cycled seasonally as part of the freezing and melting sequence [*Maykut*, 1982]. We note, however, that there are important differences in upper-ocean stratification between the Canadian and Eurasian basins, with the latter being the more weakly stratified. At the same time, the temperature gradient in the halocline of the Eurasian Basin is much larger than in the Canadian Basin, so that we might expect greater variability in the heat flux there. However, even in the strongly stratified Canadian Basin, there can be remarkable year-to-year variability in the mixed layer and the upper halocline. For example, near-surface temperatures over large areas can be above the freezing temperature even during winter [*R. Perkin*, personal communication], apparently because of anomalously warm inflow from Bering Strait and/or outflow from the shelves. Similarly, *M. McPhee* [personal communication] has observed large transient temperature maxima lying above the well-known temperature maximum near 75 m depth that originates in the Bering Sea [*Coachman and Barnes*, 1961]. Imagine, therefore, ever larger perturbations in upper ocean forcing leading to an altered vertical structure in the Arctic Ocean, perhaps through changes in continental runoff patterns or in the ice production over the shelves. Such an altered vertical structure would almost certainly also occasion major changes in the convective and diffusive fluxes of salt and sensible heat across the pycnocline and the mixed-layer interface. Ice model sensitivity studies [e.g., *Maykut and Untersteiner*, 1971] suggest that the resulting increase in the vertical heat flux would most probably lead to an ice cover that at least regionally, say over the Eurasian Basin, would be significantly reduced. The ice distribution would then more nearly resemble that of the Antarctic, where most of the ice disappears each summer, or that of the Greenland Sea, where nearly ice-free conditions can be found intermittently even during the cold season. A reduction in ice cover would in turn effect major changes in the surface and atmospheric radiation balances, and probably also in the vertical flux of carbon. The point is not so much whether any particular perturbation is realistic, but rather that a principal reason for the present ice cover and climate in the Arctic is the strong stratification of the upper ocean. We might then ask whether or not the stratification is robust, or alternatively how efficiently the ocean can be ventilated under a variety of stratified conditions.

1.2 Ocean Overturning

Turn now for a moment to the second oceanic effect, viz., the arctic influence on the large-scale thermohaline circulation of the world ocean. This circulation mode is probably intimately involved with the redistribution of heat over the globe, and in particular it appears instrumental in moderating the climate of Europe [cf., *Broecker et al.*, 1985 for a discussion]. The present evidence is that convection in the high northern latitudes of the Atlantic drives much of the global overturning cell in the ocean, but that different stable modes of this global circulation can exist, each associated with a very different climate [*Manabe and Stouffer*, 1988]. For example, in the case of a thermohaline circulation driven vigorously in the far northern Atlantic, the northern European climate is moderate, much as it is today or even warmer. On the other hand, if the North Atlantic forcing diminishes, the climate deteriorates dramatically. Weakening of the thermohaline circulation has, for example, been proposed as responsible for such major disturbances in the paleoclimatic record as the Younger Dryas interruption of the warming trend accompanying the last deglaciation [*Broecker et al.*, 1985].

We show schematically in Figure 2 how similar mode transitions might be manifested in the Arctic Ocean and the adjacent seas. At present it appears that convection in the Greenland and Iceland seas is conditioned by freshwater export from the Arctic Ocean [*Aagaard and Carmack*, 1989]. If the export were increased significantly, convection would likely diminish, while under conditions of decreased freshwater export, convection would become more vigorous. We hypothesize that at some level of increased export, convection would cease; the northward transport of Atlantic water would decrease; and the climatological ice cover would spread southward. In this scenario, the convective regions would also shift southward and the entire system would assume a new configuration. Conversely, were the supply of fresh water to the Arctic to decrease, we might expect a weakening of the stratification in the Eurasian Basin, perhaps to the point of allowing local convection to at least intermediate depths, with an attendant increase in northward heat transport and a decrease in the ice cover.

Perturbations in ocean stratification need not be confined to long time scales. For example, in the late 1960s a remarkable freshening of the upper ocean was observed in the seas north of Iceland. The perturbation propagated southward, passed through the Labrador Sea, and by the mid 1970s it had crossed the Atlantic to Europe. *Dickson et al.* [1988] have referred to this perturbation, which was most marked between 500–800 m, as "The Great Salinity Anomaly." The anomaly also appears to have quickly left its imprint on the global overturning cell in the form of a freshening and cooling of the North Atlantic deep water, first described by *Brewer et al.* [1983]. Freshwater budget calculations show that the source of the Great Salinity Anomaly almost surely was the Arctic Ocean, probably in the form of an anomalously large discharge of sea ice during the mid to late 1960s [*Aagaard and Carmack*, 1989]. Subsequent work by *Walsh and Chapman* [1990] suggests a possible mechanism for such a discharge, viz., anomalous

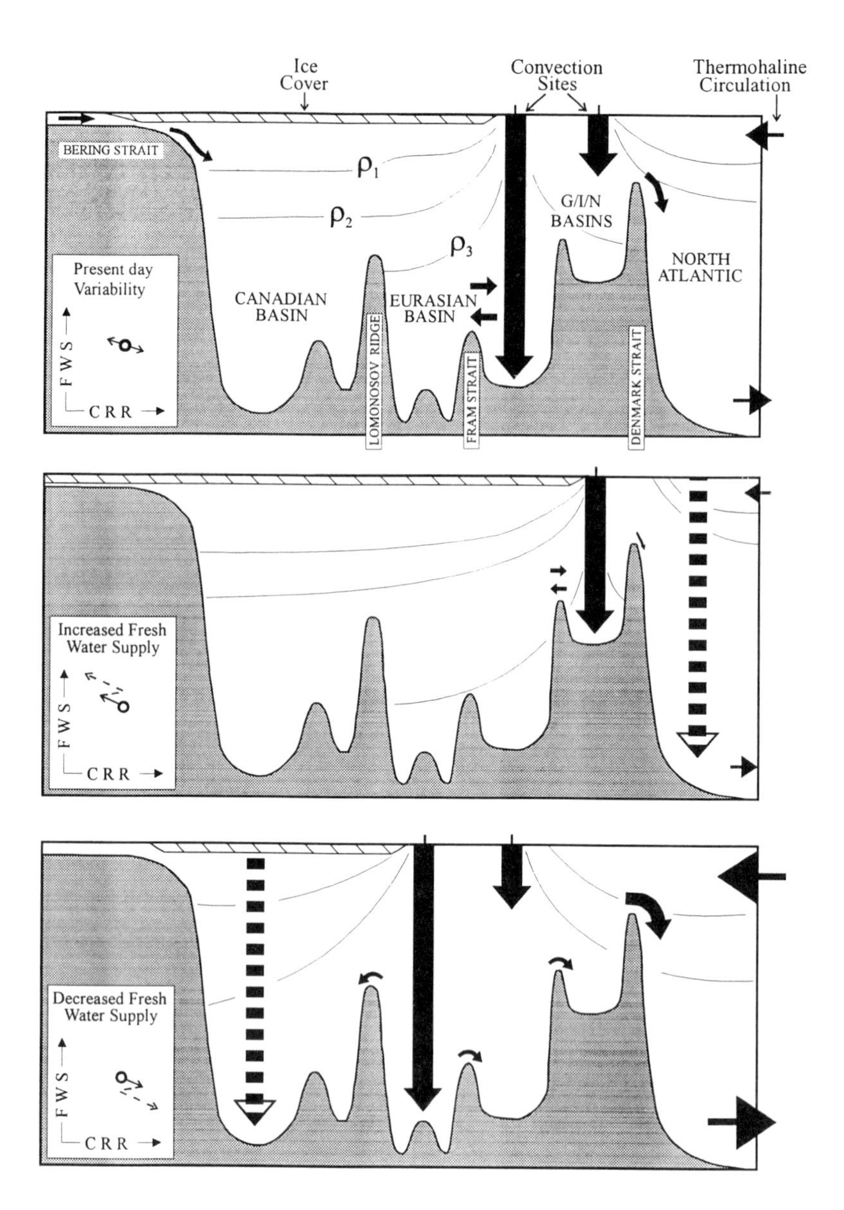

Fig. 2. Hypothesized dependence of the convective renewal rate (CRR) on freshwater supply (FWS) from the Arctic Ocean under (a) present conditions, (b) increased freshwater supply, and (c) decreased supply. The size of the arrows through the right-hand side is representative of the strength of the thermohaline circulation forced from the far northern seas. The barred arrows represent the extreme locations of convection. The solid arrows in the insets indicate the trend in convective renewal with changing freshwater supply, and the dashed arrows indicate possible transitions to different circulation modes.

winds reaching into the very thick ice pack in the northeastern Canadian Basin and incorporating it into the ice export through Fram Strait. The recent modeling study by *Häkkinen* [1993] strengthens the case for an Arctic Ocean origin of the Great Salinity Anomaly and points to other events with a similar, but somewhat reduced, impact. However, her model suggests that anomalously large ice export need not originate solely in the northeastern Canadian Basin, but can draw more generally from anywhere in the Polar Basin and is related to "...inconspicuous changes in the transport pattern...such as the widening and strengthening of the Transpolar Drift Stream" [*Häkkinen*, 1993, p. 16,400].

Certainly the Arctic Ocean stores a very large amount of buoyancy, about 100,000 km^3 of freshwater equivalent, which is enough to maintain a Great Salinity Anomaly in the Atlantic for many decades [*Aagaard and Carmack*, 1989]. Certainly also, in the discharge of this buoyancy into the convective regions in the Greenland, Iceland, and Labrador seas, nature has a means of controlling the strength of the thermohaline circulation by stabilizing the upper ocean and preventing the high-latitude convective overturn that drives the global overturning cell. This would be precisely the scenario described in the so-called "halocline catastrophes" proposed both by paleoclimatologists and by modelers of the global circulation, in which the convective regions are capped by fresh water. However, what we see now is that this capping may not be by meltwater from the continental ice sheets, nor by increased precipitation in a world warmed by carbon dioxide, but by perturbations of the freshwater component of the outflow from the Polar Basin, or by changes in the transfer of this component into the interior convective gyres from the western boundary current [*Aagaard and Carmack*, 1989]. (We should also note that increasingly the paleoclimatic record points not only toward such grand long-term events as the Younger Dryas, but also toward large high-frequency changes in the atmospheric circulation particularly. Such changes occur on time scales less than a decade [*Taylor et al.*, 1993].) While we can at present only speculate on the causes and nature of the transitions between ocean circulation modes and their associated climate, certainly perturbation of the ocean circulation by events originating in the Arctic Ocean is very much in the realm of possibility.

Before leaving the topic of freshwater control of convection, we note that because of the differential compressibility of seawater, with cold water being the more compressible (the so-called thermobaric effect) it is conceivable that within some parameter range, a small amount of salinity stratification may actually promote deep convection [*Aagaard and Carmack*, 1989]. The stratification allows the surface to be cooled significantly more than the underlying water, and when convection eventually is initiated, the very cold convective plume originating above the halocline may accelerate downward as the pressure increases during its descent [*Gill*, 1973]. Something like this happens in the Antarctic where the extremely cold ice shelf water accelerates at an angle down the continental slope, as has been modeled by *Killworth* [1977] and shown so convincingly by *Foldvik et al.* [1985]. This is a far more efficient ventilating process than the gradual deepening of the mixed layer.

We should note also that the Arctic Ocean outflow not only influences the convective process from above through the variable contribution of low-salinity water, but also from below through heating. We now know that not only low-salinity surface water, but also relatively warm and saline Arctic Ocean deep water, moves southward through Fram Strait and spreads into the interior of both the Greenland and Iceland seas [*Aagaard et al.*, 1991]. Because the effect on stability of temperature dominates in this case, the deep Arctic Ocean outflow provides a destabilizing influence on the convective gyre, warming it from below. *Aagaard et al.* [1991] have shown that the effect of this process on the Greenland Sea below 1500 m during the 1980s was to both warm and increase the salinity of the deep water, with the net effect of decreasing the deep density and thereby possibly setting the stage for later deep ventilation. The net cooling and freshening that was observed in the upper ocean during this period (cf. Figure 8 in *Aagaard et al.* [1991]) was due to increasing surface-driven convection during the late 1980s, after a long period of ineffective and shallow convection [*Schlosser et al.*, 1991]. The situation is shown schematically in Figure 3.

In effect, therefore, the Arctic Ocean appears to condition both the surface and the deep waters in the Greenland and Iceland seas, and to some extent the surface layer in the Labrador Sea [*Lazier*, 1980; 1988]. Sustained changes in ice production and in the formation rates and properties of water masses in the Arctic Ocean and its shelf seas, or in changes in the export of these products to the convective regions of the North Atlantic, might therefore alter the large-scale Atlantic thermohaline circulation cell. Whether that cell were weakened or strengthened, such a change would probably have major consequences for northern hemisphere climate in particular.

2. DENSITY STRUCTURE AND CIRCULATION

If we are to adequately understand these surface heat balance and thermohaline circulation issues, and whether or not the Arctic Ocean plays an important role on the global stage, we need to know in some detail both the density structure and the circulation of the Arctic Ocean, how they are maintained, and also something about the present variability of the system. By circulation we mean both the quasi-horizontal component and the vertical component, or convection. The density structure and the circulation are of course interactive, and they therefore at some stage have to be treated jointly for a full description of the ocean. We here first briefly point out some features of the circulation that we have in recent years come to believe are important, and then point to some new results and observations that address the issue of variability in time and space of both the density structure and the circulation.

2.1 *Circulation Features*

Figure 4 schematically shows the quasi-horizontal circulation of the Arctic Ocean. Note that:

First, the Arctic Ocean receives waters from the Pacific, and its stratification is very significantly influenced by that

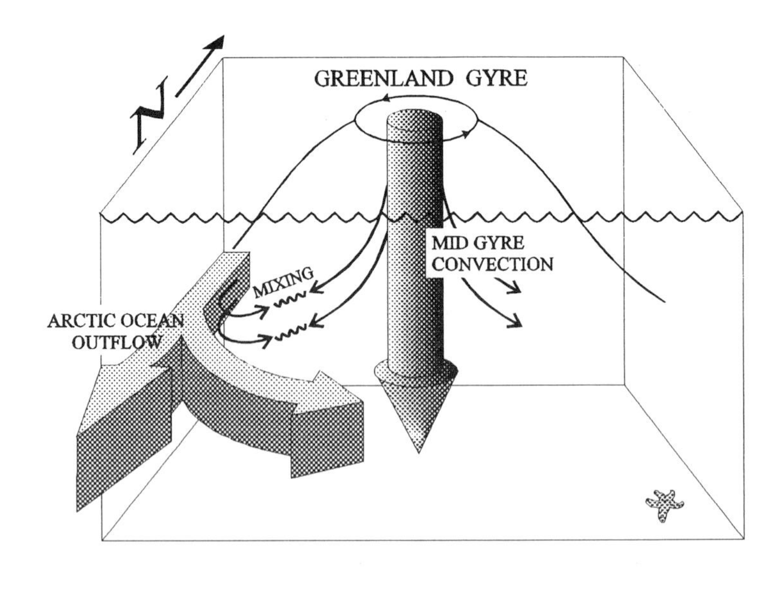

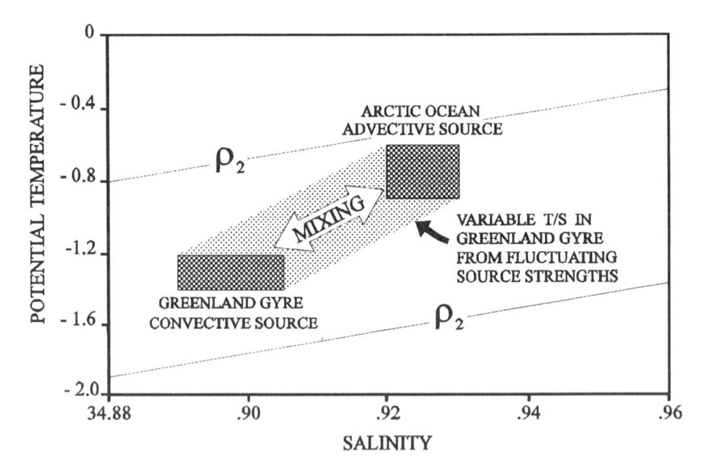

Fig. 3. Schematic representation of the competing influences on conditions in the Greenland gyre. The advection of warm and saline water from the Arctic Ocean may be relatively invariant, while the injection by convective overturn of colder and fresher water to intermediate and great depths appears to be highly variable under present climatic conditions.

inflow [*Coachman and Barnes*, 1961; *Aagaard et al.*, 1981; *Killworth and Smith*, 1984; *Björk*, 1989]; but it is genuinely interactive only with the Atlantic, with which it actually exchanges both mass and energy.

Second, there probably are boundary undercurrents along all the major topography. These boundary flows appear to be the strongest persistent currents in the otherwise generally low-velocity Arctic Ocean. They are a few tens of kilometers wide, and they are trapped over the margins of each of the major basins, including the flanks of the Lomonosov and the Nansen-Gakkel ridges. The current speed has a maximum at some intermediate depth, or even near the bottom, and the flow does not in general have a surface manifestation. Rather, these boundary currents are generally directed counterclockwise in each basin and are therefore opposite to the flow of the ice and the surface mixed layer along almost all of the continental margin surrounding the Arctic Ocean. Note, however, that our description of the boundary undercurrents is for their mean state; they characteristically show frequent reversals [*Aagaard*, 1984], possibly associated with shelf wave propagation [*Aagaard and Roach*, 1990]. We do not know how these mean boundary undercurrents arise, although *Holloway* [1987] has proposed a mechanism based on the interaction of eddies with topographic variations.

Third, intense small eddies appear to be ubiquitous in at

least the Canadian Basin. Typical eddy diameters are in the range 10–20 km, and their tangential speeds can exceed 30 cm s^{-1}. The eddies are generally embedded in the pycnocline or within the Atlantic layer, and their thickness does not usually exceed 400–500 m. The eddies have little or no surface manifestation; their water properties are generally anomalous, suggesting a distant origin; they are most probably formed near the margins of the Arctic Ocean; and they have very long lives, probably several years [*Newton et al.*, 1974; *Manley and Hunkins*, 1985; *D'Asaro*, 1988].

This brings us to an apparent inconsistency in the present view of the Arctic Ocean. On the one hand, below the surface mixed layer, the mean flow field in the interior Arctic Ocean appears to be extremely weak, and the internal wave energy is far less than the canonical value for the World Ocean [*Levine et al.*, 1987; *Padman and Dillon*, 1989], so that we might expect the ocean to transmit information very slowly. On the other hand, several modeling studies have suggested that at least horizontal stirring and mixing are rapid [*Killworth and Smith*, 1984; *Wallace et al.*, 1987]. In this case, "rapid" means the order of a decade. Since tracer data from the Eurasian Basin suggest that the upper ocean above the Atlantic layer (and removed from the boundary current) has a residence time of 7–14 years [*Schlosser et al.*, 1990], rapid mixing, in the decadal sense, has some support in the observed properties of the ocean. However, can we reconcile this view with the generally low energy levels in the interior? The only answer suggested so far has been that the transport and mixing are accomplished by the eddies: energetic rotating lenses of water that can drift long distances before slowly dissipating and discharging their contents into the surrounding ocean. We shall return to this issue.

Because of the strong stratification in the upper Arctic Ocean, there has not been a general concern that this ocean is at all delicately poised with respect to convection, the way the Greenland Sea appears to be [*Aagaard and Carmack*, 1989]. For example, after the issue of changes in the freshwater balance was raised nearly 20 years ago in connection with potential diversion of the Siberian rivers [*Aagaard and Coachman*, 1975], modeling studies by *Semtner* [1984] suggested that even the most extreme changes in the runoff could not reduce the stratification sufficiently to allow convection in the central Arctic Ocean. (However, we note that *Semtner's* calculations assume that there are no seasonal effects and that river inflow is instantly mixed uniformly across the interior ocean.)

However, there is another useful perspective on this issue. Consider an extraordinary feature of the Arctic Ocean: despite the freshwater discharge onto the shelves, it is the deep basins that are permanently stratified by a cap of low-salinity water, not the shelves. Curiously also, the Eurasian Basin is much less stratified than the Canadian Basin, despite the proximity of the former to the massive runoff from the Eurasian land mass. In contrast, the shelf waters, although stratified in summer, are relatively well mixed during the winter, and are in fact in many places homogeneous then. In effect, the central basins of the Arctic Ocean below about 50 m, which is near the maximum thickness of the surface mixed layer, are permanently insulated from local surface processes, except for the sinking of particulate matter. In contrast, the waters of the vast shelf areas along the periphery of the Arctic Ocean are thoroughly mixed on a seasonal basis, primarily by the surface density flux associated with freezing. These shelf areas are then able to communicate with the interior ocean. The result is that the Arctic Ocean is ventilated laterally from its shelves, circumventing the strong upper-ocean stratification in the interior ocean (cf. *Aagaard et al.* [1981] for an early discussion).

While we do not have a detailed understanding of the mechanisms responsible for the transfer of shelf water properties into the interior ocean, the signatures of this transfer are unmistakable throughout the Arctic Ocean, e.g., in the temperature-salinity structure of the halocline [*Aagaard et al.*, 1981]; in the nutrient and dissolved oxygen distributions [*Nikiforov et al.*, 1966; *Wallace et al.*, 1987]; and in its natural radionuclide content [*Moore and Smith*, 1986]. All these point toward large lateral influxes of shelf waters to the central basins of the Arctic Ocean. The influxes originate in the formation of dense waters on the shelves during winter, through cooling and through the brine rejection associated with freezing, and we can discern several different major sources of the shelf waters for the interior ocean. The Canadian Basin source represents low-salinity Pacific waters that have entered the Arctic Ocean through Bering Strait and have been cooled and had their salinity increased somewhat on the shelf; the Eurasian Basin source represents more saline waters from the North Atlantic that have been cooled and mixed on the western Eurasian shelves [*Jones and Anderson*, 1986; *Aagaard and Carmack*, 1989]. Tritium and helium distributions suggest that the spreading of the shelf waters within the upper Arctic Ocean is completed within about ten years of their injection [*Schlosser et al.*, in press].

Finally, we note that shelf-basin exchange is a two-way process, and that not only may near-surface waters in the deep basin move onshore to replace the denser waters injected into the halocline in winter, but deeper waters may also move onto the shelves through upwelling [*Coachman and Barnes*, 1962; *Mountain et al.*, 1976]. The potential importance of such upwelling is that it can bring deeper saline waters to shelf sites where they can be cooled by sur-

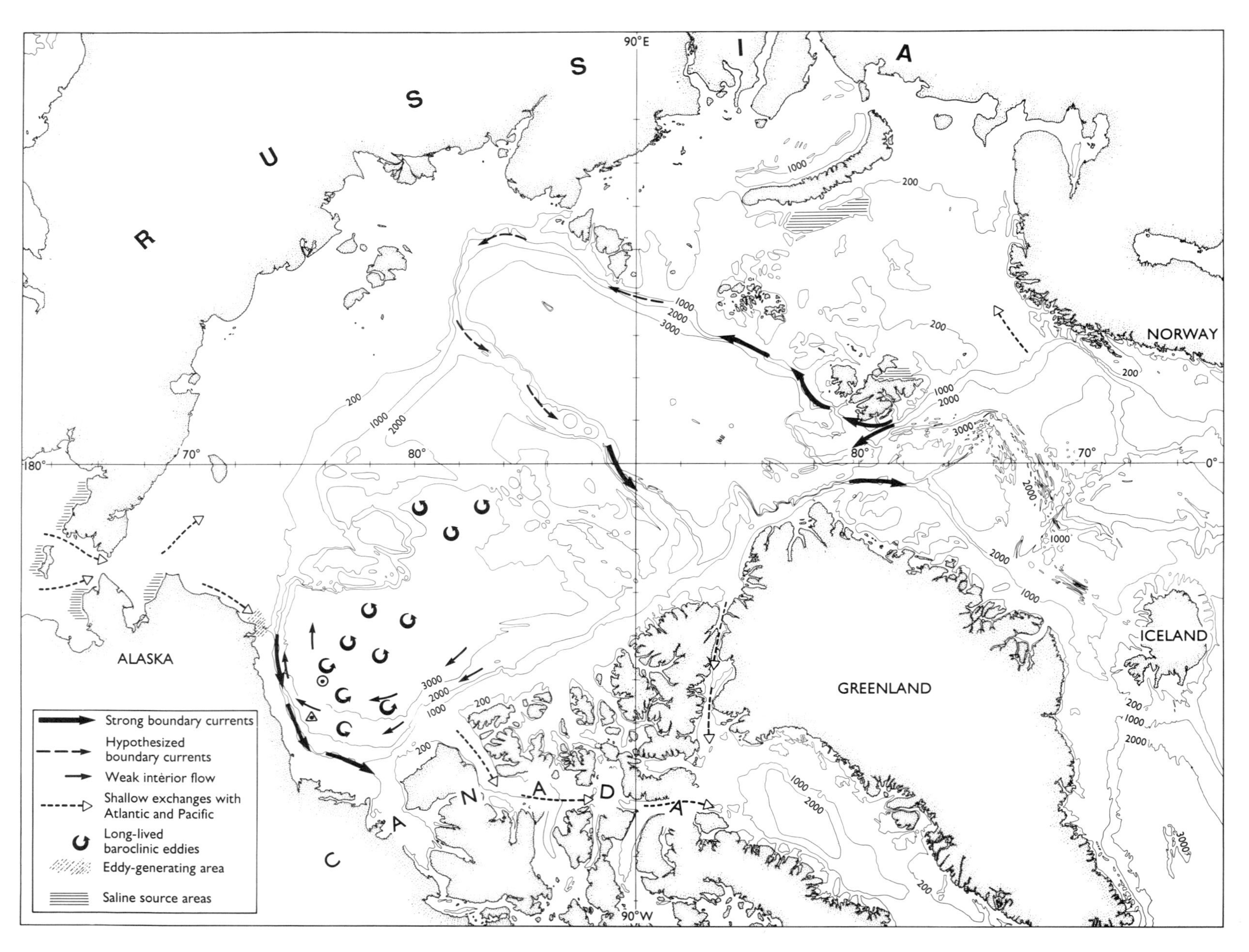

Fig. 4. Schematic subsurface circulation in the Arctic Ocean. Exchanges with the seas to the south generally extend to the sea surface. Known formation areas for saline shelf water and an eddy-generating area are also shown. Location of temperature-salinity measurements shown in Figures 6–9 is indicated by the circle. Location of current measurements shown in Figure 11 is indicated by the triangle. Adapted from *Aagaard* [1989].

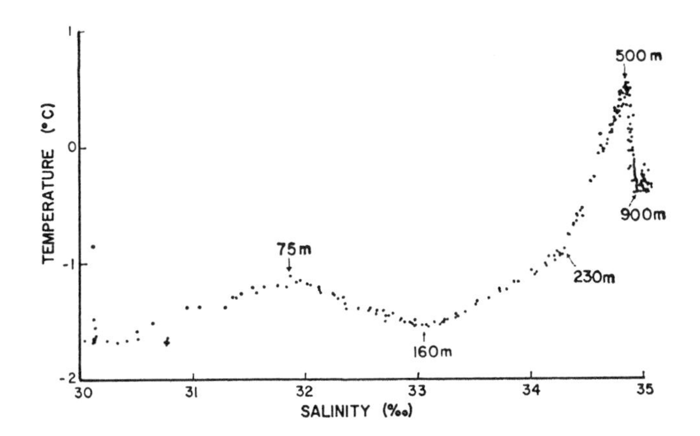

Fig. 5. Temperature-salinity diagram for the Canadian Basin based on three stations occupied from Ice Island T-3 during 1968–69. From *Kinney et al.* [1970].

face exchange and returned to the deep basins with their density increased [*Aagaard et al.*, 1981]. However, the efficacy of the process is not yet clear, and measurements by *Aagaard and Roach* [1990] suggest that the net fluxes associated with upwelling are small, except possibly in the upper reaches of selected canyons.

2.2 *Eddies and the Variability of the T/S Structure*

Figure 5, from a seminal paper by *Kinney et al.* [1970], shows the typical temperature-salinity structure of the Canadian Basin. Note that above the temperature minimum near 33.1 in salinity, which represents the Canadian Basin shelf source mode [*Aagaard and Carmack*, 1989], the T/S correlation is negative, i.e., lower temperatures are associated with higher salinities, but below the temperature minimum, the correlation is positive, with warmer water being the more saline. The change in slope near the salinity 34.2 represents the Eurasian Basin shelf source mode [*Jones and Anderson*, 1986].

If long-lived eddies are important to transport processes, we should probably expect to find considerable horizontal inhomogeneities in the tracer fields, with a length scale equivalent to that of the eddies, for the water properties of the eddies are often anomalous with respect to their surroundings [*Newton et al.*, 1974]. How horizontally uniform, then, is the interior of the Arctic Ocean? What is the variability of the vertical structure? To make a modest start on assessing the variability of the density and velocity fields, we have maintained two ocean monitoring arrays in the southern Canadian Basin during the last few years, one over the abyssal plain and the other over the continental slope.

Figure 6 shows the hourly T/S values at a nominal depth of 83 m over a 19-month period at the deep monitoring site. This instrument was sited in the upper part of the pycnocline. The T/S correlation is not at all tight, but rather is quite complex, with clouds of outlier points. The temperature excursions are particularly large, probably because for cold water, temperature has a lesser effect on density than does salinity, and temperature excursions are therefore less constrained by buoyancy forces. As an example of such an excursion, note the large, slightly curved warm anomaly at a salinity just below 32.4. Figure 7 shows the low-passed time series at this same depth and site. The temperature and salinity axes have been scaled by the T/S correlation slope from the data of *Kinney et al.* [1970], and the salinity scale has been inverted, so that parallel changes in the temperature and salinity time series simply reflect movement in T/S space along the mean correlation curve, or equivalently, the raising or lowering of a particular isopycnal surface. We see that during the first 10 months the temperature and salinity tracked each other faithfully, but that from early July onward, water mass properties were frequently anomalous. The cooling and increased salinity at this depth during the first half of the record correspond to an isopycnal being raised (and the upper layer thinning) by some 25 m. Meanwhile the large warm anomaly that we saw in Figure 6 moved past our instruments during a 3 week period in August–September 1991. The simultaneous increase in salinity caused a net increase in density, corresponding to an upward doming of the isopycnals. During the same period the velocity record from this depth (not shown) exhibits what is likely an anticyclonic eddy, with peak tangential speeds in excess of 30 cm s^{-1} being advected past the mooring in a generally westward direction (cf. *Foldvik et al.* [1988] for a discussion of eddy identification). Based on the velocity record, the center of the eddy appears to have passed very close to the mooring. Taken together, these features suggest that the large warm anomaly in Figure 6 does in fact represent a warm-core eddy. Only a very slight trace of the eddy is visible in the velocity record from the instrument 125 m deeper (again not shown), where the isopycnals are bowed down slightly. Because of the large variance in the velocity record, we cannot distinguish the mean velocity during this period from zero, and are therefore unable to estimate the horizontal scale of the warm water anomaly directly from the current records.

Figure 8 shows the T/S correlation at 207 m at this site for the same period. The instrument was located below the temperature minimum, in the depth interval in which the temperature-salinity correlation is positive. Again note the looseness of the T/S correlation, with clouds of outliers, much as recorded at the instruments 125 m higher in the

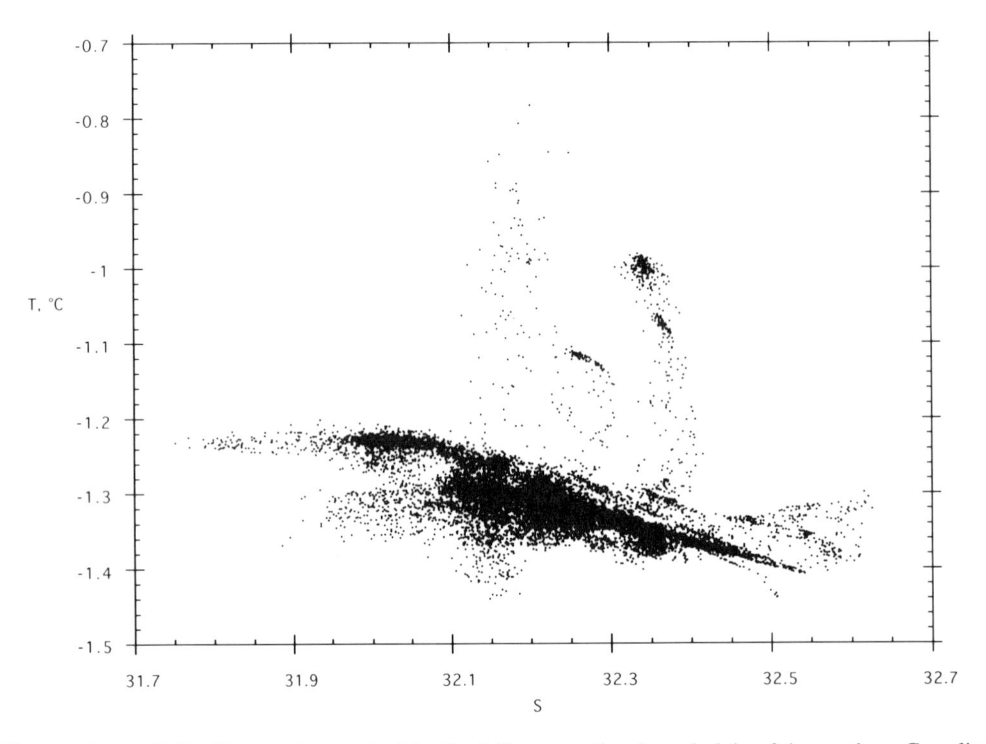

Fig. 6. Temperature-salinity diagram at a nominal depth of 83 m over the abyssal plain of the southern Canadian Basin near 72° 37'N, 143° 34'W during September 1990–April 1992. Each dot represents an hourly observation.

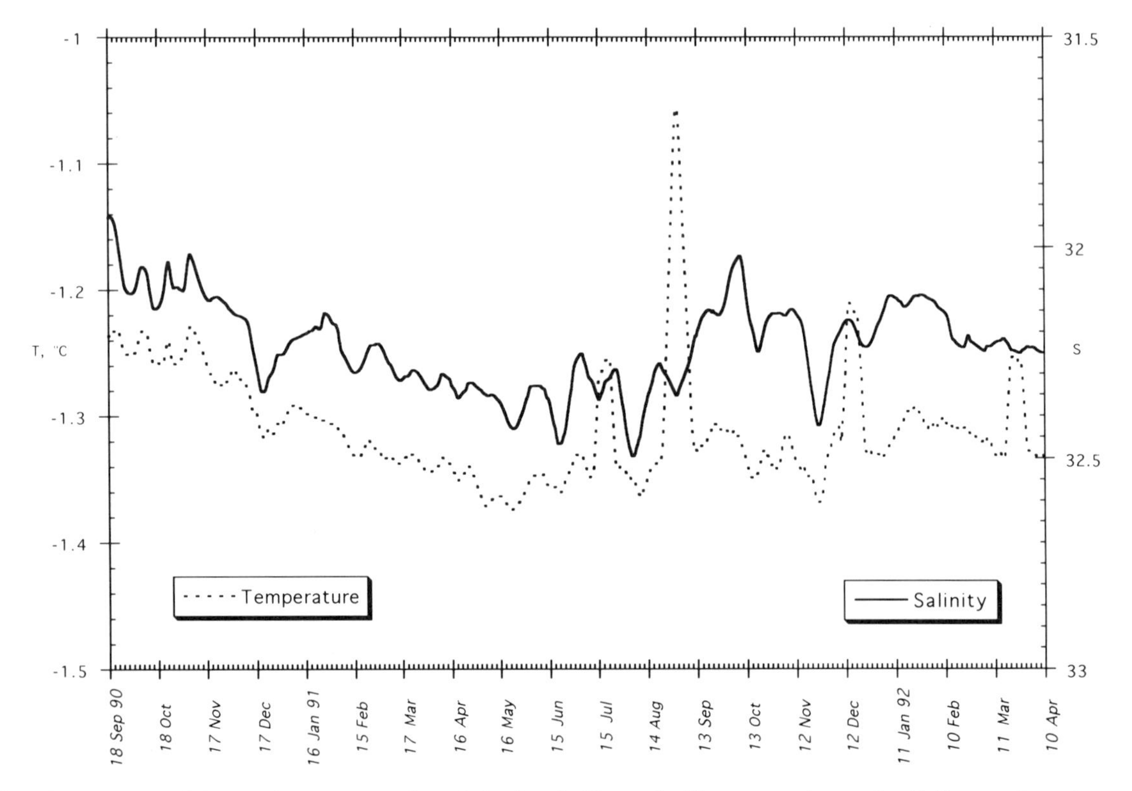

Fig. 7. Low-passed time series representation of the data in Figure 6. The temperature and salinity axes have been scaled by the slope of the correlation curve in the depth interval 75–160 m in Figure 4, and the salinity scale has been inverted.

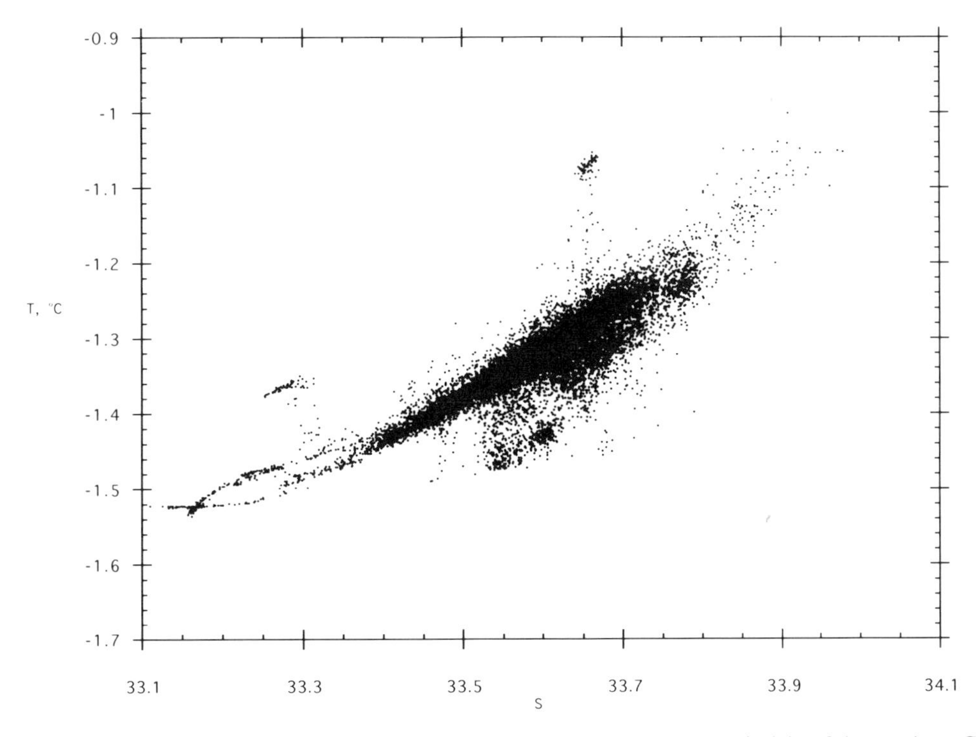

Figure 8. Temperature-salinity diagram at a nominal depth of 207 m over the abyssal plain of the southern Canadian Basin near 72° 37'N, 143° 34'W during September 1990–April 1992. Each dot represents an hourly observation.

water column. Throughout the halocline, therefore, the water mass structure appears grainy, although we do not yet know for certain the horizontal scales of these anomalous water mass structures.

The time series at this same depth (Figure 9) show the water warming and increasing in salinity during the first eight months, corresponding to the raising of an isopycnal by about 20 m, much as was the case at 83 m, or perhaps slightly less. Since the monitoring station is on the southeastern side of the Beaufort gyre, the lifting of the entire halocline over this prolonged period may represent either a northward shift of the gyre or a strengthening of the circulation, or both. We particularly note that the isopycnals did not subsequently return to the depth at which they were found at the beginning of the record.

Again, if we compare the temperature and salinity time series with that of velocity at the same depth (not shown), we see evidence of eddies. For example, the large, cold, low-salinity event in October–November 1990, corresponding to a lowering of the pycnocline at that depth, coincides with the southward drift past the mooring of an anticyclonic eddy with a peak tangential speed approaching 30 cm s^{-1}. A qualitatively similar velocity signal, but with peak speeds just under 20 cm s^{-1}, can be seen at 82 m depth, but is accompanied there by a cold high-salinity event that corresponds to a raising of the isopycnals at that depth. In other words, the isopycnals bulge upward above some intermediate depth, perhaps 150 m or so, and downward below that depth. This precisely matches the first description of these baroclinic eddies embedded in the pycnocline, provided some 20 years ago by *Newton et al.* [1974] (Figure 10).

Figure 11 shows the daily mean velocity records for 1991–92 at four depths at another site on the southern Canadian Basin abyssal plain. Peak daily mean values are as high as 40 cm s^{-1}, notably in the anticyclonic eddy (indicated by the upward doming of the isopycnals at 98 m, based on the temperature and salinity records [not shown]) that moved northward past the site during early January. This eddy is still strongly manifested in the current record from 161 m. We have temperature and salinity records from only the one depth (98 m) at this site, but if we assume that the net motion in the Atlantic layer was also northward during this same period, the current meter record from 417 m suggests a second eddy underlying the upper one. This deeper eddy is cyclonic. Vertical pairing of oppositely rotating eddies has previously been suggested by *D'Asaro* [1988]. Evidence of other eddies, both cyclonic and anticyclonic, appears throughout the records, often in counter-rotating combinations. For example, in the upper layer during mid-October a cyclonic eddy propagated westward past the mooring and was followed during late October by an anticyclonic eddy. Similarly, an anticyclonic-cyclonic

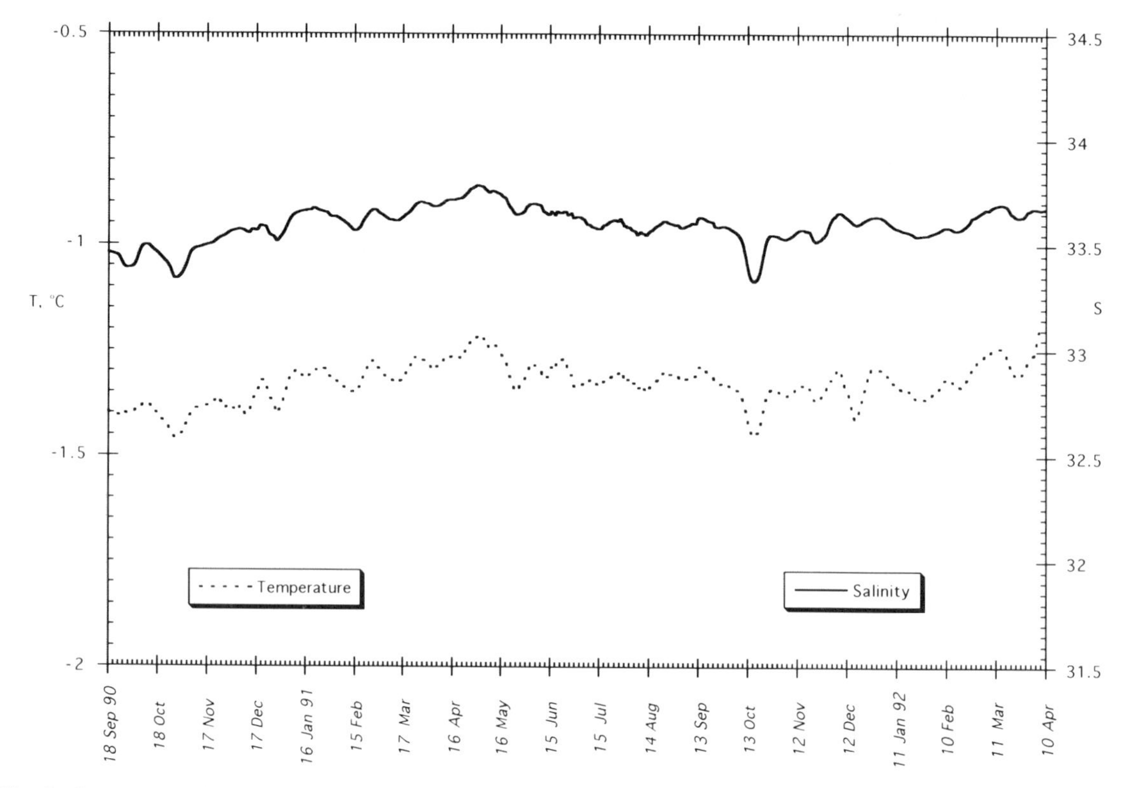

Fig. 9. Low-passed time series representation of the data in Figure 8. The temperature and salinity axes have been scaled by the slope of the correlation curve in the depth interval 160–230 m in Figure 4.

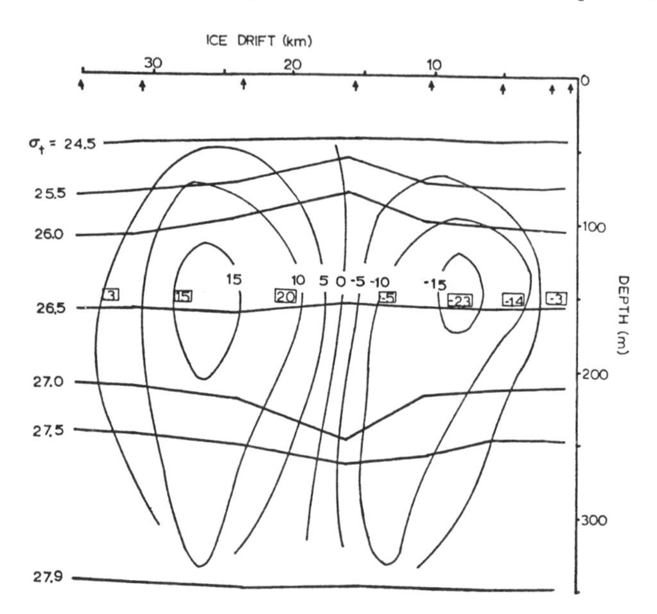

Fig. 10. Isotachs in a vertical section through an anticyclonic baroclinic eddy embedded in the Arctic Ocean pycnocline. Contours are computed geostrophic flow relative to 20 m, and numbers in boxes are measured current differences between the velocity at 150 m and 30 m. From *Newton et al.* [1974].

pair appears to have passed southward during late February–early March and a cyclonic-anticyclonic pair in a complicated translation during April. Note that speeds in the Atlantic layer, which were monitored by the lowest instrument, can be quite high, approaching 20 cm s^{-1} in the daily mean and sustained above 5 cm s^{-1} for a month or more at a time, but followed by similarly long periods of flow below the instrument threshold. On the whole, the periods of high kinetic energy tend to occur during the same general period at all depths, and are generally associated with eddy activity. The record-length mean speeds are all less than 2 cm s^{-1} and are directed northward.

What we see in the various records from these monitoring arrays, then, is an interior ocean with considerable horizontal and vertical structure, much of it associated with eddies, and therefore probably with horizontal scales of a few tens of kilometers. The eddies are frequently paired, either horizontally or vertically, and they contain most of the kinetic energy below the mixed layer. Long-term mean speeds are only a few cm s^{-1}.

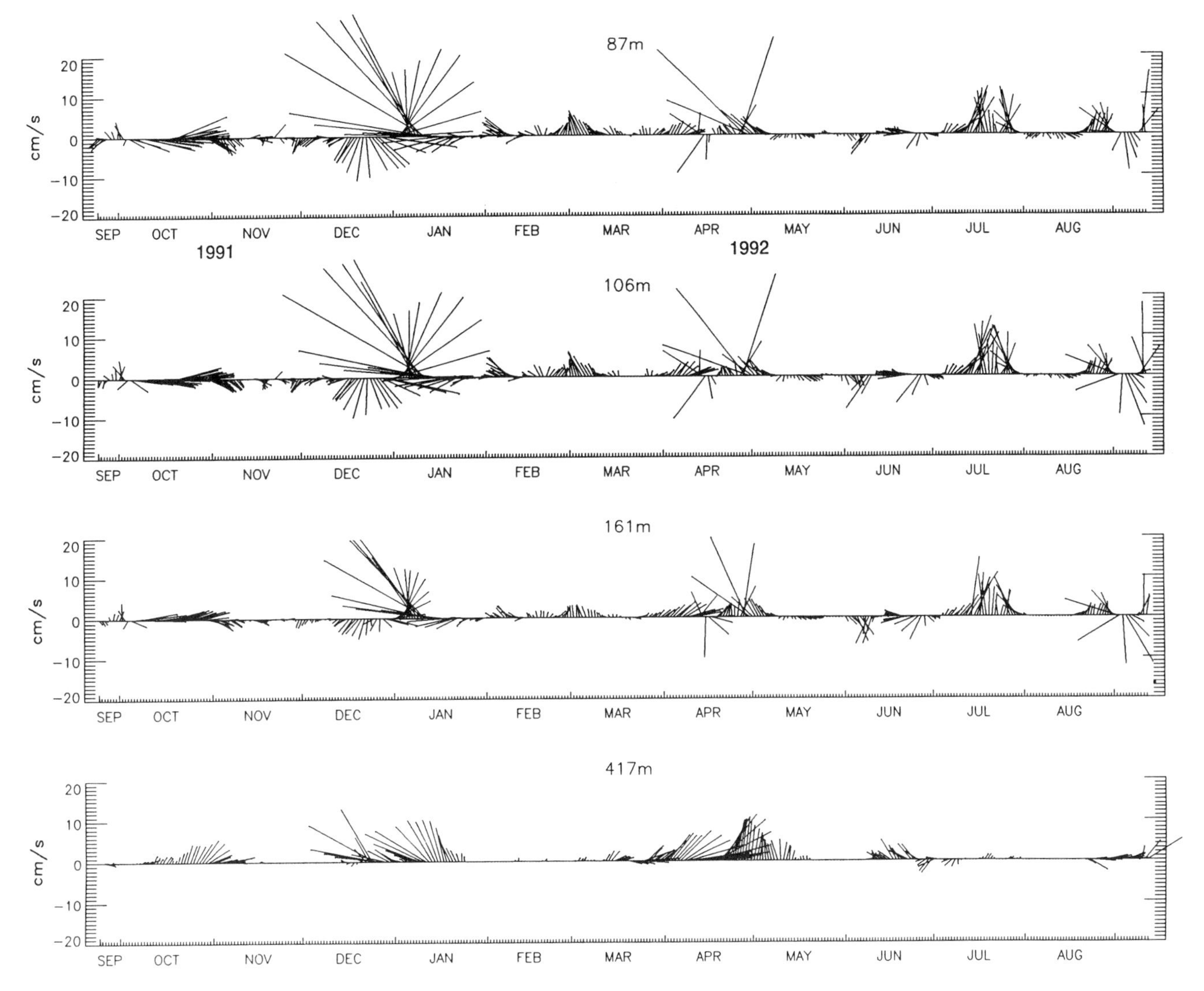

Fig. 11. Daily mean currents during September 1991–September 1992 at four depths at a site on the southern Canadian Basin abyssal plain near 71° 22'N, 141° 42'W. Vectors show direction toward which current is setting.

2.3 *Variability of Shelf-driven Ventilation*

We return now to the convection issue. The general mechanism that has been proposed to move waters off the shelves is the sinking of dense plumes along the continental slope, which eventually detach from the bottom and may then move into the interior at a level appropriate to their density [*Carmack and Killworth*, 1978; *Aagaard et al.*, 1985]. These dense outflows from the shelf appear to be episodic [*Schumacher et al.*, 1983; *Aagaard et al.*, 1985], and they may not occur every winter [*Aagaard and Roach*, 1990]. For example, in the Canadian Beaufort Sea, recent observations by *Melling* [1993] and modeling by *Omstedt et al.* [in press] suggest that a preconditioning of the shelf by prolonged wind-driven flushing the previous fall is required for dense outflow to occur. This apparent switching on and off of the shelf-generated convection, which renews the waters of the interior on a variety of time scales, indicates that the ventilation of much of the Arctic Ocean from the shelves, and the isolation of parts of the deep Arctic Ocean from these same surface processes, may not be robust, but may rather represent a "flickering switch" mode, somewhat analogous to that which has very recently been proposed for the atmospheric circulation by *Taylor et al.* [1993] on the basis of the Greenland ice core records.

Nevertheless, the various lines of evidence also suggest that on the time scale of a decade the halocline of the Arctic Ocean is presently being effectively ventilated, despite the capping of the deep Polar Basin itself by low-salinity surface waters. However, the deeper waters are another matter. In particular, the Canadian Basin provides an important counterpoint, since the very large portion of the Arctic Ocean on the North American side of the Lomonosov Ridge is either not presently being ventilated at all below sill depth [*Macdonald and Carmack*, 1991; *Macdonald et al.*, 1993], or only extremely slowly [*Östlund et al.*, 1987]. We do not know which of these is actually the case. However, if we accept the argument of *Macdonald et al.* [1993], based on a one-dimensional model of the ^{14}C distribution, that the Canadian Basin was last ventilated about 500 years ago, we are faced with an interesting question: Why did the ventilation stop, unless the shelf convection is not really robust? After all, the deep-sea climate record suggests that the Arctic Ocean has been perennially ice-covered for hundreds of thousands of years [*Herman et al.*, 1989]. What happened 500 years ago?

On occasion, human history may provide an insight into the workings of the climate machinery, and perhaps that is the case in this matter. About a thousand years ago a great migration of whale-hunting peoples began: the so-called Thule culture [*McGhee*, 1984]. These people rapidly spread eastward from northern Alaska to Greenland, following the bowhead whale, which they hunted in open water during summer [*Savelle and McCartney*, 1991]. About 500 years ago this whale hunting period came to an end. The people switched from winter houses built around whale bones to snow houses, and they abandoned whale hunting for seals [*McCartney and Savelle*, 1993]. Both the rise and the demise of the Thule culture have been attributed to changes in climate, and in particular to changes in the open water conditions during summer [*McGhee*, 1984], for the bowhead whale requires open water for its migration and feeding. We expect that this same absence of ice during summer in the Thule period would have been conducive to the generation of dense water on the shelf the following winter, both because of the reduced amount of fresh water (ice) on the shelf and because of the large potential for new ice formation. We therefore hypothesize that a decrease in open water during summer around the rim of the Canadian Basin about 500 years ago ended not only the Thule culture, but also the ventilation of the deep Canadian Basin. During the previous 500 years, the same shelf conditions that had kept the area ice free in summer and had permitted both whales and whale hunters to thrive, had also preconditioned the shelf in summer, setting it up for extensive freezing in the fall and early winter, thereby driving convection capable of ventilating the deep basin.

While our scenario is highly conjectural, it is quite in keeping with the message of change that Nansen himself preached on numerous occasions. For example, in a lecture on the Fram drift delivered in 1897 [*Nansen*, 1942], he ended with these words: "Everything is drifting, the whole ocean moves ceaselessly...a link in...Nature's never ending cycle...just as shifting and transitory as the human theories."

Acknowledgments. R. Macdonald has consistently and enthusiastically championed our mooring work in the face of competing demands for funding and ship time; C. Darnall, J. Johnson, T. Juhasz, and D. Sieberg were technically responsible for the monitoring arrays; and it all came together because of the wholehearted support of Capt. D. Johns of the Canadian Coast Guard and the officers and crew of the CGC *Henry Larsen*. C. Darnall and A. Roach processed and graphed the time series measurements. Funding for these measurements has come from the U.S. Office of Naval Research (Grants N00014-92-F-0029 and N00014-93-1-0095), the Canadian Department of Indian and Northern Affairs (NOGAP B.6 Project), and the Canadian Department of Fisheries and Oceans. Additional salary support for KA for the preparation of this manuscript came from the U.S. National Science Foundation (Grant DPP9220635) under the Arctic System Science (ARCSS) program.

REFERENCES

Aagaard. K., The Beaufort Undercurrent, in *The Alaskan Beaufort Sea: The Ecosystems and Environment*, edited by P. Barnes and E. Reimnitz, pp. 47–71, Academic Press, New York, 1984.

Aagaard, K., A synthesis of the Arctic Ocean circulation, *Rapp. P.-V. Reun. Cons. Int. Explor. Mer*, *188*, 11–22, 1989.

Aagaard, K., and E.C. Carmack, The role of sea ice and other fresh water in the arctic circulation, *J. Geophys. Res.*, *94*, 14485–14498, 1989.

Aagaard, K., and L.K. Coachman, Toward an ice-free Arctic Ocean, *Eos 56*, 484–486, 1975.

Aagaard, K., and A.T. Roach, Arctic ocean-shelf exchange: Measurements in Barrow Canyon, *J. Geophys. Res.*, *95*, 18,163–18,175, 1990.

Aagaard, K., L.K. Coachman, and E.C. Carmack, On the halocline of the Arctic Ocean, *Deep-Sea Res.*, *28*, 529–545, 1981.

Aagaard, K., J.H. Swift, and E.C. Carmack, Thermohaline circulation in the Arctic Mediterranean seas, *J. Geophys. Res.*, *90*, 4833–4846, 1985.

Aagaard, K., E. Fahrbach, J. Meincke, and J.H. Swift, Saline outflow from the Arctic Ocean: Its contribution to the deep waters of the Greenland, Norwegian, and Iceland seas. *J. Geophys. Res.*, *96*, 20,433–20,441, 1991.

Björk, G., A one-dimensional time-dependent model for the vertical stratification of the upper Arctic Ocean, *J. Phys. Ocean.*, *19*, 52–67, 1989.

Brewer, P.G., W.S. Broecker, W.J. Jenkins, P.B. Rhines, C.G. Rooth, J.H. Swift, T. Takahashi, and R.T. Williams, A climatic freshening of the deep Atlantic north of 50°N over the past 20 years, *Science, 222,* 1237–1239, 1983.

Broecker, W.S., D.M. Peteet, and D. Rind, Does the ocean–

atmosphere system have more than one stable mode of operation?, *Nature, 315*, 21–26, 1985.

Carmack, E.C., and P.D. Killworth, Formation and interleaving of abyssal water masses off Wilkes Land, Antarctica, *Deep-Sea Res., 25*, 357–369, 1978.

Coachman, L.K., and C.A. Barnes, The contribution of Bering Sea water to the Arctic Ocean, *Arctic, 14*, 147–161, 1961.

Coachman, L.K., and C.A. Barnes, Surface water in the Eurasian Basin of the Arctic Ocean, *Arctic, 15*, 251–277, 1962.

D'Asaro, E.A., Observations of small eddies in the Beaufort Sea, *J. Geophys. Res., 93*, 6669–6684, 1988.

Dickson, R.R., J. Meincke, S.-A. Malmberg, and A.J. Lee, The "Great Salinity Anomaly" in the northern North Atlantic 1968–1982, *Prog. Oceanogr., 20,* 103–151, 1988.

Foldvik, A., T. Gammelsrød, and T. Tørresen, Circulation and water masses on the southern Weddell Sea shelf, in *Oceanology of the Antarctic Continental Shelf*, Ant. Res. Ser. 43, pp. 5–20, American Geophysical Union, Washington, D.C., 1985.

Foldvik, A., K. Aagaard, and T. Tørresen, On the velocity field of the East Greenland Current, *Deep-Sea Res., 35*, 1335–1354, 1988.

Gill, A.E., Circulation and bottom water production in the Weddell Sea, *Deep-Sea Res., 20*, 111–140, 1973.

Häkkinen, S., An arctic source for The Great Salinity Anomaly: A simulation of the arctic ice–ocean system for 1955–1975, *J. Geophys. Res.*, 98, 16,397–16,410, 1993.

Herman, Y., J.K. Osmond, and B.L.K. Somayajulu, Late Neogene arctic paleoceanography: Micropaleontology, stable isotopes, and chronology, in *The Arctic Seas: Climatology, Oceanography, Geology, and Biology*, edited by Y. Herman, pp. 581–655, Van Nostrand Reinhold, New York, 1989.

Holloway, G., Systematic forcing of large-scale geophysical flows by eddy-topography interaction, *J. Fluid Mech., 184*, 463–476, 1987.

Jones, E.P., and L.G. Anderson, On the origin of the chemical properties of the Arctic Ocean halocline, *J. Geophys. Res., 91*, 10,759–10,767, 1986.

Killworth, P.D., Mixing on the Weddell Sea continental slope, *Deep-Sea Res., 24*, 427–448, 1977.

Killworth, P.D., and J.M. Smith, A one-and-a-half dimensional model of the arctic halocline, *Deep-Sea Res., 31*, 271–293, 1984.

Kinney, P., M.E. Arhelger, and D.C. Burrell, Chemical characteristics of water masses in the Amerasian Basin of the Arctic Ocean, *J. Geophys. Res., 75*, 4097–4104, 1970.

Lazier, J.R.N., Oceanographic conditions at Ocean Weather Ship *Bravo*, 1964–1974, *Atmos. Ocean, 18*, 227–238, 1980.

Lazier, J.R.N., Temperature and salinity changes in the deep Labrador Sea, 1962–1986, *Deep-Sea Res., 35*, 1247–1253, 1988.

Levine, M.D., C.A. Paulson, and J.H. Morison, Observations of internal gravity waves under the Arctic pack ice, *J. Geophys. Res., 92*, 779–782, 1987.

Macdonald, R.W., and E.C. Carmack, Age of Canada Basin deep waters: A way to estimate primary production for the Arctic Ocean, *Science, 254*, 1348–1350, 1991.

Macdonald, R.W., E.C. Carmack, and D.W.R. Wallace, Tritium and radiocarbon dating of Canada Basin deep waters, *Science, 259*, 103–104, 1993.

Manabe, S., and R.J. Stouffer, Two stable equilibria of a coupled ocean–atmosphere model, *J. Climate, 1*, 841–866, 1988.

Manley, T.O., and K. Hunkins, Mesoscale eddies of the Arctic Ocean, *J. Geophys. Res., 90*, 4911–4930, 1985.

Maykut, G.A., Large-scale heat exchange and ice production in the central Arctic, *J. Geophys. Res., 87*, 7971–7984, 1982.

Maykut, G.A., and N. Untersteiner, Some results from a time-dependent thermodynamic model of sea ice, *J. Geophys. Res., 76*, 1550–1575, 1971.

McCartney, A.P., and J.M. Savelle, Bowhead whale bones and Thule Eskimo subsistence-settlement patterns in the central Canadian Arctic, *Polar Record, 29*, 1–12, 1993.

McGhee, R., Thule prehistory of Canada, in *Handbook of North American Indians*, vol. 5 *Arctic*, edited by D. Dumas, pp. 369–376, Smithsonian Inst., Washington, D.C., 1984

Melling, H., The formation of a haline shelf front in wintertime in an ice-covered arctic sea, *Cont. Shelf Res., 13*, 1123–1147, 1993.

Moore, R.M., and J.N. Smith, Disequilibria between 226Ra, 210Pb and 210Po in the Arctic Ocean and the implications for chemical modification of the Pacific water inflow, *Earth Planet. Sci. Lett., 77*, 285–292, 1986.

Mountain, D.G., L.K. Coachman, and K. Aagaard, On the flow through Barrow Canyon, *J. Phys. Oceanogr., 6*, 461–470, 1976.

Nansen, F., The oceanography of the North Polar Basin, *Norw. N. Polar Exped., 1893–1896, Sci. Res., V* (IX), 427 pp., 1902.

Nansen, F., Foredrag om "Fram"-ferden (1897), in *Nansens Røst: Artikler og Taler av Fridtjof Nansen.1. 1884–1905*, pp. 228–252, Jacob Dybwads Forlag, Oslo, 1942.

Newton, J.L., K. Aagaard, and L.K. Coachman, Baroclinic eddies in the Arctic Ocean, *Deep-Sea Res., 21*, 707–719, 1974.

Nikiforev, Ye.G., Y.V. Belysheva, and N.I. Blinov, The structure of water masses in the eastern part of the Arctic Basin, *Oceanology, 6*, 59–64, 1966.

Omstedt, A., E.C. Carmack, and R.W. Macdonald, Modeling the seasonal cycle of salinity in the Mackenzie shelf/estuary. *J. Geophys. Res.*, in press, 1994.

Östlund, H.G., G. Possnert, and J.H. Swift, Ventilation rate of the deep Arctic Ocean from carbon 14 data, *J. Geophys. Res., 92*, 3769–3777, 1987.

Padman, L., and T.M. Dillon, Thermal microstructure and internal waves in the Canada Basin diffusive staircase, *Deep-Sea Res., 36*, 531–542, 1989.

Savelle, J.M., and A.P. McCartney, Thule Eskimo subsistence and bowhead whale procurement, in *Human Predators and Prey Mortality*, edited by M.C. Stiner, pp. 201–216, Westview Press, Boulder, CO, 1991.

Schlosser, P., G. Bönisch, B. Kromer, K.O. Münnich, and K.P. Koltermann, Ventilation rates of the waters in the Nansen Basin of the Arctic Ocean derived by a multi-tracer approach, *J. Geophys. Res., 95*, 3265–3272, 1990.

Schlosser, P., G. Bönisch, M. Rhein, and R. Bayer, Reduction of deepwater formation in the Greenland Sea during the 1980s: Evidence from tracer data, *Science*, *251*, 1054–1056, 1991.

Schlosser, P., D. Grabitz, R. Fairbanks, and G. Bönisch, Arctic river-runoff: Mean residence time on the shelves and in the halocline, *Deep-Sea Res.*, in press, 1994.

Schumacher, J.D., K. Aagaard, C.H. Pease, and R.B. Tripp, Effects of a shelf polynya on flow and water properties in the northern Bering Sea, *J. Geophys. Res.*, *88*, 2723–2732, 1983.

Semtner, A.J., The climatic response of the Arctic Ocean to Soviet river diversions, *Clim. Change*, *6*, 109–130, 1984.

Taylor, K.C., G.W. Lamorey, G.A Doyle, R.B. Alley, P.M. Grootes, P.A Mayewski, J.W.C. White, and L.K. Barlow, The 'flickering switch' of late Pleistocene climate change, *Nature*, *361*, 432–436, 1993

Wallace, D.W., R.M. Moore, and E.P. Jones, Ventilation of the Arctic Ocean cold halocline: Rates of diapycnal and isopycnal transport, oxygen utilization, and primary production inferred using chlorofluoromethane distribution, *Deep-Sea Res.*, *34*, 1957–1980, 1987.

Walsh, J.E., and W.L. Chapman, Arctic contribution to upper-ocean variability in the North Atlantic, *J. Climate*, *3*, 1462–1473, 1990.

K. Aagaard, Applied Physics Laboratory, University of Washington, HN-10, Seattle, WA 98195

E. C. Carmack, Institute of Ocean Sciences, P.O. Box 6000, Sidney, British Columbia, V8L 4B2, Canada

A Review of Coupled Ice-Ocean Models

George L. Mellor

Princeton University, Princeton New Jersey

Sirpa Häkkinen

NASA Goddard Space Flight Center, Greenbelt Maryland

This is a review of sea ice-ocean models which are used or will be used as components of global, coupled atmosphere-ocean-land-ice models. Until recent times, climate modeling has conformed to the view that the sea ice cover is a passive participant in climate change through variability in albedo. However, future climate models will couple dynamic models for atmosphere, ocean, land and ice in which sea ice and its coupling to the ocean will incorporate relatively sophisticated physics. Using numerical ocean models, research goals are to understand the numerous interactions between lateral and vertical exchanges of heat and salt that dominate the high latitude ice-ocean system and to understand how high latitude oceans interact with the global ocean or, more generally, the climate system. The ultimate goal is to predict climate change as a function of natural and anthropogenic forcing.

Our more limited objective here is to review models and modeling studies for the Arctic and its peripheral seas. We have assembled information on algorithms to model ice dynamics and thermodynamics, to model the underlying ocean and to compare various modeling strategies in an understandable manner. Furthermore, we restrict detailed discussion to three-dimensional, primitive equation numerical ocean models which acknowledge temperature and salinity variations and which have been used or are currently in preparation for use in Arctic studies in a coupled ice-ocean mode. We will focus specifically on these types of ocean models because the inclusion of thermodynamics are important for short time scale as well as climate problems where buoyancy forcing and the advection of salt and heat become integral parts of the problem. For other types of ice-ocean models, mainly limited to mesoscale dynamics, the reader is referred to a review by *Häkkinen* [1990]. A more recent review by *Barry et al* [1993] focuses mainly on observations but does include discussion of models.

1. INTRODUCTION

The Arctic Ocean, together with the Greenland, Iceland, Norwegian and Barents Seas, comprise a unique system where a large portion of the world ocean deep waters are formed. River runoff, Atlantic and Pacific inflow, ice export and surface fluxes determine the overall stratification structure. The fresh water component of the lateral flows is especially important because of its stabilizing effect; in the Arctic Ocean, it prevents heat exchange between upper and deeper parts of the water column; in the Greenland Sea, an excess fresh water cap in the form of ice can prohibit the renewal of deep waters.

Considering global climate issues, a critical element is the oceanic thermohaline circulation which is important in redistributing heat. But it is not at all clear how this thermohaline circulation is maintained. It is presently believed that the major source of the deep and bottom water is in the Greenland Sea, but water mass modification occurs everywhere at the high-latitudes and the relative contribution of these other water masses is not clear. Deep water formation is not a straightforward process of cooling of the high salinity Atlantic waters, because the deep water salinities are quite fresh compared to the North Atlantic surface salinities. Mixing of fresh water masses is essential to the small scale processes through which the deep water formation takes place; however, too much fresh water can

The Polar Oceans and Their Role in Shaping the Global Environment
Geophysical Monograph 85

impede deep convection. Fresh water masses on the other hand are carried to the Greenland Sea both as ice and seawater by the Transpolar Drift Stream/East Greenland Current.

The Arctic can also be an active participant in climate change by modifying the upper ocean properties through variability in ice formation. This variability will result in upper ocean salinity anomalies which eventually propagate to the Greenland Sea supporting larger ice cover than on average. If they are accompanied with a large ice export event, a major modification of ocean stratification occurs not only locally but even further downstream in the subpolar gyre and in midlatitudes.

A tool that has proven and undoubtedly will prove useful towards understanding and simulating the Arctic Ocean and, indeed, the climate system is numerical modeling. There has been evolutionary progress toward completely coupled, ice-ocean numerical models. At first, decoupled models were the general rule. *Semtner* [1976*a*] performed an ocean modeling study of the Arctic basin wherein ice properties were prescribed, and the general circulation was calculated by his model. On the other hand, there have been many more modeling studies where ice was explicitly modeled and where oceanic parameters were prescribed. *Maykut and Untersteiner* [1971] developed a one-dimensional, thermodynamic ice model which was simplified and shown by *Semtner* [1976*b*] to perform well even if the ice is represented by a low-resolution vertical grid. *Parkinson and Washington* [1979] used Semtner's model and a simplified ice dynamics model to simulate the yearly ice cycle in the Arctic and Antarctic. *Hibler* [1979] developed a horizontally two-dimensional, transport model of the Arctic basin which exhibited realistic properties; he used ice growth rates which were prescribed *a priori* as a function of ice thickness and time of the year. A one-dimensional, bulk mixed layer model has recently been developed by *Lemke* [1987] and coupled to Semtner's ice model. *Mellor and Kantha* [1989] coupled a one-dimensional, turbulence closure model to an ice model; the ice model also explicitly recognized the important ice to ocean heat transfer role of open leads.

The papers by *Hibler and Bryan* [1987] and *Semtner* [1987] brought forth simulation studies of the Arctic basin using a full three-dimensional, coupled ice-ocean model. It is apparent from these papers that the oceanic heat distribution does play an important role in determining the location of the marginal ice zone and the other ice cover properties such as mean ice thickness and concentration. These models did not incorporate mixed layer physics even though, as remarked by *Hibler and Bryan*, the mixed layer ought to be considered the essential coupling medium between ice and ocean. *Häkkinen and Mellor* [1992] constructed a relatively complete, three-dimensional, coupled ice-ocean model which incorporated coupling with the mixed-layer physics developed by *Mellor and Kantha* [1989].

In this paper, we assemble information on models useful for modeling the Arctic Ocean. Until the last decade or two, modeling studies have been somewhat idealized and lacking in adequate resolution. Due to the availability of increasingly affordable and powerful computer resources, numerical modeling is now entering an age where one can expect ever more realistic simulations of oceans. Observational studies will, of course, continue to be important - witness the complexity of modeling ice-ocean systems - but they will share partnership with modeling research.

We first describe separately the nuts and bolts of ice models and then ocean models in section 3 and provide rudimentary taxonomies. In section 4, we discuss ice-ocean modeling studies that have appeared in the literature.

2. ICE MODELS

We first state the governing ice equations which include the effects of the ice concentration variable, A, a statistical quantity denoting the fractional area covered by ice, whereas $(1 - A)$ is the fractional open water so that $0 \le A \le 1$. We define $\tilde{h}$ to be the local ice floe thickness whereas h_I is the average ice thickness where the average is taken only over ice covered water. Thus, Ah_I is the average thickness over the total area. This is a relatively simplistic description of an ice field, compared to studies by *Thorndike et al.* [1975], *Hibler* [1980], and *Maykut* [1982], wherein one attempts to model a thickness distribution function, $g(\tilde{h})$ where $\tilde{h}$ is the local thickness. Instead, we deal only with the concentration, A, and the moment

$$h_o = \int_0^\infty \tilde{h} g(\tilde{h}) d\tilde{h} \tag{1}$$

Nevertheless, one should keep in mind that a distribution, $g(\tilde{h})$, does exist. For example the effective ice thickness applicable to the conduction of heat through solid ice is less [*Mellor and Kantha* 1989] than that given by (1).

We also note that ice models that will be discussed below are essentially slab models where vertical structure, except that necessary to minimally model the conduction process [*Semtner*, 1976b] is neglected. Thermodynamic vertical structure was included in the pioneering studies by *Maykut and Untersteiner* [1972], but they neglected ice concentration (so that A = 1) and horizontal variability.

The Momentum Equation

At a given horizontal point, the ice velocity is characterized by a velocity vector, U_{Ii} $(i = x, y)$. The dynamic equation applicable to open water ($A = 0$), an ensemble of ice floes ($A < 1$), or a solid ice sheet ($A = 1$) is

$$\begin{aligned}\frac{\partial}{\partial t}(Ah_I U_{Ii}) + \frac{\partial}{\partial x_k}(Ah_I U_{Ik} U_{Ii}) - Ah_I \varepsilon_{ijk} f_j U_{Ik} \\ = -Agh_I \frac{\partial \zeta}{\partial x_i} + \frac{1}{\rho_I}\frac{\partial \sigma_{ij}}{\partial x_j} + \frac{A}{\rho_I}(\tau_{Ali} - \tau_{IOi})\end{aligned} \tag{2}$$

where $f_i = (0, 0, f)$ and f is the Coriolis parameter; ρ_I is the density of ice; ζ is the sea surface elevation and σ_{ij} is the internal ice stress tensor; τ_{AIi} is the atmospheric wind stress; τ_{IOi} is the ice-ocean interfacial stress; and g is the gravity constant. τ_{AIi} is calculated according to bulk drag relations presuming near surface wind speeds are available.

The ice-ocean interface stress, τ_{IOi} is obtained from a bulk drag relation in a stand-alone ice model but may be calculated according to a ocean surface layer sub-model in a more complete ice ocean model.

Current models use the rheology by *Hibler* [1977] or simpler formulations [*Häkkinen* 1987, *Semtner* 1987] to provide the internal stresses, σ_{ij}. The lowest order requirement is to ensure that the tensile stress is null. For compressive stress, a parameter describing the strength of the ice will contribute to a determination of ice thickness. Nevertheless, there remain considerable uncertainty in prescribing a constitutive stress-strain relationship to cope with the complicated internal interactions of ice sheets and floes.

The Equations for Mass and Concentration

The equation for the conservation of the mass of ice is

$$\frac{\partial}{\partial t}(Ah_I)+\frac{\partial}{\partial x_i}(Ah_I U_{Ii}) = \frac{\rho_O}{\rho_I}[A(W_{IO}-W_{AI}+(1-A)W_{AO}+W_{FR}] \tag{3}$$

where the ratio of ocean to ice density, ρ_o / ρ_I converts mass fluxes to volumetric fluxes of ice on the left side with volumetric fluxes of ocean water on the right side of (3). The various volumetric fluxes are illustrated in Figure 1; they are all positive upward and correspond to melting or freezing, depending on their location. Thus W_{AI} is melt rate (positive) on the top of the ice; it is determined by balancing the latent heat of fusion with the atmospheric fluxes (sensible, latent heat, short wave and long wave radiational heat transfer) and heat conduction through the ice and snow cover. W_{AI} can also represent a freeze rate (negative) when trapped surface water refreezes in late summer. W_{IO} is the freeze rate (positive) of congelate ice at the ice-ocean interface; it is determined by a heat of fusion balance with sensible heat transfer from the ocean and heat conduction from the ice. W_{AO} is the freeze rate (constrained to be positive) in *open water* (*open water* is defined by equation (4)); it is determined by a latent heat balance with atmospheric fluxes - which use a low albedo relative to ice and snow - and sensible heat flux from the ocean. W_{FR} is the rate of ice accretion at the surface due to frazil ice growth in the water column.

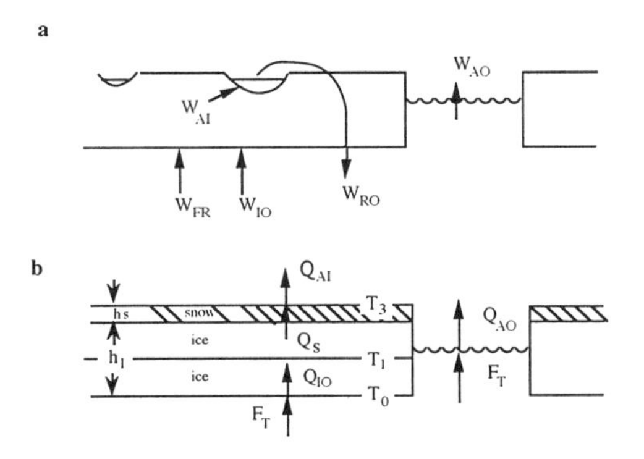

Fig. 1. a) A sketch of water to ice (or snow) volume fluxes across the ice-ocean interface, W_{IO}, the atmosphere-ice interface, W_{AI} and the atmosphere-ocean interface, W_{AO}. The W_{AI} flux can be trapped and subsequently, run off into the ocean or refrozen in late summer. W_{FR} is frazil ice flux. b) A sketch of the internal ice temperatures and the heat flux through the ice. Q_s and Q_{IO} are heat conduction fluxes at the upper snow boundary and lower ice boundary respectively. F_T is the turbulent heat flux at the upper ocean boundary; here, the fluxes under ice and open leads are assumed to be equal.

In place of the right side of (3), *Hibler* [1979] substituted a function, $S_h = f(h_I, t)\,A + f(0,t)(1 - A)$ where $f(h_I, t)$, dependent on thickness and time of year, was determined *a priori* from calculations by *Maykut and Untersteiner* [1971]. (Note that our Ah_I is equal to Hibler's h.) This is a numerically convenient procedure and appropriate to its time but it should be modified to account for greatly reduced albedo of open ocean resulting in a large amount of shortwave radiation captured by open leads.

The equation for ice concentration, A, is

$$h_I\left[\frac{\partial A}{\partial t}+\frac{\partial}{\partial x_k}\left(AU_{Ik}\right)\right]=S_A\;;\quad 0\le A\le 1 \tag{4}$$

Equation (3) is an exact conservation equation. On the other hand, equation (4), first introduced by *Nikiferov* [1957] for the null case, $S_A = 0$, is such a simple looking equation that it is necessary to keep in mind that it is empirical and is intimately associated with the (idealized) constitutive properties of ice wherein one assumes that ice cannot be horizontally dilated without creating open water. Consider a purely dynamical property of (3) and (4) for which we neglect thermodynamic forcing; i.e., the right sides of (3) and (4) are null. Then write (4) as $DA/Dt = -A\ \partial U_{Ii} / \partial x_i$. Now the divergence, $\partial U_{Ii} / \partial x_i$, is the rate of creation of new area following a Lagrangian or material surface. Consider a material ice surface where $A = 1$ when $t = 0$. After a small interval of time, Δt, equation (4), for positive divergence, yields a decrease in A according to $\Delta A = -\ \partial U_{Ii} / \partial x_i \Delta t$. Furthermore, according to a combination of (3) and (4), $Dh_I / Dt = 0$. Therefore open water is formed, but the average thickness of the ice is unchanged. On the other hand, if the divergence is negative, the constraint, $A \le 1$, overrides the differential equation, and $A = 1$ at $t = \Delta t$. Equation (3) now yields an increase in h_I according to $Dh_I / Dt = -\ h_I\ \partial U_{Ii} / \partial x_i$. This simulates ridging. Thus equation (4) is a deceptively simple but, nevertheless, remarkable relation.

Whereas the left side of (4) represents understandable physics, the right side, which must account for the role of freezing and melting on the areal size of ice floes, is problematical.

Hibler [1979] prescribed

$$S_A=\frac{(1-A)}{h_o}f(0,t)H(f(0,t))+\frac{1}{2h}S_h H(-S_h)$$

where $h_O = 0.5$ cm and the Heaviside function, $H(x) = 0$ for $x \leq 0$ and $H(x) = 1$ for $x > 0$. The first term represents the decrease in open water during freezing conditions whereas the second term accounts for the creation of open water during melting.

The prescription by *Mellor and Kantha* [1989], as modified by *Häkkinen and Mellor* [1992], is

$$S_A = 0.7A(W_{AI} - W_{IO})H(W_{AI} - W_{IO}) + 4.0(1 - A)W_{AO}$$

The constants, 0.7 and 4.0, are empirical and W_{AO} is positive definite. The first term in square brackets requires that open water be created when ice melts at the top and bottom whereas the same terms are null for freezing. The second term requires that leads be frozen over when heat is released from open water when $W_{AO} > 0$.

Thermodynamics.

Finally, an equation for the heat content of the ice is

$$\rho_I h_I \left[\frac{\partial}{\partial t} E + U_{Ik} \frac{\partial}{\partial x_k} E \right] = Q_{IO} - Q_s \qquad (5)$$

where the enthalpy, $E = E(T, r)$, is a function of temperature, T, and brine fraction, r. Following the guidance of *Semtner* [1976b] that a few grid levels adequately represented the thermal structure of ice and snow, *Mellor and Kantha* [1989] determined E based on a single average ice temperature, T_I; this is, after all, consistent with (2), (3) and (4). If one neglects the heat capacity of snow, the thermal conductivity of the snow, k_S, together with that of the ice, k_I, may be combined to yield $Q_S = k_S k_I \times (T_I - T_3)/(h_I k_S/2 + h_S k_I)$ and $Q_{IO} = k_I (T_0 - T_I)/(h_I/2)$. where h_S and h_I are snow and ice thicknesses. Some models eliminate the heat storage of the ice so that $Q_{IO} = Q_S$ and, thus, equation (5) is eliminated. As previously mentioned, by considering the ice thickness distribution, the effective thickness for conduction, h_I' is less than h_I (We estimate that $0.5 < h_I' / h_I < 0.8$) such that thin ice grows faster than thick ice. Average simulated ice thicknesses are quite sensitive to assumed values of h_I' / h_I.

The remainder of the thermodynamics is straightforward, although algebraically complicated, and involves balancing heat fluxes at snow and ice surfaces with the latent heat conversion due to melting or freezing [*Maykut and Untersteiner* 1971, *Semtner,* 1976b, *Mellor and Kantha* 1989]. Thus, the difference, $Q_{AI} - Q_S$ creates surface melting or freezing flux, W_{AI}. $Q_{IO} - F_T$ creates W_{IO}, and $Q_{AO} - F_T$ creates W_{AO}. The atmospheric fluxes, Q_{IO} and Q_{AO}, are determined from empirical formulas for sensible and latent heat fluxes and short and long wave radiation fluxes.

Although the ice thermodynamics are relatively straight forward, there is need to insert a considerable amount of empirical information such as the albedo (a function of the type of surface exposed to the atmosphere: open ocean, snow, bare ice, wet ice, etc.), the amount of water trapped in pools and the ratio of the effective heat conduction thickness to the volumetric thickness. The radiation formulas and bulk atmospheric sensible and latent heat transport all represent empirical and uncertain information. Calculated results such as ice thickness are discouragingly sensitive to variations of many of these parameters.

One-dimensional Ice Studies.

In the process of developing ice models it has been necessary to first explore the model equations and the empiricism attendant to these equations using one -dimensional, ice models; to explore *all* of parameter space using a two-dimensional ice model attached to a three-dimensional ocean model would have been a costly numerical exercise. Thus the studies due to *Maykut and Untersteiner* [1971], *Semtner* [1976b], *Lemke* [1987] and *Mellor and Kantha* [1989] have proved valuable. We wish to note here that one should not only specify oceanic heat flux - a requirement partially obviated by attaching a one-dimensional mixed layer model - but also ice divergence in these one-dimensional studies in order to approximate the mass and heat budget of the full problem. This adds to the rather ponderous list of unknown parameters. Obviously, the effect of horizontal inhomogeneity would still be missing.

A common conclusion from the one-dimensional studies is that the snow-ice system has its largest sensitivity to the surface albedo. In effect, all other variability in surface forcing components such as cloudiness, sensible and latent heat flux are secondary compared to the albedo effect. However, once one defines the albedo model, - so that, for example, the Arctic sea ice will not vanish during summer in present climate - the other forcing components, oceanic heat flux, wind stress, ice divergence, snowfall, cloudiness and air temperatures also show strong sensitivity in determining the Arctic ice mass. Considering all the uncertainties in all of the forcing components, the ice models appear to be relatively stable in giving average Arctic thickness of about 2.5 to 3.3 m when climatological radiative and turbulent heat fluxes, snow fall and oceanic heat flux representative of central Arctic are used.

We now note that, whereas the models include some vertical variability for the thermodynamics, the equations for momentum are invariably slab models; they do not allow for vertical variability in velocity, a not unreasonable simplification.

3. OCEAN MODELS

As previously stated, this discussion is restricted to three-dimensional, primitive equation numerical ocean models which include description of temperature and salinity variations; exceptions which have been used in large-scale Arctic studies in a coupled ice-ocean mode will be identified. For mesoscale ice-ocean models and ice-ocean process studies, the reader is referred to a review by *Häkkinen* [1990]. In the following we first identify model characteristics and then the specific models which adopt a cross-section of these characteristics.

Ocean Model Characteristics

We delineate the various characteristics of numerical models. These characteristics can be organized according to

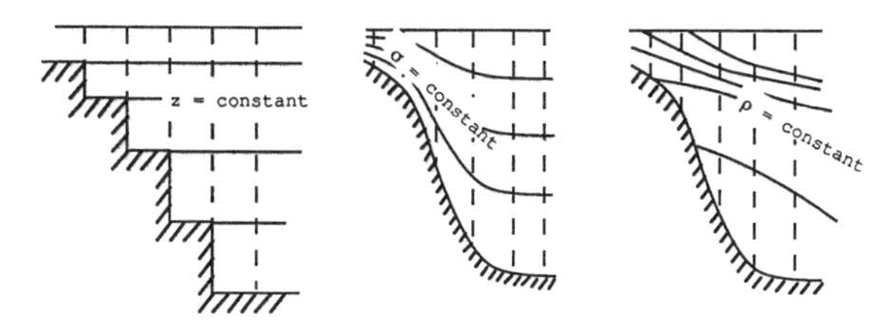

Fig. 2. Schematics of three vertical grid arrangements.

a model's vertical coordinates, horizontal coordinates, vertical and horizontal diffusional parameterizations and treatment of the sea surface.

Vertical Coordinates. Fig. 2 shows sketches of various vertical coordinate systems. The first is the so-called z-level system which is simply a rectilinear coordinate for the vertical coordinate, z. It has an apparent advantage that the coordinates are, on average, aligned with oceanic temperature, salinity and density fields. It is simple, always an advantage.

The second is the so-called sigma coordinate system where rectilinear coordinates, (x, y, z) are transformed to "sigma" coordinates, (x^*, y^*, σ) according to $x^* = x$, $y^* = y$, $\sigma = [z-\zeta(x,y)]/[H(x,y)+\zeta(x,y)]$. Here, $H(x,y)$ is the bathymetry and $\zeta(x,y)$ is the surface elevation ($\zeta = 0$ for a rigid lid model). The resulting transformed equation are not complicated. Sigma coordinates would seem to be advantageous in dealing with flows on continental shelves and whenever topographical variability is important (e.g., almost everywhere). By the same token, there is an error associated with the evaluation of horizontal density gradients in sigma coordinates. Whether or not this is a serious problem is controversial [*Haney* 1991, *Mellor et al.* 1993].

A third type of vertical coordinate is the isopycnal coordinate where instead of (x, y, z) as independent coordinates, the independent coordinates are x, y and potential density, ρ (or, equivalently, sigma-t). The advantage of this system is that high resolution is obtained near the large density gradients of important frontal regions. Conversely, vertical resolution will be poor on well mixed, continental shelves in the winter. Although we have no direct experience, a probable cost of isopycnal models is code complexity resulting from the need to cope with intersections of density surfaces with the sea surface and bottom.

Horizontal Coordinates. Fig.3 depicts a variety of horizontal coordinate systems. The first are rectilinear or spherical coordinates (both are represented by surfaces on which the two horizontal coordinates are constants). The second system is an orthogonal curvilinear coordinate which is more general and includes the rectilinear and spherical system as special cases. The added freedom of the curvilinear system allows one, for example, to follow coastlines with increased resolution (and minimize land points) and to avoid the singular behavior of spherical coordinates near the poles which is especially advantageous for the Arctic Ocean. Still more freedom is achieved with a non-orthogonal grid, sometimes called a *finite volume* system, which means that differencing is achieved by applying the integral conservation equations to the grid cell volume.

Vertical Diffusion . Mixing is a critical element of oceanic processes at high latitude oceans where a multitude of water mass modifications occur year round. Formation of deep water masses are connected to deep convection which in large-scale hydrostatic models takes place through vertical diffusion, and additionally convective adjustment maybe used. Below we discuss the several parameterizations that have been used for vertical diffusion. Generally, attention is focused on the ocean surface layer or "mixed layer" (a somewhat misleading but oceanographically ingrained term; only temperature and salinity are generally well mixed whereas momentum usually is not well mixed; we discuss this shortly) in which case, one can define two broad model categories, local models and bulk models. Local models prescribe an eddy diffusivity (we ignore here models which solve for all components of the Reynolds stress and flux tensors) and the output is the temperature, salinity and velocity profile distributions. From this information, if the eddy diffusivities are Richardson number dependent, it is possible to define the lower bound of the region of large diffusivity and therefore the depth of the mixed layer is determined diagnostically. Bulk models generally parameterize the mixed layer depth *ab initio* and assume that all properties are uniform in the mixed layer (they are mixed!) and models solve for these properties using integral

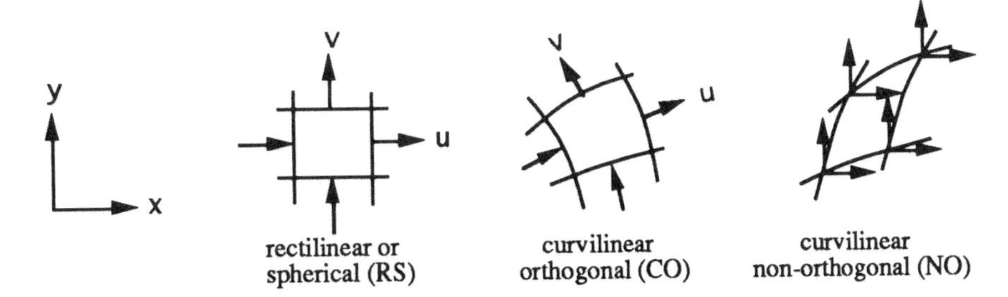

Fig. 3. Schematics of various grids. Aside from the morphology of the grid cells, the location of properties on the cells vary. Thus, the RS and CO grids illustrate "C grids" where the velocities are staggered as shown; the density and other scalars are located in the center of the cell. The NO grid is an "A or B grid" where the velocities are co-located; for A the density is co-located with velocities; for B, the density is located in the center of the cell. Note that the velocities are fixed in direction in the RS and NO grids.

conservation laws and entrainment rules. Table 1a gives the shorthand notations for the above discussed parameterizations.

Whereas temperature and salinity are generally "well mixed" in the ocean surface layers, momentum and scalars such as biochemical constituents/tracers may not be. If we assume that the diffusivity for momentum and temperature are equal, one obtains $\delta T / \delta V \approx Q / (C_p \tau)$ where δT is the change in temperature across the ocean surface layer; δV is the change in velocity; C_p is the specific heat and τ and Q are the surface wind stress and heat flux. Then, for typical values, Q = 50 w m^{-2} and τ = 1 dyne cm^{-2}, one obtains $\delta T/\delta V \approx$ 0.1 ^{o}C/(m s^{-1}). From observations or from different models one can find that $\delta V \approx$ 10 cm s^{-1} and, therefore, $\delta T \approx$ 0.01 ^{o}C. Thus, one finds that temperature is "well mixed" whereas velocity is not. Other quantities, such as carbon dioxide may similarly display important variability in the ocean surface layer. Local models [*Munk and Anderson* 1948] including the so-called turbulence closure models [*Mellor and Yamada* 1978 1982] allow one to calculate this variability and as increased computer resources enable one to afford fine vertical resolution, say 25 vertical levels or more, the ocean surface layer can be resolved with, say, five or more, near surface levels or layers.

Horizontal Diffusion. If one does a scale analysis of the basic fluid dynamic equations, terms can be deleted due to smallness of spatial variability in the vertical direction relative to the horizontal direction. This leads to the hydrostatic approximation but also to neglect of all the horizontal eddy viscosity terms. If horizontal resolution is adequate, these terms can be neglected [*Oey et al.* 1985]; vertically variable, horizontal advection together with vertical diffusion creates horizontal dispersion [Taylor 1954]. Thus, we include the very desirable null diffusivity (ND) as an option in Table 1b. However, for most present day applications with inadequate resolution, horizontal diffusion is needed to suppress numerical noise if not to prevent the model from blowing up. Thus, we also list other ways of damping the offending noise in Table 1b. Use of a constant diffusivity (CHD; which, synonymously, is a Laplacian smoother) and a velocity gradient dependent, Smagorinsky formulation which can claim a modicum of physical support. The empirical constant in the *Smagorinsky* diffusivity is at least non-dimensional and the diffusivity is proportional to the grid cell area so that it disappears as resolution is improved. The biharmonic smoother [*Holland* 1978] selectively removes small scales. Another strategy is to filter calculated fields between time steps using, say, a *Shapiro* filter [*Robinson and Walstad* 1987].

Table 1a. A listing of parameterizations for vertical eddy diffusivities.

Type	Identifier
Constant diffusivity	CVD
Convective adjustment	CA
Richardson Number modified diffusivity	RND
Turbulence Closure	TC
Bulk Mixed Layer Models	BML

Table 1b. A listing of parameterizations for horizontal eddy diffusivities.

Type	Identifier
Null Diffusivity	ND
Constant Diffusivity	CHD
Smagorinsky Diffusivity	Smag
Biharmonic Smoother	BiH
Filters	FLT

Specific Ocean Models

Table 2 is a short list of ocean models. We have constrained the list to three-dimensional, primitive equation models with full thermodynamics. The further constraint is that either the models are in use by many users or that they possess some unique characteristic. The list covers all of the modeling applications to the Arctic Ocean that are known to us. A code is assigned to each model; thus, the *Bryan-Cox-Semtner* model is denoted by BCS. The code we use, first created by *Blumberg and Mellor* [1978], is denoted POM for the Princeton Ocean Model (a label first adopted by users other than *Blumberg and Mellor*). SPEM denotes the Spectral, Primitive Equation Model since it uses a series expansion of *Chebyshev* polynomials instead of finite differences to represent vertical variability. Each of the models has a set of characteristics discussed above that define the model.

The first numerical ocean model and the present day standard is the *Bryan-Cox* model [*Bryan* 1969, *Cox*, 1984], later numerically improved by *Semtner* [1976a] so that it is sometimes called the *Bryan-Cox-Semtner* model. It is conceptually the simplest of the models, an immediate advantage, in that it uses a conventional, z-level, vertical coordinate and spherical coordinates. It is used by many large scale ocean modelers [*Bryan and Holland* 1989, *Semtner and Chervin* 1992].

The POM model by *Blumberg and Mellor* [1980], *Mellor* [1993] was initially developed for application to estuaries and coastal oceans although it is currently being applied to ocean basins. In estuarine applications, the sigma coordinate system, together with a free surface and the turbulence closure sub-model [*Mellor* 1973, *Mellor and Yamada* 1974, 1982] provided a bottom boundary which converted tidal energy into turbulence and mixing. Comparisons with current meter measurements, tide gage data and salinity intrusion into estuaries were favorable. The model's horizontal grid is an orthogonal curvilinear grid which supports rectilinear or spherical coordinates as special cases.

Table 2. A listing of three-dimensional, primitive equation models which admit full thermodynamics and density stratification.

Authors	Identifier	Vertical Grid	Horizontal Grid	Vertical Diffusivity	Horizontal Diffusivity	Surface
Blumburg-Mellor	POM	σ - C	CO/C	TC or CVD	Smag or CHD	free
Bryan-Cox-Semtner	BCS	z - C	RS/B	RND/CA[1]	CHD	rigid
Haidvogel	SPEM	σ – C/ Spect	CO/C	BML	BiH	rigid
Oberhuber	OBH	ρ- C	RS/B	BML	Smag	free

[1] The GFDL, MOM (Modular Ocean Model) version of the BCS model has an option to include a TC [Mellor and Yamada 1974 1982] vertical diffusion scheme.

A variation to the theme of sigma-coordinate models is offered by the SPEM model by *Haidvogel* [*Haidvogel et al.* 1991, *Hedstrom* 1990]. It is similar in some respects to the POM model except that it has a rigid lid and is distinguished by the fact that variables on the sigma coordinate are expanded vertically in series of Chebyshev polynomials, i.e. it is not a level model. It has been used in a number of process studies, but not yet for coupled ice-ocean studies, although such a coupled version is under development (*Hedstrom*, private communication).

The forerunner of the isopycnal layer models is the *Bleck and Boudra* model [*Bleck and Boudra* 1986, *Bleck et al.* 1992]. This particular model has not yet been coupled to sea ice, but the isopycnal formulation was adopted by *Oberhuber* [1993a] in his version of the isopycnal model with an sea ice cover using Hibler's viscous-plastic rheology and Semtner's thermodynamics. The OBH model is implicit and possibly economical for climate studies. An advantage of isopycnal coordinates is that increased resolution is automatically obtained in regions with strong density gradients. A disadvantage is the converse; for example, poor resolution is obtained in convective regions where bottom water is formed and in wintertime shelf regions. The code is complicated due to the need to cope with vanishing isopycnal layer thicknesses near the ocean surface and bottom.

4. COUPLED ICE-OCEAN, ARTIC MODELING STUDIES

In the Arctic Ocean and its peripheral seas, Greenland, Iceland, Norwegian and Barents Seas, the overall stratification structure is determined by river runoff and Atlantic and Pacific inflow and by dynamic and thermodynamic interactions with the ice cover. As a result of heat exchange with the atmosphere, deep waters are formed in this region, a process which is augmented by brine rejection due to ice formation. Alternately, since density is strongly salinity dependent, fresh water fluxes at the ocean surface are especially important because of their stabilizing effect; in the Arctic Ocean, the fresh water layer prevents heat exchange between upper and deeper parts of the water column; in the Greenland Sea, an excess fresh water cap in the form of ice can prohibit the renewal of deep waters. These processes prescribe the needs of a limited area coupled Arctic model: a good description of mixing - which poses further demands on the surface forcing - and good estimates for river runoff, lateral inflows and outflows and their spatial distribution. Either of these elements can be circumvented by using a diagnostic description for parts of the ocean model such as relaxation to climatology; however, this limits a model's usefulness for climate change studies.

Partially diagnostic models

Hibler and Bryan [1986] used the BCS ocean model and were the first to present results from a coupled ice-ocean model for the Arctic. The ocean model was partially diagnostic for the deeper ocean in that a Newtonian damping factor forced the temperatures and salinities toward climatological values, while the upper ocean could adjust prognostically to the surface forcing determined by ice freezing and melt. The main result from the simulations was to show the importance of the northward heat transport by the Norwegian Current, which is responsible for determining the ice extent in the Greenland and Barents Seas. This same heat source is responsible for year round ice melt in the Greenland Sea. The coupling also intensified the oceanic Beaufort Gyre and the East Greenland Current, which the authors described as a readjustment to the forcing because the initial salinity and temperature fields were smoothed estimates of the observations. The climatology of the Arctic ice thickness field supports high ice thicknesses north of Greenland due to the mechanical pileup of ice transported by the Transpolar Drift Stream to the vicinity. However, the model, using surface forcing from 1979, produced a highly anomalous ice thickness field with a large ice buildup along the East Siberian coast and a weak buildup north of Greenland. (A similar anomalous field resulted from 1987 surface forcing in the model by *Häkkinen and Mellor* , 1992.) Thus, the sea ice thickness field can have a large interannual variability as a response to wind forcing, and monitoring only a couple of trans-Arctic sections for ice thickness as a signal of climatic warming or cooling is unlikely to be meaningful.

This *Hibler-Bryan* model has been implemented for operational use at the Navy's Fleet Numerical Oceanography Center [*Riedlinger and Preller* 1991]. Forecasted seasonal ice trends are in good overall agreement with observations for growth and decay. However, the ice cover in the Barents and Greenland Seas appears to be somewhat excessive which the authors attribute to poor model resolution to describe narrow currents like the West Spitsbergen Current.

Another coupled ice-ocean model by *Piacsek et al.* [1991] is also partially diagnostic. The Hibler ice model is coupled to a high resolution mixed layer where turbulence is calculated according to the level 2.5, *Mellor and Yamada*, turbulence closure model. However, the deeper ocean is diagnostic with a geostrophic velocity field determined from *Levitus* [1982] climatology. Corresponding to a perpetual year, 1986 forcing, the model produces a very realistic seasonal variability, with the exception of the wintertime ice extent in the Barents Sea. They consider the inclusion of the mixed layer dynamics to give a superior ice thickness field compared to coupled models without mixed layer dynamics. The mean oceanic heat flux in the model varies in the ice-covered area from 5 to 15 W m^{-2}, attaining even larger values north-east of Spitsbergen in the area of the submerged Atlantic waters. These values are on the high side compared to the traditional view of about 2 W m^{-2}, and to even lower values, less than 0.5 W m^{-2}, as suggested by other one-dimensional modeling studies [*Mellor and Kantha* 1989, *Häkkinen and Mellor* 1990]. Comparison of observed buoy tracks and simulated drift tracks are in reasonably good agreement considering the coarse resolution of the model.

Prognostic models

The first comprehensive prognostic ocean model for the Arctic Ocean was described by *Semtner* [1976a]; he used the BCS ocean model. The model was later expanded by *Semtner* [1987] to include a dynamic-thermodynamic ice cover comprised of a three-level snow-ice system [*Semtner* 1976b] and an ice rheology simplified from the model by *Hibler* [1979]. This ice-ocean system was driven by monthly surface forcing and specified inflow-outflow fluxes at the boundaries as in Semtner's 1976 model. Overall, the results from the ice model component showed agreement with the observed ice extent but ice thicknesses were much lower compared to the generally accepted average values of 2.5 to 3m. The model predicted a modest ice growth in the Beaufort Sea Gyre and much larger growth occurring on the Eurasia coast. The model also showed that ice melt occurred in the Greenland Sea even in mid-winter. The simulated oceanic circulation produced the main circulation features in the upper ocean such as the Beaufort Gyre, Transpolar Drift Stream, East Greenland Current, Norwegian Currrent, and Barents Shelf circulation. Surface salinities were reasonably well reproduced in the model except that their gradients were not strong enough in the central Arctic. This model has been applied to the study of interannual ice variability by *Fleming and Semtner* [1991] for the period of 1971-1980. Their main conclusion is that using interannually varying forcing produces much improved sea ice cover variability for the annual cycle compared to the model forced by mean monthly climatology. The interannual forcing produced large variability in ice thickness field, with much lesser degree of variability in sea ice extent. The oceanic heat flux variability has the strongest influence in determining the monthly average ice edge positions. However, overly thin ice thickness is still a problem in the model which is. according to the authors, probably a result of an excessive melt-freeze cycle. Also, the authors consider that not having an explicit treatment of mixed layer dynamics produces inadequacies in the vertical mixing processes.

Häkkinen and Mellor [1992] used the POM ocean model together with the snow-ice model described by *Mellor and Kantha* [1989]. Using monthly climatological surface heat flux and wind stress, the seasonal variability of the model ice cover is quite realistic in that the thickest ice is located north of Greenland and the average ice thickness is about 3 meters. The largest deviation between the simulated and observed ice cover is in the Greenland Sea where oceanic conditions determine the ice edge. The monthly climatological forcing does not result in strong enough mixing to bring sufficient heat from the deep ocean to keep the central Greenland Sea gyre ice-free. The results improve for both the ice cover and the ocean structure by invoking daily wind forcing. In the ocean model, the large mixing events associated with storm passages were better resolved and, as a result, the overall oceanic structure in the Greenland Sea became more realistic. Model simulations using daily forcing for 1986 and 1987 produced quite different ice thickness fields; the 1986 ice thickness field resembled the observational climatology whereas the 1987 simulations produced anomalously thick ice in the East Siberian Sea and anomalously thin ice north of Greenland. The simulated mixed layer circulation from the model captured the main features of the upper ocean circulation such as the Beaufort Gyre and the Transpolar Drift Stream. The Atlantic waters circulate in the Arctic Basin at about the observed level, between 400 and 600 meters. However, the Atlantic water appears to weaken as it flows towards the Eurasia Basin; there, the simulated Atlantic waters are too cool by about 0.2^{o}C - 0.5^{o}C. (More recent research leads us to relate this weakening directly to a weak inflow as discussed below.)

Häkkinen [1993] used the aforementioned Arctic ice-ocean model to study the interannual variability of the sea ice during the period, 1955-1975, and to explain the large variability of the ice extent in the Greenland and Iceland Seas during the late 1960's. In particular, the model is used to test the conjecture of *Aagaard and Carmack* [1989] that the Great Salinity Anomaly was a consequence of anomalously large ice export in 1968. The model simulations explored the high latitude, ice-ocean circulation changes due to wind field changes while other forcing components such as air temperatures, cloudiness, snowfall and river runoff were climatological. The simulated ice extent in the Greenland Sea increased during the 1960's, reaching a maximum in 1968, as observed. The maxima in ice extent coincided with large pulses of ice export through the Fram Strait. The ice export event of 1968 was the largest in the simulation, being about twice as large as the simulated average for 1955-1975 and corresponding to 1600 km^3 of excess fresh water. The simulated upper water column in the Greenland Sea has a salinity minimum in the fall of 1968, followed by very low winter salinities. In addition to the above average ice export to the Greenland Sea, there was also fresh water export to support the larger than average ice cover. The total simulated fresh water input of 2500 km^3 to the Greenland Sea compare well with the estimated total fresh water excess of the Great Salinity Anomaly of about 2200 km^3 as it passed through the Labrador Sea [*Dickson et al.*, 1988].

A very recent paper by *Oberhuber* [1993b] used the OBH model for most of the Atlantic Ocean from 30°S to 90°N. Thus a sector of the Arctic Ocean was included in the model domain. The ice model was as described in section 3; however, the sensible heat storage in the ice was neglected. The model resolution is 2° by 2°. This model has been adopted into a limited area Arctic-GIN seas model by *Holland et al* (1993). Their paper presents a set of sensitivity studies of nearly all possible forcing components: They find that cloud cover is one of the least known, but most sensitive components, while snow fall (up to 2 m/year) did not appear to be crucial.

Model sensitivity to the Iceland-Norway inflow

Boundary conditions must be provided for limited model domains such as the Arctic Ocean. In the modeling study by *Häkkinen and Mellor* [1992], we used an Atlantic inflow of about 7 Sv which was converted to a constant velocity on the boundary gridpoints between Iceland and Norway. This resulted in a relatively diffuse flow at 500 m depth as shown in Fig. 4a. A consequence was that the penetration of

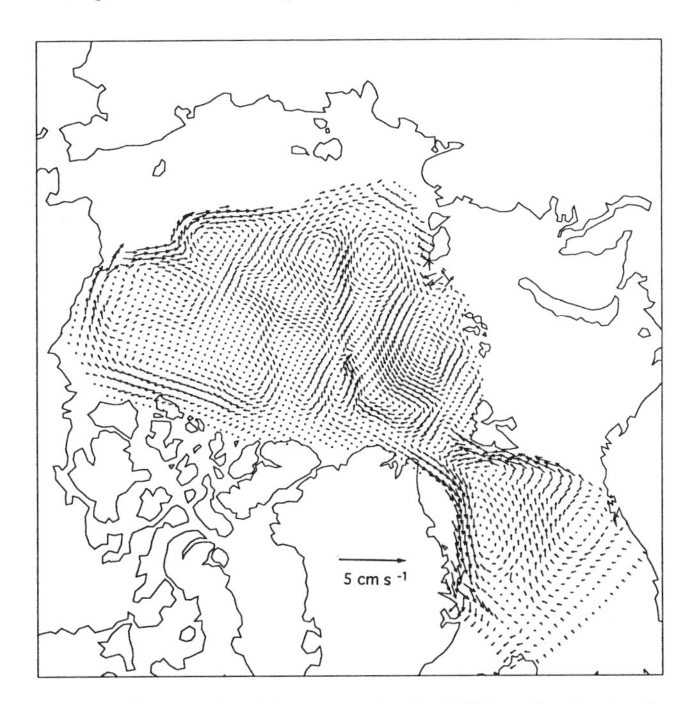

Fig. 4a. The simulated flow at a depth of 500 m for the Arctic basin wherein the inflow/outflow transport (vertically integrated velocity normal to the boundary) is uniformly distributed across the two open boundaries.

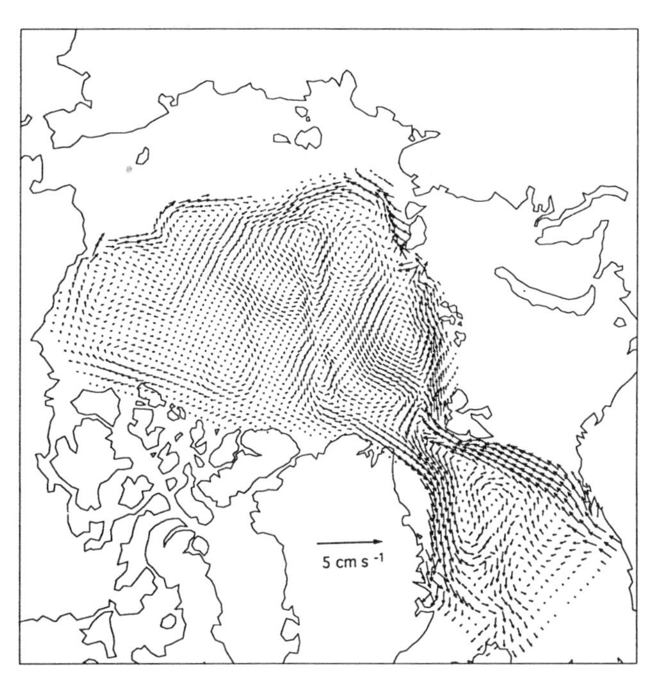

Fig. 4b. The simulated flow at a depth of 500 m for the Arctic basin wherein the inflow/outflow transport between Iceland and Norway is more concentrated along the Norwegian coast.

Atlantic water into the basin was weak. If the inflow-velocity profile (keeping the mass flow across the open boundary as 7 Sv) is specified so that it contains a strong barotropic jet at the Norwegian Coast, the results change drastically. Now, in Fig. 4b, the Atlantic flow appears quite strong north of Svalbard. It meanders into the rest of the Arctic Basin and, generally, the Atlantic water penetration is more satisfactory. A conclusion is that the limited area Arctic models are quite sensitive to the specified velocity and mass flux at the boundary to the GIN Seas area. Unfortunately, the spatial variability of the inflow velocity and mass flux between the Iceland and Norway is not known very well.

5. SUMMARY

There now exist fairly sophisticated, numerical models for the ice cover and the oceans. Presently, the ice models are conceptually complex and require a large ingestion of empirical information. Since they are basically slab models, they are not numerically or computationally demanding, however. Simulated ice thicknesses are sensitive to such parameters as snow and ice albedos, the ratio of effective ice conductivity thickness to volumetric thickness, the allowable amount of surface melt water trapped in pools, the adjustable constants in the equation for ice concentration, the way of modeling the internal ice stresses and attendant parameters and, probably, other parameters associated with processes still unknown.

The available ocean models appear to be capable of fairly realistic simulations although a lack of sufficient resolution is still evident. There are a choice of model characteristics such as the manner of discretizing the vertical coordinate. Experience has been accumulated in the use of z-level coordinates, sigma coordinates and isopycnal coordinates. The advantage of sigma coordinate system is that it allows the inclusion of shallow water and shelf processes. A somewhat separate issue is the means of coping with vertical mixing processes. The surface boundary layer is the means by which the ocean communicates with the ice and the atmosphere. Apparently, bottom water formation in the Arctic basin and in the Atlantic Ocean is a result of dense water creation at the margins of polar oceans; indeed Antarctic bottom water is formed, at rates of the order of a few Sverdrups, in the shallow regions of the Weddell Sea with possible contributions from the Ross Sea. Thus, climate modelers should be concerned with mixing processes in the bottom boundary layers as well as the surface layers. Turbulence closure, and mixing models together with sufficient vertical resolution should supply the means of coping simultaneously with surface and bottom layers. On climatic time scales, there remains the problem of specifying mixing in the interior ocean [*Gregg* 1989] which is due to energy supplied by internal waves. On the other hand, it has been hypothesized [*Armi* 1978] that interior mixing is, in reality, a manifestation of boundary mixing on sloping bottoms followed by intrusion into the interior.

Acknowledgements. The reviewers provided useful suggestions. The research was supported by NOAA's Atlantic Climate Change Program, grant N00014-93-1-0037.

REFERENCES

Aagaard, K., and E. C. Carmack, The role of sea ice and other fresh water in the Arctic circulation, *J. Geophys. Res., 94,* 14485-14498, 1989.

Aagaard, K., L.K. Coachman, and E.C. Carmack, On the halocline of the Arctic Ocean, *Deep-Sea Res., Part A, 28,* 529-545,1981.

Armi, L., Some evidence for boundary mixing in the deep ocean, *J. Geophys. Res., 83*, 1971-1979, 1978.

Barry, R.G., M.C. Serreze, J.A. Maslanik and R.H. Preller, The Arctic sea ice-climate system: observations and modeling, *Rev. Geophys., 31,* 397-422, 1993.

Bleck, R. and D.B. Boudra, Wind-driven spin-up in eddy resolving ocean models formulated in isopycnic and isobaric coordinates, *J. Geophys. Res., 91,* 7611-7621, 1986

Bleck, R., C. Rooth, D. Hu, and L.T. Smith, Salinity driven thermocline transients in a wind - and thermohaline forced isopycnic coordinate model of the North Atlantic, *J. Phys Oceanogr., 22,* 1486- 1505, 1992.

Blumberg, A.F. and G.L. Mellor, A coastal ocean numerical model, in Mathematical Modelling of Estuarine \Physics, *Proc. Int. Symp., Hamburg, 1978,* edited by J.Sunderman and K.-P. Holz, 203-214, Springer-Verlag, Berlin, 1980.

Bryan, F. and W.R. Holland, A high resolution simulation of the wind and thermohaline - driven circulation of the North Atlantic, 'Aha Huliko'a, *Proc. of the Hawaiian Winter Workshop,* University of Hawaii, 99-116, 1989.

Bryan, K., A numerical model for the study of circulation of the world oceans, *J. Comput. Phys., 4,* 347-376, 1969.

Cox, M.D., A primitive equation, 3-dimensional model of the ocean, GFDL Ocean Group Tech. Rep. No.1, GFDL,/Princeton University, 141 pp, 1984.

Dickson, R. R., J. Meincke, S.-A. Malmberg, and A. J. Lee, The "Great Salinity Anomaly" in the northern North Atlantic 1968-1982, *Prog. Oceanogr., 20,* 103-151, 1988.

Fleming, G.H. and A.J. Semtner, A numerical study of interannual ocean forcing on Arctic ice, *J. Geophys. Res., 96*, 4589 - 4603, 1991.

Gordienko, P.A., and A.F. Laktionov, Circulation and physics of the Arctic Basin waters, *Annals of the International Geophysical Year, 46*, 94-112, 1969.

Gregg, M.C., Scaling turbulent dissipation in the thermocline, J. Geophys. Res., 94, 9686-9698, 1989.

Gudkovich Z.M. and Ye.G. Nikiforov, The study of the nature of the water circulation in the Arctic Basin by means of a model, *Oceanologia 5,*52-60,1965.

Haidvogel, D.B., J.L.Wilkin and R.E.Young, A semi-spectral primitive equation ocean circulation model using vertical sigma and orthogonal curvilinear horizontal coordinates. *J. Comp. Phys. 94,* 151-185, 1991.

Haney, R.L., On the pressure gradient force over steep topography in sigma coordinate ocean models. *J. Phys. Oceanogr, 21,* 610-619,1991

Häkkinen, S., A constitutive law for sea ice and some applications, *Mathl. Modelling, 9,* 9469-9478., 1987.

Häkkinen, S., Models and their applications to Polar Oceanography, in *Polar Oceanography, part A, Physical Science*, ed W.O. Smith, Academic Press, 406pp., 1990.

Häkkinen, S. and G.L. Mellor, One hundred years of Arctic ice cover variations as simulated by a one dimensional coupled ice-ocean model, *J. Geophys. Res.,95,* 15, 15959-15969, 1990.

Häkkinen, S. and G.L. Mellor, Modeling the seasonal variability of the coupled Arctic ice-ocean system, *J. Geophys. Res., 97,* 20285-20304, 1992.

Häkkinen S., An Arctic source for the Great Salinity Anomaly: A simulation of the Arctic Ice-ocean system for 1955-1975, *J. Geophys. Res., 98*, 16397- 16410, 1993.

Hedstrom, K.S., User's manual for a semi-spectral primitive equation regional ocean circulation model. Version 3.0, Institute for Naval Oceanography Technical Note FY90-2, 90 pp, 1990.

Hibler, W. D., III, A viscous sea ice law as a stochastic average of plasticity, *J. Geophys. Res., 82*, 3932-3938, 1977.

Hibler, W.D., III, A dynamic thermodynamic sea ice model, *Mon. Wea. Rev., 108*, 1943-1973, 1979.

Hibler , W.D., III, Modeling variable thickness sea ice cover, *Mon. Wea. Rev., 108*, 1943-1973, 1980.

Hibler, W.D., III, and K. Bryan, A diagnostic ice-ocean model, *J. Phys. Oceanogr., 17*, 987-1015, 1986

Holland, W.R., The role of mesoscale eddies in the general circulation of the ocean - numerical experiments using a wind-driven quasi-geostrophic model, *J. Phys. Oceanogr., 8,* 363 - 392, 1978.

Holland, D.M., L.A. Mysak, D.K. Manak, and J.M. Oberhuber, Sensitivity study of a dynamic thermodynamic sea ice model, *J. Geophys. Res., 98*, 2561-2586, 1993.

Lemke, P., A coupled one-dimensional sea ice-mixed layer model, *J. Geophys. Res., 92*, 13164-13172, 1987.

Levitus, S., Climatological atlas of the world ocean, NOAA Publ. 13, U.S. Dept. of Commerce, Wash. D.C. 173pp, 1982.

Maykut, G.A. and N. Untersteiner, Some results from a time-dependent thermodynamic model of sea ice, *J. Geophys. Res., 76*, 1550-1575, 1972.

Maykut, G.A., Large scale heat exchange and ice production in the central Arctic, *J. Geophys. Res., 87*, 7971- 7984, 1982.

Mellor, G.L., Analytic prediction of the properties of stratified planetary surface layers, *J. Atmos. Sci., 30,* 1061-1069, 1973.

Mellor, G.L., User's guide for a three-dimensional, primitive equation, numerical ocean model, Report: AOS Program, Princeton University, Princeton NJ 08540, 1993.

Mellor , G.L., T. Ezer, and L.-Y. Oey, On the pressure gradient conundrum of sigma coordinate ocean models, *J. Atmos. Ocean. Tech., in press*, 1994.

Mellor, G.L. and L.H. Kantha, An ice-ocean coupled model, *J. Geophys. Res. , 94*, 10937-10954, 1989.

Mellor, G.L. and T. Yamada, A hierarchy of turbulence closure models for planetary boundary layers, *J. Atmos. Sci., 31*, 1791-1806, 1974.

Mellor, G.L. and T. Yamada, Development of a turbulence closure model for geophysical fluid problems, *Rev. Geophys., 20*, 851-875, 1982.

Munk, W.H., and E.R. Anderson, Notes on a theory the thermocline. *J. Marine Res., 7*, 276-295, 1948.

Nikiferov, E.G., "Variations in icecover compaction due to its dynamics", *Probl. Arktiki, 2*, Morskoi Trans., Leningrad. 1957.

Oberhuber, J.M., Simulation of the Atlantic circulation with a coupled sea ice-mixed layer-isopycnal general circulation model. Part I: Model description, *J. Phys. Oceanogr., 23*, 808-829, 1993a

Oberhuber, J.M., Simulation of the Atlantic circulation with a coupled sea ice-mixed layer-isopycnal general circulation model. Part II: Model experiment, *J. Phys. Oceanogr., 23*, 830-845, 1993b

Oey, L.-Y., G.L. Mellor and R.I. Hires, A three-dimensional simulation of the Hudson-Raritan estuary. Part I: Description of the model and model simulations, J. *Phys. Oceanogr., 15*, 1676-1692, 1985.

Parkinson, C.L., and W.M. Washington, A large-scale numerical model of sea ice, *J. Geophys. Res., 84*, 311-336, 1979.

Piacsek, S., R. Allard, and A. Warn-Varnas, Studies of the Arctic ice cover and upper ocean with a coupled ice-ocean model, *J. Geophys. Res., 96,* 4631-4650, 1991.

Riedlinger, S.H., and R. H. Preller, The development of a coupled ice-ocean model for forecasting ice conditions in the Arctic, *J. Geophys. Res., 96*, 16,955-16,977, 1991.

Robinson A.R., and L.J. Walsted, The Harvard open ocean model: calibration and application to dynamical process, forecasting, and data assimilation studies. Appl. Num. Math., 3, 89-131.

Semtner, A.J., Numerical simulation of the Arctic Ocean Circulation, *J. Phys. Oceanogr., 6*, 409-425, 1976a

Semtner, A.J., A model for the thermodynamic growth of sea ice in numerical investigation of climate, *J. Phys. Oceanogr., 6*, 379-389, 1976b.

Semtner, A.J., A numerical study of sea-ice and ocean circulation in the Arctic, J. *Phys. Oceanogr., 17*, 1077-1099,1987.

Taylor, G.I., The dispersion of matter in turbulent flow through a pipe. *Proc. Roy. Soc. London, A223*, 446-468, 1952

Thorndike, A.S., D.A. Rothrock, G.A. Maykut, and R. Colony, The thickness distribution of sea ice, *J. Geophys. Res., 80,* 4501-4513, 1975.

On the Intermediate Depth Waters of the Arctic Ocean

B. Rudels,[1] E. P. Jones,[2] L. G. Anderson,[3] and G. Kattner,[4]

The intermediate depth waters of the Arctic Ocean are supplied from the North Atlantic through Fram Strait and over the Barents Sea. Water flowing into the Arctic Ocean by these two paths results in two branches of apparent equal strength (≈2 Sv): the Fram Strait branch with relatively warm water flowing in by way of the West Spitsbergen Current, and the Barents Sea branch with water that has been cooled and made less saline in the Barents Sea. The two branches meet north of the Kara Sea, creating interleaving and inversions that diminish along their flow paths. The intermediate depth waters from these two sources may be further transformed by sinking shelf plumes, water made dense by freezing and ice growth on the shallow shelves. The descending density flows cool the Atlantic Layer and redistribute heat downwards. A calculation shows how the introduction of the dense shelf water penetrating the Atlantic Layer can explain the characteristics observed in the Canadian Basin water column. Clues to the circulation of the intermediate depth waters are provided by changes in the shapes of the Θ-*S* curves at different locations and by chlorofluorocarbon and silicate distributions. In the Eurasian Basin, the residence times of the intermediate depth waters given by chlorofluorocarbon concentrations are of the order of a decade, with the deeper layers being the oldest. In the Canadian Basin, the residence times are twice as long for similar depths. Several loops are identified in the Eurasian Basin, and similar flow pattern in the Canadian Basin is suggested based on properties of outflowing Canadian Basin water near the Morris Jesup Plateau.

INTRODUCTION

The Arctic Ocean is a strongly stratified enclosed sea with two major deep basins and extensive shallow shelves (Figure 1). Four restricted passages communicate with the world oceans. The Barents Sea, Fram Strait, and the Canadian Archipelago connect the Arctic Ocean to the North Atlantic, and Bering Strait connects it to the Pacific. Only Fram Strait, between Greenland and Svalbard, permits exchanges of deep water over a sill at about 2500 m between the Arctic Ocean and the Greenland and Norwegian seas. The communication with the North Atlantic is, however, restricted to the upper 600 m to 800 m by the Greenland-Scotland Ridge further south. Convection in the central Arctic Ocean is limited to the upper 50 m to 100 m and the deeper layers of the Arctic Ocean must be ventilated by advection, partly by water passing through Fram Strait, but also by water made dense by cooling and by brine rejection from ice formation on the shallow shelves inside the Arctic Ocean. These processes allow waters entering through the shallow Barents Sea and Bering Strait to penetrate into the intermediate and deeper layers of the central basins.

The warm temperatures (Θ > 0°C) in the Arctic Ocean associated with Atlantic Layer were observed by *Nansen* [1902], who inferred that this layer, as well as those below, were advectively renewed by water entering the Arctic Ocean in the West Spitsbergen Current west of Svalbard. Recent observations indicate that the inflow to the Arctic Ocean over the Barents Sea may be as great as, or perhaps

[1] Institut für Meereskunde, Universität Hamburg, Hamburg, Germany

[2] Department of Fisheries and Oceans, Bedford Institute of Oceanography, Dartmouth, Nova Scotia, Canada

[3] Department of Analytical and Marine Chemistry, University of Göteborg and Chalmers University of Technology, Göteborg, Sweden

[4] Alfred-Wegener-Institut für Polar-und Meeresforschung, Bremerhaven, Germany

The Polar Oceans and Their Role in Shaping the Global Environment
Geophysical Monograph 85

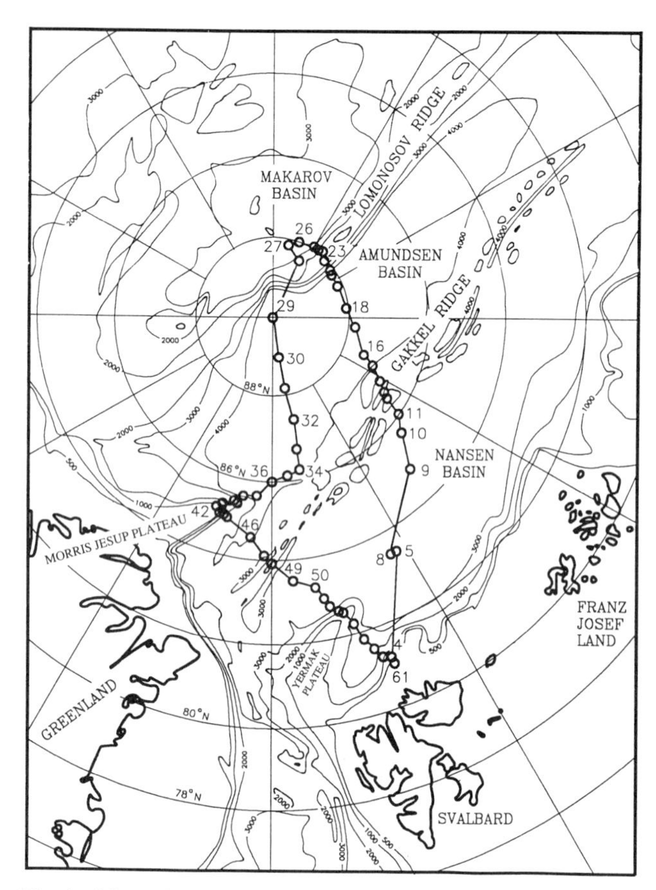

Fig. 1. Map of the Arctic Ocean indicating the main topographic features and the *Oden* 91 hydrographic stations.

greater than, the inflow carried by the west Spitsbergen current [*Blindheim*, 1989; *Rudels*, 1987].

An earlier picture of the circulation of the Atlantic Layer, based on the relatively scanty data set then available, was deduced from changes in its core temperature, which indicated a cyclonic flow around the Arctic Ocean [*Coachman and Barnes*, 1963]. The sensible heat stored in the Atlantic Layer is large and the inflow of Atlantic water through Fram Strait is considered to be the major oceanic heat source for the Arctic Ocean [*Aagaard and Greisman*, 1975]. Little has been written about the circulation of the waters below the Atlantic Layer, but their circulation is commonly assumed also to be cyclonic and mostly confined to narrow boundary currents [*Aagaard*, 1989]

Observations made during the *Oden* 91 expedition (Figure 1) allow us to add to this picture. The expedition and some general oceanographic results have been presented elsewhere [*Anderson et al.*, 1994]. Here we concentrate on the interactions between different water masses and the circulation of the waters between 200 m and 1700 m, i.e., between the bottom of the halocline and the top of the Lomonosov Ridge, which separates the two major basins of

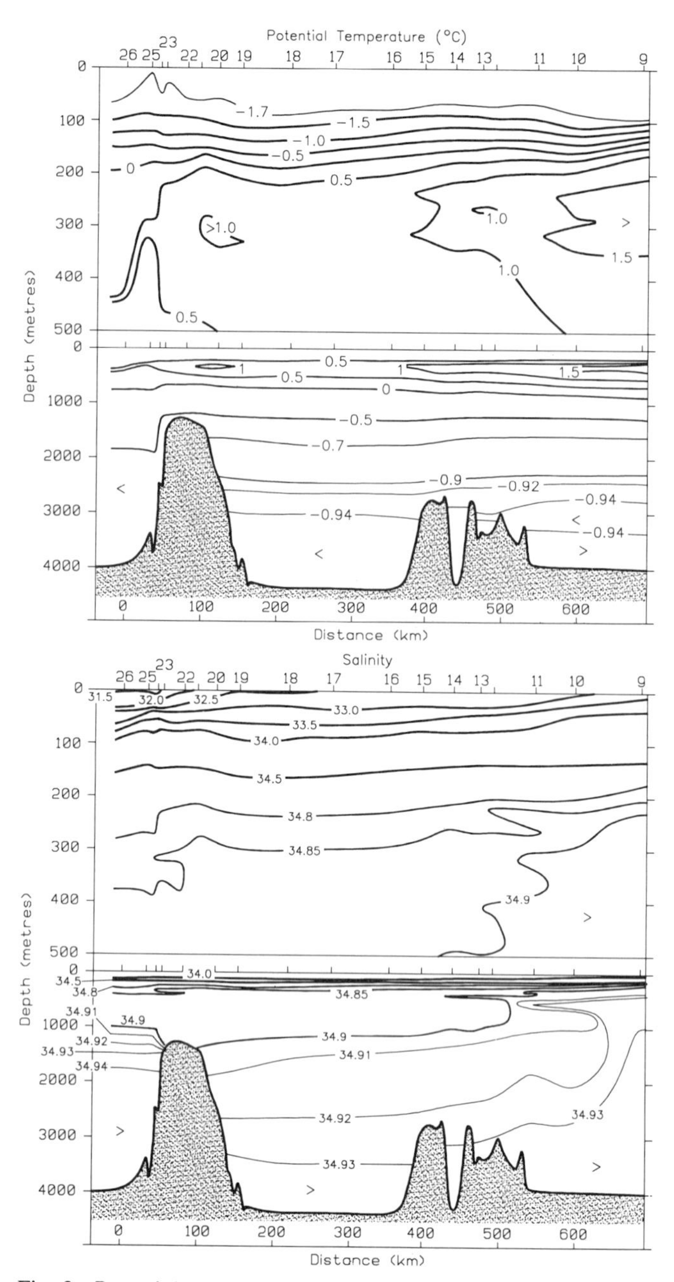

Fig. 2. Potential temperature and salinity: Section A, stations 9-26.

the Arctic Ocean. Waters in this depth range are mostly confined to the deep basins, though they may occasionally upwell onto the deeper shelves [e.g., *Aagaard et al.*, 1981].

The Atlantic Layer constitutes the upper part of these waters, historically being defined as having temperatures above 0°C, and ranges in depth from about 200 m to between 600 m and 800 m. The water below the Atlantic Layer, in the temperature range from 0°C to -0.5°C, will be called Upper Polar Deep Water (UPDW) to distinguish it from the Arctic Intermediate Water (AIW) of the same Θ-*S* range found in the Greenland and Icelandic seas [*Swift and Aagaard*, 1980]. UPDW is characterized by increasing salinity and decreasing temperature with depth, which is the result of the merging of entraining shelf plumes with the Arctic Ocean water column. This is in contrast to the AIW, which shows constant or increasing temperatures and salinities with depth, characteristics of mode waters formed by open ocean convection. The Θ-*S* features allow the UPDW to be identified also outside the Arctic Ocean. That the UPDW is created inside the Arctic Ocean is supported by the prominence of water in the -0.5 - 0°C temperature range over the Greenland slope as compared to the inflow area on the Svalbard side of Fram Strait [*Rudels,* 1986].

DISCUSSION

The temperature and salinity distribution on a section from the Nansen Basin into the Makarov Basin (Figure 2) shows a strong front at the Lomonosov Ridge. The warmer and more saline Atlantic Layer in the Eurasian Basin is separated from the cooler, less saline Atlantic Layer of the Makarov Basin. However, in the UPDW depth range the situation is reversed and the waters on the Canadian Basin side of the Lomonosov Ridge are now the warmer and more saline. The water supplying the intermediate depth layers comes primarily from the North Atlantic, some through Fram Strait and a part through the Barents Sea. To reach the Canadian Basin the Atlantic water has to cross the Lomonosov Ridge and must then pass the Eurasian Basin, probably along the continental slope.

The potential temperature-salinity diagram in Figure 3a shows the Θ-*S* characteristics of the water columns of the Fram Strait inflow and of the waters of the Nansen and Amundsen basins. The broken line indicates the characteristics of the water mass with which the Fram Strait inflow must mix to create the observed Θ-*S* curves. The water crossing the Lomonosov Ridge into the Canadian Basin should have characteristics not of the Fram Strait inflow but rather like the Θ-*S* curves in the Amundsen Basin close to the Lomonosov Ridge. Figure 3b shows two Θ-*S* curves on each side of the Lomonosov Ridge together with the Θ-*S* curve required for the hypothetical mixing water. The Θ-*S* shapes of these two, looked for water masses, one in the Eurasian and one in the Canadian Basin are distinctly different, indicating that different processes dominate in the two major basins of the Arctic Ocean. Below we shall try to explore and explain these differences.

The Eurasian Basin: Advection and Interleaving

The Atlantic Layer of the Eurasian Basin exhibits further horizontal structures. The Amundsen Basin is cooler and less saline than the Nansen Basin and cores of warmer, more saline water are observed above the Nansen-Gakkel

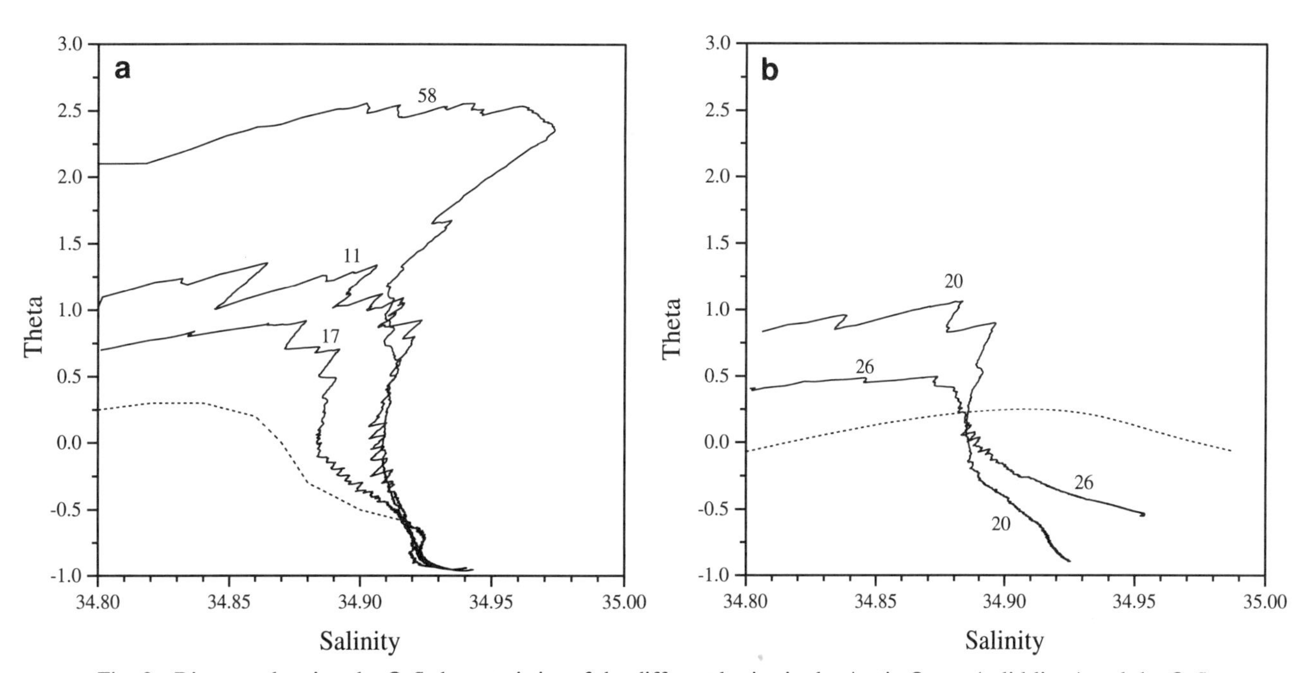

Fig. 3. Diagram showing the Θ-*S* characteristics of the different basins in the Arctic Ocean (solid lines) and the Θ-*S* characteristics (broken line) of the hypothetical water mass with which they must mix to attain the observed characteristics. a) The Fram Strait inflow (st. 58), the Nansen Basin (st. 11) and the Amundsen Basin (st. 17). b) The Lomonosov Ridge (st. 20) and the Makarov Basin (st. 26).

and Lomonosov ridges (Figure 2). The profiles (Figure 4) show sharp inversions over most of this depth interval, implying previous or ongoing strong interactions between water masses of similar densities. These inversions may supply clues as to the circulation of intermediate depth waters in the Eurasian Basin.

Similar inversions were observed in the Atlantic Layer by *Perkin and Lewis* [1984] during the Eurasian Basin Experiment, EUBEX, north of Svalbard. Their explanation was that the inversions resulted from cross-flow interactions between Atlantic water as it entered the Arctic Ocean and the older Arctic Ocean water column. The *Oden* 91 observations showed a wide distribution of inversions throughout the Eurasian Basin. The explanation given by *Perkin and Lewis* [1984] would then mean that the inversions occur as a result of the lateral displacements caused by basin scale horizontal movements or that they form locally north of the Yermak Plateau and are then advected eastward as a broad stream extending across the Nansen and the Amundsen basins. Both of these propositions are contrary to the concept of the inflowing Atlantic water progressing eastward as a boundary flow along the continental

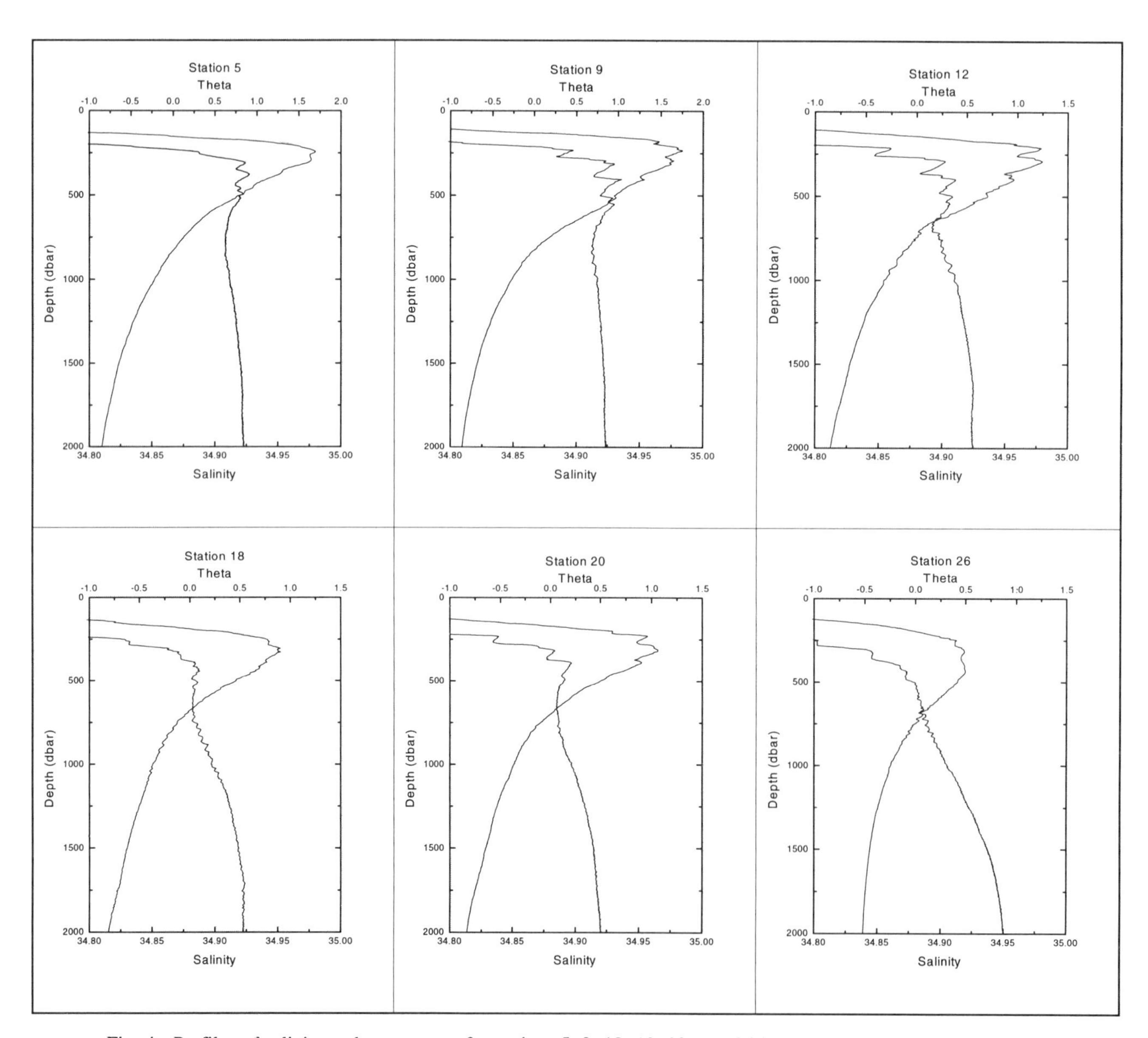

Fig. 4. Profiles of salinity and temperature for stations 5, 9, 12, 18, 20, and 26 illustrating the intrusions at several locations along section A. Note the different temperature scale for stations 5 and 9.

slope [*Aagaard*, 1989]. We therefore follow the explanation proposed by *Quadfasel et al.* [1993], who observed strong inversions in the high temperature core on most of their XBT profiles. These inversions were assumed to be created by the intrusion of dense water from the shelves, then advected along the Nansen-Gakkel Ridge towards Fram Strait. However. the regularity of the deeper lying inversions (not reached by the measurements reported by *Quadfasel et al.* [1993]) and the wider distribution of the inversions require some modification of their interpretation.

The inflow from the Atlantic entering the Arctic Ocean through Fram Strait consists of Atlantic water as well as some AIW and Norwegian Sea and Greenland Sea Deep Waters. The Atlantic water can pass over the Yermak Plateau directly into the Arctic Ocean, while the deeper layers must flow around the Yermak Plateau. In the Arctic Ocean the inflow soon becomes covered by a low salinity surface layer. This "Fram Strait branch" thus becomes shielded from the atmosphere, and direct air-sea-ice interactions are inhibited as it moves eastward along the continental slope.

The flow of the Atlantic water through the Barents Sea into the Arctic Ocean is believed to occur between Novaya Zemlya and Franz Josef Land into the Kara Sea then down the St. Anna Trough into the Arctic Ocean [*Rudels*, 1987; *Blindheim*, 1989]. In the Barents Sea, the inflowing Atlantic water becomes colder and mostly fresher. While heat loss at the sea surface can explain the reduction in temperature, a mixing with water of lower salinity is needed to reduce the salinity. The low salinity water could be supplied by the Norwegian Coastal Current, which turns northward and continues along the west coast of Novaya Zemlya. Heat loss and freezing over the shallow areas east of these islands could provide cold and dense water, brine enriched but still with lower salinity than the Atlantic water. It would then sink into and cool the Atlantic inflow passing over the Barents Sea. *Quadfasel et al.* [1992] observed indications of a process in the western Barents Sea, where dense water formed by haline convection over the Central Bank flowed off the bank and cooled the bottom water in the Hopen Deep. In addition, over shallow areas, saline, highly dense bottom water is formed by freezing and brine rejection [*Midttun*, 1985]. This water would enter beneath the Atlantic water, perhaps entraining some Atlantic water to aquire a temprature above freezing.

These mixing processes allow the initial density range of the Atlantic water entering between Norway and Bear Island to be expanded as the water moves through the Barents Sea. The time required for the Atlantic water to cross the Barents Sea is estimated to be about a year [*Rudels*, 1987] thus averaging out a seasonal fluctuation in the outflow into the Arctic Ocean. In any event, the summer ice melt and heating would lower the density of the surface layer but not affect the characteristics of the deeper outflow. If a seasonal signal exists in the Barents Sea branch, it would be expressed more in changes in volume than in Θ-S characteristics.

We propose that as the Barents Sea branch leaves the St. Anna Trough as a complete, 1000 m deep water column and becomes trapped as a narrow, eastward moving wedge between the continental slope and the Arctic Ocean water column (including the Fram Strait branch). The convergence of the Fram Strait and Barents Sea branches north of the Kara Sea creates a front with large horizontal gradients in temperature and salinity. If the water columns are perturbed laterally, inversions large enough to trigger double diffusive processes may be created, as in the laboratory experiments performed by *Ruddick and Turner* [1979]. These reinforce the disturbances and drive the intrusions across the front. This raises the temperature and salinity in the Barents Sea branch and lowers them in the Fram Strait branch.

These interactions occur, because of the expanded density range of the Barents Sea branch, not only with the Atlantic water of the Fram Strait branch but also with the layer below. These waters have in Fram Strait a fairly constant salinity and decreasing temperature and appears to consist mainly of AIW formed, perhaps in the Greenland Sea but most likely further south. Because they have to flow around the Yermak Plateau to enter the Nansen Basin we do not expect the contribution from the Fram Strait branch to be as large as that of the Barents Sea branch in this density range. This is partly born out by the cool, low salinity characteristics of the UPDW.

Dense shelf water may form and sink to intermediate depths, forming intrusions also west of Franz Josef Land, as was observed during the *Ymer* 80 expedition [Rudels, 1986]. These intrusions were not as regular as those observed on the *Oden* 91 expedition and were confined to the slope. The *Ymer* 80 stations further into the basin showed no deep interleaving and the temperature profiles in the Atlatic Layer were smooth [*Rudels*, 1989]. Also the *Oden* 91 station 5 west of Franz Josef Land (Figures 1 and 4) shows no deep interleaving, and the temperature inversions in the Atlantic Layer are smooth. This excludes a western source for the observed interleaving. The lateral velocities in the layers of an interleaving front have been estimated to be 0.1 cm/s [*Ruddick and Hebert*, 1988]. The *Oden* 91 stations in the Nansen Basin are too close to Fram Strait for the observed interleaving (Figure 4) to have penetrated northward across the eastward moving Fram Strait branch directly from the Kara Sea shelf break.

Moreover, the inversions found on station 9 (Figure 4) in the Nansen Basin are weaker, not stronger than those over the Nansen-Gakkel Ridge.

Our interpretation is that the boundary flow along the continental slope, in which the convergence of the branches and the interleaving occur, advects the inversions toward the Laptev Sea, where they are deflected into the interior of the basin. The inversions are then returned along the axis of the ridges and the Amundsen Basin towards Fram Strait. This flow pattern, which was proposed by *Quadfasel et al.* [1993] based on "fossil" intrusions and also by *Anderson et al.* [1989], is consistent with many of the observed features. The Fram Strait branch would have the "inner, basin lane", turning first and carrying the warmer, more saline part of the westward stream closest to the eastward moving boundary flow. The colder, less saline branch in the "outer, slope lane" would dominate in the Amundsen Basin and at the Lomonosov Ridge.

The difference between this interpretation and that of *Perkin and Lewis* [1984] is that the interleaving occurs to the south, not to the north, of the water entering through Fram Strait, and that the inversions seen at the *Oden* 91 stations and during EUBEX represent a return flow from the active slope area towards Fram Strait. The departure from the *Quadfasel et al.* [1993] picture is that it is not transient shelf sources but rather a continuous inflow of water from the Atlantic transformed in the Barents Sea, which merges with the Arctic Ocean water column. This causes the interleaving to occur across a sharp front that extends over a 1000 m depth range.

The persistence of the inversions in the interior of the basin indicates that double diffusive processes dominate over other disruptive processes (shears, mesoscale eddies, etc.). There is little sign of cross-interactions once the flow has been deflected into the Basin. The different cores can still be distinguished at section D between the Morris Jesup Plateau and the Yermak Plateau as is seen from the Θ-*S* diagrams showing stations 17, 32, 36 and 46 along the Amundsen Basin and stations 12 and 48 over the Nansen-Gakkel Ridge (Figure 5). The inversions at the core above the Nansen-Gakkel Ridge are still sharp, showing that double diffusive fluxes have not had time enough to smooth the profiles. This suggests that the transit time to Fram Strait above the ridge is short for the Atlantic Layer part of the water column. By comparing stations 12 and 48 it nevertheless appears as if the inversions on station 48 are less sharp and that the stability ratios over the finger and diffusive interfaces have increased.

Using flux laws such as *Kunze* [1987] for thin finger interfaces and a combination of *Linden and Shirtcliffe* [1979] and *Rudels* [1991] for the diffusive interface, and assuming only on interface between each inversion the time required to generate the observed changes can be estimated (*Rudels*, submitted). The obtained time estimate is unrealistically short, only about a week appears to be needed for the Atlantic water to flow from station 12 to station 48. This corresponds to a velocity of about 90 cm/s. This time estimate is a minimum, since it is assumed that only one interface between the layers and that the layers are well mixed. If more than one step are present the property steps are smaller and, since the flux goes as the step interval to the 4/3 power, the time increases. With three interfaces, for example, the time would be about 1 month and the velocity about 20 cm/s. This would mean, considering the time difference between the observations, that the CTD casts on station 12 and 48 were, for the Atlantic Layer, made in the same water body. These times still appear too short since it means that the Atlantic Layer, at least in the more rapid

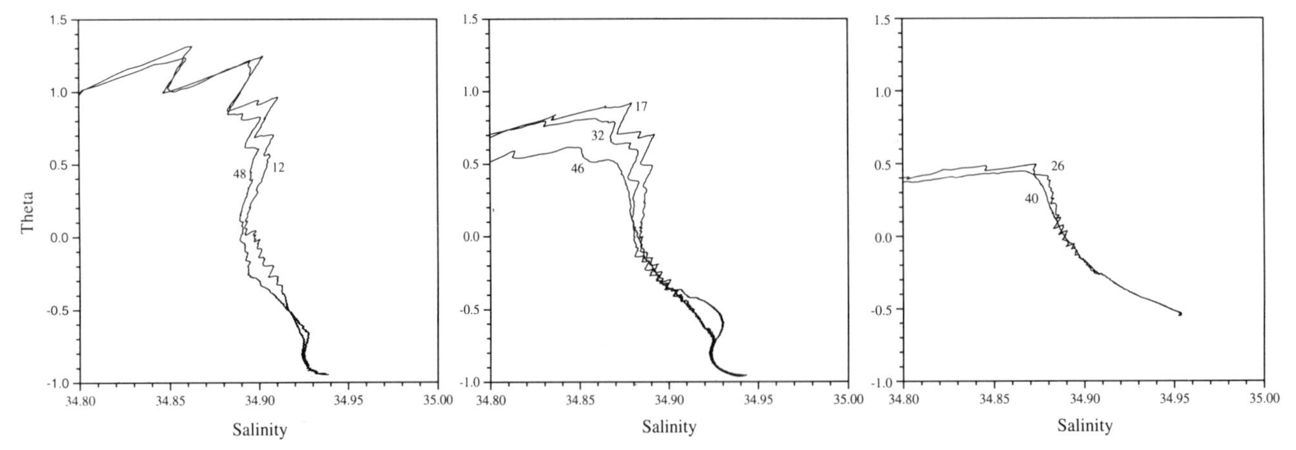

Fig. 5. Θ-*S* diagrams comparing stations 12 and 48 along the Nansen-Gakkel Ridge, stations 17, 32 and 46 along the Amundsen Basin, station 26 in the Makarov and station 40 at the Morris Jesup Plateau.

moving loops of the Eurasian Basin, is ventilated on time scales of months rather than years.

The circulation pattern, but not the ventilation times, deduced from the intrusions are supported by the chlorofluorocarbon concentrations. The CFCs are transient tracers introduced into the atmosphere and increasing monotonically in the surface of the ocean since their introduction into the atmosphere: CFC-11 and CFC-12 about five decades ago, and carbon tetrachloride (CCl_4) in the early part of this century. At intermediate depths, chlorofluorocarbon concentrations generally increase from the Nansen-Gakkel Ridge toward the Lomonosov Ridge (Figure 6). The exception is in the Atlantic Layer, at the temperature maximum, where the concentrations near the Nansen-Gakkel Ridge (stations 9-11) are higher than those along the axis of the Amundsen Basin (stations 14-17). At the temperature maximum, the lower chlorofluorocarbon values at the centre of the Amundsen Basin support the idea of a more sluggish flow. Higher concentrations near the Lomonosov Ridge compared to those over the Nansen-Gakkel Ridge would result if the flow along the Lomonosov Ridge is swifter but also if water of the Barents Sea branch were more recently ventilated as it passed over the shelves. In winter, the inflowing Atlantic water west of Svalbard is ventilated by thermal convection, and it is not far from Svalbard to the Kara Sea. If differences in chlorofluorocarbon concentrations over the Nansen-Gakkel Ridge and the Lomonosov Ridge correspond to the differences between the Fram Strait branch and the water entering over the Barents Sea, the transit time for the Fram Strait branch from where it loses contact with the atmosphere until it meets the Barents Sea branch is of the order of a year, perhaps more. Here we have assumed that the water of the two branches is saturated to the same degree and not too far from 100%, when it leaves the surface, and that the mixing can be described by a simple freight train model. Comparing concentrations at stations 59 and 60, we estimate that it takes the warm core roughly 10 years to reach station 11, flowing first along the continental slope and then along the Nansen-Gakkel Ridge. For the Barents Sea branch at station 20, the corresponding time from the St. Anna Trough along the slope and the Lomonosov Ridge is close to 6 years, with the assumption above of about the same degree of saturation for the two branches. These times are much longer than those estimated from changes in the inversions. The reason for different results from the two approaches remains to be discovered.

For the deeper layers (> 500 m), the chlorofluorocarbon concentrations increase monotonically from the Nansen-Gakkel Ridge towards the Lomonosov Ridge. The Barents Sea branch should initially be equally well ventilated at all levels and the inferred greater age of the Fram Strait branch at deeper layers could be the result of less complete winter overturning down to these depths before the water enters through Fram Strait. At these levels, there exists a difference in concentrations corresponding to a few years in the time between the last ventilation of the two branches.

Canadian Basin: Slope Convection and Entrainment

The dominance of the colder, fresher Barents Sea branch in the Amundsen Basin implies that a substantial fraction of the deeper Fram Strait branch has been deflected from the boundary current and entered the central basin before reaching the Lomonosov Ridge, forming part of a circulation confined to the Eurasian Basin. Only a part of the original Atlantic inflow, primarily from the Barents Sea branch, with a smaller fraction of the warmer, less dense part of the Fram Strait branch, appears to cross the Lomonosov Ridge to enter the Canadian Basin.

The information on the intermediate circulation in the Canadian Basin is much more fragmentary than in the Eurasian Basin. It is based on data from stations 26 and 27 in the Makarov Basin and on stations near the Morris Jesup Plateau (e.g., stations 39, 40) where water from the Canadian Basin re-enters the Eurasian Basin. The lower temperatures and salinities of the Atlantic Layer and the higher temperatures and salinities in the UPDW compared to those in the Eurasian Basin show that shelf-slope-basin interactions occurs within the Canadian Basin. The only other source water outside the Arctic Ocean basin is the less dense Pacific inflow, and any changes of temperature and salinity occurring at depths below 200 m must result from the injection of dense water from the shelves.

The freshening and cooling of the Atlantic Layer in the Canadian Basin indicate that the principal heat loss of the inflowing Atlantic water takes place by incorporating cold shelf waters. The only possibility for the deeper layers to become warmer is by vertical redistribution of heat stored in the Atlantic Layer, i.e., by entraining warmer Atlantic Layer water into density flows descending from the shelves. For a significant downward transfer of heat to occur, the core temperature of the Atlantic Layer at the shelf break must be fairly high, i.e., its maximum temperature should be comparable to what is observed on the Eurasian Basin side of the Lomonosov Ridge (station 20). Dense shelf water interacting directly with the entering (advective) Eurasian Basin waters could then form the deeper waters of the Canadian Basin.

By contrast, water in the Eurasian Basin at intermediate depths is cooled, not heated. This feature shows that the waters at these levels (below 600 m) in the Amundsen

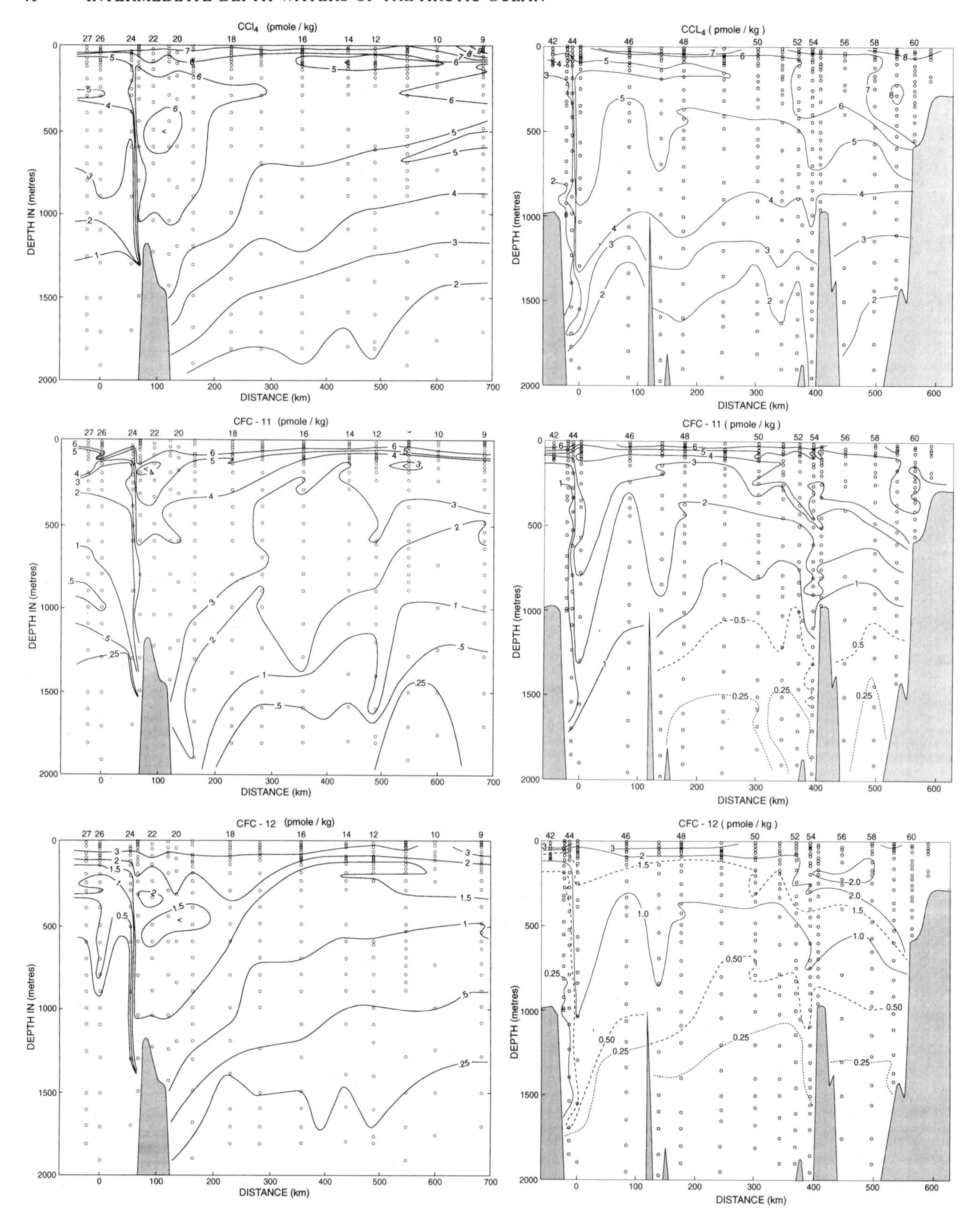

Fig. 6. CCl_4, CFC-11, and CFC-12 distributions on sections A and D between 0 and 2000 m.

Basin, which are colder than in the Nansen or Makarov Basins, cannot be formed by mixing waters from the Nansen and Makarov basins. The lower temperatures and salinities can only be explained by the entrance of the Barents Sea branch directly into the deeper layers through the St. Anna Trough.

Shelf-Slope Plume Calculation

To illustrate the effects of the shelf-slope-basin interactions in the Canadian Basin, we make a simple estimate of the dense water formation on the shelves that, together with an assumed (but realistic) inflow from the Eurasian Basin can reproduce the water column in the Makarov Basin.

The combined inflow through Fram Strait and over the Barents Sea is about 3 Sv excluding deep water exchange through Fram Strait [*Rudels*, 1987; *Aagaard and Carmack*, 1989]. About 1 Sv is transformed into lighter surface water, and, considering the indications of a strong recirculation in the Eurasian Basin already discussed, only about 0.5 to 1 Sv might cross the Lomonosov Ridge into the Canadian Basin. The subsequent Θ-S changes of the Atlantic Layer and UPDW in the Canadian Basin are results of the incorporation, and the bypassing, of entraining density flows from the shelves [*Rudels*, 1986]. The initial temperature of the density flows on the shelves must be at the freezing point and their initial densities would be determined by the content of brine released by the freezing process. The properties of the shelf contribution will then change because of entrainment of surrounding water as the plumes sink into the deep.

To assess the strength of the interactions at the slope in the Canadian Basin, the characteristics of the Eurasian Basin water as it flows over the Lomonosov Ridge into the Canadian Basin must be known. No station from the continental slope north of the Laptev Sea is available, and we shall assume that the water that crosses the Lomonosov Ridge has characteristics similar to those in the water that flows along the Lomonosov Ridge towards Greenland. The water crossing into the Canadian Basin is represented by an idealized column based on the Θ-S characteristics found on station 20. The corresponding Θ-S characteristics for the Canadian Basin water column are taken from station 26 (Table 1). The depth interval between 200 m and 1700 m is approximated by five 300 m thick homogeneous layers with constant properties.

The entrainment rate into the sinking plume is assumed constant with depth, i.e., the total amount of water entrained into a plume is a linear function of the convection depth. This is a crude parameterization, which corresponds to the entrainment into a two-dimensional free-sinking line

Table 1. Assumed characteristics of Eurasian Basin Water flowing into the Canadian Basin (Θ, S) and Canadian Basin water (Θ_i, S_i).

Layer	Depth	Θ	S	Θ_i	S_i
1	200-500	0.85	34.875	0.40	34.820
2	500-800	0.15	34.886	0.17	34.887
3	800-1100	-0.30	34.893	-0.18	34.903
4	1100-1400	-0.52	34.908	-0.36	34.923
5	1400-1700	-0.66	34.915	-0.46	34.940

plume. In reality, the entrainment depends on, among other things, density differences, ambient stratification, plume thickness and width, slope angle, and the existence of possible steering, protective canyons.

The entrainment equations then become

$$V_i = V_i' + V_{oi} \cdot \left(1 + \sum_{1}^{i-1} e\right) - \sum_{k=i+1}^{N} V_{ok} \cdot e$$

$$\Theta_i \cdot V_i = \Theta_i' \cdot V_i' + V_{oi} \cdot \left(\Theta_o + \sum_{j=1}^{i-1} \Theta_j' \cdot e\right) - \Theta_i' \cdot \sum_{k=i+1}^{N} V_{ok} \cdot e \quad (1)$$

$$S_i \cdot V_i = S_i' \cdot V_i' + V_{oi} \cdot \left(S_{oi} + \sum_{j=1}^{i-1} S_j' \cdot e\right) - S_i' \cdot \sum_{k=i+1}^{N} V_{ok} \cdot e$$

for the layers 1 to N ($N = 5$). In Equation (1), V_i, $\Theta_i \cdot V_i$, and $S_i \cdot V_i$ are the volume, heat and salt fluxes into the layer i in the interior of the Canadian Basin. V, etc., are the contributions advected along the slope from the Eurasian Basin, and V_{oi}, $\Theta_o \cdot V_{oi}$, and $S_{oi} \cdot V_{oi}$ are the fluxes from the shelves. Some entrainment must take place between the shelf break (at about 50 m) and 200 m, and the temperature of the density flows when they reach 200 m will be above freezing. The constant entrainment, e, is written as the increase in the transport of the plume as it passes through one 300 m layer. The first summation term represents the amount of entrained water entering layer i and the second summation term the corresponding loss to the underlying levels.

Initially, the entrainment suggested by *Quadfasel et al.* [1988] was used. They estimated a 500% increase in volume for a plume sinking from Storfjorden in the south of Svalbard to 2000 m in Fram Strait. However, this was found to be too small to explain the observed Θ-S charcteristics. Values of e and Θ_o were varied until, with $e = 2$ and Θ_o = -1.0°C, the observed changes of the water column could be reproduced. This value of e means that a volume

Table 2. Shelf Plume Calculation Results for the Canadian Basin. A linear velocity profile, V'_i, is assumed for inflowing Eurasian Basin Water. The flow from the shelf (entering at 200 m) is represented by V_{oi}, Θ_{oi}, and S_{oi}. The plume characteristics are represented by Θ, and S. V_i is the sum according to equation (1).

Layer	V'_i	V_{oi}	V_i	Θ_{oi}	Θ	S_{oi}	S
1	0.14	0.030	0.125	-1.0	-1.0	34.649	34.649
2	0.12	0.010	0.124	-1.0	0.233	34.920	34.890
3	0.10	0.005	0.112	-1.0	0.200	35.151	34.935
4	0.08	0.004	0.101	-1.0	0.057	35.427	34.962
5	0.06	0.003	0.099	-1.0	-0.071	35.773	34.989

compared to that leaving the shelf is entrained every 150 m. This is twice the value obtained by *Quadfasel et al.* [1988]. The temperature of the plume, Θ_0, would be attained if water with an average temperature of 0^oC were added between the shelf break at 50 m and 200 m before the sinking shelf water reaches the first layer. No further efforts were made to optimize the choices of e and Θ_0.

Equation (1) is under-determined, but since the temperature of the shelf water, Θ_0, is constant, the system can be solved for the ratio of shelf water to advective water in each layer. The ratio is first determined for the deepest layer ignoring any deeper sinking contributions. The salinity of the shelf water reaching this layer is computed. Then it is possible to work upwards step-by-step into overlying levels.

We postulate that 0.5 Sv of the boundary flow, with a linear velocity profile (Table 2), enters the Canadian Basin and remains in the intermediate layers. Since the volume entrained into deeper sinking plumes from each layer will be the same, deeper convection, penetrating below the intermediate depth layers, will not affect the calculations for the intermediate depth layers. The volume displaced downwards out of the intermediate depth layers is estimated by a similar calculation to be 0.15 Sv, making the total boundary flow crossing the Lomonosov Ridge 0.65 Sv (*Jones et al.* submitted). The results are shown in Table 2. In these tables, Θ and S represent the temperature and salinity of the descending plume, including the entrained water from the upper layers, when it interleaves into the level, ie,

$$\Theta = \frac{\Theta_{oi} + \sum_{j=1}^{i-1} e \cdot \Theta'_j}{1 + \sum_{j=1}^{i-1} e}$$

with a similar expression for S.

It should be clear that the underlying convection picture is one with several sources generating an ensamble of plumes, which sink to their own terminal depths and merge with the surrounding water column [*Rudels*, 1986]. The obtained contributions represent the integral effect of these sources.

About 0.044 Sv originating above 200 m enters the layers between 200 m and 1700 m and water is added to all layers (Table 2). The amount actually leaving the shelves could be smaller, perhaps half, because of the entrainment between the shelf break and 200 m. The amount of shelf water reaching the deepest layer (5) is about 0.002 Sv, and the salinity required for the density flows to reach this deep is high. The salinity of 35.992 (salinity of the water reaching the 1400-1700 m layer) given in Table 2 is the salinity of the shelf plume at 200 m, and if the entrainment above this depth is considered, salinities on the shelf close to 37 might be needed. These values are very high, but are perhaps not impossible to attain in wind generated polynyas over shallow waters [*Aagaard et al.*, 1985].

Quadfasel et al. [1988] found that water from Storfjorden with an initial salinity of 35.45 could sink to 2000 m in Fram Strait. The upper part of the Arctic Ocean is more strongly stratified and a higher salinity should be required. The larger entrainment rate chosen also demands higher salinities. The thickness of the outflow observed by *Quadfasel et al.* [1988] was about 10 m at 2000 m. This indicates that the entrainment leads to a widening and not to a thickening of the plume.

The dense plumes leaving the shallow Arctic Ocean shelves probably have an initial depth of 5 - 10 m and keep this thickness as they sink and spread over the slope. The large initial density difference leads to high velocities and a strong entrainment relative to the volume. In the later stages the reduced density contrast leads to smaller velocities, and the entrainment rate per unit volume will go down. However, the small thickness of the plumes keeps the turbulence level fairly high and the entrainment will not go to zero. The increase in area of the interface between the plume and the surrounding water column acts to maintain a high total entrainment. The constant entrainment rate

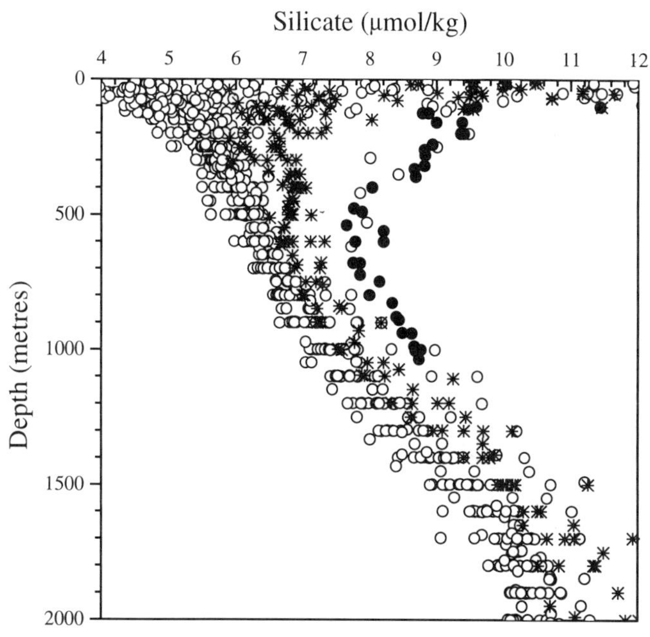

Fig. 7. Composite silicate profiles between 0 and 2000 m for several *Oden* stations. The asterisks represent stations 24 to 28 and stations 39, and 41 to 46. The filled circles represent stations 40 and 43. The open circles represent all of the remaining stations.

adopted in these calculations might therefore not be too far from reality.

The entrainment rate demanded by the requirement that the Canadian Basin water column should be reproduced is high but of the same order as that found by *Quadfasel et al.* [1988]. The difference could be due to still thinner initial plumes in the Arctic Ocean compared to Fram Strait. It is much higher than the entrainment rates implied from observations in the Weddell Sea (*Muench* pers. comm.). One reason for this could be the deeper shelves around Antarctica, which allow for thicker plumes. The entrainment per unit volume would then be smaller and the initial characteristics of the plume be more concerved. This would keep the temperature contrast between the plume and the outside water column high and the thermobaric effect can work and further facilitate the sinking of the plumes [*Killworth*, 1977; *Carmack*, 1986].

One further uncertainty in the present calculations lies in the chosen temperature and salinity value for the boundary current. A higher temperature in the entering Atlantic Layer requires a smaller entrainment rate than the one used. Furthermore, the Θ-*S* characteristics of the boundary current are assumed to hold along the entire continental slope up to the Chukchi Sea. This implies that the Canadian Basin is considered as a vertically stratified but horizontally homogenous box. Below we shall sea that this is not necessarily true.

Inferred Canadian Basin Circulation

The much lower chlorofluorocarbon concentrations (e.g., CCl_4) measured in the Makarov Basin suggest a slower circulation in the Canadian Basin compared to the Eurasian Basin (Figure 6). The "age" differences between the water columns on either side of the Lomonosov Ridge are large, the columns in the Makarov Basin at every level being a decade and more older than in the Eurasian Basin. This greater age would remove intrusions formed at the continental slope before they reach the central Makarov Basin. The small thicknesses expected for the intrusions in the Canadian Basin also allow for rapid disappearance of the anomalies. Inversions, which are probably of local origin, are observed in the Makarov Basin close to the Lomonosov Ridge (Figures 3 and 5). These interleaving structures must be formed at a near by frontal area and the sharp front at the Lomonosov Ridge is an obvious candidate for this.

Water recrossing the Lomonosov Ridge from the Canadian Basin is observed over the continental slope north of Greenland (Figures 5, 7, 8). Its Θ-*S* characteristics (Figure 5) are similar to those observed in the Makarov Basin, but the still smoother Θ-*S* curve and profiles over the Morris Jesup Plateau imply that some of the recrossing Canadian Basin water is older than that in the Makarov Basin (Figure 5).

The presence of Canadian Basin water in the Eurasian Basin near the Morris Jesup Plateau is also revealed by a composite plot of silicate concentration profiles (Figure 7). Profiles from the Makarov Basin (stations 24 to 28) show slightly higher concentrations throughout the Atlantic Layer and upper Polar Deep Water than do most profiles

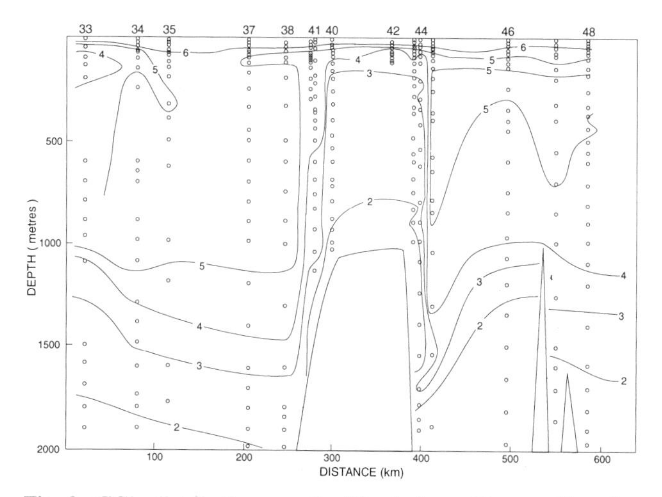

Fig. 8. CCl_4 distribution near the Morris Jesup Plateau. Note that this is a "corner" section, with stations 42 to 46 to a large degree mirroring stations 37 to 40.

from the Eurasian Basin, and correspond closely to the concentrations observed at stations 39, 41 and 46. The highest concentrations at this depth level are found at three locations (stations 40, 42, 43) in the region of the Morris Jesup Plateau.

Chlorofluorocarbon concentrations show a similar pattern. The lowest chlorofluorocarbon concentrations are found at stations 40, 42 and 43, while the stations in the Makarov Basin have concentrations not much different from those at stations 39, 41 and 46 (Figure 8). The silicate and chlorofluorocarbon data point to the existence of two sources from the Canadian Basin feeding into the Eurasian Basin near the Morris Jesup Plateau, one being the Makarov Basin and the other with higher silicate concentrations and lower chlorofluorocarbon concentrations located elsewhere in the Canadian Basin.

High silicate concentrations are the well-known signature of Upper Halocline Water formed in the Chukchi Sea region from Pacific water [*Anderson et al.*, 1994; *Jones and Anderson*, 1986]. The water with higher silicate concentrations is assumed to contain more high silicate shelf water and is likely to have passed close to the silicate source in the Chukchi Sea region. The amount of added shelf water is small and not likely to dominate the chlorofluorocarbon signal. The lower chlorofluorocarbon concentrations are therefore also indications of a longer flow path.

The two types of outflowing water from the Canadian Basin suggest that several loops also are present in the Canadian Basin, perhaps one loop in the Canada Basin transporting the high silicate Atlantic Layer and another loop in the Makarov Basin involving the water of intermediate silicate concentrations. Water from both loops cross the Lomonosov Ridge north of Greenland. All of the above arguments points towards the circulation pattern shown in Figure 9.

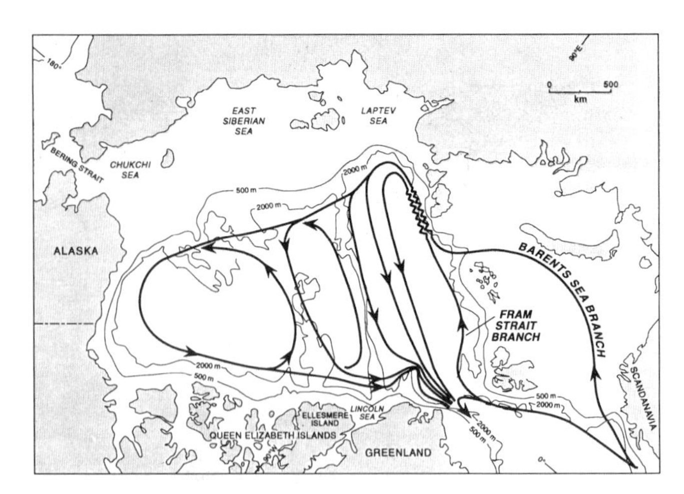

Fig. 9. Schematic diagram showing the inferred circulation in the Arctic Ocean of the Atlantic Layer and intermediate depth waters, between 200 m and 1700 m.

CONCLUDING REMARKS

The Atlantic and intermediate waters are supplied from two sources: through Fram Strait, and over the Barents Sea. The two contributions appear to be of equal strengths. The warmer, more saline Fram Strait branch dominates at lower densities, and the fresher, colder Barents Sea branch at higher densities. The main interactions between the two branches are intrusive mixing enhanced by double diffusive processes. The circulation is inferred from the distribution and amplitude of observed inversions and from chlorofluorocarbon concentrations. Cyclonic gyres with several loops are detected. In the Eurasian Basin, a tight loop brings the water back towards Fram Strait, the circulation being swifter over the Nansen-Gakkel and Lomonosov ridges than along the Amundsen Basin. Only a fraction of mainly the Barents Sea branch crosses the Lomonosov Ridge. In the Canadian Basin, the Θ-S measurements as well as the chlorofluorocarbon observations indicate a longer ventilation time. The possible existence of loops also in the Canadian Basin can be inferred from chlorofluorocarbon and silicate observations.

The heat loss from the Atlantic Layer occurs in two steps, first with the merging of the two branches north of the Kara Sea, and second by the incorporation of dense, cold descending shelf water mostly further to the east. In both cases, the heat loss to the atmosphere occurs on the shelves and the lowering of the temperature in the Atlantic Layer is caused by isopycnal mixing with and heating of the colder waters from the shelves. Another important heat loss occurs in connection with the vertical transport of entrained water downwards. Entrainment of Atlantic Layer water into sinking density flows is the main mechanism for supplying of heat to the deeper layers of the Canadian Basin. The heat loss from the Atlantic Layer to the upper layer, the ice, and the atmosphere appears small and may, to a first approximation, be neglected.

The intermediate waters of the Canadian Basin are distinct from the Eurasian Basin waters, the Atlantic Layer being colder and less saline, the layers below warmer and more saline. These features highlight the importance of the boundary convection and entrainment in the Canadian Basin. The Eurasian Basin is dominated by advection and interleaving and the intermediate waters are not laterally homogenized before they converge and become incorporated into the waters recirculating in Fram Strait.

Previous observations in Fram Strait have not revealed the multi-branched structure of the exiting Atlantic

Layer, not to mention the deeper levels, primarily because of the strong recirculation taking place in and just north of Fram Strait [*Quadfasel et al.*, 1987; *Bourke et al.*, 1988; *Rudels*, 1987]. This recirculation destroys the characteristics of the loops of the Eurasian Basin gyre, and remnants of slope-basin interleaving cannot be distinguished from the local interactions between the outflow and warm Atlantic water entering from and recirculating towards the south. The deeper inversions in particular have disappeared and, while intrusions are observed also at these levels close to Fram Strait, the regularity found in the interior of the Arctic Ocean is gone. Of the intermediate Arctic waters, only the Atlantic Layer and the upper Polar Deep Water of the Canadian Basin, both found close to the western continental shelf, appear to survive the passage through Fram Strait as distinct waters. The upper Polar Deep Water, easily distinguished by its Θ-S shape, can be followed into the Icelandic Sea and the Denmark Strait.

Acknowledgments. We wish to express our gratitude to all of our colleagues on the *Oden* who helped make the *Oden* 91 expedition a scientific success. We wish to thank M. Hingston and F. Zemlyak who were instrumental in collecting chlorofluorocarbon data, B-M. Dahlberg for skillful assistance in the nutrient determination and C. Ross and G. Björk for helpful comments on an early version of the manuscript. We wish to acknowledge support from the Deutsche Forschungsgemeinshaft SFB 318 (B.R.), the Panel on Energy Research and Development (Canada) (E.P.J.), the Swedish Natural Science Research Council (L.G.A.), and the Alfred-Wegener-Institute für Polar- und Meeresforsung (contribution no. 724) (G.K).

REFERENCES

Aagaard, K., A synthesis of the Arctic Ocean circulation, *Rapp. P.-V. Reun. Cons. Int. Explor. Mer., 188,* 11-22, 1989.

Aagaard, K., and P. Greisman, Towards new mass and heat budgets for the Arctic Ocean, *J. Geophys. Res., 80,* 3821-3827, 1975.

Aagaard, K., and E. C. Carmack, The role of sea ice and other fresh water in the Arctic circulation, J. Geophys. Res., 94, 14,485-14,498, 1989.

Aagaard, K., L. K. Coachman, and E. C. Carmack, On the halocline of the Arctic Ocean, *Deep-Sea Res., 28,* 529-545, 1981.

Aagaard, K., J. H. Swift, and E. C. Carmack, Thermohaline circulation in the Arctic Mediterranean Sea, *J. Geophys. Res., 90,* 4833-4846, 1985.

Anderson, L. G., E. P. Jones, K. P. Koltermann, P. Schlosser, J. H. Swift, and D. W. R. Wallace, The first oceanographic section across the Nansen Basin in the Arctic Ocean, *Deep-Sea Res., 36,* 475-482, 1989.

Anderson, L. G., G. Björk, O. Holby, E. P. Jones, G. Kattner, K. P. Koltermann, B. Liljeblad, R. Lindegren, B. Rudels, and J. H Swift, Water masses and circulation in the Eurasian Basin: Results from the *Oden* 91 expedition, *J. Geophys. Res., in press,* 1994.

Blindheim, J., Cascading of Barents Sea bottom water into the Norwegian Sea, *Rapp. P.-V. Reun. Cons. Int. Explor. Mer., 188,* 49-58, 1989.

Bourke, R. H., A. M. Weigel, and R. G. Paquette, The westward turning branch of the West Spitsbergen Current, *J. Geophys. Res., 93,* 14,065-14,077, 1988.

Carmack, E. C., Circulation and mixing in ice covered waters, in *The Geophysics of Sea Ice,* edited by N. Untersteiner, pp. 641-712, Plenum Press, New York, 1986.

Coachman, L. K., and C. A. Barnes, The movement of Atlantic water in the Arctic Ocean, *Arctic, 16,* 8-16, 1963.

Jones, E. P., and L. G. Anderson, On the origin of the chemical properties of the Arctic Ocean halocline, *J. Geophys. Res., 91,* 10,759-10,767, 1986.

Jones, E. P., B. Rudels and L. G. Anderson, Deep waters of the Arctic Ocean: Origin and circulation, *Deep-Sea Res.,* submitted to, 1993.

Killworth, P. D., Mixing on the Weddell Sea continental slope, *Deep-Sea Res.,* 24, 427-448, 1977.

Kunze, E., Limits of growing finite-length salt fingers. A Richardson number constraint, *J. Mar. Res., 45,* 533-556, 1987.

Linden, P. F. and T. G. L. Shirtcliffe, The diffusive interface in double diffusive convection, *J. Fluid Mech., 87,* 417-432, 1979.

Midttun, L., Formation of dense bottom water in the Barents Sea, *Deep-Sea Res., 32,* 1233-1241, 1985.

Nansen, F., Oceanography of the North Polar Basin. *The Norwegian North Polar Expedition 1893-96. Scientific Results, 2(9),* pp. 427, 1902.

Perkin, R. G., and E. L. Lewis, Mixing in the West Spitsbergen Current, *J. Phys. Ocean., 14,* 1315-1325, 1984.

Quadfasel, D., J.-C. Gascard, and P. K. Koltermann, Large-scale oceanography in Fram Strait during the 1984 Marginal Ice Zone experiment, *J. Geophys. Res., 92,* 6719-6728, 1987.

Quadfasel, D., B. Rudels, and K. Kurz, Outflow of dense water from a Svalbard fjord into the Fram Strait, *Deep-Sea Res., 35,* 1143-1150, 1988.

Quadfasel, D., B. Rudels and S. Selchow, The Central Bank vortex in the Barents Sea: water mass transformation and circulation, *ICES mar. Sci Symp.,* 195, 40-51, 1992.

Quadfasel, D., A. Sy, and B. Rudels, A ship of opportunity section to the North Pole: Upper ocean temperature observations, *Deep-Sea Res., 40,* 777-789, 1993.

Ruddick, B. and J. S. Turner, The vertical length scale of double-diffusive intrusions, *Deep-Sea Res.*, 26, 903-913, 1979.

Ruddick, B., and D. Hebert, The mixing of meddy "Sharon", in *Small-scale mixing and turbulence in the ocean,* edited by J.C.N. Nihoul and B.M. Jamart, pp. 481-507, Elsevier, Amsterdam, 1988.

Rudels, B., The Θ-S relations in the northern seas: Implications for the deep circulation, *Polar Res., 4,* 133-159, 1986.

Rudels, B., On the mass balance of the Polar Ocean with special emphasis on the Fram Strait, *Norsk Polar. Skr., 188,* 1-53, 1987.

Rudels, B., Mixing processes in the northern Barents Sea, *Rapp. P.-V Réun. Cons. Int. Explor. Mer.*, 188, 36-48, 1989.

Rudels, B., The diffusive interface at low stability: The importance of non-linearity and turbulent entrainment, *Tellus, 43A,* 153-167, 1991.

Rudels, B., On double-diffusive layering, *J. Phys. Oceangr.*, submitted to, 1993.

Swift, J. H., and K. Aagaard, Seasonal transitions and water mass formation in the Iceland and Greenland Seas, *Deep-Sea Res., 28A,* 1107-1129, 1980.

B. Rudels, Institut für Meereskunde, Universität Hamburg, Troplowitzstrasse 7, D-22529 Hamburg 54, Germany

E. P. Jones, Department of Fisheries and Oceans, Bedford Institute of Oceanography, P.O. Box 1006, Dartmouth, Nova Scotia B2Y 4A2, Canada

L. G. Anderson, Department of Analytical and Marine Chemistry, University of Goteborg and Chalmers University of Technology, S-412 96 Göteborg, Sweden

G. Kattner, Alfred-Wegener-Institute für Polar- und Meeresforsung, Postfach 12 01 61, Columbusstraße, D-275 15 Bremerhaven, Germany

Nutrient-Based Tracers in the Western Arctic: A New Lower Holocline Water Defined

David K. Salmon and C. Peter McRoy

Institute of Marine Science, University of Alaska, Fairbanks, Alaska

We re-examined historical data on oxygen, silicate, and the conservative tracers NO and PO in waters over the Chukchi and Beaufort slopes to determine the sources for waters constituting the lower halocline in the western Arctic Ocean. The lower halocline in the Chukchi and Beaufort seas is characterized by a NO minimum, as is the lower halocline in the central Arctic Ocean; however, silicate concentrations at salinities of 34.15 to 34.70 psu are higher (20 to 40 μmol/L) over the Chukchi-Beaufort slope than concentrations (<14 μmol/L) in the lower halocline at stations in the Canada, Makarov and Amundsen basins. Data from Ice Island T-3 in the Canada Basin confirm that silicate concentrations are higher (19 to 31 μmol/L) than in the lower halocline elsewhere in the Arctic Ocean. Also, slope waters are warmer than those in other Arctic regions in the salinity range of the lower halocline. Silicate data show a distinction between lower halocline water (LHW) in the Canada Basin and the other Arctic basins at the same salinities. The nutrient maximum associated with upper halocline water (UHW) in the Canada Basin spreads over a wider salinity range than in the Makarov and Amundsen basins. Also, the oxygen minimum over the Chukchi-Beaufort slopes occurs at higher salinities than elsewhere.

High silicate concentrations in the Arctic are principally derived from Pacific waters, but processes exist by which high silicate concentrations can co-occur with salinities between 34.15 and 34.70 psu (i.e., at salinities above those of Pacific waters.) The NO/PO ratio in relation to salinity suggests that high silicate concentrations in this range are the result of diapycnal mixing of upper and lower halocline waters over slope regions. However, individual distributions of NO and PO show distinct minima in the lower halocline that cannot be produced only through a mixing mechanism. Physical mechanisms and chemical properties suggest that the Chukchi, East Siberian, and Laptev Seas produce halocline waters over a wide salinity range (32.0 to 34.7 psu). However, for the Chukchi shelf, no feasible mechanism has been identified that can produce low NO water at salinities high enough to ventilate the lower halocline. We conclude that the low NO water characterizing LHW must be produced over the Barents and Kara shelves, where salinity is high and nutrient enrichment occurs as shelf-derived water interleaves with slope water as the halocline waters transit the Canada Basin, thereby producing a new water mass. Here we name this water Canada Lower Halocline Water (CLHW). The water is characterized by high salinity with high silicate (and other nutrients).

INTRODUCTION

The upper Arctic Ocean has a strong halocline that occurs between the surface mixed layer and the deeper Atlantic layer. In the Canada Basin and near the North Pole, the waters of the halocline are divided into upper halocline water (UHW) and lower halocline water (LHW) based on physical and chemical properties [Jones and Anderson, 1986]. UHW was not observed at stations occupied by the F. S. *Polarstern* in the Nansen Basin during 1987 [Anderson and Jones, 1992] although it has been found north of Fram Strait between Greenland and Svalbard [Anderson and Jones, 1986] and in western Fram Strait [Anderson and Dyrssen, 1981; Jones et al., 1991]. Nutrient concentrations are significantly higher in UHW than in LHW [Jones and Anderson, 1986; Rudels et

The Polar Oceans and Their Role in Shaping the Global Environment
Geophysical Monograph 85

al., 1991]. In UHW silicate and phosphate concentrations are nearly three and two times those of LHW, respectively [Jones and Anderson, 1986]. Nitrate concentrations in UHW and LHW are not as disparate (17 and 12 μmol/kg respectively).

A prominent feature of UHW is a nutrient maximum that coincides with salinities of 33.1 to 33.2 psu. This indicates that UHW is Pacific water that has been modified on the shelves of the Bering and Chukchi seas before entering the Arctic [Jones and Anderson, 1986; Moore et al., 1983; Wilson and Wallace, 1990]. Based on a trace metal maximum occurring in the region of the UHW nutrient maximum, Yeats [1988] argues that Pacific water has been altered only by in situ processes rather than by interaction with shelf sediments. This latter process is the mechanism for formation of the UHW nutrient maximum suggested by Moore and Smith [1986] who relied on lead and polonium isotope data.

Similar silicate concentrations in Atlantic water and LHW led Moore et al. [1983] to conclude that LHW is Atlantic water modified only by mixing with the overlying water. However, nitrate and oxygen concentrations are higher in Atlantic water than in LHW [Jones and Anderson, 1986] so the low NO signal of the LHW cannot be produced through a mixing process. LHW is characterized by a minimum in NO at salinities of 34.2 to 34.3 psu [Anderson and Jones, 1992; Jones and Anderson, 1986]. and is also delimited by a change in the slope of T/S diagrams that occurs at about 34.2 psu [Moore et al., 1983]. The source regions for LHW are usually taken to be in the Barents and Kara Seas because of the low NO and high salinities observed in these shelf-seas [Jones and Anderson, 1986; Jones et al., 1991]. Wilson and Wallace [1990] show that NO distributions alone are insufficient to identify source regions for halocline waters. They show that the occurrence of low NO water prevails in the Chukchi Sea compared to the Barents Sea. However, a crucial point overlooked by Wilson and Wallace [1990] is that the low NO water over the Chukchi shelf occurs at salinities that are too low to ventilate the lower halocline [Wilson and Wallace used summer Chukchi shelf data), and that the Chukchi shelf waters with salinities sufficient to ventilate the lower halocline have NO values that are substantially higher than those occurring in the NO minimum of LHW. The Chukchi shelf waters that might be saline enough to ventilate the lower halocline can only be produced during winter when nitrate concentrations are high throughout the water column and concomitant oxygen concentrations are close to 100% saturation [McRoy and Salmon, unpublished data]. The enigma is that the concentrations of both of these constituents are too high to produce lower halocline water with NO values that approach those observed at the NO minimum in LHW.

We show here that low NO water over the Chukchi and Beaufort slopes and in the central Canada Basin in the salinity range of LHW is characterized by nutrient concentrations (particularly silicate) higher than those in the LHW in the Amundsen, Makarov and Nansen basins. High silicate values over the Beaufort and Chukchi slopes and at Ice Island T-3 in the Canada Basin spread over a wider range of salinities than in the other Arctic basins. Silicate concentrations at salinities characterizing LHW and variations of the oxygen minimum associated with the UHW nutrient maximum distinguish lower halocline water in the Canada Basin from LHW that in the Nansen, Amundsen and Makarov basins. We propose the term Canada Lower Halocline Water (CLHW) to distinguish this water from lower halocline water occurring in other Arctic basins. This new analysis of historical data was instigated by a quasi-synoptic transect by submarine of the Canadian Basin. The transect indicated a reassessment was warranted.

DATA

This study was undertaken using historical physical and chemical data. NO and PO concentrations were computed from the nutrient and oxygen data. NO and PO are conservative tracers defined by NO= $[O_2] + 9[NO_3]$ and PO= $[O_2] + 135[PO_4]$ [Broecker, 1974]. We designate "slope" waters to be those in the depth range from 200 to 2000 m. Also, the terms "lower halocline water" and "LHW" are not used interchangeably. "LHW" refers to the water mass described by Jones and Anderson [1986, cf. their table 1], which is characterized by a NO minimum with a silicate concentration of less than 12 μmol/L at salinities of 34.2 to 34.6. The term "lower halocline water" refers to a water mass that shares the same salinity and NO characteristics as LHW, but has a silicate concentration greater than 20 μmol/L. Finally, the term "lower halocline" is used interchangeably to describe water masses with salinities between 34.2 and 34.7 psu.

The Chukchi slope data were taken during 1968 and 1969 onboard US Coast Guard icebreakers. Beaufort slope data were also acquired during icebreaker cruises in 1968 and 1969, and again in 1971 and 1972 as part of the WEBSEC study over the Beaufort shelf and slope regions (Figure 1). Physical and chemical data from Ice Island T-3 were acquired during 1967 to 1969 when the island drifted through the Canada Basin from 80°N to 85°N between 160°W and 125°W [Kinney et al., 1970b]. Other data are from the AIWEX (Arctic Internal Waves Experiment), CESAR (Canadian Expedition to Study the Alpha Ridge) and LOREX (Lomonosov Ridge Experiment) ice camps.

RESULTS

In waters over the Beaufort and Chukchi slopes, high silicate (20 to 40 μmol/L) occurs near the nutrient maximum in UHW (33.1 psu) and in the lower halocline between 34.15 and 34.70 psu (Figure 2). Twenty three of the 28 Beaufort and Chukchi slope stations (82%) with salinities of 34.15 to 34.60 psu, (which encompasses the salinity range of the LHW), have silicate concentrations in the 20 to 40 μmol/L range.

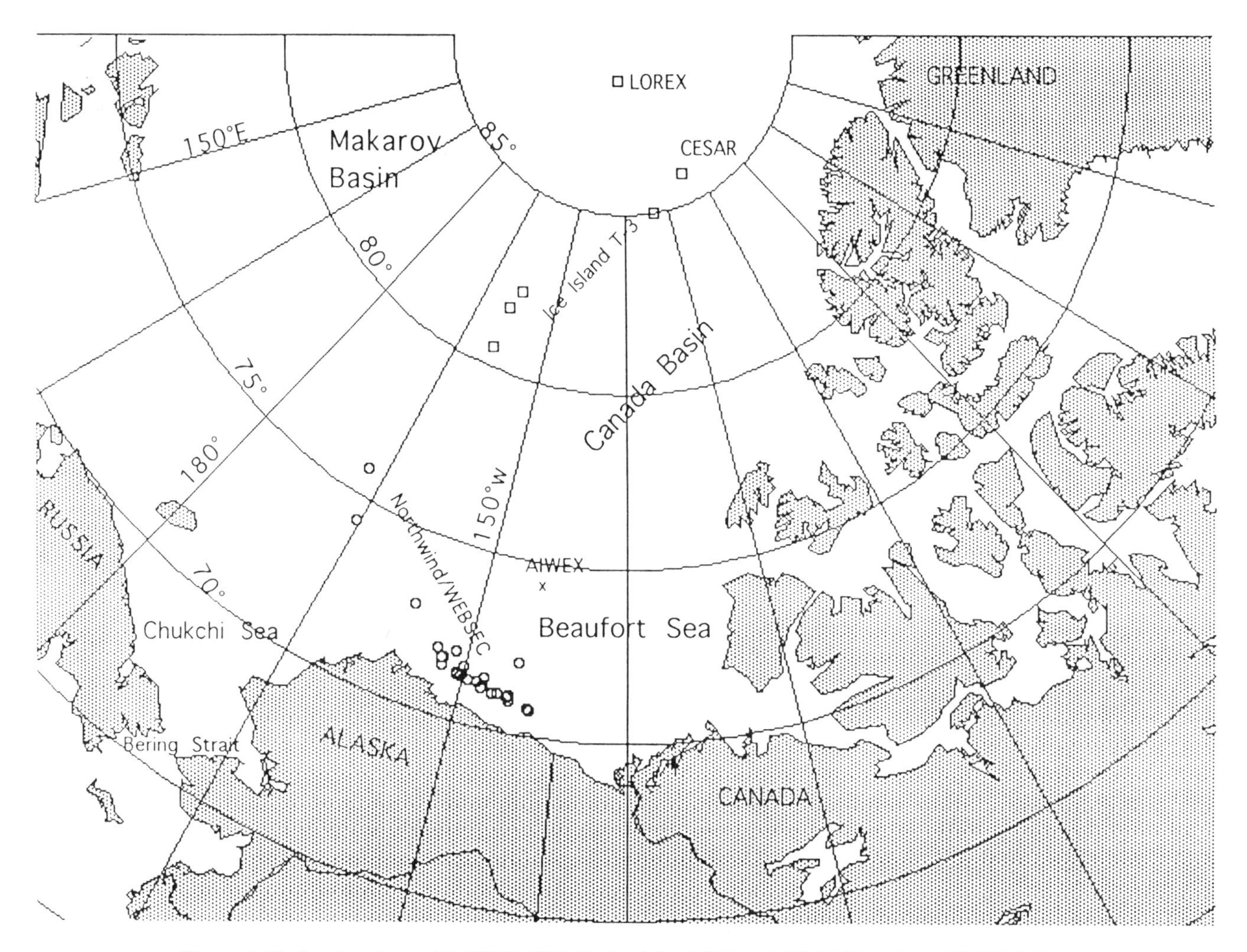

Figure 1. Station locations of LOREX, CESAR, Ice Island T-3, and USCGC Northwind/WEBSEC.

The AIWEX sites were located close to the Beaufort slope stations, and many stations have more detailed vertical resolution than other data. The AIWEX data (not shown) confirm that high silicate concentrations exist throughout the lower halocline in the Beaufort Sea. Silicate concentrations of 18 to 26 μmol/L occur at salinities from 34.2 to 34.6 psu [cf. Anderson and Swift, 1990].

High silicate occurred over a wide range of salinities at Ice Island T-3 in the central Canada Basin, including the lower region of the halocline here (Figure 3). Using data from Kinney et al. [1970 b], we examined all T-3 stations from 1967 to 1969 and found that silicates were between 19 and 31 μmol/L for salinities of 34.2 to 34.3 psu. Silicates at salinities from 34.20 to 34.50 at T-3 were higher in all but one case than those occurring in this salinity range at the LOREX and CESAR sites (Table 1). The exception is the T-3 station that was located close to the CESAR and LOREX stations (Figure 1). At this site property distributions were similar to those at CESAR and LOREX. In particular, the oxygen and silicate values at this T-3 station more closely resembled those at the CESAR and LOREX sites than they did the concentrations at T-3 stations in the central Canada Basin. The implication of these property distributions is that chemical characteristics of the halocline water of the Canada Basin are distinct from those of the Makarov and Amundsen basins (as characterized by CESAR and LOREX respectively).

A point of further clarification is necessary. Jones and Anderson [1986] reported lower silicates in the lower halocline at T-3 than we report here, but they did not have the original data of Kinney et al. [1970b] so their value was obtained through approximation in a region of very steep gradients [L. G. Anderson, 1993, personal comm.]. Although NO is the key tracer for LHW, differences in silicates between UHW and the lower halocline demonstrate that this clarification is a significant point. It suggests that the lower halocline water in the Canada Basin can be distinguished from LHW in the Makarov and Amundsen basins on the basis of

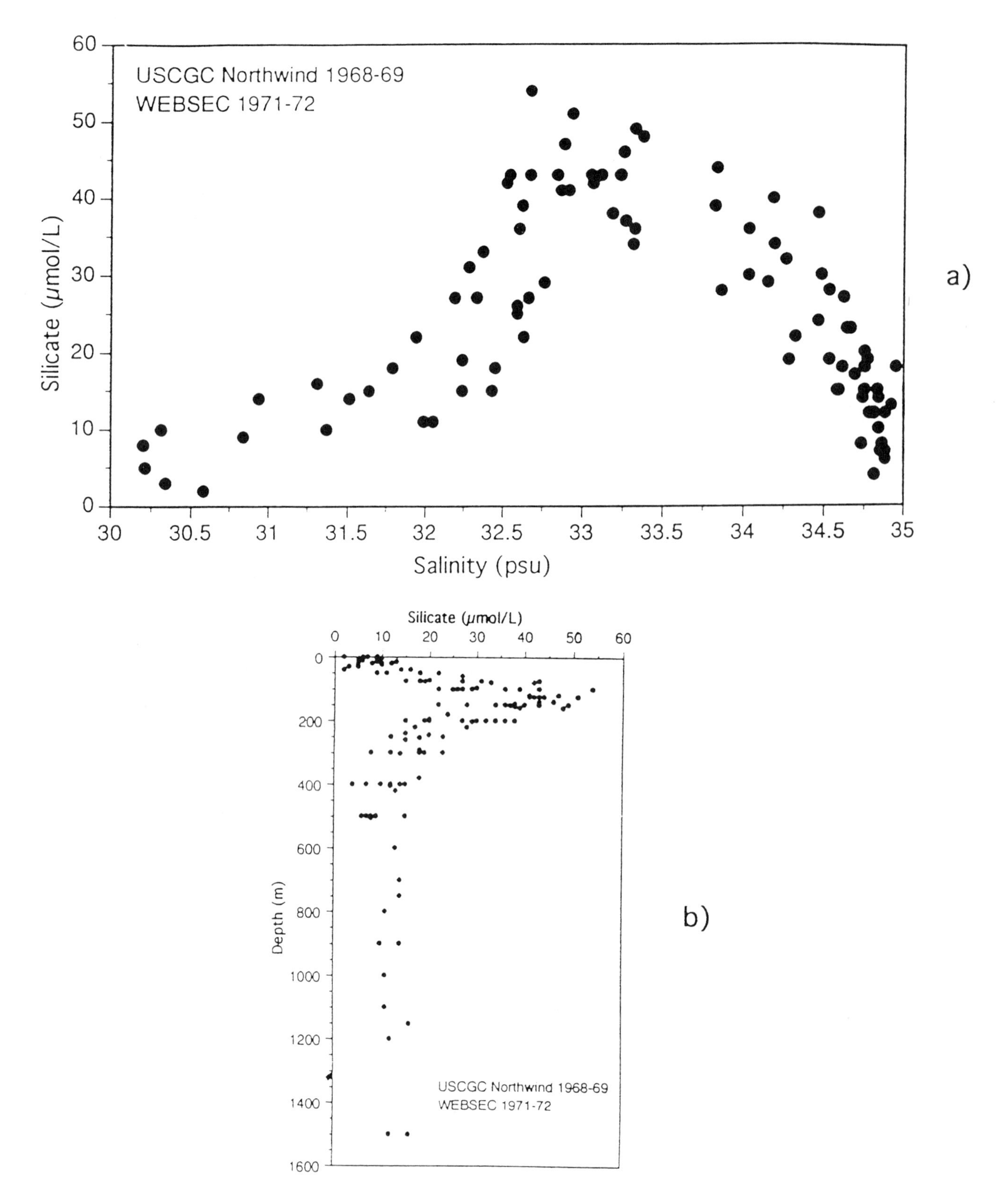

Figure 2. a) Distribution of silicate versus salinity and b) silicate versus depth at stations over the Beaufort and Chukchi slopes.

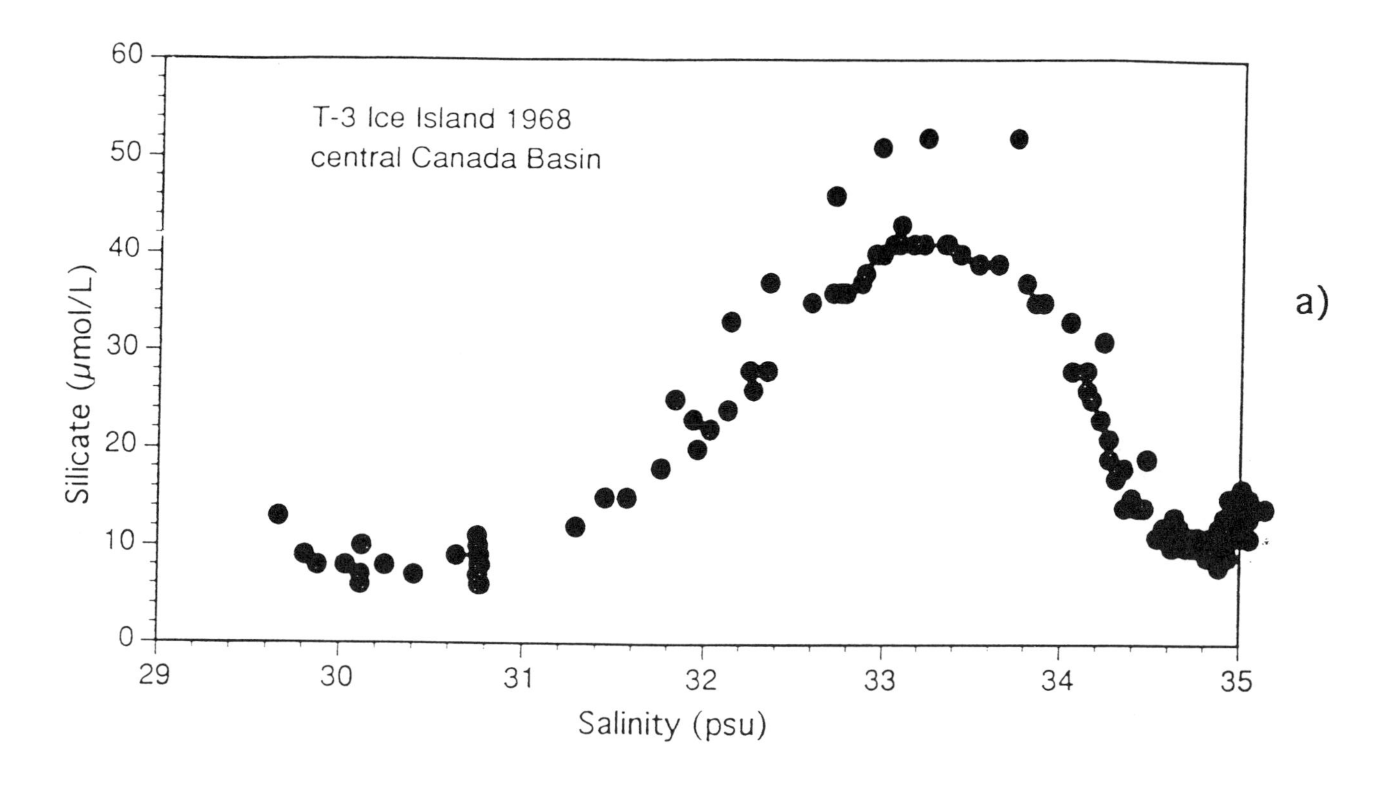

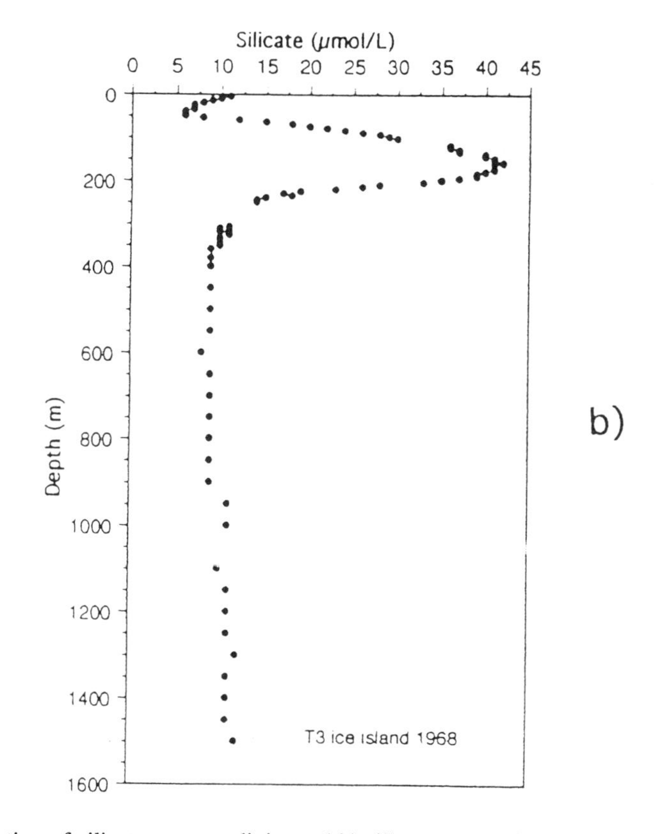

Figure 3. a) Distribution of silicate versus salinity and b) silicate versus depth at the T3 Ice Island in the central Canada Basin.

Table 1. Silicate concentrations and associated salinities in LHW at CESAR, LOREX, and Ice Island T-3.

	Salinity (psu)	Silicate (μmol/l)
CESAR		
	34.31	13
	34.38	11
	34.45	11
	34.47	10
	34.30[a]	33
LOREX (one profile)		
	34.22	16
	34.31	13
	34.37	11
	34.43	10
	34.48	10
Ice Island T-3 (six profiles)		
	34.21	23
	34.22	31
	34.22	27
	34.26	21
	34.26	19
	34.30	17
	34.34	18
	34.38	15
	34.42	16
	34.42	14
	34.45	14
	34.47	19
	34.49	12

[a] This measurement occured at a depth of 247 m, close to depth of LHW observed at T3 (230 m). The other CESAR measurements at this salinity were at depths of less than 190 m.

silicate concentrations. In other words, while the lower halocline regions in the Canada and Makarov basins (there are no NO data from LOREX stations in the Amundsen Basin) share similar characteristics in terms of their NO minima, the lower halocline waters in the Canada Basin are distinguished from LHW in the other basins by virtue of high silicates and by differences in the oxygen minimum (discussed below). On the basis of these large differences in silicates in the lower halocline regions in the Canada, Makarov and Amundsen basins we propose the term Canada Lower Halocline Water (CLHW) to distinguish this water type from LHW that occurs in the other Arctic Ocean basins.

The salinity-silicate distribution in the LHW at T-3 resembles the distributions over the Beaufort and Chukchi slopes. Silicate values are significantly lower (<12 mmol/L) at salinities of 34.15 to 34.70 at stations in the Amundsen Basin (LOREX data) and Nansen Basin [Anderson and Dyrssen, 1981] as well as for the CESAR station located along the boundary between the Canada and Makarov basins. Water near 34.2 to 34.3 psu in the central Arctic Ocean (Canada, Amundsen and Makarov basins) is classified as LHW, based largely on the observed NO minimum that occurs in this salinity range. Silicate in LHW is generally between 10 and 12 μmol/L [Jones and Anderson, 1986]. In contrast, our analysis indicates that the lower halocline of the central Arctic Ocean is not as homogeneous as previously reported by Jones and Anderson [1986].

Silicates in the lower halocline of the Canada Basin are always higher than those in the Makarov, Amundsen, and Nansen basins. Using silicate, we distinguish waters in the lower halocline in the Chukchi and Beaufort slope regions, as well as the central Canada Basin, from LHW in the

Amundsen, Makarov, and Nansen basins. This shows that the UHW nutrient maximum occurs over a wider salinity range along the Beaufort and Chukchi slopes and the Canada Basin than it does in UHW of the Amundsen and Makarov basins. Except for oxygen (and NO) concentrations in the upper and lower halocline, differences over the Beaufort slope are not as distinct as those observed in other Arctic basins (i.e. at the CESAR and LOREX stations). This result is consistent with that of Jones and Anderson [1990] who found that the UHW nutrient maximum over the shelf north of Ellesmere Island occurs over a broader salinity range than in the central Arctic.

We also find that temperatures in the lower halocline over the Beaufort and Chukchi slopes are warmer than those of LHW at LOREX, T-3, and, to some extent, CESAR (Table 2). There is considerable spatial and interannual variability, but slope waters are generally warmer (by a few tenths of a degree) at a given salinity than those in the central basins. Warmer temperatures over the slope are consistent with the results of Jones and Anderson [1990] who reported that halocline waters north of Ellesmere Island are about 0.3°C warmer than in the central Arctic Ocean (at CESAR stations).

The lower halocline in the Beaufort and Chukchi slope

Table 2. Comparison of temperatures through the LHW at CESAR, LOREX, Ice Island T-3 and Beaufort slope stations; values from multiple profiles.

	Salinity (psu)	Potential Temperature (°C)
CESAR		
	34.27	-1.11
	34.38	-0.69
	34.47	-0.51, -0.47, -0.46
LOREX		
	34.22	-1.08
	34.37	-0.84
	34.46	-0.64
Ice Island T-3		
	34.22	-0.92, -1.0
	34.28	-0.95
	34.34	-0.78, -0.78
	34.37	-1.05
	34.45	-0.55
	34.46	-0.75
	34.47	-0.62
	34.49	-0.91
Beaufort Slope		
	34.23	-0.3
	34.23[a]	-0.69
	34.24[a]	-0.63
	34.25	-0.66
	34.27	-0.31
	34.28	-0.63
	34.34[a]	-0.56, -0.47
	34.37	-0.29
	34.37	-0.60
	34.38	-0.51
	34.45	-0.34
	34.45[a]	-0.43
	34.47	-0.47
	34.47[a]	-0.36
	34.48	-0.22
	34.49	-0.30, -0.47

[a] Indicates data taken during 1959-60 by Kusunoki et al. (1962)

regions is characterized by NO and PO minima that occur at depths of 200 to 250 m (Figure 4). These minima are distinct in individual station profiles (not shown). The vertical structure of slope profiles is similar to that observed at T-3 and at CESAR (Figure 4). NO in the minimum ranges from 331 to 423 μmol/L, with most values lying in the narrower range of 360 to 410 μmol/L. NO minima are more clearly defined than PO minima and occur in conjunction with the lower halocline oxygen minimum at 200 to 250 m. The PO minimum, with values from 373 to 436 μmol/L, often lies below the NO and oxygen minima. Most NO minima occur at salinities of 34.4 to 34.6, while PO minima occur at 34.7 to 34.8 psu (Figure 5). At both CESAR and T-3 in the Canada Basin, the salinity of the NO minimum is 34.4 and the salinity of the PO minimum is 34.5 to 34.6 psu.

NO and PO decrease and NO/PO ratios increase rapidly in the lower halocline at salinities between the UHW nutrient maximum and the NO (and oxygen) minimum at CESAR, T-3 and over the Beaufort slope (Figures 5 and 6). These abrupt changes typically occur between 150 and 200 m, which is the transition zone between UHW and LHW. A minimum in NO/PO is associated with the UHW nutrient maximum. NO/PO ratios at the minimum range from about 0.70 to 0.85. These values are close to the mean of 0.78 computed by Wilson and Wallace [1990] for the UHW nutrient maximum at T-3.

The distinct PO and NO/PO ratios between UHW and LHW indicate that the water types have different sources and that the high silicate they have in common is the result of strong mixing in slope regions with Pacific water entering from the Chukchi Sea. However, in contrast to temperature-salinity and NO/PO, which suggest that mixing of UHW and Atlantic water is a feasible mechanism for the production of LHW, the NO and PO minima that occur in the lower halocline preclude the possibility of such mixing to form LHW, i.e. UHW and Atlantic Water have higher NO and PO than does the LHW, therefore LHW is not a mixing product of UHW and Atlantic Water.

A well defined oxygen minimum occurs below the nutrient maximum in the Arctic Ocean at T-3 [Kinney et al., 1970a] and CESAR [Jones and Anderson, 1986], with concentrations of 259 and 268 μmol/L and depths of 200 m and 120 m, respectively (Figure 7). The depth of the minimum in each case is related to the UHW nutrient maximum which occurs at about 160 m at T-3 and 110 m at CESAR. The oxygen minimum is deeper, 200 to 250 m, along the Beaufort slope (Figure 7). An oxygen minimum (about 260 μmol/L) has also been observed in the Mackenzie shelf region of the Beaufort Sea in 180 to 225 m at salinities of 33.8 to 34.5 psu [Kusunoki, 1962; Moore et al., 1992]. Kusunoki et al. [1962] show that the oxygen minimum in the Beaufort occurred at 200 to 250 m when the Ice Island T-3 passed through the region during 1959-60. The minimum (257 to 275 μmol/L) occurred at salinities of 34.1 to 34.6 psu.

The oxygen minimum (266 to 272 μmol/L) occurs at 34.5 psu over the Beaufort and Chukchi slopes (Figure 8a). AIWEX data confirm that this oxygen minimum occurs on or near the 34.5 isohaline [cf. Anderson and Swift, 1990]. The oxygen minimum (268 μmol/L) at CESAR occurs at 33.6 psu (Figure 8b), while the minimum (about 259 μmol/L) at T-3 occurred at 34.0 psu (Figure 8c). The exception is the T-3 station in the Alpha ridge region, where the minimum occurred at 33.6 to 33.8 psu, as at the CESAR station (Figure 8d).

The oxygen minimum over the slope occurs at a higher salinity than in the central Arctic Ocean even though the nutrient maximum occurs at the same salinity (33.1) over the slope and in the central Arctic Ocean (at T-3, LOREX and CESAR). Jones and Anderson [1986] show a close connection between the oxygen minimum and the UHW nutrient maximum. A vertical profile of apparent oxygen utilization (AOU) versus salinity shows an AOU minimum that coincides with the UHW nutrient maximum. This association again confirms that the waters of the upper halocline are spread over a wider salinity range over the slope of the Canada Basin than they are in other Arctic Ocean basins.

Oxygen in the lower halocline can be used to infer different sources for the lower halocline water in the various Arctic basins. In particular, the oxygen content at about 34.5 psu provides a clue to sources. We noted above that an oxygen minimum of 260 to 270 μmol/L exists along the Beaufort slope at 34.5 psu. In contrast, Wilson and Wallace [1990] note the existence of a local oxygen maximum (300 μmol/L) at 34.5 psu at the T-3 station near Alpha Ridge. This maximum erodes westward and does not exist in the Canada Basin at the more westward T-3 stations. They further speculate that the oxygen maximum is attributable to Atlantic waters.

At T-3 stations between the Beaufort slope and the Alpha Ridge, oxygen concentrations were 290 to 300 μmol/L at salinities near 34.5 psu. Thus a marked gradient in oxygen concentrations occurs on the 34.5 salinity surface between the slope of the Canada Basin, the central Canada Basin , and the Makarov Basin. In addition to this gradient, the oxygen minimum exists at a salinity of 34.5 psu along the Beaufort slope, while a local maximum exists at this salinity along the border of the Canada and Makarov basins. We believe that the low oxygen (< 270 μmol/L) observed along the Beaufort slope at this high salinity indicates the strong influence of Pacific water and the associated high nutrients. The existence of the minimum on less saline surfaces away from the slope is the result of the diminishing influence of the Pacific water nutrient signal away from the source region.

The intermediate oxygen content (~ 290 μmol/L) observed at a salinity 34.5 at T-3 in the Canada Basin probably reflects both the influence of Pacific and Atlantic derived waters. It is likely that Pacific and Atlantic-derived lower halocline waters mix in the boundary between the Canada and Makarov basins (i.e. near Alpha Ridge). In the Alpha Ridge region, the Pacific water signature is only observed in UHW and is lost in LHW by the stronger Atlantic water oxygen signal.

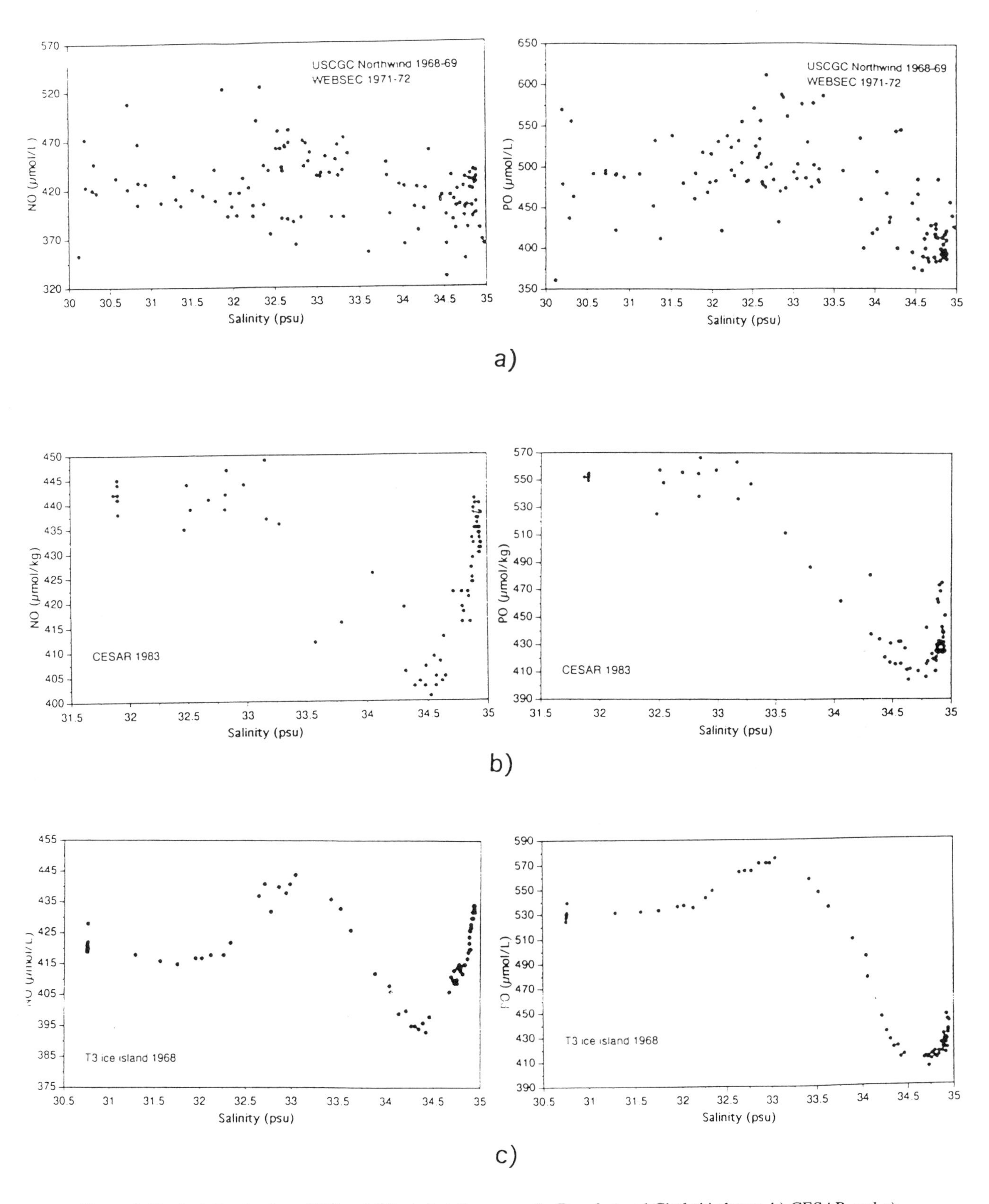

Figure 4. Vertical distribution of NO and PO at a) stations over the Beaufort and Chukchi slopes, b) CESAR and c) T3 Ice Island in the central Canada Basin.

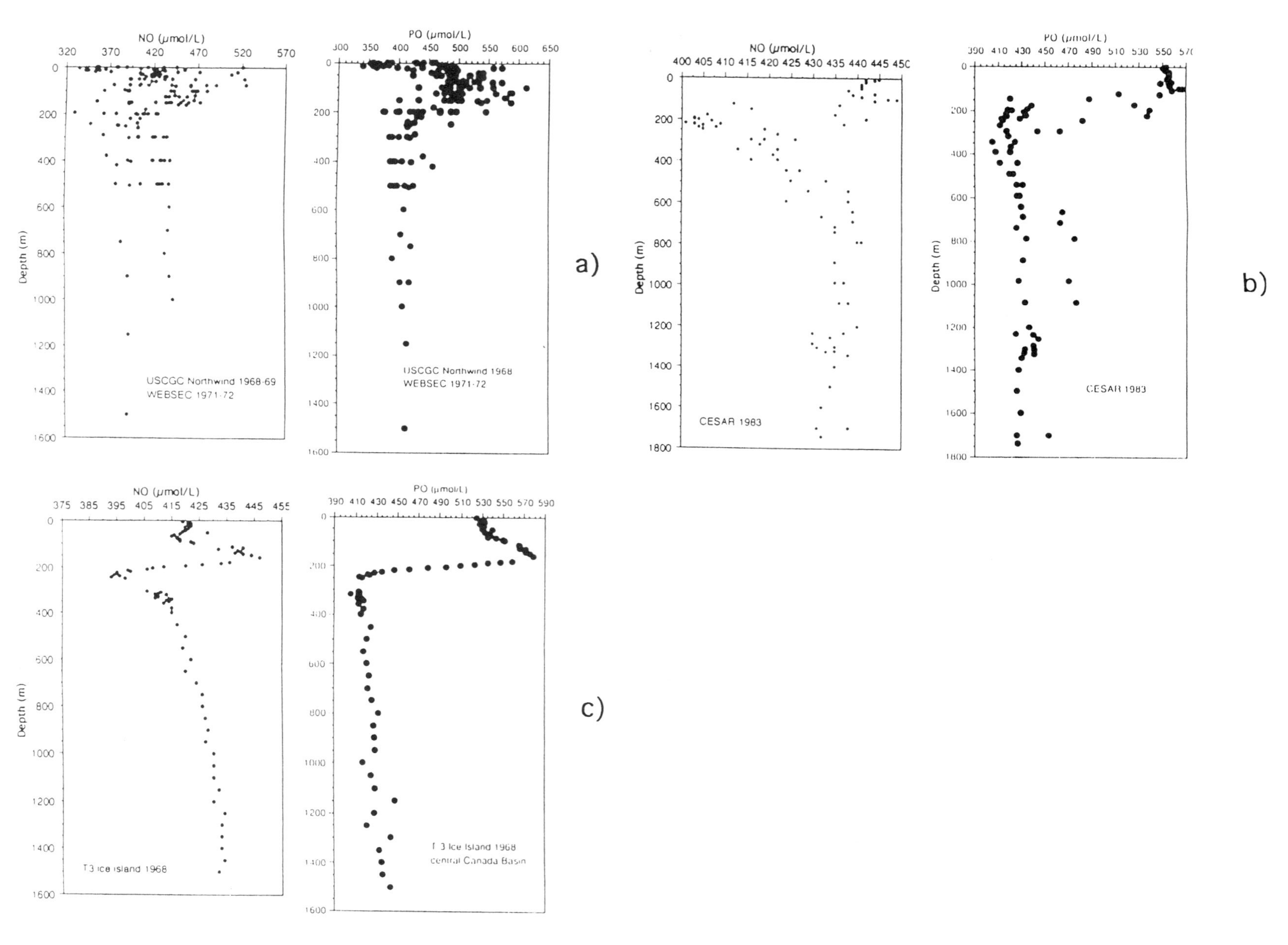

Figure 5. Distribution of NO and PO versus salinity a) over the Beaufort slope, b) at CESAR, and c) at T3 Ice Island in the central Canada Basin.

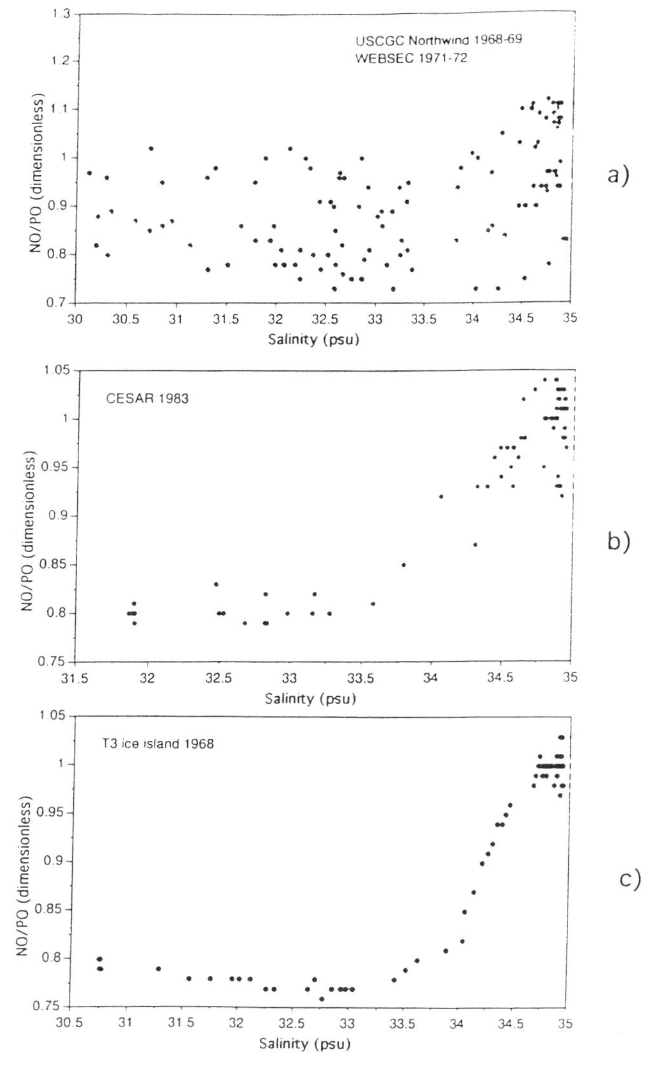

Figure 6. Distribution of NO/PO versus salinity a) over the Beaufort and Chukchi slopes, b) at CESAR, and c) at T3 Ice Island in the central Canada Basin.

Therefore, on the basis of oxygen in the lower halocline and the distinct vertical structure of oxygen profiles in the Canada and Makarov basins, we conclude that the lower halocline in the Beaufort slope region is strongly influenced by Pacific water and the lower halocline of the central Canada Basin is influenced both by Pacific and Atlantic-derived waters. The Pacific water nutrient signal in the lower halocline diminishes with distance from the Chukchi Sea source.

DISCUSSION

Jones and Anderson [1990] report that the distinction between UHW and LHW was less sharply defined over the shelf north of Ellesmere Island than at CESAR or LOREX. They also observed a spreading of the UHW nutrient maximum over a wider salinity range and to higher salinities north of Ellesmere than at CESAR or LOREX. We interpret their results in relation to ours in the following ways. First, the chemical characteristics of waters in the salinity range of the LHW are significantly modified i.e. nutrient values are increased in the shelf and slope regions of the Canada Basin through shelf-driven interleaving processes. A second possibility is that the circulation of the Beaufort Undercurrent extends farther east than 129°W (the eastern-most reported), Aagaard, [1984] and advects waters from the Chukchi and Beaufort regions all along the slope of the Canada Basin. Aagaard [1984] considers the Beaufort Undercurrent a part of the large-scale circulation of the Canada Basin that extends along the entire slope region of the Beaufort Sea. Aagaard et al. [1988] show that a major source for waters in the Beaufort Undercurrent is along the northern Chukchi margin, most likely the Hope Sea Valley and Herald Canyon. The similarity between the salinity-silicate relationships at T-3, over the Beaufort and Chukchi slopes, and north of Ellesmere Island indicates that high silicate in the lower halocline is a widely occurring feature of the Canada Basin. Furthermore, high silicate in the lower halocline of the Arctic Ocean is confined to the Canada Basin and does not occur in the LHW of the other Arctic basins. This distribution suggests that CLHW derives some of its properties (but not low NO) from the shelves surrounding the Canada Basin and is chemically distinct (in terms of silicate) from LHW that forms on the shelves of the Barents and Kara seas.

Using data collected by submarine from the first quasi-synoptic transect of the Canada Basin [McRoy, unpublished manuscript] shows that the inflow of high nutrient (particularly silicate) water from the Chukchi Sea spreads into the Canada Basin as a plume under the surface mixed layer. McRoy concludes that the inflow from the Chukchi shelf can account for the occurrence of the UHW nutrient maximum throughout the Canada Basin. From the areal distribution of nutrients mapped by McRoy it appears that the Alpha-Mendelyev ridge system exerts control over the flow by topographic steering, resulting in restriction of the plume of Pacific water largely to the Canada Basin. The plume is traceable through the Canadian archipelago and western Fram Strait. A portion of the inflow from the Chukchi shelf must be advected eastward along the slope in the Beaufort Undercurrent. The direct influence of the high silicate from Chukchi inflow appears to be responsible for maintaining chemical differences between the upper and lower halocline waters of the Canada Basin and the other Arctic basins.

We can not yet fully resolve the problem of how high silicate occurs in the waters of the lower halocline in the Canada Basin and adjacent slope waters, i.e. the apparent conflict of high silicate and high salinity. We propose the following hypotheses toward solution of the problem. First, the high silicates do not originate in the Barents or Kara seas, which are the usual proposed sources for LHW. Sections across the shelf of the Kara Sea show silicate less than 8 μmol/L in the upper 100 m and less than 6 μmol/L between

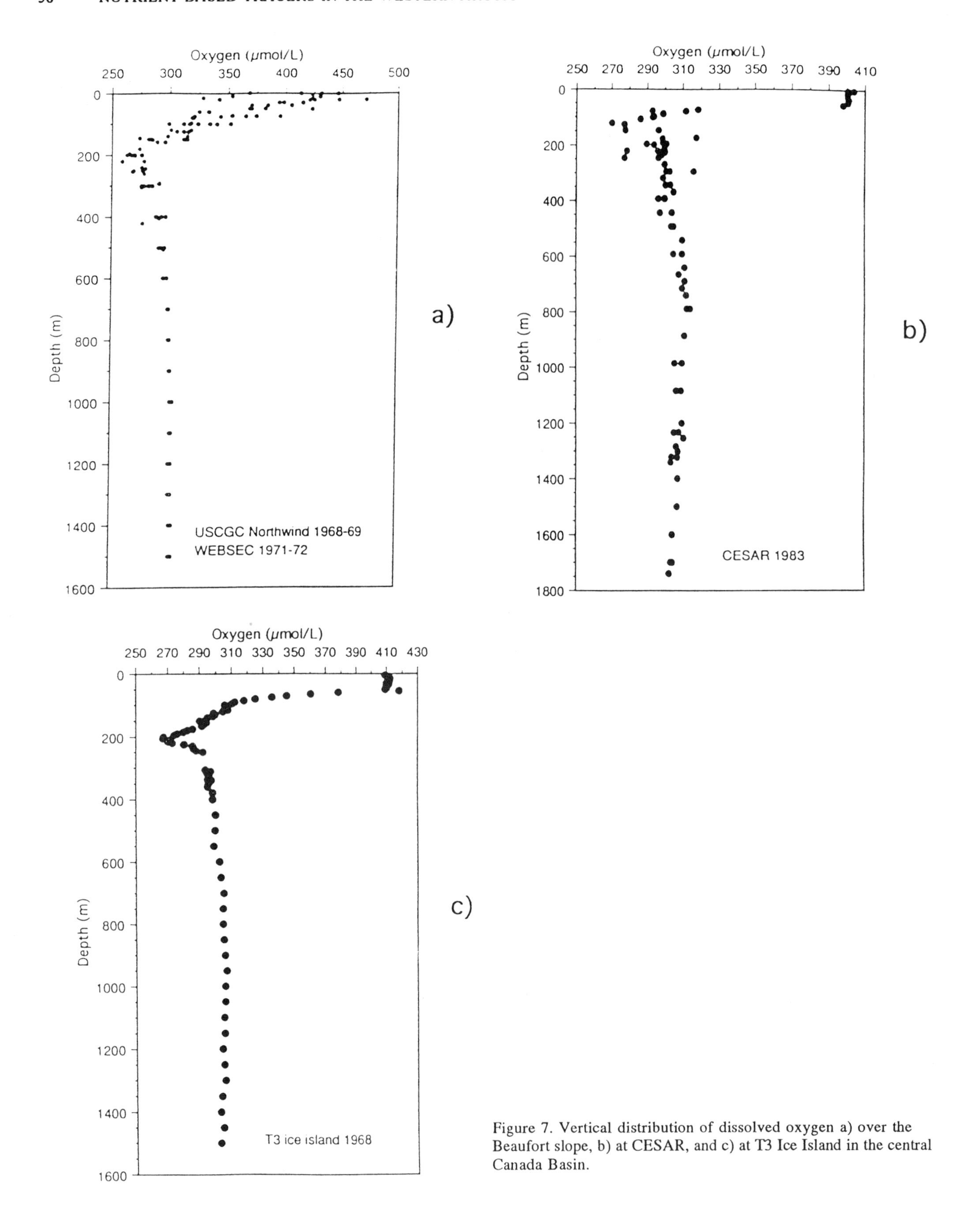

Figure 7. Vertical distribution of dissolved oxygen a) over the Beaufort slope, b) at CESAR, and c) at T3 Ice Island in the central Canada Basin.

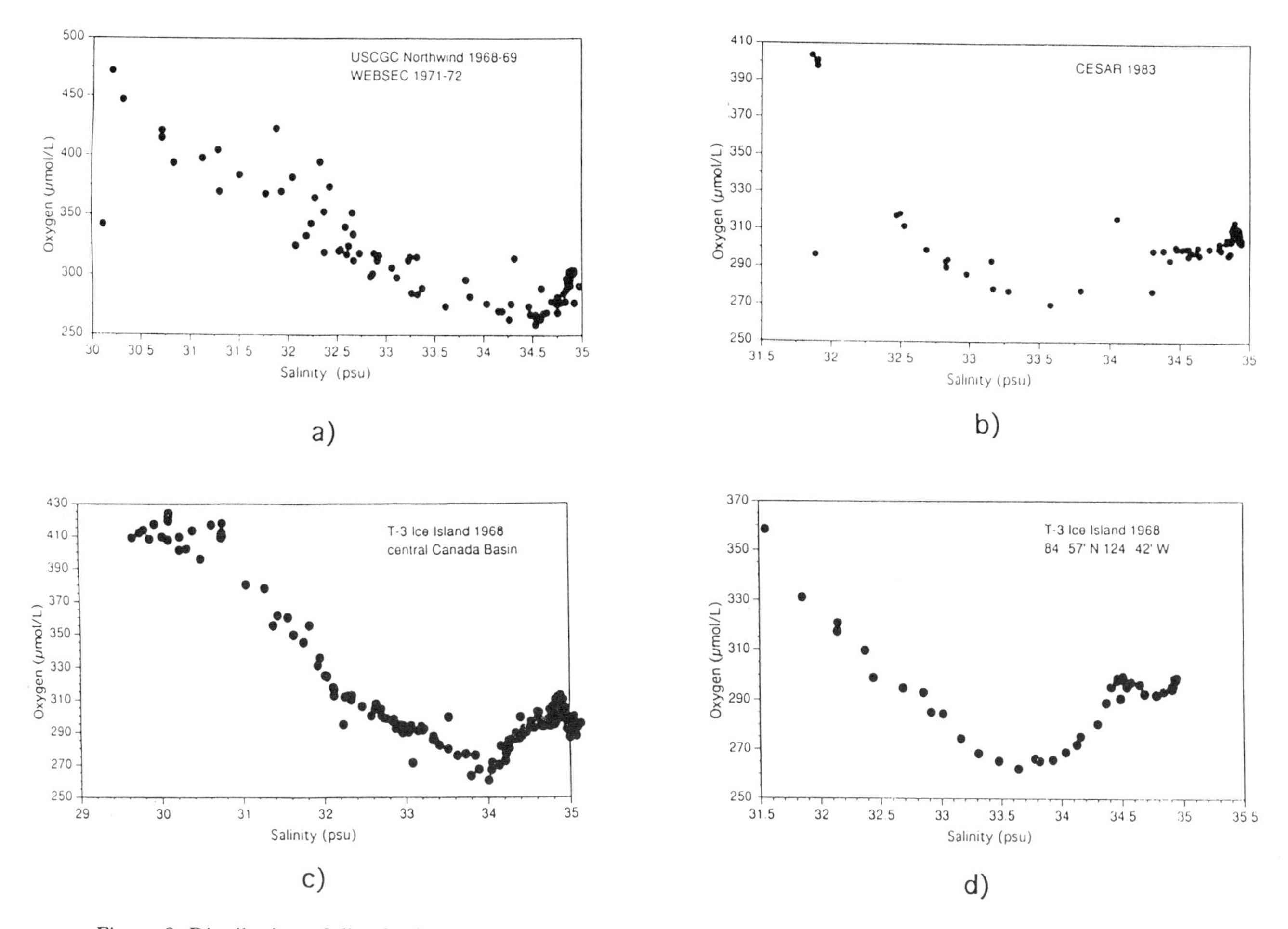

Figure 8. Distribution of dissolved oxygen versus salinity a) at stations over the Beaufort slope, b) at CESAR, c) at T3 Ice Island in the central Canada Basin, and d) at T-3 station in geographical proximity to CESAR.

100 and 600 m [Milligan, 1969]. Therefore, the high silicate in the lower halocline of the Canada Basin can not have a Kara Sea source. Similarly, data from the Barents Sea [Swift, 1984] show that silicates are too low (<6 μmol/L) to account for the observed values in the lower halocline of the Canada Basin. Moore and Smith [1986] propose that shelf water/bottom sediment interactions increase nutrients over the Barents and Kara shelves, a contention not supported by data from the Chukchi Sea [Lomstein et al., 1989]. Rusanov and Shpaikher [1979] report high silicates in the Chukchi Sea throughout the water column, especially during winter, and considered this to be the major source of silicate to both the upper and lower halocline of the Canada Basin.

Moore et al. [1983] suggest that upwelling of Atlantic water onto shelves via canyons could account for the origin of LHW. Jones and Anderson [1986] contend that the problem with this hypothesis is that NO values of Atlantic water are higher than those of LHW and there is no feasible mechanism to decrease the NO. We propose that high silicate at salinities of 34.2 to 34.7 psu is a result of further modification within the Canada Basin of LHW derived from the Barents and Kara seas, although it is not clear how the high silicates make their way into low NO water. A feasible mechanism is an interleaving of high silicate water from the Chukchi shelf during winter ventilation events [E. P. Jones, personal comm.]. Interleaving would somewhat modify NO concentrations, but would not erode the structure of the NO minimum since mixing is limited when the dense plumes sink to their appropriate densities and enter the lower halocline. Diffusive fluxes are not strong enough to erode the vertical structure in the water column. We discount simple mixing of water masses as this would act to erode the structure of the NO minimum (and PO minimum). The distinct changes in NO and PO as well as the NO/PO ratio that occur between the upper and lower halocline suggest that the waters do not share a common source. Also, as presented earlier, it is unlikely that low the NO water that characterizes LHW could have a source on the shelves of the East Siberian, Chukchi or Beaufort seas.

In the Laptev Sea, NO and NO/PO ratios are within the range of LHW [Wilson and Wallace, 1990]. Also Codispoti and Richards [1968] measured high silicate concentrations in the shallow bottom water around the Lena River delta. Again, the salinities in this region are not high enough to ventilate the lower halocline so that these conditions are an unlikely means of modifying LHW through the introduction of high silicate water. We agree it is possible that the LHW of the Canada Basin originates in the Barents and Kara Seas as hypothesized by Jones and Anderson [1986] and acquires high silicates through vertical mixing processes occurring along the slope regions. However, the NO and PO minima in the lower halocline suggest that this type of mixing does not occur to an appreciable extent, i.e. no mixing line exists in which the NO and PO minima can be formed since UHW and Atlantic water values of NO and PO are higher than those of LHW.

Finally, Jones and Anderson [1986] propose that water with salinities and NO values matching those of LHW could form in the Chukchi Sea through brine rejection. They state that "summer water with sufficient additional brine...could easily result in water with salinity and NO values matching that of the S=34.2 end member." The problem is still silicate. Since it is difficult to derive a source for high silicate other than waters that ultimately originate in the Pacific, we conclude that the lower halocline water of the Canada Basin (CLHW) originates in the Barents and Kara Sea regions (as LHW) and spreads into the central Canada Basin and along the slope of the Canada Basin in the Beaufort Undercurrent where it is modified through interleaving by high silicate Chukchi shelf waters. Canada Basin Lower Halocline Water (CLHW) is characterized by NO values that typify LHW but has silicate concentrations close to those in UHW that distinguish it from LHW and indicate additional processes in formation. Furthermore, the oxygen minimum along the Beaufort-Chukchi slopes occurs within CLHW, whereas the oxygen minimum in other regions of the Canada Basin occurs at salinities above those of LHW (i.e. above 34.2 psu).

In summary we conclude the following:

1) LHW in the Nansen, Makarov and Amundsen basins is characterized by reduced silicate compared to the lower halocline of the Canada Basin. The increase in the Canada Basin is a result of the influence on the circulation of Pacific water inflow from the Chukchi Sea.

2) The upper halocline nutrient maximum occurs over a wider salinity range along the Beaufort slope and in the Canada Basin than it does in the Amundsen and Makarov basins. The oxygen minimum associated with the UHW nutrient maximum occurs deeper and at higher salinities along the Beaufort slope than it does in other Arctic basins. The oxygen minimum occurs at salinities of LHW only along the Beaufort slope.

3) The distinction between upper and lower halocline waters is not well-defined in terms of silicate and oxygen in the Canada Basin, although NO, PO, and NO/PO ratios suggest that the water types are distinct and have different sources.

4) Canada Lower Halocline Water originates on the shelves of the Barents and Kara seas where it derives its low NO signature. This water is modified in the Canada Basin by the addition of silicate from the Chukchi Sea, probably by interleaving during winter ventilation events.

5) Oxygen concentrations and the distinct structure of vertical profiles confirm that lower halocline water of the Chukchi/Beaufort slope of the Canada Basin is strongly influenced by Pacific water. In the central Canada Basin the lower halocline is influenced by both Pacific and Atlantic-derived water as evidenced by the presence of a local oxygen maximum within the halocline.

Acknowledgments. This work was supported by the National Science Foundation under grant DPP-86-05659 from the Division of Polar Programs. We thank E.P. Jones for providing CESAR data and critical evaluation of the manuscript. We also thank R. M. Moore for LOREX data, J. Swift for AIWEX data and L. G. Anderson for critical comments.

REFERENCES

Aagaard, K., "The Beaufort undercurrent," In *The Alaskan Beaufort Sea; Ecosystems and environments*, 47-. Academic Press, Inc., 1984.

Aagaard, K., C. H. Pease, and S. A. Salo, *Beaufort Sea mesoscale circulation study. Preliminary results.*, ERL PMEL, 1988. 82.

Anderson, G. C. and J. H. Swift, *Arctic Internal Waves Experiment (AIWEX) Hydrographic Data, Scripps Institution of Oceanography*, Reference 90-10., 29 pp., 1990.

Anderson, L. G. and D. Dyrssen, "Chemical constituents of the Arctic Ocean in the Svalbard area," *Oceanologica Acta*, 4, 305-311, 1981.

Anderson, L. G. and E. P. Jones, "Water masses and their chemical constituents in the western Nansen Basin of the Arctic Ocean," *Oceanologica Acta*, 9, 227-283, 1986.

Anderson, L. G. and E. P. Jones, "Tracing upper waters of the Nansen Basin in the Arctic Ocean," *Deep-Sea Research*, 39, S425-S443, 1992.

Broecker, W. S., ""NO", A conservative water mass tracer," *Earth and Planetary Science Letters*, 23, 100-107, 1974.

Codispoti, L. A. and F. A. Richards, "Micronutrient distributions in the East Siberian and Laptev seas during summer 1963," *Arctic*, 21, 67-83, 1968.

Jones, E. P. and L. G. Anderson, "On the origin of the chemical properties of the Arctic Ocean halocline," *Journal of Geophysical Research*, 91, 10759-10767, 1986.

Jones, E. P. and L. G. Anderson, "On the origin of the properties of the Arctic Ocean halocline north of Ellesmere Island: Results from the Canadian ice island," *Continental Shelf Research*, 10, 485-498, 1990.

Jones, E. P., L. G. Anderson, and D. W. R. Wallace, "Tracers of near-surface, halocline and deep waters in the Arctic Ocean: Implications for circulation," *Journal of Marine Systems*, 2, 241-255, 1991.

Kinney, P., M. E. Arhelger, and D. C. Burrell, "Chemical characteristics of water masses in the Amerasian Basin of the Arctic Ocean," *Journal of Geophysical Research*, 75, 4097-4104, 1970a.

Kinney, P., M. E. Arhelger, and D. W. Hood, Chemical characteristics of Arctic water masses: T-3 ice island, Institute of Marine Science, University of Alaska, 1970b. Technical Report R-69-15.

Kusunoki, K., "Hydrography of the Arctic Ocean with special reference to the Beaufort Sea," *Contributions from the Institute of low Temperature Science, Hokkaido University, Series* A, 1-74, 1962.

Kusunoki, K., J. Muguruma and K. Higuchi, Oceanographic observations at Fletcher's ice island (T-3) in the Arctic Ocean in 1959-1960, Air Force Cambridge Research Laboratories, 1962. Research Paper 22. AFCRL-62-479.

Lomstein, B. A., T. H. Blackburn, and K. Henriksen, "Aspects of nitrogen and carbon cycling in the northern Bering Shelf sediment. I. The significance of urea turnover in the mineralization of NH_4^+," *Marine Ecology Progress Series*, 57, 237-247, 1989.

Milligan, D. B., Oceanographic survey results, Kara Sea, summer and fall 1965, US Naval Oceanographic Office, 1969. Technical Report TR217.

Moore, R. M., M. G. Lowings, and F. C. Tan, "Geochemical profiles in the central Arctic Ocean," *Journal of Geophysical Research*, 88, 2667-2674, 1983.

Moore, R. M., H. Melling, and K. R. Thompson, "A description of water types on the Mackenzie Shelf of the Beaufort Sea during winter," *Journal of Geophysical Research*, 97, 12,607-12,618, 1992.

Moore, R. M. and J. N. Smith, "Disequilibria between 226Ra, 210Pb, and 210Po in the Arctic Ocean and the implications for modification of the Pacific water inflow," *Earth Planetary Science Letters*, 77, 285-292, 1986.

Rudels, B., A. Larsson, and P. Sehlstedt, "Stratification and water mass formation in the Arctic Ocean: some implications for the nutrient distribution," *Polar Research*, 10, 19-31, 1991.

Rusanov, V. P. and O. Shpaikher, "Advection of dissolved silicic acid in the Chukchi Sea," *Okeanologiya*, 19, 626-631, 1979.

Swift, J. H., Preliminary data report, R/V Polarstern Cruise Arktis II/3. Physical and chemical data from rosette samples., Scripps Institution of Oceanography, 1984.

Wilson, C. and D. W. Wallace, "Using the nutrient ratio NO/PO as a tracer of continental shelf waters in the central Arctic Ocean," *Journal of Geophysical Research*, 95, 22193-22208, 1990.

Yeats, P. A., "Manganese, nickel, zinc and cadmium distributions at the Fram 3 and Cesar ice camps in the Arctic Ocean," *Oceanologica Acta*, 11, 383-388, 1988.

D. K. Salmon, Prince William Sound Science Center, Cordova, AK 99574, C. P. McRoy, Institute of Marine Science, University of Alaska, Fairbanks, AK 99775.

The Potential of Barium as a Tracer of Arctic Water Masses

K. Kenison Falkner

College of Oceanic & Atmospheric Sciences, Oregon State University, Corvallis, Oregon

R. W. Macdonald and E. C. Carmack

Institute of Ocean Sciences, Sidney, British Columbia, Canada

T. Weingartner

Institute of Marine Sciences, University of Alaska, Fairbanks, Alaska

As part of ongoing circulation studies in the Arctic, seawater samples for dissolved Ba concentrations were obtained during Sep.-Oct., 1992 at several locations in the Bering Strait, Eastern Chukchi and Southern Beaufort Seas. The results reveal a dynamic range (10 to 150 nmol kg^{-1}) for this element in the Arctic equal to or greater than that in the combined Atlantic, Indian and Pacific oceans. Lowest levels are observed in surface waters, with values tending to decrease northwards in the direction of currents generally flowing from the Bering Strait along the Alaskan coast. Low surface concentrations tend to be accompanied by relatively enriched near bottom levels. On the basis of these spatial distributions, hydrographic observations and a knowledge of its behavior in other marine settings, it appears that Ba can be significantly depleted from surface waters as a result of the highly seasonal biological activities over Arctic marginal shelves. Removal at the surface is counteracted to some extent by regeneration at depth or in the sediments and by riverine inputs. The biologically related drawdown is likely to enhance the contrast between "background" surface Ba levels in the Arctic and waters imprinted by regeneration and/or rivers. These preliminary findings suggest that Ba holds particular promise for tracing river waters and the ventilation of halocline waters by laterally sinking brines produced during ice formation over the shelves.

1. INTRODUCTION

The Arctic Ocean is often viewed as being comprised of horizontal layers which ultimately originate from Atlantic, Pacific, and fluvial inputs. A combination of biological, chemical and physical processes occurring predominantly over the Arctic shelves and in the Greenland, Iceland and Norwegian Seas modifies the properties of these source waters which are advected laterally, bearing their altered signatures. Tracer fields provide the means both to determine the nature of shelf processes and to deduce mean circulation of the Arctic; here we suggest that distributions of barium may be useful in this regard. Measurements of dissolved Ba from a Canadian Basin profile, a transect of the Bering Strait and stations along the northwestern

The Polar Oceans and Their Role in Shaping the Global Environment
Geophysical Monograph 85

Alaskan coast in the Chukchi Sea, reveal a dynamic range for this biointermediate element in the Arctic at least as great as that in all of the open oceans. Comparison with other oceanographic properties and knowledge of its behavior in other marine settings lead both to a preliminary understanding of Ba cycling in the Arctic and an indication to which aspects of circulation it might be applied.

2. BACKGROUND

Marine Geochemical Behavior of Barium

As this is the first report concerning Ba in Arctic waters, we begin our discussion with a brief summary of what is known about the cycling of this minor element elsewhere in the marine environment. In the Atlantic, Pacific and Indian oceans, as well as in marginal seas such as the Mediterranean and closed basins such as the Black sea, dissolved Ba tends to be depleted at the surface and enriched with depth and along advective flow lines, much like a hard-part nutrient such as Si or alkalinity as it reflects $CaCO_3$ cycling [Chan et al., 1977; Falkner et al., 1993; Lea, 1990]. This has been thought to be due primarily to uptake of Ba at the surface as the mineral barite ($BaSO_4$), which is formed in association with biological particulate matter and subsequently sinks and is regenerated at depth or in the sediments [Bishop, 1988; Collier and Edmond, 1984; Dehairs et al., 1980; Dehairs et al., 1987]. Oceanic waters tend to be undersaturated with respect to barite [Church and Wolgemuth, 1972; Falkner et al., 1993], thus its occurrence has been attributed to favorable microenvironments present in biological debris, although the exact mechanism of barite formation remains unclear [Bishop, 1988; Chow and Goldberg, 1960; Church and Wolgemuth, 1972; Dehairs et al., 1980].

Profiles of suspended particulate Ba tend to show a subsurface maximum comprised primarily of micron to sub-micron size barite crystals thought to be the relatively refractory remnant of organic matter decomposition in the upper water column [Bishop, 1988; Bishop, 1989; Dehairs et al., 1990; Dehairs et al., 1991]. Most barite in sinking particulate matter is apparently refractory enough to survive its journey through the water column to the sediments where a large fraction of that originating at the surface is regenerated [Dymond et al., 1992]. Recent studies indicate that sinking celestite ($SrSO_4$) shells of Acantharia, a sub-class of Radiolaria, and siliceous Radiolaria which form celestite crystals may also contribute significantly to Ba removal from surface waters [Bernstein et al., 1992]. Barium readily substitutes into celestite which, in turn, redissolves so rapidly upon the death of an organism, that it does not survive in preserved samples, nor during typical sediments trap deployments, nor is it observed in oceanic sediments [Bernstein and Betzer, 1991]. In this manner, dissolved Ba might be transferred directly and rapidly from surface waters to mid-depths.

In addition, dissolution of celestite incorporated into particulate biological debris could generate microenvironments favorable for barite formation [Bernstein et al., 1992], thus Ba transport should be enhanced in regions where Acantharia and the siliceous Radiolaria which form celestite crystals thrive. Since these organisms are typically undersampled by fine-meshed plankton nets and are subject to dissolution, their distributions in the oceans are not yet well characterized [Michaels, 1988]. It does not appear to be known, for example, to what extent such organisms thrive in the Arctic.

Since Ba cycles in a manner similar to that of hard-part constituents, dissolved Ba tends to correlate linearly with dissolved Si and alkalinity in much of the world's oceans [Bacon and Edmond, 1972; Chan et al., 1977]. The slopes and intercepts of the correlation lines vary somewhat with location, in part because these constituents display diverse surface distributions; Si is a biolimiting element that can be depleted to essentially zero concentrations in surface waters, whereas alkalinity is only slightly altered by biological activities. Barium, being biointermediate, lies between the two in that it is measurably but not fully depleted. In addition, there exist areas of the oceans where Ba, Si and alkalinity regeneration fluxes are likely decoupled, since opal, barite (and celestite), and carbonate are subject to different dissolution controls. Being more like opal in this regard, elevated barite concentrations are observed in sediments underlying the most highly productive areas of the oceans [Goldberg and Arrhenius, 1958; Revelle, 1944; Revelle et al., 1955; Turekian and Tausch, 1964]. The link between productivity and barite preservation may be obscured under anoxic conditions since the consumption of sulfate promotes the dissolution of this mineral.

The principal external sources of dissolved Ba to the world's oceans are rivers [Martin and Maybeck, 1979] and hydrothermal venting at mid-ocean ridges [Edmond et al., 1979; Von Damm et al., 1985]. Both sources tend to be elevated in Ba content over the seawater into which they arrive. Most hydrothermal Ba is probably precipitated inorganically as barite in the vicinity of hot spring sources [Von Damm, 1990]. Fluvial inputs, in contrast, are enhanced in estuaries where Ba adsorbed onto riverborne clays is desorbed in exchange for the more abundant cations of seawater [Carroll et al., 1993; Edmond et al., 1978; Hanor and Chan, 1977; Li and Chan, 1979].

Description of the Sampling Region

The region sampled spanned from the Bering Strait, through the eastern Chukchi Sea to the Beaufort Sea (Figure 1). A positive difference in sea-surface height between the Pacific and Arctic (≈0.5 m), generally causes waters to flow from the Bering into the Chukchi Sea through the shallow (≤45 m) and narrow (85 km) Bering Strait [Coachman and Aagaard, 1966]. Occasionally and more commonly in the period lasting from late August until May, sufficiently strong opposing winds out of the north generate intermittent flow reversals [Aagaard et al., 1985; Coachman, 1993; Coachman and Aagaard, 1981; Overland and Roach, 1987; and references therein]. The integrated effect of such reversals is to reduce the net inflow of Pacific water to the Arctic in winter to roughly one fourth of its summer values resulting in an annual average input to the Arctic of about 1 Sv of waters characterized by an average salinity of about 32 psu [Coachman, 1993 and references therein]. Bering Strait waters tend to be enriched in nutrients, which are derived largely from upwelling along the Gulf of Anadyr. This upwelling helps fuel intense seasonal productivity observed just north and south of the strait in the Bering and Chukchi Seas [Coachman and Hansell, 1993; Codispoti and Richards, 1971; Walsh et al., 1989].

The flow field extending northward of the Bering Strait divides into a broad current on the west that eventually veers into the central Chukchi Sea in the vicinity of Pt. Hope and a narrower and more intense current on the east which continues northeastward along the Alaskan coast toward Pt. Barrow [Coachman and Aagaard, 1988; Overland and Roach, 1987]. While the speed and direction of the Alaskan Coastal Current is variable, its net transport is in a northerly direction at a mean speed of about 0.07 ms^{-1} [Aagaard, 1984; Coachman and Aagaard, 1981; Coachman and Aagaard, 1988; T. Weingartner, unpublished data]. It is primarily this current that was transected by our RV Alpha Helix-Cruise 166 stations north of Bering Strait. These waters tend to cool through air-sea heat exchange and mixing and are influenced by strong seasonal productivity as they venture northwards [Walsh et al., 1989].

Within the deep Beaufort Sea, the water column can be divided roughly into four main layers consisting of a shallow (≈50 m) mixed layer of low salinity (30-32 psu) and high seasonal variability; a pronounced halocline maintained by lateral inputs from the Pacific and marginal shelves and having distinct upper and lower layer signatures; a relatively warm but salty deeper layer of Atlantic origin whose core occurs at about 400 m; and deep water of uncertain origin [Carmack, 1990 and references therein].

Temperature and salinity considerations suggest that the halocline is not simply a mixture of waters from above and below the halocline but arises from lateral advection along horizons of constant density of both Pacific inflow and of brines produced during ice formation over the Arctic shelves [Aagaard et al., 1981; Melling and Lewis, 1982]. Further support for this hypothesis is the occurrence a striking nutrient maximum accompanied by a not necessarily coincident oxygen minimum within the upper halocline (≈190 m); these features are thought to reflect the high nutrient content of Pacific waters and effects of organic matter regeneration to which the brines are exposed during their contact with shelf sediments [Jones and Anderson, 1986; Jones and Anderson, 1990; Kinney et al., 1970; Macdonald et al., 1987]. Spatial distributions and budgetary considerations of Ca, alkalinity and total carbonate suggest that riverine inputs primarily influence the surface layer, penetrating the halocline only to a limited extent [Anderson et al., 1990; Anderson et al., 1989]. Measurably different values of the quasi-conservative parameter NO ($NO=9x[NO_3]+[O_2]$) [Broecker, 1974] in the upper and lower halocline are thought to reflect differing brine source regions for these layers [Anderson et al., 1989; Jones and Anderson, 1986; Wilson and Wallace, 1990]. Within the Beaufort Sea, an additional subsurface temperature maximum at 50-100 m is often observed [Aagaard et al., 1981; Coachman and Barnes, 1961]. This feature results from the subsurface extension along the northern Alaskan shelf edge of the current flowing from the Bering Strait along the Alaskan coast [Hufford, 1973; Hufford, 1975; Macdonald et al., 1987; Paquette and Bourke, 1974].

3. SAMPLE COLLECTION

Alpha Helix Cruise 166

As part of a joint Japanese/Russian/American effort, the RV Alpha Helix of the University of Alaska occupied an extensive grid (Figure 1; Table 1) during Sep.-Oct., 1992 in the Bering Strait and Chukchi Sea for an ongoing general circulation study of the region. Surface and near-bottom samples for Ba (as well as for oxygen isotopes, the results of which will be reported elsewhere) were obtained at several locations in the eastern Chukchi Sea along the Alaskan coast and across the Bering Strait. Vertically detailed profiling was carried out at a few select stations (Table 1). Samples were collected using 1.7-liter Niskin bottles, outfitted with Buta-N black rubber O-rings and

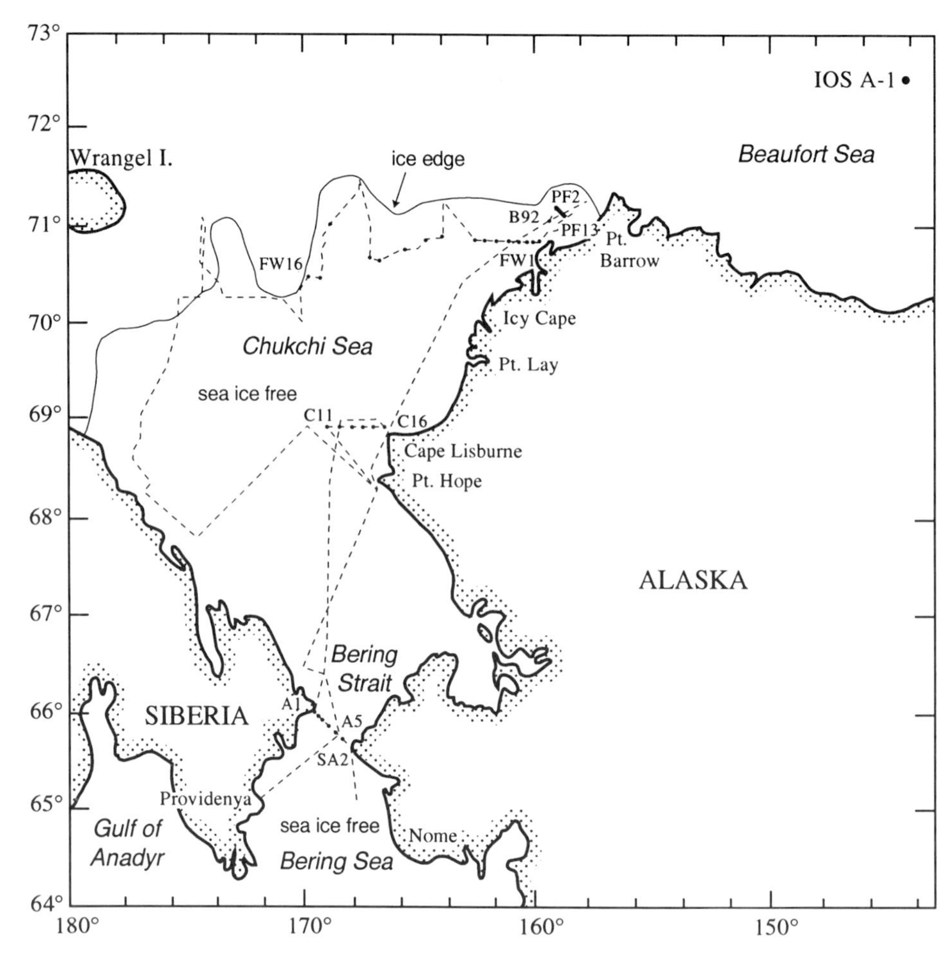

Fig. 1. Map of study area. Dashed line is RV Alpha Helix (HX166) cruise track. Dots represent stations at which Ba samples were taken along the designated lines. IOS Arctic Ocean Climate Station A-1 is indicated by larger dot in the Beaufort Sea. See Tables 1&2 for station coordinates. Solid line represents southern extent of ice edge on September 25, 1992. In general, the northern portion of the cruise track was within 2 to 15 miles of the ice edge; apparent crossings of the ice edge indicate shifting of the ice edge occurred between Sep. 25 and the time the cruise track was occupied.

internal closure by latex-tubing, cast on a stainless steel hydrowire. Continuous conductivity, temperature, depth, fluorometry (CTDF) profiles were obtained at these sites in addition to many others over the entire cruise track.

Henry Larson Cruise 92-16

A detailed vertical profile was obtained in September, 1992 from the CCGS Henry Larson at the IOS Arctic Ocean Climate Station A-1 (72°31.6'N, 143°51.9'W; 3375 m) in the Beaufort Sea (Figure 1). Samples were collected using 10-liter Niskin bottles outfitted with red-silicone rubber O-rings and epoxy-coated internal springs and mounted on a stainless steel hydrowire. Samples were also drawn for chlorofluorocarbons, oxygen, salinity, temperature, TCO_2, carbon and oxygen isotopes, tritium, total suspended solids, particulate nitrogen and carbon, chlorophyll and nutrient determinations to be discussed in further detail elsewhere when the results become available. Continuous conductivity, temperature and depth profiles were obtained at this site via a Guildline CTD.

For both the CCGS Henry Larson and RV Alpha Helix cruises, seawater samples for Ba analyses were collected into 20-ml, 0.1N HCl-washed polyethylene vials after pre-rinsing with the sample. The external joints between the caps and vials were wrapped in Parafilm to minimize evaporation; the unacidified samples were placed in cardboard racks which were enclosed in plastic bags for storage at room temperature until shipment back to the laboratory.

4. ANALYTICAL PROCEDURE

Barium concentrations were determined by isotope dilution-inductively coupled plasma quadrupole mass spectrometry (ID-ICPMS) in a manner similar to that described elsewhere [Falkner et al., 1993; Klinkhammer and Chan, 1990]. Briefly, 250 μl of seawater sample was spiked with an equal volume of ^{135}Ba-enriched (source: Oak Ridge National Laboratories) solution and diluted 34-

TABLE 1. HX166 Data.

station id	date	latitude	longitude	bottom depth m	depth m	temp °C	sal-psu	sigma-t	Ba-nM
A1*	Sep. 21, 1992	65°59' N	169°38' W	49					
A2*	Sep. 21, 1992	65°56' N	169°29' W	50					
A3*	Sep. 21, 1992	5°54' N	169°19' W	46					
A4*	Sep. 21, 1992	65°52' N	169°07' W	44					
A5*	Sep. 21, 1992	65°59' N	168°52' W	45					
SA2*	Sep. 21, 1992	65°47' N	168°37' W	50					
C11	Sep. 23, 1992	68°56' N	168°59'	52	10.2	3.419	32.230	25.637	56.9
					50.9	3.215	32.583	25.938	62.2
C12	Sep. 23, 1992	68°56' N	167°29' W	50	9.2	3.279	32.192	25.620	53.6
					48.7	3.252	32.596	25.945	77.8
C13	Sep. 23, 1992	68°56' N	167°59' W	48	5.3	3.247	31.616	25.163	56.8
					46	3.446	32.518	25.864	75.1
C14	Sep. 23, 1992	68°56' N	167°29' W	45	9.1	3.467	31.401	24.973	62.1
					45.4	3.486	32.597	25.924	78.2
C15	Sep. 23, 1992	68°56' N	166°59' W	44	10	4.044	31.233	24.787	60.4
					44.2	3.632	32.552	25.875	74.4
C16	Sep. 23, 1992	68°56' N	166°20' W	27	5	3.581	30.642	24.359	63.1
					26.2	4.466	31.704	25.119	66.1
B92	Sep. 24, 1992	71°12' N	159°43' W	73	0	-1.540	30.005	24.117	
					5	-1.538	30.013	24.124	
					10	-1.534	30.028	24.135	
					15	-1.516	30.052	24.155	
					20	-1.569	30.095	24.190	
					21	-1.475	30.134	24.220	25.5
					25	0.464	32.260	25.872	
					30	-0.054	32.461	26.060	74.5
					35	-0.066	32.467	26.065	
					40	-0.067	32.467	26.065	75.5
					45	-0.098	32.469	26.067	
					50	-0.105	32.472	26.070	71.5
					55	-0.107	32.471	26.070	
					60	-0.109	32.470	26.069	74.6
					65	-0.109	32.468	26.068	
					70	-0.113	32.468	26.067	
					72	-0.111	32.466	26.066	72.1
PF2	Sep. 24, 1992	71°06' N	159°18' W	81	80	-0.620	32.485	26.102	68.2
PF3	Sep. 24, 1992	71°05' N	159°15' W	81	10.2	-1.576	29.940	24.065	24.6
					20	-1.577	29.939	24.064	24.4
					29.9	0.584	32.379	25.962	64.1
					39.9	-0.319	32.467	26.076	66.4
					49.8	-0.374	32.468	26.079	73.5
					59.7	-0.382	32.473	26.083	66.1
					69.2	-0.377	32.471	26.081	69.7

*data for these stations shown in Fig. 2a-d

TABLE 1. HX166 Data (continued)

station id	date	latitude	longitude	bottom depth m	depth m	temp °C	sal-psu	sigma-t	Ba-nM
PF5	Sep. 24, 1992	71°05' N	159°10' W	78	10.2	-0.139	30.178	24.221	14.1
					20	0.640	30.346	24.323	13.6
					25.2	2.310	31.303	24.989	63.0
					30	0.741	32.248	25.848	76.3
					75	0.483	32.292	25.897	77.5
PF7	Sep. 25, 1992	71°04' N	159°04' W	61	10.9	-0.457	30.147	24.206	12.4
					60.5	2.858	32.157	25.628	76.6
PF9	Sep. 25, 1992	71°04' N	158°58' W	45	9.9	1.306	30.621	24.508	13.5
					45	3.165	31.570	25.134	56.6
PF11	Sep. 25, 1992	71°01' N	158°53' W	30	10	0.517	30.251	24.252	16.4
					29.8	0.553	30.263	24.260	18.4
PF13	Sep. 25, 1992	70°59' N	158°49' W	25	10.3	0.356	30.187	24.208	15.9
					24	0.368	30.192	24.211	17.8
FW1	Sep. 26, 1992	70°57' N	159°14' W	32	4.1	0.286	30.230	24.245	20.9
					28.5	1.331	30.669	24.545	36.7
FW2	Sep. 26, 1992	70°57' N	159°34' W	53	5	0.238	30.365	24.356	13.8
					50.2	3.285	32.198	25.624	70.4
FW3	Sep. 26, 1992	70°50' N	160°00' W	44	7.2	0.448	30.322	24.312	16.9
					41.2	3.515	32.103	25.528	67.7
FW4	Sep. 26, 1992	70°50' N	160°31' W	50	5.2	0.232	30.697	24.624	13.5
					48.1	2.138	32.259	25.766	93.0
FW5	Sep. 26, 1992	70°49' N	161°01' W	44	7.3	0.075	30.660	24.601	13.9
					42.1	1.856	32.266	25.792	95.4
FW6	Sep. 26, 1992	70°49' N	161°31' W	43	9.9	0.275	30.570	24.520	17.6
					40.9	2.834	32.289	25.735	100
FW7	Sep. 26, 1992	70°49' N	162°02' W	41	10.2	0.079	30.610	24.560	17.5
					40.2	2.809	32.304	25.749	151
FW8	Sep. 26, 1992	70°49' N	162°33' W	42	6.6	0.107	30.630	24.575	13.0
					41.2	2.834	32.375	25.773	91.8(93.2)†
FW9	Sep. 27, 1992	70°50' N	163°50' W	43	6.2	0.442	30.906	24.783	21.3
					41.4	1.061	32.291	25.864	91.3
FW10	Sep. 27, 1992	70°50' N	164°52' W	36	10.3	-0.564	30.432	24.440	10.8
					34.3	1.126	32.254	25.830	84.2
FW11	Sep. 27, 1992	70°43' N	165°50' W	40	9.1	-0.015	30.628	24.578	12.9
					38.9	1.109	32.370	25.924	81.2
FW12	Sep. 27, 1992	70°40' N	166°50' W	46	10	2.415	31.990	25.530	43.7
					44	2.896	32.729	26.081	94.6
FW13	Sep. 27, 1992	70°40' N	167°51' W	47	10.4	1.842	32.098	25.659	42.5
					45.2	-0.090	32.477	26.074	99.1
FW14	Sep. 28, 1992	71°00' N	168°50' W	42	5.3	0.914	31.215	25.008	27.4
					40.3	3.122	32.514	25.890	53.5(53.1)†
FW15	Sep. 29, 1992	70°26' N	169°14'W	31	4.2	0.032	31.065	24.929	25.1
					28.7	2.733	32.300	25.753	44.7
FW16	Sep. 29, 1992	70°28' N	169°41'W	31	3	-0.136	30.688	24.632	28.4
					27.8	2.333	32.363	25.834	43.8

†value in parentheses repeat determination

fold in 0.16N redistilled HNO_3. Samples were introduced into a Fisons Plasmaquad II via an autosampler-fed peristaltic pump (≈1 ml min^{-1} flow rate)-Meinhard concentric glass nebulizer system. About 1 minute was required to achieve a steady-state signal and the quadrupole was then scanned repeatedly over the 134-139 mass region for a total acquisition time of 1.5 minutes. A 4-minute washout with 0.16N HNO_3 was employed between samples to minimize sample carry over. Both a gravimetric Ba-standard, which was spiked and diluted in the same way as the samples, and a 0.16N HNO_3 blank was interspersed between every 5 samples. The gravimetric standard has been calibrated for consistency with the GEOSECS Ba data set and was used to correct for instrumental offset and drift. Based on repeated analyses, the precision is estimated to range from 2% at 100 nmol Ba kg^{-1} to 4% at 10 nmol Ba kg^{-1}.

5. RESULTS AND DISCUSSION

Alpha Helix Cruise 166

The first occupation of Bering Strait occurred when strong northerly winds (≥10 m s^{-1}) had been blowing for the previous four days and were diminishing (8 m s^{-1}). These winds are capable of forcing a reversal in the normally northward transport through the strait [Coachman, 1993]. Indeed, the sloping isohaline surfaces (which parallel the isopycnals) imply a southward baroclinic flow through Bering Strait (Sep. 21; Figure 2). A current meter just north of the eastern side of the strait registered southerly flow immediately prior to and changing flow during our occupation (K. Aagaard, personal communication). The very low salinities (28-29 psu) occurring in a 10 km wide band on the western side of the

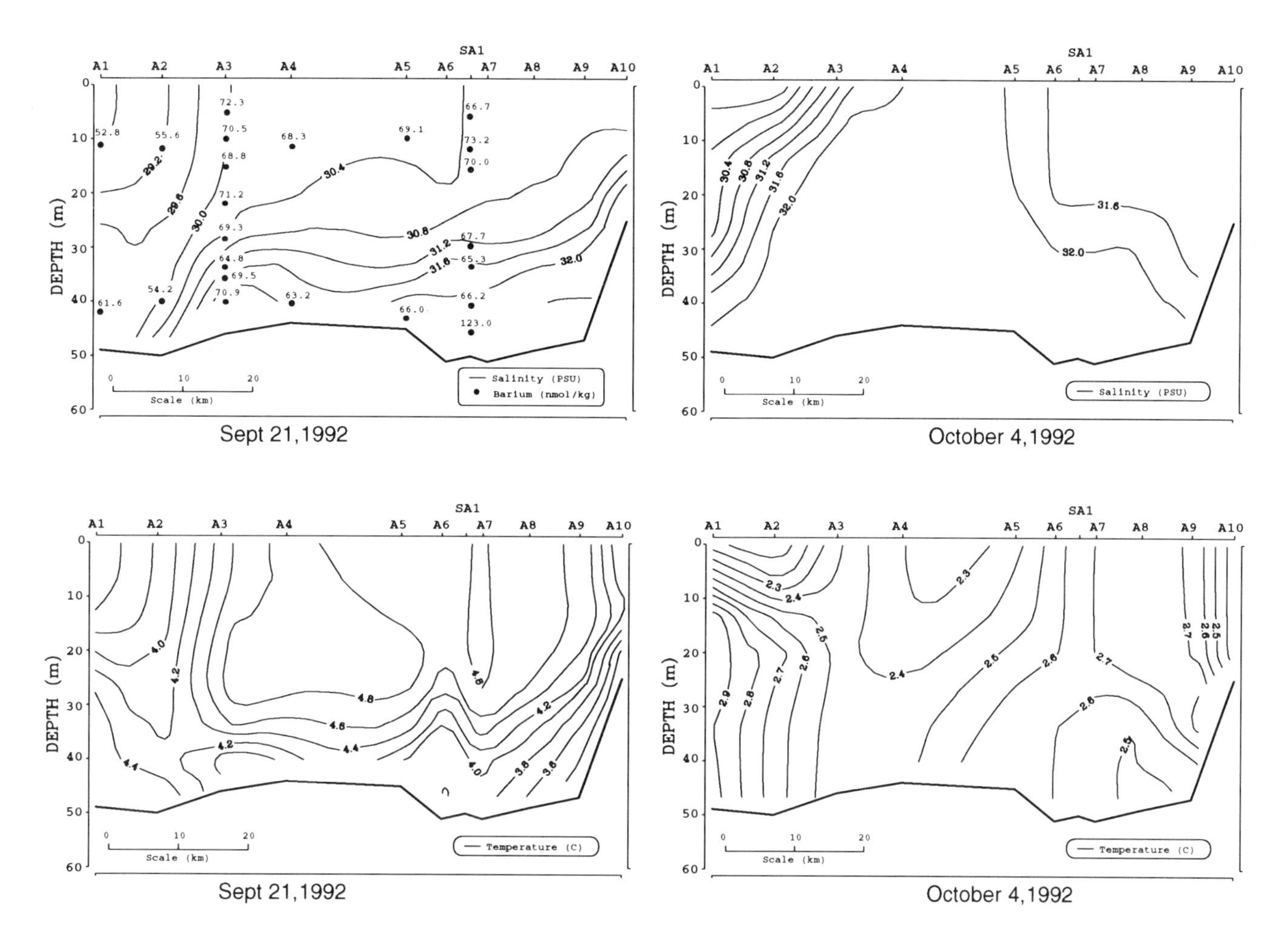

Fig. 2. Bering Strait transects. Barium concentrations superimposed on salinity contours for Sep. 21, 1993.

strait are probably reflective of river run-off and ice melt contributions from the north. In fact, quite fresh waters, with respect to an average salinity of 32 psu thought to characterize the strait [Coachman, 1993], occupied nearly the entire section. The situation had changed markedly by the time the section was occupied again (for CTDF work only) two weeks later (Oct. 4; Figure 2) and the system appeared to be in transition toward more typical flow conditions.

Barium concentrations in the Bering Strait, available only for the first occupation, generally fall within the range of 65-70 nmol Ba kg^{-1} (Figure 2). Exceptions include relatively depleted (53-55 nmol Ba kg^{-1}), lower salinity (28.8-29.8 psu) surface waters off the western coast and a single quite elevated point (123 nmol Ba kg^{-1}) associated with higher salinity (32.2 psu) waters at the deepest location on the eastern side of the strait. Unfortunately there is no Ba data for the Chukchi and Bering Sea waters just north or south of the Strait with which to compare our results. In fact, the closest available Ba measurements are for the Oct. 8, 1973 occupation of GEOSECS Station 219, located just north of the Aleutian Arc (53°6'N,177°18'W) in the southern Bering Sea [Figure 3; p. 96 in Ostlund et al., 1987]. There, Ba showed a distribution typical of the open ocean, having a depleted surface mixed layer (33.1 psu; 56-57 nmol Ba kg^{-1}) and gradually increasing concentrations with depth. Barium values comparable to those observed over most of our Bering Strait section are encountered over the upper few hundred meters but at relatively higher salinities (i.e. 348 m, 33.9 psu, 75.3 nmol Ba kg^{-1}). Concentrations equivalent to our highest Bering Strait Ba value are not encountered shallower than 1400m (or 34.5 psu); there is no obvious pathway by which salty waters of such depth could contribute to Bering Strait signals.

Southern Bering Sea surface waters represented by the GEOSECS profile are subject to considerable processing before they reach the Bering Strait. They can be freshened by a combination of riverine inputs and ice melt, be cooled via heat loss to the atmosphere, be modified by mixing, and be imprinted by upwelling and the heightened biological productivity that upwelling nutrients foster [Coachman and Hansell, 1993]. Exactly how these processes affect the distribution of Ba within the Bering Sea is the subject of studies currently underway; however, preliminary results suggest that it is highly doubtful that the concentration of Ba associated with the mean flow into the Arctic through Bering Strait is very different from the values we observed in the Strait (K. Falkner, unpublished data).

Since productivity can be quite high both north and south of the western side of the strait [Coachman and Hansell, 1993 and references therein], productivity in either

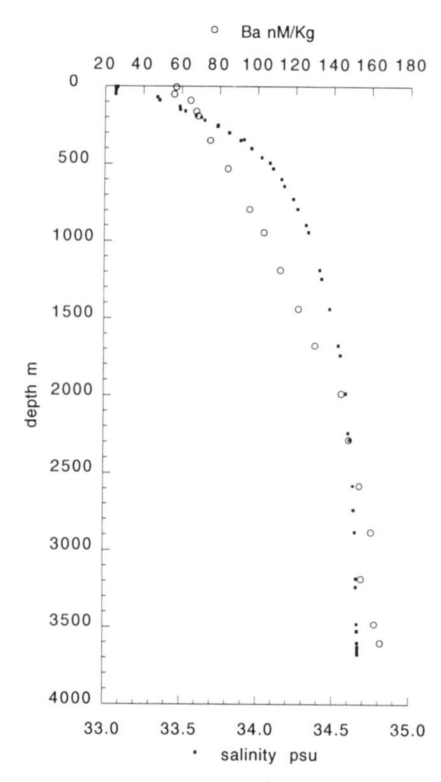

Fig. 3. GEOSECS Station 219 (Oct. 8, 1973) at 53° 6' N, 177° 18' W in the Bering Sea [Ostlund et al., 1987].

or both locations could ultimately have been responsible for the observed (20%) depletion of Ba in western surface waters with respect to the ≈70 nmol Ba kg^{-1} generally observed in the strait. As the salinity is lowered by at most 6% in the western surface waters, dilution by low Ba ice melt waters alone could not have produced such Ba depletions. (Dilution by river water would generally be expected to result in Ba concentrations greater than surface seawater levels, thus riverine inputs could partially offset the biological stripping effects particularly in waters originating from the Siberian coastal current which is fed directly by the Kolyma and Indigirka rivers.) The very high value (123 nmol Ba kg^{-1}) observed in the deep eastern strait probably represents a local or advected regeneration signature.

Transects along the northeast Chukchi Sea showed salinity distributions typical of summer stratified conditions; a 10-20 m surface-mixed layer affected by ice melt ranged in salinity from 29.8 to 32.2 psu while deeper waters had more consistent salinities of about 32.4 psu. As previously observed [Coachman and Barnes, 1961; Walsh et al., 1989], near bottom water temperatures generally decreased toward the north, ranging from 3.4°C at the C-line to 2.3°C at FW and PF sites (Figure 1; Table 1).

Surface waters displayed more dramatic temperature differences, ranging from 3.2-4.0°C at the C-line to -1.6-2.4°C northward. Stratified conditions were also evident in Ba concentrations with lower values being encountered in the mixed layer (Table 1). If it can be presumed that the surface waters along the transects originated with similar properties at the Bering Strait, i.e. Ba levels of about 68-70 nmol kg^{-1} encountered on Sep. 21, then it appears that surface waters are depleted in their Ba concentrations as they progress to the north. These trends are shown graphically by plotting composite Ba profiles constructed from surface and deep samples for each of the major sections (Figure 4). Again, the extent of these apparent depletions is far greater than can be explained by salinity changes due to ice melt.

The FW line sampling was conducted from within 2 to 15 miles from the southerly extent of the ice edge (see Figure 1). Interestingly, it is along the ice edge that the very lowest Ba concentrations are encountered. Barium removal appears to be accelerated in concert with elevated productivity known to be occurring at the ice-edge [*Walsh et al.*, 1989]. Two surface samples from stations FW 12 and 13 (10 m in Figure 4) stand out as having relatively

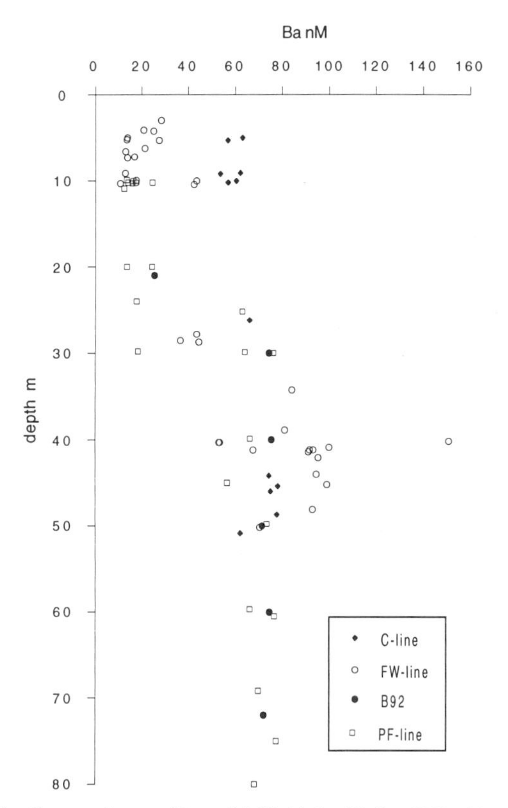

Fig. 4. Composite profiles of RV Alpha Helix (HX166) surface and deep stations; key refers to cruise lines labled in Fig. 1.

high Ba concentrations (≈40 nmol Ba kg^{-1}) and distinctly higher salinities and surface temperatures. These two stations are located over a small depression to the east of Herald shoal where current meters recorded consistently northerly flow from 1991 to 1992, including during the period leading up to the occupation of these stations (T. Weingartner, unpublished data). This probably represents a distinct band of Bering Sea water extending northwards through stations C15 and C16 which display similar hydrographic properties.

As a crude estimate, the integrated rate of depletion for the water parcels represented by our measurements can be calculated by applying the mean current speed along the coast of 0.07 m s^{-1}. From the Bering Strait to Cape Lisburne (≈350 km), 15% of surface Ba appears to have been stripped by biological productivity within a few months; from the Bering Strait to between Icy Cape and Pt. Barrow (≈350 km), up to 85% appears to have been stripped within half a year. Consistent with such a scenario, the deep waters show roughly increasing concentrations (65-150 nmol Ba kg^{-1}) toward the north. This is presumably due to regeneration from shelf sediments with greater variability in the deep values suggestive of sediment sources of variable intensity. A relatively high benthic biomass over the Chukchi shelves [Grebmeier, 1993] would be expected to process material raining from the surface relatively rapidly and in so doing augment typical Ba regeneration rates. More detailed profiles available for these sections confirm the depth trends indicated by the composite profiles (Table 1). It should be noted that the alternative scenario that the northward trends represents a rapid variation in the Ba input at the Bering Strait cannot be ruled out by these data alone although it seems less likely given the depth trends along the current path.

Henry Larson Cruise 92-16

At the IOS-A-1 Station in the Beaufort Sea (Figure 5a; Table 2), Ba generally parallels Si, displaying a pronounced maximum characteristic of nutrients the upper halocline. Barium concentrations equal to and exceeding those at the maximum in the upper halocline at IOS-A-1 were observed in near bottom waters over the Alaskan shelf. Although salinities over the shelf were not high enough for these waters to serve directly as a source to the halocline, their elevated Ba contents demonstrate accumulation does occur at depth. Thus, laterally advected brines which are believed to ventilate the halocline should display varying Ba contents depending upon their paths and residence times over the shelves. Furthermore, the amount of Ba included

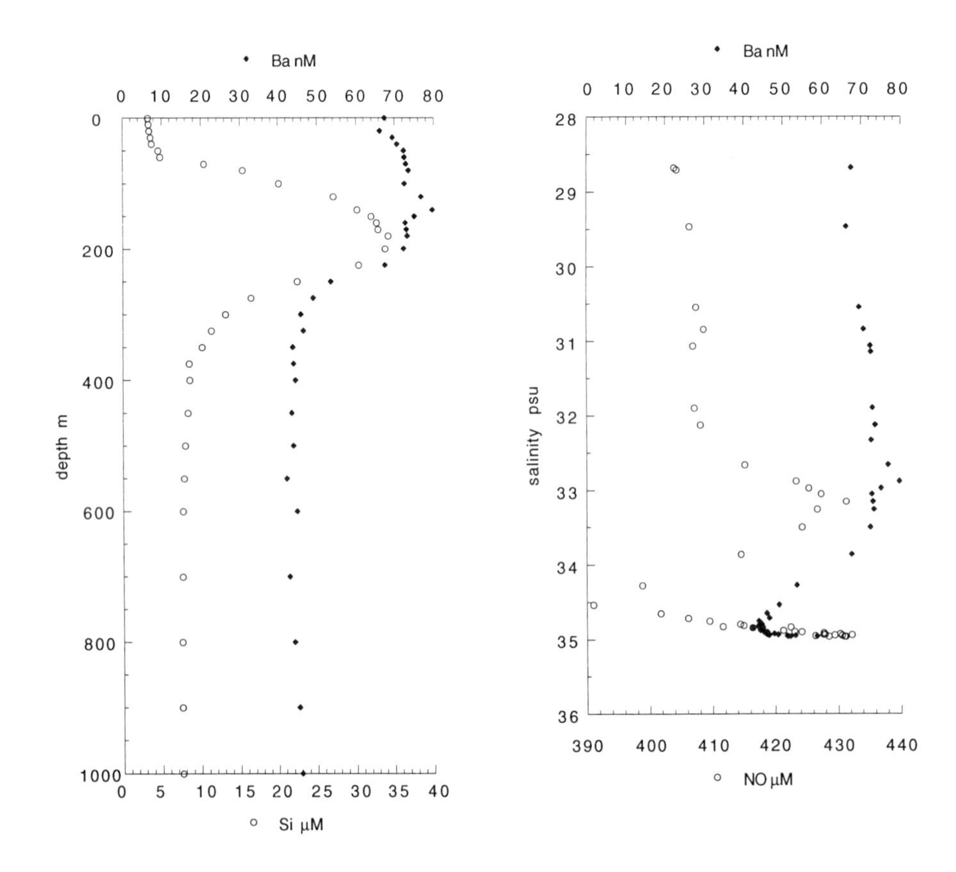

Fig. 5. IOS-A1 Arctic Ocean Climate Station (Sep. 20, 1992; 72° 32' N, 143° 54' W) profiles. For additional data see Table 2. (a) Si and Ba versus depth (b) Ba and NO (= 9[NO_3]+[O_2]) versus salinity; note samples R07 and R10 suspected to have larger than normal analytical errors in either O_2 or NO_3 and are not plotted for NO.

in the brines should be a function of local biological and sedimentary conditions and might be expected to differ over the different shelf regions and to not necessarily act coherently with the soft part nutrients and oxygen.

Although they are basically similar, it is worth noting a few distinct differences between the Ba and Si profiles. While Ba does display a broad maximum (≈75 nmol Ba kg^{-1}; 33.2 psu) in the upper halocline, a minor peak (80 nmol Ba kg^{-1}) at 32.9 psu is superimposed upon its broad peak. Since it is not observed in the nutrient profiles nor in salinity or temperature, it would be tempting to dismiss this odd feature as noise in the data except that the small peak is defined by three points and its magnitude is greater than analytical uncertainties. Furthermore, similarly unusual features, in the form of a pronounced maximum at a slightly lower salinities (32.3), appear in the quasi-conservative NO (Figure 5b) and PO parameters. Deeper in the water column, the gradients of NO, PO and Ba are in the same direction but differ in magnitude and location. From the limited data to date, it would appear that these tracers provide *similar* integrated information with regard to halocline ventilation but may track brine formation events differently. More experience gained from additional observations is needed to determine whether features such as these can aid us in delineating from where the brines originate or if they are the result of sampling artifacts.

Surface (≤50 m) Ba levels (68 nmol Ba kg^{-1}) at IOS-A-1 are not as depleted as those of Si and the other nutrients and they show marked elevation with respect to very low levels (10-20 nmol Ba kg^{-1}) in nearby Alaskan Shelf surface waters of comparable salinity. Oxygen isotope data from previous expeditions to this region suggest meteoric waters can penetrate well off shore (R. Macdonald, unpublished data). Therefore it is possible that integrated Arctic riverine Ba inputs more than compensate for biological activities at IOS-A-1. While we were unable to locate Ba values for Arctic rivers, other world river waters typically contain 100 to several hundred nmol dissolved Ba kg^{-1} with an additional 30% to over 100% of this available upon

TABLE 2. IOS-A1 Climate Station (Sep. 21, 1992; 72°32' N, 143°54' W) Data

sample id	depth m	sal-psu	O2-μM	PO4-μM	SiO4-μM	NO3-μM	Ba-nM
R01	0	28.680	403	0.93	3.30	0.07	67.5
R02*	10	28.708	404	0.93	3.36	0.07	77.8
R03	20	29.466	406	0.94	3.47	0.07	66.3
R04	30	30.548	407	0.98	3.63	0.01	69.6
R05	40	30.841	409	0.99	3.76	0.00	70.7
R06	50	31.065	405	1.03	4.59	0.20	72.4
R07	60	31.146	396	1.05	4.85	0.47	72.5
R08	70	31.899	368	1.35	10.4	4.39	72.9
R09	80	32.127	341	1.56	15.4	7.48	73.5
R10	100	32.332	299	1.74	20.1	10.0	72.5
R11	120	32.659	298	1.89	27.1	13.0	76.8
R12	140	32.880	294	1.97	30.1	14.4	79.6
R13	150	32.974	289	2.03	32.0	15.1	75.0
R14	160	33.053	288	2.04	32.7	15.5	72.7
R15	170	33.153	291	2.03	32.9	15.6	73.0
R16	180	33.259	281	2.07	34.2	16.2	73.2
R17†	190	33.374	305	2.05	34.0	16.3	72.8
R18	200	33.501	277	2.04	33.8	16.3	72.3
R19	225	33.864	268	1.86	30.3	16.3	67.5
R20	250	34.274	259	1.56	22.5	15.6	53.5
R21	275	34.533	260	1.31	16.4	14.5	49.0
R22	300	34.648	275	1.22	13.2	14.1	45.8
R23	325	34.711	281	1.17	11.4	13.9	46.5
R24	350	34.751	285	1.14	10.2	13.8	43.8
R25	375	34.812	292	1.11	8.50	13.7	44.0
R26	400	34.794	292	1.11	8.56	13.6	44.5
R27	450	34.823	289	1.11	8.35	13.7	43.5
R28	500	34.842	294	1.10	7.99	13.6	44.0
R29	550	34.846	295	1.11	7.84	13.7	42.2
R30	600	34.831	300	1.11	7.69	13.6	44.9
R31†	700	34.881	276	1.10	7.61	13.3	42.9
R32	800	34.877	301	1.12	7.51	13.4	44.1
R33	900	34.893	303	1.16	7.51	13.4	45.3
R34	1000	34.898	303	1.17	7.57	13.4	46.0
R35	1250	34.909	304	1.25	8.21	13.7	45.3
R36	1500	34.922	304	1.20	9.17	14.1	47.7
R37	1600	34.932	301	1.14	9.65	14.1	45.9
R38	1700	34.936	301	1.17	10.0	14.6	46.4
R39	1800	34.934	295	1.18	10.7	14.7	48.7
R40	1900	34.941	295	1.21	10.9	15.0	46.4
R41	2000	34.945	296	1.22	11.2	15.0	53.1
R42	2280	34.955	294	1.23	12.1	15.3	51.2
R43	2500	34.950	288	1.23	12.6	15.4	51.0
R44	3000	34.957	289	1.24	13.8	15.5	51.8
R45	3200	34.955	292	1.24	13.7	15.5	58.6

*suspected Ba contamination: not plotted in Fig.'s 5a & 5b

†suspected pre or post trips: not plotted in Fig.'s 5a & 5b

desorption from clays in the estuary [Carroll et al., 1993; Edmond et al., 1978; Hanor and Chan, 1977; Li and Chan, 1979; M. Palmer and J. Edmond, unpublished data]. The effective riverine Ba concentration contributed to the surface Arctic should be further amplified by ice formation since dissolved constituents are generally excluded during ice formation and continental run off in the Arctic is approximately balanced by ice export at the Fram Strait [Aagaard and Carmack, 1989]. Thus, in the absence of biological drawdown, the substantial combined input of river water to the Arctic (about half the input of the world's largest river, the Amazon) and limited mixing with underlying waters would be expected to result in elevated Ba levels in the surface waters of this relatively closed basin.

Superimposition of biologically driven Ba removal on edges the system should serve to accentuate the contrast between riverine and "background" Arctic surface water Ba levels. An analogous yet more extreme situation for Si renders it useful as a localized tracer of river waters for limited periods of the year [Macdonald et al., 1987]. The more biointermediate character of Ba would be expected to extend the time and space scales over which it can be applied to tracking the fate of river waters. At the same time, Ba should provide more spatial resolution for this purpose than Ca which is less impacted by removal processes and whose small but detectable excesses have been used as an integrated measure of the riverine influence on Arctic surface waters [e.g. Anderson and Dyrssen, 1981; Anderson and Jones, 1985; Jones et al., 1991; Tan et al., 1983]. Additional data will be required to determine whether the balance of dynamic forces affecting Ba are favorable in this regard.

Exactly what phase is responsible for Ba removal in the Arctic has yet to be determined. If, as in the open oceans, barite is formed in biological debris and a fraction delivered to the sediments survives burial, then it is conceivable that barite in Arctic sediments might be applied as a paleo-productivity index [Shimmield et al., 1988]. Correlated sedimentary Ba (Al-normalized) and oxygen isotope records in the Nansen Basin show promise in this regard (D. Nürnberg, personal communication, 1993). A potentially complicating factor arises from ice-rafting of sediments deposited over the shelves underlying more biologically productive waters [Pfirman et al., 1989]. High Al-normalized particulate Ba concentrations observed on several occasions in Arctic ice-cores (S. Pfirman, personal communication) may reflect transport of such material. However, Ba-enriched phases formed in association with sub-ice biological activities cannot be ruled out. While our data seem to support the former scenario, we are presently embarked on additional studies to acquire the more thorough understanding of Ba cycling in the Arctic prerequisite to exploiting its paleo-record. It is particularly important, in this regard, to determine to what extent celestite versus barite is implicated in particulate Ba transport.

6. SUMMARY

In general, the biointermediate behavior of Ba in the oceans tends to result in measurably different signatures between water masses, which once removed from the surface, estuary or sediments, should in principle be traceable by their Ba contents. The results of this initial study indicate possibilities for Ba as a water mass tracer in the Arctic. Removal of Ba from surface waters in association with biological activities and subsequent regeneration in deeper waters appear to be occurring over the eastern shelves of the Chukchi Sea. Although the temporal and spatial extents of such processes in the Arctic remain to be determined, it is likely that biological draw down accentuates the difference between riverine and "background" surface concentrations which should enhance the utility of Ba for tracking Arctic river waters. Analogous features suggest that Ba might provide information complementary to the NO and PO parameters for determining the origin of brines ventilating the upper and lower halocline. To explore these tracer capabilities and to establish the link between water column processes and the sedimentary record, additional measurements (including simultaneous determinations of the better studied tracers such as the nutrients and oxygen and its isotopes) in key regions of the Arctic and surrounding seas are presently being carried out.

Acknowledgments. E. Carmack and R. Macdonald acknowledge Rick Pearson for data processing, Doug Sieberg for shipwork, Fiona McLaughlin and Mary O'Brien for seawater sampling and the Captain, officers and crew of the CCGS Henry Larson and the support of the Coast Guard Northern, Canada for the successful 1992 field efforts. T. Weingartner was supported by the Minerals Management Service under contract #MMS01411917 to the University of Alaska. Alpha Helix ship time was provided by MMS and the Japan Marine Science and Technology Center (JAMSTEC). K. Falkner thanks Rick Pearson for his cheerful assistance with the figures in this report and Knut Aagaard for enlightening discussions. K. Falkner was supported by the Young Investigator Program of the US Office of Naval research. This manuscript was improved by the comments of two anonymous reviewers.

REFERENCES

Aagaard, K., Current, CTD, and pressure measurements in possible dispersion regions of the Chukchi sea, in *Final Report, OCSEAP RU-91,* pp. 255-333, DOI-DOC Minerals Management Service, Alaska Region, 1984.

Aagaard, K., and E. C. Carmack, The role of sea ice and other fresh water in the Arctic circulation, *J. Geophys. Res., 94,* 14,485-14,498, 1989.

Aagaard, K., L. K. Coachman, and E. C. Carmack, On the halocline of the Arctic Ocean, *Deep Sea Res., 28,* 529-545, 1981.

Aagaard, K., A. T. Roach, and J. D. Schumacher, On the wind-driven variability of the flow through the Bering Strait, *J. Geophys. Res., 90,* 7213-7221, 1985.

Anderson, L., and D. Dyrssen, Chemical constituents in the Svalbard area, *Oceanol. Acta, 4,* 305-311, 1981.

Anderson, L. G., D. Dyrssen, and E. P. Jones, An assessment of the transport of CO_2 in the Arctic Ocean, *J. Geophys. Res., 95,* 1703-1711, 1990.

Anderson, L. G., and E. P. Jones, Measurements of total alkalinity, calcium, and sulphate in natural sea ice, *J. Geophys. Res., 90,* 9194-9198, 1985.

Anderson, L. G., E. P. Jones, K. P. Koltermann, P. Schlosser, J. H. Swift, and D. W. R. Wallace, The first oceanographic section across the Nansen Basin of the Arctic Ocean, *Deep-Sea Res., 36,* 475-482, 1989.

Bacon, M. P., and J. M. Edmond, Barium at GEOSECS III in the Southwest Pacific, *Earth Planet. Sci. Lett., 16,* 66-74, 1972.

Bernstein, R. E., and P. R. Betzer, Labile phases and the ocean's strontium cycle: A method of sediment trap sampling for Acantharians, in *Marine Particles: Analysis and Characterization,* edited by D. C. Hurd and D. W. Spencer, pp. 369-374, AGU, Washington DC, 1991.

Bernstein, R. E., R. H. Byrne, P. R. Betzer, and A. M. Greco, Morphologies and transformations of celestite in seawater: The role of acantharia in strontium and barium geochemistry, *Geochim. Cosmochim. Acta, 56,* 3273-3279, 1992.

Bishop, J. K. B., The barite-opal-organic carbon association in oceanic particulate matter, *Nature, 332,* 341-343, 1988.

Bishop, J. K. B., Regional extremes in particulate matter composition and flux: Effects on the chemistry of the ocean interior, in *Productivity in the Ocean: Present and Past,* edited by W. H. Berger, V. S. Smetacek and G. Wefer, pp. 117-137, John Wiley & Sons, New York, 1989.

Broecker, W. S., "NO", a conservative water-mass tracer, *Earth Planet. Sci. Lett., 23,* 100-107, 1974.

Carmack, E. C., Large-Scale Physical Oceanography of Polar Oceans, in *Polar Oceanography, Part A: Physical Science,* edited by W. O. Smith, pp. 171-222, Academic Press, San Diego, 1990.

Carroll, J., K. K. Falkner, E. T. Brown, and W. S. Moore, The role of sediments in maintaining high dissolved ^{226}Ra and Ba in the Ganges-Bramaputra mixing zone, *Geochim. Cosmochim. Acta, 57,* 2981-2990, 1993.

Chan, L. H., D. Drummond, J. M. Edmond, and B. Grant, On the barium data from the Atlantic GEOSECS Expedition, *Deep-Sea Res., 24,* 613-649, 1977.

Chow, T. S., and E. D. Goldberg, On the marine geochemistry of barium, *Geochim. Cosmochim. Acta, 20,* 192-198, 1960.

Church, T. M., and K. Wolgemuth, Marine barite saturation, *Earth Planet. Sci. Lett., 15,* 35-44, 1972.

Coachman, L. K., On the flow field in the Chirikov Basin, *Continental Shelf Res., 13,* 481-508, 1993.

Coachman, L. K., and K. Aagaard, On the water exchange through the Bering Strait, *Limnol. Oceanogr., 11,* 44-59, 1966.

Coachman, L. K., and K. Aagaard, Reevaluation of water transports in the vicinity of Bering Strait, in *The Eastern Bering Sea Shelf: Oceanography and Resources,* edited by D. W. Hood and J. A. Calder, pp. 95-110, National Oceanic and Atmospheric Administration, Washington DC, 1981.

Coachman, L. K., and K. Aagaard, Transports through Bering Strait: Annual and interannual variability, *J. Geophys. Res., 93,* 15,535-15,539, 1988.

Coachman, L. K., and C. A. Barnes, The contribution of Bering Sea water to the Arctic Ocean, *Arctic, 14,* 147-161, 1961.

Coachman, L. K., and D. A. Hansell, ISHTAR, *Continental Shelf Res., 13,* 1-704, 1993.

Codispoti, L. A., and F. A. Richards, Oxygen supersaturations in the Chukchi and East Siberian Seas, *Deep-Sea Res., 18,* 341-351, 1971.

Collier, R., and J. Edmond, The trace element geochemistry of marine biogenic particulate matter, *Prog. Oceanog., 13,* 113-199, 1984.

Dehairs, F., R. Chesselet, and J. Jedwab, Discrete suspended particles of barite and the barium cycle in the open ocean, *Earth Planet. Sci. Lett., 49,* 528-550, 1980.

Dehairs, F., L. Goeyens, N. Stroobants, P. Bernard, C. Goyet, A. Poisson, and R. Chesselet, On suspended barite and the oxygen minimum in the southern ocean, *Biogeochem. Cycles, 4,* 85-102, 1990.

Dehairs, F., C. E. Lambert, R. Chesselet, and N. Risler, The biological production of marine suspended barite and the barium cycle in the Western Mediterranean Sea, *Biogeochem., 4,* 119-139, 1987.

Dehairs, F., N. Stroobants, and L. Goeyens, Suspended barite as a tracer of biological activity in the Southern Ocean, *Mar. Chem., 35,* 399-410, 1991.

Dymond, J., E. Suess, and M. Lyle, Barium in deep-sea sediment: A geochemical proxy for paleoproductivity, *Paleoceanogr., 7,* 163-181, 1992.

Edmond, J. M., E. D. Boyle, D. Drummond, B. Grant, and T. Mislick, Desorption of barium in the plume of the Zaire (Congo) River, *Neth. J. Sea Res., 12,* 324-328, 1978.

Edmond, J. M., C. Measures, R. E. McDuff, L. H. Chan, R. Collier, B. Grant, L. I. Gordon, and J. B. Corliss, Ridge crest hydrothermal activity and the balances of the major and minor elements in the ocean: The Galapagos data, *Earth Planet. Sci. Lett., 46,* 1-18, 1979.

Falkner, K. K., G. Klinkhammer, T. S. Bowers, J. F. Todd, B.

Lewis, W. Landing, and J. M. Edmond, The behavior of Ba in anoxic marine waters, *Geochim. Cosmochim. Acta, 57,* 537-554, 1993.

Goldberg, E. D., and G. O. S. Arrhenius, Chemistry of Pacific pelagic sediments, *Geochim. Cosmochim. Acta, 13,* 153-212, 1958.

Grebmeier, J. M., Studies of pelagic-benthic coupling extended onto the Soviet continental shelf in the northern Bering and Chukchi shelf sediments, *Continental Shelf Res., 13,* 653-668, 1993.

Hanor, J. S., and L.-H. Chan, Non-conservative behavior of barium during mixing of Mississippi River and Gulf of Mexico waters, *Earth Planet. Sci. Lett., 37,* 242-250, 1977.

Hufford, G. L., Warm water advection in the southern Beaufort Sea August-September 1971, *J. Geophys. Res., 78,* 2702-2707, 1973.

Hufford, G. L., Some characteristics of the Beaufort Sea Shelf Current, *J. Geophys. Res., 80,* 3465-3468, 1975.

Jones, E. P., and L. G. Anderson, On the origin of the chemical properties of the Arctic Ocean halocline, *J. Geophys. Res., 91,* 10,759-10,767, 1986.

Jones, E. P., and L. G. Anderson, On the origin of the properties of the Arctic Ocean halocline north of Ellesmere island: results from the Canadian Ice Island, *Continental Shelf Res., 10,* 485-498, 1990.

Jones, E. P., L. G. Anderson, and D. W. R. Wallace, Tracers of near-surface, halocline and deep waters in the Arctic Ocean: Implications for circulation, *J. Mar. Sys., 2,* 241-255, 1991.

Kinney, P., M. E. Arhelger, and D. C. Burrell, Chemical characteristics of water masses in the Amerasian Basin of the Arctic Ocean, *J. Geophys. Res., 75,* 4097-4104, 1970.

Klinkhammer, G. P., and L. H. Chan, Determination of barium in marine waters by isotope dilution inductively coupled plasma mass spectrometry, *Anal. Chim. Acta, 232,* 323-329, 1990.

Lea, D. W., Foraminiferal and coralline barium as paleoceanographic tracers, PhD Thesis, MIT/WHOI, WHOI-90-06, Cambridge, MA, 1990.

Li, Y.-H., and L. H. Chan, Desorption of Ba and ^{226}Ra from riverborne sediments in the Hudson Estuary, *Earth Planet. Sci. Lett., 43,* 343-350, 1979.

Macdonald, R. W., C. S. Wong, and P. E. Erickson, The distribution of nutrients in the southeastern Beaufort Sea: Implications for water circulation and primary production, *J. Geophys. Res., 92,* 2939-2952, 1987.

Martin, J.-M., and M. Maybeck, Elemental mass-balance of material carried by major world rivers, *Mar. Chem., 7,* 173-206, 1979.

Melling, H., and E. L. Lewis, Shelf drainage flows in the Beaufort Sea and their effect on the Arctic Ocean pycnocline, *Deep Sea Res., 29,* 967-985, 1982.

Michaels, A. F., Vertical distribution and abundance of Acantharia and their symbionts, *Mar. Bio., 97,* 559-569, 1988.

Ostlund, H. G., H. Craig, W. S. Broecker, and D. Spencer (Eds.), *Atlantic, Pacific and Indian Ocean Expeditions, Shorebased Data and Graphics,* 200 pp., National Science Foundation, Washington DC, 1987.

Overland, J. E., and A. T. Roach, Northward flow in the Bering and Chukchi Seas, *J. Geophys. Res., 92,* 7097-7105, 1987.

Paquette, R. G., and R. H. Bourke, Observations on the coastal current of Arctic Alaska, *J. Mar. Res., 32,* 195-207, 1974.

Pfirman, S., J. C. Gascard, I. Wollenburg, P. Mudie, and A. Abelmann, Particle-laden Eurasian Arctic sea ice; observations from July and August 1987, *Polar Res., 7,* 59-66, 1989.

Revelle, R. R., *Marine bottom samples collected in the Pacific Ocean by the Carnegie on its seventh cruise,* 1-133 pp., Publ. Carn. Inst., Washington, D.C., 1944.

Revelle, R. R., M. Bramlette, G. Arrhenius, and E. D. Goldberg, Pelagic sediments in the Pacific, *Spec. Paper Geol. Soc. America, 62,* 221-236, 1955.

Shimmield, G. B., N. B. Price, and A. A. Kahn, The use of Th-230 and Ba as indicators of palaeoproductivity over a 300 Kyr timescale-Evidence from the Arabian Sea, *Chem. Geol., 70,* 112, 1988.

Tan, F. C., D. Dyrssen, and P. M. Strain, Sea ice meltwater and excess alkalinity in the east Greenland currrent, *Oceanol. Acta, 6,* 283-288, 1983.

Turekian, K. K., and E. H. Tausch, Barium in deep sediments of the Atlantic Ocean, *Nature, 201,* 696-697, 1964.

Von Damm, K. L., Seafloor hydrothermal activity: Black smoker chemistry and chimneys, *Annu. Rev. Earth Planet. Sci., 18,* 173-204, 1990.

Von Damm, K. L., J. M. Edmond, B. Grant, C. I. Measures, B. Walden, and R. F. Weiss, Chemistry of submarine hydrothermal solutions at 21°N, East Pacific Rise, *Geochim. Cosmochim. Acta, 49,* 2197-2220, 1985.

Walsh, J. J., C. P. McRoy, L. K. Coachman, J. J. Goering, J. J. Nihoul, T. E. Whitledge, T. H. Blackburn, P. L. Parker, C. D. Wirick, P. G. Shuert, J. M. Grebmeier, A. M. Springer, R. D. Tripp, D. A. Hansell, S. Djenidi, E. Deleersnijder, K. Henriksen, B. A. Lund, P. Andersen, F. E. Müller-Karger, and K. Dean, Carbon and nitrogen cycling within the Bering/Chukchi Seas: Source regions for organic matter effecting AOU demands of the Arctic Ocean, *Prog. Oceanogr., 22,* 277-359, 1989.

Wilson, C., and D. W. R. Wallace, Using the nutrient ratio NO/PO as a tracer of continental shelf waters in the central Arctic Ocean, *J. Geophys. Res., 95,* 22,193-22,208, 1990.

corresponding author:
K. Kelly Falkner, COAS, OSU, Ocean. Admin. Bldg. 104, Corvallis, OR, 97331-5503

The Northern Barents Sea: Water Mass Distribution and Modification

S.L. Pfirman[1], D. Bauch[2], and T. Gammelsrød[3]

[1]Barnard College, Columbia University, New York City, New York
[2]Lamont-Doherty Earth Observatory of Columbia University, Palisades, New York
[3]Geophysical Institute, University of Bergen, Bergen, Norway

The main water masses in the northern Barents Sea are surface water, Arctic water, transformed Atlantic water, and cold bottom water. Using summer data from 1981 and 1982, the formation, distribution, modification and circulation of these water masses are discussed. Recent estimates show that about 2 Sv of Atlantic water enters the Barents Sea by the North Cape Current, balanced by a similar outflow through the strait between Novaya Zemlya and Frans Josef Land. Passing through the Barents Sea, Atlantic-derived water is modified by interaction with other water masses as well as with the atmosphere, and the end products are believed to be important contributors to the hydrographic structure of the Arctic Ocean.

INTRODUCTION

The main sources of deep waters which fill the deep basins of the global oceans are known to be located in polar regions. Here, water masses are transformed by processes that increase their density, usually by cooling and/or increasing salinity. Because deep water formation governs global oceanic circulation, it is important to determine the actual locations and processes involved, as well as to determine their sensitivity to environmental changes.

The Barents Sea is a key region for modification of water masses in the Arctic. It is one of several continental shelf seas marginal to the Arctic Ocean which influence its hydrographic structure. Water flowing over shallow regions is transformed when heat loss and brine rejection during sea-ice formation increase density in winter, while warming and addition of sea-ice meltwater and river runoff decrease surface water density during summer [e.g. *Nansen*, 1906; *Redfield and Friedman*, 1969; *Hanzlick and Aagaard*, 1980; *Midttun*, 1985; *Quadfasel et al.*, 1988; *Aagaard and Carmack*, 1989].

The Polar Oceans and Their Role in Shaping
the Global Environment
Geophysical Monograph 85

The Barents Sea is different from other Arctic shelves because it has close connections to the Norwegian Sea as well as to the Arctic Ocean. The amount of Atlantic water passing through this sea from a branch of the Norwegian Atlantic Current may equal that supplied through Fram Strait in the West Spitsbergen Current [Fig. 1; *Rudels*, 1987]. Atlantic water inflow into the Barents Sea between Norway and Bear Island appears to be about 1.6 Sv, varying between 2.1 Sv in winter and 1.4 Sv in summer (see compilation by Loeng et al., 1993). Blindheim (1989) estimates the total inflow (coastal water plus Atlantic water) through this passage to be about 3 Sv, with about 1 Sv outflow. The average mass transport leaving the Barents Sea through the strait between Novaya Zemlya and Frans Josef Land is of order 2 Sv (varying between 0.7 to 3.2 Sv; Loeng et al., 1993). Meanwhile, estimates of Atlantic water entering the Arctic Ocean via the West Spitsbergen Current vary between about 2 and 3.7 Sv [*Hanzlick*, 1983; *Rudels*, 1987].

Because of Atlantic water inflow and the very limited contribution of freshwater from river runoff [*Novitsky*, 1961; *Rudels*, 1987], the Barents Sea is also more saline

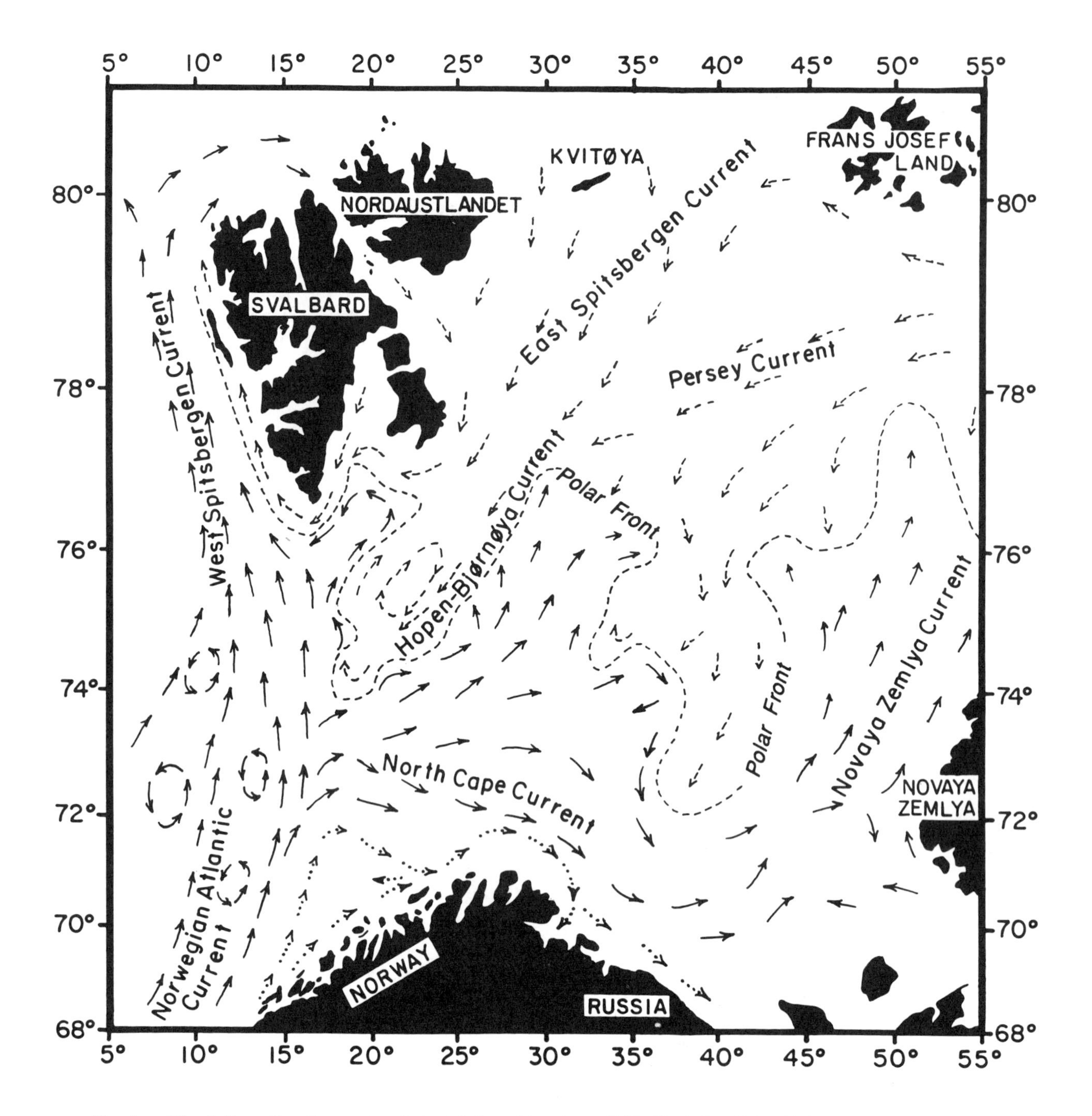

Fig. 1 Circulation of surface waters (adapted from *Tantsiura*, 1959; *Novitsky*, 1961; *Midttun and Loeng*, 1987) showing main warm currents (Atlantic water: solid arrows), cold currents (Arctic water: dashed arrows), Norwegian/Murmansk Coastal Current (dotted arrows) and location of the oceanic polar front (also called the Arctic front: dashed line).

than the other shelf seas. This means that winter cooling and additional salt contributed from sea-ice formation can easily result in formation of cold, saline water that is denser than the Atlantic layer. When such dense waters flow off the shelf, they are thought to contribute a cold, saline signal to deep waters of the Arctic Ocean [*Nansen*, 1906; *Gordienko and Laktionov*, 1969; *Aagaard et al.*, 1981; *Swift et al.*, 1983; *Midttun*, 1985; *Rudels*, 1987; *Quadfasel et al.*, 1988; *Blindheim*, 1989; *Midttun*, 1989]. Recent $\partial^{18}O$ profiles indicating that deep and bottom waters of the Arctic do not show river-runoff influence, also point to the Barents Sea as a possible source region [*Schlosser et al.*, 1994].

This paper examines the major water masses in the northern Barents Sea, using temperature and salinity data collected during summer expeditions in 1981 and 1982, to trace their distribution and modification [see also *Tantsiura*, 1959, 1973; *Novitsky*, 1961; *Pfirman*, 1985; *Midttun and Loeng*, 1987; *Loeng*, 1991].

BATHYMETRY AND HYDROGRAPHY

The main hydrographic feature of the near-surface waters of the Barents Sea is the oceanic polar front (also called the Arctic front), which divides relatively warm and saline water of Atlantic origin in the south from colder and fresher Arctic water in the north (Fig. 1). It is not as distinct in the eastern Barents Sea where a mixed water mass extends over large areas [*Midttun and Loeng*, 1987]. The polar front divides the Barents Sea roughly in half and is located in the southern portion of our study area.

Because the Barents Sea is shallow with many banks and troughs (Fig. 2), bathymetric relief guides water mass transport and exchange. Atlantic water found south of the polar front is derived from the Norwegian Atlantic Current, which flows north along the Norwegian continental slope and then east up Bjørnøyrenna (also known as the Bear Island Channel) as the North Cape Current (Fig. 1). Atlantic water transport in this current appears to be about 1.6 Sv [see compilation in Loeng et al., 1993].

The North Cape Current branches in the Barents Sea, flowing through depressions between banks. The branch to the south of Sentralbanken (also called Central Bank) continues toward Novaya Zemlya and the Kara Sea and is known as the Novaya Zemlya Current [Tantsiura, 1959; Novitsky, 1961]. Here, modified Atlantic water which has mixed with Arctic water has been called Atlantic water [e.g. Pfirman, 1985], or Barents Sea water [Loeng, 1991]. In order to clarify the origin of this water mass, we call it Barents Atlantic-derived water (BAW) in this contribution. In the Arctic Ocean and the Norwegian-Greenland Sea the modified, Atlantic-derived water is often called Arctic intermediate water [e.g. *Swift*, 1986].

Part of the Norwegian Atlantic Current also continues west and north of Bjørnøyrenna along the western flank of Spitsbergenbanken, becoming the West Spitsbergen Current. North of Spitsbergen, an easterly branch of this current flows along the northern slope of the Barents Sea. Nansen (1906) and Mosby (1938) found that Atlantic water submerges beneath Arctic water in this region and becomes an intermediate water mass. Some branches of Atlantic water enter the Barents Sea between Nordaustlandet and Frans Josef Land from the north as a near-bottom water mass [*Mosby*, 1938].

Arctic water, also called polar water or winter water, is a water mass with temperatures near the freezing point which dominates the upper part of the water column north of the polar front. It originates from convection during sea-ice formation in fall and winter [*Mosby*, 1938]. In summer, a remnant of this layer is found in the Barents Sea. Arctic water forms locally or is advected into the Barents Sea from the northern Kara Sea and the Arctic Ocean [*Novitsky*, 1961; *Tantsiura*, 1973]. There are three main currents transporting this cold water and sea ice within the Barents Sea (Fig. 1): the Persey Current flowing west across the Barents Sea towards Spitsbergenbanken; the ill-defined East Spitsbergen Current, flowing out of the Arctic Ocean into the Barents Sea between Nordaustlandet and Frans Josef Land; and the high-velocity Hopen-Bjørnøya Current flowing south along the eastern flank of Spitsbergenbanken [*Tantsiura*, 1959; *Novitsky*, 1961; *Midttun and Loeng*, 1987].

From fall to early summer, sea-ice covers the region north of the polar front. As summer progresses, the ice melts and the marginal ice zone retreats towards the north. A thin layer of surface water is formed that overlies Arctic water during mid- to late summer. This layer has a lower salinity and is warmed due to exposure to solar radiation [*Mosby*, 1938; *Loeng*, 1980; *Loeng*, 1991].

Near the sea floor, a cold, saline water mass is frequently observed. The distribution of this dense bottom water shows large seasonal and annual variations [*Tantsiura*, 1973]. It forms both by cooling of Atlantic water and by addition of salt to the water column by brine rejection during sea-ice formation [*Nansen*, 1906; *Midttun*, 1985]. Where the salinity of the cold bottom water is greater than that of Atlantic water, this is a good indication that sea ice formation contributed to development of the water mass. Following Nansen (1906), we refer to water with colder temperatures near the sea floor as cold bottom water (CBW, also called bottom water by Loeng (1991) and cold deep water by Pfirman (1985)).

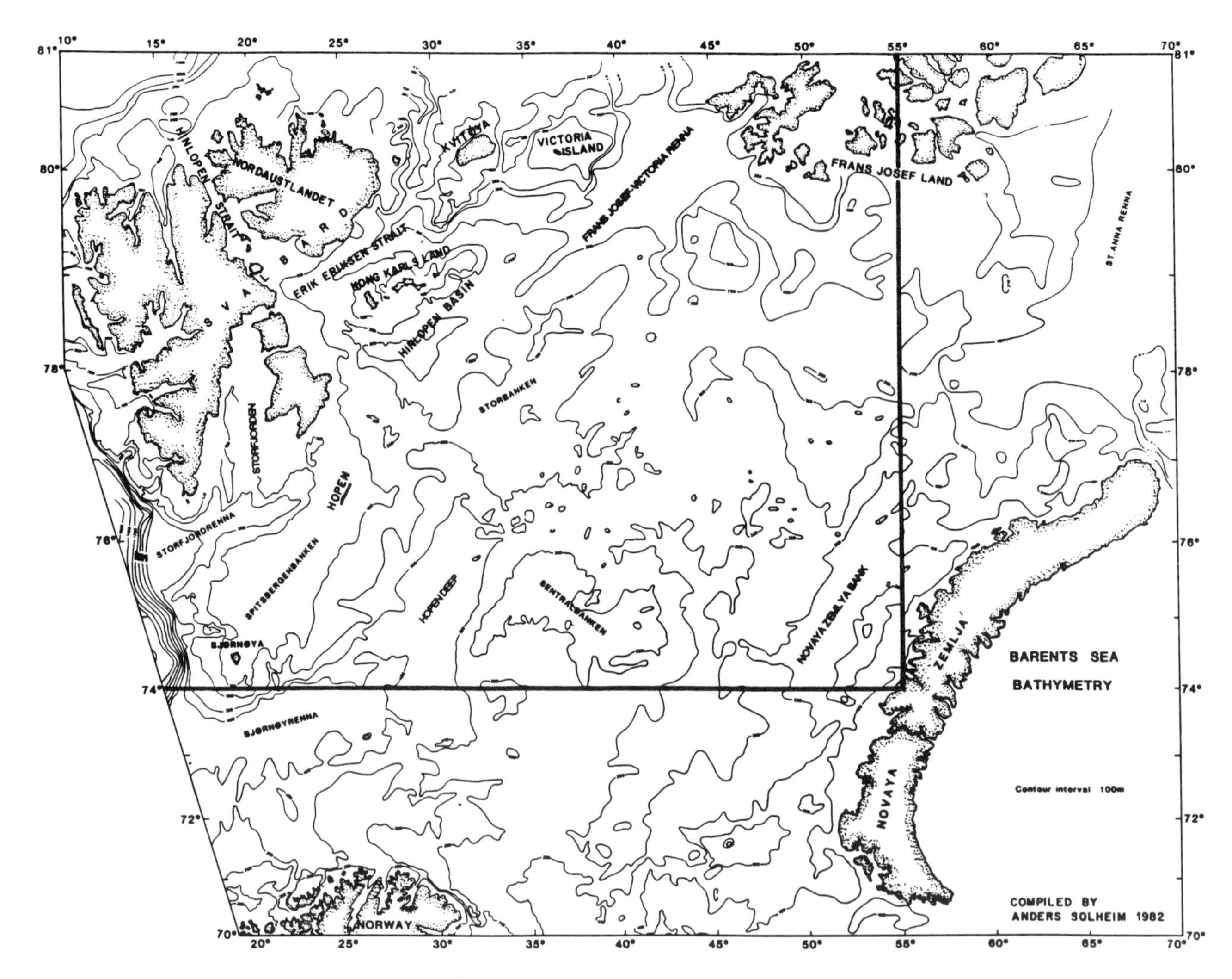

Fig. 2 Overview of Barents Sea geography and bathymetry. The study area is boxed in over the northwestern portion of the map.

During winter, sea ice formation causes vertical mixing to take place throughout the Barents Sea. Convection commonly extends down to 150 to 200m, reaching the sea floor over some banks [*Novitsky*, 1961; *Tantsiura*, 1973; *Midttun*, 1989]. Convection forms very saline and cold CBW near Novaya Zemlya Bank [Fig. 2: *Knipowitsch*, 1905; *Nansen*, 1906; *Tantsiura*, 1959; *Midttun*, 1985]. Similar cold brines form in Storfjorden [*Novitsky*, 1961; *Midttun*, 1985; *Anderson et al.*, 1988; *Quadfasel et al.*, 1988]. Less saline dense water forms over Sentralbanken [*Tantsiura*, 1959; *Midttun*, 1961; *Midttun and Loeng*, 1987; *Midttun*, 1989]. Convection over the banks also forms secondary circulation patterns, which aid in modifying adjacent Atlantic-derived water [*Tantsiura*, 1959]. For example, winter cooling of Atlantic water results in dense, but fairly warm bottom water near Spitsbergenbanken [*Tantsiura*, 1959; *Sarynina*, 1969; *Midttun and Loeng*, 1987]. Once formed, CBW seeks depressions and channels, flowing out of the Barents Sea and contributing to the intermediate, deep and bottom water characteristics of the Norwegian Sea and the Arctic Ocean [*Nansen*, 1906; *Swift et al.*, 1983; *Midttun*, 1985; *Midttun and Loeng*, 1987; *Quadfasel et al.*, 1988; *Midttun*, 1989; *Blindheim*, 1989; *Schlosser et al.*, 1994].

DATA AND METHODS

The data used in this study were obtained in the summers (July, August, September) of 1981 [*Midttun*, 1985; *Pfirman*, 1985; *Elverhøi et al.*, 1989] and 1982 (Fig. 3A). The more complete data set collected in 1981 is used as a basis of discussion. Even with this fairly extensive data set

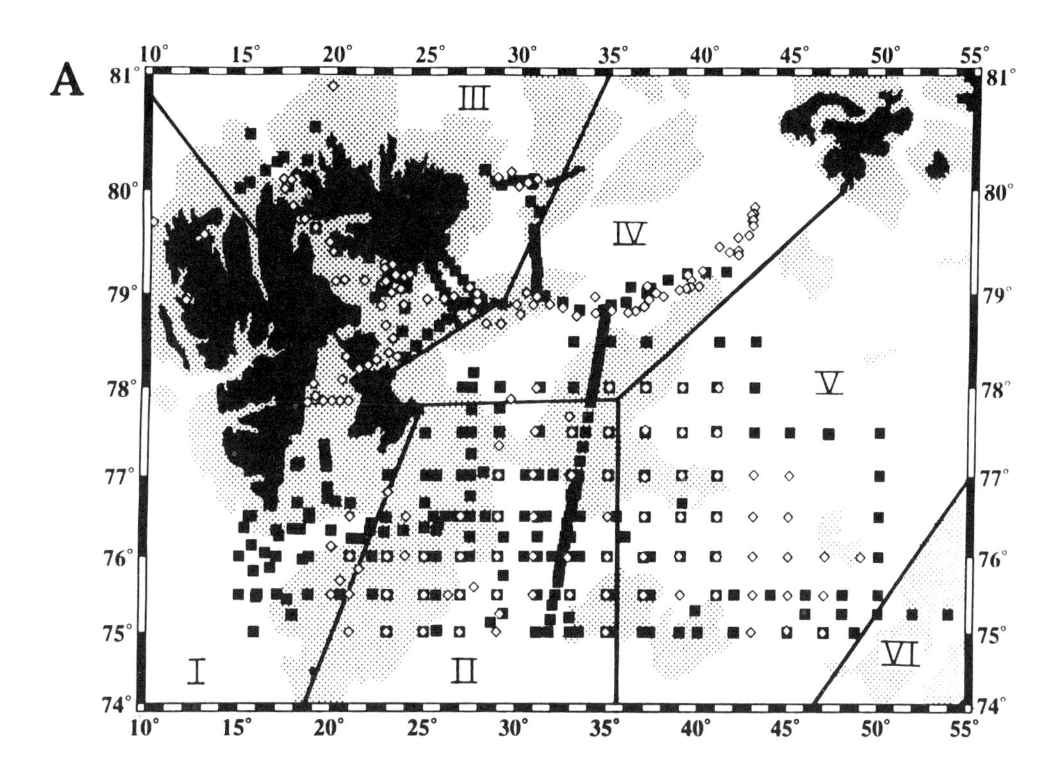

Fig. 3 A) Locations of hydrographic stations sampled in July, August, and September 1981 (squares) and 1982 (diamonds). Roman numerals indicate areas used in water mass analysis depicted in Fig. 3B, C, D, E. Stippling indicates regions where water depths are less than 200m. Potential temperature-salinity plots of: B) the temperature maximum representing Atlantic-derived water sampled in 1981, C) bottom water with salinities > 34.5°/oo sampled in 1981 (note cold, saline water in areas I (*Midttun*, 1985) and VI), D) the temperature maximum representing Atlantic-derived water sampled in 1982 (note: in area IV some stations had two Atlantic-derived water maxima. A "+" symbol is used to designate the deeper maximum, and the area is called IV*), and E) bottom water with salinities > 34.5°/oo sampled in 1982. The box labelled "EBDW" indicates the water mass characteristics of Eurasian Basin deep water of the Arctic Ocean.

it is difficult to describe the details of the water mass distribution patterns due to low sampling density, lack of synopticity, and lack of additional tracer data.

Hydrographic data were obtained from two sources: the University of Bergen and the Institute for Marine Research in Bergen (Havforskningsinstitutt). Both data sets were collected using a Neil Brown CTD calibrated by bottle data. The r.m.s. error in salinity of the University of Bergen data was ±0.008 in 1981 and ±0.004 in 1982. The precision of the temperature data was +/- 0.005°C and that of the pressure sensor was +/- 0.1%. The original CTD data was reduced to two decibar averages and the salinities were calculated from the average values. The Institute for Marine Research data was compiled from a variety of cruises and may have a slightly higher salinity error. These data were reported at standard depths (0, 5, 10, 20, 30, 40, 50, 75, 100, 125, 150, 200, 250, 300, 350, 400m).

Identification of Arctic water, Atlantic-derived water and cold bottom water is based on temperature, and to a lesser degree, salinity extrema. This method, called the core method [*Wuest*, 1935] and used in the northern Barents Sea by Mosby (1938), aids in following a water mass while it is transformed.

WATER MASS DISTRIBUTION

In late summer, north of the polar front, a typical station obtained in one of the deeper regions of the Barents Sea generally has: a relatively warm surface water layer; underlain by a temperature minimum corresponding with the Arctic water layer; underlain by a temperature and salinity maximum, which is related to the Barents Atlantic-derived water; and colder water near the bottom, here called cold bottom water (see Table 1, Fig. 4).

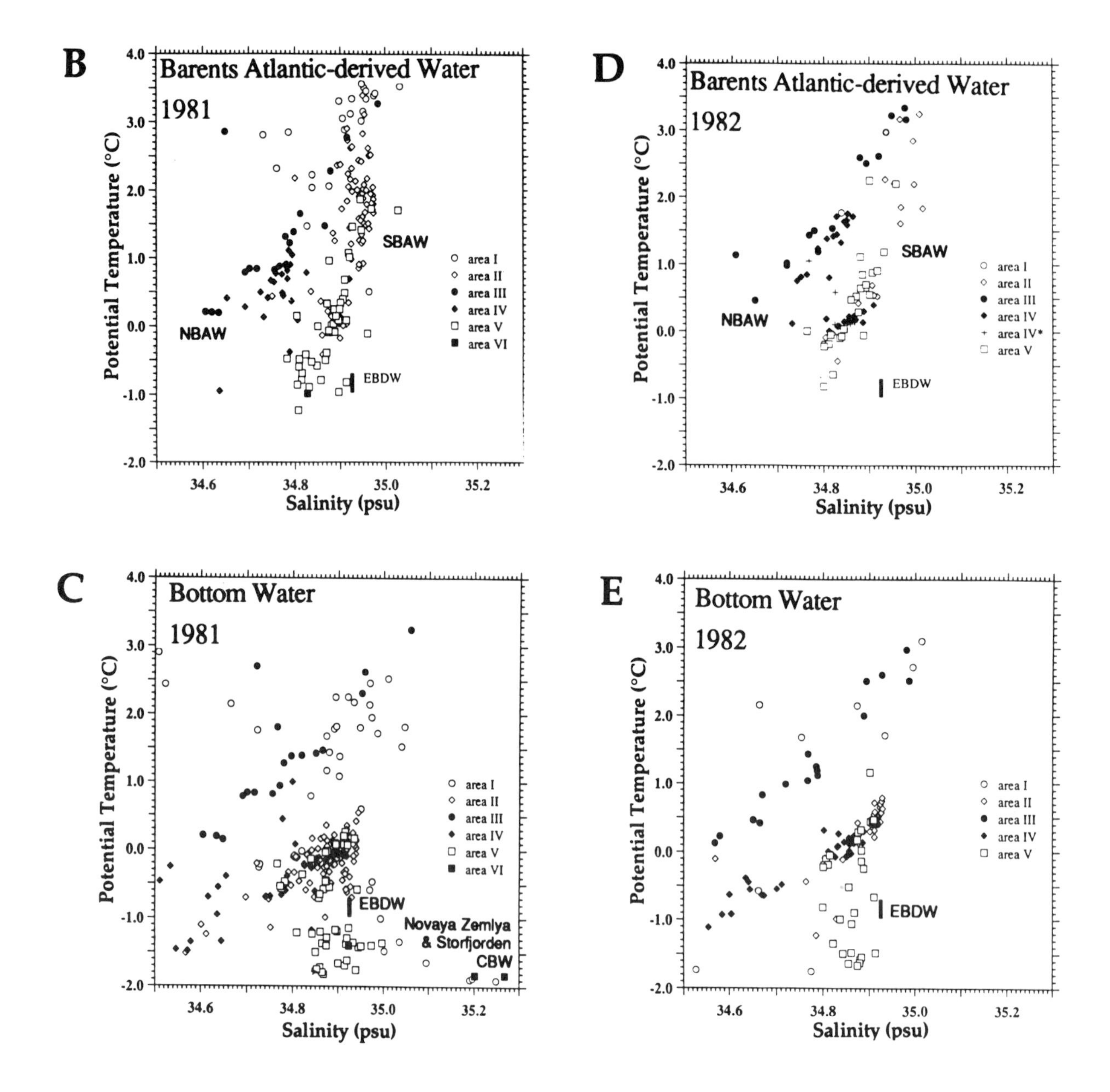

TABLE 1. Core Characteristics of Water Masses in the Northern Barents Sea in Late Summer

Water Mass	Depth [m]	Potential Temperature [°C]	Salinity [psu]	Density [σ_0]
Surface water	0 - 20	-1 to 3	32.0 - 34.0	25.0 - 27.0
Arctic water [a]	30 - 80	-1.8 to -1.0	34.3 - 34.7	27.5 - 27.8
Atlantic-derived water [b]	75 - 250	0 to - 2.0	34.75 - 34.95	27.8 - 28.0
cold bottom water		-1.8 to 0.0	34.9 - 35.2	28.0 - 28.4

[a] correlated with a temperature minimum; [b] correlated with a temperature and salinity maximum

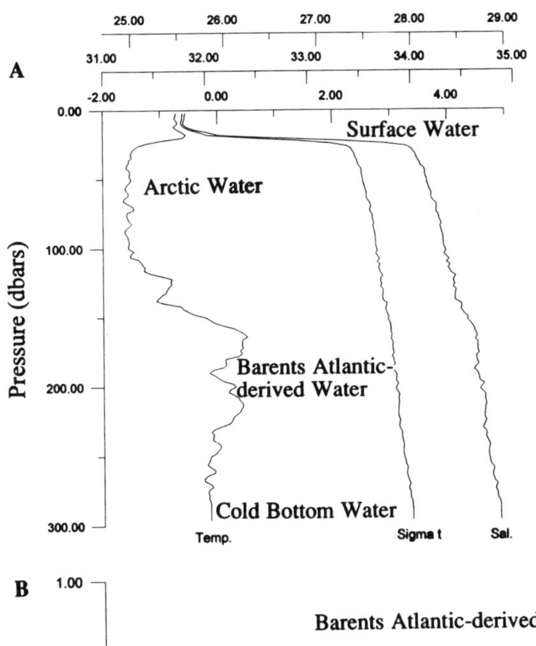

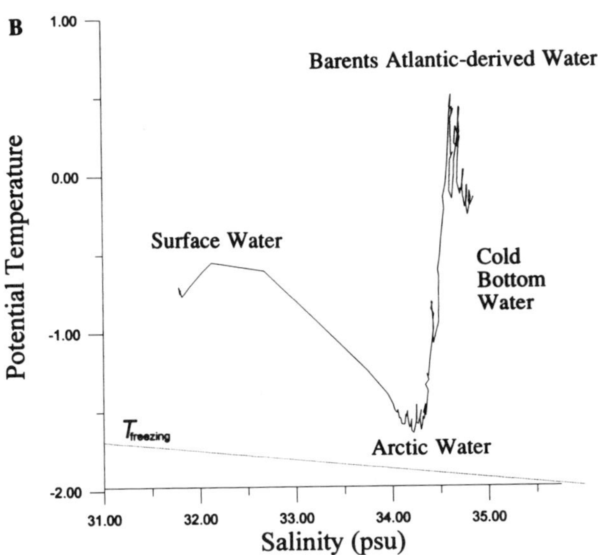

Fig. 4 A) Profile of a station north of the polar front (78°30'N and 33°11'E) in summer showing the surface water layer, the cold Arctic water, the warmer and more saline Barents Atlantic-derived water and the underlying cold bottom water. B) Potential temperature-salinity profile of the same station showing the well-defined Arctic water and Barents Atlantic-derived water masses, and the cold bottom water.

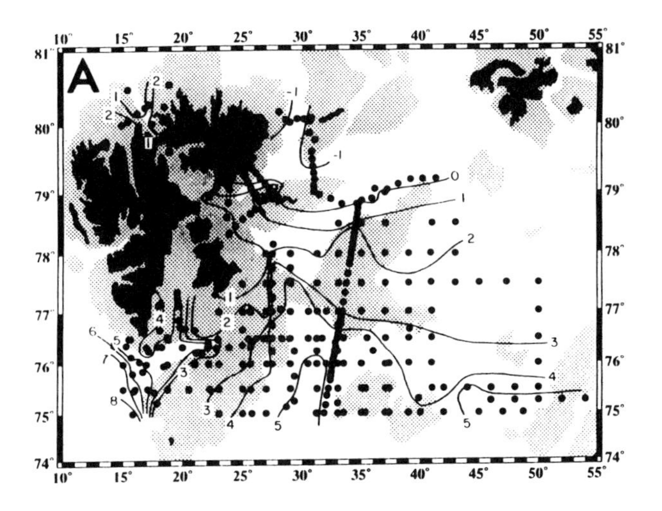

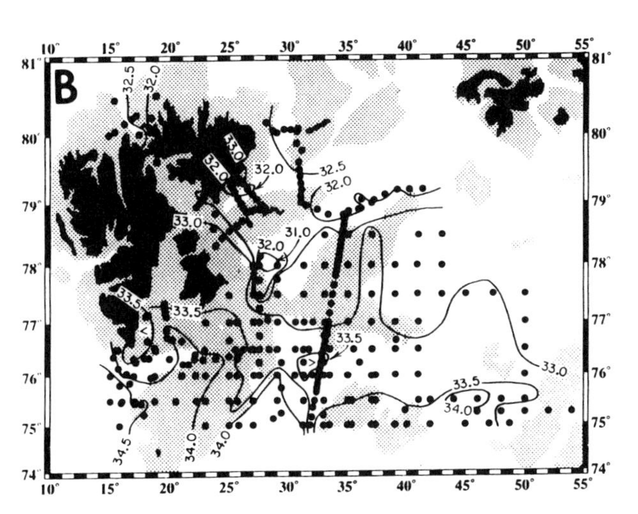

Fig. 5 Surface water characteristics: A) temperature, B) salinity. Data are from summer 1981. Stippling indicates regions with water depths less than 200m.

Surface Water

In general, during summer the surface water temperature decreases toward the north (Fig. 5A). This is due to progressive retreat of the sea ice cover by melting, with a northward decrease in the duration of exposure to solar radiation [*Mosby*, 1938; *Loeng*, 1980; *Loeng*, 1991]. Salinities also decrease toward the north, caused by admixtures of sea ice meltwater (Fig. 5B).

Along a hydrographic transect at 75.5°N (Fig. 6A) the surface layer is about 40m thick and there are lenses of low-salinity water centered over the banks (Fig. 6B). A northern hydrographic transect, from Kong Karls Land to the eastern side of Storbanken (also called the Great Bank or Persey Elevation) at 79°N (Fig. 7A), shows a thinner surface layer, less than 30m thick. Here again there is a lower salinity lens over Storbanken (Fig. 7B).

Arctic Water

The Arctic water layer typically is found between 20 to 150m [*Midttun and Loeng*, 1987], with a temperature

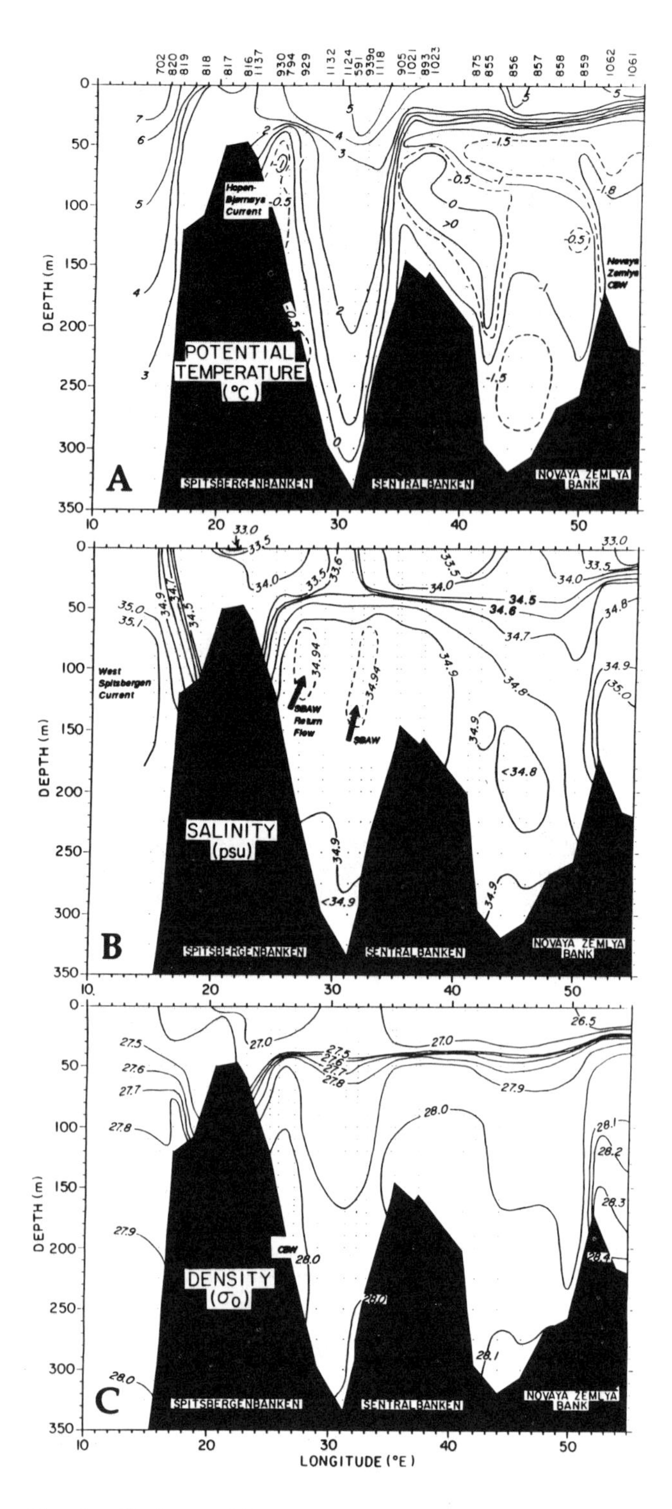

Fig. 6 Hydrographic transect along 75.5°N obtained in 1981: A) potential temperature, B) salinity, C) density.

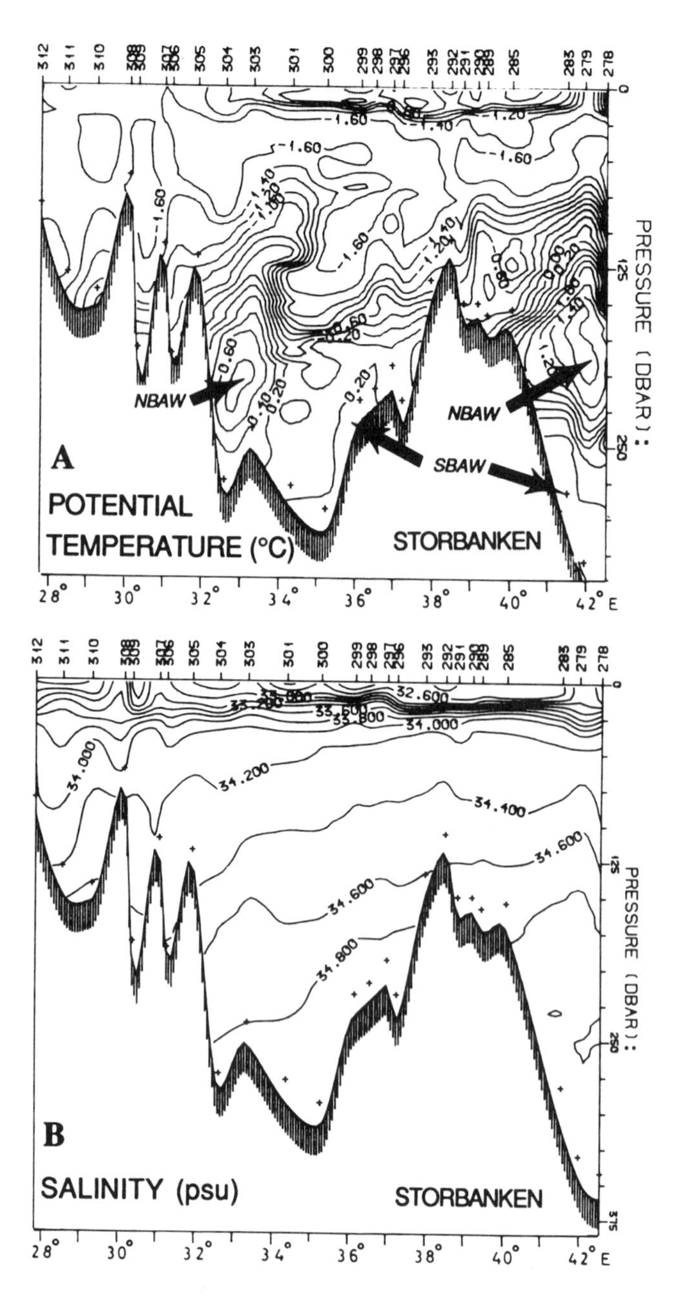

Fig. 7 Hydrographic transect along 79°N obtained in 1982: A) potential temperature, B) salinity.

minimum between 50 and 75m [Fig. 4A; *Novitsky*, 1961]. North of the polar front, there is very little variation in temperature and the water mass is between -1.6 and -1.8°C (Fig. 8A). Only over Storbanken is it below -1.8°C. Temperatures are higher in the vicinity of the polar front

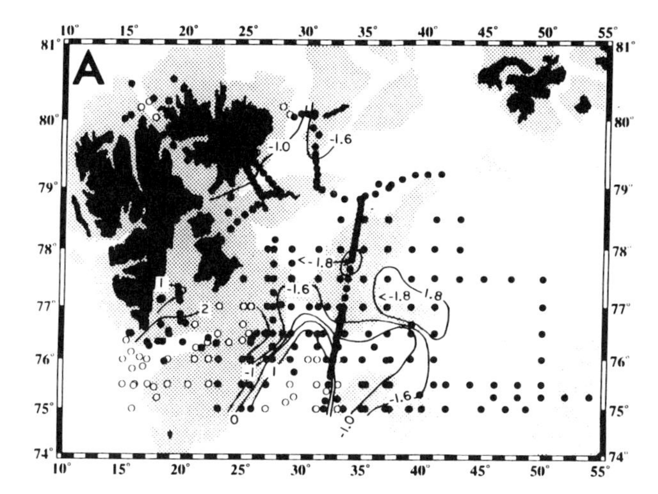

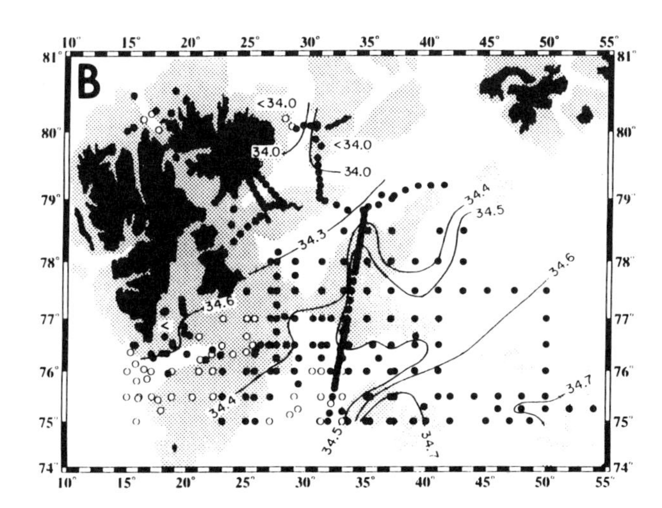

Fig. 8 Characteristics of the water column at the temperature minimum which generally corresponds with the characteristics of Arctic water: A) potential temperature, B) salinity. Data are from summer 1981. Stippling indicates regions with water depths less than 200m, open circles indicate stations where a temperature minimum was not observed.

(Fig. 1) showing the influence of nearby Atlantic-derived water. Salinity increases in this water mass from 34.0°/oo in the northwest to 34.7°/oo in the southeast (Fig. 8B, see also Fig. 6B, 7B). Corresponding to changes in the salinity distribution, the density increases from 27.6 in the north to over 27.9 σ_0 in the south (Fig. 6C).

Atlantic-derived Water

In order to distinguish Atlantic-derived water entering the Barents Sea via the North Cape Current (Fig. 1; Fig. 3, areas II and V) from that entering the Barents Sea from the north (Fig. 3, areas III and IV), we refer to the Atlantic-derived water with origin in the southern Barents Sea as southern Barents Atlantic-derived Water (SBAW) and that to the north as northern Barents Atlantic-derived Water (NBAW). These water masses are modified in the Barents Sea by cooling and mixing with Arctic water. Core characteristics fall on different mixing lines (Fig. 3B, D), with NBAW generally being less saline than SBAW in 1981 and 1982.

In Hopen Deep, south of the polar front, SBAW fills the entire water column (Fig. 3A: central part of area II; Fig. 6A, B). The salinity maximum is at about 75m depth (Figs. 6B, 9D). In northern Hopen Deep, Atlantic water flows over sills, crosses the polar front and is transformed into a submerged water mass. North and east of the polar front, SBAW is overlain by Arctic water. In this transformation region, the depth to the temperature maximum drops to > 200m (Figs. 7, 9D), while its salinity decreases by about 1.5°/oo, and its temperature decreases to 0°C (see Figs. 3B, D, area II; 8A, B). As a submerged water mass, SBAW is also observed to the north and east, following depressions between shallower banks (Fig. 3: areas IV and V).

NBAW occurs in the strait between Nordaustlandet and Kvitøya (Fig. 3: area III), with a maximum temperature greater than 1°C (Fig. 9A). Similar temperatures are also observed in: Erik Eriksen Strait, the passage south of Kvitøya and just to the northeast of Storbanken.

Bottom Water

The water overlying the sea floor has a large range in temperature and salinity (Fig. 3C, E) which is strongly dependent on the water depth (compare Figs. 2 and 10). Usually, either cold bottom water or Barents Atlantic-derived water are found at the bottom in topographic depressions, while Arctic water occurs as the bottom water mass on Spitsbergenbanken and Storbanken (Fig. 10A, B).

Densest bottom waters (Fig. 10C) are found on Novaya Zemlya Bank (Fig. 3, area VI) and in Storfjorden (Fig. 3, area I; selected stations were previously published by Midttun, 1985). Both areas are known as regions where very dense cold bottom water (CBW) with salinities > 35°/oo is formed [Figs. 3C, E, 10B; *Knipowitsch*, 1905; *Nansen*, 1906; *Novitsky*, 1961; *Midttun*, 1985; *Anderson et al.*, 1988; *Quadfasel, et al.*, 1988; *Blindheim*, 1989; *Midttun*, 1989; *Loeng*, 1991]. CBW in both of these regions is definitely more saline and colder than the deep water and bottom water of the Eurasian Basin of the Arctic Ocean (Fig. 3C). The densest water in Storfjorden is found

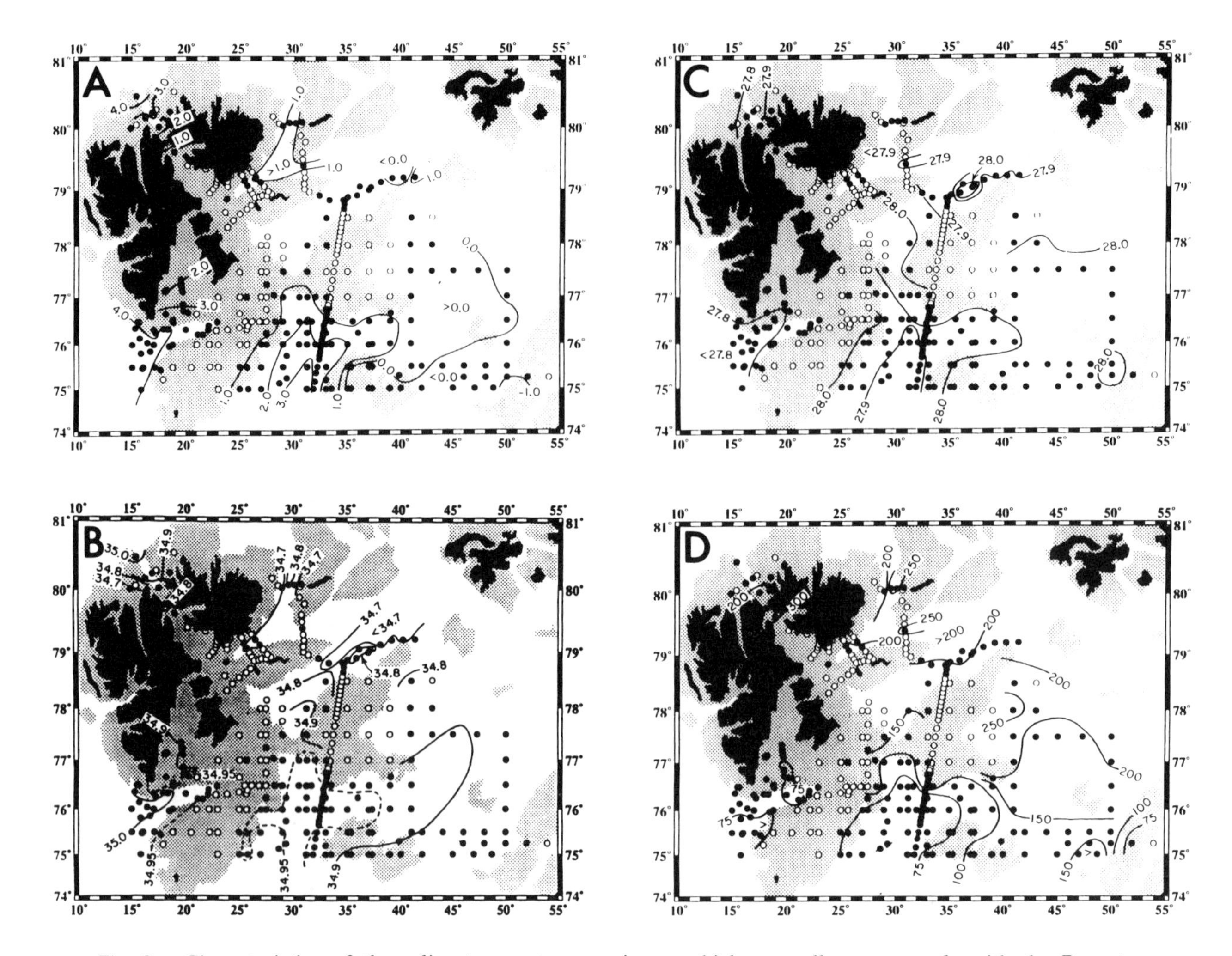

Fig. 9 Characteristics of the saline temperature maximum which generally corresponds with the Barents Atlantic-derived water: A) potential temperature, B) salinity, C) density, D) depth. Data are from summer 1981. Stippling indicates regions with water depths less than 200m, open circles indicate stations where a saline temperature maximum was not observed.

in the very north of the fjord which is separated from the southern part by a sill that is between 50 and 100m deep. The bottom water becomes continuously fresher and warmer towards the southern part of the fjord (Fig. 10). In 1982, our hydrographic stations were located north of the basin where dense water was observed in 1981 and do not show this feature (Fig. 3E).

Bottom water throughout the rest of the Barents Sea is generally less saline although it does reach 34.94°/oo in area V in 1981. Density is less than or equal to 28.1 σ_0.

CIRCULATION AND MIXING PATTERNS

Using the water mass distribution and temperature-salinity relationships observed in this investigation and analyses of hydrographic and current data by Tantsiura (1959, 1973), Novitsky (1961), Eide and Loeng (1983), Midttun (1985), Pfirman (1985), Midttun and Loeng (1987), and Loeng et al. (1993) we discuss circulation and mixing patterns in the northern Barents Sea (Figs. 1 and 11). Due to variations in winds, water exchange with adjacent seas, and continental runoff there are large seasonal and annual variations in current velocity, location and hydrographic characteristics [*Tantsiura*, 1959; *Novitsky*, 1961; *Midttun and Loeng*, 1987; *Ådlandsvik and Loeng*, 1991].

Surface Water

Although the characteristics of summer surface water is

governed by sea-ice melting and retreat, deviation from this pattern is seen along the eastern flank of Spitsbergenbanken where there is a region that is colder than that to either side (Fig. 5). This cold tongue corresponds to the southward-flowing Hopen-Bjørnøya Current (Figs. 1 and 11). Another anomaly is observed over Hopen Deep, where the surface water is warmer and more saline due to Atlantic water inflow. Similar Atlantic water influence is also observed in Storfjorden where a branch of the West Spitsbergen Current flows eastward up the trough.

The low salinity lenses observed over the banks (Figs. 5B, 6B, 7B), may represent water trapped by anticyclonic vortexes [*Midttun and Loeng*, 1987; *Midttun*, 1989].

Arctic Water

A portion of the Arctic water in the Barents Sea forms in the previous winter as a result of convection during development of the sea ice cover. The lower temperature and higher salinity Arctic water near Storbanken (Figs. 7 and 8) may represent winter water trapped by an anticyclonic vortex [*Midttun and Loeng*, 1987; *Midttun*, 1989]. However, it could also indicate advection of Arctic water into the Barents Sea from the northern Kara Sea and the region near Frans Josef Land via the westward-flowing Persey Current previously described by Tantsiura (1959) and Novitsky (1961) (Figs. 1 and 11). The Persey Current continues across the Barents Sea to the west, where it merges with the well-defined, southward-flowing Hopen-Bjørnøya Current [*Tantsiura*, 1959; *Novitsky*, 1961]. Comprised of Arctic water, with low temperatures extending from 50 down to 200m water depth [*Novitsky*, 1961], the Hopen-Bjørnøya Current, clearly observed in our data set, has a core at about 70m water depth (26°E; Figs. 7A, 8A). It has high velocities [up to 1 knot: *Tantsiura*, 1959], and generally is located over the 100 to 200m isobaths along the eastern flank of Spitsbergenbanken [*Tantsiura*, 1959]. Some Arctic water from the Arctic Ocean also contributes to this flow via the East Spitsbergen Current [*Tantsiura*, 1959; *Novitsky*, 1961].

Atlantic-derived Water

The section along 75.5°N (Fig. 6) shows Atlantic water of the West Spitsbergen Current flowing north along the western flank of Spitsbergenbanken. Maximum salinities here exceed 35.1°/oo (Fig. 3, area I). A tongue of Atlantic water enters Storfjorden with a core depth of about 75m (Fig. 9). Note that this water does not appear to continue eastward into the Barents Sea, probably because of the shallow depth of Spitsbergenbanken [*Tantsiura*, 1959: depth

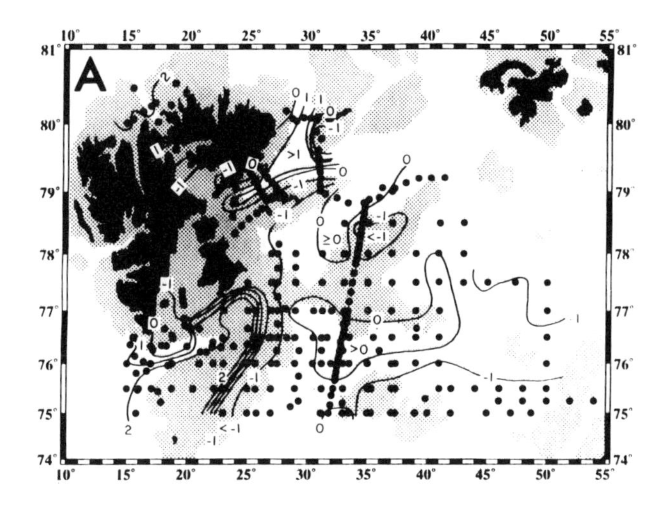

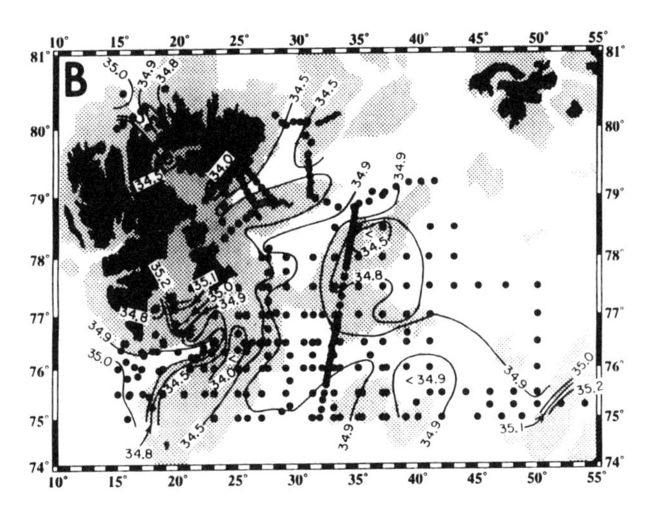

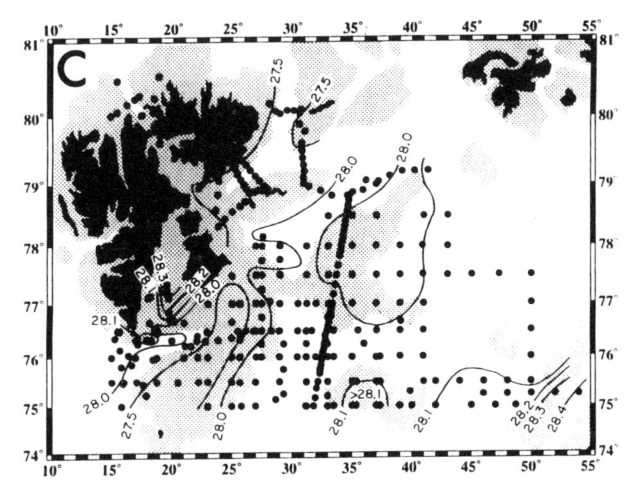

Fig. 10 Bottom water characteristics: A) potential temperature, B) salinity, C) density. Data are from summer 1981. Stippling indicates regions with water depths less than 200m.

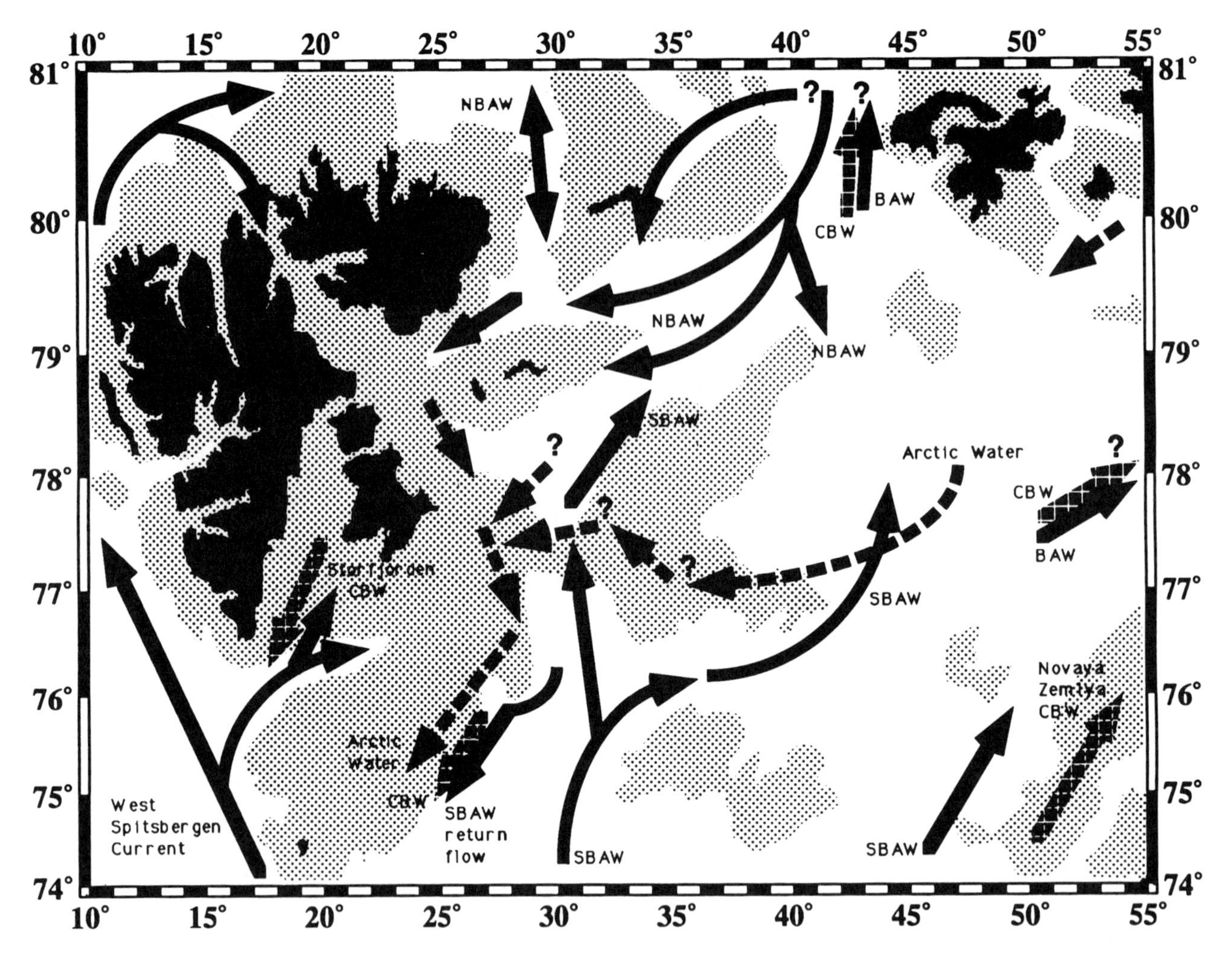

Fig. 11 Schematic of subsurface circulation of water masses observed in the study area of the northern Barents Sea using summer data from this study and information from various sources (*Tantsiura*, 1959, 1973; *Novitsky*, 1961; *Midttun*, 1985; *Pfirman*, 1985; *Quadfasel et al.*, 1988; *Loeng et al.*, 1993). Dashed arrows sketch flow of Arctic water (20-150m water depth); solid arrows sketch flow of Barents Atlantic-derived water (75-250m water depth), and cross-hatched arrows sketch flow of cold bottom water along the sea floor. Recirculation in basins is not shown. Stippling indicates regions where water depths are less than 200m.

between 60 and 80m, *Kristoffersen et al.*, 1988], which is close to the core depth in this region.

Atlantic water does enter the Barents Sea south of Spitsbergenbanken via the North Cape Current (Fig. 1). Atlantic water inflow through this passage may be about 1.6 Sv [see summary by *Loeng*, 1993]. A branch of the North Cape Current extends north along the western flank of Sentralbanken [Figs. 1 and 11; *Nansen*, 1906; *Novitsky*, 1961], forming a current of SBAW that is up to 60 miles wide [*Tantsiura*, 1959]. Current meters deployed in this region in the summer indicate surface to bottom flow to the north [*Eide and Loeng*, 1983]. A portion of this flow turns back to the south along the eastern flank of Spitsbergenbanken [*Tantsiura*, 1959; *Novitsky*, 1961; *Rudels*, 1987]. This return flow is apparently seen in the salinity data on a section through this region (Fig. 6B), as the westernmost of the two cores with salinities exceeding 34.94°/oo. The return flow is situated just to the east of the Hopen-Bjørnøya Current, and, as reported by Tantsiura (1959), it is located between the 150 to 350m isobaths. This modified Atlantic water exits the Barents Sea through Bjørnøyrenna, and returns to the Norwegian Sea [*Novitsky*, 1961].

Tongues of SBAW (Fig. 11) also extend over the sill between Storbanken and Spitsbergenbanken [sill depth between 150 and 200m, *Cherkis et al.*, 1991], as well as between Storbanken and Sentralbanken [called the Eastern Current by *Novitsky* (1961), sill depth between 200 and 250m, Cherkis et al., 1991]. In these regions, just beyond the polar front, SBAW undergoes rapid modification and submerges below the surface. Temperature-salinity relationships lie on a steep slope (Fig. 3B, D, areas II and V), indicating a drop in temperature and a slight decrease in salinity. This relationship could be explained by mixing of SBAW with a near-freezing water mass that has a salinity < 34.8°/oo. In 1981, actual Arctic water salinities were generally < 34.7°/oo in this region (Fig. 8). Rudels (1987) suggests that cooling of Atlantic water in this region occurs: locally by penetrative convection during sea ice formation, isopycnally by advection of cold water with similar densities, and/or by double diffusion. The jagged appearance of the temperature-salinity relationship observed in BAW (Fig. 4A) just to the north of this sill, indicates isopycnal interleaving in the vicinity of the polar front.

Part of the SBAW continues as a warm bottom current to the north and west of Storbanken [*Novitsky*, 1961]. In 1982, SBAW was seen again at 79°N (Fig. 7A), where temperatures along the west flank of Storbanken are greater than 0.2°C. Tantsiura (1959) and Novitsky (1961) indicate that flow in this northward branch is variable and not particularly substantial. Although neither the 1981 nor the 1982 data sets extend far enough north to trace the path of SBAW past Storbanken, there are no bathymetrical hindrances to continued northward transport into the Arctic Ocean. Therefore, some SBAW may continue flowing north into the Arctic Ocean through Frans Josef-Victoria Renna [Fig. 11: Rudels, 1987].

Note that there is no well-defined core of Atlantic water west of Novaya Zemlya in the region of the Novaya Zemlya Current (Fig. 1), as described by Nansen (1906), Tantsiura (1959), Novitsky (1961) and Loeng et al. (1993), among others. Depth to a modest temperature maximum is about 200m in the eastern portion of the study area (Fig. 9D). Apparently, in 1981, this current was not as developed as it is in other years. According to Tantsiura (1959), flow of SBAW in the Novaya Zemlya Current, joined by SBAW transported between Storbanken and Sentralbanken [Novitsky, 1961], continues eastward towards the Kara Sea [Fig. 11: see also Rudels, 1987; Loeng et al., 1993]. From here, the modified Atlantic water flows continues north, through the St. Anna Renna into the Arctic Ocean [*Rudels et al.*, this volume]. Although Cherkis et al. (1991) show depths between 150 and 200m throughout the southern part of the passage between Frans Josef Land and Novaya Zemlya which would divert flow of water deeper than that to the north, Loeng et al. (1993) indicate a passage with water depths between 200 and 300m -- sufficiently deep to allow flow of Barents Atlantic-derived water. Loeng et al. (1993) found that outflow of water through the passage between Frans Josef Land and Novaya Zemlya increases towards the bottom, and the average total outflow is about 2 Sv, with a maximum in early winter.

Turning now to the northern reaches of the Barents Sea, NBAW enters Hinlopen Strait (Figs. 2 and 11), but does not extend all the way into the Barents Sea because of the shallow water depths in the southern portion of the strait [less than 100m, *Cherkis et al.*, 1991]. NBAW does enter the Barents Sea further east through various straits between Nordaustlandet and Frans Josef Land [*Mosby*, 1938]. NBAW decreases markedly in salinity (by more than 0.3°/oo), as well as temperature, as it penetrates southward into the northern basins (Figs. 3B, D, 9). Because CBW is absent, NBAW is the bottom water mass in the northern basins (Fig. 10A), and it mixes primarily with the overlying cold and less saline Arctic water.

Both Mosby (1938) and Novitsky (1961) suggest a modest southward flow of NBAW through the strait between Nordaustlandet and Kvitøya [sill depth between 250 and 300m, *Cherkis et al.*, 1991]. However, a year of current meter measurements in the NBAW core (255m), indicate an average velocity here to the *northeast* of 0.5 cm sec^{-1} [*Aagaard et al.*, 1983]. In 1981, the maximum temperature in the strait was > 1°C and the salinity was > 34.8°/oo (Fig. 9A, B). Continuing southward, into Erik Eriksen Strait east of Svalbard (Fig. 3: area III), this branch of NBAW appears to mix with Arctic water that a salinity of about 34.4°/oo. Actual salinities at the temperature minimum of the Arctic water are generally < 34.3°/oo (Fig. 8). Because the Arctic water here is so fresh, the mixing relationship for western NBAW (Fig. 3: area III) is less steep than that for NBAW entering the Barents Sea to the east (Fig 3: area IV) where the salinity of Arctic water is greater. Novitsky (1961) indicates that this northwestern branch of NBAW is separated in some way from the main mass of NBAW that enters the Barents Sea east of Kvitøya.

According to previous analyses, some NBAW enters the Barents Sea between Kvitøya and Victoria Island, with a temperature maximum at about 200m [*Mosby*, 1938; *Pfirman*, 1985; apparent sill depth between 200 and 250m, *Cherkis et al.*, 1991], but the majority of the flow is along the eastern margin of Victoria Island through Frans Josef-Victoria Renna [Fig. 11: *Mosby*, 1938; *Novitsky*, 1961]. The sill depth of this passage appears to be between 300 and 350m [*Cherkis et al.*, 1991], making this strait the deepest, as well as the widest conduit into the northern Barents Sea from the Arctic Ocean. NBAW occurs as a bottom water mass along the western slope, influencing the sea floor from about 150 to 400m water depth [*Mosby*,

1938]. South of Victoria Island, NBAW spreads to the southwest (Fig. 11), filling the deepest part of Frans Josef-Victoria Renna, and continuing into the depression south of Kong Karls Land [*Novitsky*, 1961: see Fig. 7; the 0.8°C core with a salinity of about 34.76°/oo at 33°E]. Although some water also passes to the east of Storbanken (see Fig. 7; the 1.4°C core with a salinity of about 34.87°/oo at 42°E), according to Novitsky (1961) it does not flow far to the south because it encounters northward-flowing SBAW. NBAW mixes with overlying Arctic water during transport, with core properties falling along a mixing line that extends to Arctic water with a salinity of about 34.5°/oo (Fig. 3B, D, area IV). Actual salinities at the temperature minimum of the Arctic water in this region were between 34.3 and 34.4°/oo in 1981 (Fig. 8).

In the 1982 data set, SBAW is detected on both sides of Storbanken in this region (Fig. 3; area IV). SBAW is the bottom water mass, while NBAW, less saline and warmer (in 1981 as well as 1982 data, also noted by Rudels, 1987), is observed at shallower depths (Fig. 7A: 200m at 33°E and 175m at 42°E). Along the western flank of Storbanken, SBAW with 0.2°C and 34.86°/oo salinity is found. Another branch of SBAW, about 0.1°C colder but with the same salinity, is also observed at the very bottom on the eastern side of Storbanken. This water mass is apparently a northern extension of SBAW that passed along the east side of Storbanken, or it has circulated around Storbanken.

The depths at which NBAW and SBAW occur depend on their relative density. In 1981 and 1982, NBAW was warmer than SBAW and its maximum salinity within the confines of the Barents Sea was lower than that of SBAW (NBAW 1981: 34.8°/oo between Nordaustlandet and Kvitøya; NBAW 1982: 34.9°/oo east of Storbanken; SBAW: 34.95 to 35°/oo in Hopen Deep). In 1931, Mosby (1938) found NBAW with temperatures greater than 2°C and salinities greater than 34.9°/oo entering the Barents Sea between Kvitøya and Frans Josef Land. There could be several reasons for differences between SBAW and NBAW. One reason could be that there are variations in the currents which feed NBAW into the Barents Sea, and that sometimes the inflow is not as great [e.g. *Midttun and Loeng*, 1987]. Also, although both SBAW and NBAW originate from the Norwegian Atlantic Current, it takes longer for Atlantic water to flow around Svalbard and enter the Barents Sea from the north. Starting from the Norwegian Atlantic Current (73°N 15°E), the distance to Storbanken (79°N 40°E) is about 1650km for NBAW and about 1050km for SBAW. Assuming an average advection velocity of 1 cm/s, NBAW would take about 2 years longer than SBAW to reach Storbanken (NBAW = 5.2 years; SBAW = 3.3 years). Different travel times mean that NBAW originates from an older 'vintage' of Atlantic water than SBAW sampled during any one year. It could be that in 1981 and 1982 this vintage was less saline than that which entered the Barents Sea more recently from the south. This compares with the occurrence of the "Great Salinity Anomaly" (when the salinity of the Atlantic water was about 0.1°/oo less than average: *Dickson et al.*, 1988), which passed the North Cape and West Spitsbergen currents in 1978-1979. More generally, Atlantic water temperature and salinity characteristics are known to vary by more than ±0.5°C and ±0.06°/oo [*Mosby*, 1938; *Blindheim and Loeng*, 1981; *Midttun*, 1989]. If these were the only two reasons for the variations, then one could expect that NBAW would sometimes be denser than SBAW.

However, according to Anderson et al. (1989), along the northern slope of the Barents Sea, just north of Kvitøya, the NBAW temperature maximum lies between 200 and 250m, while the salinity maximum lies between 300 and 500m. These depths are close to the sill depths of the northern channels (250 to 300m between Nordaustlandet and Kvitøya, 200 to 250m between Kvitøya and Victoria Island, and 300 to 350m between Victoria Island and Frans Josef Land, *Cherkis et al.*, 1991). Therefore, differences in the core extrema could also be due to skimming off of the less saline fraction during entry into the northern Barents Sea. In addition, NBAW sampled in our study has had more opportunities to mix with Arctic water due to its longer travel time as a submerged water mass, and the Arctic water with which it mixes is fresher than the Arctic water with which the SBAW mixes. As a result of these processes, NBAW may generally be less dense than SBAW and be found above it when the two water masses meet in the northern Barents Sea.

Moving further east, Tantsiura (1959) and Loeng et al. (1993) indicate inflow of modified Atlantic water as a submerged water mass from the Arctic Ocean into the Kara Sea through the St. Anna Renna and then on into the northeastern Barents Sea south of Frans Josef Land. Cherkis et al. (1991), locate the deepest part of the passage between Frans Josef Land and Novaya Zemlya just south of Frans Josef Land, with a sill depth between 200 and 250m. Loeng et al. (1993) indicate another passage further south with sill depth between 200 and 300m. In warm years, this flow of Atlantic-derived water in from the Kara Sea may join the eastward-flowing SBAW of the Novaya Zemlya Current [*Novitsky*, 1961]. Our stations are located too far to the west to resolve this water mass in the present study.

Cold Bottom Water

Because data used in this study are summer values, and

CBW forms primarily in fall and winter, the distribution of CBW shown here represents only a remnant of this water mass [*Novitsky*, 1961]. CBW is modified by summer mixing and thus is probably also less dense than it originally was.

Even in mid-summer, very dense CBW is observed to accumulate over Novaya Zemlya Bank in 1981 (Fig. 6). Nansen (1906) noted that the same kind of dense, cold water is found almost everywhere on Novaya Zemlya Bank, comprising a layer more than 100m thick. Because this water is so saline, it has to have added salt from brine rejection during sea ice formation (Fig. 3, area VI). The process of dense water formation starts here in the fall, when the water column is cooled, and continues when salt is added from sea ice formation [*Loeng*, 1991]. Novaya Zemlya CBW apparently flows east into the Kara Sea (Fig. 11), and may continue on through the St. Anna Renna into the Arctic Ocean [*Nansen*, 1906; *Midttun*, 1985; *Blindheim*, 1989]. Temperature and salinity properties of this water mass are more extreme than that of Eurasian Basin deep and bottom water (Fig. 3C) and Novaya Zemlya CBW is thought to contribute to formation of these Arctic water masses [*Nansen*, 1906; *Midttun*, 1985]. Although bottom water in the trough west of the bank has slightly elevated salinity and density in proximity to the sea floor, it is much less dense and saline than the water on Novaya Zemlya Bank (Fig. 6). In 1981, a sharp front separated the two water masses and they do not appear to have been connected (also observed by *Nansen*, 1906). In Midttun's (1985) report on dense water near Novaya Zemlya Bank, CBW appeared to spread toward the west along the sea floor into the adjacent trough.

Water mass characteristics of Novaya Zemlya CBW are similar to those observed in Storfjorden [Fig. 3C, area I, *Midttun*, 1985). Dense CBW flows out of Storfjorden to the west (Fig. 11) and then spreads along the west Spitsbergen margin to the north [*Quadfasel et al.*, 1988]. During summer, the volume of cold water pooled in Storfjorden decreases, and by late fall it has disappeared or is found in only minor amounts in the deepest part of the trough [*Novitsky*, 1961]. As with Novaya Zemlya CBW, temperature and salinity properties of this water mass are more extreme than that of Eurasian Basin deep and bottom water (Fig. 3C, E), and water formed here could contribute to development of these Arctic water masses [*Midttun*, 1985; *Quadfasel et al.*, 1988].

Less extreme cold bottom water observed in Hopen Deep may contribute to waters of intermediate depth in the Norwegian Sea [*Blindheim*, 1989]. The hydrographic structure observed at 75.5°N suggests flow of such cold bottom water (Fig. 6: temperature < -0.5°C, salinity about 34.92°/oo, and density > 28.0 σ_0) southward along the eastern flank of Spitsbergenbanken at about 225m water depth (Fig. 11). This water most likely continues flowing southwestward and exits the Barents Sea through Bjørnøyrenna. Blindheim (1989) estimated about 0.8 Sv of outflowing bottom water through this passage.

CBW occurring in most of the other depressions in the Barents Sea generally is slightly warmer (Fig. 10: temperature between -0.2 and +0.2°C, salinities between 34.90 to 34.94°/oo, and densities between 28.0 and 28.1 σ_0). The highest salinities are observed east of Sentralbanken (in a region sometimes called the Central Depression; Fig. 3, area V), an area also discussed by Nansen (1906). In 1981, CBW sampled here was slightly less saline than in 1982 (Fig. 3C, E; compare salinity range in area V).

Throughout the Barents Sea there are large annual--as well as seasonal--variations in the regional accumulation and discharge of CBW [*Nansen*, 1906; *Midttun and Loeng*, 1987; *Blindheim*, 1989; *Midttun*, 1989]. For example, Midttun (1989) displays sections along 45°E collected in September of 1982 and 1983 which show a large amount of cold < -1.5°C bottom water south of 75°N in 1982 compared with only a small amount in 1983 (note: this feature is not observed in our 1982 data set, because our study area cuts off at 75°N). Also, the distribution of bottom density in September-October 1986 [*Blindheim*, 1989] showed a much larger region with densities > 28.1 σ_0 than that observed in this study (compare our Fig. 10C with Blindheim's (1989) Fig. 2). Midttun and Loeng (1987) found that the rate of dense water formation and accumulation is related to variations in its outflow, as well as variations in the salinity of the inflowing water. According to these authors, following a period of high Atlantic water influx, it takes more than a year for winter cooling to accumulate CBW with a density great enough to initiate a new outflow, which may occur as a massive discharge [*Loeng*, 1991].

Outflow of CBW could take place [*Nansen*, 1906; *Swift et al.*, 1983; *Midttun*, 1985; *Midttun and Loeng*, 1987; *Quadfasel et al.*, 1988; *Midttun*, 1989; *Blindheim*, 1989; *Loeng*, 1991]: to the west of Frans Josef Land where there is a deep conduit between this region and the Arctic Ocean (with depths between 250 and 300m, *Cherkis et al.*, 1991), between Frans Josef Land and Novaya Zemlya [the primary avenue according to *Loeng*, 1991; *Loeng et al.*, 1993], and/or to the southwest through Bjørnøyrenna.

CONCLUSIONS

Under summer conditions, the water column in the

Barents Sea, north of the polar front, is stratified: a warm, fresh surface layer is underlain by cold, relatively fresh Arctic water, which in turn is underlain by a warm, saline Atlantic-derived layer. Cold bottom water, which is colder and usually slightly more saline than the Barents Atlantic-derived layer, is found near the sea floor mostly in bathymetric depressions, but also over Novaya Zemlya Bank.

Arctic water may develop in place during sea ice formation in fall and winter. It may also be advected into the Barents Sea, primarily from the northern Kara Sea via the Persey Current and to a lesser degree from the Arctic Ocean via the East Spitsbergen Current. Arctic water is freshest in the northwestern portion of the Barents Sea, less than 34.3°/oo, and ranges up to 34.7°/oo in the southeastern portion of the study area.

Atlantic water enters the Barents Sea both from the south, as southern Barents Atlantic-derived water (SBAW) and the north as northern Barents Atlantic-derived water (NBAW). SBAW cools abruptly when it penetrates north and east of the polar front, and becomes an intermediate or bottom water mass overlain by cold Arctic water. Salinity decreases only slightly in this process (from 34.95 to 34.8°/oo). In contrast, while penetrating southward into troughs in the Barents Sea, NBAW is cooled and diluted significantly (up to 0.3°/oo) by mixing with the fresher, overlying Arctic water. North of about 79°N, during 1981 and 1982, SBAW was denser than NBAW and occurred beneath it. Some SBAW may continue flowing northward into the Arctic Ocean through the Frans-Victorica Renna west of Frans Josef Land [*Rudels*, 1987]. Here it would be located in the eastern portion of the passage, to the east of the core of NBAW which flows southward in this region. The greater volume of SBAW flows eastward into the Kara Sea south of Frans Josef Land [*Rudels*, 1987; *Loeng et al.*, 1993]. Total outflow (modified Atlantic and other water) through this passage is about 2 Sv [Loeng et al., 1993], apparently comparable to the inflow of Atlantic water through Bjørnøyrenna (1.6 Sv). This flow continues north, through the St. Anna Renna into the Arctic Ocean [*Rudels et al.*, this volume]. Therefore, while some modified Barents Atlantic-derived water is contributed directly to Arctic Ocean from the Barents Sea, most appears to be contributed indirectly through the Kara Sea.

Remnants of cold, saline water formed by brine rejection during sea ice formation are observed near the sea floor. In 1981, extremely dense water (> 28.3 σ_0) was observed both in Storfjorden and on Novaya Zemlya Bank (described earlier by *Midttun*, 1985). These waters are the densest observed in the Barents Sea. Less cold and saline dense bottom waters also occur in other regions. Following bathymetric depressions, this dense bottom water flows out of the Barents Sea and, depending on its density, influences the intermediate, deep and bottom water characteristics of the adjacent Norwegian Sea and the Arctic Ocean. Outflow occurs along the bottom and takes place:

1) into the Norwegian Sea along the northern flank of Bjørnøyrenna [*Sarynina*, 1969; *Swift*, 1986; *Midttun and Loeng*, 1987; *Blindheim*, 1989; *Midttun*, 1989],

2) from Storfjorden to the western Spitsbergen margin of the Norwegian Sea where it continues north into the Arctic Ocean [e.g. *Quadfasel et al.*, 1988],

3) directly into the Arctic Ocean west of Frans Josef Land [*Blindheim*, 1989], and

4) into the Kara Sea along the southern portion of the passage between Frans Josef Land and Novaya Zemlya, where it probably exits to the Arctic Ocean through St. Anna Renna. This is the main passageway for the extremely cold and dense Novaya Zemlya cold bottom water [*Nansen*, 1906; *Midttun*, 1985; *Midttun*, 1989; *Loeng*, 1991].

Acknowledgments. Hydrographic data from the Institute for Marine Research (Havforskningsinstitutt), in Bergen, Norway, was provided by Harald Loeng. The research was funded under U.S. Office of Naval Research contracts N00014-81-C-009 (directed by John D. Milliman) and N00014-90-J-1362, with some travel funding from the Norwegian Marshall Fund. Ship time on the M/S Lance and logistical assistance was provided by the Norwegian Polar Research Institute. Lamont-Doherty Earth Observatory contribution #5169.

REFERENCES

Aagaard, K. and Carmack, E. C., 1989, The role of sea ice and other fresh water in the Arctic circulation, *Jour. Geophys.Res.*, 94(C10), 14487-14498.

Aagaard, K., Coachman, L. K. and Carmack, E. C., 1981, On the halocline of the Arctic Ocean, *Deep-Sea Res.*, 28, 529-545.

Aagaard, K., Foldvik, A., Gammelsrød, T. and Vinje, T. (1983) One-year records of current and bottom pressure in the strait between Nordaustlandet and Kvitøya, Svalbard, 1980-81, *Polar Research In.s.*, 107-113.

Ådlandsvik, B. and Loeng, H. (1991) A study of the climate system in the Barents Sea, *Polar Res.*, 10(1), 45-49.

Anderson, L.G., Jones, E.P., Koltermann, K.P., Schlosser, P., Swift, J., and Wallace, D.W.R., 1989, The first oceanographic section across the Nansen Basin of the Arctic Ocean, *Deep-Sea Res.* 36, 475-482.

Anderson, L.G., Jones, E.P., Lindegren, R., Rudels, B., and

Sehlstedt, P.-J., 1988, Nutrient regeneration in cold, high salinity bottom water of the Arctic shelves, *Continental Shelf Res.*, 8(12), 1345-1355.

Blindheim, J., 1989, Cascading of Barents Sea bottom water into the Norwegian Sea, *Rapp.P.-v.Reun.Cons.int.Explor. Mer*, 188, 49-58.

Blindheim, J. and Loeng, H., 1981, On the variability of Atlantic influence in the Norwegian and Barents Sea, *Fisk. Dir. Skr. Ser. Hav.Unders.*, 17, 61-189.

Cherkis, N.Z., Fleming, H.S., Max, M.D., Vogt, P.R., and Czarnecki, M.F. (1991) *Bathymetry of the Barents and Kara Seas*, Geological Society of America, Inc., Boulder, Colorado.

Dickson, R.R., Meincke, J., Malmberg, S.-A. and Lee, A.J. (1988) The "Great Salinity Anomaly" in the northern North Atlantic 1968-1982, *Prog.Oceanog.*, 20, 103-151.

Eide, L.I. and Loeng, H., 1983, Environmental conditions in the Barents Sea and near Jan Mayan, *Det Norske Meteorologiske Institutt*, Oslo, August, 1983.

Gordienko, P.A. and Laktionov, A.F., 1969, Circulation and physics of the Arctic Basin waters, in Annals of the International Geophysical Year, *Oceanography*, 46, Pergamon, New York, p. 94-112.

Hanzlick, D.J., 1983, *The West Spitsbergen Current: transport, forcing, and variability*, Ph.D. thesis, University of Washington, 127 pp.

Hanzlick, D. and Aagaard, K., 1980, Freshwater and Atlantic water in the Kara Sea, *Jour.Geophys.Res.*, 85(C9), 4937-4942.

Knipowitsch, N., 1905, Hydrologische Untersuchungen in Europaeischen Eismeer, *Annalen der Hydrographie und Maritimen Meteorologie*, 33, 289-308.

Kristoffersen, Y., Sand, M., Beskow, B., and Ohta, Y., 1988, *Western Barents Sea bathymetry*, Norsk Polarinstitutt, Oslo.

Loeng, H., 1980, Physical oceanographic investigations in central parts of the Barents Sea, *Fisken Hav.*, 3, 29-60.

Loeng, H., 1991, Features of the physical oceanographic conditions of the Barents Sea, *Polar Research*, 10, 5-18.

Loeng, H., Ozhigin, V., Ådlandsvik, B. and Sagen, H. (1993) Current measurements in the northeastern Barents Sea, *ICES Statutory Meeting*, 22 p.

Midttun, L. 1961, Norwegian hydrographical investigations in the Barents Sea during the international geophysical year, *Rapp.P.-v.Reun.Cons.int.Explor.Mer*, 149, 25-30.

Midttun, L., 1985, Formation of dense bottom water in the Barents Sea, *Deep-Sea Res.*, 32(10), 1233-1241.

Midttun, L., 1989, Climatic fluctuations in the Barents Sea, *Rapp.P.-v.Reun.Cons.int.Explor.Mer*, 188, 23-35.

Midttun, L. and Loeng, H., 1987, Climatic variations in the Barents Sea, in The effect of oceanographic conditions on distribution and population dynamics of commercial fish stocks in the Barents Sea, H. Loeng, ed., *Proc. Third Soviet-Norwegian Symposium*, Murmansk, May, 1986, pp. 13-28.

Mosby, H., 1938, Svalbard Waters, *Geofysiske Publikasjoner*, 12(4), 1-85.

Nansen, F., 1906, Northern waters: Captain Roald Amundsen's oceanographic observations in the Arctic seas in 1901, *Vitenskabs-Selskapets Skrifter* 1, Mathematisk-Naturv. Klasse., 3, 145 pp.

Novitsky, V.P., 1961, Permanent currents of the Northern Barents Sea, *Trudy Gosudarstvennogo Okeanograficheskogo Instituda*, 64, 1-32, Translated by U.S.N.O. 1967, Leningrad.

Pfirman, S.L., 1985, *Modern sedimentation in the northern Barents Sea: Input, dispersal and deposition of suspended sediments from glacial meltwater*, Ph.D. Thesis, Mass. Inst.Tech./Woods Hole Oceanogr. Inst., Woods Hole Oceanogr.Inst., Tech.Rept. WHOI-85-4, 382 p.

Quadfasel, D., Rudels, B., and Kurz, K., 1988, Outflow of dense water from a Svalbard fjord into the Fram Strait, *Deep-Sea Res.*, 35(7), 1143-1150.

Redfield, A.C. and Friedman, I., 1969, The effect of meteoric water, melt water and brine on the composition of Polar Sea water and of the deep waters of the ocean, *Deep-Sea Res.*, 16, 197-214.

Rudels, B., 1987, On the mass balance of the Polar Ocean, with special emphasis on the Fram Strait, *Norsk Polarinstitutt Skrifter*, 188, 1-53.

Rudels, B., Jones, E.P., Anderson, L.G. and Kattner, G., On the intermediate depth waters of the Arctic Ocean, in *Geophysical Monograph Series*, edited by O. Johannessen, R.D. Muench, and J.E. Overland, AGU, Washington, D.C., in press, 1994.

Sarynina, R.N., 1969, Conditions of origin of cold deep-sea waters in the Bear Island Channel, *Symposium on Physical Variability in the North Atlantic*, Dublin, 23-27 September 1969.

Schlosser, P., Bauch, D., Rairbanks, R. and G. Boenisch, Arctic river-runoff: mean residence time on the shelves and in the halocline, *Deep-Sea Research*, in press, 1994.

Swift, J.H., Takahashi, T. and Livingston, H. D., 1983, The contribution of the Greenland and Barents seas to the deep water of the Arctic Ocean, *Jour.Geophys.Res.*, 88(C10), 5981-5986.

Swift, J.H., 1986, Arctic Waters, in: *The Nordic Seas*, ed. B.G. Hurdle, Springer-Verlag, New York, 129-153.

Tantsiura, A.I., 1959, On the currents of the Barents Sea, *Transactions of the Polar Scientific Research Institute of Marine Fisheries and Oceanography* - N.M. Knipovic (PINRO), 11, 35-53, in Russian, translated to English by the Norwegian Polar Research Institute, Oslo, 1983.

Tantsiura, A.I., 1973, On Seasonal changes in currents in the Barents Sea, *Transactions of the Polar Scientific Research Institute of Marine Fisheries and Oceanography* - N.M. Knipovic (PINRO), in Russian, translated to English by the Norwegian Polar Research Institute, Oslo.

Wuest, G., 1935, Die Stratosphaere. Deutsche Atlantische Expedition, Meteor 1925-1927, *Wiss. Erg.*, 6(1), 288 pp.

A Study on the Inflow of Atlantic Water to the GIN Sea using GEOSAT Altimeter Data

P. Samuel and J.A. Johannessen

Nansen Environmental and Remote Sensing Center, Bergen, Norway.

O.M. Johannessen

Nansen Environmental and Remote Sensing Center / Geophysical Institute, University of Bergen, Bergen, Norway.

Satellite altimeter data from the GEOSAT exact repeat mission were processed to obtain the sea surface height (SSH) anomalies in the GIN Sea. Improved corrections, including tidal elevations from a regional model were applied to retrieve the anomalies. The annual mean SSH from an isopycnal ocean model was added to the altimeter SSH anomalies to obtain the total dynamic sea level. Geostrophic velocities derived from the smoothed dynamic sea level field were generally small, but exhibited considerable seasonal variability. Monthly mean mixed layer depths from the same ocean model was then used together with the geostrophic velocities to estimate transports in the Norwegian Atlantic Current, which had a mean of 2.7 Sv and seasonal variability with an amplitude of 1.8 Sv. Maximum transports were in February-March and minimum in July-August. Heat fluxes estimated on the basis of these transports and average temperatures recorded by Ocean Weather Station-M had a mean of 84×10^{12}W and seasonal variability of up to 49×10^{12}W.

1 INTRODUCTION

The general features of the circulation in the Greenland-Iceland-Norwegian (GIN) Sea have been described by several investigators [*Johannessen,* 1986; *Hopkins,* 1991]. Topographic steering is seen to be a strong factor for the currents in these waters. To the south, the boundary of the GIN Sea with the Atlantic ocean is defined by a sequence of three continental ridges extending from Greenland to Scotland and known collectively as the Greenland-Scotland Ridge. This ridge is scoured by two channels, namely, the Denmark Strait between Greenland and Iceland, and the Færoe-Shetland Channel between the Færoes and Shetland.

The major part of the transport of warm, saline Atlantic water into the GIN sea is effected by the Norwegian Atlantic Current through the Færoe-Shetland Channel (see Plate 1) and over the northwestern shelf of the Færoes [*Buch et al.,* 1988]. The reverse flow of the colder, less saline GIN Sea water into the Atlantic takes place through the Denmark Strait with the East Greenland Current and as an overflow over the Greenland-Scotland Ridge [*Meincke,* 1983; *Buch et al.,* 1988].

Worthington [1970] gave estimates of the transport of water in and out of the GIN Sea, based on mass balance considerations. According to his calculations, transports into the GIN Sea were 8 Sv between Iceland and Scotland and 1 Sv to the west of Iceland. This was balanced by an outflow of 3 Sv with the East Greenland Current and overflows of 4 Sv through the Denmark Strait and 2 Sv between Iceland and Scotland.

However, estimates given by other investigators vary significantly. For instance, *Dooley and Meincke* [1981] estimated the Atlantic inflow through the Færoe-Shetland channel to be only 2 Sv or 3.3 Sv when recirculated Færoe Atlantic Water is included. Recently, *Dickson et al.* [1990] obtained values of 2.9 Sv for the overflow through the Denmark Strait and 2.7 Sv for the overflow over the Iceland-Scotland Ridge. *Buch et al.* [1988] point out that transport estimates for the exchange are correct only to within a factor of 2. Moreover, as *Meincke* [1983] mentions, there is little information on the seasonal variability of the transports. This uncertainty in the estimates for the mean flow and its variability is one of the main issues addressed by the Nordic WOCE project [*The Nordic WOCE Working Group,* 1989] of which this study forms a part.

The Polar Oceans and Their Role in Shaping the Global Environment
Geophysical Monograph 85

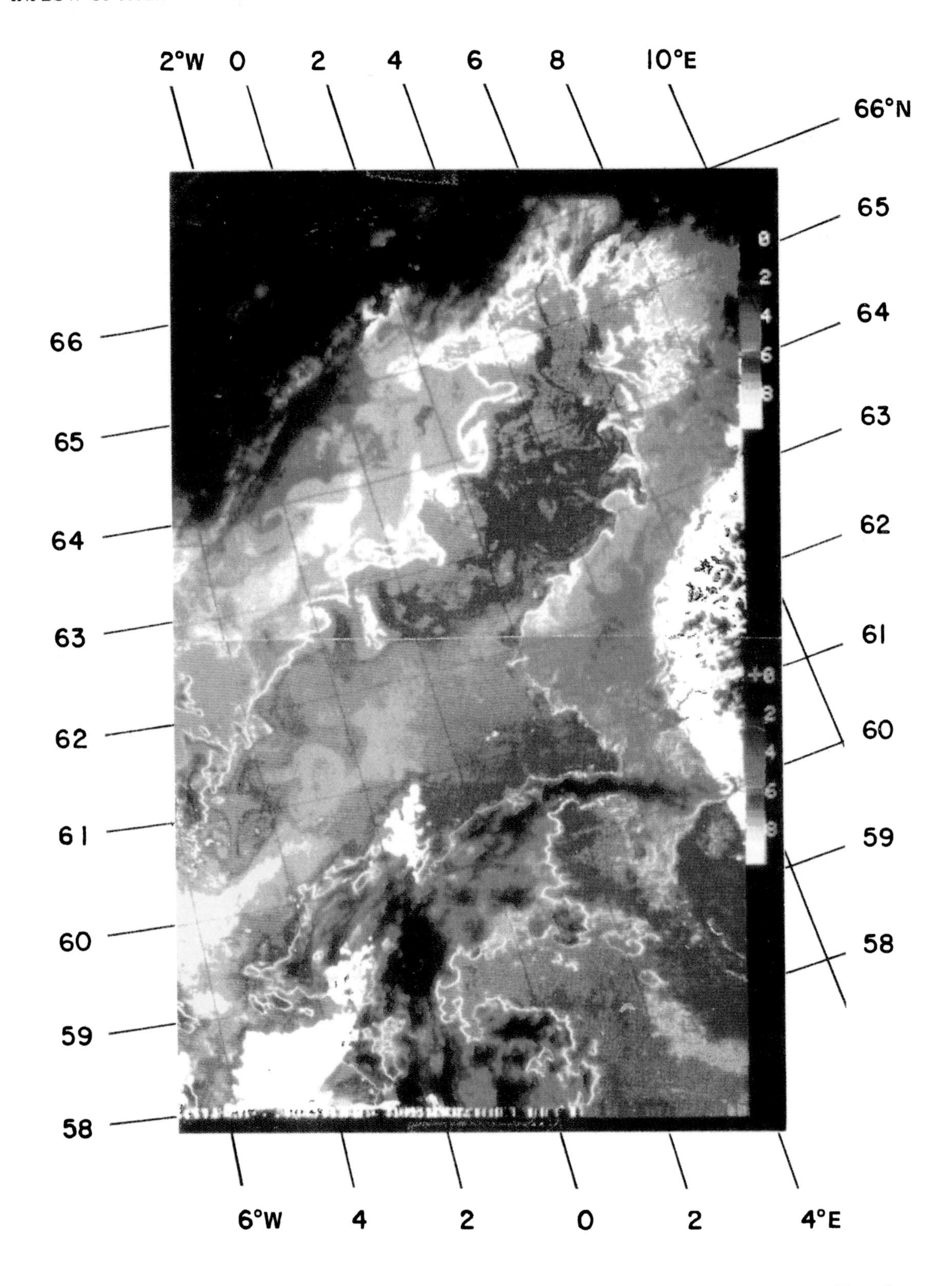

Plate 1. Infra-red image from NOAA-6, orbit 4573, 14th May, 1980 showing the North Atlantic Current. The colour coded temperature scale shows Atlantic water, at about 6°C, as red (From *Johannessen,* 1986).

Satellite altimetry is a technique that can be used to estimate upper layer circulation and transports. Several investigations have demonstrated the usefulness of altimetry in the study of strong currents like the Gulf Stream [*Cheney et al.*, 1983; *Glenn et al.*, 1991] and the Agulhas [*Wakker et al.*, 1990], where sea surface height anomalies greater than 50 cm have been observed. The altimetric signal in the Iceland-Færoe region is expected to be much smaller owing to the low energy of the ocean dynamics and special care must be taken in the processing of the altimeter data and in the interpretation of the results. *Dobson* [1988] examined some GEOSAT altimeter passes to the east of Iceland to study the movement of the Iceland-Færoe Front and obtained height anomalies below 23 cm. Similar values were also obtained by *Robinson et al.* [1989] and *Samuel* [1990]. In addition, the smaller spatial and temporal eddy scales — the baroclinic Rossby radius is of the order of 10 km, and *Meincke* [1983] found current fluctuations with periods of the order of 10 days attributable to baroclinic instabilities — could mean that scales with significant energy are not resolved by the sampling pattern. However, as *Sandwell and Zhang* [1989] point out, even though the sea surface height signals are small, mesoscale signals may still be retrievable from the sea surface slopes. GEOSAT altimeter data have been used to obtain reasonable estimates of the transport of the Norwegian Atlantic Current by combining them with hydrographic data [*Pistek and Johnson,* 1992], yielding a mean transport of 2.92 Sv with a standard deviation of 1.01 Sv.

In Section 2, the altimeter measurement principles and data processing methods will be discussed. In Section 3 the results of the study will be presented. Section 4 summarizes the results.

2 THE ALTIMETER MEASUREMENTS AND DATA PROCESSING

The radar altimeter onboard the U.S. Navy Geodetic Satellite (GEOSAT) was operational during the period from November, 1986 to October, 1989 [*Cheney et al.*, 1991]. It provided data at approximately 7 km intervals along the satellite's ground track, which formed a mesh of arcs going northwestwards (ascending tracks) and southwestwards (descending tracks). The track separation was about 50 km at 64° N and this pattern was repeated every 17.05 days (repeat period), with deviations from the nominal track confined to less than a few hundred meters. Thus, neglecting data gaps, all data points along the ground track had a sampling frequency of at least once in 17 days while the points of intersection (crossover points) between the ascending and descending tracks were sampled twice during each 17 day period. In practice, the data coverage varied from one repeat period to the next, owing to data gaps, which, in the extreme case, resulted in some of the tracks being absent altogether for some repeat periods.

The altimeter measures the time taken for an electromagnetic pulse transmitted vertically downwards to return to the antenna after reflection from the ocean surface. Given independent information about the height of the antenna above the earth ellipsoid and the velocity of light in the intervening atmospheric column, the height of the sea surface above the ellipsoid can be retrieved [*Tapley et al.*, 1982]. This idealised altimetric height may be represented as:

$$h = h_{(g)} + h_{(ssh)} + h_{(err)} \tag{1}$$

where $h_{(g)}$ is the time-invariant geoid reflecting the earth's gravity field, $h_{(ssh)}$ is the total sea-surface height component originating from the dynamic response of the ocean to currents, tidal forces and atmospheric pressure fluctuations and $h_{(err)}$ contains the errors in modelling the satellite ephemerides and the properties of the atmospheric column through which the altimetric pulse travels.

Tidal elevations and the effect of atmospheric pressure fluctuations are modelled separately to correct the altimeter height data. The errors in these models add to the residual error.

Sufficiently accurate geoid models are not currently available for the study area. However, since the altimeter data points are sampled repeatedly, the invariant part of the height measurements (the sum of the geoid and the mean dynamic sea level) can be removed by subtracting the height at a chosen reference time from each of the repeat measurements. Orbit error constitutes a major error term in the altimetric signal. However, the major part is concentrated at a wavelength of about 1 earth circumference and hence relatively simple techniques have been devised to suppress it [*Cheney et al.*, 1983; *Tai,* 1988]. We subtracted a sinusoidal function with a period of once per revolution fitted to the along-track height residuals. The corrected residuals indicate the sea level change from the reference time to the repeat measurement time. Sea surface height anomalies referred to the mean sea level may then be obtained by subtracting the mean of the residuals from each residual. We can thus write:

$$h' - \frac{1}{n}\sum_{i=1}^{n} h' = h_{(ano)} + h_{(err)} \tag{2}$$

where h' are the corrected residual heights and $h_{(ano)}$ is the sea surface height anomaly.

Given independent estimates of the sea level component due to the mean circulation, the total dynamic sea level can be computed as

$$h_{(dyn)} = h_{(ano)} + h_{(mean)} + h_{(err)} \tag{3}$$

where $h_{(dyn)}$ is the total dynamic sea level, $h_{(mean)}$ is the modelled mean dynamic sea level and $h_{(err)}$ now also contains the errors in modelling $h_{(mean)}$.

Geostrophic currents normal to the altimeter ground track (v_n) can be calculated from the total dynamic sea level using

$$v_n = \frac{g}{f}\frac{\partial h_{(dyn)}}{\partial s} \tag{4}$$

where g is the acceleration due to gravity, f is the Coriolis parameter and $\frac{\partial}{\partial s}$ is the along-track derivative.

Generally, the geostrophic velocities thus computed consist of both barotropic and baroclinic components. To a first approximation, we can assume that v_n represents the average flow within an upper layer of depth H, which is known as a function of time and position, in our case, from a circulation model. If L is the distance over which the slopes are computed, the upper layer transport, M_n^u, normal to the tracks can be estimated as:

$$M_n^u = HLv_n \tag{5}$$

These estimates are computed at each altimeter data point and integrated over any desired segment of the track.

Net heat transport (Q) can be calculated using estimates for the average temperatures of the upper-layer and lower-layer water masses (θ^u and θ^l) as

$$\begin{aligned} Q &= C_{sp}(\rho^u M_n^u \theta^u - \rho^l M^l \theta^l) \\ &= C_{sp}\rho^u M_n^u(\theta^u - \theta^l)\,, \end{aligned} \tag{6}$$

where C_{sp} is the specific heat of sea water, assumed to be equal for the upper and lower layers, and we have also assumed that $\rho^u M_n^u = \rho^l M^l$.

Global GEOSAT altimeter GDRs (Geophysical Data Records) incorporating improved T2 orbits, ocean tides and atmospheric corrections were obtained from NOAA [*Cheney et al.*, 1991]. The study area is bounded by the latitudes 55°N and 70°N and the longitudes 20°W and 15°E. The satellite's ground track within the study area is shown in Figure 1 This area excludes the Denmark Strait through which a significant part of the exchange takes place, a limitation which was dictated by two considerations. First, GEOSAT altimeter coverage of the ice-prone waters between Iceland and Greenland is relatively poor. Another reason was that the

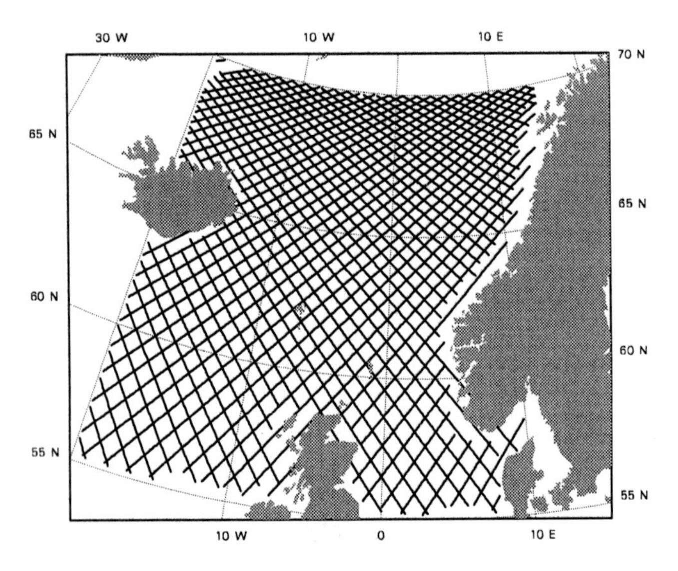

Fig. 1. GEOSAT altimeter ground track within the study area.

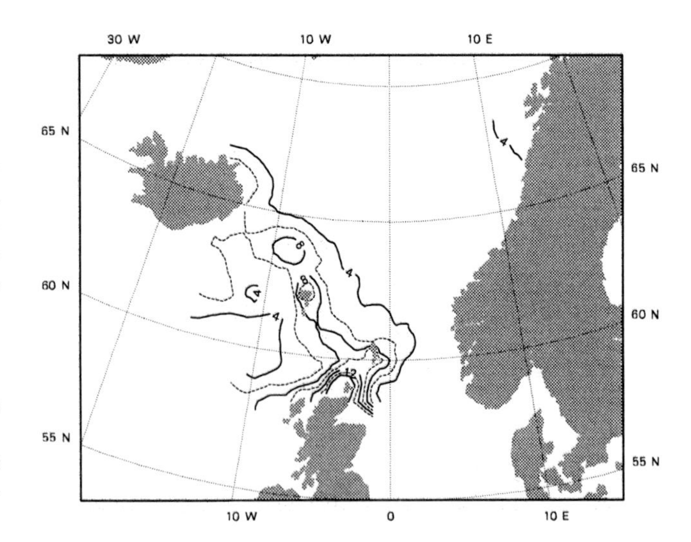

Fig. 2. Difference in GEOSAT RMS sea surface height anomalies (cm) obtained using the Schwiderski and Flather tide models.

domain of a regional geoid model [*Drottning*, 1993] which we opted to use before the repeat track processing as an attempt to reduce possible errors due to the deviations of the repeat passes from the nominal track excluded the Denmark Strait. These geoid heights were obtained by merging GEOSAT altimeter data and marine gravimetry data in a least squares collocation technique using *a priori* covariance matrices for both the geoid and the mean sea surface and were gridded on a 0.125° latitude by 0.25° longitude grid.

The tidal corrections included in the GDRs are from a global model [*Schwiderski*, 1980]. This model, with a resolution of 1° latitude by 1° longitude, is based on a scheme of hydrodynamical interpolation between sites at which tidal measurements have been made. Earlier investigations [*Tomas and Woodworth*, 1990; *Samuel et al.*, 1992] indicate that this model may be inadequate here. At least two regional models are available for the study area [*Flather*, 1981; *Gjevik and Straume*, 1989]. The North East Atlantic model [*Flather*, 1981], has a higher resolution (0.3° latitude by 0.5° longitude) and is based on considerable experience in coastal and pelagic tidal measurement and numerical modelling in this region. We used this model instead of the Schwiderski model. Tidal parameters from the model were converted to elevations using a modified version of routines developed at the Institute of Ocean Sciences [*Foreman*, 1977]. The difference in the root mean square (RMS) of dynamic sea level anomalies occurring solely due to differences in the two tide models is shown in Figure 2. Over most of the Norwegian Sea, the differences due to choice of tidal correction are small, with RMS values below 0.05 m. However, over a wide belt between Iceland and Scotland, the differences are relatively large, with RMS values of up to 0.12 m. This is comparable to the signal that we are trying to measure and underlines the need for accurate tide modelling as a prerequisite to the effective use of altimeter data in this area.

Mean dynamic sea level and mixed layer depths from the OPYC (Ocean isoPYCnal) model [*Oberhuber,* 1993] as implemented for the GIN sea [*Aukrust and Oberhuber,* Unpublished manuscript] were also used in the analysis. The model domain extends from 30°S to the North Pole and from 100°W to 50°E, with cyclic boundary conditions in the Arctic Ocean. A variable grid spacing is used, with a resolution of about 30–50 km in the GIN Sea. The model is diabatic, with prognostic temperature and salinity fields and realistic bottom topography. The surface boundary layer is parameterized by a detailed mixed layer model. A sea ice model with Hibler-type rheology is coupled to the mixed layer. Thermal forcing, wind stress and surface input of turbulent kinetic energy are determined from monthly mean values of atmospheric quantities while the surface salinity is prescribed as an annual mean. The mean sea level used is from the cyclo-stationary state attained after integrating the model over 30 years. Mixed layer depths were obtained as monthly means at the model grid points for a model year and were used as estimates for H in equation 5. Monthly mean sea level from the model were used to obtain estimates of the model upper layer transports to assess the significance of the contribution due to the altimeter SSH anomaly.

All data points falling within the study area were corrected for electromagnetic bias (2% of significant wave height), the solid earth tide, tropospheric water vapour effects (SSMI/-TOVS), dry tropospheric effects (ECMWF) and ionospheric effects. The inverse barometric effect was modelled as suggested in *Cheney et al.* [1991]. The Flather tidal elevations, the modelled geoid heights and the OPYC mean sea level were interpolated to the altimeter data positions and subtracted from the altimetric height data. If the geoid model and the mean sea level were accurate, there would have been no time-invariant component remaining in the residuals. However, it was seen that large residual mean heights remained (see Figure 3), implying that unacceptably large errors are present in either the geoid model or the ocean model or both. Hence, it was decided not to rely on the modelled geoid alone but to follow repeat track processing as outlined earlier, assuming that the errors in the mean sea level from the ocean model are smaller than those in the modelled geoid heights.

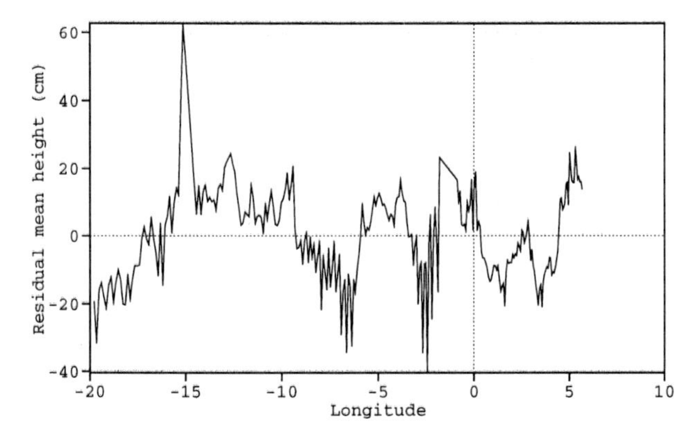

Fig. 3. Residual time-invariant component of SSH along an ascending track after subtracting the modelled geoid and mean dynamic sea level.

The resultant data set thus consisted of a space-time matrix of SSH anomalies. To obtain the total dynamic sea level, the model mean SSH could now be added back.

The root mean square (RMS) of the SSH anomalies at each data point were computed over the entire mission period. This provides an indication of the spatial distribution of mesoscale variability.

Another indicator of mesoscale variability is eddy kinetic energy (EKE). This may be computed as the variance of the geostrophic velocities computed at each data point. However, since along-track derivatives only give velocity components normal to the tracks, one would have to assume isotropy in order to compute EKE. We have exploited the sampling pattern of the altimeter to develop an interpolation technique to grid the SSH anomalies and hence to compute the zonal and meridional components of velocity for estimating EKE. Generally, adjacent tracks have time differences of 2–3 days. Thus by choosing a grid spacing of approximately 30 km and a time-step of 3 days it was possible to fill in a sufficient number of adjacent grid points at each time step. At the same time, the spatial and temporal smoothing was limited to the minimum possible. At each grid point, the SSH anomaly was computed as an exponentially weighted mean of all data points within one grid length and time step. Further details of the interpolation scheme are given in *Samuel* [1993]. Note that this interpolation was done only for computing EKE.

Transports were computed only for selected ascending tracks crossing the Norwegian Atlantic Current perpendicularly. Along-track derivatives of the total dynamic sea level were used for computing the geostrophic velocity component normal to the tracks. Since this is a high-pass filtering operation which enhances short-wavelength noise, a Gaussian low-pass filter was applied to smooth the along-track dynamic topography before computing the derivatives (see *Zhang* [1988] and *Samuel* [1993]). The resultant band-pass filter has maximum response for wavelengths in the range 30–150 km.

3 RESULTS

In this section, the results from the analysis relating to the RMS, EKE, geostrophic currents and transports are discussed.

3.1 Root Mean Square Anomalies

Plate 2 shows the RMS sea surface height anomalies computed from the GEOSAT data. RMS values observed are generally low with a maximum of about 0.15 m. These values are about half of those observed in high energy regimes

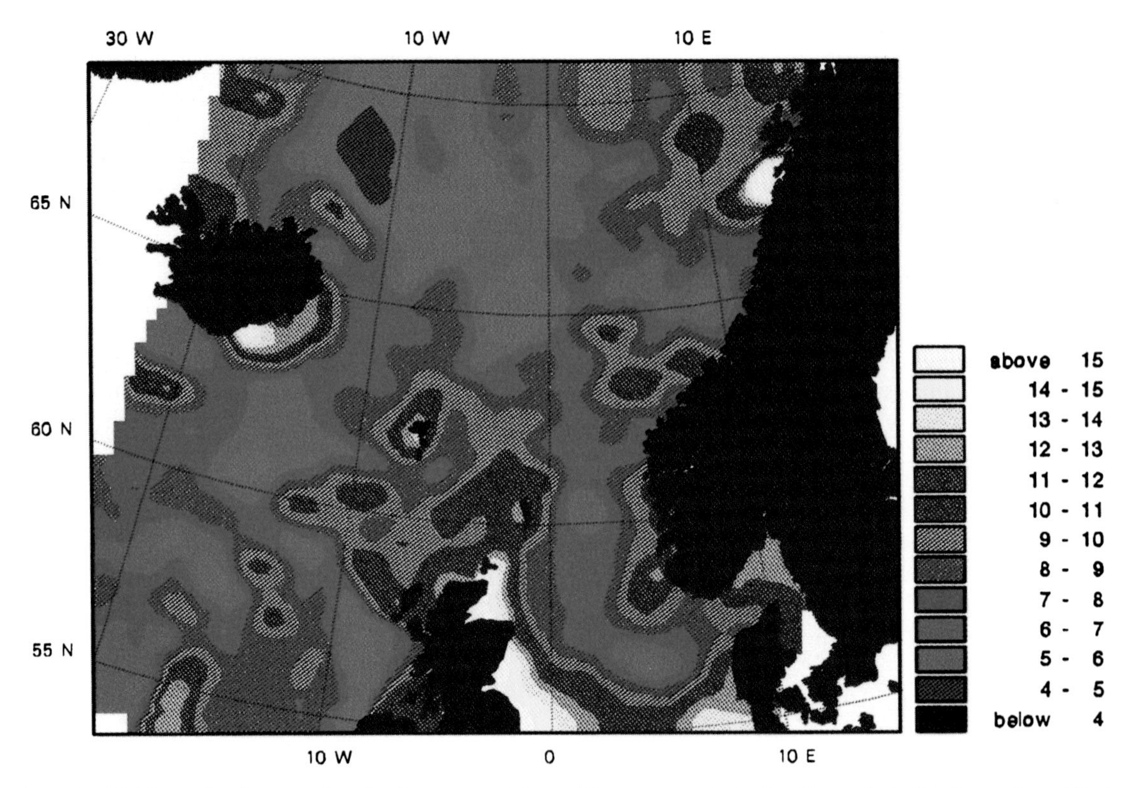

Plate 2. RMS sea level anomalies (cm) computed from 39 repeat periods for the period 8th November, '86 to 1st September, '88. Only data points with at least 11 repeat measurements were used.

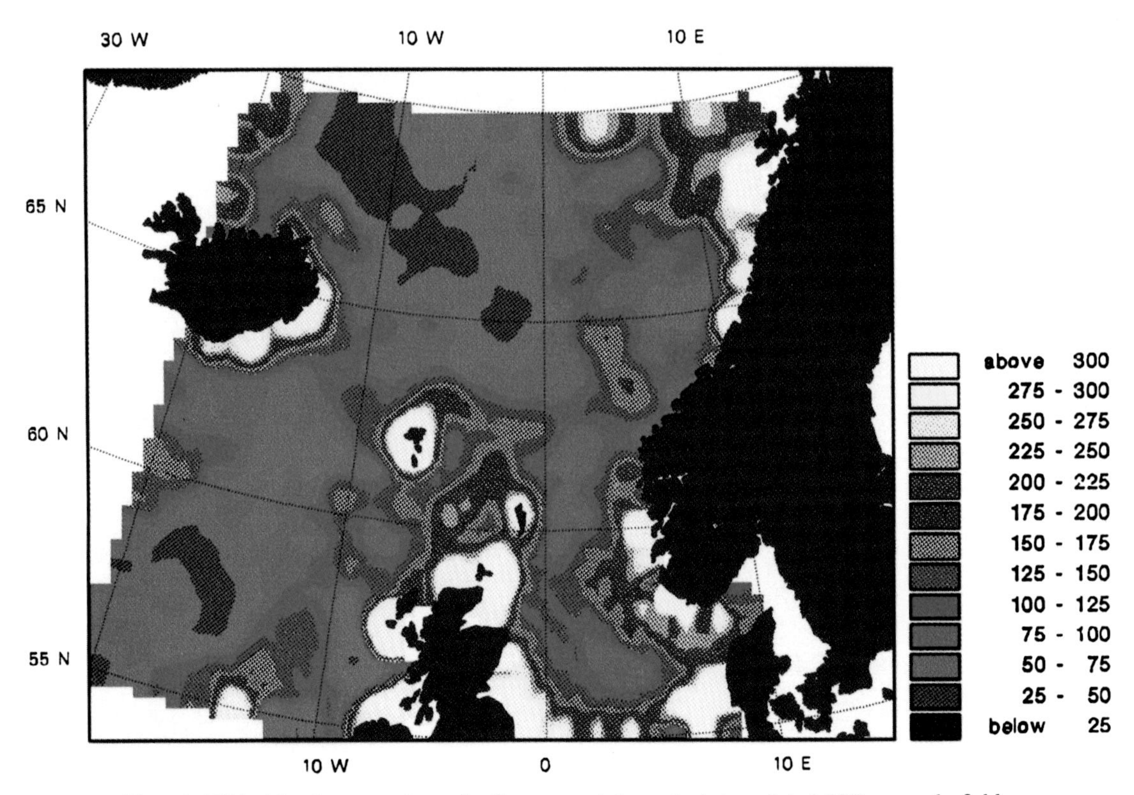

Plate 3. Eddy kinetic energy in cm^2s^{-2} computed from the interpolated SSH anomaly field.

like the Gulf Stream or the Agulhas where similar analyses have yielded RMS values of up to 0.28 m [*Le Traon and Rouquet,* 1990] and 0.32 m [*Wakker et al.,* 1990] respectively. However, an area of high RMS extends northwards from the Færoe-Shetland Channel up to about 62°N, with values above 0.10 m. This represents the variability of the Norwegian Atlantic Current through the channel as characterised by the meanders and eddies seen in Plate 1. It is also seen that a zone of high variability extends north of the Færoe Islands, possibly related to the variability in flow of Atlantic water around the east and north of the Færoes that some investigators have reported [*Buch et al.,* 1988]. RMS values up to 0.12 m are also observed off the Norwegian coast at around 63°N, where the continental slope veers westwards.

Other high variability areas are found in the northwest approaches to the British Isles, where the complicated bathymetry causes significant mesoscale activity [*Røed et al.,* Unpublished manuscript], around the coast of Scotland and around the eastern coast of Iceland. The latter could be related to the anticyclonic flow around Iceland [*Hopkins* 1991]. The areas very close to land should probably be discounted since the altimeter is known to have tracking problems when crossing from land to sea. Similarly, the high variability observed in the northwest and southeast corners of the study area could also be artifacts introduced by the detrending and interpolation operations.

Surprisingly, apart from the areas close to Iceland and the Færoes, there is little variability over the Iceland-Færoe Ridge, even though the East Iceland Polar Front which runs nearly parallel to the Iceland-Færoe Ridge is known to meander and would hence be expected to produce high anomalies [*Robinson et al.,* 1989; *Dobson,* 1988]. Part of the reason for this could be the sampling limitations of the altimeter whose temporal and spatial scales are insufficient to resolve the prevailing eddy periods of the order of 10 days [*Meincke,* 1983] and wavelengths of 10 km upwards. However, it is more likely that the strong temperature gradient across the front is offset by the salinity gradient, such that the resultant density gradient is not large enough to be registered by the altimeter. It should be mentioned in this context that RMS anomalies of up to 0.22 m were obtained in this area when the Schwiderski tidal correction was applied instead of the Flather model tidal elevations [*Samuel,* 1990].

3.2 Eddy Kinetic Energy

Plate 3 shows the distribution of EKE computed assuming isotropy. While there is a strong resemblance with the plot of RMS anomalies, the distribution of eddy kinetic energy shows a lot more detail. The energy of the Norwegian Atlantic Current flowing northwards through the Færoe-Shetland Channel and the eastward flow on the north side of the Færoes is more clearly delineated. So are the areas off the Norwegian coast further downstream.

Apart from the areas already identified by the RMS distribution, there are now additional areas of high EKE to the west of Scotland, reflecting the effects of the region's complex bathymetry on the North Atlantic Current, and another area north of the Iceland-Færoe Ridge, over the narrow Ægir Ridge.

3.3 Geostrophic Currents

In order to get an indication of the significance of the SSH anomaly signal obtained from altimetry for the upper layer current field, geostrophic velocities were computed both from the OPYC model mean SSH field and the total dynamic sea level obtained by adding the altimetric SSH anomaly field to the mean field. The geostrophic computations were carried out on a 50 km by 50 km spatial grid. Temporal interpolation was also carried out, whereby total dynamic SSH on a particular day was computed as an exponentially weighted average of data 8 days prior and 8 days after. The results are shown in Figure 4 for the mean SSH and for representative periods in autumn, winter, spring and summer.

The OPYC model geostrophic velocity vectors (Figure 4a) give a good qualitative representation of the general circulation patterns in the area [*Aukrust and Oberhuber,* Unpublished manuscript]. The inflow through the Færoe-Shetland Channel and to the northeast of the Færoes is discernible, as also the Norwegian Coastal Current and a cyclonic circulation within the Iceland-Norwegian Sea. Speeds are generally low, but relatively higher speeds are found close to the Norwegian coast, around Scotland and to the northeast of Iceland, where the East Iceland Current flows. Speeds of over 0.1 m/s are obtained near the Norwegian coast at around 62°N where the RMS anomalies and EKE from the altimeter were high. In general, the altimeter shows high RMS variability and EKE in the regions where the model geostrophic velocities are high. This is not true of the southeast coast of Iceland where the altimeter indicates high variability while the model geostrophic velocities are low. *Hopkins* [1991] also mentions a flow on the southeastern coast of Iceland which the OPYC model clearly fails to reproduce.

The upper layer current patterns derived from the total SSH preserve the main features shown in the mean flow map (Figure 4). However, as would be expected, they reveal more detail and give some indication of the seasonal variability.

The map for September indicates increased speeds in the eastern part of the Færoe-Shetland Channel. The cyclonic gyre in the GIN Sea has undergone a translation eastwards, with a seaward meander in the Norwegian Atlantic Current at around 62°N. The East Iceland Current flows closer to the Icelandic coast and penetrates further south. The high velocities seen around Scotland are suspect due to the proximity to land, as mentioned earlier in connection with the RMS anomalies.

In December, there is an overall intensification of the currents in GIN Sea and the southern arm of the gyre now flows over the Iceland-Færoe Ridge. The North Atlantic Current is also stronger than in September.

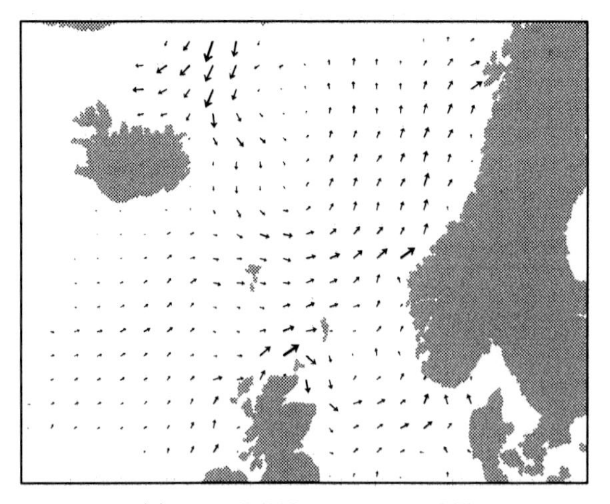

(a) From OPYC model mean SSH

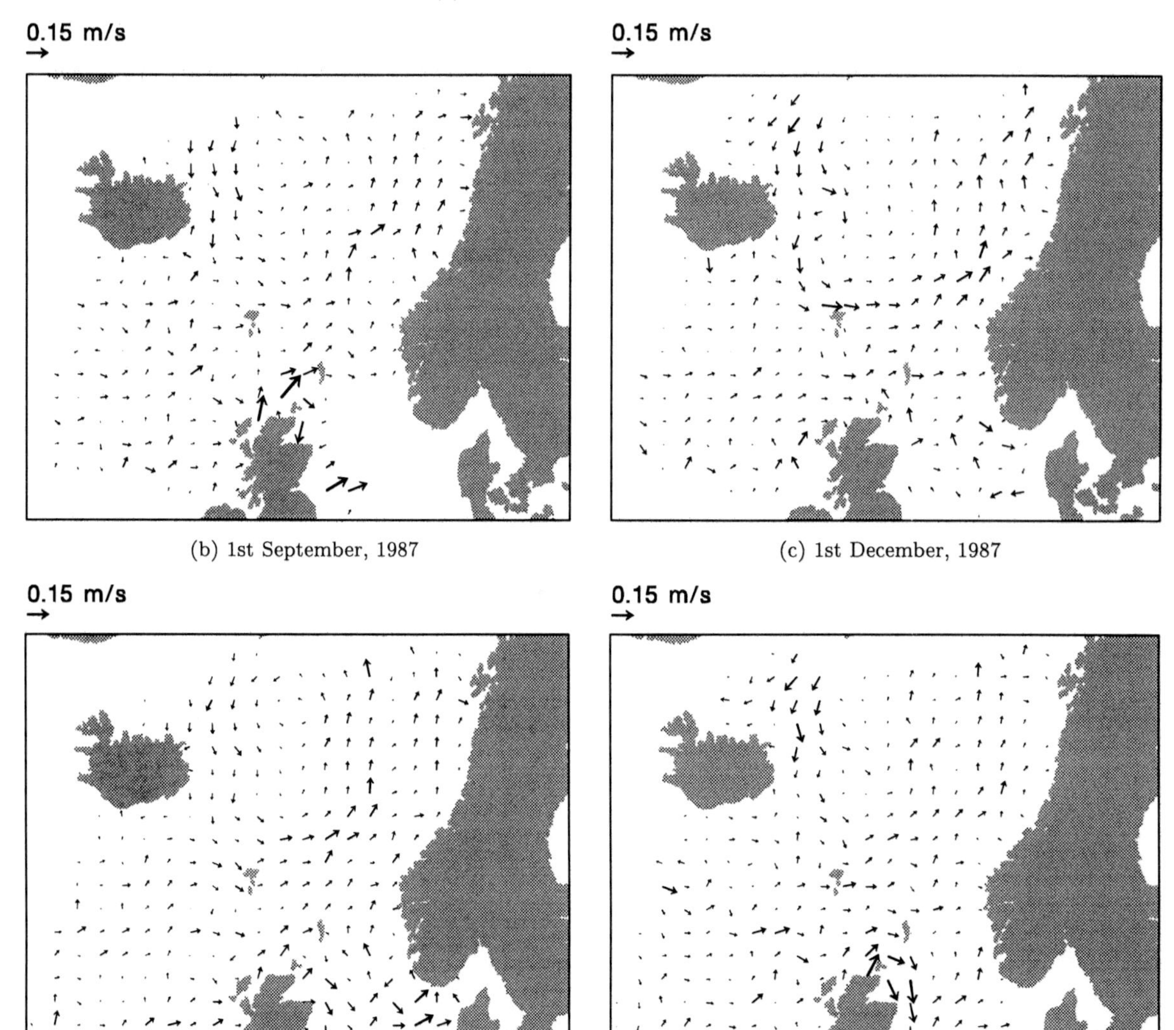

(b) 1st September, 1987

(c) 1st December, 1987

(d) 1st March, 1988

(e) 1st June, 1988

In March there is a slight weakening of the gyre except in the eastern arm which has drifted further away from the Norwegian coast. The southern arm has moved northwards closer to the mean position.

In June we observe increased speeds in the East Iceland Current but lower speeds in the Norwegian Atlantic Current. Within the Færoe-Shetland Channel, speeds are higher in the western side now.

It should be stressed that these maps only give an indication of the effect anomalous SSH values measured by the altimeter can have on the geostrophic velocity field. The interpolation over a seventeen-day period obviously filters out a large part of the mesoscale energy spectrum.

3.4 Volume Transport

To study the transport of Atlantic water through the Færoe-Shetland Channel, the ascending track segments shown in Figure 5 were used. Some of the tracks crossing the Færoe-Shetland Channel had to discarded due to lack of a sufficient number of usable data points. These tracks pass almost perpendicularly to the Norwegian Atlantic Current (see Plate 1) and are thus ideal for studying the transports. Monthly mean mixed layer depths were interpolated from the OPYC model grid points as estimates for H in Equation 5. This approximation is justified since the velocities are depth-invariant in the model mixed layer and the geostrophic velocities derived from the sea surface heights can be assumed to reflect the average velocity to at least these depths. The depths have a mean between about 200 m and 250 m in this area, with annual RMS variability of 45 m to 65 m. Transports were estimated according to Equation 5 at each data point using $L = 7$ km (along-track sample spacing) and subsequently integrated along each track segment.

The transports estimated from the model monthly mean SSH (dotted lines in Figure 6) clearly show the annual cycle, with high northward transports in winter and lower transports in summer. It should be mentioned that these are not the transports actually computed by the OPYC model, but the upper-layer transports computed under the geostrophic assumption from the model mean SSH. *Aukrust and Oberhuber* [Unpublished manuscript] give a value of 4.5 Sv as the mean inflow between Iceland and Scotland computed by the model. These estimates are averaged from transports computed at each model time step and also include ageostrophic terms.

Deviations of up to 2.5 Sv are observed when the altimetric SSH anomalies are used together with the model annual mean SSH. These differences reflect mesoscale variability of the upper-layer circulation and its significance for the Atlantic inflow. The altimeter transport is a point estimate in

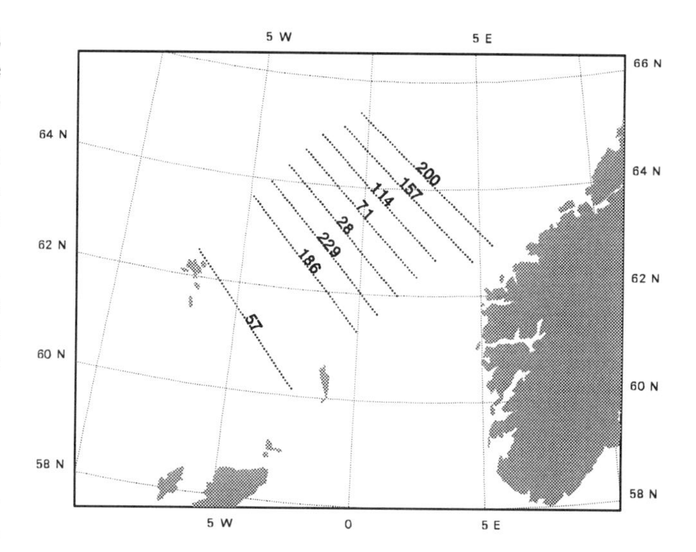

Fig. 5. Track segments used in the transport computations.

time, using the SSH on the day of the pass, albeit with a monthly mean mixed layer depth, while the model estimate represents an average for the month. Also, the model SSH used here is that from one representative year and variations from one year to the next are not considered. Some of the difference could, of course, be due to bad data points and gaps in the altimeter data. Transports were computed at each data point along a pass, summed and then scaled by the ratio of the number of data points on the ideal track to the number of points on the current pass. If the number of missing data points were large, it could have a significant effect on the transport computed for the pass. This error would be most severe for the tracks crossing the Færoe-Shetland Channel where the proximity of land reduces the number of good data points. Figure 6 shows that the discrepancy seems to be relatively high even for track 57, which was retained in the editing process.

While the mean transport for all the tracks is more or less constant at about 2.5 Sv, the amplitude of the annual cycle ranges from about 1.2 Sv for track 114 to about 2.8 Sv for track 186, and some of the individual estimates (represented by points in Figure 6) are as high as 10 Sv. Transport estimates using altimeter data, undertrack CTD measurements and climatological dynamic heights have been reported by *Pistek and Johnson* [1992]. They combined segments of tracks 114 and 200 and present a time series of transport estimates. Their results are qualitatively very similar to the results we have obtained. They give a mean transport value of 2.92 Sv for the two tracks with a standard

Fig. 4. Geostrophic velocity vectors computed from OPYC model mean SSH (a) and from total SSH, i.e. OPYC mean + smoothed altimetric anomalies (b-e).

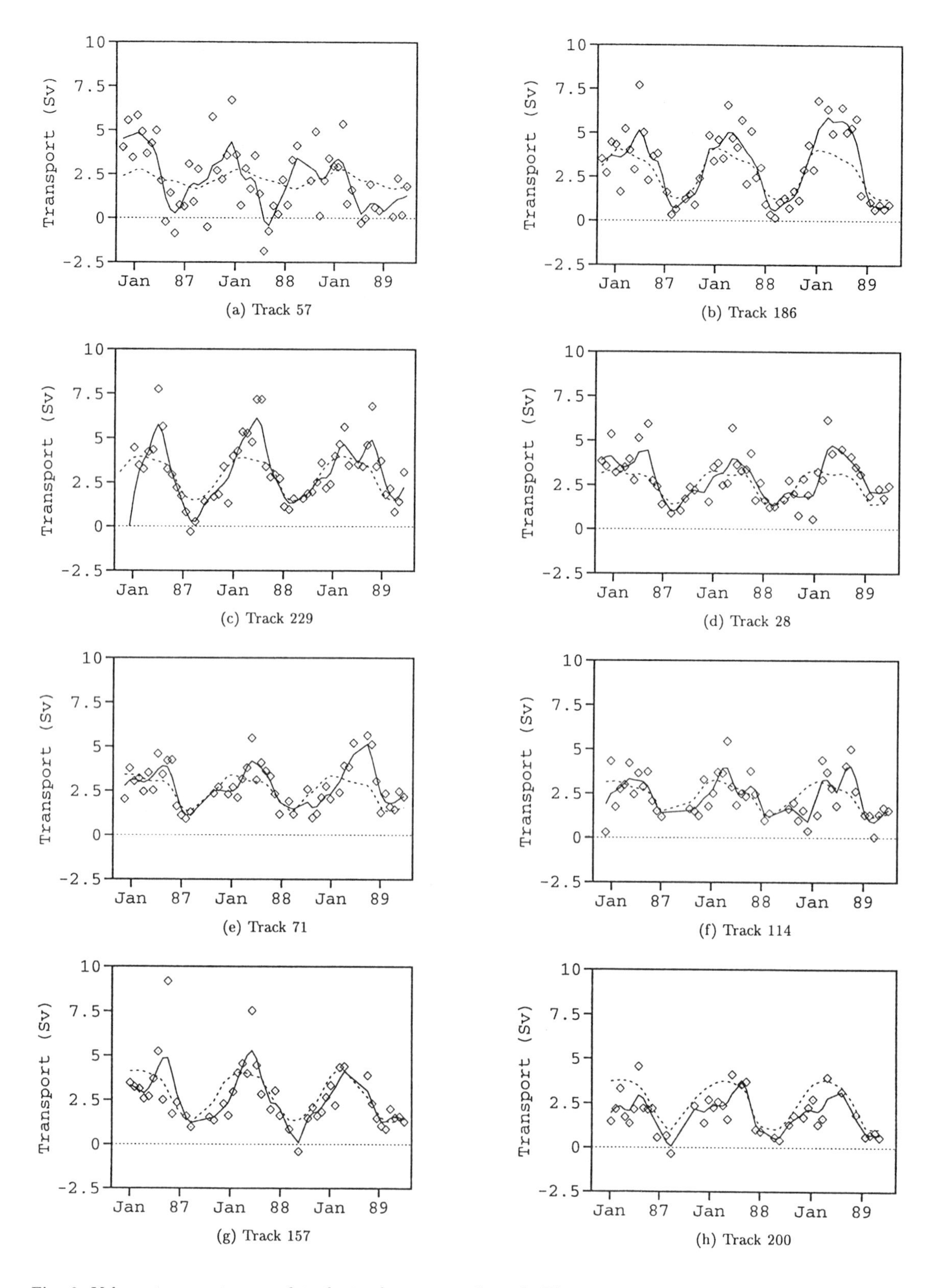

Fig. 6. Volume transports normal to the track segments shown in Figure 5. The points represent estimates for each repeat pass, the solid line connects the weighted averages for windows of width 20 days and the dotted line is a similar curve using transports estimated from the OPYC model monthly mean SSH.

deviation of 1.01 Sv and minimum and maximum values of 0.6 Sv and 4.8 Sv respectively. In our case, the mean for these two tracks is about 2.4 Sv and minimum and maximum values are −0.3 Sv and 4.6 Sv respectively. At least part of this difference can be explained by the differences in the track segment lengths and the depth of the upper layer, resulting in a smaller cross-sectional area in our case. The progression of the time series is similar, especially if track 114 is considered in isolation. The good agreement of the results from these two different methods for computing transports from altimetry justify further investigation along these lines with more rigid quality checks on the data.

Figure 7a shows the transport estimates for tracks 186, 229, 28, 71, 114 and 157 considered as one time series. Again the annual cycle is clearly visible, with a mean around 2.7 Sv and an amplitude of about 1.8 Sv. A few of the individual estimates show small negative (southward) transports, but the vast majority are in the range 1–7 Sv. Previous estimates from hydrographic sections and current meter measurements range from 3.3 Sv given by *Dooley and Meincke* [1981] to about 7.5 Sv measured by *Gould et al.* [1985]. The wide range in *in situ* measurements reflects the mesoscale variability of the transport which the altimeter data indicate. The minimum transport observed is 0.9 Sv in July-August, 1987 and the maximum is 4.5 Sv, in February-March, 1988. The vertical bars indicate the standard deviation of the weighted means calculated at each point on the solid line. It is seen to have magnitudes ranging from 0.5 Sv to 2.2 Sv. A striking feature of the transport time series is that the maximum appears in February-March rather than December-January where the model transports are highest. This observation is also true of the estimates computed by *Pistek and Johnson* [1992], but contrary to the estimates from hydrographic transects between 1927 and 1958 with a minimum in March that they present as Figure 1. In Figure 7b, where the data have been merged to a single year, this feature is even more prominent. The reasons for this discrepancy are not clear. One possible explanation could be that the tracks involved are somewhat north of the Færoe-Shetland Channel where the *in situ* estimates were obtained. Indeed, for track 57, the only track within the channel with a clear annual cycle in the transport estimates (Figure 6), we see that the maximum is in January, with a minimum in May-June. Also, it is possible that the segments considered cross the Norwegian Coastal Current so that its transport is also included in the estimates. Further, it should be remembered that the transport estimates, both from altimetry and from hydrographic transects, show large fluctuations from day to day and the agreement between the two types of estimates is well within the range of these fluctuations. Moreover, the number of hydrographic transects show a marked decrease in the winter months, casting some doubt on the reliability of the fitted curve.

3.5 Comparison with In Situ Measurements and Heat Transport Estimates

Salinity and temperature measured at the Ocean Weather Station-M [*Gammelsrød et al.,* 1992] at 66°N 2°E during the same period at 8 depths—0 m, 10 m, 25 m, 50 m, 100 m,

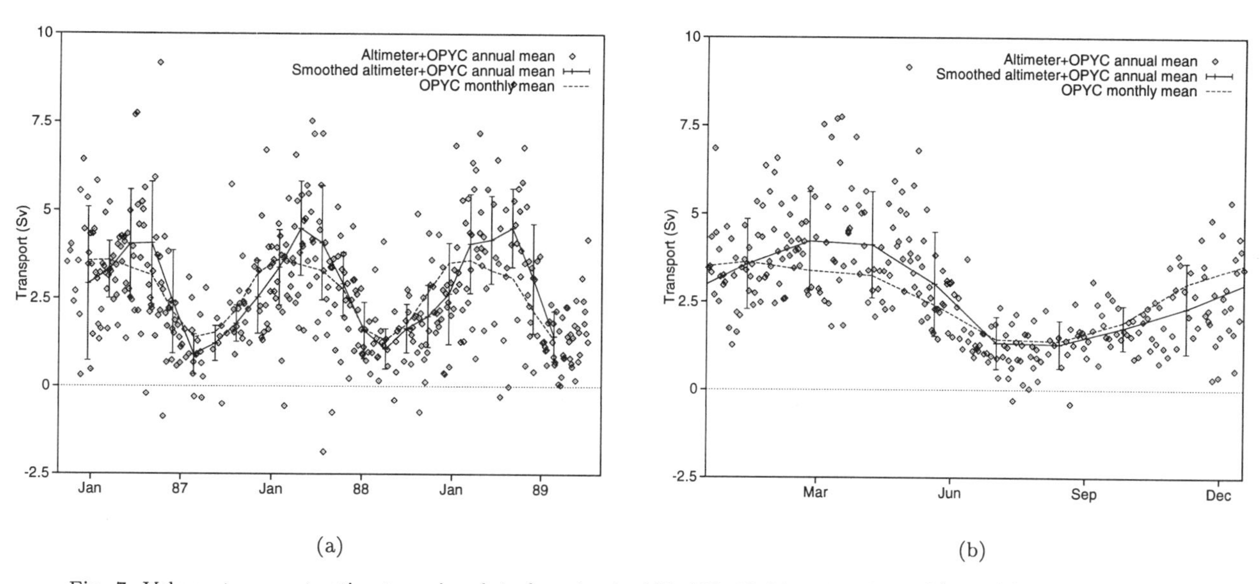

Fig. 7. Volume transport estimates using data from tracks 186, 229, 28, 71, 114 and 157 (a). In (b), the data period is merged to a single representative year. The points represent actual transport estimates for a pass segment, the solid line connects weighted means computed for windows of width 20 days with the vertical bars indicating standard deviation for the weighted means and the dotted line shows the transports estimated from the OPYC model monthly mean sea levels.

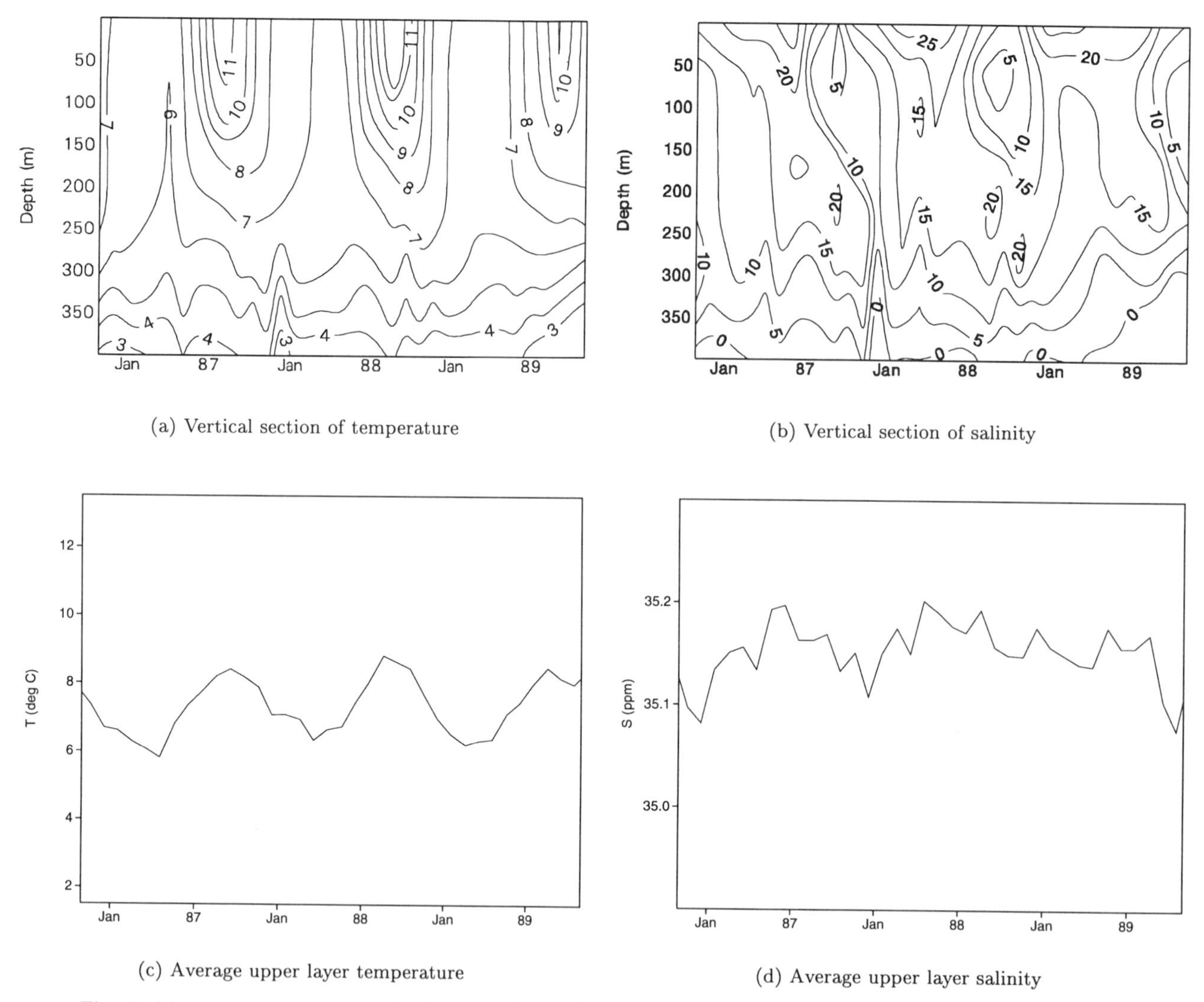

(a) Vertical section of temperature

(b) Vertical section of salinity

(c) Average upper layer temperature

(d) Average upper layer salinity

Fig. 8. Measured temperature and salinity at Ocean Weather Station-M (66°N;2°E). Temperature is given in °C. Salinity is shown as 100 times the deviation from 35 °/oo in (b) and as °/oo. in (d)

200 m, 300 m and 400 m—are shown in Figure 8. The annual cycle in both temperature and salinity is evident in the upper layer extending to a depth of about 250 m. This depth is comparable to the mixed layer depths from the OPYC model which were used in the transport computations. The vertically-averaged temperatures and salinities in this layer (from the measurements at 0 m, 10 m, 25 m, 50 m, 100 m and 200 m) are shown in Figure 8c,d. The average salinity is high in April-May. It is conceivable that this high value is a consequence of the high transports of Atlantic water (Figure 7) which mixes throughout the upper layer.

The mean temperature time series (Fig 8c) shows a well-defined annual cycle, with maximum in August-September and minimum in March-April. The phase of this time series is opposite to that of the transport (Figure 7a), with maximum temperature associated with minimum transport and *vice versa.* This would contribute to smoothing out the seasonal variations in heat transport to some extent.

An approximation to the net heat advected by the current can be found by assuming that the mass transported into the GIN Sea in the upper layer is balanced by a reverse transport to the Atlantic in the lower layers, not necessarily in the same region. This outflow will be at a lower temperature (θ^l). The choice of this lower layer temperature is critical to the heat transport derived, since the closer this temperature is to the upper layer temperature, the greater the effect of the temperature variability on the heat transport. We have chosen 0°C as the upper limit of temperature representative for the lower layers since Norwegian Arctic Intermediate Water has a temperature of *ca.* 0.5°C and Norwegian Sea Deep Water has a temperature of *ca.* −1.25°C [*Johannessen,* 1986; *Hopkins,* 1991]. The net heat trans-

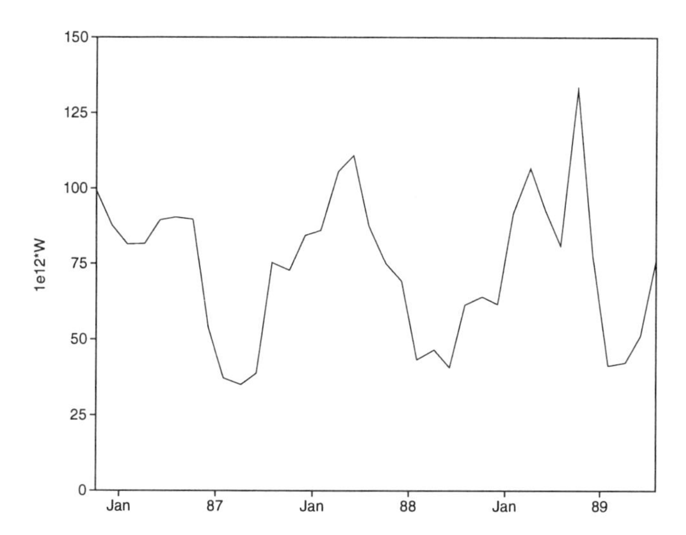

Fig. 9. Net heat flux (referred to a lower-layer temperature of 0°C) computed from the depth-averaged temperature at Station-M and the altimeter-derived volume transports.

port computed under these assumptions are shown in Figure 9. The time series of net heat transport resembles the time series of volume transport (Figure 7a) with a minimum of 35×10^{12}W in July-August, 1987 and a maximum of 133×10^{12}W in April-May, 1989. The mean value is about 84×10^{12}W with an amplitude of 49×10^{12}W. These estimates are very sensitive to the choice of a representative temperature for the outflow from the GIN Sea and caution should be exercised in interpreting them. However, they provide an indication of the significance of the Norwegian Atlantic Current in the heat balance of the GIN Sea.

For comparison, these estimates may be examined against the heat budget computed by *Worthington* [1970] for the Norwegian Sea. His estimate for the net heat lost to the atmosphere is about 264×10^{12}W. The difference is expected since his computations are based on a mean inflow of 9 Sv and different estimates for the inflow and outflow temperatures. Recent model simulations [*New et al.,* Unpublished manuscript] indicate poleward heat transports in the range 75–200×10^{12}W north of 60°N in the Atlantic, which are comparable to the estimates presented here.

4 SUMMARY

This study demonstrates the feasibility of combining model results with altimetry to study the volume transport of ocean currents. Despite the low signal to noise ratio of the altimeter data in the study area, the magnitudes of the transport estimates obtained compare well with estimates from hydrographic transects. Our results also compare well with the the results of another study using hydrographic data to estimate transports from altimetry. Transport estimates for the Norwegian Atlantic Current averaged over 20 days range between 0.9 Sv and 4.5 Sv with standard deviations between 0.5 Sv and 2.2 Sv within the averaging period. The annual cycle is pronounced, with a maximum in February-March and a minimum in July-August. The mean value is about 2.7 Sv and the amplitude of the annual cycle about 1.8 Sv. Differences between the transport computed from the model monthly mean SSH and the sum of model annual mean and altimetric anomalies are as large as 2.5 Sv, reflecting mesoscale variability.

Measured salinities during the same period at Station-M are also high in April-May, indicating the likelihood of increased inflow of Atlantic water. Temperature maxima coincide with transport minima and *vice versa*, but the transport variability dominates the net heat transport, which has a mean value of about 84×10^{12}W and shows an annual cycle with an amplitude of about 49×10^{12}W. These estimates may be examined against the globally-integrated values of northward transport of heat by the ocean given by *Vonder Haar and Oort* [1973]. They present a plot showing a transport of about 1000×10^{12}W at 50°N, decreasing polewards and vanishing at about 65°N. Our studies indicate that the northward transport is still significant at 62°N.

The distribution of the RMS of SSH anomalies identified regions of high variability, in the northwest approaches to the British Isles, in the Atlantic inflow through the Færoe-Shetland Channel, in the Norwegian Coastal Current and in the East Iceland Current. These areas are more clearly demarcated in the distribution of EKE computed from the along-track slopes, indicating that this may be a better way of studying mesoscale variability than the standard repeat track analysis which is generally used.

Satellite altimetry missions extending into the next century are planned and by the turn of the century a twenty-five year long data set will be available. The ERS-1 and TOPEX/POSEIDON satellites which are currently operational and the altimeter satellites that are to follow them will provide more accurate data than the GEOSAT which has been used in this study and will therefore greatly improve the accuracy of the estimates. Altimeter data will therefore constitute an important source of data for monitoring the oceans and its role in climate change.

Acknowledgements. This study was supported by the Norwegian Research Council (NFR). We would also like to thank Kjetil Lygre and Trond Aukrust for providing results from the OPYC model and Øystein Godøy for providing the *in situ* data. Thanks also to Peter Haugan and Moto Ikeda for useful comments.

REFERENCES

Buch, E., B. Hansen, J. A. Johannessen, S. Malmberg, and S. Østerhus, The exchange of water and heat between the Atlantic and the Nordic Seas, Document presented to the International

WOCE scientific conference, Paris, Nov. 28 – Dec. 2, 1988, 1988.

Cheney, R. E., N. S. Doyle, B. C. Douglas, R. W. Agreen, L. Miller, E. L. Timmerman, and D. C. McAdoo, *The complete Geosat altimeter GDR handbook*, National Ocean Service, NOAA, Rockville, MD, 1991.

Cheney, R. E., J. G. Marsh, and B. D. Beckley, Global mesoscale variability from collinear tracks of SEASAT altimeter data, *J. Geophys. Res.*, 88(C7):4343–4354, 1983.

Dickson, R. R., E. M. Gmitrowicz, and A. J. Watson, Deep-water renewal in the northern North Atlantic, *Nature*, 344:848–850, 1990.

Dobson, E. B., Dynamic topography as measured by the Geosat altimeter in regions of small surface height signatures, in *Proceedings of IGARSS '88 Symposium*, pp. 635–638, ESA Publications Division, 1988.

Dooley, H. D. and J. Meincke, Circulation of water masses in the Færoes channels during OVERFLOW 73, *Dtsch. Hydrogr. Z.*, 34(2):41–45, 1981.

Drottning, Å. H., *Gravity Field and Dynamic Sea Surface Topography Estimation in the Norwegian Sea using Satellite Altimeter Data and Sea-Gravity Data*, PhD thesis, Univ. of Bergen, Bergen, 1993.

Flather, R. A., Results from a tidal model of the North Atlantic ocean, in *Proceedings of the Tenth International Symposium on Earth Tides (with special sessions dedicated to ocean tides)*, edited by Vieira, R., p. 970, Madrid, Consejo Superior de Investigaciones Cientificas, 1981.

Foreman, M. G. G., *Manual for tidal heights analysis and prediction*, Institute of Ocean Sciences, Sidney, B. C., Pacific Marine Science Report 77-10, 1977.

Gammelsrød, T., S. Østerhus, and Ø. Godøy, Decadal variations of ocean climate in the Norwegian Sea observed at Ocean Station "Mike" (66° N 2° E), in *Proceedings of the ICES Mar. Sci. Symp.*, number 195, pp. 68–75, ICES, 1992.

Gjevik, B. and T. Straume, Model simulations of the M_2 and K_1 tide in the Nordic seas and the Arctic Ocean, *Tellus*, 41a:73–96, 1989.

Glenn, S. M., D. L. Porter, and A. R. Robinson, A synthetic geoid validation of Geosat mesoscale dynamic topography in the Gulf Stream region, *J. Geophys. Res.*, 96(C4):7145–7166, 1991.

Gould, W., J. Loynes, and J. Backhaus, Seasonality in slope current transports NW of Shetland, ICES, Hydrography Committee C.M. 1985/C:7, 13p, 1985.

Hopkins, T. S., The GIN Sea – A synthesis of its physical oceanography and literature review 1972–1985, *Earth-Science Reviews*, 30(3–4):318, 1991.

Johannessen, O. M., Brief overview of the physical oceanography, in *The Nordic Seas*, edited by Hurdle, B. G., chapter 4, pp. 103–127, Springer-Verlag, New York, 1986.

Le Traon, P. Y. and M. C. Rouquet, Spatial scales of variability in the North Atlantic as deduced from GEOSAT data, *J. Geophys. Res.*, 95(C11):20267–20285, 1990.

Meincke, J., The modern current regime across the Greenland-Scotland Ridge, in *Structure and development of the Greenland-Scotland Ridge*, edited by Bott, Saxov, Talwani, and Thiede, pp. 637–650, Plenum Publishing Corporation, 1983.

Oberhuber, J. M., Simulation of the Atlantic circulation with a coupled sea ice-mixed layer-isopycnal model. Part I : Model description, *J. Phys. Oceanogr.*, 23:808–829, 1993.

Pistek, P. and D. R. Johnson, Transport of the Norwegian Atlantic current as determined from satellite altimetry, *Geophys. Res. Letters*, 19(13):1379–1382, 1992.

Robinson, A., L. Walstad, J. Calman, E. Dobson, D. Denbo, S. Glenn, D. Porter, and J. Goldhirsh, Frontal signals east of Iceland from the Geosat altimeter, *Geophys. Res. Letters*, 16(1):77–80, 1989.

Samuel, P., Application of Geosat altimeter data for mesoscale variability studies in the Norwegian Sea, Master's thesis, Univ. of Bergen/NRSC, Bergen, Norway, 1990.

Samuel, P., *Applications of satellite altimeter data for studies on the water exchange between the Atlantic Ocean and the Nordic seas*, PhD thesis, Univ. of Bergen/NERSC, Bergen, Norway, 1993.

Samuel, P., J. A. Johannessen, and O. M. Johannessen, Preliminary results of altimeter data analysis for studying the water exchange between the Atlantic and the Nordic seas, in *Proceedings of the Central Symposium of the 'International Space Year' Conference*, pp. 393–396, ESA Publications Division, 1992.

Sandwell, D. T. and B. Zhang, Global mesoscale variability from the GEOSAT exact repeat mission : correlation with ocean depth, *J. Geophys. Res.*, 94(C12):17971–17984, 1989.

Schwiderski, E. W., On charting global tides, *Rev. of Geophys. and Space Phys.*, 18(1):242–268, 1980.

Tai, C.-K., Geosat crossover analysis in the tropical Pacific. 1. Constrained sinusoidal crossover adjustment, *J. Geophys. Res.*, 93(C9):10621–10629, 1988.

Tapley, B., G. Born, and M. Parker, The Seasat altimeter data and its accuracy assessment, *J. Geophys. Res.*, 87(C5):3179–3188, 1982.

The Nordic WOCE Working Group, Nordic WOCE : Project Description of the Nordic participation in the World Ocean Circulation Experiment, Technical report, 1989.

Tomas, J. and P. L. Woodworth, The influence of ocean tide model corrections on Geosat mesoscale variability maps of the North East Atlantic, *Geophys. Res. Letters*, 17(12):2389–2392, 1990.

Vonder Haar, T. and A. Oort, New estimate of annual poleward energy transport by northern hemisphere oceans, *J. Phys. Oceanogr.*, 3:169–172, 1973.

Wakker, K. F., R. C. A. Zandbergen, M. C. Naeije, and B. A. C. Ambrosius, Geosat altimeter data analysis for the oceans around South Africa, *J. Geophys. Res.*, 95(C3):2991–3006, 1990.

Worthington, L. V., The Norwegian Sea as a mediterranean basin, *Deep-Sea Res.*, 17:77–84, 1970.

Zhang, B., *GEOSAT altimeter data analysis for the determination of global oceanic mesoscale variability*, PhD thesis, Univ. of Texas at Austin, Austin, 1988.

J.A. Johannessen and P. Samuel, Nansen Environmental and Remote Sensing Center, Edv. Griegsv 3a, 5037 Solheimsvik, Norway.

O. M. Johannessen, Nansen Environmental and Remote Sensing Center, Edv. Griegsv 3a, 5037 Solheimsvik, Norway, and Geophysical Institute, University of Bergen, Allegt. 70, 5007 Bergen, Norway

Observation and Simulation of Ice Tongues and Vortex Pairs in the Marginal Ice Zone

Ola M. Johannessen[1], Stein Sandven[2], W. Paul Budgell[2]
Johnny A. Johannessen[2] and Robert A. Shuchman[3]

[1] *Nansen Environmental and Remote Sensing Center / Geophysical Institute University of Bergen, Bergen, Norway.*
[2] *Nansen Environmental and Remote Sensing Center, Bergen, Norway*
[3] *Environmental Research Institute of Michigan, Ann Arbor, USA*

Investigations in the marginal ice zone of the Greenland Sea and Fram Strait have established that ice tongues can frequently propagate from the ice edge out into open water and grow into vortex pairs. The horizontal scale of these tongues and vortex pairs is typically 30 km, and the life time of the surface signature is a few days. A two-layer primitive equation ocean model is used to simulate the ice tongues and the vortex pairs in an area of 108 by 128 km with a simplified bottom topography. Ice was treated as a passive tracer and represented by particles drifting at the sea surface. No wind forcing was applied since the observed ice tongues occurred during moderate to light wind conditions. In a series of numerical experiments, where an imposed jet current along the ice edge was perturbed, ice tongues and vortex pairs similar to the observations were generated. The study demonstrates that mixed barotropic/baroclinic instability is a sufficient mechanism to generate complex mesoscale circulation and ice edge structures in the marginal ice zone. It is estimated that about 12 % of the annual ice transport past 79° N is advected out into open water by the eddies and ice tongues along the East Greenland Current.

1. INTRODUCTION

Observations as well as numerical models of marginal ice zones indicate that jets along the ice edge can occur under favorable wind conditions [*Johannessen et al.*, 1983; *Røed and O'Brien*, 1983; *Smith and Bird,* 1991; *Johannessen et al.*, 1992]. Meanders in these jets can transport ice out into open water which is readily observed in remote sensing data [*Johannessen et al.,* 1987a]. The meanders can spin off eddies which may well grow into vortex pairs which are more or less asymmetric and propagate normal to the ice edge. This phenomenon was investigated by *Fedorov and Ginsburg* [1986, 1989] who studied what they called mushroom-like currents or vortex dipoles for several years using both satellite observations and laboratory experiments. They suggested that vortex dipoles are some of the most widespread forms of non-stationary coherent motions in the ocean. Several hypotheses have been suggested to explain the generation of the ice tongues in the marginal ice zone, such as baroclinic or barotropic frontal instability, water and ice entrainment by topographic eddies, eddies within ice covered areas leading to the squirting of a jet, and ice edge upwelling. In their laboratory experiment *Flierl, Stern and Whitehead* [1983] showed that a barotropic jet emerging from a point source in a rotating fluid deflects to the right in the northern hemisphere and develops into two counter-rotating vortices. *Couder and Basdevant* [1986] studied experimentally and numerically how vortex pairs develop in two-dimensional turbulent flow. Jets and vortex pairs were also generated

The Polar Oceans and Their Role in Shaping the Global Environment
Geophysical Monograph 85

by frontal instabilities in an experiment of coastal upwelling [*Narimousa and Maxworthy*, 1987]. Numerical simulations of the generation of mushroom-like vortices by a unidirectional upper-ocean momentum patch have been conducted by *Mied at al.* [1991].

The mesoscale activity in the East Greenland Current (EGC) has been studied by several investigators during the last few years and various generating mechanisms for this activity have been suggested. *Johannessen et al.* [1987b] proposed that barotropic and baroclinic instability as well as topographic trapping are important sources for eddies. They also argued that the eddies could be advected into the EGC from the eastern part of the Fram Strait by recirculation of Atlantic Water.

In this paper the generation and decay of ice tongues and vortex pairs observed by time series of airborne Synthetic Aperture Radar (SAR) imagery combined with drifting buoys and hydrographic sections obtained in the Marginal Ice Zone Experiment of 1987 (MIZEX 87) are described. Then the results of several numerical experiments of mesoscale ocean circulation are presented, and finally, the generation of ice tongues and vortex pairs is discussed.

2. OBSERVATIONS DURING MIZEX 87

During MIZEX 87, which was located in the EGC between 76° and 79° N, mesoscale processes along the ice edge were intensively investigated during winter conditions (Figure 1). The summer experiments in 1983 and 1984 [*MIZEX Group*, 1986] had shown the importance of having SAR coverage of the same area every day in order to map the rapid changes of the ice edge processes. Therefore, a unique 12-day time series of SAR images from the Greenland Sea, combined with Seasoar and Conductivity -Temperature - Depth (CTD) sections, and data from drifting buoys, were obtained during MIZEX 87, providing detailed information of the temporal and spatial variability of ice edge features [*MIZEX Group*, 1989]. The real-time downlink of the SAR images from the aircraft to the research vessel was invaluable in the planning of buoy deployment and the CTD/Seasoar sections. Because of its high resolution and independence from light and cloud conditions, SAR is the most useful sensor for mapping of mesoscale sea ice variables such as ice edge location, detailed ice edge morphology, ice concentration, ice motion, ice types and ice edge processes [*Johannessen et al.,* 1992]. Moreover, ice floes in SAR imagery are useful tracers of upper ocean circulation if the wind conditions are moderate. During most of the MIZEX 87 experiment the winds were moderate (3 - 12 m/s) which allowed the mesoscale circulation to be mirrored by the ice floes observed by sequential SAR images.

In the X-band SAR images presented in Figure 2, 3 and 4, the sea ice of different types appear as bright signatures in contrast to open water which has a very dark signature. This is because the SAR was set for optimum ice contrast [*Johannessen et al.*, 1992]. The SAR images showed two distinct ice regimes: 1) the ice edge zone, and 2) the interior of the ice pack with large multiyear floes.

The 20 - 40 km wide ice edge zone is characterized by numerous eddies, meanders, ice tongues and vortex pairs. In this zone, the ice has a uniform and high backscatter indicating many small floes of size from 20 - 100 m. Inside the ice edge zone, large multiyear floes and clusters of such floes are seen surrounded by thin ice (dark signature). The multiyear ice clusters can be recognized from one day to

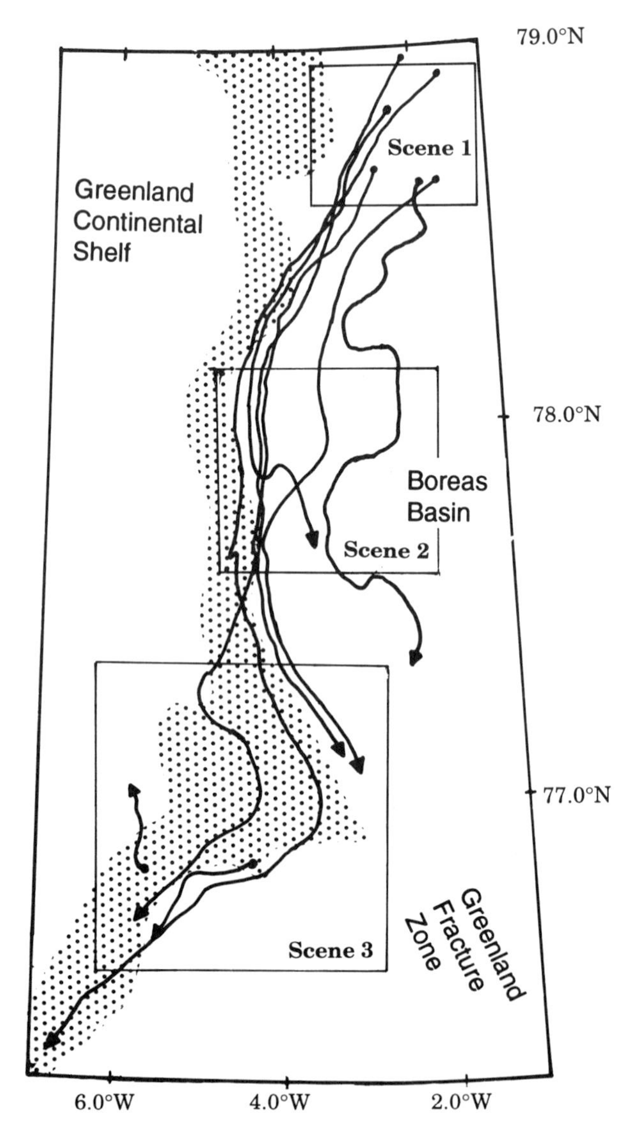

Fig.1. Trajectories of drifting buoys deployed in the ice edge region of the Greenland Sea during MIZEX 87. The shaded area indicates the continental shelf break (bottom depths from 1200 to 2000 m). The boxes mark the location of the scenes described in section 2.

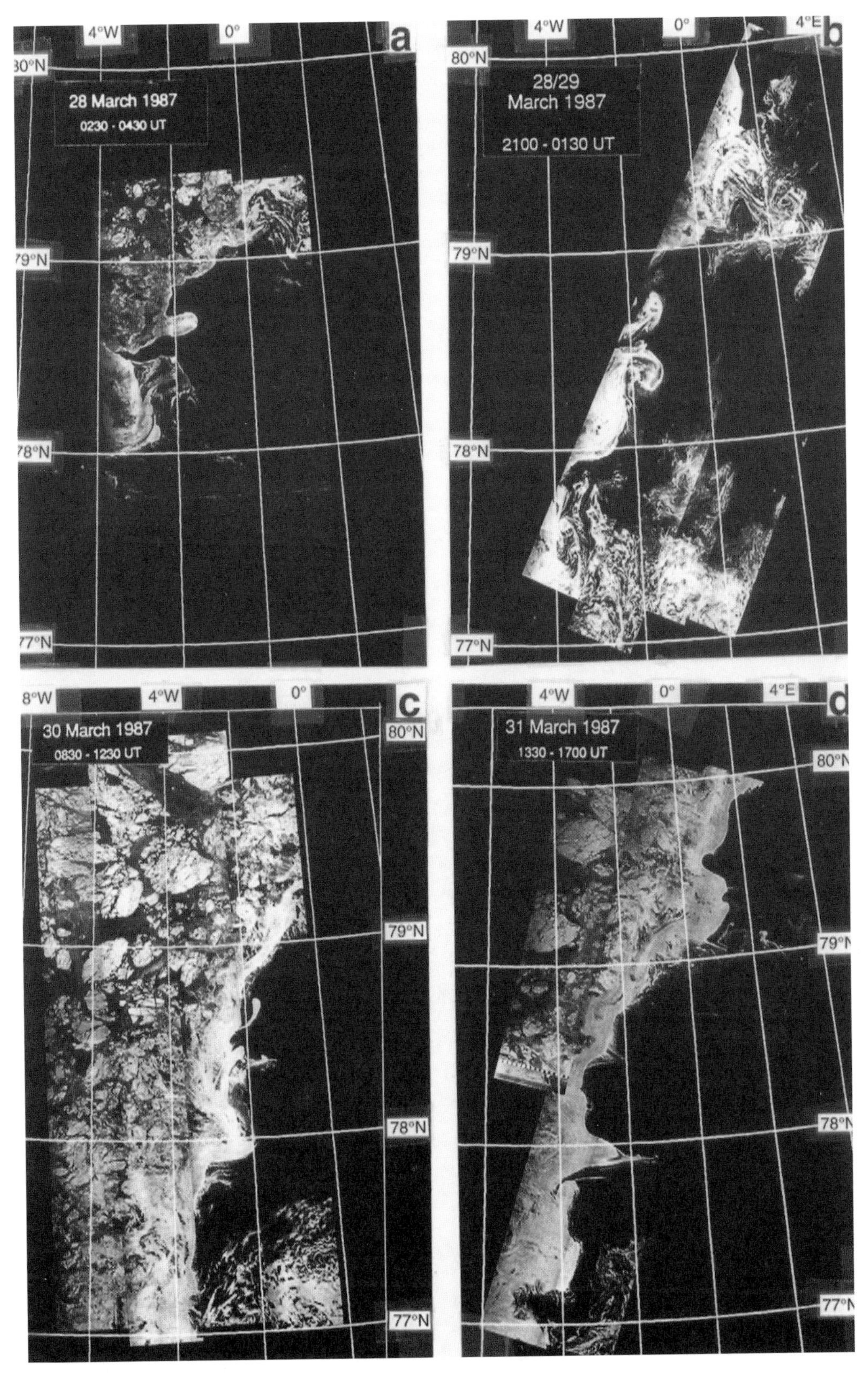

Fig. 2. Airborne X-band SAR mosaics obtained from March 28 to 31 1987.

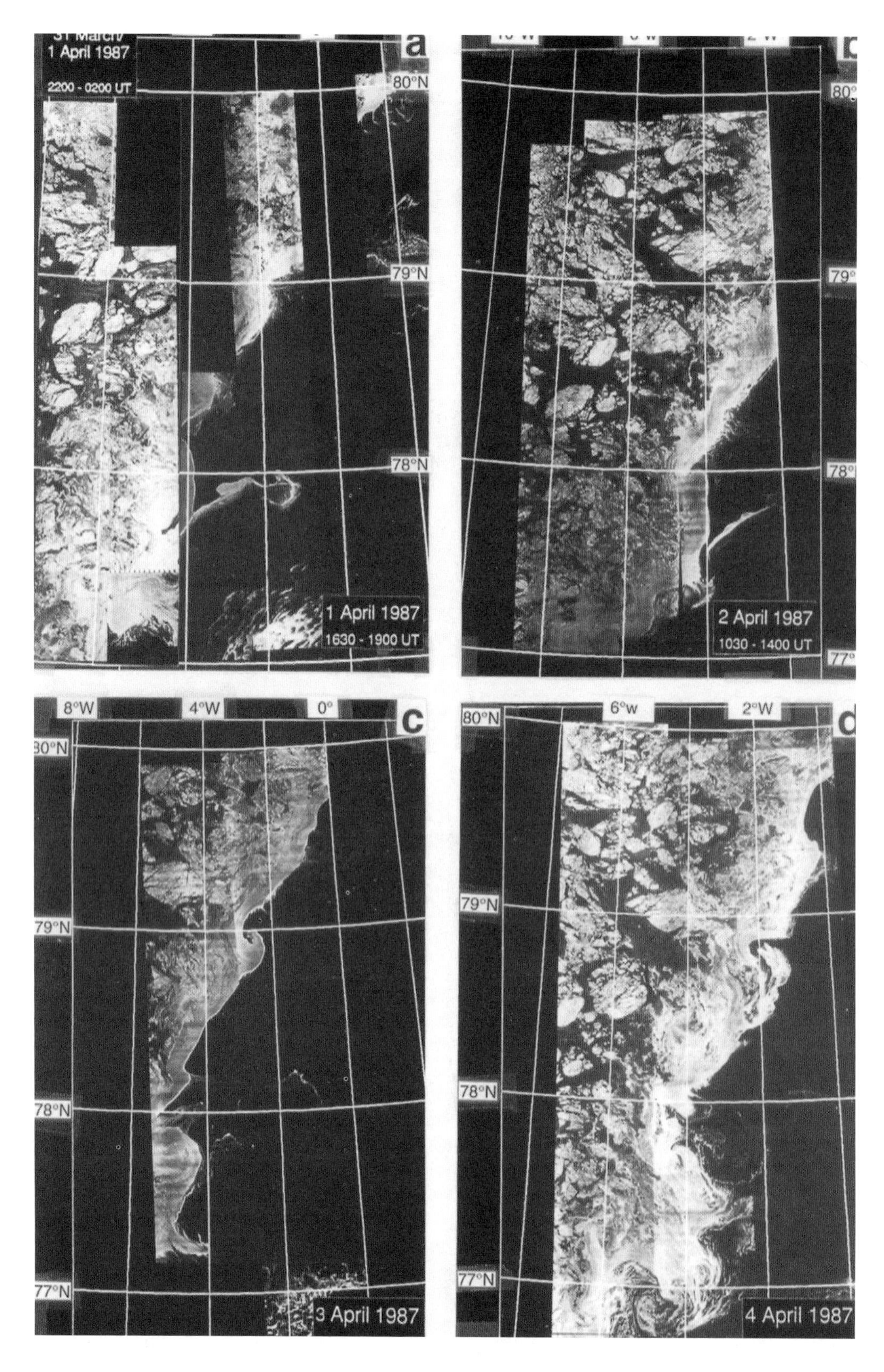

Fig. 3. Airborne X-band SAR mosaics obtained from April 1 to 4 1987.

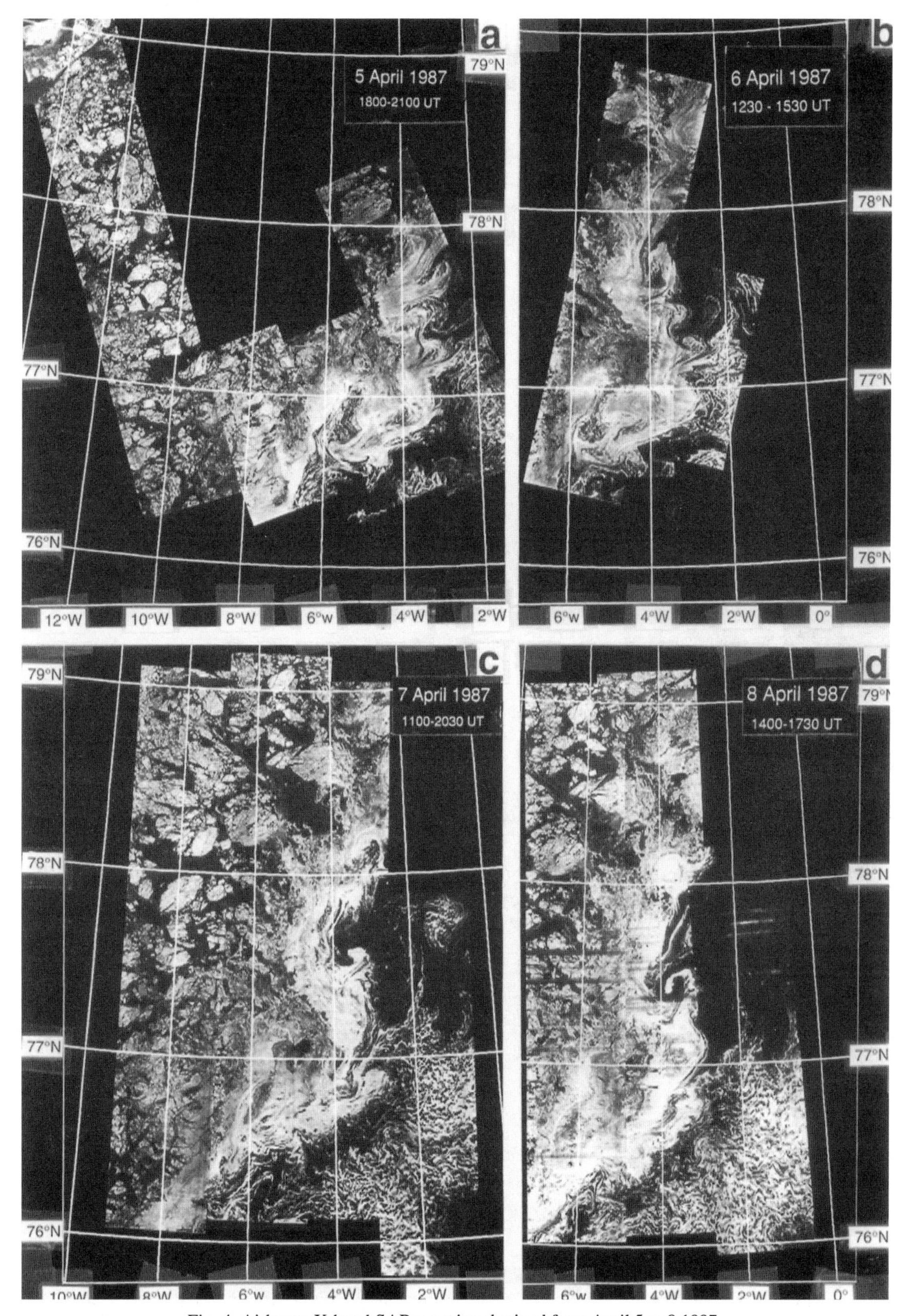

Fig. 4. Airborne X-band SAR mosaics obtained from April 5 to 8 1987.

another, thus time series of SAR images can be used to derive ice velocity vectors from the area. The sequence of SAR images display several episodes where ice eddies and tongues evolve and decay. A few of these events are selected for further examination.

An array of 10 drifting Argos buoys, three of which had current meter strings, were deployed in the ice edge region, providing ice and ocean velocity at various depths down to 250 m (Figure 1 and 5). Zonal velocity profiles across the ECG were obtained from the array, while the current meter strings provided vertical velocity profiles.

2.1. *Scene 1: Ice tongues and vortex-pair north of 78°30' N.*

In the period from March 28 to 31, the growth and decay of two ice tongues were observed in a sequence of four SAR images (Figure 6) and documented by CTD and Seasoar sections. The image obtained early in the morning on March 28 showed two tongues of ice shooting out from the ice edge. The northern tongue was characterized by a compact and well-defined ice edge. Later in the morning, as the icegoing vessel "Polar Circle" obtained a CTD section along 78° 50'N, the northern tongue was also observed from helicopter. The southern tongue which had the signature of an anticyclonic eddy was more diffuse with lower ice concentration, but the SAR image showed clear ice bands extending out into the open ocean. The wind was 5 m/s decreasing to 2 m/s from northeast during the day.

One day later the northern tongue had developed into a well-defined mushroom-shaped vortex-pair with horizontal scale of about 20 km (Figure 6 b). The front of the tongue did not show any eastward or southward advection. Approximately 8 hours after the SAR acquisition time a helicopter survey of the northern ice tongue showed that it was growing into a cyclone. Meanwhile, the southern tongue turned into an anticyclone about 30 km in diameter.

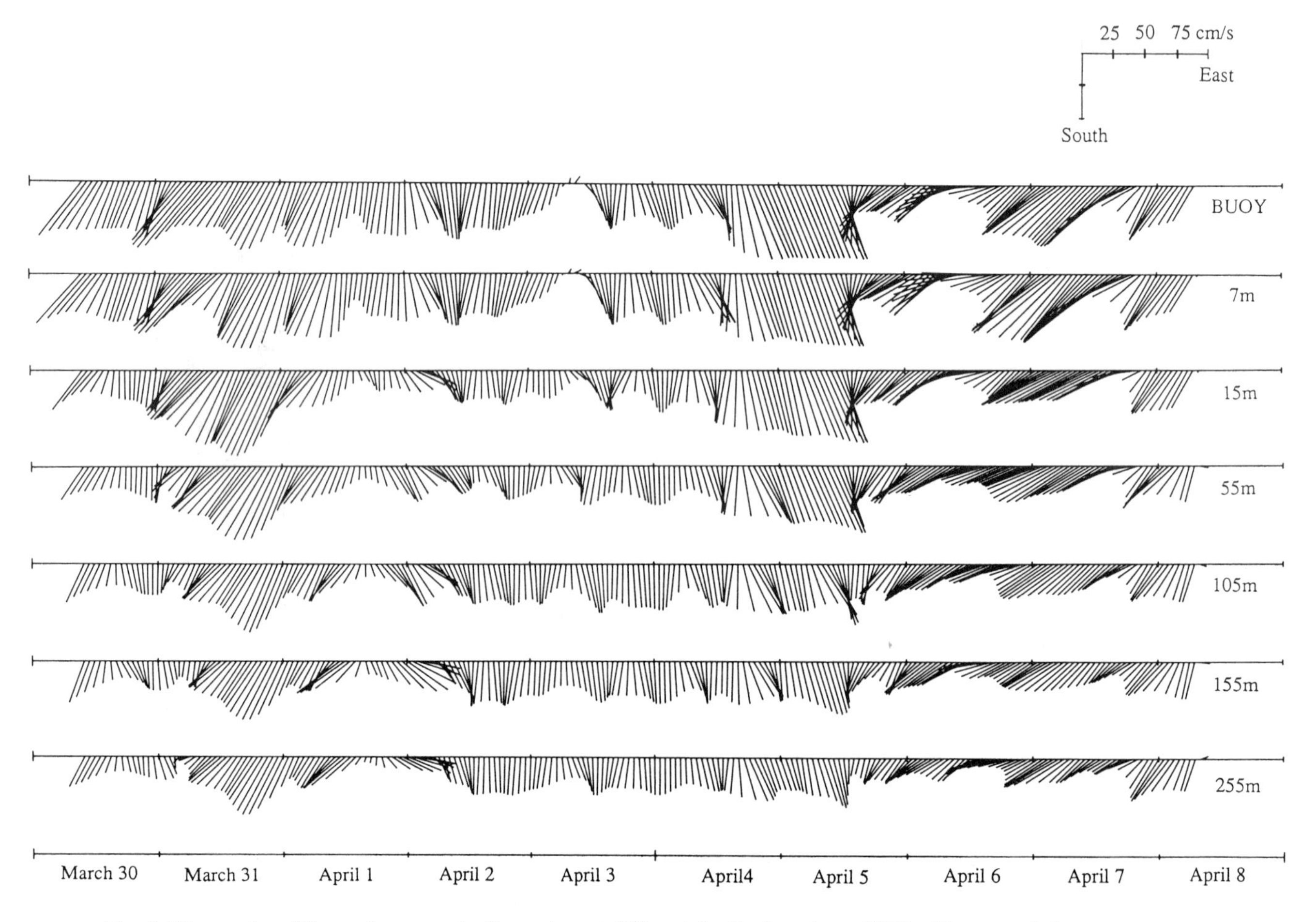

Fig. 5. Time series of ice and ocean velocity vectors at different depths from buoy 5064 with suspended current meters. The ice velocity is calculated from Argos positions interpolated to hourly values. The absolute ocean velocity is estimated by adding relative ocean velocity to the ice velocity. All vectors are plotted every hour and high frequency noise (cutoff period of 3 hour) has beeen removed from the data.

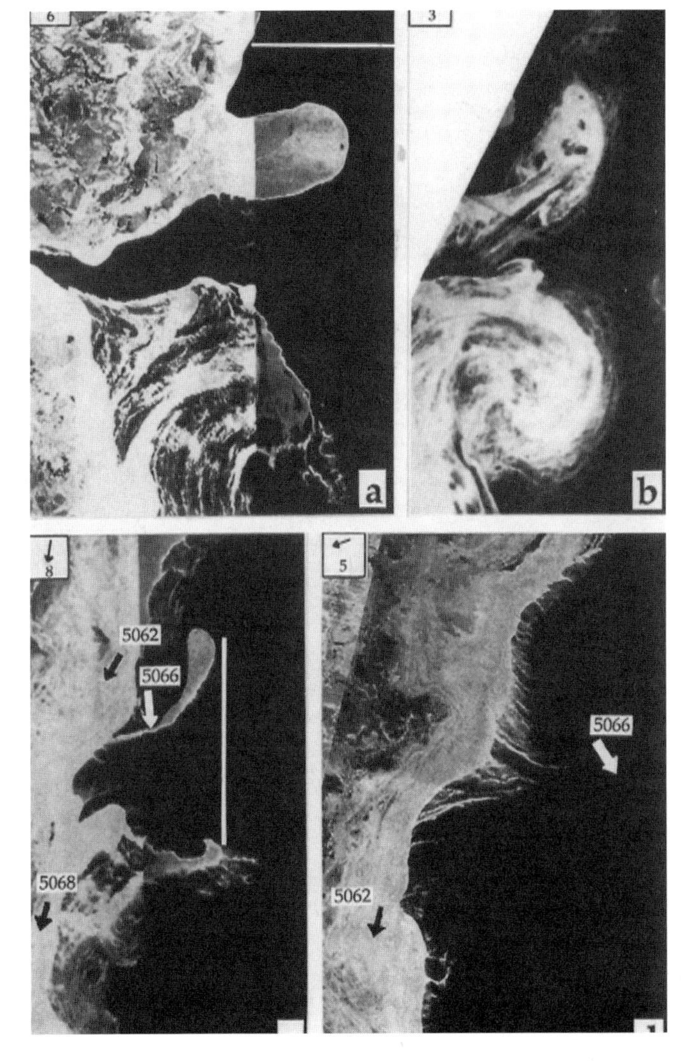

Fig. 6. Zoomed SAR images of the scene 1 area showing growth and decay of two ice tongues and a vortex pair between 78° and 79° N. The images were obtained on (a) March 28 0300 Z, (b) March 29 0100 Z; (c) March 30 1100 Z and (d) March 31 1400 Z. White/grey is ice while black is open water. White lines in (a) and (c) mark the location of CTD and Seasoar sections. At the time of each SAR image the buoy drift is indicated by bold arrows and the wind (m/s) is inserted in the upper left corner.

On March 29 several Argos buoys were deployed in the ice edge region. Buoy (5066) was dropped in open water to the north of the cyclonic center, while another (5068) was deployed on a floe south of the cyclone´s center (Figure 7). The trajectory of 5066 indicated that it was caught in the cyclonic eddy; first it drifted westwards and then towards south along the main ice edge, before it turned southeastward into open ocean. Buoy 5068 drifted towards southwest and into the main ice pack. A third buoy, 5062, deployed north of the eddy, remained inside the ice edge throughout the experiment. The Argos buoys and the SAR images showed that the core of the EGC as well as the mean ice edge was located along the slope of the continental shelf.

The zonal velocity profile (from the Argos buoys) across the EGC had a maximum of 0.40 - 0.60 m/s at the ice edge and decreased to about 0.20 m/s further into the ice. In the open ocean the velocity field was dominated by eddies and the mean southward current was weaker (below 0.10 m/s).

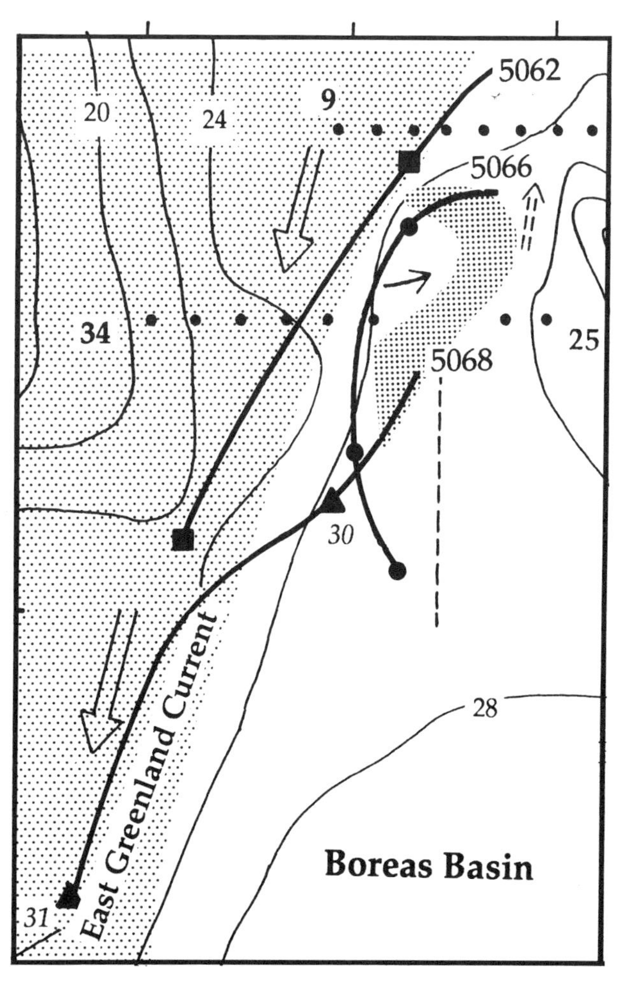

Fig. 7. Bathymetric chart of scene 1 area, the northern Boreas Basin including the East Greenland Current (bold arrows) and the mean ice-covered area (shaded area). The shape of the cyclonic ice tongue (shaded area off the ice edge) was observed from helicopter 8 hours after the image in Fig. 6 b. Small dots indicate two deep CTD sections oriented in east-west direction, and the dashed line mark the north-south Seasoar section. The vertical temperature structure in these sections are shown in Fig. 8. Bold lines are the trajectories of buoys 5062, 5066 and 5068 from March 29 to 31. A northerly current (dashed bold arrow) east of the ice tongue is derived from geostrophic calculations of the CTD sections. Bottom contours are in hectometer.

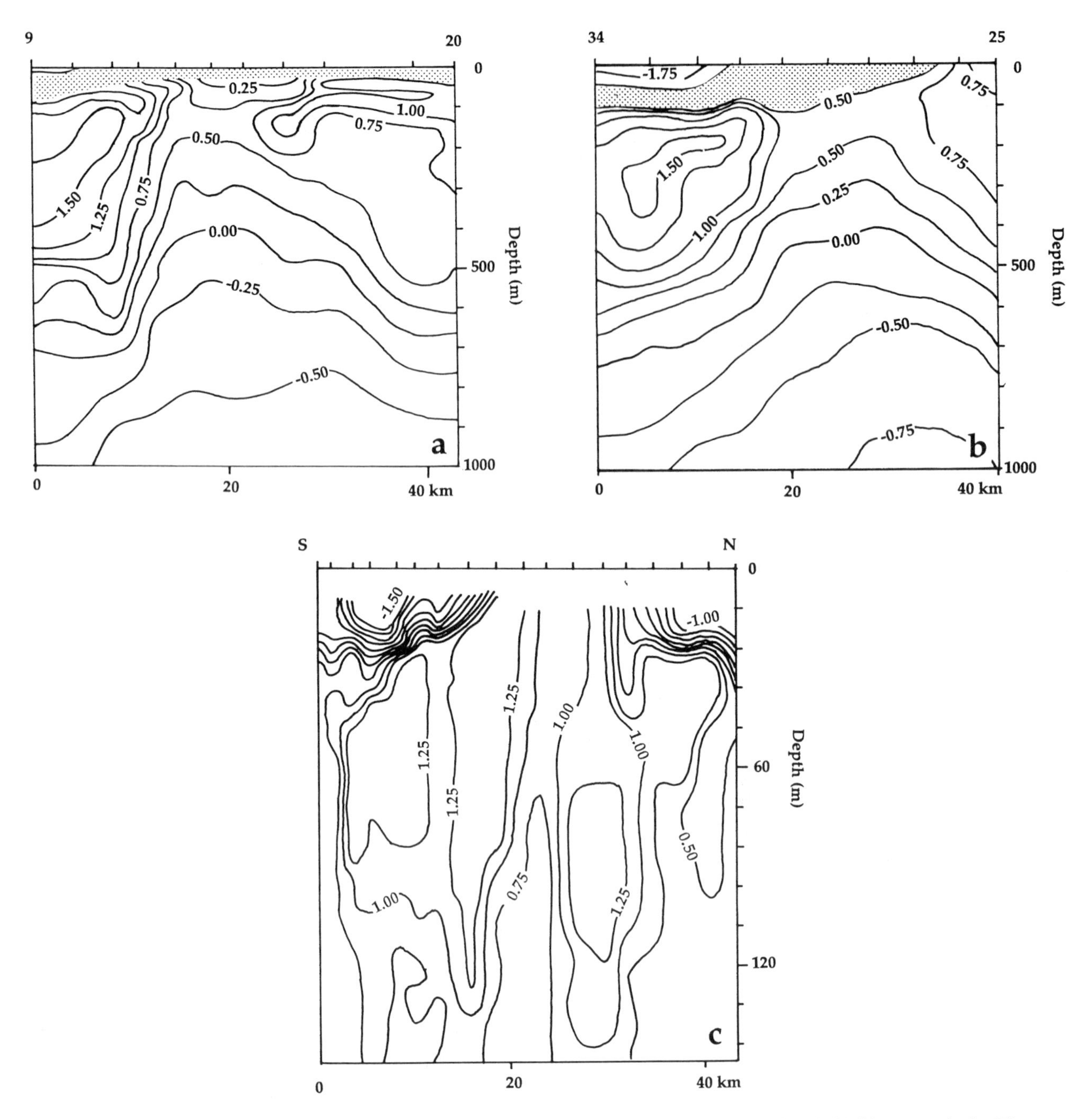

Fig. 8. Vertical sections of temperature obtained by the "Polar Circle" on (a) March 28 and (b) 31, respectively. The thermocline (from -1.5° to 0.5° C) is shaded. The Seasoar temperature section (c) was obtained by the "Håkon Mosby" on March 30 across the front of two ice tongues. Station spacing is approximately 2 km. The location of the three sections is shown in Fig. 7.

On March 30, the third SAR image (Figure 6 c) showed that the vortex pair had been reduced to a smaller tongue of ice which was oriented towards the north. The surface signature of the cyclone, which was clearly seen the previous day, was about to collapse due to an increasing northeasterly wind of 10 - 12 m/s. Also, the anticyclonic tongue was diminishing. The fourth SAR image obtained on March 31 showed no sign of any ice tongues. After a period of easterly winds 6 - 12 m/s the remnants of the tongues had been pushed back into the main ice edge, indicating that the wind was the dominant ice forcing mechanism on this day, thus masking the ocean circulation. Melting could also play a role to diminish the ice tongue.

Based on the real-time SAR images downlinked to the "Polar Circle" a CTD section was obtained across the ice edge (Figure 7 and 8 a), while the northern ice tongue developed into a vortex-pair. The EGC is seen in the western part of the section as a sloping boundary separating the relatively cold and fresh surface water from the warmer and more saline Atlantic Water between 50 and 300 m. In the center of the section deep water masses dome towards the surface surrounded by Atlantic water. In the western part of the section the southerly geostrophic velocity relative to 1000 m is up to 0.20 m/s, while it is 0.05 - 0.10 m/s to the north in the eastern part. These observations confirm the presence of a cyclonic eddy with a diameter of 30 - 40 km extending at least down to 1000 m. A second east-west CTD section, 15 km to the south, was obtained by the "Polar Circle" three days later, on March 31, (Figure 8 b) and showed the same characteristics as the first section. The trajectory of buoy 5066, shown in Figure 7, also indicated the cyclonic circulation.

On March 30 the research vessel "Håkon Mosby" obtained a Seasoar section in front of both the ice tongues while they were in the decaying phase. This section showed that the ice tongues were associated with a 20 m thick surface layer of cold Polar Water on top of a core of Atlantic Water between 30 and 120 m (Figure 8 c). This water mass structure is characteristic for the core of the EGC. As this water propagates to the east the surface layer becomes shallower and the isotherms rise to the surface.

2.2. *Scene 2: Ice tongue and vortex-pair at 78° 00' N*

The second occurence of ice tongues was observed from March 30 to April 2. The growth of this tongue is illustrated schematically in a composite of the ice edge from four SAR images (Figure 9). The tongue started on March 30 as a meander of the main ice edge at 78°00' N. The day after a well-defined tongue had formed and moved about 10 km towards the east. In the following two days the tongue was advected southwards and changed its orientation due to the zonal velocity shear across the EGC. In spite of an on-ice wind of up to 12 m/s the tongue grew eastwards and retained its structure. Examination of the 31 March SAR image (Figure 10 a) clearly shows an ice distribution that traces the velocity structure within the tongue.

One of the Argos buoys (5068) was drifting in the tongue, providing quantitative information on the ice velocity. As the buoy approached the center of the tongue on its northern side it drifted east-southeast at 0.30 - 0.40 m/s. Instead of being advected further east and out into open water it suddenly turned southwest and increased its speed to 0.70 - 0.80 m/s in the southern part of the tongue (Figure 11). During the next day the tongue was stretched

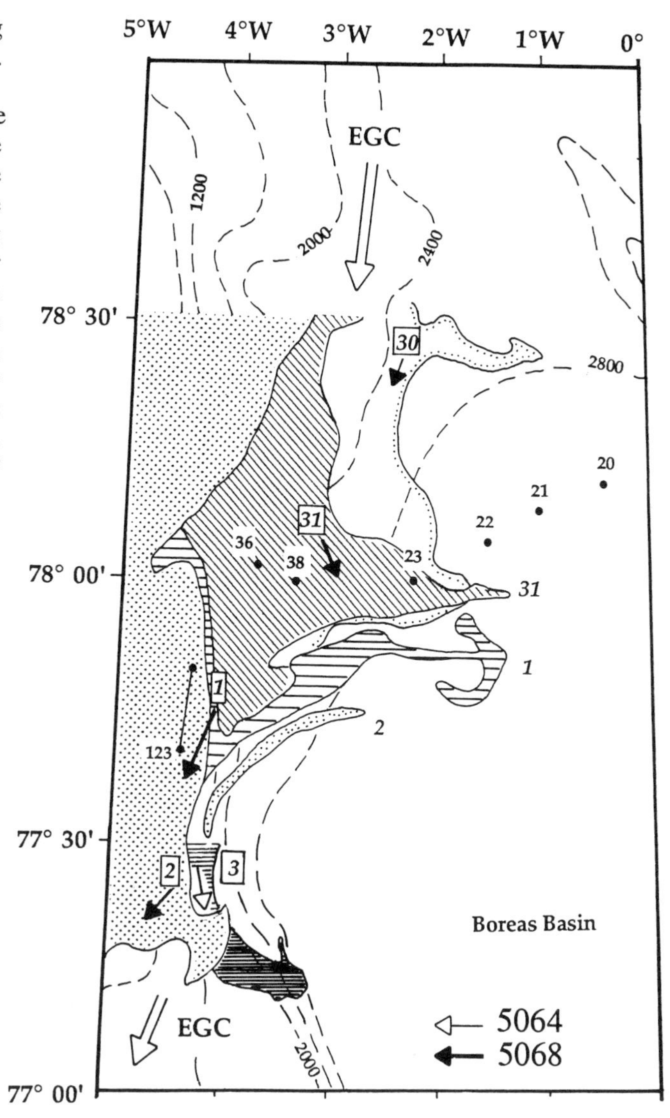

Fig. 9. Composite map of ice edge from daily SAR images between March 30 and April 3 superimposed on a bathymetric chart. Dates are shown in italics and black dots indicate CTD stations. Thin arrows mark position and drift velocity of ice buoys at the time of the SAR images. Bold arrows indicate the core of the East Greenland Current (EGC).

as it was advected southwards. In front of the tongue a vortex pair was formed with a horizontal scale of 20 km (Figure 10 b). Both the cyclonic and anticyclonic parts of the vortex pair were clearly developed. One day later the surface signature of the vortex pair had vanished. The lifetime of the tongue is estimated to be 2 - 3 days and the length was up to 50 km. The two ships attempted to characterize the subsurface structure of the water below the ice tongue but did not succeed due to the rapid growth and decay of the tongue.

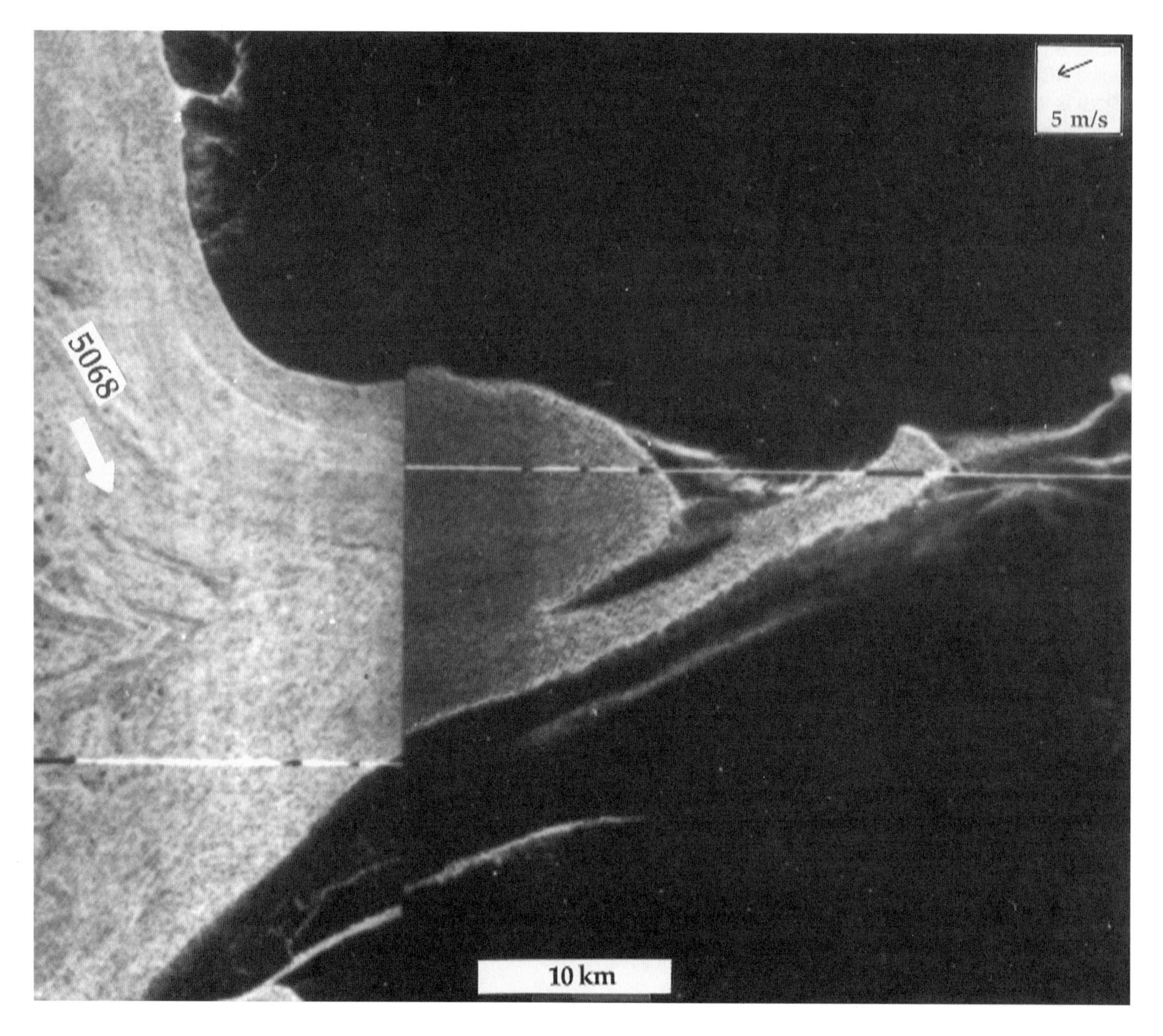

Fig. 10. (a) A zoomed SAR image with resolution of 15 m obtained on March 31 at 78° 00' N where a tongue of ice is evolving. The bold white arrow indicates drift of buoy 5068. (b) SAR image from the same area obtained one day later (April 1). The isecond mage shows how the tongue is stretched and the development of a vortex-pair.

2.3. *Scene 3: Southerly jet and anticyclonic eddy.*

In the SAR image of April 2, a fan-shaped ice tongue is seen to develop in a southeasterly direction normal to the ice edge at 77° 20' N 4° W (Figure 12 a). The tongue is seen as scattered ice advected out from the main ice edge. The wind was northerly 6 m/s. During the next day, when the wind was 8 m/s from northwest, a well-developed tongue of more compact ice was formed and advected out from the main ice edge in a southeasterly direction (Figure 12 b). The SAR image also shows filaments of ice which extend out from the ice tongue as well as stripes of lower concentration inside the tongue, having the form of a vortex pair at the other end of the tongue. The width of the main tongue is about 10 km and the length is approximately 20 - 25 km at the time when the SAR image was obtained. In contrast to the tongues in scene 1 and 2, which propagated against the wind, this jet developed in the same direction as the wind. The front of the tongue was advected about 20 km in one day to the southeast (Figure 9).

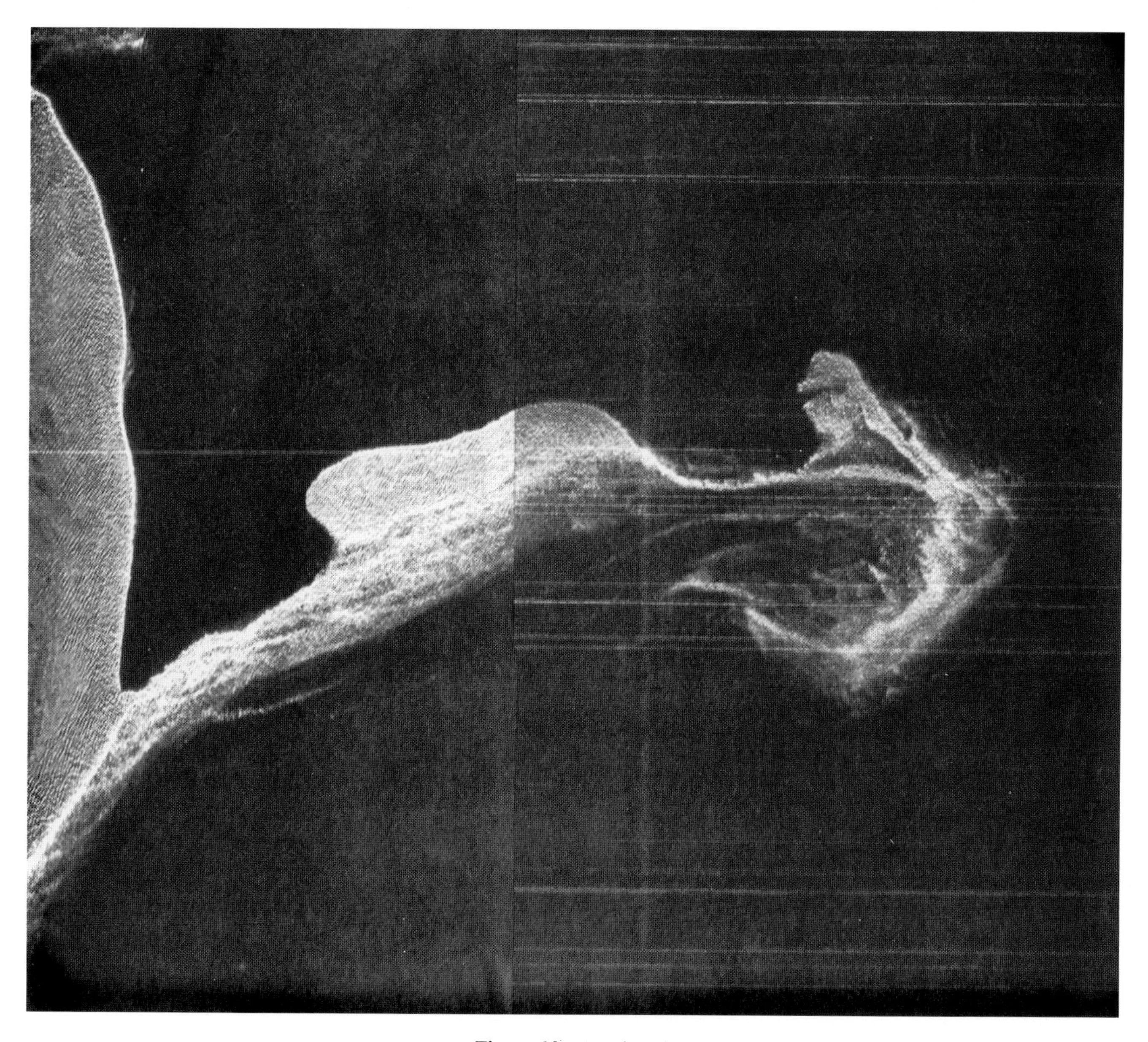

Figure 10. (continued)

The drifting buoys showed that this tongue was associated with a strong oceanic jet parallel to the isobaths. On April 4 the easternmost buoys (5064 was one of them) were caught by the jet and advected southeastward out into open water and had to be recovered in the afternoon (Figure 13). Buoy 5063, which was initially located only two miles further into the ice compared to 5064, was initially caught in the jet but was detached from it towards the end of the day. Buoy 5068 was located 10 - 15 km into the ice pack and drifted southwards at a slower speed, about 0.20 m/s. The surface speed of the jet, as measured by the drift velocity of 5064, exceeded 1.0 m/s and the horizontal shear between 5064 and 5063 was as strong as 10^{-4} s^{-1}, on the same order as the Coriolis parameter.

Mean vertical profiles of velocity and temperature of the jet as well as the core of the EGC, estimated for April 4, are shown in Figure 14. The velocity profile of buoy 5063, which followed the core of the EGC, had a maximum of 0.60 m/s at the surface and decreased to about 0.30 m/s at 250 m. The jet, which branched off towards the southeast along the isobaths, had a similar profile but with 0.30 - 0.40 m/s higher values in the upper 250 m of the water column. The temperature profiles were similar at the two buoys, with cold polar water in the surface layer and warmer

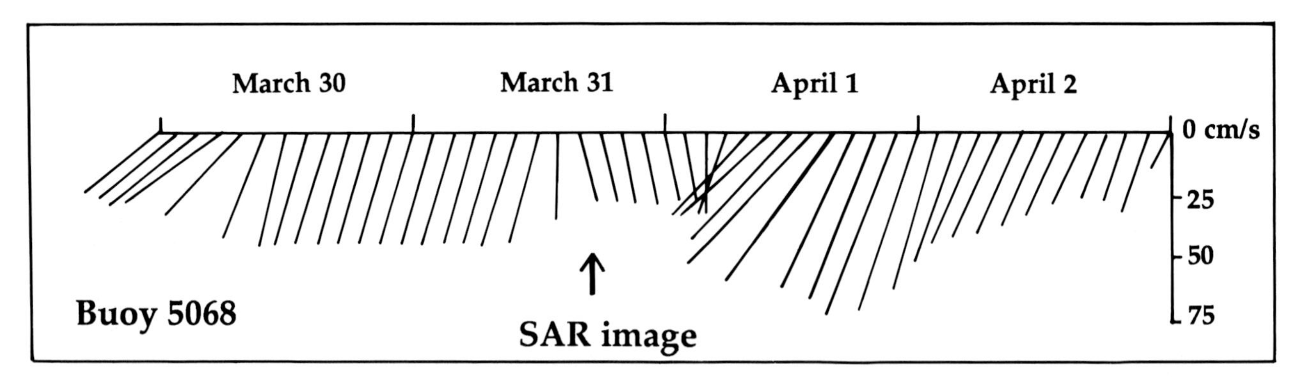

Fig. 11. Velocity vectors of buoy 5068 from March 30 to April 2.

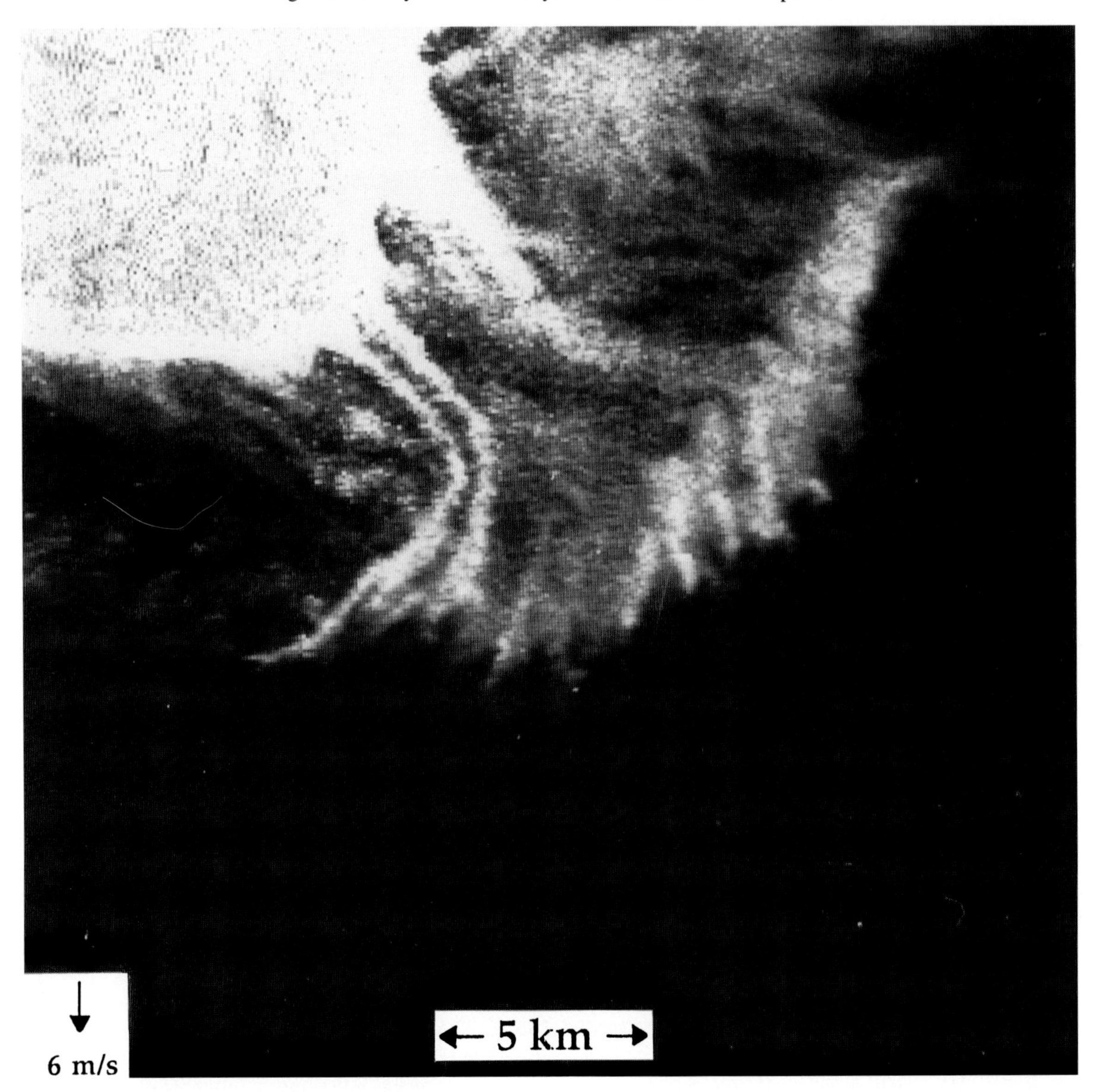

Fig. 12. SAR images showing the evolution of a south-easterly ice tongue on (a) April 2 and (b) April 3. The location of the tongue is shown in Fig. 9 at about 77° 15' N.

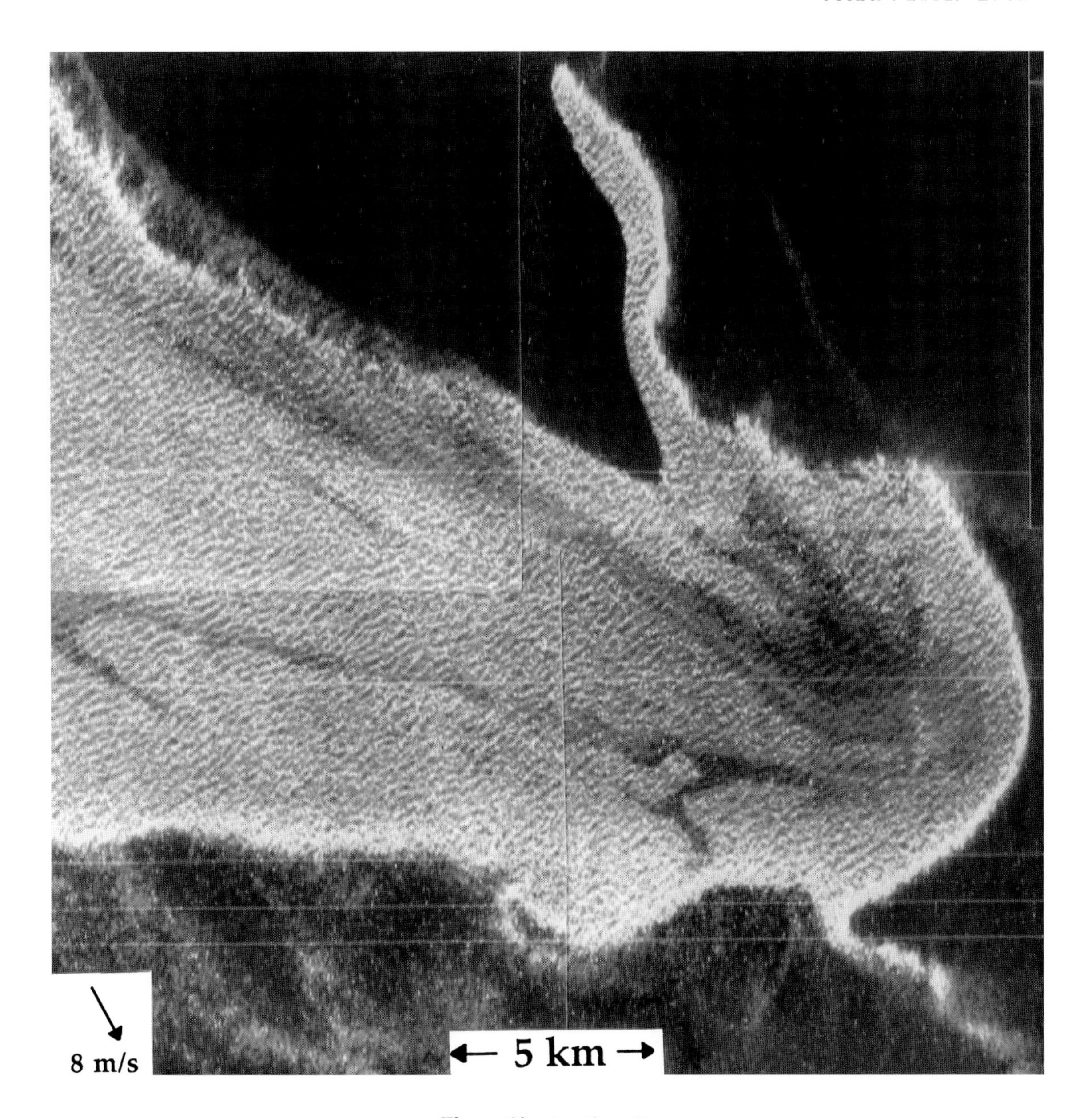

Figure 12. (continued)

Atlantic Water below 150 m. Characteristic density profiles from CTD data are used to estimate a stratification parameter to be used in the two-layer model described in section 3. Upper and lower layer density of 1027.2 kgm^{-3} and 1027.9 kgm^{-3}, respectively, yield a reduced gravity $g' = 0.007\ ms^{-2}$. Similarly, upper and lower layer velocity values to be used in the model were obtained from the current meter data.

In contrast to the upper layer short-lived ice tongues described in scene 1 and 2, this jet penetrated deeper than 250 m (Figure 14) and propagated along the isobaths which are directed to the southeast in this area because of the Greenland Fracture Zone. A likely explanation is that this topographic feature can branch off a part of the EGC which is entrained in gyre scale cyclonic circulation outside the ice edge in the Boreas Basin [*Quadfasel and Meincke*, 1987; *MIZEX Group*, 1989]. The observed jet could therefore be an indication of this circulation along the pronounced bottom topography of Greenland Fracture Zone.

On April 4 the ice jets had developed into a large anticyclonic eddy near the Greenland Fracture Zone at 77° N (Figure 13). As the array of buoys drifted through the anticyclone they documented the velocity of the ice as well as the upper ocean current. The ice and upper ocean velocity was 0.50 m/s, decreasing to about 0.30 m/s at 250

Fig. 13. The displacement of three buoys (5068 west, 5063 center, and 5064 east) from April 3 1800 Z to April 4 1800 Z superimposed on the SAR image of April 4. The length of the velocity vector of 5064 is 50 km.

m. A composite of SAR ice edge positions between April 4 and 6 combined with ice velocity vectors from the buoys shows the surface characteristics of the anticyclone (Figure 15). A deep CTD section was successfully obtained through the center of the eddy by using real time SAR images (Figure 16). The section documented the anticyclonic rotation by a characteristic downward doming of the isopycnals in the eddy center. In addition to the surface ice transport, a dominant feature was a core of Atlantic water circulating at a radius of 20 km from the eddy center at a depth between 200 and 400 m. The geostrophic velocity estimates showed that the southward flow in the upper 100 m was above 0.40 m/s, in good agreement with the buoy velocities. From this eddy, which had a diameter of about 50 km, several smaller jets and vortex-pairs were pinched off (Figure 4 a, b). The interaction of the EGC with the abrupt topographic change at 77° N is a probable generating mechanism for this anticyclone.

2.4. *Other observations*

The spatial distribution of ice tongues over a larger area was analyzed using an AVHRR satellite image of July 1 1984. The image shows a cloudfree ice edge from about 75° to 79° N with at least eight ice tongues or eddy features (Plate 1). The ice tongue at about 78° (# 3) is remarkable

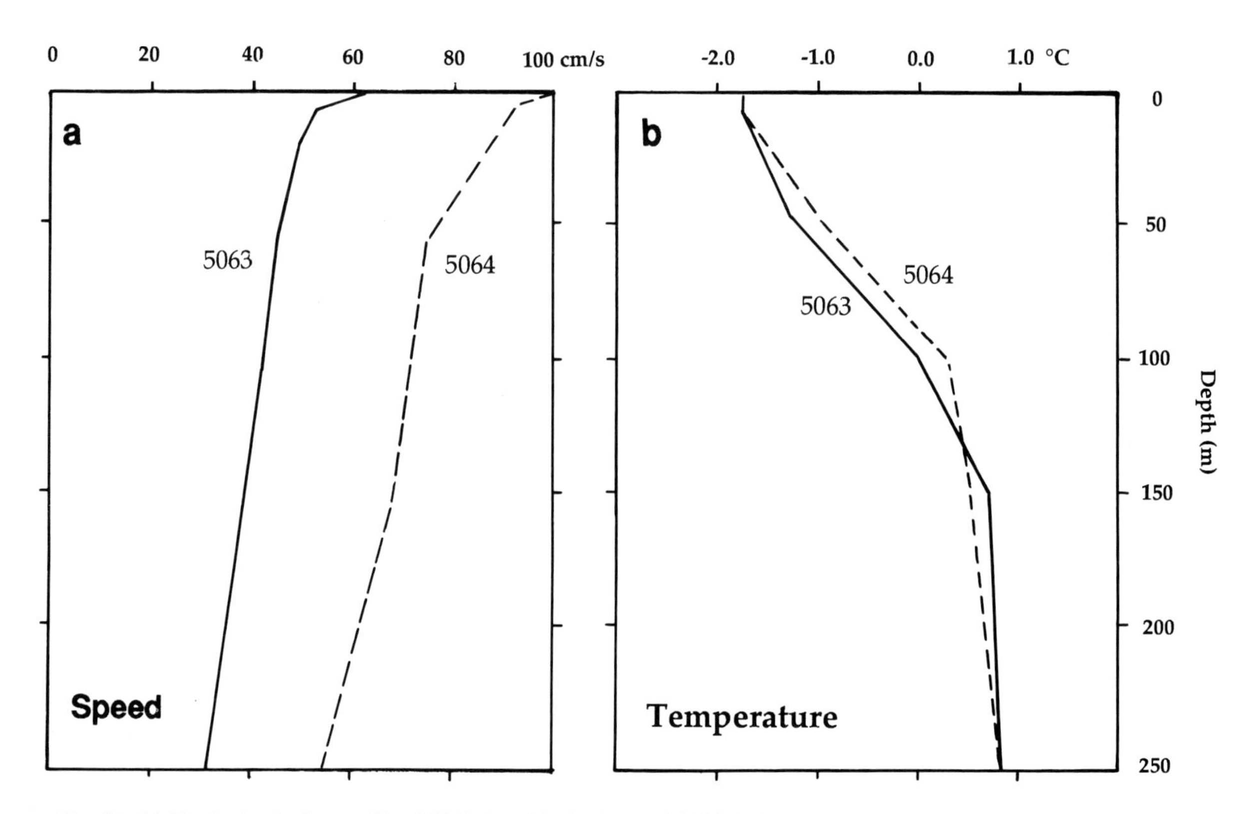

Fig. 14. (a) Vertical velocity profile of 5063 (outside the jet) and 5064 (inside the jet) on April 4. (b) Vertical profile of temperature from the same buoys.

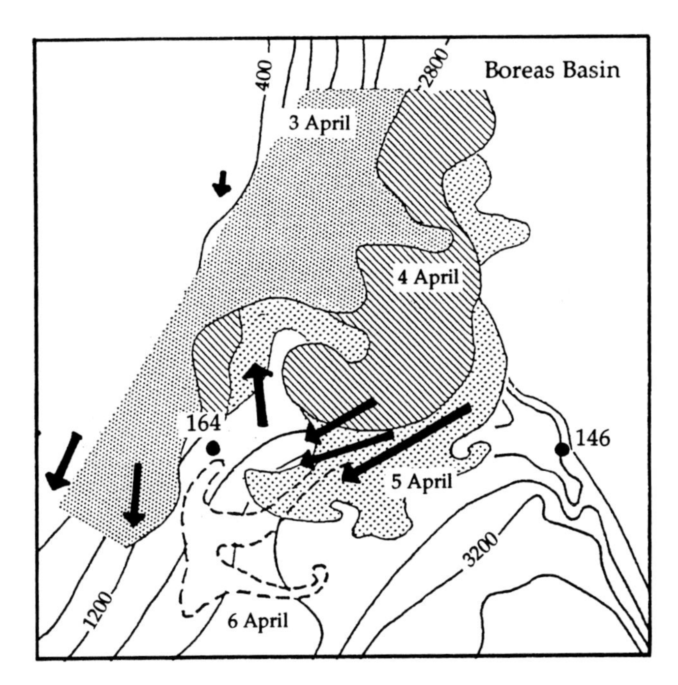

Fig. 15. A composite of the ice edge obtained from daily SAR images between April 3 and 6 superimposed on the bottom topography. Included are also ice velocity vectors from drifting buoys (bold arrows). The two bold dots (146 and 164) mark the end points of the CTD section taken across the anticyclone.

similar to Figure 10 a. At 76° 30' N a tongue of cold water and ice (# 5) is advected to the east, probably caused by the topography, which was disussed in section 2.3. Further south two tongues of ice (# 6 and 7) grow out in open water and develop into vortex pairs. The horizontal scale of all these eddy features is 30 - 50 km, similar to the observations in 1987. The image also shows a characteristic separation between each ice tongue of 50 km.

In January 1992 SAR images from the ERS-1 satellite were obtained during another field experiment by the "Håkon Mosby" in the same area as MIZEX 87. These SAR images covered 600 km long and 100 km wide swaths in the ice edge region between 75° and 80° N (Plate 2, a). On January 13 an anticyclone was observed in same location as in MIZEX 87 (77° N 4° W). The anticyclone had a vortex-pair directed in a southwesterly direction. In addition the ERS-1 SAR image showed several tongues of ice extending out from the main ice edge. Since the ERS-1 had a repeat cycle of three days, daily mapping of the growth and decay of the tongues could not be obtained. However, a 300 km long Seasoar section, measuring ocean temperature, salinity and density in the upper 100 m was taken parallel to the ice edge (Plate 2 b, c, and d). This section ran across several cold plumes of water associated with tongues of ice extending out from the ice edge. The section showed a complex horizontal structure with typical scales of 10 km or less. The plumes of cold low salinity water associated with the ice tongues were deeper than 100 m. The data from 1992 confirm several of the features observed in the MIZEX 87 experiment, demonstrating that eddies, ice tongues and vortex-pairs occur frequently in the area and that the anticyclone at 77° N may be a more permanent feature.

2.5. Summary of observations

The interpretation of the ice velocity field from SAR or other images (AVHRR) is not a straightforward procedure. The problem is to separate the Lagrangian from the Eulerian description of the velocity field. Therefore, ice features seen in the SAR images cannot be used to estimate velocities directly. Another problem is that the ice may not be a passive tracer which drifts freely. It can have its own dynamics, in contrast to the filaments off California and dyed streamers in tank experiments. With this in mind the observations of eddy features in the marginal ice zone of the Greenland Sea can be analyzed.

A typical length scale of the ice tongues and vortex pairs along the ice edge is 30 km. The lifetime is more difficult to estimate, but the ice signature from a time series of SAR images combined with CTD data suggests that in the upper structure (< 30 m depth) it can be as short as 3 - 4 days, while the deeper structures last much longer (> 10 days). The SAR images showed several cases where a filament of ice shoots off the ice edge one day and grows into a fully developed vortex pair the next day. One or two days later it has degenerated. This apparent degeneration may be due to ice which diffuses and melts or ice advection by the wind. Thus the surface ice signature may disappear and obscure the subsurface structure of the circulation which can persist for a longer period. In situ data from the ships and the drifting buoys were therefore of vital importance to supplement the SAR data in the interpretation of the circulation.

The background circulation for the ice tongues and vortex pair observations described in this paper is a boundary current flowing along a continental shelf break as well as a front separating colder and fresher Polar Water from warmer and more saline water masses of Atlantic origin. In addition, there is an ice edge separating open water from pack ice which has its own dynamics. Boundary currents and frontal jets, as described in section 2.3, are deeper and more permanent than the ice tongues extending out from the ice edge. These tongues or filaments, often with a vortex-pair in the front, are shallow (30 - 100 m) and have a lifetime of a few days. It is hypothesized that local dynamic instability of synoptic scale motions such as meanders, eddies and frontal jets plays an important role in their formation.

3. NUMERICAL MODEL

In this section, a model for describing the generation of ice eddies, tongues and vortex pairs through mixed

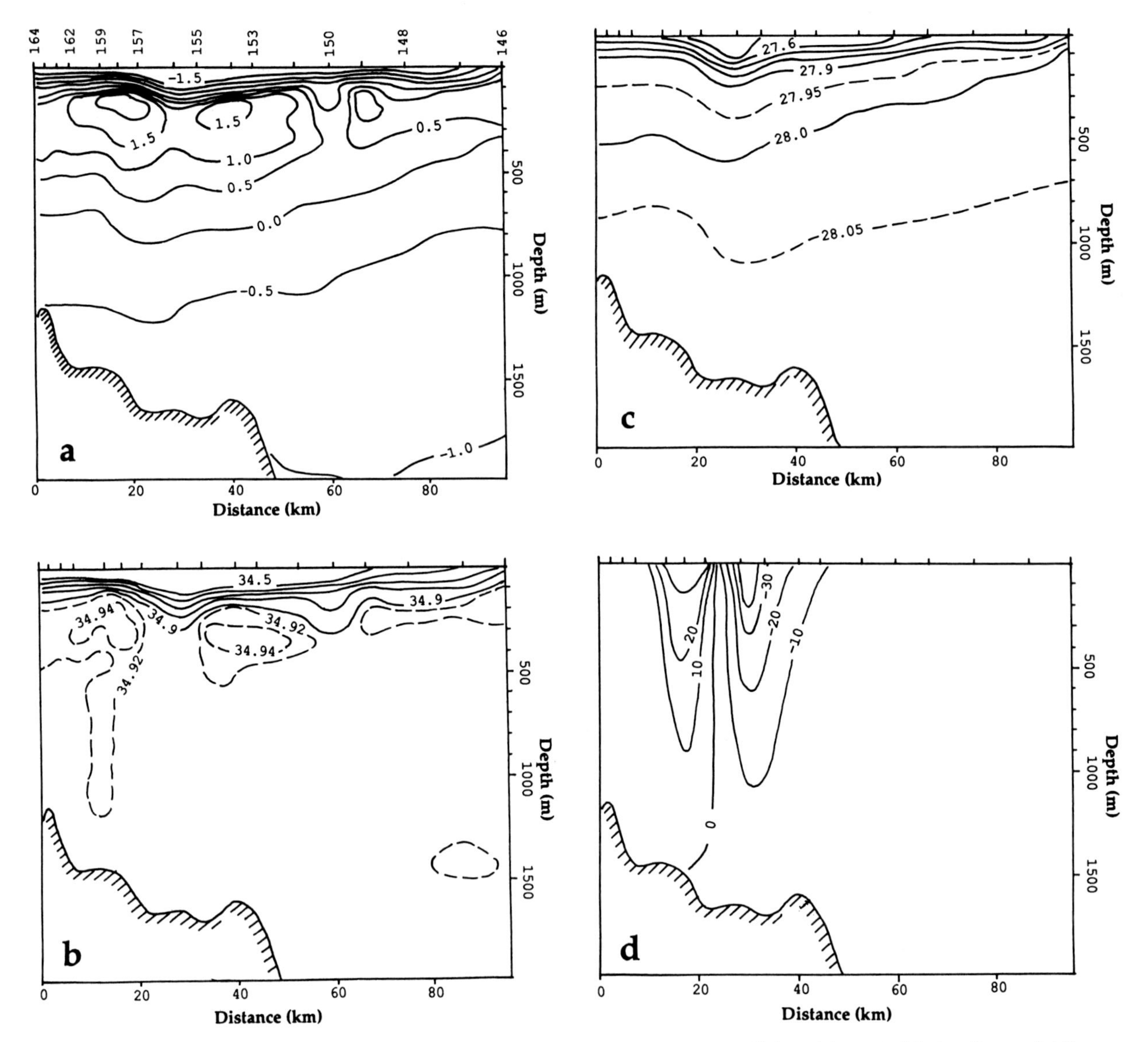

Fig. 16. CTD section across the anticyclone: (a) potential temperature, (b) salinity, (c) potential density, and (d) geostrophic velocity.

barotropic-baroclinic instability is defined. In the study, a two-layer primitive equation representation of the ocean is employed with typical parameters derived from the MIZEX 87 experiment.

3.1. Governing Equations

The governing equations for the model consist of statements of momentum and mass conservation for a two-layer ocean:

$$\frac{\partial \mathbf{V}_i}{\partial t} + \left(\nabla \cdot \mathbf{V}_i + \mathbf{V}_i \cdot \nabla\right)\mathbf{v}_i + f \times \mathbf{V}_i = -h_i \nabla P_i - A_{bh} \nabla^4 \mathbf{V}_i \qquad (1)$$

$$\frac{\partial h_i}{\partial t} + \nabla \cdot \mathbf{V}_i = 0 \qquad (2)$$

The symbols are defined in Table 1. Sea ice is treated as a passive tracer in this study, and is represented by Lagrangian particles drifting on top of the ocean surface. This means that no ice dynamics are used implying that wind forcing as well as internal ice stress are absent. The only mechanism for ice motion is the surface velocity of the ocean. The neglect of wind forcing is normally not realistic, but can be justfied in these experiments because the wind conditions during the field experiment were moderate and most of the mesoscale activity was caused by the ocean dynamics. The neglect of internal ice stress is assumed to

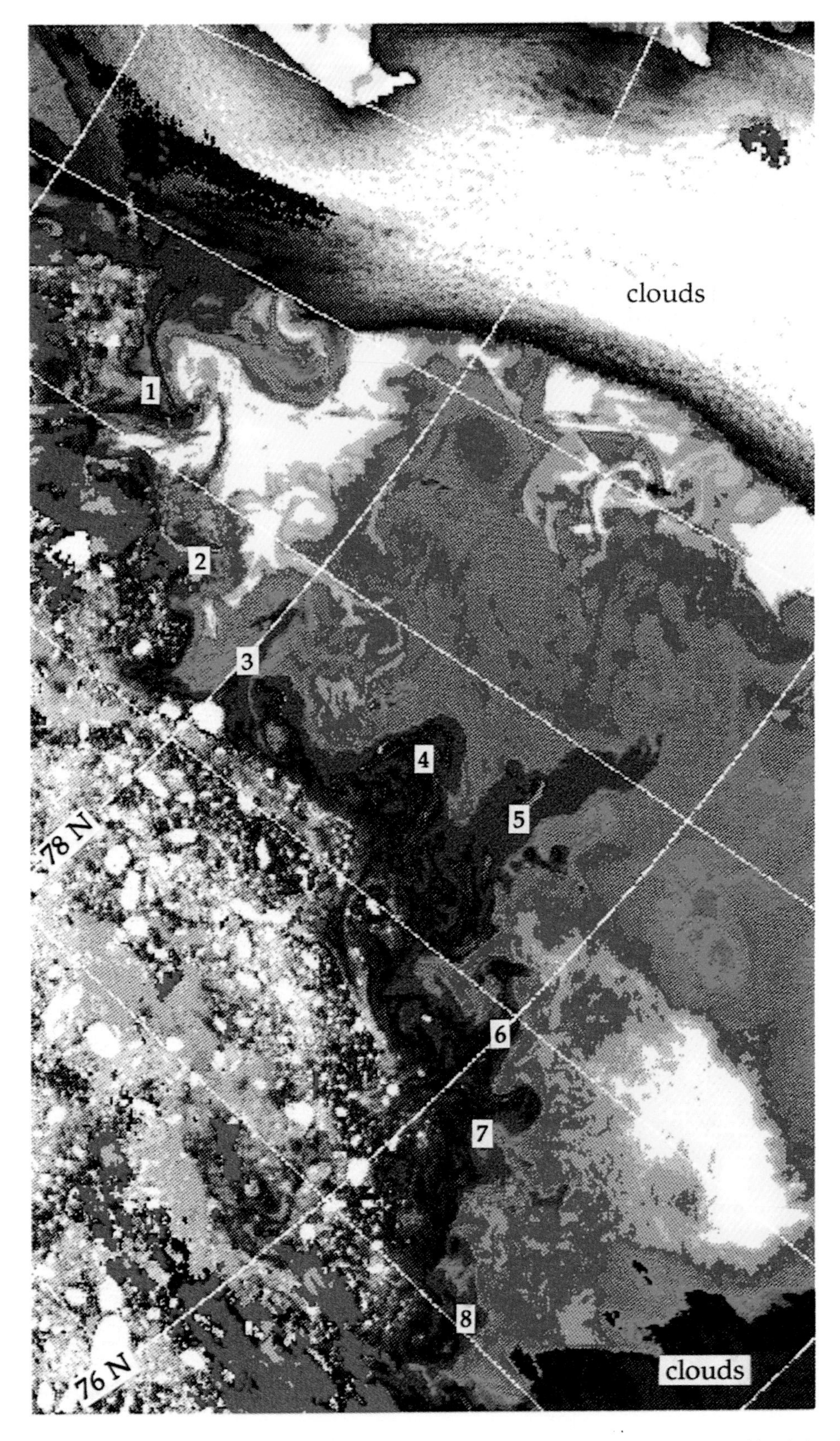

Plate 1. NOAA AVHRR image from July 1 1984 during the MIZEX 84 experiment. The image combined visual (channel 2) data over ice with IR data (channel 4) in open water. The dark blue indicates cold water along the ice edge, while light blue, red and yellow shows warmer water (> 2°C). The numbers indicate ice tongues similar to the SAR observations in MIZEX 87.

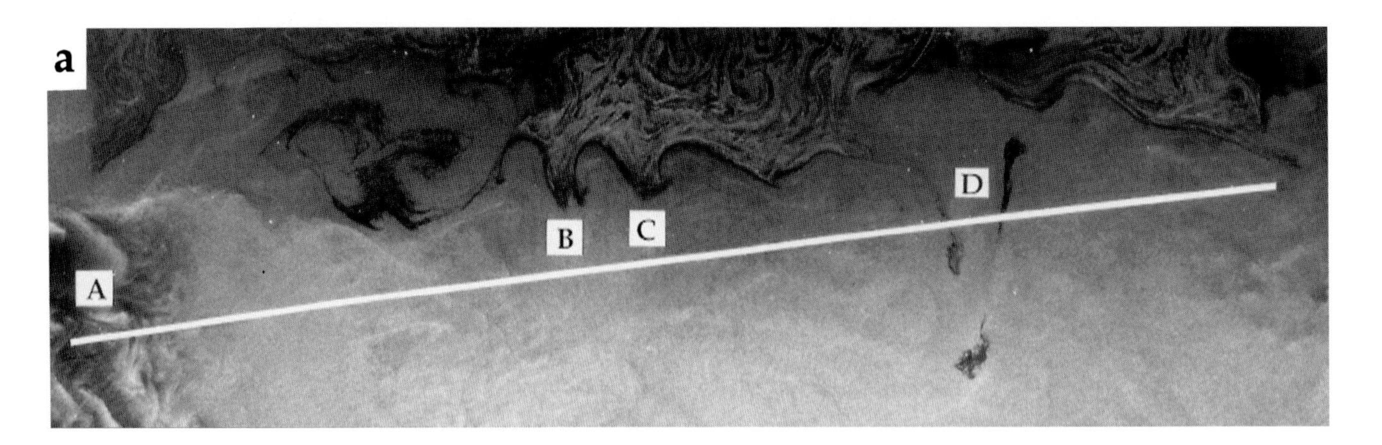
a
A
B
C
D

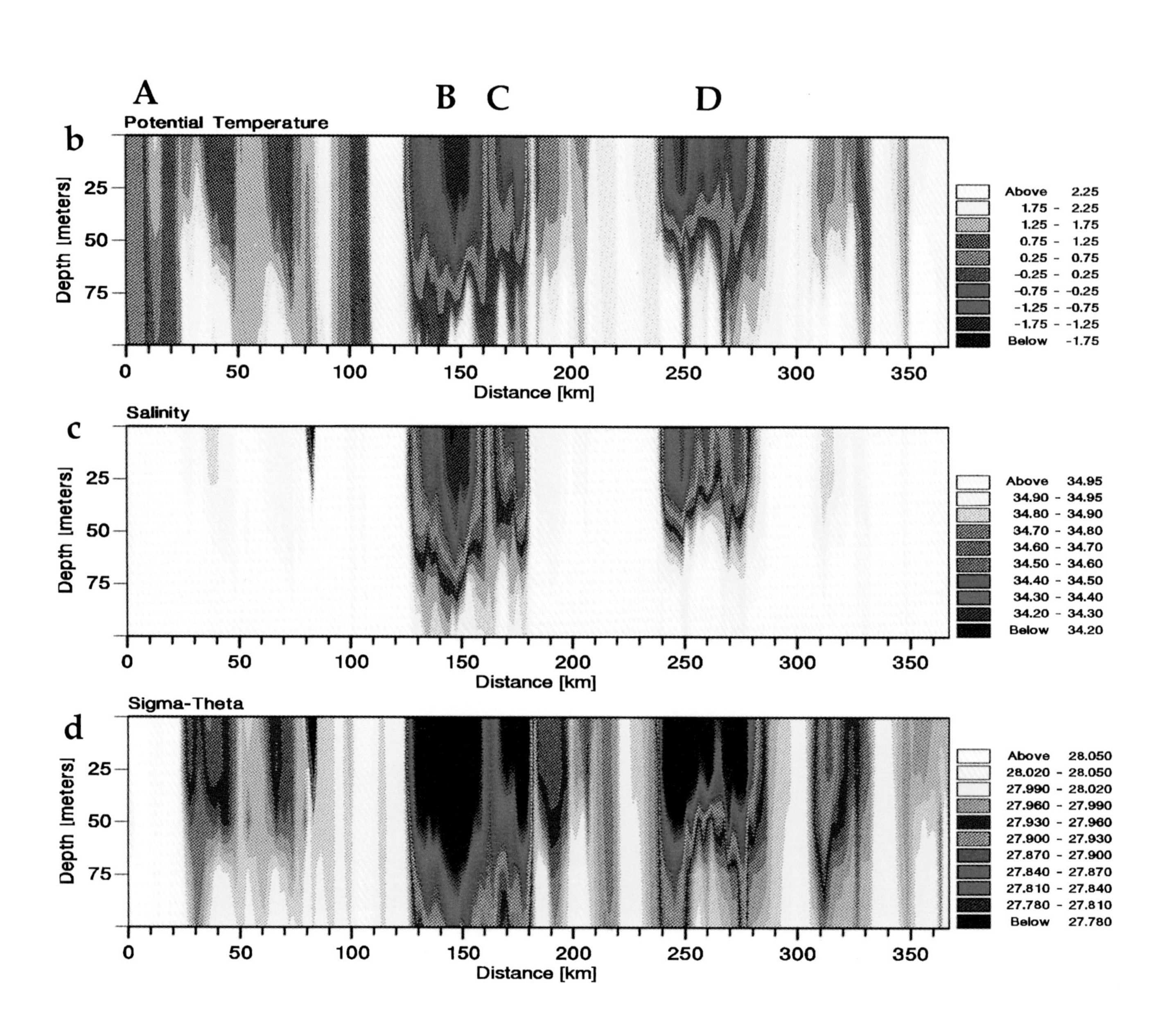
A
B
C
D
b
Potential Temperature
Depth [meters]
25
50
75
0
50
100
150
200
250
300
350
Distance [km]
Above 2.25
1.75 - 2.25
1.25 - 1.75
0.75 - 1.25
0.25 - 0.75
-0.25 - 0.25
-0.75 - -0.25
-1.25 - -0.75
-1.75 - -1.25
Below -1.75
c
Salinity
Above 34.95
34.90 - 34.95
34.80 - 34.90
34.70 - 34.80
34.60 - 34.70
34.50 - 34.60
34.40 - 34.50
34.30 - 34.40
34.20 - 34.30
Below 34.20
d
Sigma-Theta
Above 28.050
28.020 - 28.050
27.990 - 28.020
27.960 - 27.990
27.930 - 27.960
27.900 - 27.930
27.870 - 27.900
27.840 - 27.870
27.810 - 27.840
27.780 - 27.810
Below 27.780

TABLE 1. List of symbols

$\mathbf{V}_i$	Transport in the ocean layer i
$\mathbf{v}_i$	Velocity in the ocean layer i
$\mathbf{f}$	$\mathbf{f} = f_0\mathbf{k}$, $f_0 = 1.427 \times 10^{-4}\ s^{-1}$, Coriolis parameter
h_i	Thickness of ocean layer i
P_i	Pressure normalized by density in ocean layer i $P_i = g\eta - \delta_{i2} g'H_1$
A_{bh}	Biharmonic friction factor, $A_{bh} = 1 \times 10^7\ m^4s^{-1}$
$\mathbf{X}$	Particle position
$\mathbf{v}_{1L}$	Lagrangian velocity in upper layer

be a valid approximation outside the main ice edge where the eddies and tongues were observed. However, further inside the ice edge internal ice stress must be invoked in realistic simulations.

3.2. Boundary and Initial Conditions

The model domain is a 128 by 108 km channel with periodic boundary conditions in the along-channel direction and no-slip side walls. The model schematization is as shown in Figure 17. The model bottom topography is a simplification of the East Greenland continental shelf break between 78° N and 79° N (we are not yet including the Greenland Fracture Zone). The topography is constant in the along-channel direction with flat-bottomed shelf and abyssal zones of depth 400 m and 2600 m, respectively. The two zones are connected by a slope of 0.11. The initial particle distribution is uniform over the ice covered area (the western half of the model area) with a density of 16 particles per km^2 for a total of over 54000 particles in a 27 km wide strip across the model domain. To simulate the core of the EGC the ocean is initialized with a Gaussian jet with a pressure field of the form

$$P_i = \left(\alpha_1 g\eta - \delta_{i2}\alpha_2 g' H_1\right)\exp\left(-\frac{y^2}{2L^2}\right) \tag{3}$$

where α_1 and α_2 are constants defining the strength of the upper and lower layer jets, respectively, and L is the e-folding scale of the jet. The ocean velocity field is initially in geostrophic balance. The upper and lower layer jets are displaced from their axes with random perturbations. The jet displacements have an along-channel wavenumber spectrum of the form exp($-\beta\kappa$), where $\beta = 5000/(2\pi)$ and κ is the wavenumber.

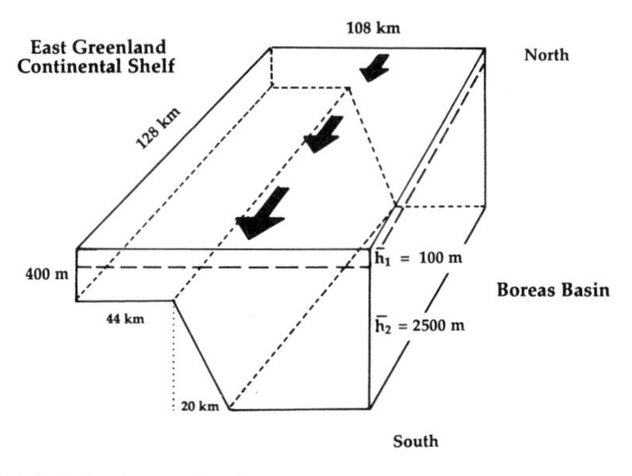

Fig. 17. Model schematization.

3.3. Numerical Scheme

A semi-implicit numerical scheme as described by *Hurlburt and Thompson* [1980] and *Smith and O'Brien* [1983] is used to perform the time integration of the primitive equations ocean model. An Arakawa "C" grid is used for the spatial discretization of the model. A time step of 300 s and a grid size of 1000 m between like variables are used in the study. The particle trajectories are computed by integrating the Lagrangian equation

$$\frac{d\mathbf{X}}{dt} = \mathbf{v}_{iL}(\mathbf{x}, \mathbf{t}) \tag{4}$$

where $\mathbf{X}$ is the particle position and $\mathbf{v}_{iL}$ is the Lagrangian velocity, using a modified midpoint method [*Press et al.*, 1986] with four subintervals in each 300 s time step. Velocities were obtained through bilinear interpolation in space and linear interpolation in time.

4. NUMERICAL EXPERIMENTS

A series of numerical experiments were conducted with the eddy-resolving ocean model to examine the role of mixed barotropic/baroclinic instability in the formation of the observed ice edge eddies, tongues and vortex pairs described in section 2. The experiments are designed to determine the influence of jet width and intensity, topography and baroclinicity upon the generation and structure of the resulting ice tongues and vortex-pairs. The experiments conducted are summarized in Table 2. This series of experiments represents the first systematic study of ice tongues and vortex-pairs in the EGC.

Plate 2. (a) 100 km wide SAR image from the ERS-1 satellite obtained in January 1992 in the same area as MIZEX 87. The white line shows the location of a Seasoar section obtained by "Håkon Mosby" at the same time as the SAR image. Vertical structure of (b) temperature, (c) salinity and (d) density from the Seasoar section are shown in the upper 100 m.

TABLE 2. Summary of numerical experiments

Experiment number	Jet amplitude (m/s)	U_2/U_1	Topography	Jet width L (km)	H_1 (m)	H_2 (m)
1	0.4	0.33	yes	5	100	2500
2	0.4	0.33	no	5	100	2500
3	0.4	0	yes	5	100	2500
4	0.4	1	yes	5	100	2500
5	0.4	0.5	no	10	100	2500
6	0.4	0.5	yes	10	100	2500
7	0.4	0	no	5	100	2500
8	0.4	0.33	no	5	100	300

Case 1. Baseline Experiment

The baseline experiment consists of an upper layer jet representing the core of the EGC with amplitude 0.4 m/s situated over a lower layer jet with strength 0.13 m/s. Both jets flow from right to left (from north to south) in the along-channel direction centered over the shelf slope with shallower water on the right (Figure 17). The lateral e-folding scale, L, of both jets is 5 km for a total jet width of approximately 15 km. The upper layer mean thickness is 100 m, while the lower layer mean thickness is 2500 m. The reduced gravity parameter g', is 0.007 ms^{-2} with a resulting internal Rossby radius of deformation of approximately 5 km. These parameters were obtained from the observations described in section 2 and represented the marginal ice zone of the EGC north of about 77° N. The parameters are summarized in Table 2.

The results from the baseline experiment are depicted in Figure 18. In all panels the orientation is north on the right and south on the left. The contour interval for all sea level contour plots is 0.005 m. Negative sea level displacements are denoted by dashed contours. The displacement of the sea level provides a description of the circulation of the upper layer, since the upper layer pressure $p_1 = gh_1$ and the velocity field are nearly in geostrophic balance. By day 5 of the simulation undulations have appeared in the ice edge with a wavelength of 25 km. By day 7, the meanders have intensified and ice tongues have formed. The dominant wavelength has increased to about 32 km. By day 10, the ice tongues have evolved into 30 - 35 km long streamers. The series of graphs in Figure 18 shows how the jet instability process can serve to transport ice into open water. The change in orientation in the ice tongues is caused by shear in the jet, as discussed in section 2.2. The extents and growth rates of the tongues are summarized for each experiment in Table 3.

Case 2. Baseline Experiment with Flat Bottom.

To examine the role of the continental shelf, a simulation is performed with the same parameters as in Case 1 but with a flat bottom at a depth of 2600 m. By day 7 of this experiment, the velocity jet is broader than at the corresponding time in the baseline case where bottom topography was included (Figure 19). Furthermore, the excursions of the ice edge are larger, the ice tongues are broader and the displacements in the ice edge have longer wavelengths than in the baseline experiment. These results are consistent with the absence of the stabilizing effect of bottom slope and could therefore represent a situation where the ice edge is located further east over a flat deep bottom. In Case 1 the cross-isobath movement is suppressed somewhat by the presence of the shelf slope. In the experiment with a flat bottom the filaments are wider and the eddies are larger (30 - 40 km in diameter by day 7), but weaker compared to the baseline experiment.

Case 3. Baseline Experiment with Upper Layer Jet

In this experiment the ocean jet is initially constrained to exist only in the upper layer. Thus, the effects of enhanced baroclinicity can be examined. The stabilizing effect of the topography will also be lessened somewhat. After 9 days, the structure of the ice concentration (Figure 20) is similar to that of the baseline experiment except that the scale of the features has been reduced due to the enhanced baroclinicity. This experiment seems to be the most realistic of the 7 cases because the results are the most similar to the observations. The small vortex-pair evolving at the front of one of the ice filaments is similar to the observation in the SAR image in Figure 10 b. The distance between the simulated ice filaments is about 50 km. A vortex-pair which propagates into the ice (Figure 20 b) has a scale of about 30 km and resembles the vortex-pair seen in the SAR image in Figure 4 a, b. In the simulations it takes typically 5 days to build up instability which then creates tongues and vortex-pairs in 2 - 3 days. This is the same growth rate as was observed in the SAR images.

Case 4. Barotropic Jet

In the first experiment, conducted with a barotropic jet on a flat bottom, no development of instability occurred until

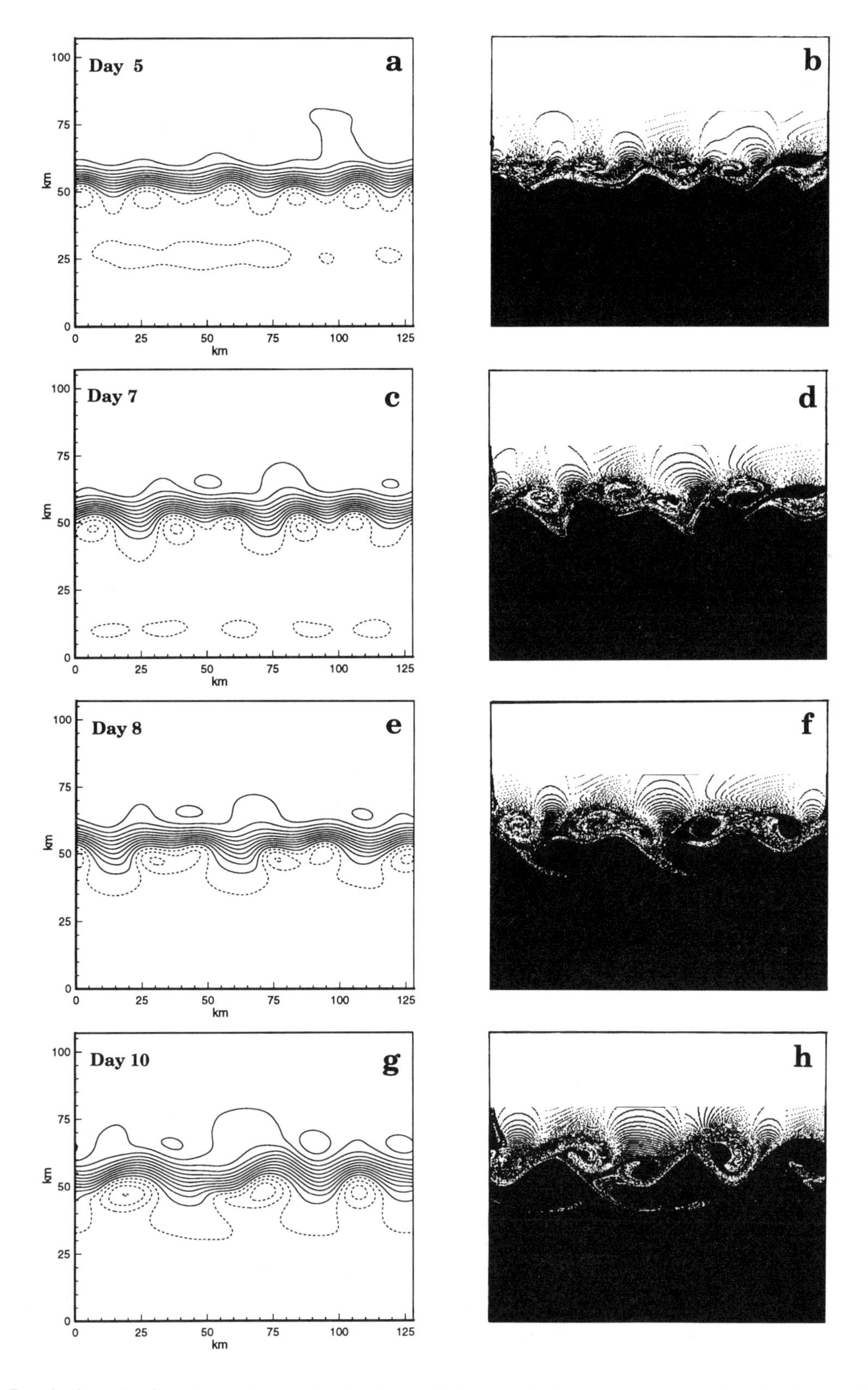

Fig. 18. Results from the Case 1 experiment after 5, 7, 8 and 10 days of simulations, (a, c, e, and g) surface height; (b, d, f, and h): corresponding particle distributions.

TABLE 3. Properties of computed ice filaments on day 9

Experiment number	Filament extent (km)	Growth rate (km/day)
1	33	15
2	26	7
3	33	11
4	49	14
5	20	6
7	29	30
8	60	10

after 18 days. By day 22, shown in Figure 21, the particle distribution developed a large sinusoidal meander pattern with a wavelength of about 40 km. The regular sinusoidal particle distribution is not representative of the typical ice patterns observed by the SAR images. In the second experiment, where shelf slope topography was included, a stable jet was produced. No instability developed after 40 days of simulation. The second experiment is a more realistic simulation of the EGC flowing along the shelf break and suggests that the core of the EGC is not necessarily unstable.

Case 5. Broader Baroclinic Jet with Flat Bottom

In this case there is a flat bottom, the lower layer jet velocity amplitude is increased to 0.2 m/s, the upper layer

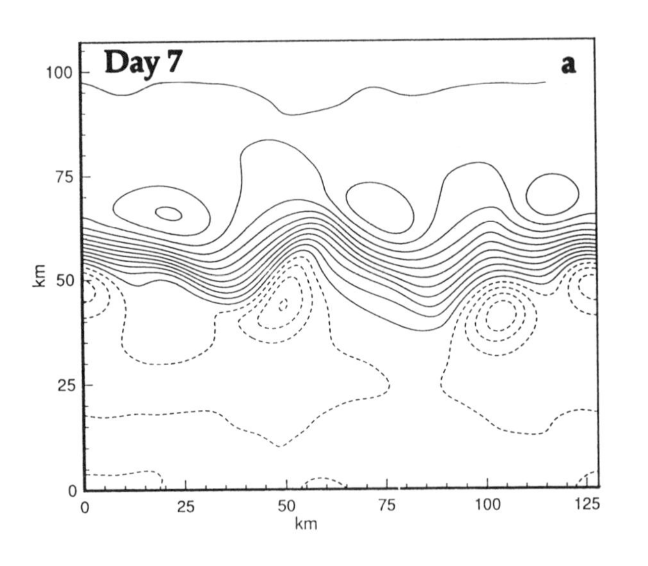

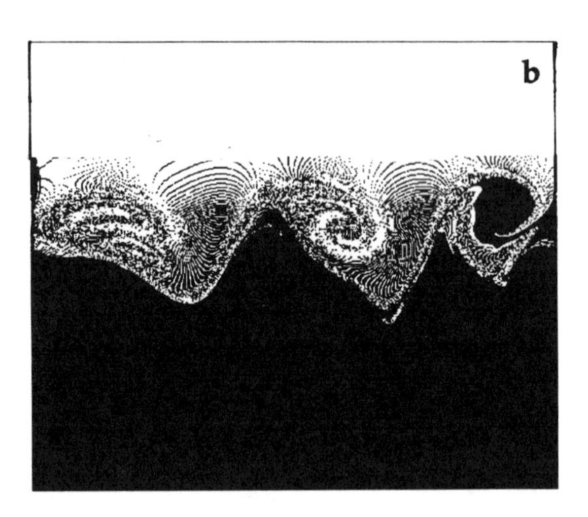

Fig. 19. Results from the Case 2 experiment after 7 days. (a) Surface height and (b) particle distribution.

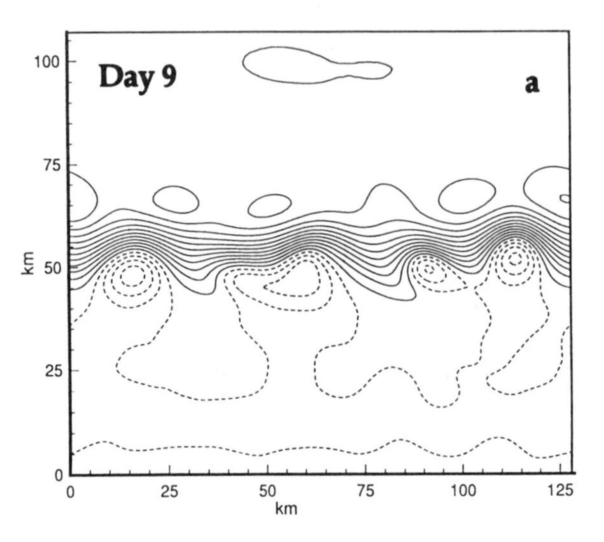

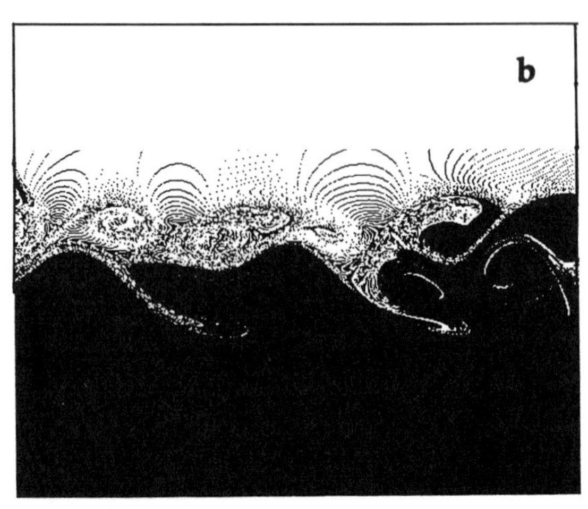

Fig. 20. Results from the Case 3 experiment after 9 days. (a) Surface height and (b) particle distribution.

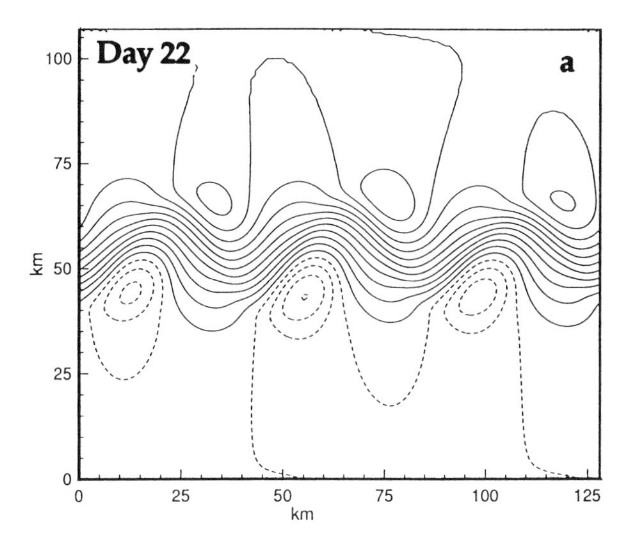

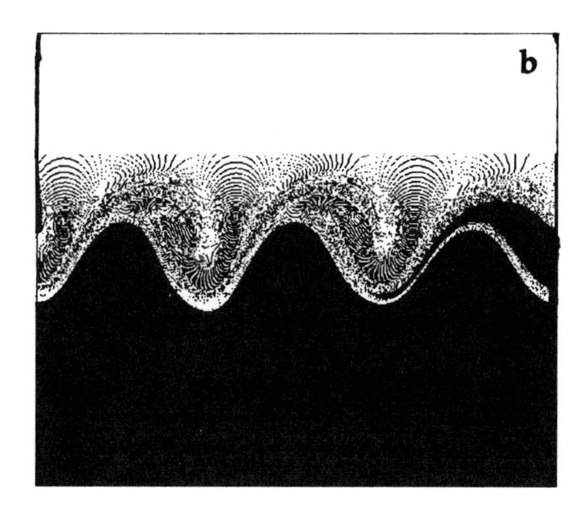

Fig. 21. Results from the Case 4 experiment after 22 days. (a) Surface height and (b) particle distribution.

amplitude remains at 0.40 m/s, and the jet width scale, L, is increased to 10 km. The growth rate for this case is less than half that of the baseline case. Two ice tongues associated with anticyclonic eddies of diameter 30 - 40 km have evolved separated by 60 - 70 km (Figure 22). Larger scale disturbances evolve in this experiment than in the previous cases.

Case 6. Broad Baroclinic Jet with Topography

This case is similar to the baseline experiment but with a wider jet and increased lower layer velocity. When shelf slope topography is included, there is no instability even after 30 days. The topography again stabilizes the jet. This case may be most representative of typical conditions when the EGC flows as a stable jet locked to the shelf break.

Case 7. Upper Layer jet with Flat Bottom

This experiment is similar to Case 3 but with a flat bottom at a depth of 2600 m. It represents the case where an intense upper layer jet is advected into deep water. From Figure 23 it can be seen that the characteristic length scale is somewhat larger than in Case 3. The instability in this case is initially stronger. This is due to the absence of the stabilizing effect of the bottom topography. After 6 days the instability has slowed to a very low growth rate.

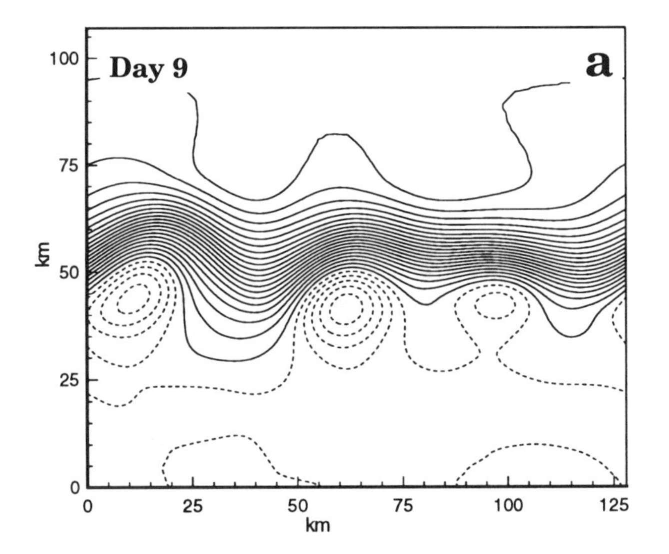

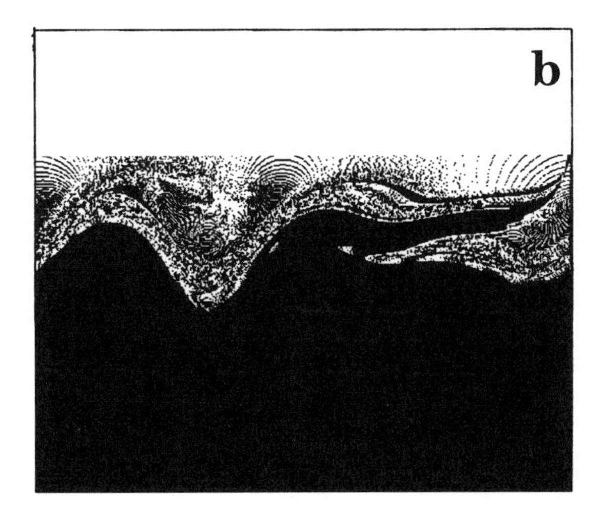

Fig. 22. Results from the Case 5 experiment after 9 days. (a) Surface height and (b) particle distribution.

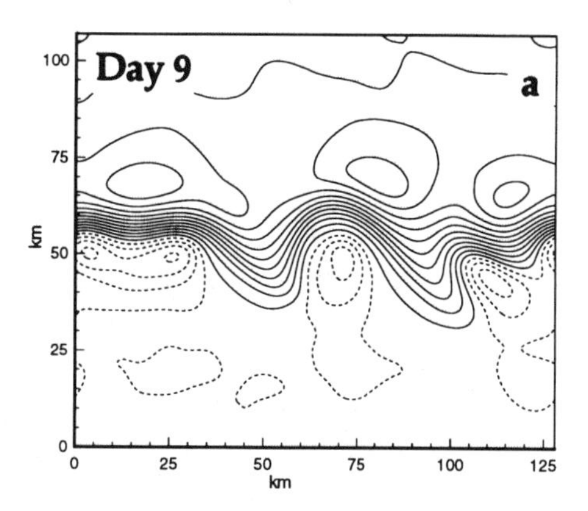

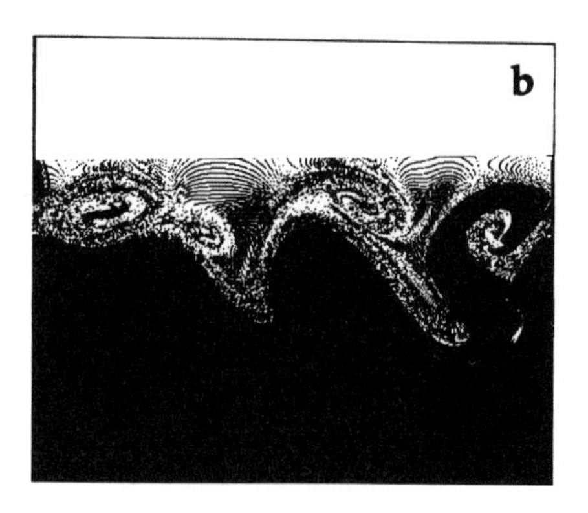

Fig. 23. Results from the Case 7 experiment after 9 days. (a) Surface height and (b) particle distribution.

Case 8. Baseline Experiment with Flat Shelf Topography

In this experiment there is a flat bottom at a depth of 400 m. This case simulates the response of an ice edge jet over the continental shelf. This experiment is similar to Case 2, but with a shallower depth. From Figure 24 it can be seen that the instability is more energetic than in Case 1 and 2.

4.2. DIAGNOSTIC ANALYSES

For a more quantitative description of the model simulation results, linear stability analyses and power spectral analyses were performed on the model output for selected experiments.

The linear stability analysis is based upon a two-layer quasi-geostrophic model. The analysis method is that of *Holland and Haidvogel* [1980], with the exception that the topographic-beta term has been included to model the effects of the bottom topography. The results of the analyses are the wavelength and e-folding scale of the growth rate of the most unstable wave. At days 0, 3, 6 and 9 the mean cross-channel profiles are computed from the model output and used in the stability analysis.

Shown in Table 4 are the linear stability results for cases 1, 2, 3, 5, 7, and 8, which are the key experiments in this study. It can be seen that the general tendency is for the wavelength of the most unstable mode and the e-folding time of the growth rate to increase over time. This is due to the jet-eddy interactions producing a broadening and weakening of the jet. The effect of eddy dynamics is to stabilize the system over time. It should be noted that several unstable modes are present in the system at any given time. Different modes will be the fastest growing at various times. Thus, in Case 8, for example, a mode with a shorter wavelength becomes dominant at day 9. The e-folding times of 1 - 3 days during days 3 - 6 for the most realistic cases, 1 and 3, are consistent with the field observations of the evolution of the ice tongues. The characteristic wavelengths of 28 - 38 km for these cases from Table 4, corresponding to eddy diameters of 14 - 19

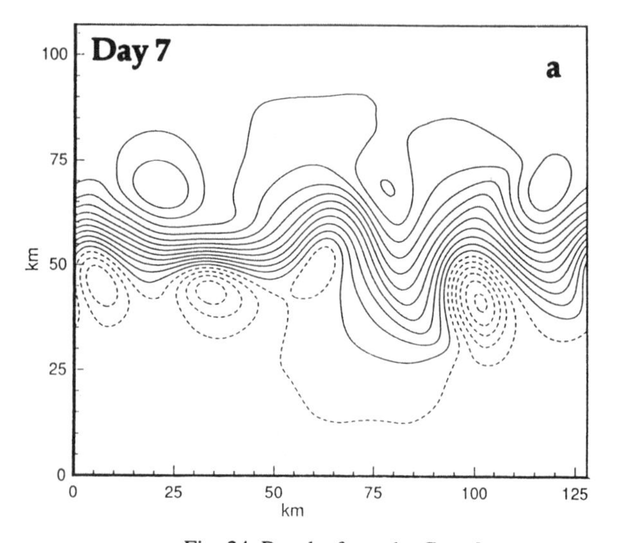

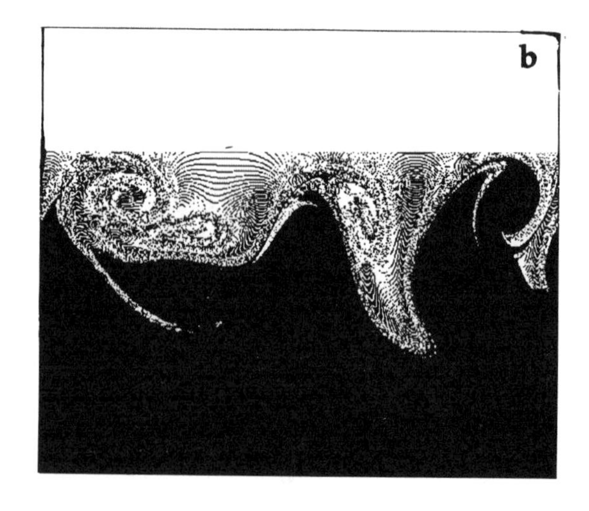

Fig. 24. Results from the Case 8 experiment after 7 days. (a) Surface height and (b) particle distribution.

TABLE 4. Linear stability analysis results

Experiment number	Wavelength (km) Day 0	Day 3	Day 6	Day 9	E-folding time (days) Day 0	Day 3	Day 6	Day 9
1	21	30	38	48	0.56	1.1	2.2	3.0
2	21	34	49	56	0.56	1.6	4.5	6.1
3	19	28	38	38	0.55	1.1	3.1	3.6
5	41	55	61	76	1.60	2.2	2.6	3.8
7	19	32	50	86	0.55	1.7	6.8	11.8
8	21	36	70	45	0.56	1.8	4.9	4.1

km, are within the observed range of eddy scales in the marginal ice zone of the Greenland Sea.

To obtain estimates of the dominant zonal length scales produced in the model simulations, a maximum entropy spectral analysis [*Press et al.*, 1986] was performed on the model results. The analysis was based on 30 coefficients and computed for a 10 km wide, along-channel strip which was centered over the initial jet position. The results were checked for consistency against periodograms computed from Fast Fourier Transforms. The results from the analysis are illustrated in Figure 25 a, b where the kinetic energy (KE) spectral density is plotted against wavelength for Case 1 and 3. The evolution of the KE spectrum over the course of the simulations is depicted in the graphs. In both cases the KE distribution is bimodal, with peaks at 24 - 26 km and 42 - 45 km. Initially, the peak at wavelength 24 - 26 km builds up rapidly, reaching its maximum at day 5 in Case 1 and day 7 in Case 3, but as time progresses energy is transferred to the 42 - 45 km scale. From Figure 18 and 20, these results can be interpreted as eddy events, where the eddies have typical diameters of 12 - 13 km and are separated by 42 - 45 km.

Cases 2 and 5 also exhibit a second clear spectral peak at a larger wavelength, at 44 - 51 km and 53 - 55 km, respectively. The low wavenumber portion of the spectra from Cases 7 and 8 varied too much over time to permit a simple characterization of a larger length scale.

These results are summarized in Table 5. Also shown in this table are the growth rates for the velocity amplitudes at the dominant scales. These values are compared to the growth rates determined from the linear stability analysis (LSA) at the same wavelengths.

A comparison of Tables 4 and 5 is instructive. Scale 1 in Table 5, which corresponds to twice a typical eddy diameter, is consistently smaller than the wavelength of the most unstable mode predicted from the LSA for day 3, as shown in Table 4. The model Scale 2 values for Case 1 and 3, which have topography, are larger than the LSA-predicted values at day 6. However, the model Scale 2 values for the non-topography Case 2 and 5 are longer than

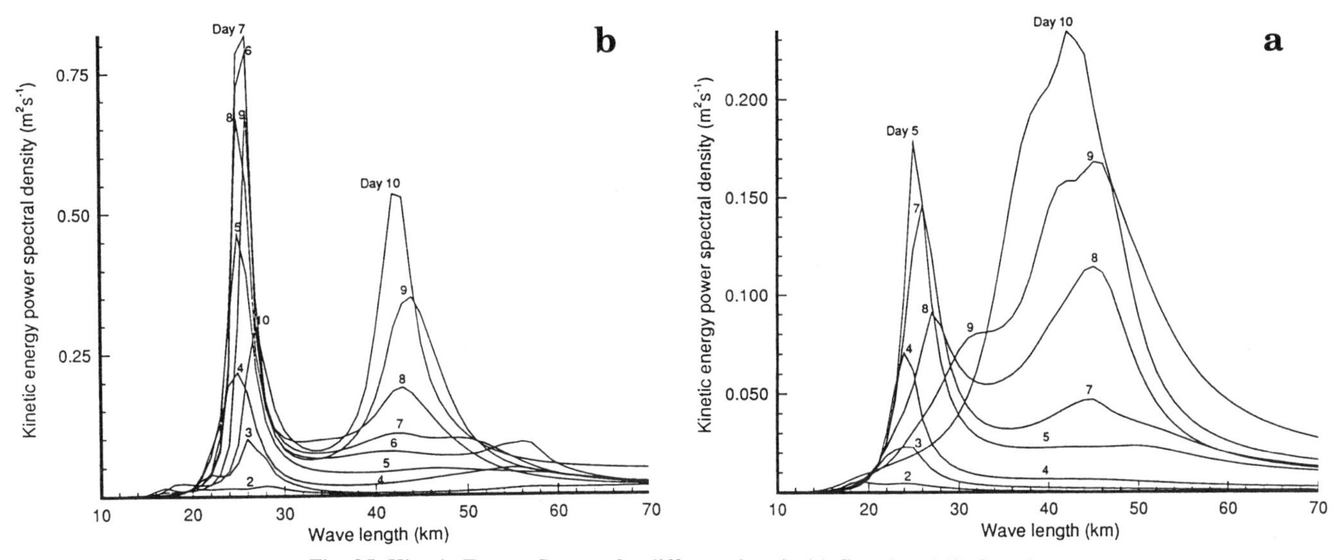

Fig. 25. Kinetic Energy Spectra for different days in (a) Case 1 and (2) Case 3.

TABLE 5. Dominant length scales and associated growth rates

Case	Scale (km)	Growth rate ($x10^{-6}s^{-1}$)	LSA growth rate ($x10^{-6}s^{-1}$)
	Scale 1		
1	24 - 25	5.7	6.1
2	34 - 35	2.0	4.0
3	25 - 26	4.4	6.7
5	33 - 35	3.0	0.5
7	27 - 30	1.4	3.1
8	29 - 32	6.6	6.1
	Scale 2		
1	42 - 45	3.7	3.6
2	44 - 51	1.2	2.8
3	41 - 44	3.4	3.3
5	53 - 55	2.7	9.8

those of the day 6 LSA. These differences in the larger dominant length scales may be partially attributable to the differences in the representation of bottom topography in the two-layer primitive equation and the quasigeostrophic models.

The model and LSA growth rates are in good agreement for cases 1, 3 and 8. The e-folding growth rate time scales of 2.0 days for Case 1 and 2.6 days for Case 3 for the eddy-containing 24 - 26 km length scale are consistent with the field observations discussed in section 2.

5. DISCUSSION

Ice tongues and vortex pairs are most readily observed along the ice edge where the images show a sharp contrast between ice and water. Vortex-pairs may also occur within the ice and in the open ocean, but here they are more difficult to detect in the SAR images as well as in other remote sensing data. But do they also occur frequently elsewhere in the ocean? What physical conditions are necessary to generate ice tongues and vortex pairs and what is their characteristic temporal and spatial distribution? Based on a large number of observations, *Fedorov and Ginsburg* [1986] argue that filaments with attached vortex pairs occur frequently in the ocean. The problem is to observe them systematically. So far only remote sensing data and simulation experiments have been able to catch the phenomenon, and then only in cases where a tracer is present.

The evolution of two counter-rotating vortices can be considered to be a consequence of the conservation of the initial angular momentum of the jet. This has been demonstrated in a series of laboratory experiments in which a variety of modes of forcing were applied to a rotating fluid [*Flierl et al.*, 1983]. The underlying generating mechanisms are frontal instabilities associated both with the ice edge and the ocean front which are present in the area. Such instabilities can produce all sorts of meanders, tongues and eddies. However, it is not clear whether vortex-pairs are just random manifestations of the turbulent eddy field, or if they are caused by specific mechanisms, as shown in this study. Another possibility is that the structures can be produced by the action of an impulsive surface stress as shown by *Mied et al.* [1991] for the open ocean.

A series of numerical experiments was carried out to examine the sensitivity of the evolution of ice edge eddies, tongues and vortex pairs to variations in model parameters. A general result from all the experiments is that although the ice distribution fields evolve into complicated structures within a few days, the velocity fields remain relatively simple with the along-channel jet slightly perturbed by weak eddies. Meanders in the velocity field are sufficient to generate off-ice tongues and filaments similar to the observations described in section 2.1 and 2.2. One must be careful to avoid confusing a Lagrangian tracer distribution with an instantaneous Eulerian circulation pattern.

An encouraging aspect of the numerical investigation has been that the spatial structure of the observed ice filaments and vortex pairs is very similar to the computed fields in Case 1 (Figure 18) and Case 3 (Figure 20). In these cases the most realistic parameter settings were used.

In a comparison of the observed and modelled ice jet evolution, summarized in Table 3, it is seen that the modelled ice tongue growth rate and extent in Case 1 and 3 corresponds well to the observed values. Furthermore, the structure of the sea ice is also similar to the observations in the SAR images.

In order to assess the significance of the instability processes on the transport of sea ice in the Greenland Sea, lateral flux calculations using particle distributions were conducted. The number of particles positioned seaward (away from the ice edge) from the initial ice edge (center of the jet) after 10 days was tabulated for each of the simulations in cases 1, 3, 5, 7, and 8. If each particle is assumed to cover a surface area of 250 x 250 m, with ice concentration of 0.8 and thickness of 1 m [*Shuchman et al.*, 1994], it is possible to compute an ice volume flux associated with eddy activity. The ice mass flux is estimated by examining the transport of particles seaward averaged over a two - day period of the most intense eddy activity. It is assumed that these conditions are representative of the marginal ice zone for 150 days per year. The seaward ice flux extrapolated from the 128 km long model domain for the 2000 km extent of the MIZ is given in Table 6.

If it is assumed that each of the particles (ice floes) advected seaward melts, then the Case 1 and 3 estimates of

TABLE 6. Seaward ice flux between 76° and 80°N

Experiment number	Ice mass flux (km^3/year)
1	273
2	109
3	120
5	121
7	264
8	442

273 and 120 $km^3 year^{-1}$ respectively, correspond to an ice edge melt rate of 0.9 and 0.4 km per day. These values are comparable to the estimates based on direct melt observations by *Johannessen et al.* [1987] for eddy-induced ice edge melt of 1 - 2 km per day.

The melt rate caused by eddies can also be estimated directly by assuming that ice edge eddies are present 150 days of the year along 2000 km of the EGC. This eddy melting mechanism of 1 km per day in 150 days will cause an annual melt of 300 km^3 assuming an ice thickness of 1 m. This estimate is in agreement with the results from the model simulations presented in Table 6. It is approximately 12 % of the annual transport of 2600 km^3 per year past 79° N [*Kvambekk and Vinje*, 1993], implying that the ice edge eddies are important and should be included in regional modelling of the ice edge location. This "eddy-melt" process will also contribute to the freshwater budget of the Greenland Sea, causing increased stabilization of the surface layer and reduced winter convection.

Wind forcing was not applied in the simulations since the wind conditions during the MIZEX 87 experiment were moderate to light and the main driving mechanism on the ice motion was the ocean circulation. When the wind is above 12 - 15 m/s, the ice is mainly driven by the wind forcing and the surface signature of the ocean circulation is masked. Under such conditons the SAR images cannot be used to study the mesoscale ocean features. In the most pronounced wind events observed during the MIZEX 87 experiment (10 - 12 m/s) the wind acted in the opposite direction of the ice tongues and was therefore not considered to be the driving force for these tongues.

Acknowledgements. This study has been supported by Office of Naval Research, University of Bergen and Environmental Research Institute of Michigan. Computing for this study was conducted at the IBM Bergen Environmental Services and Solutions Centre. Special thanks go to the SAR aircraft crews from INTERA Technologies Ltd., and to the crews onboard R/V "Polar Circle" and R/V "Håkon Mosby" for excellent cooperation in the field experiment.

REFERENCES

Couder, Y. and C. Basdevant, Experimental and numerical study of vortex couples in two-dimensional flows, *J. Fluid Mech, 173*, 225-251, 1986.

Fedorov, K. N. and A. I. Ginsburg, "Mushroom-like" currents (vortex dipoles) in the ocean and in a laboratory tank, *Annales Geophysicae, 4*, B, 5, 507-515, 1986.

Fedorov, K. N. and A. I. Ginsburg, Mushroom-like currents (vortex dipoles): One of the most widespread forms of non-stationary coherent motions in the ocean, in *Mesoscale/Synoptic Coherent Structures in Geophysical Turbulence*, edited by J. C. J. Nihoul and B. M. Jamart, pp. 1-14, 1989.

Flierl, G. R., M. E. Stern, and J. A. Whitehead, The physical significance of modons: laboratory experiments and general integral constraints, *Dyn. of Atmos. and Oceans, 7*, 233-263, 1983.

Holland and Haidvogel, A parameter study of the mixed instability of ideal ocean currents. *Dyn. Atmos. Oceans, 4*, 185 - 215, 1980.

Hurlburt, H. E. and J. D. Thompson, A numerical study of Loop Current intrusions and eddy shedding, *J. Phys. Oceanogr., 9*, 1611-1651, 1980.

Johannessen, O. M., J. A. Johannessen, J. Morison, B. A. Farrelly and E. A. S. Svendsen, Oceanographic conditions in the marginal ice zone north of Svalbard in early fall 1979 with an emphasis on mesoscale processes. *J. Geophys. Res., 88*, 2755-2769, 1983.

Johannessen, O. M., J. A. Johannessen, E. Svendsen, R. A. Shuchman, W. J. Campbell, and E. Josbergeer, Ice-edge eddies in the Fram Strait Marginal Ice Zone. *Science, 236*, 427-729, 1987a.

Johannessen, J. A., O. M. Johannessen, E. Svendsen, R. Shuchman, T. Manley, W. J. Campbell, E. G. Josberger, S. Sandven, J. C. Gascard, T. Olaussen, K. Davidson, and J. Van Leer, Mesoscale eddies in the Fram Strait marginal ice zone during 1983 and 1984 Marginal Ice Zone Experiments, *J. Geophys. Res., 97*, 6754-6772, 1987b.

Johannessen, O. M., W. J. Campbell, R. Shuchman, S. Sandven, P. Gloersen, J. A. Johannessen, E. G. Josberger, and P. M. Haugan. Microwave study programs of air-ice-ocean interactive processes in the seasonal ice zone of the Greenland and Barents seas, in *Microwave Remote Sensing of Sea Ice* Geophys. Monogr. Ser., vol. 68, edited by F. Carsey, pp. 261 - 289, AGU, Washington, 1992.

Kvambekk, Å. S., and T. Vinje, The ice transport through the Fram Strait. (abstract) Nansen Centennial Symposium, June 21 - 25, Nansen Environmental and Remote Sensing Center, 1993.

Mied, R. P., J. C. McWilliams and G. J. Lindemann. The generation and evolution of Mushroom-like Vortices. *J. Phys. Ocean.ogr 21*, 489 - 510, 1991.

MIZEX Group, MIZEX East 83/84: The summer Marginal Ice Zone Program in the Fram Strait/Greenland Sea. *EOS, Trans. AGU 67*, No.23, 513-517, 1986.

MIZEX Group, MIZEX East 1987: The winter Marginal Ice Zone Program in the Fram Strait/Greenland Sea. *EOS, Trans. AGU 70*, No.17, 545, 1989.

Narimousa, S., and T. Maxworthy, Coastal upwelling on a sloping bottom: the formation of plumes, jets and pinched-off cyclones, *J. Fluid Mech. 176*, 169-190, 1987.

Press, W. H., B. P. Flannery, S. A. Tekolsky, and W. T. Vetting, *Numerical Recipes. The Art of Scientific Computing,* 818 pp., Cambridge, University Press, Cambridge, 1986.

Quadfasel, D., and J. Meincke, Note on the thermal structure of the Greenland Sea. *Deep Sea Res., 34,* 1883-1888, 1987.

Røed, L. P., and J. J. O'Brien, A coupled ice-ocean model of upwelling in the marginal ice zone, *J. Geophys. Res., 88,* 2863-2872, 1983.

Shuchman, R. A., C. L. Rufenach, and O. M. Johannessen, The extraction of marginal ice zone thickness using gravity wave imagery. *J. Geophys. Res.* (in press) 1994.

Smith, D. C. IV, and J. J. O'Brien, The interaction of a two-layer isolated mesoscale eddy with topography, *J. Phys. Oceanogr., 13,* 1681-1697, 1983.

Smith, D. C. IV, and A. A. Bird, The interaction of an ocean eddy with an ice edge ocean jet in a marginal ice zone, *J. Geophys. Res., 96,* 4675-4690, 1991.

O. M. Johannessen, S. Sandven, W. P. Budgell, J. A. Johannessen, NERSC, Edv. Griegsvei 3a, N-5037 Solheimsvik, Norway.

R. A. Shuchman, ERIM, P. O. Box 134001, Ann Arbor, MI 48113-4001.

The Arctic Ocean Tides

Z. Kowalik and A. Y. Proshutinsky

Institute of Marine Science, University of Alaska, Fairbanks

To study tides in the Arctic Ocean a set of equations describing dynamical interactions in the ice-water system is solved numerically with a space grid of about 14 km. Four major tidal waves have been calculated, namely: semidiurnal constituents M_2 and S_2, and diurnal constituents K_1 and O_1. The distribution of amplitude, phase, current ellipses and ice motion related to these constituents is computed and depicted in figures. Residual motion in water due to nonlinear tidal dynamics is also studied. Experiments with non-linear internal ice stresses reveal concomitant tide-induced residual ice motion caused by the non-linear ice floe interactions. The latter is of transient character in time because the ice compactness changes due to both residual motion and seasonal conditions. The information on tidal harmonic constants from more than 300 tide stations and more than 400 Geosat altimetry measurements are used as boundary conditions and as data for comparison against the model results.

1. INTRODUCTION

Perpetual tidal motion in the Arctic Ocean influences ice distribution and generates periodic and permanent leads in the pack ice. These leads are caused by the periodic divergence and convergence of the tidal flow over the tidal cycles and residual water currents and ice drift. Periodic and permanent leads in the pack ice cover influence both heat exchange between the ocean and the atmosphere and the rate of ice formation. Tides are also responsible for maintaining the basic level of turbulence in the ice, in the boundary layers beneath the ice cover, and in proximity to the bottom. The mixing and stirring due to tides play an important role in the deep Arctic Ocean because the general circulation there appears to be extremely weak; they also play a very important role in the ice cover close to the shore line and shore-fast ice.

Periodic changes and strong ice shear have been observed by early northern travelers in the Barents Sea and White Sea region, where strong tide motion is present. Explanations were suggested by *Litke* [1844] and *Nansen* [1902].

The first important contribution toward understanding the Arctic tides was made by *Harris* [1911] who drew a rough chart showing a tidal wave entering from the Atlantic Ocean and traveling across the Arctic in 20 hours.

Mathematical calculations of tides in the Arctic were first performed by *Goldsbrough* [1913, 1927]. He considered a closed polar basin and concluded that semidiurnal tides should be small because they do not meet the resonance condition. The motion of a tide wave in the polar basin of a regular form was also considered by *Sretensky* [1937] through application of analytical solution.

The usefulness of the numerical approach to tide calculations had been demonstrated by *Defant* [1918, 1924]. In the course of numerical experiments, he established that 1) the semidiurnal tides in the Arctic Ocean are caused by the Atlantic tides, and 2) the diurnal tides are formed directly in the Arctic Ocean by astronomical forces. Many characteristics of the tides on the Siberian shelf were explained by *Sverdrup* [1926]. For example, he was able to explain the vertical profile of velocity by including a variable eddy viscosity coefficient and density stratification. *Fjeldstad* [1929] con-

The Polar Oceans and Their Role in Shaping the Global Environment
Geophysical Monograph 85

tinued Sverdrup's research and showed that the semidiurnal waves cross the Arctic basin in 12 hours.

In 1928 *Sterneck* published a new tidal chart for the semidiurnal tide in the central Arctic. Further studies of polar ocean tides have advanced mainly by using improved numerical models for various geographical regions [*Nekrasov*, 1962; *Kagan*, 1968; *Dvorkin*, 1970; *Godin*, 1980].

The global ocean models, which included the Arctic Ocean [*Zahel*, 1977; *Schwiderski*, 1980] have contributed to understanding interaction with the Atlantic Ocean although the modeling accuracy did not increase in polar regions. Local Arctic Ocean models indicated satisfactory agreement with observed data by using improved resolution of 75 km to 37 km [*Kowalik and Untersteiner*, 1978; *Kowalik*, 1981; *Gjevik and Straume*, 1989; *Proshutinsky and Polyakov*, 1991; *Proshutinsky*, 1991].

Measurements and tidal theories in the Arctic Ocean are primarily concerned with the semidiurnal tides (M_2 and S_2) whose amplitudes generally dominate all tidal constituents. *Mooers and Smith* [1968] on the Oregon shelf and *Cartwright* [1969] off the west coast of Scotland found that in the field of velocity the reverse situation can occur, i.e., diurnal tidal currents can dominate over semidiurnal currents. In the Arctic Ocean the local response to tidal forcing is especially conspicuous in the diurnal range of velocity field. *Hunkins* [1986] reported a significant topographic amplification of the diurnal tide over the Yermak Plateau with high velocities and enhanced mixing. *Huthnance* [1981] described high velocities in the vicinity of Bear Island. *Aagaard et al.* [1990] observed unexpected behavior of tide amplitudes and velocities of the different tidal constituents. For example, at the Beaufort Sea shelf, the amplitudes are dominated by the semidiurnal tide and, in contrast, the flow is usually dominated by the diurnal tide. *Muench et al.* [1992] demonstrated that the tidal flow maintains the unusually thick mixed surface layer over the Yermak Plateau and increases the vertical exchange of heat between the Atlantic water and the ocean surface. Recent measurements and tidal current analysis over the Yermak Plateau has been summarized by *Padman et al.* [1992]. Their measurements clearly show that Atlantic layer mixing is enhanced by tidally-generated wave packets. We have applied a high-resolution model to study diurnal tides in the Arctic Ocean [*Kowalik*, 1994; *Kowalik and Proshutinsky*, 1993] and have shown that current enhancement is related to shelf waves of tidal origin.

Ice-tide interaction not only affects ice distribution but also changes amplitude and phase of the tide. This change, although small in deep water, can be significant in the shallow water under shore-fast ice. In the very shallow water, shore-fast ice alters water depth and influences tide propagation [*Murty*, 1985]. Harmonic constants for these locations change from summer to winter and it is quite difficult to calculate the tidal constituents.

Recent investigations and observations in the Arctic Ocean have shown that tides and tide-related ice motion play an important role in formation of polynyas, openings and leads. *Martin and Cavalieri* [1989], *Kozo* [1991] and *Deming et al.* [1993] suggested that the enhanced mixing observed in polynyas along the Siberian shelf, in the Lincoln Sea, and in the Greenland Sea is caused by tidal currents.

Nonlinear dynamics of the tidal wave leads to formation of residual tidal circulation over the tide period. Although the overall residual currents are rather small, they may influence water properties and move suspended particles. Residual tidal motion is a frequent source of eddies in shallow areas and straits [*Proshutinsky*, 1988]. Concomitant with residual tide motion, residual ice motion occurs due to both the momentum transferred from the ocean and the non-linear interaction of ice floes [*Kowalik*, 1981].

Tidal models with low spatial resolution failed to demonstrate any significant non-linear and resonance effects. In this paper, a numerical model with spatial resolution of about 14 km is used primarily to obtain charts of tidal amplitude, phase, currents and ice motion for the M_2, S_2, K_1 and O_1 tidal constituents. Moreover, we have investigated secondary effects such as: ice and water residual circulation, ice production due to convergence/divergence of the tidal motion, and the role of tides in the generation of leads.

To study tides we apply two-dimensional, non-linear equations for water and ice dynamics. The model domain includes the Arctic Ocean, all the Nordic Seas, Hudson Bay and Baffin Bay. It is implemented on a stereographic map projection with a spatial grid of about 14 km (Figure 1). The forcing of tide is done through the open boundaries of the computational domain and by astronomical forcing over the whole domain. The astronomical forcing includes not only tide-generating potential but also various corrections due to earth tide and ocean loading.

This paper is organized in the following manner. Basic equations and boundary conditions are described in Section 2. The data base is discussed in Section 3. The model results are given in Section 4, and Section 5 contains summary and discussion.

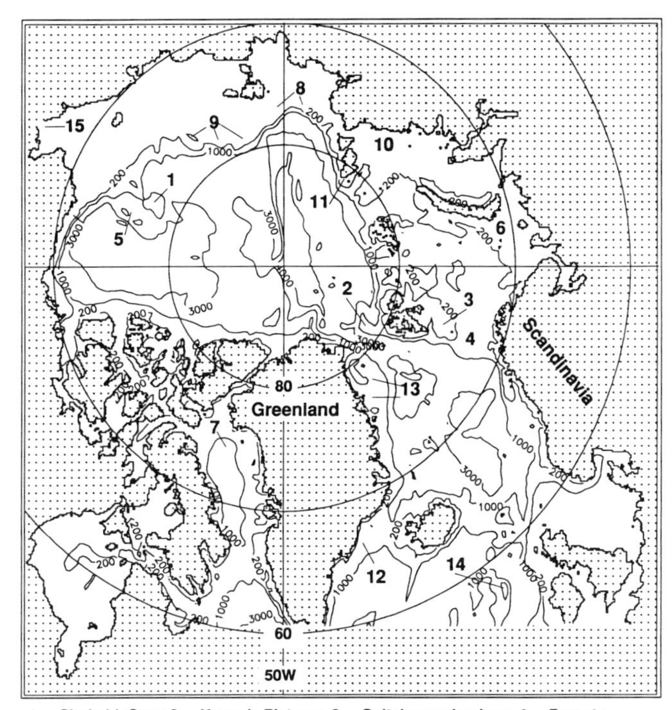

1 - Chukchi Cap; 2 - Yermak Plateau; 3 - Spitsbergenbanken; 4 - Barents-Norwegian Seas shelf; 5 - Arlis Plateau; 6 - Gusinaya Bank; 7 - Baffin Bay; 8 - Laptev Sea shelf edge; 9 - East-Siberian shelf edge; 10 - Kara Sea; 11- Severnaya Zemlia shelf edge; 12 - South Greenland Atlantic shelf edge; 13 - Greenland Sea shelf; 14 - Rockal Bank edge; 15 - Bering Strait

Fig. 1. Computational domain. Numbers denote major bathymetric features.

2. EQUATIONS AND BOUNDARY CONDITIONS

We shall use a system of equations of motion and continuity written in the vector notation and in the stereographic polar coordinate system

$$\frac{d\mathbf{U}}{dt} + f\mathbf{k} \times \mathbf{U} =$$

$$-gDm\nabla(\alpha\zeta - \beta\zeta_0) + N_h m^2 \nabla^2 \mathbf{U} + \frac{c\mathbf{T}_i - \mathbf{T}_b}{\rho} \quad (1)$$

$$\frac{\partial \zeta}{\partial t} = -m^2 \nabla(\mathbf{U}/m) \quad (2)$$

and the equations of motion and continuity for ice,

$$\frac{d\mathbf{u}_i}{dt} + f\mathbf{k} \times \mathbf{u}_i = -gm\nabla\zeta + \frac{\mathbf{T}_i}{\rho_i h_i} + \mathbf{F}_i \quad (3)$$

$$\frac{\partial c}{\partial t} = -m^2 \nabla(\mathbf{u}_i c/m) \quad (4)$$

Here, x, y are the lateral coordinates, with their origin at the North Pole; t is time; ζ denotes free surface elevation; $\mathbf{U}$ is a vector of volume transport with components U, V along x and y- directions; $\mathbf{u}_i$ is a vector of ice velocity; $\mathbf{T}_i$ is a vector of ice stress between water and ice; $\mathbf{T}_b$ is a vector of bottom stress; $\mathbf{F}_i$ is a vector of internal ice forces; ρ_i is ice density; ρ is water density; c is ice compactness , $0 \le c \le 1$; h_i is ice thickness; N_h is horizontal eddy viscosity ($=5 \cdot 10^7 cm^2/s$); D is total depth ($= H + \zeta$); f denotes Coriolis parameter; ζ_0 is equilibrium tide; α, β are parameters accounting for tidal potential perturbations; $\mathbf{k}$ is unit vector along vertical direction; m denotes map coefficient.

$$\frac{d}{dt} = \frac{\partial}{\partial t} + m[\frac{U}{D}\frac{\partial}{\partial x} + \frac{V}{D}\frac{\partial}{\partial y}]$$

In the above operator only the first order nonlinear terms are retained. The second order terms have been neglected [*Kowalik*, 1994].

The stereographic polar coordinate system used in (1) – (4) is very close to the rectangular system of coordinates. The difference is due to the map coefficient m. It describes map correction from spherical projection to polar stereographic projection. Its value changes from 1 at 90^0N to 1.071 at 60^0N.

Bottom stress is described by

$$\mathbf{T}_b = \rho R_b \frac{|\mathbf{U}|\mathbf{U}}{D^2} \quad (5)$$

where R_b is a bottom friction coefficient ($= 2.6 \cdot 10^{-3}$).

The interaction of ice and water is described by

$$\mathbf{T}_i = \rho R_i |\mathbf{u}_i - \frac{\mathbf{U}}{D}|(\mathbf{u}_i - \frac{\mathbf{U}}{D}) \quad (6)$$

The ice-water drag coefficient R_i ($= 5.5 \cdot 10^{-3}$) was estimated by *McPhee* [1980].

To describe internal ice forces the nonlinear viscous constitutive law proposed by *Rothrock* [1975] is taken:

$$\mathbf{F}_i = \eta m^2 \nabla^2 \mathbf{u}_i + \Lambda m^2 \nabla(\nabla \mathbf{u}_i) - m\nabla p \quad (7)$$

Here

$$\nabla(\nabla \mathbf{u}_i) = grad\ (div \mathbf{u}_i)$$

Rothrock [1975] suggested that the tensile stress in ice is negligible compared to compressive stress. Pressure (p) in (7) is given by:

$$p = -A_p m \nabla \mathbf{u}_i \quad \text{if} \quad \nabla \mathbf{u}_i < 0$$

$$p = 0 \quad \text{if} \quad \nabla \mathbf{u}_i \ge 0 \quad (8)$$

In the above formulas both bulk (Λ) and shear (η) viscosity coefficients are taken to be equal; A_p is the ice pressure coefficient.

The magnitude of the frictional coefficients used in the above equations should result in numerical stability and reasonable reproduction of the turbulent processes in the water and ice. Horizontal eddy viscosity $N_h = 5.\cdot 10^7 cm^2 s^{-1}$ is taken close to the threshold of numerical stability $N_h = 5.\cdot 10^6 cm^2 s^{-1}$ [*Kowalik*, 1981]. Variable level of the vertically generated turbulence is controlled through the ice-water stress. In the ice-free areas this term is set to zero and in the fast ice regions the friction is expressed in similar manner to the bottom stress.

Tidal forcing is described in (1) through the terms which are multiplied by coefficients α and β. These terms include tide-generating potential but they also contain various corrections due to earth tide and ocean loading, [e.g., *Schwiderski*, 1979, 1981 a, b, c]. Coefficient α defines ocean loading and self-attraction of the ocean tides. Its value ranges from 0.940 (diurnal) to 0.953 (semi-diurnal) according to *Ray and Sanchez* [1989]. The higher-order correction for the loading effect can be implemented as well, [e.g., *Francis and Mazzega*, 1990]. The term $\beta\zeta_0$ includes both the tide–generating potential and correction due to the earth tide. It is usually expressed as

$$\beta\zeta_0 = (1 + k - h)\zeta_0 \tag{9}$$

Here k and h denote Love numbers, which are equal to 0.302 and 0.602 for the semi-diurnal tides respectively. These values for the individual diurnal constituent may differ and according to *Wahr* [1981] for K_1 constituent $k = 0.256$ and $h = 0.520$, and for O_1 constituent $k = 0.298$ and $h = 0.603$.

The equilibrium tide for the diurnal constituents is:

$$\zeta_0 = H_n \sin 2\phi \cos(\sigma_n t + \lambda) \tag{10}$$

Here $H_n = 14.565$ cm, $\sigma_n = 0.729221 \cdot 10^{-4}$ s^{-1} for K_1, and $H_n = 10.0514$ cm, $\sigma_n = 0.675981 \cdot 10^{-4}$ s^{-1} for O_1; λ denotes the longitude angle; ϕ is a latitude.

The equilibrium tide for the semi-diurnal constituents is:

$$\zeta_0 = H_n \cos^2\phi \cos(\sigma_n t + 2\lambda) \tag{11}$$

Here $H_n = 24.2334$ cm, $\sigma_n = 1.495189 \cdot 10^{-4}$ s^{-1} for M_2, and $H_n = 11.2841$ cm, $\sigma_n = 1.454410 \cdot 10^{-4}$ s^{-1} for S_2.

To obtain a unique solution to the above system in the domain of integration it is sufficient to specify normal and tangential water transport and ice drift velocities everywhere along the boundaries [*Marchuk and Kagan*, 1989].

Initially, the dependent variables in the integration domain are taken as zero:

$$\zeta(x,y)_{t=0} = 0 \tag{12}$$

$$\mathbf{U}(x,y)_{t=0} = 0 \tag{13}$$

$$\mathbf{u}_i(x,y)_{t=0} = 0 \tag{14}$$

Along the solid boundary (S) we assume a no-slip condition for water transport and ice velocity

$$\mathbf{U}(x,y,t)_S = 0 \tag{15}$$

$$\mathbf{u}_i(x,y,t)_S = 0 \tag{16}$$

On the open boundary (O) of the domain, often only the sea level is known. In such cases, according to *Marchuk and Kagan* [1989], a unique solution cannot be obtained to (1) and (2). Therefore, we proceed as follows: in the vicinity of the open boundary (along the first line parallel to the open boundary), the linear hyperbolic problem is solved (horizontal friction and advective terms in (1) are omitted). This procedure yields a unique solution for the volume transport with sea level defined at the open boundary, [e.g., *Kowalik and Murty* 1993]. When transport is specified along the line parallel to the open boundary, the solution process for the full set of equations (parabolic problem) can be extended into the integration domain. With the above restriction, the boundary conditions pertinent to equation (1) to (2) can be stated as follows:

$$\mathbf{U}(x,y,t)_S = 0 \tag{17}$$

$$\mathbf{u}_i(x,y,t)_S = 0 \tag{18}$$

$$\zeta_O = \zeta(x,y,t) \tag{19}$$

For ice cover at the open boundary the following conditions are prescribed

$$\frac{\partial \mathbf{u}_i}{\partial n} = 0 \tag{20}$$

$$\frac{\partial c}{\partial n} = 0 \tag{21}$$

where n is a normal to the open boundary.

Along the open boundaries, amplitudes and phases for every tidal constituent are specified. This is done from the model results obtained by *Schwiderski* [1979, 1981 a, b, c], *Zahel* [1977], satellite observations [*Cartwright et al.*, 1991], and from observations at coastal tide stations. We performed an extensive comparison between satellite observations and Schwiderski's results and found that both data sets are practically identical. Along the open boundaries we use Schwiderski's data, and recent observations are blended into these data.

3. DATA BASE

To cover the Arctic Ocean with a 14-km spatial grid step required a thorough examination of the data base on depth distribution. In the Barents Sea and the Kara Sea, depth was compiled from the Norwegian [*Norsk Polarinstitutt*, 1989] and U.S. charts [*Geophysical Society of America*, 1991]. Large-scale bottom features in the Greenland, Laptev, and Beaufort Seas were compiled from *Naval Research Laboratory* [1986].

The model results are verified through comparison with observations. Four kinds of observations are available to perform this comparison:

a) Sea level observations are available from more than 300 sites around the Arctic Ocean. These are mainly data from tide gauges located at the shore, while some data are from the shelf area (pelagic stations). A one-year record of bottom pressure in the strait between Nordaustlandet and Kvitoya, Svalbard, was obtained by *Aagaard et al.* [1983], and a series of about three years of pressure data was recorded at 568-m close to Spitsbergen [*Morison*, 1991]. These are probably the farthest-north recorded sea level data.

The principal source for the pelagic tidal constants is an IAPSO publication by *Cartwright and Zetler* [1985]. The tidal constituents information for the tide gauges located at the shore is taken from *Gidrographicheskoe Upravlenie VMF SSSR* [1941] and *Hydrographer of the Navy* [1987]. Some tide data were received from the Canadian Hydrographic Service and the International Hydrographic Bureau in Monaco.

As an additional source of sea level information in the open ocean, we used results from the *Schwiderski* [1979, 1981a, b, c] model. A comparison against Schwiderski's model is made at more than 6000 grid points. The purpose of this comparison is to locate differences between the low-resolution Schwiderski model (1^0 grid space) and the high resolution model with about a 14-km grid space.

b) More then 400 points of Geosat altimetry measurements [*Cartwright and Ray*, 1990; *Cartwright et al.*, 1991] are used as data for comparison against model results.

c) Current meter data from 12 locations are available for comparison against the model results. Most measurements were taken in the Norwegian, Barents, and Beaufort Seas and in Hudson Bay. The current meter data were published by *Huggett et al.* [1975], *Huthnance* [1981], *Aagaard et al.* [1983], *Aagaard et al.* [1985], *Drinkwater* [1988], and *Aagaard et al.* [1990].

d) Ice motion data in the tidal range of periods are mainly available from ice drifting stations and buoys. There are no direct eulerian measurements of tidal ice motion in a fixed point in the Arctic Ocean. Only with upward looking Doppler sonar will it be possible to obtain velocity and direction of ice motion in fixed coordinate frame during long periods of time. Presently, ice motion induced by tides is estimated from ice drift data using a special technique of data processing and analysis [*McPhee*, 1986; *Pease et al.*, 1994]. From these lagrangian data, it is possible to estimate ice current due to various tidal constituents over 14-day period in a fixed location [*Pease et al.*, 1994].

Satellite images of ice cover are primary sources of indirect information on ice-tide interaction. Ice motion and ice concentration obtained by SAR may be used to calibrate and to validate the model. Moreover, one can use high-quality satellite imagery in regions where elliptically-shaped traces formed by grounded icebergs in the ice fields are observed [*Dmitriev et al.*, 1991]. From these measurements, it is possible to determine length, form, and orientation of the tide ellipses.

Measurements of internal ice stresses are also very good sources of data on ice-tide interactions. Numerous measurements were made around Spitsbergen during several experiments at the ice drift stations [*GSP Group*, 1990; *CEAREX Drift Group*, 1990].

It is interesting to note that ambient noise measured by hydrophone could be a good index of ice floe interactions in the tidal range of periods. For example, during CEAREX [*CEAREX Drift Group*, 1990], ambient noise was dominated by 20- to 25-dB oscillations of about 12-hr period.

4. RESULTS

The semidiurnal (M_2, S_2) and diurnal (K_1, O_1) tidal wave simulations have been carried out with different versions of the model in order to assess the effects of tide-generating forces, horizontal eddy viscosity, bottom friction, non-linear terms and tide-ice interactions.

4.1. *Structure of the co-tidal and co-range charts*

The tidal charts of K_1, O_1 diurnal waves and M_2, S_2 semidiurnal waves are shown in Figures 2, 3, 4 and 5, respectively. The numbers on the co-range lines in Figures 2–5 are given in cm. Phase angle on the co-

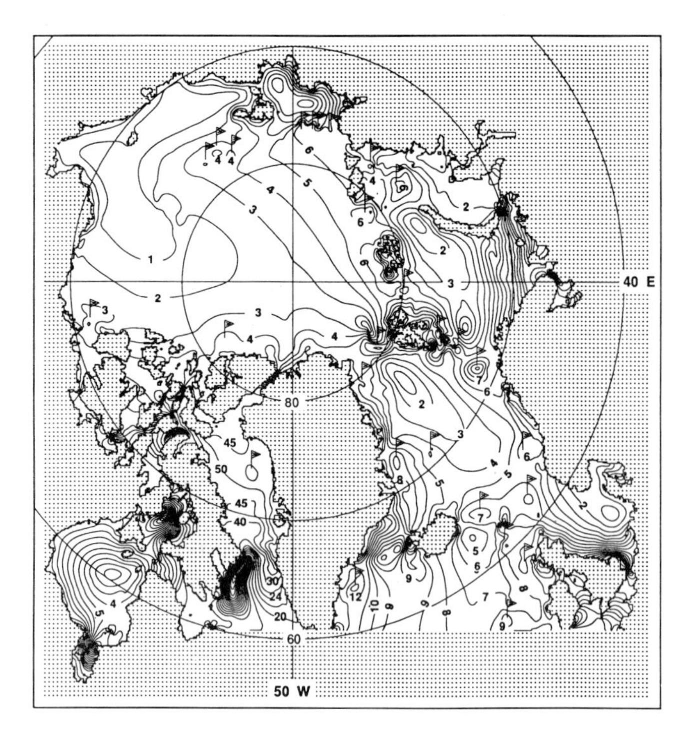

Fig. 2a. Computed amplitude (cm) of surface elevation for the diurnal constituent K_1. Flags denote shelf wave regions.

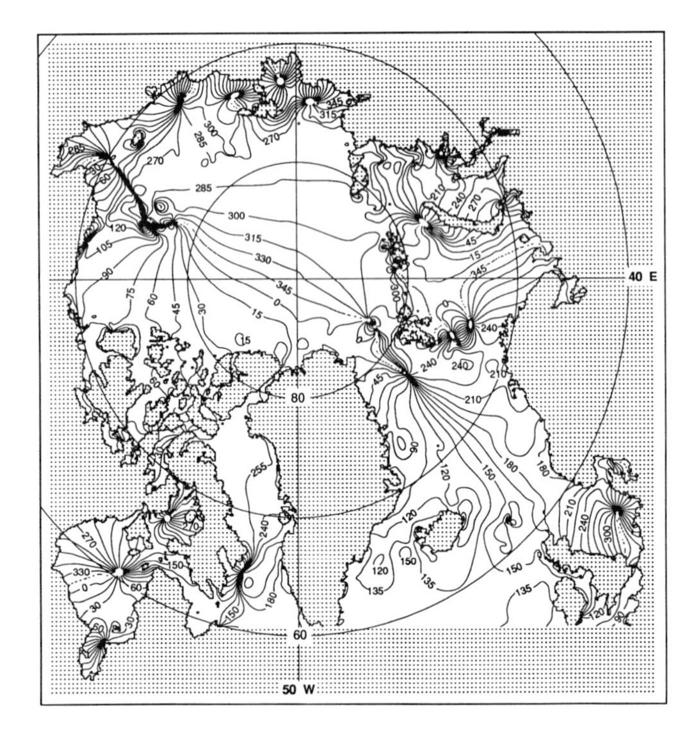

Fig. 2b. Computed phase (deg) of surface elevation for the diurnal constituent K_1.

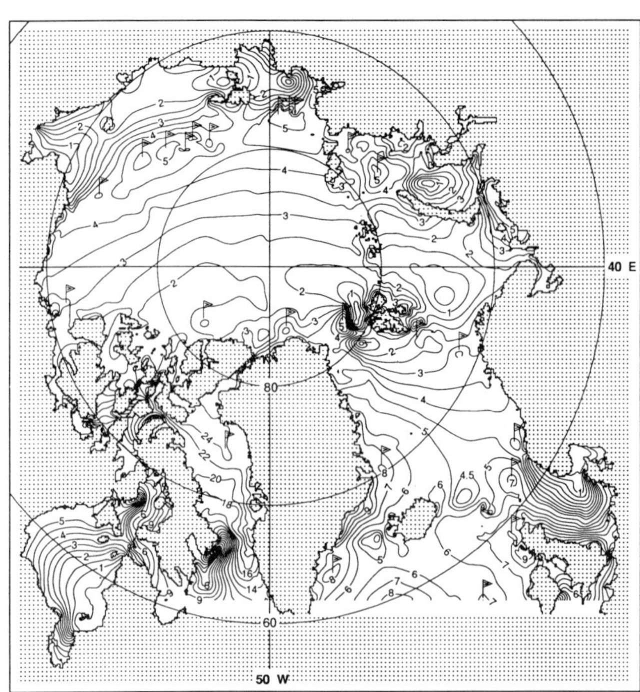

Fig. 3a. Computed amplitude (cm) of surface elevation for the diurnal constituent O_1. Flags denote shelf wave regions.

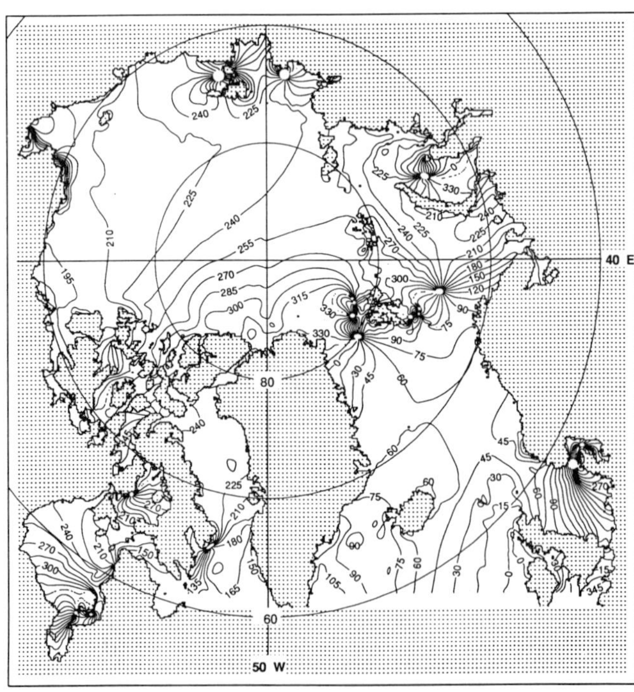

Fig. 3b. Computed phase (deg) of surface elevation for the diurnal constituent O_1.

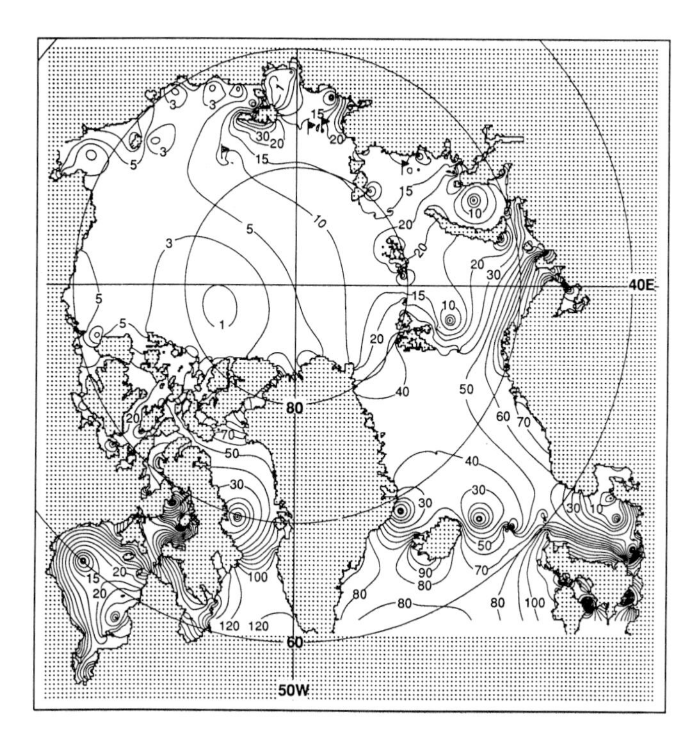

Fig. 4a. Computed amplitude (cm) of surface elevation for the semidiurnal constituent M_2. Flags denote shelf wave regions.

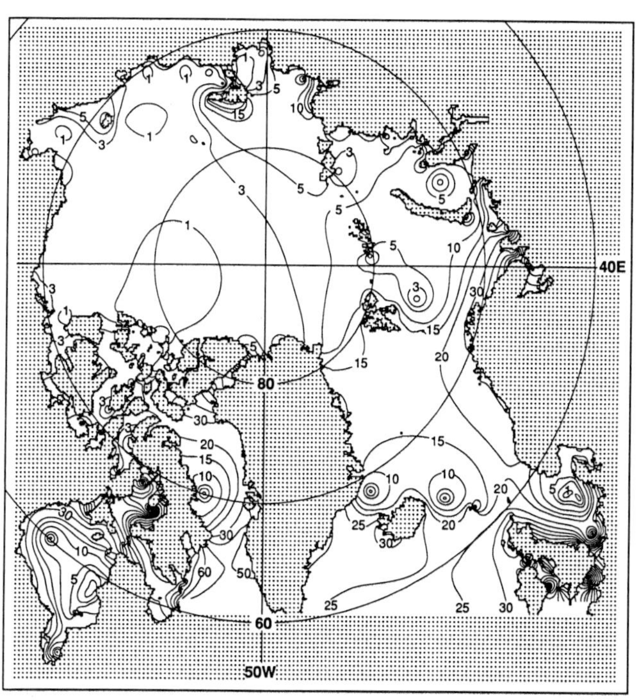

Fig. 5a. Computed amplitude (cm) of surface elevation for the semidiurnal constituent S_2.

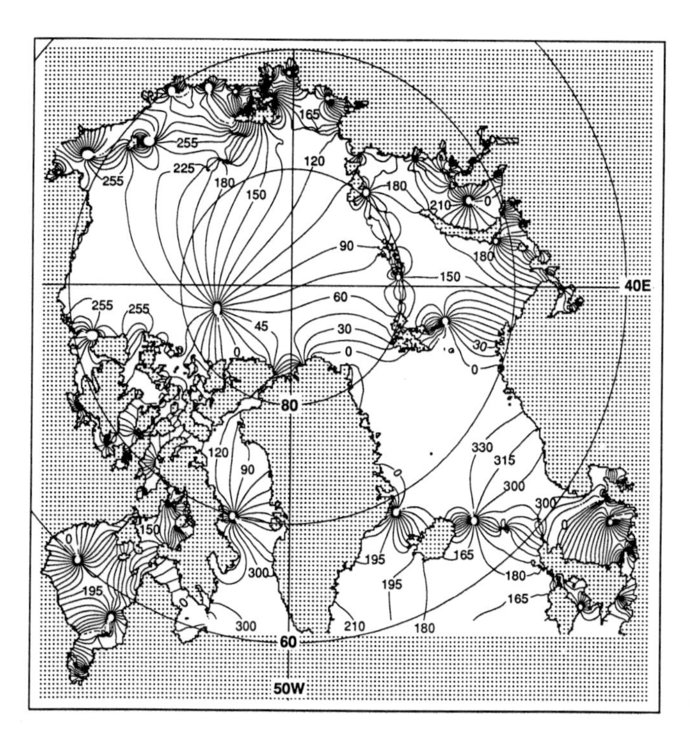

Fig. 4b. Computed phase (deg) of surface elevation for the semidiurnal constituent M_2.

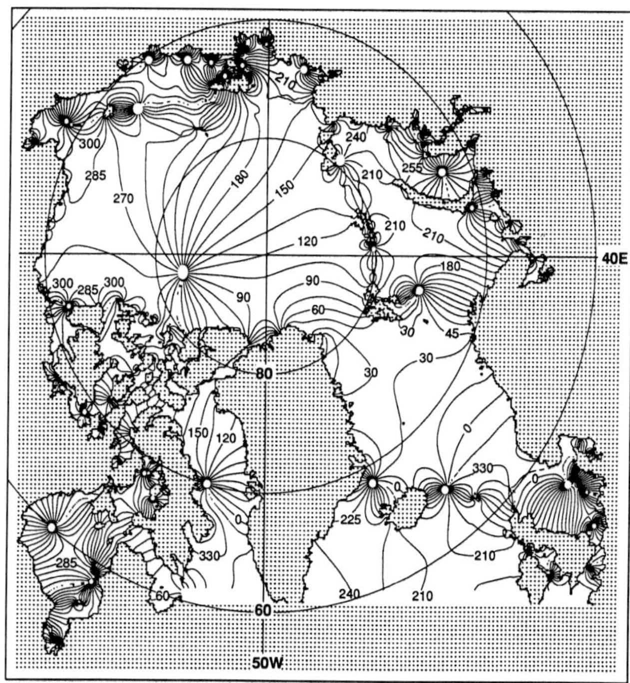

Fig. 5b. Computed phase (deg) of surface elevation for the semidiurnal constituent S_2.

tidal charts is referred to Greenwich and is expressed in degrees.

Distribution of the diurnal tides in the Arctic Ocean have been described in [*Kowalik and Proshutinsky*, 1993]. Here, for the sake of completeness, we depict only tidal charts with brief conclusions. The K_1 and O_1 tidal charts (Figures 2 and 3) derived with a 14-km space step possess in many locations an unexpected pattern. Huthnance [1981] measured enhanced diurnal currents near the Bear Island and suggested the influence of the trapped diurnal waves. Along the Barents Sea shelf break this pattern was discovered by *Gjevik* [1990] and was attributed to the shelf wave components in the diurnal tide. The homogeneous structure of the classical tidal pattern is often broken by the local regions of an increased amplitude and variable phase (see Figures 2 and 3). In the same regions, enhanced tidal currents occur. We have also noted that these regions of variable amplitude and phase are often confined to the shelf edge with a nonuniform bottom topography. We believe that this shorter wavelength phenomena occurs over seamounts, submarine peninsulas, and bays.

The M_2 and S_2 tidal maps are given in Figures 4 and 5, respectively. The calculated results (see Tables 1 and 2) turned out to be the closest to observational data on the Scandinavian coast, the southern part of the Barents Sea (including the entrance to the White Sea) where observed amplitude reaches 221 cm and computed value is 224 cm. It is interesting to note that three observations taken in the open water are close to the model results. Comparing these new charts against *Schwiderski* [1980], *Kowalik and Untersteiner* [1978], *Kowalik* [1981] and *Zahel* [1977] computations, one can note several new amphidromic points on the shelf of the Arctic seas.

Comparison model results against Geosat altimetry (Table 2) shows that both data sets are practically identical in the North Atlantic, Norwegian and Greenland Seas. There is only one line of 15 points of Geosat altimetry measurements along 69.5^0 N latitude in the southern Barents Sea. A scattering diagram (Figure 6, left portion) depicts comparison between Geosat altimetry and computed M_2 amplitudes in this area. Unfortunately, comparison against satellite data in the Chukchi Sea shows large differences (Figure 6, right portion). Computed amplitudes in this region are close to the tide gage data. One must conclude that satellite data were distorted by the frequent occurrence of ice cover in the Chukchi Sea.

Tidal maps of the semidiurnal waves M_2 and S_2 are, in general, quite similar. The phase difference of these waves of 50^0-80^0 restricts the age of semiduirnal tides to 2-3 days.

The largest discrepancy among various M_2 tidal charts obtained by the different authors occurs in the Barents Sea. *Proshutinsky and Polyakov* [1991] showed that the periods of natural oscillations in the southern part of the Barents Sea and especially in the entrance to the White Sea are close to the periods of semidiurnal tidal constituents. The rate of resonance amplification of the tide can be determined as the amplitude ratio between observed and equilibrium tides. For the southern part of the Barents Sea, this ratio at the frequency of the M_2 wave is equal to 58, and for S_2 is equal to 45.4. Tide computation in the Arctic Ocean without the White Sea leads to considerable redistribution of the tidal energy not only in the Barents sea but in the adjacent areas as well.

Local resonance is also important in other regions of the Arctic Ocean. Natural periods close to the 12-hr period occur in: Baydaratskaya Guba in the Kara Sea, Khatanga Gulf in the Laptev Sea, the regions north of the New Siberian Islands, the Wrangel Island area, and a number of straits of the Canadian Arctic Archipelago.

Proshutinsky and Polyakov [1991] investigated natural oscillations of the Arctic Ocean through the energy spectra of these oscillations. In the diurnal band of periods, the energy for all the Arctic Ocean, except the Norwegian and Greenland seas, and Baffin and Hudson bays was estimated to be very small. Therefore they concluded that the diurnal tides in the Arctic Ocean are very small due to anti-resonance conditions of generation and propagation. Recent numerical experiments with spatial resolution of about 14 km confirmed an absence of large scale resonance in the diurnal band but at least 26 local regions were found in the Arctic Ocean, including the continental slope of the Siberian shelf, where resonance occurs [*Kowalik and Proshutinsky*, 1993]. In these local areas, natural oscillation periods ranged from 23 to 25 hours.

The difference in the periods between diurnal and semidiurnal tidal waves results in the different dynamics of tide propagation. To elucidate at least some of these differences we consider energy fluxes for the M_2 and K_1 constituents in the area of the Yermak Plateau and Spitsbergen (Figure 7).

One can glean from Figure 7a that the M_2 wave does not interact with the Yermak Plateau. Energy flux of this wave is trapped by the system of Spitsbergen Islands and Spitsbergenbanken. Energy flux of the diurnal tidal wave K_1 is trapped by Yermak Plateau (Figure 7b) and partly by Spitsbergenbanken and Spitsber-

TABLE 1. Comparison of Computed (*Comp.*) and Observed (*Obs.*) M_2/K_1 Amplitude h and Phase g at Coastal and Offshore Tide Gauges

Gauge	*Latitude*/*Longitude*	h, cm *Comp.*	h, cm *Obs.*	g, deg *Comp.*	g, deg *Obs.*
Open sea 1	80 00′N/030 00′E	020.2/007.0	021.0/007.0	091.2/321.8	097.0/305.0
Open sea 2	74 32′N/030 58′E	016.2/003.8	15.8/004.0	085.7/341.6	092.0/322.0
Open sea 3	79 52′N/008 07′E	035.2/006.9	34.2/006.9	014.4/274.1	047.4/233.0
North Star Bay	76 33′N/068 53′W	077.1/041.0	078.9/037.7	093.2/256.3	103.1/249.8
Wrangel Bay	82 00′N/062 30′W	043.3/009.5	044.0/010.2	098.9/298.5	105.4/311.2
Maxwell Bay	74 41′N/088 54′W	070.4/30.6	067.5/028.1	176.3/309.7	178.5/319.8
Arctic Bay	73 02′N/085 10′W	059.2/018.8	061.2/015.8	153.3/269.3	153.0/280.0
Sachs Harbour	71 58′N/125 15′W	004.6/003.1	004.6/004.0	161.4/107.2	142.4/150.5
Grise Fiord	76 25′N/083 05′W	096.0/037.5	096.3/032.0	140.5/269.9	137.6/265.0
Cape Flora	79 57′N/049 59′E	013.2/006.4	013.3/006.8	184.4/352.1	175.7/340.2
Vardo	70 20′N/031 06′W	100.9/010.9	100.7/011.9	103.3/260.8	100.4/252.1
Bear Island	74 29′N/019 12′E	033.9/005.5	034.2/005.4	005.7/180.0	011.6/210.8
Hammerfest	70 40′N/023 24′E	087.9/007.4	088.3/008.1	025.8/217.5	018.6/214.0
Alesund	62 29′N/006 09′E	061.2/006.1	060.9/006.0	295.9/167.1	302.0/155.0
Longyearbyen	78 13′N/015 38′E	052.4/007.5	052.2/006.9	013.6/243.0	027.0/236.0
Ribachiy	69 55′N/032 40′E	112.6/014.8	112.0/015.0	120.0/267.9	120.0/270.0
Sosnovka	67 14′N/041 00′E	212.7/020.6	211.0/020.0	232.1/320.0	234.0/325.0
Amderma	69 45′N/061 42′E	020.0/002.5	020.0/004.0	351.7/161.6	348.0/121.0
B. Solnechnaya	7813′N/103 06′E	015.7/006.0	016.0/006.0	127.4/289.9	129.0/299.0
B. Pronchishevoy	75 38′N/113 25′E	044.1/008.0	043.0/008.0	191.2/306.8	177.0/313.0
B. Tiksi	71 39′N/128 53E	011.5/005.4	013.0/004.0	059.7/281.2	069.0/312.0
Tempa	75 47′N/137 25′E	014.4/002.8	015.0/002.0	080.0/178.8	093.0/163.0
Live River	59 15′N/069 25′W	400.8/020.0	429.0/19.0	002.7/165.2	014.5/188.0
Lake Harbour	62 47′N/069 50′W	335.0/017.9	350.0/015.0	007.7/137.9	017.5/177.0
Frederikshab	61 52′N/051 25′W	102.4/17.1	103.0/15.0	284.9/165.0	266.7/160.0

gen Islands. Diurnal constituents depict strong interaction with bottom irregularities whose dimensions are in the range 150 km to 300 km. *Kowalik and Proshutinsky* [1993] showed that the energy of the diurnal tides flows to the shelf wave regions and, afterwards, this energy is strongly dissipated in the same regions. The semidiurnal tidal energy flux, with the source of energy in the North Atlantic, propagates mainly to the north along the eastern boundaries of the Arctic Ocean and dissipates in regions of the semidiurnal tide resonance.

Part of the Arctic Ocean is located above critical latitude ($\phi = 74.48^0$) for the M_2 tide wave. In this area, a topographically amplified semidiurnal shelf wave should occur and, indeed, we found three regions

TABLE 2. Mean Absolute Error (MAE), Intercept (I), Coefficient (B), Standard Deviation (S) of a Linear Regression Line, and Correlation Coefficient (R) Between Observed and Computed Amplitudes of M_2/S_2 Waves

Data Base		MAE, cm	I, cm	B	S, cm	R
Coastal stations	(217)	3.33 / 1.28	−2.20 / 0.00	1.02 / 0.99	±4.69 / ±1.74	0.992 / 0.993
Pelagic stations	(24)	2.94 / 1.25	0.00 / 1.00	1.01 / 1.00	±3.40 / ±1.50	0.995 / 0.995
Geosat altimetry	(421)	2.98 / –	1.10 / –	0.97 / –	±3.44 / –	0.993 / –
Schwiderski	(6350)	2.13 / 1.35	−0.90 / −0.90	0.98 / 0.95	±2.88 / ±2.08	0.993 / 0.990

on the Siberian shelf where these phenomena exist (Figure 4). It will require further investigation to understand how these shelf waves were generated.

4.2. *Tidal currents*

In Plate 1, the maximum of the tidal currents in the Arctic Ocean is depicted. These currents are due to M_2, S_2, O_1 and K_1 tidal constituents. The largest currents in the Arctic Ocean occur at the entrance to the White Sea and at the Spitsbergenbanken, south from Spitsbergen. Relatively large currents also occur along the continental slope of the Arctic Ocean from Spitsbergen to the New Siberian Islands. Investigation of the enhanced velocity along the continental slope shows that these are caused by near-resonant amplification of the diurnal tides by the continental slope topography [*Kowalik and Proshutinsky*, 1993]. The topographically-amplified diurnal tidal currents are especially strong over the Yermak Plateau, located northeast from Spitsbergen [*Kowalik*, 1994].

Comparison of computed M_2 currents against observed ones is presented in Table 3. In Table 4 the maximum M_2 currents observed in the Barents Sea are presented (from *Gjevik* [1990]) and compared against computed values.

Together with oscillatory tidal currents, a residual tidal motion occurs due to nonlinear terms. This weak motion may influence water properties, ice concentration and suspended sediments, and is also a frequent

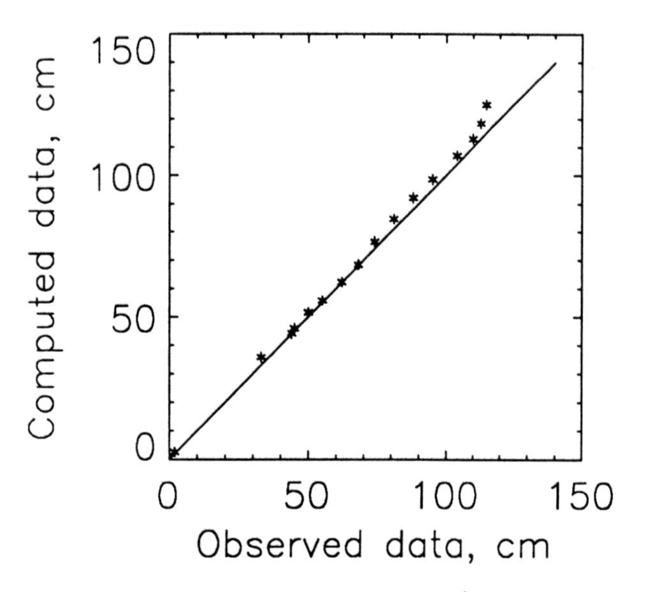

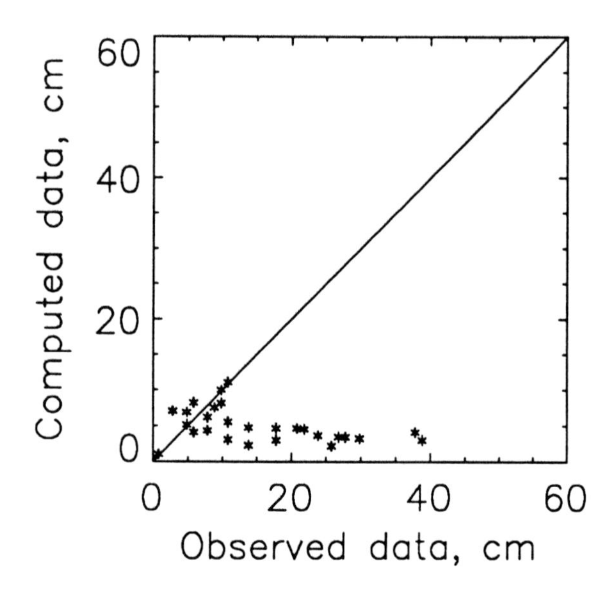

Fig. 6. M_2 amplitude along 69.5^{0}N in the Barents Sea (left) and in the Chukchi Sea (right). Observed data courtesy R. D. Ray (personal communication), computed data obtained from a 14-km resolution tide model.

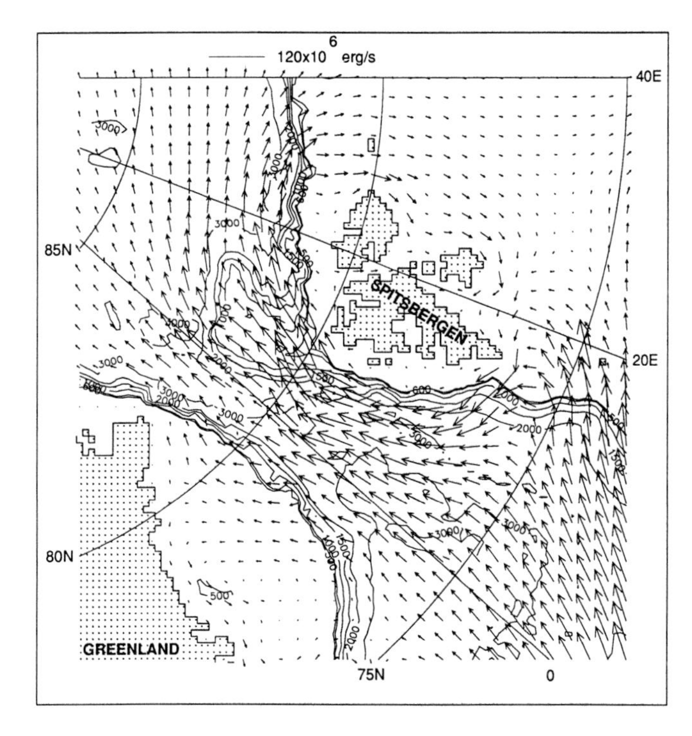

Fig. 7a. Tidal energy flux for M_2 wave in the Fram Strait region. Continuous lines denote depth contours (m).

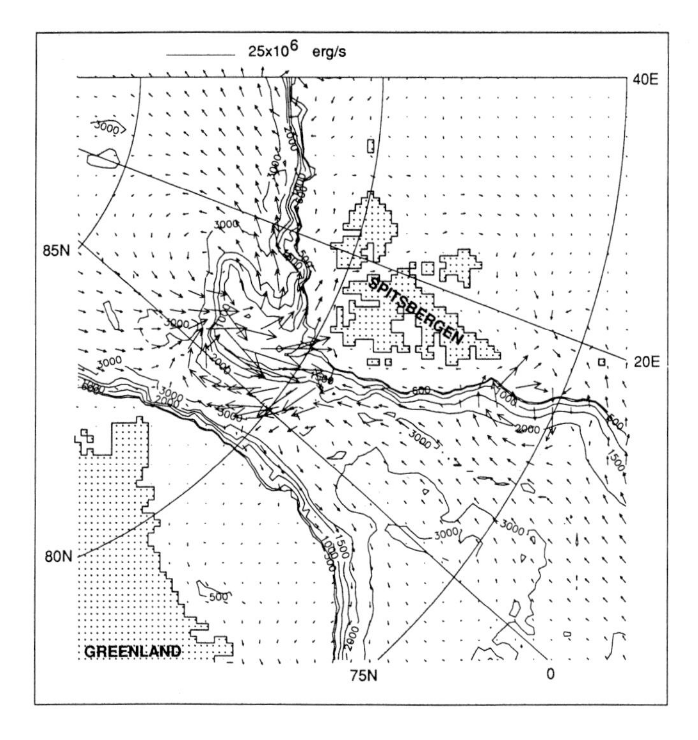

Fig. 7b. Tidal energy flux for K_1 wave in the Fram Strait region. Continuous lines denote depth contours (m).

source of eddies. The residual tidal currents were obtained by averaging the tidal currents over the tidal period. In Figures 8a, b, c, and d, residual motion at the Siberian shelf of the Laptev Sea is depicted. The anticyclonic residual eddies form over elevations of the sea floor, whereas cyclonic eddies develop in areas of depressed bottom relief with considerable slopes. It is interesting to note that residual tidal eddies are quite similar for different tidal constituents. Anticyclonic eddies develop over the same sea bottom irregularities and approximately at the same places for different tidal waves (Figure 8). We have also noticed that residual tidal currents are often observed in the regions where shelf waves are induced by diurnal tides. One can therefore assume, that through nonlinear terms in the equations, due to the variable bathymetry, the oscillating currents transfer vorticity to the mean field of motion.

The major regions of the Arctic Ocean with significant residual tidal currents are located at the entrance to the White Sea, Spitsbergenbanken, and continental slopes of the East-Siberian and Laptev Seas.

4.3. *Tide–ice interaction*

At least four different methods have previously been used for the tidal ice drift estimation. *Zubov* [1945] and *Legenkov* [1968] assumed that ice motion could be completely determined by surface tidal currents. *Kagan* [1968], *Kheysin and Ivchenko* [1973], *Kowalik* [1981] and *Proshutinsky* [1991] took into account the dynamic interaction of ice cover and formulated coupled water-ice models.

Constitutive law describing internal ice forces (7) was tested over the tidal range of periods by *Kowalik* [1981]. In this law, three empirical parameters, i.e., coefficients of shear (η) and bulk (Λ) viscosity and pressure coefficient (A_p), ought to be specified. Moreover, the transfer of energy between ice and water depends on the magnitude of the ice-water drag coefficient (R_i). The magnitude of this drag coefficient changes from summer to winter, due mainly to the change of ice roughness. An increase in the drag coefficient will result in the increased energy transfer between ice and water. Under the pack ice we use $R_i=5.5\cdot10^{-3}$, the value proposed by *McPhee* [1980]. Under the fast ice, drag coefficient is increased to $R_i=7.8\cdot10^{-3}$ due to increased ice roughness beneath the fast ice.

Internal ice forces in the tidal range of periods, measured as internal ice stresses (force per unit surface), display wide variability. A mean diurnal cycle of the internal ice stress in the Fram Strait from 7 May to 31 May 1988 displays variability in the range of 0.01

Maximum of the tidal currents, cm/s

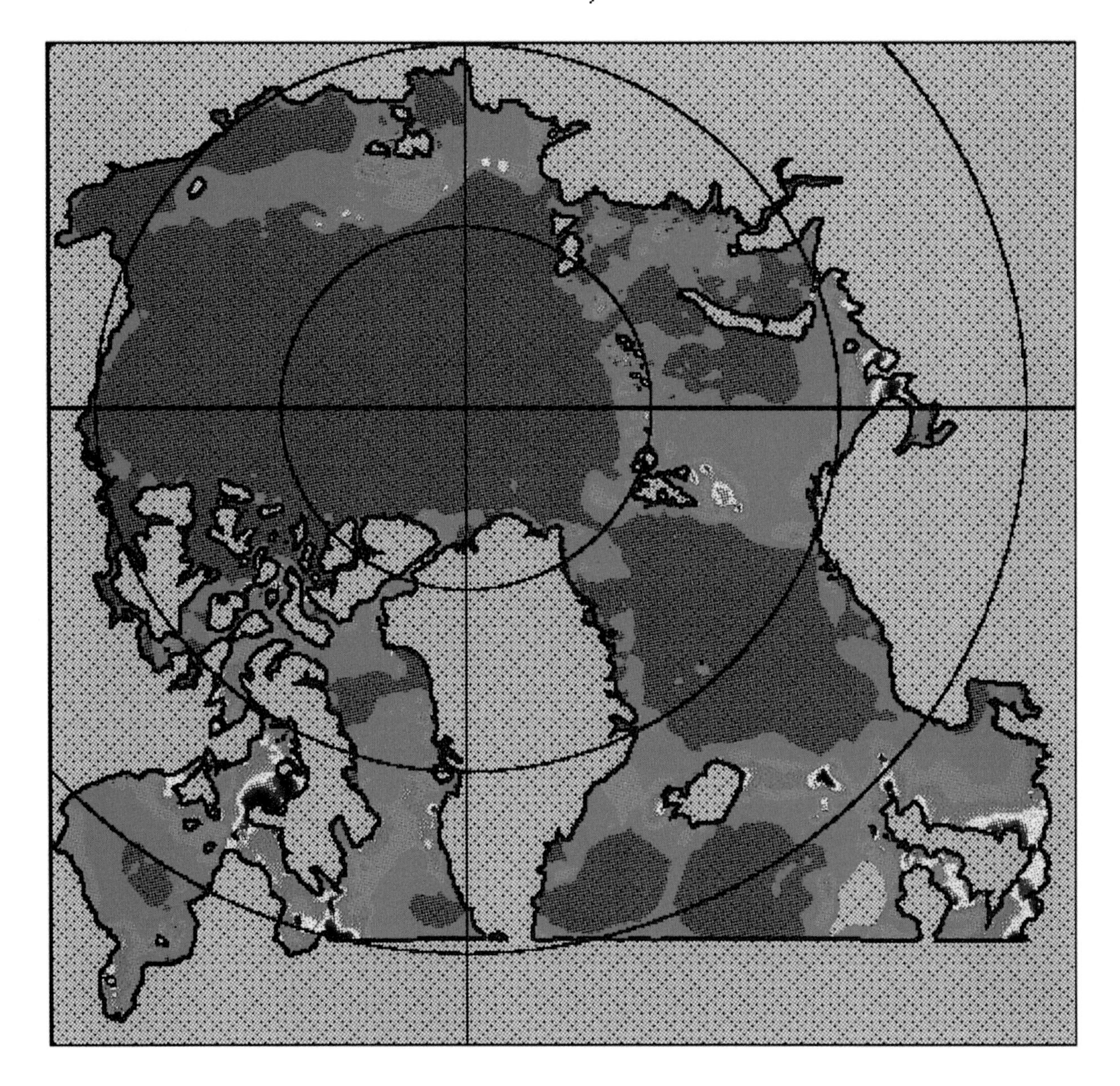

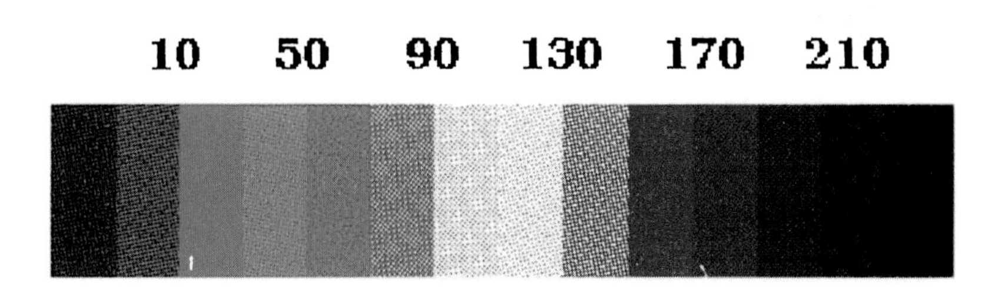

Plate 1. Maximum tidal currents due to four tidal constituents M_2, S_2, K_1 and O_1.

TABLE 3. Comparison of Observed (1) and Computed (2) M_2 Ellipse Parameters

		Major Half-Axis A, cm s^{-1}		Minor Half-Axis B, cm s^{-1}		Azimuth Angle of A, degrees		Rotation (+ clockwise)	
Latitude	Longitude	1	2	1	2	1	2	1	2
78° 59N	05° 15E	2.5	4.0	0.4	0.2	18	-7	+	+
79° 00N	04° 25E	2.5	3.7	0.3	0.3	14	-11	+	+
78° 55N	03° 18E	2.5	3.6	0.4	0.3	8	-6	+	+
65° 03N	07° 35E	5.6	5.7	0.7	0.3	74	76	+	+
66° 18N	09° 31E	3.7	3.7	2.1	2.2	137	145	-	-
66° 02N	07° 32E	3.5	3.4	1.7	1.8	108	84	-	-
61° 10N	69° 37W	28.1	28.8	5.1	5.4	180	184	+	+
62° 07N	68° 25W	40.0	41.0	2.8	0.8	224	213	+	+

to 0.02 N/m^2 [*GSP Group*, 1990]. Ice stress recorded in late fall of 1988 to the east of Spitsbergen [*Tucker III and Perovich*, 1992] shows twice-daily oscillations in the range of $25 \cdot 10^3$ to $50 \cdot 10^3$ N/m^2. This large variability, expressed through coefficient of viscosity (η) or pressure coefficient (A_p), will place these coefficients in the range of 10^7 cm^2/s – 10^8 cm^2/s to 10^{11} cm^2/s – 10^{12} cm^2/s.

To study the tide-ice interaction, we consider a multi-year average ice cover distribution in the Arctic Ocean for winter and summer seasons. In Figure 9, average winter ice cover is shown. Three areas are clearly delineated in this figure: the ice-free region in the North Atlantic, Greenland and Barents Seas; the pack ice region in the Arctic Ocean; and shore-fast ice regions along the coasts. A set of equations expressing ice-water interaction is taken according to location of a numerical grid in the above areas.

In the ice-free areas, equations (1) and (2) are solved subject to the boundary condition at the bottom (5), $\mathbf{T}_i = 0$, and $c = 0$. Under the shore-fast ice the same equations are used. Ice velocity is equated to zero and compactness is $c = 1$. In the areas covered by pack ice, the full system of equations (1)-(8) were used to obtain a solution. A series of numerical experiments have been carried out with different versions of the model in order to study the effects of internal ice stresses, ice distribution, ice-water friction and ice thickness. To test influence of the variable coefficients in the constitutive law, we used a friction coefficient from a range of 10^7 – 10^9 cm^2/s and pressure coefficient from a range of 10^8 – 10^{10} cm^2/s. These numerical experiments show that ice cover in the Arctic Ocean is of minor importance for tidal wave propagation. Due to ice cover, the amplitudes of sea level oscillations decreased about 3% and the phase is lagging about 4.5° to 5°. Locally, these differences occur in a wider range. Tides in the Arctic Ocean generally propagate in the same fashion as in ice-free water bodies. The long tide wave of several hundred kilometers in length has very small surface slopes. On such surfaces, pack ice cover behaves like a flexible membrane which weakly damps the vertical

TABLE 4. Comparison of Computed (1) and Observed (2) Maximum M_2 Currents

Latitude		Longitude		Maximum velocity, cms^{-1} 1	Maximum velocity, cms^{-1} 2	Rotation (+ clockwise) 1	Rotation (+ clockwise) 2
74°	15N	19°	00E	42.0	41.3	+	+
74°	00N	20°	00E	21.0	21.0	+	+
73°	12N	19°	30E	10.0	10.5	+	+
71°	27N	19°	20E	19.0	18.8	+	+

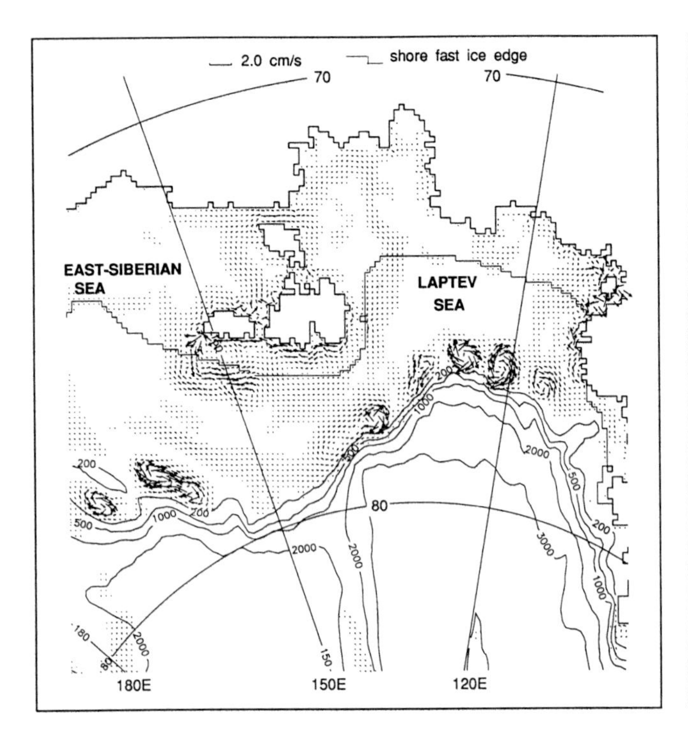

Fig. 8a. Residual tidal circulation in the East Siberian and Laptev Seas due to M_2 constituent. Notice the extent of shore-fast ice.

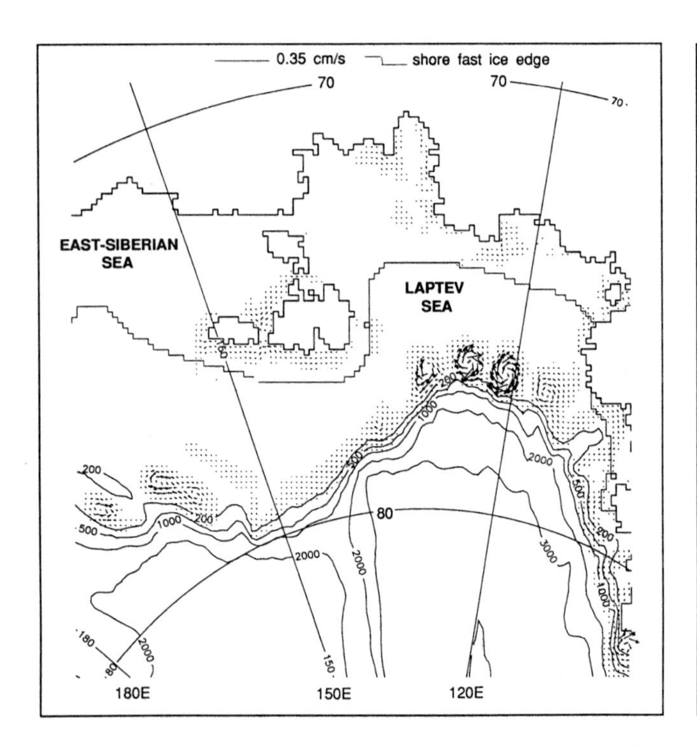

Fig. 8b. Residual tidal circulation in the East Siberian and Laptev Seas due to K_1 constituent. Notice the extent of shore-fast ice.

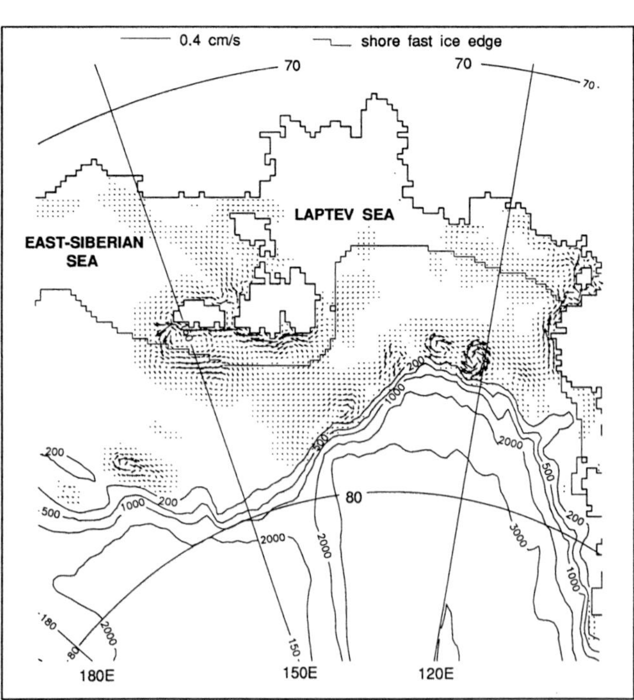

Fig. 8c. Residual tidal circulation in the East Siberian and Laptev Seas due to S_2 constituent. Notice the extent of shore-fast ice.

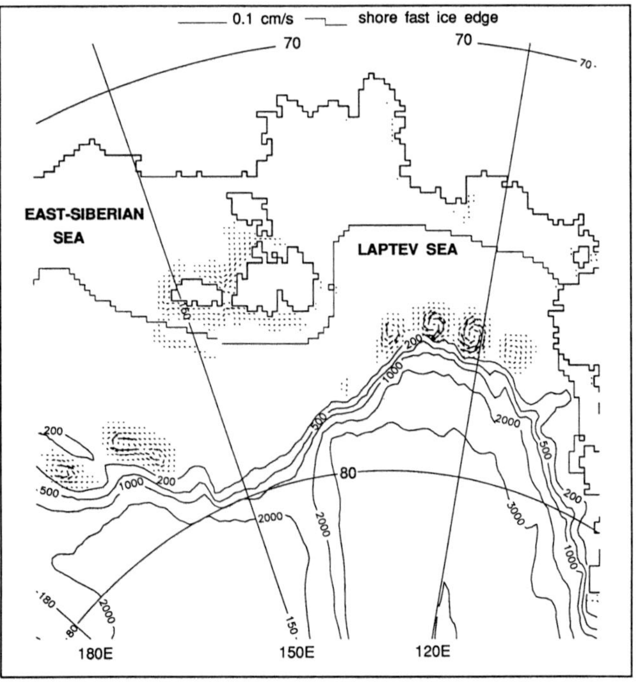

Fig. 8d. Residual tidal circulation in the East Siberian and Laptev Seas due to O_1 constituent. Notice the extent of shore-fast ice.

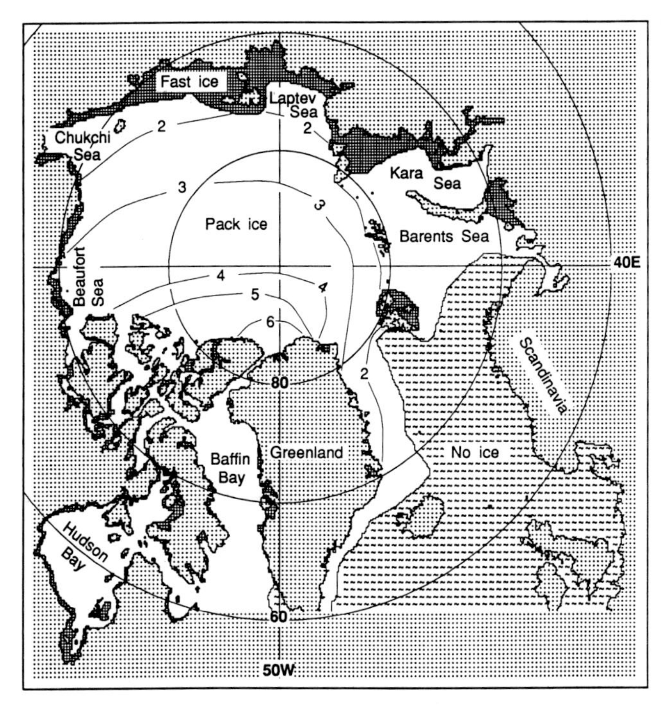

Fig. 9. Ice distribution in the Arctic Ocean for the average winter conditions. Ice thickness is given in meters.

mode of motion but strongly resists horizontal motion.

We should stress that we were unable to reproduce the phase lag and amplitude change observed during winter in the very shallow water under shore-fast ice [*Kowalik and Murty*, 1993]. First, due to the 14-km grid step we could not reproduce the bathymetry of small and narrow bays; and second, we did not have information to estimate changes in bathymetry caused by the shore fast-ice. Observations taken at Tuktoyaktuk, located on the southern part of the Beaufort Sea, show that the average depth of 2 m could be reduced to 1 m by ice cover or increased to 3 m by a storm surge [*Murty*, 1985].

To demonstrate the model performance in Figure 10, we compare mean diurnal ice stresses measured 7-31 May 1988 at an ice floe station in the Fram Strait [*GSP Group*, 1990] and calculated K_1 internal ice stresses in the same region. In this experiment $\eta = 10^7$ cm^2/s and $A_p = 10^8$ cm^2/s, $\Lambda = \eta$.

Ice motion due to tides usually repeats elliptical motion in water. This observation can be derived by comparing the pack ice motion from Figure 11a and water particle motion from Figure 11b caused by the M_2 tidal wave. Tidal motion measured by the buoys deployed on the drifting ice during the CEAREX experiment and the motion of water under the ice was practically identical [*Pease et al.*, 1994]. One can conclude that the ice-tide interaction is only moderately affected by internal stresses and in the first approximation it can be considered through an approach suggested by *Zubov* [1945] and *Legenkov* [1968].

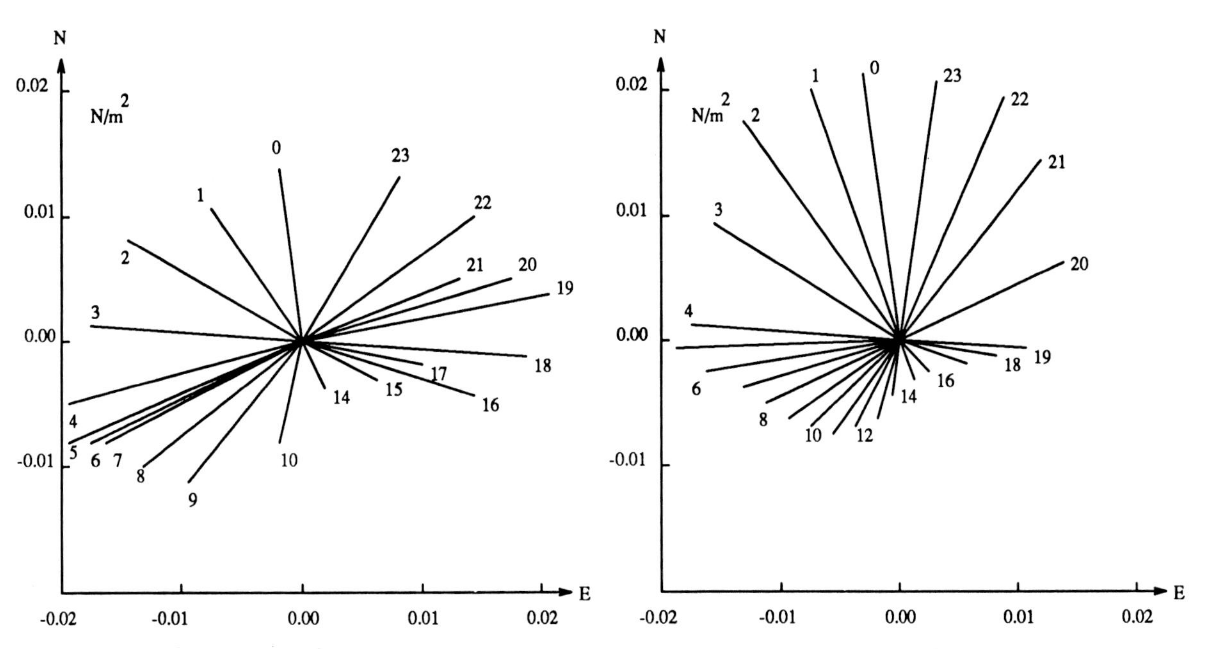

Fig. 10. Diurnal march of internal ice stress in the Yermak Plateau region. Measured at an ice floe station May 1988 during ARKTIS experiment [*GSP Group*, 1990] (left) and computed for the K_1 constituent (right).

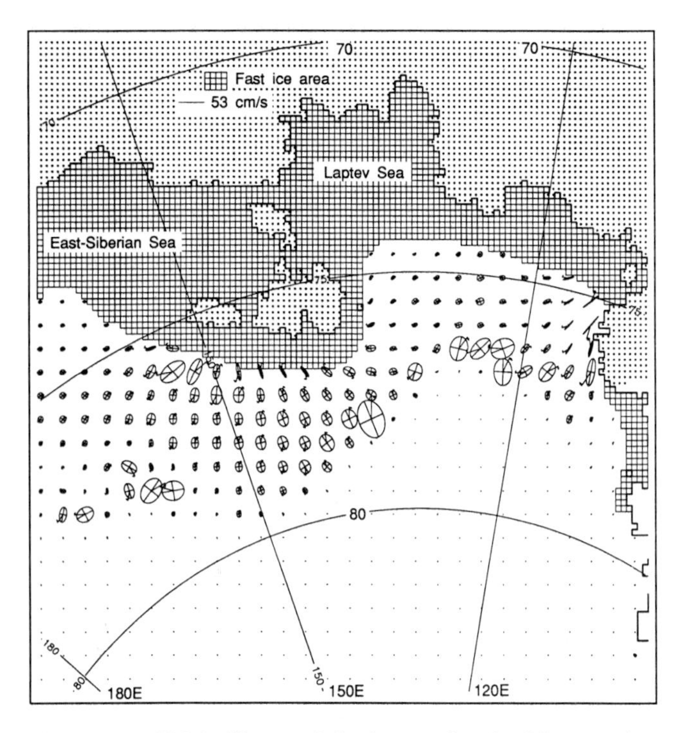

Fig. 11a. Tidal ellipses of the ice motion for M_2 wave in the East Siberian and Laptev Seas.

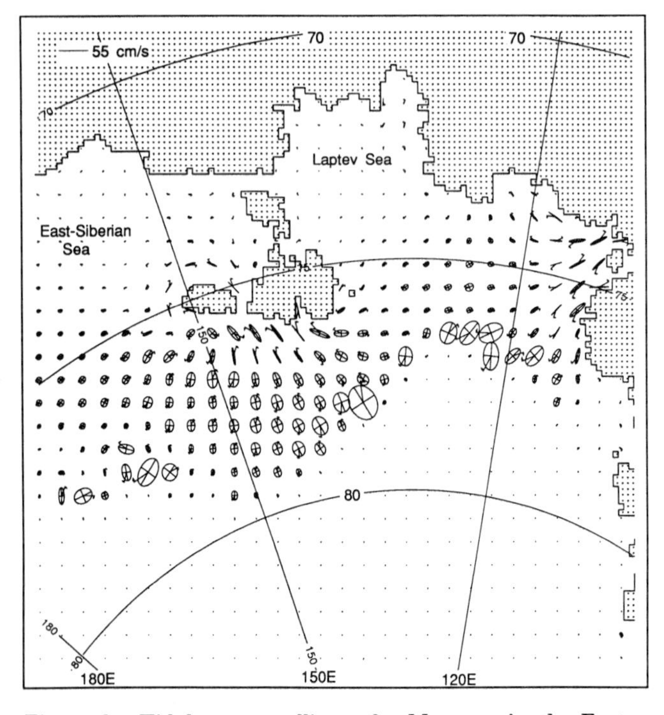

Fig. 11b. Tidal current ellipses for M_2 wave in the East Siberian and Laptev Seas.

Compressive non-linear stress (8) incorporated into equation (7) for the internal ice forces relates stress to ice motion. This stress should suppress, in the case of convergence, the ice movement. Time dependence of the internal ice pressure displays a strong nonlinear process. An example of the temporal behavior of internal ice pressure in the Fram Strait is depicted in Figure 12. Pressure in the ice occurs only for part of the tidal period. Stress recorded by *Tucker III and Perovich* [1992] in drifting pack ice shows similar temporal variations.

The nonlinear effects manifest their presence in the calculated spectra of the internal ice pressure. Calculation in the Yermak Plateau region depicts both in the diurnal component (Figure 13, top) and in the semidiurnal component (Figure 13, bottom), secondary peaks due to nonlinear interaction.

We have already described residual motion in water caused by nonlinear tidal dynamics. This motion will transfer momentum to the ice through the ice-water stress. Therefore, one can assume that similar motion will occur in the ice cover as well. Moreover, if the ice motion is averaged over the tidal period a residual motion superposed on the residual water motion will occur due to non-linear internal ice pressure. According to (7), residual motion is a function of ice velocity convergence and magnitude of the pressure coefficient. It is obvious that the residual ice motion generated through nonlinear ice pressure is of transient character in time because: 1) ice compactness is changing seasonally and thus influencing interactions between ice floes, and 2) residual ice motion will change ice compactness, thus causing change in ice floe interactions. An example of residual ice motion on the shelf of the Laptev Sea is depicted in Figures 14a and 14b. Due to non-linear effects in the ice cover, residual ice motion close to the shore (or shore-fast ice) is directed from the solid boundary into the open ocean. Owing to the high space resolution we obtained a number of very interesting off-

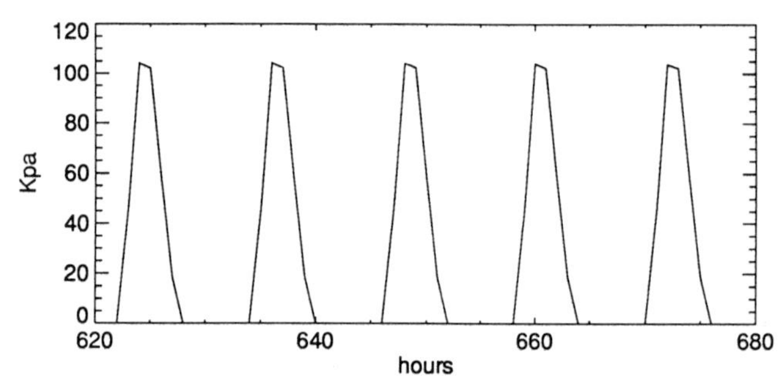

Fig. 12. Time dependence of the internal ice pressure in the Yermak Plateau region due to M_2 wave.

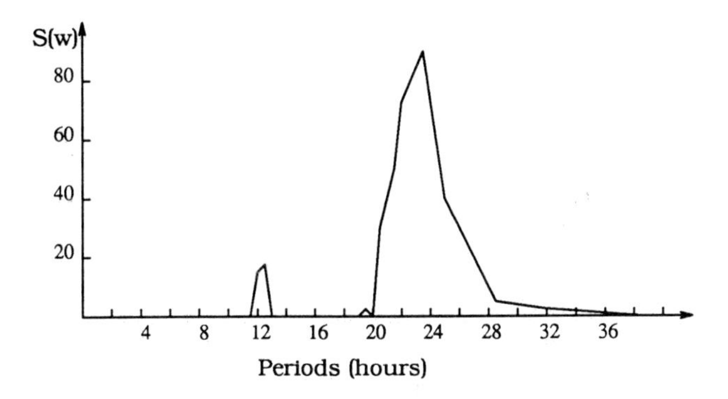

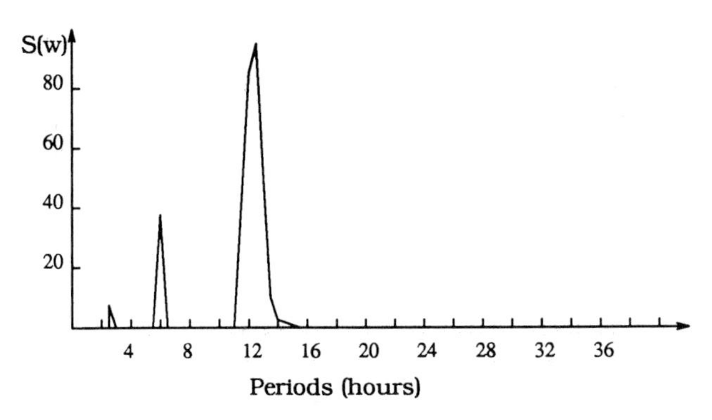

Fig. 13. Spectra of the internal ice pressure in the Yermak Plateau region. Due to K_1 wave (top) and due to M_2 wave (bottom).

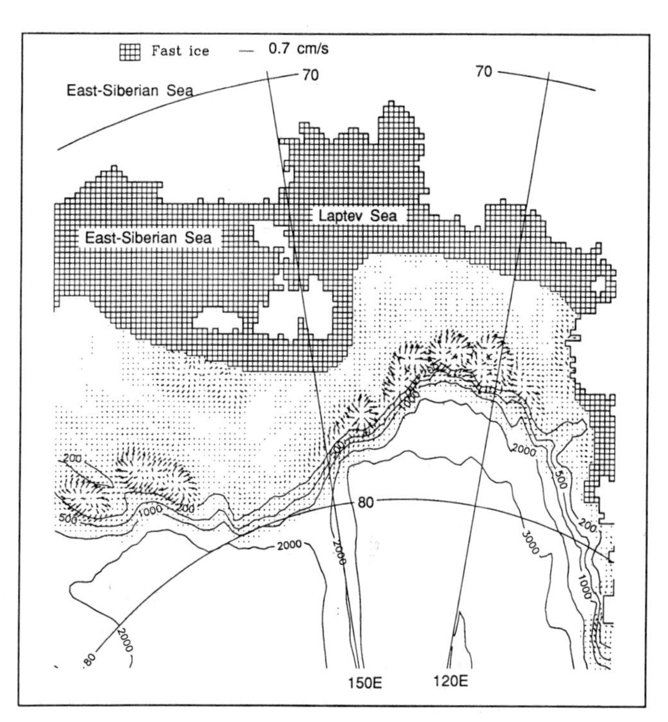

Fig. 14a. Residual ice motion in the East Siberian and Laptev Seas caused by M_2 wave. Ice viscosity coefficient is taken as $\eta = 10^7\text{cm}^2/\text{s}$ and ice pressure coefficient is $A_p = 10^8\text{cm}^2/\text{s}$.

shore patterns in the tidal residual ice motion (Figure 14a). We have considered two different ice conditions described by the different sets of the coefficient: an average condition given by the coefficients $\eta = 10^7$ cm^2/s and $A_p = 10^8$ cm^2/s, and a winter condition with pressure coefficient increased to $A_p = 10^{10}$ cm^2/s. For the average ice condition, at least seven regions of residual ice motion are depicted in Figure 14a. In Figure 14b, the results for winter are shown. Due to very high ice pressure one can see strong enhancement of the residual ice motion directed from the solid boundary into the open sea and from the centers of the ice convergence. This residual ice motion is related to the ice convergence and divergence. During half of the tidal cycle, ice motion is suppressed by the high ice compactness or by the solid boundary. During the second part of the tidal cycle, ice moves in the opposite direction, the compactness of the ice decreases and regions of open water appear. Therefore, averaging ice motion over the tidal cycle one obtains residual offshore propulsion of ice in the vicinity of the coastline, and in the open ocean, residual ice motion directed from the centers of high convergence.

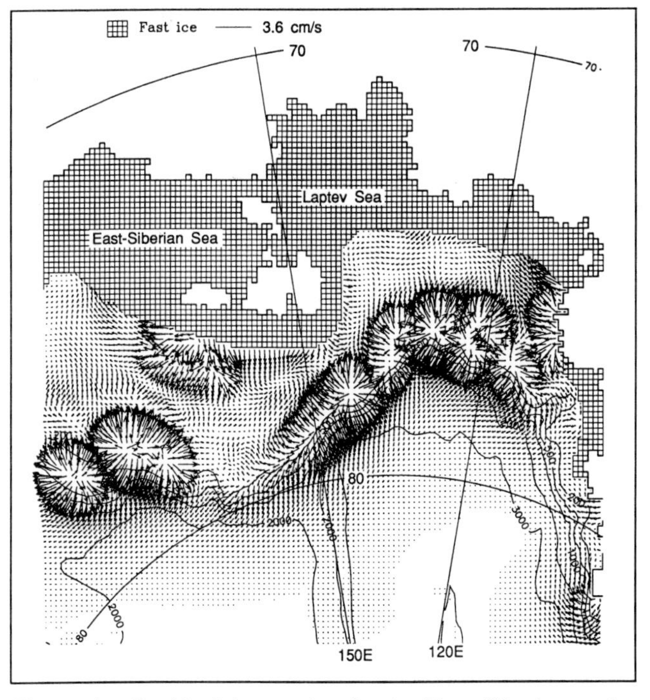

Fig. 14b. Residual ice motion in the East Siberian and Laptev Seas caused by M_2 wave. Ice viscosity coefficient is taken as $\eta = 10^7\text{cm}^2/\text{s}$ and ice pressure coefficient is $A_p = 10^{10}\text{cm}^2/\text{s}$.

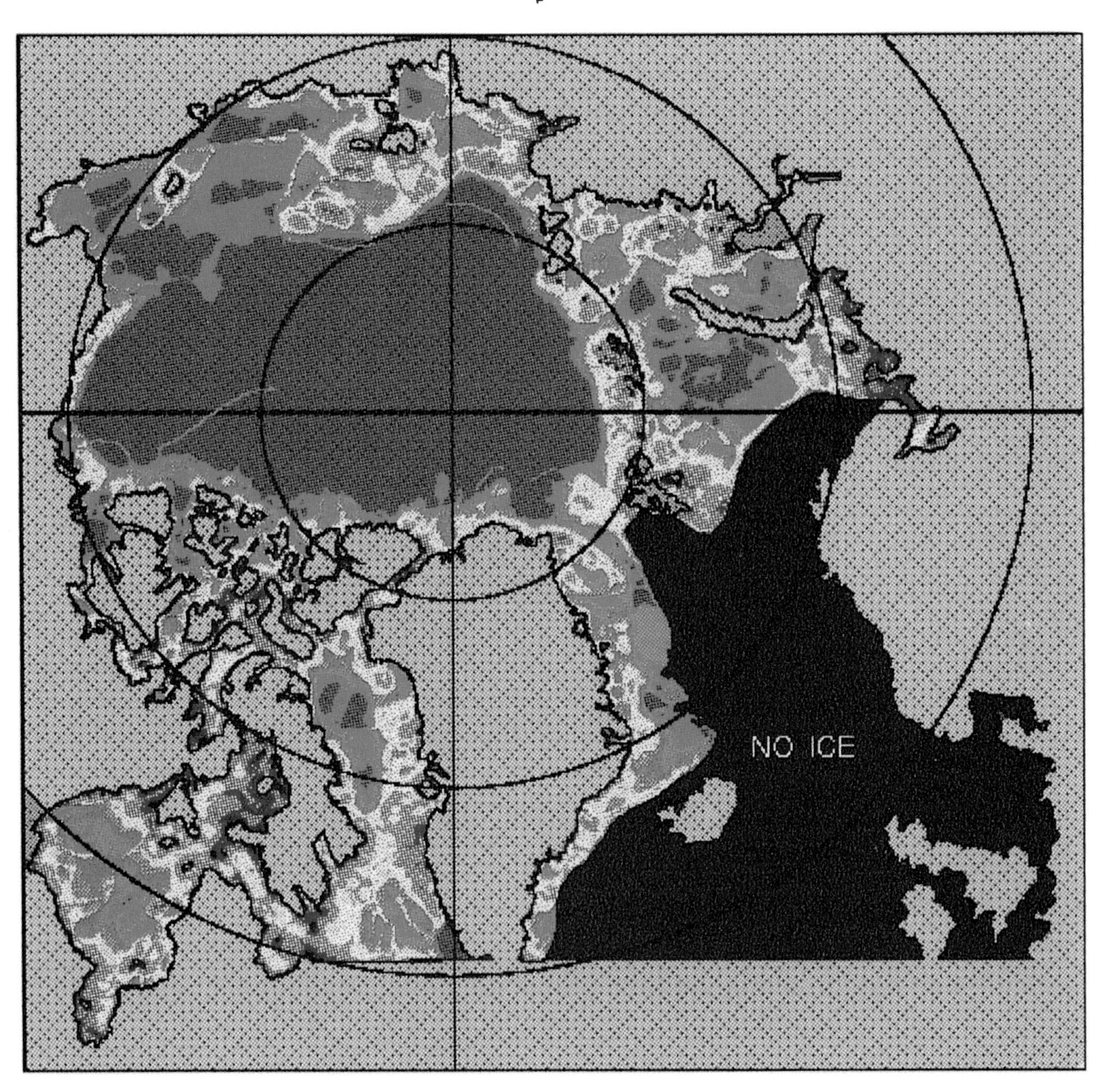

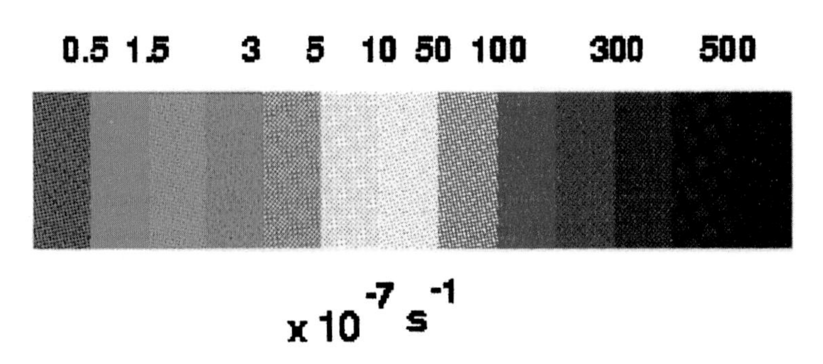

Plate 2. Distribution of maximal shear of ice velocity in the Arctic Ocean caused by four tidal waves (M_2, S_2, K_1 and O_1).

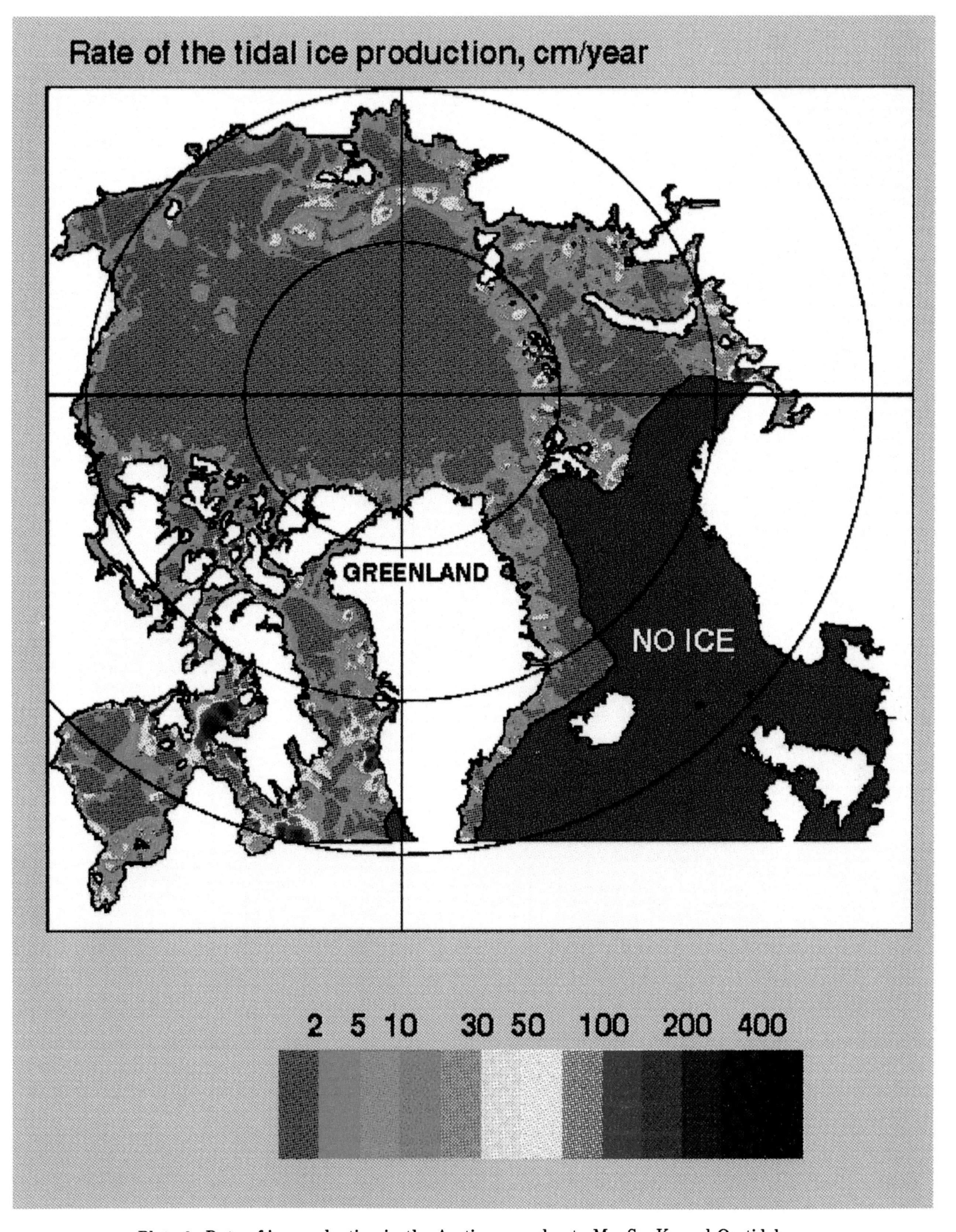

Plate 3. Rate of ice production in the Arctic ocean due to M_2, S_2, K_1 and O_1 tidal waves.

5. SUMMARY AND DISCUSSION

We have constructed a tidal model with resolution of about 14 km, obtained solutions, and analyzed charts of amplitude, phase currents, and ice motion in the Arctic Ocean. Earlier [*Kowalik and Proshutinsky*, 1993], we thoroughly analyzed diurnal tidal constituents (K_1 and O_1). Here, we investigate semidiurnal components (M_2 and S_2) and compare them with diurnal components. The difference has been elucidated by a concomitant pattern of energy flux in the Yermak Plateau – Spitsbergen for the M_2 and K_1 constituents. While the diurnal component tends to be trapped by the irregular bottom topography, semidiurnal tides propagate over such region without any change.

We have also introduced ice cover into the present model because tidal motion in the Arctic Ocean is frequently observed through pack ice. The process of tide propagation can be of practical significance to navigation in ice if it generates opening and closing of cracks in the ice cover. We use a maximal shear of ice velocity as a measure of ice cover deformation [*Kowalik*, 1981],

$$V_s = [(\frac{\partial u_i}{\partial x} - \frac{\partial v_i}{\partial y})^2 + (\frac{\partial u_i}{\partial y} + \frac{\partial v_i}{\partial x})^2]^{0.5}$$

Distribution of V_s in the Arctic Ocean is plotted in Plate 2. The most probable areas for the occurrence of ice leads, where V_s exceeds 10^{-6} s^{-1}, are: the entrance of the White Sea, the Spitsbergenbanken area, the northern part of the western Greenland continental slope, the continental slope of the Laptev and East Siberian Seas, and the Wrangel Island region (Plate 2). In this plate, one can locate a number of narrow, elongated areas of increased ice deformation. These areas must be a subject of special attention during satellite data processing.

Tidally-induced, periodic and residual motion of the ice interact with other motions of the ice cover caused by wind and permanent currents. These interactions may result in a number of interesting phenomena in the Arctic Ocean. For example, along the continental slope of the East Siberian and Laptev Sea, the "Great Siberian Polynya" develops. There are three major factors which influence the development of polynyas: mean seasonal wind directed to the north, heat advection of the Atlantic water, and tides and residual tidal ice drift. Most favorable conditions for development of the Siberian polynya start with southerly winds so that the pack ice moves into the open sea, Atlantic water is upwelled in the area of the continental slope and, finally, strong tidal currents mix surface Arctic and subsurface Atlantic waters and transfer heat to the surface. Additionally, a residual tidal ice drift transports ice away from the continental slope regions. Similar phenomena may occur over the Yermak Plateau, the Chukchi Cap, in the Lincoln Sea, and in regions of the Northeast Water polynya. Tidal mixing over the Arctic seamounts transfers heat from the warm Atlantic water to the surface. In the Norwegian Sea and North Atlantic, because the surface layer is warmer than the subsurface layers, the opposite situation occurs: tidal mixing transfers colder water to the surface. These colder and warmer spots can probably be observed in satellite imagery. *Proshutinsky* [1988] described, both through measurements and modeling, the cold water area (cold eddy) resulting from tides in the Faroe-Shetland Strait.

The periodic tidal currents, through divergence and convergence processes, generate the periodic change in the ice compactness distribution. To demonstrate the role of tide-generated openings, one may consider a hypothetical process of new ice formation in the Arctic Ocean due to the M_2, S_2, K_1 and O_1 constituents. We shall assume that for half of the tide cycle (during diverging ice motion when compactness decreases) young ice is formed in leads at a rate of 3 cm/day and that for the second half of the cycle (during converging motion) ice ridging occurs. The increase of ice volume caused by this mechanism will take place at a rate of $8 \cdot 10^{11}$ m^3/year (Plate 3). The ice will form mainly along the Eurasian shelf. This process, through salt brine rejection, will also result in increased salinity of the water and weaker thermohaline circulation.

Acknowledgments. We would like to express our gratitude to R. D. Ray from NASA, Goddard Space Flight Center for offering Geosat altimetry data and discussions. We wish to thank N. Z. Cherkis, Naval Research Laboratory, for his advice and data on the bathymetry charts of the Arctic Ocean. T. Vinje, R. Korsnes and T. Eiken, Norsk Polarinstitutt, Norway, kindly offered their advice and data on the bathymetry and tidal constituents in the Spistbergen area. We extend our appreciation to reviewers for very valuable comments.

We gratefully acknowledge support for this work from the National Aeronautic and Space Administration under grant NAGW-2972 and from National Science Foundation under grant DPP 9114549. Numerical computations were carried out using CRAY Y-MP8/464 from NASA, Goddard Space Flight Center.

REFERENCES

Aagaard, K., A. Foldvik, T. Gammelsrod, and T. Vinje, One-year records of current and bottom pressure in the strait between Nordaustlandet and Kvitoya, Svalbard, 1980-1981, *Polar Res., 1 (2)*, 107-113, 1983.

Aagaard, K., L. Darnall, A. Foldvik, and T. Torresen, Fram Strait current measurements 1984-1985, *Report no. 63*, Dept. of Oceanography, University of Bergen, 49 pp., 1985.

Aagaard, K., C. H. Pease, A. T. Roach, and S. A. Salo, Beaufort Sea mesoscale study, *Outer Continental Shelf Environmental Assessment Program, v. 65*, 1–136, 1990.

Cartwright, D. E., Extraordinary tidal currents near St. Kilda, *Nature, 223*, 928–932, 1969.

Cartwright, D. E., and B. D. Zetler, Pelagic tidal constants, *IAPSO Publication Scientifique No. 33*, Compiled by IAPSO Advisory committee on tides and mean sea level, 59 pp., 1985.

Cartwright, D. E., and R. D. Ray, Oceanic tides from GEOSAT altimetry, *Geophys. Res. Lett., 17*, 619-622, 1990.

Cartwright, D. E., R. D. Ray, and B. V. Sanchez, Oceanic tide maps and spherical harmonic coefficients from GEOSAT altimetry, *NASA Technical Memorandum 104544*, 75 pp., 1991.

CEAREX Drift Group, CEAREX Drift Experiment, *EOS Trans., AGU, 71, 40*, 1115–1118, 1990.

Defant, A., Neue Methode zur Ermittlung der Eigenschwingungen (seiches) von abgeschlossenen Wassermassen. (Seen, Buchten usw.), *Ann. Hydr. Mar. Met., 46*, 78–85, 1918.

Defant, A., Die Gezeiten des Atlantischen Ozeans und des Arctischen Meeres, *Ann. Hydr. Mar. Met., 52 (8–9)*, 153–166, 177–184, 1924.

Deming, J., et al., Northeast water polynya: Polar Sea cruise results, *EOS Trans., AGU, 74, (16)*, 185–196, 1993.

Dmitriev, N. E., A. Y. Proshutinsky, T. B. Loyning, T. Vinje, Tidal ice dynamics in the area of Svalbard and Frans Josef Land, *Polar Res. 9(2)*, 193–205, 1991.

Drinkwater, K. F., On the mean and tidal currents in Hudson Strait, *Atmosphere-Ocean, 26(2)*, 252–266, 1988.

Dvorkin, E. N., Tides, in: *Soviet Arktik*, Izd. Nauka, Moskwa, pp. 191-197, 1970.

Fjeldstad, J. E., Contributions to the dynamics of the free progressive tidal waves, *Sci. Res. Norweg. North Polar Exped. "Maud" 1918-1925*, Bergen, *4(3)*, 3–80, 1929.

Francis, O., and P. Mazzega, Global charts of ocean loading effects, *J. Geophys. Res., 95*, 11,411–11,424, 1990.

Geophysical Society of America, *Bathymetry of the Barents and Kara Seas*, Boulder, Colo., 1991.

Gidrographicheckoe Upravlenie VMF SSSR, *Tables of tides*, (in Russian), vol. 2, Harmonic constants for tide prediction, 295 pp., Leningrad, 1941.

Gjevik, B., Model simulations of tides and shelf waves along the shelves of the Norwegian-Greenland-Barents Sea, in *Modeling Marine Systems*, vol. 1, ed. A. M. Davies. CRC Press Inc, 187–219, 1990

Gjevik, B., and T. Straume, Model simulations of the M_2 and the K_1 tide in the Nordic Seas and the Arctic Ocean, *Tellus, 41*, 73–96, 1989.

Goldsbrough, G. R., The dynamical theory of the tides in a polar basin, *Proc. Lond. Math. Soc. 14(2)*, 31–66, 1913.

Goldsbrough, G. R., The tides in oceans on a rotating globe, *Proc. Roy. Soc. Lond., A, 117*, 692–718, 1927.

Godin, G., Cotidal charts for Canada, *Manuscript Rep. Ser. 55*, 91 pp., 1980.

GSP Group, Greenland Sea Project, *EOS Trans. 71(24)*, 1990

Harris, R. A., *Arctic Tides*, U.S. Coast and Geodetic Surv. Rep., 103 pp., 1911.

Huggett, W. S., M. J. Woodward, F. Stephenson, F. V. Hermiston, and A. Douglas, Near bottom currents and offshore tides, *Beaufort Sea Proj. Tech. Rep., 16*, Ocean and Aquatic Sciences, Dept. of Environment, Victoria, B.C., 38 pp., 1975.

Hunkins, K., Anomalous diurnal tidal currents on the Yermak Plateau, *J. Mar. Sci., 44*, 51–69, 1986.

Huthnance, J. M., Large tidal currents near Bear Island and related tidal energy losses from the North Atlantic, *Deep-Sea Res., 28A*, 51–70, 1981.

Hydrographer of the Navy, *Admiralty Tide Tables*, vol. 2, 300 pp., London, 1987.

Kagan, B. A., *Hydrodynamical models of tidal motion in the seas*, Gidrometeoizdat, Leningrad, 231 pp., 1968.

Kheysin, D. E., and V. O. Ivchenko, A numerical model of tidal ice drift with the interaction between floes, *Izv. Acad. Sci. USSR: Atmospheric and Oceanic Physics, 9*, 420–429, 1973.

Kowalik, Z., A study of the M_2 tide in the ice-covered Arctic Ocean, *Modeling, Identification and Control, 2(4)*, 201–223, 1981.

Kowalik, Z., Modeling of topographically-amplified diurnal tides in the Nordic Seas, *in press J. Phys. Oceanogr.*, 1994.

Kowalik, Z., and T. S. Murty, *Numerical modeling of ocean dynamics*, World Scientific Publ., 481 pp., 1993.

Kowalik, Z., and A. Yu. Proshutinsky, Diurnal tides in the Arctic Ocean, *J. Geophys. Res. 98*, 16,449–16,468, 1993.

Kowalik, Z., and N. Untersteiner, A study of the M_2 tide in the Arctic Ocean, *Deutsh. Hydr. Zeit., 31(6)*, 216–229, 1978.

Kozo, T. L., The hybrid polynya at the northern end of Nares Strait, *Geophys. Res. Lett., 18*, 2059–2062, 1991.

Legenkov, A. P., Determination of ice concentration, dispersion and compaction using tidal currents. *Trudy Arkticheskogo i Antarkticheskogo Instituta, 285*, 215–222, 1968.

Litke, F., On tides in the Great North Sea and Ice–Covered Sea, *Zapiski Hydrograficheskogo Departamenta Morskogo Ministerstva*, ch. 2, p. 353–376, 1844.

Marchuk, G. I., and B. A. Kagan, *Dynamics of Ocean Tides*, Kluwer Academic Publishers, 327 pp., 1989.

Martin, S., and D. J. Cavalieri, Contributions of the Siberian shelf polynyas to the Arctic Ocean intermediate and deep water, *J. Geophys. Res., 94*, 12,725–12,738, 1989.

McPhee, M. G., An analysis of pack ice drift in summer, *Proceedings of the AIDJEX Symposium*, Univ. of Washington Press, 62–75, 1980.

McPhee, M. G., Analysis and prediction of short-term ice drift, *Proc. 5th Int. Offshore Mechanics and Arctic Engineering Symposium, 4*, ASME, 385–393, 1986.

Morison, J. H., Seasonal variations in the West Spitsbergen Current estimated from bottom pressure measurements, *J. Geophys. Res., 96*, 18,381–18,395, 1991.

Mooers, C. N. K., and R. L. Smith, Continental shelf waves off Oregon, *J. Geophys. Res., 73*, 549–557, 1968.

Muench, R. D., M. G. McPhee, C. A. Paulson, and J. H. Morison, Winter oceanographic conditions in the Fram Strait–Yermak Plateau region, *J. Geophys. Res. 97*, 3,469–3,483, 1992.

Murty, T. S., Modification of hydrographic characteristics, tides, and normal modes by ice cover, *Marine Geodesy, 9(4)*, 451–468, 1985.

Nansen, F. (ed.), The oceanography of the north polar basin. *The Norwegian North Polar Expedition, 1893–1896, Scientific Results*, Publ. by Longmans, Green and Co., vol. 3, 422 p., 1902.

Naval Research Laboratory, *Bathymetry of the Arctic Ocean*, Washington, D. C., 1986.

Nekrasov, A. V., Computation and construction of M_2 tidal charts in the Norwegian and Greenland Seas by the Hansen's method, *Trudy Leningrad Gidromet. Inst., 16*, 49–57, 1962.

Norsk Polarinstitutt, *Western Barents Sea Bathymetry*, Oslo, 1989.

Padman, L., A. J. Plueddemann, R. D. Muench, and R. Pinkel, Diurnal tides near the Yermak Plateau, *J. Geophys. Res., 97*, 12,639–12,652, 1992.

Pease, C. H., P. Turet, R. S. Pritchard, and J. E. Overland, Barents Sea Tidal, Wind Drift, and Inertial Motion fron ARGOS Ice Buoys During CEAREX, *in press J. Geophys. Res.*, 1994.

Proshutinsky, A. Y., Generation of eddy structures in the Faroe-Shetland Strait by tidal currents, *Oceanology, 28*, *5*, 567–571, 1988.

Proshutinsky, A. Y., Tidal water and ice dynamics in the Arctic Ocean, in: *Proc. Int. Conf. on the Role of the Polar Regions in Global Change, Vol. 1*, 296–303, Geophysical Institute, University of Alaska Press, Fairbanks, 1991.

Proshutinsky, A. Y. and I. V. Polyakov, The Arctic Ocean eigen oscillations, in: *Proc. Int. Conf. on the Role of the Polar Regions in Global Change, Vol. 1*, 347-354, Geophysical Institute, University of Alaska Press, Fairbanks, 1991.

Ray, R. D., and B. V. Sanchez, Radial deformation of the earth by oceanic tide loading, *NASA Technical Memorandum 100743*, 51 pp., 1989.

Rothrock, D. A., The mechanical behavior of pack ice, *Annual Review of Earth Planetary Sciences, 3*, 317–342, 1975.

Schwiderski, E. W., *Global Ocean Tides, Part II: The Semidiurnal Principal Lunar Tide (M_2), Atlas of Tidal Charts and Maps*, Naval Surface Weapon Center, Dahlgren, VI 22248, 87 pp., 1979.

Schwiderski, E. W., On charting global ocean tides, *Rev. Geoph. Space Phys., 18*, 243–268, 1980.

Schwiderski, E. W., *Global Ocean Tides, Part III: The Semidiurnal Principal Solar Tide (S_2), Atlas of Tidal Charts and Maps*, Naval Surface Weapon Center, Dahlgren, VI 22248, 96 pp., 1981a.

Schwiderski, E. W., *Global Ocean Tides, Part IV: The Diurnal Luni-Solar Declination Tide (K_1)*, Naval Surface Weapon Center, Dahlgren, VI 22248, 87 pp., 1981b.

Schwiderski, E. W., *Global Ocean Tides, Part V: The Diurnal Principal Lunar Tide (O_1)*. Naval Surface Weapon Center, Dahlgren, VI 22248, 85 pp., 1981c.

Sretensky, L. N., On motion of free tidal wave inside polar basin, *Izv. AN SSSR, Ser. Geogr. i Geof., 3*, 383–402, 1937.

Sterneck, R., Die Gezeiten im nördlichen Eismeer, *Ann. Hydr. Mar. Met., 56 (21)*, 51–72, 1928.

Sverdrup, H. U., Dynamics of tides on the northern Siberian shelf, *Geophys. Publ., 4,5*, Oslo, Norway, 76 pp., 1926.

Tucker, W. B. III., and D. K. Perovich, Stress measurements in drifting pack ice, *Cold Regions Science and Technology, 20*, 119–139, 1992.

Wahr, J., Body tides on an elliptical, rotating, elastic and oceanless earth. *Geophys. J. R. Ast. Soc., 64*, 677–703, 1981.

Zahel, W., A global hydrodynamic–numerical 1^0-model of the ocean tides; the oscillation system of the M_2–tide and its distribution of energy dissipation, *Ann. Geophys., 33*, 31–40, 1977.

Zubov, N. N., *Arctic Ice*, Izd. Glavsevmorputi, Moscow, 350 pp., 1945.

Z. Kowalik and A.Y. Proshutinsky, Institute of Marine Science, University of Alaska, Fairbanks, AK 99775-1080.

Distribution of Water Masses on the Continental Shelf in the Southern Weddell Sea

T. Gammelsrød, A. Foldvik, O.A. Nøst, and Ø. Skagseth[1]

L.G. Anderson, E. Fogelqvist, K. Olsson, and T. Tanhua[2]

E.P. Jones[3]

S. Østerhus[4]

Abstract. Water properties on the continental shelf in the southern Weddell Sea observed during NARP 92/93 are presented. The station distribution includes a section close to the floating ice shelf from the Filchner Depression to the Antarctic Peninsula. Temperature, salinity, oxygen, silicate, CFC-11 and CFC-12 distributions are shown. Melting under the ice shelves, circulation systems, residence times, sediment/water interactions and bottom water formation are discussed. Ice Shelf Water (ISW), which is formed by cooling and melting below the floating ice shelf, seems to be about 10 years older than its parent water mass, which indicates the residence time below the ice shelf. The average melting rate below the Filchner Ronne ice shelf, based on the volume flux of ISW in the Filchner Depression is estimated to be 0.1 m/year. Compared with earlier observations considerable changes were found in the water characteristics and distribution: The temperature of the Weddell Deep Water has increased 0.7°C since 1977. Western Shelf Water, usually dominating the bottom layers in the Filchner Depression and on the Berkner Shelf, was found only in the Ronne Depression.

1. INTRODUCTION

The southern Weddell Sea is of particular interest in oceanography and climate research because of the presence of an immense (~$5 \cdot 10^5$ km^2 [*Robin et al*, 1983]) floating ice shelf, the Filchner-Ronne Ice shelf. The thickness of this ice shelf is typically about 300m at the seaward edge, and up to 1600m at the grounding line [*Jenkins and Doake*, 1991].

The interactions between atmosphere, ocean, sea ice and ice shelf give rise to several processes which are relevant to the earth's climate. High density water that ventilates the world ocean abyss is formed in this region, [*Gill*, 1973; *Carmack and Foster*, 1975a; *Foster and Carmack*, 1976; *Foldvik and Gammelsrød,* 1988] The uptake of anthropogenic CO_2 (and other greenhouse gases) by this high density water has been estimated [*Anderson et al.*, 1991]. A burning issue in climate research today is the stability of the West Antarctic ice sheet, changes in which could possibly have profound consequences for the sea level. It is therefore of vital interest to study the mass balance of the floating ice shelves, and much can be learned about the processes below the ice shelf by studying the water masses outside it. Accordingly, during the Nordic Antarctic Research Programme 1992/93 (NARP 92/93) we

[1]Geophysical Institute, University of Bergen, Bergen, Norway
[2]Department of Analytical and Marine Chemistry, University of Göteborg and Chalmers University of Technology, Göteborg, Sweden
[3]Department of Fisheries and Oceans, Bedford Institute of Oceanography, Dartmouth, Nova Scotia, Canada
[4]Norwegian Polar Institute, Oslo, Norway

The Polar Oceans and Their Role in Shaping the Global Environment
Geophysical Monograph 84

focused on studies of the circulation, residence times, melting processes below the ice shelves, formation of bottom water, and long term changes in the hydrographic conditions.

Several investigators have addressed these questions before: Direct measurements of the hydrography below the Ronne Ice Shelf were obtained by drilling through the ice shelf [*Nicholls et al.*, 1991]. *MacAyeal* [1984, 1985] formulated mathematical models for the tidal currents and buoyancy driven plumes under the Ross Ice Shelf. The thermohaline circulation beneath ice shelves has been modelled by *Hellmer and Olbers* [1989] and *Jenkins* [1991]. Residence times below the Ross Ice Shelf were discussed by *Trumbore et al.* [1991] using chlorofluorocarbons (CFC´s), but similar studies below the Filchner Ronne Ice Shelf have not been possible due to the lack of knowledge of the chemical constituents of the WSW found in the Ronne Depression [*Schlosser et al.*, 1990]. For a recent discussion of melting rates and mass balance of the Antarctic Ice Shelves see *Jacobs et al.* [1992].

There are basically two hypotheses concerning the formation of bottom water in the Weddell Sea: *Foster and Carmack* [1976] suggested a mechanism for the mixing of several water masses at the continental shelf break, whereas *Foldvik et al.* [1985c] showed that ISW mixes with the overlying Warm Deep Water (WDW) on its way down the continental slope. Studies of the chemical tracer distribution *Schlosser et al.* [1990]; *Schlosser et al.* [1991]; *Bayer and Schlosser* [1991] confirm that the latter process is of importance north of the sill of the Filchner Depression. Recent observations show that one or both of these processes takes place in the western part of the Weddell Sea shelf [*Gordon et al.*, 1993]

Access to the area in the south-western Weddell Sea is normally difficult because of heavy sea ice conditions. During the survey presented here we succeeded in obtaining an oceanographic section along the Filchner-Ronne Ice Shelf all the way to the Antarctic Peninsula, (see Figure 1 for station positions). This section was first occupied in 1980 [*Gammelsrød and Slotsvik*, 1981], but in addition to the traditional CTD-data obtained in 1980, this expedition carried out a major chemical observation program that included oxygen, nutrients, the carbonate system, CFC´s, tritium, helium-3 and oxygen-18. The section was obtained

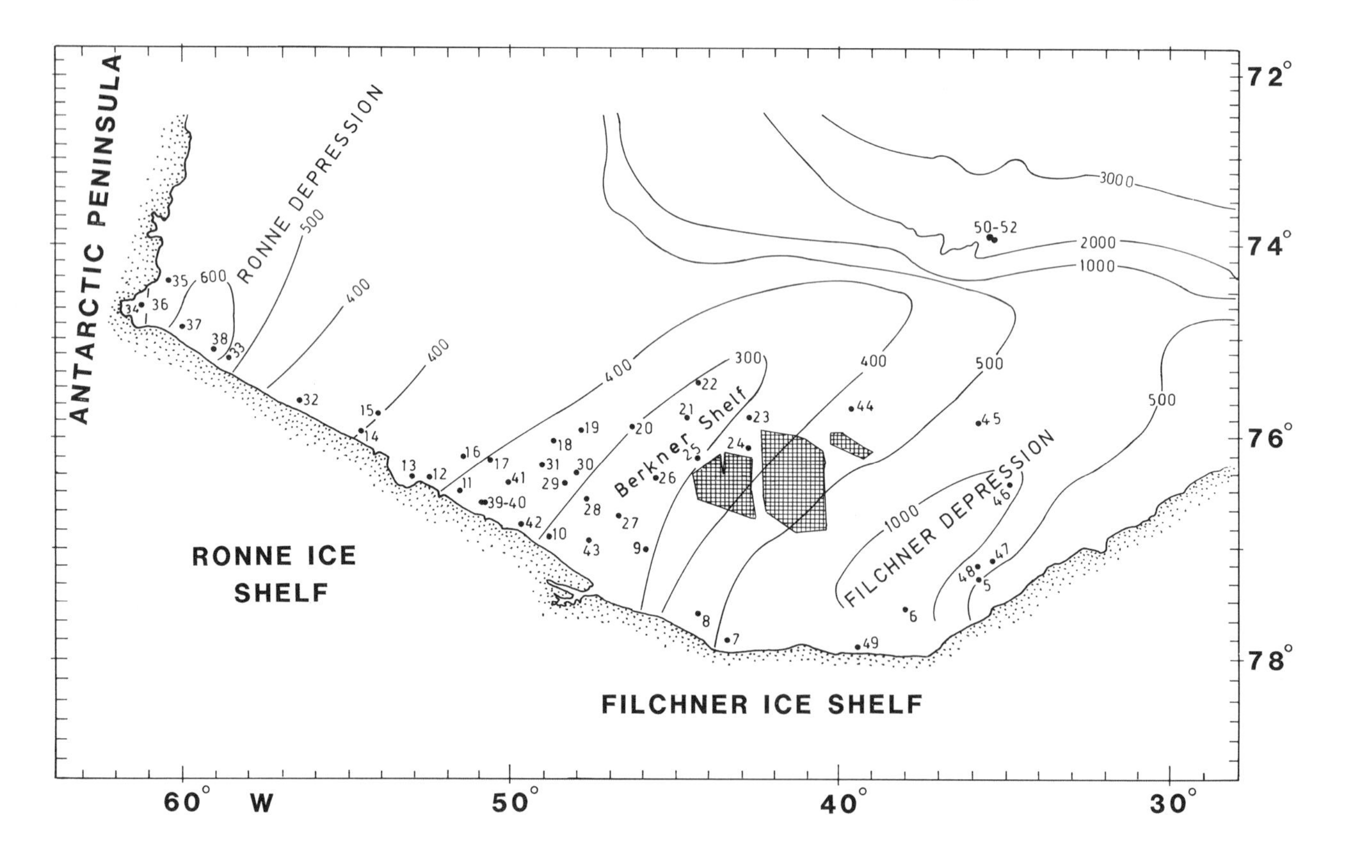

Fig. 1. Map of the southern Weddell Sea showing the positions of the CTD-stations occupied during NARP 92/93. The positions of the three grounded icebergs on the western slope of the Filchner Depression are also shown

during February 1993, and the results from the CTD, oxygen, silicate, and CFC-11 and -12 observations are presented here. In addition to the observations close to the ice barrier another section was taken further north across the shallow Berkner Shelf. Also a few deep stations (more than 2000 m) were occupied at the continental slope, around 36°W.

2. INSTRUMENTS AND METHODS

Vertical profiles of temperature and salinity were collected using a Neil Brown Instrument System Mark IIIB CTD. The CTD was combined with a General Oceanic Rosette sampler holding 24 ten litre Niskin type bottles. Samples were drawn from the rosette bottles immediately after they were brought on board in order to minimise contamination. The CTD temperature sensor was calibrated before and after the cruise. During the cruise it was checked against two SIS (Sensoren Instrument System) reversing thermometers. The accuracy was found to be better than 0.003°C. The pressure sensor was checked against a SIS 6000H reversing pressure meter. The pressures reported here are believed to be accurate to better than 5 dbars. The CTD-salinity was checked against 350 water samples analysed on board using a Guildline Portasal 8410 salinometer. When obvious outliers were removed, 297 samples were accepted for calibration yielding an accuracy better than ±0.002 psu.

Oxygen concentrations were determined by an automated Winkler titration [*Anderson et al.*, 1992]. Duplicate samples were regularly collected from the same Niskin bottles and the differences between individual pairs were always better than ±0.4 μmol/kg. The accuracy was set by titration of a standard potassium iodate solution. The error in the concentration of this standard solution is considerably less than the precision of the titration, thus the accuracy is expected to be of the same order of magnitude as the precision.

Silicate concentrations were determined with a Technicon AutoAnalyzer, using their standard methods. During each run working standards, prepared on board, were used as references. The accuracy of the working standards was compared with CSK standards. The temperature in the laboratory fluctuated quite severely both from day to day as well as within each separate run, thus affecting both accuracy and precision. The precision was estimated to be of the order ±0.5 μ*M*.

The CFC concentrations were measured using a method designed for the simultaneous determination of the chlorofluorocarbons (CFC-11, CFC-12, CFC-113 and carbon tetrachloride) and for a large set of other low molecular weight halogenated substances, anthropogenic (e.g., tri- and tetrachloroethylene) as well as biogenic (e.g., methyl iodide and bromoform). They were measured by a gas chromatographic method after extraction from the water matrix by purging with nitrogen and pre-concentration in a cold trap. Detection and quantification were accomplished by electron capture detection. A total of 25 compounds were detected, 15 of them quantified in the sea water. Air samples were analysed for a total of 13 compounds.

For the CFC measurements, sea water samples were drawn from the Niskin bottles with 100 ml glass syringes avoiding contact between the water sample and ambient air. The syringes were stored in a bucket of sea water prior to the analysis, which generally took place within 4 hours and never later than 12 hours after sampling. Air samples were taken in the same kind of syringes and analysed directly.

3. RESULTS

3.1. Water mass definitions and origins

The 297 samples analysed on board for salinity are presented with the corresponding temperature values in a θ-S diagram (Figure 2). Water mass definitions according to *Foster and Carmack* [1976] are also shown. Winter Water (WW), which is formed during winter by cooling, has

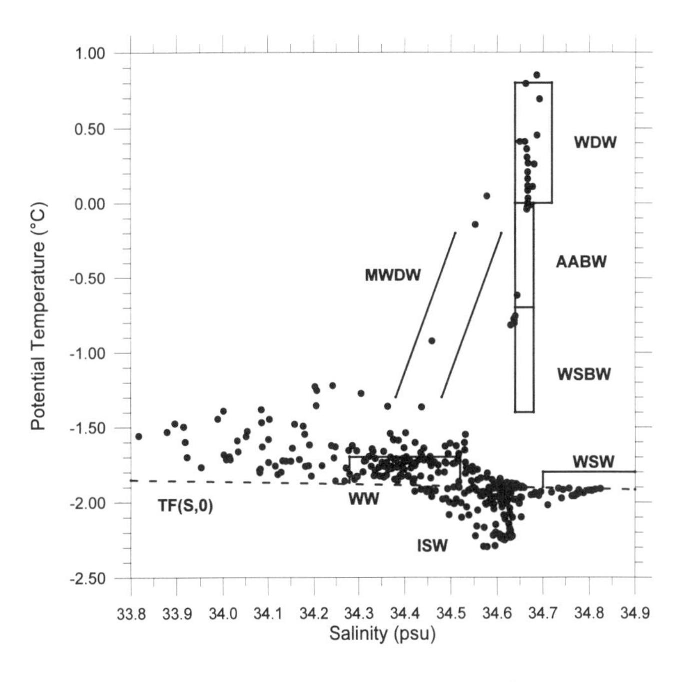

Fig. 2. θ-*S* diagram for the water samples used for calibration. The freezing point and typical water masses are also marked.

temperatures between -1.7°C and -1.9°C (the surface freezing point), and is usually found above 200 m depth. Western Shelf Water (WSW) has salinity above 34.7 psu, and a temperature close to the freezing point. It is formed from WW by cooling, freezing and brine release. Ice Shelf Water (ISW) is formed by the cooling of WSW and melting below the ice shelf. Due to the lowering of the freezing point with increasing pressure [*Millero,* 1978], ISW is easy to identify, because the temperature will be lower than the surface freezing point (-1.9°C). Warm Deep Water (WDW) which is derived from the warm (θ>0.5°C) Circumpolar Deep Water, is usually defined by salinity 34.64 psu<S<34.72 psu and temperature 0°C<θ<0.8°C. Figure 2 shows that the WDW observed this year had a maximum temperature near 0.9°C, so the upper temperature limit of WDW should be revised accordingly. The Modified Warm Deep Water (MWDW) is formed as a mixture of WW and WDW at the shelf break. The two remaining water masses indicated in Figure 2 are the Weddell Sea Bottom Water (WSBW) and the Antarctic Bottom Water (AABW), with the separation being indicated by the θ=-0.7°C isotherm [*Carmack and Foster*, 1975a].

3.2. Distribution of properties along the Filchner-Ronne Ice Shelf

Figure 3 shows the distribution of the various parameters observed close to the ice barrier, for station positions, see Figure 1. The section is shown projected onto a line of constant latitude, except for station 5, which is dislocated along the isobath towards the north-east to fit into the general bottom slope. As may be observed from Figure 1, the distance of the stations from the ice edge varies considerably, so caution should be taken in interpretation of the data because the water characteristics may change rapidly with distance from the barrier [*Foldvik et al.*, 1985b]. Also note that the observation density may be different for the various parameters.

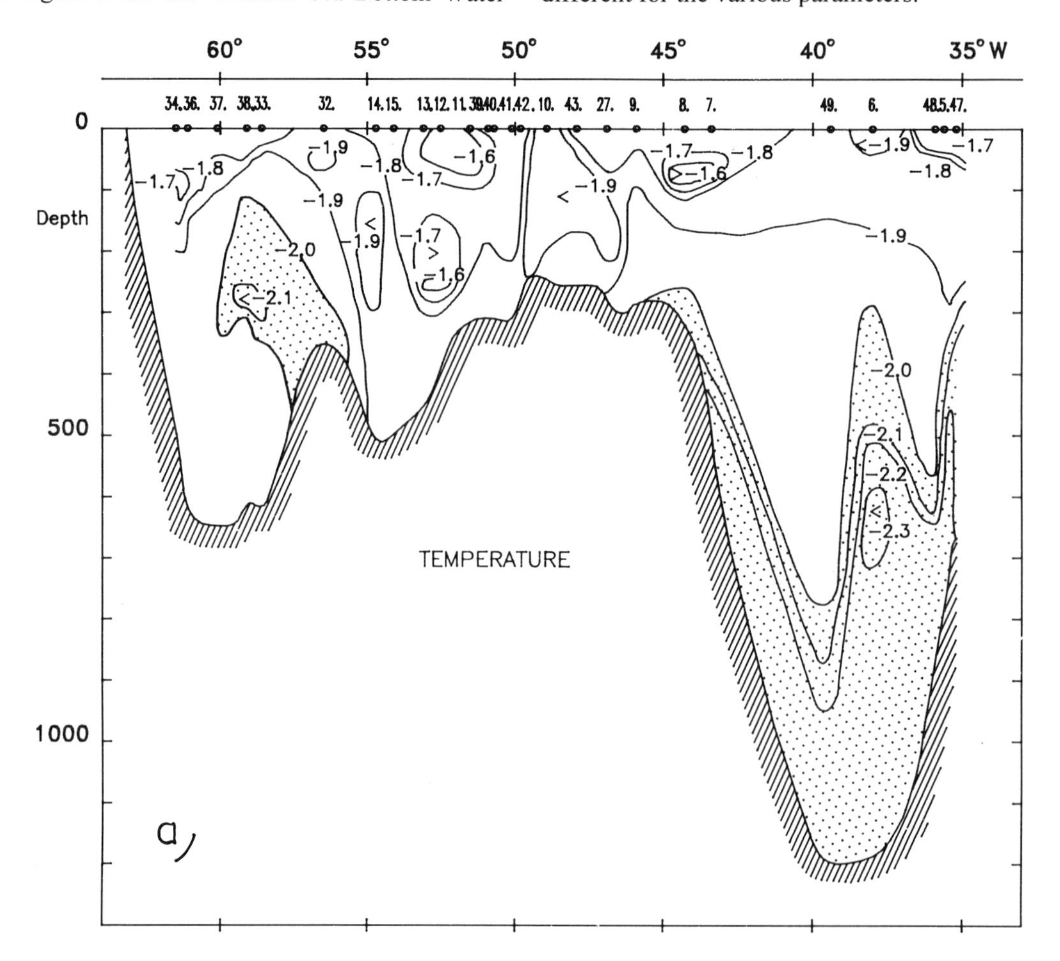

Fig. 3. Section along the Filchner-Ronne Ice shelves showing the distribution of a) Potential Temperature, b) Salinity, c) Oxygen, d) Silicate, e) CFC-11, and f) CFC-12. Note the different information density for the various parameters. The hatched areas in all the figures correspond to temperatures lower than -2.0°C (Figure 3a).

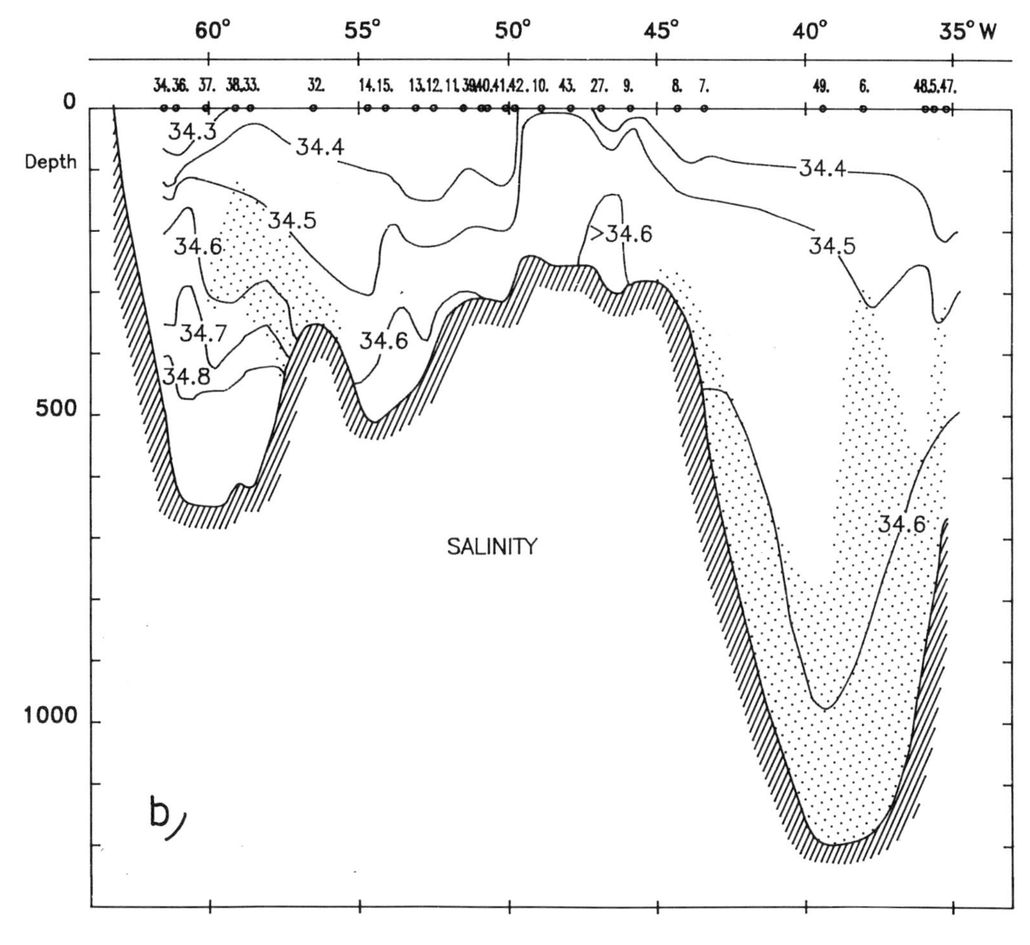

Figure 3. (continued)

The temperature distribution is displayed in Figure 3a, and shows that below about 200 m the Filchner Depression was entirely occupied by ISW. The lowest temperature (θ <-2.3°C) was observed at station 6 at a depth of about 700 m. For convenience we have hatched the areas where the temperature is below -2.0°C. To simplify the discussion these areas are also marked on the rest of Figure 3. Note that the wave-like form of the isotherms on the eastern slope of the Filchner Depression may be an artefact, because stations 47-49 were sampled about two weeks later than stations 5 and 6. Also note the coarse station spacing at the western slope of the Filchner Depression where the ISW is usually more pronounced [*Carmack and Foster*, 1975b; *Foldvik et al.*, 1985c]. The maximum salinity (Figure 3b) in the Filchner Depression was only 34.64 psu.

At the Berkner Shelf the temperature was close to the surface freezing point (-1.9°C), but the maximum salinity was only 34.65 psu, hence the salinity was too low for this water mass to be classified as WSW. Note the weak stratification at the Berkner Shelf, especially at station 10. On the western slope of the Berkner Shelf we note a temperature maximum (θ >-1.6 °C) at about 250 m depth centred around station 12. In the Ronne Depression WSW was found below 300 m depth with maximum salinity above 34.84 psu at the westernmost station (station 34). A core of ISW was observed at the eastern slope.

The most striking features in the oxygen distribution (Figure 3c) are the high surface values, caused by biological primary production, and the minimum centred around station 13, associated with the deep temperature maximum. The bottom water in the western part, in the region with high salinity, also had lower values compared with the water on the Berkner Shelf. In the Filchner Depression the values were between 320 and 330 μmol/kg below 150 m, except for a slight drop in concentration near the bottom.

The silicate distribution (Figure 3d) generally shows almost constant concentration throughout the water column, but with slightly depleted surface concentrations. On top of

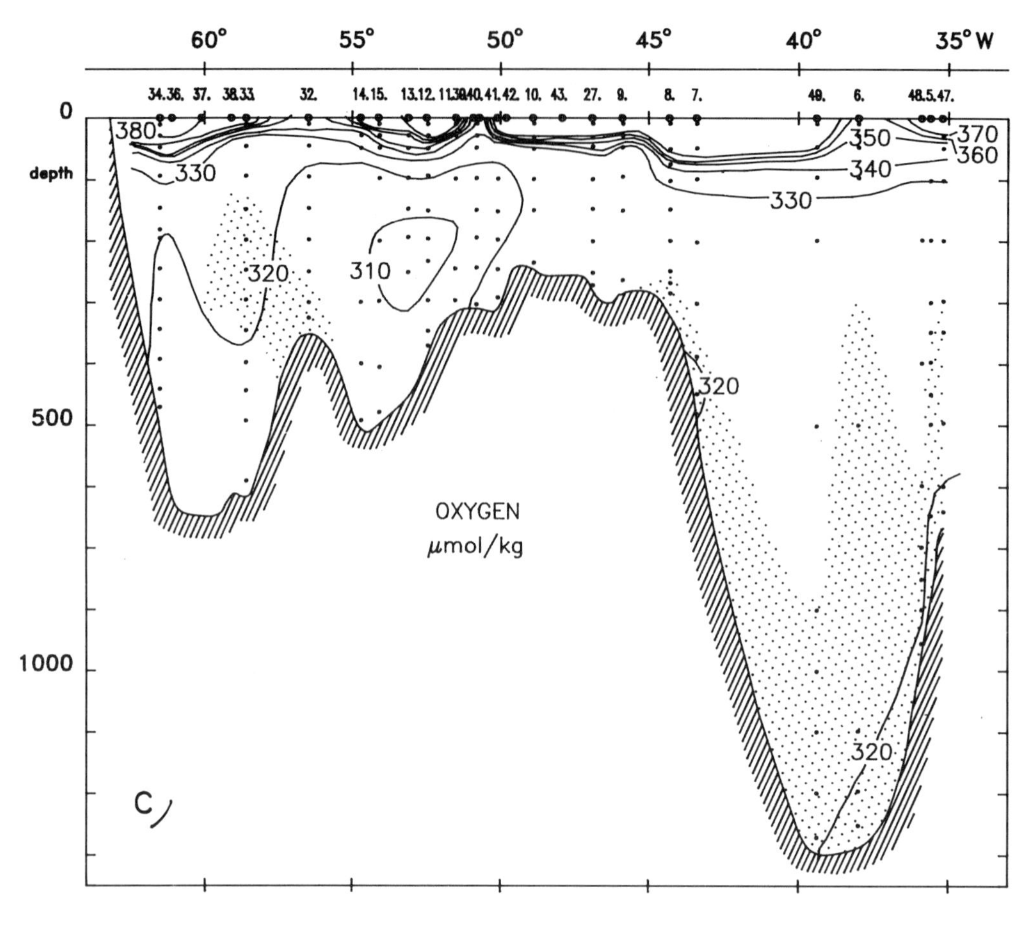

Figure 3. (continued)

this general picture we found elevated concentrations in the high temperature region of stations 11 - 14, in the waters associated with the WSW to the west and in the deep Filchner Depression to the east.

The concentrations of CFC-11 and CFC-12 are presented in the Figures 3e and f. Because of the better resolution of the CFC-11 values, more features are seen here (Figure 3e) than in the CFC-12 distribution. The WSW at station 37 in the Ronne Depression is associated with relatively high CFC values, except at the very bottom. At the eastern slope of the Ronne Depression where ISW was observed, relatively low CFC values were found. The deep, high temperature region at the western slope of the Berkner Shelf is associated with low CFC values. The vertical homogeneity at the Berkner Shelf (station 10) is also confirmed by the CFC-11 and CFC-12 distributions. The ISW in the Filchner Depression had CFC values comparable with those of the ISW in the Ronne Depression. The lowest values were found at the eastern slope below 700 m.

3.3. Distribution of properties on the Berkner shelf

A section starting at about 52°W, running east-north-east from the Ronne Ice Shelf across the Berkner Shelf up to the central Filchner Depression, is shown in Figure 4 (for positions see Figure 1). As in the Filchner-Ronne Ice Shelf section, we note a temperature maximum at about 250 m depth over the western slope (Figure 4a), but here it was even more distinct than close to the barrier, with temperatures above -1.4°C at station 31. The bottom salinity (Figure 4b) was below 34.65 psu in the whole section, so, as in the southern part, no WSW was observed on the central Berkner Shelf. ISW was observed below 300 m in the Filchner Depression.

The oxygen concentrations were highest at the surface, with a strong gradient at about 50 m and minimum values occurring at a depth of about of 200 m all along this section (Figure 4c). The minimum is more distinct in the western part, again associated with the temperature maximum. The

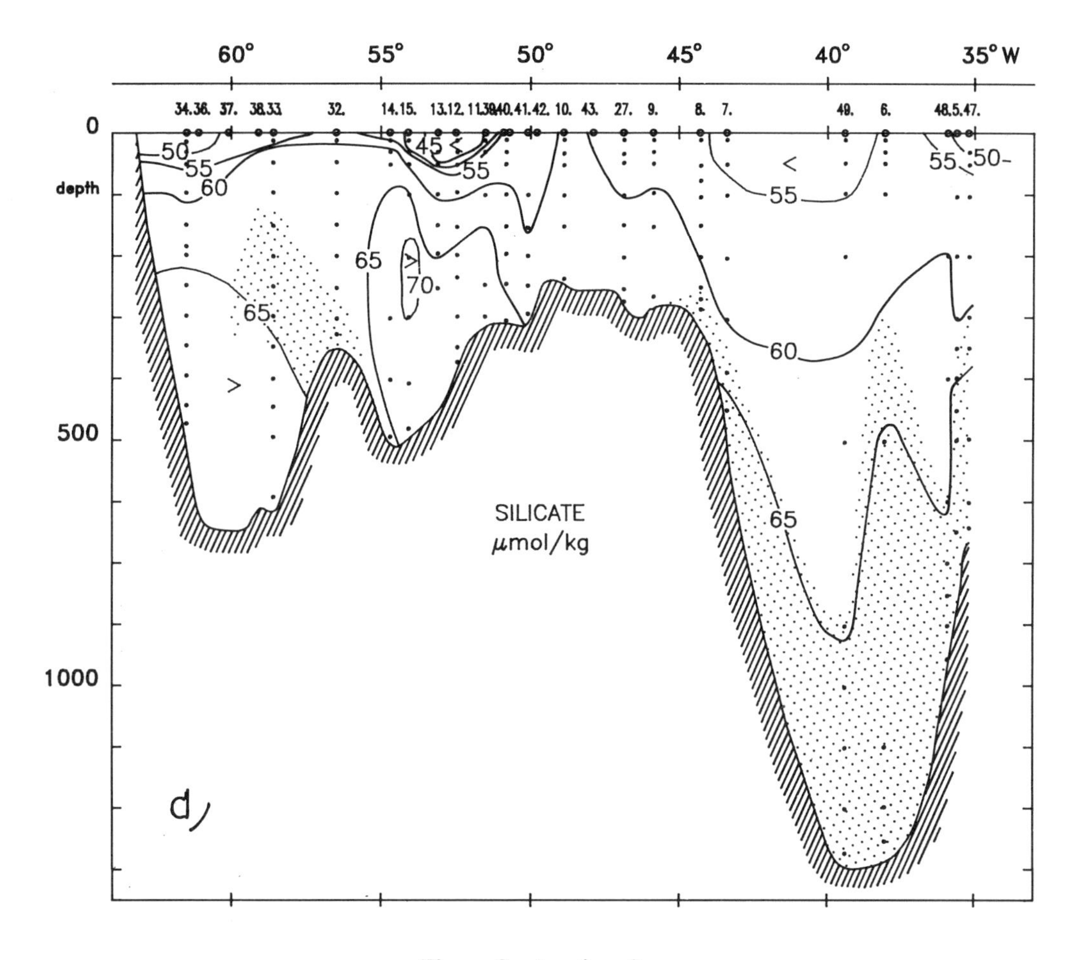

Figure 3. (continued)

oxygen concentrations in the Filchner Depression were between 330 and 320 µmol/kg with a few values even below 320 µmol/kg , the same as found further to the south in this depression.

The silicate concentration showed a general trend towards low values at the surface and high values at the bottom (Figure 4d). As with oxygen, the strongest gradient was at about 50 m depth. East of station 23, the maximum silicate concentrations were found at the bottom, but west of station 23, the maximum was found between 200m and 300m. No concentration above 65 µmol/kg was found in the Filchner Depression at this latitude.

The CFC-11 and CFC-12 distributions along the Berkner Shelf showed the same general features as the Filchner-Ronne Ice Shelf transect; characterised by weak vertical stratification and low values associated with the high temperature region at the western slope of the Berkner Shelf (Figures 4e and f).

4. DISCUSSION

4.1 The intrusion of WDW on the Continental Shelf

The deep temperature maximum at the western slope of the Berkner Shelf is interpreted as MWDW (Figure 2), formed by mixing WW and WDW. Low oxygen, high silicate and low CFC-11 and 12 concentrations support this view (Figures 3 and 4). Note that neither the exact position nor the shape of the core for the different properties is quite equal. This may stem from the somewhat different sampling frequency, from the variations in the relative accuracy of the different parameters, and/or from the fact that not all of the parameters are conservative. An examination of the individual temperature profiles shows that the deep temperature maximum may be traced all over the western slope of the Berkner shelf. The fact that the -1.8°C isotherm

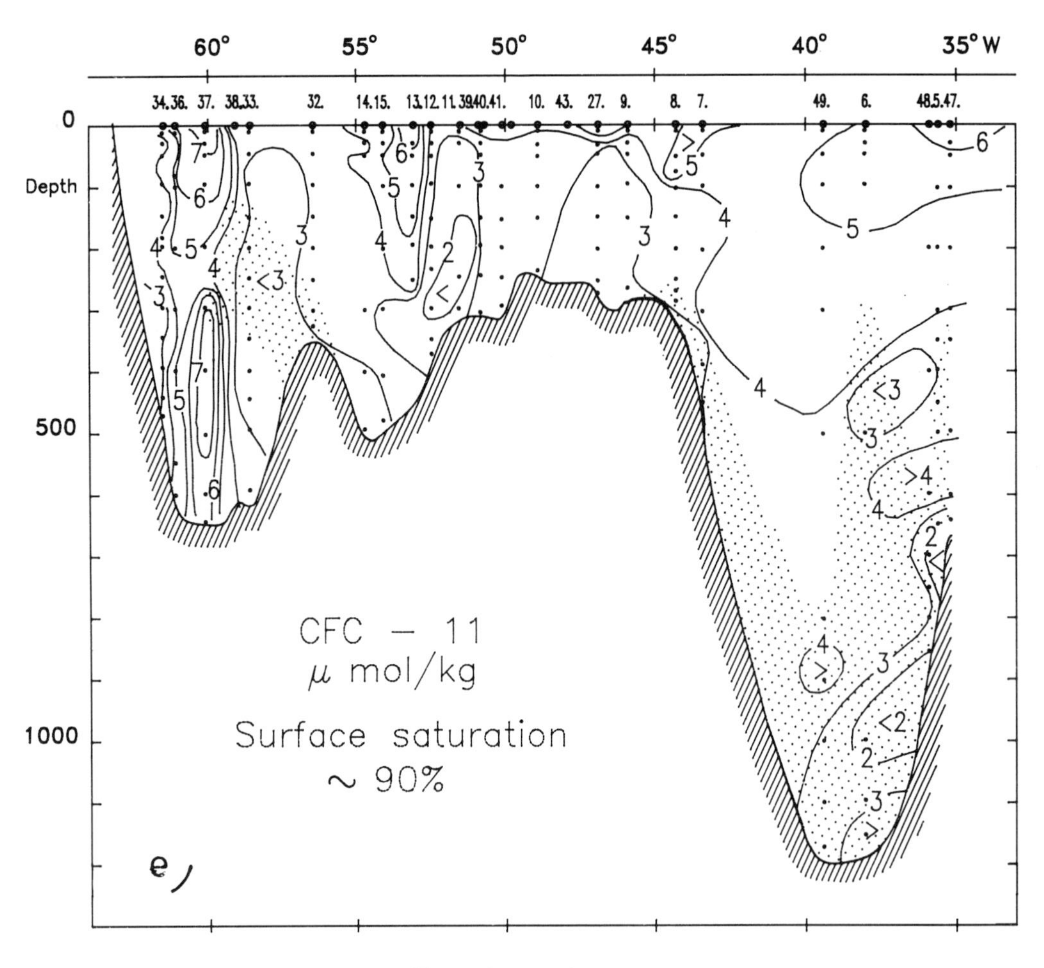

Figure 3. (continued)

at the central Berkner Shelf is as deep as 200 m (Figure 4a), may be a result of intrusion of WDW. The oxygen minimum and low CFC-12 values (less than 1.5 pmol/kg) found all along the Berkner Shelf section are associated with this temperature maximum. This connection is less obvious in the CFC-11 distribution.

Foster and Carmack [1976] have demonstrated how WDW intrudes southwards onto the continental shelf, especially at 40°W where it could be traced forming MWDW up to 100 km from the shelf break. It is somewhat surprising to find this water as a permanent feature so far from the continental shelf break as the Ronne Ice Shelf (~ 500 km). A persistent SW current must exist in the area transporting the MWDW southwards. This current is probably geostrophically balanced, and steered by the bottom topography. The forcing of this current may be related to melting processes below the ice shelf, which provides a buoyancy source at great depths, driving ISW plumes up and northwards. This water has to be replaced by a southward volume flux of the same magnitude, thus providing a possible energy source for the southward current of MWDW. A similar flow of warm water is observed in the Ross Sea [*Jacobs et al.*, 1985]. The MWDW is the only water mass on the southern Weddell Sea shelf which has a temperature above the surface freezing point in winter. Therefore it may have an important influence on the heat budget and melting processes near and under the ice shelf.

4.2 Indication of upwelling at the Berkner Shelf

The stratification at the Berkner Shelf is much weaker close to the ice shelf than further out (compare Figures 3 and 4), indicating that the water mass here was not liable to be subject to vertical overturning. The relative low CFC values indicate that upwelling related to the ice barrier takes place at the Berkner Shelf. *Gammelsrød and Slotsvik* [1981] reported upwelling amplitudes of more than 150 m in the area related to the tidal currents.

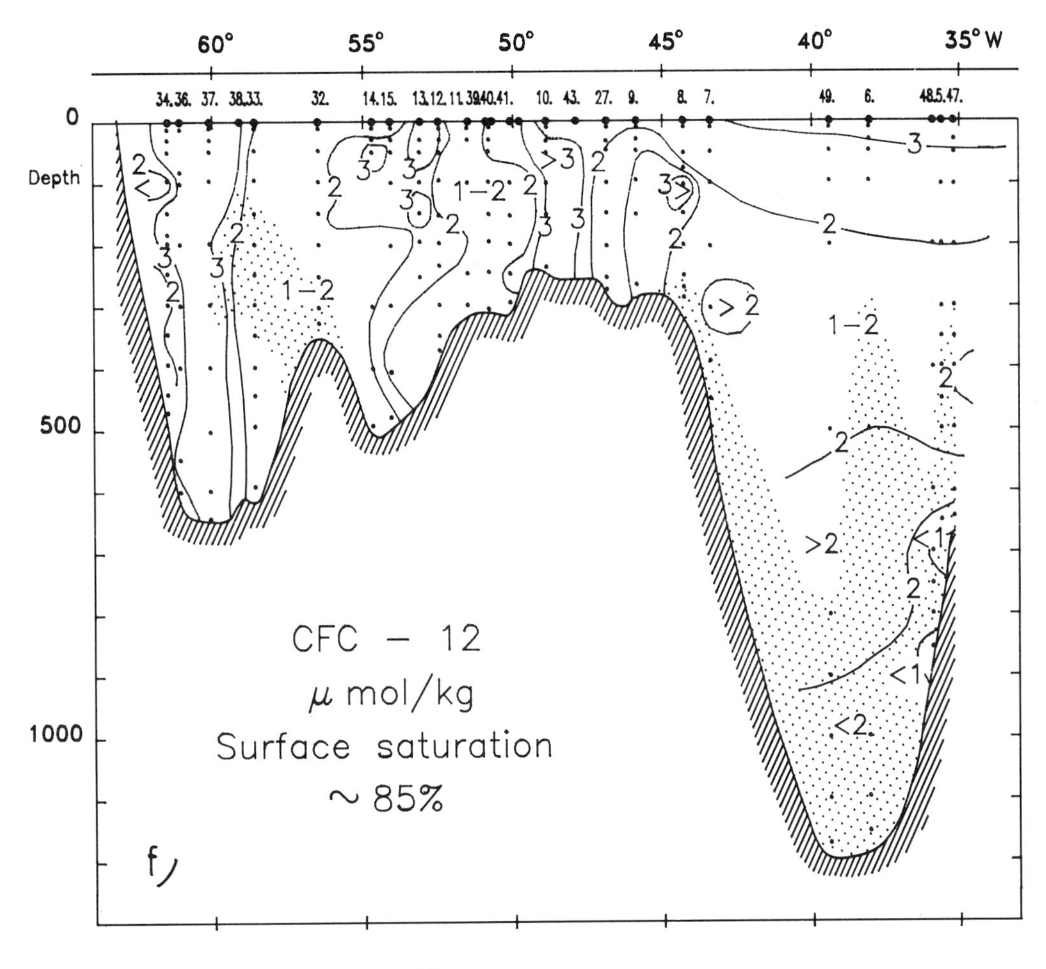

Figure 3. (continued)

The stratification is particularly weak at station 10 (Figure 3), and the CFC-11 and CFC-12 concentrations are higher at station 10 than elsewhere on the Berkner Shelf, consistent with a recent ventilation.

4.3. Melting rates beneath the ice shelf

To estimate the content of meltwater in the ISW-overflow at the sill of the Filchner Depression, we need to know the potential temperature of the source water (θ_W), the meltwater content of the source water and the potential temperature of the ISW (θ). Following *Nøst and Foldvik* (submitted), the relation between meltwater and source water in ISW is given by:

$$\frac{m_s}{m_e} = \frac{c}{L}(\theta_w - \theta)$$

m_S is the mass of meltwater, m_e is the mass of source water, c is the specific heat capacity of sea water and L is the latent heat of fusion for ice.

From the mass and energy balance of a water mass interacting with the ice shelf, *Nøst and Foldvik* (submitted) found that ISW in the Filchner Depression is formed from the WSW observed in the Ronne Depression. The WSW in the Ronne Depression has potential temperature almost exactly equal to the surface freezing point (Figure 2 and 3a). This means that the temperature of the source water is $\theta_w = -1.9°C$. This WSW most probably does not contain any meltwater, because an even warmer and more saline source water mass would then be needed near the Ronne Depression. Such a water mass is not found anywhere in front of the ice shelf (Figure 3a and 3b).

The annual mean temperature of the ISW overflowing at the sill of the Filchner Depression is $-2.0°C$ [*Foldvik et al.*, 1985a]. Using the equation above with $\theta_w = -1.9°C$ and $\theta = -2.0°C$ we obtain the relation between

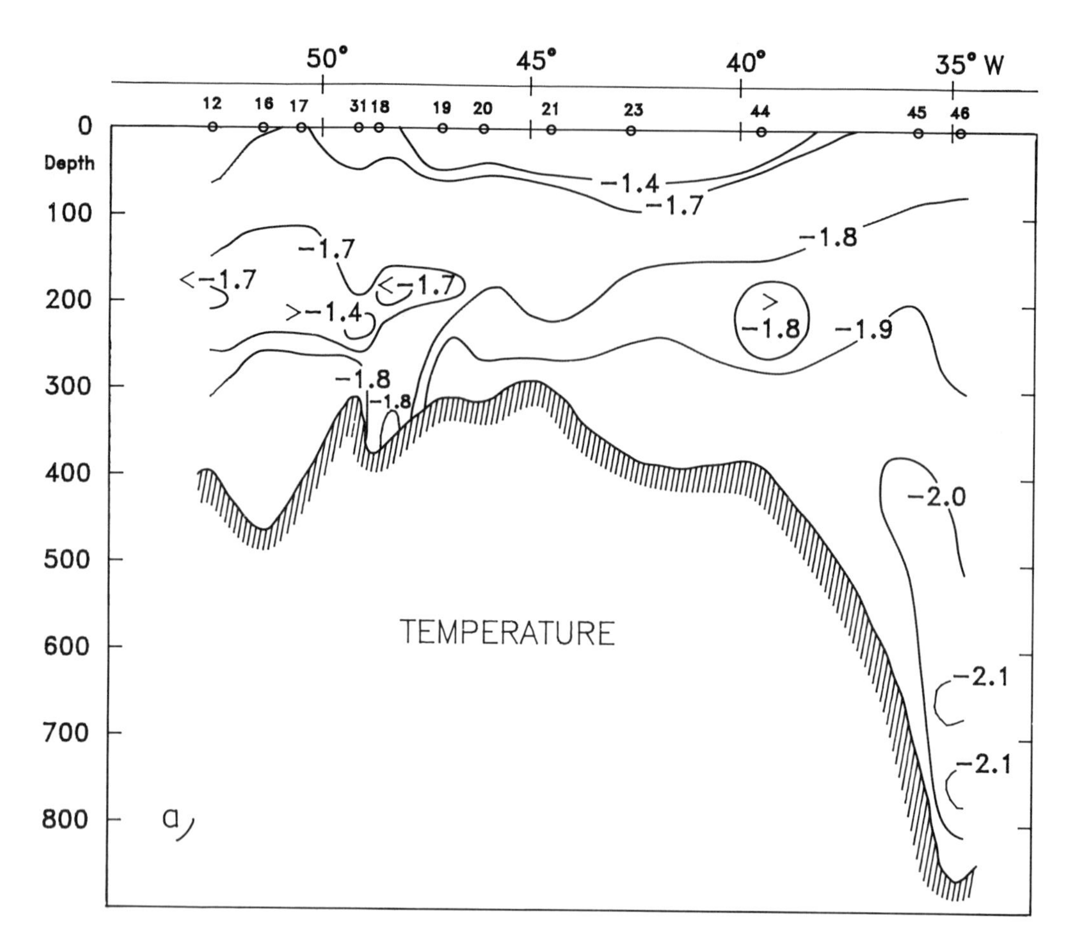

Fig. 4. Section of the central Berkner Shelf showing the distribution of a) Potential Temperature, b) Salinity, c) Oxygen, d) Silicate, e) CFC-11, and f) CFC-12. Note the different information density for the various parameters.

meltwater and source water in the ISW overflow to be $m_s/m_e \approx 1 \cdot 10^{-3}$.

Taking the volume transport of this ISW overflow to be 1 Sv (10^6 m^3/s) [*Foldvik et al.*, 1985c], the volume transport of glacial meltwater is 10^3 m^3/s. If this melting is spread over the total area of the ice shelf, which is about $5 \cdot 10^5$ km^2 [*Robin et al.*, 1983], the corresponding melt rate is ~0.1 m/year. *Nøst and Foldvik* (submitted) suggest that the flow from the Ronne Depression to the Filchner Depression takes place in a deep trough close to the grounding line. The melting caused by the water which overflows at the sill of the Filchner Depression then probably takes place along this trough. The area along this trough is about half of the total area of the ice shelf, and the corresponding melt rate for this area, caused by the flow from Ronne to Filchner Depressions, would then be ~0.2 m/year.

The net melting rates presented above are about one third of those calculated by *Schlosser et al.* [1990], who used the $\delta^{18}O$ distribution to estimate the content of meltwater in the ISW. The $\delta^{18}O$ concentration of the ISW is only marginally affected by freezing at the underside of the ice shelf, meaning that the water which freezes on to the ice shelf base will not be taken into account in a calculation based on the $\delta^{18}O$ distribution. Hence their calculations give the total melt, while ours give the net melt.

Using the salinity difference between WSW and the overflowing ISW one obtains a meltwater fraction of about 4‰. However, between the Filchner Ice Shelf front and the sill, the ISW is mixing with overlaying water which has salinities lower than WSW (Figures 3b and 4b). Hence, the difference in salinity between WSW and the overflowing ISW is not only due to input of meltwater.

The mixing with overlaying water will not influence the calculations of meltwater content based on the mass and energy balance, as long as the overlaying water is either water at the surface freezing point and with no meltwater

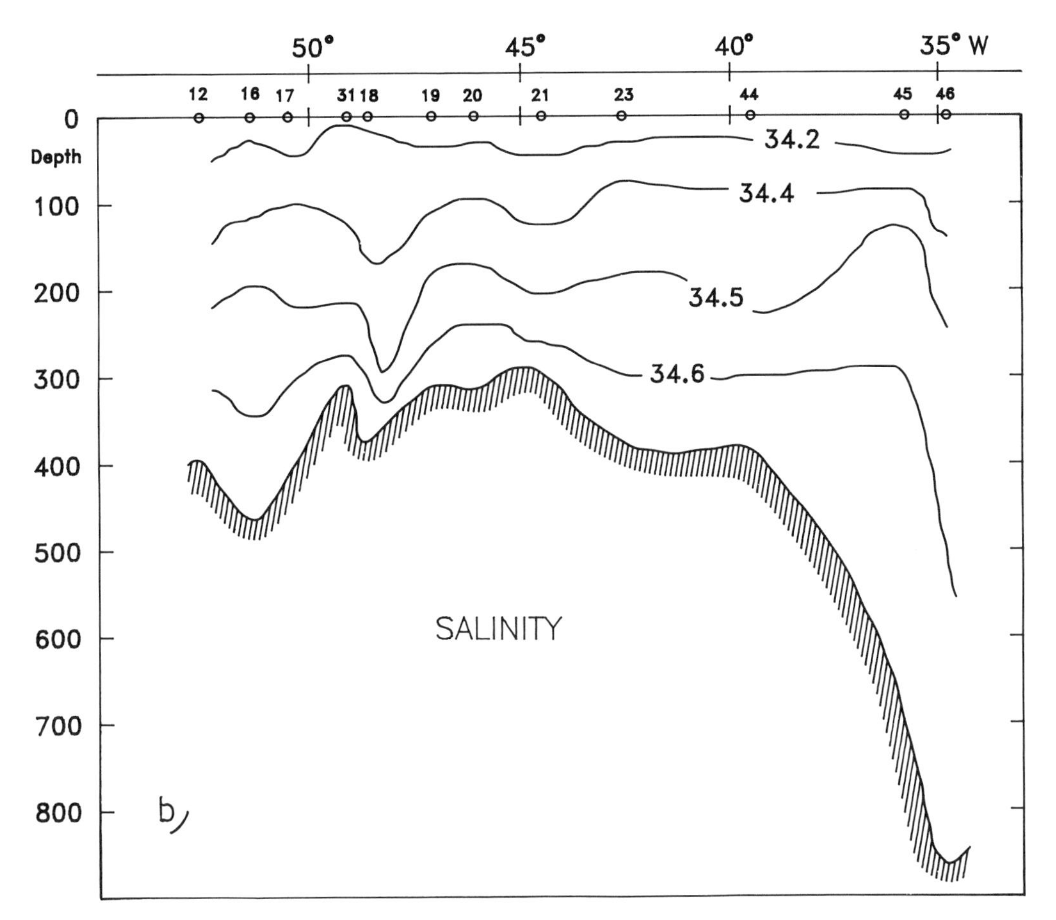

Figure 4. (continued)

content, or ISW with source water at the surface freezing point. This is probably a good assumption, because Figure (3a) and (4a) show that the temperature of the water surrounding the ISW in the Filchner Depression is at the surface freezing point or colder.

4.4. Residence times below the ice shelf

The concentrations of CFC's in the various water masses can be used to assign an apparent age to the water, here defined as the time elapsed since the water was last in contact with the atmosphere. The estimated surface water concentration of the two CFC's as a function of time is illustrated in Fig. 5. These curves are based on the atmospheric history of CFC's 11 and 12 from estimates done within the ALE (Atmospheric Lifetime Experiment) programme [Cunnold et al. 1986, and references therein], solubility data for the same CFC's at various temperatures and salinities [Warner and Weiss, 1985] and finally the relative saturation in surface water measured during the cruise. It follows that the youngest water mass at depth along the Filchner-Ronne section was found in the Ronne Depression, where concentration levels of CFC's at station 37 below 300 m were highest, and the water column seems recently ventilated. The CFC-11 concentrations higher than 7 pmol/kg and CFC-12 values higher than 3 pmol/kg indicate that the water has been equilibrated with the atmosphere during the last 1 to 3 years, i.e. 1990-92. The most saline WSW found at the western slope is more than 10 years older, it might be a water mass which is too heavy to escape from the Ronne Depression.

The ISW observed at the eastern slope of the Ronne Depression and in the Filchner Depression both have a CFC age of roughly 18-20 years, indicating similar residence times for these two water masses. The assumption that ISW is formed from a mixture of the WSW found in the Ronne Depression with a CFC age of about 5-10 years, leads to a rough estimate on ISW residence times under the ice shelf of the order of 10 years.

Assuming complete mixing an estimate of the maximum residence time (t) may be calculated as $t = V/(dV/dt)$, where V is the volume under the ice shelf and dV/dt is the

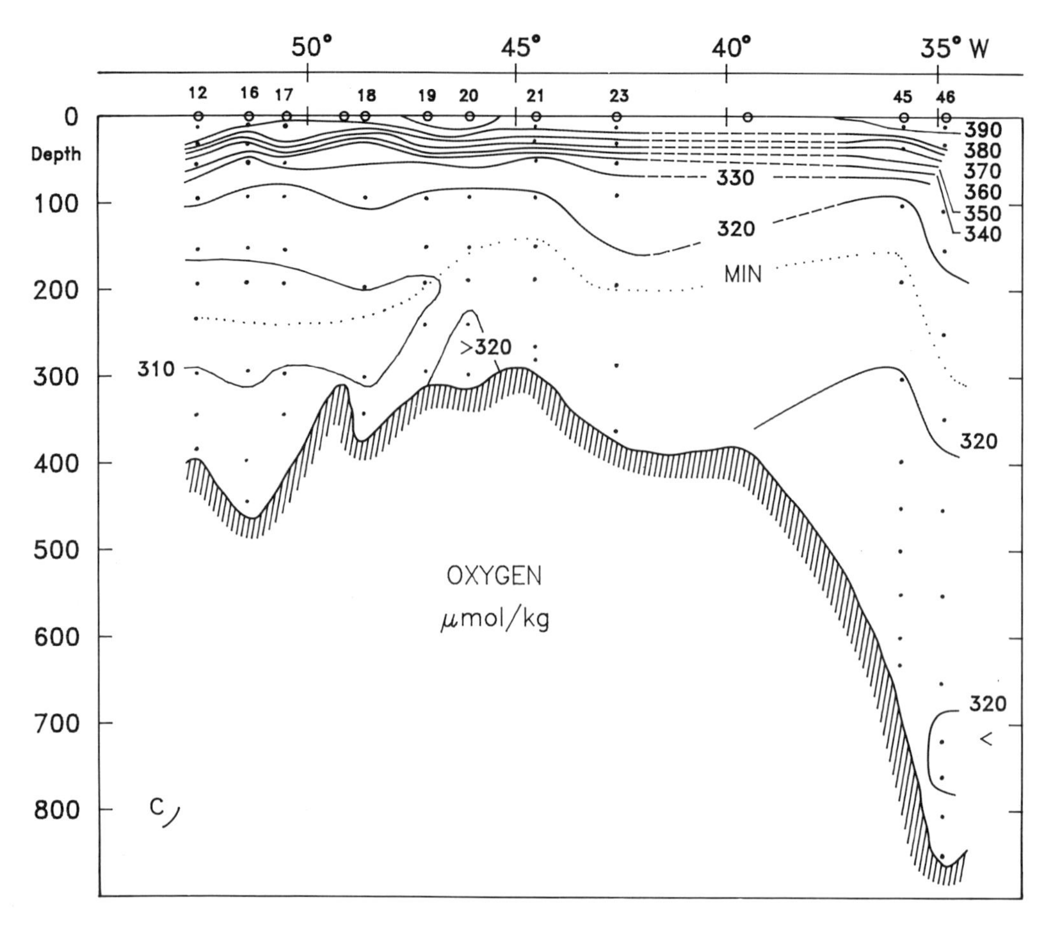

Figure 4. (continued)

volume transport out of the ice shelf cavity. A crude estimate of the volume along the trough between the Ronne and Filchner Depressions gives V ~ 10^{14} m^3. The ISW overflow over the sill in the Filchner Depression is found to be about 1 Sv, giving a residence time in the order of 3 years. However, due to entrainment into the ISW plume between the Filchner Ice Front and the sill, 1 Sv is obviously an overestimate for actual ISW produced under the Filchner-Ronne Ice Shelf. A smaller rate of this ISW would give a better agreement with the CFC ages obtained.

4.5 Water mass transformations and bottom water formation

The modification and mixing of the different water masses is clearly seen in a plot of properties versus salinity for station 6 (containing ISW), station 34 (containing WSW) and station 50 obtained at 2300 m depth at the continental slope (containing WW, WDW, AABW and WSBW), for station positions, see Figure 1. In the θ-*S* plot (Figure 6a) symbols mark every 50 m increase in depth, thus showing the thickness of the different layers. When a mixing line is extended from the WDW through the WSBW found at the bottom of station 50, it merges with the data of station 6 at a salinity lower than the maximum (Figure 6a). Hence, given that WSBW may be formed as a mixture of ISW and WDW [*Foldvik et al.*, 1985c, *Schlosser et al.*, 1990] the ISW that flows down the continental margin and mixes with the overlying WDW is not the most dense found at the bottom of the Filchner Depression, but rather the coldest ISW found at depths corresponding to the sill depth. This may have relevance to the dynamics of the plume of ISW flowing down the continental slope, as cold water is more compressible than warmer water (the thermobaric effect).

The θ-*S* distribution below the temperature minimum at station 6 is very close to the theoretical line (dotted in Figure 6a) obtained when a water parcel, with high salinity and freezing temperature at atmospheric pressure, interacts with and melts the underside of the ice shelf, resulting in

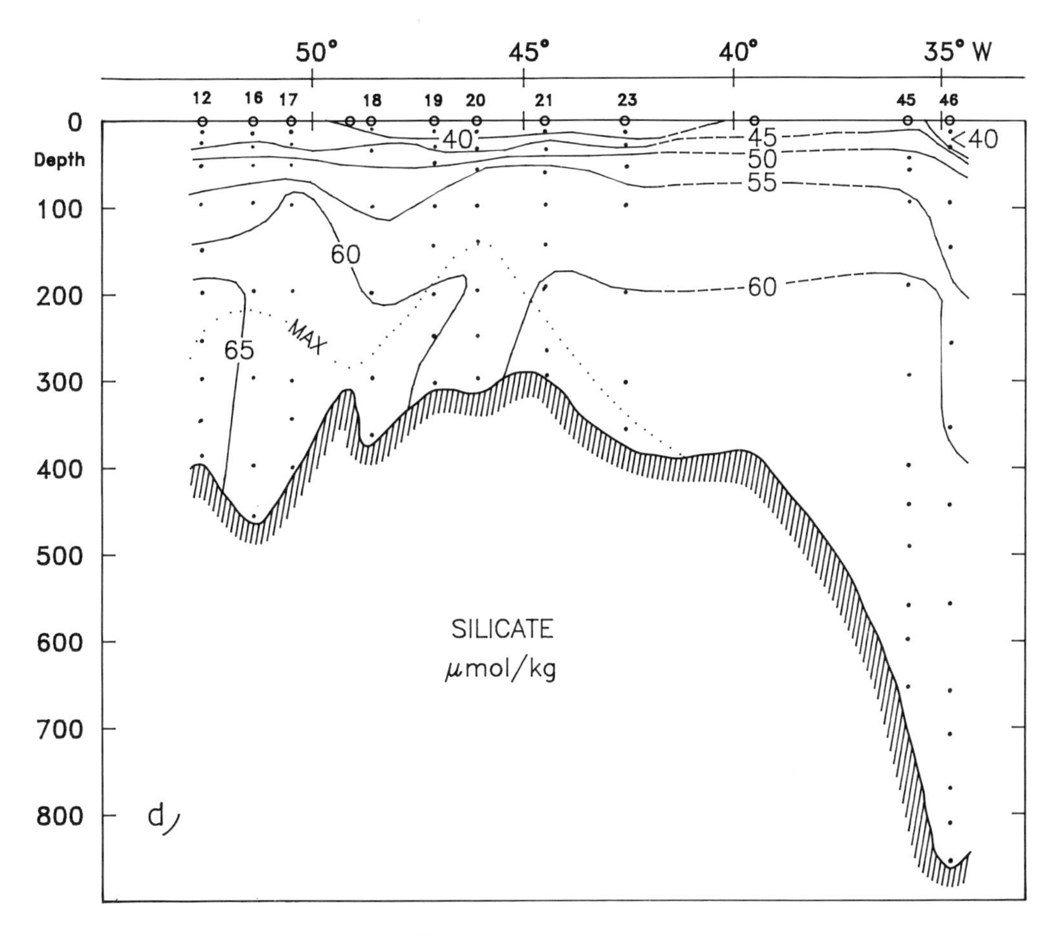

Figure 4. (continued)

loss of heat and reduced salinity (*Nøst and Foldvik*, submitted). The theoretical line that best fits the data of station 6 does not pass through the highest salinity encountered in the WSW at station 34, indicating either that the most dense WSW does not circulate below the ice shelf, or that it is mixing with less saline WSW on its way towards the Filchner Depression.

In Figure 6 we summarise the formation processes and mixing of the water masses participating in the production of bottom water in the Weddell Sea with the aid of θ-*S*, O_2-*S* and *Si-S* curves from station 6 in the Filchner Depression, station 34 in the Ronne Depression and station 50 at the continental slope. The processes are indicated by numbered arrows in the figures and explained in paragraphs 1 - 3 below.

1. WSW is formed from WW by cooling, freezing and brine release (Figure 6a). When WSW loses contact with the surface layer, its oxygen content decreases as a result of the decay of organic soft matter (Figure 6b), and the silicate concentration increases as a result of the dissolution of biogenic silica over the continental shelf as illustrated in Figure 6c.

2. The WSW flowing south underneath the ice shelf is cooled and mixed with melt water (Figure 6a). During this process the oxygen content is in fact increasing (Figure 6b). It has earlier been pointed out that such enriching of oxygen takes place when glacial ice melts because air bubbles are trapped in the ice [*Jacobs et al.*, 1970; *Gow and Williamson*, 1975]. Also the silicate concentration is higher in the ISW than the WSW (Figure 6c). The WSW is believed to first come in contact with the ice at the grounding line some 500 km from the ice edge [*Jenkins*, 1991], and during this journey the WSW may be enriched with silicate by interaction with the sediments before the melting process forming the ISW starts.

The main difference between the oxygen and silicate plots is the bend in the WW - WDW - AABW mixing line. It is obvious that dissolution of biogenic silica occurs deeper than the salinity maximum in the water column, and that it is only the mixing with down flowing ISW that causes the

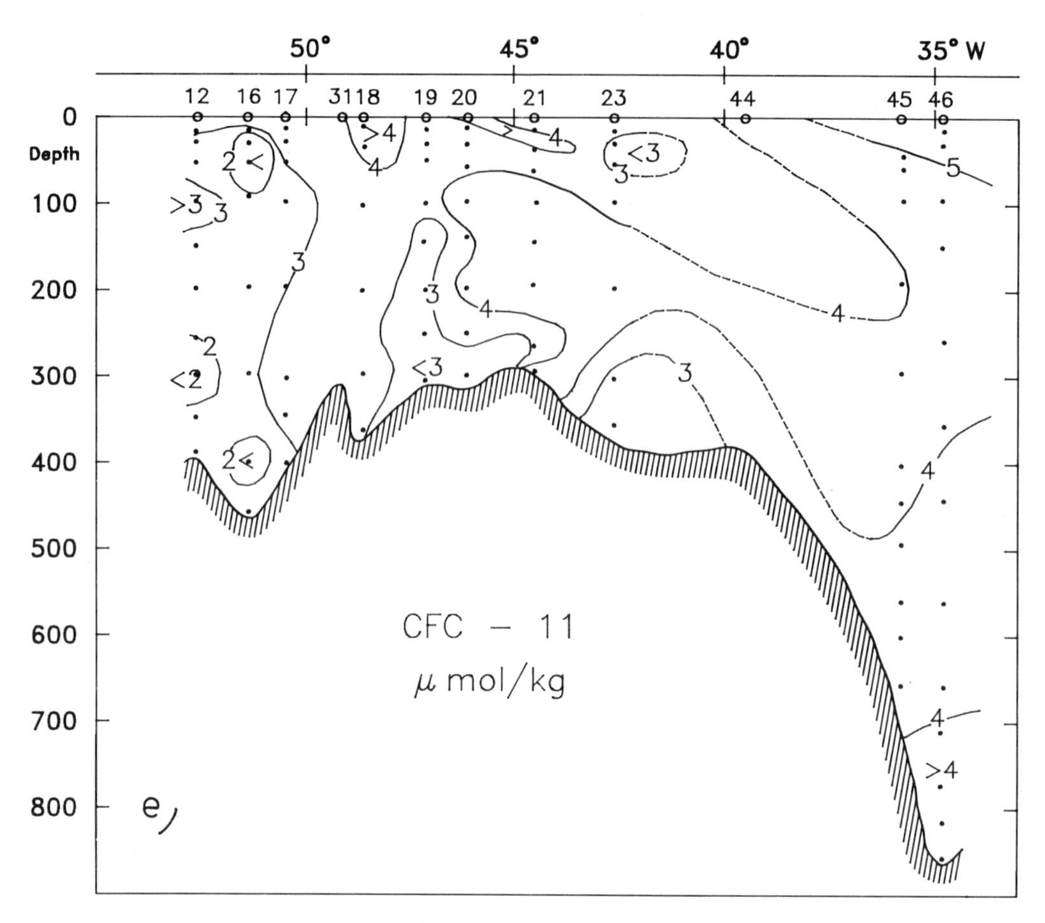

Figure 4. (continued)

drastic decrease in the bottom layer at station 50. This is in agreement with earlier investigations, which have shown that the biogenic silica particles to a large degree sink to the bottom before dissolving [e.g. *Pudsey et al.*, 1988]. Decay of organic soft matter takes place largely at shallower depths in the water column hence the consumption of oxygen (Figure 6b).

3. Finally, WSBW may be formed as a mixture of ISW and WDW. If ISW is mixed with a core of WDW, about equal parts of the two water masses are needed to form WSBW. We observe from Figure 6a, however, that the core of the WDW is situated about 1000m above the down flowing ISW. It has been argued [*Foldvik et al.*, 1985a] that WSBW is formed by admixture of WDW with temperatures near 0°C, which means that the estimate of the production rate of WSBW must be increased by about 50% if the simple mixing of ISW and WDW is accepted.

Note that the extension of the mixing lines between WDW and WSBW meet the ISW at almost the same intermediate salinity for all of the properties discussed. This indicates that the mixing time at the continental slope is much shorter than the decay time of organic matter, and oxygen can be considered as a conservative tracer in this instance.

4.6. Long term variability

Although observations from the area are scarce, it is possible to see changes in the hydrographic situation that have occurred since earlier expeditions. The most striking change is the lack of WSW both in the Filchner Depression and on the Berkner Shelf. At the Berkner Shelf WSW was observed in the bottom layer both in 1980 [*Foldvik et al.*, 1985b] and in 1984 [*Rohardt*, 1984], while WSW has been found in the Filchner Depression abyss each time it has been visited since 1973 to 1987 [*Carmack and Foster* 1975a; *Foldvik et al.*, 1985d; *Foldvik et al.*, 1985b; *Rohardt*, 1984]. However, in 1989 no WSW was observed in the Filchner Depression, but only ISW [Anderson et al., personal communication].

Three huge icebergs grounded on the Berkner Shelf in 1987 (their positions in February 1993 are indicated on

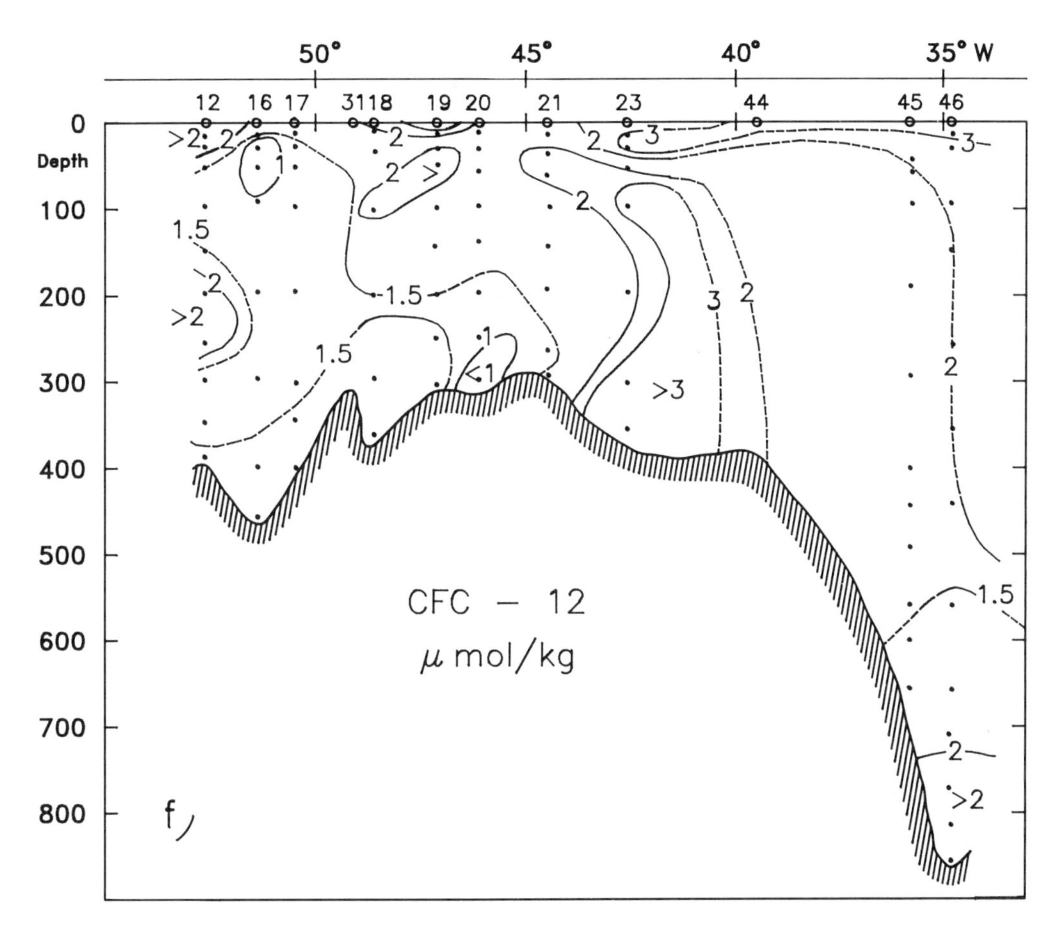

Figure 4. (continued)

Figure 1) have caused a large polynya on their western side and much more ice on their eastern side. This significant change in sea ice conditions might be the cause of the observed changes in water conditions. More open water on the Berkner Shelf may explain the lack of WSW found during this cruise, as the stations were worked rather late in the season, and an increased solar heating may have increased the surface layer temperature sufficiently to inhibit ice formation even during cold nights. After the stations on the Berkner Shelf were occupied (February 13) we observed that new ice started to form in the Weddell Sea. During winter time one should expect an increased production of WSW on the Berkner Shelf, which is in contrast to the disappearance of WSW from the Filchner Depression.

The intrusion of MWDW from the north seems to be a permanent feature, as it was observed both in 1980, [*Gammelsrød and Slotsvik, 1981*] and in 1984 [*Rohardt*, 1984]. The temperature of the warm parent water mass (WDW) has increased by some 0.7°C since 1977 [Foldvik et al., [1985d]), probably due to the disappearance of the Weddell Polynya [*Gordon*, 1982].

4.7. Concluding remarks

For the first time a substantial chemical oceanographic observation program was carried out in the Ronne Depression. Combined with the other data obtained on this cruise, both along the Filchner-Ronne Ice Shelf, on the Berkner Shelf and on the continental slope, much can be learned about the processes taking place in the Southern Weddell Sea, both outside and under the ice shelves. Based on the data presented here we have arrived at the following preliminary conclusions:

Active convection takes place in the Ronne Depression, where relatively young WSW is found down to 500 m depth.

A warm deep current (MWDW) follows the western slope of the Berkner shelf all the way from the continental slope to the ice barrier, and seems to be a permanent feature.

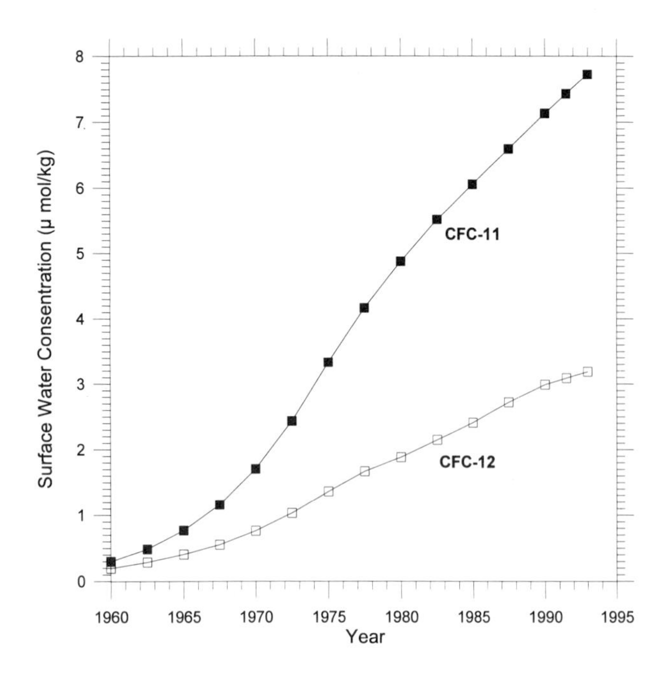

Fig.5 Estimated surface water concentrations of CFC-11 at 90% saturation, and CFC-12 at 85% saturation for T=-1.8°C and S=34.4 psu as a function of time.

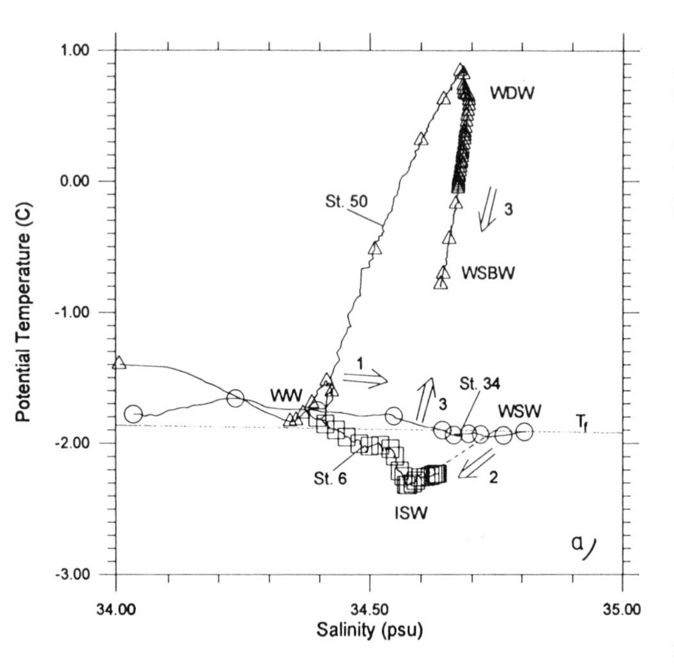

Fig. 6. Property versus salinity of station 6 in the Filchner Depression, station 34 in the Ronne Depression, and station 50 at the continental slope. a) θ-S, b) O_2-S,and c) Si-S. Symbols are annotated every 50 m. Arrows indicate water mass formation processes (see text).

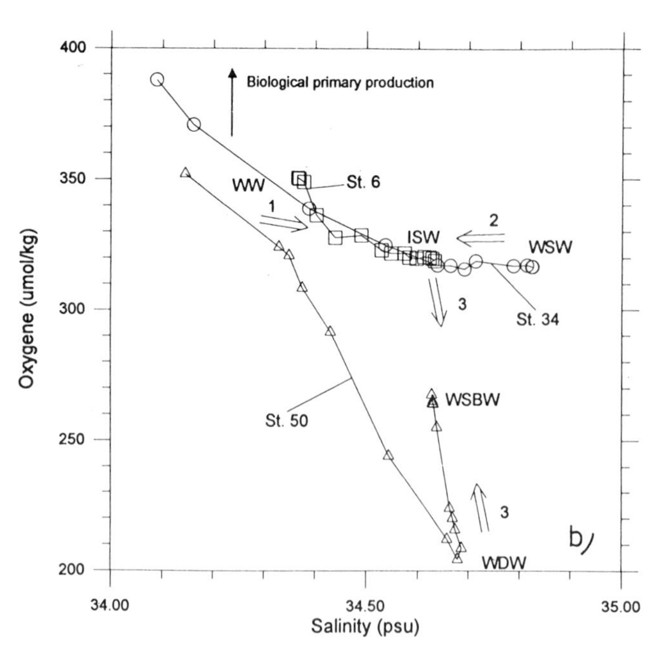

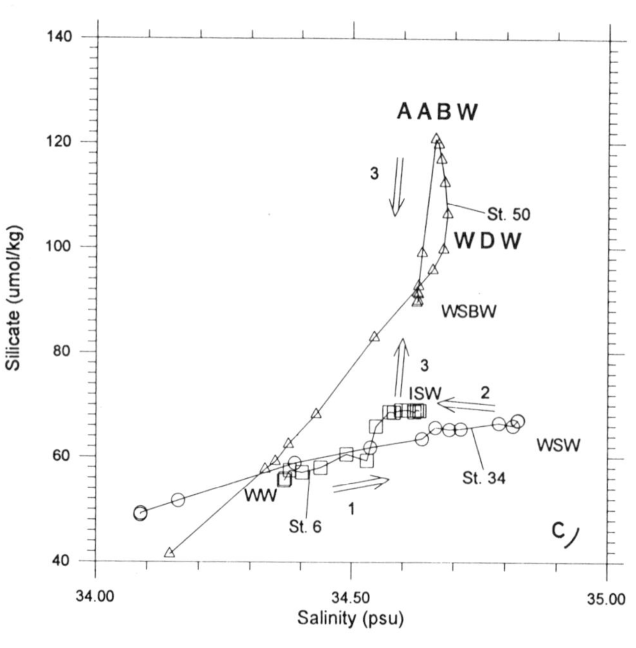

Figure 6. (continued)

The Berkner Shelf close to the barrier is found to be a region of upwelling and vertical convection.

Melting rates and residence times below the ice shelves have been calculated on the basis of difference in CFC ages and measured transport of meltwater spilling over the sill of the Filchner Depression. ISW is found to have residence times of about 10 years. Average melting rates, due to ISW overflowing the sill of the Filchner Depression, is calculated to be of order 10 cm/year.

It is confirmed that bottom water is formed as a mixture of WDW and ISW on the continental slope north of the Filchner Depression.

The core temperature of the WDW in the southern Weddell Sea was found to be close to 0.9°C, which is the highest temperature ever recorded of this water mass. This is 0.7°C warmer than observed in 1977 [*Foldvik et al.* 1985d], and thus represent a substantial climatic change. This warming may be related to the cessation of the Weddell Polynya activity in the late seventies [*Gordon*, 1982].

WSW seems to have disappeared from the Berkner Shelf and the Filchner Depression. This may be related to the break-off of three huge icebergs from the Filchner ice shelf in 1986. These bergs later grounded on the western slope of the Filchner Depression, where they still are found (see Figure 1). This has resulted in substantial changes in the sea ice distribution, with more sea ice in the Filchner Depression and less on the Berkner Shelf, due to the general cyclonic circulation in the area. This sea ice redistribution will influence the freezing and brine release, and thus the WSW production.

Acknowledgements. Our sincere thanks to Captain J. Olsen and his crew on board R/V *Lance* for their efficient and skilful performance under difficult conditions. We wish to thank Mr. Frank Zemlyak for his dedication and skill in performing the nutrient analyses. The Swedish group was supported by the Swedish Natural Science Research Council. Canadian participation was supported by the Panel on Research and Development (Canada). The Norwegian group was supported by the Norwegian Research Council (NAVF).

REFERENCES

Anderson, L.G., O. Holby, R. Lindegren, and M. Ohlson, The transport of anthropogenic carbon dioxide into the Weddell Sea. *J. Geophys. Res.*, *96*, 16,679-16,687, 1991.

Anderson, L.G., C. Haraldsson, and R. Lindegren, Gran linearization of potentiometric Winkler titration, *Mar. Chem.*, *37*, 179-190, 1992.

Bayer, R., and P. Schlosser, Tritium profiles in the Weddell Sea. *Mar. Chem.*, *35*, 123-126, 1991.

Carmack, E.C. and T. D. Foster, On the flow of water out of the Weddell Sea, *Deep Sea Res.*, *22*,711-724, 1975a.

Carmack, E. C., and T. D. Foster, Circulation and distribution of oceanographic properties near the Filchner Ice Shelf, *Deep Sea Res.*, *22*, 77-90, 1975b.

Cunnold, D.M., Prinn, R.G., Rasmussen, R.A., Simmonds, P.G., Alyea, F.N., Cardelino, C.A., Crawford, A.J., Fraser, P.J., and Rosen, R.D., Atmospheric Lifetime and Annual Release Estimates for $CFCl_3$ and CF_2Cl_2 From 5 Years of ALE Data, *J. Geophys. Res.*, *91(D10)*, 10,797-10,817, 1986.

Foldvik, A., and T. Gammelsrød, Notes on Southern Ocean Hydrography, sea-ice and bottom water formation. *Palaeogeogr., Palaeoclimatol., Palaeoecol.*, *67*, 3-17, 1988.

Foldvik, A., T. Gammelsrød, and T. Tørresen. Physical oceanography studies in the Weddell Sea during the Norwegian Antarctic Research Expedition 1978/79. *Polar Res.*, *3*, 195-207, 1985a.

Foldvik, A., T. Gammelsrød, N. Slotsvik, and T. Tørresen, Oceanographic observations on the Weddell Sea Shelf during the German Antarctic Expedition 1979/80, *Polar Res.*, *3*, 209-226, 1985b.

Foldvik, A., T. Gammelsrød, and T. Tørresen, Circulation and water masses on the Southern Weddell Sea Shelf. *Oceanology of the Antarctic Continental Shelf, Antarct. Res. Ser., Vol 43*, 5-20, 1985c.

Foldvik, A., T. Gammelsrød, and T. Tørresen. Hydrographic observations from the Weddell Sea during the Norwegian Antarctic Research Expedition 1976/77, *Polar Res.*, *3*,177-193, 1985d.

Foster, T. D., and Carmack, E.C., Frontal zone mixing and Antarctic Bottom Water formation in the southern Weddell Sea, *Deep Sea Res.*, *Vol 23*, 301-317, 1976.

Gammelsrød,T., and N. Slotsvik, Hydrographic and current measurements in the Southern Weddell Sea 1979/80, *Polarforschung*, *51(1)*, 101-111, 1981.

Gill, A. E., Circulation and bottom water production in the Weddell Sea, *Deep Sea Research, Vol. 20*, 111-140, 1973.

Gordon,A.L., Weddell deep water variability, *J. Mar. Res.*, *Vol50*, suppl. 199-217, 1982.

Gordon, A.L., B.A. Huber, H.H. Hellmer and A.Field, Deep and bottom water of the Weddell Sea's western rim,. *Science Vol 262*,95-97, 1993.

Gow, A.J., and T. Williamson, Gas Inclusion in the Antarctic Ice Sheet and Their Glaciological Significance, *J. Geophys. Res.*, *80*, 5101-5108, 1975.

Hellmer, H. H., and D. J. Olbers, A two-dimensional model for the thermohaline circulation under an ice shelf, *Antarct. Sci.*, *1*, 325-336, 1989.

Jacobs, S.S., A.F. Amos, and P. Bruchhausen, Ross Sea oceanography and Antarctic Bottom Water formation, *Deep Sea Res, Vol. 17, 935-962*, 1970.

Jacobs, S.S., and R.G. Fairbanks, Origin and evolution of water masses near the Antarctic continental margin: evidence from $H_2{}^{18}O/H_2{}^{16}O$ ratios in sea water, in *Oceanology of the Antarctic Continental Shelf, Antarctic res.ser. 43*, 59-85, 1985.

Jacobs, S.S., H.H. Hellmer, C.S.M. Doake, A. Jenkins, and R.M. Frolich, Melting of ice shelves and the mass balance of Antarctica, *J. of Glaciology, 38*, 375-386,1992.

Jenkins, A., A One-Dimensional Model of Ice Shelf Ocean Interaction, *J. Geophys. Res., 96,* 791-813, 1991.

Jenkins, A., and C.S.M. Doake, Ice-Ocean Interaction on Ronne Ice Shelf, Antarctica, *J. Geophys. Res., 96*, 791-813, 1991.

MacAyeal, D. R., Numerical Simulation of the Ross Sea Tides, *J. Geophys. Res., 89,* 607-615, 1984.

MacAyeal, D. R., Evolution of tidally triggered meltwater plumes below ice shelves, in *Oceanology of the Antarctic Continental Shelf, Antarct. Res. Ser., vol. 43,* edited by S. S. Jacobs, pp. 133- 143, AGU, Washington, D. C., 1985.

Millero, F. J., Freezing point of seawater, Eighth Report of the Joint Panel of Oceanographic Tables and Standards, Appendix 6, *UNESCO Tech. Pap. Mar. Sci., 28,* 29-31, 1978.

Nicholls, K.W., K. Makinson, and A. V. Robinson, Ocean circulation beneath the Ronne ice shelf, *Nature, Vol. 354,* 221-223, 1991.

Nøst, O.A., and A. Foldvik, A model of ice shelf-ocean interaction with applications to the Filchner-Ronne and the Ross Ice shelf, *J. Geophys. Res.*, in press, 1994.

Pudsey, C.J., P.F. Barker, and N. Hamilton, Weddell Sea abyssal sediments: A record of Antarctic bottom water flow, *Mar. Geol., 81*, 289-314, 1988.

Robin, G.Q., C.S.M. Doake, H. Kohnen, R.D. Crabtree, S.R. Jordan, and D. Möller, Regime of the Filchner-Ronne Ice Shelves, Antarctica., *Nature, Vol 302, 582-586,* 1983.

Rohardt, G., Hydrographische Untersuchungen am Rand des Filchner Schelfeises, *Ber. Zur Polarforschung, 19,* 141-144, 1984.

Schlosser, P., R. Bayer, A. Foldvik, T.Gammelsrød, G. Rohardt, and K.O. Munnich, Oxygen 18 and Helium as tracers of Ice Shelf Water and Water/Ice Interaction in the Weddell Sea, *J. Geoph..Res., 95,* 3253-3263, 1990.

Schlosser, P., J.L. Bullister, and R. Bayer, Studies of deep water formation and circulation in the Weddell Sea using natural and anthropogenic tracers, *Mar. Chem., 35*, 97-122, 1991.

Trumbore, S.E., S.S. Jacobs, and W.M. Smethie Jr. Chlorofluorocarbons evidence for rapid ventilation of the Ross Sea, *Deep Sea Res., C4,* 3583 3590, 1991.

Warner, M.J. and Weiss, R.F., Solubilities of chlorofluorocarbons 11 and 12 in water and seawater, *Deep-Sea Res. 32,* 1485-1497, 1985.

^{228}Ra and ^{228}Th in the Weddell Sea

Michiel M. Rutgers van der Loeff

Alfred Wegener Institute for Polar and Marine Research, Bremerhaven, Germany

^{228}Ra and its granddaughter ^{228}Th were measured on a N-S transect from 45°S to the Antarctic continent across the Antarctic Circumpolar Current (ACC) and the Weddell Sea. The distributions of ^{230}Th, ^{228}Th and ^{228}Ra show that southward transport across the ACC of Circumpolar Deep Water (CDW), the source of Warm Deep Water (WDW) in the Weddell Sea, occurs on a time scale between 8 and 30 years, in qualitative agreement with estimates of the upwelling rate of WDW. The distribution of ^{228}Ra in deep waters is controlled by advection and isopycnal mixing rather than diapycnal mixing. In the Weddell Sea, deep-water ^{228}Ra activities reach 15-20 dpm.m^{-3}. Enrichment in deep water is controlled by the production in the deep-sea floor, favoured by low biogenic sediment accumulation rates and consequently high ^{232}Th contents in the surface sediment (3 to 5 dpm.g^{-1}). The highest ^{228}Ra value (73 dpm.m^{-3}) was observed near the sea floor in a channel where an eastern outflow of Weddell Sea Bottom Water (WSBW) is suspected. It is not yet known whether this value is produced *in-situ* by accumulation in the stratified bottom water, or contains a signal of enrichment in shelf- and Ice Shelf Water. High ^{228}Ra activities on the south-eastern shelf (22 dpm.m^{-3}) and low activities offshore yield an estimated residence time of 1.5 years on this shelf and imply slow exchange with offshore waters.

INTRODUCTION

^{228}Ra is produced by radioactive decay of ^{232}Th. As ^{232}Th has very low activities in seawater [*Moore*, 1981; *Bacon and Anderson*, 1982], whereas it is present in all marine sediments, the major source of ^{228}Ra in the ocean is the sea floor. The distribution in the ocean is controlled by mixing and decay (5.8 y half-life). The highest activities have been reported in coastal seas where the flux from the continental shelf sediments is diluted in a shallow water column [*Moore*, 1969; *Kaufman et al.*, 1973]. The contribution from rivers is significant but smaller than the shelf source [*Key et al.*, 1985; *Moore*, 1987].

In principle, the oceanic distribution of ^{228}Ra, as mapped in the GEOSECS and TTO programs, contains information on rates of mixing processes in the ocean [*Moore*, 1969]. It has been used to derive rates of horizontal mixing between the shelf and the open ocean [*Kaufman et al.*, 1973, *Moore*, 1987] and in the deep sea [*Sarmiento et al.*, 1982], and rates of vertical mixing in the thermocline [*Li et al.*, 1980; *Sarmiento et al.*, 1990] and in the deep sea [*Sarmiento et al.*, 1976]. In many studies, however, the distribution of ^{228}Ra alone was not sufficient to distinguish between mixing and advection of water masses [*Kaufman et al.*, 1973; *Moore and Santschi*, 1986].

New results from the polar oceans show that ^{228}Ra can serve here as a welcome additional tracer. In the Arctic surface waters high ^{228}Ra activities have been reported far from the shelf source [*Kaufman et al.*, 1973, *Bacon et al.*, 1989]. The distribution is controlled by contact with the extensive Siberian shelves and may be used together with other tracers to derive transport and mixing rates [*Rutgers van der Loeff et al.*, in preparation].

The few available data from the Southern Ocean show very low activities of both isotopes in surface waters south of the Antarctic Convergence [*Kaufman et al.*, 1973; *Li et al.*, 1980], with to my knowledge only one observation of an increase in ^{228}Ra activity near the Antarctic continent to 5 dpm.m^{-3} [*Kaufman et al.*, 1973]. Near-bottom ^{228}Ra activities in the Indian sector of the Southern Ocean reach 12-22 dpm.m^{-3} [*Moore and Santschi*, 1986.]

This paper presents the distribution of ^{228}Ra and ^{228}Th on a transect across the Antarctic Circumpolar Current and the Weddell Sea from the Agulhas Basin to the Antarctic continent. It discusses what these tracers can tell us about mixing and transport of water masses in the Weddell Sea.

The Polar Oceans and Their Role in Shaping
the Global Environment
Geophysical Monograph 85

MATERIAL AND METHODS

During Polarstern expedition ANT VIII/3 (November 1989) five stations were sampled on a transect from the Agulhas Basin to the northern edge of the Weddell Sea, and three more further west (stations PS1751 to PS1785, Figure 1) . On expedition ANT IX/3 (Jan-March 1991) samples were collected in the Weddell Sea proper (stations 126 to 227), and during expedition ANT X/6 (October-November 1992) six deep stations were sampled in the ACC (862-917). Methods of sampling and thorium analysis have been given in *Rutgers van der Loeff and Berger* [1993]. Shortly, 300-1200 L of water were filtered *in-situ* (COSS pumps) through a 1-μ filter (293mm diameter) and two MnO_2-coated cartridges. Thorium was isolated from filter and cartridges and counted by alpha spectrometry. Natural ^{234}Th served as a yield tracer, and was assumed to be in equilibrium with ^{238}U below 100m water depth. The ^{234}Th activity in surface waters was determined from the ratio of the activities on the two cartridges in series (ANT VIII/3) or with discrete samples collected with Gerard bottles (other expeditions). Total ^{228}Th was determined as the sum of particulate and dissolved ^{228}Th.

Radium was analysed only during expedition ANT IX/3 an ANT X/6. After separation of Th from the 3-Liter cartridge leaches by $Fe(OH)_3$ precipitation, radium was precipitated from the supernatant as $BaSO_4$ by dropwise adding, under stirring, 30ml of a 0.3M solution of $BaCl_2$. The ^{228}Ra/^{226}Ra activity ratio was determined by gamma spectrometry [*Moore*, 1984; *Moore et al.*, 1985] in a well-type germanium detector. The ^{226}Ra activity was estimated from the silicate concentrations according to the nearly identical relationships given by *Chung and Applequist* [1980] for the Weddell Sea and by *Ku and Lin* [1976] for circumpolar waters. The resulting ^{226}Ra activities are close to 220 dpm.m^{-3} in all major water masses studied, decreasing to about 170 dpm.m^{-3} in North Atlantic Deep Water and in shelf waters.

^{228}TH AS ANALOGUE FOR ^{228}RA

The relevant part of the ^{232}Th decay series, with the corresponding half-lifes, can be represented as

$$^{232}\text{Th} \rightarrow {}^{228}\text{Ra} \rightarrow {}^{228}\text{Ac} \rightarrow {}^{228}\text{Th} \rightarrow \rightarrow {}^{208}\text{Pb}$$
$$(1.4\ 10^{10}\text{yr})\quad (5.8\ \text{yr})\quad (6.1\ \text{hr})\quad (1.9\ \text{yr})\quad (\text{stable})$$

^{228}Ra is a more suitable water-mass tracer than ^{228}Th because Ra is much less particle-reactive than Th. For

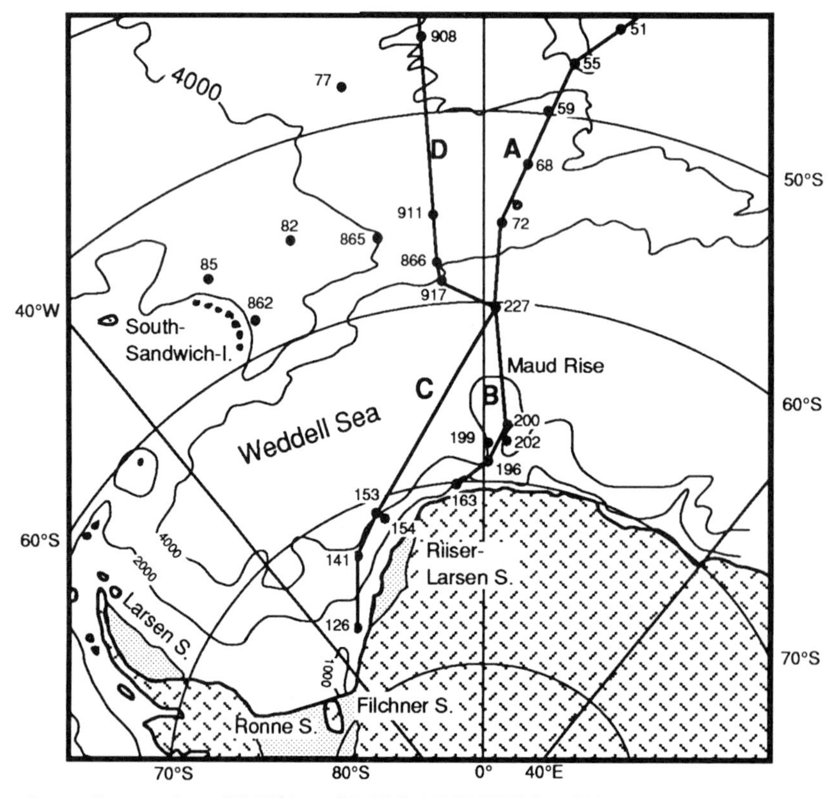

Figure 1. Map of sampling stations PS1751 to PS1785 (ANT VIII/3, 1989, indicated by last two digits), 126 to 227 (ANT IX/3, 1991), and 862 to 917 (ANT X/6, 1992), and the position of transects A, B, C and D.

^{228}Ra, removal by uptake on particles can be disregarded as the oceanic residence time of Ra is much greater than the half-life of ^{228}Ra. As the sampling and analytical procedures used here yield more precise ^{228}Th than ^{228}Ra data, and as ^{228}Ra data are not available from the first expedition, the question arises to what extent ^{228}Ra can be measured through its granddaughter ^{228}Th.

In the absence of scavenging the $^{228}Th/^{228}Ra$ ratio in a water mass depends on the hydrographical conditions: In a well-mixed water mass in contact with a ^{228}Ra source we would expect a secular equilibrium to develop with an activity ratio of 1. In a water mass that detaches from the ^{228}Ra source, e.g. by upwelling, ^{228}Ra is now unsupported and follows simple radioactive decay. A transient equilibrium, where $^{228}Th/^{228}Ra = \lambda(Ra)/(\lambda(Ra)-\lambda(Th)) \approx 1.5$, would develop, in which the activity of both isotopes declines with the halflife of ^{228}Ra.

$^{228}Th/^{228}Ra$ ratios below unity indicate that scavenging is significant, as in the surface ocean [*Li et al.*, 1980] and near the seafloor [*Broecker and Peng*, 1982; this work]. Ratios above 1.5 have been explained as resulting from the release of ^{228}Th associated with mineralization [*Li et al.*, 1980].

In the present study the ratio of total ^{228}Th to ^{228}Ra was below 1 in all bottom and shelf waters (Table 1). The distribution of ^{228}Th is clearly influenced by scavenging, and consequently, ^{228}Ra cannot be measured through ^{228}Th near the coast and near the seafloor. In intermediate waters with low ^{228}Ra contents, the large counting error in the ^{228}Ra values precludes a determination of an eventual depletion of ^{228}Th relative to ^{228}Ra. As ^{230}Th data from section A [*Rutgers van der Loeff and Berger*, 1993] show that the scavenging residence time of Th is too long (≥30y) to affect the distribution of ^{228}Th (1.9 y half-life), it can be expected that ^{228}Th serves as a proxy for ^{228}Ra in the intermediate waters of this section.

RESULTS

The distribution of total (dissolved + particulate) ^{228}Th along sections A and B, from the Agulhas Basin to the Antarctic Continent (Figure 2), shows the same general features as the distribution of ^{228}Ra, measured along sections D and B (Figure 3). ^{228}Ra sources at the seafloor are apparent. Bottom waters are enriched to 5-10 dpm.m^{-3} in the ACC, and up to 23 dpm.m^{-3} in the Weddell Sea. Total ^{228}Th activities reach from 1.5-3.2 dpm.m^{-3} in the ACC and 7-14 dpm.m^{-3} in the Weddell Sea, or about 30-70% of the ^{228}Ra values.

This significant depletion of ^{228}Th with respect to ^{228}Ra in deep waters is observed in all sections. *Broecker and Peng* [1982] mention $^{228}Th/^{228}Ra$ ratios of 0.5-1.0 in deep waters from other oceans. The strong depletions in the bottom waters of the ACC may be related to the existence of a well-developed nepheloid layer [*Biscaye and Eittreim*, 1977], which enhances scavenging of Th isotopes (*Bacon and Rutgers van der Loeff*, 1989).

Figures 2 and 3 and Table 2 show how sediment characteristics control the distribution of ^{228}Ra (and by consequence of ^{228}Th) in the deep waters of ACC and Weddell Basin. The deep-sea sediments can be subdivided into three zones with markedly different ^{232}Th contents (Table 2): First, the abyssal Weddell Sea, where ^{232}Th contents range from 3-5 dpm.g^{-1}. Second, the opal belt underlying the ACC, characterized by very high opal contents (50-95%, *Bohrmann et al.*, personal communication) and correspondingly low ^{232}Th activities, and third, the area north of the Polar Front with an opal content of only 3-5% and ^{232}Th activities around 1 dpm.g^{-1}. The higher ^{232}Th activities in Weddell Sea sediments can explain the higher ^{228}Ra and ^{228}Th values in deep waters in the Weddell Sea compared with the deep waters from the Agulhas Basin (Table 2).

As deep waters from the ACC exchange with shallow waters in the Weddell Sea along sloping isopycnal surfaces, the tracer data have been plotted in Figures 2 and 3 on potential density contours. The data sets are mutually supportive in depicting the behaviour of ^{228}Ra: the ^{228}Th data are more precise, but in surface and bottom waters they are depleted relative to ^{228}Ra as a result of scavenging; the ^{228}Ra values have relatively large counting errors, but are unaffected by scavenging.

The main point shown by these plots is, that in contrast to the distribution of ^{230}Th, the distributions of ^{228}Ra and of ^{228}Th do not follow the density surfaces. The activities of ^{228}Ra and of its granddaughter ^{228}Th decrease along the upwelling path from the enhanced bottom water values in the ACC to very low values in shallower waters of the same potential density, including the Warm Deep Water (WDW), at station 227. Further south, the temperature maximum layer in the WDW remains low in ^{228}Ra throughout the transect, until station 163 where it touches the slope at 750m depth.

The gradient of ^{228}Th in shallow waters across the Polar Front agrees with the data of *Kaufman et al.* [1973] and *Li et al.* [1980], and is related to the basin-wide distribution of ^{228}Ra in surface waters with apparently stronger sources from the American and African continents than from Antarctica.

The influence of sources on the shelf is shown by enhanced ^{228}Ra values in the surface water near the Antarctic continent. However, in the Coastal Current near Kap Norvergia (Station 163, Figure 3) the shelf water is not highly enriched with ^{228}Ra. The narrow shelf and strong currents do not allow ^{228}Ra to accumulate above 6.4 dpm.m^{-3} in the surface water.

On the relatively wide southern shelf (Section C, Figure 4), ^{228}Ra accumulates to 22 dpm.m^{-3}. The influence of shelf and slope extends to the entire water column at station 141. As in section B, bottom waters are enriched up to about 20 dpm.m^{-3}. The extreme value at station 153 is discussed below.

Table 1. ^{228}Ra and ^{228}Th activities with 1-sigma counting errors

Expedition ANT VIII/3

depth (m)	total 228Th (dpm/m3)
Station PS1751 (a)	
10	4.29 ± 0.24
200	3.69 ± 0.07
1000	1.57 ± 0.03
2500	1.23 ± 0.03
4700	3.23 ± 0.04
Station PS1755	
10	0.98 ± 0.11
500	0.54 ± 0.04
1000	0.35 ± 0.03
2500	0.97 ± 0.03
4000	2.28 ± 0.05
Station PS1759	
10	2.20 ± 0.10
300	0.62 ± 0.03
800	0.48 ± 0.02
2200	2.43 ± 0.04
3600	1.51 ± 0.03
Station PS1768	
10	2.24 ± 0.15
300	0.50 ± 0.03
600	0.50 ± 0.02
1900	1.02 ± 0.02
3200	2.81 ± 0.04
Station PS1772	
400	0.95 ± 0.02
1600	1.08 ± 0.02
2800	1.89 ± 0.03
4000	3.78 ± 0.04
Station PS1777	
10	1.25 ± 0.10
400	0.76 ± 0.02
900	0.70 ± 0.02
1650	0.74 ± 0.02
2400	1.23 ± 0.02
Station PS1782	
10	2.07 ± 0.14
600	0.78 ± 0.01
2000	2.62 ± 0.03
3500	3.58 ± 0.04
5000	6.05 ± 0.05
Station PS1785	
600	0.80 ± 0.02
2500	2.39 ± 0.06
4000	4.26 ± 0.05
5500	6.35 ± 0.04

Expedition ANT IX/3

depth (m)	total 228Th (dpm/m3)	Ra228 (dpm/m3)	Th/Ra ratio
St 126	*76° 25.5′ S*	*30° 28.0′ W*	*322m.*
20	7.63 ± 0.18	22.4 ± 4.0	0.34
200	6.74 ± 0.08	21.3 ± 2.6	0.32
305	5.91 ± 0.14	21.1 ± 4.3	0.28
360	8.04 ± 0.09	22.4 ± 1.3	0.36
St 127	*76° 37.0′ S*	*31° 25.7′ W*	*307m.*
20	7.07 ± 0.15	23.2 ± 5.2	0.30
100	8.26 ± 0.13	20.7 ± 1.8	0.40
270	7.51 ± 0.39	32.3 ± 5.7	0.23
St 141	*73° 36.4′ S*	*26° 6.5′ W*	*3379m.*
25	4.61 ± 0.18	13.5 ± 2.8	0.34
300	3.23 ± 0.08	10.4 ± 2.2	0.31
1100	1.19 ± 0.05	5.7 ± 2.5	0.21
2100	2.92 ± 0.06	6.1 ± 1.0	0.48
3100	4.64 ± 0.09	15.1 ± 3.0	0.31
3280	5.49 ± 0.10	16.2 ± 2.1	0.34
St 153	*71° 5.4′ S*	*20° 47.4′ W*	*4416m.*
30	1.00 ± 0.05	<6.0 b	
470	1.53 ± 0.05	2.4 ± 1.7	0.63
1700	1.54 ± 0.04	<5.3 b	
3000	5.65 ± 0.09	13.7 ± 4.0	0.41
4250	7.27 ± 0.13	19.7 ± 3.9	0.37
4400	10.69 ± 0.18	73.1 ± 6.9	0.15
St 163	*70° 5.6′ S*	*4° 0.1′ W*	*1139m.*
13	2.20 ± 0.21	6.4 ± 2.3	0.35
100	3.15 ± 0.06	ND	
350	3.15 ± 0.04	ND	
750	1.47 ± 0.04	<2.6 b	
920	1.45 ± 0.03	5.2 ± 2.0	0.28
St 196	*69° 4.9′ S*	*0° 53.3′ E*	*3324m.*
35	2.84 ± 0.25	ND	
85	1.66 ± 0.72	ND	
275	2.01 ± 0.08	ND	
1200	1.31 ± 0.04	ND	
2200	2.33 ± 0.27	ND	
3150	5.42 ± 0.10	13.8 ± 1.7	0.39
St 199	*68° 14.0′ S*	*0° 59.1′ E*	*4465m.*
35	2.36 ± 0.15	<8.8 b	
95	0.63 ± 0.03	<3.5 b	
275	0.72 ± 0.03	<4.6 b	
2900	2.70 ± 0.04	5.8 ± 1.7	0.46
4200	5.76 ± 0.05	14.8 ± 3.6	0.39
St 200	*66° 48.2′ S*	*6° 15.3′ E*	*4321m.*
145	0.93 ± 0.05	<3.6 b	
240	1.13 ± 0.08	<3.8 b	
1000	0.96 ± 0.04	<4.6 b	
2550	2.18 ± 0.06	3.7 ± 1.9	0.59
4100	7.59 ± 0.12	16.2 ± 1.5	0.47
St 202	*67° 59.9′ S*	*6° 16.0′ E*	*3896m.*
90	0.52 ± 0.03	<3.3 b	
295	1.35 ± 0.04	<2.9 b	
1000	1.04 ± 0.03	<2.2 b	
2335	2.85 ± 0.04	2.2 ± 1.1	1.29
3670	7.10 ± 0.08	22.4 ± 2.1	0.32
St 227	*60°34.2′ S*	*3° 57.8′ E*	*5373m.*
160	0.67 ± 0.03	<2.1 b	
1300	0.69 ± 0.14	<2.1 b	
2600	1.37 ± 0.03	3.7 ± 1.3	0.37
3900	5.49 ± 0.07	7.0 ± 2.2	0.78
5200	13.78 ± 0.10	21.6 ± 3.5	0.64

Expedition ANT X/6

depth (m)	total 228Th (dpm/m3)	Ra228 (dpm/m3)	Th/Ra ratio
St 862	*57° 28′ S*	*23° 23′ W*	*5007m.*
3000	1.76 ± 0.03	7.2 ± 1.2	0.24
4000	3.72 ± 0.06	9.3 ± 2.3	0.40
4650	5.32 ± 0.06	18.4 ± 2.1	0.29
4850	6.86 ± 0.09	19.7 ± 1.6	0.35
4880	6.59 ± 0.08	15.0 ± 1.2	0.44
St 865	*56° 10′ S*	*12° 23′ W*	*4860m.*
3000	1.82 ± 0.03	6.6 ± 1.3	0.28
4000	2.05 ± 0.05	9.4 ± 1.5	0.22
4610		9.3 ± 1.3	
4820	23.83 ± 0.05	14.0 ± 1.8	0.20
4850	3.27 ± 0.04	11.3 ± 1.6	0.29
St 866	*57° 39′ S*	*6° 23′ W*	*3400m.*
800	0.43 ± 0.04	<3.7 b	
2000	1.24 ± 0.04	4.6 ± 3.3	0.27
3000	1.41 ± 0.05	10.4 ± 3.3	0.14
3370	1.67 ± 0.04	6.1 ± 2.5	0.28
3420	1.96 ± 0.04	6.9 ± 1.8	0.28
3450	1.47 ± 0.03	7.3 ± 1.5	0.20
St 908	*46° 52′ S*	*5° 44′ W*	*3615m.*
1600	0.55 ± 0.02	<2.3 b	
3400	1.74 ± 0.05	6.1 ± 1.3	0.29
3600	1.80 ± 0.06	6.1 ± 1.9	0.30
3650	1.89 ± 0.07	5.4 ± 1.7	0.35
St 911	*55° 51′ S*	*5° 56′ W*	*3995m.*
2500	1.28 ± 0.03	5.6 ± 1.6	0.23
3600	2.24 ± 0.04	8.4 ± 1.4	0.27
3800	2.13 ± 0.05	6.1 ± 1.6	0.35
3900	2.19 ± 0.05	10.8 ± 1.5	0.20
3950	2.38 ± 0.07	10.4 ± 1.9	0.23
St 917	*58° 23′ S*	*5° 55′ W*	*5078m.*
2500	2.23 ± 0.07	8.2 ± 1.3	0.27
4000	4.96 ± 0.12	16.9 ± 1.9	0.29
4600	6.36 ± 0.10	14.3 ± 1.3	0.44
4800	8.10 ± 0.13	23.2 ± 1.7	0.35
4950	8.73 ± 0.11	16.5 ± 1.6	0.53

a: positions in Rutgers van der Loeff and Berger, 1993
b: detection limit based on 2 times background noise
ND: not determined

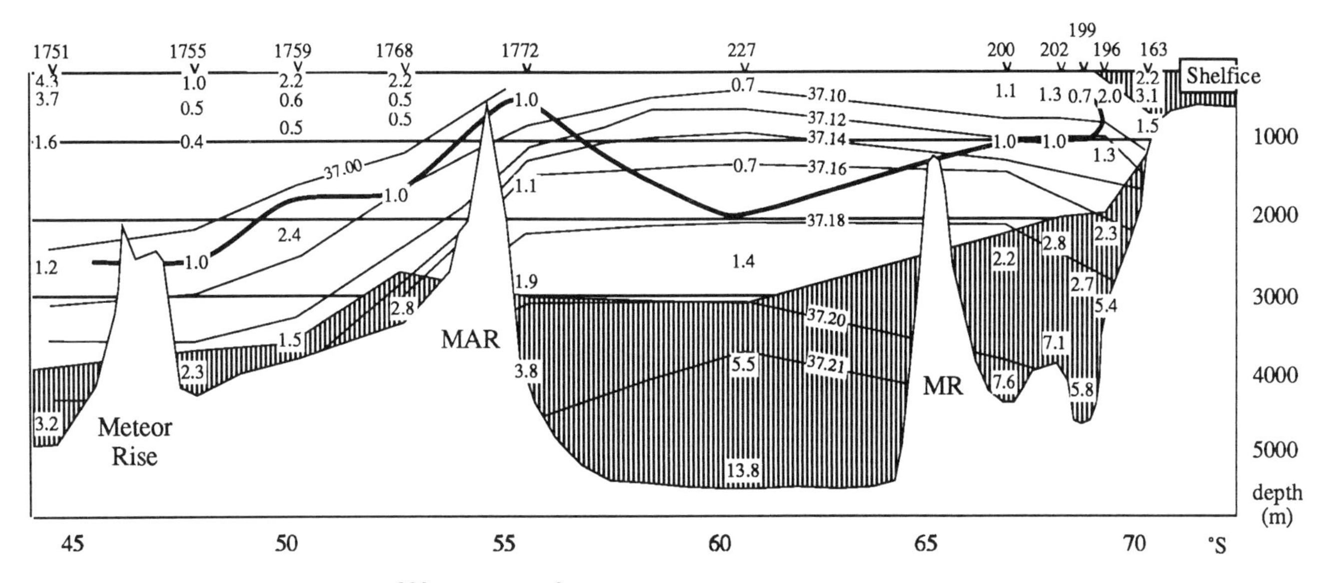

Figure 2. The distribution of ^{228}Th (dpm.m^{-3}) in section A + B, superimposed on the distribution of potential density anomaly σ_2 (kg.m^{-3}). Activities of 1 dpm.m^{-3} and above 1.5 and 2 dpm.m^{-3} indicated by heavy line, light and heavy hatching, respectively. MAR, Mid Atlantic Ridge; MR, Maud Rise.

DISCUSSION

^{228}Ra is produced wherever water masses come into contact with the seafloor. In the following discussion, we distinguish between the ^{228}Ra enrichment in surface waters resulting from contact with the shelf sediments, and in deep waters by release from the deep-sea floor.

Shelf Waters

We expect ^{228}Ra to accumulate to high activities where thin water layers have a long residence time over a ^{232}Th-containing sediment. These conditions are met on the shelf, where the water depth is only about 400m. We can estimate the residence time τ of water on the shelves

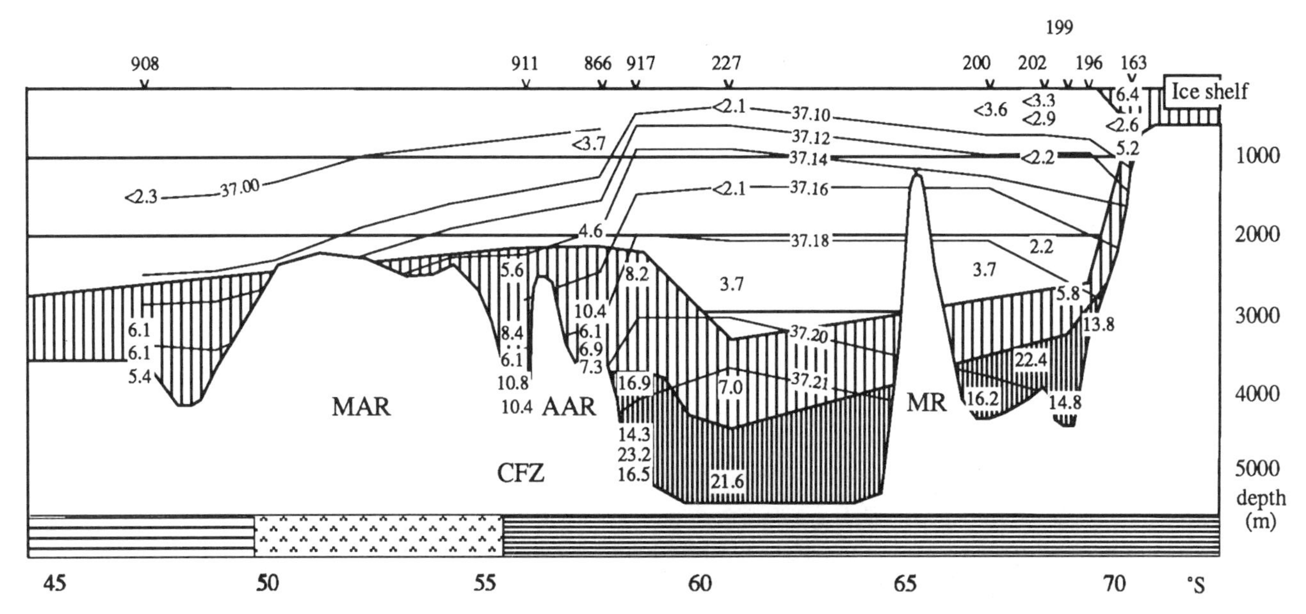

Figure 3. The distribution of ^{228}Ra (dpm.m^{-3}) in section D + B, superimposed on the distribution of potential density anomaly σ_2 (kg.m^{-3}). Activities above 5 and 10 dpm.m^{-3} indicated by light and heavy hatching, respectively. Differences in the composition of bottom sediments (Table 2) have been indicated by shading. MAR, Mid Atlantic Ridge; CFZ, Conrad Fracture Zone; AAR, America Antarctic Ridge; MR, Maud Rise.

Table 2. Deep-sea and shelf sediment sources of ^{228}Ra, and the range of ^{228}Ra and ^{228}Th values observed in the overlying waters

Source area	^{232}Th sediment ($dpm.g^{-1}$)	^{228}Ra bottom water ($dpm.m^{-3}$)	^{228}Th bottom water ($dpm.m^{-3}$)
Deep-sea			
South Atlantic N of 50°S	1	5	2 - 3
ACC 50-56°S	0.5	5 - 10	3 - 6
deep Weddell Sea	3 - 5	14 - 73	5 - 14
Shelf			
Lazarev shelf	3 - 5	6	3
Southeastern Weddell shelf	3 - 5	21 - 32	6 - 8

required to account for the observed enrichments in ^{228}Ra, using the simple model illustrated in Figure 5. The ingrowth of ^{228}Ra activity C_s on the shelf follows from

$$\frac{\partial C_s}{\partial t} = \frac{F}{H} - \lambda C_s + 1/\tau (C_o - C_s) \qquad (1)$$

where C_o is the activity of offshore surface water exchanging with the shelf at a rate $1/\tau$, F is the ^{228}Ra flux from the sediment, H the water depth and λ the decay constant of ^{228}Ra. At steady state, this yields

$$C_s = \frac{1}{\lambda\tau+1} \left(\frac{F\tau}{H} + C_o\right) \qquad (2)$$

In the absence of data for the flux F from Antarctic shelf sediments, we have to base our estimate on reported ^{228}Ra fluxes from other continental shelves. The major variables controlling the flux F are the ^{232}Th content in the sediment, giving the production rate, and biological and physical mixing, controlling the release rate. The average ^{228}Ra flux from Atlantic continental shelves to the water column has been estimated by *Li et al.* [1980] to be 0.6 ± 0.1 dpm $cm^{-2}.yr^{-1}$, but reported values range from 0.3 to 3 dpm $cm^{-2}.yr^{-1}$, the higher values probably reflecting higher proportions of fine-grained sediment and perhaps more extensive sediment mixing [*Moore*, 1987 and references therein]. Even higher fluxes are observed in the Amazon delta related to intensive and deep sediment reworking [*Moore et al.*, 1986], an extreme situation which cannot be compared to the moderate bioturbation observed on the Antarctic shelf [*DeMaster et al.*, 1991].

The ^{228}Ra activity of 6.4 ± 2.3 $dpm.m^{-3}$ on the shelf of the Lazarev Sea (4°W) is similar to the value observed in the Coastal Current at 130°E by *Kaufman et al.* [1973]. With the estimated flux of 0.6 dpm $cm^{-2}.yr^{-1}$, and setting C_o=0, this observed ^{228}Ra enrichment would require a

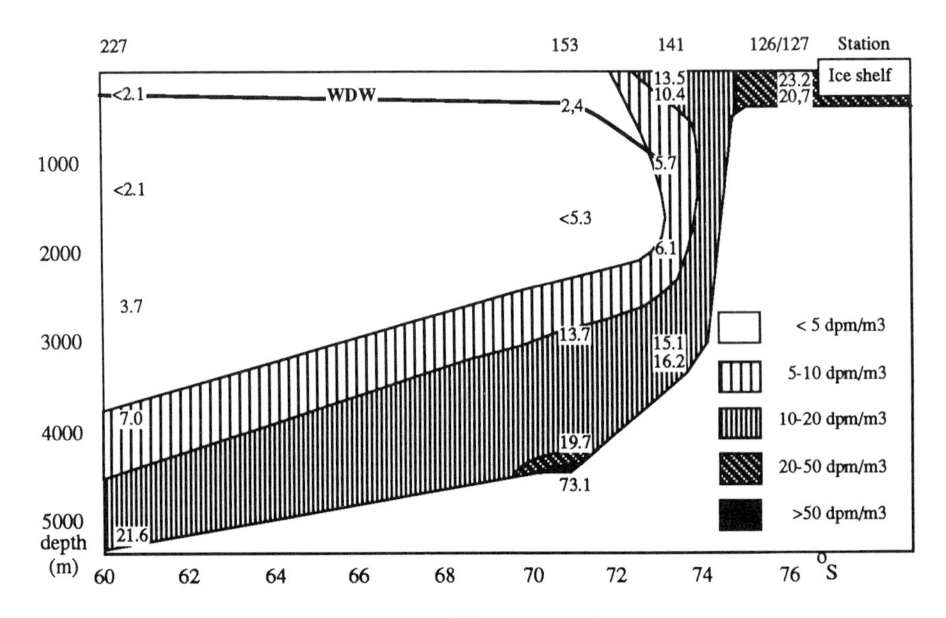

Figure 4. Distribution of ^{228}Ra ($dpm.m^{-3}$) on transect C.

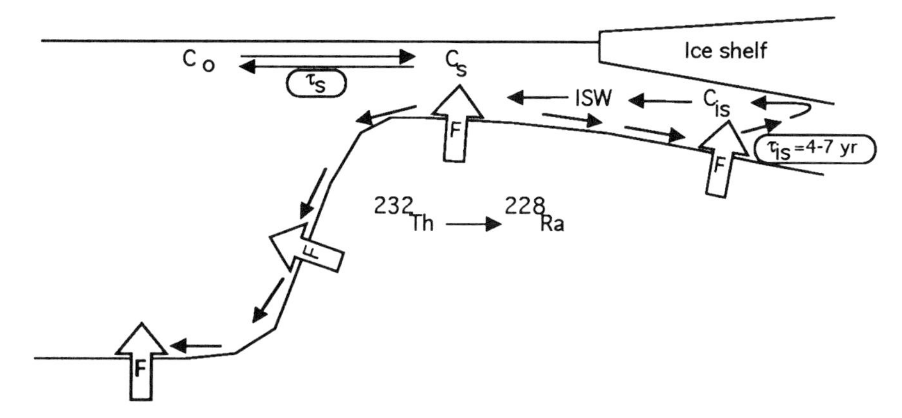

Figure 5. Schematic diagram of the exchange of shelf water (^{228}Ra activity C_s) with offshore water (activity C_o) and with water under the ice shelf (activity C_{is}), with residence time on the shelf (τ_s) and under the shelf ice (τ_{is}) and the origin and flow pattern of Ice Shelf Water (ISW).

residence time on the around 300m deep shelf in the order of 4 months. A similar calculation yields a residence time of about 1.5 year on the south-eastern shelf (average water depth 350m) to account for the 22 dpm.m^{-3} observed in the Eastern Shelf Water at station 126/127. As a result of the uncertainty in the sediment flux, these are only rough estimates. It is clear, however, that the residence time on the southeastern shelf is much longer than on the narrow shelf of the Lazarev Sea.

It is remarkable that horizontal mixing in the surface water does not bring the shelf influence far offshore (Figures 3 and 4). The only deep station with clearly enhanced surface values is station 141. Surface water in the central Weddell Sea does not exchange with shelf waters on the time scale of ^{228}Ra decay.

^{228}Th is depleted throughout in the surface waters. In the Coastal Current at station 163, where ^{234}Th was depleted by 27% relative to ^{238}U [*Bathmann et al.*, in preparation], the ^{228}Th/^{228}Ra ratio was 0.35. At the southeastern shelf, under a closed ice cover where ^{234}Th was not depleted relative to ^{238}U, the ^{228}Th/^{228}Ra ratio was similar (Table 1). This larger ^{228}Th depletion at station 126 compared to station 163 is explained by the longer residence time of water on the southeastern shelf.

Deep-Sea:

1. upwelling and ^{228}Ra distribution in the Antarctic Circumpolar Current (ACC). The ^{228}Ra distribution in transects A, B and D (Figures 2 and 3) sets a maximum to the upwelling rate of CDW toward the Weddell Basin. The Weddell Basin is flushed with Circumpolar Deep Water, including a core of North Atlantic Deep Water (NADW). After entering the Weddell Basin CDW is called Warm Deep Water (WDW), which mixes with Weddell Sea Bottom Water (WSBW) to form Weddell Sea Deep Water (WSDW), which then leaves the Basin as Antarctic Bottom Water (AABW). If we take the transitions between AABW, CDW, and northward flowing Antarctic Intermediate Water (AAIW) at a potential temperature of 0°C and 2°C, respectively, the CDW has a range in σ_2 of about 37.00 to 37.155. The temperature maximum in the WDW in the central Weddell Sea is derived from Lower Circumpolar Deep Water with a potential density anomaly around 37.12 [*Whitworth and Nowlin*, 1987]. The source of the WDW around Maud Rise is somewhat shallower [σ_2 around 37.03, *Whitworth and Nowlin*, 1987].

The distribution of ^{230}Th closely follows the density distribution [*Rutgers van der Loeff and Berger*, 1993]. This implies that the southward transport of CDW must be rapid compared to the scavenging residence time of Th, which is about 30 years in the southern Atlantic. Although the scavenging residence time is longer in the Weddell Sea [*Rutgers van der Loeff and Berger*, 1993], this is unlikely in the area around the Polar Front based on the relatively high particulate fluxes recovered here in sediment traps [*Wefer and Fischer*, 1991].

In contrast to the distribution of ^{230}Th, ^{228}Ra does not follow the density distribution. The ^{228}Th data of transects A+B (Figure 2) suggest an effect of upwelling, similar to the ^{230}Th data, in the somewhat enhanced concentrations at station 1772 at 400 and 1600m. Further south however, very low ^{228}Th values are found in the WDW at this density, indicating the gradual decay of its parent ^{228}Ra. As it cannot be excluded that the distribution of the particle-reactive ^{228}Th in this shallow water mass would to some extent be influenced by scavenging or mineralization, it is reassuring that the behaviour concluded from the ^{228}Th data, is supported by the actual ^{228}Ra data (Figure 3). It can be concluded that the upwelling of Circumpolar Deep Water is slow relative to the mean life of ^{228}Ra (8.3y).

We can compare these two estimates with hydrographical information. *Gordon and Huber* [1990] estimate the upwelling rate of WDW in the full circumpolar 60-70°S belt at 24 Sv. This upwelling must be compensated by a southward component of the CDW transport below the Polar Front. It should be noted that the meridional transport of dissolved substances and heat across the Polar Front is largely transient, the mean advective transport being negligible [*De Szoeke and Levine*, 1981]. This is confirmed by calculations with the Fine Resolution Antarctic Model (FRAM), which show highly variable meridional velocities at all depths of the ACC [*Webb et al.*, 1991], but a negligible mean flow [*Thompson*, 1993]. As the estimate of *Gordon and Huber* [1990] is based on heat flux and not on mass transport, it appears legitimate to estimate the meridional transport of dissolved radionuclides in analogy with their advective model of heat transport.

The southward flowing CDW with a range of σ_2 of about 37.00 to 37.155 has a depth range in the south-eastern Atlantic of about 2000m at the Polar Front at 50°S, thinning to about 1000m at 60°S [*Gordon and Molinelli*, 1982; *Whitworth and Nowlin*, 1987; Figures 2 and 3]. Thus the southward flow averaged over all longitudes increases from 0.04 cm.s^{-1} at 50°S to 0.06 cm.s^{-1} at 55°S, with a total transit time in the ACC of 34 years. As the influx of WDW is concentrated in the eastern edge of the Gyre [*Deacon*, 1979; *Gouretski and Danilov*, 1993], the velocities are probably higher and the transit time correspondingly shorter. This result is in qualitative agreement with the radionuclide data, which tell us that the residence time of CDW in the ACC should be between the life time of ^{228}Ra (8.3 years) and the scavenging residence time of ^{230}Th (30 years).

2. *^{228}Ra in the deep Weddell Sea.* The vertical distribution of ^{228}Ra and ^{228}Th in the Weddell Basin is characterized by a gradual increase from the minimum in the WDW downwards. WSDW, the major water mass, is produced by the mixing of cold, oxygen- and ^{228}Ra-rich bottom waters (WSBW) with oxygen- and ^{228}Ra-poor Warm Deep Water (WDW). The oxygen-temperature relationship is usually linear, showing a conservative mixing [*Rutgers van der Loeff and van Bennekom*, 1990]. The highly non-linear mixing lines of ^{228}Ra can be used to estimate the rate of vertical eddy diffusion in the deep water.

A steady state is established when the supply of ^{228}Ra by mixing balances radioactive decay. If this mixing is vertical (approximately diapycnal), the ^{228}Ra distribution is described by a simple one-dimensional model [e.g., *Sarmiento et al.*, 1976]:

$$\frac{\partial C}{\partial t} = K \frac{\partial_2 C}{\partial z^2} - \lambda C = 0 \qquad (3)$$

$$C = C_{z=0}\, e^{-\sqrt{\frac{\lambda}{K}}\, z} \qquad (4)$$

where K is the eddy diffusion coefficient and z is the height above seafloor. The high penetration of ^{228}Ra above the seabed (Table 1, Figures 3 and 4) would require values for K on the order of 50-70 cm^{2}s^{-1}. Such values are unrealistically high [*cf. Sarmiento et al.*, 1976; *Sarmiento et al.*, 1982] and show that in the central Weddell Sea, as at the more coastal stations 141 and 153, the distribution of ^{228}Ra is not controlled by diapycnal mixing but rather by horizontal advection and isopycnal mixing. This result agrees with observations in the Atlantic [*Sarmiento et al.*, 1982] and Indian Oceans [*Moore and Santschi*, 1986], and has two causes:

First, water circulation brings waters at intermediate depth in contact with shallower areas like ridges and the continental slope. In the Atlantic sector of the Southern Ocean, the ACC is forced to flow through many ridges and seamounts. Horizontal extension of inputs of ^{228}Ra is enhanced by isopycnal mixing. A special case is the wide Explora Escarpment. This large plateau at 2000-3000m depth is a likely source for ^{228}Ra enrichment at intermediate depths even far offshore, as observed at station 141.

Second, newly formed bottom water has a strong shelf component and must consequently be enriched in ^{228}Ra (no data of ^{228}Ra in newly formed bottom water or ISW are yet available). This water is not only introduced at abyssal depth as the classical WSBW, but also interleaves at intermediate depths along the western slopes of the Weddell Basin [*Gordon*, 1993; *Fahrbach et al.*, in preparation].

3. *enrichment at station 153.* The highest ^{228}Ra activity in this study was observed in the bottom water at station 153, in a channel which is believed to constitute an eastern pathway of newly formed Weddell Sea Bottom Water (WSBW) from the Filchner Depression [*Kuhn and Weber*, 1993]. Until recently, it was generally believed that WSBW flows westward from the overflow area, to enter the deep circulation along the southern and western slope of the Weddell Basin. However, both the distribution of surface sediment types [*Kuhn and Weber*, 1993] and recent hydrographic observations based on detailed CTD sections and current meter deployments [*Fahrbach et al.*, in preparation] make us believe that there exists an eastern short-cut for WSBW. Topographic features prevent the Coriolis accelaration to divert the northward flowing WSBW to the west. Although no WSBW (Θ<-0.7°C) was encountered at station 153, there was a clear temperature stratification (even more pronounced at nearby station 154) which allowed a strong gradient in ^{228}Ra (Figure 6).

It has yet to be investigated whether this ^{228}Ra signal is produced at the deep-sea floor or contains a signature from the Filchner shelf. No ^{228}Ra data exist to my knowledge from the formation area of WSBW. One of the processes responsible for the formation of Weddell Sea Bottom Water is the circulation of Western Shelf Water under the shelf ice, thus producing undercooled Ice Shelf Water (Figure 5).

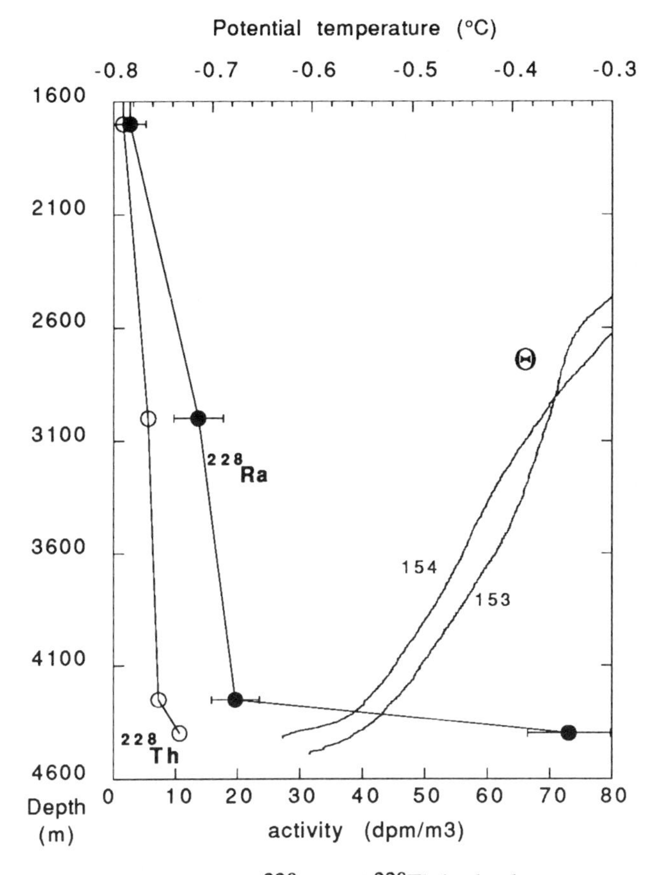

Figure 6. Distribution of ^{228}Ra and ^{228}Th in the deep water at station 153, with potential temperature at stations 153 and 154.

The residence time of this water under the shelf ice (τ_{is}) has been estimated from tritium and CFC data to be about 4-7 years [*Bayer and Schlosser*, 1991; *Gammelsrød et al.*, this issue]. The ^{228}Ra enrichment in an approximately 400 m thick water mass under the ice shelf can again be described with equations (1) and (2), if we substitute C_s for C_o, and τ_{is} for τ_s. C_s now represents the activity of the shelf water that flows under the ice shelf and thus forms the source of ISW. With τ_{is} = 5 yr and C_s = 22 dpm.m^{-3}, the resulting activity C_{is} of the ISW would be 38 to 250 dpm.m^{-3}, depending on the ^{228}Ra flux out of the sediment within the reported range given above. Mixed with two parts of ^{228}Ra-free WDW to a potential temperature of -0.6°C, and neglecting radioactive decay during transport down the slope, one third of this enrichment would be left in the bottom water of station 153. Although we have no information on the actual sediment-water exchange of solutes under the ice shelf, this admittedly rather speculative calculation shows that a strong ^{228}Ra signal could be present in newly formed ISW.

CONCLUSIONS

High ^{228}Ra activities occur on the southeastern Weddell shelf and in stratified bottom waters.

In surface waters, ^{228}Ra can trace the dispersion of shelf water. There is little exchange of shelf water with surface water in the central Weddell Sea. This contrasts with the situation in the Arctic, where the deep basins are covered with a layer of ^{228}Ra-rich water (up to over 100 dpm.m^{-3}) derived from the Siberian shelves.

Enrichment of ^{228}Ra in water circulating under floating shelves might be a tracer of residence time under these shelves, irrespective of melting or freezing processes, provided that the average ^{228}Ra flux from the sediment can be determined.

The distribution of ^{228}Ra in deep waters is controlled by isopycnal rather than diapycnal mixing. The distributions of ^{230}Th, ^{228}Th and ^{228}Ra in a transect across the Polar Front show that transport of dissolved matter within the CDW towards the Weddell Sea occurs on a time scale between 8 and 30 years, in agreement with an estimate based on heat flux considerations.

As the silicate flux from deep-sea sediments in the Weddell Sea is negligible, ^{228}Ra provides the most suitable tracer for bottom contact of water masses in this Basin.

Acknowledgments. I wish to thank Gijs Berger, Andreas Michel and Heike Höltzen for pleasant cooperation during the expeditions. Andreas Wysotzki, Michael Schröder and Kees Veth kindly provided the CTD data of expeditions ANT VIII/3, ANT IX/3 and ANT X/6, respectively. Silicate determinations were made by Gerhard Kattner, Marthi Stürcken-Rodewald, Peter Fritsche and Karel Bakker. I thank Arnold Gordon for his advise, Eberhard Fahrbach for discussions, and Peter Killworth and Berverly De Cuevas for providing the FRAM output data. This is AWI publication nr. 719 and SFB 261 contribution nr. 80.

REFERENCES

Bacon, M. P., and R. F. Anderson, Distribution of thorium isotopes between dissolved and particulate forms in the deep sea., *J. Geophys. Res.*, *87*, 2045-2056, 1982.

Bacon, M. P., and M. M. Rutgers van der Loeff, Removal of Thorium-234 by scavenging in the bottom nepheloid layer of the ocean., *Earth Planet. Sci. Lett.*, *92*, 157-164, 1989.

Bayer, R., and P. Schlosser, Tritium profiles in the Weddell Sea, *Mar. Chem.*, *35*, 123-136, 1993.

Biscaye, P. E., and S. L. Eittreim, Suspended particulate loads and transport in the nepholoid layer of the abyssal Atlantic Ocean, *Mar. Geol.*, *23*, 155-172, 1977.

Broecker, W. S., and T. -H Peng, Tracers in the Sea, Lamont-Doherty Geol. Obs., Columbia University, 1-690, 1982.

Chung, Y., and M. D. Applequist, ^{226}Ra and ^{210}Pb in the Weddell Sea., *Earth Planet.Sci.Lett.*, *49*, 401-410, 1980.

De Szoeke, R. A., and M. D. Levine, The advective flux of heat by mean geostrophic motions in the Southern Ocean, *Deep-Sea Res., 28A*, 1057-1085, 1981.

Deacon, G. E. R., The Weddell Gyre, *Deep-Sea Res., 26A*, 981-995, 1979.

DeMaster, D. J., T. M. Nelson, S. J. Harden, and C. A. Nittrouer, The cycling and accumulation of biogenic silica and organic carbon in Antarctic deep-sea and continental margin environments., *Mar. Chem., 35*, (1-4), 489-502, 1991.

Gammelsrød, T., L. F. Anderson, E. Fogelkvist, A. Foldvik, E. P. Jones, O. A. Nøst, K. Olsson, Ø. Skagseth, T. Tanhua, and S. Østerhus, Distribution of water masses over the Continental Shelf in the Southern Weddell Sea, *J. Geophys. Res.*, (Nansen Centennial Symposium volume), this issue, 1994.

Gordon, A. L., Weddell Sea Exploration from Ice Station, *EOS, 74*, (11), 121-126, 1993.

Gordon, A. L., and B. A. Huber, Southern Ocean winter mixed layer, *J. Geophys. Res., 95*, (C7), 11655-11672, 1990.

Gordon, A. L., and E. Molinelli, Southern Ocean Atlas: Thermohaline and chemical distributions, Columbia University Press, New York, 1982.

Gouretzki, V. V., and A. I. Danilov, Weddell Gyre: Structure of the Eastern Boundary, *Berichte aus dem Fachbereich Physik. AWI*, (21), 1-18, 1991.

Kaufman, A., R. M. Trier, W. S. Broecker, and H. W. Feely, Distribution of ^{228}Ra in the World Ocean, *J. Geophys. Res., 78*, (36), 8827-8848, 1973.

Key, R. M., R. F. Stallard, W. S. Moore, and J. L. Sarmiento, Distribution and flux of ^{226}Ra and ^{228}Ra in the Amazon River Estuary, *J. Geophys. Res., 90*, (C4), 6995-7004, 1985.

Ku, T. L., and M. C. Lin, ^{226}Ra distribution in the Antarctic Ocean, *Earth Planet. Sci. Lett., 32*, 236, 1976.

Kuhn, G., and M. E. Weber, Acoustical characterization of sediments by *Parasound* and 3.5 kHz systems: Related sedimentary processes on the southeastern Weddell Sea continental slope, Antarctica, *Mar. Geol., 113*, 201-217, 1993.

Li, Y. -H, H. W. Feely, and J. R. Toggweiler, ^{228}Ra and ^{228}Th concentrations in GEOSECS Atlantic surface waters, *Deep-Sea Res., 27A*, 545-555, 1980.

Moore, W. S., Oceanic concentrations of ^{228}Ra, *Earth Planet.Sci.Lett., 6*, 437-446, 1969.

Moore, W. S., Radium isotope measurements using germanium detectors, *Nucl. Instrum. Meth. Phys. Res., 223*, 407-411, 1984.

Moore, W. S., Radium 228 in the South Atlantic Bight, *J. Geophys. Res., 92*, (C5), 5177-5190, 1987.

Moore, W. S., The thorium isotope content of ocean waters, *Earth and Planetary Science Letters, 53*, 419-426, 1981.

Moore, W. S., R. M. Key, and J. L. Sarmiento, Techniques for precise mapping of ^{226}Ra and ^{228}Ra in the ocean, *J. Geophys. Res., 90*, (C4), 6983-6994, 1985.

Moore, W. S., and P. H. Santschi, Ra-228 in the deep Indian Ocean, *Deep-Sea Res., 33*, (1), 107-120, 1986.

Moore, W. S., J. L. Sarmiento, and R. M. Key, Tracing the Amazon component of surface Atlantic water using ^{228}Ra, salinity and silica, *J. Geophys. Res., 91*, (C2), 2574-2580, 1986.

Rutgers van der Loeff, M. M., and A. J. van Bennekom, Weddell Sea contributes little to silicate enrichment in Antarctic Bottom Water, *Deep-Sea Res., 36*, (9), 1341-1357, 1989.

Rutgers van der Loeff, M. M., and G. W. Berger, Scavenging of ^{230}Th and ^{231}Pa near the Antarctic Polar Front in the South Atlantic, *Deep-Sea Res.I, 40*, (2), 339-357, 1993.

Sarmiento, J. L., H. W. Feely, W. S. Moore, and A. E. Bainbridge, The relationship between vertical eddy diffusion and buoyancy gradient in the deep sea., *Earth Planet. Sci. Lett., 32*, 357-370, 1976.

Sarmiento, J. L., C. G. H. Rooth, and W. S. Broecker, Radium 228 as a tracer of basin wide processes in the abyssal ocean, *J. Geophys. Res., 87*, (C12), 9694-9698, 1982.

Sarmiento, J. L., G. Thiele, R. M. Key, and W. S. Moore, Oxygen and Nitrate New Production and Remineralization in the north Atlantic subtropical Gyre, *J. Geophys. Res., 95*, (C10), 18303-18315, 1990.

Thompson, S. R., Estimation of the transport of heat in the Southern Ocean using a fine-resolution numerical model, *J. Phys. Oceanogr., 23*, 2493-2497, 1993.

Webb, D. J., P. D. Kilworth, A. C. Coward, and S. R. Thompson, *The FRAM atlas of the Southern Ocean*, Natural Environmental Research Council, Swindon, 1991.

Wefer, G., and G. Fischer, Annual primary production and export flux in the Southern Ocean from sediment trap data, *Mar. Chem., 35*, (1-4), 597-614, 1991.

Whitworth, T., and W. D. Nowlin, Water masses and currents of the Southern Ocean at the Greenwich meridian, *J. Geophys. Res., 92*, (C6), 6462-6476, 1987.

M.M. Rutgers van der Loeff, Alfred-Wegener-Institut für Polar- und Meeresforschung, P.O. Box 120161, D27515 Bremerhaven, Germany.

Modelling the Extent of Sea Ice Ridging in the Weddell Sea

Markus Harder

Alfred-Wegener-Institut für Polar- und Meeresforschung, Bremerhaven, Germany

Peter Lemke

Alfred-Wegener-Institut für Polar- und Meeresforschung, Bremerhaven, Germany

and Institut für Umweltphysik, Universität Bremen, Germany

A dynamic-thermodynamic sea ice model is modified to distinguish level and ridged ice and is applied to the Weddell Sea. The model explicitly states different prognostic thicknesses for ridged and unridged ice, using four extended continuity equations which include terms to transform level ice into ridged ice due to deformation. Significantly different temporal evolution and spatial distribution for the two ice classes are predicted. Another important extension of the sea ice model, the age of both level and ridged ice, is introduced additionally through two prognostically determined variables, described by two extended continuity equations.

1. INTRODUCTION

The spatial and temporal evolution of sea ice is determined by the coupling of dynamic and thermodynamic processes. Freezing of sea water dominantly takes place in leads of open water, produced by divergent ice drift as well as due to formation of pressure ridges in shear zones. This thermodynamic freezing process is accompanied by accumulation of ice in regions with convergent ice drift, where thick ice (pressure ridges) is built mainly due to compression of ice floes formerly produced in other regions. Thus, the observed thickness distribution of the ice cover is a result of the coupled mechanism of thermodynamic ice growth especially in regions with lower ice thicknesses and compactnesses and of ice dynamics due to wind stress and ocean currents which strongly modify the areal distribution of sea ice.

The Polar Oceans and Their Role in Shaping the Global Environment
Geophysical Monograph 85

Simulations with a coupled dynamic-thermodynamic, one-ice class sea ice model applied to the Weddell Sea (described by *Fischer* and *Lemke*, this issue) yield a prognostic ice cover and ice drift in good agreement with observations obtained from satellite sensors and drift buoys. To explicitly identify regions and seasons where deformation of the ice cover and formation of thick, ridged ice occurs, and to estimate the extent of these ridging processes, this sea ice model has been extended to distinguish two classes of ice, here referred to as "level" and "ridged" ice. Originally, all new ice formed by freezing of open water is defined to be level ice. Under condition of convergent drift or strong shearing deformation, a portion of the level ice is assumed to be transformed into ridged ice. Thermodynamic growth and decay as well as advection are separately calculated for both ice classes. This scheme allows the distinct areas to be identified where thin new ice is frozen, where level ice is compressed to thick ridged ice, where the thick ice drifts to and reaches its maximum thickness, and where it finally melts.

Modelling the roughness of sea ice provides an additional prognostic variable that may be compared with observed sea ice roughness to validate the quality of the model predictions. Another aspect is that the prognostic roughness may be used to improve the model, e. g. by making the drag coefficients dependent on the surface roughness. However, it must be emphasised that the definition of ridged versus unridged ice used in the model is essentially different from the distinction of ridged and level ice as used in observations. This model regards all ice subject to deformation (ridging and rafting) as becoming ridged and staying ridged until this ridged ice totally melts. New, very thin ice compressed in leads is considered as ridged ice as well as pressure ridges several meters thick created by collisions of thick ice floes. Therefore, the observable, thick ridges are only a portion of all the ice regarded as "ridged" here. In this sense, the spatial and temporal distributions of roughness predicted by this model are of major concern, whereas numerical values of ridged ice fractions are related to but not identical with geometrical and dynamic properties of the sea ice cover.

Another important extension of the model is the introduction of the age of level and ridged ice as two additional prognostic variables. Here the age is defined as an average age of all ice of one class in a model grid cell, weighing different ice portions with different ages by their respective volumes. This average has to be calculated with regard to advection mixing ice produced in different regions as well as thermodynamic freezing adding new ice to the old ice that already exists.

2. DESCRIPTION OF THE MODEL

The two-dimensional sea ice model consists of two ice classes representing level and ridged ice, and a snow layer assumed to be equally distributed on the surfaces of level and ridged ice. The model is an extension of *Hibler's* model [1979] with the thermodynamic part following *Semtner* [1976] and *Parkinson* and *Washington* [1979]. It is coupled with a one-dimensional prognostic oceanic mixed layer developed by *Lemke et al.* (1990).

The ice momentum balance is

$$m \frac{D\mathbf{u}}{Dt} = -m f \mathbf{k} \times \mathbf{u} + \tau_a + \tau_o - mg\nabla H + \mathbf{F} \qquad (1)$$

where m is the total sea ice mass per unit area, **u** is the ice velocity, f is the Coriolis parameter, **k** is the vertical upright unit vector, and g is the gravitational acceleration. Equation (1) includes inertial terms, Coriolis force, wind (τ_a) and water (τ_o) stresses, the tilt of the sea surface H, and internal forces **F**. Wind and water stresses are determined from integral boundary layer theory [e. g. *McPhee*, 1979] assuming a constant turning angle of 25° for the ocean and no turning angle for the atmosphere since surface winds are used. The surface tilt is derived from the prescribed geostrophic ocean currents using geostrophy.

The internal forces due to ice interaction are given as the divergence of the two-dimensional stress tensor, $\mathbf{F} = \nabla \cdot \boldsymbol{\sigma}$. Following *Hibler* [1979], the ice is considered to obey the non-linear viscous-plastic constitutive law

$$\sigma_{ij} = 2\eta\left(\dot{\varepsilon}_{ij}, P\right)\dot{\varepsilon}_{ij} + \left[\left(\zeta\left(\dot{\varepsilon}_{ij}, P\right) - \eta\left(\dot{\varepsilon}_{ij}, P\right)\right)\dot{\varepsilon}_{kk} - P/2\right]\delta_{ij} \qquad (2)$$

$\dot{\varepsilon}_{ij}$ is the strain rate tensor, P is the ice strength, and ζ and η are the non-linear bulk and shear viscosities. Using an elliptical yield curve gives

$$\zeta = \frac{P}{2 \cdot \Delta(\dot{\varepsilon})} \qquad \eta = \frac{\zeta}{e^2} \qquad (3)$$

where

$$\Delta = \sqrt{\left(\dot{\varepsilon}_{11}^2 + \dot{\varepsilon}_{22}^2\right)\left(1 + e^{-2}\right) + 4\dot{\varepsilon}_{12}^2 e^{-2} + 2\dot{\varepsilon}_{11}^2\dot{\varepsilon}_{22}^2\left(1 - e^{-2}\right)} \qquad (4)$$

and e is the eccentricity of the elliptical yield curve. In case of very small strain rates the viscosities are limited to be less then

$$\zeta_{\max} = P \cdot \left(2.5 \times 10^8 s\right) \qquad \eta_{\max} = \frac{\zeta_{\max}}{e^2} \qquad (5)$$

The ice strength is coupled to compactness and thickness of both level and ridged ice by

$$P = P^* \cdot h \cdot C(A) \qquad (6)$$

where

$$C(A) = \exp\left[-C^*(1 - A)\right] \qquad (7)$$

is a dimensionless factor making the ice strength strongly dependent on the compactness A, $C^* = 20$ and $P^* = 20{,}000$ N/m^2 are fixed empirical constants, and

$$h = h_l + h_r \qquad (8a)$$

$$A = A_l + A_r \qquad (8b)$$

h_l and h_r are the volumes per unit area of level and ridged ice, respectively, given as an average (in meters) over a whole grid cell. The compactnesses A_l and A_r represent the fractions of a grid cell area covered by level and ridged ice respectively. The actual thicknesses are (h_l / A_l) for level ice and (h_r / A_r) for ridged ice. Previous investigations of sea ice ridging presented by *Flato* and *Hibler* [1991] distinguished level and ridged ice with respect to their volume per unit area using equation (8a) but did not use separate compactnesses for level and ridged ice as expressed in equation (8b). The model presented here can be understood as a further extension of *Hibler*'s [1979] and *Flato* and *Hibler*'s [1991] approaches.

Four continuity equations linking dynamic and thermodynamic effects describe the local changes in ice compactnesses and thicknesses

$$\frac{\partial A_l}{\partial t} = -\nabla(\mathbf{u}A_l) + G_{A_l} + G_{A_{ow}} - Q_A - R_A T_A \tag{9a}$$

$$\frac{\partial A_r}{\partial t} = -\nabla(\mathbf{u}A_r) + G_{A_r} + R_A T_A \tag{9b}$$

$$\frac{\partial h_l}{\partial t} = -\nabla(\mathbf{u}h_l) + G_{h_l} + G_{h_{ow}} - T_A \frac{h_l}{A_l} \tag{9c}$$

$$\frac{\partial h_r}{\partial t} = -\nabla(\mathbf{u}h_r) + G_{h_r} + T_A \frac{h_l}{A_l} \tag{9d}$$

The velocity of the ice drift, **u**, as given by equation (1), is used for the advection terms of both level and ridged ice since all ice in a grid cell is assumed to move with the same velocity. G_{Al}, G_{hl}, G_{Ar}, G_{hr}, G_{Aow} and G_{how} are the thermodynamic growth rates for the compactnesses and thicknesses of level ice, ridged ice, and freezing of open water respectively. The growth rates G_{Al}, G_{hl}, G_{Ar}, and G_{hr} are determined from conductive heat fluxes which arise from the conservation of energy at the upper ice surface following *Semtner* [1976] and *Parkinson* and *Washington* [1979]. The growth rates are separately calculated for level and ridged ice taking into account that their thicknesses are significantly different. The growth rates for both the ice classes are estimated from equal, seven-level distributions between zero and twice the average thickness of the ice class concerned, so altogether 14 different ice thicknesses are employed to compute the thermodynamic growth rates. Formation of new thin ice due to freezing of open water (G_{Aow} and G_{how}) is considered to contribute solely to the growth of level ice. Q_A is the rate of production of open water due to ridging as a result of shearing deformation. T_A is the rate of transformation of level ice into ridged ice, given as the level ice area per time transformed to ridged ice. (Areas here are expressed in terms of compactness, i. e. they are regarded as areal fractions of the grid cell represented by dimensionless numbers between 0 and 1.) The increase in area covered with ridged ice is assumed to be proportional to the area involved in ridging processes and thus stated as $R_A T_A$. The ridging parameter R_A is a dimensionless positive number smaller than 1 defined as the ratio of the area of newly produced ridged ice to the former area of level ice these ridges have been built of. This empirical constant reflects the fact that the thickness of newly produced ridged ice is larger than the thickness (h_l / A_l) of the level ice. Multiplying the area of level ice transformed to ridged ice per time by the thickness of this level ice yields $T_A (h_l / A_l)$ as the decrease in the level ice volume per unit area, h_l. Assuming the total ice volume being conserved during the ridging process, this is identical with the increase in the ridged ice volume per unit area, h_r. Following *Flato* and *Hibler* [1991],

$$Q_A = \psi_s(\dot{\varepsilon}_{ij}) \cdot C(A) \tag{10}$$

and

$$T_A = \left[\psi_s(\dot{\varepsilon}_{ij}) + \psi_c(\dot{\varepsilon}_{ij})\right] \cdot C(A) \tag{11}$$

where the ice strength is described by *C(A)* given by equation (7). The contribution to ridging due to shearing deformation is

$$\psi_s(\dot{\varepsilon}_{ij}) = 0.5 \cdot \left(\Delta(\dot{\varepsilon}) - \left|\dot{\varepsilon}_{11} + \dot{\varepsilon}_{22}\right|\right) \tag{12}$$

with $\Delta(\dot{\varepsilon})$ defined by equation (4), and ridging due to convergent ice drift is given by

$$\psi_c(\dot{\varepsilon}_{ij}) = -(\dot{\varepsilon}_{11} + \dot{\varepsilon}_{22}) = -\operatorname{div}\mathbf{u} \quad \text{if} \quad \operatorname{div}\mathbf{u} < 0 \tag{13a}$$

$$\psi_c(\dot{\varepsilon}_{ij}) = 0 \quad \text{if} \quad \operatorname{div}\mathbf{u} \geq 0 \tag{13b}$$

The evolution of the age of level ice, a_l, and of ridged ice, a_r, is given by the extended continuity equations

$$\frac{\partial(a_l h_l)}{\partial t} = -\nabla(\mathbf{u}a_l h_l) + h_l + a_l M_{h_l} - a_l T_A \frac{h_l}{A_l} \tag{14a}$$

$$\frac{\partial(a_r h_r)}{\partial t} = -\nabla(\mathbf{u} a_r h_r) + h_r + a_r M_{h_r} + a_l T_A \frac{h_l}{A_l} \quad (14b)$$

where the melting rates M_{hl} and M_{hr} are identical with the thermodynamic growth rates G_{hl} and G_{hr} in case of melting conditions (G_{hl}, $G_{hr} < 0$), and equal to 0 for freezing conditions. The terms on the right hand side of equations (14a) and (14b) represent advection, aging, thermodynamic growth, and transformation of level to ridged ice. The aging process is described by h_l and h_r which provide that

$$\frac{\partial a_l}{\partial t} = 1 \quad \text{and} \quad \frac{\partial a_r}{\partial t} = 1 \quad (15)$$

The thermodynamic growth has to distinguish between melting and freezing, explained here taking the example of level ice. If melting occurs, the age of the ice, $a_l = (a_l h_l) / h_l$, is assumed not to be affected. This is provided for by the melting term M_{hl} making $(a_l h_l)$ vary as h_l. In contrast, freezing decreases the average age of the ice, $a_l = (a_l h_l) / h_l$, since new ice with an age of zero is added to the old ice. This effect is included in the model by letting h_l grow with the growth rate G_{hl} whereas $(a_l h_l)$ is not affected by freezing.

3. MODEL DOMAIN AND FORCING DATA

The model is applied to a spherical grid with a latitudinal and longitudinal resolution of 2.5° covering the Weddell Sea region, here taken to extend from 60° W to 60° E and 80° S to 47.5° S. The boundary conditions are "no slip" at continental boundaries and open outflow at water boundaries. The time step is one day, and the integration time is 7 years consisting of 6 spin-up years employing forcing data of 1986 to reach a cyclostationary state and a final seventh year with 1987 forcing to obtain the simulation results displayed here. Wind field, air temperature and humidity are taken from global analyses of the European Center for Medium Range Weather Forecasts (ECMWF) averaged to daily mean values. The 1000 hPa wind field is assumed to be the surface (2 m) wind [*Stössel*, 1992], whereas air temperature and humidity have been interpolated from the 850 hPa and 1000 hPa pressure levels to the surface following the *ECMWF Research Department* [1986]. Geostrophic ocean currents forming the well known Weddell Gyre have been defined as the annual mean of an integration with the Princeton General Circulation Model [*Bryan*, 1969; *Cox*, 1984] carried out by *Olbers* and *Wübbers* [1991]. Cloudiness is described as a function of latitude by climatological annual means taken from *van Loon* [1972]. A precipitation rate of 35 cm/year constant in space and time is prescribed.

The numerical solution of the simultaneous momentum, energy and continuity equations as an initial-value problem uses finite-difference techniqes on a spatial grid of 750 cells. This numerical scheme is similar to that of *Hibler* [1979, Appendix A], who gives a detailed description of this topic. The program requires about 2 Megabyte of computer memory and 700 seconds of CPU time on a CRAY-YMP for the 7-year standard simulation.

4. RESULTS

4.1. Standard simulation

The ridging parameter is set to $R_A = 0.5$ to yield an average ratio of ridged ice thickness to level ice thickness of about 2.5. This is in agreement with field observations obtained during the Winter Weddell Gyre Study (WWGS) 1989 (*Eicken et al.*, 1994, in press) which have shown that the typical thickness of pressure ridges is about 2 to 3 times the thickness of level ice. Sensitivity studies on the effect of variations of R_A have been carried out and are shown below (chapter 4.2.). The ice strength parameter $P^* = 20\,000$ N/m^2 and the drag coefficients $c_a = 1.5 \times 10^{-3}$ for the atmosphere and $c_o = 3.0 \times 10^{-3}$ for the ocean have been adjusted to give best agreement between simulated velocities and data of 7 ARGOS drift buoys in 1986 and 1987. A more detailed description of this comparison, including an illustration showing observed and simulated buoy trajectories, is given by *Fischer* and *Lemke* (this volume). Simulations with various parameter settings gave the result that the simulated ice drift is dominated by the ratio of the drag coefficients c_a / c_o rather than by their absolute values. The constants $C^* = 20$ and $e = 2$ are the same as used by *Hibler* [1979]. The initial conditions are defined as the entire region being covered with level ice with a thickness $h_l = 0.5$m, a compactness $A_l = 100\%$, a velocity $\mathbf{u} = (0,0)$, and an age $a_l = 0$. Tests with different initial conditions have shown that after reaching a cyclostationary state the model results do not significantly depend on the choice of the initial conditions.

The spatial distribution of ice thickness h (i. e. total ice volume per unit area) is shown in Figure 1 for September 1987, when the Antarctic sea ice cover reaches its annual maximum extent. The total ice thicknesses $h = h_l + h_r$ and

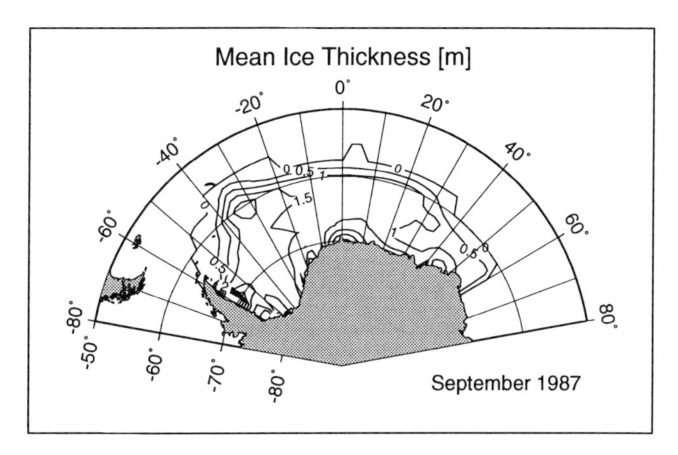

Fig. 1. Total ice thickness (m) for September 1987.

velocities **u** predicted by this model employing level and ridged ice are not significantly different from those obtained by a similar model with one ice class only (*Fischer* and *Lemke*, this volume). Due to the introduction of two ice classes with different thicknesses, this model provides more areas with thinner, level ice in most regions, causing a moderate increase in ice growth rates which yield thicknesses generally 10 to 20 cm higher than those of the one-class model. The spatial distribution of ridged ice in September 1987, expressed as ridged ice volume per unit area (m), is shown in Figure 2. High thicknesses of ridged ice of about 3 m are predicted east of the Antarctic Peninsula, accompanied by considerable ridged ice thicknesses up to 2 m northeast of the tip of the peninsula, both being results of the ice drift following the Weddell Gyre circulation pattern. The spatial distribution of the volumetric ridged ice fraction h_r / h (given in %) is shown in Figure 3. Lower values are found in the Eastern

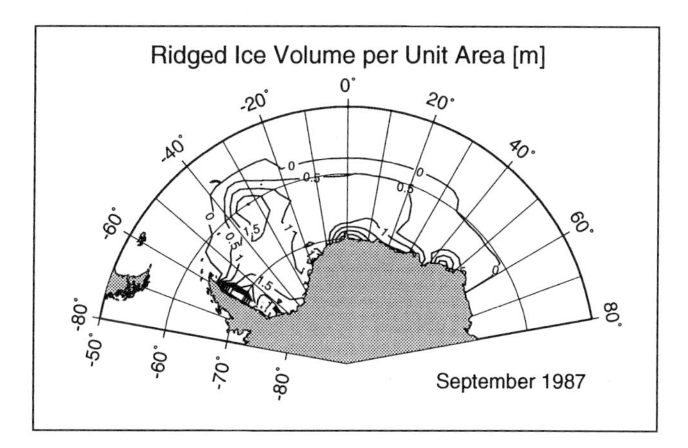

Fig. 2. Ridged ice volume per unit area (m) for September 1987. Maximum values of 3.5 m occur east of the Antarctic Peninsula.

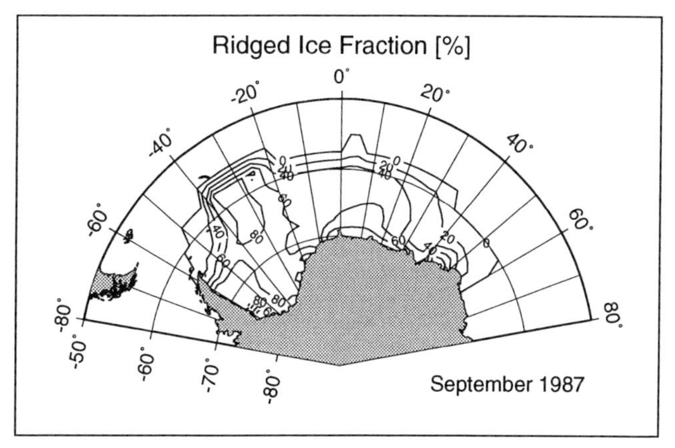

Fig. 3. Volumetric fraction of ridged ice (%) for September 1987. Ratios higher than 95% are found in coastal regions.

Weddell Sea, where larger portions of the area are covered with young, fast-growing level ice, whereas in the Western Weddell Sea the fraction of older, heavily ridged ice reaches up to about 95%. As discussed in the introduction, these ridged ice fractions include thin ice deformed during its first days as well as thick pressure ridges. Therefore, the fraction of observable pressure ridges is much smaller than the ridged ice fraction given by this simulation. However, regions with high prognostic ridged ice fractions correspond with regions of large amounts of observable ridges (see chapter 4.3).

Another important result is that sea ice deformation plays a major role in the ice build-up as stated by *Hibler* [1979]. Comparing Figure 1 with Figures 2 and 3, it is found that regions with thick ice show high fractions and thicknesses of ridged ice. This is an essential feature of the model since most of the ice in the Weddell Sea region is first or second year ice, rarely older than several months, due to the realistic, prognostic velocity field transporting the Weddell Sea ice out to the Antarctic Circumpolar Current during a period of less than one year. Since the ice cannot thermodynamically grow to a thickness of more than about 1 m during this short lifetime, almost all the thick ice found in the Western Weddell Sea must have participated in a deformation process at least once in its lifetime even though this is not obviously found as a pronounced surface roughness. This is what this model predicts when ridged ice fractions of more than 80% are modelled for the north-western Weddell Sea.

The regions where ridged ice is created can be identified in Figure 4 showing the annual production of ridged ice volume per unit area (m). Highest values up to 3 m ridged ice per year are found in coastal regions. However, the

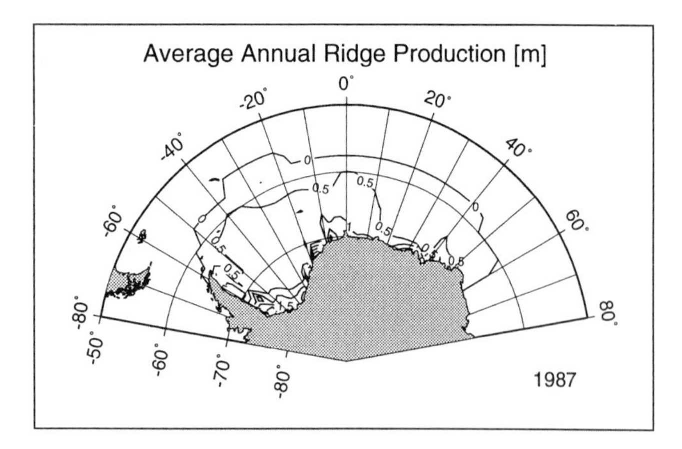

Fig. 4. Annual production of ridged ice volume per unit area (m) for 1987. Maximum values up to 3 m per year occur near the coast, whereas the ridge production in large regions of the Central Weddell Sea is between 0.5 and 1 m per year.

entire ice cover in the Central Weddell Sea also considerably contributes to the ridge formation with values between 0.5 and 1 m per year, which is a result of the daily wind forcing yielding strong convergences and divergences with high variability on short time scales of the order of days. Large amounts of ridged ice are produced in the southern Weddell Sea along the Filchner-Ronne ice shelf (Figure 4), whereas the thick ice is accumulated in the north-western Weddell Sea along the Antarctic Peninsula. This is due to the ice drift following the Weddell Gyre and emphasises the important role of advection, since it shows that the regions where the ice characteristics are formed are not the same as those where these ice masses are finally found.

Figure 5 shows the annual cycle for 1987 of the ice volumes in 10^3 km^3 of level ice (thin line), ridged ice (dashed line) and the total ice volume (thick line) integrated over the whole model domain. The interesting feature is the annual variation in these ice volumes and in their relative proportion. The model predicts that only ridged ice survives the austral summer (with the minimum ice extent around March 1st, i. e. day 60), giving a ridged ice fraction h_r / h close to 100% in most regions. The thinner, level ice melts almost totally in summer and freezes nearly constantly from March to end of August (about day 240). The increase of ridged ice volume, also starting in March, is caused by thermodynamic freezing as well as ridging. The latter effect causes the peak of the ridged ice volume to occur at the end of October (day 300) about two months after the maximum level ice volume, since when thermodynamic freezing stops at the end of the winter and turns to net melting, ridging processes still continue to transform level ice to ridged ice. This is illustrated by Figure 6 showing the increase in total ice volume in 10^3 km^3 per day caused by thermodynamic freezing (or melting for negative values), and the ice volume per day being transformed from level to ridged ice. During the first two months of net freezing (March

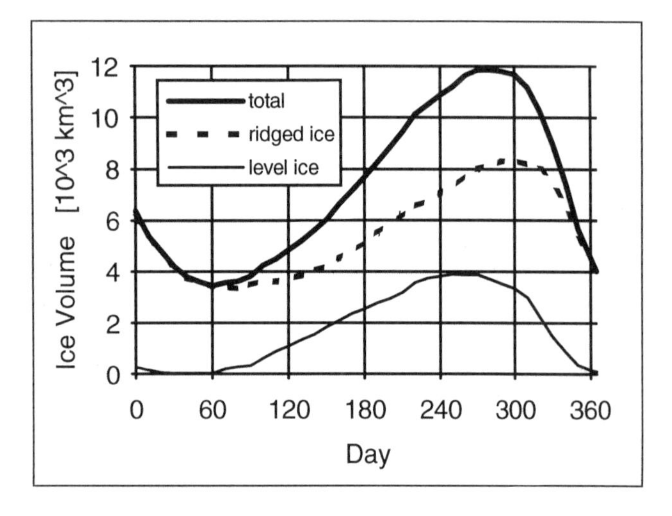

Fig. 5. Time series of ice volumes (10^3 km^3) spatially integrated over the whole model domain for level ice (thin solid line), ridged ice (dashed line), and total (thick solid line), for 1987. The ridged ice volume reaches its maximum several weeks after the peak in level ice volume.

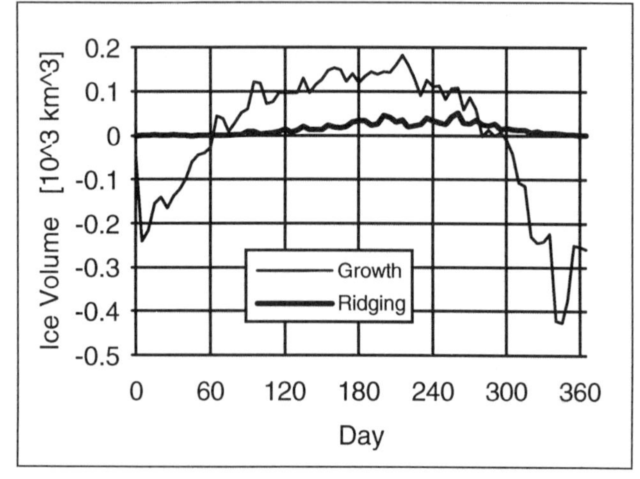

Fig. 6. Time series of ice volumes (10^3 km^3) per day produced by thermodynamic growth (thin line) and transformed from level to ridged ice (thick line), spatially integrated over the whole model domain for 1987. Negative ice growth volumes per day indicate melting.

and April, days 60 to 120) ridging effects are small since the ice cover has not yet closed and thus large amounts of open leads allow almost free drift. As the ice cover closes in May and later, ridging increases considerably, with the daily created ridge volume increasing as the level ice thickness increases. In October and November (days 270 to 330) ridging decreases but still continues while the thermodynamic growth turns to net melting.

The annual cycle of the average age of ridged ice (dashed line), level ice (thin solid line), and the volume-weighted average of both yielding the average age of the total ice mass (thick solid line) as a mean value of the whole model domain is shown in Figure 7. The minimum age of level ice, occuring around March 1 (day 60), indicates the beginning of extensive freezing of new, thin ice. Comparing figures 5 and 7, the minimum in the average age of level ice in March and the constant increase in subsequent months coincide with the increase in the level ice volume from March to September. During the freezing period, when large amounts of new ice are added to the level ice volume, the average age of level ice increases more slowly than after the end of the freezing period in September, because at that time no more growth of young ice counteracts the aging of the old ice that already exists. The result shown in Figure 5, that almost no level ice survives the summer, is expressed in Figure 7 by the fact that the maximum age of level ice, about 140 days, is significantly less than one year. In contrast, the ridged ice that in a considerable amount survives the summer melting, shows a significantly higher age of about 280 to 380 days, i. e. in the order of one year. This

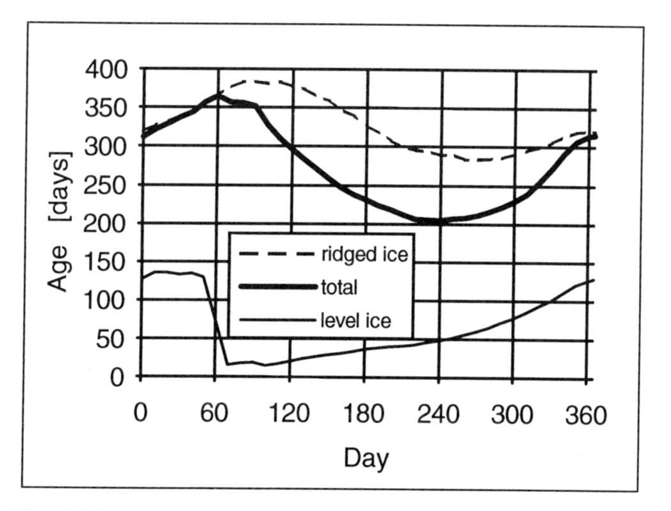

Fig. 7. Time series of the age (days) of ridged ice (dashed line), of level ice (thin solid line), and of the total ice mass (thick solid line), given as the volume-weighted average age of the whole model domain for 1987.

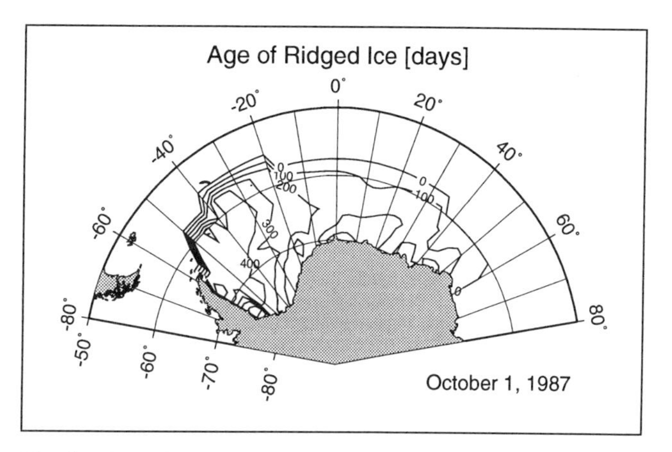

Fig. 8a. Age of ridged ice (days) on October 1, 1987. The oldest ridged ice with an age of 500 days is found in the north-western Weddell Sea along the Antarctic Peninsula.

indicates that the total ridged ice volume consists of first-year ice as well as multi-year ice. Whereas the minimum of the age of level ice is found at the beginning of the freezing period in March, the minimum of the age of ridged ice occurs at the end of the freezing period in September. This is a consequence of ridging processes continually transforming young level ice to ridged ice even if the thermodynamic freezing stops in September. Due to the significantly higher thickness of ridged ice as compared with level ice, thermodynamic growth of ridged ice is much slower and has only a minor effect on its average age. At the summer minimum ice extent, when almost all ice is ridged, the average age of ice is nearly identical with the age of ridged ice. In winter, the age of the total ice volume is significantly lower than the age of the ridged ice, the difference being determined by the volume of young, level ice formed in the freezing period.

Figures 8a and 8b show the geographical pattern of the age of ridged and level ice, respectively. A striking difference between the two ice classes is stated: In the north-western Weddell Sea, where the oldest ridged ice is predicted with an age of up to 500 days, the level ice is relatively young with an age of about 25 days. In contrast, in the Antarctic Circumpolar Current (ACC) in the north-eastern Weddell Sea the oldest level ice is found with an age up to 100 days, whereas the age of the ridged ice is of the order of only 200 days in that region. As a general result, a high age of level ice corresponds with a relatively low age and a low fraction of ridged ice (see Figure 3), and vice versa. This indicates that level ice has the possibility to grow old only in regions where ridging effects are small, whereas ice masses subject to strong deformation undergo a rapid transformation of level ice into ridged ice. The spatial pattern of the age of ridged ice

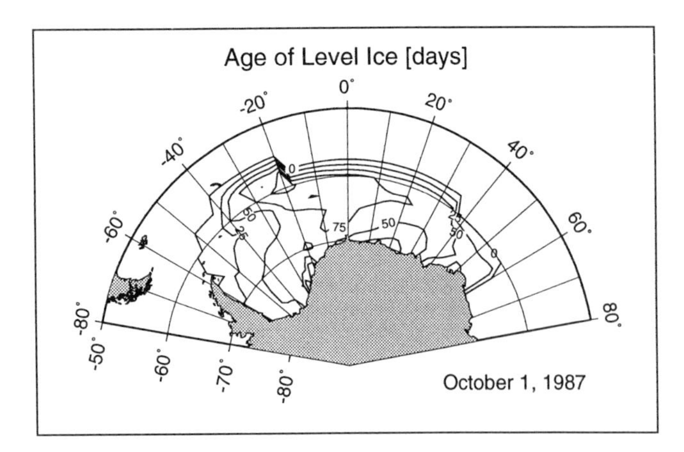

Fig. 8b. Age of level ice (days) on October 1, 1987. The oldest level ice with an age of 100 days is found in the Antarctic Circumpolar Current in the north-eastern Weddell Sea.

reflects the high ridge production in the southern Weddell Sea, yielding a low age of ridged ice in the south and an increasing age northwards along the Antarctic peninsula. This shows that the ridges found in the thick ice in the north-western Weddell Sea are mainly not locally produced, but are advected to that region following the Weddell Gyre circulation. The spatial distribution of the average age of the total ice mass, calculated as the volume-weighted average of the age of both ridged and level ice, is dominated by the ridged ice and yields a similar pattern as displayed in Figure 8a.

4.2. *Investigation of the ridging parameter R_A*

To investigate the influence of the ridging parameter R_A on the prognostic model variables, different simulations with R_A varying between 0.05 and 1.0 have been carried out. Figure 9 gives the annual mean values of the areal and volumetric fractions (A_r / A) and h_r / h of ridged ice in % as a function of R_A , taken as averages over the whole model domain. The volumetric fraction of ridged ice shows only a minor dependence on R_A, varying from about 65% to 85% with increasing R_A . Opposed to this, the areal fraction of ridged ice strongly depends on R_A . It is approximately proportional to R_A with a lower increase for $R_A > 0.5$. As the ridged ice volume is not strongly affected by variations in R_A, the increase in the areal fraction covered by ridged ice with increasing R_A corresponds to a simultaneous decrease of the thickness of ridged ice. Smaller values of R_A directly induce higher thickness of ridged ice, approximately following the linear relation

$$(h_r / A_r) \sim 1 / R_A \tag{16}$$

Therefore, the (average) thickness of ridged ice is the appropriate observable variable that should be used to tune the ridging parameter R_A to give best agreement between model and observations. Other prognostic model variables, such as ice concentration, average ice thickness and ice drift velocity are affected much less and in a more indirect manner by variations of R_A. This leads to the satisfying conclusion that the scheme of treating level and ridged ice separately, coupled with the uncertain empirical ridging parameter, R_A , has not introduced an additional, artificial parameter allowing arbitrary tuning of the prognostic model variables. Although this two-ice class model generates local effects moderately depending on the ridging parameter, the general features of the one-ice class model are preserved. Additionally, the basic features of this two-ice class model, i. e. the areal distribution pattern of roughness and the annual cycle of level and ridged ice volumes, do not considerably depend on the choice of R_A . To get a general view of where ridged ice is produced, where it drifts to, and how it evolves during the cycle of seasons, any choice of R_A within a reasonable range (about 0.3 to 0.7 for the Weddell Sea) will basically yield the same results in a qualitative sense.

4.3 *Comparison of simulation results with observations*

Estimates of the ice thickness are available from more than 5000 drill-hole measurements carried out during the

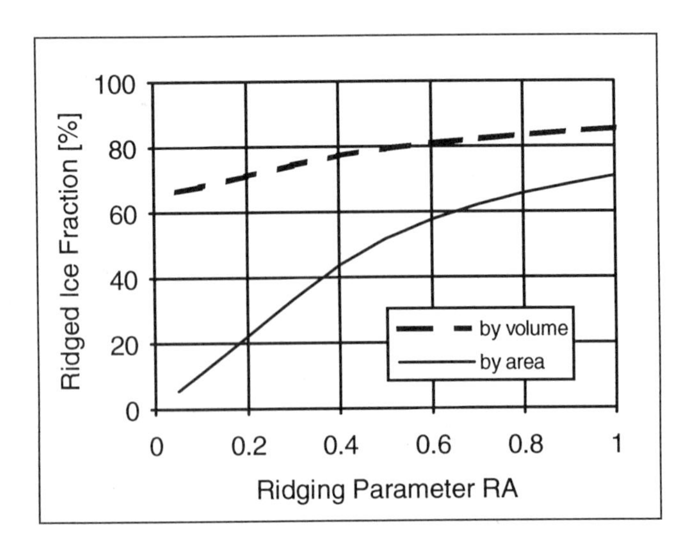

Fig. 9. Effect of the ridging parameter R_A on the areal (solid line) and volumetric (dashed line) fractions of ridged ice (%). Shown are the annual mean values, spatially averaged over the whole model domain for 1987.

Winter Weddell Gyre Study (WWGS) in September and October 1989 (*Eicken et al.* [1994, in press]). The ice cores were taken along the cruise track of *RV Polarstern* from Kapp Norwegia at the north-eastern coast of the Weddell Sea to the northern tip of the Antarctic Peninsula. A distinction between level and ridged (or rafted) ice and between first-year (FY) and multi-year (MY) ice was performed. The mean thicknesses were found to be 0.6 m (FY, level ice), 1.0 m (FY, ridged ice), 1.2 m (MY, level ice) and 2.5 m (MY, ridged ice). These observations agree with the simulation results of typical ice thicknesses of 0.5 to 1 m for level ice and 2 to 3 times higher thicknesses of ridged ice. About 50% of the sampled ice cores showed evidence that they had participated in deformation. This confirms that deformation is an important process in the build-up of the ice cover as predicted by the model.

Figure 10 gives the mean ice thickness at 29 locations along the cruise track of WWGS 1989 as a function of the geographical longitude. Each data point represents the average of more than 100 measurements, including ice of all classes (level and ridged, FY and MY). A striking difference between the western and the eastern Weddell Sea is found. The ice thicknesses in the eastern region rarely exceed 1 m, whereas significantly higher ice thicknesses of 2 m and more are found in the western region. According to *Eicken et al.* [1994], "ice core analysis and shipboard ice observations indicate that this is the transition between the second-year and first-year ice regimes. Thus, the largest fraction of the floes west of 35° W had survived the previous summer, whereas all of the ice sampled east of this line was in its first year of growth". These observed ice characteristics agree with the model predictions that thick, ridged ice older than a year is found in the western Weddell Sea, whereas the ice in the eastern region is predicted to be younger, thinner and with a lower fraction of ridged ice.

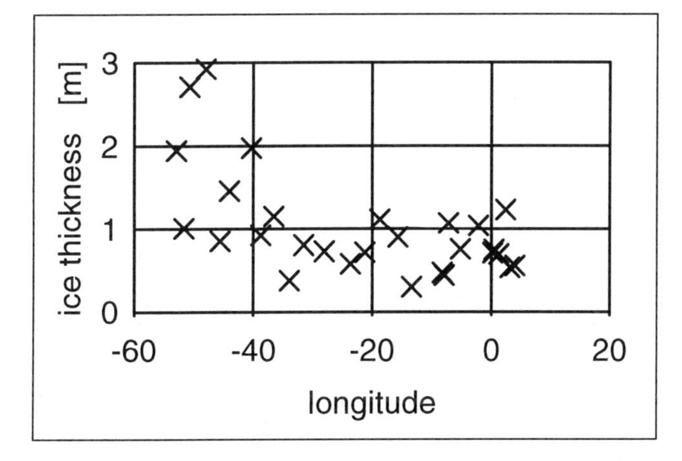

Fig. 10. Observed average ice thickness (m) as a function of geographical longitude, based on data obtained by *Eicken et al.* during WWGS 1989. Each data point represents the mean thickness of more than 100 drill-hole measurements.

Observations of the surface roughness were obtained by *Dierking* (1993, personal communication; "Laser profiling of the ice surface topography during WWGS 1992" submitted to JGR) using 800 km of helicopter-mounted laser profiling of the ice surface during WWGS 1992. The cruise track was similar to WWGS 1989. The average thickness of ridged ice was determined to be about 3.1 to 3.7 m. This is in good accord with the model predictions as well as with the measurements of *Eicken et al.* [1994] during WWGS 1989. Since small ridges with a height of less then 0.8 m have been excluded from the evaluation of the altimeter profiling, the average height of ridged ice given by *Dierking* is slightly higher than that predicted by the model and observed for the ice core samples.

Investigating the WWGS 1992 laser altimeter data, *Dierking* estimated the ridging intensity according to

$$R_1 = \frac{\mu_h}{\mu_s} \tag{17}$$

where μ_h is the mean height of pressure ridge sails, and μ_s is the mean spacing (horizontal distance) between two ridges. R_1 is proportional to the aerodynamic form drag of ridges. High values of R_1 indicate ice floes with non-uniform, rough surfaces and high numbers of ridges. However, one should keep in mind that not all types of deformed ice (i. e. rafted floes) can be identified by observable surface roughness, thus ridging indices like R_1 show only the fraction of deformed ice consisting of large, obviously ridged ice structures.

Figure 11 displays the spatial pattern of the ridging intensity R_1 (multiplied by 100 for convenient numbers) measured during WWGS 1992. Three regions can be distinguished, each showing good agreement with the model predictions. Low ridging intensities (0.4 to 0.5) were observed in the central Weddell Sea, in accordance with the simulated low but still noticeable ridged ice production in that region. Increased ridging intensities (values of 1.2 and 3.0) were found westward toward the tip of the Antarctic Peninsula, where the model predicts an outflow of relatively old, thick, ridged ice. Highest ridging intensities (2.2 and 4.0) were observed close to the coast near Kapp Norwegia at the eastern boundary of the Weddell Sea. This confirms the simulation result that the

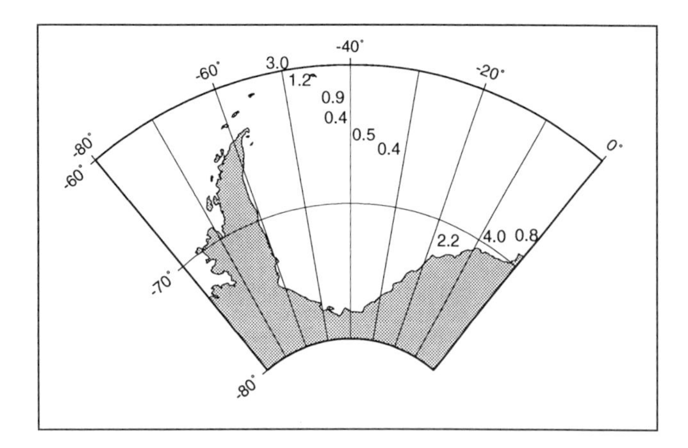

Fig. 11. Spatial distribution of the observed ridging intensity R_1. Numbers give R_1 multiplied by 100 at the approximated locations of measurements. Data were obtained from a helicopter-mounted laser altimeter by *Dierking* during WWGS 1992.

coastal regions show significantly higher ridge production rates than the central Weddell Sea.

5. DISCUSSION AND CONCLUSIONS

Experiments with a dynamic-thermodynamic sea ice model which distinguishes level and ridged ice, applied to the Weddell Sea region, show that these different ice classes evolve in a significantly different way, concerning the areal distribution as well as the annual cycle. Regarding the prognostic properties of sea ice predicted by the model, a significant relationship between ice thickness, ridged ice fraction, and the age of the ice is found. The southeastern Weddell Sea, as an area with a high ice production, is characterised by young ice with a low ridged ice fraction, whereas in the north-western Weddell Sea and the adjacent Antarctic Circumpolar Current (ACC) thick, old ice with a high fraction of ridges is predicted. The coastal regions in the southern Weddell Sea are identified as the major sources of ridged ice, although ridging is found to play an important role for the ice build-up in the whole ice covered region.

Level ice and ridged ice are predicted to evolve in qualitatively different ways during the cycle of seasons. The variation of the level ice volume, integrated over the whole region, is determined by the annual cycle of thermodynamic freezing and melting, supplemented by ridging processes transforming level to ridged ice acting as a sink of level ice. Opposed to this, the build-up of ridged ice is mainly determined by ridging, whereas the contribution of thermodynamic freezing of thick, ridged ice is not as dominant for the ice build-up as it is for level ice. As a consequence, the evolution of the ridged ice volume shows a delay of several weeks, compared with the annual cycle of level ice.

The age of the ice, here described separately for level and ridged ice by two extended continuity equations, has turned out to be a useful, additional prognostic variable for sea ice models, reflecting both dynamic and thermodynamic effects which determine the evolution of the ice cover. Since the annual cycles of the average age are qualitatively different for level and ridged ice, the distinction between level and ridged ice has shown evidence to be helpful for the investigation of the different processes participating in the build-up of the ice cover.

The most important additional, prognostic variables of the enhanced sea ice model described here, the surface roughness, and the age and the thickness of both level and ridged ice, are reasonably predicted and accord with observations of ice core samples and laser altimeter profiles as shown in chapter 4.3. The application of the sea ice model displayed here, probably with further improvements, to other polar regions seems to be a promising future work, which may yield improved methods of sea ice modelling as well as a better understanding of the physics involved in the build-up of sea ice due to the coupling of dynamic and thermodynamic processes.

Acknowledgements. The authors would like to thank H. Fischer for valuable discussion of the model features, H. Eicken and W. Dierking for advice and disposal of data concerning the observed properties of sea ice in the Weddell Sea, and R. Schnur for preprocessing the ECMWF forcing data delivered by Deutscher Wetterdienst (DWD), Offenbach. This is AWI publication number 733.

REFERENCES

Bryan, K., A numerical method for the study of the circulation of the world ocean, *J. of Comp. Phys.*, 4, 347-376, 1969

Cox, M. D., A primitive equation, 3 dimensional model of the ocean, *GFDL Ocean Group Technical Report No. 1*, 1978

ECMWF Research Department 1986, Research manual 1, ECMWF data assimilation, scientific documentation. *Meteorological Bulletin M1.5/1(1)* (eds. P. Lönnberg and D. Shaw). Reading/UK: ECMWF

Eicken, H., M. A. Lange, H.-W. Hubberten, and Peter Wadhams, Characteristics and distribution patterns of snow and meteoric ice in the Weddell Sea and their contribution to the mass balance of sea ice, *Ann. Geophys.*, 1994, in press

Fischer, H., and P. Lemke, On the required accuracy of

atmospheric forcing fields for driving dynamic-thermodynamic sea ice models , this volume

Flato, G. M., and W. D. Hibler, III, An initial numerical investigation of the extent of sea ice ridging, *Ann. Glaciol.*, 15, 31-36, 1991

Hibler, W. D., III, A dynamic thermodynamic sea ice model, *J. Phys. Oceonogr.,* 9(4), 815-846, 1979

Lemke, P., W. B. Owens, and W. B. Hibler, III, A coupled sea ice-mixed layer-pycnocline model for the Weddell Sea, *J. Geophys. Res.*, 95, 9527-9538, 1990

McPhee, M. G., The effect of the oceanic boundary layer on the mean drift of pack ice: Application of a simple model, *J. Phys. Oceonogr.*, 9, 388-400, 1979

Olbers, D. and C. Wübber, The role of wind and buoyancy forcing of the Antarctic Circumpolar Current, *Berichte aus dem Fachbereich Physik, Report No. 22,* AWI, Bremerhaven, FRG, 1991

Parkinson, C. L., and W. M. Washington, A large-scale numerical model of sea ice, *J. Phys. Oceonogr.*, 84, 311-337, 1979

Semtner, A. J., Jr., A model for the thermodynamic growth of sea ice in numerical investigations of climate, *J. Phys. Oceonogr.,* 6, 379-389, 1976

Stössel, A., Sensitivity of Southern Ocean sea ice simulations to different atmospheric forcing algorithms, *Tellus*, 44A, 395-413, 1992

van Loon, H., Cloudiness and precipitation in the southern hemisphere, *Meteorol. Monogr.*, 13, 101-111, 1972

M. Harder and P. Lemke, Department of Physics 1, Alfred-Wegener-Institut für Polar- und Meeresforschung, Am Handelshafen 12, D-27515 Bremerhaven, Federal Republic of Germany

Thermobaric Convection

Roland W. Garwood, Jr., Shirley M. Isakari

Naval Postgraduate School, Monterey, California, USA

and

Patrick C. Gallacher

NRL, Stennis Space Center, Mississippi, USA

Large-eddy simulation of two cases of free convection in the polar seas reveal the three-dimensional structure of thermobaric-enhanced turbulence, with and without salinity stratification. In the first case, with no initial stratification, 3.6 km-deep free thermal convection produces anticyclonic cells of rising warmer water at the surface, with largest cell diameters of about 1 km. Narrow linear shear zones of colder near-surface water between these warm cells are the source of sinking cyclonic thermobaric plumes that provide the energy to power the convective system. The prediction of a mid-depth maximum in the turbulent kinetic energy caused by thermobaricity is corroborated numerically. In the second case, thermal convection in a mixed layer overlying salinity-stratified warmer water may generate two kinds of conditional instabilities. In a *thermobaric parcel instability*, detraining parcels of mixed-layer water may penetrate the pycnocline without significant mixing of the stable surrounding water. In a *thermobaric layer instability*, a nonturbulent layer may become statically unstable and turbulent if advected below a predicted critical depth. For either kind of instability, the thermobaric increase in density of either a parcel or a layer of cold water may cause plumes of near-surface water to penetrate deep into the pycnocline and possibly to the bottom as "cumulus towers" of the polar seas.

1. INTRODUCTION

Predictions of vertical convection and deep water formation in the polar seas need to include the nonlinear equation of state effect called "thermobaricity." The term thermobaricity was coined by McDougall [1984] and generally refers to phenomena related to the pressure dependence of the thermal expansion coefficient for the density (ρ) of seawater, $\alpha = -\rho^{-1}\partial\rho/\partial T$. The magnitude of α increases with pressure (p) and may be approximated by the first two terms of a Taylor series expansion in p,

$$\alpha \cong \alpha_0\left(1 + \frac{p}{\rho_0 g H_\alpha}\right) \tag{1}$$

The Polar Oceans and Their Role in Shaping the Global Environment
Geophysical Monograph 85

where α_0 and ρ_0 are the surface values for thermal expansion and density, T is temperature, g is gravity, and H_α is the thermobaric depth scale,

$$H_\alpha = \alpha_0 / \left.\frac{\partial\alpha}{\partial z}\right|_{z=0}, \tag{2}$$

with the vertical coordinate (z) being positive up. For sea water near freezing, H_α is about 1 km.

As first pointed out by Gill [1973], this nonlinearity in the equation of state should cause a cold plume of saline shelf water produced by freezing to experience an additional decrease in stability as it flows down-shelf if the potential temperature of the surrounding water exceeds that of the plume by a finite amount, $\delta\theta$. This buoyancy reduction (δb) with depth (z) is explained by the thermobaric term of (3) that reduces the static stability for finite vertical displacement, δz:

$$\delta b = \left(N^2 - \frac{\alpha_0 g \delta\theta}{H_\alpha}\right)\delta z \tag{3}$$

where $N = (\alpha g \partial\theta/\partial z - \beta g \partial S/\partial z)^{1/2}$ is the buoyancy frequency for infinitesimal parcel displacement, and $\theta(z)$ and $S(z)$ are the ambient profiles of potential temperature and salinity.

A possible open-ocean role for thermobaricity was first considered by Killworth [1979], following the discoveries of chimneys in the Weddell [Gordon, 1978] and Iceland Seas [Foldvik, personal communication]. Killworth included thermobaricity in the calculation of hydrostatic stability for the Weddell Sea chimney. Farmer and Carmack [1981] and Carmack and Farmer [1982] showed its importance in freshwater lakes. Although other deep convection numerical studies [Brugge et al., 1991; Jones and Marshall, 1993] have considered nonhydrostatic effects, none have included thermobaricity until recently [Garwood, 1991; 1992a,b; 1993]. The dynamic nature of thermobaric instabilities seems to have been overlooked.

1.1. Dynamic Effects of Thermobaricity Hypothesized

Earlier mixed-layer models for polar-sea application have not included thermobaricity either [e.g. Lemke, 1987; Martinson, 1990]. However, searching for mechanisms to explain deep mixed-layer entrainment, Garwood [1991] showed a significant increase in the buoyancy flux (Figure 1) and in the predicted mixed-layer entrainment rate when thermobaricity was included in the turbulent kinetic energy (TKE) budget. A mid-depth maximum in the TKE was predicted for the deepest polar sea free convection because of thermobaricity.

In spite of the enhanced mixed-layer buoyancy flux, when salinity stratification is included, mixed-layer physics can not explain recent small-scale deep convection observations. CTD sections by Rudels et al. [1989], towed thermistor chain observations reported by Scott and Killworth [1991], and moored thermistor/conductivity time series of Schott et al. [1993] provide compelling evidence of small scale vertical convection events with horizontal widths of order 2 km and less in the Iceland and Greenland Seas. None of these observations of vertical plumes included evidence of horizontal homogenization that would be predicted by traditional mixed-layer physics.

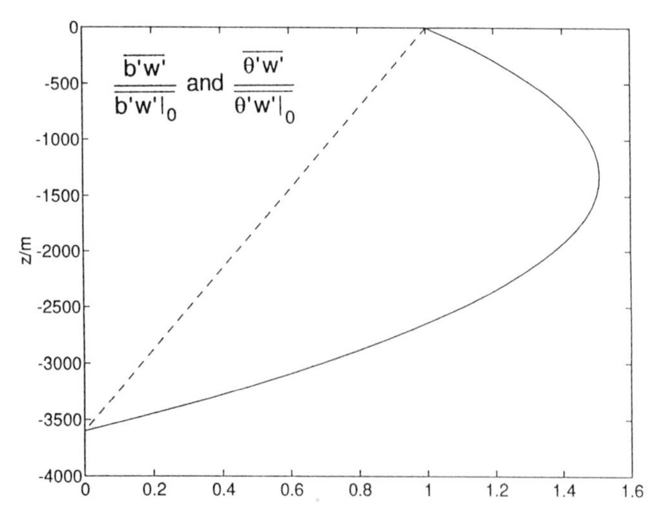

Fig. 1. Normalized buoyancy flux (solid) versus depth associated with a linear turbulent heat flux (dashed) in deep sea water with surface temperature near freezing [Garwood, 1991].

To explain these plumes that extend below the surface mixed layer into the halocline, Garwood [1992a,b] has suggested that deep penetrative convection in the polar seas may have dynamic and thermodynamic similarities to atmospheric cumulus convection, in that buoyancy is not conserved in either system. Two kinds of oceanic conditional instabilities were hypothesized. These processes, a *parcel instability* and a *layer instability*, are analogous to atmospheric conditional instabilities of the first and second kind [Holton, 1972].

1.2. Large-Eddy Simulation (LES): a Tool to Test Hypotheses

A nonhydrostatic numerical model for high Reynolds number turbulent flow is used to test the above hypotheses. The LES model of Moeng [1984], which was adapted to shallow ocean mixing by McWilliams et al. [1993], was modified for application to oceanic deep convection by adding a prognostic equation for salinity and including the pressure dependence (1) in the equation of state. The Boussinesq equations plus heat and salinity budgets are used to explicitly calculate the three-dimensional large-eddy velocity, salinity, and potential temperature fields:

$$\frac{du}{dt} = -\frac{1}{\rho}\frac{\partial p}{\partial x} + fv - 2\Omega_y w + \frac{\partial}{\partial x}\left(K_M\frac{\partial u}{\partial x}\right) + \frac{\partial}{\partial y}\left(K_M\frac{\partial u}{\partial y}\right) + \frac{\partial}{\partial z}\left(K_M\frac{\partial u}{\partial z}\right) \quad (4)$$

$$\frac{dv}{dt} = -\frac{1}{\rho}\frac{\partial p}{\partial y} - fu + \frac{\partial}{\partial x}\left(K_M\frac{\partial v}{\partial x}\right) + \frac{\partial}{\partial y}\left(K_M\frac{\partial v}{\partial y}\right) + \frac{\partial}{\partial z}\left(K_M\frac{\partial v}{\partial z}\right) \quad (5)$$

$$\frac{dw}{dt} = -\frac{1}{\rho}\frac{\partial p}{\partial z} + \alpha g(\theta - \theta_0) - \beta g(S - S_0) + 2\Omega_y u + \frac{\partial}{\partial x}\left(K_M\frac{\partial w}{\partial x}\right) + \frac{\partial}{\partial y}\left(K_M\frac{\partial w}{\partial y}\right) + \frac{\partial}{\partial z}\left(K_M\frac{\partial w}{\partial z}\right) \quad (6)$$

$$\frac{\partial u}{\partial x} + \frac{\partial v}{\partial y} + \frac{\partial w}{\partial z} = 0 \quad (7)$$

$$\frac{dS}{dt} = \frac{\partial}{\partial x}(K_S\frac{\partial S}{\partial x}) + \frac{\partial}{\partial y}(K_S\frac{\partial S}{\partial y}) + \frac{\partial}{\partial z}(K_S\frac{\partial S}{\partial z}) \quad (8)$$

$$\frac{d\theta}{dt} = \frac{\partial}{\partial x}(K_\theta\frac{\partial \theta}{\partial x}) + \frac{\partial}{\partial y}(K_\theta\frac{\partial \theta}{\partial y}) + \frac{\partial}{\partial z}(K_\theta\frac{\partial \theta}{\partial z}) \quad (9)$$

Here u, v and w are the easterly, northerly and vertical velocity components, respectively, f is the vertical Coriolis parameter, and $2\Omega_y$ is the horizontal Coriolis parameter. In (6), β is the salinity contraction coefficient for buoyancy, and thermobaricity is introduced by the pressure dependence (1) for α. The total derivative in (4)-(6), (8) and (9) is

$$\frac{d}{dt} = \frac{\partial}{\partial t} + u\frac{\partial}{\partial x} + v\frac{\partial}{\partial y} + w\frac{\partial}{\partial z}.$$

The prognostic equations (4-9) for resolved scale momentum, salinity and potential temperature are solved using second order, centered finite differencing in the vertical and the pseudospectral method of Fox and Orszag [1973] in the horizontal. This pseudospectral method allows use of a high-wavenumber cutoff filter to define the resolved scales and to remove the small-scale noise without artificially damping the resolved scale motions [Moeng and Wyngaard, 1988]. Time advancement is accomplished using the Adams-Bashforth scheme.

The subgrid scale fluxes are parameterized with eddy mixing coefficients ($K_{M,S,\theta}$) that are time- and space-dependent and calculated with second order turbulence closure as functions of the subgrid TKE (e), following Smagorinsky [1963], with

$$K_M = 0.1\lambda\sqrt{e} \quad (10)$$

and

$$K_S = K_\theta = [1 + (2\lambda)/L]K_M \quad (11)$$

The subgrid TKE length scale λ is equal to the grid scale L,

$$\lambda = L = (\Delta x \Delta y \Delta z)^{1/3} \quad (12)$$

unless the stratification is stable, when it is

$$\lambda = \lambda_s = 0.76\sqrt{e}N \quad (13)$$

if $\lambda_s<L$, where N is the buoyancy frequency. The unresolved subgrid TKE (e) is computed by

$$\frac{de}{dt} = K_M\left[(\frac{\partial u}{\partial z})^2 + (\frac{\partial v}{\partial z})^2\right] - gK_\theta(\alpha\frac{\partial \theta}{\partial z} - \beta\frac{\partial S}{\partial z}) + \frac{\partial}{\partial z}(2K_M\frac{\partial e}{\partial z}) - \varepsilon \quad (14)$$

where the four terms on the right of (14) are subgrid shear production, buoyancy flux, turbulent transport, and viscous dissipation. Subgrid dissipation (ε) is modeled as a function of the subgrid TKE,

$$\varepsilon = (0.19 + 0.51\lambda/L)\varepsilon^{1.5}/\lambda \quad (15)$$

More extensive details concerning the subgrid scale fluxes and the numerical method are provided by Moeng (1984).

For the numerical solutions here, the predicted eddy viscosity is on the order of 0.1 m^2s^{-1} or less, and the LES Reynolds number is of order 10^3 or larger and sufficient to cause a robust turbulence spectrum having the correct -5/3 slope at high wavenumbers [Gallacher, 1990].

In the following sections, thermobaric convection is examined and numerically simulated, first for purely thermal free convection and then for conditional instabilities in which salinity stratification allows for the build up and subsequent release of thermobaric potential energy.

2. THERMOBARIC FREE CONVECTION

To demonstrate the power magnification attributable to thermobaricity in the generation of turbulence, the vertical integral over the water column of the buoyancy flux is considered,

$$\frac{\text{Power}}{(\text{Area})\rho} = \int_{-h}^{0} (\alpha g\overline{\theta w} - \beta g\overline{Sw})\,dz \equiv w^{*3}, \quad (16)$$

where h is mixed-layer depth, overbars denote horizontal averages in a system that is approximately horizontally homogeneous in the mean, and w is the vertical turbulent velocity. The velocity scale w* that is defined by (16) is expected to be representative of the magnitude of the turbulent velocity present in the system at equilibrium when dissipation balances buoyant production, and storage of TKE is negligible. For no salinity flux, $\overline{Sw} = 0$, and a heat flux, $Q = \rho C_p\overline{\theta w}$, that decreases linearly from the surface value (Q_0) to zero at a depth z=-h, (1) and (16) yield

$$w^* = \{0.5\alpha_0 gh\frac{Q_0}{\rho C_p}[1 + h/(3H_\alpha)]\}^{1/3}, \quad (17)$$

where C_p=3990 J/Kg/C and ρ=1028 Kg/m^3 are representative mixed-layer values for specific heat and density.

For typical surface winter polar sea conditions (Q_0 =200 watts/m^2; S=34.5 psu, T(0)=-1.8 C), Figure 2 shows that w* increases from 0.7 cm/s for a convection depth h = 50 m to

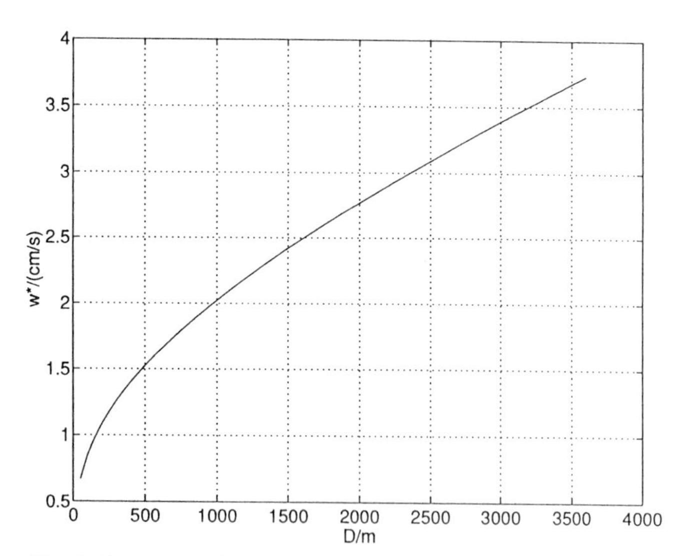

Fig. 2. Free convection velocity scale (w*) as a function of mixing depth (h) for a fixed value of surface heat flux, Q_0 =100 watts/m^2.

2.4 cm/s for h = 1500 m, and to more than 3.5 cm/s if convection penetrates below 3500 m.

2.1. Large-Eddy-Simulation of Thermally-Driven Deep Free Convection Without Salinity

The first of two numerical experiments is for thermally-driven deep polar sea free convection, without salinity. The purpose is to verify the TKE budget estimates and the prediction that thermobaricity should cause a mid-depth maximum in TKE.

The model domain is cubical, with each side 3.6 km, and periodic lateral boundary conditions are prescribed (see Figure 3). The ocean was assumed initially quiescent and homogeneous with the surface temperature at -1.8 C, just above freezing and typical of the western Greenland gyre during winter. Convection was initiated with application of a constant upward surface heat flux of 200 watts/m^2. With no wind stress in these free convection experiments, a slip condition was prescribed for the surface velocity, allowing the surface temperature field to be freely advected by the buoyancy-driven convection. Without an underlying salinity stratification, there was no loss of TKE to entrainment damping or to radiating internal waves. The simulation was continued for several days, until turbulence filled the model domain and a statistical equilibrium was approximated.

Figure 4 shows representative instantaneous horizontal averages of both the turbulent buoyancy flux and the subgrid scale conductive buoyancy flux versus depth. Thermobaricity causes the turbulent buoyancy flux to be greatest near the 2500-m depth. The peak buoyancy flux in Figure 4 exceeds that predicted in Figure 1 because of the unsteadiness in the

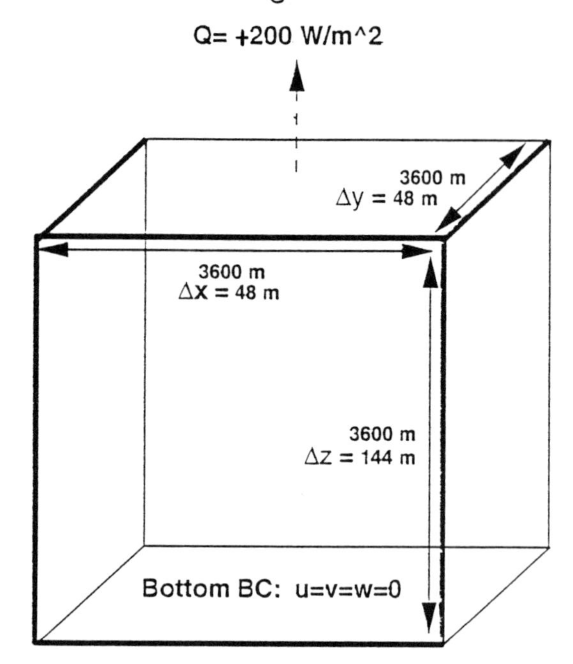

Fig. 3. Schematic diagram of LES model for first numerical experiment.

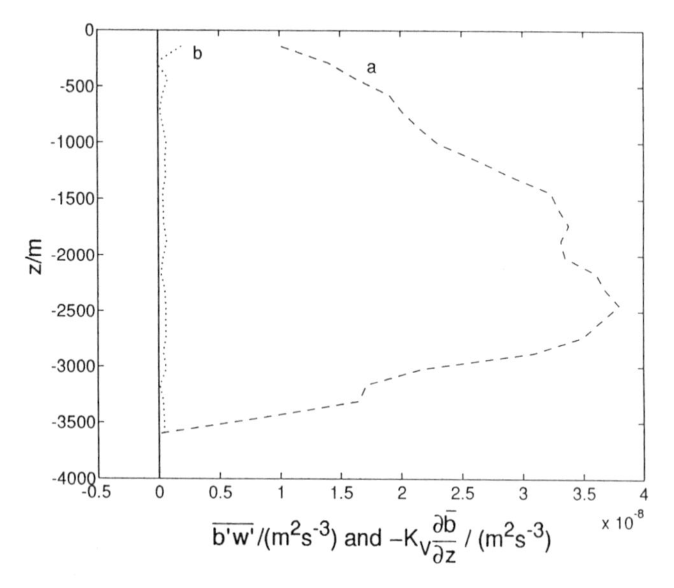

Fig. 4. Vertical profiles of horizontally-averaged buoyancy fluxes versus depth for (a) resolved turbulent motion (dashed), and (b) subgrid-scale diffusion (dotted). Averages are over the x-y domain of (3.6 km)^2 and computed for an arbitrary time during the fifth day after model initiation when the convective turbulence has achieved an approximate statistical equilibrium.

LES plume field and because individual large plumes can penetrate to depth without being significantly dissipated. The buoyancy flux profile in Figure 4 was during one of these plume events. The subgrid-scale conductive buoyancy flux also shown is considerably smaller in magnitude than the resolved flux, indicating that the energetics of the large eddies are independent of the subgrid-scale parameterization.

Figure 5 confirms the expectation of a maximum vertical TKE near mid-depth, with peak root-mean-square vertical velocities of about 0.03 m/s. Near the 2000-m depth there is a broad maximum in the total TKE, with $\overline{u^2+v^2+w^2}=3.2\times10^{-3}m^2/s^2$. There is also a peak in the horizontal TKE at the surface because of rotation and inertial effects. Turbulent transport and pressure redistribution together tend to reduce local maxima in total TKE and in each of the components.

The structure of the convective elements throughout the three-dimensional model domain has been studied by Garwood [1993] using hundreds of vertical and horizontal sections of contoured temperature together with time series of parcel trajectories. Video recording of time series of these sections showed that the longevity of vertical plumes seldom exceeds the time for a parcel of surface-cooled water to sink to the bottom. Because of the intensity of the three-dimensional vorticity, the loss of energy from the largest plumes to smaller eddies prevents the most energetic vertical plumes from lasting much longer than several hours, the integral time scale for the turbulence. The effects of thermobaricity are evident in Plate 1a, with the lower portions of downward-accelerating plumes appearing to separate from the upper part.

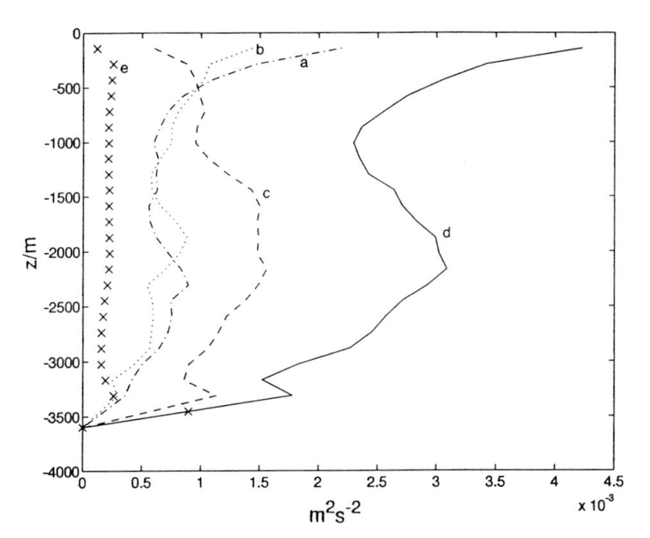

Fig. 5. For the same period as for Figure 4, vertical profiles of horizontally-averaged values of TKE components: (a) $\overline{u^2}$, (b) $\overline{v^2}$, (c) $\overline{w^2}$, (d) total TKE, and (e) subgrid TKE.

Horizontal sections showing the surface temperature field have a considerable degree of organization. See Plate 1b. Particularly noteworthy are the organized mesoscale features that somewhat resemble Rayleigh-Benard cells [Carsey and Garwood, 1993]. Unlike Rayleigh-Benard cells, however, these are nonstationary cells that are influenced by both planetary rotation and the smaller-scale three-dimensional turbulent vortices.

The large cells with three to six sides are warmer (red) than the cold (blue) areas of the field, by about 0.02 C. The largest cells are divergent and rotate anticyclonically. These warmer regions are fed by rising water that diverges at the surface and begins to spin under the influence of Coriolis. With a maximum horizontal speed of about 0.06 m/s and a horizontal scale size (D) of about 1 km, the largest cells have a local Rossby number (Ro=w*/fD) of about 0.5.

The coldest near-surface water lies in linear convergence lines between the expanding warm cells and has a large cyclonic vorticity that is accentuated by the vertical stretching induced by sinking. The local Rossby number of the sinking plumes is therefore much greater than unity.

Carsey and Garwood [1993] believe that they see similar surface features to those predicted by the LES in ERS-1 SAR data from the Greenland Sea during the winter. A functional relationship was suggested between the horizontal scale size (D) of the mesoscale features at the surface and the interior state of the ocean and surface forcing as measured by the depth of mixing, h, the thermobaric depth, H_α, and the free and forced convection velocities, w* and $u^*=\tau^{0.5}$, where τ is the wind stress. For a statistical steady state, this function should have the form,

$$D/h = \Phi[Ro, h/H_\alpha, w^*/u^*] \quad (18)$$

The function Φ needs to be evaluated from future observations, both in situ and remote, together with model simulations. Then Q_0, τ, and h could potentially be diagnosed from satellite observations alone.

3. THERMOBARIC CONDITIONAL INSTABILITIES

Thermobaric effects in the presence of salinity stratification are now considered. For typical polar sea conditions near freezing, the surface value of the thermal expansion coefficient is $\alpha_0 = 2.75\times10^{-5}C^{-1}$, the thermobaric depth defined by (2) is H_α=990 m, and the salinity contraction coefficient, $\beta=0.791\times10^{-3}psu^{-1}$, is nearly depth independent. Under these conditions, the magnitude of the buoyancy

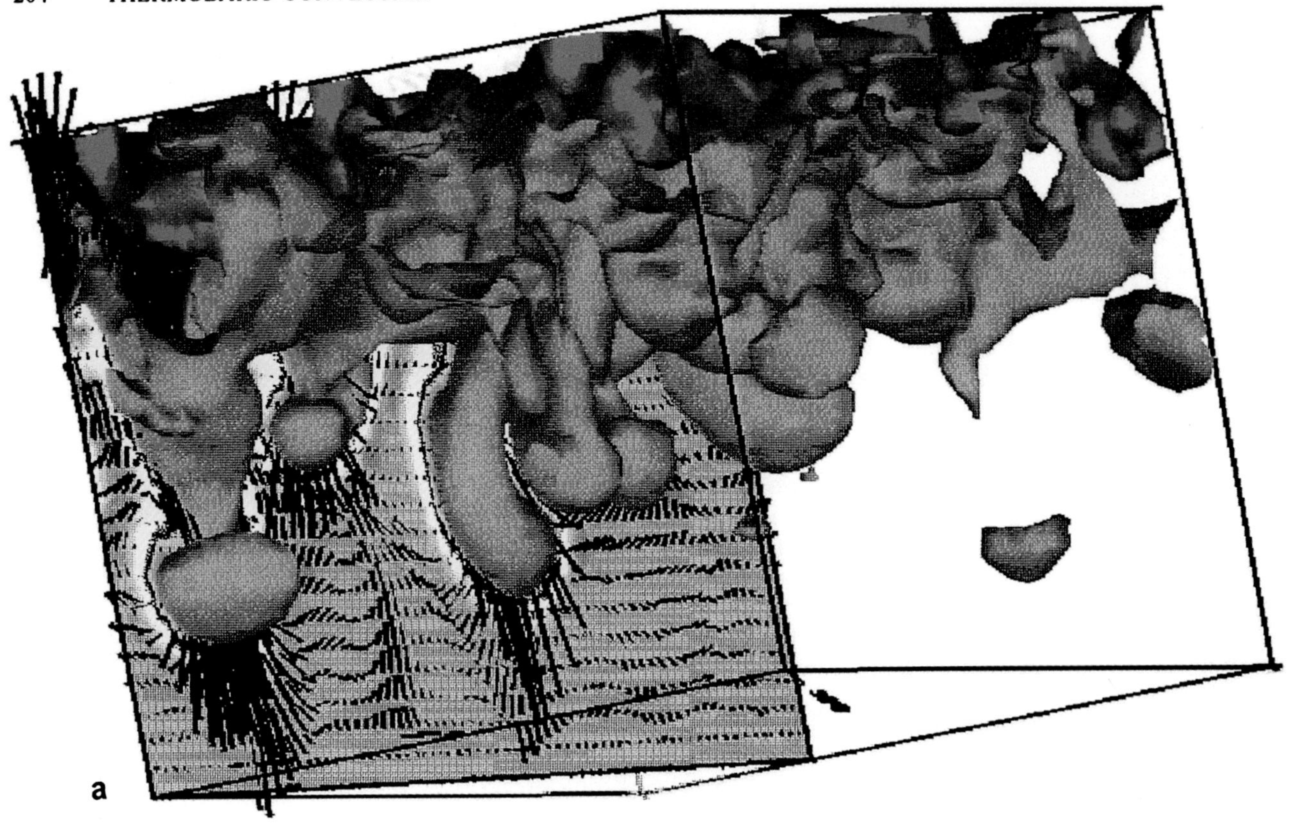

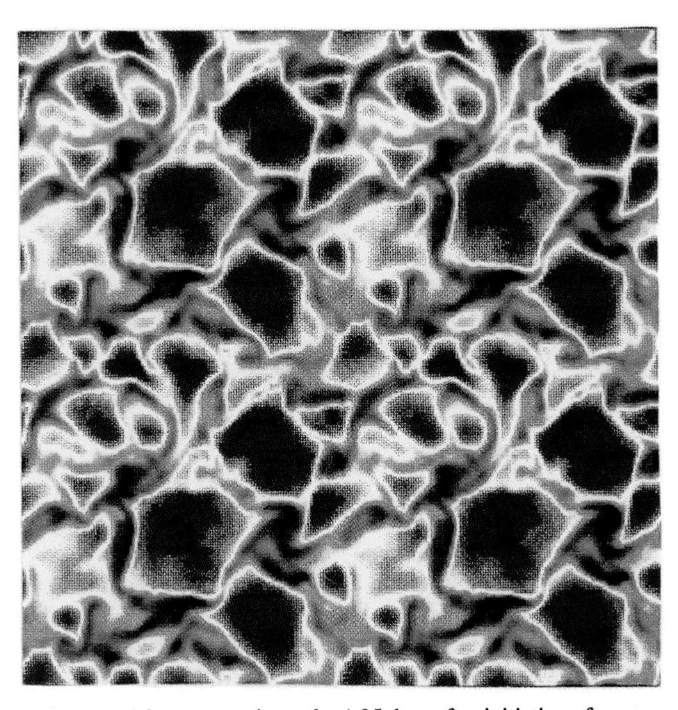

Plate 1. First LES experiment with thermal free convection only, 1.25 days after initiation of an upward surface heat flux of 200 watts/m^2. (a) Three-dimensional depiction of constant-temperature surfaces in a 3.6-km cube. Blue is -1.806 C, and green is -1.802 C. (b) Cellular mesoscale convection cells predicted for the surface temperature field. Area shown is 7.2 km x 7.2 km. Cyclical horizontal boundary conditions enable the depicted area to be expanded four-fold for visual effect.

change attributable to temperature difference increases a little over 0.1% with each meter of downward displacement, while the buoyancy change attributable to salinity difference remains constant. In other words, if two parcels of equal density in the mixed layer that have different θ-S properties are displaced downward together, the colder and fresher parcel will be compressed more than is the warmer and more saline parcel.

Garwood [1992a,b] hypothesized that an upside-down physical analog to cumulus convection may occur in polar seas due to the combined effects of salinity and temperature stratification in the nonlinear equation of state. In this thermodynamic analog between the ocean and atmosphere, the halocline plays the same stabilizing role in the polar seas as does the potential temperature profile in the tropical atmosphere, and low oceanic mixed-layer temperature plays a destabilizing role as does high humidity in the atmospheric boundary layer.

Perhaps the simplest case demonstrating thermobaric conditional instability is a two-layer system with a surface mixed layer of colder and fresher water overlying a more saline and warmer deeper layer. If mixed-layer water is displaced a short distance downward into the deeper layer, it will be buoyant and lifted back to the interface. However, analogous to the release of latent heat by a rising parcel of air from the marine atmospheric surface layer, the increase in thermal expansion coefficient with depth may make temperature differences overcome salinity differences in determining buoyancy if the ocean mixed-layer parcel is displaced to sufficient depth and differentially compressed by high pressure. Therefore, if a mixed-layer parcel or the interface itself is given sufficient downward motion by turbulence or internal waves, then the displaced water may become more dense than the ambient deeper water, with buoyancy not conserved. Parcels or plumes could then break away from the interface and be accelerated downward through the deeper layer. If the deeper layer is itself nonturbulent, such downward moving parcels or plumes may be convected to the bottom, or until they reach water of greater in situ density, without appreciable mixing with the intermediate water.

A critical depth (h_{cr}) is predicted for the thermobaric instability to occur,

$$h_{cr} = [\beta \Delta S/(\alpha_0 \Delta\theta) - 1]H_\alpha \qquad (19)$$

where ΔS and $\Delta\theta$ are the increases in salinity and potential temperature for the underlying layer relative to the surface mixed layer. The critical condition (19) applies to vertical displacement, but is similar in form to Aagaard et al.'s [1985] critical pressure level applied to horizontal interleaving of water masses.

3.1. *Parcel Instability Criterion and Large-Eddy Simulation*

Whenever the depth (h) of the surface turbulent boundary layer approaches h_{cr} because of entrainment in response to surface forcing, parcels of turbulent fluid may be detrained, escaping from the mixed layer. Such a parcel of surface-cooled water will accelerate downward until it meets the underlying more saline and warmer water. Upon entering the pycnocline, the parcel will at first be positively buoyant, slowing its penetration. If the parcel's initial downward speed is small, it may return to the mixed layer. However, if the speed is large enough, the parcel may pass below the critical depth before its downward speed is lost, and a *thermobaric parcel instability* should occur. The parcel will then fall until it meets the bottom or another layer of greater density, or until it is diffused by turbulent mixing. As an example, if H_α = 904 m, and the magnitude of the stratification due to salinity exceeds by 24% the magnitude of the compensating stratification due to temperature, then the critical depth would be (1.24-1)x(904 m), or 217 m. If the mixed-layer depth were 200 m, then a plume of mixed-layer water that penetrated more than 17 m into the lower layer would initiate a parcel instability. Of course, partial mixing between the parcel and the surrounding water will influence these tendencies, but the basic premise is expected to apply because the water column being penetrated is not expected to be turbulent and mixing should be minimized.

Figure 6 from observations by Quadfasel and Ungewiß [1988] shows that the lower layer of such a system need not be homogeneous. It may be stably stratified. The dashed profiles are for a neutral parcel: if a parcel of mixed-layer water is displaced vertically into the lower layer, it will have buoyancy identical to that of the surrounding water regardless of depth. The thermal expansion increase with depth exactly compensates the in situ temperature gradient. As can be seen, only a minor increase in surface salinity, equivalent to freezing 25 cm of water, will make the water column susceptible to parcel instability, potentially allowing parcel penetration deep into the thermocline without a concomitant mixing of the thermocline. In observations during the following winter, Rudels et al. [1989] observed such a narrow "pipe," with vertical penetration at a single site to 1250 m, while the surrounding deep water remained unmixed. Although Rudels et al. suggest a double-diffusion mechanism for the density anomaly to develop, we suggest the parcel instability mechanism. However, what initiates the parcel displacement?

If the initial conditions for a parcel ejected from a mixed layer of thickness h are

$$\frac{\partial}{\partial t} z|_{t=0} = -w_0 \text{ and } z|_{t=0} = -h,$$

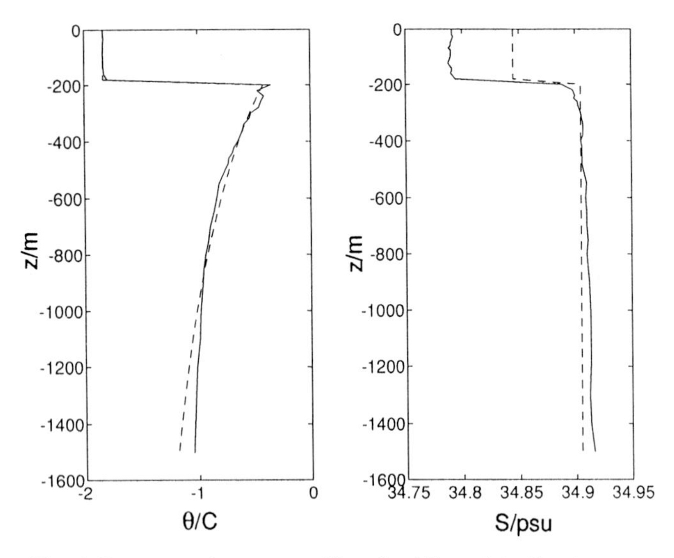

Fig. 6. Representative western Greenland Sea wintertime temperature and salinity profiles, from observations by Quadfasel and Ungewiβ [1988] (solid curves). Dashed curves are hypothetical profiles that approximate the change in the observations that would accompany a mixed-layer salinity increase associated with the freezing of 25 cm of surface water.

then the subsequent vertical displacement of the parcel can be predicted from the vertical momentum balance. For the period during which the detrained parcel is contained within the underlying layer, $z < -h$, the acceleration of the parcel is balanced approximately by its buoyancy,

$$\frac{d^2z}{dt^2} = \beta g \Delta S - \alpha_0 g \Delta\theta\left(1 - \frac{z}{H_\alpha}\right) \qquad (20)$$

where $g=9.83\ m/s^2$ is gravity. Using (19) the solution to (20) may be written as

$$z(t) = -h_{cr} + 0.5C_1 \exp\left[\left(\frac{\alpha_0 g \Delta\theta}{H_\alpha}\right)^{0.5} t\right]$$

$$+ 0.5C_2 \exp\left[-\left(\frac{\alpha_0 g \Delta\theta}{H_\alpha}\right)^{0.5} t\right] \qquad (21)$$

where $C_1 = h_{cr} - h - w_0\left(\frac{H_\alpha}{\alpha_0 g \Delta\theta}\right)^{0.5}$ and

$$C_2 = h_{cr} - h + w_0\left(\frac{H_\alpha}{\alpha_0 g \Delta\theta}\right)^{0.5}$$

Evaluating (21) as t approaches infinity, and requiring z to approach the critical depth gives the critical initial parcel speed w_0 to be

$$w_{cr} = (h_{cr} - h)\left(\frac{\alpha_0 g \Delta\theta}{H_\alpha}\right)^{0.5} \qquad (22)$$

If the initial parcel speed is assumed to be provided by mixed-layer free convection so that $w_0 = w^*$, then (17) and (22) may be combined to predict the minimum necessary surface heat flux to give a parcel instability,

$$Q_0 > 2\rho C_p \frac{(h_{cr} - h)^3}{h\left(1 + \frac{h}{3H_\alpha}\right)} (\alpha_0 g)^{0.5} \left(\frac{\Delta\theta}{H_\alpha}\right)^{1.5} \qquad (23)$$

The dashed profiles of Figure 6 provide an example. With $\Delta\theta = 1.50$ C, $\Delta S = 0.0636$ psu, and $h = 200$ m, (19) gives $h_{cr} = 217.5$ m. Then (23) predicts that $Q_0 > 200$ watts/m^2 would lead to a parcel instability by causing a sufficiently large initial velocity of $w_0 > w_{cr} = 0.011$ m/s to propel mixed-layer parcels below the critical depth.

With these analytical results as a guide, an LES numerical experiment was initiated with these values for h, $\Delta\theta$ and ΔS and forced by $Q_0 = 200$ watts/m^2. Plate 2 illustrates a sequence of x-z sections from the LES model. Four hours after initializing convection in an initially quiescent and stable ocean, the density interface near $z = -200$m was deformed and experienced some entrainment mixing. The first parcel instability to break through the interface and continue below the critical depth occurred some 5 hours after initialization. In the second picture in Plate 2, an 80-m wide pipe of mixed-layer water has penetrated the lower layer below the critical depth, and detrainment has occurred.

3.2. Layer Instability Criterion: Possible Explanation for Scott and Killworth [1991] Chimneys

While mixed-layer turbulence may provide the initial energy to generate a parcel instability, downwelling by the larger-scale circulation may lead to a related instability. If h is made to exceed h_{cr} in (19), either by Ekman pumping or other downwelling process, then the entire upper layer will become hydrostatically unstable, and a *thermobaric layer instability* may occur. This may have been the case for the chimney features having a 2-3 km width reported by Scott and Killworth [1991]. Although there were no local salinity measurements for verification in Scott and Killworth's observations, an originally hydrostatically stable layer with stable vertical gradients, $(\partial S/\partial z)/(\partial\theta/\partial z) > \alpha/\beta$, could have been made unstable if advected below a critical depth,

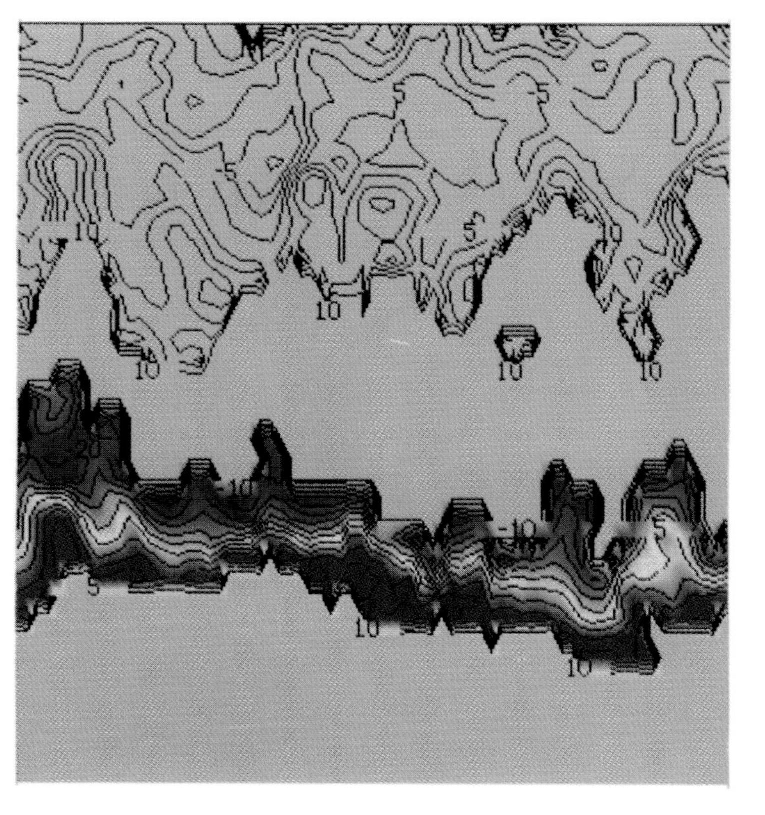

Plate 2. Second LES experiment including salinity stratification, (a) 4 hours, and (b) 5 hours after initiation of surface cooling. Parcel instability occurs in the second of these two sequential vertical temperature sections. The first section shows significant interface deformation by mixed-layer turbulence. The second picture shows the onset of the parcel instability with an 80-m wide plume penetrating the pycnocline. Contours of relative temperature in the upper 100 m are in millidegrees, and contours of relative temperature near the interface at z=-200 m are in centidegrees.

$$z_{cr} = -\left(\frac{\beta}{\alpha_0}\frac{\partial S}{\partial \theta} - 1\right)H_\alpha . \qquad (24)$$

Thus (24) is a generalization of the critical depth defined by (19) and applies to both stratified and well-mixed layers. In any case, advection of a layer with vertical temperature and salinity gradients below z_{cr} could conceivably lead to the draining of either a surface or an intermediate layer down a pipe or chimney created in the region that was advected below the critical depth.

4. CONCLUSIONS AND FUTURE NUMERICAL INVESTIGATION

Numerical simulations of thermobaric convection appear to verify two earlier hypotheses:

- There is a mid-depth maximum in the total TKE for three-dimensional simulations of deep thermally-driven free convection in the polar seas.
- A conditional instability thermodynamically similar to atmospheric cumulus convection can occur for the right combination of cold mixed-layer water overlying warmer more saline water. A thermobaric parcel instability was numerically simulated here for representative winter polar-sea conditions, but many polar sea thermobaric conditional instabilities may be of the layer-type.

The numerical solutions revealed details of the thermobaric convection that had not been anticipated:

- The most observable aspect of the surface during free convection may be the "ragged-net" field of larger cells of rising warmer water separated by linear shear zones of colder sinking water, as reported by Carsey and Garwood [1993]. Remote sensing of this surface manifestation of convection may be used potentially to diagnose the state of the ocean interior as well as the surface forcing.
- With regard to the vertical structure, the largest sinking plumes of colder water may be broken into discontinuous parcels because of both the thermobaric-enhanced downward acceleration of their leading edges and the interaction with the more isotropic smaller-scale turbulence.

Additional numerical experiments are required to answer key questions:

- How dissipative are the convection cells?
- How important is the tendency to conserve entropy [Garwood and Isakari, 1993] and not mix the intermediate water column that is penetrated by plumes that extend deep into the intermediate waters and possibly to the bottom?
- Finally, LES computations may guide parameterization of both deep mixed-layer dynamics and subgrid scale penetrative convection. A major goal is to include the effects of deep convection in ocean circulation models without having to resolve the plumes themselves.

Acknowledgments. This research was sponsored by the National Science Foundation (DPP 91-14161) and the Office of Naval Research (Code 112L). We thank Arlene Guest for valuable discussions and a careful reading of the manuscript. The numerical simulations were performed on the Primary Oceanographic Prediction System CRAY Y-MP 8/8128 at the Naval Oceanographic Office with funds provided by ONR and NRL. Support for PCG was provided by the Naval Research Laboratory.

REFERENCES

Aagaard, K., J. H. Swift, and D. C. Carmack, Thermohaline circulation in the Arctic Mediterranean Seas, *J. Geophys. Res.*, 90, 4833-4846, 1985.

Brugge, R., H. L. Jones, and J. C. Marshall, Non-hydrostatic ocean modeling for studies of open-ocean deep convection, In *Deep Convection and Deep Water Formation in the Oceans*, Ed. by P. C. Chu and J. C. Gascard, 325-340, Elsevier Science, 1991.

Carmack, E. C., and D. M. Farmer, Cooling processes in deep, temperate lakes: A review with examples from two lakes in British Columbia, *J. Mar. Res.*, 40, Supp., 85-111, 1982.

Carsey, F. D. and R. W. Garwood, Jr., Identification of modeled ocean plumes in Greenland Gyre ERS-1 SAR data, *Geophys. Res. Lett.*, 20, 2207-2210, 1993.

Farmer, D. M., and E. C. Carmack, Wind mixing and restratification in a lake near the temperature of maximum density, *J. Phys. Oceanogr.*, 11, 1516-1533, 1981.

Fox, D. G., and S. A. Orszag, Pseudospectral approximation to two-dimensional turbulence, *J. Comput. Phys.*, 11, 612-619, 1973.

Gallacher, P. C., Large eddy simulation of the turbulent boundary layer in the upper ocean (abstract), *Trans. Am. Geophys. Union,* 71, 1354, 1990.

Garwood, R. W., Jr., Enhancements to deep turbulent entrainment, In *Deep Convection and Deep Water Formation in the Oceans,* Ed. by P. C. Chu and J. C. Gascard, 197-213, Elsevier Science, 1991.

Garwood, R. W., Jr., Missing physics for deep convection? Arctic System Science Ocean-Atmosphere-Ice Interactions Modeling Workshop, Pacific Grove, Report No. 1, 49-54, 1992a.

Garwood, R. W., Jr., Oceanic convective instabilities hypothesized, *ARCSS-OAII Newsletter,* 2, 4-5, 1992b.

Garwood, R. W., Jr., Parcel and layer instability convection, Nansen Centennial Symposium, Solstrand-Bergen, Norway, 21-25 June 1993.

Garwood, R. W., and S. M. Isakari, Entropy Conserving deep convection in the Weddell Sea, *Proceedings* Fourth International Conf. Southern Hemisphere Meteorol. and Oceanogr., Hobart, 29 March - 2 April, 503-504, 1993.

Gill, A. E., Circulation and bottom water formation in the Weddell Sea, *Deep-Sea Res.*, 20,111-140, 1973.

Gordon, A. L., Deep Antarctic convection west of Maud Rise, *J. Phys. Oceanogr.*, 8, 600-612, 1978.

Holton, J. R., *An Introduction to Dynamic Meteorology,* Academic Press, New York, 319 pp., 1972.

Jones, H. and J. Marshall, Convection with rotation in a neutral ocean: A study of open-ocean deep convection, *J. Phys. Oceanogr.*, 23, 1009-1039, 1993.

Killworth, P. D., On "chimney" formations in the deep ocean, *J. Phys. Oceanogr.*, 9, 531-554, 1979.

Lemke, P., A coupled one-dimensional sea ice model, *J. Geophys. Res.*, 92, 13164-13172, 1987.

Martinson, D. G., Evolution of the southern ocean winter mixed layer and sea ice: Open ocean deepwater formation and ventilation, *J. Geophys. Res.*, 95, 11641-11654, 1990.

McDougall, T. J., The relative roles of diapycnal and isopycnal mixing on subsurface watermass conversion, *J. Phys. Oceanogr.*, 14, 1577-1589, 1984.

McWilliams, J. C., P. C. Gallacher, C.-H. Moeng, and J. C. Wyngaard, Modeling the oceanic planetary boundary layer, in *Large Eddy Simulation of Complex Engineering and Geophysical Flows*, Cambridge University Press, 441-454, 1993.

Moeng, C.-H., A large-eddy simulation model for the study of planetary boundary layer problems, *J. Atmos. Sci.*, 41, 2052-2062, 1984.

Moeng, C.-H., and J. C. Wyngaard, Spectral analysis of large-eddy simulations of the convective boundary layer, *J. Atmos. Sci.*, 45, 3573-3587, 1988.

Quadfasel, D. and M. Ungewiβ, MIZEX 87 - RV Valdivia Cruise 54, CTD Observations in the Greenland Sea, *Technical Report 5-88*, Institut fur Meereskunde der Universitat Hamburg, 1988.

Rudels, B., D. Quadfasel, H. Friedrich, and M.-N. Houssais, Greenland Sea convection in the winter of 1987-1988, *J. Geophys. Res.*, 94, 3223-3227,1989.

Schott, F., M. Visbeck and J. Fischer, Observations of vertical currents and convection in the central Greenland Sea during the winter of 1988-1989, *J. Geophys. Res.,* 98, 14401-14422, 1993.

Scott, J. C. and P. D. Killworth, Upper ocean structures in the southwestern Iceland Sea: A preliminary report, In *Deep Convection and Deep Water Formation in the Oceans*, Ed. by P. C. Chu and J. C. Gascard, 107-122, Elsevier Science, 1991.

Smagorinsky, J., General circulation experiments with the primitive equations, *Mon. Wea. Rev.*, 91, 99-165, 1963.

R. W. Garwood, Jr. and S. M. Isakari, Department of Oceanography, Naval Postgraduate School, Code OC/Gd, Monterey, Ca 93943, USA.

P. C. Gallacher, Naval Research Laboratory, Stennis Space Center, MS 39529, USA.

Oceanic Convection in the Greenland Sea Odden Region as Interpreted in Satellite Data

Frank D. Carsey
California Institute of Technology, Jet Propulsion Laboratory, Pasadena CA

and

Andrew T. Roach
Applied Physics Laboratory, University of Washington, Seattle, WA

Satellite and in-situ data for the "Odden" region of the Greenland Sea are discussed with respect to describing regions of convection. The convection is discussed in terms of regional ice retreat, observed in passive microwave data, that has previously been associated with convection observed in ocean mooring data. These regions are tentatively identified in SAR data which shows plumes of about 300 m separation in an area about 20 km by 90 km immediately north of the rapidly retreating ice edge at the southern end of an ice edge embayment, taken to be the consequence of the flow of warm, saline Arctic Intermediate Water to the surface during convection. Although there is no way to determine the depth of the convection it is assumed to be to intermediate depths. The embayment in 1989 is seen in passive microwave data to expand downwind at the rate expected of wind-forcing of either ice or surface water, but a propagation along a salinity gradient is also possible; the actual mechanism at work is not known.

1. INTRODUCTION

1.1 Background

Ocean convection is seen as a globally important process in which air-sea interactions influence oceanic circulation through the production and ventilation of deep and intermediate waters. A key site of convection, active at least some winters, is in the Greenland gyre, and the convection seems to be related to the development of an ice feature called Odden ("the Icy Cape" in Norwegian), an eastward extension of the ice edge in the latitude range 71° to 75°N as shown in Figure 1 for the winter of 1989. Previously published results from the 1988-89 winter [*Roach et al*, 1993] showed that convection near the Odden ice edge at 75°N, 4°W immediately preceded the formation of the Nordbukta ("North Bay"), the large embayment or central retreat in Odden, occurring nearly every winter.

The key dynamic elements of oceanic convection are taken to be the individual plumes, the clusters of plumes called chimneys, the eddies that are the consequence of chimneys aging in a rotating frame, and, in the Greenland Sea, the embayments and polynyas resulting from the outcropping and spread of intermediate-level convective-return water on the surface. Recent numerical work, which has not included surface wind driving, has suggested that the chimneys, of scale 10-60 km, should grow through increase in plume numbers, decay through baroclinic instability, and circulate cyclonically with the gyre. Plumes are expected to have dimensions in the range of 100-1000 m; chimneys in the range 20-60 km; and eddies in the range 5-60 km [*Jones and Marshall*, 1993; *Garwood*, 1991; *Gascard*, 1991]. The processes of plume formation and convection are not well understood, and other mechanisms have been proposed for initialization of convection, notably that of ice-edge upwelling [*Häkkinen*, 1987; *Johannessen et al*, this issue]. Additionally, eddies have been observed in other parts of the Greenland Sea, and numerous dynamic origins have been proposed for them [*Johannessen et al*, 1987].

In this paper we discuss features seen in satellite data; these features are hypothesized to be the surface signals of the plumes, and possibly the eddies, which have been numerically simulated. Specifically we present an interpretation of passive microwave data for 1989 and 1992 in the context of a simple model of convection-driven ice-edge retreat causing Nordbukta growth;

The Polar Oceans and Their Role in Shaping
the Global Environment
Geophysical Monograph 85

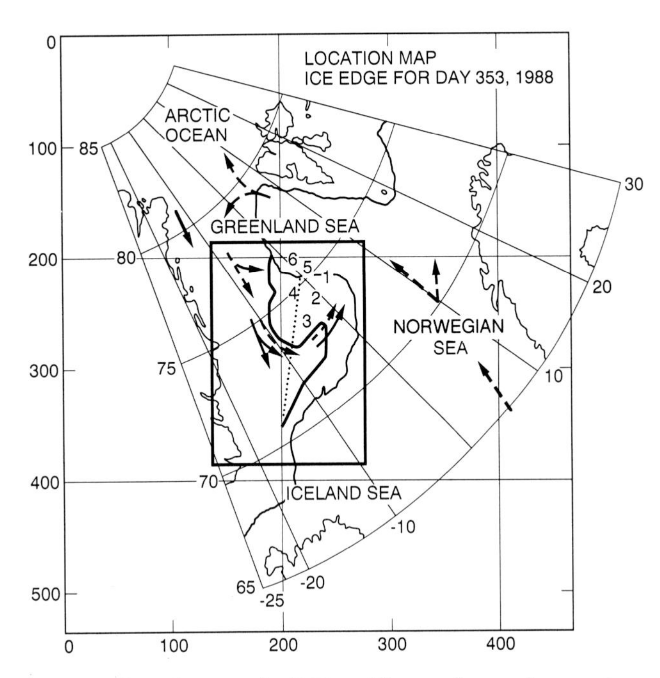

Figure 1: Location map for Odden. The numbers at the margin are row and column numbers for a 5 km grid. The numbers represent Greenland Sea Program Moorings as discussed in *Roach et al* [1993]. The box encloses the region known as Odden, and the satellite data discussed here are within this box. The dashed line that runs nearly up the box is approximately along the section of *Bourke et al* [1992] and the solid hooking line outlines their JMC 1.5°C boundary.

constraints are imposed by processes associated with the formation and migration of a small polynya. We also present imaging radar data which seems to describe the surface structure resulting from convecting plumes and supports our hypothesis that convective action in an area at the ice edge controls the ice edge motion in this region. The satellite data interpretation, especially of the radar image, is conjectural; there has been insufficient in-situ oceanic and ice-cover data collection for conclusive interpretation of either the satellite data or the processes of the upper ocean.

1.2 The Odden Region

The oceanography of the Greenland-Iceland-Norwegian Seas has received much attention throughout the century, and here only a very quick overview is supplied; a comprehensive discussion of the region can be found in *Hurdle* [1986]. The principal water masses involved in the Odden processes are the upper Arctic Intermediate Water (uAIW), a warmer saltier water of Atlantic origin, and the Polar Water (PW), a cooler fresher water of Arctic origin [*Johannessen*, 1986]. The uAIW approaches the Greenland gyre from the northeast after moving north past Norway and recirculating west beneath Spitzbergen. Some of this flow is bathymetrically turned again to the east and forms the Greenland gyre with its center approximately at the location of GSP-4 in Figure 1. PW flows along the northwest side of the gyre moving in a southeasterly direction along Greenland as the East Greenland Current (EGC). Filaments of uAIW and PW forming the lower and upper branches of the Jan Mayen Current (JMC) flow to the south approximately along the ice edge before turning with the gyre to the east, as shown in Figure 1, to cut across the southern half of the gyre , and PW, freshened by seasonal sea ice melt, also forms the surface water of the Odden area [*Bourke et al*, 1992].

The uAIW is underlain, below about 500 m, with deeper waters that are cooler and slightly fresher than the uAIW, the Greenland Sea Deep Water (GSDW) and Norwegian Sea Deep Water (NSDW). These are the end-point waters for the Greenland Sea deep convection; their salinity ranges from 34.88 to 34.94 psu, and their temperatures from -0.5°C to -1.3°C, with the GSDW the fresher and cooler [see e.g. *Johannessen*, 1986].

1.3 Winter Processes in the Odden region

In winter the waters of the Greenland Sea are cooled by cold, mostly northerly and northeasterly winds but with some strong northwesterly cold air outbreaks [*Roach et al*, 1993]. The waters at the surface of what will be Odden are buoyant with a mixed layer of uncertain depth, but in the range 50-100 m at end of summer [Bourke et al, 1992], and this area will be cooled enough, in most winters, to form an ice cover. The uAIW to the north has enough sensible heat that while it cools through the winter (*Worcester et al*, 1993), it does not form an ice cover. As Odden ice grows, brine is injected into the upper water increasing its salinity and density. This water will, if enough cooling and brine are supplied, convect into and sometimes through the uAIW water, and this process brings up convective return water with enough heat to stop ice formation or even melt ice, thus liberating fresh water [*Killworth*, 1979]. If conditions are right convection will continue and deepen until it extends to the bottom, but this step is apparently not as simple as it sounds. As a consequence of greater compressibility of lower salinity waters, surface waters fresher relative to the uAIW probably convect through a progressive mixed-layer deepening while surface waters more saline can undergo abrupt deep convection [see *Aagaard and Carmack*, 1989]. For deep convection, dynamical constraints must be met [*Gascard*, 1991]. The issues of convection and deep water formation are examined in detail by *Chu and Gascard* [1991].

There is a temptation to think that the Odden ice growth simply converts the fresher surface PW to uAIW by brine generation, but this is not the case. The *Bourke et al* [1992] section shows a surface salinity change of about 1 psu/100 km; to remove this layer by brine from ice growth with the surface fluxes available is not practical. Specifically if we use a mean flux of 200 w m^{-2} and a mixed layer of 50 m mean thickness, the retreat of the edge of meltwater front (lighter than the warm intermediate water below), and thus of the ice edge, would occur at a maximum (if all heat lost at the surface is latent heat, which it is not) of only 3 km d^{-1}, 20 % of the observed rate [*Roach et al*, 1993, and below]. Thus, the heat and salt in the convective-return uAIW is required. The fact that most of the surface heat loss goes into ocean cooling is also apparent in the salination record in figure 2 of *Roach et al* [1993] which shows a brine change during winter of about 0.004 psu d^{-1}, appropriate for about 1 cm d^{-1} of ice growth (on a 50 m mixed layer) or a heat flux of only 30 w m^{-2}; this result is consistent with those of *Schott et al*, [1993]. This seems to indicate that, in part, Odden ice growth may serve to generate negative buoyancy to stir the uAIW up into the mixed layer.

1.4 The Convective Events of 1989

As part of the Greenland Sea Project, oceanographic data from the upper 200 m at two locations were examined [*Roach et al* 1993; see also *Schott et al*, 1993]. The growth of sea ice in the central Greenland gyre injected brine into the upper water column locally and reduced the vertical stability profile during December and January. Subsequent cooling increased the density of the surface waters to a critical point and a cold air outbreak in late January 1989 provided enough buoyancy loss to convectively overturn at least the upper 200 m. Replacement water then rose from the warmer pool of intermediate water at mid-depth causing an increase in the heat available in the upper layer. The surface signature of that warming was the retreat of the ice cover near GSP-4 and then along the Greenland shelf edge to the southwest, where the warmer water apparently was advected. As will be discussed below the meltback rate estimated from sequential satellite images is about 11 km d^{-1} (or 13 cm s^{-1}), comparable to both the 10 cm s^{-1} mean current in the EGC at 79°N and the wind forcing of ice or surface water [*McPhee*, 1990].

2 THE SATELLITE DATA

2.1 Data

To generate regional sea ice distribution we used image data from the Special Sensor Microwave Imager (SSM/I), an instrument designed to make a variety of oceanic and terrestrial observations [*Hollinger et al*, 1990], including the concentration and type of sea ice. SSM/I brightness temperatures (T_b) are acquired at 19, 37, 22 and 85.5 GHz at both polarizations except at 22 GHz. To obtain fine-scale data on ocean surface processes we used SAR data from the AMI on the ESA ERS-1 [*Carsey*, 1992]; these radar images have resolution of about 30 m with swaths of 100 km and have been used in other air-sea-ice studies in this region [*Johannessen et al*, this issue].

2.2 SSM/I Interpretation

The interpretation of microwave radiance for new and young ice types is complex, and is different from the interpretation for thick first-year ice or older ice. The observed T_b for new and young ice depends on thickness according to whether the ice grows in calm or rough conditions. This situation has been examined using surface and satellite data [*Grenfell et al*, 1992] and with satellite data alone [*Steffen and Maslanik*, 1988]. Essentially, nilas growth in calm water is characterized by a change from the low open sea T_b to the high ice T_b as a consequence of growth to only a few millimeters of thickness [*Wensnahan et al*, 1993], depending on frequency, while pancake ice growth in rough conditions is characterized by a nearly linear (albeit noisy) increase in T_b with ice thickness over the range 0 to about 15 cm, largely independent of frequency. *Grenfell et al* [1992] also noted that the vertically polarized radiance was enhanced for pancakes. Pancake ice has previously been indicated as the dominant form in the regions of the ice edge in the Odden region [see e.g., *Tucker et al*, 1991; *Sutherland et al*, 1989], and the meteorological conditions of this site are appropriate for this kind of growth [*Weeks and Ackley*, 1986]. We assume that the Odden ice is principally pancake form. In a microscopic sense the microwave signal from pancake ice [as discussed in *Grenfell et al* 1992] may be the consequence of some variable, e.g., pancake wetness, that is correlated to ice thickness rather than the consequence of ice thickness itself; on this there is no definitive data set or applicable model.

2.3 SSM/I Variables

To further examine the ice condition record we form the Polarization Ratio (PR) and the Gradient Ratio (GR), these variables have proven useful in the examination of the major ice types of the polar seas [*Cavalieri et al*, 1984].

$$PR(freq) = (T_bV(freq)-T_bH(freq))/(T_bV(freq)+T_bH(freq))$$
$$GR = ((T_bV(37)-(T_bV(19)))/(T_bV(37)+T_bV(19)) \quad (1)$$

where T_bV and T_bH are the vertically and horizontally polarized microwave brightness temperatures at the frequency indicated in GHz.

In the formation of PR and GR we have variables that have reduced sensitivity to surface temperature and weather, and we have also generated somewhat "tuned" variables as PR is more sensitive to open water fractional coverage or pancake thickness while GR is more sensitive to the presence of old ice although it is sensitive to open water as well [*Cavalieri et al*, 1984]. From the definition of PR and GR their sensitivity to weather is reduced approximately by half, but it still can be appreciable. PR and GR are shown in Figure 2 for January 17, 1993 for a data set in which the 37 GHz data have been expressed on a 5 km grid which preserves the resolution at about 30 km. These data sets are consistent with the data for the entire winter, and they show that PR and GR are redundant variables of this ice cover, and we will use only PR to describe ice conditions.

2.4 The SSM/I Ice Edge

There are several interpretation schemes for Odden ice PR values. We could use the traditional approach in which thick ice

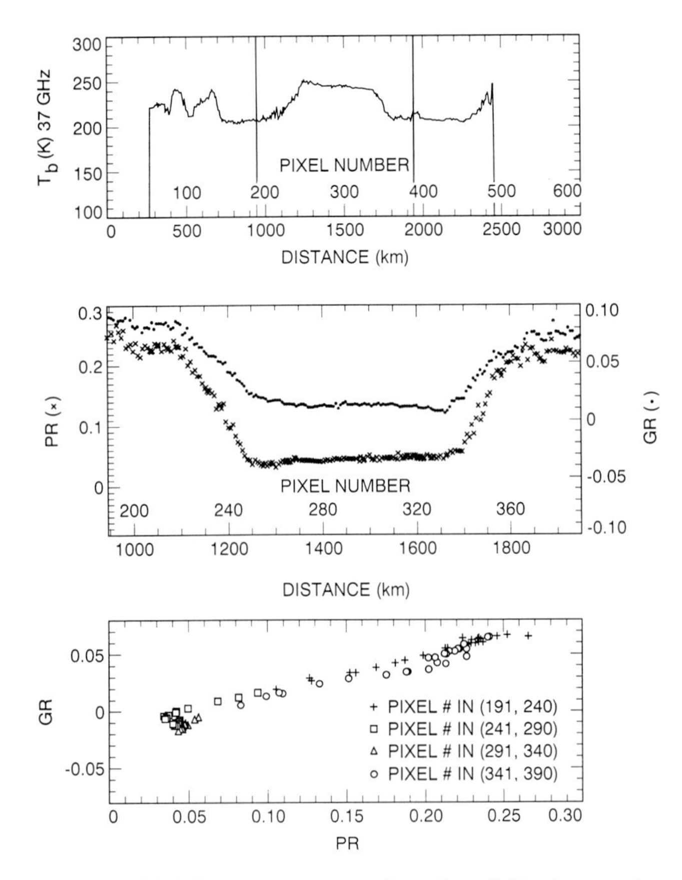

Figure 2. SSM/I Data for the centerline of the Odden box on day 17 of 1989. In the upper frame is a plot of T_b vs distance down the box along column 208 of Figure 1. The heavy lines indicate the top and bottom of the Odden box. In the center frame the calculated PR and GR are shown for the profile through the box only. Both of these variables have minimum values in thick consolidated ice. In the bottom frame PR and GR in the box are plotted against each other with different symbols used for different parts of the profile. The presence of old ice would draw GR down; the variation shown is due to either ice concentration or thickness of pancakes.

is assumed and ice concentration is solved for [*Steffen et al*, 1992]; we could utilize the *Grenfell et al* [1992] result and interpret PR as a pancake thickness for an area covered to some concentration by uniform-thickness pancakes; or we could acknowledge that there is a concentration ranging from 0-100 % of variable thickness pancakes. While none of these is wholly satisfactory, the last category is doubtless more correct, but we do not have the information to pursue it quantitatively. The approach open to us is to assume, strictly for purposes of locating the ice edge, that the actual concentration and pancake thickness profiles along lines normal to the ice edge are essentially constant over the winter so that a given PR value is the locus of ice of essentially invariant spatial relationship to the ice edge by any definition.

3. THE 1989 and 1992 ODDEN EXTENT RECORDS

3.1 *The 1989 Odden*

The timing and strength of development of Odden is different in every year, and, in fact, it has not formed in some years [see *Sutherland et al*, 1989; *Wadhams*, 1986]. The Odden in 1988-89 formed in November, reached maximum extent in December [Figure 1], and began to retreat in late January. The retreat took on its typical pattern as the formation of the Nordbukta at its northern edge near 75°N, 4°W. In the late winter-early spring time frame the Nordbukta can, as it did in 1989, separate Odden from the EGC ice. Ice conditions in the spring were highly variable in 1989, and this is also typical.

3.2 *Ice Retreat in 1989*

In Plate 1 the PR(37GHz) is shown for the Odden box of Figure 1 for the entire winter. According to *Roach et al* [1993] the convective events start about January 20. Shortly thereafter an embayment forms at the ice edge in the upper center of the Odden box, and the embayment grows by steady ice retreat of 10-15 km d^{-1} to the southwest (down the box). the retreat continues steadily until about day 66 when there is some episodic alternation of PR increase and decrease. The tongue of ice that is the most persistent is seen to lie along the axis of the Jan Mayen Current (see Figure 1) as observed by *Bourke et al* [1992]. In our analysis we assume that the Nordbukta growth (the ice retreat) is the consequence of convection, to at least intermediate depth, beginning near the center of the Greenland gyre, occurring essentially annually.

Near day 30, three small scallop-shaped embayments appear in the ice edge, and these features migrate down the box, approximately to the southwest. We initially interpreted them to be the chimney features as discussed by both theoretical [*Jones and Marshall*, 1993] and observational [*Gascard*, 1991] investigators, although the scale of the scallops is larger than has been discussed at 60-100 km (the uncertainty arising from SSM/I resolution). Another difficulty in the identification of the scallops as chimneys is that chimney drifts should retain the cyclonic sense of the gyre, but the scallops simply move to the southwest at the rate of the Nordbukta retreat. Finally, embayments on all sorts of scales are common in ice edges, and this particular geometry is not always due to convective events.

The scallop which appears on the eastern side of the convective embayment does not simply move SW on the embayment fringe as the others do; it closes into a migrating open-water feature which we argue is a convective sensible-heat polynya, in Plate 1 a yellow spot of 60-90 km diameter. When it has moved some 120 km to the SW of its formation another scallop forms at its origination site. The migrating polynyas "fill" with ice at approximately the northern edge of the JMC. It may be that the upper PW filament of the JMC supplies enough fresh water to terminate the convection and permit formation of an ice cover once the heat pulse brought to the surface by the convection has been lost to the air. These polynyas are large

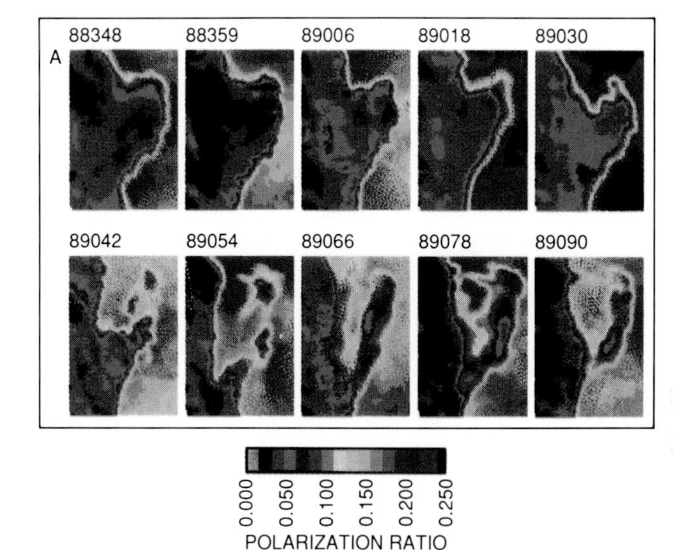

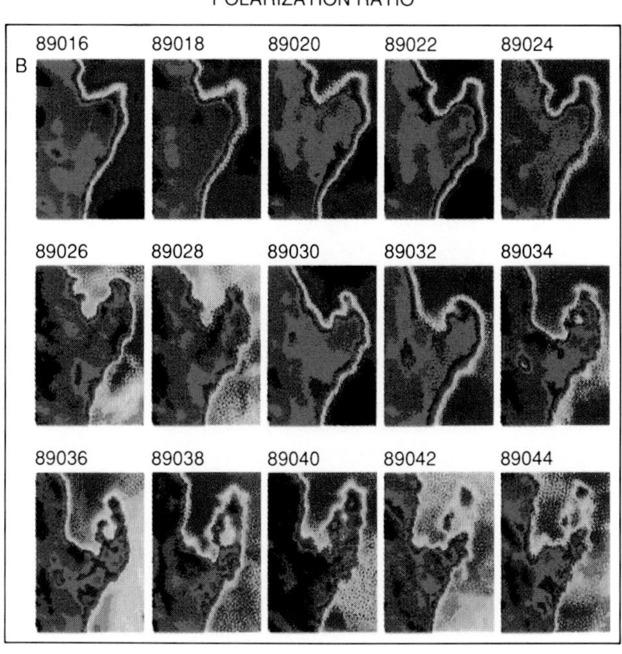

Plate 1. SSM/I values of 37 GHz PR {Where PR = $(T_{bV}-T_{bH})/(T_{bV}+T_{bH})$} for the winter of 1989; day of year numbers are shown above each map. In part A the entire winter is summarized with PR data on 12 day separation. In part B the period of rapid evolution of the Nordbukta [the central ice retreat] is shown with PR data on 2 day separation. In these maps the reds and oranges are open water (with weather), and the blues and greens are ice while yellow is a transition color which can be open water if heavy clouds are present. The yellow spot to the east of the Nordbukta in days 89034 and following is hypothesized to be a convective sensible-heat polynya.

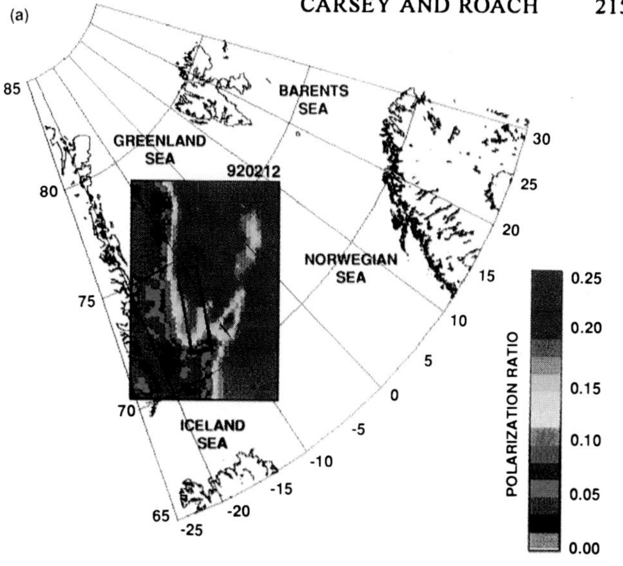

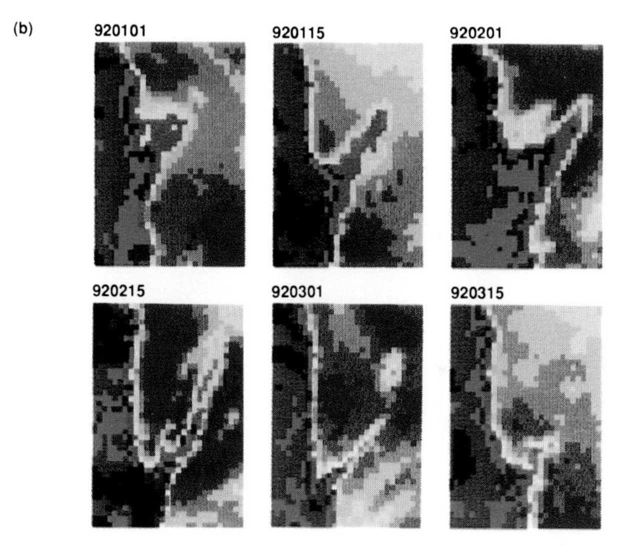

Plate 2. SSMI/I data for 1992 for the Odden region, as in Figure 1 and Plate 1. a) Outline of the region shown in Plate 4 ERS-1 SAR data.

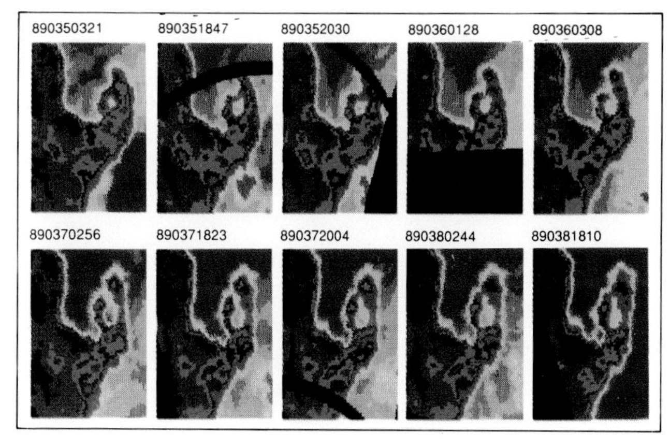

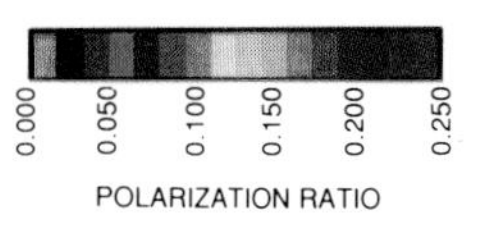

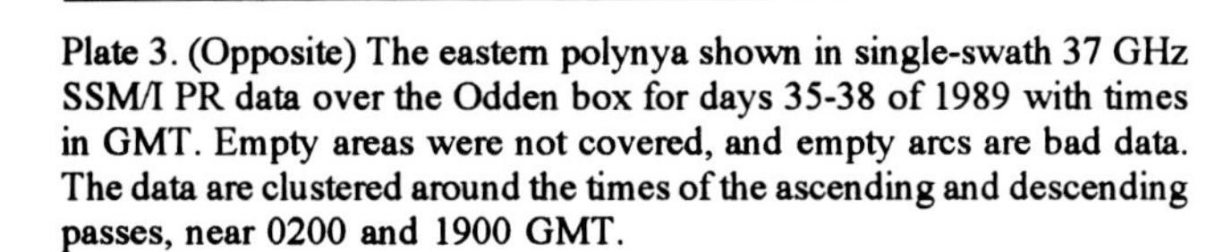

Plate 3. (Opposite) The eastern polynya shown in single-swath 37 GHz SSM/I PR data over the Odden box for days 35-38 of 1989 with times in GMT. Empty areas were not covered, and empty arcs are bad data. The data are clustered around the times of the ascending and descending passes, near 0200 and 1900 GMT.

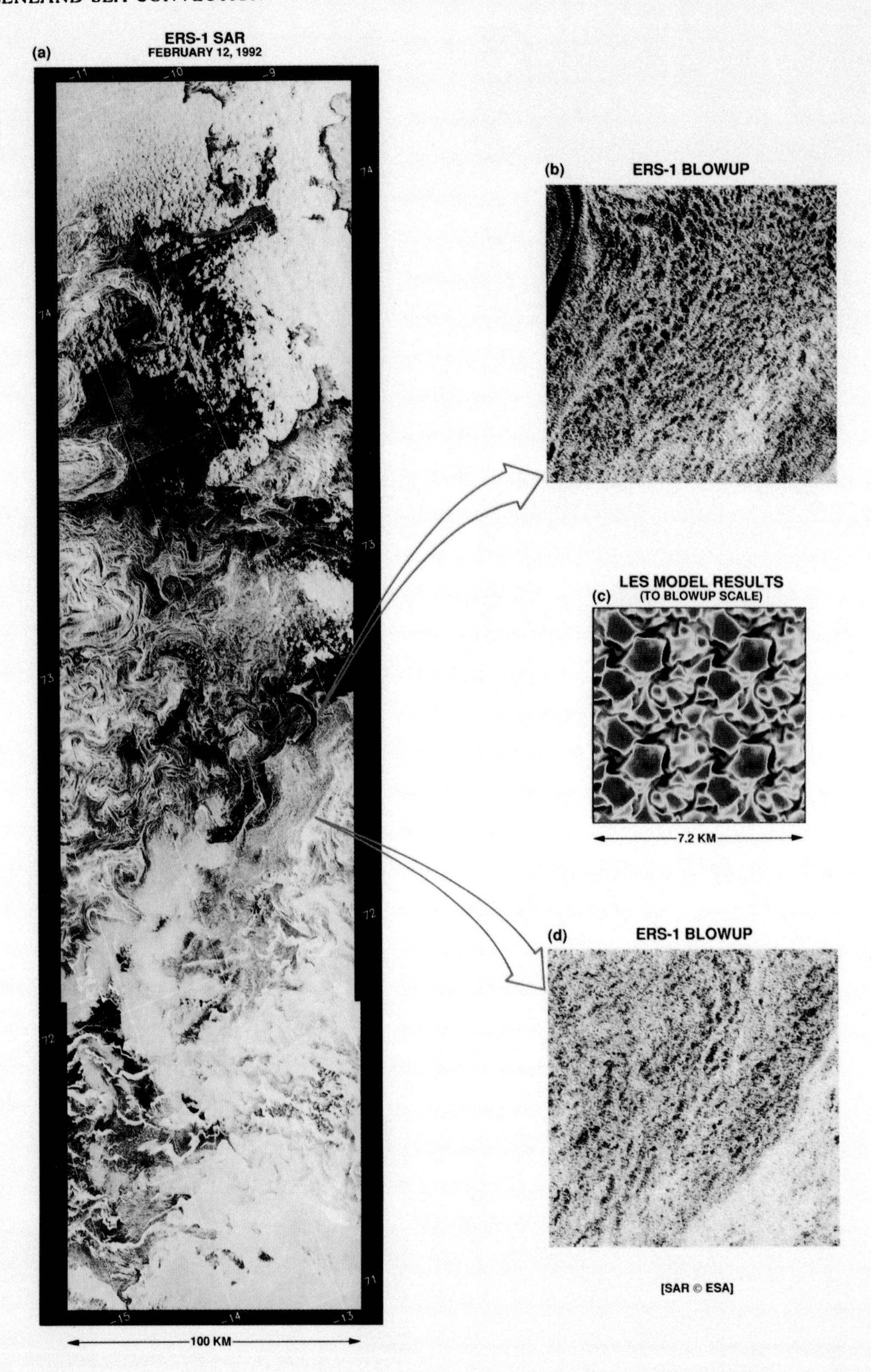

Plate 4. Plumes in the Greenland Sea as modeled [*Garwood*, 1991] and observed in ERS-1 SAR images [*Carsey and Garwood*, 1993]. 4a is the ERS-1 SAR data at nominal swath of 100 km and reduced resolution, about 100 m; in the blowups of 4b and 4d the resolution is shown at 30 m. The model result only covers an area 3.6 km on a side; four identical such regions are grouped in 4c.

enough that the SSM/I data should reasonably resolve their character, and the T_b data indicate a polarization intermediate between open water and a high concentration of thick pancakes; thus, this eastern polynya may be partially or completely filled with thin pancakes and may be difficult or impossible to observe in radar or visible-light data sets or even visually from a ship's deck.

All the scallops move at within 10 % of the rate of the Nordbukta growth and are thus likely to be controlled by a common mechanism. At the same time, this tendency to move to the southwest is not universal to Odden-area features; the far northeast tip of Odden, for example, moves to the north-east; its behavior is uncorrelated with the embayment features.

3.3 The 1992 Odden

Plate 2 shows the 1992 Odden from SSM/I data. This was a winter in which the Odden formed late and was small, but there was the formation of Nordbukta a bit later than 1989. This data set is being used for comparison with ERS-1 SAR data, and in what follows SAR data for the outlined rectangular area will be discussed.

4. ODDEN ICE COVER BEHAVIOR

4.1 A Simple Model of Ice Retreat

We would like to develop a quantitative picture of the processes at work in Greenland Sea convection. From the discussion of *Roach et al* [1993], it seems that the rapid ice retreat is a signal feature of the convection and that this process must be the consequence of the sensible heat brought to the surface by the convective-return uAIW. The first question to address deals with the rate of ice retreat. From visual inspection of Plate 1 the ice retreat has a rate of about 12 km d^{-1}. The chimney growth rates suggested by theoretical analyses are 2-3 km d^{-1} [*Legg and Marshall*, 1993], and the currents of the region are negligible [*Roach et al*, 1993]. Thus we have only the wind as external source for the rapid ice edge motion. If we speculate that the convective-return water terminates ice growth exactly, i.e., no ice at the ice edge is formed or melted after the convection begins, then the last ice that formed will move under simple wind forcing, and the open-water area will grow at that rate. To examine that prospect we use the Norwegian Hindcast winds for the location of GSP-4.

According to *Moritz* [1988], Greenland Sea ice moves according to

$$U - c = BG \tag{2}$$

where U is the [complex] ice velocity, c is the current, G is the geostrophic wind, and B is a complex constant which contains the drag coefficient and Coriolis turning. We will use c = 0 for the Odden area Further, we will concern ourselves only with the ice motion component down the center of the Odden box. Following *Moritz* [1988] we use

$$|B| = 1.21 \times 10^{-2}$$
$$\Theta = \arg(B) = -3° \tag{3}$$

In the above we are specifically modeling ice motion, but the modeling of the motion of warm surface uAIW would use equivalent terms [see *McPhee*, 1990]. Thus, we are examining the motion of the ice edge to see if it is controlled by wind-driven properties, but we are not specifying what is being driven.

Figure 3 shows the geometry including a sample of geostrophic wind and ice motion down the box. Positive x-component of wind and ice motion are taken to be down the box. We generated daily average wind-forced and observed ice edge positions where the observed ice edge is the location, on the line down the box center, of PR = 0.12. Both resulting displacement series were smoothed for 7 days over the whole season. Figure 4 shows the resulting ice edge retreat velocity component by both SSM/I and the wind-forced calculation.

In interpreting Figure 4 errors must be considered. there are errors in the winds; there are geophysical variations in the parameters of 1 and 2; there may be local surface currents so that

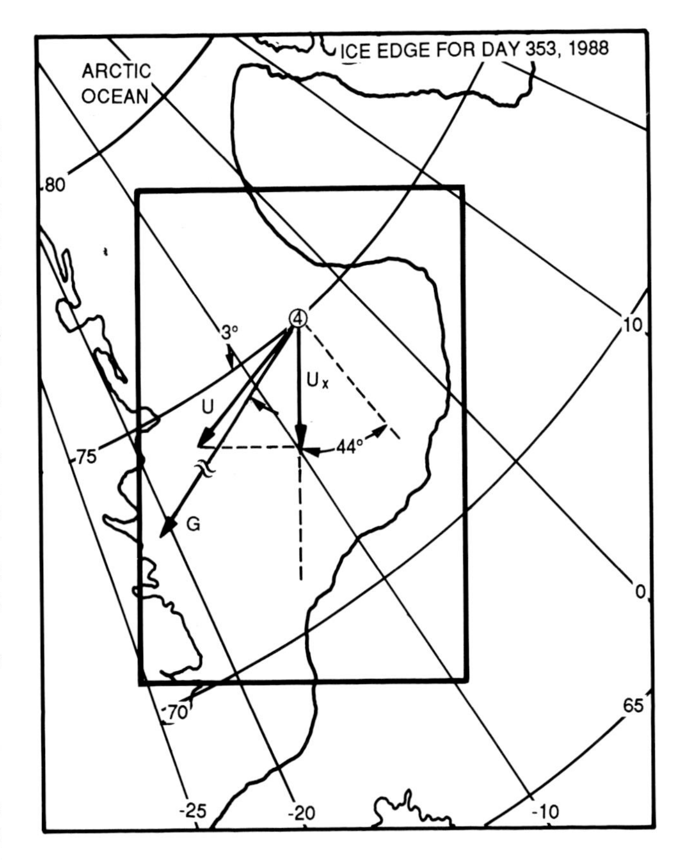

Figure 3. The geometry for the ice edge model. The centerline of the box is seen to have an offset from true north of 44°. G is the geostrophic wind, U is the modeled ice motion and Ux is the component of ice motion down the box centerline.

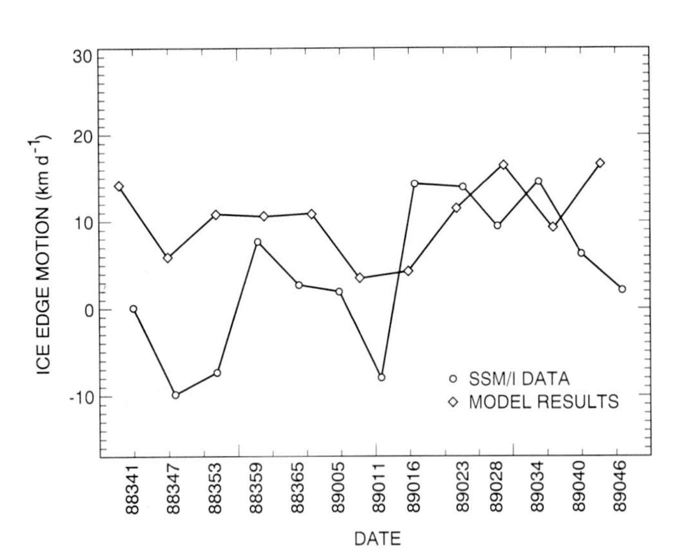

Figure 4. Ice retreat rates down the box of Fig. 3 in 1988-89 as measured in the SSM/I data and as predicted with a simple wind-driven model. The errors estimated for the model result are about 40 % so that a nominal agreement is found.

$c \neq 0$; and there are errors in Earth-location for the SSM/I data. For and uncorrelated uncertainty of 25 % for each, a fairly conservative estimate [*Brown*, 1990], there is an uncertainly of about 45 % in the comparison. Thus, the specifics of the curves cannot be interpreted closely. In Figure 4 the ice edge is seen to be going upwind to NE early in the winter, consistent with expected early-season thermodynamic ice advance. Beginning at about the time of convection onset the two indications of ice edge motion agree within the error estimates. During this period there is a tendency for negatively correlated departures in the two curves; an increase in predicted wind-driven retreat occurs with a decrease in SSM/I retreat, and vice-versa, suggesting an additional loss of ice cover by wind-induced mixing. In the late winter the SSM/I ice edge becomes erratic as ice covers and retreats from large areas very quickly. In this situation the wind-forced model is too simple.

This calculation fails to disprove that wind advection is a control of ice retreat in mid and later winter, but it is neither strong enough to confirm wind-forcing nor to specify whether the ice at the edge is moving a bit faster than or more slowly than the edge itself, or if the ice itself might be growing or becoming thinner near the edge. These considerations are important to the surface salinity budget.

4.2 *Refinements of Odden Behavior*

Two features of the Odden are thus arguably convective, the Nordbukta and the eastern polynya. It would be possible for the Nordbukta to be formed entirely by a convective-return water source limited in geographic extent to an area at the gyre center immediately around GSP-4; the convective-return water would simply be blown downwind to lose heat to the air while it mixes with local surface water. However, the eastern polynya has to bring its source of convective-return water with it as it moves to the southwest down the box. Thus we speculate that the convection in both the Nordbukta and the eastern polynya is confined to a region near the ice edge. Additionally, since ice is found all around the eastern polynya, in particular to the northeast, there must be, to stabilize the column, surface water with reduced salinity after the convection has moved on. This fresh water could be residual PW, or it could be the consequence of the melt of ice advected from the NE. The ice edge may move more slowly that the ice, so that ice is always forming at the lower edge of the polynya, providing brine for convection, and melting at the upper edge. Thus, the data lead us to hypothesize that the convective water is located in, and confined to the southern end of, the open water area; the remainder of the Nordbukta and the eastern polynya are modified uAIW of convective-return origin.

Plate 1 permits a rough but useful calculation. From the rate of motion of the polynya, a given spot on the ocean is in the polynya for 8-10 days as the polynya is 100 -120 km across and is moving at about 12 km d^{-1}. For mean winter fluxes of about 200 w m^{-2}, a 1°C change in temperature would indicate a mixed layer of 40 m. The actual change is not known and by water characteristics may be smaller that 1° by as much as half, and the mixed layer thickness is not well defined, but is in the range of 50-100 m [*Bourke et al*, 1992]. Thus the heat loss by the ocean is consistent with the cooling of the mixed layer that has been warmed by mixing with uAIW as long as the uAIW-PW resultant water is fresh (light) enough. At the point of origin of the eastern polynya the process of initiation of convection seems to be cyclical on approximately a 10 day period.

Although the Greenland Sea situation is explicitly not covered by the simulation, *Killworth* [1979] has examined convection in the Weddell Sea, and has suggested that convection in the presence of ice can take on a form in which the ice cover is intermittent. In this mode surface cooling causes the sea surface to freeze, and ice grows until the stability is destroyed, and then convection begins. The convection brings up warm water from depth; the warm water melts some or all the ice, and the convection is terminated until the melt-induced buoyancy is destroyed by ice growth whereupon the convection restarts. In *Killworth* [1979] this sequence is called ABCDA. For the Weddell Sea data the predicted ice cover is cyclic with a frequency about 1.2 d^{-1}. In Plate 3 we show a sample of individual passes of SSM/I over Odden; there are about 3 per day that cover the region reasonably well, and this rate spans frequencies adequately, considering that no intermittency is visible in the daily data. The polynya is not seen to be changing in size or shape on the time scale of the satellite revisit schedule which suggests that the polynya is responding to mechanisms other than the intermittent-ice mode suggested by Killworth although the possibility of chimney-scale (5-10 km) intermittency on the polynya edge is not ruled out at the SSM/I resolution. The capability of the point of origin of the polynya to generate another polynya in about 10 days may be related to the Killworth processes in that the water at the ice edge has increasing salinity due to ice growth until convection is triggered.

5. POSSIBLE PLUMES IN THE SAR DATA

5.1 SAR interpretation

Plate 4 a, b, and d show ERS-1 Synthetic-Aperture Radar [SAR] data for the area in the Odden region outlined in Plate 2, and 4c shows model results for deep convection [*Carsey and Garwood*, 1993]. This image is complex, and here we will present a brief and qualitative interpretation of the features as this is all that is needed for our argument, and a more concrete interpretation of the image calls for in-situ observations.

The interpretation to follow is reasonable, in our view, but is necessarily somewhat speculative. For other treatments of open-ocean SAR data see *Johannessen et al* [1993a and 1993b], *Johannessen et al* [1992], *Johannessen et al* [this issue], *Johannessen et al* [1983], and *Tucker et al*, 1991. In the upper portion are white puffy features interpreted as low winds on open ocean; wind speeds of about 3 m s^{-1} would generate enough backscatter to make the image bright [see *Donelan and Pierson*, 1987], and ECMWF-analyzed winds at this time were about 2 m s^{-1}; thus one might expect visual evidence of regions of low [dark] and higher [bright] winds like these. Below the wind-puffs there is a zone of dark water followed by a zone of dark water with eddy-like features etched in narrow bright lines. We interpret the dark area to be open water and the bright lines to be ice streamers advected by the eddy currents. In the upper right near 73.5°N and again near 74°N there are scallops 15-20 km across; these appear to be waves. Below the eddy field there is on the right a textured gray area; this is hypothesized to contain the plumes and will be discussed further below. To the left of that is a bright region that we interpret to be open water roughened by wind. Below this area (below 72°N) there is a region that we interpret to be pancake ice, on the right, and ice bands, on the left. Pancake ice has been reported to be common in this region [*Tucker et al*, 1991 and *Wadhams et al*, 1993]. The pancakes show structure similar to the eddies to the north but less intense. In the bottom right is wind-roughened open water to the southwest of Odden. This interpretation is made upon inspection of the SAR image and is consistent, in general, with the SSM/I data of Plate 2 which shows ice, either thin pancakes or first-year ice in a concentration of about 50 %, in a triangular area on the east side of the SAR image frame and in the Odden ice tongue. Open water roughened by higher winds are at the edge of the frame at the southeast.

Quantitatively, there is only limited analysis that can be done to substantiate the interpretations; pancakes have inherently a wide range of possible backscatter as does the wind-roughened sea [*Tucker et al*, 1991; *Donelan and Pierson*, 1987], and winds derived from even the best analyses have expected errors in the range of 2 m s^{-1}, large enough for ambiguous interpretations. Thus, with respect to the SAR image alone, no concrete conclusions are possible. The geometry of the plumes and eddies is concrete, but there has been no in-situ verification of the oceanic processes hypothesized to be at work.

5.2 The Hypothesized Plumes

The key element of oceanic convection, the active plume, has as yet not been convincingly observed or simulated. Some plume data have been acquired. From the temperature series on GSP 4 during the convective period, an upper bound on the vertical velocity of 3.1 cm s^{-1} was noted [*Roach et al*, 1993]. This is in general agreement with vertical velocities measured directly by Doppler profilers in the Mediterranean Sea and in the Greenland Sea [*Schott and Leaman*, 1991; *Schott et al*, 1993]. The modeling community has recently predicted what one can expect to find in ocean convection. In particular, *Jones and Marshall* [1993] and *Garwood* [1991], through scaling arguments, have found the important terms in the convective process are buoyancy flux, Coriolis force and ocean depth. The *Jones and Marshall* [1993] calculations applied to our data with a nominal 500 W m^{-2} peak heat loss [for initiation of convection] will yield a plume of about 160 m diameter with a vertical velocity of 2.2 cm s^{-1} while the Garwood approach finds that the plume array should have spacing dependent on convection depth and ranging up to 2 km for deep convection in the Greenland Sea.

ERS-1 SAR data, discussed above, and modeling results are shown in Plate 4 [see also *Carsey and Garwood*, 1993]; these figures are hypothesized to represent modeled and observed plume surfaces. The model result is from *Garwood* [1991] for deep convection; thus the plume spacing in the SAR data would indicate intermediate convection to about 1000 m, consistent with the results of *Worcester et al* [1993]. In the blowups the dark regions are interpreted to be convective return water, and the bright regions are interpreted to be pancakes growing on the plumes. We recognize that the bright regions may be concentrations of small-scale surface waves herded onto the plume tops by surface convergence, and it is clear that no conclusive argument can be made from SAR data alone. The (hypothesized) plume-filled region is seen to be about 20 km by 90 km and to be located directly north of the Odden ice edge. As discussed above, the ECMWF wind analysis for this area indicated a day of very low winds, about 2 m s^{-1}; in more usual high wind conditions the plumes might have a very different appearance, as well as surface structure, and might not be visible at all.

We argue that the "ragged net" appearance is not commonplace and may well be the consequence of convection. In blowups of the other parts of the image we could find no zones possessing this appearance although we note that an amazing variety of features can occur in SAR images of the ocean. Finally, the ragged-net features we have hypothesized to be due to plumes may be visible in other SAR data of the ocean (and convection is certainly present in other parts of the ocean), but we have not observed them in examination of well over 100 ERS-1 SAR images of this area, and we find no reference to them in literature [see e.g. the summary in *Johannessen et al*, 1992, p283].

5.3 Convective Regions

Some properties of convective regions in this area can be derived from these data sets. Since the images of Plates 1 and 3 for 1989 indicate that the scallop-shaped embayments and the polynya are most likely all the same phenomenon at work, we can count embayments, e.g., on day 38, to estimate that there are 3 or perhaps 4 regions of convection in Odden. In the images of Plate 2 for 1992 the scallop-shaped features are not as clearly shown, but one could argue that there are 2 such features present. In the 1989 data the points of origin of the embayments are apparently near the upper edge of Odden about 75°N, 4°W, and the site of origination of the eastern polynya is capable of sequential generation of the transient convecting features. We note that the SSM/I data are ineffective in monitoring any convection processes not at the ice edge, e.g., at GSP4 after the ice has started to retreat.

The data discussed here indicate an association of a type of convective behavior with the ice edge, and seem to argue that convection in other locations would have to have a different evolution. An interesting issue with respect to comparison of the 1989 data with the model results of *Legg and Marshall* [1993] is whether the 3 embayments clearly visible on day 28 started off as one convective event which grew and subdivided into three as in their Figure 9.

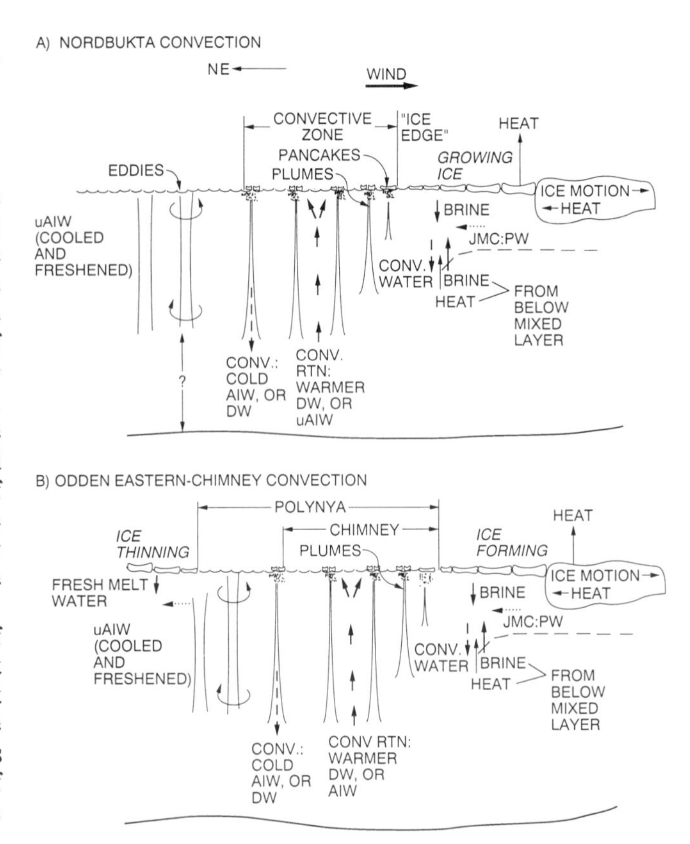

Figure 5. Hypothesized processes in the Nordbukta (A) and in the eastern polynya (B) in cartoon form, not to scale. DW stands for Deep Water, and a dotted line represents a virtual current due to the propagation of the convection. The role of ice is shown as providing brine for convection; this assumes that the convecting density is approached by adding brine to cold water so that the sinking water will be a bit fresher and colder than the surrounding water. An analysis of heat and salt fluxes indicates that some heat and brine supplied to the upper ocean has to come from intermediate depths.

6. CONCLUSIONS

In short, we conclude that the satellite data, taken together with the interpretation of mooring data from 1989 [*Roach et al*, 1993; *Schott et al*, 1993], strongly suggest that the Nordbukta as well as polynya-like features in the Odden ice cover are the result of convection which is confined to an area smaller than 100 km across located at the retreating ice edge of the Odden or the leading edge of a migrating polynya. Inspection of ERS-1 SAR data for this ice-edge region yields features which strongly resemble modeled plumes, and we hypothesize that these structures are the surface signatures of plumes. Clearly, these conclusions are tentative; further in-situ and satellite observations of ice and upper ocean are required.

Modeled and observed convection behavior are integrated in Figure 5 which shows in cartoon form the oceanic structure suggested by the satellite and ocean data for the central chimney and eastern polynya. The key features are the plumes, chimneys (aggregates of plumes), and open-water. The key difference between the eastern polynya and the Nordbukta is that the eastern polynya is advected as a closed, partially ice-covered, chimney-like feature while the Nordbukta (the central retreat of Odden) is an embayment, a difference probably arising from lower initial surface salinity in the polynya area.

We still have no model or strongly-indicative data on the mechanism for the observed propagation or wind-advection of convection although we hypothesize that the same mechanism is at work in both the Nordbukta and the eastern polynya. Also, we tentatively conclude that the Odden region of the Greenland Sea has convection at work only in a few small (≤ 100 km) regions near the ice edge, and is characterized over most of the open-water area by a well mixed layer at least 200 m and possibly 500 m deep.

We speculate that the eastern polynya is a convective, sensible-heat polynya that may be partly filled with ice; in-situ observations of features of this sort would be interesting and useful.

The plumes that we argue are observed in the Nordbukta seem to be small structures, about 100 m across, organized in a "ragged-net" array with separation about 300 m, consistent with model results for intermediate convection. We would expect structurally similar plumes to be active in the eastern polynya, but the actual dimensions could vary from those of Plate 4, and there may be, in addition, convective plumes in numerous other areas of the Greenland Sea.

These tentative conclusions call for more data, from both satellite and in-situ platforms. To verify the presence, densities, and scales of plumes, horizontal profiling over a range of depth is called for; this data acquisition is quite challenging in seas which are partially to fully ice covered and will require significant efforts.

Acknowledgments. SSM/I data are from NSIDC in boulder, and ERS-1 data are from ESA through ESRIN and the UK-PAF. We thank them for their work and cooperative spirit. John Crawford and I-Lin Tang at JPL performed analysis tasks on the data sets. Knut Aagaard and Johnny Johannessen have been sources of good advice in numerous areas. The modeling data of Plate 4 are from Roland Garwood. We appreciate the useful and insightful comments provided by the reviewers. The research at California Institute of Technology Jet Propulsion Laboratory was supported by NASA under contract, and the work at the University of Washington was supported by NASA and ONR.

7. REFERENCES

Aagaard, K., and E. Carmack, 1992, The role fo sea ice and other fresh water in the Arctic circulation, *J. Geophys. Res., 94,* 14485-14,498.

Bourke, R., R. Paquette and R. Blythe, 1992, The Jan Mayan Current, J. Geophys. Res., 97, 7241-7250.

Brown, R., 1990, Meteorology, in Smith, ed., *Polar Oceanography, Part A, Physical Science*, Academic Press, San Diego, 1-46.

Carsey, F. and R. Garwood, 1993, Identification of modeled ocean plumes in Greenland gyre ERS-1 SAR data, *Geophys. Res. Lett.*, 2207-2210.

Carsey, F., 1992, ed. *Microwave Remote Sensing of Sea Ice*, AGU Monograph 68, 474p.

Cavalieri, D., J. Crawford, M. Drinkwater, D. Eppler, L. Farmer, R. Jentz, and C. Wackerman, 1991, Aircraft active and passive microwave validation of sea ice concentration from the Defense Meteorological Satellite Program Special Sensor Microwave Imager, *J. Geophys Res., 96,* 21,989-22,008.

Chu, P. and J.-C. Gascard, 1991, eds., *Deep Convection and Deep Water Formation in the Oceans*, Elsevier, 382 p.

Donelan, M. and W. Pierson, jr, 1987, Radar scattering and equilibrium ranges in wind-generated waves with application to scatterometry, *J. Geophys. Res.* 92, 4971-5029.

Garwood, R., 1991, Enhancements to Deep Turbulent Entrainment, in Chu and Gascard, eds., *Deep Convection and Deep Water Formation in the Oceans,* Elsevier Science Pub., 197-213.

Gascard, J.-C., 1991, Open ocean convection and deep water formation revisited in the Mediterranean, Labrador, Greenland, and Weddell Seas, in Chu, P. and J.-C. Gascard, eds., *Deep Convection and Deep Water Formation in the Oceans*, Elsevier, 382 p.

Grenfell, T., D. Cavalieri, J. Comiso, M. Drinkwater, R. Onstott, I. Rubinstein, K. Steffen, and D. Winebrenner, 1992, Chapter 14. considerations for microwave remote sensing of thin sea ice, in Carsey, F., ed., 1992, *Microwave Remote Sensing of Sea Ice*, AGU Monograph 68, 291-301.

Häkkinen, S., 1987, Upwelling at the ice edge: A mechanism for deep water formation?, *J. Geophys. Res., 92*, 5031-5034.

Hollinger, J., J. Pierce and G. Poe, 1990, SSM/I instrument evaluation, *IEEE Trans on Geosci. and Remote Sens., 28,* 781-790.

Hurdle, B., 1986, ed., *The Nordic Seas*, Springer-Verlag, New York, 777p.

Johannessen, O. M., J. A. Johannessen, J. Morison, B. Farrelly, and E. Svendsen, 1983, Oceanographic conditions in the marginal ice zone north of Svalbard in early fall 1979 with and emphasis on mesoscale processes, *J. Geophys. Res., 88,* 2755-2769.

Johannessen, O., 1986, Brief overview of the physical oceanography, in Hurdle, B., ed., *The Nordic Seas*, Springer-Verlag, New York, 103-127.

Johannessen, J. A., O. M. Johannessen, E. Svendsen, R. Shuchman, T. Manley, W. Campbell, E. Josberger, S. Sandven, J. Gascard, T. Olaussen, K. Davidson and J. van Leer, 1987, Mesoscale eddies in the Fram Strait marginal ice zone during the 1983 and 1984 marginal ice zone experiments, *J. Geophys. Res., 92,* 6754-6772.

Johannessen, O. M., W. Campbell, R. Shuchman, S. Sandven, P. Gloersen, J. Johannessen, E, Josberger, and P. Haugen, 1992, Microwave study programs of air-sea-interactive processes in the s seasonal ice zone of the Greenland and Barents Seas, in Carsey, ed. *Microwave Remote Sensing of Sea Ice*, AGU Monograph 68, 261-289.

Johannessen, J. A., R. A. Shuchman, K. Davidson, Ø. Frette, G. Dagranes, O. M. Johannessen, 1993a, Coastal ocean studies with ERS-1 SAR during NORCSEX'91, *Proc. First ERS-1 Symp., Space at the service of our environment*, Cannes, France, ESA-SP-359, 113-117.

Johannessen, O. M., S. Sandven, W. J. Campbell, and R. Shuchman, 1993b, Ice studies in the Barents Sea by the ERS-1 SAR during SIZEX'92, *Proc. First ERS-1 Symp., Space at the service of our e environment*, Cannes, France, ESA-SP-359, 277-282.

Johannessen, O. M., S. Sandven, W. P. Budgell, J A. Johannessen and P. M. Haugen, 1994, Observations and simulations of ice tongues and vortex pairs in the marginal ice zone This issue].

Jones, H. and J. Marshall, 1993, Convection with rotation in a neutral ocean: A study of open ocean deep convection, *J. Phys. Oceanogr., 23*, in press.

Killworth, P., 1979, On chimney formation in the ocean, *J. Phys. Oceanogr.*, 9,. 531-554.

Legg, S. and J. Marshall, 1993, A heaton model of the spreading phase of open ocean convection, *J. Phys. Oceanogr., 23*, in press.

McPhee, M., 1990, Small-scale processes, in Smith, W., ed., *Polar Oceanography, part A Physical Science*, Academic Press, San Diego, 406p.

Moritz, R., 1988, *The ice budget of the Greenland Sea*, Tech. Report, Applied Physics Laboratory of the Univ. of Washington, Seattle, APL-UW TR 8812, 117p.

Roach, A. T., K. Aagaard and F. Carsey, Coupled Ice-Ocean Variability in the Greenland Sea, *Atmosphere-Ocean, 31*, 319-337.

Shott, F., and K. Leaman, 1991, Observations with moored acoustic Doppler profilers in the convection regime in the Gulf of Lions, winter 1987, J. Phys. Oceanogr., 221, 558-574.

Schott, F., M. Visbeck and J. Fischer, 1993, Observations of vertical currents and convection in the central Greenland Sea during the winter of 1988/89, *J. Geophys. Res.*, in press.

Steffen, K. and J. Maslanik, 1988, Comparison of Nimbus 7 Scanning Multichannel Microwave Radiometer radiance and derived sea ice concentrations with Landsat imagery for the North Water area of Baffin Bay, *J. Geophys. Res., 93*, 10769-10781.

Steffen, K, J. Key, D. Cavalieri, J. Comiso, P. Gloersen, K. St. Germain and I. Rubinstein, 1992 in Carsey [ed.] *Microwave remote Sensing of Sea Ice*, AGU Monograph 68, 201-231.

Sutherland, L., R. Shuchman, P. Gloersen, J. Johannessen, and O. Johannessen, 1989, SAR and passive microwave observations of the Odden during MIZEX '87, *IGARSS 1989 Digest*, 1539-1544.

Tucker, W., T. Grenfell, R. Onstott, D. Perovich, A. Gow, R.Shuchman, and L. Sutherland, 1991, Microwave and physical properties of sea ice in the winter marginal ice zone, *J. Geophys. Res., 96*, 4573-4587.

Wadhams, P., 1986, The ice cover, in Hurdle, B., ed., *The Nordic Seas*, Springer-Verlag, New York, 21-87.

Wadhams, P., J. Comiso, E. Prussen, S. Wells, D. Carne, M. Brandon and E. Aldworth, 1993, Active and passive microwave imagery of the Odden ice tongue in the Greenland Sea compared with filed observations, *EOS 74*, 323.

Weeks, W. and S. Ackley, 1986, The growth, structure and properties of sea ice, in Untersteiner, N., ed., *The Geophysics of Sea Ice*, NATO ASI Series B: Physics vol. 146, Plenum Press, New York, 9-164.

Wensnahan, M., T. Grenfell, D. Winebrenner, G. Maykut, 1993, Observations and theoretical studies of microwave emission from thin saline ice, *J. Geophys. Res., 98*, 8531-8546.

Worcester, P., J. Lynch, W. Morawitz, R. Pawlowicz, P. Sutton, B. Cornuelle, O. Johannessen, W. Munk, W. Owens, R. Shuchman, and R. Spindel, 1993, Evolution of the large-scale temperature field in the Greenland Sea during 1988-89 from tomographic measurements, *Geophys. Res. Lett., 20*, 2211-2214.

Corresponding addresses:
Frank D. Carsey
California Institute of Technology, Jet Propulsion Laboratory, Pasadena CA 91109; 818-354-8163, F.Carsey/Omnet, fdc@pacific.jpl.nasa.gov
and
Andrew T. Roach
Applied Physics Laboratory, University of Washington, Seattle WA 98105; 206-685-7911; A.Roach/Omnet

Transport of Biogenic Particulate Matter to Depth Within the Greenland Sea

T. O. Manley
Marine Research Corp., Middlebury, VT
Department of Geology, Middlebury College, Middlebury, VT
and
W. O. Smith, Jr.
Botany Department and Graduate Program in Ecology, University of Tennessee, Knoxville, TN

During a study of the biological-physical interactions within the northern Greenland Sea and Fram Strait during spring, 1989, substantial biogenic material was observed to depths of 750 m. The proximity of significantly elevated isopycnals to the deep fluorescence maxima suggests that the outcropping density surfaces represented paths along which particulate matter was transported. In one specific case, a relict-chimney (observed during a three-dimensional survey near the ice edge) appeared to have transported cooled, saline water of Atlantic origin and its associated biomass to mid-depths along inclined isopycnals. It is postulated that enhanced surface cooling necessary to establish deep isopycnal outcropping within the euphotic zone coupled with the high rate of particulate matter generation in the spring represents the primary mechanism by which the observed biogenic material was vertically transported to depth. The spatial and temporal scales over which biogenic transport occurs range from several km (chimneys) to hundreds of km (northern front of the Greenland Sea) and days to weeks.

1. INTRODUCTION

The removal of atmospheric carbon through the creation of biological particulate material at the ocean's surface as well as its subsequent fate (recycling to the surface or incorporation into sediments) are climatically important issues needed to be addressed. The rate at which this material is transported into the deep ocean, and consequently removed from ocean-atmosphere exchange processes, is a function of both particle size and ocean dynamics. A substantial increase in the vertical transport rate decreases the transit time over which deeper foraging can occur while concurrently minimizing the dissolution effects on the skeletal remains. Even though aggregation and biological packaging provide substantial increases in settling rates over phytoplankton sinking rates by one or two orders of magnitude (i.e., meters to 10s of meters per day; Longhurst and Harrison [1989]), ocean dynamics can accelerate this downward transport. A recent study concluded that fluorescence maxima observed between the euphotic zone and 200 m were a result of biomass being downwardly advected along inclined isopycnals of a frontal boundary observed during the 1988 Coastal Transition Zone Program [Washburn et al., 1991; Brink and Cowles, 1991]. Quantification of the vertical subsidence rate at the front of 25 m/d was obtained from analysis of the temperature, salinity, ^{222}Rn, ^{226}Ra, and O_2 data [Kadko et al., 1991]. In high-latitude regions, the combined effect of spring blooms [Smith et al., 1991] and isopycnal sinking of cooled, potentially brine enriched surface waters (as first proposed by Metcalf [1960] for the replenishment of deep and bottom waters of the Greenland Sea) may provide conditions suitable for largescale biogenic transport to depth.

The processes which produce deep and bottom water on continental shelves are different from those occurring in the deep ocean [Killworth, 1979]. On the shelf, these processes are primarily driven by the creation of brine during ice formation [Foldvik et al., 1985; Midttun, 1985; Jacobs, 1987; Quadfasel et al., 1988]. In the central ocean, however, complex interactions between ice, ocean, and atmosphere are re-

The Polar Oceans and Their Role in Shaping the Global Environment
Geophysical Monograph 85

quired. These open-ocean interactions result in the eventual destruction of the pycnocline, as well as the maintenance of destabilized conditions during active-convection [Gordon, 1978; Killworth, 1979; Martinson et al., 1981]. We report on the role of frontal structure observed along the northern Greenland Sea gyre, and what is interpreted to be a relict open-ocean convective feature (chimney), in controlling both the production and enhancement of the vertical transport of biogenic matter from the euphotic zone of Fram Strait.

2. FRAM STRAIT CIRCULATION

Although complex in its dynamics, circulation within the northern Greenland Sea and Fram Strait (between northeast Greenland and the islands of Svalbard) can be represented as two opposing flow fields (Figure 1a) [Carmack, 1990; Muench, 1990]. First, along the eastern side of the basin is the northerly inflow of warm, saline Atlantic Water (AW) of the West Spitsbergen Current (WSC). The second is a combined southerly flow of sea-ice, cold and relatively fresh Polar Water (PW) and Atlantic Intermediate Water (AIW) (preferred over arctic intermediate water) that exits the Arctic Ocean as the East Greenland Current (EGC). The East Greenland Polar Front (EGPF), represents the boundary between the PW and AW/AIW, and is typically associated with the transition between the open and ice-covered ocean (the marginal ice zone or MIZ). Along its northern path, the WSC has been observed to split into separate filaments. One branch is the Return Atlantic Current (RAC), which flows westward across Fram Strait and eventually becomes associated with the southerly flowing EGC [Paquette et al., 1985; Bourke et al., 1988]. The remaining flow field of the WSC enters the Arctic Ocean as three separate branches controlled by bottom topography [Perkins and Lewis, 1984; Aagaard et al., 1987; Manley et al., 1992]. The dynamic complexity of the region is evidenced by the rich environment of mesoscale activity such as eddies, squirts, jets, dipoles, frontal instability, chimneys, and meandering ice-edges [Hanzlick, 1983; Smith et al., 1984; Häkkinen, 1987; Johannessen et al., 1987; Manley et al., 1987a,b; Gascard et al., 1988; MIZEX Group, 1989; Sandven and Johannessen, 1989].

Compared with other mesoscale features, investigations of chimneys are infrequent. They have been observed in the Weddell Sea [Gordon,1978; Foldvik (reported by Killworth, 1979], the Labrador Sea [Lazier, 1973; Gascard and Clarke, 1983], the Iceland Sea [Scott and Killworth, 1991], the Greenland Sea [Rudels et al., 1989], and Fram Strait [Johannessen et al., 1989; MIZEX Group, 1989; Sandven et al., 1991]. Chimneys that have been studied within the adjacent seas of the Arctic Ocean have not been observed repeatedly through time, and hence their age and evolutionary history remain uncertain. In a relative sense, a majority of these features could be defined as either juvenile (early stages of development) or relict (post active-convection) in that both have elevated isopycnals of only a few hundred meters and weakly stratified central cores. Two chimneys, one reported by the MIZEX Group [1989] and the second by Rudels et al. [1989], were at or close to a mature state (potentially active deep-convection). These features possessed elevated isopycnals in excess of 1000 m and central cores that were virtually isohaline and isothermal. Although the origin and mechanism(s) of formation of the Greenland Sea chimneys are still unknown, T-S properties within the central cores of these features led Johannessen et al. [1991] to suggest that a cyclonic eddy was the most likely pre-conditioning mechanism, whereas Rudels et al. [1989] postulated repetitive surface cooling events.

3. OBSERVATIONS

During the spring of 1989, the distribution of density and particulate matter in the Greenland Sea were determined as part of CEAREX (Coordinated Eastern Arctic Experiment). The field program was designed to gain insight into the coupled physical-biological processes of mesoscale eddies located near the MIZ. From 10 April to 17 May 1989, 212 hydrographic stations and 224 XBT profiles were completed (Figure 1).

The distribution of temperature, salinity and fluorescence were obtained using a Neil Brown CTD equipped with a Sea Tech fluorometer and twelve 10-liter Niskin bottles mounted on a rosette. Samples for chlorophyll-a and other biological parameters [Smith et al., 1991] were collected for discrete analyses. The CTD sensors had both pre- and post-cruise calibrations along with salinity samples taken at the top and bottom of all profiles. Salinity samples were analyzed at sea on a Guideline Autosal. Final processing of temperature and salinity yielded average accuracies of 0.003 °C and 0.006 PSU, with the largest errors being 0.02 PSU in ca. 5% of the data. The Sea Tech fluorometer was calibrated by regressing the chlorophyll values obtained from discrete sample analyses with the fluorescence of values obtained from the same depths. The discrete chlorophyll samples were quantified by filtering known volumes of seawater through 25 mm Whatman GF/F filters (nominal pore size 0.7 μm) under low pressure, extracting the filters in 90% acetone, sonicating for 15 minutes at 0 °C, extracting for 15 minutes more in darkness, and measuring the fluorescence before and after acidification on a Turner Designs Model 10 fluorometer. The fluorometer was calibrated before and after the cruise with commercially purified chlorophyll a (Sigma Chemical). The resulting 803

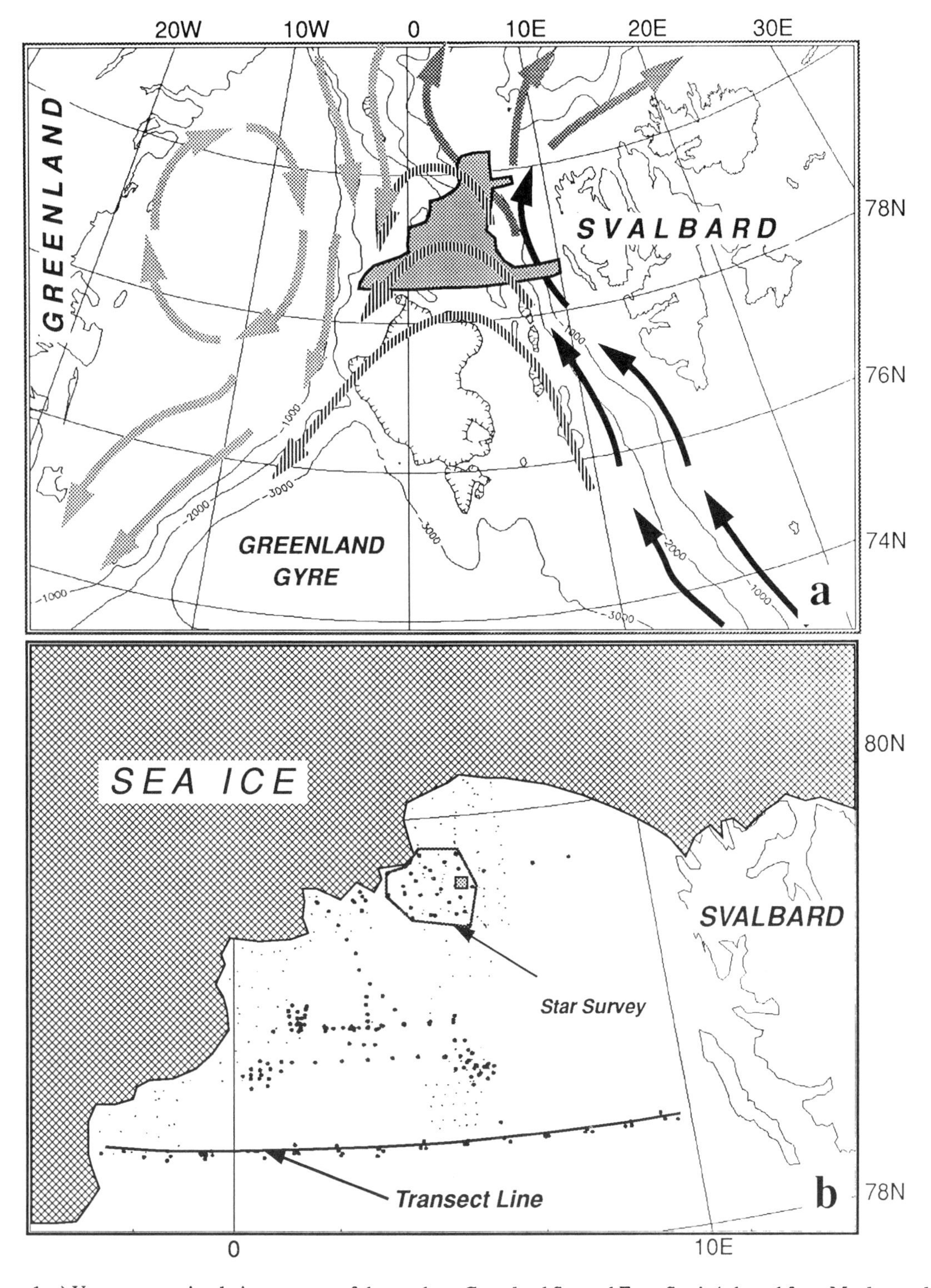

Figure 1. a) Upper ocean circulation patterns of the northern Greenland Sea and Fram Strait (adapted from Manley et al., 1991). Bathymetry is in meters. Darkest to lightest arrows represent the WSC, the branching of the WSC north of Svalbard, and the outflowing PW of the EGC, respectively. Barred arrows denote the RAC filaments. The study region is shown as a stippled pattern. b) Location of the 212 CTD (solid circles) and 254 XBT (small dots) stations taken during the six-week cruise. The region denoted as the star survey contained the relict-chimney (square), while the transect along 78.5° N was occupied approximately every two weeks. The extent of sea ice for the study period is also shown.

fluorescence (fl) and chlorophyll (chl) concentrations were linearly correlated by the equation chl = 0.735*fl + 0.109 (r^2 = 0.6). Estimated accuracy of the sensor was 0.02 µg/l with a threshold of 0.08 µg/l.

4. RESULTS

Significant amounts of fluorescence were found below the euphotic zone (the depth to which 0.1% of the surface photosynthetically active radiation penetrates) to depths of 750 m (e.g., Station 113; Figure 2). Because the mean depth of the euphotic zone was 58 m (minimum and maximum of 28 and 130 m, respectively), the observed deep biomass could not be the result of *in situ* production. To illustrate the large number of stations which had substantial amounts of fluorescence at sub-euphotic depths, the maximum fluorescence value at meter intervals for all stations, as well as the number of stations in which detectable fluorescence values occurred were plotted (Figure 3). These results indicate that the occurrence of substantial fluorescence peaks below the euphotic zone was a regular feature in the study region.

Cross-sections of potential temperature, salinity, potential density, and fluorescence taken from a transect across Fram Strait (occupied from May 12-15) illustrate the large vertical displacement of isotherms and isopycnals (and to a lesser extent, isohalines) which occurred in the region (Figure 4). The 30 - 60 km horizontal scale of the warmer lenses are similar to other mesoscale features observed within this region [Johannessen et al., 1987; Manley et al., 1987a,b; Muench, 1990]. The most developed of these features (centered at station 205) had the largest vertical displacement of isopycnals while the least developed feature was centered at station 208. This basic structure of the 12-15 May transect was also observed during the previous two crossings (10-12 and 25-27, April), yet none of the features could be tracked from one transect to the next with any degree of reliability. Geostrophic calculations along the transect, using a reference level of 700 db, showed oscillating patterns of cyclonic and anticyclonic flow (centered around the axes of the upwarped and downwarped isopycnals, respectively). If these cyclonic features are to be considered isolated (i.e., eddies; chimneys being a sub-division), then they are atypical of the warm-core cyclonic eddies observed within the region. Although they may represent chimneys created from warm-core mesoscale eddies [Johannessen et al., 1991], the number of complex eddy to chimney conversions required to account for the observations appears unlikely. In contrast, an unstable front along the northern Greenland Sea gyre (between the Greenland Sea and the AW/AIW of the RAC) would be capable of producing meanders that could account for the observed hydrographic characteristics and predominance of mesoscale activity. If any meander did become unstable and break away from the front (towards Fram Strait) as an isolated cold-core cyclonic eddy, it could account for the observation of chimney-like features by Johannessen et al. [1991] without the requirement of warm-core eddy conversion. Additionally, a cold-core cyclonic eddy would provide a stronger pre-conditioning field for the creation of a chimney [Killworth, 1979] than that of a warm-core eddy.

The term chimney is defined as a deep-reaching actively-convective mesoscale feature having a homogeneous internal core extending from the surface to the deep or bottom water. This term is not associated with the smaller convective plumes (~1 km) which theoretically operate within the chimney. By classifying feature as chimney-like, it is acknowledged that the feature is not actively convective, yet possesses the characteristics of either a 'juvenile' or 'relict' chimney. 'Juvenile' and 'relict', in turn, reflect the 'pre-conditioning' and 'sinking and spreading' phases defined by the MEDOC Group [1970].

The interior of these cold-core meanders (or chimney-like features) found along the transects possessed only small quantities of fluorescence. In fact, the deep fluorescence maxima were consistently located along the most inclined sections of the front (between the warm- and cold-core meanders). It is proposed that this correlation can be explained as accelerated sinking of biomass from the euphotic zone along isopycnal surfaces (i.e., in a similar fashion as that of Washburn et al. [1991] and Kadko et al. [1991]). Although some of the isopycnals bounding the deeper fluorescence showed no penetration into the euphotic zone at the time of observation, the spatial and temporal variability of the front was probably sufficient to inhibit the proper mapping of all isopycnal surfaces at any given period of time.

A mesoscale feature, believed to be an isolated relict-chimney, was discovered while mapping a surface-defined warm-core eddy located near the ice-edge approximately 160 km north of the transect line ('Star Survey' in Figure 1b). During the 64-hour mapping period from 17-20 April, 29 hydrographic stations were completed within a 40 km x 40 km x 500 m region. Of these stations, only one (station 43), defined the relict-chimney. Potential-temperature (using three-dimensional minimum-tension gridding; Manley and Tallet [1990]) showed the warm-core eddy occupying the northwestern portion of the surveyed region with surface and 100 m temperatures approaching 3.0 °C and 3.5 °C, respectively (Plate 1a). The proximity of the ice edge can be inferred from the surface layer of PW with temperatures less than 0 °C (Plate 1b). Although surface temperatures suggested that the feature was a discrete warm-core eddy, it was part of a more extensive sub-surface filament of AW/AIW.

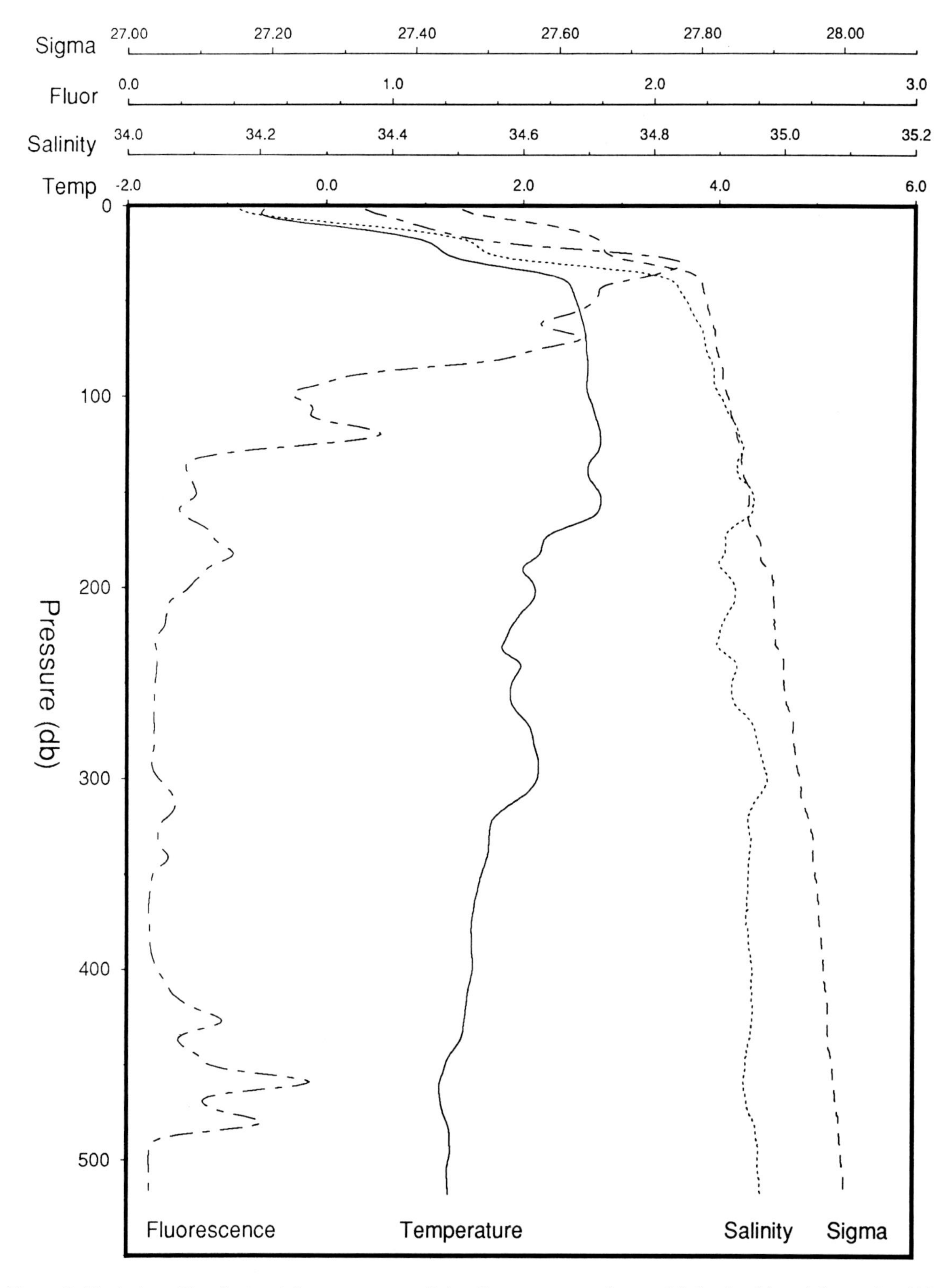

Figure 2. Vertical profile of potential temperature, salinity, fluorescence, and potential density (sigma) for station 113 (May 5, 1989). Note the elevated levels of fluorescence below 400 m.

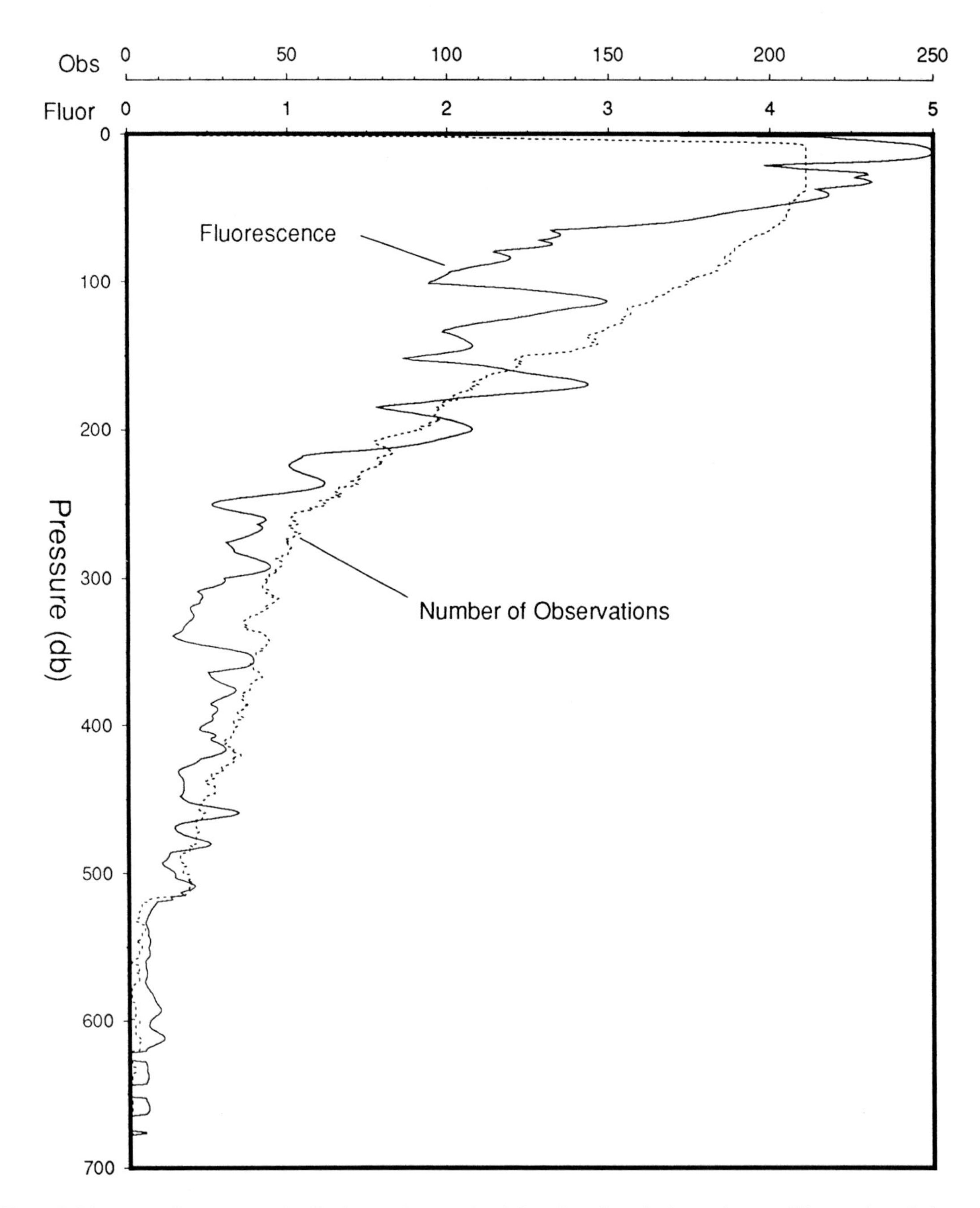

Figure 3. Maximum fluorescence (μg/l) observed at any depth for all stations in the study area. The number of observations where fluorescence exceed the fluorometer's threshold value (0.08 μg/l) is also provided. Fluorescence is indicative of the absolute concentration of pigments within the water column.

The cold and vertically-coherent thermal structure of the relict-chimney was, for the most part, encompassed by AW/AIW. Temperatures less than 0.75 °C define the central part of the relict-chimney's core; however, it is the continuous nature of the 0.75 to 1.5 °C isothermal shells (Plate 1b) that clearly define its structure. The coldest (-1.00 °C to -.75 °C) and least saline waters (34.25 to 34.65 PSU) in the upper 70 m reflect PW that was most likely entrained from the EGC during the initial stages of collapse. The lower portion of the relict-chimney's core consisted of potential temperatures and salinities ranging from -0.25 to 0.25 °C and 34.90 to 34.95 PSU (Plate 2b), respectively. These properties encompass

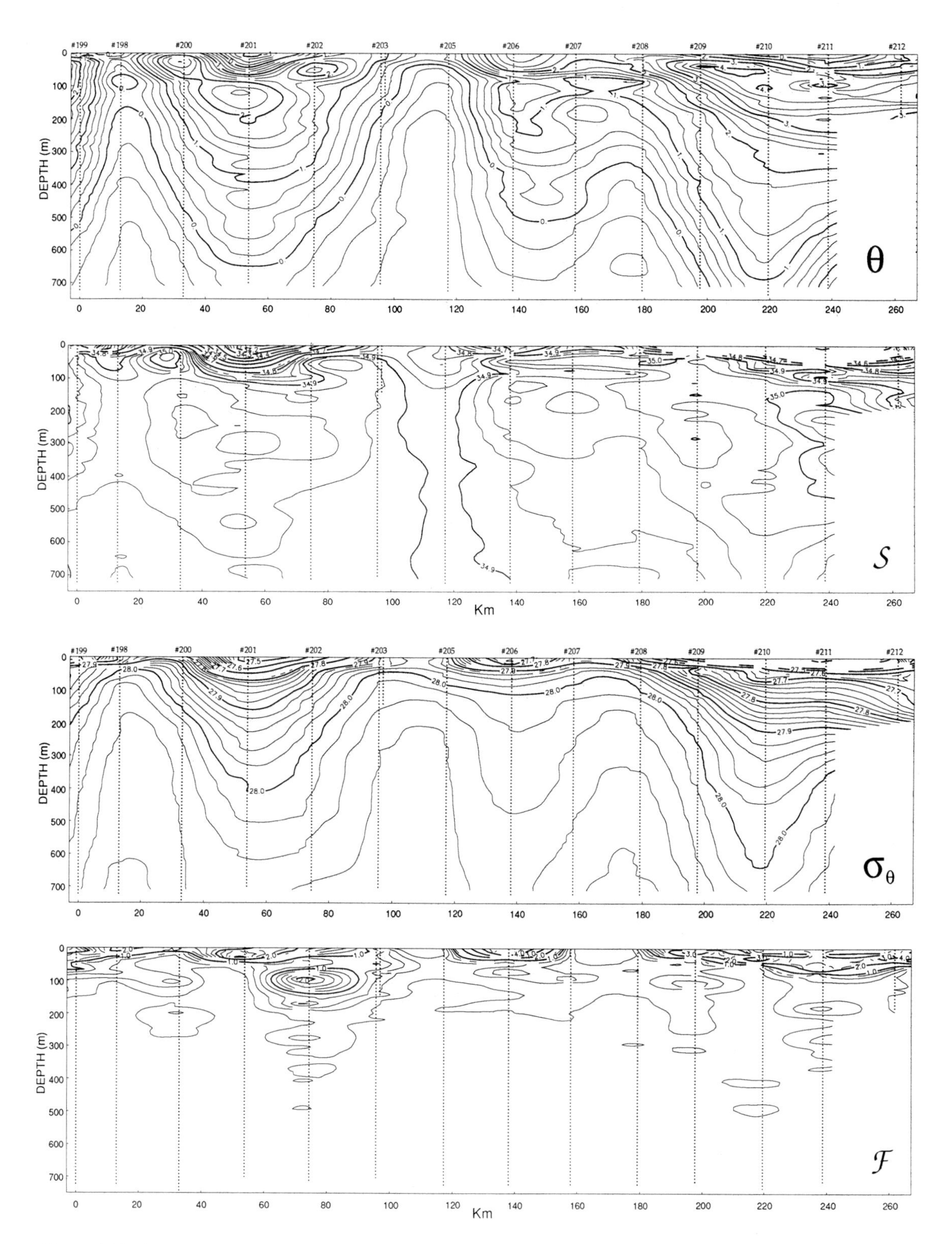

Figure 4. Vertical sections of a) potential temperature (Θ), b) salinity (S), c) potential density (σ_Θ), and d) fluorescence (F) at 78.5° N from the May 15-17, 1989 Fram Strait transect. Vertical lines define hydrographic stations (numbers are provided at the top of the diagram). Contour intervals for potential temperature, salinity, potential density, and fluorescence, are 0.25, 0.025, 0.02, and 0.2, respectively.

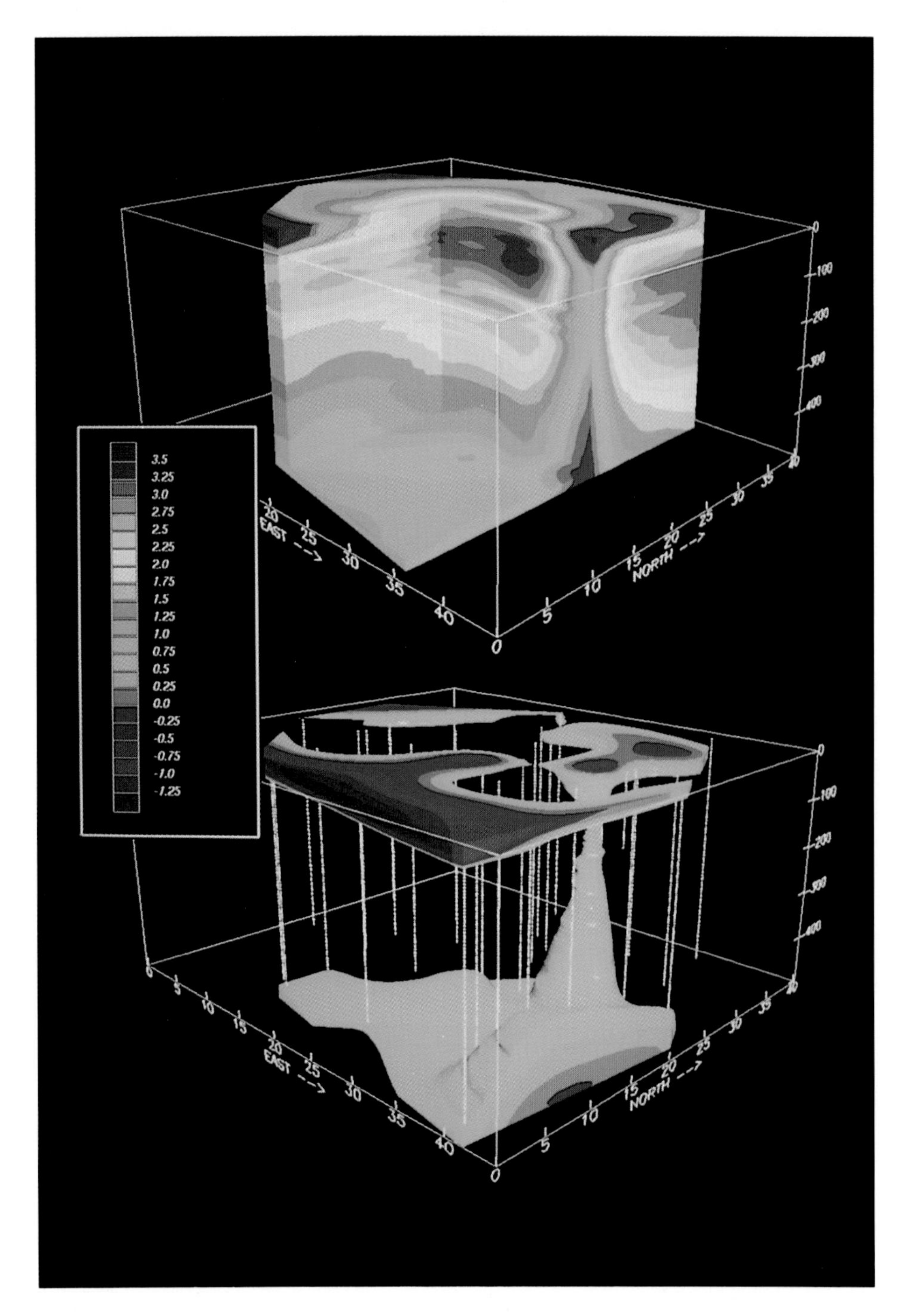

Plate 1. a) Volumetric rendering of the relict-chimney's thermal structure based on data taken during the 'star survey'. The volume is 40 km x 40 km x 500 m. Local coordinate axes have major tick marks every 5 km in the horizontal and 100 m in the vertical. Color key changes are every 0.25 °C. North and east are indicated below the axes. All thermal layers are retained, but model is cut to show the interior structure of the relict-chimney. b) Only temperatures less than 0.75 °C are shown. The relict-chimney's structure is best observed in this manner. Station locations are indicated by vertical lines within the volume.

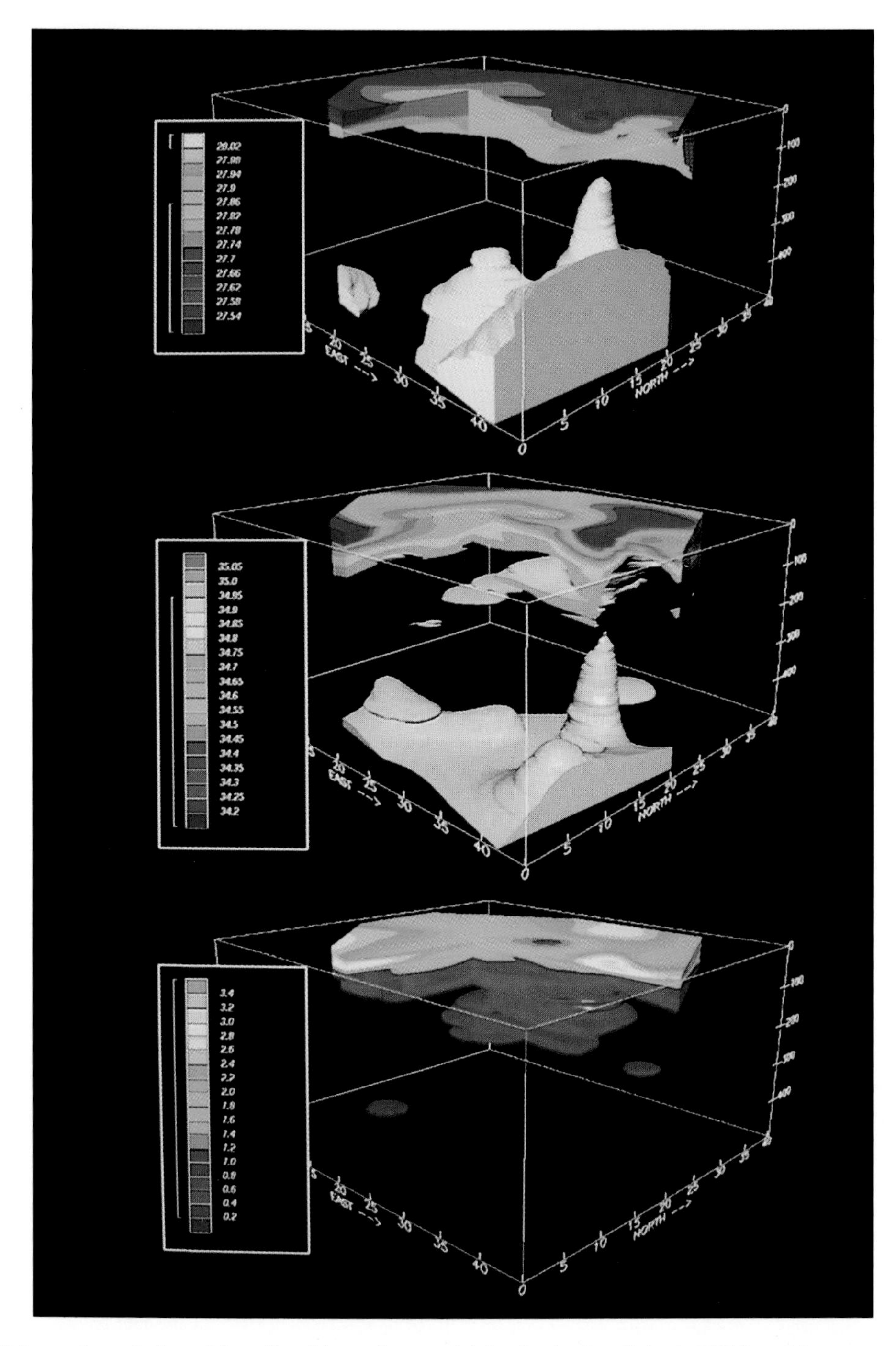

Plate 2. Volumetric rendering of the relict-chimney's potential density (top), salinity (middle), and fluorescence (bottom) based on data taken during the 'star survey'. The format of each image is identical to Plate 1. For potential density, values less than 27.86, which define the near surface waters, and greater than 28.02 (defining the deep water) are shown. For salinity, only values less than 34.95 PSU are displayed. This removes the more saline AW/AIW from the image and provides better view of the upwarped salinity surface. Fluorescence values greater than 0.2 μg/l are displayed. Although biomass shows no chimney-like structure (as do T, S, and potential density), the visual correlation between the inclined westward trend of biomass and the similar trending low-salinity lenses is apparent. For potential density, salinity, and fluorescence, the color steps are 0.04, 0.05 and 0.2, respectively. See text for further details.

the densest component of AIW to that of Norwegian Sea Deep Water (NSDW; which is found around the periphery of the Greenland Sea [Swift, 1986]). The largest elevation of potential density, to within 100 m of the surface, was found within the core of the relict-chimney using the 28.02 iso-surface (Plate 2a). As previously mentioned, the least dense waters (σ_Θ < 27.54) were found directly over the relict-chimney as a capping lens of PW. Maximum surface geostrophic velocities of 23 cm/s (0/500) indicate that the relict-chimney was still dynamically active.

Elevated levels of fluorescence were found at depth along the inclined isopycnals between the relict-chimney and the eddy along their western and northern sectors, respectively (Plate 2c). The highest fluorescence values within the region were associated with the strongest density gradient of the PW cap, and therefore, the most likely site for most rapid phytoplankton growth and accumulation [Smith and Sakshaug, 1990]. Correlated with the deep fluorescence signature and trending along a similar downward-trending northwestern path were discrete lower-salinity (relative to the surrounding water) lenses (Plate 2b). These lenses, observed from 130-290 m, suggest the recent cooling of warmer, saline AW or AIW and mixing with fresher near-surface water prior to subsequent subduction along isopycnals (ranging from 27.88 to 27.98). Based on spatial orientation of the salinity lenses within the density field, an origin can be traced back to within 30-140 m of the sea surface (at the base of the euphotic zone) and in close proximity of the relict-chimney core. This observation, in conjunction with the fact that the isopycnal surfaces of the eddy were only weakly inclined and oppositely tilted to that of the relict-chimney, supports the hypothesis that dynamics related to the relict-chimney, and not the eddy, were the causative factor of the deep salinity and fluorescence lenses.

The discrete nature of salinity and fluorescent lenses suggest that they were formed by either episodic cooling events or continuous cooling with episodic sinking of localized water parcels. The fact that these lenses were observed only in the northern quadrant and not around the complete circumference of the relict-chimney may reinforce issues of complex chimney dynamics [Martinson et al., 1981; Gascard and Clarke, 1983; Rudels et al., 1989; Madec and Crepon, 1991; Chu, 1991a,b], or that the observed complexity was a result of the combined proximal effects of the surface-defined cyclonic eddy (to the west) and the sub-filament of AW/AIW.

5. DISCUSSION

The elevated levels of fluorescence associated with the salinity lenses suggest a unique relationship between chimney dynamics, water mass modification, and removal of biogenic material from the euphotic zone. The lack of appreciable fluorescence in the central (100-500 m) core of the relict-chimney suggests that either a mature convective state was never reached or the biomass was rapidly transported below 500 m due to large vertical velocities expected during active convection [Gascard, personal communication]. Concurrent with active convection, phytoplankton productivity would be near zero in the central core due to a lack of stratification and adequate irradiance to drive net primary production [Sverdrup, 1953]. As such, *additional* deep transport of biomass from the euphotic zone must be dependent on processes other than active convection. For example, particulate matter from a stratified surface layer could be incorporated in the modified surface water during the pre-conditioning phase. Similarly, vertical collapse of the chimney would force a lateral advection of the peripherally located, less-dense near-surface water (and its associated elevated biomass) towards the core. Continued cooling could still produce sinking along intermediate to shallow isopycnals still outcropping within the euphotic zone.

The rates of biomass transfer into the deep ocean can be estimated from knowledge of the photosynthetic responses of the surface and deep-water phytoplankton assemblages [Gallegos et al., 1983]. If an assemblage had been recently exposed to surface irradiance (i.e., within the past 48 hours), then its photosynthetic response would be expected to be similar to that of the surface assemblage. Conversely, if the assemblage had been maintained at low light levels for some time, its photosynthetic response would reflect adaptation to a lower irradiance (e.g., large photoinhibition, elevated initial (light-limited) rates of photosynthesis, and slightly lowered maximal (light-saturated) rates of photosynthesis; Gallegos et al. [1983]; Smith and Sakshaug [1990]). Since Cota et al. [1993] observed that the mean photosynthetic response of assemblages for this cruise was similar throughout the water column, we would expect that the surface phytoplankton assemblage had been transported to depth within the past two weeks.

The volumes of the discrete salinity lenses shown (Plate 2b) were calculated to estimate the quantity of water that was removed from the surface. The analysis shows that the salinity-defined lenses (located between 130-290 m and with salinities between 34.85-34.96 PSU) comprised a volume of 15.1 km^3. This calculation was limited to the northern section of the volume to eliminate interference from the warm-core eddy. A slightly larger volume of 18.7 km^3 was calculated using fluorescence criteria of 0.2 to 1.0 μg/l over the same depth range and sector. If the 5.95 metric tons of pigment within the 18.7 km^3 of sub-euphotic fluorescence maxima is converted to carbon, using the mean carbon/chlorophyll ratio observed within this depth range (196;

Smith, unpublished), then approximately 1,200 metric tons of carbon were removed from the surface by convective processes. It should be noted that the single deep-fluorescence peak below 290 m (centered at 430 m) was not considered in the volumetric determination. Its relatively small volume when compared to those at the intermediate depths, would not appreciably influence the volumetric or particulate carbon calculations.

Other explanations for the creation of deep fluorescence layers, such as passive sinking of individual particles, accelerated sinking of aggregates, resuspension of bottom sediments with subsequent entrainment into the water column, and lateral advection from a distant region were also investigated. The depth of the water column (2-3 km) within central Fram Strait, combined with the maximum depth of observation (750 m), precludes resuspension of sediments as a realistic hypothesis unless the water column had been in recent contact with the Norwegian, Spitsbergen or Barents Sea continental shelves. Analysis of biogenic and lithogenic silica concentrations within the upper 150 m indicate that particulate resuspension from the bordering continental shelves was insignificant [D.M. Nelson, personal communication]. Furthermore, there was no correlation between high concentrations of lithogenic silica and fluorescence as might be expected in waters influenced by sediment resuspension events.

Passive sinking of individual particles is also an unlikely cause of the observed deep fluorescence maxima. Sinking rates for individual phytoplankton cells are much less than 1 m/d [Culver and Smith, 1989] and cannot transport particles to depth rapidly enough to maintain the observed photosynthetic capacities [Cota et al., 1993]. Particle flux from the surface layer via passive sinking would also tend to produce a more uniform biomass distribution with depth, rather than discrete layers that range from 5 to 100 m thick. In addition, the primary species of phytoplankton observed within our study area, *Phaeocystis pouchetii* [Smith et al., 1991], occurred as colonies rather than solitary cells. Although such colonies would have higher settling rates, it is unlikely that these rates would be an order of magnitude greater or more. Aggregate formation and subsequent accelerated sinking (as high as 100 m/d; Smetacek [1985]), has been suggested to be an important mechanism by which particles can be transported to depth within short periods of time [Jackson, 1991]. Thus, aggregate formation is a possible mechanism by which biogenic particles might be transported to depth within the required time frame. Although rates of 100 m/d would be adequate to account for the appearance of material at 200-400 m, the net sinking rate from a *Phaeocystis pouchetii* bloom in the Barents Sea was estimated to be a maximum of 3.5 m/d (calculated from the mean particulate matter concentration divided by the measured flux rate; Wassmann et al., [1990]). Sinking rates also decreased markedly with increasing depth as a result of zooplankton grazing on aggregates, disintegration and remineralization. Even though aggregate sinking rates of *Phaeocystis pouchetii* and their environmental controls are not well known, there is no reason to expect them to be dramatically different from those (1-2 m/d) found within other productive regions [Bienfang, 1981; Smith and Barber, 1987].

By using the maximal sinking rate (3.5 m/d) of a *Phaeocystis* colony in conjunction with a 14 day settling period (starting at the mean depth of the euphotic zone; 58 m), only fluorescence observations less than 120 m could be a result of passive sinking. Similarly, elevated fluorescence at 500 m would require a rate of 32 m/d. Since this 2-week period represents a conservative estimate of biomass removal from the photic zone, the 32 m/d decent rate reflects a lower limit of the convective process, particularly when considering that the initial depth was not assumed to be at the surface. A potential upper limit using a 1 week duration and a starting depth of 0 m would result in a net sinking rate of approximately 70 m/d. These values are bounded by those of Kadko et al. [1991] (25 m/d) and Smetacek [1985] (100 m/d).

Hydrodynamic measurements of vertical velocities associated with chimneys were observed from 0.5 mm/s to 10 cm/s [Gascard, 1991]. The value of 10 cm/s was associated with small scale (~1 km) convective cells within the upper 1 km of the surface during active convection. This may also explain why the upper 1 km of a chimney's core would be devoid of biomass shortly after the onset of convection. The lower values of 0.5 - 2 mm/s were more typical of baroclinic circulation associated with frontal boundaries. Converting the estimated descent rates of 32 and 70 m/d yields values of 0.4 and 0.8 mm/s, respectively. Although the 0.4 mm/s decent rate is at the lower end of the calculated ranges, it should be reiterated that this value is conservative. The more realistic 70 m/d (0.8 mm/s) descent rate falls within the expected range of values and tends to support the conclusion of frontal/chimney forced removal of biomass from the euphotic zone.

Lateral advection also does not appear to be important in the production of deep fluorescence maxima. The average horizontal velocity of 5 cm/s (~5 km/d) within the deep basin would be too slow to transport particles into the central and western reaches of the Greenland Sea within a two week period. Although the WSC has higher velocities, they are primarily confined to the shelf/slope region and do not significantly affect the interior circulation of Fram Strait. The EGC may be a source region for the deep biomass layers located on the western edge of the strait; however, its mean south-southwesterly flow and westward sloping isopycnals suggest that the EGC/EGPF should not be considered a source of biomass for the central deep-basin of Fram Strait. Furthermore, the EGC has very low levels of fluorescence [Smith, unpublished]. On a smaller scale, Häkkinen's [1988] model of ice-edge chimney formation (which has not, as yet, been confirmed by direct observations) may have a potential to transport biomass located near the EGPF into the central basin. Thus, the lateral advection of biogenic particulate matter into the central and western portions of the Greenland Sea from distant sources can not explain the observations.

Meteorological observations obtained from the ship confirm the presence of relatively high ocean-atmosphere heat fluxes necessary for surface water modification. By assuming an average sea surface temperature of 2 °C, and that shipboard measurements were representative of synoptic conditions within Fram Strait, sensible heat fluxes from 200 to 600 W/m^2 could be obtained. These fluxes, only associated with polar outbreaks of cold air (-10 to -18 °C) having wind velocities from 8 to 16 m/s and lasting for more than 2 days, were typically observed within 7 days prior to the Fram Strait transects and the 'star survey'. Only the third transect was preceded by a longer time interval of 14 days. Since heat losses are only partially controlled by sensible heat [Bunker, 1976, 1980; Bunker and Worthington, 1976], the addition of latent heat losses could increase the total heat flux to 400 - 1200 W/m^2. This range of ocean-atmosphere fluxes would be sufficient to drive vertical convection of the surface waters while still being well within the requisite 14 day time frame imposed by the photosynthetic response criteria.

Therefore, the deep-reaching cyclonic nature of the Greenland Sea gyre offers a temporally consistent 'pre-conditioned' structure of properly inclined isopycnal surfaces (sloping downwards into Fram Strait along its northern border) required for accelerated biogenic transport. Although, largescale atmospheric forcing during the late fall, winter, and early spring increase the dynamic range of isopycnal surfaces intersecting the euphotic zone of the Greenland Sea frontal zone, its large areal extent and temporal permanency augment the potential for significant carbon input into the deep ocean. The instability of the northern Greenland Sea front is also of interest, in that such a boundary could produce cold-core cyclonic eddies possessing strongly upwarped isopycnal surfaces relative to the background mean (i.e., chimney-like structures). These eddies may, in turn, continue the process of deep biomass transport as they move northward into central Fram Strait, while concurrently providing a 'pre-conditioned' structure that is theoretically required for the development of chimneys (given proper atmospheric forcing).

The effectiveness of chimneys in enhancing the vertical flux of biogenic material, as well as the depth to which the biogenic material is transported, is not only a function of frontal dynamics, the number of chimneys formed, and their developmental stage (e.g., juvenile, mature), but also the season in which they are created. Deep convection dynamics, from either frontal activity or chimneys, and phytoplankton growth are processes that may have their strongest interaction during the spring, when primary productivity and biomass accumulation are maximal. Deep convective processes would be expected to be greatest during winter due to large atmospheric heat fluxes, whereas biomass generation is maximal during periods of stratification of previously mixed water columns (i.e., those which are nutrient replete). Hence, the largest amounts of biogenic material which can be removed by convective processes would be during spring to early summer. Although the degree of oceanic heat loss decreases during this time period, thereby resulting in shallower convective penetration, the removal of biomass to levels of only a few hundred meters may effectively remove carbon from the upper ocean/atmospheric system.

6. CONCLUSIONS

During the CEAREX cruise in central Fram Strait, a consistent relationship between deep fluorescence layers and the isopycnal structure was observed within the upper 750 m of the ocean. These layers were located within strongly upwarped density fields having typical horizontal scales of 5 to 30 km. Lateral advection, individual and aggregate particle sinking, or resuspension of sediments cannot account for the observed deep biomass. This suggests that processes associated with frontal structure along the northern boundary of the Greenland Sea gyre and deep-convective chimneys may play an active role in vertical transport of water and particulate matter.

The observation of a relict-chimney during a survey at the marginal ice zone provided a unique view of its structure. Temperature and salinity fields within the relict-chimney indicated that it was no longer undergoing active convection and probably was in the process of vertical collapse. Discrete layers of biomass, associated with lenses of high salinity water, were observed along inclined isopycnal surfaces from 130-290 m at the periphery of the relict-chimney. These lenses were indicative of one or more convective events that transported cooled, slightly fresher AW/AIW, along with its particulate constituents, to depth. The volume of the subducted salinity-lenses was 15.1 km^3 while the associated carbon transport was calculated to be 1,200 metric tons.

The depth to which biomass is transported is primarily dependent on the depth from which isopycnals are displaced. For example, isopycnals located on the peripheral rim of a chimney originated from a shallower depth than do those that are proximal to the central core. The amount of biomass that is removed to depth is, however, dependent on the amount of primary production and the period over which vertical convection remains active. It is believed that heightened frontal structure of the northern Greenland Sea gyre, as well as chimneys created during the spring, provide the greatest removal of biogenic material to the deep ocean, even though these springtime processes may not provide the deepest convective penetration. Although speculative, it is felt that the frontal region of the northern Greenland Sea gyre, due to its temporal and spatial consistency, may represent a larger sink for biomass than those of isolated chimneys or chimney-like features.

Acknowledgments. This work was funded by the Office of Naval Research under contracts N00014-92-C-0074 (T. O. Manley) and N00014-87-K-0022 (W. O. Smith). We would like to extend our appreciation to the three anonymous reviewers who provided valuable comments on this paper. The two- and three-dimensional visualization of the scattered data was accomplished using the *EarthVisions* software of Dynamic Graphics, Inc., Alameda, CA.

REFERENCES

Aagaard, K., A. Foldvik, and S. R. Hillman, The West Spitsbergen Current: disposition and water mass transformation, *J. Geophys. Res., 92 (C4)*, 3778-3784, 1987.

Bienfang, P. K., Sinking rate dynamics of Cricosphaera Carterae Braarud. 1. Effects of growth rate, limiting substrate, and diurnal variations in steady-state populations, *J. Exp. Mar. Ecol., 49*, 217-233, 1981.

Bourke, R. H., A. M. Weigel, and R. G. Paquette, The westward turning branch of the West Spitsbergen Current, *J. Geophys. Res., 93 (C11)*, 14065-14077, 1988.

Brink, K. H. and T. J. Cowles, The Coastal Transition Zone, J. Geophys. Res., 96, 14637-14647, 1991.

Bunker, A. F., Computations of surface energy flux and air-sea interaction cycles of the North Atlantic Ocean, *Mon. Weather Rev., 104*, 1122-1140, 1976.

Bunker, A. F., Trends of variables and energy fluxes of the Atlantic Ocean from 1948 to 1972, *Mon. Weather Rev., 108*, 720-732, 1980.

Bunker, A. F. and L. V. Worthington, Energy exchange charts of the North Atlantic Ocean, *Bull. Am. Meteorol. Soc., 57*, 670-678, 1976.

Carmack, E. C., Large-scale physical oceanography of polar oceans. In *Polar Oceanography - Part A* (W. O. Smith, Jr., ed.), Academic Press, Inc., 406 pp, 1990.

Chu, P. C., Geophysics of deep convection and deep water formation in oceans, in *Deep Convection and Deep Water Formation* (P. C. Chu and J. C. Gascard, eds.), Elsevier Oceanography Series, 3-16, 1991a.

Chu, P. C., Vertical cells driven by vortices - a possible mechanism for the preconditioning of open-ocean deep convection, in *Deep Convection and Deep Water Formation* (P. C. Chu and J. C. Gascard, eds.), Elsevier Oceanography Series, 267-282, 1991b.

Cota, G. F., W. Smith, Jr., and B. G. Mitchell, Photosynthesis of *Phaeocystis* in the Greenland Sea, *Lim. and Ocean.*, in press, 1993.

Culver, M. E. & W. O. Smith, Jr., Effects of environmental variation on sinking rates of marine phytoplankton. *J. Phycol., 25*: 262-270, 1989

Foldvik, A., T. K. Kvinge, and T. Torresen, Bottom current near the continental shelf break in the Weddell Sea, *Antarc. Res. Ser., 43*, 21-24, 1985.

Gallegos, C. L., T. Platt, W. G. Harrison & B. Irwin, Photosynthetic parameters of arctic marine phytoplankton: vertical variations and time scales of adaptation, *Limnol. Oceanogr. 28:* 698-708, 1983.

Gascard, J. C., and R. A. Clarke, The formation of Labrador Sea water. Part II: Mesoscale and smaller-scale processes, *J. Phys. Oceanogr., 13*, 1779-1797, 1983.

Gascard, J. C., C. Kergomard, P. Jeannin and M. Fily, Diagnostic study of the Fram Strait MIZ and circulation during summer from MIZEX 83-84 lagrangian observations, *J. Geophys. Res., 93(C4)*, 3613-3641, 1988.

Gascard, J. C., Open ocean convection and deep water formation revisited in the mediterranean, Labrador, Greenland and Weddell Seas, in *Deep Convection and Deep Water Formation* (P. C. Chu and J. C. Gascard, eds.), Elsevier Oceanography Series, 157-181, 1991.

Gordon, A. L., Deep Antarctic convection west of Maud Rise, *J. Phys. Ocean., 8*, 600-612, 1978.

Häkkinen, S., A coupled dynamic-thermodynamic model of an ice-ocean system in the marginal ice-zone, *J. Geophys. Res., 92*, 9469-9478, 1987.

Häkkinen, S., A note on deep water formation via a "chimney" formation in the ice edge regions, *J. Geophys. Res., 93*, 8279-8282, 1988.

Häkkinen, S. and D. J. Cavalieri, A study of oceanic surface heat fluxes in the Greenland, Norwegian, and Barents Seas, *J. Geophys. Res., 94*, 6146-6157, 1989.

Hanzlick, D. J., The West Spitsbergen Current: transport, forcing, and variability, Ph.D. thesis, Univ. of Washington, Seattle, 1983.

Jackson, G., A model of formation of marine algol flocs by physical coagulation processes, in press, *Deep-Sea Res.*, 1991.

Jacobs, S. S., Injecting Ice-Shelf Water and air into the deep Antarctic Oceans, *Nature, 321*, 196-197, 1987.

Johannessen, J. A., O. M. Johannessen, E. Svendsen, R. Shuchman, T. O. Manley, W. Campbell, E. Josberger, S. Sandven, J. C. Gascard, T. Olaussen, K. Davidson, and J. Van Leer, Mesoscale eddies in the Fram Strait marginal ice zone during MIZEX 1983 and 1984, *J.Geophys. Res., 92(C7)*, 6754-6772, 1987.

Johannessen, J. A., O. M. Johannessen and S. Sandven, Chimneys in the marginal ice zone, In *Proceedings from the Workshop on the Regional and Mesoscale Modelling of Ice Covered Oceans*, Bergen, Norway, 23-27 October, 1989.

Johannessen, O. M., S. Sandven, and J. A. Johannessen, Eddy-related winter convection in the Boreas Basin, in *Deep Convection and Deep Water Formation* (P. C. Chu and J. C. Gascard, eds.), Elsevier Oceanography Series, 87-105, 1991.

Kadko, D. C., L. Washburn, and B. H. Jones, Evidence of subduction within cold filaments of the northern california coastal transition zone, *J. Geophys. Res.*, 96(C8), 14909-14926, 1991.

Killworth, P. D., On "Chimney" formation in the ocean, *J. Phys. Oceanogr, 9*, 531-554, 1979.

Lazier, J., The renewal of Labrador Sea water, *Deep-Sea Res., 20*, 341-353, 1973.

Longhurst, A. R. and W. G. Harrison, The biological pump: profiles of plankton production and consumption in the upper ocean, *Prog. Oceanogr, 22*, 47-123, 1989.

Madec, G. and M. Crepon, Thermohaline-driven deep water formation in the northwestern mediterranean sea, in *Deep Convection and Deep Water Formation* (P. C. Chu and J. C. Gascard, eds.), Elsevier Oceanography Series, 241-265, 1991.

Manley, T. O., J. C. Gascard, J. Villanueva, P. F. Jeannin, K. L. Hunkins and J. Van Leer, Mesoscale variability within the region of the Yermak Plateau and East Greenland Polar Front, *Science, 236*, 432-434, 1987a.

Manley, T. O., K. L. Hunkins and R. D. Muench, Current regimes across the East Greenland Polar Front at 78'40' north latitude during summer 1984, *J. Geophys. Res., 92*, 6741-6753, 1987b.

Manley, T. O. and J. A. Tallet, Volumetric visualization: an effective use of GIS technology in the field of oceanography, *Oceanography, 3(1)*, 23-29, 1990.

Manley, T. O., R. H. Bourke and K. L. Hunkins, Near-surface circulation over the Yermak Plateau in northern Fram Strait, in press, *J. Marine Sys.*, 3, 107-125, 1992.

Martinson, D. G., P. D. Killworth, and A. L. Gordon, A convective model for the Weddell Polynya, *J. Phys. Oceanogr., 11*, 466-488, 1981.

MEDOC Group, Observation of deep water formation in the Mediterranean Sea, *Nature, 227*, 1037-1040, 1970.

Metcalf, W. G., A note on water movement in the Greenland-Norwegian Sea, *Deep-Sea Res., 8(3)*, 190-200, 1960.

MIZEX Group, MIZEX East 1987: The winter marginal ice zone experiment in Fram Strait/Greenland Sea, *EOS, Trans. Am. Geophys. Union, 70(17)*, 1989.

Midttun, L., Formation of dense bottom water in the Barents Sea, *Deep-Sea Res., 32, 1233-1241, 1985.*

Muench, R. D., Mesoscale phenomena in the polar oceans in *Polar Oceanography - Part A* (W. O. Smith, Jr., ed.), Academic Press, Inc., 406 pp, 1990.

Nelson, D. M., W. O. Smith, Jr., L. I. Gordon, and B. A. Huber, Spring distributions of density, nutrients, and phytoplankton biomass in the ice edge zone of the Weddell-Scotia Sea, *J. Geophys. Res., 92(C7)*, 7181--7190, 1987

Paquette, R. G., R. Bourke, J. F. Newton and W. Perdue, The East Greenland Polar Front in autumn, *J. Geophys. Res., 90*, 4866-4882, 1985.

Perkin, R. G. and E. L. Lewis, Mixing in the West Spitsbergen Current, *J. Phys. Oceanog., 14*, 1315-1325, 1984.

Quadfasel, D., B. Rudels, and K. Kurz, Outflow of dense water from a Svalbard fjord into Fram Strait, *Deep-Sea Res., 35*, 1143-1150, 1988.

Rudels, B., D. Quadfasel, H. Friedrich and M.N. Houssais, Greenland Sea convection in the winter of 1987-1988, *J. Geophys. Res., 94(C3)*, 3223-3227, 1989.

Sandven, S. and O. M. Johannessen, Jets and vortex pairs in the marginal ice zone, In *Proceedings from the Workshop on the Regional and Mesoscale Modelling of Ice Covered Oceans*, Bergen, Norway, 23-27 October, 1989.

Sandven, S., O. M. Johannessen, and J. A. Johannessen, Mesoscale eddies and chimneys in the marginal ice zone, *J. Marine Systems, 2, 195-208*, 1991.

Scott, J. C. and P. D. Killworth, Upper ocean structures in the south-western Iceland Sea - a preliminary report, in *Deep Convection and Deep Water Formation* (P. C. Chu and J. C. Gascard, eds.), Elsevier Oceanography Series, 107-121, 1991.

Smetacek, V. S., Role of sinking in diatom life-history cycles: ecological, evolutionary, and geological significance, *Mar. Bio., 84*, 239-251, 1985.

Smith, D. C.,IV, J. H. Morison, J. A. Johannessen, and N. Untersteiner, Topographic generation of an eddy at the edge of the East Greenland Current, *J. Geophys. Res., 89(C5)*, 8205-8208, 1984.

Smith, W. O., Jr. and R. T. Barber, The influence of horizontal and vertical displacements on phytoplankton assemblages in tropical upwelling systems, *Oceanologica Acta, 6:* 137-144, 1987.

Smith, W. O., Jr. and E. Sakshaug, Polar phytoplankton. In: *Polar Oceanography, Part B* (W.O. Smith, Jr., ed.), Academic Press, San Diego. pp. 477-525, 1990.

Smith, W. O., Jr., L. A. Codispoti, D. M. Nelson, T. Manley, E.J. Buskey, H. J. Niebauer, and G. Cota, Importance of Phaeocystis blooms in the high-latitude ocean carbon cycle, *Nature, 352*, 514-516, 1991.

Sverdrup, H. U., On conditions for the vernal blooming of phytoplankton, *J. Cons., Cons. Intl. Explor. Mer, 18*, 287-295, 1953.

Swift, J. H., The arctic waters, in The Nordic Seas, B. G. Hurdle, Ed., Chapter 5, Springer-Verlag, 128-153, 1986.

Washburn, L., D. C. Kadko, B. H. Jones, T. Hayward, P. M. Kosro, T. P. Stanton, S. Ramp, and T. Cowles, Water mass subduction and the transport of phytoplankton in a coastal upwelling system, *J. Geophys. Res.*, 96(C8), 14927-14946, 1991.

Wassmann, P., M. Vernet, B. G. Mitchell, and F. Rey, Mass sedimentation of *Phaeocystis pouchetii* in the Barents Sea, Marine Ecology Prog. Series, 66, 183-195, 1990.

Modelling of Deep-Sea Gravity Currents Using an Integrated Plume Model

Guttorm Alendal, Helge Drange and Peter M. Haugan

Nansen Environmental and Remote Sensing Center, Bergen, Norway

Abstract

An integrated plume model is used to describe large scale gravity currents in the ocean. The model describes competing effects of (negative) buoyancy, friction, entrainment and Coriolis force, as well as a pressure term due to variable plume thickness, on the flux, speed and flow direction of the plume. Equations for conservation of salt and internal energy (temperature) and a full equation of state for seawater is included in the model.

The entrainment of ambient water is parameterized with support in empirical data, and a drag coefficient consistent with the entrainment is introduced.

The model is tested against the overflow through the Denmark Strait, the flow down the Weddell Sea continental slope, and the outflow of saline water through the Gibraltar Strait and from the Spencer Gulf, Australia. The former gain an extra driving mechanism due to the thermobaric effect, while in the two latter cases the initial density difference is so large that this effect is not essential.

Order of magnitude fit with measurements requires drag coefficient between 0.01 and 0.1. Conditions susceptible to meander behaviour and a singularity arising from the pressure dependency on the current thickness variations are briefly discussed.

Introduction

Buoyancy is the driving force for atmospheric and oceanic phenomena such as snow avalanches, volcanic eruption columns, chimneys, deep water formation, deep sea gravity currents, turbidity currents and thermal vents [Rudnicki and Elderfield, 1992; Simpson, 1987; Turner, 1973; Woods, 1988]. Since the pioneering work of Morton and co-workers [Morton *et al.*, 1956], it has been shown that a set of simple conservation equations can be used to describe the main dynamics of the above mentioned features. This paper elaborates on the dynamics of deep-sea gravity currents, and a dynamic model is derived that represents a slight extension to the gravity current models set up by Smith [1975] and Killworth [1977]. The model also represents an extension of the model used by Speer and Tziperman [1990] since we are treating the salinity and temperature explicitly and we use Richardson number dependent entrainment parameterization. Understanding the dynamics of such currents is also important in view of shallow injection of CO_2 enriched water in the ocean [Haugan and Drange, 1992; Drange *et al.*, 1993].

The Polar Oceans and Their Role in Shaping the Global Environment
Geophysical Monograph 85

The model is developed by integrating the local momentum and continuity equations over a cross-section of the current. This means that interior structures are averaged out and we are describing the currents overall, or bulk, behaviour. The model is steady state excluding time varying phenomena such as upwelling and tidal currents and the influence such events might have. First we give a description of the model introducing a drag coefficient consistent with the Richardson number dependent entrainment parameter. A thorough derivation of the model is given in the appendix. The occurrence of a singularity in the model originating from the inclusion of effects from variation in current thickness is briefly discussed.

We have used measurements from four different sites to test the model: the overflow through the Denmark Strait between Iceland and Greenland, the flow down the Weddell Sea continental slope, the outflow from the Mediterranean through the Gibraltar Strait and the outflow from Spencer Gulf, Australia. The two former are important sites for formation of deep-sea bottom water [Killworth, 1983], and the density difference is mainly caused by the current water being colder than the environment. This gives rise to the thermobaric effect, i.e., cold water has higher compressibility than warmer water. For the two latter currents the excess density is set up by higher salinity in the current relative to the environment.

The model runs are performed as a sensitivity study of the

drag coefficient, since this is the most sensitive parameter. The testing is rough in the sense that we are not performing detailed comparisons of the solutions against measurements. The events are time dependent and the measurements sparse in time and space. Therefore only order of magnitude comparisons with our simple steady state model is justified.

THE INTEGRATED PLUME MODEL

The Physical Configuration

The current is assumed to move down an inclined bottom and we distinguish between three different coordinate systems as shown in Fig. 1. The (x, y, z) system has z in the vertical direction and the x- and y-directions in the horizontal plane, while the (x', y', z') system has z' perpendicular to the plane and with x' and y' directions lying in the inclined plane. The x- and x'-directions are similar giving that y- and y'- directions forms an angle θ with each other. We then have the following relations between the marked and the unmarked co-ordinate systems:

$$\mathbf{i}' = \mathbf{i},\ \mathbf{j}' = \cos\theta\,\mathbf{j} - \sin\theta\,\mathbf{k},\ \mathbf{k}' = \sin\theta\,\mathbf{j} + \cos\theta\,\mathbf{k}, \tag{1}$$

where $\mathbf{i}$, $\mathbf{j}$ and $\mathbf{k}$ are the unit vectors in x-, y-, and z-directions, respectively, and $\mathbf{i}'$, $\mathbf{j}'$ and $\mathbf{k}'$ are the corresponding unit vectors in the (x', y', z') system.

The model is developed in the curvi-linear coordinate system (ξ, η, z') which has ξ in the along-stream direction and η perpendicular to both the ξ and the z' direction. The η direction is to the left of the current looking downstream. β is the angle between the along-stream direction and the x-direction and this relates the two coordinate systems fixed to the slope as follows:

$$\mathbf{e}_\xi = \cos\beta\,\mathbf{i} + \sin\beta\,\mathbf{j}', \qquad \mathbf{e}_\eta = -\sin\beta\,\mathbf{i} + \cos\beta\,\mathbf{j}', \tag{2}$$

where $\mathbf{e}_\xi$ is the unit vector in along-stream direction and $\mathbf{e}_\eta$ the unit vector in η direction. Together with $\mathbf{k}'$ these vectors form an orthogonal span of the 3D space. From equations (1) and (2) it follows that

$$\mathbf{k} = \cos\theta\,\mathbf{k}' - \sin\theta\sin\beta\,\mathbf{e}_\xi - \sin\theta\cos\beta\,\mathbf{e}_\eta. \tag{3}$$

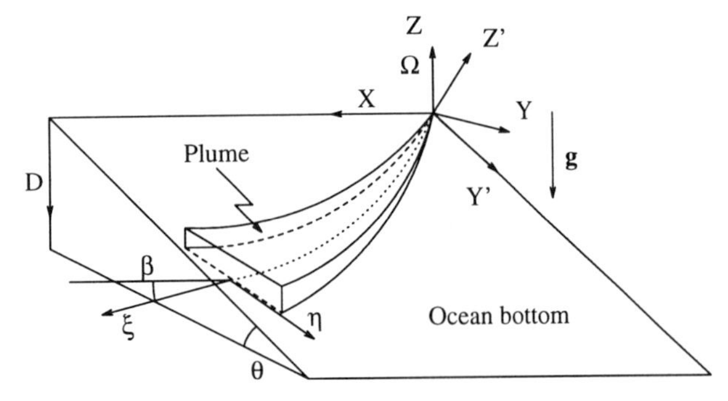

Fig. 1. The physical configuration. Defining the coordinate systems.

and since the (ξ, η, z') system is accelerated relative to the bottom fixed coordinate systems, the fictive centrifugal force

$$\frac{\partial}{\partial\xi}\mathbf{e}_\xi = \frac{\partial}{\partial\xi}\left(\cos\beta\mathbf{i} + \sin\beta\mathbf{j}'\right) = \left(\frac{\partial}{\partial\xi}\beta\right)\mathbf{e}_\eta \tag{4}$$

arises.

Averaging over a cross-section of the current with area $A = w \cdot h$, where w (m) and h (m) are, respectively, the width and the height of the current, gives the model equations (subscript ξ means $\partial/\partial\xi$)

$$(\rho A U)_\xi = \rho_e E(Ri)\, w\, U, \tag{5}$$

$$(T\rho A U)_\xi = \rho_e T_e\, E(Ri) w U, \tag{6}$$

$$(S\rho A U)_\xi = \rho_e S_e\, E(Ri) w\, U, \tag{7}$$

$$\left(\rho A U^2\right)_\xi = A g(\rho - \rho_e)(\sin\beta\sin\theta - h_\xi) - C_D\,\rho\, w\, U^2, \tag{8}$$

$$\rho U^2 \beta_\xi = g(\rho - \rho_e)\sin\theta\cos\beta - fU. \tag{9}$$

Here ρ (kg m^{-3}), S (psu) and T (K) is the density, salinity and temperature of the current, respectively, while the corresponding variables with subscript e are connected to the ambient water. U (m s^{-1}) denotes the current mean velocity, E (dimensionless) the entrainment parameter and C_D (dimensionless) the friction parameter, or the drag coefficient.

The Coriolis parameter is given by $f = 2\Omega(\sin\varphi\cos\theta - \cos\varphi\sin\theta\sin\zeta)$ (s^{-1}), φ (rad) being the latitude and ζ (rad) being the orientation of the slope. For $\zeta = 0$ the slope sinks in the eastward direction, while $\zeta = \pi/2$ means that the slope is in the north direction [Alendal *et al.*, 1993].

Equations (5)-(7) are conservation equations for mass, internal energy (heat) and salt, respectively, while the two latter are momentum equations in the along stream direction (8) and in the cross-stream-direction (9). A rigorous derivation of these equations is given in the appendix.

The x and y positions together with the total depth enter the differential system through the geometric relations

$$\frac{\partial}{\partial\xi}x = \cos\beta, \quad \frac{\partial}{\partial\xi}y = \sin\beta\cos\theta, \quad \frac{\partial}{\partial\xi}D = \sin\beta\sin\theta. \tag{10}$$

The equation of state of seawater [UNESCO, 1981] relates the density of seawater to the salinity, temperature and pressure (depth);

$$\rho = \rho(S, T, p). \tag{11}$$

This system of ordinary differential equations is not closed and thereby not solvable until we have an extra relation between some of the variables in the model. We have assumed constant width over thickness ratio

$$\frac{w}{h} = constant. \tag{12}$$

In our numerical study this parameter is set by the initial values of volume flux $F_0 = w_0\, h_0\, U_0$, (m^3 s^{-1}), velocity U_0 and height h_0 (m) of the current.

The Entrainment Parameterization and The Friction Parameter

The entrainment parameter $E(Ri)$ is strongly related to the overall Richardson number for a wall bounded current [Turner, 1973]

$$Ri = \frac{gh(\rho - \rho_e)\cos\theta}{\rho U^2} = \frac{g' h \cos\theta}{U^2}, \qquad (13)$$

where $g' = g\,h\,(\rho - \rho_e)/\rho$, is the reduced gravity. With support in laboratory experiments, Christodoulou [1986] found general laws for this dependency:

$$E(Ri) = \begin{cases} 0.07 & \text{for } Ri < 10^{-2} \\ 0.007 Ri^{-1/2} & \text{for } 10^{-2} < Ri < 1 \\ 0.007 Ri^{-3/2} & \text{for } 1 < Ri < 10^2 \end{cases}, \qquad (14)$$

and

$$E(Ri) = 0.002\,Ri \quad \text{for} \quad 0.08163 < Ri < 12.25. \qquad (15)$$

Throughout this paper equation (14) has been used for the entrainment parameter, with the mean value of (14) and (15) for $0.08163 < Ri < 12.25$.

The friction parameter is usually defined by

$$u_*^2 = C_D u^2, \qquad (16)$$

where u_* ($\mathrm{m\,s^{-1}}$) is the friction velocity. Adopting the Kato-Phillips like equation [Stigebrandt, 1987]

$$E\,u = \left(2\,m_0\,u_*^3\right) / \left(g'\,h\,cos\theta\right), \qquad (17)$$

where m_0 is a constant of order 1 [Stigebrandt, 1987; Oberhuber, 1993] and solving (16) and (17) with respect to C_D gives

$$C_D^{3/2} = \frac{Ri\,E(Ri)}{2\,m_0}. \qquad (18)$$

Plotting C_D as function of Ri, Fig. 2, shows a peak at $Ri = 1$, or for $U^2 = g\,h\,\cos\theta$. When the current moves with this velocity it is in resonance with long-wavelength waves on the boundary between the current and the ambient water. This resonance draws energy from the current to the waves and it is therefore reasonable that the current experiences higher friction around $Ri = 1$. Notice also that this friction parameter is of the order 10^{-3} for large Richardson number which is of the order used in tidal modelling [Gjevik and Straume, 1989].

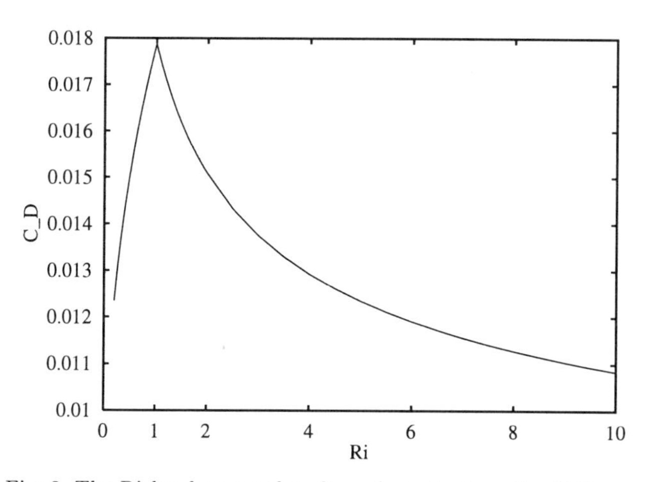

Fig. 2. The Richardson number dependent C_D from Eq. (18).

Price *et al.* [1993], reporting on measurements of the outflow from the Mediterranean through the Gibraltar Strait, give empirical evidence of high friction parameter values. They define bottom stress τ_b (Pa) in the usual manner

$$\tau_b = \rho_0\,C_D\,U^2, \qquad (19)$$

and at the top of the continental slope, which is the site for high acceleration and thereby high entrainment [McCartney, 1992], they found bottom stress of 13 Pa connected to a maximum velocity of 1.3 $\mathrm{m\,s^{-1}}$. With $\rho_0 \sim 10^3$ kg $\mathrm{m^{-3}}$, this gives $C_D = 0.007$. But a mean velocity of the current at say 1 $\mathrm{m\,s^{-1}}$, which is a high estimate, gives $C_D = 0.013$. In our use of the model we have used both the variable and the constant friction parameter approach.

The Appearance of a Singularity in the Model

The inclusion of the pressure term due to variations in the plume thickness, hereafter called the h_ξ-term, gives rise to a singularity. To see this we use the constraint $w/h = constant$ in (5) and (8). Assuming that the variation in density is negligible this results in the following equation for the plume thickness

$$\left(g'\,h - 2\,U^2\right)\,h_\xi = g' h \sin\theta\,\sin\beta - (C_D + 2\frac{\rho_e}{\rho}E)U^2. \qquad (20)$$

The singularity occurs when

$$U^2 = \frac{1}{2}g' h,$$

i.e. when the velocity equals the phase- and group-velocity of internal waves on a current with the constraint that $w/h = constant$ (see Stoker [1957] for analysis of a similar case with current in a channel of constant width).

If during the numerical integration the singularity is approached, the derivative of the current thickness h_ξ will become plus or minus infinity, depending on whether we approach from the subcritical ($U^2 < g'h/2$) or the supercritical ($U^2 > g'h/2$) side, unless the right-hand side is equal to zero simultaneously. In the latter case

$$2\,\sin\theta\,\sin\beta = (C_D + 2\,\frac{\rho_e}{\rho}E), \qquad (21)$$

which gives a critical slope θ. Solutions that pass through this singularity go through a hydraulic jump from subcritical to supercritical flow, or vice versa. This is somewhat similar to the control point analysis for obstacles in a current [Pratt, 1986; Wajsowicz, 1993].

Furthermore, the h_ξ-term includes effects from waves in the momentum equations and thereby the possibility of triggering instabilities. The steady state model cannot treat unsteady waves, indicating that a time dependency should be present when the h_ξ-term is used.

In cases where the singularity has no effect the solutions show no major differences whether the h_ξ-term is present or not. The h_ξ term is therefore neglected in the numerical studies given in this paper. The problem with non-vanishing h_ξ-term will be subject of further studies.

Modelling the Outflow through the Denmark Strait

The Environment and the Numerical Procedure

Dense water formed in the Nordic Seas, i.e., the Greenland, Icelandic and Norwegian Seas, overflow the ridge between Greenland and Iceland (the Denmark Strait), the Iceland-Faroe ridge, and the Faroe-Shetland Channel forming the North Atlantic Bottom Water (NABW) [Dickson *et al.*, 1990].

Our objective is to model the overflow through the Denmark Strait. The surface water contains warm saline water of the Irminger Current and cold but low salinity Polar Water. The intermediate water column is the Arctic Intermediate Water (AIW) with salinity between 34.7 psu and 34.9 psu and with temperature from 0 ℃ to 1 ℃, and Polar Intermediate Water (PIW) with salinity below 34.7 psu and temperatures below 0 ℃. At the bottom there is the Norwegian Sea Bottom Water (NSBW) with temperature below 0 ℃, typically −0.4 ℃, and salinity higher than 34.9 psu, typically 34.94 psu [Malmberg, 1985].

There is some dispute over which water mass contributes to the formation of NABW. Measurements show overflow through the Denmark Strait that experiences an increase in volume transport from a depth of 500m to 2 000m from 2.9 Sv to slightly less than 6 Sv (1 Sv equals 10^6 m^3s^{-1}) [Dickson *et al.*, 1990; Swift *et al.*, 1980]. The initial width over height ratio is 1500. We do not discuss this any further but use these properties of NSBW as initial values for the gravity current.

Using data from Swift *et al.* [1980], we set the ambient potential density σ_{T_θ} to 1027.95 kg m^{-3} and the salinity S to 34.9 psu. The potential density is defined by the expression

$$\sigma_{T_\theta} = \rho(S, T_\theta, P = 0) \qquad (22)$$

where T_θ and P are the potential temperature and the pressure, respectively, and ρ is given by Eq. (11). Our model uses *in-situ* quantities so we have to solve equation (22) with respect to the potential temperature T_θ whereafter we find the absolute temperature from the potential temperature using standard routines [UNESCO, 1983]. When this conversion is done we find the *in-situ* density from the equation of state of seawater.

Equation (22) gives that the potential temperature stays constant but that the *in-situ* temperature increases with depth. This follows from the fact that *in-situ* temperature takes the increasing pressure into account and thereby the compressibility. This also gives rise to the extra driving force, the thermobaric effect, since the current water is colder than the ambient water and therefore is more compressible.

The model is integrated as an initial value problem for S, T, U and β (the initial density is then given from the equation of state) using the *Livermore Solver for Ordinary Differential Equations* [Hindmarsh, 1980]. Relative to the sensitivity for the drag coefficient the solutions are not sensitive to perturbations in the initial values and the numerical study is therefore performed as a sensitivity study for the drag coefficient. The integration is stopped if the plume velocity or the excess density become lower than 10^{-2} m s^{-1} and 10^{-2} kg m^{-3}, respectively.

The Modelled Volume Transport

We estimate the order of magnitude required for the friction parameter in order to obtain an increase in volume transport from 2.9 Sv at 500 m, to slightly less than 6 Sv at 2 000 m [Dickson *et al.*, 1990]. We set the slope angle θ to the value 0.01, and integrate the model for three different constant drag coefficients, $C_D = 0.1$, $C_D = 0.01$ and $C_D = 0.001$, in addition to the Richardson number dependent drag coefficient given by Eq. (18). This gives the volume fluxes shown in Fig. 3. The high drag coefficient solution gives a too small volume flux at 2 000 m, while the two solutions with small C_D have entrained too much ambient water giving too high volume transport. The reason for this is that the current has higher acceleration for the smaller friction cases, so the velocity and the entrainment increase, cf. Eq. (5). The model requires C_D between 0.01 and 0.1 in order to fit measurements. This is less than the requirement of Smith [1975] of $C_D = 0.15$, which was also used by Killworth [1977]. In addition, $C_D = 0.11$ was used by Speer and Tziperman [1990]. The Richardson number dependent friction parameter stays near the $C_D = 0.01$ solution and gives slightly too

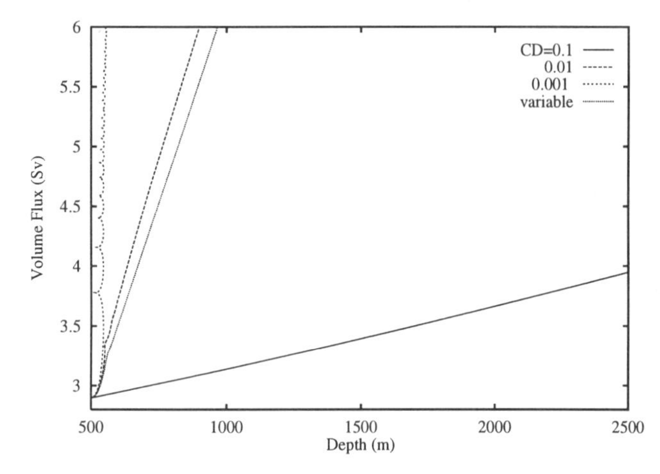

Fig. 3. The modelled volume transport through the Denmark Strait for different friction parameters. Measurements indicates increase in volume transport from 2.9 Sv at 500 m to 6 Sv at 2000 m [Dickson *et al.*, 1990].

much entrainment. Notice also the meander behaviour of the $C_D = 0.001$ solution to be discussed below.

Variable Bottom Topography and a Discussion of Meandering Behaviour

Topographic maps show that the inclination is steeper at 500 meters than at say 1 500 meters. The model can take this into account and we set a steep inclination at 500 m decreasing to zero as indicated in Fig. 4 . This gives the remarkable results shown in Figs. 5 and 6.

When the slope is steep, the gravity term is dominant in the momentum equations, (8) and (9). The low drag solution gains higher velocity than the solution with higher friction parameter. The relative importance of the Coriolis force compared to gravity becomes larger with increasing velocity and with decreasing slope angle θ, which causes the low drag solution to bend fastest of the two cases.

When θ has become smaller the low drag solution oscillates with decreasing amplitude over an equilibrium solution where

$$g(\rho - \rho_e)\sin\theta\cos\beta = \rho f U.$$

This may be explained using conservation of energy arguments. When the plume has high velocity, i.e., high kinetic energy, the Coriolis force bends the current until it moves upwards along the sloping bottom. But as it moves higher, the kinetic energy is lost to potential energy causing the plume to stop the upward motion. When the plume has lost enough kinetic energy for the gravity to overcome the Coriolis force, the current is forced downslope again. This continues until the drag has retarded the current enough to gain force balance. The current then moves in geostrophic balance.

One remark on this oscillatory behaviour has to be made. If the radius of curvature $(\partial\beta/\partial\xi)^{-1}$ become less than half of the plume width, then the model is no longer valid because the plume would cross itself. Nevertheless, it seems that the oscillations do not affect the solutions after the oscillations have come to rest. For instance if the initial velocity is low, the gravity will accelerate the plume causing similar oscillations as those present in Fig. 5. Another case with higher initial velocity may show similar oscillation but with lower amplitude, but when the oscillations have been damped out the two solutions appear to be equal.

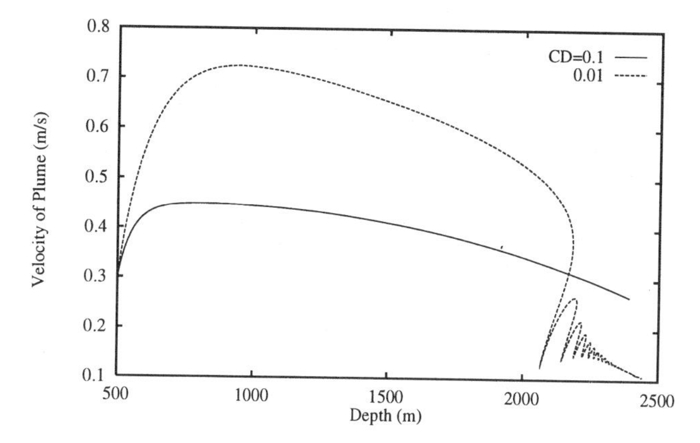

Fig. 5. The plume velocity when the inclined angle varies as shown in Fig. 4. Denmark Strait environment and initial values.

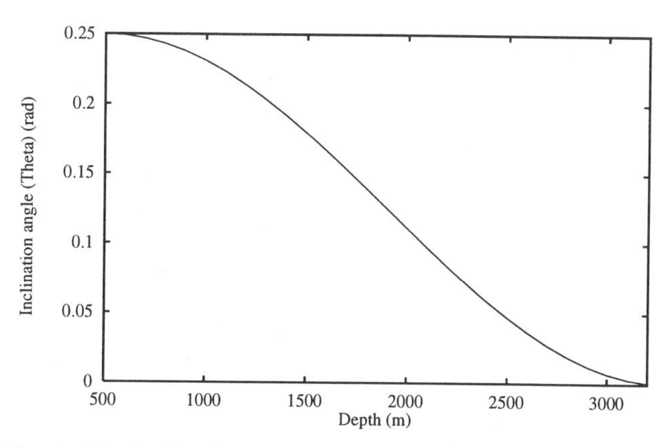

Fig. 4. The inclination angle θ used to study meandering behaviour.

Modelling of Other Deep-Sea gravity Currents

The Weddell Sea Continental Slope

The water in this gravity current, Ice Shelf Water (ISW), is produced from melting under the Ronne Ice Shelf where there exists an upward current with smaller density than the environment [Jenkins, 1991]. This water flows over the Filchner Depression before it flows down the continental slope where it forms the Weddell Sea Bottom Water (WSBW). Killworth [1977] has modelled this gravity current with a model similar to ours, but used gradients instead of *in-situ* quantities for the the ambient water and he also used constant entrainment parameter and drag coefficient.

Using data from Foldvik and Gammelsrød [1988] and Killworth [1977] gives the following quantities: The ambient wa-

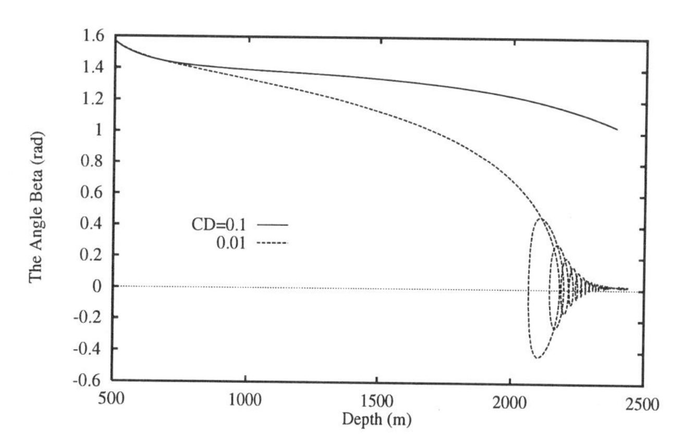

Fig. 6. The angle β as a function of depth when the slope angle decreases as in Fig. 4. Denmark Strait environment and initial values.

ter has potential temperature ranging from 0.0 ℃ at 500 m to -0.2 ℃ at 2 000 meters depth. The ambient salinity is between 34.64 psu and 34.72 psu; we use the mean value 34.68 psu. The current is initialized with a volume flux of 0.5 Sv, velocity 0.01 $\mathrm{m\,s^{-1}}$, potential temperature -1.8℃, and salinity 34.61 psu. The slope is $\theta = 10^{-2}$ rad in the mean directed northward and the plume moves straight downslope at the starting point, i.e. $\beta = \pi/2$ rad. This current is initially very wide and not too high which gives $w\,h^{-1}$ equal to 4 900.

At 2 000 meters depth the volume transport ($A\,U$) should increase to 200-300% of the initial value [Killworth, 1977]. The resulting volume transport for different friction parameter is given in Fig. 7. As for the Denmark Strait case, we must have drag coefficient somewhere between $C_D = 0.01$ and $C_D = 0.1$ in order to get the wanted increase in volume transport. In this case however, the variable C_D gives higher entrainment than the $C_D = 0.01$ solution.

Outflow of High Salinity Water from Spencer Gulf, Australia

In the two previous examples of gravity currents the compressibility due to lower temperature inside the current than outside gives rise to an extra driving mechanism. The initial density differences are respectively 0.1 kg/m^3 and 0.01 kg/m^3 for the Denmark Strait and for the Weddell Sea overflows. This effect Killworth [1977] found necessary to include in his plume model in order to let the gravity current down the Weddell Sea continental slope reach as deep as it should according to measurements.

In Spencer Gulf, South Australia, initial excess density in the plume is much higher, 0.6 $\mathrm{kg\,m^{-3}}$, and the compressibility effect has little influence on the current [Bowers and Lennon, 1987; Lennon *et al.*, 1987].

Due to evaporation the Spencer Gulf water becomes saline during summer, and when the high salinity water cools in the autumn it sinks to the bottom and moves out of the gulf along the bottom while lighter water moves inwards on top. The volume flux is 0.042 Sv which is much smaller than the corresponding values for the currents treated earlier. The current width over height ratio is set to 1 500.

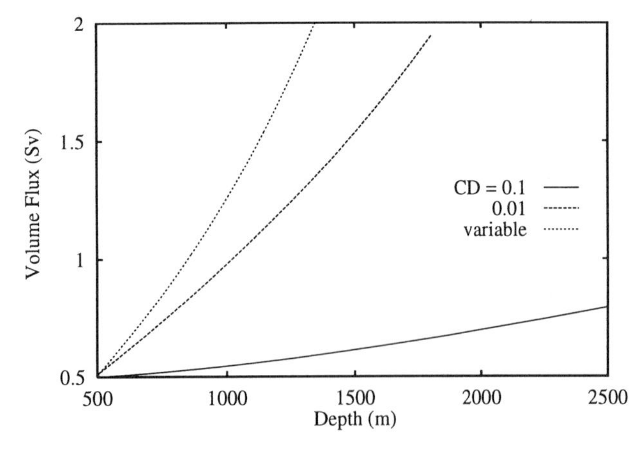

Fig. 7. The modelled volume flux in the Weddell Sea as a function of depth for different drag coefficients. According to Killworth [1977] the volume transport at 2000m should be 1-1.5 Sv.

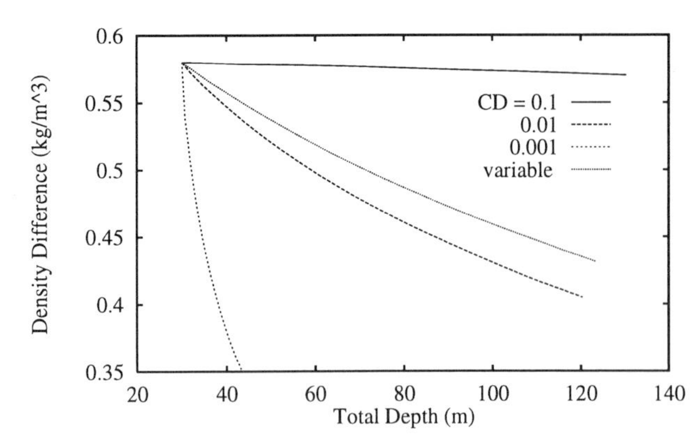

Fig. 8. The density differences for different drag coefficients, in Spencer Gulf.

We have set the ambient temperature to 17 ℃ and salinity to 36.0 psu. Together with initial values for the current, temperature 16.5 ℃ and salinity 36.6 psu, this gives the desired density difference of 0.6 $\mathrm{kg\,m^{-3}}$. The slope angle is set to $\theta = 10^{-2}$ rad, and initially the plume moves with angle $\beta \sim \pi/4$ rad [Bowers and Lennon, 1987].

If there was little entrainment, as stated by Bowers and Lennon [1987], then the density differences should stay nearly constant. Fig. 8 shows the modelled density differences for different friction parameters. Notice that the solution with $C_D = 10^{-3}$ gives too much entrainment so in order to fit the measurements we must have higher friction parameter and thereby lower entrainment, as for the previous examples. This is in contrast to the analyzes of Bowers and Lennon [1987] and Lennon *et al.* [1987].

The Gibraltar Strait

Another current mainly driven by high density difference due to high salinity difference is the Mediterranean outflow through the Gibraltar Strait. Again, due to evaporation in the Mediterranean the water gains high salinity and flows out through the Gibraltar Strait with less saline Atlantic water flowing in on top. We start our modelling at 400 meters depth when the current has left the canyon in the strait. The bottom topography is complex in the area, and the current splits into two cores [Ambar and Howe, 1979]. We are therefore not modelling deeper than 900 meter.

The ambient water has salinity 35.8 psu and potential density 1027 $\mathrm{kg\,m^{-3}}$, giving a temperature of 13.1 ℃. The initial volume flux of the currents is set to 2 Sv with salinity 38.0 psu and temperature 13.2 ℃[Price *et al.*, 1993; Ochoa and Bray, 1991]. This gives density difference $\sim$ 1.68 $\mathrm{kg\,m^{-3}}$ which is somewhat higher than used by Smith [1975] (1.25 $\mathrm{kg\,m^{-3}}$). The inclination is set to 0.015 rad and the width to thickness ratio is estimated to 47 [Smith, 1975]. At 700 - 900

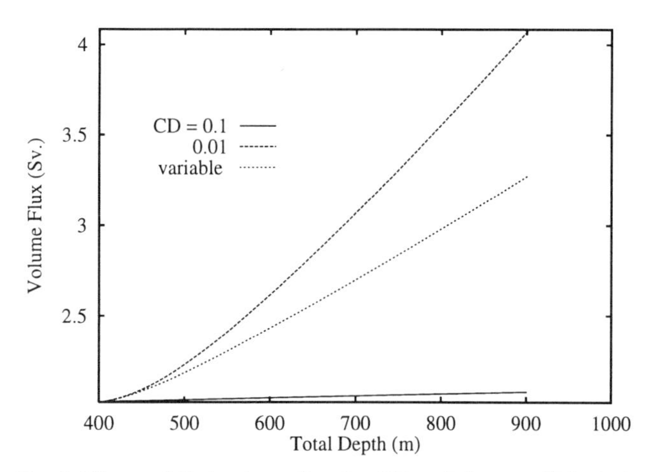

Fig. 9. The modelled volume flux for different drag coefficients in the Gibraltar Strait. Measurements indicates 2 Sv at 400m with an increase of approx. 0.6 Sv at 900 m [Price *et al.*, 1993].

m depth the observed volume flux increases by 0.6 Sv and the density anomaly decreases to 0.7 $\mathrm{kg\,m^{-3}}$ [Price *et al.*, 1993].

The variable friction parameter in this case turns out like 0.015. Results from our modelling shows that this friction parameter fits the observed volume flux (Fig. 9) best, while the low constant friction parameter 0.01 fits the observed density difference best (Fig. 10). Thus our model does not identify a unique friction parameter for this case. This indicates that there may be missing physics in the model. One possibility may be variable bottom friction. The way we defined the dependency in Eq. (18) it represents friction on the interface between the current and the ambient water. Price *et al.* [1993] shows with support in measurements that the bottom friction can be higher than the interfacial friction.

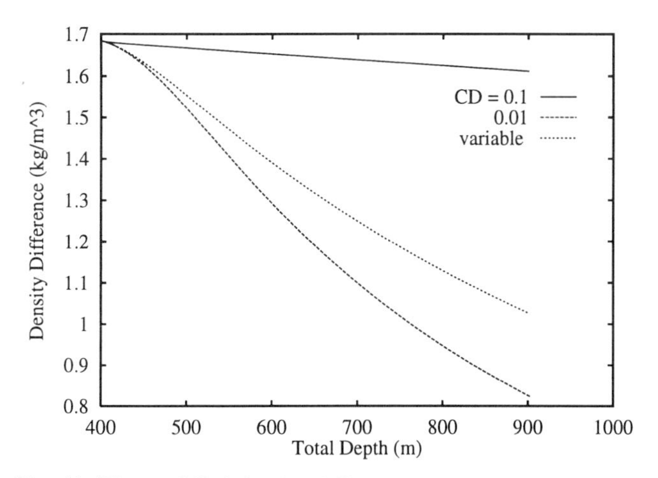

Fig. 10. The modelled density differences for different drag coefficients in the Gibraltar Strait. According to measurements the density anomal should decrease from $\sim$ 1.7 kg $\mathrm{m^{-3}}$ at 400 m to 0.7 kg $\mathrm{m^{-3}}$ at 700 -900 m [Price *et al.*, 1993; Ochoa and Bray, 1991].

Fig. 11 shows the resulting friction parameter from Eq. (18) for three of the cases reported in this paper. In the shallower Spencer Gulf case case the friction parameter starts at 8×10^{-3} decreasing to approximately 5×10^{-3}.

Conclusions

An extensive description of an integrated plume model is given and derivation of the pressure term due to variations in plume thickness is included. This term includes waves in the momentum equations and thereby the possibility of net energy loss to wave energy and instabilities. At low friction, meandering behaviour is to be expected. This may occur e.g. in the downstream part of the overflow from the Denmark Strait. Further study of waves and meanders would require a time dependent model.

We used an entrainment parameterization based on empirical data, and introduced a drag coefficient consistent with the entrainment parameter. Within the limitation of the steady state model, the numerical results shows that the friction parameter should be between 0.01 and 0.1 in order to provide an order of magnitude fit to measurements. The consistent Richardson dependent drag coefficient gives solutions that give too little entrainment but this drag coefficient parameterizes only the interfacial friction so additional bottom friction should perhaps be added. This study indicates that gravity current modelling should use friction values approximately a factor ten higher than used for instance in tidal modelling. This may be important for the representation of such currents in large scale climate models.

Appendix: Derivation of The Integrated Plume Model

The configuration is as given in Fig. 1.

The Local Governing Equations

An infitesimal volume inside the current follows the local

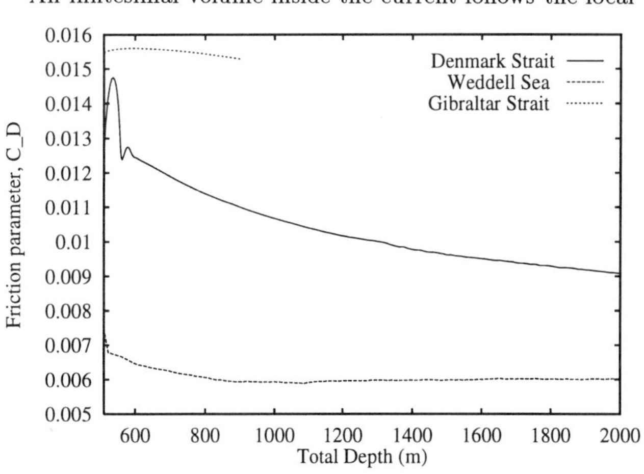

Fig. 11. The resulting drag coefficient, C_D, for three of the currents. In Spencer Gulf, not shown, the frictrion parameter starts at approx. 0.008 decreasing to approx. 0.005

continuity equation and the equation of motion which reads for steady state:

$$\nabla \cdot \rho_l \, \mathbf{u} = 0, \tag{23}$$

$$\rho_l \mathbf{u} \cdot \nabla \mathbf{u} = \mathcal{F} + \nabla \cdot \tilde{\tau}, \tag{24}$$

where ρ_l (kg m^{-3}) is the local density, $\mathbf{u}$ (m s^{-1}) the local velocity, $\mathcal{F}$ is the volume force, and $\tilde{\tau} = \tau_{ij}\mathbf{e}_i\mathbf{e}_j$ ($\mathbf{e}_i$ and $\mathbf{e}_j$ being unit vectors), the stress tensor, represents forces acting on the boundary of the volume. Volume forces in our study are gravity ($\mathcal{F}_{grav} = -g\rho_l\mathbf{k}$), and Coriolis force ($\mathcal{F}_{coriolis} = -2\rho_l\boldsymbol{\Omega} \times \mathbf{u}$), where $\boldsymbol{\Omega}$ (s^{-1}) is the angular velocity of the earth.

The first subscript of the stress tensor indicate the direction in which the normal to the plane is pointing while the second subscript indicates the direction in which the force is acting. This means that the diagonal elements (τ_{ii}) represent forces which are acting normal to the surface while those elements with different subscripts (τ_{ij}, $i \neq j$) are shear forces.

Let $-p = \tau_{11} + \tau_{22} + \tau_{33} = trace(\tilde{\tau})$, then the local steady state momentum equation may be written:

$$\rho_l \mathbf{u} \cdot \nabla \mathbf{u} = \mathcal{F} - \nabla p + \nabla \cdot (\tilde{\tau} - trace(\tau)\tilde{\mathcal{I}}) \tag{25}$$

where $\tilde{\mathcal{I}}$ is the unit dyad.

The pressure is assumed to be hydrostatic. Fig. 12 illustrates a section through the current in the along stream direction. A reference particle moving a distance $\Delta\xi$ in the along-stream direction experiences an increase in pressure given by the expression

$$p(\xi + \Delta\xi) = p(\xi) + g\rho_e\,(\Delta\xi \sin\phi - r) + g\rho r, \tag{26}$$

where ρ is the mean density of the current to be defined.

Simple geometric relations and Taylor expansion to second order in $(\Delta\xi - s)$ gives:

$$\begin{aligned} r\sin\phi &= s, \\ r\cos\phi &= h\,(\xi + \Delta\xi - s) - h\,(\xi) \\ &= \tfrac{dh}{d\xi}\,(\Delta\xi - s) + O\left((\Delta\xi - s)^2\right). \end{aligned} \tag{27}$$

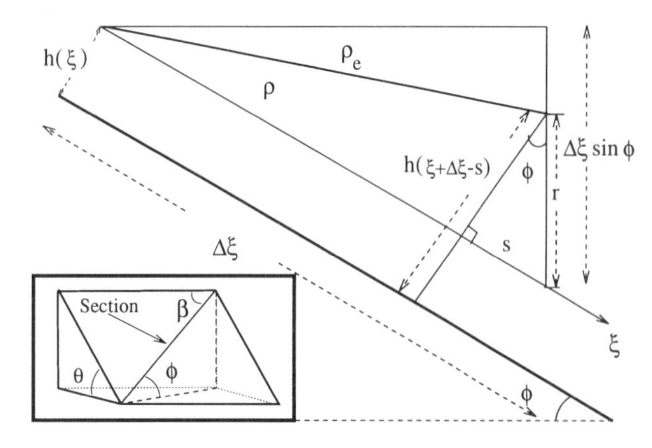

Fig. 12. Illustration of the pressure term.

For small ϕ we have that

$$r = \Delta\xi \, \frac{\partial}{\partial \xi} h. \tag{28}$$

Using (28) in (26) and letting $\Delta\xi \to 0$ gives that

$$\frac{\partial}{\partial \xi} p = g\,\rho_e\,\sin\theta\sin\beta + g\,(\rho - \rho_e)\,\frac{\partial}{\partial \xi} h, \tag{29}$$

where the identity $\sin\phi = \sin\theta\sin\beta$ has been used (see inset in Fig. 12).

In the other directions the variation in plume thickness has no effect, so

$$\frac{\partial}{\partial \eta} p = g\,\rho_e\,\sin\theta\,\cos\beta, \quad \frac{\partial}{\partial z'} p = g\,\rho_e\,\cos\theta. \tag{30}$$

Integration of the Local Equations

The plume model represents the overall behaviour of the current, so the mean values of velocity, density, salinity, etc., are studied while interior structures from turbulence are neglected. The model equations are obtained by integration of the local governing equations over a cross-section of the current normal to the along-stream direction. The cross-section has area A, width w and thickness h; $A = w \cdot h$. The mean over the cross section is defined as

$$B = \frac{1}{A}\int_0^h \int_{-w/2}^{w/2} b\,d\eta\,dz', \tag{31}$$

where b is an arbitrary local scalar or vector quantity.

Conservation of mass. The continuity equation integrates and gives

$$\frac{\partial}{\partial \xi}(A\,\rho\,U) = \rho_e w v_e = \rho_e w E U, \tag{32}$$

where ρ is the mean density of the current and where the entraining velocity v_e (m s^{-1}) is set proportional to the mean velocity U of the current, [Smith, 1975]. For a wall bounded currents the entrainment parameter E is strongly related to the overall Richardson number, Eq. (13) [Turner, 1973].

Conservation of salt and heat. In the same manner as for the total mass conservation, continuity equations have to be valid for the salinity and for the internal energy (heat). The salinity S is defined as gram salt per kilogram water so, similar to the conservation of total mass, the conservation equation of salt reads

$$\frac{\partial}{\partial \xi}(S\,\rho\,A\,U) = \rho_e S_e E(Ri_0) w U. \tag{33}$$

Neglecting energy sources such as heat production due to friction gives in a same manner the internal energy conservation equation

$$\frac{\partial}{\partial \xi}(\rho\,C_p\,T_K\,A\,U) = \rho_e C_p T_{K,e} E(Ri_0) w\,U. \tag{34}$$

Here T_K (K) is the absolute temperature and C_p (J m^{-3} K^{-1}) the heat capacity of the fluids. Since C_p for the ambient water approximates the heat capacity for the current water and remains almost constant with ξ, we have that

$$\frac{\partial}{\partial \xi}(\rho T_K A U) = \rho_e T_{K,e} E(Ri_0) w U. \qquad (35)$$

The equations of motion. Neglecting turbulent fluctuations perpendicular to the along stream direction (assuming that they sum up to zero) means that the left hand side of (24) becomes

$$\rho_l u \frac{\partial}{\partial \xi} u \mathbf{e}_\xi + \rho_l u^2 (\frac{\partial}{\partial \xi}\beta)\mathbf{e}_\eta. \qquad (36)$$

The local continuity equation gives that $\partial(\rho_l u)/\partial\xi = 0$, and assuming that $u \to 0$ at the boundary between the current and the ambient water, integration of (36) over the cross-section gives

$$\frac{\partial}{\partial \xi}\left(A\rho U^2\right)\mathbf{e}_\xi + A\rho U^2\left(\frac{\partial}{\partial \xi}\beta\right)\mathbf{e}_\eta. \qquad (37)$$

Here we have assumed that the mean of ρU is equal to the mean density times the mean velocity.

Integrating the divergence of the stress tensor gives

$$\int\int_A \nabla\cdot\left(\tilde{\tau} - trace(\tau)\tilde{\mathcal{I}}\right) dA \equiv \int\int_A \nabla\cdot\tilde{\mathcal{D}}\, dA = \int_\Gamma \tilde{\mathcal{D}}\, d\Gamma, \qquad (38)$$

where Γ is the boundary of the cross-section. We distinguish between stresses normal and tangential to the boundary. The normal stresses are due to a force from the bottom responding on the weight of the plume and the hydrostatic pressure acting on the top of the plume, we call this N and it points in the z' direction.

The shear stress must oppose the movement of the current. We assume the usual drag law, proportional to the square of the velocity and to the width of the current, so

$$\int_\Gamma \tilde{\mathcal{D}}\, d\Gamma = N\mathbf{k}' - C_D w U^2 \mathbf{e}_\xi. \qquad (39)$$

Integration of the gravity and Coriolis terms follows trivially so the equations of motion read:

$$\frac{d}{d\xi}\left(\rho A U^2\right) = Ag(\rho - \rho_e)(\sin\theta\sin\beta - \frac{dh}{d\xi}) - C_D w U^2, \qquad (40)$$

$$\rho U^2 \frac{d}{d\xi}\beta = g(\rho - \rho_e)\sin\theta\cos\beta - \rho f U, \qquad (41)$$

$$0 = -g(\rho - \rho_e)\cos\theta - \rho f_k U + N. \qquad (42)$$

Here $f = 2\Omega(\sin\varphi\cos\theta - \cos\varphi\sin\theta\sin\zeta)$ is the Coriolis parameter, φ being the latitude and ζ being the orientation of the slope; for $\zeta = 0$ the slope sinks in the eastward direction, while $\zeta = \pi/2$ means that the slope is in the north direction [Alendal *et al.*, 1993]. The Coriolis parameter f_k has a lengthier expression but since the normal force N acting from the bottom on the current opposes the weight of the plume, it responds in such a manner that equation (42) is always satisfied and this equation is not of interest.

Acknowledgments. This work originated from a project supported by Den Norske Stats Olje Selskap (Statoil), Norsk Hydro a.s., the Norwegian State Pollution Authority (SFT) and the Norwegian Research Council. GA is presently supported by the Norwegian Research Council and HD is supported by Saga Petroleum a.s.

References

Alendal, G., H. Drange, and P. M. Haugan, Injection of CO_2 in the ocean: Modelling bottom gravity currents with CO_2-enriched seawater, Technical Report 66A, Nansen Environmental and Remote Sensing Center, 1993.

Ambar, I. and M. R. Howe, Observation of the Mediterranean outflow - I: Mixing in the Mediterranean outflow, *Deep-Sea Res.*, **26A**:535–554, 1979.

Bowers, D. G. and G. W. Lennon, Observations of stratified flow over a bottom gradient in a coastal sea, *Cont. Shelf Res.*, **7**(9):1105–1121, 1987.

Christodoulou, G. C., Interfacial mixing in stratified flows, *J. Hydr. Res.*, **24**(2):77–92, 1986.

Dickson, R. R., E. M. Gmitrowicz, and A. J. Watson, Deep-water renewal in the northern North Atlantic, *Nature*, **344**:848–850 1990.

Drange, H., G. Alendal, and P. M. Haugan, P. M. A bottom gravity current model for carbon-dioxide enriched sinking seawater, *Energy Conv. Mgmt.*, **34**(9-11):1065–1072, 1993.

Foldvik, A. and T. Gammelsrød, Notes on southern hydrography, sea-ice and bottom water formation, *Palaeogeogr. Palaeoclimatol. Palaeoecol.*, **67**:3–17, 1988.

Gjevik, B. and T. Straume, Model simulations of the m_2 and the k_1 tide in the Nordic seas and the Artic ocean, *Tellus*, **41 A**:73–76, 1989.

Haugan, P. M. and H. Drange, Sequestration of CO_2 in the deep ocean by shallow injection, *Nature*, **357**:318–320, 1992.

Hindmarsh, A. C., LSODE and LSODI, two new initial value ordinary differential equation solvers, *ACM-Sign. Newsl.*, **15**:10–11, 1980.

Jenkins, A., A one-dimensional model of ice shelf-ocean interaction, *J. Geophys. Res.*, **96**(C11):20,671–20,677, 1991.

Killworth, P. D., Mixing on the Weddell sea continental slope, *Deep-Sea Res.*, **24**:427–448, 1977.

Killworth, P. D., Deep convection in the world ocean, *Rew. Geophys. Space Phys.*, **21**(1):1–26, 1983.

Lennon, G. W. *et al.*, Gravity currents and the release of salt from an inverse estuary., *Nature*, **327**:695–697, 1987.

Malmberg, S.-A., The water masses between Iceland and Greenland, in *Chemical tracers for studying water masses and physical processes in the sea*, edited by Stefánsson, U., volume IX, RIT Fiskideildar, 1985.

McCartney, M. S., Recirculating components to the deep boundary current of the northern North Atlantic, *Prog. Oceanog.*, **29**:283–383, 1992.

Morton, B. R., G. I. Taylor, and J. S. Turner, Turbulent gravitational convection from maintained and instantaneous sources, *Proc. Roy. Soc. A*, **234**:1–23, 1956.

Oberhuber, J. M. Simulation of the Atlantic Circulation with a coupled Sea Ice -Mixed Layer - Isopycnal General Circulation Model, Part I: Model Description, *J. Phys. Oceanogr.*, **23**(5):808–829, 1993.

Ochoa, J. and N. A. Bray, Water mass exchange in the Gulf of Cadiz, *Deep-Sea Res.*, **38**(Suppl 1):S465–S503, 1991.

Pratt, L. J., Hydraulic control of sill flow with bottom friction, *J. Phys. Ocanogr.*, **16**:1970–1980, 1986.

Price, J. F., M. O'Neill Baringer, R. G. Lueck, G. C. Johnson, I. Ambar, G. Parilla, A. Cantos, M. A. Kennelly, and T. B. Sanford, Mediterranean outflow mixing and dynamics, *Science*, **259**:1277–1282, 1993.

Rudnicki, M. D. and H. Elderfield, Theory applied to the Mid-Atlantic Ridge hydrothermal plumes; the finite-difference approach, *J. Volc. and Geoth. Res.*, **50**:161–172, 1992.

Simpson, J. E., *Gravity currents in the environment and the laboratory*, Halsted Press, Chichester, 1987.

Smith, P. C., A streamtube model for bottom boundary currents in the ocean, *Deep-Sea Res.*, **22**:853–873, 1975.

Speer, K. and E. Tziperman, Convection from a source in an ocean basin, *Deep-Sea Res.*, **37**(3):431–446, 1990.

Stigebrandt, A. A model for the vertical circulation of the Baltic deep water, *J. Phys. Oceanogr.*, **17**:1772–1785, 1987.

Stoker, J. J., *Water Waves*, Interscience Publishers Inc, New York, 1957.

Swift, J. H., K. Aagaard, and S.-A. Malmberg, The contribution of the Denmark Strait overflow to the deep north atlantic, *Deep-Sea Res.*, **27A**:29–42, 1980.

Turner, J. S., *Buoyancy effects in fluids*, Cambridge University Press, Cambridge, 1973.

UNESCO, Tenth report of the joint panel on oceanographic tables and standards, UNESCO Tecnical Papers in Marine Sci. 36, Paris, 1981.

UNESCO, Algoritm for computation of fundamental properties of seawater, UNESCO tecnical papers in marine science 44, Paris, 1983.

Wajsowicz, R. C., Dissipative effects in inertial flows over a sill, *Dyn. Atmosph. Oceans*, **17**:257–301, 1993.

Woods, A. W., The fluid dynamics and thermodynamics of eruption columns, *Bull. Volcanol.*, **50**:169–193, 1988.

Guttorm Alendal, Helge Drange and Peter M. Haugan, Nansen Environmental and Remote Sensing Center, Edvard Griegsvei 3a, N–5037 Solheimsviken/Bergen, Norway.

Net Primary Production & Stratification in the Ocean

Trevor Platt

Bedford Institute of Oceanography, Dartmouth, N.S., Canada.

John D. Woods

NERC Oceanography Unit, Physics Department, University of Oxford, Oxford, U.K.

Shubha Sathyendranath[1]

Oceanography Department, Dalhousie University, Halifax, N.S., Canada.

Wolfgang Barkmann

Oceanography Department, University of Southampton, Southampton, U.K.

The current status of the Sverdrup theory for the initiation of plankton blooms is examined. A prescription is given for the computation of the Sverdrup critical depth, using recently-published algorithms for mixed-layer primary production and a generalised loss term. Using no further information, the intrinsic rate of increase of phytoplankton biomass in the mixed layer can also be found. This rate, compared against the local frequency of storm occurrence, provides an alternative criterion for the initiation of blooms. The Eulerian (bulk property) methods used to derive these results are contrasted with the Lagrangian Ensemble method. The Lagrangian approach provides one avenue to the elaboration of the Sverdrup criterion to include the effect of processes with characteristic timescales small compared to one day. The incidence of blooms in the apparent absence of vertical stratification is reviewed: it is concluded that these observations do not undermine the basic logic of the Sverdrup theory. However, they do provoke interest in a re-examination of the feedbacks between the physical and biological dynamics in the mixed layer: an example is given. Finally, suggestions are made for further work in this subject area.

1. INTRODUCTION

Nansen was an eclectic scientist, with a range of interests from the nervous system of fish to the circulation of the polar ocean. Indeed, Norway has been blessed with a disproportionate share of oceanographers of great intellectual breadth. One example where this breadth of vision has allowed Norwegian scientists to make fundamental contributions concerns our understanding of the relation between primary production and local, physical oceanographic conditions. As described by *Mills* [1989], the pioneers in this field were Gran and Braarud. Their conclusions, won by hard work at sea over many years, were formalised in a famous theoretical paper published by Sverdrup in 1953. So if 1993 is the centenary of the Nansen voyage, it is also the 40th anniversary of another important Norwegian contribution, one of the landmarks in modern, biological oceanography. In this paper we shall examine what Sverdrup's theory has to tell us after forty years, and where it might lead us in the years ahead.

[1] Also at Bedford Institute of Oceanography, Dartmouth, N.S., Canada.

The Polar Oceans and Their Role in Shaping the Global Environment
Geophysical Monograph 85

2. THE BASIC THEORY

The Sverdrup theory refers to watercolumn properties, not to properties at discrete depths. In other words, it refers to quantities integrated over depth. The lower limit of integration will always be the surface (depth $z = 0$, where z is positive downwards). The upper limit of integration will be allowed to vary: the object of the analysis is to find the value of the limit that satisfies the conditions of the problem. Initially, let Z_a be the arbitrary depth to which the integration is made.

For the layer from surface to Z_a, we calculate the vertically-averaged, daily rates of primary production and losses by respiration. Let $P_{Z_a,T}$ be the daily primary production for the layer. Then the vertically-averaged production will be $(Z_a)^{-1}P_{Z_a,T}$. It is an essential feature of the theory that the layer is assumed to be well mixed. Thus we can assume that the biomass is distributed uniformly with depth and that the loss terms are independent of depth. Let B be the autotrophic biomass and L^B the hourly, biomass-specific, loss rate [*Platt et al.*, 1991]. Then the vertically-averaged daily loss is simply $24BL^B$. As pointed out by *Smetacek and Passow* [1990], Sverdrup intended the loss rate to embrace more than just the dark respiration of phytoplankton. In fact, we should interpret it in its broadest biogeochemical sense to mean any process by which the products of photosynthesis are remineralised or removed from the mixed layer [*Platt et al.*, 1991]. Thus L^B is a generalised loss rate for the entire pelagic community, consistent with the time and space scales under discussion.

The Sverdrup theory is concerned directly with the significance of vertical mixing for the local primary production. Let us therefore identify Z_a with the depth of the surface mixed layer Z_m. The theory now asserts that development of a phytoplankton bloom is possible in the layer if

$$(Z_m)^{-1}P_{Z_m,T} \geq 24BL^B. \quad (1)$$

Observe that, because primary production is forced by irradiance, an exponentially-decreasing function of depth, the average primary production for the layer will decrease as Z_m increases. On the other hand, the average loss rate $24BL^B$ is independent of depth. Hence, the possibility that the condition expressed in (1) is true will be strengthened as Z_m decreases. From its maximum value in the winter, Z_m begins to decrease with the increase of surface heating in Spring. Eventually, Z_m becomes sufficiently shallow that the equality in (1) is satisfied. The depth for which the equality holds is called the critical depth Z_m^{cr}. The Sverdrup criterion for the initiation of a phytoplankton bloom is $Z_m < Z_m^{cr}$.

3. ESTIMATION OF CRITICAL DEPTH

Computation of $P_{Z_m^{cr},T}$ is simplified by the (justifiable) assumption of a uniform biomass profile in the mixed layer. *Sverdrup* [1953] ignored the wavelength dependence of primary production, and all subsequent authors have followed suit in the estimation of critical depth. For simplicity, we suppress spectral effects in this paper, noting at the same time that systematic relationships exist between the spectral and non-spectral results for vertically-uniform models [*Platt and Sathyendranath*, 1991].

The available non-spectral models have been reviewed recently [*Platt and Sathyendranath*, 1993]. Here, we choose the model of *Platt et al.* [1990]. In its simplest form, this model can be written as

$$P_{Z_m^{cr},T} = \frac{BP_m^B D}{K}\sum_{x=1}^{5}\Omega_x\,(I_*^m)^x\left(1 - e^{-xKZ_m^{cr}}\right), \quad (2)$$

where P_m^B is the assimilation number, D is the day-length in hours, K is the vertical attenuation coefficient for photosynthetically-active radiation, I_*^m is the noon surface irradiance scaled to the photoadaptation parameter I_k, and the Ω_x are weights whose magnitudes are known (when the scale factor is taken to be $BP_m^B D/K$, as here, the weights differ by a factor of order π from those given in *Platt et al.* [1990]: see *Platt and Sathyendranath* [1993]).

We can then state the equality in (1) as

$$\frac{P_m^B D}{K}\sum_{x=1}^{5}\Omega_x\,(I_*^m)^x\left(1 - e^{-xKZ_m^{cr}}\right) = 24L^B Z_m^{cr}. \quad (3)$$

This equation may be solved, by iteration, for Z_m^{cr}. The most difficult step is the assignment of a value for the generalised loss term L^B: a detailed discussion of this step is given in *Platt et al.* [1991].

4. REALISED RATE OF INCREASE

The Sverdrup criterion tells us only whether net growth is possible in the mixed layer. It does not tell us how rapid the growth will be. In other words, it is a necessary but not sufficient condition for the initiation of a bloom.

Without invoking any new principles or any additional information, we can estimate the rate of biomass increase in the mixed layer [*Platt et al.*, 1991]. The easiest way to see this is to simplify, as Sverdrup did, the calculation of daily primary production. He assumed that the mixed layer was forced with a constant, depth-, and time-averaged irradiance $\langle I_* \rangle_{Z_m,T}$. With a sinusoidal variation of irradiance through the day, $\langle I_* \rangle_{Z_m,T} = (2I_*^m/\pi K Z_m)(1 - e^{-KZ_m})$. Then, the daily production for the mixed layer is

$$P_{Z_m,T} = (2/\pi)(BP_m^B D/K)I_*^m(1 - e^{-KZ_m}) \quad (4)$$

Now suppose that the biomass increase in the mixed layer can be described by exponential growth in the standard notation of population dynamics; see also *Mitchell and Holm-Hansen* [1991] and *Mitchell et al.* [1991]. That is

$$B(t) = B(0)e^{(g-l)t}, \quad (5)$$

where t is the time, and where g and l are, respectively, the exponential growth and loss rates. Their algebraic sum $r = g - l$ is the realised rate of increase.

It can be shown [*Platt et al.*, 1991] that equation (5) leads to an expression for g of the form

$$g = \log_e \left(1 + \frac{2\eta}{\pi K Z_m}(1 - e^{-KZ_m}) \right). \quad (6)$$

In this equation, η is the dimensionless quantity $\eta = P_m^B D I_*^m / \chi$, where χ is the carbon-to-chlorophyll ratio. Thus, g is maximal for low values of the dimensionless product KZ_m. A high value of the mixed-layer depth tends to lower the magnitude of g but this effect is partially offset by the initially-low value of K in pre-bloom conditions: observe that K and Z_m occur everywhere as a product in equation (6). To the extent that changes in K and Z_m do not offset each other so as to keep their product roughly constant, g and l will be a functions of time, and equation (5) should be applied with caution.

The corresponding equation for the realised rate of increase is

$$r = \log_e \left(1 + \frac{2\eta}{\pi K Z_m}(1 - e^{-KZ_m}) - \frac{24L^B}{\chi} \right). \quad (7)$$

Hence, rather than taking the Sverdrup theory as providing a condition for the possibility of positive net growth, we can see it as leading to a nominal timescale $r^{-1} \log_e 2$ or $0.69r^{-1}$ for the biomass in the mixed layer to double. Let us define the occurrence of a bloom as an increase in biomass by a factor of ten, or some 3.3 doublings. This will take a time $\sim 2.27r^{-1}$. We can then see that a bloom is likely if the surface mixed layer remains undisturbed by storms for a time period of order a few multiples of r.

5. EULERIAN AND LAGRANGIAN METHODS

So far we have employed strictly Eulerian (bulk-property) methods of calculation. Because the phytoplankton cells are suspended in a turbulent medium, because the problem under study relates to the effect of vertical mixing on local phytoplankton cells, and because the forcing variable for primary production (irradiance) is an exponential function of depth, we should consider whether a Lagrangian approach would be desirable. Some of the important differences between the two approaches are listed in Table 1.

The Lagrangian Ensemble (LE) method of integrating a mathematical model of the upper ocean has been developed by Woods and his colleagues over the last twelve years [*Woods and Onken*, 1982; *Wolf and Woods*, 1989; *Woods and Barkmann*, 1993*a*]. In the LE method, the phytoplankton biomass is represented by an ensemble of a few thousand particles, each of which contains a population of cells, which changes as the cells divide or are lost to grazing. Each particle follows a trajectory computed from the rules governing displacement of a single phytoplankton cell by the combination of its own sinking through the water and mixing by turbulence. As the particle progresses along its trajectory it experiences a unique variation of ambient environment. The biological equations are integrated in the LE method independently for each particle under the influence of its own time-varying ambient environment, including irradiance, temperature, etc. The net energetics of plankton in the mixed layer is computed by summing the separate energy balance of every particle in the ensemble that lies in that layer.

Woods and Barkmann [1993*b*] have recently used the LE method to integrate a model of the upper-ocean ecosystem that incorporates one guild (functional group) of phytoplankton (like diatoms) and one guild of herbivorous zooplankton (copepods like *Calanus*), in a physical environment described by solar radiation in 27 spectral bands, turbulence and temperature, and a

Table 1. Approaches to primary production models.

Bulk-property models	Ensemble models
Eulerian	Lagrangian
Biomass concentration	Individual cells
Variances ignored	Variances significant
Analytic	Monte Carlo
Computationally-efficient	Computationally-intensive
Less rigorous	More rigorous

chemical environment described by dissolved nitrate, ammonium and inorganic plus particulate nitrogen and carbon in phytoplankton detritus and faecal pellets. The simulated environment included biofeedback due to the optical effects of plankton biomass, and nitrate and carbon chemistry. They used the model to compare the seasonal variations of compensation depth and mixed-layer depth under conditions of diurnal variation of insolation and monthly-mean surface climatology. The same model has since been used to compute the seasonal variation of critical depth [*Woods and Barkmann* 1993*b*].

6. BLOOMS IN APPARENT ABSENCE OF STRATIFICATION

In recent years, several reports have appeared of spring phytoplankton blooms in the apparent absence of vertical stratification [*Eilertsen et al.*, 1981; *Eilertsen et al.*, 1989; *Townsend et al.*, 1992; *Eilertsen*, 1993]. At first sight, these reports might be taken to mean that the Sverdrup theory is violated. We now examine this impression in the light of the formalism outlined above.

Dealing with the loss terms first, we note from equation (1) that large values of Z_m^{cr} are possible if L^B is low. This may very well be true, especially early in the Spring, and particularly for the grazing contribution to L^B. In any case, to our knowledge, L^B has not been measured directly in the context of any study of critical depth. Indeed, it is doubtful whether all the components of the generalised respiration term L^B have ever been measured at the same station in any context. So the first possibility is that, at the stations where blooms have been reported in the apparent absence of stratification (mixed layer depth in exess, say, of 100 m), the critical depth is deeper than the mixed layer depth by virtue of a very low loss term.

Turning now to the growth term, consider the exponential rate g as determined by equation (6). The dimensionless quantity η increases with increasing daylength. Especially in high latitudes, daylength increases rapidly in the Spring: for example at 75°latitude, D increases by five hours in the first ten days after the sun shows above the horizon. The magnitude of η is also favoured by low values of the carbon-to-chlorophyll ratio, significant for the blooming of diatom assemblages [*Langdon*, 1988]. Thus, from the viewpoint of the influence of the growth terms on critical depth, high values of the critical depth are at least partially accounted for by the inevitable increase in D every Spring, a point documented so well by *Eilertsen* [1993].

The final point to make about blooms in the apparent absence of stratification is that the conventional way of judging stratification, inspection of a conductivity-temperature-depth profile, is not a secure way of diagnosing the extent of the layer in which active vertical mixing occurs [*Oakey and Elliott*, 1982; *Woods and Barkmann*, 1986; *Federov and Ginzburg*, 1992]. This can only be determined by direct estimation of turbulent energy dissipation rates from observations of temperature or velocity microstructure. From these data it will usually be found that vertical mixing extends over a smaller vertical interval than would be deduced from inspection of a standard hydrographic profile. This will increase the probability, at any station, that the mixed layer depth is less than the critical depth, and that a bloom is therefore possible.

We conclude that the recent observations of phytoplankton blooms in the apparent absence of stratification do not undermine the utility of the Sverdrup criterion. They may be an indication that the mixed-layer depth has been misjudged; they may simply be a result of a low magnitude for the loss term L^B; they may be a manifestation of the abrupt increase in the rate of biomass increase in the mixed layer associated with the increasing daylength early in the year. It is likely that more than one, or all of these factors, are implicated in the typical case.

7. FEEDBACK BETWEEN THE PHYSICAL AND BIOLOGICAL DYNAMICS

In the previous section we referred to the difficulty of judging the extent of the layer of active mixing given only a hydrographic profile. A theoretical approach allows to diagnose the lower boundary of this layer (the turbocline *sensu Woods and Barkmann* [1986]). The turbocline depth is determined by the balance between the heat gain and heat loss terms for the water column. Whereas the heat loss terms are localised at the surface, the heat gain terms are distributed through the water column to an extent that depends on the optical attenuation coefficient: greater attenuation implies reduced turbocline depth. Because K depends on the biomass (indeed is often controlled by the biomass), the turbocline depth will be a function of the biomass, a point made earlier by *Denman* [1973]. At the same time, the growth rate of the phytoplankton is, by equation (6), a strong function of the product of turbocline depth and attenuation coefficient. Hence, we have the possibility of a tight coupling and feedback between the physical and biological dynamics in the mixed layer. Is this feedback of potential significance? Is it strong enough that phytoplankton assemblages can create the stratification that will favour their growth (a possibility raised by *Platt et al.* [1991] and by *Townsend et al.* [1992])? This is the subject of the present section.

The Sverdrup model of critical depth can be used to evaluate whether the stratification of the water column is sufficient to favour net growth of phytoplankton. The same formalism can easily be extended to examine the reciprocal effect of phytoplankton growth on stratification.

It is easy to see that the net growth of phytoplankton in the mixed layer, initiated when the mixed-layer depth becomes shallower than the critical depth, would lead to an increase in the absorption of sunlight within the mixed layer, compared to a clear-water case. Since only a small fraction of the incident radiation (usually less than 1%) is used for photosynthesis, the bulk of the absorbed energy is dissipated as heat, leading to an increase in the local temperature, and a shoaling of the mixed layer depth.

To assess the significance of this effect, we carried out a simple simulation for a spring bloom at 60°N. In the simulation, the total short-wave radiation reaching the sea-surface was computed using a clear-sky model, as in *Platt et al.* [1990]. Typical values of monthly-mean, surface-heat fluxes at this latitude for the period from January to April were taken from *Isemer and Hasse* [1987], and fitted to a smooth function. The computations were begun on the 75th day of the year, when the net heat transfer at the surface became positive, and the mixed-layer began to shallow. The mixed-layer depth was computed as the equilibrium depth, according to the model of *Denman* [1973]. The initial mixed-layer depth was 210 m, and the initial biomass in the water column was assumed to be $0.1\,\mathrm{mg\,Chl\,m^{-3}}$. The attenuation coefficient K for the short-wave radiation was assumed to be a linear function of biomass B.

In one simulation, the decrease in mixed-layer depth was computed from day 75 to day 120, without allowing for changes in the attenuation coefficient due to phytoplankton growth. In the second simulation, we allowed the changes in biomass to modify the value of K. To compute the phytoplankton growth in the second simulation we used the mixed-layer production model of *Platt and Sathyendranath* [1991], assigning a value of $3\,\mathrm{mg\,C\,(mg\,Chl)^{-1}\,h^{-1}}$ to the assimilation number P_m^B and a value of $0.1\,\mathrm{mg\,C\,(mg\,Chl)^{-1}\,(W\,m^{-2})^{-1}\,h^{-1}}$ to α^B, the initial slope of the photosynthesis-light curve. We also assumed that net losses (L^B) were 90% of the mixed-layer production. In this simulation, the biomass increased to $7\,\mathrm{mg\,Chl\,m^{-3}}$ by day 120, and K increased to $0.43\,\mathrm{m^{-1}}$. Both the simulations were forced with a steady wind of $10\,\mathrm{m\,s^{-1}}$. In neither simulation did we account for changes in back radiation or in sensible heat transfer with changes in the sea-surface temperature.

The results of the two simulations are shown in Figure 1. In the first simulation (no bio-feedback), the mixed layer shallowed to 70 m by day 120, whereas in the second case, the mixed-layer shallowing was faster, reaching 38 m by the same date. These preliminary results are sufficiently interesting to encourage further work with more realistic conditions, such as variable winds, and accounting for changes in nutrient availability and air-sea fluxes with changes in stratification. For the present, we are encouraged in the view that the biofeedback in mixed-layer dynamics is a plausible mechanism.

8. DIRECTIONS FOR FUTURE WORK

Sverdrup's model outlines, in a simple and elegant manner, how physical conditions control the growth of phytoplankton in the upper well-mixed layer of the ocean. His insights have been a stimulus to research in biological oceanography during the last forty years. Yet, the potential of Sverdrup-type models is still not fully explored. One fruitful direction for future research could be the study of biofeedback mechanisms that could modify the physical properties of the upper ocean. It is interesting to note that, although models of atmospheric circulation or heat budget include the effect of clouds as a matter of course, models of ocean dynamics that include the effect of phytoplankton (which may be considered in this context merely as clouds of particles in the ocean) are still the exception (see, for example, *Simonot et al.* [1988]) rather than the rule.

One reason could be our failure to realise in the past that biologically-induced variations in optical properties could be significant in open-ocean waters: let us recall that, according to the classification of *Jerlov* [1968], most open-ocean waters, with the exception of major upwelling zones, would fall into a single optical class. Satellite-derived images of ocean colour have discredited this simple view and opened our eyes to the highly dynamic and variable nature of phytoplankton distribution in the ocean. It is now generally accepted that, at least in the open-ocean waters, light transmission characteristics can be modelled as a function of phytoplankton concentration; such models would include, either explicitly or implicitly, the role of other (dissolved and particulate) material that co-exists with phytoplankton. But the development of coupled biological-physical models also requires information on the rate constants for phytoplankton growth and loss. Although considerable progress has been made towards understanding of the natural variability in phytoplankton growth rates, there is still much to be learned about the factors that affect these rates in natural environments. This state-

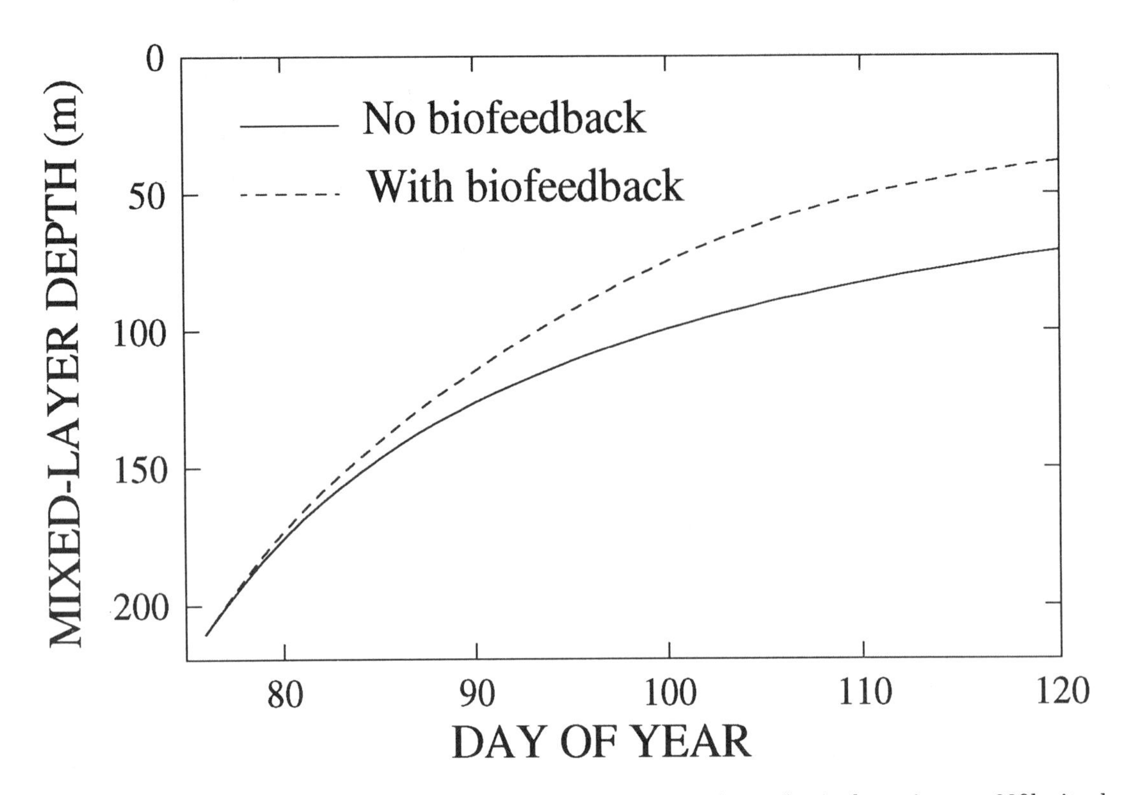

Figure 1. Evolution of the mixed-layer depth with time at a hypothetical station at 60°latitude, computed using the model of *Denman* [1973], with parameters as given in the text. Results of two simulations are given. In the simulation with no biofeedback, the attenuation coefficient does not change with time. In the case with biofeedback, the attenuation coefficient changes as a function of biomass, which increases during the spring bloom.

ment would be even more true for the phytoplankton loss terms.

Coupled biological-physical models for the oceanic mixed layer, of the type discussed in the previous section, could be refined to include spectral effects in light transmission and in photosynthesis. More realistic models would also include the effect of variable winds, nutrients, and air-sea fluxes. The model discussed here also assumes (reasonably) that the biomass distribution in the mixed-layer is uniform. But the evolution of the vertical stratification is also affected by light absorption below the mixed layer, and this depends, in turn, on the vertical biomass profile. Although satellites could provide information on biomass variability in the surface layer, information on vertical profile of biomass would still have to come from in situ observations.

For definitive field work in this area, it will be necessary to estimate the loss terms for biomass as well as the growth terms (including the parameters of the photosynthesis-light curve). Further, confusion about the extent of the layer subject to active mixing should be eliminated by direct measurements of turbulent microstructure. Indeed, it will be of interest to study the influence of phytoplankton on the temperature microstructure itself, and the feedbacks between the microstructure in the physical and biological fields.

If the definition of the mixed layer were taken to be the layer in which properties are vertically uniform, it could be that the biomass profile itself would be a better indicator than temperature and salinity (see, for example, Figure 1 in *Eilertsen* [1993]). To marry the field results with theory, it will be preferable to compute the turbocline depth, imposing the requirement to measure also the wind stress and the surface heat balance.

Finally, detailed numerical simulations of phytoplankton dynamics by the LE method should be continued and extended, with the objective of reconciling the results of the Lagrangian approach with those from the bulk-property approach.

9. STATUS OF THE SVERDRUP THEORY

Notwithstanding recent criticisms of it, the Sverdrup theory is based principles simple enough to be irrefutable. It is a necessary but not sufficient condition for the development of a phytoplankton bloom. Less generally appreciated is its value as a criterion for the breakdown of a bloom, for example as a consequence of the increased optical attenuation associated with the blooming algae, even in the presence of nutrients. At times, the Sverdrup criterion for bloom initiation is so weak that it is of little interest: it leads to a test with low resolving power. Of more universal value will be the exponential growth rates (positive or negative) expected in the mixed layer. Beyond telling us whether a bloom is possible, these will forecast how rapidly the biomass can be expected to accumulate. We have seen how this information can be used to help diagnose the evolution of the turbocline depth.

There is a sense in which the Sverdrup theory has outlived its initial purpose. But it has pointed us in the right direction to study one of the leading problems of the day: to understand the detailed dynamics of the physical and biological fields, and their mutual interactions, in the mixed layer of the ocean.

Acknowledgments. Norwegian scientists have played a fundamental role in our understanding of the dependence of primary production on the physical structure of the ocean water column. Gran and Braarud were pioneers in this field. Sverdrup crystallised their thinking in a justly-famous paper published in 1953. Forty years later, this subject area is still a rich one for research and discussion. The Nansen symposium is a suitable occasion to assess the present status of critical-depth theory: it provides a vantage point from which, in the spirit of these early Norwegian oceanographers, we can seek the way ahead to further understanding.

The work presented in this paper was supported, in part, by the Office of Naval Research, the National Aeronautics and Space Administration, and the Natural Sciences and Engineering Research Council, Canada. This work was carried out as part of the Canadian contribution to the Joint Global Ocean Flux Study.

REFERENCES

Denman, K. L., A time-dependent model of the upper ocean, *J. Phys. Oceanogr.*, *3*, 173–184, 1973.

Denman, K. L., and M. Miyake, Upper layer modification at ocean station *Papa*: observations and simulation, *J. Phys. Oceanogr.*, *3*, 185–196, 1973.

Eilertsen, H. C., Spring blooms and stratification, *Nature*, *363*, 24, 1993.

Eilertsen, H. C., B. Schei, and J. P. Taasen, Investigations on the plankton community of Balsfjorden, northern Norway. The phytoplankton 1976-1978. Abundance, species composition, and succession, *Sarsia*, *66*, 129–141, 1981.

Eilertsen, H. C., J. P. Taasen, and J. M. Węslawski, Phytoplankton studies in the fjords of West Spitsbergen: physical environment and production in spring and summer, *J. Plankton Res.*, *11*, 1245–1260, 1989.

Fedorov, K. N., and A. I. Ginzburg, The near-surface layer of the ocean, p. 259, Zeist, VSP, 1992.

Isemer, H.-J., and L. Hasse, *The Bunker Climate Atlas of the North Atlantic Ocean. 2. Air-Sea Interactions*, p. 252, Springer-Verlag, Berlin, New York, 1987.

Jerlov, N. G., *Optical Oceanography*, p. 194, Elsevier Publishing Company, Amsterdam, London, New York, 1968.

Langdon, C., On the causes of interspecific differences in the growth-irradiance relationship for phytoplankton. II. A general review, *J. Plankton Res.*, *10*, 1291–1312, 1988.

Mills, E. L., *Biological Oceanography. An Early History, 1870-1960*, p. 378, Cornell University Press, Ithaca, London, 1989.

Mitchell, B. G., and O. Holm-Hansen, Observations and modeling of the Antarctic phytoplankton crop in relation to mixing depth, *Deep Sea Res.*, *38*, 981–1007, 1991.

Mitchell, B. G., E. A. Brody, O. Holm-Hansen, C. McClain, and J. Bishop, Light limitation of phytoplankton biomass and macronutrient utilization in the Southern Ocean, *Limnol. Oceanogr.*, *36*, 1662–1677, 1991.

Oakey, N. S., and Elliott, J. A., Dissipation within the surface layer, *J. Phys. Oceanogr.*, *12*, 171–185, 1982.

Platt, T., and S. Sathyendranath, Biological production models as elements of coupled, atmosphere-ocean models for climate research, *J. Geophys. Res.*, *96*, 2585–2592, 1991.

Platt, T., and S. Sathyendranath, Estimators of primary production for interpretation of remotely sensed data on ocean colour, *J. Geophys. Res.*, *98*, 14561–14576, 1993.

Platt, T., S. Sathyendranath, and P. Ravindran, Primary production by phytoplankton: analytic solutions for daily rates per unit area of water surface, *Proc. R. Soc. Lond. Ser. B*, *241*, 101–111, 1990.

Platt, T., D. F. Bird, and S. Sathyendranath, Critical depth and marine primary production, *Proc. R. Soc. Lond. Ser. B*, *246*, 205–217, 1991.

Simonot, J.-Y., E. Dollinger, and H. Le Treut, Thermodynamic-biological-optical coupling in the oceanic mixed layer, *J. Geophys. Res.*, *93*, 8193–8202, 1988.

Smetacek, V., and U. Passow, Spring bloom initiation and Sverdrup's critical-depth model, *Limnol. Oceanogr.*, *35*, 228–234, 1990.

Sverdrup, H. U., On conditions for the vernal blooming of phytoplankton, *Journal du Conseil*, *18*, 287–295, 1953.

Townsend, D. W., M. D. Keller, M. E. Sieracki, and S. G. Ackleson, Spring phytoplankton blooms in the absence of vertical water column stratification, *Nature*, *360*, 59–62, 1992.

Wolf, U. and J. D. Woods, Lagrangian simulation of primary production in the physical environment – the deep chlorophyll maximum, in *Towards a Theory of Biological-Physical Interactions in the World Ocean*, edited by B. J. Rothschild, pp. 51–70, Kluver, Dordrecht, 1988.

Woods, J. D., and W. Barkmann, The response of the upper ocean to solar heating. I: The mixed layer, *Quart. J. R. Meteorol. Soc.*, *112*, 1–27, 1986.

Woods, J. D., and W. Barkmann, The phytoplankton multiplier, positive feedback in the greenhouse, *J. Plankton Res.*, *15*, 1053–1074, 1993*a*.

Woods, J. D., and W. Barkmann, Diatom demography in winter – simulated by the Lagrangian Ensemble method, *Fisheries Oceanography, 2*, 202–222, 1993*b*.

Woods, J. D., and R. Onken, Diurnal variation and primary production in the ocean – preliminary results of a Lagrangian Ensemble model, *J. Plankton Res., 4*, 735–756, 1982.

Wolfgang Barkmann, Oceanography Department, University of Southampton, Southampton, U.K.

Trevor Platt and Shubha Sathyendranath, Department of Fisheries and Oceans, Biological Oceanography Division, Bedford Institute of Oceanography, Dartmouth, N.S., Canada B2Y 4A2.

John D. Woods, NERC Oceanography Unit, Physics Department, University of Oxford, Oxford, U.K.

Distribution of Dissolved Inorganic and Organic Carbon in the Eurasian Basin of the Arctic Ocean

Leif G. Anderson, Kristina Olsson and Annelie Skoog

Department of Analytical and Marine Chemistry, University of Göteborg and Chalmers University of Technology, Göteborg, Sweden

Measurements of total carbonate (total inorganic carbon) and total organic carbon were carried out during the International Arctic Ocean Expedition 91 on three sections across the Nansen and Amundsen basins, and into the Makarov Basin. In the surface mixed layer a distinct front was observed to the north, identified by elevated total carbonate concentrations, reflecting the signature of Siberian river runoff. This front forms the southern border of the Siberian Branch of the Transpolar Drift, which is located over the northern part of the Nansen Basin, at the eastern part of the investigated area (about 30°E). To the west, the front is located over the tip of the Yermak Plateau, showing the very wide extent of the Transpolar Drift just north of the Fram Strait. Within the Siberian Branch, about 200 km north of the front, a maximum in the total organic carbon concentration may indicate a shorter residence time since the water left the river mouth, possibly related to a higher flow rate. At the tip of the Morris Jesup Plateau, clear signs of outflowing Canadian Basin waters are seen. In the surface layer, high total carbonate concentrations were present in the water of salinity 33.1, indicating upper halocline water. At intermediate depths (1500 - 2000 m), elevated total carbonate concentrations demonstrate the existence of the outflowing Canadian Basin intermediate water. Part of this outflowing water follows the continental slope to the south while part is injected into the Makarov Basin.

1. INTRODUCTION

River runoff into the Arctic Ocean may hold a key to global climate through its influence on Greenland Sea deep convection [*Aagaard and Carmack*, 1989; *Rudels*, 1989]. Deep water convection could be inhibited by stratification in the upper water layers as a result of outflowing freshwater from the Arctic Ocean via Fram Strait. Changes in global temperature may affect the extent of sea-ice meltwater as well as the river runoff contribution. Not only could this affect the convection gyres and thereby the global deep water ventilation but also the flow of chemical constituents in the area. Salinity and ice anomalies in the Greenland, Norwegian and Iceland seas have been detected in the 60´s, 70´s and 80´s [*Swift*, 1984] and the influence of freshwater input has been studied [*Schlosser et al.*, 1991]. The deep waters of the Arctic Ocean are renewed by advection through Fram Strait and by dense plumes originating on the shelves. The cause of the high density is the brine that is released during sea-ice formation over the large shelf areas. The detailed circulation of the deep waters are not well resolved yet, but it is assumed that the general circulation is topographically driven with several cyclonical gyres guided by the ridges. The distribution of both the organic and inorganic carbon can help to deduce the distribution and circulation of the Arctic Ocean water masses, especially the surface waters which show the strongest signals.

Carbon enters the Arctic Ocean by three different routes; the atmosphere via air-sea interactions, river run-off, and inflow of different water masses to the Arctic Ocean. The amount of carbon transported depends on (1) biological fixation of carbon dioxide into organic matter, which takes place on the tundra as well as in the ocean, (2) dissolution

The Polar Oceans and Their Role in Shaping
the Global Environment
Geophysical Monograph 85

of calcium carbonate, which also occurs both on the tundra and in the ocean, and (3) concentrations of carbon in the water masses flowing into and out of the Arctic Ocean.

The river runoff transports organic matter as well as carbonates into the Arctic Ocean. This in turn leads to higher total carbonate, total organic carbon as well as total alkalinity concentrations in the river runoff [*Anderson and Dyrssen*, 1981]. This offers a means to distinguish the riverine freshwater from that of sea-ice melt water. Rivers transport about 2,500 km^3 of freshwater each year (about 0.1 Sv) [*Aagaard and Carmack,* 1989] to the Arctic Ocean, of which approximately 2,200 km^3 come from Siberia. The river runoff thus acts as an important link between the terrestrial river basin carbon pool in Siberia and the dissolved carbon pool of the Arctic Ocean.

Carbon may also be redistributed within the Arctic Ocean by sea ice formation and melting in combination with biological and hydrographical processes on the shelves. Variations in sea-ice extent over the continental shelves are seasonal and affect the Arctic Ocean through the release of brine during ice formation, producing a dense water that sinks towards the shelf bottom. This water parcel follows the bottom towards the deep interior where it interleaves at its appropriate density level [*Aagaard and Coachman*, 1981; *Melling and Lewis,* 1982; *Rudels et al*., 1994]. The melting of sea-ice in spring is important for the biological primary production in that the meltwater increases stratification and because the absence of ice allows sunlight to reach the water without additional attenuation. At the end of the season, the bulk of the dead organic matter ends up at the sediment surface, where it slowly decays. The decay products diffuse back to the overlaying bottom water [*Anderson et al*., 1988]. The melting sea-ice can also have a high nutrient concentration, but as the volume is small the total amount added is of minor importance compared to other sources of nutrients.

The ice cover hampers gas exchange between the atmosphere and ocean, thus restricting gas transfer to the ice free shelf regions (including polynyas) and to leads within the ice pack. Gas exchange over the air-sea boundary is driven by differences in partial pressure at the interface. This difference in turn is due to the atmospheric carbon dioxide content, and the salinity, temperature, alkalinity and total carbonate content of the surface water. The total carbonate concentration in particular, but also alkalinity, are very sensitive to biological activity, which is mostly constrained to the shelf areas that are ice free during the summer periods. The continental shelves within the Arctic Ocean comprise more than 1/3 of the total area, and being shallow, they allow high productivity during the ice-free periods followed by slow decay of organic matter at the sediment surface in the winter [*Anderson and Jones*, 1991].

2. METHOD

2.1. *Cruise Description*

The International Arctic Ocean Expedition with the Swedish ice breaker *Oden* took place between the 1 August and the 14 October 1991 and covered the western parts of the Nansen and Amundsen basins as well as a few stations in the Makarov Basin; all together 57 stations were sampled (Figure 1). For the chemical and physical oceanography program, water samples were drawn from 10-L Niskin bottles mounted on 24- and 12-bottle rosette samplers (General Oceanic), respectively. Underneath the sampler units a CTD instrument (Neil Brown) was positioned. In order to achieve good coverage and surface resolution, the 24-bottle rosette was normally used from the surface down to bottom while the 12-bottle rosette overlapped in the upper 500 meters.

Sampling procedures started directly after the rosette had been transferred into a shelter to avoid freezing in the bottles. Altogether, a total of 41 variables were measured, including total carbonate and total organic carbon.

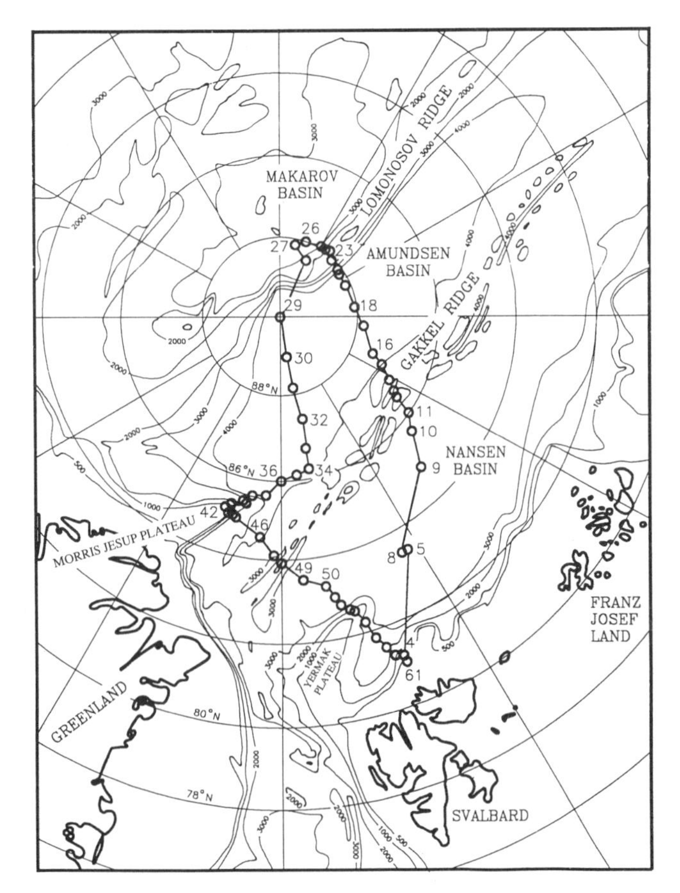

Fig. 1. Map of the Eurasian Basin and Fram Strait showing the oceanographic stations occupied during the *Oden* 91 expedition.

2.2. *Analytical Methods*

2.2.1. *Sampling techniques.* The seawater samples were drawn into 100-mL and 250-mL plastic capped Pyrex glass bottles for analysis of total organic carbon and total carbonate, respectively, and stored under dark and cooled conditions before analyses. For total organic carbon 5 mL of the sample was transferred to a quartz tube and acidified to approximately pH 3 with 2 m*M* HCl (pro analysis, Kebo) prior to analysis. Measurements of total carbonate took place at 10°C to avoid temperature effects on the cell solution as well as to minimize air bubble formation in the sampling loop, thus keeping the error of the volumetric determination to a minimum. All measurements were performed on board the ship within 24 hours of sampling.

2.2.2. *Sample analysis of total organic carbon (TOC).* The measurements of total organic carbon were performed with a Shimadzu TOC-5000 equipped with a Shimadzu auto sampler. Combustion of organic carbon to carbon dioxide takes place on a platinum surface at 670°C utilizing a normal-sensitivity aluminum oxide-based Shimadzu catalyst.

A standard stock solution of 100 m*M* potassium hydrogen phtalate (Kanto Chemical Company Inc.) in milli-RO water was prepared every ten days. The working standard solutions were prepared by successive dilution with milli-RO every day. A standard with a concentration of 80 μ*M* was used for calibration during the run. The built in system for calibration in the TOC-5000 was not used since this does not allow sparging of the standard solution. Instead, one set of measurements was organized in loops starting with initial background measurement followed by analyses of one standard and four samples respectively. All samples, including background and standards, were acidified and sparged for five minutes to drive off CO_2. A 100-μL volume of solution was injected and a background correction made assuming that the drift of the background was linear between every background measurement. Background values ranged between 30-60 μ*M* and the standard deviation related to the mean concentration was 8-10 %. For the background it was < 20%.

No prefiltration of the samples was done but the needle inlet of the instrument provided a direct physical separation of particles > 0.18 mm. In general, with the exception of a few surface samples, the amount of particulate organic matter in this region was extremely low and justifies why the results can be interpreted as dissolved organic carbon (*DOC*).

The ice-breaking motions of the vessel caused vibrations that interfered with the IR detector of the Schimazu TOC-5000. The analyses of total organic carbon therefore had to be carried out when the ship was on station. Because of the vibrations, the analytical precision was less than that obtained under laboratory conditions.

2.2.3. *Sample analysis of total carbonate (C_t)* For the determinations of inorganic carbon, a UIC 5011 Coulometer unit was used in combination with a gas extraction unit. After acidification of the sample (10% ortho-phosphoric acid, pro analysis, Merck) to a pH of 2-3, the carbon dioxide gas was extracted from the sample by the carrier gas ($N_2(g)$). The evolved carbon dioxide gas was then coulometrically titrated in a dimethylsulphoxide solution according to *Johnson et al.* [1985]. Duplicate analysis of samples and standards was performed. Standard solutions were made up from preweighed dry bisodium carbonate [*Dickson and Goyet*, 1991] and analyzed before changing the cell solution, i.e. in between each station. Background values were 10-16 μ*M* with the extraction time used. The standard deviation related to the mean concentration was 0.05-0.15% for samples and standards and < 8% for the background. The total carbonate concentrations were normalized to a salinity of 35 to remove the variation of total carbonate with salinity. The bisodium carbonate solutions used for calibration of the inorganic carbonate determinations showed day-to-day variations, therefore the sample concentrations were normalized against the Eurasian Basin deep water carbonate concentrations of 2190 μmol/kg [*Anderson et al.*, 1989].

3. RESULTS AND DISCUSSION

The most obvious features in the distribution of the normalized total carbonate (C_t-35) are the fairly constant concentrations in the deep waters (below about 250 meters), and the 150 μmol/kg higher surface concentrations. These features are seen in all three sections (Figure 2) except for the southern ends of sections A and D. A similar trend is also seen in the *TOC* distribution (Figure 3). The salinity of the surface water is below 33.5 all along the sections shown, except for stations 60 and 61. The general pattern of both the C_t-35 and *TOC* concentrations are a result of the large inflow of river runoff, high in both inorganic and organic carbon content as a result of the large bioactivity on the tundra, in combination with the flow pattern of the surface water. The C_t-35 front in the surface layer at about station 10 (section A) and Station 55 (section D) thus represents the border between (1) seawater where river runoff has been added as the fresh water source and (2) seawater where sea ice melt water has been added. Hence, the surface water from the Siberian shelves, rich in runoff, follows the Gakkel Ridge towards Fram Strait and the "front" represents the southern border of the Siberian Branch of the Transpolar

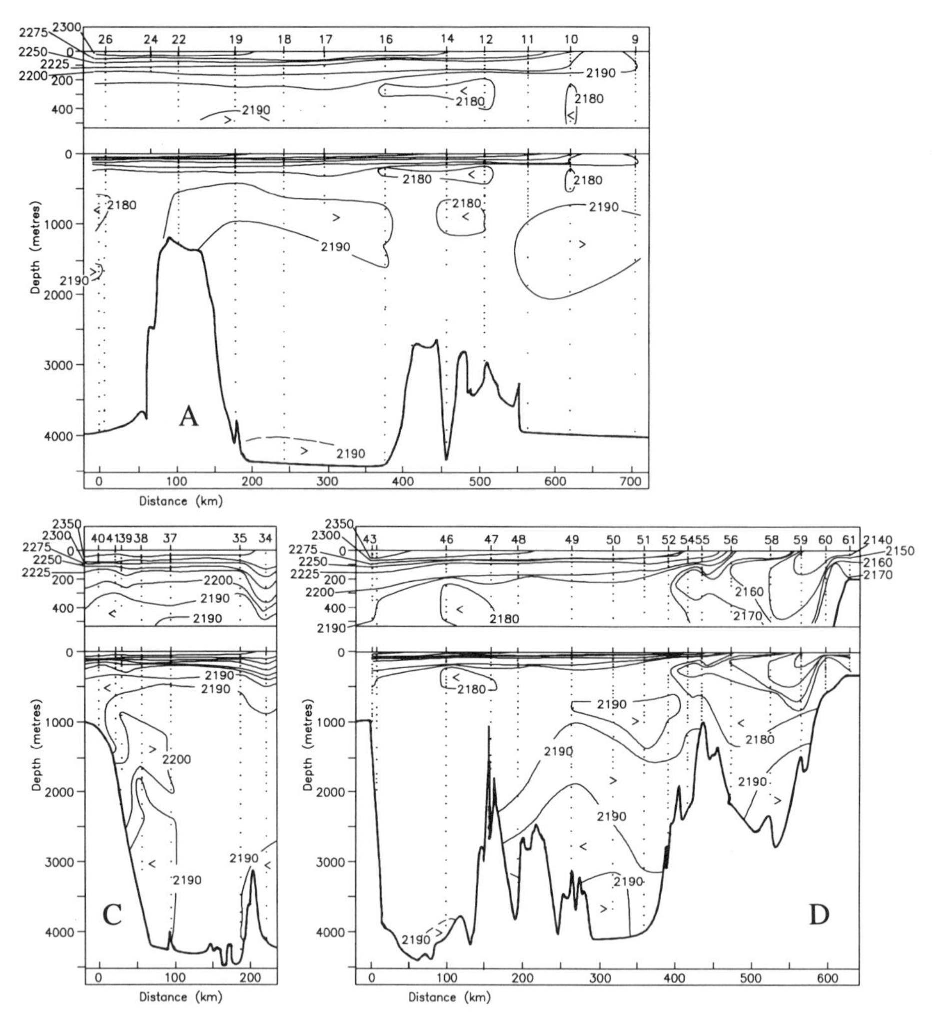

Fig. 2. The distribution of normalized total carbonate (µmol/kg seawater) along the four sections, A (stations 9 to 27), C (stations 34 to 41) and D (stations 43 to 61).

Drift [*Anderson and Jones*, 1992; *Schlosser et al.*, 1993]. The front in section A is situated just south of the Gakkel ridge, while it in section D is close to the Yermak Plateau (see Figure 1).

The very low C_t-35 concentration in the surface water of stations 59 - 61 coincides with the inflowing Atlantic water. Because primary production consumes total carbonate, these low values probably reflect that the bulk of the primary production took place before the water mass entered the ice-covered regions. High biological activity normally also results in the release of dissolved organic carbon, which probably is why no *TOC* minimum was observed in the same region. The elevated concentrations of *TOC* in the surface water showed the highest value in a maximum at station 49 (section D), over the southern flank of the Gakkel Ridge. In section A, the mean *TOC* concentration was about the same as in section D, except for a surface minimum north of the maximum at station 16. This maximum, however, was located over the northern flank of the Gakkel Ridge. The combination of the maxima in sections A and D might indicate a higher flow rate of the surface water along the ridge, if we assume a significant river runoff contribution and a decay rate of *TOC* that is in the right relationship to the transport time of water from the Laptev Sea to the investigated area, i.e. in the order of years [*Schlosser et al.*, 1993]. A significant river runoff component is seen from the C_t-35 data, which for both stations show concentrations from 2275 - 2300 µmol/kg. The decay rate of the *TOC*, however, is difficult to evaluate because it depends on the type of *TOC* (in the present study, more likely *DOC*). River run-off normally has a significant content of quite stable organic compounds, e.g. humic

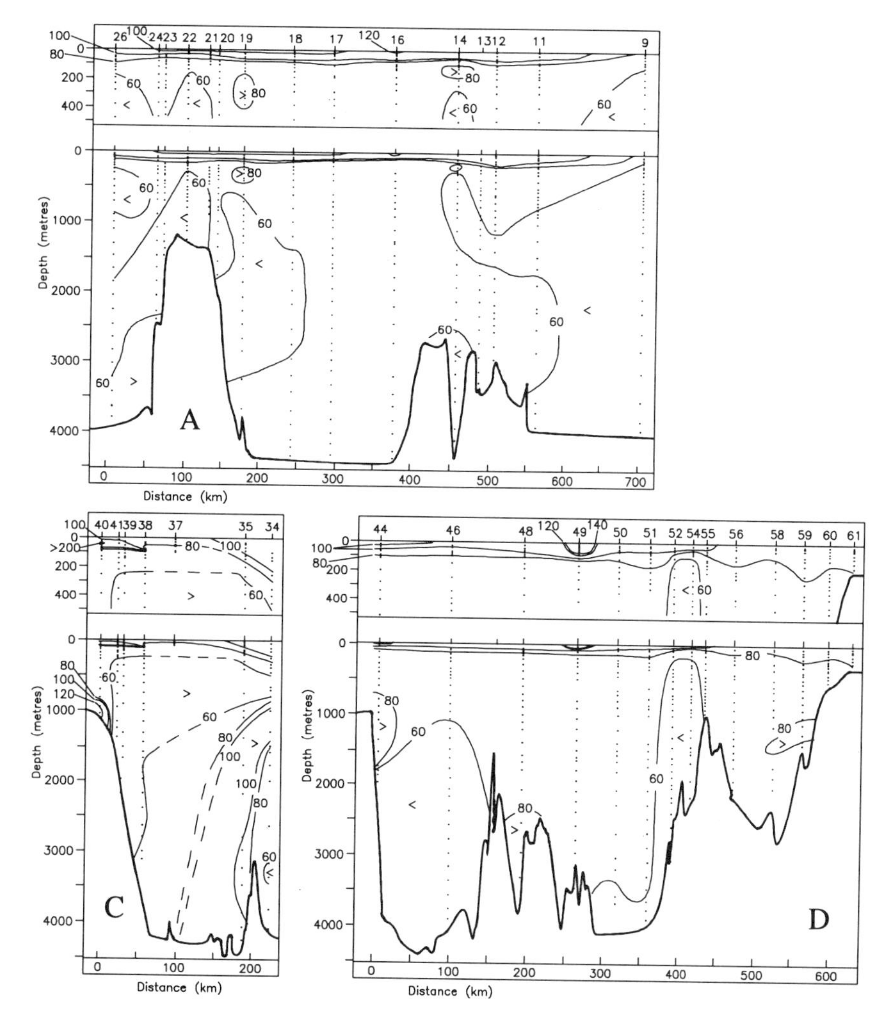

Fig. 3. The distribution of Total Organic Carbon (µmol/kg seawater) along sections A, C and D.

substances, and in a water parcel with a higher flow rate (that is, shorter time) these compounds may not have been degraded to the same extent as in waters of lower flow rate (that is, longer time). The combination of maxima situated over the northern and southern flank of the ridge could either indicate that the faster flowing surface water is meandering along the ridge, or that it is located at about the same distance from the runoff front. In both sections the surface waters with high *TOC* are found about 200 km from the river runoff front.

There was a small but significant increase in C_f-35 towards the bottom in the center of the basins (Figure 2), however, it is difficult to draw any firm conclusions regarding the total carbonate structure in the deep water. This increase is illustrated by an isoline for 2190 µmol/kg and applies to all sections except C, for which most of the water column is above 2190 µmol/kg. The elevated bottom water concentrations can be explained by the general circulation pattern [*Anderson et al.*, 1989; *Jones et al.*, 1991], which yields the oldest water along the bottom in the central area of the basins. The oldest water would be enriched in inorganic carbon as a result of decay of organic matter, which also decreases the oxygen concentration, and/or dissolution of metal carbonates. The oxygen distribution in the deep water (not shown here) have a consistent trend, but with a smaller gradient than what is expected from the decay of organic matter according to *Redfield et al.* [1963]. Hence, the elevated C_f-35 concentrations at larger depths presumably are a result of both decay of organic matter and dissolution of metal carbonates.

The reason for the major fraction of the water column in section C having C_f-35 values above 2190 µmol/kg could be that Canadian Basin intermediate water (the oldest water encountered during the cruise) flows along the shelf slope, at 1500 to 2000 m depth north of Greenland, and part of it is steered outwards by the Morris Jesup Plateau (stations 37-41) and into of the Amundsen Basin. This influence of Canadian Basin intermediate water has been seen in other constituents of the seawater [e.g. *Anderson et al.*, 1994].

The region above and north of the Morris Jesup Plateau exhibited large variability both at great depths and in the shallow layers (Figure 2C). The water situated deeper than 1000 m, with C_t-35 values above 2200 μmol/kg, comprises the core of the outflowing Canadian Basin intermediate water, as described above. The outflow from the Canadian Basin is also evident in shallower layers, where the nutrient signal of the upper halocline water is most apparent [*Jones et al.*, 1991]. Thus, elevated concentrations of C_t are located at 50 to 100 m depth at all stations but 41, on top of the Morris Jesup Plateau (Figure 4). This shows that the shallow outflow from the Canadian Basin forms a very narrow boundary current. Nowhere else in the Eurasian Basin was the C_t signature of the upper halocline water found, which puts a limit to the extent of this water mass. The lack of excess C_t at salinities around 33.1 and the increased concentrations of silicate at similar salinities in the Makarov Basin and over the Lomonosov Ridge [*Anderson et al.*, 1994] lead us to believe that the upper halocline water is confined to the Canadian Basin. The region where upper halocline water is found is more restricted than was assumed by *Anderson et al.* [1990] who calculated the input of total carbonate to different depth layers of the Arctic Ocean. They based their calculation of the input, on an extent of the upper halocline water equal to the deep Arctic Ocean basins. With the more restricted area of that water mass, found in this investigation, their estimate

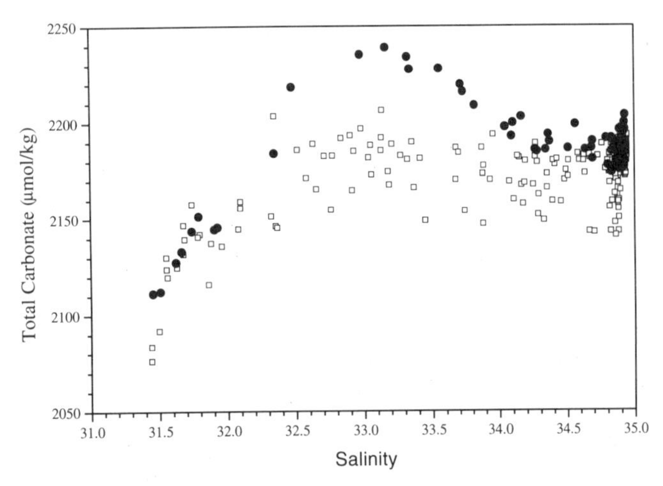

Fig. 4. Salinity versus normalized total carbonate for stations 29 to 39, 41 and 46 (open squares) and stations 39, 40, 43 and 44 (filled circles).

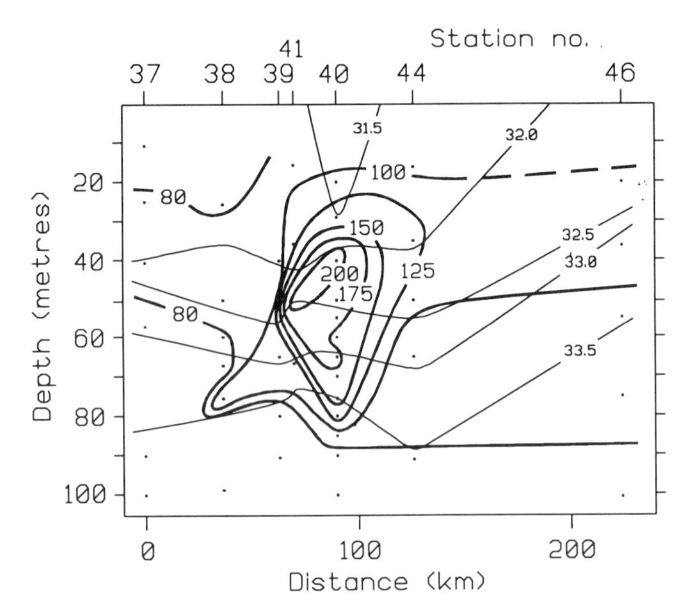

Fig. 5. The distribution of TOC in the top 100 meters (thick lines), over the tip of the Morris Jesup Plateau. The thin lines represent the salinity distribution as recorded by the CTD.

of the mean new production in the Laptev, East Siberian and Chukchi seas has to be decreased from about 45 to about 25 g C m^{-2} y^{-1}. This is an estimate of the mean new production over the whole of these three seas, however, it is likely that the bulk of the primary production is restricted to some local regions.

The *TOC* distribution in the top 100 meters at the stations over the Morris Jesup Plateau (Figure 5) shows a distinct maximum at a depth of about 50 meter, centered around stations 40 and 41. This *TOC* maximum is located at a lower salinity than the C_t maximum. Therefor it is not possible to directly relate them to each other. Moreover it is unlikely that dissolved organic carbon that has been added to the water over the shelves, and thereafter incorporated into the upper halocline water, could have survived the relatively long transit time (several decades according to *Wallace and More* [1985]). Furthermore, *Gordon and Cranford* [1985] did not report any elevated *DOC* concentrations within the upper halocline water at ice station CESAR. The source of this maximum is hence not obvious, but the distinct maximum at the tip of the Morris Jesup Plateau is yet another indication of the importance of boundary currents in the Arctic Ocean.

A less obvious feature is the low values for C_t-35 in the maximum temperature Atlantic layer, depth 200 - 500 m (Figure 2). The low values form this distinct minima with concentrations from 2180 - 2190 μmol/kg. One of them is in section A over the Gakkel Ridge (stations 12 - 16) and another in section D at the northern rim of the Gakkel Ridge (stations 46 and 47). Except for the surface waters, the Atlantic layer constitutes the youngest water mass and would therefore not be as much enriched in C_t because of decaying organic matter and dissolution of metal carbonates. Furthermore, the lowest C_t-35 values found in the investigated region occur in the inflowing Atlantic Layer (southern part of section D) and could be explained by recent high primary production. As primary production varies strongly seasonally it could be expected that the C_t concentration in the Atlantic layer within the investigated area should be fairly variable if mixing is limited. Indications of limited mixing is seen by the very sharp intrusions of the two Atlantic source waters, the Fram Strait branch and the Barents Sea branch, that are largely maintained within the whole investigated region [*Rudels et al.*, 1994].

4. SUMMARY

The distribution of normalized total carbonate in the surface mixed layer along the *Oden* 91 cruise track clearly shows a front between river runoff and sea-ice meltwater as the fresh water source. This front represents the southern border of the Siberian Branch of the Transpolar Drift and it is situated at about 30°E, over the northern part of the Amundsen Basin, and just north of Svalbard, over the tip of the Yermak Plateau. Hence, the Transpolar Drift is very wide just north of Fram Strait, and is forced to the west only by the inflowing Atlantic water, making it more narrow further south. Within the Siberian Branch, about 200 km north of the front, there is a core of surface water with high total organic carbon content which could indicate a higher flow rate.

The outflowing Canadian Basin intermediate water is distinguished by slightly elevated normalized total carbonate concentrations in the Amundsen Basin, to the east of the Morris Jesup Plateau. Over the tip of the Morris Jesup Plateau higher concentrations of total carbonate was present, at salinities around 33.1, showing the existence of upper halocline water. This is the only place where this signal was found, indicating that the upper halocline water is confined to the Canadian Basin. Bearing this in mind, we reduce the estimated mean new production over the Laptev, East Siberian and Chukchi Seas of *Anderson et al.* [1990] from 45 to 25 g C m^{-2} y^{-1}. A marked maximum in total organic carbon is also present over the Morris Jesup Plateau, but restricted to a salinity range 32 to 33. This is lower than the salinity of the upper halocline water and there is no obvious source of this maximum.

In conclusion, the distribution of both dissolved inorganic and organic carbon is a useful tool to evaluate the distribution of different water masses in the upper water column, and thereby also a mean to deduce the circulation pattern of these waters. Furthermore, the high *TOC* concentrations in a core of surface water following the Gakkel Ridge, adds information that has not been reported before. To explain the distribution in the total carbonate concentration between the different water masses investigated, we suggest some processes that likely have occurred during their history. In order to make a more detailed discussion of the general carbon cycle useful, we need increased information from the Canadian Basin and the Siberian shelf seas.

Acknowledgments. We express our gratitude to our colleagues on board *Oden* who made this investigation possible. Financial support from the Swedish Natural Science Research Council, the and Alice Wallenberg Foundation and the Marianne and Marcus Wallenberg Foundation is gratefully acknowledged.

6. REFERENCES

Aagaard, K., and E. C. Carmack, The role of sea ice and other fresh water in the Arctic circulation, *J. Geophys. Res., 94*, 14,485-14,498, 1989.

Aagaard, K., and L. K. Coachman, On the halocline of the Arctic Ocean, *Deep-Sea Res., 28*, 529-545, 1981.

Anderson, L., and D. Dyrssen, Chemical constituents of the Arctic Ocean in the Svalbard area, *Oceanol. Acta, 4*, 305-311, 1981.

Anderson, L. G., and E. P. Jones, The transport of CO_2 into Arctic and Antarctic seas: Similarities and differences in the driving processes, *J. Mar. Syst., 2*, 81-95, 1991.

Anderson, L. G., and E. P. Jones, Tracing upper waters of the Nansen Basin in the Arctic Ocean, *Deep-Sea Res., 39*, S45-S433, 1992.

Anderson, L. G., D. Dyrssen, and E. P. Jones, An assessment of the transport of atmospheric CO_2 into the Arctic Ocean,. *J. Geophys. Res.*, *95*, 1703-1711, 1990.

Anderson, L. G., E. P. Jones, R. Lindegren, B. Rudels, and P-I. Sehlstedt, Nutrient regeneration in cold, high salinity bottom water of the Arctic shelves, *Cont. Shelf Res., 8 (12)*, 1345-1355, 1988.

Anderson, L. G., E. P. Jones, K. P. Koltermann, P. Schlosser, J. H. Swift, and D. W. R. Wallace, The first oceanographic section across the Nansen Basin in the Arctic Ocean, *Deep-Sea Res., 36*, 475-482, 1989.

Anderson, L. G., G. Björk, O. Holby, E, P. Jones, G. Kattner, P. K. Koltermann, B. Liljeblad, R. Lindegren, B. Rudels, and J. H. Swift, Water masses and circulation in the Eurasian Basin: Results from the *Oden* 91 expedition, *J. Geophys. Res.*, in press, 1994.

Dickson, A. G., and C. Goyet, Handbook of methods for the analysis of the various parameters of the carbon dioxide system in the sea water, *DOE CO_2 analysis handbook*, U.S.DOE SRGP-89-7A, 1991.

Gordon, D. C., and P. J. Cranford, Vertical distribution of dissolved and particulate organic matter in the Arctic Ocean at

86°N and comparison with other regions, *Deep-Sea Res., 32*, 1221-1232, 1985.

Johnson, K. M., A. E. King, and J. McN. Sieburth, Coulometric TCO2 analyses for marine studies; an introduction, *Mar. Chem., 16*, 61-82, 1985.

Jones, E. P., L. G. Anderson, and D. W. R. Wallace, Tracers of near-surface, halocline and deep waters in the Arctic Ocean: Implications for circulation, *J. Mar. Syst.*, *2*, 241-255, 1991.

Melling, H., and E. L. Lewis, Shelf drainage flows in the Beaufort Sea and their effects on the Arctic Ocean pycnocline, *Deep-Sea Res., 29*, 967-985, 1982.

Redfield, A. C., B. H. Ketchum, and F. A. Richards, The influence of organisms on the composition of seawater, in *The Sea*, Vol 2, edited by M. N. Hill, pp. 26-77, Wiley, New York, 1963.

Rudels, B., The formation of polar surface water, the ice export and exchanges through the Fram Strait, *Prog. Oceanogr., 22*, 205-248, 1989.

Rudels, B., E. P. Jones, L. G. Anderson and G. Kattner, On the origin and circulation of Atlantic layer and intermediate depth waters in the Arctic Ocean, in *The Role of the Polar Oceans in Shaping the Global Environment*, edited by R. Muench and O. M. Johannessen, pp. ??-??, American Geophysical Union, Washington, D.C., 1994.

Schlosser, P., G. Bönisch, M. Rhein, and R. Bayer, Reduction of deep water formation in the Greenland Sea during the 1980's: Evidens from tracer data, *Science 251*, 1054-1056, 1991.

Schlosser, P., D. Grabitz, R. Fairbanks and G. Bönisch, Mean residence time of river-runoff in the halocline and on the shelves of the Arctic Ocean derived from salinity, tritium, ^{3}He and ^{18}O data, *Deep-Sea Res.*, in press, 1993.

Swift, J. H., A recent Θ-S shift in the deep water of the northern north Atlantic, in *Climate Processes and Climate Sensitivity, Geophysical Monograph 29* (Maurice Ewing Vol. 5), edited by J. E. Hansen and T. Takahashi, American Geophysical Union, Washington, D.C., 1984.

Wallace, D. W. R., and R. M. More, Vertical profiles of CCL_3F (F-11) and CCl_2F_2 (F-12) in the central Arctic Ocean basin, *J. Geophys. Res., 90*, 1155-1166, 1985.

L. G. Anderson, K. Olsson, and A. Skoog, Department of Analytical and Marine Chemistry, University of Göteborg and Chalmers University of Technology, S-412 96 Göteborg, Sweden.

Primary Productivity of a Phaeocystis Bloom in the Greenland Sea During Spring, 1989

Walker O. Smith, Jr.

Botany Department and Graduate Program in Ecology, University of Tennessee, Knoxville, Tennessee

During April-May, 1989 the development of a massive bloom of the prymnesiophyte Phaeocystis pouchetii was observed in the Greenland Sea. The population was highly productive, with the average primary productivity being 2.06 g C m^{-2} d^{-1} (± 1.80; n = 25) and the maximum observed rate being 8.06 g C m^{-2} d^{-1}. Biomass and rate process results from assemblages near the ice edge were similar to those from the central portion of the basin, suggesting that ice-edge processes such as meltwater-induced stratification played a minor role in bloom promotion. Light-saturated chlorophyll-specific productivity did not vary with time for surface populations, which suggests that the Phaeocystis assemblage was adapted to the surface incident radiation. Rates of chlorophyll-specific photosynthesis from the surface layer averaged 3.47 mg C (mg chl)$^{-1}$ h^{-1}, which are substantially higher than those typical of Arctic diatom populations. Growth rates averaged 0.31 ± 0.20 d^{-1}. The mean carbon:chlorophyll ratio, as determined by regression of results from the upper 50 m, was 81.0 near the ice edge and 73.7 for the entire study. Particulate carbon:nitrogen ratios were moderate, averaging 4.92 for ice-edge stations. There was little particulate detritus in the surface layer. The characteristics of the particulate matter in the euphotic zone suggest that the polysaccharide mucoid sheath of Phaeocystis did not contribute a substantial amount of carbon-rich (and nitrogen-poor) material to the particulate pool. The productivity of this Phaeocystis bloom was exceptional, and if such blooms are reoccurring features of the region, then they are quantitatively important components of the carbon cycle of the Greenland Sea. A complete understanding of the region's carbon cycle will be dependent on understanding the initiation and fate of these blooms.

1. INTRODUCTION

The Greenland Sea is an important region for the entire ocean, in that it is the site of deep-water formation which drives much of the ocean's thermohaline circulation [*Broecker*, 1991]. As a result of this deep-water formation, it has the potential for injecting substantial amounts of inorganic carbon into the deep-water circulation and effectively removing it from atmospheric exchange [*Peng et al.*, 1987]. Hence during winter when deep convection occurs, the region can act as a sink for atmospheric carbon. However, during summer the productivity is generally moderate, although it can be elevated along the ice edge [*Smith et al.*, 1987] and during blooms of the prymnesiophyte Phaeocystis pouchetii [*Smith et al.*, 1991]. Vertical flux of organic matter from the surface layer is generally low [*Hebbeln and Wefer*, 1991] and the amount of particulate organic material reaching the benthos is small [*Honjo*, 1990]. As a result, the "biological pump" [*Longhurst and Harrison*, 1989] is not considered to be a significant means of removing carbon relative to non-biological processes in the Greenland Sea [*Honjo*, 1990]. Blooms of Phaeocystis have been observed in the North At-

The Polar Oceans and Their Role in Shaping the Global Environment
Geophysical Monograph 85

lantic directly and indirectly for many years. For example, *Steffansson and Olafsson* [1991] concluded from nitrate:silica ratios that Phaeocystis has occurred in detectable amounts for 56% of the years from 1959-1989. Phaeocystis has also been observed at or near the ice edge in six of seven cruises to the area in the last decade [*Smith*, unpublished], and it regularly occurs in the Barents Sea and Norwegian fjords [*Wassmann et al.*, 1990; *Wassmann*, 1993] as well as the Antarctic [e.g., *El-Sayed et al.*, 1983; *Palmisano et al.*, 1986]. In a study of the Phaeocystis bloom encountered in the Greenland Sea during spring, 1989, *Smith et al.* [1991] found particulate carbon concentrations of more than 1 g C m^{-3} and a mean primary productivity based on ^{14}C-uptake measurements to be ca. 2.1 g C m^{-2} d^{-1}. These rates, along with the associated nitrogen removal rates, are as large as have ever been observed in the Greenland Sea, and suggest that Phaeocystis blooms have the potential for quantitatively influencing the carbon cycle of the Greenland Sea.

There appear to be a number of aspects concerning the constituent ratios and metabolism of Phaeocystis that are highly unusual. For example, *Lancelot and Mathot* [1985] reported that as much as 65% of the total photosynthate is released extracellularly, probably as polysaccharide material in the mucilaginous sheath. Non-colonial forms, however, do not release such a high proportion of photosynthate [*Guillard and Hellebust*, 1971]. Because of this large reservoir of organic carbon which is not associated with cells *per se*, the carbon:nitrogen ratio of particulate matter in Phaeocystis blooms can be high (near 10, which is similar to detrital material; *Verity et al.*, 1988; *Wassmann et al.*, 1990) and may increase substantially upon the depletion of available ambient nitrogen. Protein synthesis has been shown to continue at night, using the energy derived from the oxidation of the polysaccharide sheath [*Lancelot and Mathot*, 1985]. Carbon:chlorophyll ratios also can be high [*Verity et al.*, 1988], although the degree to which this is controlled by the unusual metabolic features of Phaeocystis and/or its irradiance response is uncertain. We do know, however, that phytoplankton assemblages dominated by Phaeocystis often have unusual biochemical characteristics and responses to environmental conditions.

In spring, 1989 a cruise was conducted to the Greenland Sea to investigate the physical-biological coupling at the ice edge. During the cruise a massive bloom of Phaeocystis was encountered [*Smith* et al., 1991]. The bloom developed over a period of six weeks, during which nitrate concentrations were reduced from ca. 12 µM to undetectable levels at a number of stations [*Codispoti et al.*, 1991]. Phaeocystis was also found to have extremely high values for its photosynthetic parameters (P_{max}, α, and I_k) and its photosynthetic efficiency approached the theoretical maximum at the temperatures encountered [*Cota et al.*, 1993]. The bloom also had very rapid nitrogen uptake rates, and *f*-ratios averaged 0.56, which suggested that a large amount of biogenic production was available for export [*Smith*, 1993]. Additionally, a mass balance for nitrogen for the central Greenland Sea indicated that a large amount of material was exported from the euphotic zone [*Smith*, 1993]. In this paper the primary productivity of the Phaeocystis bloom encountered in spring, 1989 is described in detail, as well as the particulate matter characteristics which were observed during this bloom. The data used for these analyses are from sections normal to the ice edge, and the results are related to the dynamics of the marginal ice zone.

2. MATERIALS AND METHODS

Vertical profiles of nutrients and particulate matter were determined at a series of transects in the Greenland Sea as part of CEAREX (Coordinated Eastern Arctic Experiment). In general transects were occupied across the basin to the ice edge [*Smith et al.*, 1991; Figure 1). A total of 22 stations made up the ice-edge transects. In addition to the routine hydrographic and particulate measurements, samples for the determination of productivity and nitrogen uptake were taken. Continuous profiles of temperature, salinity and fluorescence were made using a Neil Brown Mark III CTD and a Sea-Tech fluorometer. Water samples were collected using a rosette system equipped with Niskin bottles fitted with teflon-coated stainless steel closing springs. Nutrients were analyzed using automated techniques [*Codispoti et al.*, 1991] and chlorophyll was determined fluorometrically after extraction in 90% acetone on a Turner Designs fluorometer before and after acidification. Particulate carbon and nitrogen concentrations were determined by filtration of a known volume of seawater through precombusted GF/F filters followed by pyrolysis on a Carlo Erba elemental analyzer. Details of particulate matter determinations are given in *Nelson et al.* [1987].

Primary productivity was quantified by using simulated *in situ* ^{14}C-uptake measurements. Approximately 20 µCi of radioactive sodium bicarbonate were added to 280 ml polycarbonate bottles and incubated for 24 h in on-deck incubators cooled with running seawater. Incubations were terminated by filtration of the entire volume through GF/F filters and rinsing with 0.01N HCl prior to the completion of the filtration. The uptake of radioactivity was quantified using a Beckman liquid scintillation counter. The amount

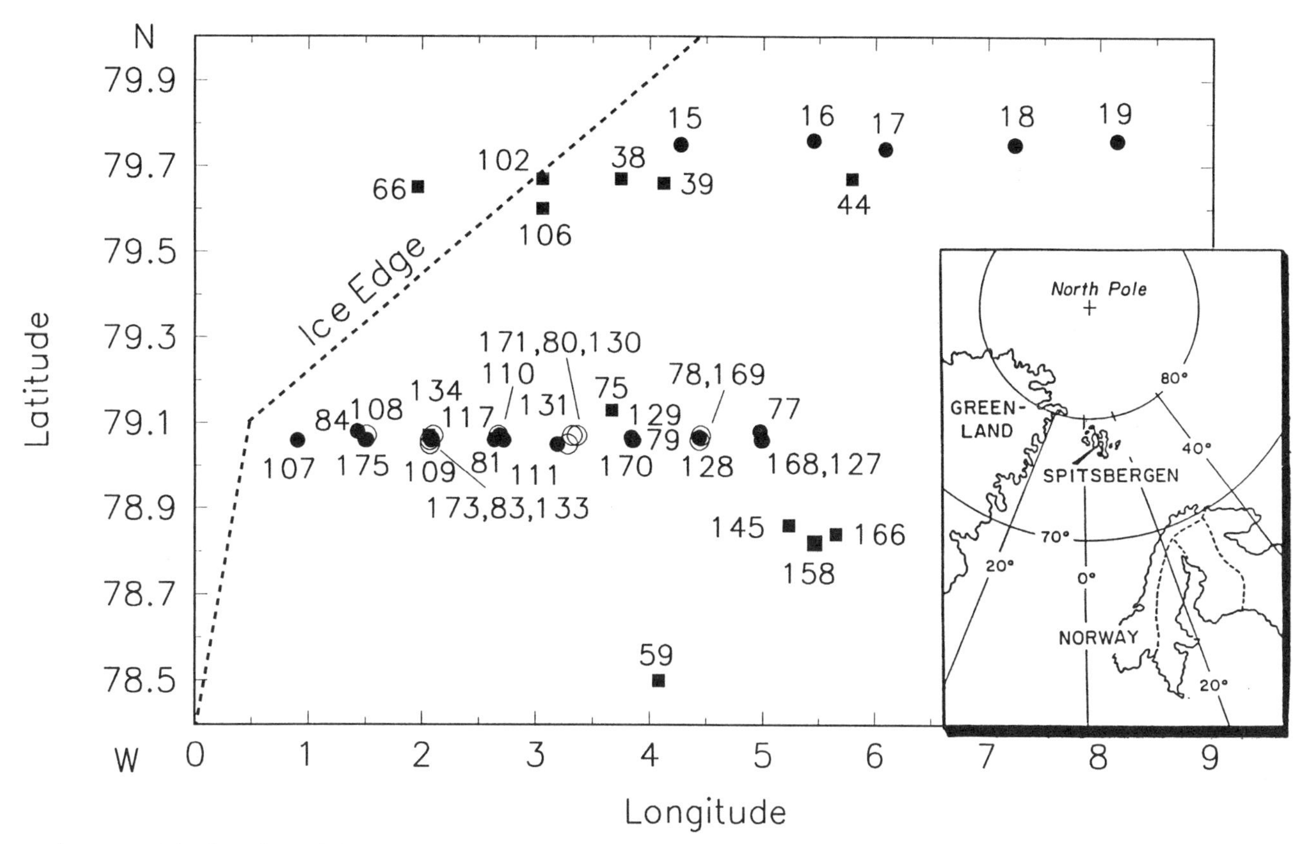

Figure 1. The location of the stations within the ice-edge transects. O symbolizes those stations not on ice-edge transects from which the ^{14}C data are included, ■ symbolizes those stations where only chlorophyll data were collected, and ● those stations along transects where ^{14}C data were collected.

of available isotope was determined by counting a 0.5 ml aliquot of unacidified, unflitered sample directly [*Smith and Nelson*, 1990]. Growth rates were calculated using the equation

$$\mu = \ln[[\Delta C + POC]/POC] \quad (1)$$

where μ is the growth rate (d^{-1}), ΔC is the ^{14}C-productivity, and POC is the particulate carbon concentration ($\mu g\ l^{-1}$) [*Eppley*, 1968]. ^{15}N-tracer techniques were used to quantify nitrogen uptake [*Smith*, 1993]. It should be noted that all filtrations (particulate carbon and nitrogen, chlorophyll, nitrogen uptake, and primary productivity) used GF/F filters, and hence particles were retained to the same degree during all filtrations.

3. RESULTS

1.1. *Biomass Distributions*

Phytoplankton biomass at the ice edge increased with time, in that the chlorophyll and particulate matter levels observed at Stations 15-19 (April 16) and Stations 77-84 (April 30; Figure 2a-c) were less than those observed at Stations 168-175 (May 12; Figure 2d-f). This was similar to the temporal pattern observed in the northern Greenland Sea basin in areas removed from the ice edge [*Smith et al.*, 1991]. Biogenic silica concentrations (a measure of diatom biomass) were low throughout the marginal ice zone and did not increase with time, similar to the entire northern Greenland Sea region [*Nelson*, unpublished]. The phytoplankton at the ice edge was dominated by the prymesiophyte Phaeocystis pouchetii, as it was over most of the Greenland Sea. No strong spatial correlation between reduced salinity from melting ice and phytoplankton biomass was observed (Figure 2a-f).

Particulate carbon:nitrogen ratios were determined by regression analysis, which gave the following equation:

$$POC = 4.92PON + 40.8 \quad (2)$$

($n = 122$; $R^2 = 0.868$; $p < 0.001$) where PON is particulate organic nitrogen ($\mu g\ l^{-1}$). The C/N ratio (4.92) is similar to

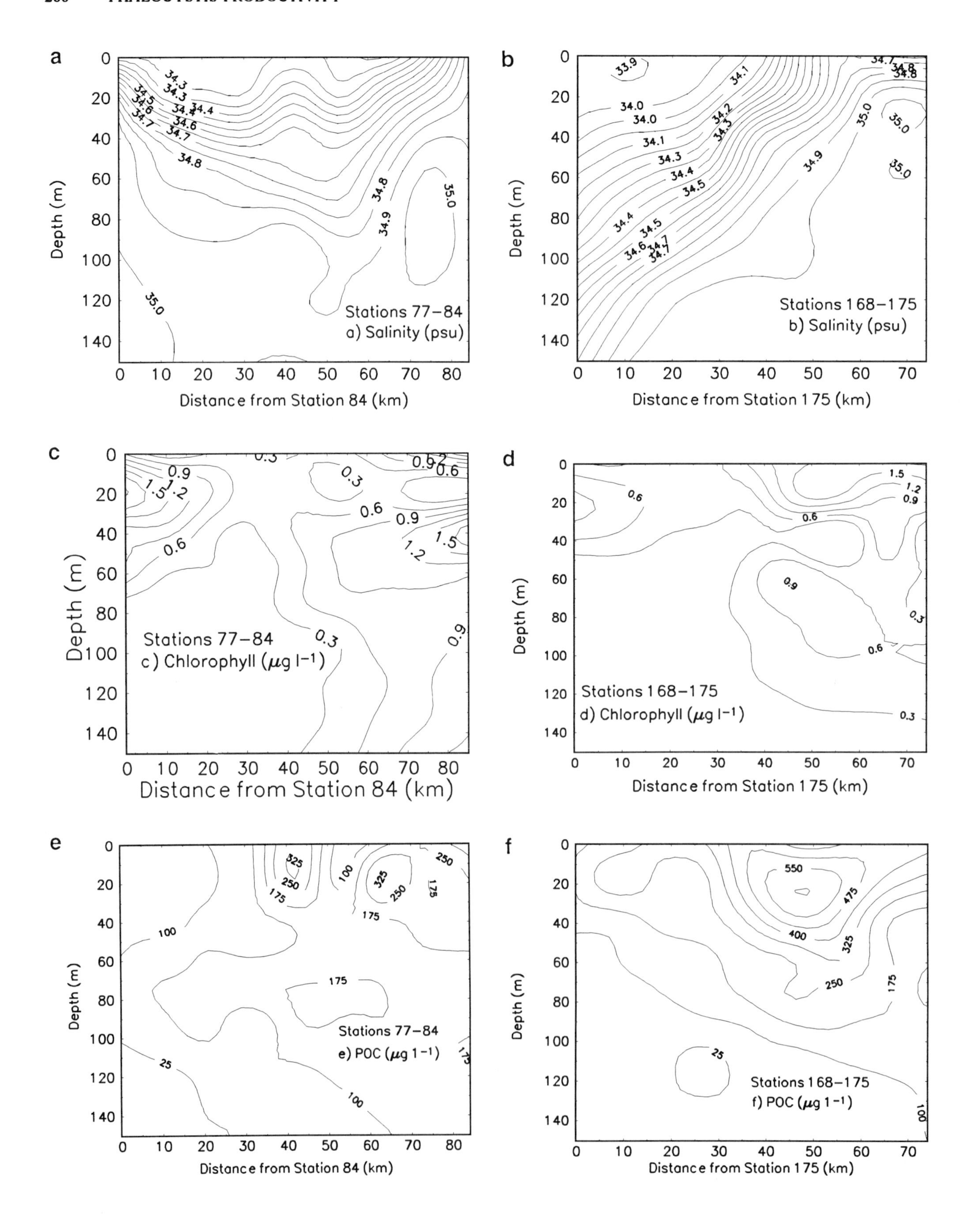
a
b
c
d
e
f
Depth (m)
Distance from Station 84 (km)
Distance from Station 175 (km)
Stations 77–84
a) Salinity (psu)
Stations 168–175
b) Salinity (psu)
Stations 77–84
c) Chlorophyll (μg l⁻¹)
Stations 168–175
d) Chlorophyll (μg l⁻¹)
Stations 77–84
e) POC (μg 1⁻¹)
Stations 168–175
f) POC (μg 1⁻¹)

those found in other polar regions and in cultures of actively growing phytoplankton [*Smith and Sakshaug*, 1990]. Particulate carbon and nitrogen were highly correlated with each other throughout the study as would be expected (Figure 3). A positive intercept was noted, which implies that some detrital material (or at least material not associated with particulate nitrogen) was present, but the absolute amount was not much greater than observed at the ice edge (25.7) in 1984 [*Smith et al.*, 1987]. Similarly, the carbon:chlorophyll regression equation for the same samples was

$$POC = 81.0CHL + 134 \quad (3)$$

($n = 122$; $R^2 = 0.273$; $p < 0.005$) where CHL is chlorophyll ($\mu g\ l^{-1}$). The C/Chl ratio is greater than what is usually considered to be average for the world's oceans (Table 1), but polar regions, with their extreme variations in solar irradiance, tend to have elevated values [*Smith and Sakshaug*, 1990]. The particulate matter relationships do not suggest that an extensive nitrogen-poor polysaccharide mucoid sheath greatly altered the ratios of cellular constituents when Phaeocystis was present.

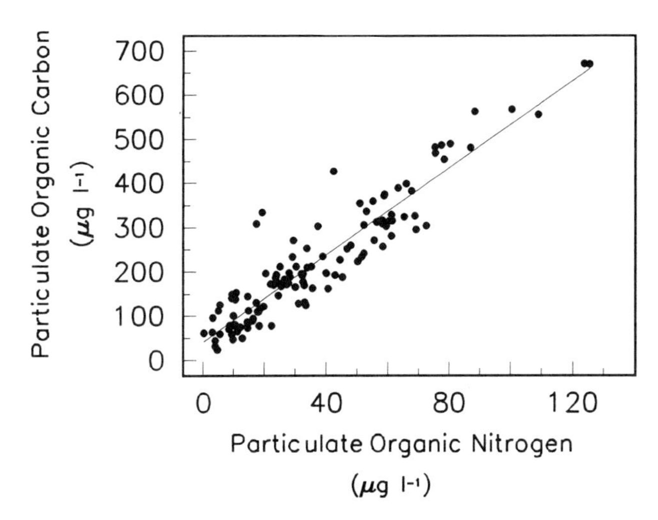

Figure 3. The relationship of particulate carbon and particulate nitrogen concentrations in the upper 50 m at the ice edge. The two are highly correlated [$PC = 4.92PN + 40.8$; $n = 122$, $R^2 = 0.868$; $p < 0.001$].

TABLE 1. Integrated values of particulate carbon, particulate nitrogen and chlorophyll and their ratios for those stations within the ice-edge transects. Integrations are through 150 m. Means of the ratios are ratios of the separate means.

Station Number	Integrated POC ($mg\ m^{-2}$)	Integrated Chlorophyll ($mg\ m^{-2}$)	Integrated PN ($mg\ m^{-2}$)	PC/ CHL	C/N
15	7,650	224	917	34.1	8.34
16	19,100	131	2,999	146	6.38
17	10,400	70.7	1,450	147	7.17
18	22,600	147	4,614	153	4.90
19	30,900	71.0	3,830	435	8.06
77	24,100	132	5,556	182	4.34
79	24,900	91.7	5,012	271	4.96
81	13,100	17.7	1,958	738	6.69
84	13,500	116	3,428	116	3.94
107	20,000	182	2,851	109	7.03
109	13,300	97.5	1,297	136	10.2
111	18,600	78.3	3,296	236	5.63
127	22,000	102	4,400	216	5.00
129	31,100	125	4,731	249	6.58
131	22,900	49.5	3,027	463	7.57
134	12,900	74.0	2,149	174	6.00
141	25,300	80.5	2,917	313	8.66
168	24,900	83.6	2,627	297	9.47
170	41,800	97.9	6,372	426	6.55
172	20,400	24.3	3,937	841	5.19
175	13,000	40.4	3,096	321	4.19
Mean	20,590	86.7	3,356	237	6.14
σ	7,890	42.5	1,386	-	-

3.2. *Primary Productivity*

Primary productivity was successfully measured at 25 stations during the seven-week period and averaged 2.06 g C m^{-2} d^{-1} (± 1.76; range 0.44 to 8.06 g C m^{-2} d^{-1}; Table 2). Productivity of just those stations in the ice-edge sections averaged 2.00 g C m^{-2} d^{-1} ($n = 13$), and hence was not significantly different from the entire cruise mean. Chlorophyll-specific productivity was also high, and the

Figure 2. Vertical sections of a) salinity (Stations 77-84), b) salinity (Stations 168-175), c) chlorophyll (Stations 77-84), d) chlorophyll (Stations 168-175), e) particulate organic carbon (POC; Stations 77-84), and f) particulate organic carbon (POC; Stations 168-175) along a transect normal to the ice edge. Biomass increased through time throughout the marginal ice zone.

TABLE 2. Biomass, productivity and growth indices for stations in the northern Greenland Sea. Productivity integrated from the surface through the 0.1% isolume. POC = particulate organic carbon; ND = no data.

Station Number	Surface Chlorophyll ($\mu g\ l^{-1}$)	Surface Productivity ($mg\ C\ m^{-3}\ h^{-1}$)	Chl-specific Productivity ($mg\ C\ (mg\ chl)^{-1}\ h^{-1}$)	Integrated Productivity ($mg\ C\ m^{-2} d^{-1}$)	POC ($\mu g\ l^{-1}$)	Growth Rate (d^{-1})
18*	3.32	7.58	2.28	2254	303	0.47
19*	1.33	1.19	0.90	1329	197	0.14
38	1.72	ND	ND	1210	284	ND
39	1.40	6.39	4.58	1311	296	0.42
44	2.14	17.9	8.36	4646	233	1.05
59	0.80	1.23	1.53	643	171	0.16
66	1.34	1.85	1.39	4421	221	0.18
75	2.53	ND	ND	2361	309	ND
77*	1.99	2.46	1.23	1264	372	0.15
79*	0.50	0.87	1.75	738	77.4	0.24
102	5.31	3.58	0.67	1864	253	0.29
106	1.73	3.81	2.21	3404	250	0.31
117	0.26	0.18	0.71	2230	124	0.03
127*	2.27	4.77	2.10	8058	562	0.19
129*	2.11	3.52	1.67	1031	382	0.20
131*	0.07	2.07	1.73	1381	149	0.29
134*	0.19	0.70	3.62	692	82.0	0.19
141*	0.42	0.95	2.26	454	183	0.12
145	1.31	2.14	1.64	893	348	0.14
158	1.48	2.70	1.82	1882	247	0.23
166	0.59	1.76	3.01	537	307	0.13
170*	1.57	3.24	2.06	1182	389	0.18
172*	0.46	3.98	8.59	1309	324	0.26
175*	0.37	3.62	9.74	5505	163	0.43
Mean	2.74	3.47	2.92	2058	266	0.25
σ	2.46	3.62	2.60	1797	111	0.20

*: Ice-edge transect stations.

light-saturated average rate was 3.47 mg C (mg chl)$^{-1}$ h^{-1} (Figure 4). No obvious temporal trend was note for light-saturated chlorophyll-specific productivity (Figure 5), which suggests that the Phaeocystis assemblage was well adapted to surface irradiance conditions throughout the entire length of the study (36 d), despite the fact that both the day length and absolute daily irradiance were increasing rapidly. Carbon growth rates averaged 0.25 d^{-1} at the surface (Table 2), which is less than the temperature-limited maximum.

3.3. *Productivity Calculations*

By comparing the increases over time in particulate matter concentrations at one point in space, a net production can be calculated. Similarly, the production can be estimated by calculating the net nutrient removal over time and converting the nutrient deficit to particulate matter production using the appropriate elemental ratios [*Smith*, 1993]. Finally, the export of particulate matter can be approximated by comparing the two. This procedure was used on a location in the marginal ice zone in which two stations (Station 79 and 170) were occupied 10.9 d apart. Nitrate levels decreased by ca. 8 μM at the surface, and particulate nitrogen concentrations increased by ca. 3.5 μmol l^{-1} (Figure 6). Particulate carbon concentrations increased by approximately 25 μmol l^{-1} at the surface and chlorophyll levels also increased throughout the water column (Figure 6). The nutrient deficit equaled 227.8 mmol m^{-2} from the surface

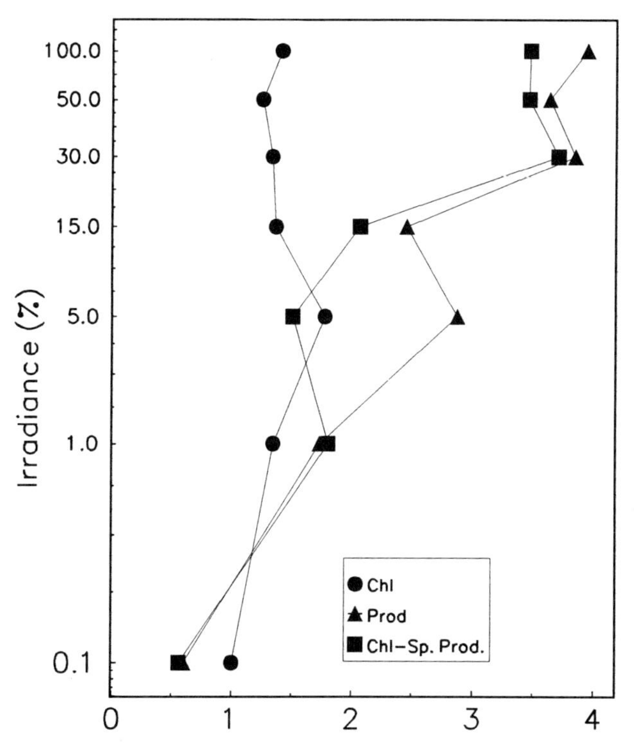

Figure 4. Vertical profiles of the average chlorophyll concentrations, productivity, and chlorophyll-specific productivity as a function of irradiance where ^{14}C-incubations were conducted.

through 75 m, whereas the particulate nitrogen increase in the same depth range equaled 256.6 mmol m^{-2}. The increase in particulate carbon was 1859 mmol m^{-2}, which corresponds to a C/N ratio during growth of 6.21 (w/w), similar to the value determined by regression.

If the nitrate disappearance (227.8 mmol m^{-2}) is converted to moles of carbon weight by multiplying by the Redfield ratio (106/16) and then multiplying by the atomic weight (14) to convert to weight units, then a new production can be estimated (Table 3). If that production is divided by the total time interval, then the estimated new production is 1.66 g C m^{-2} d^{-1}. Similarly, if the new produdtion from the particulate nitrogen appearance is calculated in a similar manner, the calculated rate is 1.87 g C m^{-2} d^{-1}. Given that the C/N ratio determined both from regression analysis and from the particulate matter budget at these two stations is lower than the Redfield ratio, these calculated productivities are overestimates to the extent that the nitrogen to carbon conversion is an overestimate. Finally, if the

TABLE 3. The mass balance of particulate material and nitrate at Stations 79 and 170, which were occupied at the same location 10.9 days apart. Integrations are through 75 m.

Variable	St. 79		St. 170
Nitrate (mmol m^{-2})	655.3		427.5
Particulate Nitrogen (mmol m^{-2})	77.3		385.3
Chlorophyll (mg m^{-2})	53.5		63.9
Particulate carbon (mmol m^{-2})	874.2		2732
ΔNO_3 (mmol m^{-2})		227.8	
ΔPN (mmol m^{-2})		256.6	
ΔPC (mmol m^{-2})		1859	
Productivity-NO_3 (mg C m^{-2} d^{-1})		1.66[1]	
Productivity-PN (mg C m^{-2} d^{-1})		1.87[1]	
Productivity-PC (mg C m^{-2} d^{-1})		2.05[2]	

[1]: Calculated by converting the change in nitrogen to carbon units by multiplying by the Redfield ratio (6.63) and dividing by the time differential

[2]: Calculated by dividing the change in particulate carbon units by the time differential

productivity is calculated from particulate carbon appearance data, then the productivity is estimated to be 2.05 mg C m^{-2} d^{-1}. If these values can be used to calculate *f*-ratios [*Eppley and Peterson*, 1979] to estimate the degree of dependence on regenerated nitrogen, then the estimates are 0.81 and 0.91. It is clear that for this one location at the ice edge, all productivity estimates converge to reveal the exceptionally high production of the Phaeocystis bloom and its dependence on nitrate.

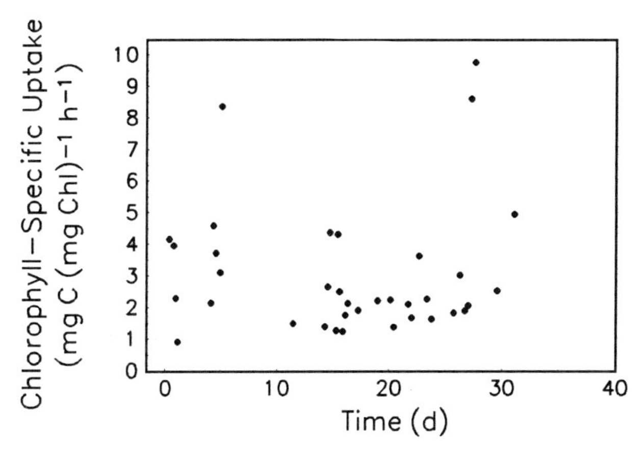

Figure 5. The temporal variations of chlorophyll-specific productivity of samples at the surface (0 m) throughout the entire northern Greenland Sea. No significant change was noted.

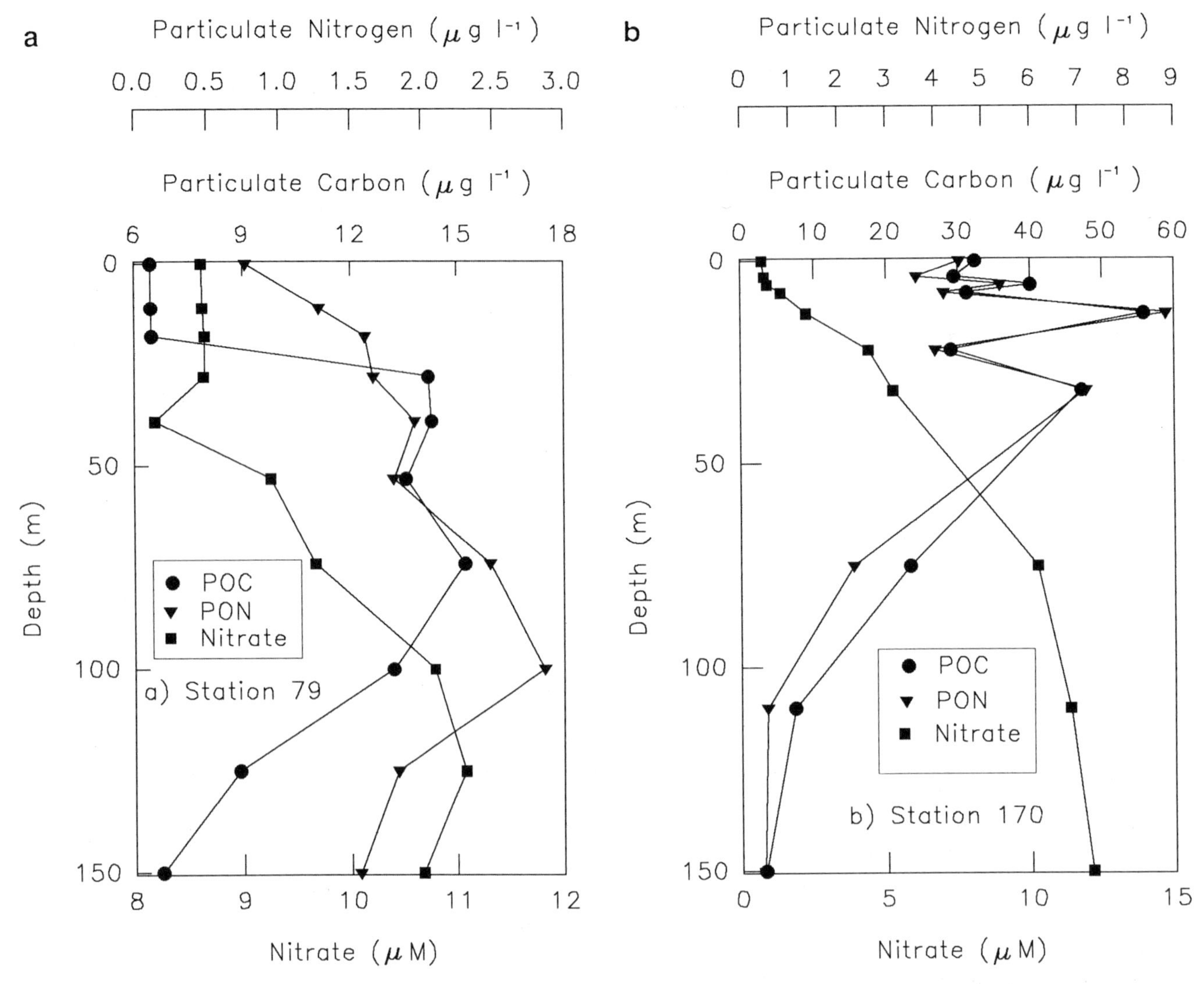

Figure 6. The distribution of particulate carbon, particulate nitrogen and nitrate at Stations a) 79 and b) 170. The difference between integrated nitrate concentrations is an estimate of net new production; the difference between particulate matter levels is an estimate of particulate matter production (see Table 3).

4. DISCUSSION

Most of the characteristics of the phytoplankton (species composition, biomass, productivity, growth rates) investigated within the marginal ice zone were similar to those found in waters outside the MIZ during spring, 1989 in the Greenland Sea. It does appear that growth and accumulation of phytoplankton at the ice-edge did begin earlier than in the central portion of the Greenland Sea, in that elevated particulate matter concentrations were observed within the first ice-edge transect (Stations 15-19) but not in a transect from Spitzbergen to the ice edge five days earlier [*Smith et al.*, 1991]. This difference between the MIZ and the northern Greenland Sea in early spring is not surprising, because waters near the ice edge are thought to be sites of production earlier than those in open waters uninfluenced by meltwater and increased vertical stratification [e.g., *Smith and Sakshaug*, 1990]. What is surprising is that the difference between the ice edge and the rest of the Greenland Sea is not greater and does not persist for longer periods.

Cota et al. [1993] found that Phaeocystis was exceptionally well adapted to low temperatures and all irradiances encountered. Specifically, Phaeocystis had high a values and very high P_{max} values, and when compared to diatoms from the same temperatures, Phaeocystis would photosynthesize much more rapidly than diatoms *at all irradiances*. It is possible that the massive bloom in 1989 was a result of the rapid growth of Phaeocystis, which in turn allowed it to utilize the available nutrients present during this period. It also suggests that the reason that only slight differences between the MIZ and the open-ocean waters were noted was that the photophysiological adaptations of Phaeocystis were such that the increased stratification found at the ice edge did not provide any sustained advantage to photosynthesis, growth and accumulation.

Another somewhat surprising feature of the Greenland Sea Phaeocystis bloom was that ratios of the cellular consituents within the particulate matter were not greatly different from diatom-based communities in other polar regions. It has been shown for other areas in which Phaeocystis dominated that elevated C/N ratios result from the polysaccharide mucoid sheath. *Verity et al.* [1988] showed that C/N ratios were ca. 5-8 when nitrate was non-limiting, but that ratios increased to values of 15 upon nutrient limitation. *Wassmann et al.* [1990] also found C/N ratios in a bloom of Phaeocystis to be 9.6 and 10.4 when nitrate was less than 1 μM in the surface layer. Such high values were not normally encountered in our samples, even at stations where nutrients were undetectable. Carbon:chlorophyll ratios were somewhat elevated, with the mean ratio as determined by regression being 81.0. An arithmetic mean from all the stations gives an average of 237, but reflects the influence of the non-phytoplanktonic carbon material (reflected in the non-zero intercept in the regression). Given that the slope of the regression (and assuming that it is a linear relationship) represents growth of the population, the regression slope gives the best indication of the true C/Chl ratio (or any ratio of cellular constituents) within a particular area and minimizes interference of detrital and non-cellular material.

The average productivity from radiocarbon uptake measurements was 2.06 g C m^{-2} d^{-1} [*Smith et al.*, 1991]. This is a substantial productivity, especially since it has been shown that this production occurred at this rate throughout the northern Greenland Sea [*Smith et al.*, 1991]. It is among the highest rates ever observed in polar regions and appears to be the largest ever reported in the Greenland Sea. It is uncertain if or why 1989 was such an exceptional year, and how Phaeocystis became established to the extent that it did. Certainly based on the nitrogen uptake [*Muggli and Smith*, 1993; *Smith*, 1993] and photosynthetic [*Cota et al.*, 1993] characteristics, it was highly adapted to the environment and grew to the near exclusion of other species. Its growth was clearly not controlled by grazing, although the quantitative effect of grazer removal of Phaeocystis by the dominant copepods is uncertain. Finally, its control by aggregation and rapid sinking has been suggested [*Smith et al.*, 1991; *Wassmann*, 1993], but because of the absence of sediment trap data and time-series deep-water profiles, it is impossible to know the extent of the vertical flux of Phaeocystis-derived organic material or the extent to which it was remineralized in the water column during this study.

A nutrient and particulate matter budget at an ice-edge station showed that the new production based on nutrient removal was nearly identical to the production calculated from particulate matter appearance. Therefore, for the period between the two stations, little vertical flux (export) of organic matter occurred. This is in marked contrast to the observations in a transect across the Greenland Sea, where a large fraction (ca. 62%) of the particulate matter had been removed [*Smith*, 1993]. The difference may have resulted from the longer time period of analysis (11 vs. 35 days), in that at the ice edge nutrients were depleted over a much shorter period. If nutrients disappeared in the non-MIZ areas as rapidly, and nutrient depletion lasted for significant time periods, then growth rates may have decreased substantially, and aggregate formation may have been enhanced and resulted in greater vertical flux rates. *Kiorboe et al.* [1990] experimentally showed that nutrient depletion induces an increase in cellular "stickiness" in a diatom, and if this finding can be extrapolated to Phaeocystis, then increased stickiness and aggregate formation might have occurred throughout the northern Greenland Sea. However, without a time-series of stations occupied at closer intervals, the temporal dynamics of loss processes cannot be resolved.

Based on the results of this investigation, we know that the observed Phaeocystis bloom was one of the most productive and widespread blooms ever observed in the Greenland Sea. However, many aspects of the bloom (such as controls of initiation, grazing and bacterial regeneration rates, vertical flux rates) remain poorly known. We do know, however, that Phaeocystis can greatly structure the local food web dynamics and biogeochemical cycles in the Greenland Sea. Without a more complete appreciation of the dynamics of Phaeocystis blooms, our understanding of the region's carbon cycle will remain incomplete.

Acknowledgments. This research was supported by the Office of Naval Research as part of the CEAREX project. I thank G. Cota, H. Kelly, and D. Muggli for assistance at sea and L.A. Monty for data analysis.

REFERENCES

Broecker, W. S., The great ocean conveyor, *Oceanography, 4,* 79-88, 1991.

Codispoti, L.A., G.E. Friederich, C.M. Sakamoto, and L.I. Gordon, Nutrient cycling and primary production in the Arctic and Antarctic: contrasts and similarities. *J. Mar. Syst., 2,* 359-384, 1991.

Cota, G.F., W.O. Smith, Jr., and B.G. Mitchell, Photosynthesis of Phaeocystis in the Greenland Sea, *Limnol. Oceanogr.* (in press), 1993.

El-Sayed, S.Z., D.C. Biggs, and O. Holm-Hansen, Phytoplankton standing crop, primary productivity, and near-surface nitrogenous nutrient fields in the Ross Sea, Antarctica. *Deep-Sea Res., 30,* 871-886, 1983.

Eppley, R.W., An incubation method for estimating the carbon content of phytoplankton in natural samples, *Limnol. Oceanogr., 13,* 574-582, 1968.

Eppley, R.W., and B.J. Peterson, Particulate organic matter flux and planktonic new production in the deep ocean, *Nature, 282,* 677-680, 1979.

Guillard, R.R.L., and J.E. Hellebust, Growth and the production of extracellular substances of two strains of Phaeocystis pouchetii, *J. Phycol., 7,* 330-338, 1971.

Hebbeln, D., and G. Wefer, Effects of ice coverage and ice-rafted material on sedimentation in the Fram Strait, *Nature, 350,* 409-411, 1991.

Honjo, S., Particle fluxes and modern sedimentation in the polar oceans, in *Polar Oceanography,* Part B, edited by W.O. Smith, Jr., pp. 687-740, Academic, San Diego, Calif., 1990.

Kiorboe, T., K.P. Anderson, and H.G. Dam, Coagulation efficiency and aggregate formation in marine phytoplankton, *Mar. Biol., 107,* 235-245, 1990.

Lancelot, C., and S. Mathot, Biochemical fractionation of primary production by phytoplankton in Belgian coastal waters during short- and long-term incubations with ^{14}C-bicarbonate, II, Phaeocystis pouchetii in colonial populations, *Mar. Biol., 86,* 227-232, 1985.

Longhurst, A.R., and W.G. Harrison, The biological pump: profiles of plankton production and consumption in the upper ocean, *Prog. Oceanogr., 22,* 47-123, 1989.

Muggli, D.L., and W.O. Smith, Jr., Regulation of nitrate and ammonium uptake in the Greenland Sea, *Mar. Biol., 115,* 199-208, 1993.

Nelson, D.M., W.O. Smith, Jr., L.I. Gordon, and B.A. Huber, Spring distributions of density, nutrients, and phytoplankton biomass in the ice edge zone of the Weddell-Scotia Sea, *J. Geophys. Res., 92,* 7181-7190, 1987.

Palmisano, A.C., J.B. SooHoo, S.L. SooHoo, S.T. Kottmeier, L.L. Craft, and C.W. Sullivan, Photoadaptation in Phaeocystis pouchetii advected beneath annual sea ice in McMurdo Sound, Antarctica, *J. Plankton Res., 8,* 891-906, 1986.

Peng, T.-H., T. Takahashi, W.S. Broecker, and J. Olafsson, Seasonal variability of carbon dioxide, nutrients and oxygen in the northern North Atlantic surface water: observations and a model, *Tellus, 39,* 439-458, 1987.

Smith, W.O., Jr., Nitrogen uptake and new production in the Greenland Sea: the spring Phaeocystis bloom, *J. Geophys. Res., 98,* 4681-4688, 1993.

Smith, W.O., Jr., M.E. Baumann, D.L. Wilson, and L. Aletsee, Phytoplankton biomass and productivity in the marginal ice zone of the Fram Strait during summer 1984, *J. Geophys. Res., 92,* 6777-6786, 1987.

Smith, W.O., Jr., L.A. Codispoti, D.M. Nelson, T. Manley, E.J. Buskey, H.J. Niebauer, and G.F. Cota, Importance of Phaeocystis blooms in the high-latitude ocean carbon cycle, *Nature, 352,* 514-516, 1991.

Smith, W.O., Jr., and D.M. Nelson, Phytoplankton growth and new production in the Weddell Sea marginal ice zone in the austral spring and autmn, *Limnol. Oceanogr., 35,* 809-821, 1990.

Smith, W.O., Jr., and E. Sakshaug, Polar phytoplankton, in *Polar Oceanography,* Part B, edited by W.O. Smith, Jr., pp. 477-525, Academic, San Diego, Calif., 1990.

Steffanson, U., and J. Olafsson, Nutrients and fertility of Icelandic waters, *J. RitFisk., 12,* 1-56, 1991.

Verity, P.G., T.A. Villareal, and T.J. Smayda, Ecological investigations of blooms of colonial Phaeocystis pouchetti (*sic*)-I. Abundance, biochemical composition, and metabolism, *J. Plankton Res., 10,* 219-248, 1988.

Wassmann, P., Significance of sedimentation for dynamics of Phaeocystis blooms, *J. Mar. Syst.,* (in press), 1993.

Wassmann, P., M. Vernet, B.G. Mitchell, and F. Rey, Mass sedimentation of Phaeocystis pouchetii in the Barents Sea, *Mar. Ecol. Prog. Ser., 66,* 183-195, 1990.

Air-Sea CO_2 Fluxes in the Southern Ocean Between 25°E and 85°E

Alain Poisson, Nicolas Metzl, Xavier Danet, Ferial Louanchi, Christian Brunet, Bernard Schauer, Bernard Brès, and Diana Ruiz-Pino

Laboratoire de Physique et Chimie Marines, Université Pierre et Marie Curie, Paris.

The variations of ΔfCO_2 (difference of sea surface and atmospheric carbon dioxide fugacities) and air-sea CO_2 fluxes are described in the western Indian sector of the Southern Ocean. The study is based on measurements made continuously with an Infrared technique during two cruises in January-April 1991 and one recent cruise in February 1993. We also compare ΔfCO_2 distributions observed in February 1987 and February 1993 at 30°E. Large ΔfCO_2 variations have been observed, most of which can be related to dynamical variability and/or local biological activity, near the ice edge, the antarctic divergence, the Polar Front Zone or local change in topography (e.g. Kerguelen Plateau). Near the pack ice, clear relationships are found between low ΔfCO_2 (CO_2 sink area) and high chorophyll-a concentrations. A significant CO_2 source region has been found in February 1993 centered on the antarctic divergence. South of the Polar Front Zone, fCO_2 decreases strongly; this signal was also found in 1987 and 1991 and can be related to high chlorophyll signal. In the region of the Kerguelen Plateau ΔfCO_2 spatial variability is very high. Neutral, source or sink conditions can be found at small scale (10 km). With such horizontal gradients, large-scale integrated air-sea CO_2 exchange flux calculation is not possible. Nevertheless, as an indication, we compute the average of the CO_2 fluxes along the cruises tracks, using observed ΔfCO_2 and ship winds; three gas transfer velocities versus wind speed relationships are also used. During March 1991, in the 60°E-90°E band, local fluxes vary from -30 to 30 $mmol.m^{-2}.d^{-1}$; the average flux of 1.1 (±4.0) $mmol.m^{-2}.d^{-1}$ or 2.0 (±8.0) $mmol.m^{-2}.d^{-1}$ (depending of the gas transfer used) suggests an ocean source of CO_2. During February 1993, in the 30°E-80°E region, the average flux of -1.4 (±2.9) $mmol.m^{-2}.d^{-1}$ or -2.5 (±5.3) $mmol.m^{-2}.d^{-1}$ suggests an ocean sink. Comparing year 1987 and 1993 it is shown that when considering interannual varibility (mainly related to SST differences between the two periods) the change in ΔfCO_2 has been low during 6 years.

INTRODUCTION

The role of the Southern Ocean in global evaluation of air-sea CO_2 exchange is poorly known. This is mainly because this ocean was and is still under-sampled on spatial as well as on seasonal scales. The difficulty is also related to high spatial variability of sea surface CO_2 fugacity (or CO_2 partial pressure, pCO_2) that has been always observed in the Southern Ocean. [Keeling, 1968; Miyake and Sugimura, 1969; Miyake et al., 1974; Inoue and Sugimura, 1988; Takahashi, 1989; Metzl et al., 1991; Murphy et al., 1991; Poisson et al., 1993]. Furthermore, the winds that control the rate of air-sea gas exchange are not very well known in this area. It is clear that for constraining air-sea CO_2 flux overall in the Southern Ocean, much more fCO_2 data are needed [Tans et al., 1990] as well as a better estimation of sea surface winds (there are few merchand ship and wind observations in the Southern Ocean). In addition, relationship between gas transfer velocity and the wind speed is not well established [e.g. Wanninkhof, 1992];

The Polar Oceans and Their Role in Shaping the Global Environment
Geophysical Monograph 85

this increases the difficulty when concluding on air-sea CO_2 fluxes. This paper aims to describe distributions of sea surface carbon dioxide fugacity (fCO_2) and air-sea CO_2 fluxes in the western Indian sector of the Southern Ocean. The study is focused on the measurements made continuously with an Infrared technique during two cruises made in January-April 1991 (MINERVE 07 and 08) in the region of the Kerguelen Plateau and one more recent cruise made in February 1993 (MINERVE 23) between Kerguelen archipelago, Antarctic continent and Africa (figure 1). Comparison with older data (cruise INDIVAT V, February 1987) in the same region is also presented.

METHOD

A continuous Infrared technique was used for the oceanic and atmospheric fCO_2 measurements. It has been set up on board the N.O. Marion-Dufresne. The system consists of an IR analyser, a thermostated equilibrium cell, a thermosalinograph, and a fluorimeter. Seawater and a

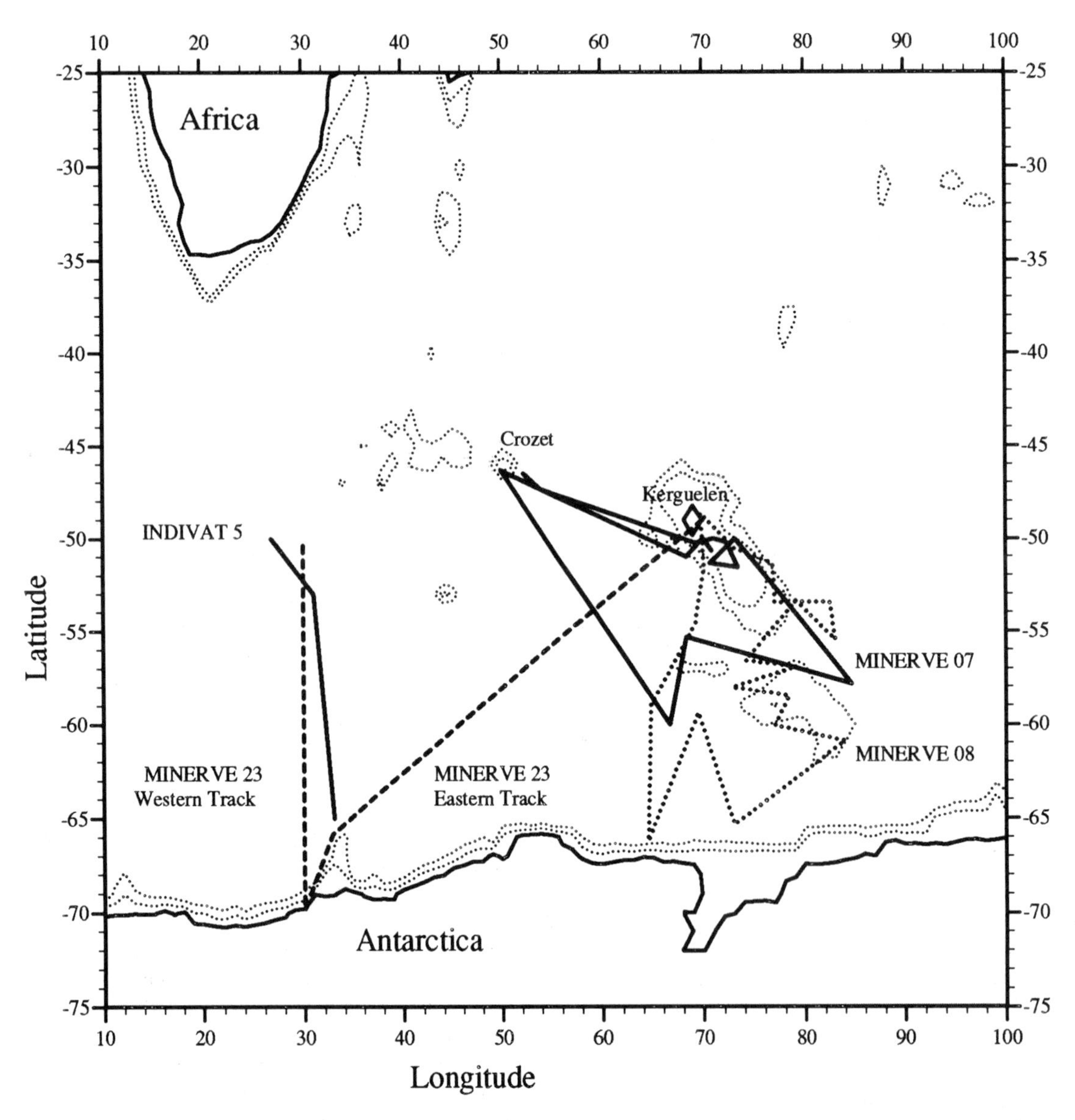

Fig 1. Tracks of the cruises performed in the Southern Ocean: INDIVAT 5 (February 1987, line), MINERVE 07 (January-February 1991, line) and MINERVE 08 (March-April 1991, dotted line) and MINERVE 23 (February 1993, dashed line). 1000m and 2000m isobaths are shown.

closed loop of air circulate in the equilibrium cell in a counter current design. Then equilibrated air goes through an automatic cold trap (-35°C) and the IR analyser. In the same time, sea-water circulates through the fluorimeter and the thermosalinographer. Instrumentation and calibrations used during MINERVE cruises are described in detail in Poisson et al. [1993]. The fCO_2 measurements are corrected to in-situ temperature by using Copin-Montégut polynomials [1988, 1989]; because the equilibrium cell is thermostated with sea surface water, the temperature correction does not exceed 1°C. For all the parameters, we average the continuous measurements recorded during 10 minutes (corresponding to about 3 kilometers and 50 data). Except when the measurements are located near or in a frontal zone, the standard deviation of fCO2 measured over 10 minutes is lower than 0.3% (about 1 μatm). Fluorescence records were made only during the most recent cruise (MINERVE 23 in February 1993) and have been converted to Chlorophyll-a by fitting Chlorophyll-a concentrations measured on discrete samples taken during the cruise (6,3 litres of seawater filtered on Whatman GF/F filter) and measured on board with a spectrophotometer (Spectronic 301 Milton Roy).

Air-sea CO_2 fluxes are computed following the relation

$$F = k.s. (fCO_{2ocean} - fCO_{2atmosphere}) = k.s. \Delta fCO_2 \quad (1)$$

where k, the gas transfer velocity, depends on wind speed and on temperature by the Schmidt coefficient [Jähne, 1980]. The CO_2 solubility (s) of a non-ideal gas depends on temperature and salinity [Weiss, 1974]. Average atmospheric fCO_2 measured was 353.2 μatm during the 1991 cruises and 354.8 μatm during the 1993 ones. We use these values for ΔfCO_2 calculations. Natural variability observed during the cruises lead to an average standard deviation smaller than 0.5 μatm on atmospheric fCO_2. The wind speed was measured each hour during the cruises. To link k with oceanic parameters in equation (1), we combine the wind speed ship measurements with the continuous measurements of fCO_2, SST and salinity recorded in the following hour. Because the relationship between k and the wind speed is still uncertain [for a review of this problem, see Wannhinkof, 1992] and having no definite answer of which relation to use, we have selected three of them that can be found in the literature: Liss and Merlivat [1986], Tans et al [1990] and Wannhinkof [1992] (the gas transfer coefficients will be refered respectively as kLM, kT and kW in the following). As an example of the range of kLM, kW and kT encountoured during the cruises, figure 2 shows the results of the three computations of k with measured wind speed and sea surface temperature during the cruise MINERVE 23. Clearly, kT and kW are higher than kLM by a factor of about 2, as expected. For low wind speed (figure 2a), kW is higher than kT and kLM because it includes chemical enhancement [see equation 8 in Wanninkhof, 1992] and kT is forced to zero at wind speed lower than 3m/s [equation 2 in Tans et al.,1990]. The air-sea CO_2 fluxes computed by using these three gas transfer velocities will then be described as extreme ranges of oceanic sources and sinks when one uses ship wind speed observations and when averaging over the cruises periods (about one month).

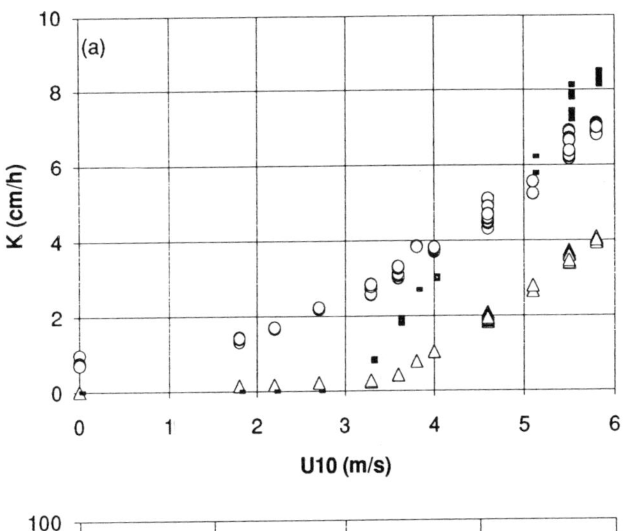

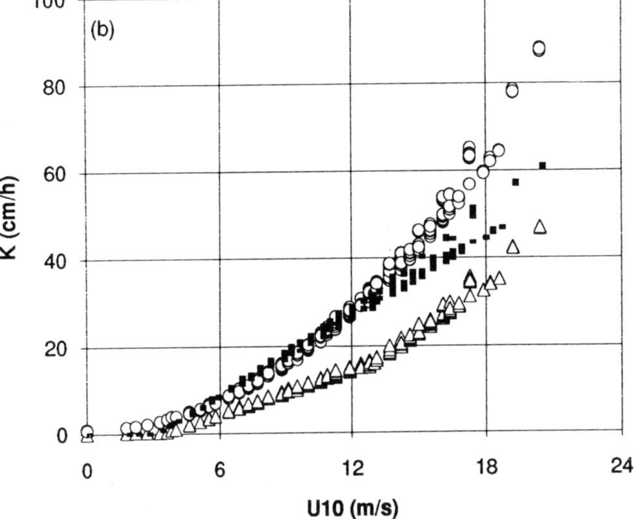

Fig. 2. CO2 transfer velocity versus wind speed as computed by using Liss and Merlivat, 1986 (Δ), Wanninkhof, 1992 (O) and Tans et al., 1990 (■) relations, (a) for low wind speeds. (b) for all wind speeds and SST measurements made during MINERVE 23 cruise. Transfer velocities are not normalised here (they are usually presented at SST=20°C); for the same wind speed, differences in transfer velocity calculation for each relation correspond to different SST.

MERIDIONAL PROFILES IN THE WESTERN AREA

Sea surface observations made in February 1993 are presented in figure 3. Along the both tracks the distribution

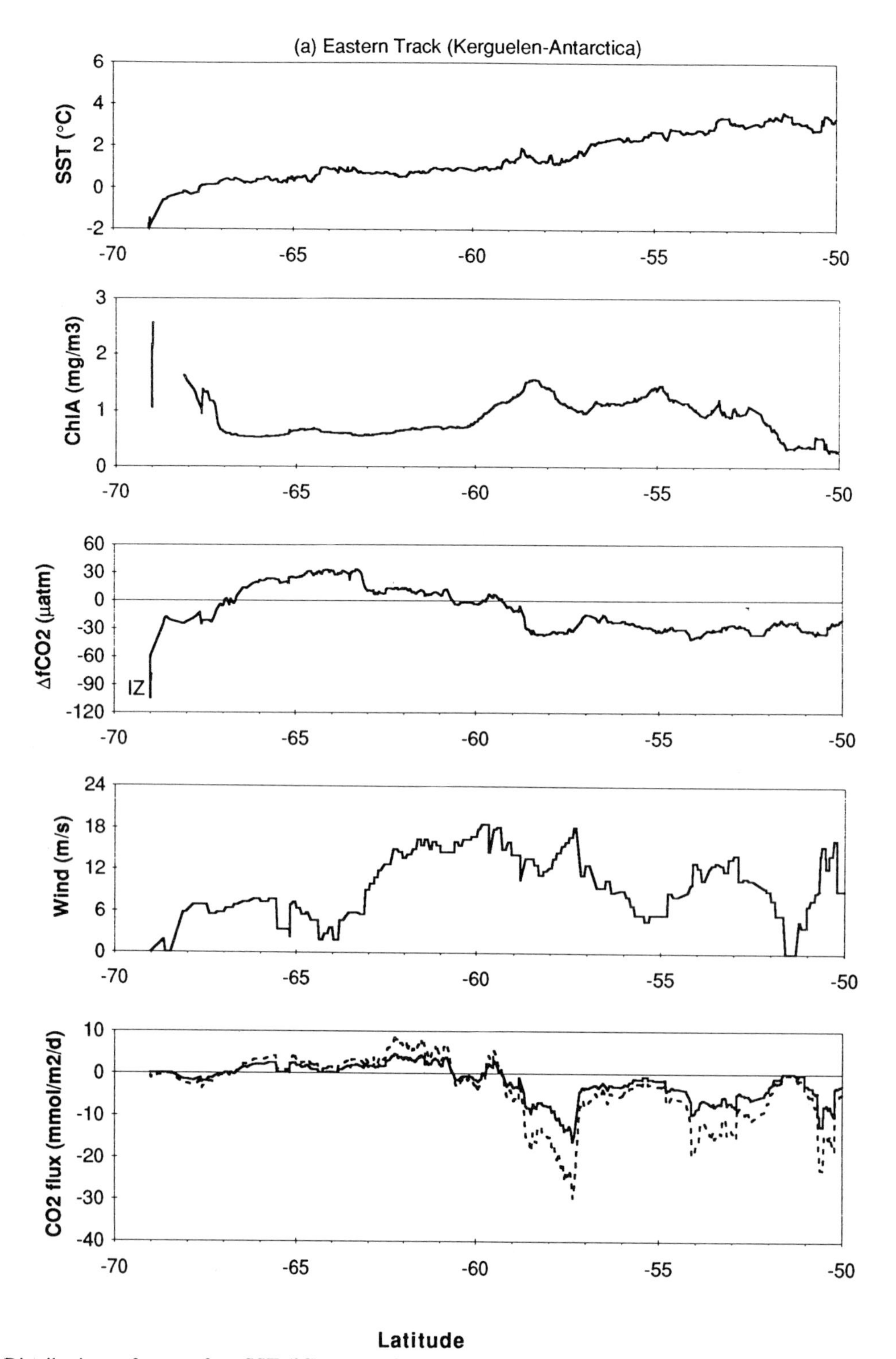

Fig. 3. Distributions of sea surface SST (°C), sea surface chlorophyll-a (mg/m3), ΔfCO_2 (μatm), wind speed ($m.s^{-1}$) and air-sea CO_2 fluxes ($mmol.m^{-2}.d^{-1}$) calculated using kLM (bold line) and kW (dashed line) along the two tracks performed in February 1993:

(a) the eastern track from Kerguelen to Antarctic continent.

(b) the western track from Antarctic continent to 50°S.

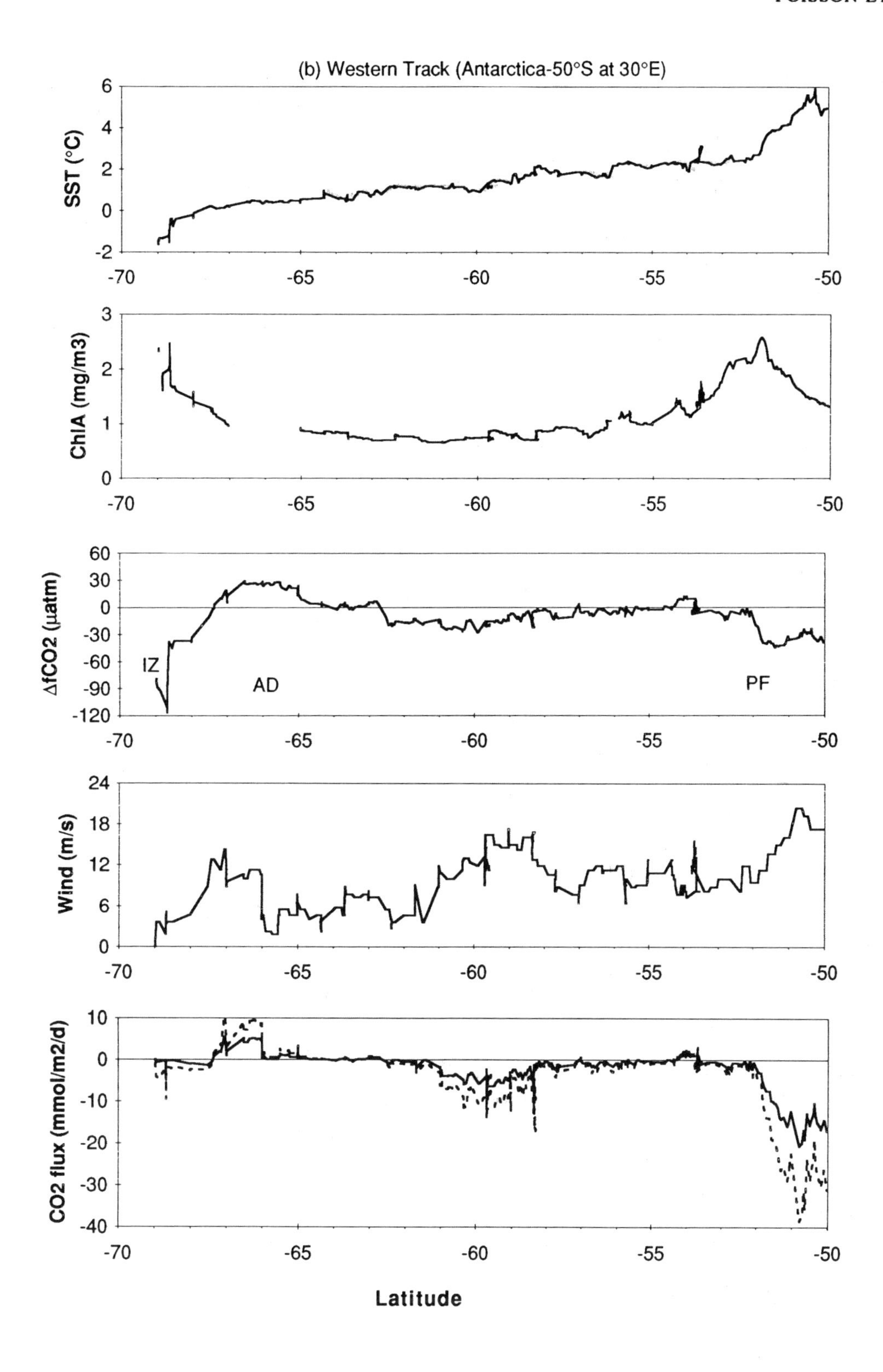

Figure 3. (continued)

can be described following four domains: the pack ice zone (IZ), the antarctic divergence (AD), the region between the AD and the Polar Front (PF), and the PF zone.

Near the pack ice very low ΔfCO_2 were observed (minimum of -115 µatm at 69°S). This large potential CO_2 sink is clearly related to biological activity as shown by high fluorescence records. Such strongly negative Chlorophyll-a/fCO_2 correlations have been previously encountered in this region during the same season [Metzl et al., 1991]. During the cruise period, the winds were very low near the Antarctic continent; consequently, instantaneous CO_2 fluxes are not dramatically negative although ΔfCO_2 is very low.

In the Antarctic Divergence (AD) zone (67°S-63°S), CO_2 sources are observed along both tracks. The source zone presents a larger extension along the Eastern track (figure 3a) than along the Western one (figure 3b). This is certainly due to a diagonal crossing of the AD zone along the Eastern track. Chlorophyll-a decreases from the South, and is low in the AD zone. On the Western track, the location of ΔfCO_2 maximum at 66°S was linked to the shallower position of the upper circumpolar deep water as shown by temperature, salinity or oxygen sections (figure 4) obtained during the CIVA-1/WOCE cruise (the MINERVE 23 program was coupled to CIVA-1/WOCE cruise). We believe this is also true for the position of the ΔfCO_2 maximum around 65°S along the Eastern track. In the AD zone, winds were stronger than at higher latitude; the CO_2 sources are significant (fluxes around 5 mmol. $m^{-2}.d^{-1}$ with kLM or 10 $mmol.m^{-2}.d^{-1}$ with kW) and as high as in equatorial areas [e.g. Feely et al., 1987; Lefevre and Dandonneau, 1992].

The domain between the AD and the PF zone is a CO_2 sink (please, note that the Polar Front was not crossed along the Eastern track). On both tracks, there is a sharp horizontal gradient of ΔfCO_2 (63°S and 65°S on Eastern and Western tracks respectively); this position separates the surface waters influenced (or not) by the Antarctic Divergence. Along the Eastern track there is also sharp decrease of ΔfCO_2 at 58°S related to higher Chlorophyll-a concentrations; the CO_2 exchange is from atmosphere to ocean (-15 $mmol.m^{-2}.d^{-1}$ with kLM or -30 $mmol.m^{-2}.d^{-1}$ with kW); from there, ΔfCO_2 is constant (-30 µatm) and the CO_2 flux variabilities are controlled by the wind variations. Along the Western track, the maximal undersaturation is -15 µatm at 60°S and reaches slowly the air-sea equilibrium at 55°S (ΔfCO_2=0 µatm); from 58°S to 53°S, the CO_2 flux is small compared to the one along the Eastern track.

The Polar Front was crossed at 52°S-30°E only along the Western track. It is located at 52°S according to the sub-surface definition: the northern position of the 2°C isotherm measured during CIVA-1/WOCE cruise (figure 4). Because the front is generally shifted northward between Crozet and Kerguelen [Park et al., 1991], it was not crossed on the Eastern track which starts south of Kerguelen. At the Polar Front there is an abrupt decrease of ΔfCO_2; this is clearly related to higher Chlorophyll-a. The CO_2 undersaturation is around -40 µatm. High wind speeds (up to 20 $m.s^{-1}$) lead to a strong CO_2 sink (minimum at 50°40'S of -21, -27 or -39 $mmol.m^{-2}.d^{-1}$ by using kLM, kT and kW respectively). In term of integrated air-sea CO_2 exchanges, the differences between these three gas transfer velocities formulations is significant in this CO_2 undersaturated and high wind speeds oceanic region.

The meridional distribution described above, appears to be permanent during austral summer. In February 1987 we also made sea surface CO_2 measurements near 30°E in the Southern Ocean during the INDIVAT 5 cruise (see the track on figure 1). Here, we compare the meridional ΔfCO_2 distribution in February 1987 and February 1993 (figure 5). ΔfCO_2 in figure 5 are referred to measured atmospheric CO_2 concentrations for each period. In the 66°S-62°S band, ΔfCO_2 have the same order of magnitude; one can conclude that in this area, atmospheric and oceanic fCO_2 have the same annual increase. Both distributions show a northward increase of ΔfCO_2 from 62°S to 54°S but the measurements made in 1987 show higher ΔfCO_2. In the Polar Front Zone, there is a rapid decrease which was associated with high Chlorophyll-a in 1987 [Metzl et al., 1991]; this was also the case in 1993 (figure 3b) and as in 1987, this occurs to the south of the Polar Front. Note that the fCO_2 measurements techniques were different (semi-continuous by gas chromatography in 1987, continuous by Infrared analyser in 1993); according to the precisions (1% in 1987 and 0.3% in 1993) the 6 years ΔfCO_2 change between the two cruises cannot be explained by technical differences only. Temporal change in both atmospheric and oceanic fCO_2 have to be taken into account. In addition, interannual variations have to be considered. As a matter of fact, in the region where ΔfCO_2 are significantly lower in 1993 (58°S-50S) the SST was also lower than in February 1987 (figure 5a). Being a uniform atmospheric increase of about 10 µatm in the band 58°S-50°S from 1987 to 1993, one has to considere an average of -14 µatm change in oceanic fCO_2 between the two data sets; this can be explained by SST interannual variations of 1.5°C on average between the two legs. Therefore, ΔfCO_2 change has been low from 1987 to 1993.

It is interesting to note that 20 years ago, in austral summer and along the same transect, sea surface CO_2 presented similar distribution [Miyake and Sugimura, 1969], but at the period of measurements, 1967, ΔfCO_2 was larger because of much lower atmospheric CO_2 concentration.

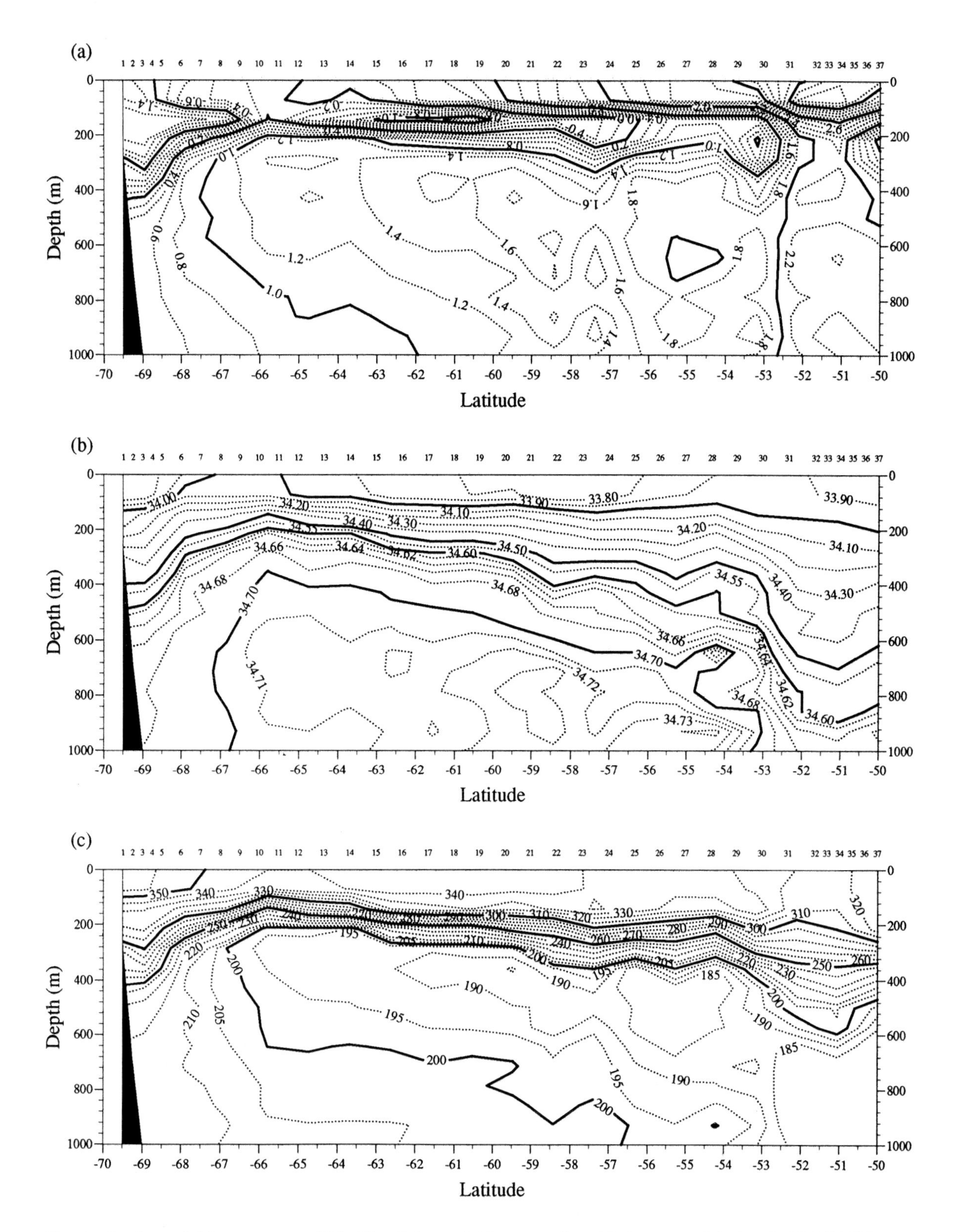

Fig. 4. Sections of (a) temperature (°C), (b) salinity and (c) oxygen (µmol/kg) measured during the WOCE/CIVA 1 cruise (February 1993, stations between 69°S and 50°S). We relate the positive ΔfCO_2 in figure 3b with the position of the upwelling zone around 66°S.

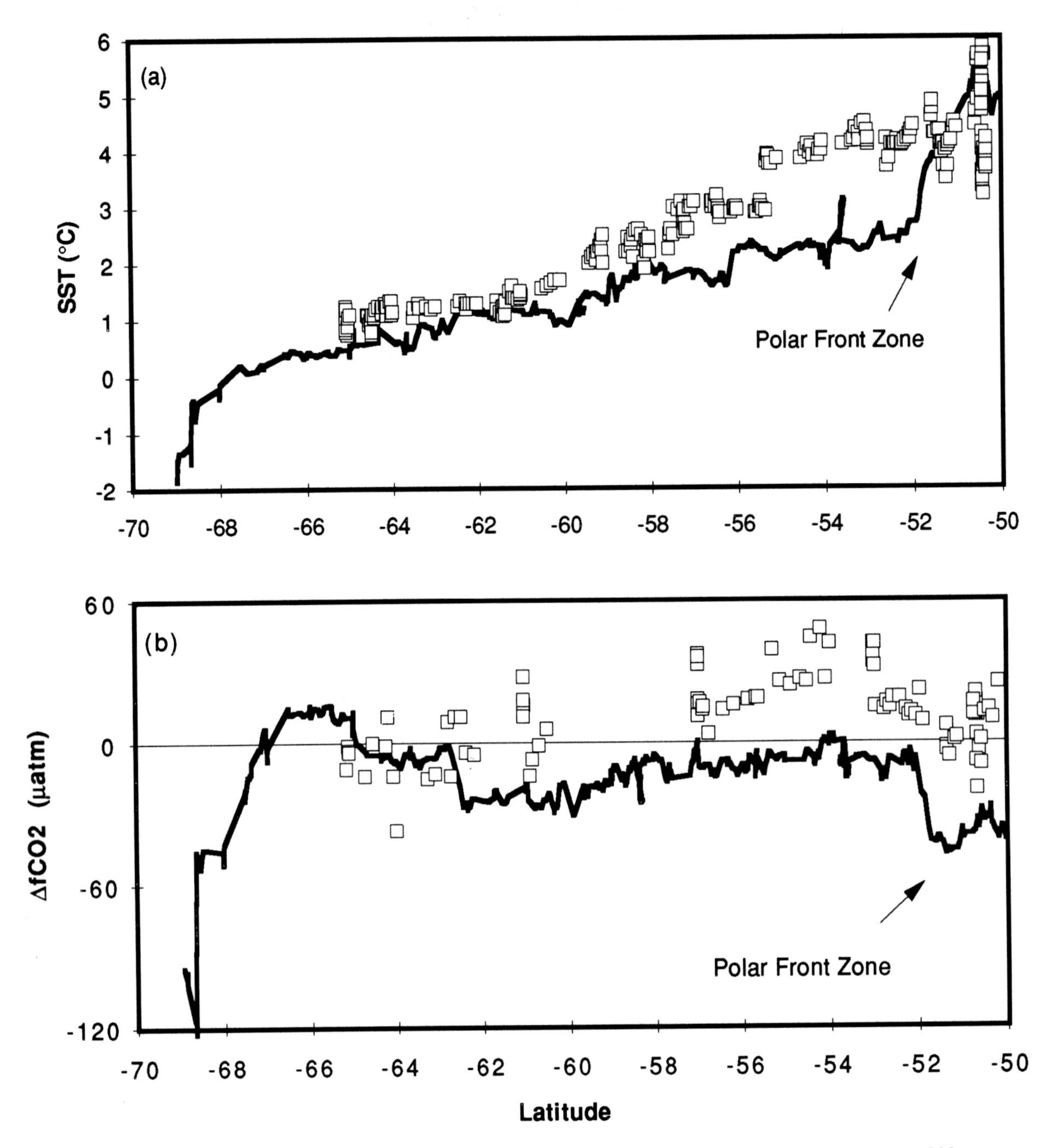

Fig. 5. Meridional SST (a) and ΔfCO2 (b) distributions around 30°E for February 1987 and February 1993.

ON THE WESTERN SIDE OF THE KERGUELEN PLATEAU

ΔfCO_2 measurements made in January-April 1991 near the Kerguelen Plateau are presented in Plate 1. Very high ΔfCO_2 spatial variabilities were observed: the domain composes a mozaic of CO_2 sinks, sources and near equilibrium values. This high ΔfCO_2 variability in the region of the Kerguelen Plateau is probably associated to the high dynamical variability described at meso-scale by Altimeter data [Cheney et al., 1983; Chelton et al., 1990] and high resolution model [Webb et al., 1991]. This complex dynamics may also govern sea surface meso-scale variability in chlorophyll-a [Krey and Babenerd, 1976; plate 23]. Offshore the Kerguelen Plateau, ΔfCO_2 distribution is much less variable (see the track from 46°S-50°E to 60°S-

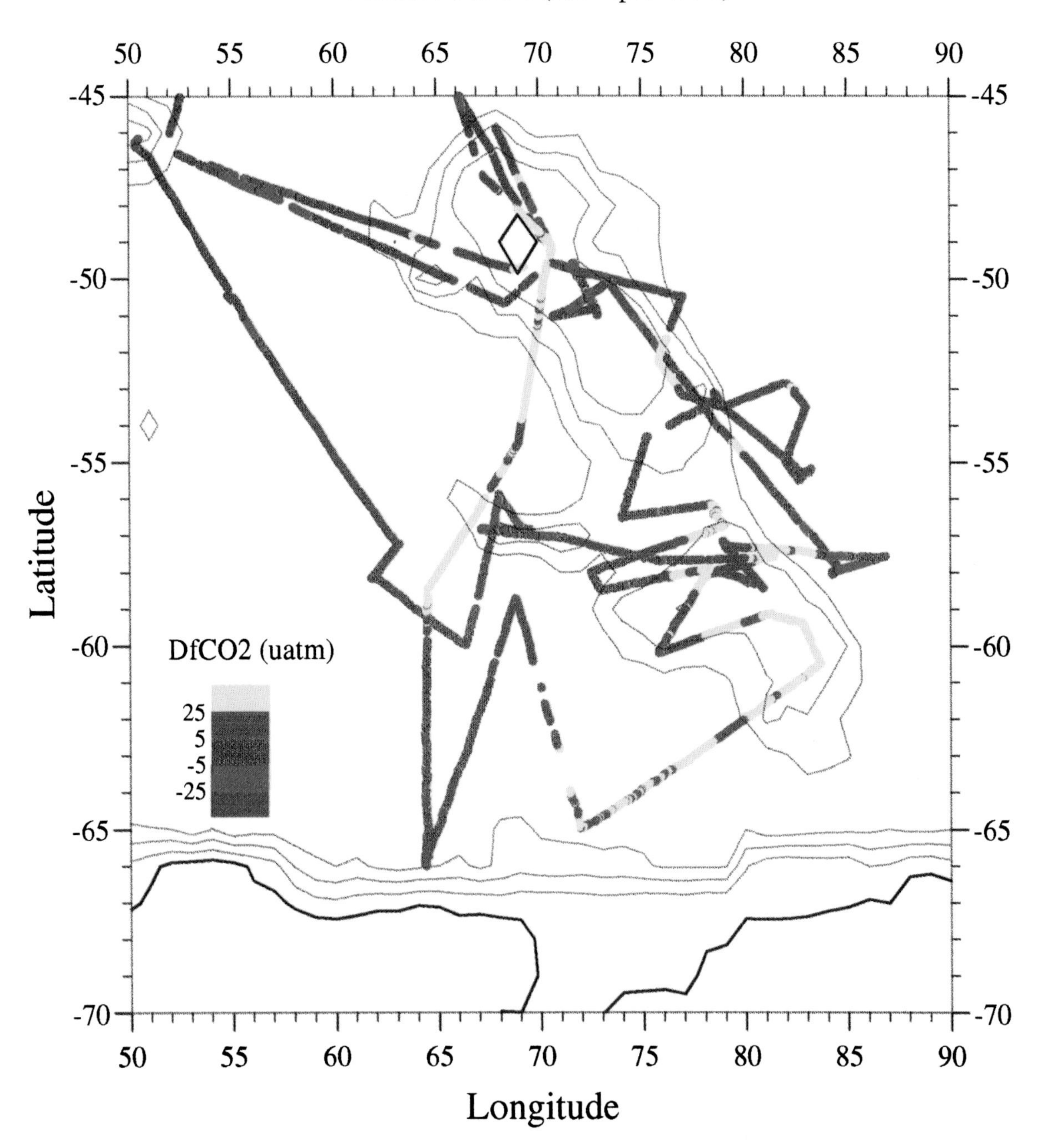

Plate 1. ΔfCO_2 maps obtained during MINERVE 07 and 08 cruises (January-April 1991). The color scale corresponds to CO_2 sources zones (yellow and red), near equilibrium (brown) and CO_2 sink zones (green and blue). Isobaths 0, 1000, 2000 and 3000m are also shown, indicating the extension of the Kerguelen Plateau.

67°E in Plate 1). The large ΔfCO_2 variability leads to a high variability in the air-sea CO_2 fluxes. For example, in March 1991, the local CO_2 flux varies from -30 mmol. $m^{-2}.d^{-1}$ (the ocean is a CO_2 sink) to 30 $mmol.m^{-2}.d^{-1}$ (the ocean is a CO_2 source) by using kW. Although the number of measurements is quite high for such a domain, it is not possible to construct objectively a continuous field for fCO_2 and for air-sea fluxes.

			Transfer Velocities				Fluxes		
Cruise	Period	Wind (m/s)	kLM (cm/h)	kW (cm/h)	kT (cm/h)	ΔfCO2 (μatm)	FLM	FW (mmol/m2/d)	FT
Minerve07	Jan 91	9.5 (3.2)	11.5 (7.3)	21.0 (14.0)	20.9 (10.3)	-6.3 (15.7)	-0.9 (2.3)	-1.7 (4.2)	-1.7 (4.2)
Minerve08	Mar 91	10.7 (3.4)	13.5 (7.2)	24.9 (13.4)	23.6 (10.3)	5.2 (18.9)	1.1 (4.0)	2.1 (7.3)	2.0 (6.8)
Minerve23 (E)	Feb 93	10.0 (4.5)	11.8 (8.5)	21.8 (15.7)	20.6 (12.8)	-11.2 (25.2)	-1.8 (3.3)	-3.3 (6.2)	-3.3 (6.1)
Minerve23 (W)	Feb 93	8.7 (3.7)	9.7 (6.7)	18.1 (12.2)	17.9 (10.8)	-8.1 (24.5)	-0.8 (2.2)	-1.7 (4.0)	-1.4 (3.8)
Minerve23	Feb 93	9.3 (4.0)	11.0 (7.7)	20.4 (14.2)	19.7 (11.8)	-8.7 (22.9)	-1.4 (2.9)	-2.6 (5.4)	-2.5 (5.3)

Table 1: Average of wind speed, gas transfer velocity (kLM, kW and kT, see text for a description of these computations), ΔfCO_2 and air-sea CO_2 flux (FLM, FW, and FT) for the Southern Ocean cruises MINERVE 7, 8 and 23. For MINERVE 23, calculations are presented for the Eastern track (E), Western track (W) and the whole cruise (see figure 1). Number in paranthesis are standard-deviations as an indicator of spatial variability (the period of each cruise is about one month).

MEAN AIR-SEA CO_2 FLUXES IN THE SOUTHERN OCEAN

As mentioned above, we cannot compute integrated air-sea CO_2 fluxes over the domains visited in 1991 and 1993. We thus present these fluxes in term of averages along the observed tracks. As shown previously (see figure 3), there exist areas of CO_2 source (in the AD zone) and sink (North of the AD zone); mean calculations over the Southern Ocean domain have to be taken with caution. For each fCO_2 measurement, intantaneous air-sea flux has been computed following equation (1) and the three gas transfert velocity relationships (kLM, kW and kT). For each cruise we calculated the average of these calculations made south of the PF (considering the Southern Ocean south of the PF). Therefore, the large sink zone between 52°S and 50°S along the Western track of the MINERVE 23 cruise (figure 3b) is not included in the mean computation. Results are presented in Table 1. Average of wind speed was about the same for the three cruises (from 9 to 10 m/s) with about the same spatio-temporal variability (3 to 4 m/s). On the contrary, average ΔfCO_2 are different. Thus, one could expect that differences in average of air-sea CO_2 fluxes will be mainly controlled by oceanic fCO_2 variations rathern than wind speed. Comparing the average of transfer velocities, kLM are always 1.8 smaller than kW and kT. This is the same for the fluxes. For the MINERVE 08 cruise which occured near the Kerguelen Plateau, the region was a CO_2 source. The average is comparable to CO_2 fluxes estimated in a region of the Pacific sector of the Southern Ocean reported by Murphy et al. [1991] who also used both kLM and kT formulations. The large CO_2 flux standard deviation is related to observed ΔfCO_2 spatial variability (5 ±19 μatm; see Table 1). We conclude that during March 1991 the 60°E-90°E zone of the Southern Ocean was a weak CO_2 source region on average, but with a large spatial variability. During the other cruises and on average, the western Indian sector of the Southern Ocean is a CO_2 sink; again, the spatial variability is high. In February 1993, the region in the 40°E-70°E band (Eastern track) was a stronger sink than at 30°E (Western track) because wind speed was higher and ΔfCO_2 lower especially in the zone 58°S-52°S (see figure 3).

CONCLUSIONS

The sea surface CO_2 observations made in austral summers 1991 and 1993 in the Western Indian sector of the Southern Ocean show that variabilities of air-sea CO_2 exchange exist at various time and space scales in this poorly sampled ocean. Very large ΔfCO_2 variations at small-scale have been found near the pack ice and in the region of the Kerguelen Plateau. This leads to high variability in air-sea CO_2 fluxes. In the AD zone and between the AD and the PF zone, ΔfCO_2 distribution is much less variable; there, the spatial variations of the air-sea exchanges are mainly controlled by wind variability.

A comparison with older data (1987 versus 1993) collected also during austral summer in the same area indicates that the meridional large-scale structure of ΔfCO_2 presents permanent characteristics (e.g. minima near the pack-ice, maxima in the AD zone, minima in the PF zone). This distribution can be related to permanent oceanic processes: biological activity near the pack-ice leading to very low ΔfCO_2; large-scale circumpolar upwelling that brings CO_2 rich waters from the bottom and governs CO_2 source in the AD zone. On the other hand, the magnitude of ΔfCO_2 are different; this temporal change of ΔfCO_2 level can be related to the decadal CO_2 atmospheric increase but oceanic processes, such as interannual SST variations, have to be taken into account for explaining the non uniformity of ΔfCO_2 changes at 30°E.

Average air-sea CO_2 flux computations for each observed period show both CO_2 sinks and sources for the atmosphere. The calculation of the fluxes performed by using different relationships between gas transfer velocity and wind speed leads to an uncertainty of a factor of about 2, but the sign of the air-sea fluxes is indeed always preserved. Keeping that in mind, we conclude that a sink domain was found in the 30°E-60°E austral region; sources were found in the area of the Kerguelen Plateau which corresponds to high eddy energy region. Because of high variabilities for both sea surface fCO_2 and wind speed, it is impossible, at present, to establish an objective continuous field of ΔfCO_2 and integrated air-sea CO_2 fluxes in these regions. From results obtained in averaging fluxes for each cruise it is clear that a single cruise at one period is not enough for estimating the role of the Southern Ocean in term of CO_2 source and sink. More in-situ studies, especially during austral winter, are needed to compute yearly integrated flux over the whole Southern Ocean.

Acknowledgements . The cruises of the R. V. Marion-Dufresne were supported by the Institut Français pour la Recherche et la Technologie Polaire (IFRTP); we thank Y. Balut for his assistance in the organisation of the MINERVE program and B. Ollivier for his continuous help during the cruises. We wish to thank all the captains and crew members of the R. V. Marion-Dufresne who recorded hourly the meterological parameters during these cruises. We thank also Philippe Laurent and Christine Blanc for their very active participation during MINERVE 07, 08 and 23 cruises.

REFERENCES

Chelton, D.B., M.G. Schlax, D.L. Witter, and J.G. Richman, Geosat Altimeter Observations of the surface Circulation of the Southern Ocean. *J. Geophy. Res., 95*, C10, 17,877-17,903, 1990.

Cheney, R.E., J.G. Marsh and B.D. Beckley, Global Mesoscale Variability from Collinear Tracks of SEASAT Altimeter Data. *J. Geophy. Res., 88*, C7, 4343-4354, 1983.

Copin-Montégut, C., A new formula for the effect of temperature on the partial pressure of CO2 in seawater. *Mar. Chem., 25*, 29-37, 1988.

Copin-Montégut, C., A new formula for the effect of temperature on the partial pressure of CO2 in seawater. Corrigendum. *Mar. Chem., 27*, 143-144, 1989.

Feely, R.A., R.H. Gammon, B.A. Taft, P.E. Pullen, L.S. Waterman, T.J. Conway, J.F. Gendron, and D.P. Wisegraver. Distribution of Chemical Tracers in the Eastern Equatorial Pacific During and After the 1982-1983 El Niño/Southern Oscillation Event. *J. Geophys. Res., 92,* C6, 6545-6558, 1987.

Inoue, H. and Y.Sugimura, Distribution and variations of oceanic carbon dioxide in the western North Pacific, eastern Indian, and Southern Ocean south of Australia. *Tellus, 40B*, 308-320, 1988.

Jähne, B., Zur Parametrisierung des Gas antaushes mit hilfe von Laborexperimenten. Doct. Dissertation, Univ. Heidelberg, 124 pp, 1980.

Keeling, C.D., Carbon Dioxide in Surface Ocean Waters. 4-Global Distribution. *J. Geophys. Res., 73*, 14, 4543-4553, 1968.

Krey, J., and B. Babenerd, Phytoplankton production, Atlas of the International Indian Ocean Expedition, Institut für Meereskunde and der Universität Kiel, *Int. Ocean. Comm. UNESCO*, 70 pp., 1976.

Lefevre, N. and Y. Dandonneau. Air-Sea CO2 Fluxes in the Equatorial Pacific in January-March 1991. *Geophysical Res. Let.*, 19, 22, 2223-2226, 1992.

Liss, P. and L. Merlivat, Air-sea exchange rates, introduction and synthesis. In *The Role of Air-Sea Exchange in Geochemical Cycling*, (ed. P.Buat-Ménard), NATO/ASI Series, D.Reidel, Dordrecht, 113-127, 1986.

Metzl, N., C. Beauverger, C. Brunet, C. Goyet and A. Poisson, Surface water pCO2 in the Western Indian Sector of the Southern Ocean: a highly variable CO2 source/sink region during the austral summer. *Mar. Chem. 35*, 85-95, 1991.

Miyake, Y. and Y. Sugimura, Carbon Dioxide in the Surface Water and the Atmosphere in the Pacific, the Indian and the Antarctic Ocean Areas. Rec. of Ocean. *Works in Japan, 10*, N°1, 28-33, 1969.

Miyake, Y., Sugimura, Y. and K. Saruhashi, The Carbon Dioxide Content in the Surface Waters in the Pacific Ocean. Rec. of Ocean. *Works in Japan, 12*, N°2, 45-52, 1974.

Murphy, P.P., R.A. Feely, R.H. Gammon, D.E. Harrison, K.C. Kelly and L.S. Waterman, Assesment of the Air-Sea Exchange of CO2 in the South Pacific During Austral Autumn. *J. Geophys. Res., 96*, C11, 20,455-20,465,1991.

Park, Y.-H., L. Gambéroni and E. Charriaud, Frontal structure and transport of the Antarcic Circumpolar Current in the south Indian Ocean sector, 40-80°E. *Mar. Chem., 35*, 1-4, 45-62, 1991.

Poisson, A., N. Metzl, C. Brunet, B. Schauer, B. Brès, D. Ruiz-Pino and F. Louanchi. Variability of sources and sinks of CO2 and in the Western Indian and Southern Oceans during the year 1991. *J. Geophys. Res.* (in press), 1993

Takahashi T., The carbon dioxide puzzle. *Oceanus, 32*, 2, 22-29, 1989.

Tans, P.P., I.Y. Fung and T. Takahashi, Observational Constraints on the Global Atmospheric CO2 Budget. *Science, 247*, 1431-1438, 1990.

Wanninkhov, R. relationship Between Wind Speed and Gas Exchange Over the Ocean. *J. Geophys. Res.* C5, 7373-7382, 1992.

Webb, D.J., P.D. Killworth, A.C. Coward, and S.R. Thompson, in The FRAM Atlas of the Southern Ocean, *NERC Pub.*, 67 pp., Nat. Environ. Res. Council, Swindon, England, 1991.

Weiss, R.F., Carbon dioxide in water and seawater: the solubility of a non-ideal gas. *Mar. Chem., 2*, 203-215, 1974.

B. Brès, C. Brunet, F. Louanchi, N. Metzl, A. Poisson, D. Ruiz-Pino, and B. Schauer, Laboratoire de physique et Chimie Marines, Université Pierre et Marie Curie, case 134, 4 place Jussieu, 75252, Paris Cedex 05, France.

Arctic Structural Evolution: Relationship to Paleoceanography

G. Leonard Johnson
Office of Naval Research, Arlington, Virginia, U. S. A.

Julian Pogrebitsky
VNII Okean Geologie and Dynam. Geol. Depart., St. Petersburg, RUSSIA

Ron Macnab
Atlantic Geoscience Centre, Bedford Inst., Dartmouth, N. S., CANADA

This paper summarizes major events in the tectonic evolution of the Arctic from a paleoceanographic perspective focused on the formation of oceanic basins in the Mesozoic and Cenozoic. Despite its present cold climate, the Arctic land masses once lay much farther to the south and were gradually carried northward by plate tectonic motions. For example, in the Late Carboniferous northern Greenland, which was part of Pangea, was situated at only 30°N with the Panthalassa Sea to the north. By Late Jurassic-Early Cretaceous, a series of allochthonous blocks had collided with both the North American and Siberian plates. The Pacific Ocean was separated from the Arctic Basin by rotation of the North-Slope Chukotka block counterclockwise away from the Canadian Arctic Islands by Mid Cretaceous. It is speculated, based on the magnetic patterns, that the Alpha-Mendeleev Ridge complex may either have been a separate precursor block or may represent oceanic crust modified by a hot spot. Sea floor spreading in Late Cretaceous or Early Paleocene rifted the present Makarov Basin in a wedge-shaped manner, with the Siberian end wider than the Canadian. The Eurasia Basin has a well-ordered set of magnetic anomalies, and thus this basin can be confidently dated to have been created just after the Cretaceous-Tertiary boundary. Creation of the Eurasia Basin eventually opened the pathway for exchange of waters between the Arctic and Atlantic oceans. The cold-oxygen rich waters from the Arctic ventilate the world's oceans and provide an important mechanism for global heat transfer and for the cycling of nutrients and carbon.

INTRODUCTION

Only in the last several decades has the morphology of the major sea floor structural features of the Arctic been defined, and detailed bathymetric data is still largely lacking. The Arctic Ocean is unique among the oceans of the world in that 49% of its area is underlain by continental shelf, primarily the wide European and Siberian continental shelves. These shelves act as ice factories, producing the Arctic sea ice and the cold residual brines that are formed by the freezing process. These waters sink as an oxygen rich dense water mass contributing to the renewal of water masses of the global ocean and to the cycling of nutrients and carbon [Broecker, 1987]. Thus the polar and subpolar seas have a global impact on the entire marine environment [Nansen Arctic Drilling Program Science Committee, 1992]. An additional Arctic characteristic of potential importance to global change is gas hydrates. These have the potential to release large quantities of methane, a greenhouse gas, if they become unstable [Kvenvolden and Grantz, 1990]. The complex sea floor topography of the Arctic Ocean is pivotal in its strong influence over the world ocean, and evolution of this sea floor must correspondingly be related to paleoclimate changes. Accordingly, this paper will focus on the paleogeography and time of formation of the Arctic Ocean and its two contained basins: the Amerasia and Eurasia

The Polar Oceans and Their Role in Shaping
the Global Environment
Geophysical Monograph 85

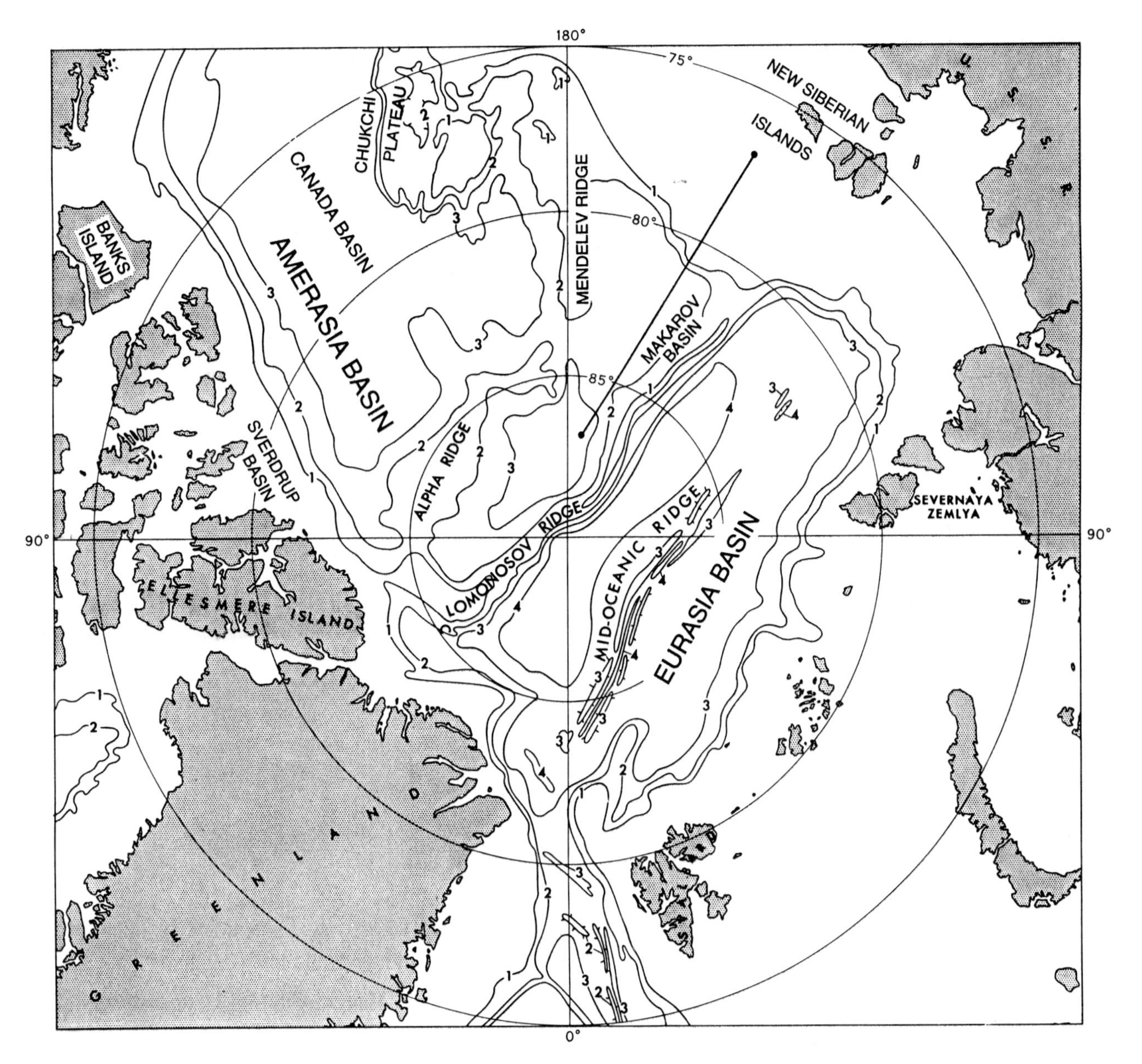

Figure 1. Physiographic Provinces of the Arctic modified from Johnson [1990]. Heavy bar denotes location of refraction line shown in Figure 2.

(Figure 1). For a morphological description of the Arctic see Johnson [1990 and references contained within], and for a plate tectonic summary see Lawver et al. [1990 and references contained within].

The Cenozoic history of the Eurasian Basin is well understood because it involves the Eurasia and North America Plates (Figure 1) and is therefore constrained by data from more southerly regions. Additionally, it contains a readily decipherable sea floor magnetic pattern (Plate 1). The oldest positive magnetic lineation that can be identified with certainty in the Eurasia Basin is anomaly 24, similar to the Norwegian Sea. Other lineations at the base of the Lomonosov Ridge suggest that spreading may have occurred as early as 54-64 million years ago (Ma) [Jackson et al., 1993; Jackson and Johnson, 1986 and references contained therein]. The source of Amerasia Basin however is problematic, and there are as many hypotheses on its origin as there are investigators. An excellent summary is contained in Lawver and Scotese [1990]. In this paper we will present our ideas, based on aeromagnetic data, and relate the tectonic history to the paleogeography.

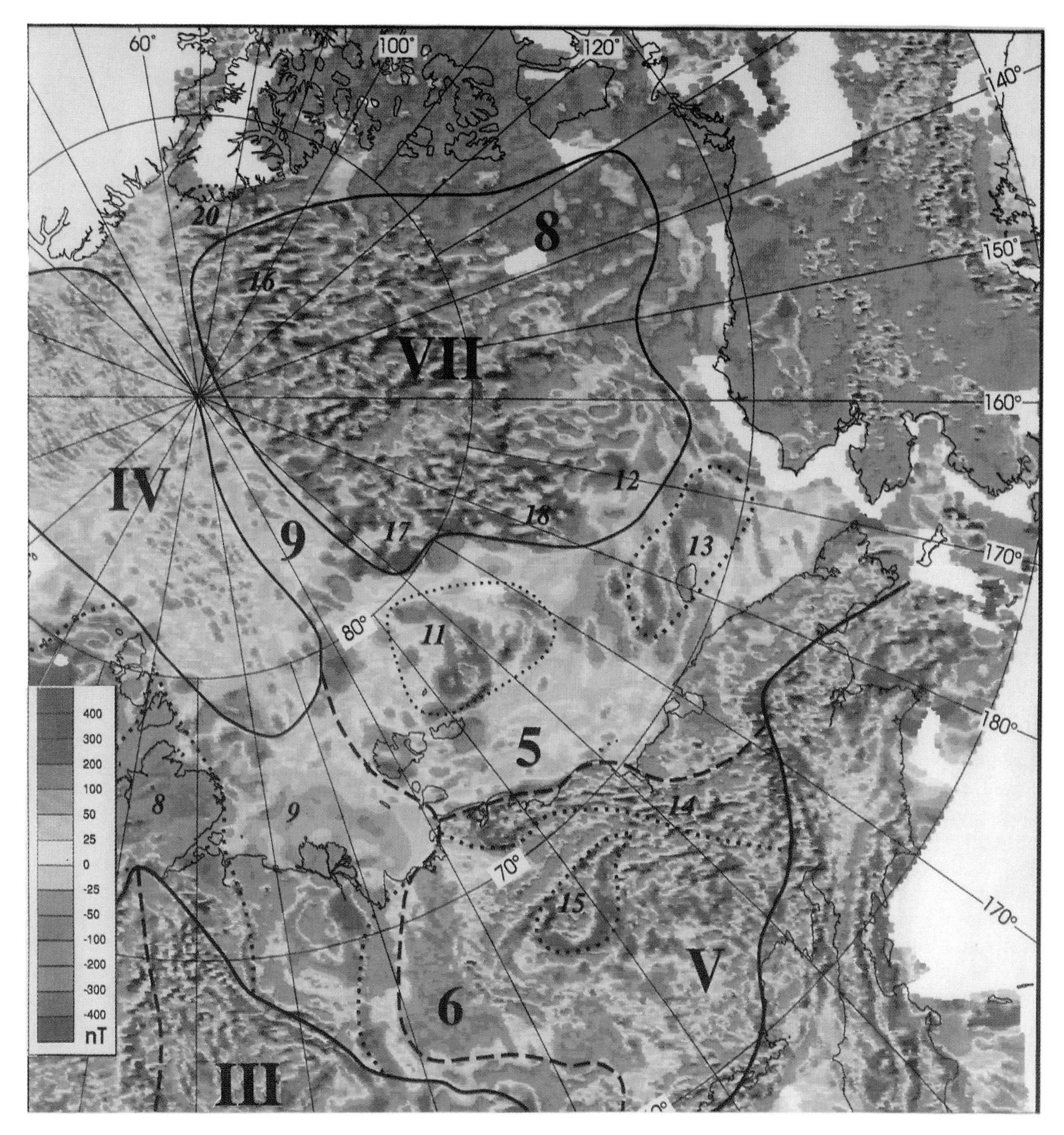

Plate 1. Magnetic Field and Tectonic structures of the Arctic Region (Macnab et al. 1992a, b; Macnab, 1993). Magnetic data over mainland Russia were collected by the Ministry of Energy of the former USSR and digitized by the U. S. Naval Oceanographic Office. Portions of the data over the Arctic Ocean were assembled at Sevmorgeologia and digitized at the Atlantic Geoscience Centre; most of the remaining offshore data were collected by the U. S. Naval Research Laboratory. Data over the North American mainland were assembled in part during the DNAG Project. Tectonic structures are in part based on the work of Y. Pogrebitsky and L. Zonenshain (Pogrebitsky et al., 1993). III Siberian Craton, IV Eurasia Basin, V Northeast Asiatic Orogenic Belt, VII Amerasia Basin; 5 Nova Sibir-Chukchi Fold Belt, 6 Verkoyansk-Kolymia Fold Belt, 8 Canada Basin, 9 Lomonosov Ridge; *8* Taimyr-Nova Zemlya Suture, *9* Laptev Rift Zone, 11 De Long Rise, *12* Chukchi Cap, *13* Wrangel Zone, 16 Alpha Ridge, *17* Makarov Basin, *18* Mendeleev Ridge, 20 Pearya Terrane.

AMERASIA BASIN EVOLUTION

The following discussion is devoted to the Amerasia Basin; however it is recognized that the major paleoceanographic effect in the northern hemisphere was related to the opening of the Eurasia Basin. Three basic models have been proposed for the origin of the Amerasia Basin and the tectonic structures contained within: 1. creation of the basin by sea floor spreading, with possible modification by "hot spot" activity; 2. entrapment of continental crust with subsequent crustal thinning; and 3. a former region of subduction or compression. See Lawver and Scotese [1990] for a synopsis of the three models. For specific details and an expanded bibliography see Churkin and Trexler [1980, 1981]; Green et al. [1984]; Jackson et al. [1990]; Pogrebitsky et al. [1993a, b]; Sweeney et al. [1978]; Taylor et al. [1981]; and Vogt et al. [1984].

The key to the problem is the origin and nature of the Alpha-Mendeleev Ridge, which is the largest single submarine feature in the Arctic Ocean. In areal extent it exceeds the Alps, and, in addition to the massive exposed structure, large portions are buried beneath the Canada Abyssal Plain. The Alpha Ridge is covered for the most part by a sedimentary sequence up to 1 km in thick and has yielded Cretaceous marine sediments [Clark et al., 1986; Nansen Arctic Drilling Program Science Committee, 1992, and references contained within the latter]. The basement material on which the sedimentary cover was deposited has a sound velocity of 5.3 km/s, typical of oceanic layer 2 and also of indurated sedimentary rocks. A continental origin based on aeromagnetic data from Pogrebitsky et al. [1993a, b] is worth serious consideration and is supported by the magnetic anomaly patterns (Plate 1). It should be noted that King et al. [1966] had reached the same conclusion using fewer data. Dredged material during the Canadian Expedition for Study of the Alpha Ridge (CESAR), from exposed basement of the ridge yielded a weathered alkaline volcanic [Van Wagoner and Robinson, 1985] consistent with a volcanic origin. The sound velocity in the layer below ranged from 6.45 to 6.8 km/s and at a depth of 20 km a velocity of 7.3 km/s was measured [Forsyth et al., 1986]. This velocity structure is similar to oceanic plateaus [Carlson et al., 1980]. The measured depth of the crust mantle boundary is 38 km at the CESAR site. This thickness could be interpreted as either continental or thickened oceanic crust due to hot spot activity such as beneath the Iceland-Faeroe Plateau. Figure 2 from Gramberg et al. [1993] shows a seismic refraction line coincident with the axis of the Makarov Basin. Moho lies at a depth of 15-20 km, which is intermediate between normal oceanic depths of about 8-10 km in the Eurasian Basin, and the Alpha Ridge with depths in excess of 30 km [Jackson and Johnson, 1986].

The magnetic anomaly pattern of the Alpha Ridge is extremely variable with peak to trough anomalies of up to 1500 nT and wave lengths of 20-75 km. While some anomalies are traceable for some distance and have a lineation pattern which is predominately NE-SW in the Canada Basin [Jackson et al., 1993], they do not exhibit a regular pattern consistent with sea floor spreading. This is obvious by comparing the pattern in the Eurasia Basin with that of the Alpha Ridge province (Plate 1). Plate 1 extends the area covered by the Alpha Ridge pattern to encompass the Canadian side of the Makarov Basin and a large section of the Canada Abyssal Plain, suggesting that this feature exists at depth there and has been covered by sediments. Riddihough et al. [1973] first noticed that the intense short-wave length magnetic anomaly pattern associated with the Alpha Ridge appears to extend across the shelf of northwest Ellesmere Island to include the Pearya terrane. Trettin [1987] defines the Pearya terrane as an amalgam of four largely shallow marine sequences which he suggests may have originated from the Caledonides of Svalbard. The rocks in this geologic provenance range from Proterozoic to Upper Silurian [Sweeney et al., 1990]; Macnab et al. [1992a] recognized this as an "exotic terrane." It should also be noted that this pattern is similar to the Siberian craton [Zonenshain et al., 1990; Pogrebitsky et al., 1993a, b] (Plate 1). In general the magnetic pattern is similar to the Iceland-Faeroe hot spot province except that sea floor spreading magnetic anomalies are totally

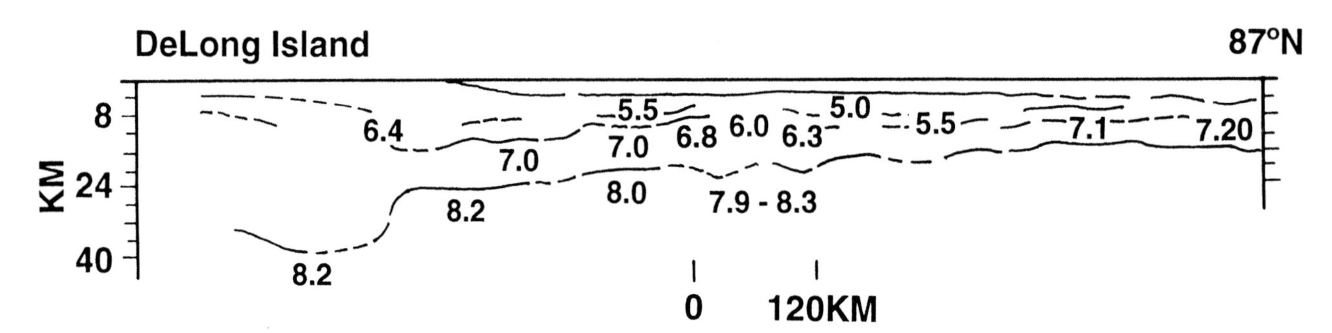

Figure 2. Seismic refraction profile along the strike of Makarov Basin based on Gramberg et al. [1993]. Location is shown in Figure 1.

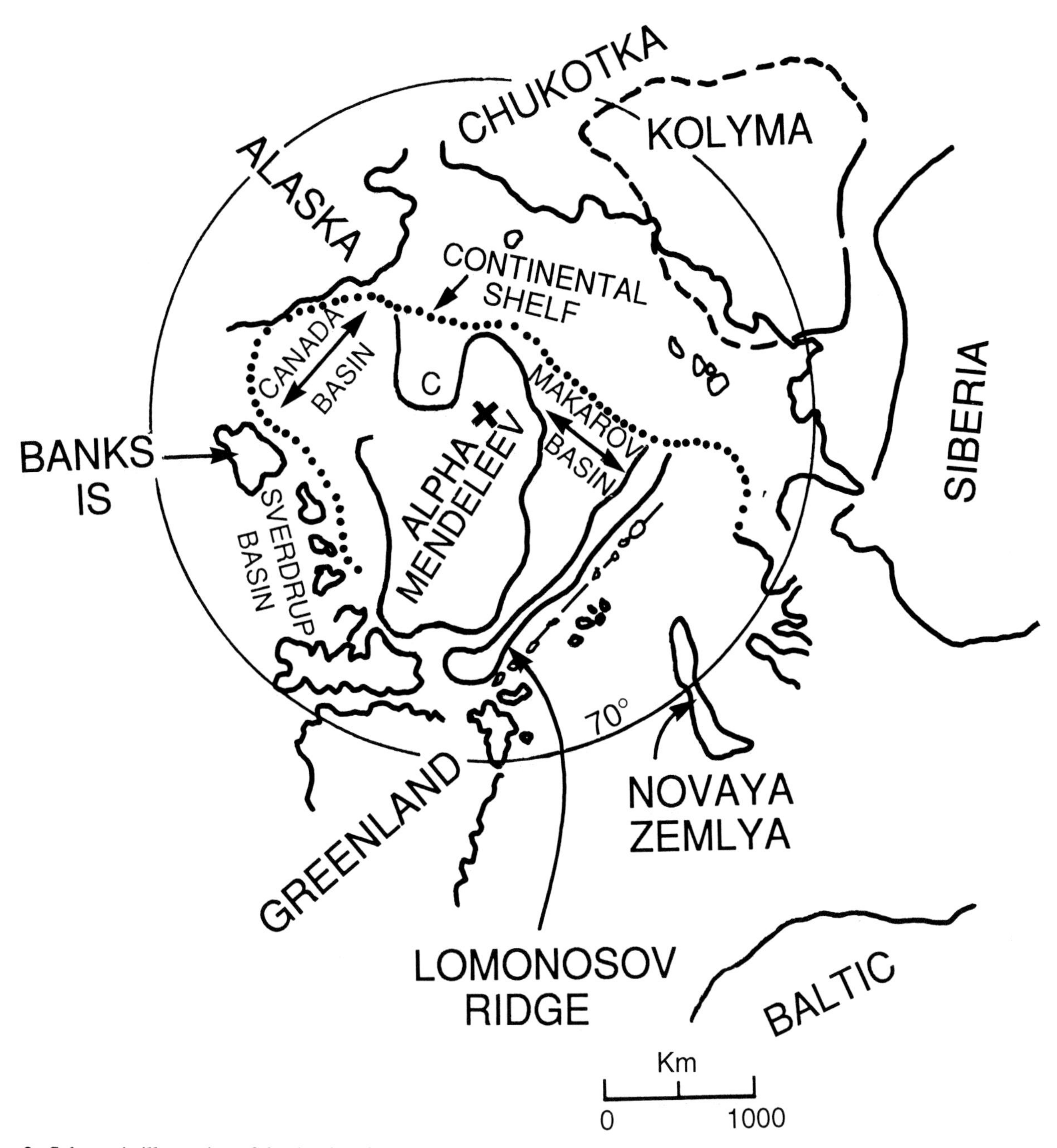

Figure 3. Schematic illustration of the Arctic prior to sea floor spreading in the Eurasia Basin. It shows limited sea floor creation in the Makarov and southern Canada Basin by sea floor spreading. C is the Chukchi Plateau region which is assumed to be continental and rifted towards the NE from the adjacent shelf (arrow). Map is not intended to show rigorous geographic locations but relative locations.

obscured [Johnson and Tanner, 1971]. Other analogous terranes in terms of depth and size include the Campbell Plateau, Lord Howe Rise and Kerguelen Plateau (M. Langseth and K. Crook, personal communication, 1993). This suggests that a continental origin for Alpha Ridge must be considered. A possible scenario is that it is an exotic terrane that entered the Arctic prior to the closing of the Arctic by the rotation of the Alaska-Chukotka block.

The magnetic pattern in the southern Canada Basin is quite different from farther north and it exhibits a subdued pattern consistent with sea floor spreading (Plate 1) [Taylor et al., 1981]. It has been suggested that these anomalies date to

M-12, or 127 Ma of the Early Cretaceous, to M-25 or 153 Ma, Oxfordian (Upper Jurassic) [Jackson and Johnson, 1986 and references contained within], consistent with dates based on other geologic and geophysical data [Sweeney, 1985]. This rifting event would have been responsible for rotation of the North Slope-Chukotka block away from the Canadian Islands to form the southern Canada Basin. It is apparent in Plate 1 that the rifting did not completely pierce the Alpha Ridge, however, it may well have caused magmatic intrusions and thinned the presumed continental crust causing subsidence of the Alpha Ridge. The limited seismic reflection data over the complex shows many extensional features such as step faults and grabens [Jackson et al., 1990].

Figure 3 shows the hypothesized fit of the blocks at the end of the Mesozoic. The region between the Kolyma and Alpha-Mendeleev blocks is presumed to be stretched continental crust morphologically represented by the wide continental shelf of the East Siberian Sea (Figure 1) and on Plate 1 by the Nova Sibir-Chukchi Foldbelt (NSCF). The magnetic field in the NSCF is subdued, with wide linear northeast striking anomalies of 200-500 nT (Plate 1). These may represent sea floor spreading flow lines and be related to northeasterly rifting of the Chukchi Plateau region as postulated by Grantz et al. [1979]. The shelf region consists of a number of east-west striking basins which are very poorly documented but which are assumed to represent a Cretaceous tensional event [Okulitch et al., 1989]. It is assumed, in this event, that the continental crust was stretched and thinned and broke into a series of fault blocks in the upper, brittle section. The thinning did not, however, pass a threshold point (20%) and the continental crust thus did not completely rupture.

CHRONOLOGY OF THE ARCTIC BASIN

In the earliest Mesozoic the present day Arctic was either a low lying shallow marginal sea or dry land at low paleolatitudes (Figure 4). The Panthalassa Sea lay to the north [Green et al., 1984], however, its extent was probably quite restricted allowing free exchange of terrestrial and marine fauna and flora between the Canadian Territories and Siberia [Kos'ko, 1993]. Initial rifting occurred during the Carboniferous to Late Jurassic in the Sverdrup Basin [Balkwill and Fox, 1982], and along the continental margin from Banks Island to the continental slope west of Point Barrow dating from the Early Jurassic (Table 1) [Grantz et al., 1990]. The continental margin of the East Siberian Sea was strongly deformed in Late Kimmerian time [Sekretov, 1993] perhaps as a result of compression from the initial Canada Basin spreading. The initial Early Cretaceous has been associated with an episode of crustal dilation, nearly normal to the present continental margin, that may have generated the NE striking system of faults and dikes within the Sverdrup Basin [Sweeney et al., 1990]. Rifting of the paleo-Canada Basin created an oceanic basin shown which comprises the southern Canada Basin of today (Figure 3). By Middle Cretaceous time, this rifting had rotated the North Slope -Chukotka block counterclockwise and moved it along transform faults nearly 70 degrees. In late Lower Cretaceous time, large extensional basins formed in the continental shelf of Siberia [Kos'ko et al., 1993]. Northward drift carried northern Greenland to 60°N by Albian time. The Nova Sibir-Chukchi Foldbelt also formed in the Early-Mid Cretaceous along the Amerasia margin. In late Maastrichtian time the Sverdrup rim began to collapse, which too may reflect tensional conditions in the Canada Basin.

Sea floor spreading in the Late Cretaceous or Early Paleogene rifted the present Makarov Basin in a wedge shaped manner, with the Siberian end wider than the Canadian. On northernmost Ellesmere Island mafic flows and intrusions have isotope ages of 93-88 Ma, and there may be a chronological and structural relationship with the opening of the Makarov Basin. The Lomonosov Ridge is commonly accepted to be a sliver of the Barents/Kara Shelf split off by initiation of sea floor spreading in the Eurasia Basin by the present mid-ocean ridge at about the Cretaceous-Tertiary boundary.

Grantz et al. [1992] and Grantz and shipboard party [1993] suggest that the Chukchi Borderland was separated from the continental margin by east-west extension of the East Siberian Sea in the Tertiary. As noted earlier Grantz et al. [1979] suggested a more northerly motion. They further indicate that the basin to the west of the ridge is underlain by continental crust that has been thinned in an east-west direction sufficiently intensively to have created the Chukchi Abyssal Plain by block rotation and listric faulting. This tectonic event completed the formation of the present day Amerasia Basin. This model, as do all others, requires large strike slip faults to accommodate the creation of new crust created by the opening of the Canada Basin and later the Makarov Basin. The location of these faults is still shrouded beneath the sea ice; however Kos'ko et al. [1993] note that large strike slip displacements up to 600 km may be present north of Wrangel Island. Sea floor spreading continues today at the slow half rate of 0.2 to 1 cm/yr in the Eurasia Basin.

PALEOCEANOGRAPHY

The opening of the Eurasia Basin and the Greenland/Norwegian Seas was a major event having profound effects on the world's oceans as it allowed the fresh ice-laden polar waters a pathway to the south. The low salinity upper layers, conditioned by river inputs, enter the Greenland/Norwegian Seas and the North Atlantic and affect the entire upper ocean by strengthening stratification. The cooled, brine-enriched deeper waters sink to contribute to the renewal of the deep and bottom water masses of the global ocean. Variations in the

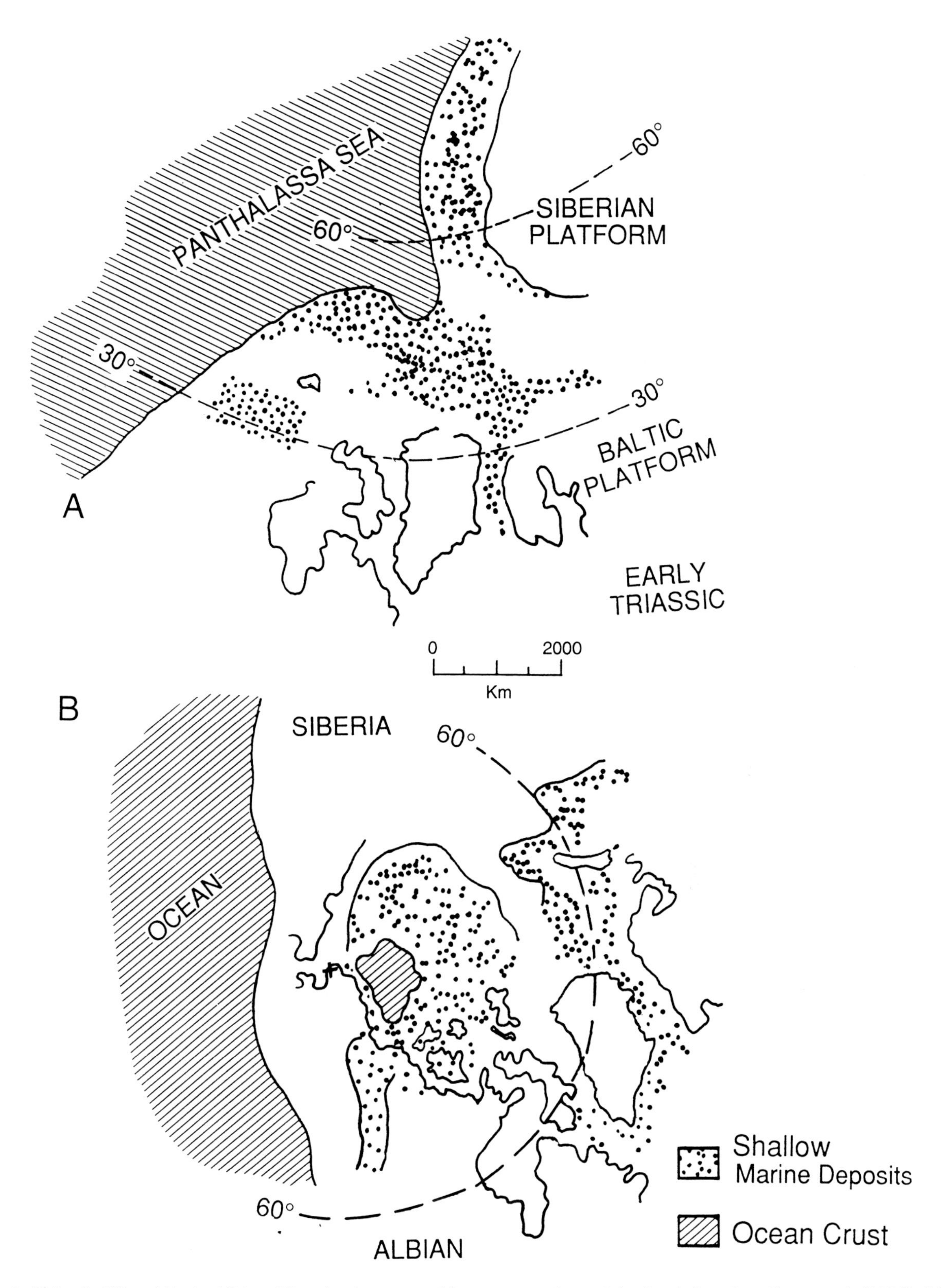

Figure 4. Triassic (A) and Early Albian (B) and paleogeographic reconstructions of the Arctic based on Green et al. [1984]. Maps are not intended to show rigorous geographic locations but rather relative location of the major structural units.

Table 1 Major Arctic Ocean Structural Events

Ma	Era	Period	Events
0	Cenozoic	Tertiary	From Strait becomes an ocean passage Greenland Sea commences opening Eurasia Basin commences spreading
100	Mesozoic	Cretaceous	Makarov Basin formed Makarov Basin commences opening S. Canada Basin formed Connection to Pacific Ocean closed Initial spreading Canada Basin
		Jurassic	East Siberian Sea Margin deformed Margin rifting continues
250		Triassic	Early rifting continental Margin Canada, Alaska

nature and flux of these deep waters have a profound impact on global temperature distribution and on ocean and atmospheric chemistry [Nansen Arctic Drilling Program Science Committee, 1992]. The Fram Strait is the major gateway for exchange of waters between the Arctic and the Atlantic. Sea floor spreading began in the Greenland Sea at anomaly 13 (approximately 35 Ma) but the adjacent Yermak and Morris Jesup plateaus inhibited water mass interchange until much later, possibly until the Miocene and/or Pliocene time [Kristoffersen and Husebye, 1985]. Also within the Eurasia Basin, the Lomonosov Ridge is a present day barrier to all water masses deeper than 1600 m. In the past it must have lain at shallower depths, however, it is uncertain at what time it started to subside.

In the Amerasia Basin a small oceanic basin (present southern Canada Basin) was created by partial rifting of the Canada Basin about 153 Ma. The Makarov Basin was formed by a later separate rifting event in the Late Cretaceous-Early Tertiary. Shallow marine conditions existed, however, over larger areas prior to the initial rifting (Figures 3, 4). During the late Albian through early Maastrichtian (Late lower Cretaceous-late Upper Cretaceous) there was a shallow water connection between the Arctic and the Gulf of Mexico through the western interior seaway, likewise, another connection existed to the region of Baffin Bay [Marincovich et al., 1990]. The northern migration of the land masses during the Mesozoic gradually moved them into high latitudes, establishing the present day climate. The obstruction by land masses to free meridianal circulation of oceanic water masses since Early Cretaceous led to the isolation of the marine Arctic from the world's oceans.

SUMMARY

Several investigators have suggested that the Alpha-Mendeleev Ridge originated by rifting from the Lomonosov Ridge while both were attached to the Barents-Kara Shelf [Zonenshain et al., 1990; and Green et al., 1984]. Green et al. [1984] postulated that the Alpha Ridge was a thinned, intruded continental fragment that was rifted away from the Barents Platform. Based on the magnetic fabric, we suggest rather that the Alpha Ridge is more nearly similar to the Siberian sector. It may well be a terrane which preceded the Kolyma block and docked up against the Canadian Ellesmere Island. This may be related to the Late Devonian-Early Carboniferous regional folding and faulting that was most intense near the continental margin of NW Canada [Sweeney et al., 1990]. Based on magnetics the Alpha Ridge province has a much greater areal extent than is evident morphologically. Also, the magnetic pattern does not resemble that of the Barents Shelf. However, based on the geologic affinities of Pearya and Svalbard, a European origin is possible. We suggest that it is possible that the Alpha-Mendeleev block is a continental terrane which was emplaced prior to the opening of the Canada Basin and subsequent rifting of the Chukchi Plateau. A major problem is the subsidence of this block to oceanic depths. It is suggested that it is not normal "continent," but rather denser material with extensive intrusions [Pogrebitsky, 1976] perhaps as the result of hot spot activities as postulated by Lawver [1993].

It is hoped that analysis of the recently collected Carboniferous-Permian limestones from the Northwind Escarpment [Grantz and shipboard party, 1993] will shed further light on their affinities. They may be related to the shallow-water carbonate deposits of the Omolon Massif which is part of the South Anyui suture [Zonenshain et al., 1990] (Plate 1) or similar deposits in the NW Territories. From Early Cretaceous until the Miocene/Pliocene the Arctic has been isolated from deep water mass exchange with the world's oceans by tectonic events which have additionally moved it northward to its present high latitude and present day key role in global change processes.

Acknowledgments. Great credit is due to Jacob Verhoef and the project team that assembled the magnetic data which made this analysis possible. Likewise L. C. Kovacs was instrumental in both the collection of part of these data with a U. S. Navy program and the subsequent analysis. Many helpful suggestions were received from Marc Langseth and Larry Lawver for which the authors are most appreciative. The manuscript greatly benefited from reviews by Ruth Jackson, Art Grantz and Robin Muench.

REFERENCES

Balkwill, H. R., and F. G. Fox, Incipient rift zone, western Sverdrup Basin, Arctic Canada, *Arctic Geology and Geophysics, 8,* edited by A. F. Embry, and H. R. Balkwill, 171-187, Canadian Soc. of Petrol. Geol. Memoir, 1982.

Broecker, W. S., The Biggest Chill, *Nat. History, 96,* 74-82, 1987.

Carlson, R. L., N. I. Christensen, and R. P. Moore, Anomalous Crustal Structures in Ocean Basins: Continental Fragments and Oceanic Plateaus, *Earth and Planet. Sci. Letters, 51,* 171-180, 1980.

Churkin, M. and J. H. Trexler, Circum-Arctic plate accretion-isolating part of a Pacific plate to form the nucleus of the Arctic Basin, *Earth and Planet. Sci. Letters, 48* (2), 336-362, 1980.

Churkin, M. and J. H. Trexler, Continental plates and accreted oceanic terranes in the Arctic, in *The Oceans Basins and Margins Vol. 5,* edited by A. E. M. Nairn, M. Churkin, and F. G. Stehli, 1-16, Plenum Press, New York, 1981.

Clark, D. L., C. W. Byers, and L. M. Pratt, Cretaceous black mud from the central Arctic Ocean, *Paleoceanography, 1,* 265-271, 1986.

Forsyth, D. A., P. Morel-a-l'Huissier, I. Asudeh, and A. G. Green, Alpha Ridge and Iceland-Products of the same plume? *J. Geodyn., Vol. 6,* edited by G. L. Johnson, and K. Kaminuma, 197-214, 1986.

Gramberg, I. S., V. V. Verba, G. A. Kudryavtzev, M. J. Sorokin, and L. J. Charitonova, Crustal structure of the Arctic Ocean along the D'Long Island-Makarov Basin Transect, *Reports of the Russian National Academy of Sciences,* Vol. 376, No. 4, 484-486 (in Russian), 1993.

Grantz, A., S. Eittreim, and D. A. Dinter, Geology and tectonic development of the continental margin north of Alaska, in Crustal Properties across Passive Margins, edited by C. E. Keen, *Tectonophysics, 59,* 263-291, 1979.

Grantz, A., S. D. May, and P. E. Hart, Geology of the Arctic Continental Margin of Alaska, in *The Arctic Ocean Region: The Geology of North America, Vol. L,* edited by A. Grantz, G. L. Johnson, and J. F. Sweeney, 257-288, Geol. Soc. of Amer., Boulder, Co., 1990.

Grantz, A., M. W. Mullen, R. L. Philips, and P. E. Hart, Significance of extension, convergence, and transform faulting in the Chukchi Borderland for the Cenozoic tectonic development of the Arctic Basin (abs.), *Geol. Assoc. of Canada Annual Mtg.,* Wolfville, Nova Scotia, Canada, *17,* 42, 1992.

Grantz A. and shipboard party, Cruise to the Chukchi Borderland, Arctic Ocean, *EOS, Tran. AGU, 74,* 250, 1993.

Green, A. R., A. A. Kaplan, and R. C. Vierbuchen, Circum-Arctic petroleum potential, *W. E. Pratt Mem. Conf., American Assoc. Petrol. Geology,* 44 pp., EXXON Production Res., Houston, Tech. Rept., 1984.

Jackson, H. R. and G. L. Johnson, Summary of Arctic geophysics, In Polar geophysics., *Jour. of Geodynamics, 6,* edited by G. L. Johnson and K. Kaminuma, 246-262, 1986.

Jackson, H. R., C. C. Currie, and J. F. Sweeney, Arctic Ocean: Alaska-Norway, *International Lithosphere Program: Global Geoscience Transects, Atlantic,* Geoscience Center, Bedford Inst., Halifax, Nova Scotia, 1993.

Jackson, H. R., D. A. Forsyth, J. K. Hall, and A. Overton, Seismic reflection and refraction, in *The Arctic Ocean Region: The Geology of North America, Vol. L,* edited by A. Grantz, G. L. Johnson, and J. F. Sweeney, 153-170, Geol. Soc. of Amer., Boulder, Co., 1990.

Johnson, G. L., and B. Tanner, Geophysical Observations on the Iceland-Faeroe Ridge, *Jokull, 21,* 45-52, 1971.

Johnson, G. L., Morphology and Plate Tectonics: The Modern Polar Oceans, Geological history of the *Polar oceans: Arctic versus Antarctic,* edited by U. Bleil and J. Thiede, *Vol. 308,* 11-28, Kluwer Acad. Press, ASI series, NATO Dordrecht, The Netherlands, 1990.

King, E. R., I. Zeitz, and L. R. Alldredge, Magnetic data on the structure of the central Arctic region, *Geol. Soc. America Bull., 77,* 619-646, 1966.

Kos'ko, M. K., Geology of the New Siberian Islands: Constraints on Paleogeodynamic Reconstructions (abs.), *L. P. Zonenshain Memorial Conference on Plate Tectonics,* Institute of Oceanology, 84, 1993.

Kos'ko, M. K., M. P. Cecile, J. C. Harrison, V. G. Ganelin, N. V. Khandoshko, and B. G. Lopatin, Geology of Wrangel Island between Chukchi and East Siberian Seas, Northeastern Russia, *Canad. Geol. Surv. Bull., 461,* 101 pp., 1993.

Kvenvolden, K. A., and A. Grantz, Gas Hydrates of the Arctic Ocean region, in *The Arctic Ocean Region: The Geology of North*

America, Vol. L, edited by A. Grantz, G. L. Johnson, and J. F. Sweeney, 539-549. Geol. Soc. of Amer., Boulder, Co., 1990.

Kristoffersen, Y., and E. S. Husebye, Multichannel seismic reflection measurements in the Eurasian Basin, Arctic Ocean, from U. S. ice station Fram-IV, in *Geophysics of the Polar Regions*, edited by E. S. Husebye, G. L. Johnson, and Y. Kristoffersen, Tectonophysics, 114, 103-115, 1985.

Lawver, L. A., The Iceland Hotspot: Tail of the Siberian Traps? (abs.), *L. P. Zonenshain Memorial Conference on Plate Tectonics,* Institute of Oceanology, 94, 1993.

Lawver, L. A., R. D. Muller, S. P. Srivastava, and W. Roest, The opening of the Arctic Ocean, *Geological history of the Polar Oceans: Arctic versus Antarctica*, edited by U. Bleil and J. Thiede, *Vol. 308*, 29-62, Kluwer Acad. Press, ASI series, NATO Dordrecht, The Netherlands, 1990.

Lawver, L. A., and C. R. Scotese, A review of tectonic models for the evolution of the Canadian Basin, in *The Arctic Ocean Region: The Geology of North America, Vol. L,* edited by A. Grantz, G. L. Johnson, and J. F. Sweeney, 593-618, Geol. Soc. of Amer., Boulder, Co., 1990.

Macnab, R., Russia and the Arctic Ocean, Magnetic Field and Tectonic Structures, Provisional diagram based on preliminary data and interpretations. *Atlantic Geoscience Centre chart*, 1993.

Macnab, R., S. Le'vesque, G. Oakley, K. G. Shih, S. P. Srivasta, A. Stark, K. Usow, and J. Verhoef, Magnetic anomaly map of the Arctic: a new resource for tectonic investigations, *EOS Trans. Amer. Geophys. Un., Vol. 73, n. 14,* (suppl.), 279, 1992a.

Macnab, R., D. A. Forsyth, D. Hardwick, D. Marcotte, B. Nelson, A. V. Okulitch, and D. J. Tesky, Detailed Aeromagnetic Mapping of the Continental Margin of Arctic Canada: Initial Geologic and Tectonic Interpretations, *EOS, Trans. Amer. Geophys. Un., Vol. 73, n. 43,* (suppl.), 144, 1992b.

Marincovich, L., E. M. Browers, D. M. Hopkins, and M. C. McKenna, Late Mesozoic and Cenozoic paleographic and paleoclimate history of the Arctic Ocean Basin, based on shallow-water marine faunas and terrestrial vertebrates, in *The Arctic Ocean Region: The Geology of North America, Vol. L,* edited by A. Grantz, G. L. Johnson, and J. F. Sweeney, 403-426, Geol. Soc. of Amer., Boulder, Co., 1990.

Nansen Arctic Drilling Program, The Arctic Ocean Record: Key to Global Change (Initial Science Plan), *Polarforschung, 61/1*, 1-102, 1992.

Okulitch, A. V., B. G. Lopatin, and H. R. Jackson, Circumpolar Geologic map of the Arctic, *Geological Survey of Canada, Map 1765A*, Scale 1:6000000, 1989.

Pogrebitsky, Y. E., The geodynamic system of the Arctic Ocean and its structural evolution, *International Geology Review, 20*, 1251-1266, 1976.

Pogrebitsky, Y. E., V. Verba, V. Volk, V. Shimaraev, L. Zonenshain, L. C. Kovacs, S. Levesgne, R. Macnab, G. Oakley, S. P. Srivastava, A. Stark, and J. Verhoef, USSR Onshore and Offshore Magnetic Anomalies: Compilation and Tectonic Interpretation, *Proc. of the Internat. Confer. on Arctic margins*, edited by D. Thuston, Anchorage, Alaska, in press, 1993a.

Pogrebitsky, Y. E., V. N. Shimaraev, V. V. Verba, J. Verhoef, R. Macnab, J. Kisabeth, and G. Jorgensen, Magnetic Anomaly Map of Russia and Adjacent Land and Marine Areas, Atlantic Geoscience Center, Geological Survey, Dartmouth, Canada, in press, 1993b.

Riddihough, R. P., G. V. Haines, and W. Hannaford, Regional Marine anomalies of the Canadian Arctic, *Canad. Jour. of Earth Sci., 10*, 147-163, 1973.

Sekretov, S. B., The Continental Margin North of East Siberian Sea: Some Geological Results and conclusions on the base of CDP seismic-reflection data (abs.), *L. P. Zonenshain Memorial Conference on Plate Tectonics*, Institute of Oceanology, 124, 1993.

Sweeney, J. F., Comments about the age of Canada Basin, *Geophysics of the Polar Seas,* edited by E. S. Husebye, G. L. Johnson, Y. Kristoffersen, 1-10, Elsevier Pub. Co., New York, 1985.

Sweeney, J. F., E. Irving, and J. W. Geuer, Evolution of the Arctic Basin, Arctic Geophysical Review, edited by J. F. Sweeney, *Publ. of the Earth Physics Branch,* Ottawa, Canada, *45, 4*, 91 pp., 1978.

Sweeney, J. F., L. W. Sobczak, and D. A. Forsyth, The continental margin northwest of the Queen Elizabeth Islands, in *The Arctic Ocean Region: The Geology of North America, Vol. L,* edited by A. Grantz, G. L. Johnson, and J. F. Sweeney, 227-238, Geol. Soc. of Amer., Boulder, Co., 1990.

Taylor, P. T., L. C. Kovacs, P. R. Vogt, and G. L. Johnson, Detailed aeromagnetic investigations of the Arctic Basin, *J. Geophys. Res., 86*, 6323-6333, 1981.

Trettin, H. P., Pearya-a composite terrane with Caledonian affinities in northern Ellesmere Island, Canad. *Jour. of Earth Sci., 24*, 224-245, 1987.

Van Wagoner, N. A., and P. T. Robinson, Petrology and Geochemistry of a CESAR bedrock sample; implication for the origin of the Alpha Ridge, Initial Geological Report on CESAR, *Geol. Survey of Canada, 84-22,* 47-58, 1985.

Vogt, P. R., R. K. Perry, and P. T. Taylor, Amerasian Basin, Arctic Ocean: Magnetic Anomalies and their decipherment, *Arctic Geology*, 27th Inter. Geol. Congress, *4*, 152-161, 1984.

Zonenshain, L. P., Kuzmin, M. I., and L. V. Natapov, Geology of the USSR: A Plate-Tectonic Synthesis, *Geodynamics Series, 21*, 242 pp., edited by B. M. Page, Amer. Geophys. Union, Washington, D. C., 1990.

G. L. Johnson, Code 324, Office of Naval Research, Arlington, VA 22217-5660, (703-696-4119, FAX: 703-696-4884), L.Johnson/ Omnet, Johnson@tomcot.onr.navy

R. Macnab, Atlantic Geosciences Centre, Box 1006, Dartmouth, Nova Scotia, CANADA, B2Y 4A2, (902-426-5687, FAX: 902-426-6152) macnab@agcrr.bio.ns.ca

J. Pogrebitsky, Scientific and Technical Complex, Moika 120, St. Petersburg, RUSSIA 190121, (Telex: 121430 onies su) ocean@sovamsce.sovusa.com

Mid-Pleistocene Climate Shift - The Nansen Connection

Wolfgang H. Berger[1] and Eystein Jansen[2]

1) Scripps Institution of Oceanography University of California, San Diego La Jolla, California 92093-0215, USA

2) Department of Geology, University of Bergen, Allégaten 41, N-5007 Bergen, Norway

The Mid-Pleistocene Revolution (MPR), a climate shift dated at 920 to 900 kyr, involves a substantial change in mean state, in amplitude of ice-mass buildup and decay, and in the frequency of variations of ice-mass. Pre-shift and post-shift time differ in that precession-driven variations become more important in the late Quaternary; at the same time the 100 kyr cycle becomes dominant. A number of possible explanations have been put forward to explain the change in character of climatic fluctuations, calling on non-linear feedback mechanisms within a steady-state system, among which are ice-mass responses to extreme size increase (such as basal melting) and extension of lowland ice over shelves, where marine downdraw becomes possible. Here we emphasize the role of the unidirectional modifications of land morphology by ice erosion (as proposed by Nansen) as a factor in generating deep shelves (e.g., Barents Sea), valleys for channelling of ice, and uplift providing for mountain peaks where a new ice buildup can begin after an interglacial has run its course. Also, we consider that increased positive feedback from variations in albedo and atmospheric carbon dioxide is important, and we present evidence for a significant role of ocean circulation (Nordic heat pump) for the time after the Mid-Pleistocene Revolution.

1. INTRODUCTION

When contemplating the course of sealevel variations during the Quaternary, as seen in the $\delta^{18}O$ record in deep-sea sediments, it is immediately obvious that there is a striking difference in the nature of fluctuations between the early and late part of this period (Fig. 1a). The early Quaternary is characterized by roughly equally spaced sinusoidal variations, with a period of ca. 41,000 years (41 kyr). The late Quaternary shows seemingly less regular fluctuations, with larger amplitudes. In fact, here the mixture of frequencies contains a dominant component with a period near 100 kyr. The mid-Quaternary switch in climatic variability from an obliquity-dominated mode to an eccentricity-dominated one has been recognized for some time, and has been the subject of much discussion. It represents one of the major puzzles in Quaternary research [Pisias and Moore, 1981; Prell, 1982; Start and Prell, 1984; Ruddiman et al., 1986; 1989; Ruddiman and Raymo, 1988; DeBlonde and Peltier, 1991].

Questions arising concern the appearance of the transition, as well as its dynamics. Is it gradual [Ruddiman et al., 1989], or is it abrupt [Berger and Wefer, 1992]? What is the origin of the 100 kyr cycles - is it positive feedback on the (minor) direct effect on insolation that stems from eccentricity? Or is it a beat phenomenon [Wigley, 1976]? Or does it result from non-linear ice dynamics [Oerlemans, 1982]? What is the role of basal melting? Of marine incursions during deglaciation? Of isostatic adjustment, high latitude topography, ice albedo-temperature feedback, ocean circulation... - the list goes on (see articles in Berger et al. [1984], Ruddiman and Wright [1987], Kukla and Went [1992]). Discussions of these and related questions have attained increasing complexity on a formal level [Le Treut and Ghil, 1983; Hagelberg and Pisias, 1991], in physical modelling [Manabe and Broccoli, 1985; Rind et

The Polar Oceans and Their Role in Shaping the Global Environment
Geophysical Monograph 84

al., 1989; Maasch and Saltzman, 1990], and in data treatment [Raymo et al., 1990; Imbrie et al., 1992].

Our argument is comparatively simple, traditional, and geological: the nature of the MPR (Mid-Pleistocene Revolution) is similar to the onset of the ice ages, which may be understood as a threshold phenomenon. A long-term cooling trend, from orogenesis and uplift, reaches a point where ice sheets can grow in high northern latitudes [Flint, 1947; Emiliani and Geiss, 1957; Ruddiman and Raymo, 1988]. The result is a "sudden" onset of ice-buildup (through albedo amplification, largely), from a "gradual" shift in boundary conditions. In analogy, the cumulative effects of glacial erosion and deposition, combined with continued uplift, provide for conditions favorable for the growth of large ice sheets subject to channelling and marine downdraw. The Nansen connection appears in several aspects: Nansen quite generally proposed an important role of glaciations in shaping land and shelf morphology, and specifically that the Barents Sea resulted from subsidence of a subaerally exposed region [Nansen, 1904, 1922]. Also, his description, with Helland Hansen, of the northward penetration of warm water in the Nordic Seas emphasize the importance of the circulation in providing heat and moisture to high latitudes [Nansen, 1906, 1912; Helland-Hansen and Nansen, 1909].

2. MORPHOLOGY OF THE TRANSITION

The record here discussed (Fig. 1a) is from the western equatorial Pacific (Ontong Java Plateau) where Quaternary temperature fluctuations are thought to be of minor importance. Between 80 and 90 percent of the $\delta^{18}O$-signal is ascribed to ice-mass variation [Berger et al., 1993]. Records from elswehere (e.g. the North Atlantic) may have a stronger temperature component, where spectral properties may differ from those of the ice-mass component. We consider that the Ontong-Java record is the best proxy of Quaternary ice-mass fluctuations available at present. Dating is by tuning to astronomical forcing [see Berger et al., 1993]. In the Ontong Java record (Fig. 1a) the transition appears to be quite abrupt: after a long series of rather evenly spaced fluctuations of moderate amplitude in the earlier portion of the record, there is a failure, at 920 kyr, to return to the interglacial state. Instead, we see a drop of $\delta^{18}O$ values well beyond 2 standard deviations of the accustomed glacial excursions. This event, marked as "MPR" on Fig. 1a, is dated by counting obliquity cycles backward from the present [Berger et al., 1993]. After the MPR, the previous mode of fluctuations is never again in evidence.

The Mid-Pleistocene Revolution involves a substantial change in mean state and in amplitude of ice-mass buildup and decay. Of course, this change in mean state, and also the amplitude change, are smoothed when observed in a 300 kyr window (Fig. 1b). However, this does not necessarily mean that the change is in fact transitional. Seen against the entire Pliocene-Pleistocene background, the special

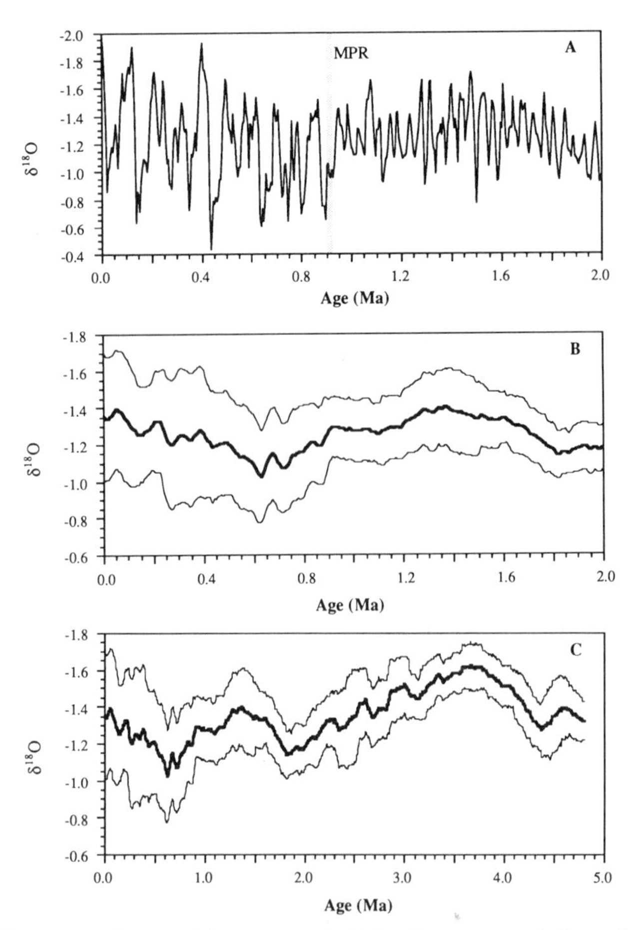

Figure 1. Oxygen isotope record of the Quaternary. a) Record of *Globigerinoides sacculifer* of Hole 806B, last 2 million years, as given in Berger et al. [1993]. b) Same record, with running mean only (200 kyr window) and with 2 standard deviations to either side of the mean. c) Same record seen within the frame of 5 million years of long-term trends (5-2 Ma isotopes from Hole 806B as given in Jansen et al. [1993]).

character of the late Quaternary as a high-amplitude climatic regime becomes especially obvious (Fig. 1c). For the Quaternary, there are 3 important trends in the mean state: a rise in the oldest third, a fall in the second, and another rise (but now associated with a large amplitude) in the youngest. This pattern supports a threefold partitioning of the Quaternary. The MPR is centered on the middle portion, that is, the one showing the overall cooling. It is interesting that the subsequent warming trend does not lead to a reversal of conditions toward the earlier configuration.

A similar three-fold partitioning emerges when studying the frequency pattern. We have already noted, from inspection, that pre-shift and post-shift time differ in terms of regularity and spacing of glaciation events: a ca-100 kyr cycle becomes strong after the Mid-Pleistocene Revolution (counting, we find nine major glaciations since the MPR; Fig. 1a). Fourier analysis provides detailed information on the periodicities contained in the record (Fig. 2). Starting with the last 400 kyr (bottom of the graph), we note the presence of peaks, within the spectrum, near 100 kyr (a rather broad maximum), near 41 kyr, and at 23 kyr, confirming the presence of the orbital frequencies, as expected from previous work [Hays et al., 1976; Imbrie et al., 1984]. Moving the window backward (in the manner illustrated by Mayer [1991]), a distinct change in spectral properties is first noticed for 0.8 to 1.2 Ma, where the power near 41 kyr attains equality with the one near 100 kyr. In the next earlier window 100 kyr power is still important, but essentially disappears before that. At the same time, power in the precessional part of the spectrum no longer rises above background, suggesting that precession and eccentricity work together (as is proper, since the precessional effect on seasonal insolation depends on eccentricity, and since much of the 100 kyr cycle may in fact represent a beating of interacting precessional frequencies; see Wigley [1976], Pollard [1982, 1984]. The techniques here used for spectral analysis are exceedingly

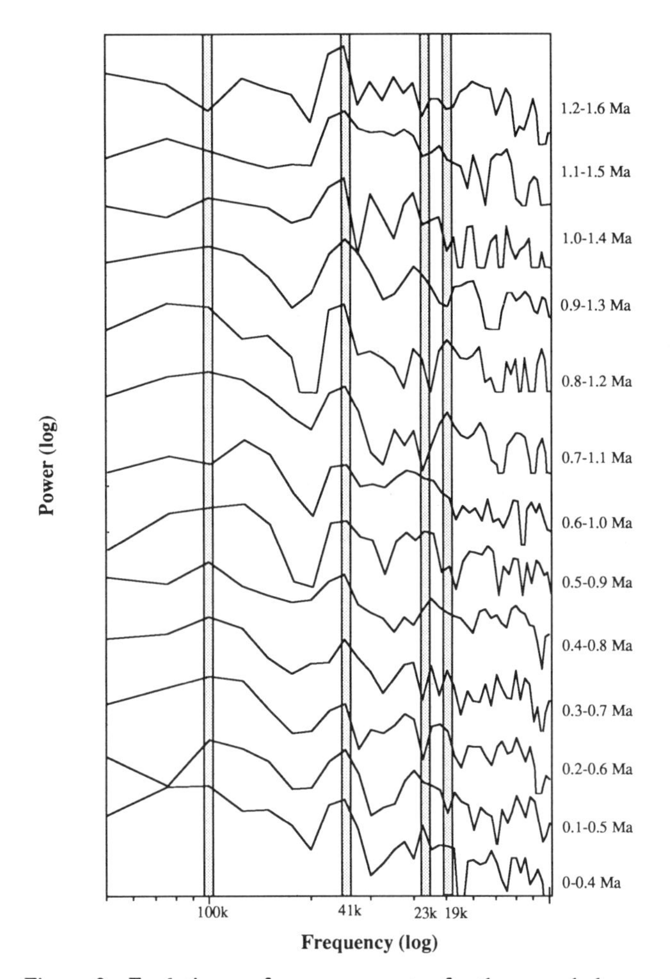

Figure 2. Evolutionary frequency spectra for the record shown in Fig. 1a. Each spectral line represents a 400 kyrs interval, with a 300 kyr overlap with neighbouring intervals. Power is the square-root of the sum of squares of the Fourier coefficients.

simple, and the window quite short. Nevertheless, the trends identified appears reliable: the record shows a three-fold partitioning for the Quaternary, the lower third dominated by obliquity-related fluctuations, the last third by eccentricity, and the middle by a transition between the two. These divisions were identified as Fourier chrons in Berger and Wefer [1992], and named "Milankovitch", "Croll", and "Laplace" chrons, in sections of 15 obliquity cycles each. The MPR is part of the transitional Croll chron; it occurs on a long-term (since 4 Ma) as well as on a short-term (1.2 to 0.7 Ma) cooling trend (Fig. 1b,c). The change in the nature of fluctuations is not reversed despite a warming trend in the Milankovitch chron.

This last point, that amplitude keeps increasing despite (slight) warming would seem to be an important observation. If we attribute the increasing amplitude in the late Quaternary to cooling during the Croll chron we must admit that conditions change in such a fashion as to impede reversal of the process responsible. Of course, we do not know the exact significance of the $\delta^{18}O$ trends: if oxygen isotopes in the glacial ice sheets were gradually less fractionated in the Milankovitch chron (for example, because of secular lowering of the crust they sit on, and a greater access for vapor from the sea), the trend would not be related to warming at all.

In any case, the post-MPR conditions are orthogonal to the trends within the Pliocene and early Quaternary (Fig. 3). The unique nature of the late Quaternary (= post-MPR, <0.9Ma), and the surprising excursion toward ever greater amplitudes, is strikingly evident. The main event at the MPR is the buildup of unusually large amounts of glacial ice. The ice-mass which goes beyond the usual amount experienced all through the early Quaternary, the "extra" or "surplus" ice, becomes part of a feedback mechanism stimulating long-period, large-amplitude fluctuations [cf. Oerlemans, 1982]. The extra ice of the super-glaciations, somehow, helps produce more complete deglaciations and longer interglacials. We call this the "ice surplus effect" (Fig. 3B, inset). It appears, from the long record, that such an effect had one earlier opportunity to take hold of the system: at 2.5 Ma. There is indeed evidence for large fluctuations following this unusually severe glaciation. However, after 300 kyr or so these fluctuations ceased. Presumably, the "correct" boundary conditions (low-lying shelves, deeply eroded hinterland, among others) had not yet materialized. Unfortunately for more detailed analysis of the period following the post-Gauss event, the section between 2 Ma and 2.1 Ma is very uncertain: this is the unadjusted patch between the records of Berger et al. [1993] and Jansen et al. [1993].

3. A TEST FOR ABRUPTNESS OF CHANGE

We have suggested, on inspection of the record (Figs. 1a, 3a), that the change from early to late Quaternary climatic style is abrupt; yet we have presented no statistical evidence for this assertion. Clearly, when using a window to depict changes in mean state, mean amplitudes, or spectral properties of sections, changes in these various properties cannot appear as being abrupt, even if they are. When the window straddles the event, some compound state for the time before and after the event will be delivered. To define a change, an expectation must be compared with the observation, and this expectation must be based on experience away from the event, not on information from the event itself. We must find a way, therefore, to extrapolate the record forward or backward in time. To make the extrapolation, we use two types of information: the nature of the cycles in the record (as shown in the

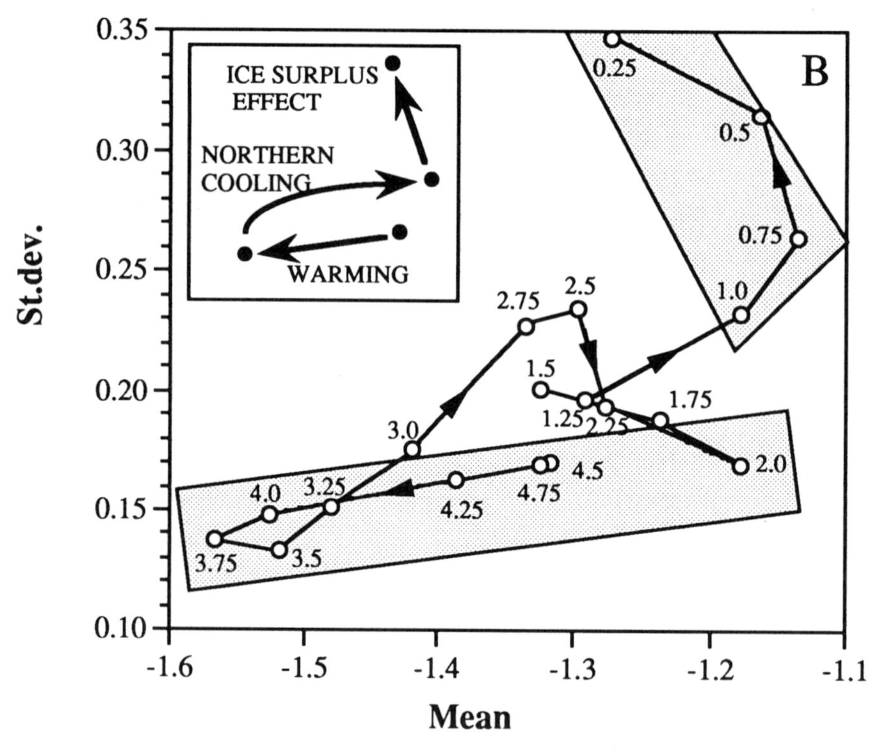

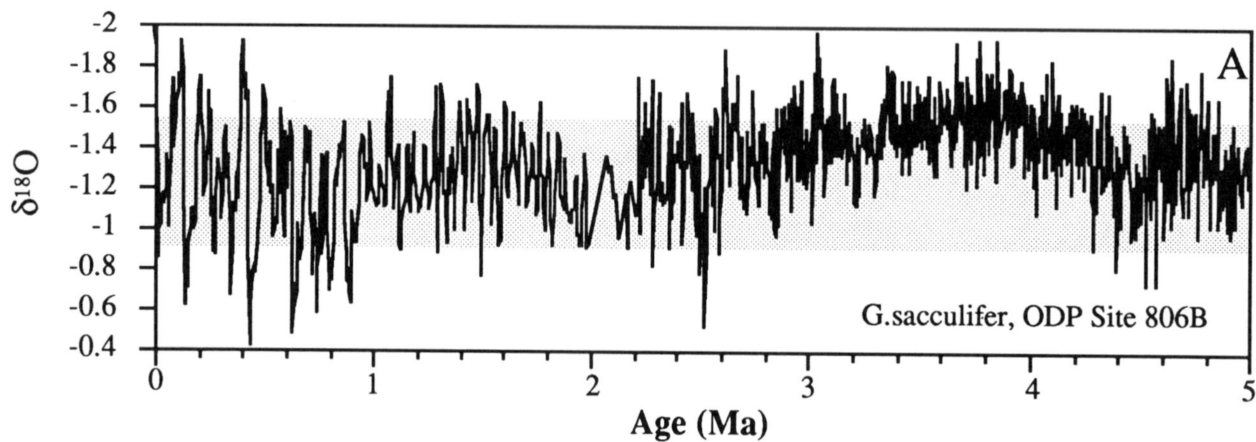

Figure 3. Evolution of system state, as seen in the $\delta^{18}O$ record of the last 5 Ma. A. $\delta^{18}O$ record of the planktonic foraminifer *Globigerinoides sacculifer* in ODP Site 806, Hole B, from Berger et al. [1993] (0-2 Ma) and Jansen et al. [1993] (2-5 Ma). Note the break between 2 Ma and 2.1 Ma, which probably reflects mismatches from dating. B. Evolution of conditions within the dimensions of mean state and amplitude of variation for consecutive 500 kyr intervals with 250 kyr overlap over the time period 0-5 Ma. Each point on the graph represents the mean value and one st.dev. of $\delta^{18}O$-values from a 500 kyr interval. The associated numbers show mid-point age of each 500 kyr period. Inset shows interpretation. Shaded areas show contrasting trends.

appropriate Fourier spectrum in Fig. 2), and the fact that orbital forcing (that is, summer insolation in high latitudes) can be readily related to the oxygen isotope record using simple conceptual models [Calder, 1976; Imbrie and Imbrie, 1980]. The task is to make a "model", based on orbital information, which fits the "sample record" as tightly as feasible, so that the "target record" can be predicted using the same convolution for the orbital series as for the "sample". The "sample" represents the experience on which expectation is based, the "target" includes the change to be tested for abruptness.

The conceptual model we use [Berger et al., in press] calculates changes in $\delta^{18}O$ (expressed as sea level: for conversion use 0.1 permil for 0.08 units of sea level in normalized space) from the following algorithm which balances ice melting against ice growth:

$$dS/dt = ins^a*mass^b*mem^c - gro \quad (1)$$

Here ins is the insolation in July at 65°N (as calculated by Berger and Loutre, [1991]), mass is the current ice-mass, and mem is the average ice-mass for the last x ky. Calculations are in normalized space, ranging from 0 to 1; thus, ins, mass and mem are less than one. The exponents a, b, c are adjustable; when large they suppress all but the maximum values of the parameters responsible for melting. The parameter mem (for memory) is a filter. For the last 1 Ma it is set at 40 kyr. The parameter gro (ice growth) is set constant, reflecting the observation that snow falls every winter, whether it be severe or mild. (In fact, in Norway, more snow falls in the milder climate on the westward facing mountains close to the coast than in the colder interior.) It will be appreciated that the melting term in equation (1) represents lagged negative feedback to ice-growth, and that melting potential is modulated by summer insolation. This basic balance is common to ice-sheet models (it is not important to the outcome whether ice-growth itself is modulated also). The model is not very sensitive to the exact setting: a from 1 to 4, b and c from 1 to 2, and gro from 0.1 to 0.2 per step (= 4 kyr), give similar results in all combinations. (This suggests that physics is less important than dynamics of the model system, in providing for goodness of fit between model output and $\delta^{18}O$ record.)

The performance of the model in turning the insolation series into a series resembling the $\delta^{18}O$ record (in the following called "template") may be judged from the fit shown in Fig. 4a. We submit that the major features of the $\delta^{18}O$ record are captured rather well. One important exception is Stage 11, for which summer insolation is too low to produce the impressively warm interglacial actually indicated in the data. This discrepancy contains an important message about the inertia of the system: once melting of a large ice sheet starts, the process tends to go to completion. This feature is not captured in our model. Nevertheless, we are reasonably confident that the phase relationships of the various periods contained in the model output are well suited to describe the record, even if amplitudes do not match exactly (as is evident also in stages other than Stage 11).

We next employ the $\delta^{18}O$ record of the first core taken above the MPR-event to find the amplitudes of the various cycles contained in post-MPR fluctuations, by Fourier analysis. The use of a single core (806B-H2) eliminates disturbance from core breaks. We then use the amplitudes of the various periods found for the core's record to adjust the amplitudes of the same periods in the Fourier matrix of that portion of the template that contains the section for which a comparison is to be made between expectation and observation (here: 300 to 1300 kyr).

Results of the exercise described (Fig. 4b) produce a prediction (or rather "post-diction"), from the point of view of an observer having the experience represented by Core 806B-H2, regarding the probable course of events in the previous 500 kyr, based on summer insolation at 65°N. Clearly, the indication is that the time around 900 kyr saw unusual change. The severity of the (mid-Croll) glacial Stage 22 is evident. The mildness of pre-MPR glaciations stands out. Thus, the abruptness of the change is verified, in agreement with previous findings based on statistics [Maasch, 1988].

4. THE QUESTION OF CAUSE

A considerable number of hypotheses have been put forward to explain the change in character of climatic fluctuations from early to late Quaternary. In essence, the problem is one of explaining the 100 kyr cycles of post-MPR time, using a mechanism that is not activated before the climate shift. Many simple (and not so simple) ice-mass models, conceptual and physical, do quite well in producing 100 kyr cycles from orbital forcing. Thus, there is little or no question that ice-sheet dynamics are at the heart of the system whose response must be studied [Calder, 1976; Birchfield and Weertman, 1978; Imbrie and Imbrie, 1980; Ghil and Le Treut, 1981; Birchfield et al., 1981; Oerlemans, 1982; Pollard, 1982, 1984; Le Treut and Ghil, 1983; Saltzman et al., 1984; Birchfield and Grumbine, 1985; Hyde and Peltier, 1985; DeBlonde and Peltier, 1991]. The exact source of the 100 kyr oscillation (that is, the precise physical interaction of the various components of the system) is under discussion, as well as the degree to which they are due to internal resonance phenomena, to beating of precessional periods, and to forcing from the (small) direct effects of eccentricity. Many investigators agree, however, that the instabilities introduced by growth of ice-sheets beyond a certain size provide a crucial element in enabling the system to respond sporadically to slightly uneven short-cycle forcing, and thereby to move power from precessional forcing into the 100 kyr band. Basal melting, isostatic subsidence under ice-

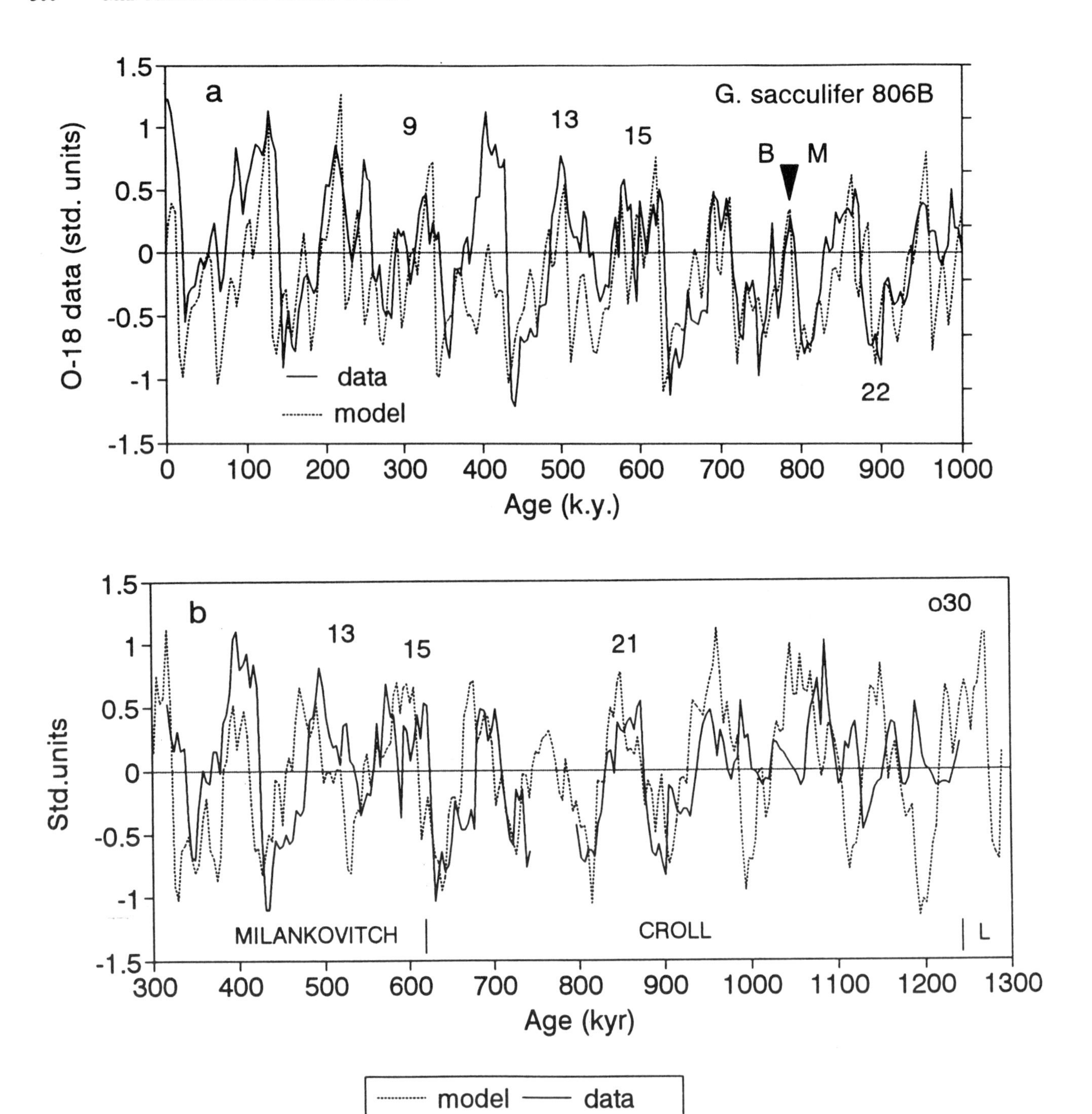

Figure 4. Test for abruptness of the Mid-Pleistocene Revolution. a) Verification of efficiency of orbital template, based on equation (1) (text) with a setting of a=3, b=2, c=2, mem=40 kyr, gro=0.14 per 4 kyr. b) Post-diction of orbitally controlled $\delta^{18}O$ fluctuations, based on experience in Core 806B-2H, for convolution of the template for 300 to 1300 k.y.

loading and interaction of ice with marine invasion are among the more important physical processes with a non-linear dependency on ice-sheet size [Wilson, 1964; Andrews and Peltier, 1976; Denton and Hughes, 1983; also see articles in Denton and Hughes, 1981; Ruddiman and Wright, 1987; Fjeldskaar and Cathles, 1991].

In agreement with these various concepts, the obvious candidate for the switch from one mode to the other is "extra" ice, as proposed in Fig. 3. If this is so, we must ask: Whence came the surplus ice? And why did it build up again after major interglacials? Given unchanging boundary conditions, one might expect the system to return to the pre-shift mode eventually, unless *something else* changed.

The alternative to postulating a shift in boundary conditions at the time of the climate shift is to call on non-linear response of ice buildup to a general cooling trend - which, however, should be reversible. Another possibility is that indeed a cooling trend is responsible for initiating the shift, but that the occurrence of the first super-glaciation (Stage 22) changes boundary conditions for the subsequent developments. On the basis of the arguments concerning the strange trend of post-shift time (Fig. 3), we assume that boundary conditions changed during the Croll chron, both as a result of the cumulative effects of earlier glaciations, and of the onset of super-glaciations. We suggest that the changes are lodged in topographic boundary conditions, and in a permanent and continuing readjustment of the prime positive feedback mechanisms, albedo and CO_2.

We propose (in agreement with many other authors) that changes in topography, increased sensitivity to albedo, and a lowering of atmospheric CO_2 are the ultimate causes for the mid-Pleistocene climate shift. The system resists adjustments to these changes in boundary conditions thanks to negative feedback. However, upon reaching a threshold between 0.9 and 1.0 million years ago, the system responds by building larger ice sheets, which in turn introduces instabilities and threshold phenomena that change periodicities and amplitudes of ice-mass fluctuations.

5. EFFECTS OF TOPOGRAPHIC CHANGES

Fridtjof Nansen (1861-1930) gave much thought to the evolution of continental margins and their morphology, and especially to the origin of the continental shelf. He interpreted this feature (as marine geologists still do) as resulting from a combination of processes of erosion, deposition, and vertical motion (Fig. 5). As a result of observations made on his transits along the Norwegian and Siberian coasts during his North Pole expedition and later cruises, he developed concepts stressing the importance of glacial processes in shaping high latitude shelves and coastal areas [Nansen, 1904, 1922] (Fig. 5b). He pointed out that glacial erosion would lower land areas to below sea level, hence producing shelf seas. Further erosion, in his view, would deepen and channelize the shelf floor, and deposition of the debris from erosion along the shelf break would enlarge the shelf. Glacial processes and frost action were seen by Nansen as important elements in producing a highly dissected coastal zone which produced many conduits

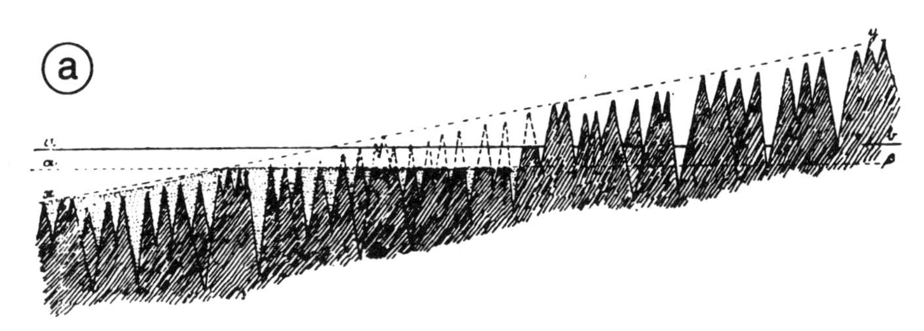

1 *ab*, Present sea-level. *αβ*, Mean ancient sea-level. *xy*, Ancient land-slope.

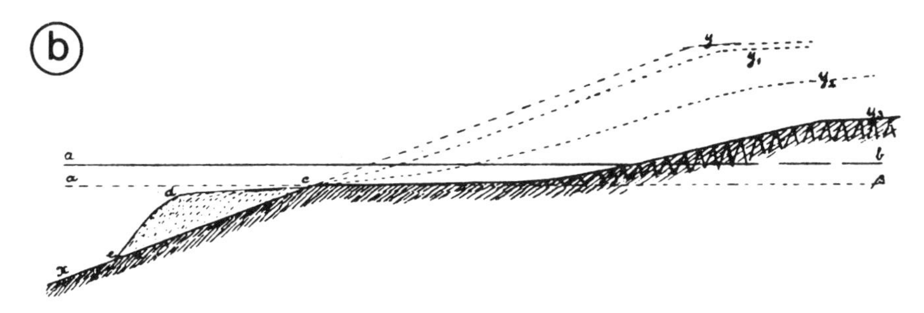

ab, Present sea-level. *αβ*, Ancient sea-level. *xy*, Original land-slope.
cy_1, cy_2, cy_3. Land-slope at different stages of denudation. *cde*, Shelf formed by waste.

Figure 5. Conceptual models of Nansen [1904] for the formation of continental shelves. Upper panel: Shelf buildup by progradation. Lower panel: Shelf buildup by glacial erosion, deposition and adjacent fjord formation. From Holtedahl [1993].

for ice-masses moving out from the hinterland toward the sea.

In the present context, several elements of the evolution of glaciated shelves and their hinterland areas are of importance: (1) The denudation of regions adjacent to shelves, through removal of overburden by glacial action, which produces new shelf seas and deepens existing ones. Thus, additional areas become available for low-lying, marine-based ice-sheets, which are especially susceptible to rapid deglaciation. (2) The delivery of enormous amounts of sediment to the outer shelves and upper slope, causing shelves to build outward, further increasing shelf areas. (3) Focussed erosion, both in coastal areas and on shelves, producing deep troughs and fjords, which provides channels for ice flow, and also access for marine invasion, during deglaciation. Channelling of glacier tongues descending from ice sheets makes for ice-streams where ice can flow quickly and efficiently drain the source region [Hughes, 1987]. This is another mechanism (besides marine downdraw) that favors ice sheet collapse upon deglaciation: The information from the shelf (i.e., rise of sealevel) can be rapidly transmitted to the interior of the large ice sheets, along the ice-stream channels. (4) Deep focussed erosion in the mountains, which removes materials while forming valleys, thus causing isostatic uplift with increased summit heights acting as nuclei for reglaciation [Fjeldskaar and Riis, 1992; see Fig. 6]. The valleys, as mentioned, provide for channelled ice flow, while the uplift changes the albedo potential and provides areas for glacial inception [cf. Molnar and England 1990].

6. BARENTS SEA SHELF AND GLACIATION

The Barents Sea and Kara Sea shelves are strategically located for receiving large ice sheets during glaciation (Fig. 7). They are, on the whole, closer to the pole than Greenland, and their combined size roughly matches that of that island. The bottom morphology and the glacial deposits at the edges of the shelves show that they harbored large ice-masses in the past [Elverhøi and Solheim, 1983; Solheim and Kristoffersen, 1984; Vorren et al., 1988; Eidvin and Riis, 1989; Eidvin et al., 1993; Solheim et al., 1990; Vorren et al., 1991]. The potential of this large shelf area as a receptacle for marine-based (and hence potentially unstable) grounded ice-masses is well recognized [e.g., Kvasov, 1978; Kvasov and Blazhchishin, 1978; Jones and Keigwin, 1988]. The question in the present context is whether there is evidence that this potential was realized at the time of the MPR, and whether it is important enough to influence the subsequent course of climatic variation.

There is little doubt about its potential importance. Because of the size involved, a sealevel change of around 10 m could be readily envisaged, from melting a Barents/Kara ice sheet. This is already a substantial portion of the "surplus" ice seen in Fig. 1a. An abrupt rise of sea level early during deglaciation, even just by several meters, would have had considerable ramifications for other marine-based ice-sheets, in Canada and even in Antarctica [Denton and Hughes, 1981, 1983]. The question about the timing of the glacial history of the Barents Sea shelf is a more difficult matter. In the last decade or so, however,

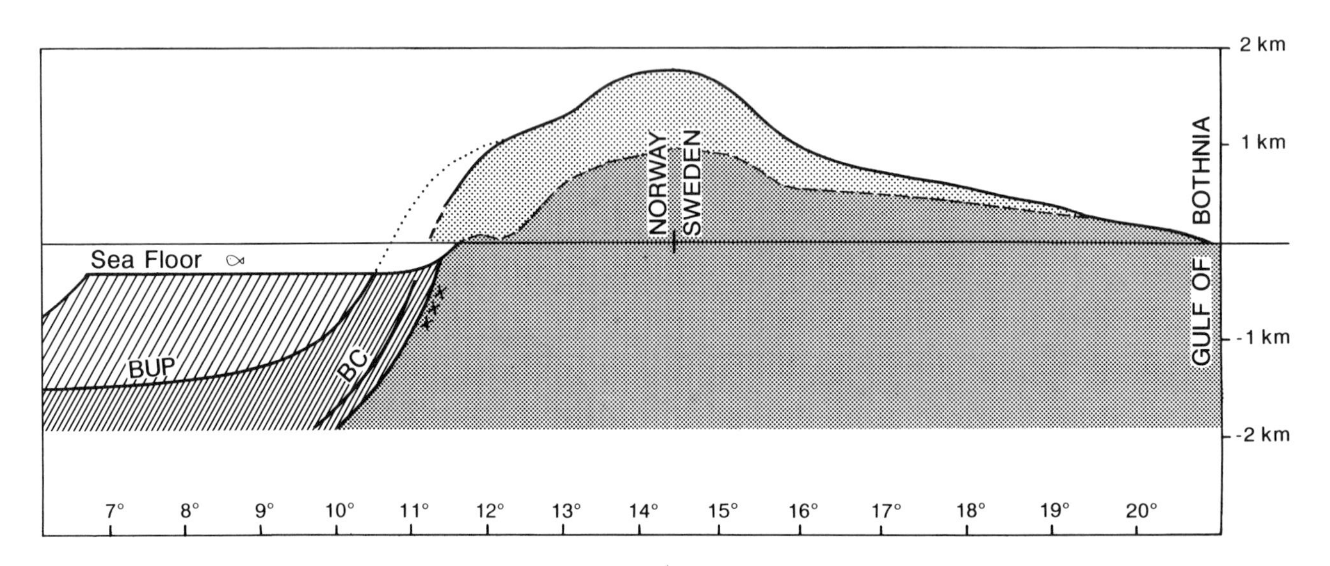

Figure 6. Modelled effects of glacial erosion on uplift and summit heights in a W-E transect across mid Scandinavia (modified from Fjeldskaar and Riis [1992]. The model results show the effects of glacial erosion on shelf progradation and on uplift in the hinterland. Light shading shows the observed summit heights and the modelled net uplift due to erosion. The dark shading shows residual uplift of Scandinavia due to other causes than erosion. BUP=Base Upper Pliocene. Light hachures above BUP indicates the volume of erosion products from glacial erosion in the Plio-Pleostocene which forms the prograding shelf. BC= Base Cenozoic sequence.

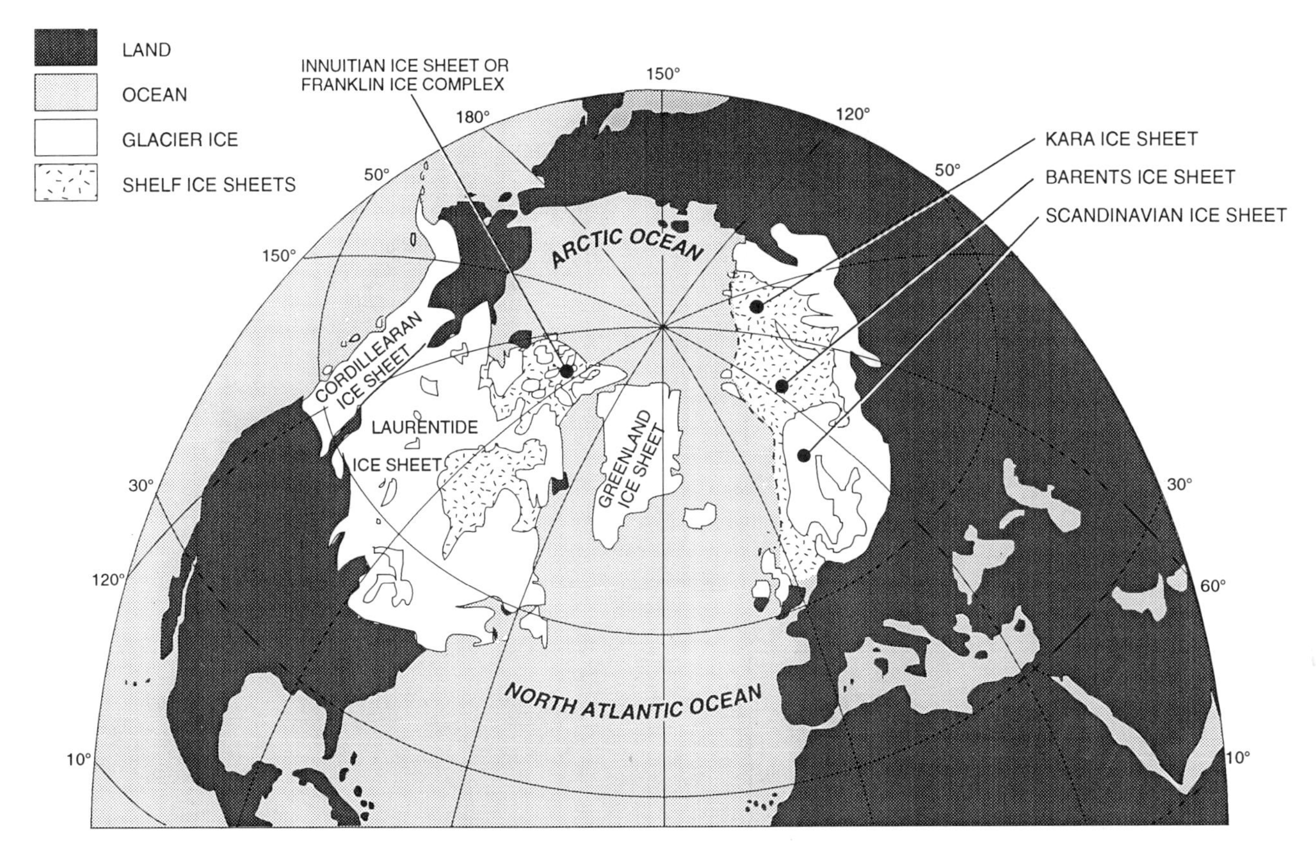

Figure 7. Distribution of glacial ice in the northern hemisphere, and the extent of shelf ice sheets on the Kara, Barents and Norwegian Sea shelves. Modified from Denton and Hughes [1981].

considerable information on this point has become available [Vorren et al., 1991; Eidvin et al., 1993].

There are several important aspects to the glaciation history of the Barents Shelf: (1) removal of overburden during the Tertiary, especially the latest Tertiary, (2) subsidence (or uplift, or both), and (3) the development of grounded ice-sheets. As to denudation, estimates of total thickness of material removed typically range from 600 to more than 1500 m. These estimates are based on the morphology and seismic stratigraphy of sediment wedges beyond the edge of the shelf [e.g. Vorren et al., 1990, 1991]. The timing assumed for the accumulation of the sediment masses (and hence of the denudation activity) varies widely, but has recently been constrained by drilling results [Eidvin et al., 1993]. Indications are that much or most of the erosion of the shelf took place in the last few million years, that is, with the onset of the northern ice ages [probably at or after 2.8 Ma; Jansen et al., 1990; Jansen and Sjøholm, 1991]. This exact view was already expressed by Fridtjof Nansen [1904], who considered the fans on the continental slope (of which he had but rough morphological knowledge) as a product of glacial erosion and deposition. If this view is correct, the major denudation of the Barents Shelf becomes part of an ice-driven morphological preparation of Arctic shelves for receiving marine-based ice sheets during the late Quaternary.

The timing of the growth of grounded ice sheets on the Barents Shelf is uncertain, of course, although a mid-Quaternary beginning is now widely assumed. The fact that this assumption leans heavily on paleoclimatic results of deep-sea stratigraphy in the Norwegian Sea [e.g., Vorren et al., 1991, p. 337; Eidvin et al., 1993] introduces a certain circularity to the argument. However, it does seem reasonable to correlate the marked increase of ice-rafted debris, as observed near the mid-Quaternary climate shift, to a marked increase in ice-mass on shelves able to provide that debris. As long as the argument is a glaciologic-sedimentologic one (rather than generally based on presumed climatic correlation) circularity is largely avoided.

There is yet another important aspect to the geographic setting of the Barents Shelf: its relationship to the influx of warm water from the south (Fig. 7, arrow). The access of (relatively) warm water is probably crucial both for ice buildup (delivery of vapor for snowfall) and for deglaciation (delivery of heat for melting). The basic concepts are schematically depicted in Fig. 8. The slope in elevation, of

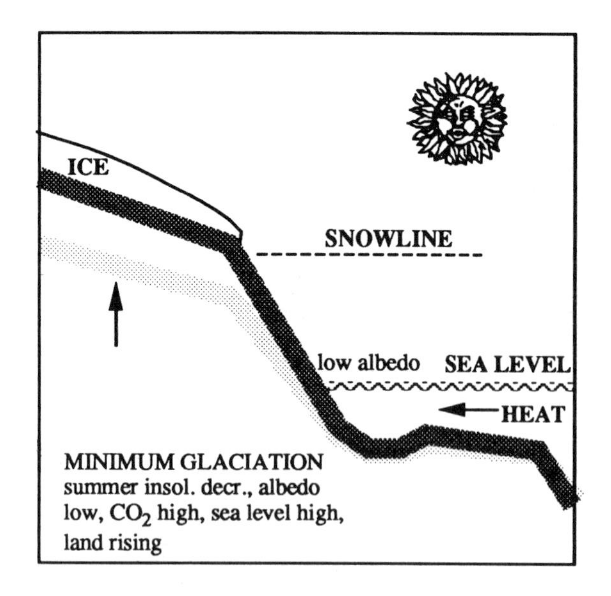

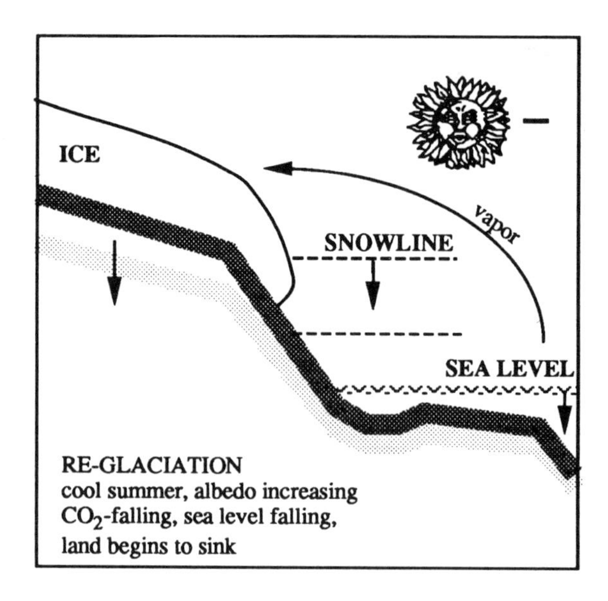

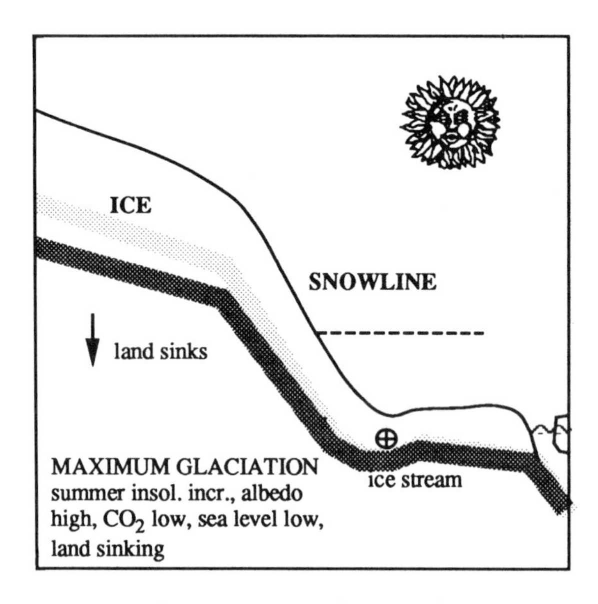

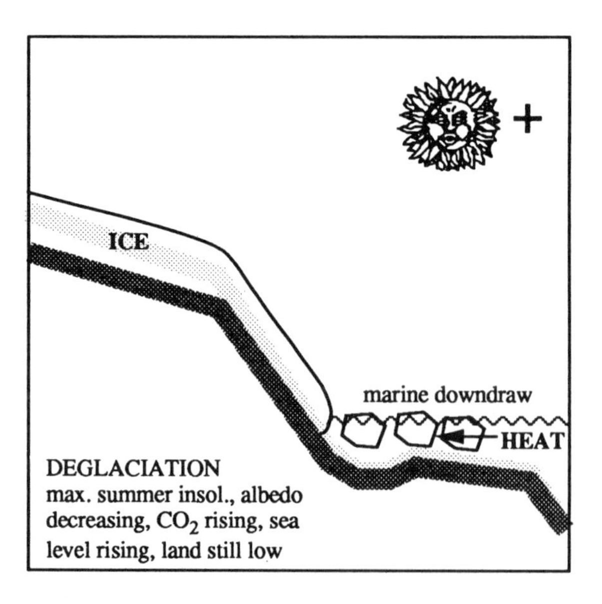

Figure 8. Conceptual diagram of ice growth and ice decay, involving differences in topography, isostatic adjustment, and marine invasion over deep shelves. The snowline moves by roughly 1 km in the vertical (not to scale).

the topography above sealevel, is exaggerated to make the point that buildup of ice is necessarily a non-linear process: when reaching the lowlands and especially when reaching the shelves, the rules must change. The isostatic adjustment is likewise exaggerated, to emphasize that the weight of glacial ice, by decreasing elevations at its base, makes the ice sheet progressively more vulnerable to marine invasion.

In summary, we believe that the Barents-Kara Ice Sheet played a major role in modulating climatic variations, beginning with the mid-Quaternary climatic shift at 0.9 Ma. The area is large enough to be important in terms of ice-mass. It is deep enough in major troughs (exceeding 300 m) to provide for marine-based ice, and it has sufficiently deep channels to allow for ice-streams and deep access of the sea during deglaciation. Also, it is situated favorably for benefitting from open waters, at least seasonally or sporadically. To this latter aspect we turn next.

As Fridtjof Nansen pointed out [Nansen, 1906; Helland Hansen and Nansen, 1909], the heat budget of the Arctic Ocean, and especially the region around Svalbard, is much influenced by the advection of warm water with the Norwegian Current (Fig. 9). A substantial portion of this water, after yielding its heat to Norway and adjacent areas, partakes in vertical convection and contributes to the North Atlantic Deep Water. The remainder moves north and branches off northern Norway, with one arm heading toward Fram Strait, the other into the Barents Sea. As a result, there is a distinct retreat of the polar front toward higher latitudes in both areas. Bottom water formation [first hypothesized by Nansen, 1912] is an important aspect of the heating process on the Barents Sea Shelf [Midttun, 1985]. At present, 1 to 3 sv flow off the shelf as cold saline bottom water [Loeng et al., 1993]. With a temperature difference of ca. 3°C between incoming surface water and outgoing bottom water, the entire shelf is warmed markedly by this process.

The advection of heat is of significance both for reglaciation and for deglaciation, as has been well recognized for some time [Ewing and Donn, 1956; Olausson and Jonasson, 1969; Boulton, 1979; Hebbeln and Wefer, 1991]. There is now evidence that the Norwegian Sea was open even during a glacial maximum (at least seasonally), so that advection of warm Atlantic water reached as far north as Spitsbergen, where it could provide a source of moisture for growth of glaciers [Vorren et al., 1988a; Veum et al., 1992; Dokken and Hald, 1993,

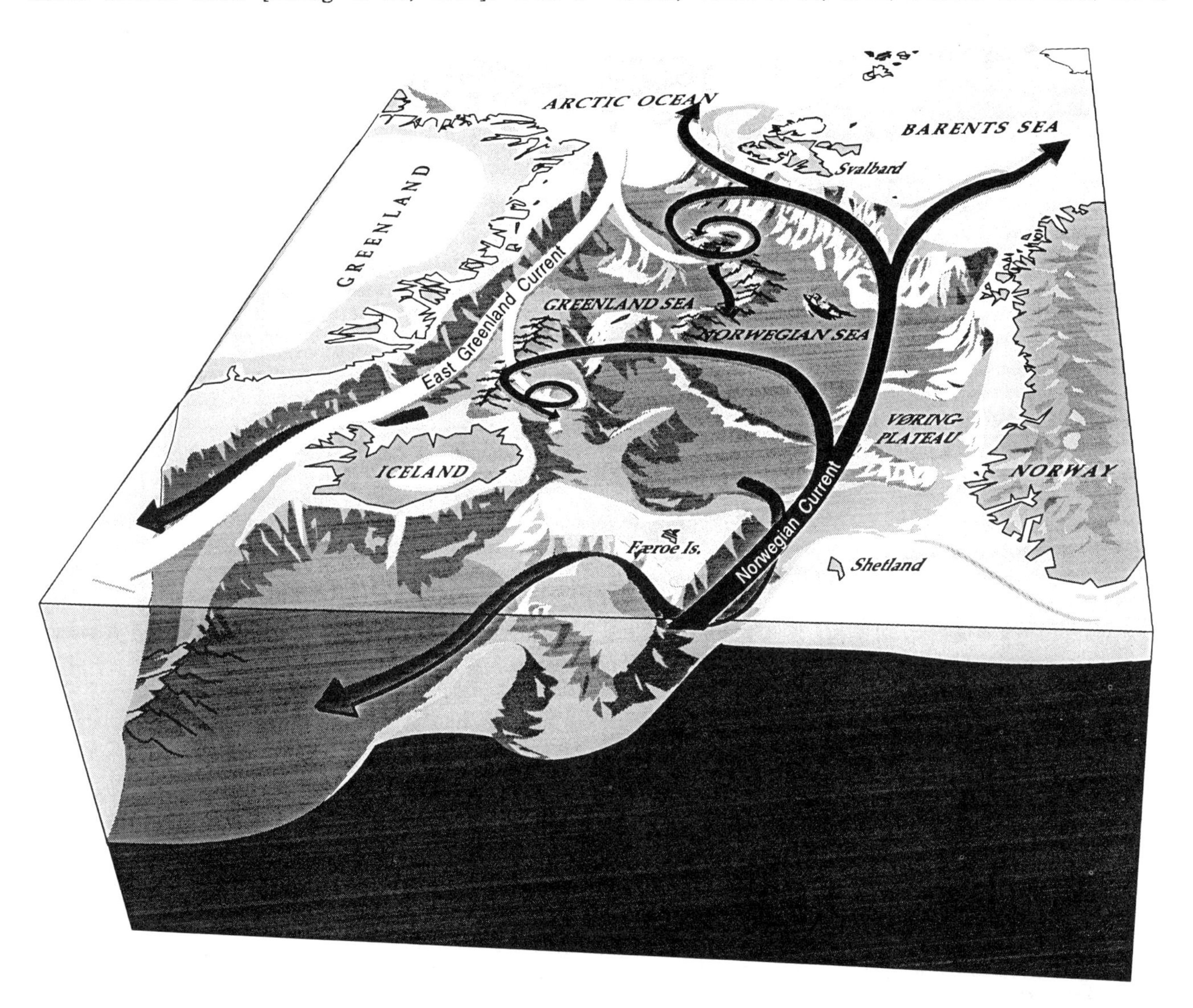

Figure 9. Main features of the circulation in the Nordic Sea. Note the formation of deep water and the advection of warm water into Fram Strait and onto the Barents Sea Shelf. This heat influx is important in maintaining the status of an interglacial, and (albeit with modified circulation) in deglaciation. During glaciation also there is advection from the south, at least seasonally or sporadically (see text). Computer graphics by Masaoki Adachi, Bergen Univ.

Johannessen et al., in press]. Studies on the last deglaciation in the region [Jansen and Erlenkeuser, 1985; Jones and Keigwin, 1988; Hald et al., 1991; Lehman, 1991; Weinelt et al., 1991; Sarnthein et al., 1992] suggest that it occurred in pulses, with an early onset, in particular documented by early strong meltwater fluxes off the Barents Sea [Sarnthein et al., 1992], and that ocean circulation may have been considerably modified by the process [summarized in Jansen, 1992]. These pulses were part of the large-scale pulsing charactizing deglaciation [Fairbanks, 1989; Jansen and Veum, 1990].

7. OCEAN CIRCULATION AND THE NORWEGIAN SEA

Helland-Hansen and Nansen [1909] in their classic study of the Nordic Sea's circulation made a series of important observations on the nature and effects of the oceanic circulation in the region. These are still considered essentially correct. They noted, for instance, the importance of the northward heat transport with the inflowing Atlantic water in the Norwegian Current, and that this water mass also reached, in parts, the Barents Sea and the Arctic Ocean. Nansen [1906, 1912] and Helland-Hansen and Nansen [1909] also were the first to describe the deep water formation in the Nordic Sea, a process depending on cooling of inflowing salt-rich water, and which releases substantial amounts of heat to the region. Nansen [1912] even considered deep water formation in the Barents Sea, from brine formation due to sea ice formation, as a contributor to the overall convection in the Nordic Seas - a proposition that has been verified recently by direct observations [Midttun, 1985].

Modification of deep ocean circulation during deglaciation [see summary in Crowley and North, 1991; articles in Bard and Broecker, 1992], and in fact over the course of the entire glacial cycle [Boyle and Keigwin, 1982; Shackleton et al., 1983; Duplessy et al., 1988; Broecker and Denton, 1989] has become a major item of discussion, since it was first proposed [Olausson, 1965; Weyl, 1968; Newell, 1974]. An important aspect of the NADW-modification mechanism is the downstream interaction with the heat and salt budget of the Southern Ocean, since the NADW feeds the Antarctic Circumpolar Current [Gordon, 1991; Martinson, 1991; also see articles in Bleil and Thiede, 1990]. Thus, NADW production influences sea-ice formation around Antarctica, and hence Southern Ocean albedo. Perhaps the most important aspect of the deep circulation in the present context, is that its response to relatively small changes in climate can be drastic and self-stabilizing [Bryan, 1986].

Without attempting to explore the various ramifications of deep circulation patterns for climatic change [which is complicated and can only be addressed by global modelling; e.g., Maier-Reimer and Mikolajewicz, 1989], we can be

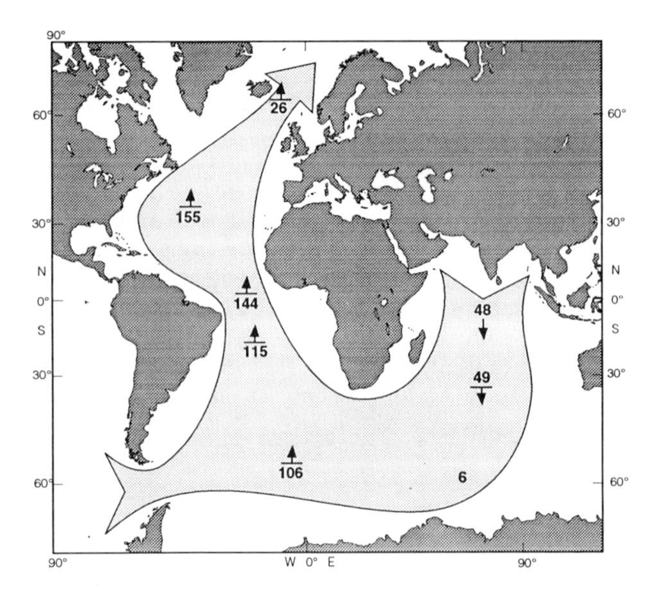

Figure 10. The North Atlantic heat trap: meridional component of heat flow (in 10^{13} watts) in Atlantic and surrounding regions. After Stommel [1980] and Woods [1981], from Berger and Labeyrie (1987; p.285).

quite certain that deep circulation plays a significant role in ice-sheet dynamics, because of its contribution to the overall asymmetry of heat flow from the southern to the northern hemisphere, and specifically into the northern North Atlantic [Crowley and Kim, 1992]. The warm Norwegian Current is one component (and the most important one, we think, in the present context) of the "North Atlantic Heat Trap". It is of great interest, therefore, that warm water advection increased substantially at the time of the Mid-Pleistocene Revolution, at the same time as ice-rafting of sediments eroded from the shelf became conspicuous (Fig. 11).

Contemporaneously with the increased warm water influx during warm stages, as documented by the appearance of strong $CaCO_3$-peaks at the MPR (Fig. 11) [Henrich and Baumann, in press], there is clear evidence for (a) much higher rates of delivery of ice-rafted debris to the Norwegian Sea [interpreted as increase in ice-bergs from increased contact of ice-mass with the ocean: Jansen and Sjøholm, 1991], and (b) frequent input of Cretaceous and Tertiary fossils into deep-sea sediments. Occurrence of these fossils (Fig. 11), which can only be eroded from shelf deposits, documents that ice sheets now entered the shelf region more frequently than before. In essence this shows that the MPR, as documented by records from sediments originating from the Scandinavian ice sheet, was characterized by both more vigorous influx of oceanic heat, and by more extensive glaciations with large ice-masses typically located on the shelf in glacial periods.

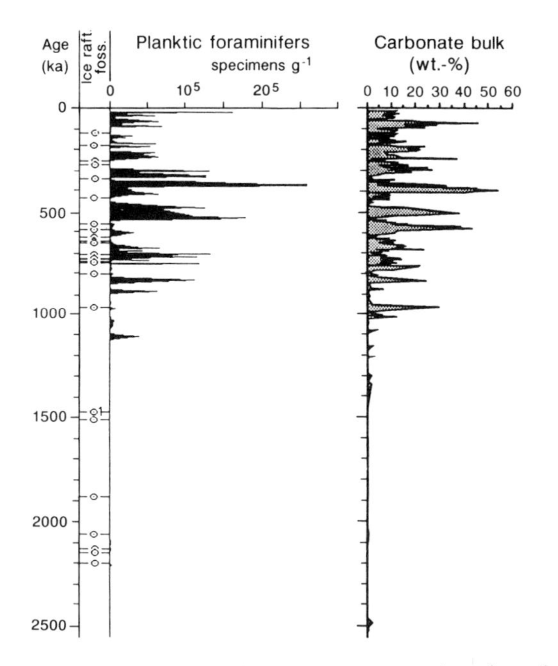

Figure 11. Evidence that warm-water advection during interglacials and supply of ice-rafted debris during glacials greatly intensified at 0.9 Ma in the Norwegian Sea. Left: abundance of planktonic foraminifers and abundance of ice-rafted fossils with origin from the continental shelf in ODP Site 643, Vøring Plateau (data from Spiegler [1989]). Right: abundance of carbonate in the same sediments (data from Henrich [1989]).

8. CIRCULATION, ALBEDO AND CO_2: THE FEEDBACK MACHINE

Considering the complexity of the feedback machine that makes up the climate system, we have so far not even listed (and certainly not discussed) all the important components. Most of these components make their influence felt through modifications of albedo and of atmospheric CO_2. The latitudinal distribution of albedo patterns and their sensitivity to seasonal insolation acts as a geographic "filter" for orbital forcing as proposed by Short et al. [1991]. Numerical experiments by these authors suggest that in the real record the temperature variation is much larger than expected from Milankovitch forcing, and has frequencies derived from the response of low latitudes to forcing [c.f. Imbrie et al., 1992]. Clearly, the long-distance effects of oceanic circulation, as well as the carbon cycle, could provide the necessary cross-latitude links.

The long-term aspects of changing albedo and atmospheric CO_2 have recently been the subject of much discussion, in connection with new evidence for the traditional concept that uplift and mountain building increases albedo and decreases CO_2 [Ruddiman and Kutzbach, 1989, Raymo, 1991]. Here we are interested in that type of feedback that is able, in principle, of changing the framework of climatic conditions, through continued glaciations. Such a feedback is given, as mentioned, by the uplift that follows erosion [Molnar and England, 1990; Fjedskaar and Riis, 1992]. For the carbon connection, also, we can hypothesize this type of "ruinous feedback", that is, positive feedback responsible for changing the boundary conditions in such a fashion as to produce ever larger amplitudes in the externally stimulated system. We note, in Fig. 11, that such ruinous feedback is indicated in the overall increase of both carbonate sedimentation in the Norwegian Sea (warm water influx) and ice-rafting (glaciers reaching the sea). The process begins at the MPR, and increases from there [Jansen et al., 1988]. Is is striking that at the same time as the North Atlantic circulation becomes more anti-estuarine (reflected in the carbonate trap aspect), circulation in the northern North Pacific becomes more estuarine [Sancetta and Silvestri, 1986], emphasizing the well-known see-saw aspect of these end-points of circulation.

It is reasonable to postulate an overall drop in the carbon dioxide content resulting from increased wind-driven upwelling which sequesters additional organic carbon in marginal marine sediments, during the large glaciations after the MPR [Sarnthein et al., 1988]. In contrast, during interglacials, the upwelling would be greatly reduced, allowing CO_2 to build up. Such a process would constitute positive feedback. Alternatively, or in addition, the large drop in sea level associated with the MPR for the first time bared shelf areas which could provide much loose carbonate for dissolution and for redeposition. The resulting increase in alkalinity also would have reduced CO_2 content in the atmosphere. A strong pulse of eolian loess carried to the ocean would have had a similar effect. At a lower overall value, the system would have been somewhat more sensitive to changes in CO_2 from that time on, with an increased contribution to the amplitudes of climatic variation. During deglaciations, the buildup of carbonate on the shelves would have released CO_2 from the ocean to the atmosphere, enhancing the greenhouse effect [Berger and Keir, 1984; Opdyke and Walker, 1992]. Concerning albedo, the increased area of shelves involved in post-MPR sealevel fluctuations would have increased the contrast in albedo from these regions: when covered with water, albedo of the shelves is considerably reduced.

9. CONCLUSIONS

The Mid-Pleistocene Revolution (MPR), comprises a climate shift that involves a substantial change in mean state and in amplitude of climatic variation, as well as in the frequency of variations. The shift may be dated rather more precisely than usually admitted, at 0.90-0.9.2 Ma [in agreement with Berger and Wefer, 1992]. Pre-shift and post-shift time differ in that precession-driven variations become more important in the late Quaternary; at the same time the

100 kyr cycle becomes strong. There is no substantial decrease in obliquity-related variation. The normal relationship between mean state and amplitude seen in the Pliocene and early Pleistocene (which shows increasing amplitude for increasing average ice-mass and vice versa) is not strictly followed after the MPR. In the last million years, amplitude of fluctuations increases regardless, suggesting the operation of some kind of "ruinous positive feedback". We suggest that this feedback must be sought in long-term topographic changes wrought by repeated glaciations, especially in the region of the Arctic. At the same time, this same topographic trend favors a shift to long-period fluctuations, through involvement of marine shelves as receptacles of "surplus" ice. The Barents-Kara Shelf is the key area in this development, providing a large receptacle for marine-based ice. Circulation, albedo and atmospheric CO_2 provide the mechanisms by which the amplitude-increasing process is translated into climatic fluctuations. The role of the unidirectional modifications of land morphology by ice erosion (as proposed by Nansen) as a factor in generating deep shelves (e.g., Barents Sea) has been recognized for some time [Denton and Hughes, 1981]. In addition, the carving of deep valleys must be considered, which results in channelling of ice (ice streams), and in uplift providing for mountain peaks at higher elevations than before where new ice buildup can begin.

A striking and counter-intuitive feature of post-MPR climate is the strenghtening of the Nordic heat pump during interglacials. Why, when there is more maximum ice-mass, should interglacials increase in length, and more heat be advected into the Norwegian Sea during the warm intervals, compared with pre-MPR time? Since many interglacials reach higher peaks during the early Pleistocene than during the early part of post-MPR time (when carbonate was deposited on Vøring Plateau), it is not the amplitude that is responsible, but more likely the length of the interglacials. If so, it cannot be a simple readjustment of deep circulation, which works on short time-scales. We suspect, instead, that the carbon cycle is involved (cumulative buildup of carbon dioxide), which amplifies fluctuations in meridional heat transport, the Pacific-Atlantic asymmetry and NADW formation. The complexity of the climatic system, with positive feedback from variations in albedo and atmospheric carbon dioxide, and from ocean circulation (Nordic heat pump), combined with the relative ease with which orbitally derived insolation series can be made to mimick ice-mass, makes it difficult to assess the relative importance of the various competing and cooperating processes. Thus, while we regard the ice-erosion hypothesis here offered as a reasonable explanation for the Mid-Pleistocene Revolution, we are aware that it will be most difficult to put it to a test.

Acknowledgments. The present study was undertaken as a follow-up to ODP Leg 130 results [Berger et al., 1993] at the University of Bremen, and continued during a sabbatical stay of W.H.B. in Bergen, at the department of Geology, incorporating the older portion of the record for geological perspective [Jansen et al., 1993]. We thank our shipmates and co-workers who made this study possible. Discussions with O. Eldholm, A. Elverhøi, M. Hald, J. Mangerud, H. Loeng, A. Solheim, T.O. Vorren, and G. Wefer, as well as other colleagues in Bremen, Bergen, Oslo, Trondheim, and Tromsø were very helpful and are much appreciated. We are grateful to M. Adachi and J. Ellingsen for drafting assistance.

REFERENCES

Andrews, J. T., and W. R. Peltier, Collapse of the Hudson Bay ice center and glacio-isostatic rebound. *Geology 4*, 73-75, 1976.

Bard, E., and W. S. Broecker (Eds.), The last deglaciation, absolute and radiocarbon chronologies, Springer Verlag, Heidelberg, XXpp,, 1992.

Berger, A., and M. F. Loutre, Insolation values for the climate of the last 10 million years, *Quat.Sci.Rev. 10*, 297-317, 1991.

Berger, A., J. Imbrie, J. Hays, G. Kukla, and B. Saltzman (Eds.), Milankovitch and Climate. 2 vols. Reidel, Dordrecht, Holland, 1984.

Berger, W. H., and R. S. Keir, Glacial-Holocene changes in atmospheric CO_2 and the deep-sea record. In J. E. Hansen and T. Takahashi (Eds.) Climate Processes and Climate Sensitivity. *Geophys. Monogr. 29*, 337-351. AGU, Washington, D. C. 1984.

Berger, W. H., and L. D. Labeyrie (Eds.), Abrupt Climatic Change: Evidence and Implications. D. Reidel, Dordrecht., Holland, 1987.

Berger, W. H., and G. Wefer, Klimageschichte aus Tiefseesedimenten: Neues vom Ontong Java Plateau (Westpazifik). *Naturwissenschaften, 79*, 541-550, 1992.

Berger, W. H., T. Bickert, H. Schmidt, G. Wefer, Quaternary oxygen isotope record of pelagic foraminifers: Site 806, Ontong Java Plateau. In Berger, W. H., Kroenke, L. W., and Mayer, L. A. (Eds.), *Proc. ODP, Sci. Results* (pp. 381-395). College Station, Texas: Ocean Drilling Program, 1993.

Berger, W. H., M. Yasuda, T. Bickert, and G. Wefer, Brunhes-Matuyama boundary: 790 k.y. date confirmed by ODP leg 130 oxygen isotope records using a new ice-mass template, *Geophysical Research Letters* (in press), 1994.

Birchfield, G. E., and R. W. Grumbine, "Slow" physics of large continental ice sheets and underlying bedrock, and its relation to the Pleistocene ice ages. *J. Geophys. Res., 90*,11,294-11,302, 1985.

Birchfield, G. E., and J. Weertman, A note on the spectral response of a model continental ice sheet. *J. Geophys. Res., 83(C8)*, 4123-4125, 1978.

Birchfield, G. E., J. Weertman, and A. T. Lunde, A paleoclimatic model of northern hemisphere ice sheets. *Quaternary Res., 15*, 126-142, 1981.

Bleil, U., and J. Thiede (Eds.), *Geological History of the Polar Oceans: Arctic Versus Antarctic*. Kluwer Academic, Dordrecht, Holland, 1990.

Boulton, G. S., A model of Weichselian glacier variation in the North Atlantic region. *Boreas 8*, 373-395, 1979.

Boyle, E. A., and L. D. Keigwin, Deep circulation of the North Atlantic over the last 200,000 years: Geochemical evidence. *Science 218*, 784-787, 1982.

Broecker, W. S., and G. H. Denton, The role of ocean-atmosphere reorganizations in glacial cycles. *Geochim. Cosmochim. Acta 53:* 2465-2501, 1989.

Bryan, F., High-latitude salinity effects and interhemispheric thermohaline circulations. *Nature 323*, 301-304, 1986.

Crowley, T. J., and G. R. North, *Paleoclimatology*. Oxford University Press, New York, 339pp, 1991.

Calder, N., Arithmetic of ice ages. *Nature, 252*, 216-218, 1974.

Crowley, T. J., and K.-Y. Kim, Complementary roles of orbital insolation and North Atlantic Deep Water during late Pleistocene interglacials. *Paleoceanography, 7*, 521-528, 1992.

DeBlonde, G., and W. R. Peltier, A one-dimensional model of continental ice volume fluctuations through the Pleistocene: implications for the origin of the Mid-Pleistocene climate transition. *Journal of Climate, 4*, 318-344, 1991.

Denton, G. H., and T. J. Hughes (Eds.), *The Last Great Ice Sheets*. Wiley, New York, 484pp., 1981.

Denton, G. H., and T. J. Hughes, Milankovitch theory of ice ages: hypothesis of ice-sheet linkage between regional insolation and global climate, *Quat. Res., 20*, 125-144, 1983.

Dokken, T., and M. Hald, Late Weichselian growth and decay of the Barents Sea Ice Sheet reflected in sediments at the margin of Spitsbergen/Barents Sea - paleoceanographic implications. *Geonytt, 29(1)*, 20 [Abstract], Tromsø, 1993.

Duplessy, J. C., N. J. Shackleton, R. G. Fairbanks, L. D. Labeyrie, D. W. Oppo, and N. Kallel, Deepwater source variations during the last climatic cycle and their impact on the global deepwater circulation, *Paleoceanography 3*, 343-360, 1988.

Eidvin, T., E. Jansen, and F. Riis, Chronology of Tertiary fan deposits off the western Barents Sea: implications for the uplift and erosion history of the Barents Shelf, *Marine Geology 112*, 109-131, 1993.

Eidvin, T., and F. Riis, Nye dateringer av de tre vestligste borehullene i Barentshavet. Resultater og konsekvenser for den tertiære hevingen. *Norwegian Petroleum Directorate Contribution No. 27*, 44pp., 1989.

Elverhøi, A., and A. Solheim, The Barents Sea ice sheet, a sedimentological discussion. *Polar Research, 1*, 23-42, 1983.

Emiliani, C., and J. Geiss, On glaciations and their causes. *Geol. Rundschau, 46*: 576-601, 1958.

Ewing, M., and W . L. Donn, A theory of ice ages. *Science, 123*, 1061-1066, 1956.

Fairbanks, R. G., A 17,000-year glacio-eustatic sea level record: Influence of glacial melting rates on Younger Dryas event and deep-ocean circulation. *Nature 342*, 637-642, 1989.

Fjeldskaar, W., and L. Cathles, Rheology of mantle and lithosphere inferred from post-glacial uplift in Fennoscandia. In: R. Sabadini, K. Lambeck, and E. Buschi (Eds.) *Glacial Isostasy, Sea Level and Mantle Rheology*. Kluwer, Dordrecht, Holland, pp. 1-19, 1991.

Fjeldskaar, W., and Riis, F., On the magnitude of the late Tertiary and Quaternary erosion and its sgnificance for the uplift of Scandinavia and the Barents Sea. In: R. M. Larsen, H. Brekke, B. T. Larsen and N. E. Tallerås (Eds.) *Structural and tectonic modelling and its application to petroleum geology*. Norwegian Petroleum Society (NPF) Spec. Publ. 1, Elsevier, Amsterdam, 163-185, 1992.

Flint, R. F., *Glacial and Quaternary Geology*. J. Wiley and Sons., N.Y., 1971.

Ghil, M., and H. Le Trout, A climate model with cryodynamics and geodynamics. *J. Geophys. Res. 86*, 5262-5270, 1991.

Gordon, A.L., The Southern Ocean: Its involvement in global change. In G.Weller, C.L. Wilson, and B.A.B. Severin (Eds) *Proc. International Conference on the Role of the Polar Regions in Global Change*, Fairbanks, Alaska, pp. 249-255, 1991.

Hagelberg, T., N. G. Pisias, and S. Elgar, Linear and nonlinear couplings between orbital forcing and the marine $\delta^{18}O$ record during the late Neogene, *Paleoceanography 6*, 729-746, 1991.

Hald, M., L. D. Labeyrie, D. A. R. Poole, P. I. Steinsund, and T. O. Vorren, Late Quaternary paleoceanography in the southern Barents Sea, *Norsk Geologisk Tidsskrift, 71*, 141-144, 1991.

Hays, J. D., J. Imbrie, and N. J. Shackleton, Variations in the Earth's orbit: Pacemaker of the ice ages, *Science, 194*, 1121-1132, 1976.

Hebbeln, D., and G. Wefer, Effects of ice coverage and ice-rafted material on sedimentation in the Fram Strait, *Nature, 350*, 409-411, 1991.

Helland-Hansen, B., and F. Nansen, The Norwegian Sea, Its Physical Oceanography Based Upon the Norwegian Researches 1900-1904. *Report on Norwegian Fishery and Marine Investigations, 2, Pt. 1, N. 2*, Mallingske, Christiania [Oslo], 1909.

Henrich, R., Glacial/Interglacial cycles in the Norwegian Sea: Sedimentology, paleoceanography, and evolution of late Pliocene to Quaternary northern hemisphere climate. In Eldholm, O., J. Thiede, E. Taylor, et al., *Proc. Ocean Drilling Program, Scientific Results, V. 104* (pp. 189-232). College Station, Texas, 1989.

Henrich, R. and K.H. Baumann, Evolution of Norwegian Current and the Scandinavian Ice Sheets during the past 2.6 Myrs.: Evidence from ODP leg 104 biogenic carbonate and terrigenous records, *Palaeogeography, Palaeoclimatology, Palaeoecology*, in press. 1994.

Holtedahl, H., Marine geology of the Norwegian continental margin, *Geological Survey of Norway Spec. Publ. 6*, 150pp., 1993.

Hughes, T. J., Ice dynamics and deglaciation models when ice sheets collapsed. In: W. F. Ruddiman and H. E. Wright Jr. (Eds.) *North America and adjacent oceans during the last deglaciation. The Geology of North America, v. K-3*. Geol. Soc. Am., Boulder, Col. 183-220, 1987.

Hyde, W. T., and W. R. Peltier, Sensitivity experiments with a model of the ice age cycle. The response of harmonic forcing. *J. Atmos. Sciences, 42*, 2170-2188, 1985.

Imbrie, J., and J. Z. Imbrie, Modeling the climatic response to orbital variations, *Science, 207*, 943-953, 1980.

Imbrie, J., J. D. Hays, D. G. Martinson, A. McIntyre, A. C. Mix, J. J. Morley, N. G. Pisias, W. L. Prell, and N. J. Shackleton, The orbital theory of Pleistocene climate: support from a revised chronology of the marine $\delta^{18}O$

record. In Berger, A. et al. (Eds.) *Milankovitch and Climate* (Part I, pp. 269-305). Reidel Publ. Comp., 1984.

Imbrie, J., E. A. Boyle, S. C. Clemens, A. Duffy, W. R. Howard, G. Kukla, J. Kutzbach, D. G. Martinson, A. McIntyre, A. C. Mix, B. Molfino, J. J. Morley, L. C. Peterson, N. G. Pisias, W. L. Prell, M. E. Raymo, N. J. Shackleton, and J. R. Toggweiler, On the structure and origin of major glaciation cycles 1. Linear responses to Milankovitch forcing. *Paleoceanography, 7*, 701-738, 1992.

Jansen, E., Deglaciation and its impact on ocean circulation. In Nierenberg, E. (ed) *Encyclopedia of Earth System Science 2*, 35-46. Acad. Press, San Diego, 1992.

Jansen, E., and H. Erlenkeuser, Ocean circulation in the Norwegian Sea during the last deglaciation: isotopic evidence. *Palaeogeography, Palaeoclimatology, Palaeoecology, 49*, 189-206, 1985.

Jansen, E., and T. Veum, Evidence for two step deglaciation and its impact on North Atlantic deep-water circulation. *Nature 343*, 612-616, 1992.

Jansen, E., and J. Sjøholm, Reconstruction of glaciation over the past 6 Myr from ice-borne deposits in the Norwegian Sea. *Nature 349*, 600-603, 1991.

Jansen, E., U. Bleil, R. Henrich, L. Kringstad, and B. Slettemark, Paleoenvironmental changes in the Norwegian Sea and the northeast Atlantic during the last 2.8 m.y.: Deep Sea Drilling Project/Ocean Drilling Program Sites 610, 642, 643 and 644. *Paleoceanography, 3*: 563-581, 1988.

Jansen, E., L. A. Mayer, J. Backman, R. M. Leckie, and T. Takayama, Evolution of Pliocene climate cyclicity at Hole 806B (5-2 Ma): Oxygen isotope record. *Proceedings Ocean Drill. Program, Scient. Res. 130*, 349-362, 1993.

Johannessen, T., E. Jansen, A. Flatøy, and A. C. Ravelo, The relationship between surface water masses, oceanographic fronts and paleoclimatic proxies in surface sediments of the Greenland, Iceland Norwegian Seas. In.: Zahn, R. and M. Kaminski (Eds.), *Carbon cycling in the glacial ocean*. NATO ASI Series, Springer Verlag, Heidelberg, in press.

Jones, G. A., and L. D. Keigwin, Evidence from Fram Strait (78°N) for early deglaciation. *Nature, 336*, 56-59, 1988.

Kvasov, D. D., The Barents Sea Ice Sheet as a relay regulator of glacial-interglacial alternation. *Quaternary Res. 9*, 288-299, 1978.

Kvasov, D. D., and A. I. Blazhchishin, The key to sources of the Pliocene and Pleistcene glaciation is at the bottom of the Barents Sea. *Nature, 273*, 138-140, 1978.

Kukla, G., and E. Went, (Eds.), *Start of a Glacial*. Springer-Verlag Berlin, 1992.

Lehman, S. J., Initiation of Fennoscandian ice-sheet retreat during the last deglaciation. *Nature 349*, 513-516, 1991.

Le Treut, H., and M. Ghil, Orbital forcing, climatic interactions and glaciation cycles. *J. Geophys. Res., 88(C9)*, 5167-5190, 1983..

Loeng, H., V. Ozhigin, B. Ådlandsvik, and H. Sagen, Water transport from the Barents Sea to the Polar Ocean. Poster presented at the Nansen Centennial Symposium, June 1993, Solstrand, Bergen, 1993.

Maasch, K. A., Statistical detection of the mid-Pleistocene transition. *Clim. Dyn. 2*, 133-143, 1988.

Maasch, K. A., and B. Saltzman, A low-order dynamical model of global climatic variability over the full Pleistocene. *Journal of Geophysical Research, 95(D)*, 1955-1964, 1990.

Maier-Reimer, E. T., and U. Mikolajewicz, Experiments with an OGCM on the cause of the Younger Dryas. *Rept. No. 39. Max-Planck-Institut f. Meteorologie, Hamburg*, 1989.

Manabe, S., and A. J. Broccoli, The influence of continental ice sheets on the climate of an ice age. *J. Geophys. Res., 90*, 2167-2190, 1985.

Martinson, D. G., The role of the Southern Ocean/sea ice interaction in global climate change. In G. Weller, C. L. Wilson, and B. A. B. Severin (Eds.), *Proc. International Conference on the Role of the Polar Regions in Global Change*, Fairbanks, Alaska, pp. 269-274, 1991.

Mayer, L. A., Extraction of high-resolution carbonate data for paleoclimate reonstruction. *Nature 352*, 148-150, 1991.

Midttun, L., Formation of dense bottom water in the Barents Sea. *Deep Sea Res. 32*, 1233-1241, 1985.

Milankovitch, M., Mathematische Klimalehre und astronomische Theorie der Klimaschwankungen. *Handbuch der Klimatologie, Bd 1, Teil A*. Bornträger, Berlin, 176pp., 1930.

Molnar, P., and P. England, Late Cenozoic uplift of mountain ranges and global climate change: Chicken or egg? *Nature, 346*, 29-34, 1990.

Nansen, F., The bathymetrical features of the North-Polar Seas. *Norwegian North Polar Expedition 1893-1896 Scientific Results 4 (13)*, 232pp. Christiania (Oslo), 1904.

Nansen, F., Northern Waters: Captain Roald Amundsen's Oceanographic Observations in the Arctic Seas in 1901. *Vid. Selsk. Skr. Mat.-Nat. Kl. 1, 3*, 1-313, 1906.

Nansen, F., Das Bodenwasser und die Abkühlung des Meeres. *Int. Rev. d. ges. Hydrobiol. Hydrograph. 5, 1*, 1-42, 1912.

Nansen, F., The strandflat and isostasy. *Vid. Selsk. Skr. Mat.-Nat. Kl. 1921, 11*, 1-313, 1922.

Newell, R.E., Changes in poleward energy flux by the atmosphere and ocean as a possible cause for ice ages. *Quat. Res. 4*, 117-127, 1974.

Oerlemans, J., Glacial cycles and ice-sheet modelling. *Climatic Change, 4*, 353-374, 1982.

Olausson, E., Evidence of climatic changes in North Atlantic deep-sea cores, with remarks on isotopic paleotemperature analysis, *Prog. Oceanogr. 3*, 221-252, 1965.

Olausson, E., and U. C. Jonasson, The Arctic Ocean during the Wurm and early Flandrian. *Geol. Foren. Stockholm Forhand., 91*, 185-200, 1969.

Opdyke, B. N., and J. C. G. Walker, Return of the coral reef hypothesis: Basin to shelf partitioning of $CaCO_3$ and its effect on atmospheric CO_2. *Geology, 20*, 733-736, 1992.

Pisias, N. G., and T. C. Moore, The evolution of Pleistocene climate: a time series approach. *Earth and Planetary Science Letters, 52*: 450-456, 1981.

Pollard, D., A simple ice-sheet model yields realistic 100 kyr glacial cycles. *Nature, 272*, 233-235, 1982.

Pollard, D., A coupled climate-ice sheet model applied to the Quaternary ice ages. *Journal of Climate, 1*, 965-997, 1983.

Prell, W. L., Oxygen and carbon isotopic stratigraphy for the Quaternary of Hole 502B: evidence for two modes of isotopic variability. *Init. Repts. Deep Sea Drilling Project, 68*, 455-464, 1982.

Raymo, M. E., Geochemical evidence supporting T. C. Chamberlin's theory of glaciation. *Geology, 19*, 344-347, 1991.

Raymo, M. E., W. F. Ruddiman, N. J. Shackleton, and D. W.

Oppo, Evolution of Atlantic-Pacific $\delta^{13}C$ gradients over the last 2.5 m.y. *Earth and Planet. Sci. Lett. 97*, 353-368, 1990.

Ruddiman, W. F., and J. E. Kutzbach, Forcing of late Cenozoic northern hemisphere climate by plateau uplift in southeast Asia and the American Southwest. *J. Geophys. Res. 94*, 409-427, 1989.

Ruddiman, W. F., and M. E. Raymo, Northern hemisphere climate regimes during the past 3 Ma: possible tectonic connections. *Phil. Trans. R. soc. Lond. B 318*, 411-430, 1988.

Ruddiman, W. F., and H. E. Wright (Eds.), *North America and Adjacent Oceans During the Last Deglaciation.* Geological Society of America, Boulder, Colorado, 1987.

Ruddiman, W. F., A. McIntyre, and M. E. Raymo, Matuyama 41,000-year cycles: North Atlantic and northern hemisphere ice sheets. *Earth and Planetary Science Letters, 80,* 117-129, 1986.

Ruddiman, W. F., M. E. Raymo, D. G. Martinson, B. Clement, and J. Backman, Pleistocene evolution: Northern Hemisphere ice sheets and North Atlantic Ocean. *Paleoceanography, 4,* 353-412, 1989.

Saltzman, B., A. R. Hansen, and K. A. Maasch, The late Quaternary glaciations as the response of a three-component feedback system to earth-orbital forcing. *J. Atmos. Sciences, 41,* 3380-3389, 1984.

Sancetta, C., and S. M. Silvestri, Pliocene-Pleistocene evolution of the North Pacific ocean-atmosphere system, interpreted from fossil diatoms. *Paleoceanography, 1,* 163-180, 1986.

Sarnthein, M., E. Jansen, M. Arnold, J.C. Duplessy, H. Erlenkeuser, A. Flatøy, M. Hahn, T. Veum, and E. Vogelsang, $\delta^{18}O$ time-slice reconstruction of meltwater anomalies at Termination I in the North Atlantic between 50 and 80°N. In: Bard, E., and W. S. Broecker (Eds.): *The last deglaciation, absolute and radiocarbon chronologies.* Springer Verlag., 183-191, 1992.

Sarnthein, M., K. Winn, J. C. Duplessy, and M. R. Fontugne, Global variations of surface ocean productivity in low and mid latitudes: Influence on CO_2 reservoirs of the deep ocean and atmosphere during the last 21,000 years. *Paleoceanography 3*, 361-399, 1988.

Shackleton, N. J., M. A. Hall, J. Line, and C. Shuxi, Carbon isotope data in core V19-30 confirm reduced carbon dioxide concentration of the ice age atmosphere. *Nature 306*, 319-322, 1983.

Short, D.A., J. G. Mengel, T. J. Crowley, W. T. Hyde, and G. R. North, Filtering of Milankovitch cycles by Earth's geography. *Quaternary Res. 35*, 157-173, 1991.

Solheim, A., and Y. Kristoffersen, The physical environment, western Barents Sea, 1:1 500 000: Sediment distribution and glacial history of the western Barents Sea. *Norsk Polarinstitutt, Skrifter, 179B*, 1-26, 1984.

Solheim, A., L. Russwurm, A. Elverhøi, and M. Nyland Berg, Glacial geomorphic features in the northern Barents Sea: direct evidence for grounded ice and implications for the pattern of deglaciation and late glacial sedimentation. In J. Dowdeswell and J. D. Scourse (Eds.) Glacimarine Environments: Processes and Sediments. *Geol. Soc. Spec. Publ. 53*, 253-268, 1990.

Spiegler, D., Ice rafted Cretaceous and Tertiary fossils in Pleistocene/Pliocene Leg 104 Sites. In Eldholm, O., Thiede, J., Taylor, E., et al. *Proc. Ocean Drilling Program, Scientific Results, V. 104* (pp. 739-744). College Station, Texas, 1989.

Start, G. G., and W. L. Prell, Evidence for two Pleistocene climatic modes: data from DSDP Site 502. In A. L. Berger and C. Nicolis (Eds.) *New Perspectives in Climate Modeling.* Elsevier, New York, pp. 3-22, 1984.

Stommel, H., Asymmetry of interoceanic fresh-water and heat fluxes. *Proc. Natl. Acad. Sci. USA, Geophys., 77,* 2377-2381, 1980.

Veum, T., E. Jansen, M. Arnold, I. Beyer, and J. C. Duplessy, Water mass exchange between the North Atlantic and the Norwegian Sea during the last 28.000 years. *Nature 356,* 783-785, 1992.

Vorren, T. O., K.-D. Vorren, T. Alm, S. S. Gulliksen, and R. Løvlie, The last deglaciation (20,000 to 11,000 BP) on Andøya, northern Norway. *Boreas 17,* 41-77, 1988a.

Vorren, T. O., M. Hald, and E. Lebesbye, Late Cenozoic environments in the Barents Sea. *Paleoceanography, 3,* 601-612, 1988b..

Vorren, T. O., G. Richardsen, S.-M. Knutsen, and E. Henriksen, The western Barents Sea during the Cenozoic. In: Bleil, U., and J.Thiede (Eds.) *Geological History of the Polar Oceans: Arctic Versus Antarctic.* Kluwer Academic, Dordrecht, Holland, 95-118, 1990.

Vorren, T. O., G. Richardsen, S.-M. Knutsen, and E. Henriksen, Cenozoic erosion and sedimentation in the western Barents Sea. *Marine and Petroleum Geology, 8,* 317-340, 1991.

Weertman, J., Rate of growth or shrinkage of nonequilibrium ice sheets. *J. Glaciol., 38,* 145-158, 1964.

Weertman, J., Milankovitch solar radiation variation and ice age sheet sizes. *Nature, 261,* 17-20, 1976. .

Weinelt, M. S., M. Sarnthein, E. Vogelsang, and H. Erlenkeuser, Early decay of the Barents Shelf Ice Sheet - spread of stable isotope signals across the eastern Norwegian Sea. *Norsk geologisk Tidsskrift, 71,* 137-140, 1991.

Weyl, P. K., The role of the ocean in climatic change: A theory of the ice ages. *Meteorol. Monogr. 8,* 37-62, 1968.

Wigley, T. M. L., Spectral analysis: astronomical theory of climatic change. *Nature, 264,* 629-631, 1976.

Wilson, A. T., Origin of ice ages: an ice shelf theory for Pleistocene glaciation. *Nature, 201,* 147-149, 1964.

Woods, J., The memory of the ocean. In: A. Berger (Ed.) *Climatic Variations and Variability: Facts and Theories.* Reidel, Dordrecht, Holland, pp. 63-83, 1981.

Wolfgang. H. Berger[1] and Eystein Jansen[2]

1) Scripps Institution of OceanographyUniversity of California, San Diego
La Jolla, California 92093-0215, USA

2) Department of Geology, University of Bergen, Allégaten 41, N-5007 Bergen, Norway

Variability of the Atmospheric Energy Flux Across 70°N Computed from the GFDL Data Set

James E. Overland

Pacific Marine Environmental Laboratory/NOAA, Seattle, Washington

Philip Turet

Joint Institute for the Study of the Atmosphere and Ocean, University of Washington, Seattle

The primary energy balance for the arctic atmosphere is through northward advection of moist static energy—sensible heat, potential energy, and latent heat—balanced by long wave radiation to space. Energy flux from sea ice and marginal seas contributes perhaps 20–30% of the outgoing radiation north of 70°N in winter and absorbs a nearly equal amount during summer. Thorndike's toy model shows that extreme climate states with no ice growth or melt can occur by changing the latitudinal energy flux by ±20–30% out of an annual mean flux of 100 W m^{-2}. We extend the previous work on latitudinal energy flux by Nakamura and Oort (NO) to a 25-year record and investigate temporal variability. Our annual latitudinal energy flux was 103 W m^{-2} compared to the NO value of 98 W m^{-2}; this difference was from greater fluxes during the winter. We found that mean winter (NDJFM) energy flux was 121 W m^{-2} with a standard deviation of 11 W m^{-2}. There were no large outliers in any year. An analysis of variance showed that interannual variability does not contribute towards explaining monthly variability of northward energy transport for the winter, summer or annual periods. Transient eddy flux of sensible heat into the arctic basin was the largest component of the total energy flux and is concentrated near the longitudes of the Greenland Sea (~10°W) and the Bering and Chukchi Seas (180°). There is a minimum in atmospheric heating north of Greenland, a known region of thick ice. While there was little interannual variability of energy flux across 70°N, there was considerable month-to-month variability and regional variability in poleward energy flux. Sea ice may play a role in storage and redistribution of energy in the arctic climate.

1. INTRODUCTION

The primary energy balance in the Arctic is advection of moist static energy, i.e., sensible heat, potential energy and latent heat, from lower latitudes and net radiation of energy outward at the top of the atmosphere. These fluxes are positive in all seasons of the year as the Arctic provides the northern hemisphere sink region for the atmospheric thermodynamic engine. In response to the large annual cycle in solar radiation, sea ice and the upper ocean absorbs heat energy during the summer season and are a net energy source during the long winter season.

The Polar Oceans and Their Role in Shaping the Global Environment
Geophysical Monograph 84

The annual cycle of this energy flux was estimated by Nakamura and Oort [*NO*, 1988], who calculated radiation fluxes from satellite data and latitudinal fluxes across 70°N from atmospheric rawinsonde observations over a ten year period, 1963 to 1973. Energy flux across the earth's surface was estimated as a residual. Figure 1 summarizes their results. F_{wall}, the latitudinal flux of moist static energy across 70°N, is normalized by dividing by the surface area north of 70°N, so that units in Figure 1 are all in W m^{-2}, with a value of 108 W m^{-2} during the three winter months (DJF) for example. All latitudinal energy fluxes calculated in this paper will be normalized in such a manner. The NO calculated net outward flux is 157 W m^{-2} in winter (DJF) and reduced to 15 W m^{-2} during summer (JJA). On an annual basis the net surface flux is small but there is a large annual signal. The 48 W m^{-2} at the surface for winter includes marginal seas; the winter upward surface flux over the

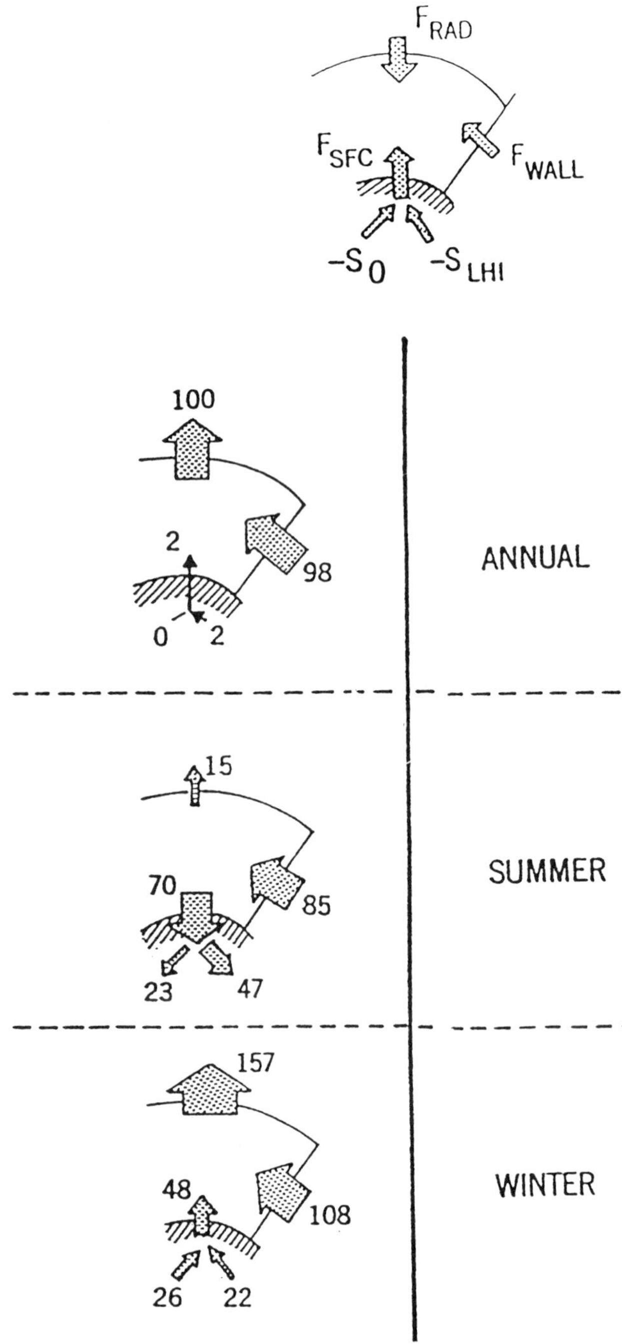

Fig. 1. Schematic breakdown of the observed poleward energy budget during typical annual-mean, summer and winter conditions for the 70°–90° polar caps (after NO). The seasonal values represent 3-month averages of June, July, and August for summer and of December, January, and February for winter. Units are in W m $^{-2}$ which is the meridional flux divided by the area north of 70°N. F_{rad} is the radiative flux through the top of the atmosphere and F_{wall} is the lateral flux into the arctic basin. The surface flux F_S is composed of S_0, the rate of storage of sensible heat in the ocean, and S_{LHI}, the rate of storage of latent heat in the form of snow and ice.

central arctic pack ice is closer to 21 W m^{-2} [*Maykut*, 1982].

More recently *Overland and Guest* [1991] and *Moritz et al.* [1992] have coupled an atmospheric radiation model to a one-dimensional sea-ice model. These models produce a net cooling of the atmosphere of about 0.5° day^{-1} in winter over the central Arctic, which can be balanced by the lateral energy fluxes suggested by NO. In a similar calculation with the atmosphere acting as a grey body, *Thorndike* [1992] showed that a simple model is enough to produce an annual cycle of temperature and ice thickness not unlike today's conditions. However, reducing the atmospheric poleward heat transport by 20 W m^{-2} is enough to make the ice so thick (>12 m) that it never warms up to the melting point in summer. Contrariwise, increasing the poleward flux by 30 W m^{-2} is enough to prevent any ice forming in the winter. Because of the potential sensitivity of polar climate shown by *Thorndike* [1992] to changes in latitudinal fluxes, it is important to look more closely at the variability of moist static energy flux into the Arctic. With an expanded GFDL data set of 25 years (1965–1989) we investigate the temporal variability of the atmospheric energy flux across 70°N.

2. DATA AND METHODS: ESTIMATING F_{WALL}

We continue with analysis techniques similar to *Oort* [1974, 1983] and NO. These studies considered fluxes across 60°N and 70°N for 10 years. Based on an expanded GFDL data set provided to us by A. Oort, we are able to expand the record to 25 years and investigate interannual variability. The GFDL data set of monthly mean variables and flux terms is based on rawinsonde data at eleven levels from 1000 mb to 50 mb. Approximate station coverage for the northern hemisphere is shown in Figure 2 after NO; lack of stations is an issue for the analyses north of 80°N in the western Arctic from 135°E to 135°W, but there is reasonable coverage at 70°N. Several error checks are performed; data beyond plus or minus four standard deviations from the mean for each level and season are excluded. Horizontal fields are developed at each level by objective analyses techniques applied on a 2.5° latitude by 5° longitude grid. Zonal averages of the monthly climatology are used as first guess fields. Only grid points above topography contribute to the statistical fields. The energy flux into the Arctic across a latitude circle can be expressed as

$$F_{wall} = \iint Cp[\overline{vT}] \frac{dxdp}{g} + \iint g[\overline{vz}] \frac{dxdp}{g} + \iint L[\overline{vq}] \frac{dxdp}{g} \quad (1)$$

where operator

$$[A] = (1/2\pi) \int_0^{2\pi} A d\lambda$$

indicates zonal mean and

$$\overline{A} = \int_0^{\tau} A dt/\tau$$

indicates a time mean, and v is the northward wind component. Other terms are defined in Table 1. The terms on the right hand side of (1) are fluxes of sensible heat, potential energy and latent heat, collectively termed the moist static energy. Each term in (1) can be expanded into four components: the transient eddy flux (TE), the stationary eddy flux (SE), the mean meridional circulation flux (MMC) and the net mass flow (NMF). For example,

$$\{[\overline{vT}]\} = \underset{TE}{\{[\overline{v'T'}]\}} + \underset{SE}{\{[\overline{v}^*\overline{T}^*]\}} + \underset{MMC}{\{[\overline{v}]''[\overline{T}]''\}} + \underset{NMF}{\{[\overline{v}]\}\{[\overline{T}]\}} \quad (2)$$

where $A' \equiv A - \overline{A}$ departure from time mean
$A^* \equiv A - [A]$ departure from zonal mean
$\{A\} \equiv \int A dp/g$ mass weighted vertical average
$A'' \equiv A - \{A\}$ departure from vertical average

For monthly time averages we assume that $\{[\overline{v}]\} \approx 0$. This physical constraint is usually not obeyed exactly when using real data. Therefore we subtract the computed $\{[\overline{v}]\}$ from the vertical profile of the original $[\overline{v}]$ data. The estimation of MMC is still sensitive to values of $[\overline{v}]$, which are generally small and subject to error. One approach is to calculate $[\overline{v}]$ from a angular momentum balance [*Oort*, 1983] or the vorticity balance [*Savijärvi*, 1988], but this was not done in NO. We found that most difficulties were with the value of the 50-mb wind. Observed values for $[\overline{v}]$ are used except at 50 mb as noted below.

TABLE 1. Definition of Variables and Parameters.

Cp	specific heat at constant pressure
g	gravity
L	latent heat of evaporation
T	temperature
z	geopotential height
q	specific humidity
p	pressure
x	unit distance (m)
λ	longitude
t	time
τ	time averaging interval

We have visually inspected the fields and variance of parameters at different levels. A major issue is the contribution of $[\overline{v}]$ at 50 mb to the variability of the fluxes compared to its contribution to the mean flux (Figure 3a). The vertical profiles of the mean values show a thermally direct circulation in MMC with equatorward flow in the lower atmosphere. Calculations of F_{wall} using the observed values of winds at 50 mb had twice the variance of F_{wall} computed with each monthly value of the 50 mb wind replaced by 25-year mean value (Figure 3b). The difference in the means for F_{wall} is small for the 5-month winter period, NDJFM, 121 W m^{-2} compared to 123 W m^{-2}, but the standard deviations are 34 W m^{-2} and 47 W m^{-2}. Although we may miss some variance, we based our calculations on replacing the 50-mb wind by the 25-year monthly mean values for $[\overline{v}]$. We do not have a clear understanding why the contribution to the total variance is so much greater at 50 mb than at other levels in the stratosphere.

3. TEMPORAL VARIABILITY OF F_{WALL} AND ITS COMPONENTS

a. *The Annual Cycle*

Figure 4 shows the annual cycle of F_{wall} at 70°N; Table 2 provides monthly values. Table 3 summarizes the mean fluxes for the annual, 3-month winter (DJF), 5-month winter (NDJFM), and 3-month summer (JJA) periods and compares them to the 10-year sample of NO. Our annual flux northward of 103 W m^{-2} is greater than NO by ~5 W m^{-2}. This is due to an increase in flux during winter (Table 3). Following NO we compute the 95% confidence limits as twice the standard error of the mean given by $2\sigma/\sqrt{N}$ where σ is the standard deviation (Table 4) and N = 25. For the winter (NDJFM) F_{wall} value of 121 W m^{-2} the confidence limits are

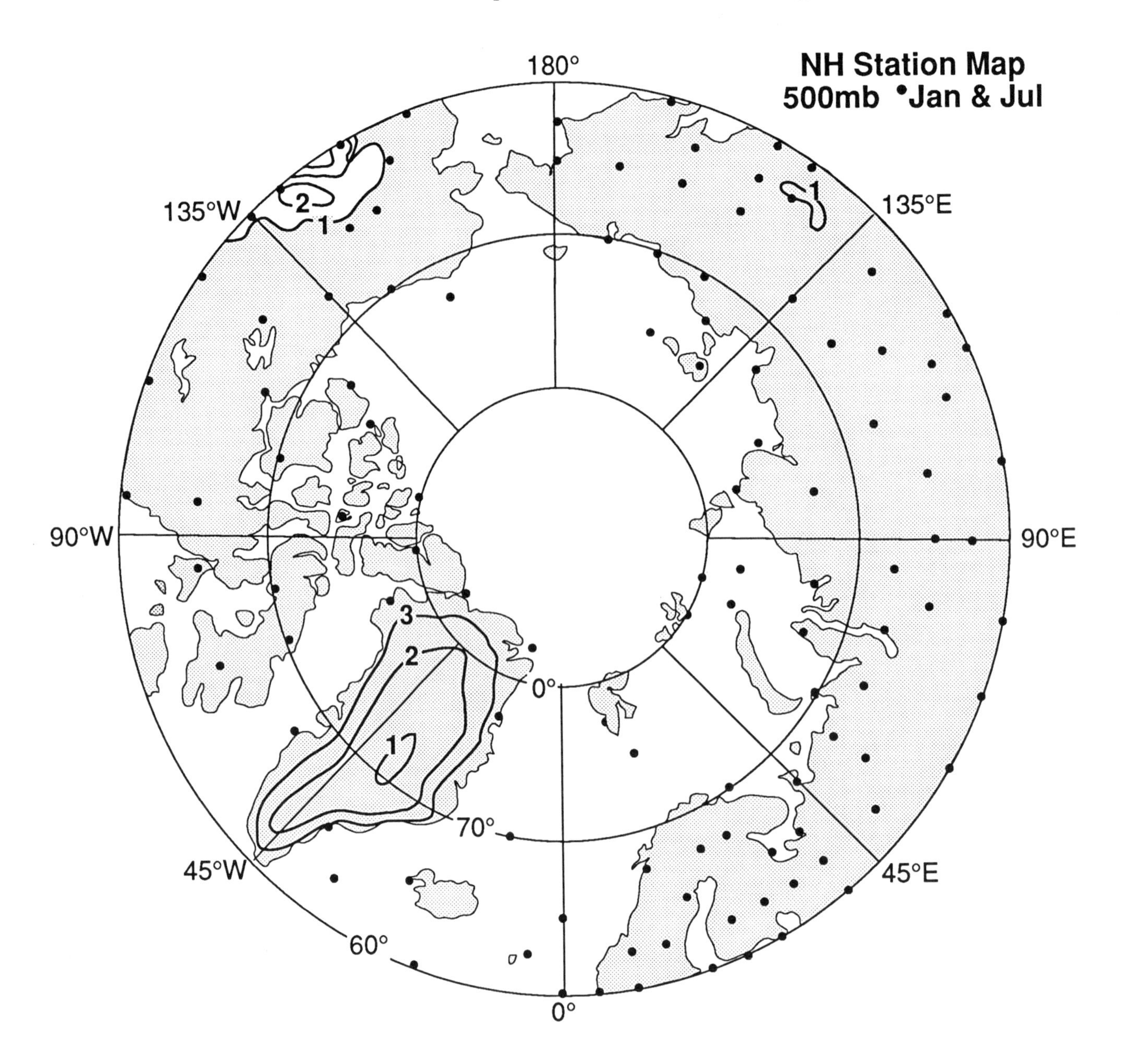

Fig. 2. Distribution of rawinsonde stations for the north polar region. The stations are plotted at the nearest grid point of a 2.5° latitude by 5° longitude grid. Height contours of surface topography are indicated in kilometers. After NO.

±4.6 W m^{-2}. Confidence limits on individual monthly values are ±13.5 W m^{-2}.

Transient eddy sensible heat flux has a small annual cycle and is the largest component in F_{wall} in both summer and winter. Stationary eddies play a greater relative role during winter. Latent heat is small in winter but is nearly 17 W m^{-2} during summer. Figure 5*a* shows the vertical profiles of annual total energy flux and flux by component; figure 5*b* and 5*c* show the same plots for winter and summer seasons. Transport by transient

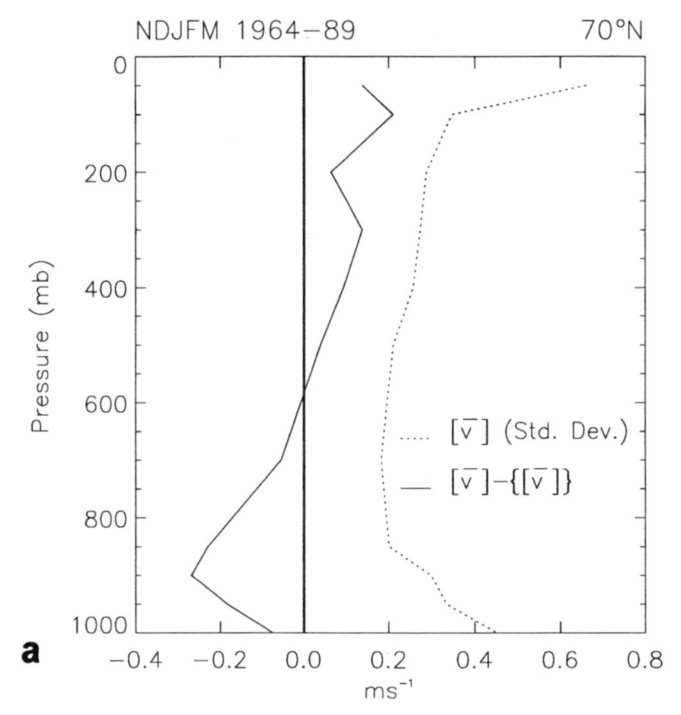

eddies, both from sensible heat and geopotential, are the major components below 300 mb. Northward transport of sensible heat is large at mid-altitudes, from 900 mb to 400 mb, but is smaller at the surface.

b. *Interannual Variability*

Figure 6 shows all nine components of the energy flux for winter (NDJFM) and summer (JJA) plotted for the 25-year period. Two major components are the sensible heat flux by transient eddies and the sensible heat by

---

Fig. 3a. Vertical variation of the mean and standard deviation of the mean meridional velocity, $[\bar{v}] - \{[\bar{v}]\}$. The standard deviation at 50 mb is two to three times greater than all values except for 1000 mb.

Fig. 3b. Vertical variation of the standard deviation of the components to F_{wall} using the $[\bar{v}]$ value at 50 mb. The largest contribution to F_{wall} is from MMC at 50 mb.

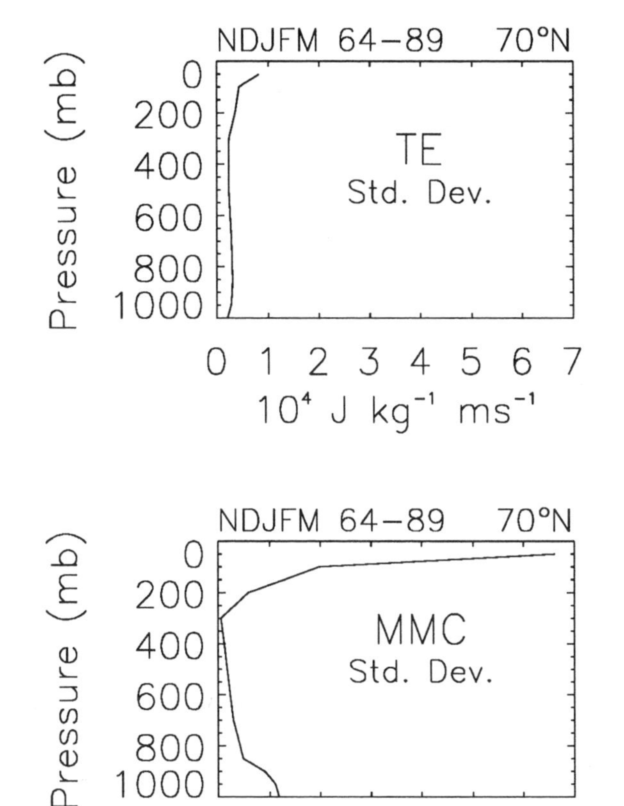

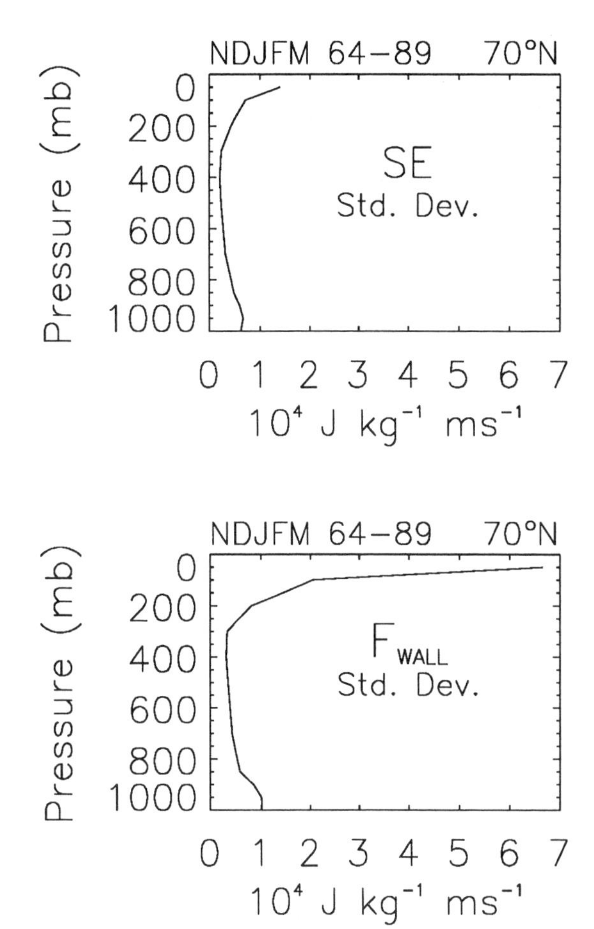

TABLE 2. Annual cycle of monthly means and standard deviations (in parentheses) at 70°N based on a 25-year data base. Also listed are the monthly values from NO. NO applied a 1/4:1/2:1/4 smoother to neighboring monthly values. Units are W m^{-2}.

	TE		SE		MMC		F_{wall}		F_{wall} (NO)
Jan	54.1	(20.8)	34.1	(25.6)	39.1	(24.7)	127.2	(30.7)	113
Feb	62.2	(22.0)	40.7	(24.9)	6.8	(27.4)	109.8	(38.8)	94
Mar	53.0	(9.9)	34.2	(27.5)	32.5	(34.3)	119.7	(37.1)	94
Apr	59.0	(12.2)	14.9	(15.0)	32.3	(22.0)	106.2	(25.8)	96
May	46.2	(10.0)	10.4	(7.6)	16.0	(12.0)	72.6	(15.4)	82
Jun	51.4	(7.8)	13.2	(7.1)	13.8	(15.0)	78.4	(15.7)	81
Jul	52.2	(11.4)	10.8	(8.2)	24.8	(13.7)	87.8	(14.3)	85
Aug	56.0	(8.8)	9.6	(7.0)	22.8	(15.2)	88.4	(15.0)	88
Sep	60.8	(8.3)	16.0	(8.8)	16.6	(16.6)	93.5	(17.9)	102
Oct	59.9	(14.2)	19.2	(14.2)	29.0	(17.2)	108.1	(21.9)	116
Nov	59.6	(17.7)	25.6	(20.6)	36.2	(24.0)	121.4	(27.2)	114
Dec	58.1	(22.2)	32.1	(27.3)	32.4	(18.7)	122.7	(31.1)	116
	SH		**GP**		**LH**		**F_{wall}**		
Jan	58.8	(30.4)	60.6	(45.9)	7.8	(3.1)	127.2	(30.7)	
Feb	86.4	(38.1)	14.5	(47.6)	8.9	(2.8)	109.8	(38.8)	
Mar	62.3	(35.1)	49.2	(51.5)	8.1	(2.3)	119.7	(37.1)	
Apr	44.7	(27.6)	52.6	(36.6)	8.9	(2.9)	106.2	(25.8)	
May	18.3	(13.7)	45.1	(17.9)	9.2	(2.3)	72.6	(15.4)	
Jun	23.3	(17.4)	39.7	(28.1)	15.4	(3.7)	78.4	(15.7)	
Jul	14.0	(16.9)	57.1	(23.1)	16.7	(4.7)	87.8	(14.3)	
Aug	25.7	(24.5)	45.5	(36.7)	17.2	(5.9)	88.4	(15.0)	
Sep	41.0	(24.3)	37.0	(40.2)	15.4	(4.1)	93.5	(17.9)	
Oct	34.3	(21.6)	62.1	(35.9)	11.7	(3.9)	108.1	(21.9)	
Nov	54.0	(27.2)	57.9	(40.8)	9.5	(3.3)	121.4	(27.2)	
Dec	61.5	(30.3)	52.8	(33.0)	8.4	(3.1)	122.7	(31.1)	

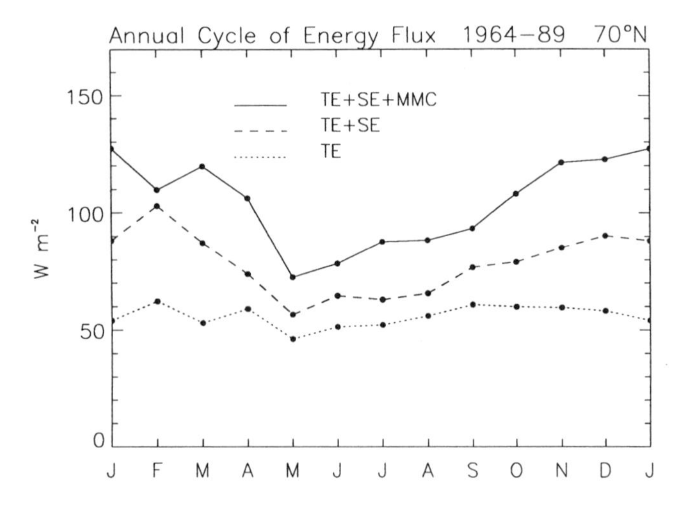

Fig. 4. Annual cycle of F_{wall}. Units of F_{wall} are W m^{-2}, energy flux through a latitude circle at 70°N divided by the enclosed surface area.

standing eddies. The MMC has contributions from both the sensible heat flux and geopotential flux, which are out of phase. Figure 7 summarizes the balance showing the monthly flux and seasonal values of the total flux, and transports by TE, SE, and MMC. In all winters and summers TE is the largest component.

With reference to Table 4 we provide the standard deviation and range of winter and summer monthly values. For winter $\sigma = 34$ W m^{-2} and for summer $\sigma = 16$ W m^{-2}. We also compute the interannual standard deviation from the 25 winter and 25 summer mean values. The interannual variability during winter as given by the interannual standard deviation, σ_I, is 11.4 W m^{-2} compared to the mean energy flux of 121 W m^{-2}; the method of calculating σ_I is given in the caption for Table 4.

Analysis of variance (Table 5) of monthly and seasonal energy flux values showed that interannual variability was not a significant component of total monthly

TABLE 3. Mean Fluxes for the annual, 3-month winter (DJF), 5-month winter (NDJFM), and 3-month summer (JJA) periods. Values from Nakamura and Oort (NO), Table B1, are also listed below in parentheses. Mean monthly 50-mb winds are used to calculate fluxes in this table. Units are W m^{-2}.

	Ann	DJF	NDJFM	JJA
F_{wall}	103	121	121	85
(NO)	(98)	(108)	(106)	(85)
TE	56	58	58	53
(NO)	(53)	(53)	(54)	(50)
SE	22	35	33	11
(NO)	(13)	(30)	(29)	(3)
MMC	25	27	31	20
(NO)	(32)	(21)	(29)	(31)
SH	44	68	63	21
TE	45	52	51	35
SE	19	33	31	8
MMC	−21	−17	−18	−22
GP	48	44	49	47
TE	−1	−1	−1	0
SE	0	0	0	0
MMC	48	45	50	48
LH	11	8	9	17
TE	12	7	8	18
SE	2	2	2	3
MMC	−3	−1	−1	−5

TABLE 4. Standard deviation, σ, and range, R, of the monthly values and the interannual standard deviation, σ_I, and range, R_I, of the seasonal averages for each year. The interannual standard deviation is defined as $\sigma_I^2 = \Sigma(\bar{A}_{i.} - \bar{A}_{..})^2 / (K-1)$ and the interannual range is defined as $R_I = [\min(\bar{A}_{i.} - \bar{A}_{..}), \max(\bar{A}_{i.} - \bar{A}_{..})]$, where $\bar{A}_{i.}$ is the seasonal average of monthly values for each year and $\bar{A}_{..}$ is the average value of all months and all years for winter, NDJFM, or summer, JJA, seasons; K=25. Units are W m^{-2}.

	NDJFM						JJA					
	σ	R		σ_I	R_I		σ	R		σ_I	R_I	
F_{wall}	33.7	1	188	11.4	95	138	15.6	53	120	9.7	68	103
TE	19.4	12	130	8.0	46	77	9.7	30	75	6.4	42	68
SE	25.1	−18	108	9.9	13	51	7.4	−3	29	3.8	4	21
MMC	27.9	−49	84	9.7	18	54	15.2	−18	52	9.0	1	38
SH	33.9	−4	166	13.0	37	85	20.5	−35	75	13.9	−16	54
TE	17.2	10	115	7.2	40	67	7.2	14	52	4.8	27	46
SE	22.4	−16	99	8.7	13	48	5.6	−2	21	2.8	3	15
MMC	17.8	−57	33	8.0	−32	0	16.8	−63	19	12.3	−50	7
GP	46.4	−90	138	18.0	22	90	30.7	−25	121	21.5	7	94
TE	2.0	−6	7	0.8	−2	1	1.2	−3	3	0.7	−2	1
SE	3.4	−8	11	1.4	−3	3	0.9	−2	2	0.4	−1	1
MMC	45.3	−80	139	17.4	22	88	30.4	−23	122	21.3	8	94
LH	2.9	1	16	1.3	6	12	4.9	3	30	3.5	8	25
TE	2.1	1	13	0.9	6	10	2.9	13	26	1.8	16	22
SE	2.7	−3	9	1.2	−1	3	2.2	−1	10	1.3	1	6
MMC	1.0	−3	2	0.5	−2	0	3.4	−16	3	2.4	−11	1

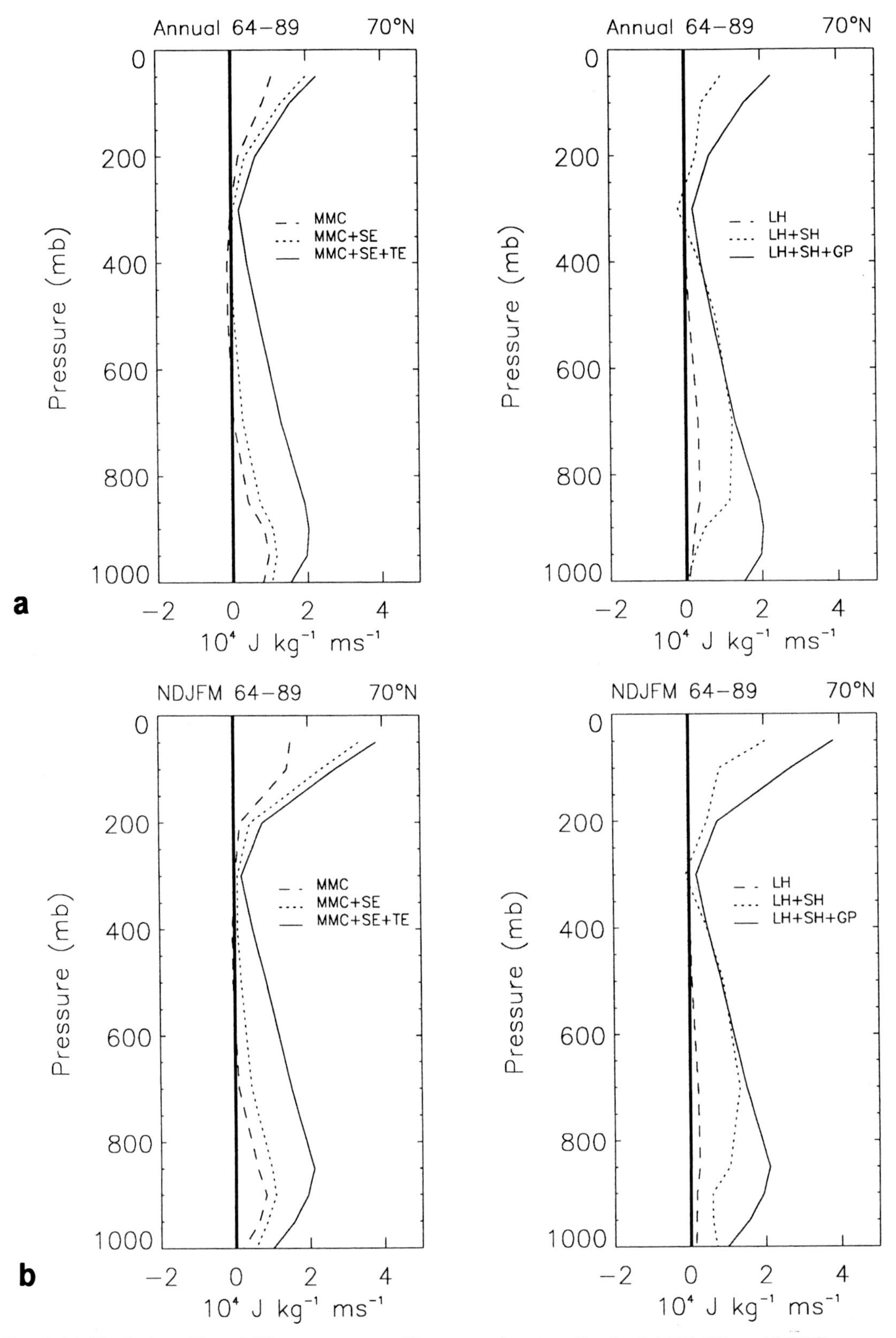

Fig. 5. (*a*) Vertical profiles of 25-year mean zonally averaged energy flux by (*a*) TE, SE, and MMC components and by LH, SH, and GP at 70°N. Units are in W m^{-1} Pa^{-1}. Integrate vertically and around the 70° latitude circle to get a total F_{wall}. (*b*) Same as 3*a* for winter months, (*c*) same as 3*a* for summer months.

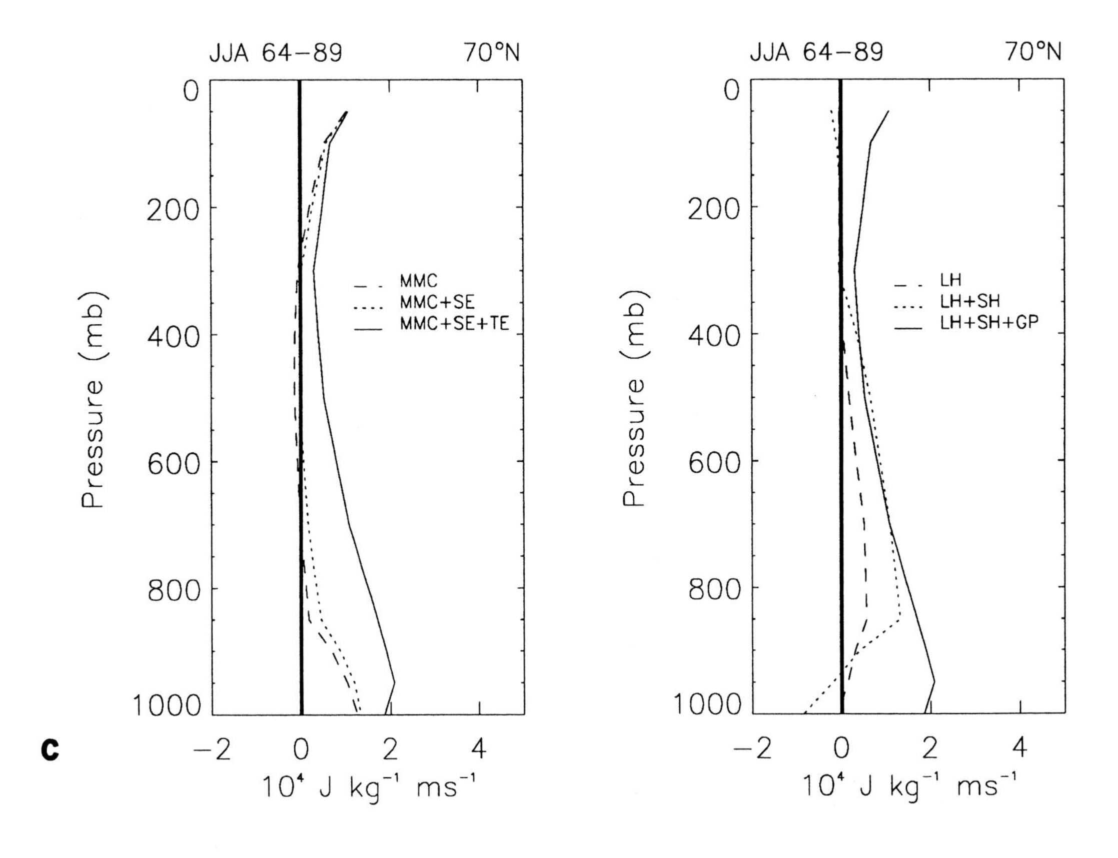

Figure 5. (continued)

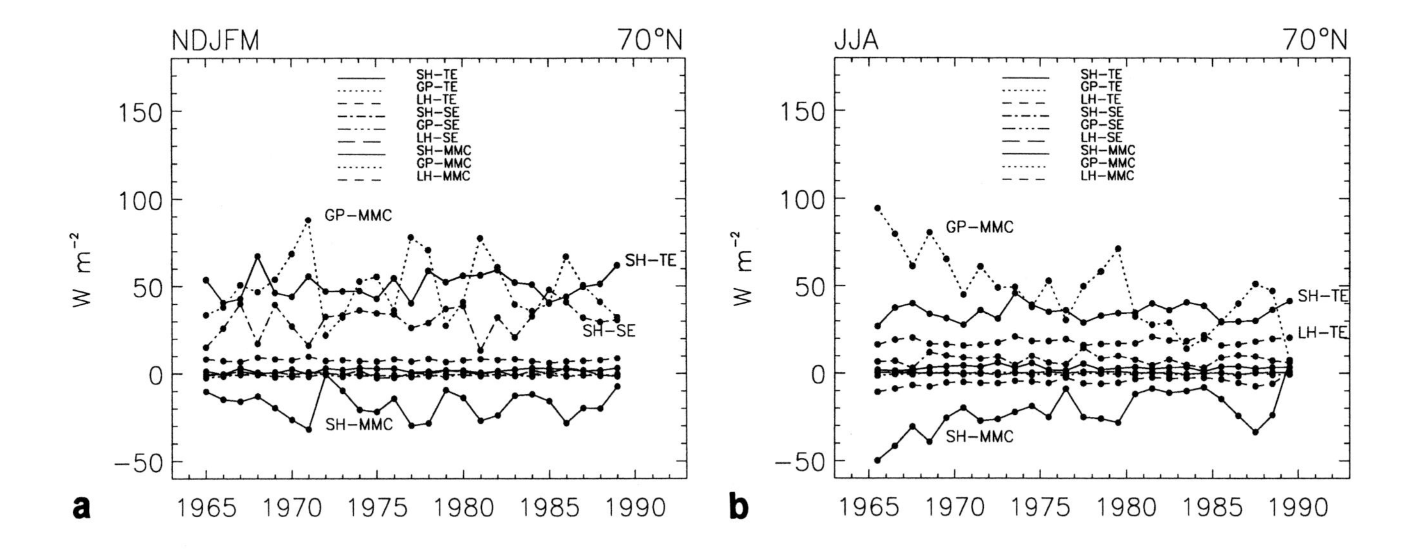

Fig. 6a–b. Winter (NDJFM) and summer (JJA) mean fluxes by components. Fluxes are listed by sensible heat (SH), geopotential (GP), and latent heat (LH), for transient eddies (TE), stationary eddies (SE), and mean meridional circulation (MMC). Units are W m^{-2}, energy flux through a latitude wall at 70°N divided by the surface area of the polar cap at that latitude.

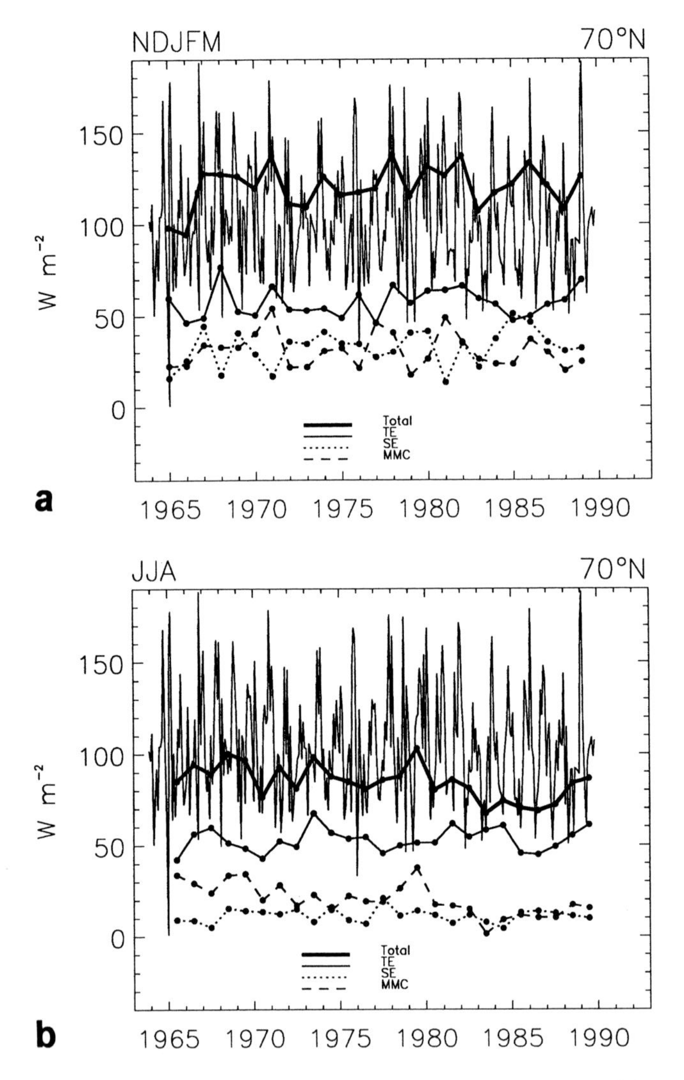

Fig. 7a–b. Seasonal mean flux by component (Total, TE, SE, MMC) for winter and summer and the monthly values.

variability in the annual series or in either the winter or summer seasons. F statistics (Table 5) based on the ratio of the interannual to within-year mean squares were used to measure the significance of the contribution of the interannual to the total monthly variability in the energy flux series [*Seber*, 1977]. Lack of significance in this statistic indicated that we could not say that interannual variability contributed to the variability in any month compared to chance; for example, if any given winter month had a higher than average value of F_{wall} we could not infer that the seasonally averaged F_{wall} would also be high.

From inspection of Figure 5, no individual year represents a major outlier. The monthly variability, σ, is substantial, however, suggesting that there is considerable compensation from month to month. There is also compensation between different types of energy transport. *Gloersen and Campbell* [1988] note that there is little interannual variability in total surface area of sea ice.

TABLE 5. Analysis of Variance of F_{wall}. Following is a table of sums of squares (SS), degrees of freedom (DF), and mean squares (MS) for within-year and interannual variability. The F statistic is the ratio of the interannual to the within-year mean squares. Interannual variability is not a significant component of total variability in any season or for the annual series. An F statistic of 1.288 on 24 and 50 degrees of freedom (JJA) is significant at about the 80% level.

Source	SS	DF	MS	F
		Annual		
Interannual:	13525.50	25	541.02	0.550
Within Year:	281529.25	286	984.37	
Total:	295054.75			
		NDJFM		
Interannual:	15633.00	24	651.38	0.519
Within Year:	125477.12	100	1254.77	
Total:	141110.12			
		JJA		
Interannual:	6835.75	24	284.82	1.288
Within Year:	11059.25	50	221.18	
Total:	17895.00			

4. SPATIAL VARIABILITY OF TRANSIENT EDDY TRANSPORT

Figure 8 shows the height/longitude plot of the winter (NDJFM) moist static energy flux across 70°N. Figure 9 shows the total energy flux by transient eddies (arrows) and flux divergence (contours) for the 5-month winter period. The major pathways of energy flux into the Arctic sea-ice region are through the Greenland and Barents Seas and the vicinity of Bering Strait in the Pacific sector similar to Figure 8.

In regions of red shading in Figure 9 the local atmosphere is warmed through the convergence of lateral heating. There is strong flux convergence in the eastern Arctic between Greenland and Scandinavia. This is a major winter cyclone track into the Arctic. Net heating is the case for most of the Arctic, however, there are two interesting exceptions. The first is north of Greenland. Here is a region of zero or small net cooling. One would expect greater ice thicknesses in this region.

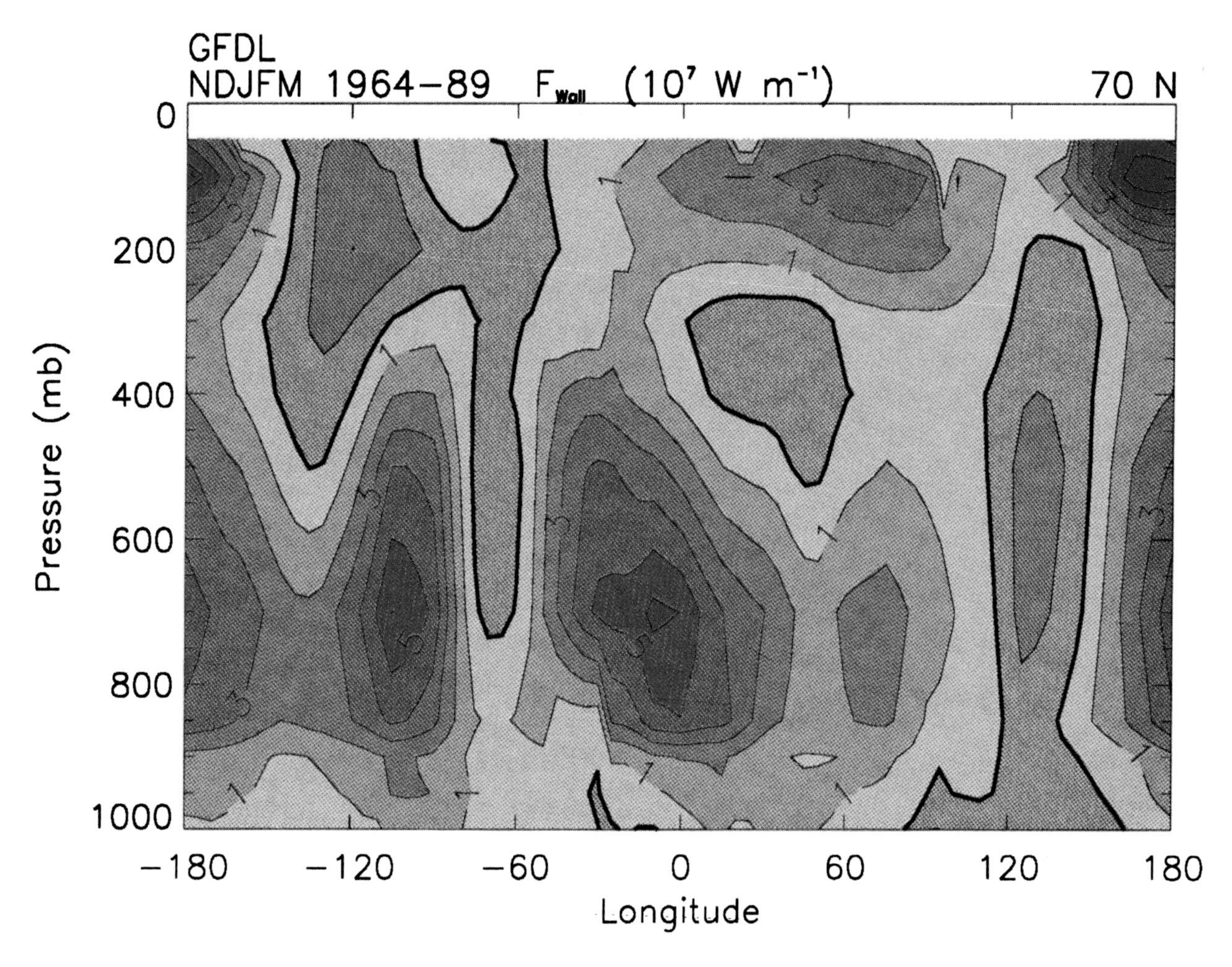

Fig. 8. Longitudinal and vertical variation of F_{wall}. Major northward components (Red) are near 0° longitude (Greenland Sea sector) and 180° (Bering Strait).

Greater ice thicknesses are observed in this region, and the primary explanation has been that the greater thicknesses are due to the pattern of ice advection in the central Arctic. We extend this explanation by suggesting that there is less advective heating and thus lower temperatures at the sea ice surface in this region compared to the rest of the Arctic basin. The sensible heat flux makes up the predominant contribution to the total TE flux over the Arctic basin; however, in the region north of Greenland, both the sensible heat flux and the geopotential flux make negative contributions to the flux convergence. The flux convergence from the geopotential flux has a wave number 2 structure with a minimum north of Canada and Greenland. The latent heat flux convergence is small and positive everywhere in the central Arctic.

A second area of minimum flux convergence is near Bering Strait. These are regions with polynyas and ice production due to ice divergence near coastlines. The minimum flux convergence in this area also supports ice production. Although we cannot be sure how robust these flux values are near 80°N based on station coverage (Figure 2), it is clear that there is considerable spatial variability of F_{wall} within the arctic basin.

5. SUMMARY

We compute a northward transport of moist static energy of 121 W m^{-2} during winter (NDJFM) and 85 W m^{-2} during summer (JJA), as well as monthly means and variances. The standard deviation of the interannual variability was 11 W m^{-2} based on 25 years. This observed interannual variability of 11 W m^{-2} is considerably less than the Thorndike thresholds of ±20–30 W m^{-2}; in fact, interannual variability is not significant as a predictor of monthly variability. This is

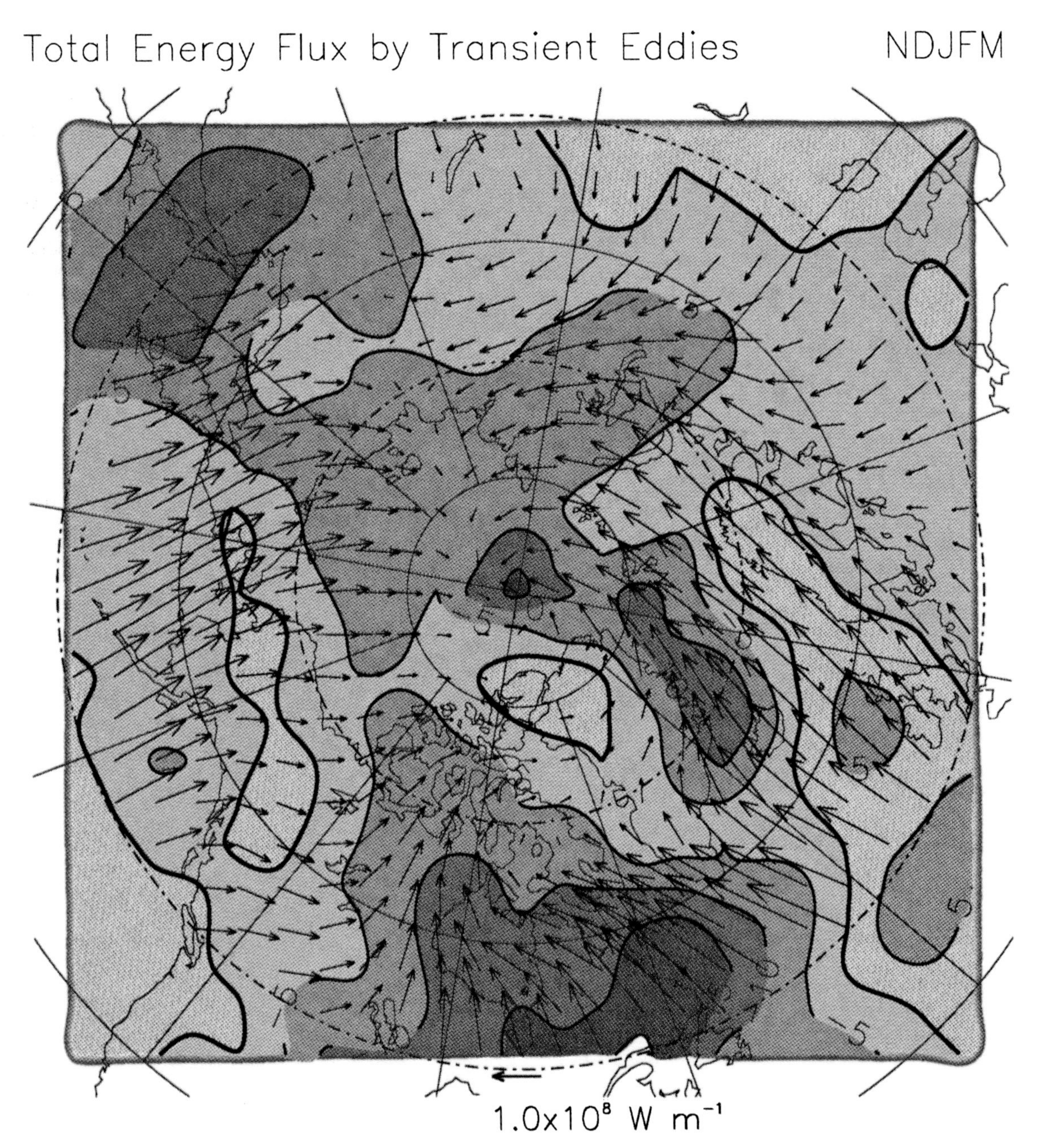

Fig. 9. Total energy flux by transient eddies (arrows) and flux divergence (contours) for the 5-month winter season (NDJFM) averaged for 25 years. The scale arrow near the bottom of the plot indicates a transport of 10^8 W m^{-1}. Flux divergence is contoured in units of 10 W m^{-2}.

true for individual maximum and minimum years as well as from the statistics. There is however considerable month-to-month variability and regional variability. Sea ice may play a role in short term energy storage and redistribution within the arctic basin. Although Thorndike's model is based on a steady state analysis and F_{wall} may have to change for many consecutive years to obtain the large shift in predicted ice conditions, it does provide a magnitude estimate of the sensitivity of Arctic climate to the northward transport of moist static energy flux. That the interannual variability of F_{wall} is small compared to Thorndike thresholds argues for the present stability of the arctic climate system.

Acknowledgments. We thank A. Oort for graciously providing access to his expanded data sets and providing comments on this manuscript. We thank R. Whitney and V. Vose for typing the manuscript. This paper is a contribution to Arctic Programs, Office of Naval Research. Contribution 1453 from NOAA Pacific Marine Environmental Laboratory and contribution 231 from the University of Washington Joint Institute for the Study of Atmosphere and Ocean.

REFERENCES

Gloersen, P., and W. J. Campbell, Variations in the Arctic, Antarctic, and global sea ice covers during 1978–1987 as observed with the Nimbus 7 Scanning Multichannel Microwave, *J. Geophys. Res., 93*, 10,666–10,674.

Maykut, G. A., Large-scale heat exchange and ice production in the central Arctic, *J. Geophys. Res., 87*, 7971–7984, 1982.

Moritz, R. E., J. A. Beesley, and K. Runciman-Moore, Modeling the interactions of sea ice and climate in the central arctic, *Third Conference Polar Meteorology and Oceanography*, AMS, 105–108, 1992.

Nakamura, N., and A. H. Oort, Atmospheric heat budgets of the polar regions, *J. Geophys. Res., 93*, 9510–9524, 1988.

Oort, A. H., Year-to-year variations in the energy balance of the Arctic atmosphere, *J. Geophys. Res., 79*, 1253–1260, 1974.

Oort, A. H., Global atmosphere circulation statistics, 1958–73, *NOAA Prof. Pap., 14*, 180 pp., Natl. Oceanic and Atmos. Admin., Washington, D.C., 1983.

Overland, J. E., and P. S. Guest, The Arctic snow and air temperature budget over sea ice during winter, *J. Geophys. Res., 96*, 4651–4662, 1991.

Savijärvi, H. I., Global energy and moisture budgets from rawinsonde data. *Mon. Weather Rev., 116*, 417–430.

Seber, G.A.F., *Linear Regression Analysis*, Wiley, 465 pp., 1977.

Thorndike, A., S., A toy model linking atmospheric thermal radiation and sea ice growth, *J. Geophys. Res., 97*, 9401–9410, 1992.

J. E. Overland and P. Turet, Pacific Marine Environmental Laboratory, National Oceanic and Atmospheric Administration, 7600 Sand Point Way NE, Seattle, WA 98115-0070.

The Influence of Polar Oceans on Interannual Climate Variations

Genrikh Alekseev

Arctic and Antarctic Research Institute St.Petersburg, Russia

Influence of oceans in high and moderate latitudes on the interannual climate variations is examined on the base of climate data analysis as well as estimates from energy- balance models. In particular, that influence may be detected in empirical data and manifests itself via strong interannual variations of mean surface air temperature of the Northern hemisphere in winter. Those variations are suggested to be forced by fluctuations of zonal heat transport in the atmosphere with ocean as a passive source of heat. It is found also that variations in latitudinal heat exchange act stronger on those of mean surface air temperature with periods being more than 10 years. Although anomalies of mean surface air temperature in high and low latitudes are synchronous on low frequencies, their full correlation is negative. Redistribution of heat in the "ocean- atmosphere- land" system leads also to the so- called advective- radiative fluctuations of mean temperature which are believed to be caused by a nonlinear relationship between surface temperature and outgoing longwave radiation (OLR). Impact of the ocean into those fluctuations may be also realized via interseasonal redistribution of heat. That redistribution causes increase of annual mean surface air temperature in the system as proportional to product of heat flux to (from) the ocean in summer (winter), relative surface of the ocean in the system, and difference between reciprocals of temperature sensitivities of OLR for winter and summer. The polar oceans affect on the global climate variability through deep winter convection processes and evolution of upper fresh layer and sea ice. Field data and fresh water balance estimates suggest the important role of saline water advection for deep winter convection development and that of the Arctic river run-off for heat surface balance in the Arctic ocean.

INTRODUCTION

Investigation of the ocean's role in climate is known as an important component of the global climate research. Strong influence of the ocean on climate have been determined both theoretically and empirically. The existing estimates indicate changes in water masses and global ocean circulation with the periods from decades to millennia [*Broecker et al.*,1985; *Delworth et al.*, 1992; *Weaver et al.*, 1993] which are governed by physical processes in oceans at high latitudes with sea ice on surface and strong water masses stratification. That is why the polar ocean are now the subject of special interest of scientists (*WCRP-72*, 1992).

One of directions in studying the above processes is connected with the global fresh water cycle whose reverse high latitudes' branch is formed in the polar oceans as outgoing cold fresh water and ice. Part of that water reaches deep layers due to deep winter convection (Figure 1). That process is fairly sensitive to the upper layer salinity in areas of convection [*Aagaard and Carmack*, 1989] which is controlled by the competitive transports of fresh and saline water from high and moderate latitudes respectively.

More short-period oceanic component of climate variability is produced at high and moderate latitudes in winter through the atmosphere circulation fluctuations with ocean as a passive source of heat and moisture. Those fluctuations manifest themselves as stable regimes of the

The Polar Oceans and Their Role in Shaping
the Global Environment
Geophysical Monograph 85

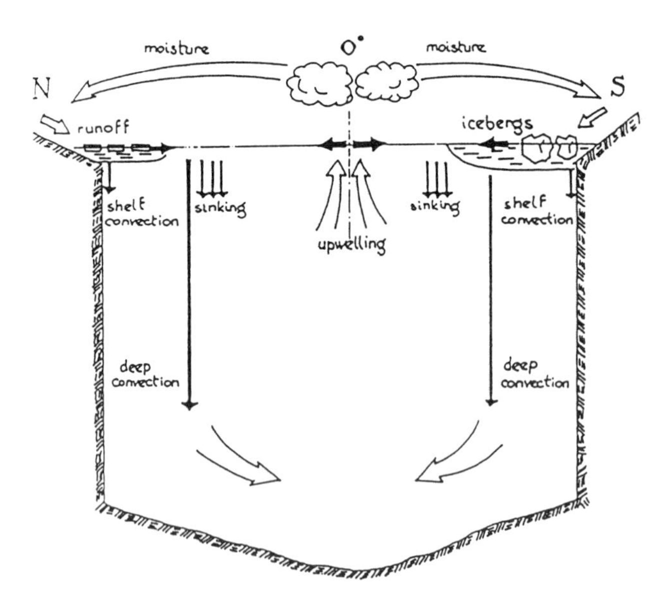

Fig.1. A diagram of the interaction of polar and extrapolar processes.

atmosphere circulation [*Dole,* 1983], as anomalously cold or warm winters in the Northern hemisphere, and mean air temperature variations with time scales up to 10 years which are detected with the use of the global climate models [*Bretherton et al.*, 1990].

This paper represents empirical and simple model estimates of oceanic component contribution to mean surface air temperature variability and discuss related mechanisms. A special attention is paid to advective-radiative fluctuations of mean temperature in the "ocean-atmosphere-land" system in result of nonlinear sensitivity of outgoing longwave radiation (OLR) to the surface - temperature.

In connection with the polar oceans influence on climate, the role of the Arctic ocean fresh water in its surface heat balance is examined. The advection of saline water into the Greenland convective gyre in deep winter convection development, and transformation of Intermediate Atlantic Water are discussed.

THE INFLUENCE OF THE OCEANS ON MEAN AIR TEMPERATURE VARIATIONS IN THE "OCEAN-ATMOSPHERE- LAND SYSTEM"

Heating effect of the ocean on climate is thought to be a result of two factors. The first one leads to short time climate variations and is caused by the fluctuation of the atmosphere circulation, whereas the second one is due to the variations of the ocean circulation and hence to those of poleward heat transport. In winter, fluctuations of the atmosphere circulation cause variability of mean surface air temperature in the "ocean-atmosphere-land" system with a broad spectrum. A part of that spectrum is suggested to be due to strong interannual variations in winter atmospheric heat and moisture transport from the low to high latitudes as well as from the ocean to land areas. So, wintertime intensification of the atmosphere circulation is accompanied by significant increase of air temperature over continents and polar regions and by a weak temperature decrease over oceans and in the low latitudes. That is, evidently, a consequence of large thermal capacity of the "ocean-atmosphere" system comparing to those of the "sea ice-atmosphere" or "atmosphere-land" system. As a result, strong internnual hemisphere- mean temperature (T) variations take place, which are strong correlated with those of temperature contrasts (D) in winter (see Figure 2).

Spectral analysis of relationship between anomalies of zonal (D^λ) and meridional (D^φ) contrasts of air temperature distribution in January with those of mean air temperature shows (Figure 3) that short period (less than 10 years) fluctuations of T are strongly correlated with D^λ . At the same time, fluctuations of longer period are correlated with those of D^φ . In the above estimates the values of T, D, D^λ and D^φ have been obtained as follows:

$$T_t=<T_{\varphi t}>_\varphi,\ T_{\varphi t}=<T_{\varphi\lambda t}>_\lambda,\ T_t'=T_t-\overline{T},$$

$$\overline{T}=\frac{1}{N}\sum_{t=1}^{N}T_t,\ N=96,$$

$$D_t^\lambda=<<(T_{\varphi\lambda t}-T_{\varphi t})^2>_\lambda>_\varphi,$$

$$D_t^{\lambda'}=(D_t^\lambda)^{\frac{1}{2}}-\overline{(D^\lambda)^{\frac{1}{2}}},$$

$$D_\varphi^t=<(T_{\varphi t}-T_t)^2>_\varphi,$$

$$D_\lambda^{\varphi'}=(D_t^\varphi)^{\frac{1}{2}}-\overline{(D_t^\varphi)^{\frac{1}{2}}},$$

$$D_t=D_t^\varphi+D_t^\lambda,\ D_t'=D_t^{\frac{1}{2}}-\overline{D^{\frac{1}{2}}},$$

$$<.>_\varphi=\frac{1}{\sin\varphi_2-\sin\varphi_1}\int_{\varphi_1}^{\varphi_2}.\cos\varphi d\varphi,$$

$$<.>_\lambda=\frac{1}{2\pi}\int_0^{2\pi}.d\lambda,\ \varphi_1=25^0\ N,\ \varphi_2=85^0\ N$$

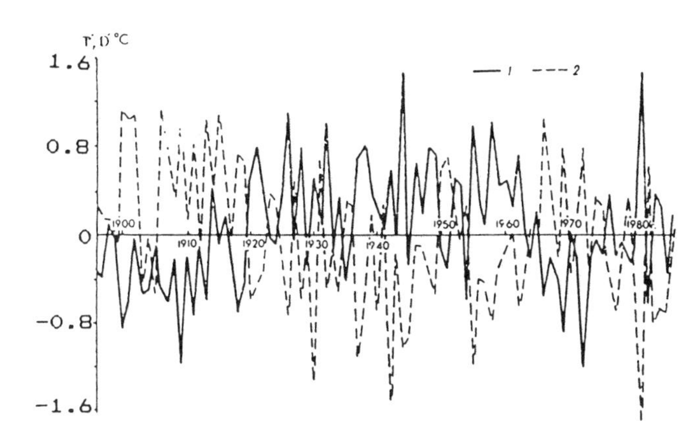

Fig.2. Anomalies of mean air temperature T' (1) and one of a spatial inhomogeneity of temperature distribution D' (2) over 25 - 85 N in January 1891 - 1986.

The second cause of correlation between T' and D'is believed to be a nonlinear relationship between surface air temperature and outgoing longwave radiation (OLR). The latter must lead to a decreased (increased) heat loss from the system when inhomogeneity in temperature distribution is increasing (decreasing). Accordingly, mean temperature of the system must change. Relevant component of mean temperature variability in the Earth's climate system has been called advective-radiative climate fluctuations [*Alekseev and Podgorny,* 1991].

Let us estimate an advective-radiative mean temperature change with the use of a simple box model that consists "warm" (T_1) and"cold" (T_2) atmospheric boxes.

$$R_1 - A - \Delta A = F_1 + b_1 \Delta T_1,$$

$$R_2 + A + \Delta A = F_2 + b_2 \Delta T_2 \qquad (1)$$

When advection between the two boxes is changed by ΔA, full energy balance in the system may be written as follows: Here $R_{1,2}$ - solar radiation flux into a box; $F_{1,2}$ - outgoing longwave radiation; A - heat advection from the warm box to the cold one; $b_{1,2}=4\delta T^3_{1,2}$ - temperature sensitivity of OLR; δ - integral transparency of the atmosphere; $\Delta T_{1,2}$ - temperature change in a box caused by that of advection.

Mean temperature change in the system may be found from Eq.(1) as

$$<\Delta T> = \frac{\Delta T_1 + \Delta T_2}{2} = \frac{\Delta A}{2}(b_2^{-1} - b_1^{-1})$$

Since $T_1 > T_2$, b_1 must be greater that b_2. Therefore, the sign of <T> depends on that of ΔA. For example, when $T_1 = 255K$, $T_2 = 270K$, $\delta = 5x10^{-8}$ wm^{-2} K^{-4}, $(b_2^{-1} - b_1^{-1})$ = 0.08w^{-1} m^2 K, ΔA=10wm^{-2}, the value of <T> is equal to 0.4K. Besides, advective-radiative component of mean air temperature change in the "ocean-atmosphere-land" system may arise due to interseasonal redistribution of the summer heat influx

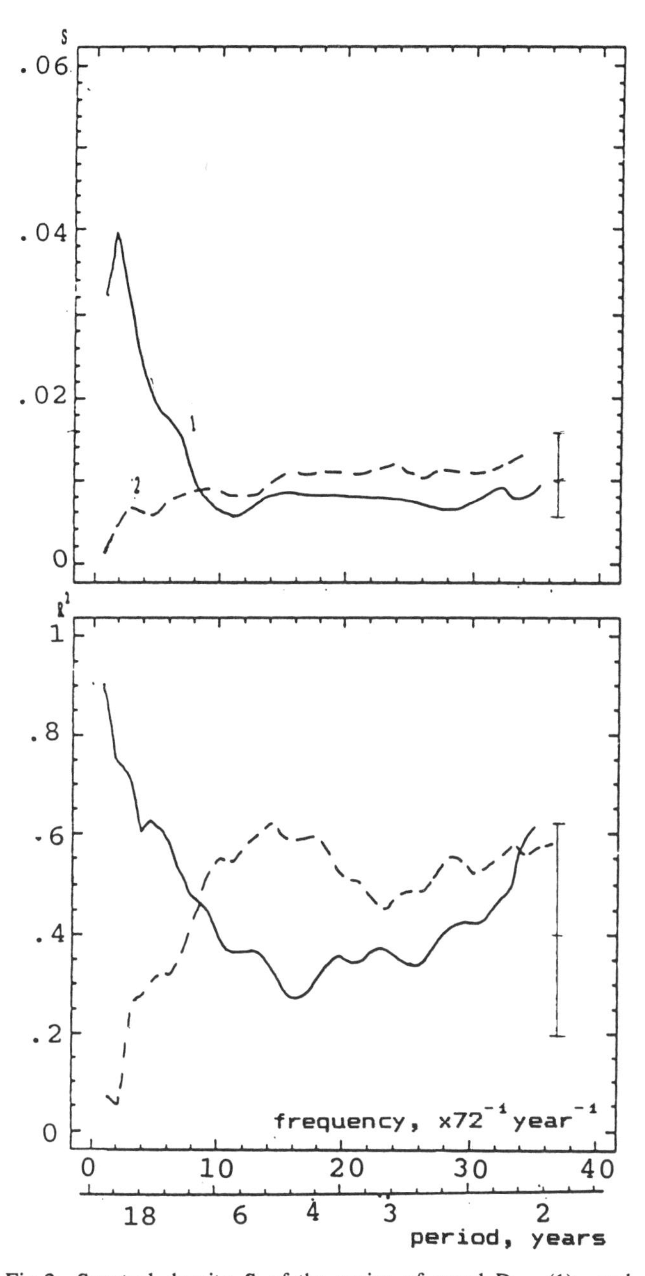

Fig.3. Spectral density S of the series of zonal D (1) and meridional D (2) contrasts and their coherency R with the mean temperature anomalies T' over 25 - 85 N in January (11- point running means. Vertical bars show 90 % confidence interval of estimates).

produced by the ocean. Let us estimate the magnitude of that component using two simple box models. In one of them the ocean consists a part of underlying surface for atmosphere where as the
other model is unoceanic. The energy balance equations on upper boundary of unoceanic model for summer and winter are:

$$R_1=F_1,$$

$$R_2=F_2,$$

and for model with ocean:

$$R_1-\alpha Q=F_1+b_1\theta_1,$$

$$R_2+\alpha Q=F_2+b_2\theta_2, \qquad (2)$$

where indices 1 and 2 correspond to summer and winter, R and F are incoming and outgoing radiation, Q is summer influx (winter outflow) of the heat to (from) the ocean, $\theta_{1,2}$ are resulting temperature changes in the systems, and $b_{1,2}$ are temperature sensitivities of OLR.

It follows from (2) taking into account the former equations:

$$\theta_1=-\frac{\alpha Q}{b_1};\ \theta_2=\frac{\alpha Q}{b_2}$$

Hence, for the mean annual temperature we obtain on addition in the model with ocean relatively the unoceanic model:

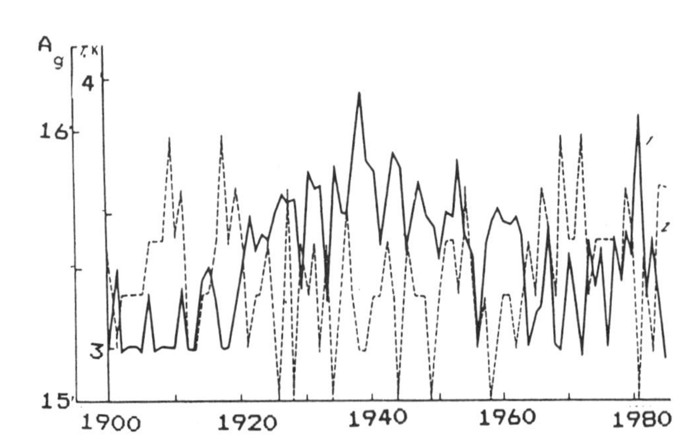

Fig.4. Mean annual air temperature T (1) and estimates of its annual variation amplitude Ag (2) over 25 - 85 N for 1900 - 1986.

$$\bar{\theta}=-\frac{\alpha Q}{2}(b_2^{-1}-b_1^{-1}) \qquad (3)$$

When T_1 = 290K, T_2 = 250K (summer and winter temperatures in unoceanic model), δ= 5×10^{-8} wm^{-2}K , $10 \le Q \le 100$ wm^{-2} , α=0.6, we have $0.4 \le \theta \le 4$K.

A confirmation of that relationship between mean air temperature change and intensity of interseasonal heat redistribution may be revealed in a strong inverse dependence between annual mean air temperature in the Northern hemisphere and characteristics of its interseasonal variability (Figure 4). The amplitude of mean temperature seasonal variation (A_g) is estimated from the expression:

$$A_g=2\sigma_g,$$

$$\sigma_g=\left(\frac{1}{12}\sum_{m=1}^{12}(T_{mg}-\bar{T}_m)^2\right)^{\frac{1}{2}},$$

$$g=1891-1986.$$

It is well known that annual mean air temperature in the Northern hemisphere is more than that of the Southern one. One of the reasons for that difference is suggested to be the ocean heat transport from the Southern to Northern hemisphere [*Hastenras*, 1980]. Relative importance of that factor on ocean's heating effect in both hemispheres may be approximately estimated using Eqs.(2-3) where the interhemispheric heat transport is introduced as Q_0. In this case, an increase of annual mean temperature in the Northern (N) or Southern (S) hemisphere is given by:

$$\theta_N=\frac{\alpha_1 Q}{2}(b_2^{-1}-b_1^{-1})+\frac{Q_0}{2b_2},$$

$$\theta_s=\frac{\alpha_2 Q}{2}(b_2^{-1}-b_1^{-1})-\frac{Q_0}{2b_2},$$

where α_1 and α_2 are relative parts of ocean surface in Northern and Southern hemispheres. So, relevant temperature difference is as follows:

$$\theta_N-\theta_S=\frac{Q}{2}\left(\frac{2Q_0}{b_2Q}-(\alpha_2-\alpha_1)(b_2^{-1}-b_1^{-1})\right)$$

When α_1 =0.6 and α_2 =0.8, inequality $\theta_N \ge \theta_S$ holds for $Q_0/Q \ge 0.1(1-(T_1/T_2))$. For example, $\theta_N \ge \theta_S$ for $Q_0/Q \ge 0.036$ when T_1 =290K, T_2 =250K.

An analysis of empirical data shows typical air temperatures in the high and low latitudes to be strongly negatively correlated in winter and in July (Table 1). In other months, correlations are weak and positive. However, one may reveal using spectral analysis that slow components of the high and low latitudes air temperature variability in winter (January-March) are correlated positively (Table 2). At the same time, relevant temperature contrasts and mean temperature are negatively correlated on all the frequencies. So, taking into account the above phenomena, a conclusion may be drawn about the need for an additional factor to explain them.

One of such factors may be the change of atmospheric transparency to longwave radiation which has an influence on both high and low latitudes air temperatures. Let us use the theory of the planetary atmospheres similarity by Golitsyn (1973) to illustrate relevant mechanism.

The rate of kinetic energy production (ϵ) in the atmosphere is determined by:

$$\varepsilon = k\frac{\Delta TQ}{T_1 M} \equiv \gamma \frac{\Delta T}{T_1}, \qquad (4)$$

where k is a dimensionless constant; $\Delta T=(T_1 - T_2)$; $T_{1,2}$ - are the equator and pole temperatures correspondingly; Q is the solar heat flux and M is the mass of the atmosphere. Change of $\delta(\Delta\delta)$ results in relevant changes of T_1 (ΔT_1) and T_2 (ΔT_2) which are given by:

$$\Delta T_1 = -\frac{1}{4}T_1^3\frac{\Delta\delta}{\delta}, \quad \Delta T_2 = -\frac{1}{4}T_2^3\frac{\Delta\delta}{\delta}$$

Accordingly

$$\Delta T_1 - \Delta T_2 = -\frac{\Delta\delta}{4\delta}(T_1^3 - T_2^3)$$

i.e. the difference of temperature changes increases when $\Delta\delta < 0$, and vice versa. According to Eq.(4), the rate of kinetic energy production has a change of:

$$\Delta_\delta \varepsilon = \frac{\gamma T_2}{4\delta_1 T}(T_2^2 - T_1^2)\Delta\delta,$$

i.e. it increases when $\Delta\delta < 0$, since $T_1 > T_2$. It may be seen from the above expressions that decrease in atmospheric transparency leads to increased temperatures on the equator and poles as well as their difference. Consequently, e also increases and atmospheric circulation becomes more intense. That leads to decreasing equator-pole temperature contrast due to an increased meridional heat transport. It is understood that the magnitude relevant temperature change on equator is considerably less than that on the pole due to significant difference in heat capacities. As a result, dependence of above mentioned correlation on periods of variations may be obtained.

CONTRIBUTION ESTIMATES OF THE FREEZING PROCESS IN THE HEAT BUDGET OF THE ARCTIC BASIN

Among water masses of the polar oceans the upper freshened layer is characterized by the largest contribution of regional processes into its formation, being along with sea ice the most active oceanic part of the polar climatic system. At the same time is still no consensus on the heat sources in the upper layer of the Arctic basin,which compensate for its losses from the surface of the basin on the whole for a year. Some investigators consider ocean advection of heat to be the major source. Others believe the main sources to be related with the internal thermal processes in the upper layer,first of all with phase transitions water-sea ice.

The estimates of heat flux components at the surface of the Arctic basin, obtained from changes in heat content of the water layer of the Atlantic or Pacific origin,or from experimental data on the coefficient of vertical turbulent heat diffusion from the Atlantic water through the upper layer [*Treshnikov and Baranov*, 1972] are quite consistent and sometimes even exceed the estimates of total heat flux at the surface of the Arctic basin (Table 3) from data of

TABLE 1. Correlation coefficients (r) between amomalies of the mean monthiy air temperature in the areas 85 - 65 and 40 - 25 n for 1891-1986. ($|R| \geq 0.20$)

Month	1	2	3	4	5	6	7	8	9	10	11	12
r	-0.17	-0.24	-0.24	0.01	0.14	0.12	-0.17	-0.01	0.12	0.23	0.10	0.02

Table 2. Phases (rad) of the correlation components in different months from table 1. (It is estimated by dint of cross spectral analysis).

Period (years)	8	96	48	32	24	16	14	12	8	6	4	3
January	6.28	5.62	5.60	2.64	3.17	3.56	3.05	3.26	1.77	4.56	2.87	0.27
February	6.28	5.71	5.18	3.20	3.25	3.72	2.52	2.31	6.08	5.96	3.24	3.56
March	6.28	6.17	5.65	4.68	5.06	5.85	3.24	4.43	0.10	2.54	3.93	5.94

climatic calculations [*Khrol*,1992] and measurements [*Makshtas*, 1984].

On the other hand,the estimates of the contribution of the internal heat sources also appear to be close to climatic values of total heat losses [*Treshnikov*, 1959; *Timofeev*, 1960]. Here these estimates have been made by means of observational data at the drifting station "North-Pole-16" in 1970-1971 in the area between 85°N,140°W and 86°N,120°W [*Alekseev and Buzuyev*, 1973]. These observations included detailed measurements of the upper ocean active layer, meteorological and actinometric measurements, measurements of ice growth and upper,lower and lateral melting. According to these data the value of the upper layer salivation due to freezing was determined to be equal to 0.60 psu.,which corresponds to the freezing out of the fresh water layer with the thickness of 80 cm. The inflow of the fresh water into the upper layer, estimated by direct measurements of melting of snow ice and precipitation and by summer freshening of water was equal 50 and 53 cm.(Table 4). The excess of winter freezing over the fresh water inflow in summer is equal 27 cm. It is close to the value 29 cm. of the mean layer of fresh water at the surface unit of the Arctic basin,corresponding to the annual ice outflow through the Fram strait [*Aagaard and Carmack*, 1989].

Measurements of multi-year ice growth in the winter period have shown simultaneously with the estimates of the upper layer salivation that the process of fresh water freezing out is connected, to a greater extent, with young ice growth in leads, than with the increase of thickness of old ice. Finally, the ice mass increase in winter is accompanied by heat release, corresponding to the flux of 12 W/m^2 , mean over this period, from which about 7 W/m^2 is due to the freeze out of local (summer) fresh water and 5 W/m^2 - to the freezing of advective fresh water. Summer heat influx along with fresh water advection to the upper layer are the main source for winter heat loss compensation by Arctic basin in its central part.

There is a different relationship between the oceanic and fresh water components of the heat influx to the upper layer in the Antarctic zone of the Southern Ocean. According to the estimates [*Gordon and Huber*, 1990] the main role here in maintaining the heat balance of

TABLE 3. Estimates of total heat flux from the ocean at the surface ofthe Arctic Basin.

Estimate	Value (W/m^2)	Mean over the period
1. By fresh water and salt balance in this paper	12.0	September-May, 1970-71
2. By meteorological observations		
(Makshtas,1984)	12.1	December-April, 1976-77
(Nazintsev,1964)	11.1	December-April, 1956-57
3. Climatic calculations (Khrol,1992)		
Central part	12.6	September-May, 1950-70
Western part	14.0	September-May, 1950-70

TABLE 4. *Components of the fresh water local balance in the upper layer in the Drift area.*

	Direct estimates	Indirect estimates	
Source	Fresh water inflow (sm)	Freshening in the layer (m)	Fresh water inflow (sm)
Ice melt from the top	2.0	0 - 3	21.0
Snowmelt	20.0	3 - 5	12.0
Ice melt from the bottom	7.0	5 - 10	12.0
Lateral ice melt	15.0	10 - 25	4.0
Liquid precipitation	6.0	In puddels on the ice	4.0
Total	50.0		53.0

the surface antarctic water belongs to the entrainment of the lower more warm and saline waters. The layer of the entrained water is estimated to be equal to 45 m over the year, which corresponds to the mean annual heat influx of 16 W/m^2 . At the same time the available estimates of the total heat flux of the ocean over the year are in this zone from 13.3 W/m^2 to 15.9 W/m^2. The annual excess of fresh water in the Antarctic is being formed mainly due to the iceberg discharge from the continent (about 2000 km^3 of fresh water over a year [*Romanov,* 1984] and from precipitation exceeding the evaporation, the fairly approximate estimates of which are 300 - 500 mm annually or 10.5 + 19.5 thousand km^3 of fresh water. To make up for such an excess of fresh water (not taking into account a possible ice and iceberg outflow outside antarctic waters) 30 + 47 meters of the lower water layer are necessary, which are accompanied by mean annual heat fluxes within 10.0 - 17.6 W/m^2 , which is close to the estimates of Gordon and Huber (1990). These estimates one should, probably, consider as the upper limit for the values of the oceanic heat influx to the upper layer, as the iceberg and ice outflow outside the Antarctic zone during the period of maximum extent of the antarctic ice cover appears to be quite possible. The main reason for a different heat contribution of the fresh water inflow to the heat balance of the surface layers of the Arctic basin and the antarctic zone of the Southern Ocean with close relative values of the influxes, is that largest past the inflow to the Arctic basin includes liquid river run-off and to the antarctic waters - solid precipitation and icebergs. That is why the latent heat of the phase transition water-ice is being transported to the Arctic basin, and the antarctic waters this heat is taken for a reverse process. And one can account for the differences between the estimates of the annual mean heat flux entrained in the upper layer and lost to the atmosphere in the Antarctic.

ATLANTIC WATER ROLE IN HEAT AND SALT BUDGET AND IN WATER MASS FORMATION OF THE ARCTIC OCEAN

The conclusion about main contribution of the local sources into upper layer heat budget at first glance does not agree with the established understanding of the great significance of the advection of the Atlantic water in the Arctic Ocean for its thermohaline structure. Actually, the estimates of the volumes of the incoming Atlantic water to the Arctic basin through the Fram Strait and the amount heat brought appear to be quite significant [*Nikiforov and Shpayher,* 1980., *Aagard et.al.*, 1987].

The heat balance of the Atlantic water is consider here, basing on the estimates of their unquestionable contribution to the salt balance of the Arctic basin and its upper layer, which is being continuously freshened by a river run-off and precipitation. For this purpose let us use the method for calculating salt balance by Knudsen equation [*Nikiforov and Spayher,* 1980] and estimate the volumes of the Atlantic water, necessary to neutralize the annual fresh water inflow to the upper layer and also that entrained to the surface and deep water layer (Figure 5).

Summing up the estimates of the heat effect due to the participation of the Atlantic water in the formation of surface and deep water of the Arctic basin and the estimate of the return heat flux of the Atlantic water outcoming from the Arctic basin, we receive the total heat effect within 1.4 - 2.0 x 10^{20} cal per year (or 1.9 - 2.7 x 10^{9} W), which includes the known estimates of the general heat inflow with the Atlantic water to the Arctic basin.

Thus, indirect estimates of the heat contribution of the Atlantic water on the base of the salt balance equation also indicate two or three- fold excess of the heat influx to the upper layer of the Arctic basin due to the freeze out of the incoming fresh water, as compared with the heat influx from the Atlantic water. Taking into account, that major portion of this heat incomes to the upper layer at the periphery of the Arctic basin, one can conclude, that the obtained estimates do not contradict the suggestion on a rather small thermal contribution of the heat of the Atlantic water to the heat balance of the upper layer directly in the Arctic basin.

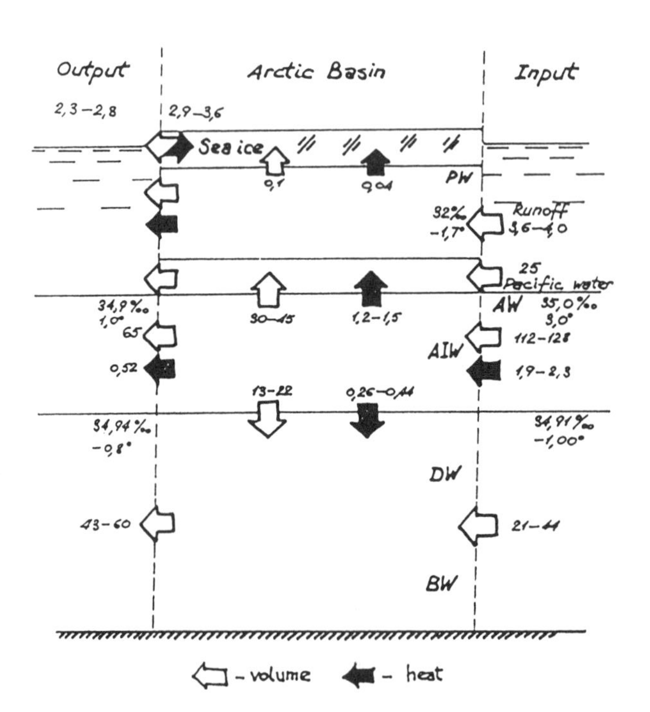

Fig.5. A general diagram of the entrainment of volume (thousands of cubic km) and heat (x 10 cal/year) of the Atlantic water in the upper deep layer of the Arctic Ocean.

Another problem to be discussed with regard to the Atlantic water is to determine the mechanisms and localize its entrainment in the upper and lower layers. The earliest view point on this problem belongs to F.Nansen (1902). Later it was shared by Timofeyev (1960), Treshnikov and Baranov (1972). The above mentioned authors consider the Atlantic water to mix with the river water and partly with the Pacific water (in the east) at the shelves of the Arctic seas, forming polar water mass, which is then transported with ice to the North Atlantic. The main volume of the Atlantic water is entrained into the deep and bottom layers, apparently in the vicinity of the Fram strait, where the transformation of the Atlantic water, (incoming with the West-Spitsbergen current through the strait) into the intermediate Atlantic water of the Arctic basin occurs [*Aagaard et al.*, 1987]. Then this three water masses spread partially eastward, being gradually transformed and selected by the system of underwater ridges and fill the deep-sea valleys of the Arctic basin. Such understanding of the formation of the large-scale water mass structure is also consistent with recent observations from icebreakers [*Anderson et al.*, 1989; 1991].

In addition to the warming effect on climate the inflow of more saline and warmer waters from low latitudes to the polar oceans also plays an important role in the production of deep and bottom waters for the World Ocean. Due to the mixing of this water with the desalinated polar waters an unstable thermohaline structure of the water masses develops in high- latitudinal gyres of the Greenland and Labrador seas in the Northern Polar Region and the gyres of the Weddell and Cosmonauts Seas in the Southern Polar Region, which under the influence of winter cooling can mix down to great depths up to the bottom [*Nansen*, 1928; *Gordon*, 1978; *Clarke and Gascard*, 1983; *Killworth*, 1983].

The increase of the discharge of desalinated waters and ice from the internal parts of the polar oceans blocks the processes of the deep convection in polar gyres [*Aagaard, and Carmack*, 1989], and vice versa, penetration of more saline waters into the polar gyres [*Koltermann*, 1991; *Bourke et al.*, 1992; *Alekseev et al.*, 1994] and into the internal regions of the polar oceans enhances vertical mixing, formation of deep and bottom waters, and thus, the large-scale vertical ocean circulation and oceanic heat transfer into high latitudes.

The second source of the deep and bottom water formation is considered to be the cooling and salinization of the surface polar waters with their subsequent mixing with lower more saline waters in some regions over the shelves of the polar seas as a result of winter divergence of sea ice near fast ice or the shores [*Nikiforov, and Shpayher,* 1980; *Aagaard et al.*, 1981]. Both sources depend to a great extent on the ratios between the inflow of saline and discharge of fresh water and sea ice, on large- scale oceanic circulation, on fluctuations of the atmospheric circulation, constituting finally the most sensitive and vulnerable links in the chain of global climate- forming processes, which, in addition, are the least adequately taken into account in global coupled sea- air circulation models. All this results in a particular attention, given to the studies of the oceans in the polar climatic system, first of all, the arctic one, the studies of which become a topic of international program ACSYS [*WCRP-72,* 1992].

CONCLUSION

Interannual mean air temperature fluctuations in the "ocean-atmosphere-land" system are thought to be a result of free variations of both ocean and atmosphere circulation under condition of inhomogeneity in the climate system key parameters and heat fluxes into the system. In this connection, the empirically found inverse dependence between anomalies of mean temperature and inhomogeneity of its spatial distribution in the winter half-year is suggested to be a consequence of non-stationary fluctuations of ocean-atmosphere heat fluxes together with the so-called advective-radiative mean air temperature fluctuations, although global scale factors should also be taken into account. Advective-radiative climate fluctuations arise due to a nonlinear relationship between surface temperature and outgoing longwave radiation. Variations of the ocean's heat impact on climate is a result of joint action of fluctuation of the atmosphere circulation, those of equator-pole heat transport as well as change of the ocean in high and moderate latitudes capability to interseasonal solar heat redistribution.

Influence of the polar oceans on climate change is connected with their key role in regulating the World ocean vertical circulation and with thermodynamical properties of the upper layer. Fresh water inflow to the Arctic basin, first, provides stability of the upper layer and, second, supply heat to compensate its loss to the atmosphere due to winter ice formation. On the contrary, the Antarctic area of the Southern ocean is supplied by fresh water in the form of ice which provides cooling effect on the average. Besides of the heating effect, important role of the polar oceans consists in production of deep and bottom water. Although regional scale conditions for deep winter convection development in the polar and subpolar seas are controlled by the large scale dynamics of the upper layer and sea ice, its local conditions arise due to sporadical advection of saline water to the convective gyres.

Acknowledgments. This work has been supported by the Ministry of Science of Russia and by the Russian Fund of Fundamental Research (Grant N 93-05-14106). The publication of the paper has been sponsored by the organisers of the Nansen Centennial Symposium. The author is grateful to Drs. I.Podgorny, P.Svyastchennikov and P.Bogorodsky for data processing and preparation of the manuscript, and to Drs. A.Buzuev, V.Zakharov for fruitful discussions.

REFERENCES

Aagaard,K., L.K.Coachman, and E.C.Carmack, On the halocline of the Arctic Ocean, *Deep-Sea Res.*, 28, 529-545, 1981.

Aagaard,K., A.Foldvik and S.R.Hillman, The West Spitsbergen Current: Disposition and water mass transformation, *J.Geophys.Res.*, 92, 3778-3784, 1987.

Aagaard,K., and E.C.Carmack, The role of sea ice and other fresh water in the arctic circulation, *J.Geophys. Res.*, 94, 14485-14498, 1989.

Alekseev,G.V., and A.Y. Buzuyev, On the evolution of the system "ice-surface ocean layer" in the drift area of the station "North Pole-16". *Problems of the Arctic and Antarctic,* 42, 37-43, 1973.

Alekseev, G.V., and Y.A.Podgorny, Advective-radiation fluctuations of the climate in the atmosphere-land -ocean system, Izv.Acad.Sci.USSR, *Atmosphere and ocean physics,* 37, 1120-1130, 1991.

Alekseev,G.V., V.V.Ivanov and A.A.Korablev, Interannual variability of thermohaline structure in the convective gyre of the Greenland Sea, *J. Geophys.Res.*, in press, 4, 1-8, 1994.

Anderson, L., E.P.Jones, K.P.Koltermann, P.Schlosser, J.H.Swift and D.W.R.Wallce, The first oceanographic section across the Nansen Basin in the Arctic Ocean, *Deep-Sea Res.*, 36, 475-482, 1989.

Bretherton, F., K.Bryan, J.D.Woods, Time-Dependent Greenhouse -Gas -Induced Climate Change in Scientific Assessment of Climate Change, *Report Prepared for IPCC by Working Group*, 175-194, 1990.

Broecker,W.S., D.M.Pettet, and D.Rind, Does the ocean-atmosphere system have more that one stable mode of operation?, *Nature,* 315, 21-26, 1985.

Bourke,R.H., R.G.Paquette and R.F.Blythe, The Jan Mayen Current of Greenland Sea, *J.Geophys.Res.*, 97, 7241- 7250, 1992.

Clarke,R.A., and J.-C.Cascard, The formation of Labrador Sea Water, Part 1, Large-Scale Processes, *J.Phys.Oceanogr.*, 13,

1764-1778, 1983.

T.Delworth, S.Manabe and R.J.Stouffer, Interdecadal Variability of the Thermohaline Circulation in a Coupled Ocean Atmospheric Model, Submitted to *J. of Climate,* 1992.

Gordon,A.L., Deep Antarctic Convection west of Maud Rise, *J.Phys.oceanogr.,* 8, 600-612, 1978.

Gordon,A.L. and B.A.Huber, Southern Ocean Winter Mixed Layer, *J.Geophys.Res.,* 95, 11655-11672, 1990.

Khrol,V.P., *Atlas of the energy balance of the Northern Polar Area* (in Russian), 72 pp., Gidrometeoizdat, St.-Petersburg, 1992.

Koltermann K.P., The deep circulation of the Greenland Sea as a consequence of the thermohaline system of the Europen Polar Seas, *Deutsche Hydrograph.Zeit.,* 23, 181 pp., 1991.

Makshtas,A.P., *The heat budget of Arctic ice in the winter* (in Russian), 67 pp., Gidrometeoizdat, Leningrad, 1984.

Nansen F., The oceanography of North Polar Basin, The Norv.North.Pol.Exp. 1893-1896, *Sci.Res.,* 1902.

Nansen F., The oceanographic problems of the still unknown Arctic Regions, Problems of Polar Research, Amer. Geogr.Ser, *Spec.Publ.,* 7,1928.

Nazintsev Yu.L., Thermal balance of the surface of the perennial ice cover in the central Arctic, *Tr.Arkt.Antarkt.Iust.,* 267, 110-126, 1964.

Nikiforov Ye.G. and A.O.Shpayher, *Features of the formation of hydrological regime large-scale variations in the Arctic Ocean* (in Russian), 269 pp, Gigrometeoizdat, Leningrad, 1980.

Romanov,A.A., *Southern Ocean Sea ice and the shipping condition* (in Russian), 88 pp., Gridrometeoizdat, Leningrad, 1984.

Swedish Arctic Research Programme 1991, International Arctic Ocean Expedition 1991, *Icebreaker "Oden"- A Cruise Report,* edited by L.C.Anderson and M.L.Carlsson, 128, pp., *Swedish Polar Research Secretariat,* Stockholm, 1991.

Timofeev,V.T., *Water masses of the Arctic basin* (in Russian), 190 pp, Gidrometeoizdat, Moscow, 1960.

Treshnikov,A.F., The surface water in the Arctic basin, *Problems of the Arctic,* 7, 5-14, 1959.

Treshnikov,A.F. and G.I.Baranov, *Structure of the water circulation in the Arctic basin* (in Russian), 158 pp., Gidrometeoizdat, Leningrad, 1972.

WCRP-72, *Scientific Concept of the Arctic Climate System Study* (ACSYS). WMO/TD-No. 486, 1992.

Weaver A.J., J.Marotzke, P.F.Cummins, E.S.Sarachik, Stability and variability of the thermohaline circulation, *J.Phys. Oceanogr.,* 23, 39-60, 1993.

G.V.Alekseev, Arctic and Antarctic Research Institute, Bering st.,38,St.Petersburg,199397,Russia

Sea Ice Thickness Changes and Their Relation to Climate

Peter Wadhams

Scott Polar Research Institute, University of Cambridge, Cambridge , England.

We review the current state of knowledge of sea ice thickness variability in the Arctic and Antarctic, and examine to what extent measurements made to date provide evidence for the impact of climate change.

We begin by examining the statistical properties of the ice thickness distribution and the mechanisms underlying its development. The shape of the distribution and the statistics of pressure ridge keel depths, slopes, widths and spacings are discussed. We look at how the thickness distribution is measured. Methods presently in use are: submarine sonar profiling; moored upward sonars; airborne laser profilometry; airborne electromagnetic techniques; and drilling. New techniques which show promise include: sonar mounted on an AUV or neutrally buoyant floats; acoustic tomography or thermometry (thickness inference from travel time changes); and inference from a combination of microwave sensors.

In relation to climate change in the Arctic, the measurements that have been most relevant to date have been submarine sonar profiles in identical parts of the Arctic Basin repeated on an interannual or interseasonal basis. These are reviewed: significant interannual changes in mean ice draft have been observed in the Canada Basin, at the North Pole, and in the region north of Greenland, while there is evidence of high stability in the Trans Polar Drift Stream between the Pole and Fram Strait. Measurements to date do not unequivocally demonstrate a trend due to climate change, although results both from the North Pole and from the Eurasian Basin do show a decline in mean thickness between measurements made during the 1970s and the 1980s.

In the Antarctic useful results to date on ice thickness distribution have come mainly from drilling, which greatly limits the generality of the conclusions that can be drawn. Data are far too sparse to provide evidence for climate-related changes. Some interesting aspects of Antarctic sea ice are the fact that first-year ice is very thin (about 60 cm); that the deep snow cover causes a significant fraction of the ice to undergo flooding leading to snow-ice formation; and that pressure ridges tend to be shallow with drafts of less than 6 m.

In assessing likely climate change impacts we conclude that the fast ice zone (shallow water close to shore) should show a definitive thinning in response to warming because ice thickness here is chiefly determined by air temperature. In moving pack ice the effects of dynamics and deformation are important and difficult to assess because of positive and negative feedback mechanisms. An area of special climatic importance is the Odden ice tongue in the Greenland Sea, where convection occurs in winter associated with local production of young ice, especially pancake; if ice production were to cease, we might expect convection in turn to slow down or cease, resulting in reduced oceanic absorption of anthropogenic CO_2.

1. THE ICE THICKNESS DISTRIBUTION

The thickness distribution of sea ice is one of the fundamental attributes defining the character of an ice regime. Together with the ice velocity, it defines the mass flux of sea ice, which in key regions such as Fram Strait is a major component of the overall energy and fresh water exchange between the polar oceans and sub-polar seas. The thickness distribution determines ocean-atmosphere heat exchange and the shape of the distribution is a measure of the degree of deformation of an icefield. Together with multi-year fraction, it defines the strength and other mechanical properties of the ice cover that are important for ice-structure and ice-vessel interaction. Finally, and possibly of greatest importance, long-term trends in ice

The Polar Oceans and Their Role in Shaping the Global Environment
Geophysical Monograph 85

thickness can be interpreted as a response of sea ice to climatic change.

At present most large-scale ocean-atmosphere models and climate models assume that sea ice can be approximated by a uniform sheet. Such treatments describe a thermodynamic equilibrium in which the ice grows thick enough that heat conducted up through the ice balances the net radiative loss to the atmosphere from the upper surface. In global warming simulations, increases in CO_2 levels cause a strongly amplified warming in polar regions, partly because of a feedback due to snowline retreat on land, but also because of a thinning and retreat of sea ice. The validity of these model experiments depends on the validity of the sea ice model used.

Of course we know that sea ice is not a uniform slab in equilibrium. It is densely fractured by open leads, even in winter, due to the differential wind and current stresses acting on the ice surface. These leads rapidly refreeze in winter, and are easily crushed by convergent or shear stresses to form pressure ridges, which consolidate by inter-block freezing to become semi-permanent features of the ice cover. The ridges are transported with the surrounding undeformed ice in which they are embedded, and partake of the later melt processes by which the ice gradually loses thickness in sub-polar seas such as the Greenland Sea. The fresh water flux into the Greenland Sea, for instance, is only partially due to thermodynamically grown ice, the rest being ice which has been produced by mechanical means. Pressure ridging is of importance in other contexts: it increases the hydrodynamic and aerodynamic drag coefficients of the ice surface; deep ridges can generate internal waves in the pycnocline; in shallow water deeper ridges scour the seabed and are responsible for a coastal "stamukhi zone" of highly deformed fast ice; ridges are the determining factor in strength calculations for offshore platforms designed for Arctic use [*Sanderson*, 1988] and in icebreaker hull design; and ridges cause scattering of underwater sound such that long-range propagation can only be achieved at very low frequencies.

Definitions

The thickness distribution can be defined as follows [*Thorndike et al.*, 1992]. Let R define a finite area within the ice cover, centred on a point **x**. Let dA(h, h + dh) represent the area within R covered by ice of thickness between h and (h + dh). Then the probability density function of ice thickness g(h) is given by

$$g(h)\, dh = dA(h, h + dh) / R \qquad (1)$$

and has dimensions L^{-1}. Fig. 1 [from *Wadhams*, 1981] shows some typical distributions. g(h) can have a delta-function at h=0 if open water is present.

When we consider how thickness changes with time, the governing equation is

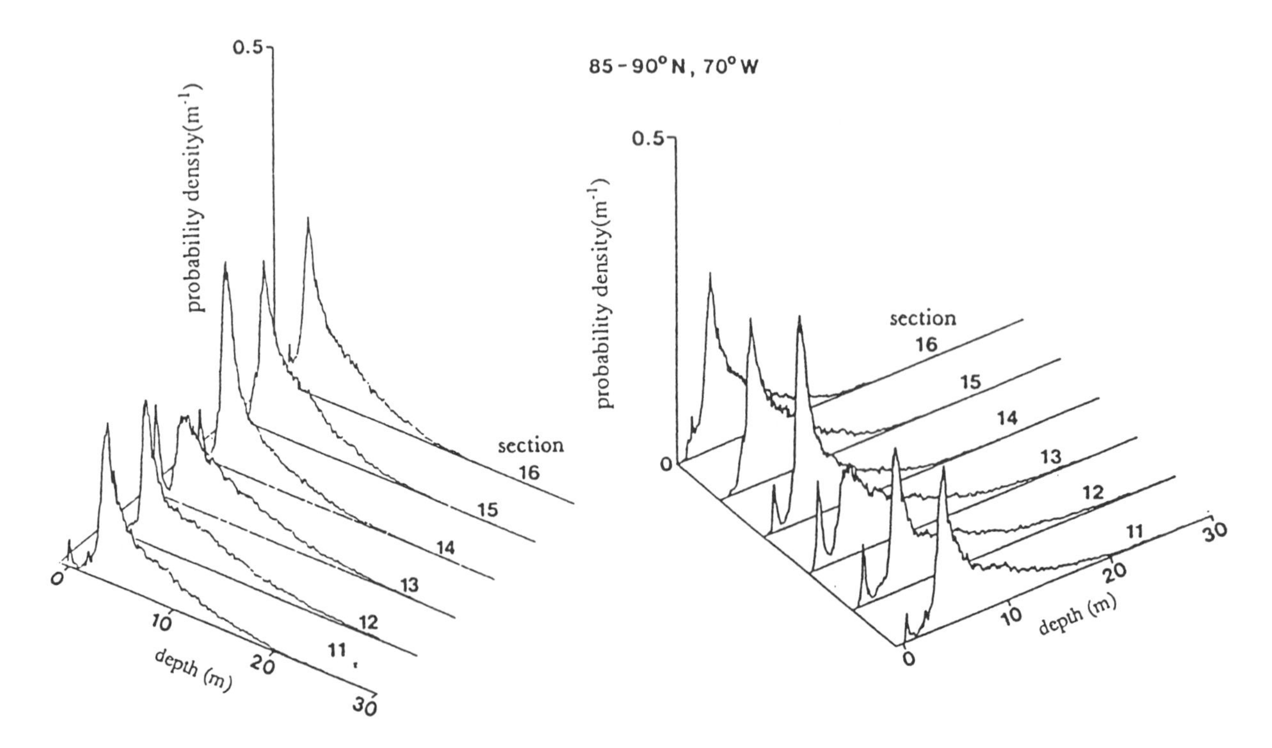

Fig. 1. Some typical probability density functions of ice draft from the Arctic Basin [after *Wadhams*, 1981].

$$\partial g/\partial t = \mathrm{div}(\mathbf{v}g) - \partial(fg)/\partial h + \Omega \quad (2)$$

where $f(h,\mathbf{x},t) = dh/dt$ is the thermodynamic growth rate of ice of thickness h at time t and at point $\mathbf{x}$; $\mathbf{v}$ is the horizontal velocity vector and Ω is a function that mechanically redistributes ice from one thickness to another, modelling the formation of leads and pressure ridges.

The first term in (2), divergence within ice, is a source of open water and a sink of ice-covered area. The second, thermodynamic, term causes thin ice to grow thicker and thick ice to grow thinner when averaged over an annual cycle. The third, and least understood term, simultaneously produces open leads and squeezes thin ice together to form pressure ridges. Thus it is at the same time a source of open water, a sink of thin ice, and a source of thick ice. The relative effects of the three processes on g(h) are shown schematically in fig. 2 [modified from *Thorndike et al*, 1992]. It is likely that the g(h) resulting from these processes never reaches an equilibrium state.

The term Ω is the real mystery. In current ice models which take account of ice thickness variations, Ω is treated in a mainly empirical way [e.g. *Walsh and Chapman*, 1991], and it seems probable that if the physics and mechanics of the ridge-building process can be more adequately described, it will be possible to radically improve such models to provide a better description of air-ice-ocean interaction and its role in global climate. We also might expect better predictions of the effect of global warming upon Arctic sea ice. It is not obvious *a priori* that warming will cause a decrease in the mean ice thickness, since thermodynamic thinning may be offset by a greater ease of pressure ridge production.

Figure 1 demonstrates the typical features of an Arctic Basin ice thickness distribution. Firstly, there are one or more small peaks in the thickness range 0-1 m. These correspond to refrozen leads at different stages of development, and young ice which has grown in polynyas during periods of ice divergence. There may also be a peak at zero thickness due to open water. There is usually a

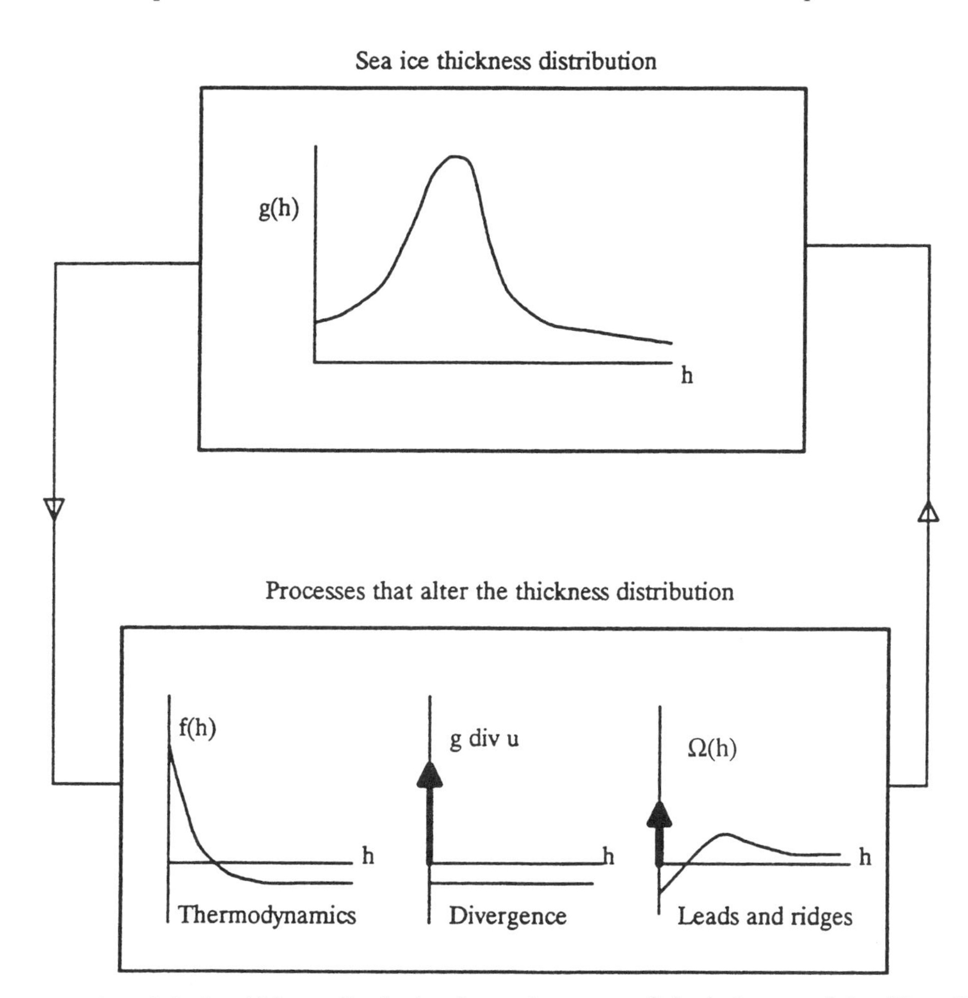

Fig. 2. The evolution of the ice thickness distribution due to three types of physical process [after *Thorndike et al.*, 1992]. For explanation see text.

relative lack of ice in the 1-1.5 m thickness range, followed by one or more main peaks. The widths and relative heights of these peaks provide important information about the undeformed ice types (first-year and multi-year of different ages) that are present in the ice regime and their relative abundances. Typically the first-year peak occurs in the vicinity of 2 m and the multi-year peak or peaks in the range 3-5 m. Finally there is a tail which contains ice which is thicker than can be attained by thermodynamic growth (> 5 m), and thus represents the flanks and crests of pressure ridges and hummocks. Deformed ice is, of course, found at lower thicknesses as well, but at thicknesses beyond about 5 m it comprises the whole of g(h).

2. STATISTICAL PROPERTIES OF THE THICKNESS DISTRIBUTION

In considering how the character of an icefield is revealed by the thickness characteristics of the ice, we have to deal not only with g(h) but also with the distribution of thicknesses and spacings of pressure ridges and refrozen leads, i.e. we must consider the thickness of *features* in the ice as well as the ice as a whole.

The shape of g(h) as shown in fig. 1 has the interesting statistical property that the tail of the distribution usually gives a good fit to a negative exponential distribution. Figure 3 demonstrates this for the case of a set of ice thickness distributions obtained at different latitudes in the Greenland Sea, plotted on a semilogarithmic scale. Although the slope of the tail increases as the latitude decreases (evidence that thick ice is melting preferentially as the ice stream moves southward), each distribution is a good approximation to a negative exponential. Since the tail of g(h) represents the ice contained in pressure ridges, we must seek an explanation for why ridged ice takes up this particular form of distribution.

There have been a number of attempts to model ridge building forces in order to predict sail and keel dimensions of pressure ridges under different conditions, but none of the models yet proposed adequately describes the complete process. The earlier models described limits to size and rates of development by equating kinetic energy loss of ice floes with the potential energy of the ice debris to calculate the ridge-building force [*Zubov, 1945*; *Parmerter and Coon*, 1972]. Later models have attempted to equate the limits of ridge development with the weight, density and plasticity of the ridge building material and the maximum possible ridge

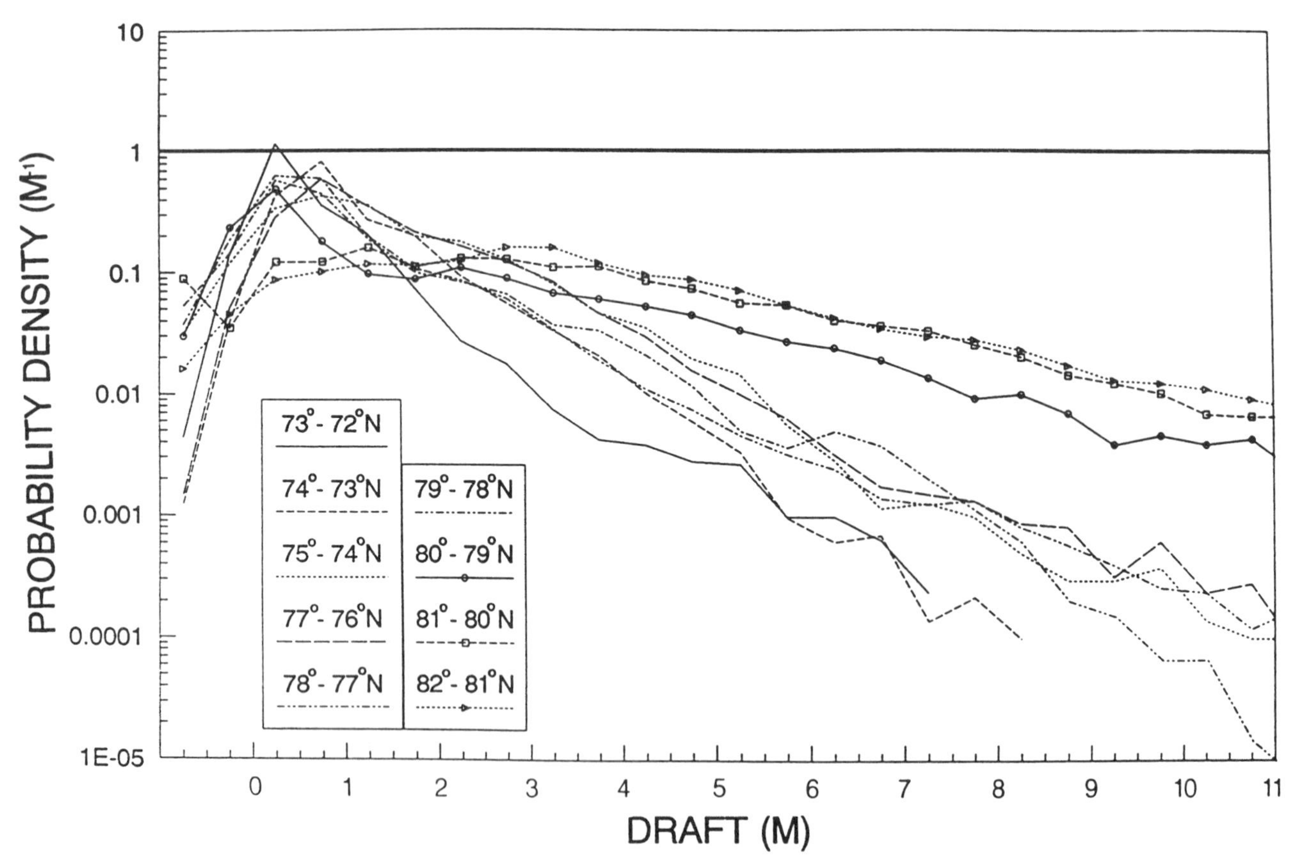

Fig. 3. Probability density functions of ice draft in 1° latitude bins in the Greenland Sea, from 82°N to 72°N, drawn on a semilogarithmic scale [after *Wadhams*, 1992].

slope [*Kovacs and Sodhi*, 1980; *Sayed and Frederking*, 1986]. However, it is apparent from these studies that ridge development is limited by the availability of deformable material and the momentum of the converging ice floes; these two conditions will be dependent on variable meteorological and oceanographic parameters.

Pressure ridges have been observed to range over a wide variety of shapes and sizes [*Wadhams*, 1977; *Weeks et al.*, 1971; *Rigby and Hanson*, 1976]. It is difficult to define pressure ridges in terms of dimensions such as ridge width or ridge slope (angle of repose) when multiyear keels have been eroded to semi-elliptical or even semi-circular profiles by seasonal melt and ocean currents, and sails have been sculpted by the wind or obscured by snow fall. The simplest representation of the ridge cross section that has been seriously considered is an isosceles triangle with rounded crest, and many ridges fall into this category. In addition, many authors have assumed that pressure ridge cross-sections are symmetrical, but an examination of published ridge profiles appears to refute this assumption and asymmetry has been reported ([*Kovacs*, 1971; *Dickins and Wetzel*, 1981].

Pressure ridges have keel draft to sail height ratios of the order 3-4:1, with larger ratios for first-year ice [*Tucker*, 1989]. The deepest ridge keel recorded had a draft of 47 m [*Lyon*, 1961] and the highest recorded free-floating sail was 13 m high [*Kovacs et al.*, 1973].The draft:height ratio is lower than the volume ratio because keels are usually wider than sails. Studies of the composition of the blocks in first year ice [e.g. *Tucker*, 1989] indicate a relationship between the thickness of the parent ice blocks and ridge height, implying that there is a limiting height to which ice of a certain thickness will build, given sufficient force. In first year ice, keel slopes have been measured in the range 20°-55°, averaging at 33° with sail slopes ranging from 15°-30° and averaging 24° [*Kovacs*, 1971]. For multiyear ice, keel slopes from almost zero to 51° have been recorded [*Wadhams*, 1977] with a concentration around 16°-28° averaging at 23.9°, whereas sail slopes are of the order of 14°-19° [*Weeks et al.*, 1971]. The Parmerter-Coon model [1972] predicts typical ridge slopes of 25° and 35° for sails and keels respectively, which they claim to be representative of pressure ridges from a variety of Arctic environments; these are in reasonable agreement with the findings of Kovacs [1971] for first year ice, but not so well matched by the multiyear ridges reported by *Wadhams* [1977].

Our knowledge of ridge distributions is thus empirical and derived from observations of ice profiles. The definition of individually identifiable ridges is commonly based on the Rayleigh criterion in optics, stating that an independent ridge must have a crest elevation or draft relative to the local undeformed ice which is more than double that of the troughs which bracket it. Using this method for the identification of ridges in the analysis of laser profiles [*Leppäranta*, 1981; *Tucker et al.*, 1979; *Wadhams*, 1976, 1981; *Wadhams and Lowry*, 1977], the distribution of sail heights was shown to fit a negative exponential. Sonar profiles of the ice underside obtained using a narrow beam instrument [*McLaren et al.*, 1984; *Wadhams and Horne*, 1980; *Wadhams et al.*, 1985; *Wadhams*, 1992], show a similar distribution such that

$$n(h)\,dh = B \exp(-bh)\,dh \;\Big|_{h>h0} \tag{3}$$

where n(h) is the number of keels per km of track per m of draught increment, and B,b are derived in terms of the experimentally observed mean keel draft (h_m), the mean number of keels per unit distance (n_k) and a low level cutoff draft (h_0).

$$b = (h_m - h_0)^{-1} \tag{4}$$

$$B = n_k b \exp(bh_0) \tag{5}$$

There is presently no physical explanation for this empirical result.

The mean slope angle of pressure ridges can be inferred from g(h) if it is assumed that pressure ridges of all depths are, on average, geometrically similar and can be approximated in shape by an isosceles triangle. Each ridge then contributes equal quantities of ice to g(h) down to its maximum depth. *Wadhams and Horne* [1980] showed that the relative slopes of log n(h) and of the tail of log g(h) are related to the mean along-track slope angle ∂ of pressure ridges by the relationship

$$g(h)\,dh_{\,[h>5]} = [2 B \cot\partial \exp(-b h) / b]\,dh \tag{6}$$

The best values for ∂ calculated from these relationships appear to lie in the range 11-14°. Thus the problems of why the pressure ridge draft distribution and the tail of the ice draft distribution are both negative exponentials can be reduced to a single unsolved problem in ice mechanics if it is assumed that ridges are geometrically congruent.

The spacing of pressure ridges was also initially reported as fitting a negative exponential distribution [*Mock et al.*, 1972], such that

$$P_r(x)\,dx = n_k \exp(-n_k x)\,dx \tag{7}$$

This relationship is expected if the generation of ridges is assumed to occur randomly over an initially undeformed ice cover. However, this may not necessarily be the case, and the work of *Wadhams and Davy* [1986] showed that a log normal distribution offers a more accurate description of the spacing distribution. Again, this is not predicted by any model.

We have recently completed a new study of ridge keel morphology, using upward looking (UL) and sidescan (SS) sonar records obtained simultaneously during a 1987 cruise, focusing on pressure ridges from six geographically distinct Arctic regions [*Davis and Wadhams*, 1992]. We examined 729 independent ridges which could be identified on the UL and also on SS, and thus for the first time it was possible to convert the observed keel slopes (from UL data) into real keel slopes, using the ridge orientation information provided by the SS. Thus we have the first very large dataset on real underwater ridge shapes, enabling us to draw statistically valid conclusions about the morphology of ridges.

The analysis revealed a low mean keel draft near the North Pole and in the Greenland Sea, with a higher mean draft in the heavily ridged zone off north Greenland. Following the Trans Polar Drift Stream downstream into the East Greenland Current revealed a decline in mean keel draft at a greater rate than the downstream decline in mean ice draft reported by *Wadhams* [1992], suggesting a faster melt rate for ridge keel bottoms than for undeformed ice. Keel spacing fitted a log normal distribution.

A wide range of keel widths (defined as the distance between the troughs which bracket the keel crest) was seen, clustered around 50 - 150 m. The widths fitted a log normal distribution. Variations with latitude were comparatively small with mean total widths around 70 m and a modal value of 65 m suggesting that keel widths are distributed around some theoretical optimum, which is dependent on the mechanism of formation. The widths fitted a log normal distribution at all latitudes with correlation coefficients better than 0.95.

Keel slopes were in good agreement with the earlier study of *Wadhams* [1977], and there was a similar good fit to a log normal distribution (correlation coefficient 0.991) with no significant difference between leading and trailing slopes. The ratios of half width to draft were similarly distributed (correlation coefficient 0.994) and variations of slope with latitude appeared to be slight.

A tentative relationship between slope and half width/draft ratio was found, of the form

$$\text{Slope} = a + b.c^{\text{Ratio}} \quad (R = 0.93) \tag{8}$$

where a,b,c are parameters which are best fitted by a = 8.71, b = 76.3, c = 0.585 when the keel slope is expressed in degrees. This relationship requires further investigation.

Summarising, we can say that what was known about ridge statistics until recently was:

- Keel drafts and sail heights fit negative exponential
- Ridge spacings fit log normal
- Ridge slopes are variable; keels are wider than sails.

The additional knowledge given by the new study is:

- Keel widths fit log normal
- Keel slopes fit log normal
- There is a relationship between slope and (width/draft) ratio.

Finally, as another piece of the puzzle of ice deformation mechanisms we must mention leads, which when refrozen are the raw material out of which ridges are created. It has been found [*Wadhams*, 1992] that the distribution of lead widths fits a power law, i.e.

$$P(w) = K\, w^{-n} \tag{9}$$

where P(w) is the probability of width w per unit increment. The best fit to an exponent [*Wadhams*, 1992] was 1.45 for leads less than 100 m wide and 2.50 for wider leads, using the criterion that ice in a refrozen lead cannot exceed 1 m in draft (1.67 and 2.76 for a 0.5 m criterion). Lead spacings fit a negative exponential at moderate spacings (400-1500 m), but with an excess of lead pairs at small and very large spacings.

3. MEASURING ICE THICKNESS DISTRIBUTION

Current techniques

Five techniques have been employed for measuring ice thickness distribution. In decreasing order of total data quantity, they are:-

1. Submarine sonar profiling
2. Moored upward sonars
3. Airborne laser profilometry
4. Airborne electromagnetic techniques
5. Drilling.

Most synoptic data to be published so far have been obtained by upward sonar profiling from submarine. Beginning with the 1958 voyage of *Nautilus* [*Lyon*, 1961; *McLaren*, 1988] many tens of thousands of km of profile have been obtained in the Arctic by US and British submarines, and our present knowledge of Arctic ice thickness distributions derives largely from the analysis and publication of data from these cruises. Problems include the necessity of removing the effect of beamwidth where a wide-beam sonar has been employed [*Wadhams*, 1981], and the fact that the data are obtained during military operations, which necessitates restrictions on the publication of exact track lines. For the same reason the dataset is not systematic in time or space

The second technique is the use of upward sonar mounted on moorings, so as to obtain a time series of g(h) at a fixed location. Early experiments were carried out in shallow water in the Beaufort Sea [*Hudson*, 1990; *Pilkington and Wright*, 1991], but the most recent studies have been in the Chukchi Sea [*Moritz*, 1991], Fram Strait and the southern Greenland Sea [*Vinje*, 1989]. More work of this kind has been taking place since 1991 and will continue until 1996, using lines of sonars which span the East Greenland

Current in Fram Strait and at 75°N (P. Lemke, T. Vinje, P. Wadhams). In conjunction with the use of AVHRR or ERS-1 SAR imagery to yield ice velocity vectors, this technique permits the time dependence of ice mass flux to be measured, and hence the fresh water input to the Greenland Sea. The observations involve mounting the sonars at 50 m depth (together with a current meter) in water up to 2500 m deep on a taut wire mooring.

The chief advantage of submarine sonar surveys as a way of generating ice thickness distributions is that upward sonar (mobile or moored) is still the only direct and accurate means of measuring the draft of sea ice, from which the thickness distribution can be inferred with very little error, while submarine-mounted (as opposed to moored) sonar allows basin-scale surveys to be carried out on a single cruise, giving the geographical variation in ice thickness characteristics. Submarine-mounted sonar also permits the shape of the ice bottom to be determined more accurately than moored sonar, including pressure ridges and the roughness of undeformed multi-year ice, allowing spectral and fractal studies to be undertaken and an understanding of the mechanics of the ridge-building process to be gained. By the use of additional sensors and concurrent airborne studies, a submarine can be used as a powerful vehicle for validating remote sensing techniques, including laser, passive and active microwave. The chief drawbacks of submarines are that, unlike moored sonar, a submarine profile cannot generate a systematic time series of ice thickness at a point in space. Submarines also cannot carry out surveys safely in very shallow water, so that many interesting aspects of ice deformation near shore cannot be studied; for instance, the dataset obtained by USS *Gurnard* north of Alaska in 1976 and analysed by *Wadhams and Horne* [1980] began at the 100 m isobath and so did not cover the whole of the Alaskan shear zone. Finally, it is unlikely that a military submarine would be available in the Antarctic, both because of remoteness and because the Antarctic Treaty requires that military vessels used in the Antarctic must be available for international inspection. The Antarctic is, however, a very suitable region for the use of sonar on an AUV.

Laser profiling of sea ice in the Arctic Ocean has been carried out extensively during the last two decades [e.g. *Ketchum*, 1971; *Weeks et al.*, 1971; *Wadhams*, 1976; *Tucker et al.*, 1979; *Krabill et al.*, 1990], while limited studies have also been carried out in the Antarctic [*Weeks et al*, 1989]. The aim has been to delineate the frequency and height distributions of pressure ridge sails and the spatial distribution of surface roughness. On only two occasions has it been possible to match a laser profile against a coincident profile of ice draft over substantial lengths of joint track. The first was a joint aircraft-submarine experiment [*Wadhams and Lowry*, 1977; *Lowry and Wadhams*, 1979; *Wadhams*, 1980, 1981] while the second was a similar experiment in May 1987 which involved a NASA P-3A aircraft equipped with an Airborne Oceanographic Lidar (AOL) and a British submarine equipped with narrow-beam upward-looking sonar [*Wadhams*, 1990a; *Comiso et al.*, 1991; *Wadhams et al.*, 1991]. Only the more recent experiment permitted a direct comparison of the PDFs of draft and freeboard to be made, since the AOL has a superior capability over earlier lasers in the removal of the sea level datum from the record. *Comiso et al.* [1991] found from this experiment that over a 60 km sample of track the overall PDFs of ice freeboard and draft could be brought to a close match across the entire range of data by a simple co-ordinate transform of the AOL data based on the ratio of mean densities of ice and water. Specifically, they showed that if R is the ratio of mean draft to mean freeboard, then matching of the freeboard PDF with the draft PDF is achieved by expanding the elevation scale of the freeboard PDF by a factor of R, and diminishing the magnitude of the PDF per m by the same factor. This is equivalent to saying that if a fraction F(h) of the ice cover has an elevation in the range h to (h + dh), then the same fraction F(h) will have a draft in the range R h to R(h + dh). R is related to mean material density (ice plus snow) ρ_m and near-surface water density ρ_w by

$$R = \rho_m / (\rho_w - \rho_m) \tag{10}$$

The success of this correlation prompted an analysis of the entire 300 km of coincident track [*Wadhams et al.*, 1992] divided into six 50 km sections, all from north of Greenland, within the zone 80.5 - 85°N, 2-35°W. The results of the analysis were as follows:-

(i) Despite variations in mean draft from 3.6 to 6 m, the six values of R all lay within a narrow range, of mean 8.04 ± 0.19. This corresponds to a mean material density of 910.7 ± 2.3 kg m^{-3}.

(ii) When each section was subjected to a co-ordinate transform based on its own value of R, the PDFs matched the sonar PDFs extremely well when plotted on a semilogarithmic scale (figure 4).

(iii) When plotted on a linear scale, the agreement was less good, in that mid-range depth probabilities are enhanced by the transformation, while very thin and very thick ice probabilities are reduced (figure 5). This is comprehensible on the basis of considering a uniform snow cover, which would give a low value of R for thin ice and a high value for thick ice. The use of a single average value for the transformation causes thin ice to be moved into thicker categories, and thick ice into thinner, thus making the converted distribution more narrow and peaked than the real ice draft distribution. In principle, a freeboard-dependent R could be used for the transform, but this would involve a sacrifice of simplicity .

The question arises of how R might vary with time of year and location. We developed a simple model for the seasonal variation of R, based on the best available data on seasonal snow thickness and the mean densities of ice, overlying snow, and near-surface water displaced by the ice

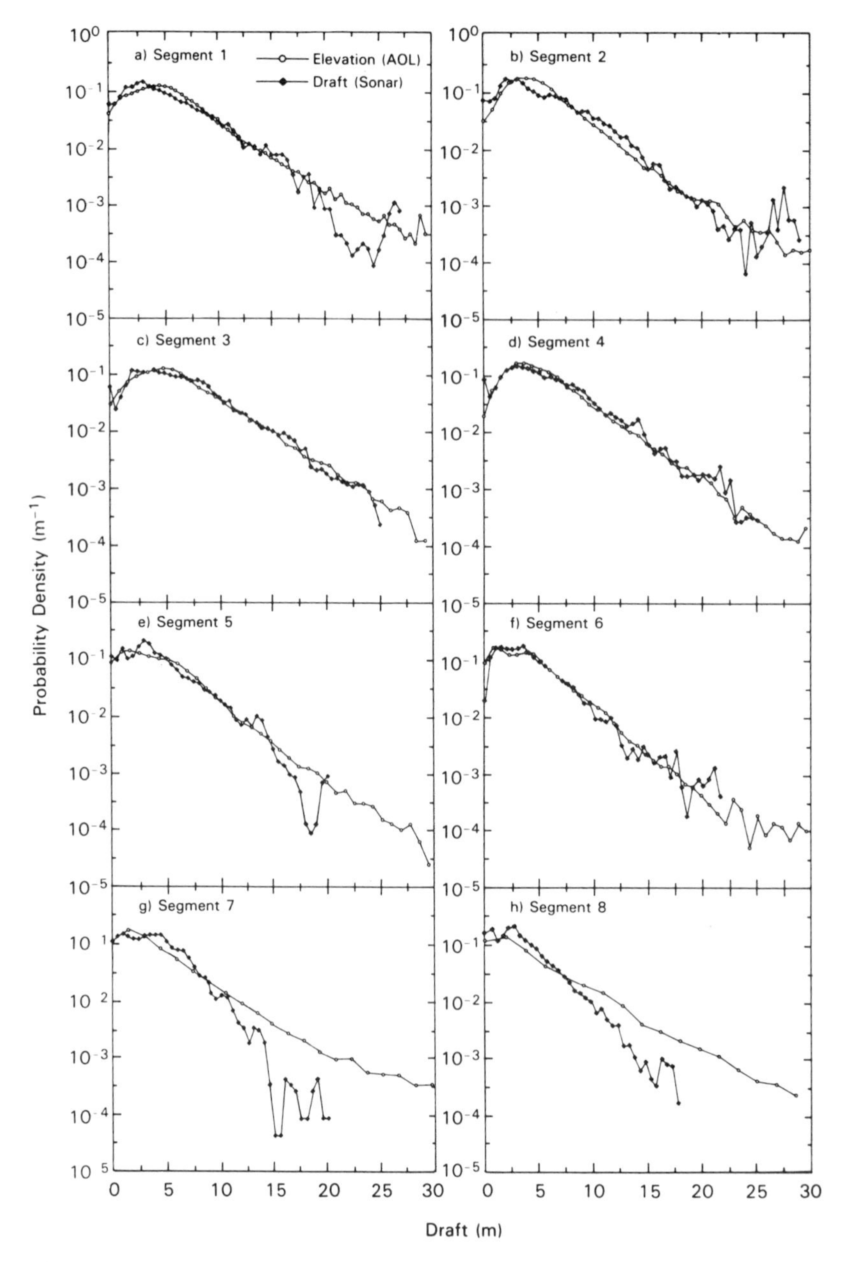

Fig. 4. Results of carrying out a co-ordinate transformation on eight 50 km sections of coincident laser and sonar track, using the mean draft-elevation ratio. Segments 7 and 8, with a poor fit, did not comprise coincident data [after *Wadhams et al.*, 1992].

cover. We found that knowledge of some key parameters is inadequate. Seasonal snow thickness appears to be overestimated in the source used for our model [*Maykut and Untersteiner*, 1971], with few more recent data available. There are few systematic measurements of ice density, especially of any fundamental difference between the densities of first- and multi-year ice. Near-surface water density is not known with the accuracy that is required for our model (since R depends on the difference between water and ice density), although we do know that it diminishes during summer because of dilution by meltwater. One set of results is shown in figure 6. We see that there is a large seasonal variation in R, mainly due to snow load. R is at its highest value on August 20, at the start of the snow season, with bare ice. R falls rapidly during the autumn snow falls of September and October, then diminishes only very slowly between November and the end of April, when little snow falls. A further onset of spring snow brings R

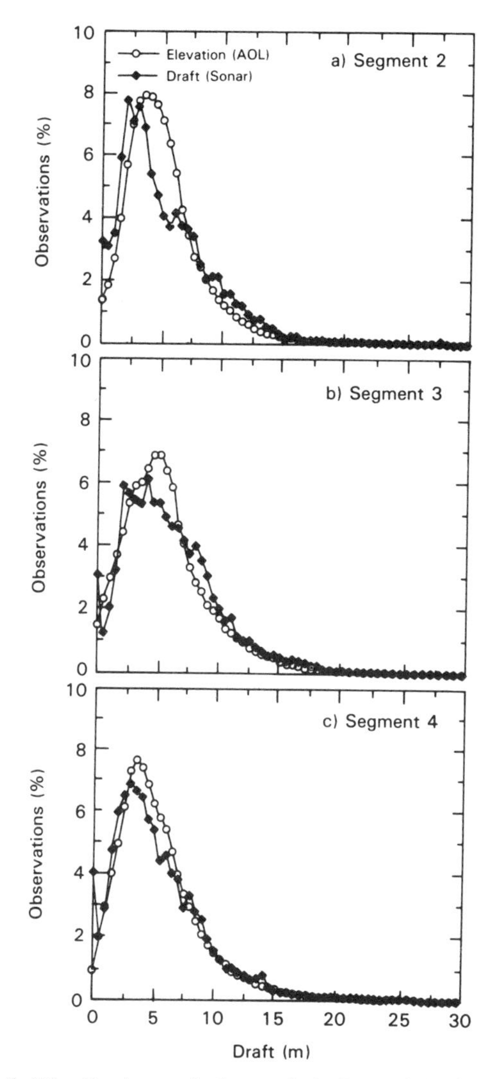

Fig. 5. Distributions of observed drafts and drafts predicted from laser elevations using a mean draft-elevation ratio of 7.89 [after *Wadhams et al.*, 1992].

to its lowest level at the beginning of June. As soon as surface snow melt begins R rises rapidly until by the end of June it has risen again almost to its August value. The final slow drift is due to surface water dilution. In fact there will be one or more higher peaks for R during the summer period, as meltwater pools form on the ice surface (increasing R), then drain (decreasing R), but no useful data exist on mean meltwater pool coverage, mean depth, or draining dates. In any case the results show that snow load is a critical parameter, and that the best time to carry out surveys is when dR/dt is at its lowest value, i.e. between November and the end of April.

Our conclusions from this study and the model were that it is indeed promising to consider the use of an airborne laser operating alone as a way of surveying both the mean ice thickness and the thickness distribution over the Arctic. The model suggests that a survey during early spring would be most useful, with annual repetition in order to examine interannual fluctuations and trends (possibly climate-induced) in the mean ice thickness or the form of the distribution. Certainly better data are required on the seasonal and spatial variability of snow load, ice density and surface water density, and it would be valuable to make at least one more attempt at a coincident airborne-submarine profile, covering as many different ice regimes as possible, for validation purposes. However, even in the absence of improved background data, the present results suggest that in regions with mean ice thickness in the range 4-6 m, the

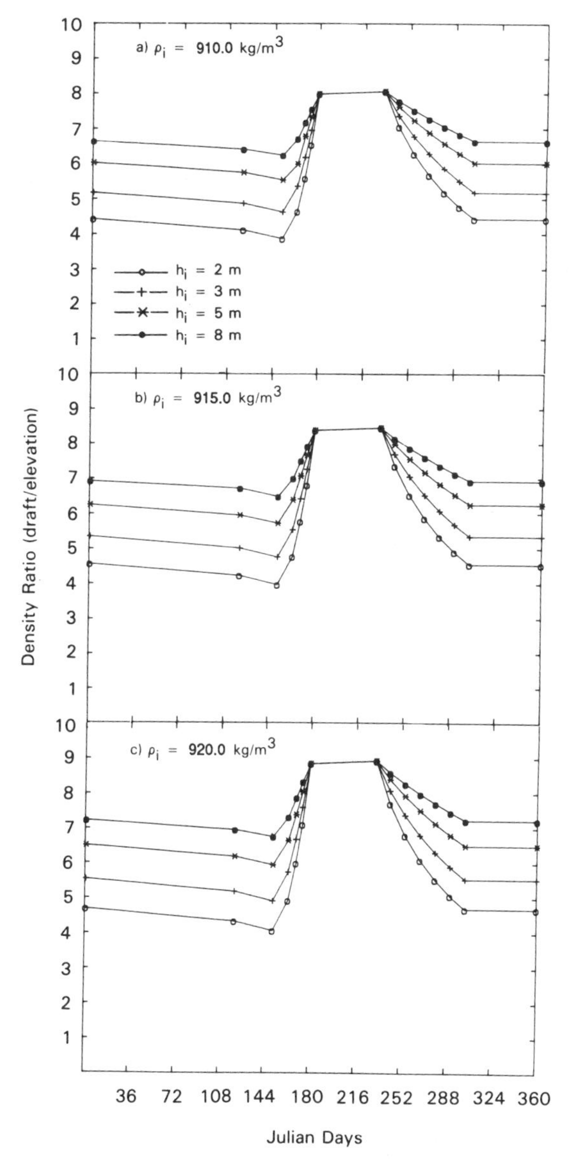

Fig. 6. Results of a model of the variation of draft/freeboard ratio R with season [after *Wadhams et al.*, 1992].

ratio R will lie in the vicinity of 8.0 in spring, and can be estimated to an accuracy of ± 2.4% in 300 km of track. This yields an accuracy of about ± 12 cm in mean thickness over 300 km of track (±30 cm in 50 km), neglecting other sources of error. This is an acceptable accuracy for detecting and mapping variability. In the Antarctic much more validation would be needed before a laser could be employed in this way, since snow has a much stronger influence on mean density than in the Arctic, and there is a major difference between snow load on first-year and multi-year ice (since the snow cover does not all melt during summer). However, if such validation were done, this technique could be a highly effective mapping tool for the vast expanse of sea ice cover in winter.

Moving now to electromagnetic techniques, the first method to be applied to sea ice was impulse radar [*Kovacs and Morey*, 1986], in which a nanosecond pulse (center frequency about 100 MHz) is applied to the ice via a paraboloidal antenna on a low- and slow-flying helicopter. The technique was found to have many limitations besides the slowness of data gathering. Superimposition of surface and bottom echoes means that it does not function well over ice less than 1 m thick, while absorption and scattering by brine cells means that better results are obtained over multi-year than first-year ice, with in any case a fading of the return signal at depths beyond 10-12 m. Thus the full profile of deeper pressure ridges is not obtained. Such instruments can therefore be regarded as of restricted application to local surveys in ice regimes of a favourable kind.

A more recent development is the use of electromagnetic induction. The technique was devised by Aerodat Ltd. of Toronto, and has subsequently been developed by CRREL [*Kovacs and Holladay*, 1989, 1990] and by Canpolar, Toronto [*Holladay et al.*, 1990]. The technique involves towing a "bird" behind a helicopter flying at normal speeds. The bird contains a coil which emits an electromagnetic field in the frequency range 900 Hz - 33kHz, inducing eddy currents in the water under the ice, which in turn generate secondary EM fields. The secondary fields are detected by a receiver in the bird; their strength depends on the depth of the ice-water interface below the bird. The bird also carries a laser profilometer to measure the depth of the ice-air interface below the bird, and the difference gives the absolute thickness of the ice. The method appears promising but requires further validation. Figure 7 illustrates the principle and shows some results obtained in the Labrador Sea during the LIMEX experiment [*Holladay et al.*, 1990].

Drilling as a technique was considered statistically by *Rothrock* [1986], who estimated that 62 independent random holes would give a mean thickness with a standard deviation of 30 cm, while 560 holes would give 10 cm. Thus the technique is not bad as a means of estimating mean ice thickness, but is very poor as a way of giving the shape of g(h).

(a)

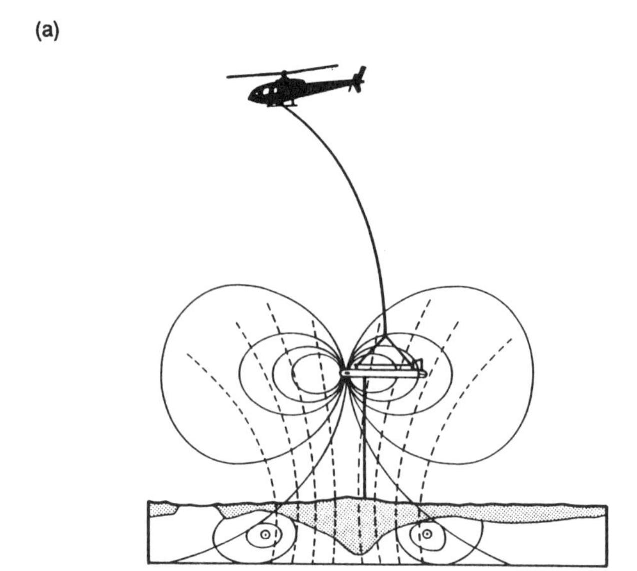

(b)

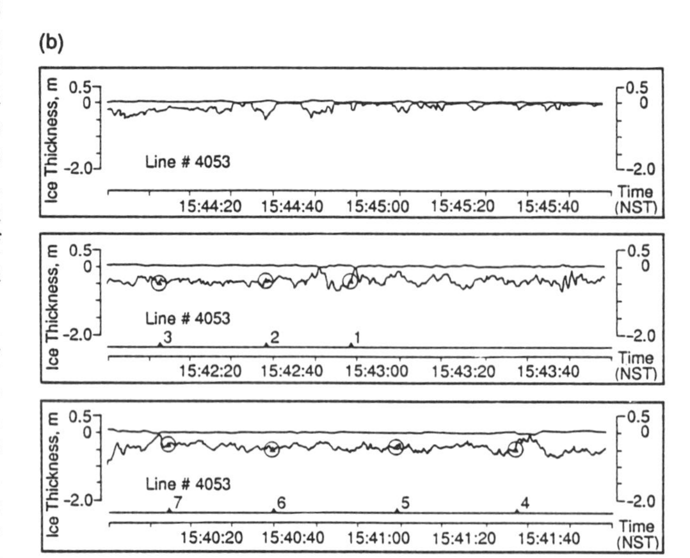

Fig. 7. (*a*) Principle of electromagnetic induction sounding of ice. (*b*) Results from electromagnetic induction sounding over first-year ice in the Labrador Sea [after *Holladay et al.*, 1990].

Possible future techniques

There are several new techniques which show some promise for measuring g(h), or parts of it, under certain conditions. We may mention the following:-

1. Sonar mounted on AUVs or neutrally buoyant floats
2. Acoustic tomography or thermometry
3. Inference from a combination of microwave sensors.

The purpose of mounting upward sonar systems on mobile platforms other than military submarines is to obtain systematic datasets along a repeatable grid of survey tracks, enabling interseasonal and interannual comparisons to be carried out for identical geographical locations. Possible platforms with synoptic potential include autonomous underwater vehicles (AUVs), of which several are under development including long-range systems with basinwide capability; long-range civilian manned submersibles, of which the Canadian-French "Saga I" is the only currently available vehicle [*Grandvaux and Drogou*, 1989]; and neutrally buoyant floats, in order to accumulate and store a PDF which can then be transmitted acoustically to an Argos readout station on a floe (J-C Gascard, personal commun.).

Acoustic tomography is a technique for monitoring the structure of the ocean within an area of order 10^6 km^2 by measuring acoustic travel times between the elements of a transducer array enclosing that area [*Munk and Wunsch*, 1979]. *Guoliang and Wadhams* [1989] showed in a theoretical study that the presence of a sea ice cover should measurably decrease travel times by an amount which is dependent on the modal ice thickness. Data obtained during the 1988-9 Scripps-WHOI tomography experiment in the Greenland Sea are being examined at present [*Jin et al.*, 1993], and this technique may have a more general application within the Arctic Ocean itself, although the additional scattering due to under-ice roughness has been found to have an important limiting effect on resolution. Proposals exist for an Arctic ATOC experiment (acoustic thermometry of ocean climate) using much lower frequencies (57 Hz) to achieve greater transmission lengths [*Johannessen et al.*, 1993], and this may allow the technique to be used to estimate a modal ice draft for the Arctic Basin.

Possible techniques involving microwave sensors have been reviewed by *Wadhams and Comiso* [1992]. In principle, after sufficient validation has been done, we might hope to find empirical relationships between the distribution of passive microwave brightness temperatures, of SAR backscatter levels, and of aspects of g(h) such as the mean thickness. To date, however, the only quantitative validation of a microwave sensor against ice thickness has been a 1987 joint survey between a submarine, an aircraft equipped with the STAR-2 X-band synthetic aperture radar system, and a second aircraft equipped with a laser profilometer and passive microwave radiometers. The SAR system operates at HH polarisation and 9.6 GHz, with a swath width of 63 km and resolution of 16.8 m. This provided an opportunity to examine correlations between ice draft as measured by the sonar, and backscatter level along the same track measured by the SAR. The question which can be addressed is to what extent can SAR brightness variability be used to infer the shape of the ice thickness distribution. The results are discussed in *Comiso et al.* [1991] and *Wadhams et al.* [1991], and are summarised here.

Firstly, a qualitative examination of the profile of SAR backscatter along the tracks of the submarine and aircraft suggested that there was a clear positive correlation with both the draft and elevation profiles. This is to be expected since pressure ridges in particular give strong returns on account of their geometry [*Onstott et al*, 1987; *Livingstone*, 1989; *Burns et al.*, 1987]. Next we examined 125 km of matched SAR and laser data. We found that there was poor correlation between SAR backscatter and laser elevation as averaged within a single SAR pixel (16.8 m square). Since this could well be due to mismatch between the two aircraft tracks, we considered longer averaging windows for both SAR backscatter and laser elevation (figure 8) obtaining the best correlation at window lengths of 1-2 km. The correlation of means exceeded the correlation of standard deviations. At a window length of 2 km the correlation coefficient between elevation and SAR backscatter reached 0.53. It should be noted that the SAR backscatter is not calibrated absolutely: calibration tests showed that surfaces differing in reflectivity by 1 dB yield backscatter values differing by 12.24 units (on a recorded scale of 0 to 255), but the offset is not accurately known.

Figure 9 shows the mean elevation and mean backscatter using a 1 km averaging window. Clearly there were regions of both good and bad correlation, the best correlation often appearing to occur in areas of low elevation. This is probably because these correspond to open water, young ice and first-year ice, which on X-band SAR offer a high contrast with multi-year and ridged ice. The correlation at this window length is 0.51, so only 26% of the variance of the backscatter can be explained by elevation differences. An imperfect correlation is to be expected, since the SAR is responding to features other than the elevation of the surface, such as brine and air content of the near-surface ice as well as the small-scale surface roughness. Even ridged areas do not offer a perfect correlation. Evidence indicates

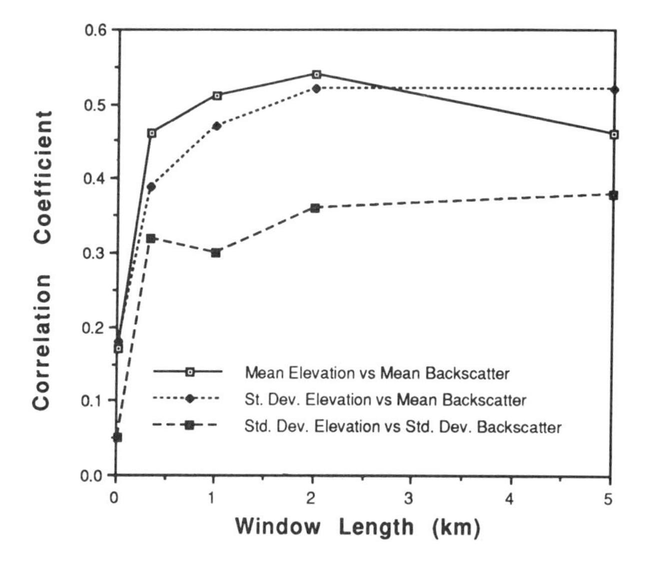

Fig. 8. Cross-correlation coefficients for mean ice elevation (from laser) versus mean SAR backscatter using different window lengths [after *Comiso et al.*, 1991].

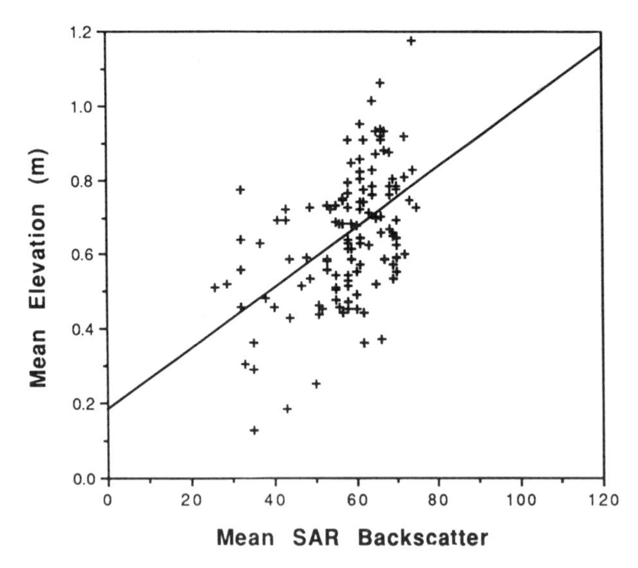

Fig. 9. Mean elevation versus mean SAR backscatter with best fit regression line for 1-km averaging window [after *Comiso et al.*,1991].

that SAR backscatter for ridges is dependent on look-angle and angle of incidence [*Leppäranta and Thompson*, 1989], while other results [*Holt et al.*, 1990] show that lower-frequency SAR (0.44 GHz) may well be superior in its resolution of ridges to high-frequency SAR such as X-band.

Next we examined the correlation between SAR backscatter and sonar ice draft. This was carried out over a more restricted 22 km section of track [*Wadhams et al.*, 1991] where we were sure of excellent matching between tracks. Once again it was found that the correlation coefficient depends on the window length, but in fact the highest correlation was obtained with a lower window length. At an averaging length of 252 m (15 SAR pixels) the correlation between draft and backscatter reached 0.68, which is better than the best correlation with elevation and which implies that 46% of the backscatter variance can be explained by draft differences. Figure 10 shows the scatter plot of draft versus SAR backscatter.

It is unexpected that SAR backscatter correlates better with ice draft than with elevation, since the SAR is responding mainly to surface features and to the upper part of the ice sheet. However, it is useful that this is the case, since the draft distribution is close to the thickness distribution g(h) (it is conventional to multiply drafts by a density factor of 1.112 to obtain thicknesses, without otherwise altering the shape of the PDF). However, it is clear that SAR alone cannot be used to infer the complete shape of the PDF. Firstly, only 46% of the variance of the SAR can be explained by draft variations, so it cannot be used alone as a predictor. Secondly, this correlation is developed over averaging lengths of 252 m, indicating that to some extent we are relating a mean ice draft to a mean SAR backscatter, rather than obtaining an algorithm which can generate a fine-resolution PDF as in the case of the laser technique. Thirdly, the experiment described above was specific to X-band SAR, whereas the types of satellite SAR which are producing routine synoptic data from the polar oceans are of different frequencies (e.g. C-band, 5.3 GHz, on ERS-1). A new and more thorough set of SAR-underside validation experiments will be needed for the SAR frequencies employed on ERS-1, JERS-1 and Radarsat. Nevertheless the results show some promise for the use of SAR, in conjunction with other sensors, as one element in an empirical scheme for determining ice thickness by microwave remote sensing.

4. SUMMARY OF OBSERVATIONS OF VARIABILITY

Arctic

Knowledge of the regional variability of g(h) in the Arctic comes almost entirely from upward sonar profiling by submarines. The ice in Baffin Bay is largely thin first-year ice with a modal thickness of 0.5-1.5 m [*Wadhams et al.*, 1985]. In the southern Greenland Sea, too, the ice, although composed largely of partly melted multi-year ice, also has a modal thickness of about 1 m [*Vinje*, 1989; *Wadhams*, 1992], with the decline in mean thickness from Fram Strait giving a measure of the fresh water input to the Greenland Sea at different latitudes. Over the Arctic Basin itself there is a gradation in mean ice thickness from the Soviet Arctic, across the Pole and towards the coasts of north Greenland and the Canadian Arctic Archipelago, where the highest mean thicknesses of some 7-8 m are observed [*LeSchack*, 1980; *Wadhams*, 1981, 1992; *Bourke and McLaren*, 1992]. These overall variations are in accord with the predictions of numerical models [*Hibler* 1979, 1980] which take account of ice dynamics and deformation as well as ice thermodynamics.

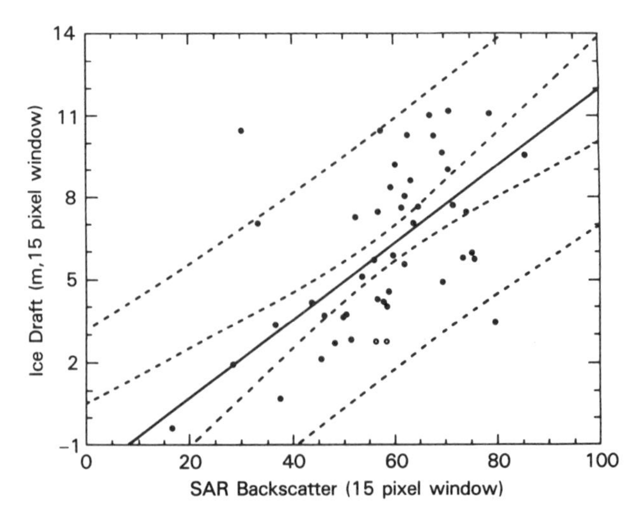

Fig. 10. Scatter plot of SAR backscatter versus draft from sonar with 95% and 99% confidence limits [after *Comiso et al.*, 1991].

The temporal variability of g(h) has been much less extensively measured. On only a few occasions have submarine sonar tracks been obtained over similar regions in different years or different times of year. Results show that in the Eurasian Basin far from land, in the region between the Pole and Svalbard, the mean ice draft is remarkably stable between different seasons and different years [*Wadhams*, 1989, 1990b; *McLaren*, 1989], although *McLaren et al.* [1992] have found considerable interannual variability at the Pole itself. In the region between the Pole and Greenland, where a build-up of pressure ridging normally occurs due to the motion of the ice cover towards a downstream land boundary, very considerable differences in P(h) and in mean draft (more than 15%) were observed between October 1976 and May 1987 [*Wadhams*, 1990a]. In the Beaufort Sea in summer, also, considerable differences were observed between records taken several years apart [*McLaren*, 1989]. *Bourke and Garrett* [1987] used all available data to attempt to construct seasonal contour maps of mean ice draft. Figure 11 shows the interannual variability observed by Wadhams, while figure 12 shows Bourke and Garrett's maps for summer and winter. It should be noted that the data used for the Bourke and Garrett map did not include open water, so these maps are over-estimates of the mean ice draft. The maps were updated by *Bourke and McLaren* [1992], who also gave contour maps of standard deviation of draft, and mean pressure ridge frequencies and drafts for summer and winter, based on 12 submarine cruises.

In order to assess reliably whether ice thickness changes are occurring in the Arctic it is necessary to obtain area-averaged observations of mean ice thickness over the same region using the same equipment at different seasons or in different years. Ideally the region should be as large as possible, to allow us to assess whether changes are basin-wide or simply regional. Also the measurements should be repeated annually in order to distinguish between a fluctuation and a trend. Because of the unsystematic nature of Arctic submarine deployments this goal has not yet been achieved, but a number of comparisons between pairs of datasets have been carried out.

McLaren [1989] compared data from two US Navy submarine transects of the Arctic Ocean in August 1958 and August 1970, stretching from Bering Strait to the North Pole and down to Fram Strait. He found similar conditions prevailing in each year in the Eurasian Basin and North Pole area, but significantly milder conditions in the Canada Basin in 1970. The difference is possibly due to anomalous cyclonic activity as observed in the region in recent summers [*Serreze et al.*, 1989]. Another possibility is that since August is the month of greatest ice retreat in the Beaufort Sea, the difference is simply due to differences between the extent to which the ice retreated in the Chukchi and southern Beaufort Seas during the respective summers.

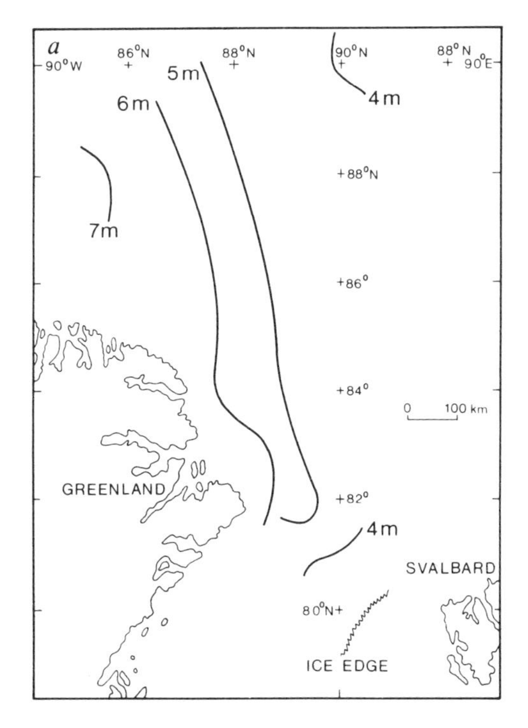

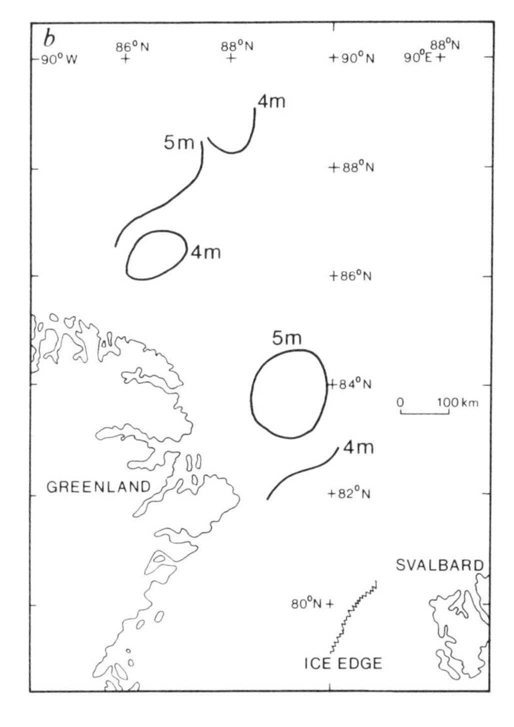

Fig. 11. Contour maps of mean ice drafts from Eurasian Basin, October 1976 and May 1987 [after *Wadhams*, 1990a].

(a)

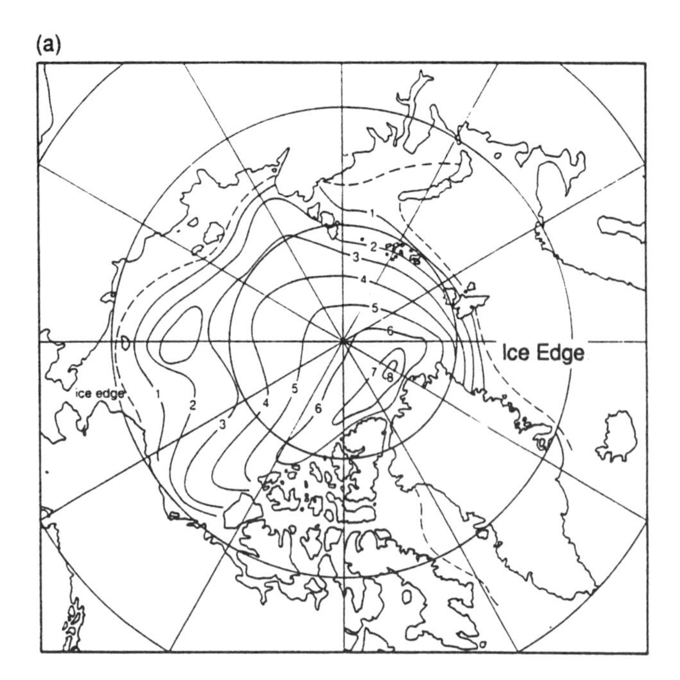

(b)

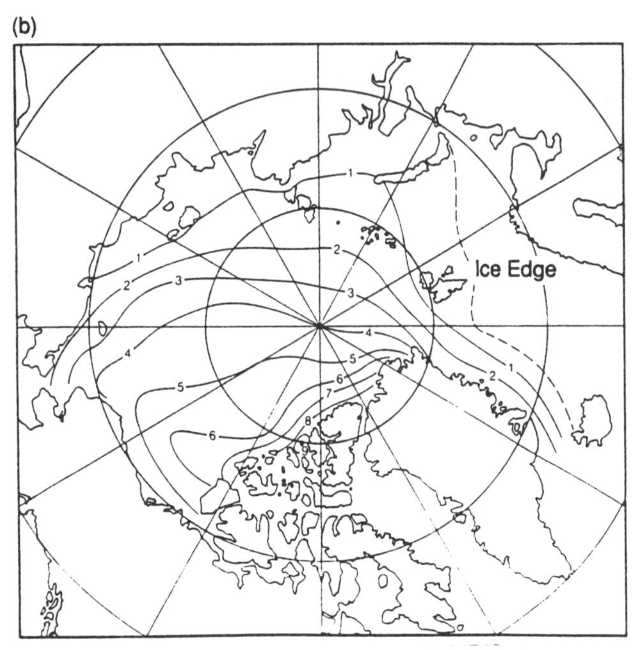

Fig. 12. Contour maps of estimated mean ice drafts for (a) summer and (b) winter in the Arctic Basin [after *Bourke and Garrett*, 1987].

The extent and duration of the open water season in the Beaufort Sea are known to have a high interannual variation, and an unusually open southern Beaufort Sea would lead to more open conditions within the pack itself.

Wadhams [1989] compared mean ice thicknesses for a region of the Eurasian Basin lying north of Fram Strait, from British Navy cruises carried out in October 1976, April-May 1979 and June-July 1985. All three datasets were recorded using similar sonar equipment. It was found that a box extending from 83°30'N to 84°30'N and from 0° to 10°E had an especially high track density from the three cruises (400 km in 1976, 400 km in 1979 and 1800 km in 1985), and this was selected for the comparison. It is a region far from any downstream boundary, and represents typical conditions in the Trans Polar Drift Stream prior to the acceleration and narrowing of the ice stream which occurs as it prepares to enter Fram Strait. The mean thicknesses from the three cruises were remarkably similar: 4.60 m in 1976; 4.75 m in 1979; and 4.85 m in 1985. It should be remembered that these datasets were recorded in different seasons as well as different years.

More recently Wadhams was able to compare data from a triangular region extending from north of Greenland to the North Pole, recorded in October 1976 and May 1987 [*Wadhams*, 1990a]. Mean drafts were computed over 50 km sections, and each value was positioned at the centroid of the section concerned; the results were contoured to give the maps shown in figure 11. There was a decrease of 15% in mean draft averaged over the whole area (300,000 km^2), from 5.34 m in 1976 to 4.55 m in 1987. Profiles along individual matching track lines (figure 13) show that the decrease was concentrated in the region south of 88°N and between 30° and 50°W. From fig. 11 it appears that the build-up of pressure ridging which gave the high mean drafts near the Greenland coast in 1976 was simply absent in 1987, but in fact the situation is not that simple. Table 1 shows a comparison between the probability density functions of ice thickness from the pairs of profiles shown in fig. 13 (strictly, two 300 km sections from the southern part of the N-S transect and two 200 km sections from 40-50°W in the E-W transect). In 1987 there was more ice present in the form of young ice in refrozen leads (coherent stretches of ice with draft less than 1 m) and as first-year ice (draft less than 2 m). There was less multi-year ice (interpreted as ice 2-5 m thick) and less ridging (ice more than 5 m thick) in 1987 in the E-W transect, although slightly more ridged ice in the N-S transect. The main contribution to the loss of volume appears to be, then, the replacement of multi-year and ridged ice by young and first-year ice.

To determine how this may have occurred the tracks of drifting buoys from the Arctic Ocean Buoy Program [*Colony et al.*, 1991; R.L. Colony, pers. commun.] were examined. Four buoys were in the region during the months prior to the 1987 cruise (fig. 14). The three in the westernmost positions, corresponding to a portion of the Beaufort Gyre, remained almost stationary during the period January - May 1987, while buoy 1897, in the Trans Polar Drift Stream, moved towards Fram Strait at an average speed of 2 km per day. The result of this anomalous halting of the motion of part of the Beaufort Gyre would be a divergence within the experimental region, as the Trans Polar Drift Stream ice continued its motion towards the SE.

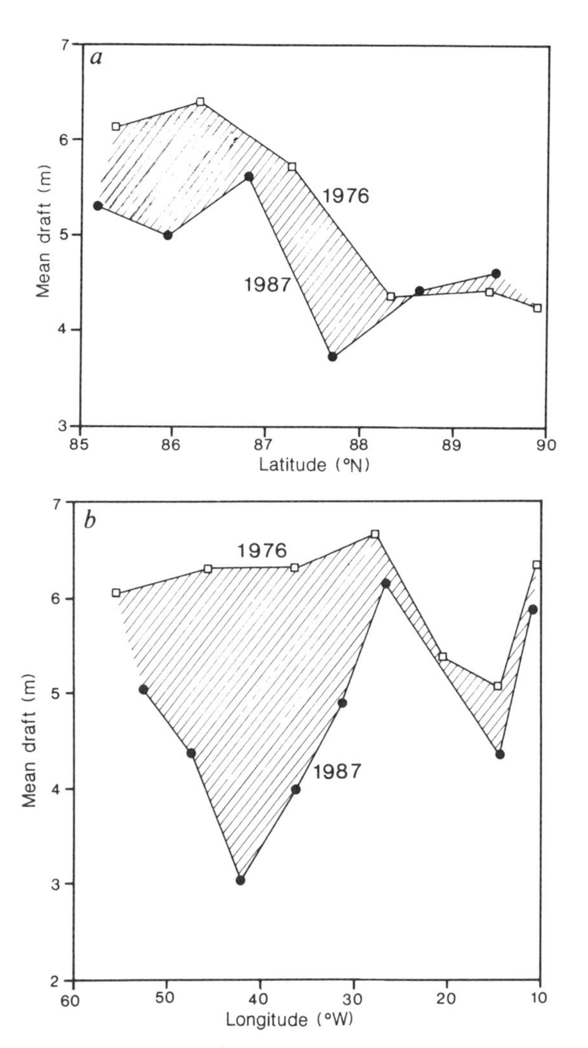

Fig. 13. (*a*) Comparison of mean ice drafts from 1976 and 1987 along a N-S transect from North Pole to 85°N. (*b*) Comparison of mean ice drafts for transect across north of Greenland from 60°W to 10°W [after *Wadhams*, 1990a].

This would lead to the opening up of the pack, creating areas of young and first-year ice. Thus the indications are that it is an ice motion anomaly rather than, or as well as, an ice growth anomaly that is responsible for the observed decrease in mean ice draft. This supposition is supported by modelling results from J. Walsh (personal commun.), who ran the Hibler ice model using daily wind forcing and monthly thermodynamic forcing, both varying interannually. He found that the simulated thicknesses showed negative anomalies in this region in May 1987 and positive anomalies in October 1976.

Further observations are clearly necessary from other years (past, recent and future), and a closer study should be made of pressure fields over the Arctic and the variations that they may be capable of creating in the geographical distribution of mean ice drafts. One of the surprising aspects of the results described above is that the build-up of mean ice draft towards Greenland was thought to be a very stable aspect of the ice climatology of the Arctic, appearing consistently in the model predictions of *Hibler* [1980] and in the tentative seasonal climatology of *Bourke and Garrett* [1987]. It appears that the ice cover, like the ocean, possesses a weather as well as a climate, and it is vital that sufficient data be examined to resolve details of this weather such that the underlying trend in basin-wide mean ice thickness, if any, can be revealed. In this respect the author is now examining more recent datasets.

A final regional comparison which the 1987 dataset has made possible occurs in the region immediately north of Fram Strait, between 82°N and 80°N (fig. 15). Here it was possible to compare data from the four years 1976, 1979, 1985 and 1987. This is a region which is ice-covered in most years, and where mixing occurs between the various ice streams preparing to enter Fram Strait, notably the streams of old, deformed ice moving S from the North Pole region and SE from the region north of Greenland; and the stream of younger, less heavily deformed ice moving SW from the seas north of the USSR. Figure 15 shows all available mean drafts from 50 km sections [from *Wadhams*, 1981, 1983, 1989 and current analyses]. There is very good consistency among these four datasets, regardless of year or season of generation; fluctuations appear to be random in character, and where centroids from different experiments lie close to one another, the mean drafts are usually similar. Only the 1976 data points appear somewhat thicker than their neighbours.

The statistical validity of the 15% decrease in mean ice draft in the Eurasian Basin observed by *Wadhams* [1990a] between 1976 and 1987 has been questioned by *McLaren et al.* [1992], but using reasoning which this author considers invalid. *McLaren et al.* analysed 50 km and 100 km sections of ice profile centred on the North Pole from 6 cruises from 1977 to 1990. They found that the mean ice draft from 50 km sections in the late 1970s (1977, 1979) was 4.1 m (4.2, 4.0 m respectively), while the mean draft for the late 1980s was 3.45 m (2.8, 4.1, 3.3, 3.6 m for

TABLE 1. Comparisons of ice statistics from 1976 and 1987 datasets

	N-S transect 1976	N-S transect 1987	E-W transect 1976	E-W transect 1987
Mean draft (m)	6.09	5.31	6.32	4.07
Ice< 2 m draft (%)	11.6	16.7	7.9	29.9
Ice2-5m draft (%)	48.7	38.6	46.5	39.6
Ice>5m draft (%)	39.7	44.7	46.0	30.5
Refrozen leads (%)	4.0	7.9	3.7	15.6

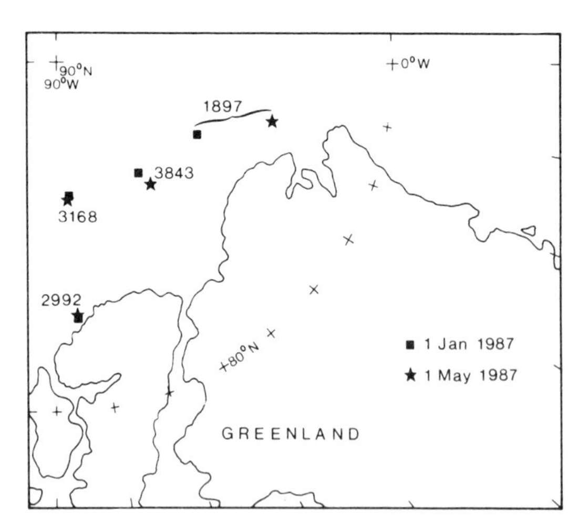

Fig. 14. Positions of drifting buoys north of Greenland on 1 January and 1 May 1987 [after *Wadhams,* 1990a].

1986, 1987, 1988, 1990). The difference of 0.65 m is 15%. Using a t-test they showed that this difference of means is significant only at the 20% level, i.e. non-significant. They then went on to claim that this implies that the 15% decline in mean drafts found by Wadhams between 1976 and 1987 is similarly non-significant. The claim is false because the McLaren *et al.* comparison is between two datasets of total length 100 km and 200 km, while the Wadhams comparison is between two datasets of length 3900 km (1976) and 2200 km (1987), giving a much greater statistical stability to the mean values.

An alternative way of interpreting McLaren *et al's* results in fact suggests that they are more significant than their authors suppose. When mean ice draft is computed from a dataset of finite length it is subject to a statistical variability which is independent of variations in ice climatology and which is merely due to the fact that the profile is sampling only a finite number of ice floes, leads and pressure ridges. This variability may be termed the sampling error. There is no *a priori* way of estimating what the sampling error should be. The only valid method is an experimental one: to examine the statistical stability of sections taken from long ice profiles obtained within a homogeneous ice regime at a single instant. *Wadhams et al.* [1992] discussed two such experimental datasets and found them very compatible. The first comprised 23 50-km sections obtained from a restricted area of the Beaufort Sea [*Wadhams and Horne*, 1980] while the second comprised 18 100-km sections from a homogeneous part of the central Arctic Basin [*Wadhams*, 1981]. If we assume that the ice regime was homogeneous, then the standard error in the means of the 50- or 100-km sections is the sampling error which we seek. The results from the two experimental datasets were that the 50-km sections gave 3.67 ± 0.19 m as the overall mean draft (a 5.2% sampling error) while the 100-km sections gave 4.51 ± 0.18 m (a 3.9% sampling error). The difference in mean is due to the different ice regimes, but the sampling errors are almost identical, given that the sampling length is double in the second case (if they were completely identical, a 5.2% error in 50 km would become a 3.7% error in 100 km). Assuming that a 3.9% sampling error per 100 km is indeed characteristic of all Arctic sea ice, and applying it to the McLaren *et al.* data listed above, we see that on the hypothesis that all 1970s data and all 1980s data were drawn from only two populations the means and sampling errors should be 4.1 ±

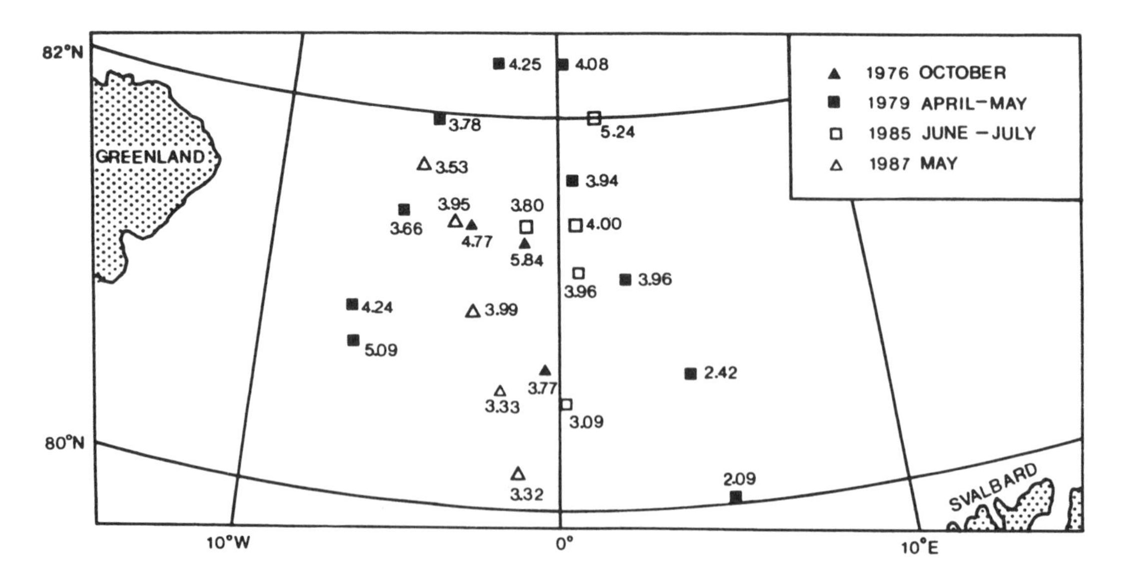

Fig. 15. Comparisons of mean ice drafts measured in the region north of Fram Strait [after *Wadhams,* 1992].

0.16 m (1970s) and 3.45 ± 0.10 m (1980s). The appropriate test for the significance of the difference between the means is to compute the d-statistic of the Fisher-Behrens distribution [*Campbell*, 1974], which gives d=3.444, significant at better than the 1% level. The data presented therefore do show evidence of a difference between 1970s and 1980s data; there is also evidence of significant interannual variability for some years, notably 1986. Thus these observations are more significant than supposed by the authors in relation to interdecadal ice thickness changes; also they do not cast valid doubts on the significance of the *Wadhams* [1990a] findings.

From the data comparisons made so far in the Arctic, we can thus draw the following tentative conclusions:-

1. Ice reaching Fram Strait via the Trans Polar Drift Stream along routes where it is not heavily influenced by a downstream land boundary shows great consistency in its mean thickness from season to season and from year to year, at latitudes from 84°30'N to 80°N and longitudes in the vicinity of 0°.
2. Ice upstream of the land boundary of Greenland can show great changes in its mean ice draft, notably a significant decline between 1976 and 1987, but it is possible to trace these to anomalies in the balance between pressure ridge formation through convergence (the normal source of a high ice draft in the region) and open water formation through divergence (the anomalous situation). A deeper knowledge of wind field anomalies is needed to understand these changes fully, together with more adequate datasets.
3. Data obtained from the North Pole region in 1977-90 show evidence of a decline in mean ice thickness in the late 1980s relative to the late 1970s.
4. Ice in the Canada Basin in summer also appears to show great variability in mean ice draft, but here there is a free boundary with ice-free marginal seas, permitting relaxation of the ice cover into a less concentrated state under certain wind conditions.

We cannot yet conclude that there is conclusive evidence of systematic thermodynamic thinning of the sea ice cover, as would be caused by the impact of the greenhouse effect, but the data are suggestive enough to make further profiling on an annual basis a very desirable aim.

Antarctic

In the Antarctic our knowledge of g(h) is much less extensive. Systematic data have been obtained only by repetitive drilling. Figure 16 shows the ice thickness distribution obtained from drilling 4400 holes in a region of first-year ice (the eastern Weddell-Enderby Basin) in mid-winter of 1986 [*Wadhams et al.*, 1987], showing that the modal ice thickness is only 50-60 cm, that maximum observed keel drafts are only about 6 m, and that the snow cover (measured independently) is sufficient in many cases to push the ice surface below water level and cause flooding and the formation of a new ice type, snow-ice [*Lange et al.*, 1989]. These surprising results have important implications

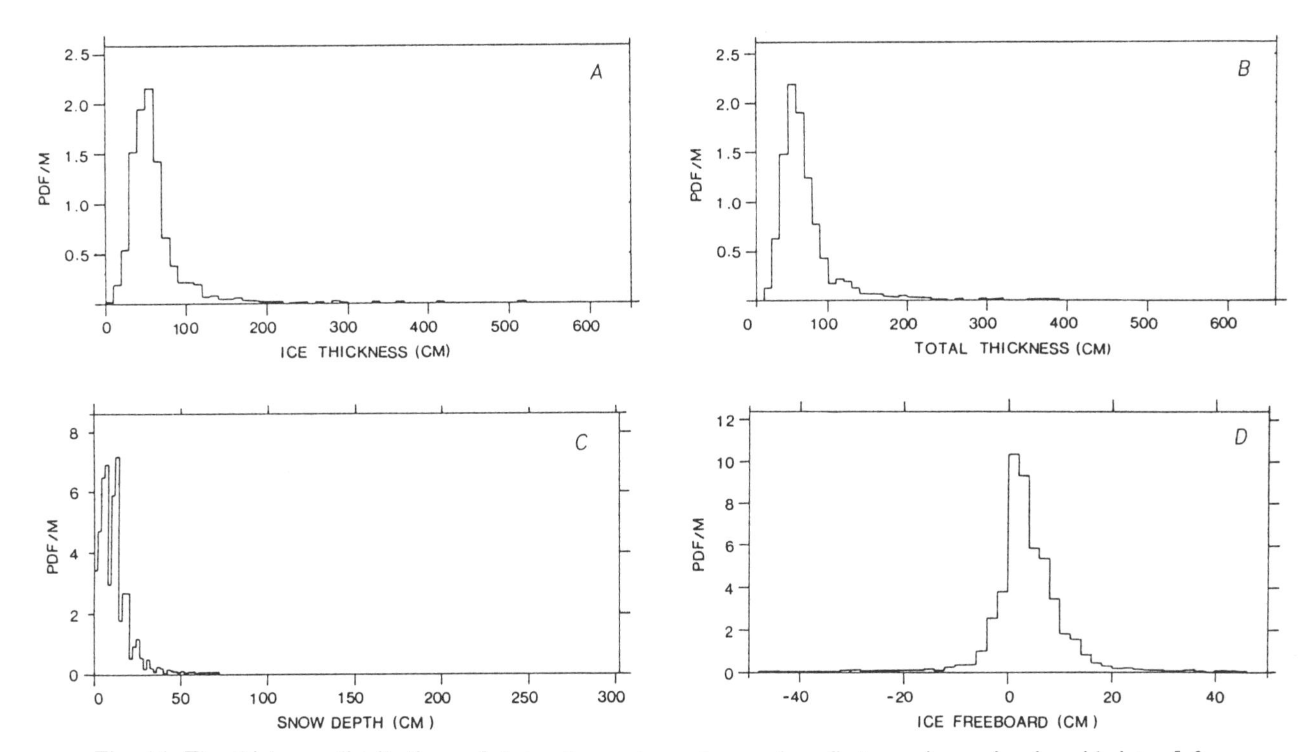

Fig. 16. The thickness distributions of Antarctic sea ice and snow in a first-year ice region in midwinter [after *Wadhams et al.*, 1987].

for energy exchanges in the Antarctic sea ice zone, and for the likely stability of the ice cover under climatic warming. In the limited regions of the Antarctic where multi-year ice occurs, there is a preferred ice thickness of about 1.4 m (western Weddell Sea, winter) for this ice type, but with no apparent increase in maximum ridge drafts [*Wadhams and Crane*, 1991; *Lange and Eicken*, 1991]. Given the paucity of data, it is in the Antarctic that the development of new techniques for synoptically monitoring g(h) would yield the greatest impact.

The only profiling data of any kind from the Antarctic are airborne laser profiles of the upper surface [*Weeks et al.*, 1989] and some limited thickness profiling by impulse radar from a helicopter [*Wadhams et al.*, 1987]. Laser data are valuable in delineating the occurrence of pressure ridging and in assessing its likely contribution to the overall thickness distribution, but cannot be used reliably to infer ice thickness directly by isostasy, largely because of the effect of the thick and variable snow cover. The usefulness of impulse radar in the Antarctic is limited by the fact that the first-year ice is thin and has a high salinity, and by the unconsolidated nature of ridges, which present misleading water horizons to the electromagnetic radiation; thus, such radar has only been used as an auxiliary to drilling efforts.

The winter pack ice in the Antarctic is of real global importance. The annual variation in sea ice area in the Antarctic is from 4 x 10^6 km^2 in February to 20 x 10^6 km^2 in September [*Zwally et al.*, 1983; *Gloersen et al.*, 1992], and so it is of the greatest importance to know the thickness of the ice which covers such a vast area of ocean for part of the year. The first penetration of the pack during Antarctic spring was the cruise of the *Mikhail Somov* in October-November 1981, which reached 62.5°S in the eastern Weddell Sea, but it was not until 1986 that the first deep penetration into the circumpolar Antarctic pack during early winter, the time of ice edge advance, took place. This was the Winter Weddell Sea Project (WWSP) cruise of FS *Polarstern*, during which systematic ice thickness measurements were made throughout the eastern part of the Weddell-Enderby Basin, from the ice edge to the coast, covering Maud Rise and representing a typical cross-section of the first year circumpolar Antarctic pack during the season of advance [*Wadhams et al.*, 1987; *Lange et al.*, 1989]. After a spring cruise in 1988 [*Lange and Eicken*, 1991] a second winter cruise was carried out in 1989: the Winter Weddell Gyre Study (WWGS) of Alfred Wegener Institute involved a crossing of the Weddell Sea from the tip of the Antarctic Peninsula to Kap Norvegia in the east during September-October, and thus allowed the multi-year ice regime of the western Weddell Sea to be studied in midwinter [*Wadhams and Crane*, 1991]. During the later part of this cruise, and in collaborative work carried out by *Akademik Fedorov*, ice conditions were studied in the same region of the eastern Weddell Sea as was covered by *Polarstern* in 1986, permitting the first interannual comparison of winter ice thicknesses in the same region of the seasonal pack to be carried out.

Ice thickness measurements in the cruises were made by drilling holes at 1 m intervals along lines of about 100 holes. Typically, during a daily ice station, two such lines would be laid out at right angles to one another, with an attempt at a random choice of location so that ridges were properly represented in the results. Clearly very thin ice and open water could not be sampled, but their contribution to the overall PDF was estimated from the results of aerial photography and video recording. Ice thickness, snow thickness and freeboard (hence ice draft) were measured in each hole. Cores were taken at each site to examine ice composition and character.

In the 1986 experiment our first discovery was a new mechanism for the formation of ice at the advancing edge, the so-called pancake-frazil cycle [*Lange et al.*, 1989]. We found that the ice which first forms at such an edge, in a region of high turbulence and wave activity, does so in the form of frazil ice, a suspension of small ice crystals in water. As the suspension thickens, it congeals into small cakes called pancakes, probably by wave-induced compression as described by *Martin and Kauffman* [1981]. At the ice margin the pancakes are only a few cm in diameter, but with increasing penetration into the pack they grow to reach 3-5 m in diameter and about 50 cm in thickness. The growth occurs by accretion from the frazil, facilitated by the continued presence of an open water surface which allows latent heat to be dissipated. The pancakes begin to freeze together in groups as one proceeds further from the margin, but it required (in the 1986 experiment) some 270 km of penetration before the wave field was damped down sufficiently to permit the pancakes to freeze together fully, the frazil acting as "glue". The final product was an ice type, consolidated pancake, in which the typical thickness was that of one of the original pancakes (50-60 cm), but with a rough bottom caused by rafting of two or three pancakes over one another at the time of freeze-up. It was found that the closure of the sea surface cut down the subsequent growth rate to a very low value (estimated at 0.4 cm/day) so that across the entire remainder of the pack to the coast the mean thickness increased by only a few cm. Within this first-year pack ice new smoother ice would form by freezing in the calm waters of polynyas. Ridging was scarce, except very close to the coast where episodes of wind-induced shoreward motion would have led to the observed increase. Most ridges were also very shallow, less than 6 m in draft. Examination of the structure of ridges showed that they are formed by the buckling of the floe itself, rather than the crushing of refrozen leads between thicker floes. Thus a compressive stress can be relieved by the creation of a number of small ridges rather than one large ridge.

The end product of these processes was the ice thickness distribution shown in fig. 16 [after *Wadhams et al.*, 1987]. Note the peak at the very low value of 50-60 cm, with a peak in snow cover thickness at 14-16 cm. The snow cover was sufficient to push the ice surface below water level in some 17% of holes drilled, and this leads to water infiltration into the snow layer and the possibility of formation of a new type of ice, snow-ice, at the boundary between ice and overlying snow. The mechanism of formation of most of the ice out of frazil and pancake explains structural observations [*Gow et al.*, 1987] which showed most of the thickness of ice cores from the Antarctic to be composed of small randomly oriented frazil-like crystals, rather than the long columnar crystals characteristic of freezing onto the bottom.

It is reasonable to suppose that these mechanisms are typical of the entire circumpolar advancing ice edge in winter (neglecting embayments such as the Ross and Weddell Seas). This implies that the first-year ice, which makes up the bulk of the winter pack, has a low mean draft of less than 1 m.

Multi-year ice was virtually absent in the region covered by the 1986 cruise. Only a small number of thick "islands" were seen (up to 11 m thick), which are thought to be very old fast ice broken out from sites along ice shelf fronts. Only in 1989 were we able to sample both first- and multi-year ice in an intimate combination, in the western Weddell Sea. The Weddell Gyre carries ice from the eastern Weddell Sea deep into high southern latitudes in the southern Weddell Sea off the Filchner-Ronne Ice Shelf, and then northward up the eastern side of the Peninsula. This journey takes about 18 months, and so permits much of the ice to mature into multi-year (strictly, second-year) ice. Such ice was seen only west of 40°W in our crossing of the Weddell Sea, the zone which experiences the northward drift regime.

Multi-year ice could be identified by its structure in cores, and by the very thick snow cover which it acquires, which is almost always sufficient to depress the ice surface below the waterline. Ice profile lines were divided into four types on the basis of this identification: undeformed first-year ice; first-year ice profiles containing deformed ice; undeformed multi-year ice; and multi-year profiles containing deformed ice. Table 2 shows the mean thickness results from these classes. It can be seen that :-

1. The mean thickness of undeformed multi-year ice (1.17 m), is about double that of first-year ice (0.60 m), indicating conditions in the southern Weddell Sea which permit much more rapid growth for an existing ice sheet than those experienced in the eastern Weddell Sea in 1986;
2. The mean thickness of first-year ice throughout the Weddell Sea is very similar to the thickness observed in 1986 in the Maud Rise area;
3. The presence of ridging roughly doubles the mean draft of the 100-m floe sections in which it occurs (0.60 to 1.03 m in first-year; 1.17 to 2.51 m in multi-year);
4. Snow is very much deeper on multi-year ice (0.63-0.79 m) than on first-year (0.16 -0.23 m).

If we consider only those ice stations in the passage northwards out of the eastern Weddell Sea, i.e. the region covered in the 1986 results, we find mean ice thicknesses as shown in fig. 17. The results have been plotted over similar data from 1986. The plots were made to examine the thicknesses reached by the natural growth process; profiles consisting largely of ridging were excluded, and those with some ridging are represented with brackets around them. At first sight it appears that the thicknesses tend to be greater in 1989. However, we should note that the decline in mean thickness in the latitude range 62° to 58° during the 1986 homeward leg represented the approach to a stationary, but compact, ice edge. In 1989 the ice edge lay much further north (the last ice was seen at 53°44'S), although its outer regions were diffuse and composed of wide bands, characteristic of off-ice winds. Consolidated pack ice could be studied as far north as 57°, at which point fig. 17 shows that there were some signs of melting. The six ice stations north of 62° in which undeformed ice was profiled can therefore be seen as reproducing the trend of ice thickness against latitude seen in 1986, but with a bodily northward shift of about 3°.

We conclude that there is no clear evidence of a change in the thickness to which first-year ice grows by congelation in the eastern Weddell Sea between the 1986 and 1989 winters. The difference lay in the fact that the ice edge at this longitude lay much further north in 1989.

Thus, from the limited data available so far (4441 holes in 1986 and 5339 in 1989) we cannot detect a change in the thickness of Antarctic sea ice. The datasets are much too sparse, however, and meaningful results must await the development of a profiling method, from above or below, which will generate genuinely synoptic maps of mean thickness.

5. LIKELY CLIMATE CHANGE IMPACTS

Fast ice

The simplest effect of warming is likely to be on the thickness of fast ice, which grows in fjords, bays and inlets in the Arctic, along the open coast in shallow water, and in channels of restricted dimensions. Because this type of ice generally forms in shallow water, oceanic heat flux is small or non-existent, and the thickness of the ice is determined almost entirely by air temperature history (modified by the thickness of the snow cover, which alters the growth rate). Empirical relationships have been successfully developed relating thickness achieved to the number of degree days of freezing since the beginning of winter [e.g. *Bilello*, 1961]. We can easily see from such curves that if the average daily air temperature increases by a known amount, the ultimate ice thickness will diminish by a calculable amount, and the ice-free season will lengthen. Using this technique, *Wadhams* [1990c] predicted that in the Northwest Passage and Northern Sea Route an air temperature rise of 8°C (equivalent to about a century of warming) will lead to a decline in the winter fast ice thickness from 1.8-2.5 m

TABLE 2. Mean ice thicknesses for different categories of ice floe drilled during 1989 Winder Weddell Gyre Study (courtesy of M.A.Lange).

Class	Holes		Mean thicknesses Snow	Ice	Draft	Total
Undeformed First-year	2034	Mean	0.16	0.60	0.60	0.76
		St Dev	0.10	0.21	0.20	0.25
Deformed First-year	2195	Mean	0.23	1.03	1.00	1.26
		St Dev	0.17	0.60	0.55	0.64
Undeformed Multi-year	349	Mean	0.63	1.17	1.25	1.80
		St Dev	0.18	0.35	0.36	0.45
Deformed Multi-year	282	Mean	0.79	2.51	2.48	3.30
		St Dev	0.23	1.08	1.05	1.09
All ice	5339	Mean	0.26	0.97	0.96	1.23
		St Dev	0.23	0.73	0.70	0.86

(depending on snow thickness) to 1.4-1.8 m and an increase in the ice-free season from 41 days to 100 days. This effect would be of great importance for the extension of the navigation season.

Even in this comparatively simple case, however, there is a possible feedback with snow thickness. If Arctic warming produces increased open water area and thus increased atmospheric water vapour content, it will lead to increased precipitation. Thicker snow cover decreases the growth rate of fast ice, as has been directly observed [*Brown and Cole*, 1992], except if the thickness is increased to the point where the snow does not all melt in summer, in which case the protection that it offers the ice surface from summer melt leads to a large increase in equilibrium ice thickness [*Maykut and Untersteiner*, 1971].

Moving pack ice

This is a much more difficult problem because of the importance of dynamics. Thermodynamic growth and decay rates no longer determine the area-averaged mean thickness of the ice cover. Wind stress acting on the ice surface causes the ice cover to open up to form leads, and later under a convergent stress refrozen leads and thicker ice elements are crushed to form pressure ridges. One result is a redistribution of ice from thinner to thicker categories, with the accompanying creation of open water areas. Another is to make the ice cover as a whole more resistant to convergent than to divergent stresses, and this causes its motion field under wind stress to differ from that of the surface water. Thus the exchanges of heat, salt and momentum are all different from those that would occur in a fast ice cover. We can only indicate a few of the interactions and how they may be changed by global warming.

Firstly, the effects of variable thickness are very important. On account of the creation of leads by the ice dynamics and the very high growth rates which occur in thin ice, it has been found that the overall area-averaged growth rate of ice is dominated (especially in autumn and early winter when much lead and ridge creation take place) by the small fraction of the sea surface occupied by ice less than 1 m thick [*Hibler*, 1980]. In fast ice, climatic warming increases sea-air heat transfer by reducing ice growth rates. However, over open leads a warming would decrease the sea-air heat transfer, so the area-averaged change in this quantity over moving ice (and hence its feedback effect on climatic change itself) depends on the change in the rate of creation of new lead area, which is itself a function of a change in the ice dynamics, either driving forces (wind field) or response (ice rheology).

Already we see that ice dynamics are important through this thickness feedback. But they also have other effects. In the Eurasian Basin the average surface ice drift pattern is a current (the Trans Polar Drift Stream) which transports ice across the Basin, out through Fram Strait, and south via the East Greenland Current into the Greenland and Iceland Seas where it melts. This implies that a typical parcel of ice forms by freezing in the Basin, the latent heat being transferred to the atmosphere; is then transported southward (which is equivalent to a northward heat transport); and then when it melts in the Greenland or Iceland Sea it absorbs the latent heat required from the ocean. The net result is a heat transfer from the upper ocean in sub-Arctic seas into the atmosphere above the Arctic Basin. A change in area-

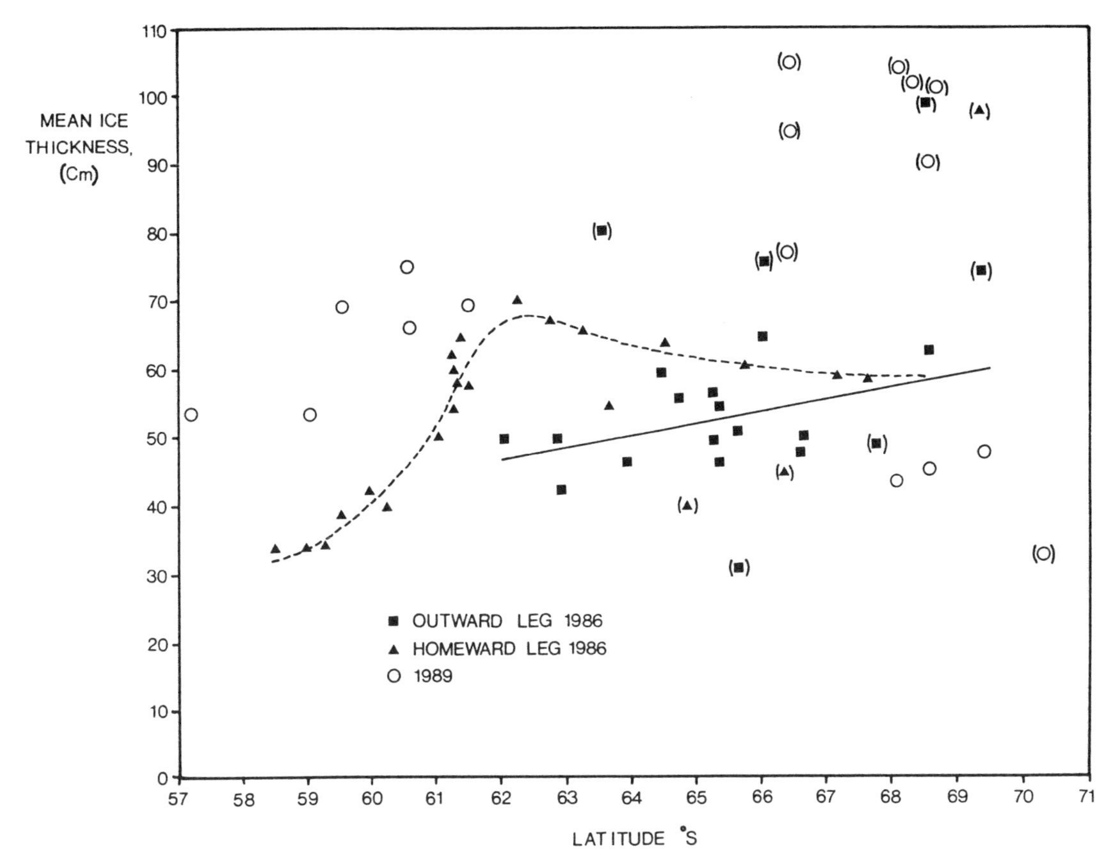

Fig. 17. Mean ice thicknesses from drilling sites in first-year ice during winter 1986 and 1989, as a function of latitude. Points in brackets included pressure ridging or young ice in the sections drilled. The continuous and dotted lines are apparent trends for the outward and homeward legs of the 1986 cruise.

averaged freezing rate in the Basin would thus cause a change of similar sign to the magnitude of this long-range heat transport. An identical argument applies to salt flux, which is positive into the upper ocean in ice growth areas and negative in melt areas. Thus salt is also transported northward via the southward ice drift. A relative increase in area-averaged melt would cause increased stabilisation of the upper layer of polar surface water, and hence a reduction in heat flux by mixing across the pycnocline, while a relative increase in freezing would cause destabilisation and possible overturning and convection.

Finally, *Hibler* [1989] has drawn attention to the role of ice deformation in reducing the sensitivity of ice thickness to global warming in areas of net convergence. The largest mean ice thicknesses in the Arctic - 7 m or more - occur off the Canadian Arctic Archipelago [*Hibler*, 1979; see also fig.12] where the wind stress drives ice against a downstream land boundary. Here the mean ice thickness is determined by mechanical factors, largely the strength of the ice, which set a limit to the amount of deformation by crushing that can occur. In this area the thickness is likely to be insensitive to atmospheric temperature changes. The main sensitivity would be to a change in the overall pattern of winds over the Arctic.

Given the complexity of these interactions and feedbacks, it is not at all clear what the overall effect of an air temperature increase on the Arctic ice cover and upper ocean would be. Sensitivity studies using coupled ocean-ice-atmosphere models are required, but results are not yet available.

In the Antarctic where the ice cover is divergent and where land boundaries are less important, it is more reasonable to suppose that the main effect of global warming will be a simple retreat of the ice edge southward. However, *Martinson* [1990] has demonstrated that a complex set of feedback mechanisms comes into play if the air temperature changes. The balance of lead concentration, upper ocean structure and pycnocline depth will adjust itself to minimise the impact of changes, tending to preserve an ice cover even though it may be thinner and more diffuse.

Ice edge regions and convection

In the central Greenland Sea, just south of the main gyre centre, an ice tongue usually develops during winter. It grows eastwards from the main East Greenland ice edge in the vicinity of 72-74°N latitude and often curves round to the northeast until it reaches east of the prime meridian. It is called Odden, and in its curvature it embraces a bay of open water, centred on the gyre centre, which is known as Nordbukta. It is believed to form mainly by local ice production, since the tongue-shaped region corresponds to the region of influence of the Jan Mayen Polar Current (the southern part of the Greenland Sea Gyre) which maintains cold surface water. There is currently great interest in Odden, for two reasons. Firstly, the contribution of local ice production as opposed to advection of older ice from the East Greenland Current is not known for certain. Secondly, and related to this, the local ice production is believed to occur largely in the form of frazil and pancake, since the intense wave field in the winter Greenland Sea inhibits ice sheet formation, while remote sensing imagery shows rapid changes in the size, shape and position of Odden, such as would not occur if it were composed of large floes. Again, frazil and pancake production implies high growth rates and high salt fluxes, possibly on a cyclic basis related to cold air outbreaks from Greenland. In recent models of winter convection in the central Greenland Sea (e.g. *Rudels*, 1990; J. Backhaus, personal commun.), this periodic salt flux plays an important role in triggering narrow convective plumes.

One of the objectives of the current EC-supported ESOP (European Subpolar Ocean Programme) is to understand winter convection and the role of sea ice-ocean interactions in it. Field operations during 1993 involved the direct study of convective plumes and the developing water structure in the central Gyre region associated with convection, and investigations of the ice characteristics, thickness and variability within the convection region. Work began in early February using the vessels *Valdivia* and *Northern Horizon* and a BAC 1-11 research aircraft of the Royal Aerospace Establishment, Farnborough. This established that Odden was developing in 1993 as a tongue of dense pancake ice, with little or no older ice present. From February-April 1993 FS *Polarstern* operated in the Greenland Sea, with intensive studies of the Odden in early April. Passive microwave and SAR images had showed that the Odden ice tongue completely separated from its root in March and became a long island of ice oriented N-S centred on the prime meridian. This island later broke into separate north and south islands, and (on passive microwave) the north island appeared to dissolve and reform late in March while the south island steadily shrank.

The ice physics programme carried out in April by *Polarstern* [*Wadhams and Viehoff*, 1993] showed that the southern island was composed of a dense mass of pancake ice. The pancakes had a mean diameter of 0.96 m (sample of 2501 pancakes in an aerial photograph) with a maximum of 4.8 m. Pancakes recovered and brought on board ranged in thickness from 9 cm to 40 cm, and the larger pancakes had low salinities (2.8 - 5.5 ‰) normally regarded as characteristic of first-year ice, implying that they had been stable components of the ice cover for some time. Frazil sampled between the pancakes had a thickness corresponding to 4-20 cm fresh water equivalent. Our preliminary conclusion from direct physical examination and a concurrent study of SAR imagery (which distinguishes frazil from pancake by the dark return from the former and bright return from the latter) is that in the centre, or core region, of the Odden, ice remains all winter as pancake. It grows into thicker and larger cakes with salinities and brine drainage systems characteristic of first-year ice, but the high ambient wave energy precludes the cakes freezing into larger floes or ice sheets. In the outer fringes of Odden the ice cover consists of much thinner pancakes and frazil, which are subject to rapid freezing and melting episodes, depending on local air temperature and wind or wave conditions. This gives the shape of the Odden a rapidly changing appearance on SSM/I images. However the absolute volume changes of ice involved in these fluctuations are small, because the parts of the Odden which fluctuate never possess ice more than a few centimetres thick, as opposed to the stable core region of the ice tongue or island. This raises questions as to whether the salt flux which can be obtained from Odden fluctuations is adequate to perform the role required of it in convective plume models. The question can only be answered when analysis of the joint field and remote sensing dataset from 1993 is complete. Further work is planned for 1994.

The reason why the role of ice in the convection mechanism is important to assess is that if the ice were to retreat from the central Gyre region on account of climatic change it may cause deep convection to cease. Already there is evidence from tracer studies [*Schlosser et al.*, 1991] of a severe reduction in the volume of convection during the last decade. If convection were to cease it would have a positive feedback effect on global warming, since the ability of the world ocean to sequestrate CO_2 through convection would be reduced.

REFERENCES

Bilello, M.A., Formation, growth and decay of sea ice. *Arctic, 14*(1), 3-24, 1961.

Bourke, R.H. and R.P. Garrett, Sea ice thickness distribution in the Arctic Ocean. *Cold Regions Sci. Technol., 13,* 259-280, 1987.

Bourke, R.H. and A.S. McLaren, Contour mapping of Arctic Basin ice draft and roughness parameters. *J. Geophys. Res., 97*(C11), 17715-17728, 1992.

Brown, R.D. and P. Cole, Interannual variability of landfast ice thickness in the Canadian High Arctic, 1950-1989. *Arctic, 45*(3), 273-284, 1992.

Burns, B.A., D.J. Cavalieri, M.R. Keller, W.J. Campbell, T.C. Grenfell, G.A. Maykut and P. Gloersen, Multisensor comparison of ice concentration estimates in the marginal ice zone. *J. Geophys. Res., 92*, 6843-6856, 1987.

Campbell, R.C., *Statistics for Biologists*. 2nd Ed. Cambridge Univ. Press, 385pp., 1974.

Colony, R.L., I. Rigor and K. Runciman-Moore, A summary of observed ice motion and analysed atmospheric pressure in the Arctic Basin, 1979-1990. Appl. Phys. Lab., Univ. Washington, Seattle, Tech. Rept., APL-UW TR9112, 106pp., 1991.

Comiso, J.C., P. Wadhams, W.B. Krabill, R.N. Swift, J.P. Crawford, and W.B. Tucker III, Top/bottom multisensor remote sensing of Arctic sea ice. *J. Geophys. Res.*, *96*(C2), 2693-2709, 1991.

Davis, N.R. and P. Wadhams, Structure and distribution of Arctic pressure ridges. SPRI Report (92-2), Scott Polar Research Institute, Technical report series, 1992.

Dickins D.F.and V.F. Wetzel, Multiyear pressure ridge study, Queen Elizabeth Islands. *POAC 81*, Proceedings of the Sixth International Conference on Port and Ocean Engineering under Arctic Conditions, II, pp 765 - 775, 1981.

Gloersen, P., W.J. Campbell, D.J. Cavalieri, J.C. Comiso, C.L. Parkinson and H.J. Zwally, *Arctic and Antarctic Sea Ice, 1978-1987: Satellite Passive-Microwave Observations and Analysis*. NASA, Washington D.C., SP-511, 290pp., 1992.

Gow, A.J., S.F. Ackley, K.R. Buck and K.M., Golden, Physical and structural characteristics of Weddell Sea pack ice. CRREL Rept. 87-14, US Army Cold Regions Res. & Eng. Lab., Hanover N.H., 1987.

Grandvaux, B. and J.-F. Drogou, Saga 1, une première étape vers les sous-marins autonomes d'intervention. In *Arctic Technology and Economy - Present Situation and Problems, Future Issues*. Bureau Veritas, Paris., 1989.

Guoliang, J. and P. Wadhams, Travel time changes in a tomography array caused by a sea ice cover. *Prog. Oceanogr.*, *22*(3), 249-275, 1989.

Hibler, W.D. III , A dynamic thermodynamic sea ice model. *J. Phys. Oceanogr.*, *9*(4), 815-846, 1979.

Hibler, W.D. III, Modeling a variable thickness sea ice cover. *Mon. Weather Rev.*, *108*, 1943-1973, 1980.

Hibler, W.D. III Arctic ice-ocean dynamics. In *The Arctic Seas. Climatology, Oceanography, Geology, and Biology* (Y. Herman, ed.). Van Nostrand Reinhold, New York, 47-91, 1989.

Holladay, J.S., J.R. Rossiter and A. Kovacs, Airborne measurement of sea ice thickness using electromagnetic induction sounding. *Proc. 9th Int. Conf. Offshore Mech. and Arctic Engng.* (ed. O.A. Ayorinde, N.K. Sinha, D.S. Sodhi), Am. Soc. Mech. Engrs., 309-315, 1990.

Holt, B., J. Crawford and F. Carsey, Characteristics of sea ice during the Arctic winter using multifrequency aircraft radar imagery. In *Sea Ice Properties and Processes* (ed. S.F. Ackley and W.F. Weeks), Monograph 90-1, US Army Cold Regions Res. & Engng. Lab., Hanover, N.H., 224 (abstract), 1990.

Hudson, R., Annual measurement of sea-ice thickness using an upward-looking sonar. *Nature, Lond.*, *344*, 135-137, 1990.

Jin, Guoliang, J.F. Lynch, R. Pawlowicz, P. Wadhams and P. Worcester, Effects of sea ice cover on acoustic ray travel times, with applications to the Greenland Sea Tomography Experiment. *J. Acoust. Soc. Am.*, *94*(12), 1044-1056, 1993.

Johannessen, O.M., L. Bjørnø, G. Bienvenu, K. Hasselmann, P.M. Haugan, J. Johnsen, U. Lie, J. Papadakis, S. Sandven, P. Wadhams and M. Zakharia, ATOC - Arctic: acoustic thermometry of the ocean climate in the Arctic Ocean. Nansen Envtl. and Remote Sensing Centre, Bergen, rept., 1993.

Ketchum, R.D., Airborne laser profiling of the Arctic pack ice. *Remote Sensing Environ.*, *2*, 41-52, 1971.

Kovacs, A., On pressured sea ice. In *Sea Ice* (ed. T. Karlsson), Proceedings of International Sea Ice Conference, Reykjavik, Iceland, May 10-13 1971, 276 - 295, 1971.

Kovacs, A. and J.S. Holladay, Development of an airborne sea ice thickness measurement system and field test results. CRREL Rept. 89-19, 1989.

Kovacs, A. and J.S. Holladay, Airborne sea ice thickness sounding. In *Sea Ice Properties and Processes* (ed. S.F. Ackley, W.F. Weeks), US Army Cold Regions Res. & Engng. Lab., Hanover, N.H., Monograph 90-1, 225-229, 1990.

Kovacs, A. and R.M. Morey, Electromagnetic measurements of multiyear sea ice using impulse radar. *Cold Regions Sci. Technol.*, *12*, 67-93, 1986.

Kovacs, A. and D.S. Sodhi, Shore ice pile-up and ride-up: field observations, models and theoretical analyses. *Cold Regions. Sci. Technol.*, *2*, 209 - 288, 1980.

Kovacs, A., W.F. Weeks, S.F. Ackley and W.D. Hibler, Structure of a multiyear pressure ridge. *Arctic*, *26*, 22 - 32, 1973.

Krabill, W.B., R.N. Swift and W.B. Tucker III Recent measurements of sea ice topography in the Eastern Arctic. In *Sea Ice Properties and Processes* (ed. S.F. Ackley and W.F. Weeks), US Army Cold Regions Res. & Engng. Lab., Hanover N.H., Monograph 90-1, 132-136, 1990.

Lange, M.A. and H. Eicken, The sea ice thickness distribution in the north-western Weddell Sea. *J. Geophys. Res.*, *96*(C3), 4821-4837, 1991.

Lange, M.A., S.F. Ackley, G.S. Dieckmann, H. Eicken and P. Wadhams, Development of sea ice in the Weddell Sea Antarctica. *Ann. Glaciol.*, *12*, 92-96, 1989.

Leppäranta, M., Statistical features of sea ice ridging in the Gulf of Bothnia. Winter Navigation Research Board, Helsinki, Tech. Rep., 32, 46p., 1981.

Leppäranta, M. and T. Thompson, BEPERS-88 sea ice remote sensing with synthetic aperture radar in the Baltic Sea. *Eos, Trans. Am. Geophys. U.*, *70*(28), 698-699, 708-709, 1989.

LeSchack, L.A., Arctic Ocean sea ice statistics derived from the upward-looking sonar data recorded during five nuclear submarine cruises. LeSchack Associates Ltd., 116-1111 University Blvd. W., Silver Spring, Md., Tech. Rept., 1980.

Livingstone, C.E., Combined active/passive microwave classification of sea ice. *Proc. IGARSS-89*, *1*, 376-380, 1989.

Lowry, R.T. and P. Wadhams, On the statistical distribution of pressure ridges in sea ice. *J. Geophys. Res.*, *84*, 2487-2494, 1979.

Lyon, W.K., Ocean and sea-ice research in the Arctic Ocean via submarine. *Trans. N.Y. Acad. Sci.*, srs. 2, *23*, 662-674, 1961.

McLaren, A.S., Analysis of the under-ice topography in the

Arctic Basin as recorded by the USS *Nautilus* during August 1958. *Arctic, 41(2),* 117-126, 1988.

McLaren, A.S., The under-ice thickness distribution of the Arctic basin as recorded in 1958 and 1970. *J. Geophys. Res., 94,* 4971-4983, 1989.

McLaren, A.S., P. Wadhams and R. Weintraub, The sea ice topography of M'Clure Strait in winter and summer of 1960 from submarine profiles. *Arctic, 37,* 110-120, 1984.

Martin, S. and P. Kauffman, A field and laboratory study of wave damping by grease ice. *J. Glaciol., 27*(96), 283-313, 1981.

Martinson, D.G., Evolution of the Southern Ocean winter mixed layer and sea ice: open ocean deepwater formation and ventilation. *J. Geophys. Res., 95*(C7), 11641-11654, 1990.

Maykut, G.A. and N. Untersteiner, Some results from a time-dependent thermodynamic model of Arctic sea ice. *J. Geophys Res., 76,* 1550-1575, 1971.

Mock, S.J., A.D. Hartwell and W.D. Hibler, Spatial aspects of pressure ridge statistics. *J. Geophys. Res., 77,* 5945 - 5953, 1972.

Moritz, R.E., Sampling the temporal variability of sea ice draft distribution. *Eos* supplement, Fall AGU Meeting, 237-238 (abstract), 1991.

Munk, W.H. and C. Wunsch, Ocean acoustic tomography: a scheme for large scale monitoring. *Deep-Sea Res., 26,* 123-161, 1979.

Onstott, R.G., T.C. Grenfell, C. Maetzler, C.A. Luther and E.A. Svendsen, Evolution of microwave sea ice signatures during early and mid summer in the marginal ice zone. *J. Geophys. Res., 92,* 6825-6837, 1987.

Parmerter, R.R. and M. Coon, Model of pressure ridge formation in sea ice. *J. Geophys. Res., 77*(33), 6565 - 6575, 1972.

Pilkington, G.R. and B.D. Wright, Beaufort Sea ice thickness measurements from an acoustic, under ice, upward looking ice keel profiler. *Proc. 1st Intl. Offshore & Polar Engng Conf., Edinburgh, 11-16 Aug 1991,* 1991.

Rigby, F.A. and A. Hanson, Evolution of a large Arctic pressure ridge. *AIDJEX Bull. 34,* 43 - 71. Div. Marine Resources, Univ. Washington, Seattle, 1976.

Rothrock, D.A., Ice thickness distribution - measurement and theory. In *The Geophysics of Sea Ice* (ed. N. Untersteiner), Plenum, New York, 551-575, 1986.

Rudels, B. Haline convection in the Greenland Sea. *Deep-Sea Res., 37*(9), 1491-1511, 1990.

Sanderson, T.J.O., *Ice Mechanics. Risks to Offshore Structures.* Graham and Trotman, London., 1988.

Sayed, M. and R. Frederking, On modelling of ice ridge formation. International Association for Hydraulic Research Ice Symposium, 603 - 614, 1986.

Schlosser, P., G. Bönisch, M. Rhein and R. Bayer, Reduction of deepwater formation in the Greenland Sea during the 1980s: evidence from tracer data. *Science, 251,* 1054-1056, 1991.

Serreze, M.C., R.G. Barry and A.S. McLaren, Summertime reversals of the Beaufort Gyre and its effects on ice concentration in the Canada Basin. *J. Geophys. Res., 94,* 10955-10970, 1989.

Thorndike, A.S., C. Parkinson and D.A. Rothrock (eds), Report of the Sea Ice Thickness Workshop, 19-21 November 1991, New Carrollton, Maryland. Polar Science Center, Applied Physics Lab., Univ. Washington., Seattle, 1992.

Tucker, W.B. III, An overview of the physical properties of sea ice. Proceedings of Workshop on Ice Properties, Associate Committee on Geotechnical Research, National Research Council Canada (June 21-22, 1988, St. Johns, Newfoundland). Technical Memorandum No. 144, NRCC 30358, 1989.

Tucker, W.B. III, W.F. Weeks and M. Frank, Sea ice ridging over the Alaskan continental shelf. *J. Geophys. Res., 84,* 4885-4897, 1979.

Vinje, T.E., An upward looking sonar ice draft series. In *Proc. 10th Intl. Conf. Port & Ocean Engng under Arctic Condns.* (ed. K.B.E. Axelsson, L.A. Fransson). Lulea Univ. Technology, *1,* 178-187, 1989.

Wadhams, P., Sea ice topography in the Beaufort Sea and its effect on oil containment. *AIDJEX Bull., 33,* 1-52. Div. Marine Resources, Univ. Washington, Seattle, 1976.

Wadhams, P., Characteristics of deep pressure ridges in the Arctic Ocean. *POAC 77,* Proceedings of the Fourth International Conference on Port and Ocean Engineering under Arctic Conditions, St. John's, *1,* 544 - 555. Memorial Univ., St. John's, Nfld., 1977.

Wadhams, P., A comparison of sonar and laser profiles along corresponding tracks in the Arctic Ocean. In *Sea Ice Processes and Models* (ed. R.S. Pritchard), Univ. Washington Press, Seattle, 283-299, 1980.

Wadhams, P., Sea ice topography of the Arctic Ocean in the region 70°W to 25°E, *Phil. Trans. Roy. Soc., Lond., A302*(1464), 45-85, 1981.

Wadhams, P., Sea ice thickness distribution in Fram Strait. *Nature, Lond., 305,* 108-111, 1983.

Wadhams, P., Sea-ice thickness in the Trans Polar Drift Stream. *Rapp. P-v Reun Cons. Int. Explor. Mer, 188,* 59-65, 1989.

Wadhams, P., Evidence for thinning of the Arctic ice cover north of Greenland. *Nature, Lond., 345,* 795-797, 1990a.

Wadhams, P., Ice thickness distribution in the Arctic Ocean. In *Ice Technology for Polar Operations* (T.K.S. Murthy, J.G. Paren, W.M. Sackinger, P. Wadhams, eds.), Computational Mechanics Publns, Southampton, 3-20, 1990b.

Wadhams, P., Sea ice and economic development in the Arctic Ocean - a glaciologist's experience. In *Arctic Technology and Economy. Present situation and problems, future issues.* Bureau Veritas, Paris, 1-23, 1990c.

Wadhams, P. Sea ice thickness distribution in the Greenland Sea and Eurasian Basin, May 1987. *J. Geophys. Res., 97*(C4), 5331 - 5348, 1992.

Wadhams, P. and J.C. Comiso, The ice thickness distribution inferred using remote sensing techniques. In *Microwave Remote Sensing of Sea Ice* (ed. F. Carsey), Geophysical Monograph 68, Amer. Geophys. U., Washington, ch. 21, 375-383, 1992.

Wadhams, P. and D.R. Crane, SPRI participation in the Winter Weddell Gyre Study 1989. *Polar Record, 27,* 29-38, 1991.

Wadhams, P. and T. Davy, On the spacing and draft distributions for pressure ridge keels. *J. Geophys. Res., 91*(C9), 10697-10708, 1986.

Wadhams, P. and R.J. Horne, An analysis of ice profiles obtained by submarine sonar in the Beaufort Sea. *J. Glaciol.*, *25*(93), 401-424, 1980.

Wadhams, P. and R.T. Lowry, A joint topside-bottomside remote sensing experiment on Arctic sea ice. *Proc. 4th Canadian Symp. on Remote Sensing, Quebec, 16-18 May 1977*. Canadian Remote Sensing Soc., 407-423, 1977.

Wadhams, P. and T. Viehoff, The Odden ice tongue in the Greenland Sea: SAR imagery and field observations of its development in 1993. *Proc. 2nd ERS-1 Symp. - Space at the Service of our Environment, Hamburg, 11-14 October 1993*. ESA, 1993, in press.

Wadhams, P., A.S. McLaren and R. Weintraub, Ice thickness distribution in Davis Strait in February from submarine sonar profiles. *J. Geophys. Res.*, *90*(C1), 1069-1077, 1985.

Wadhams, P., M.A. Lange and S.F. Ackley, The ice thickness distribution across the Atlantic sector of the Antarctic Ocean in midwinter. *J. Geophys. Res.*, *92*(C13), 14535-14552, 1987.

Wadhams, P., J.C. Comiso, J. Crawford, G. Jackson, W. Krabill, R. Kutz, C.B. Sear, R. Swift, W.B. Tucker and N. R. Davis, Concurrent remote sensing of Arctic sea ice from submarine and aircraft. *Int. J. Remote Sensing*, *12(9)*, 1829-1840, 1991.

Wadhams, P., W.B. Tucker III, W.B. Krabill, R.N. Swift, J.C. Comiso and N.R. Davis, Relationship between sea ice freeboard and draft in the Arctic Basin, and implications for ice thickness monitoring. *J. Geophys. Res.*, *97*(C12), 20325-20334, 1992.

Walsh, J.E. and W.L. Chapman, Model simulation of changes in Arctic sea ice thickness, 1960-1989. Proc. 5th AMS Conf. on Climate Variations, Denver, October 1991, 1991.

Weeks, W.F., A. Kovacs and W.D. Hibler III, Pressure ridge characteristics in the Arctic coastal environment. *Proc. 1st Intl. Conf. Port & Ocean Engng. under Arctic Condns* (ed. S.S. Wetteland, P. Bruun), Tech. Univ. Norway, Trondheim, 152-183, 1971.

Weeks, W.F., S.F. Ackley and J. Govoni, Sea ice ridging in the Ross Sea, Antarctica, as compared with sites in the Arctic. *J. Geophys. Res.*, *94*(C4), 4984-4988, 1989.

Zubov N.N., *Arctic Ice*. Izdatel'stvo Glavsermorputi, Moscow. (Translation AD426972, National Technical Information Service, Springfield, Va.), 1945.

Zwally, H.J., J.C. Comiso, C.L. Parkinson, W.J. Campbell, F.D. Carsey and P. Gloersen, *Antarctic Sea Ice 1973-1976: Satellite Passive Microwave Observations*. NASA, Washington D.C., Rept. SP-459, 1983.

Variability in Sea-Ice Thickness Over the North Pole From 1958 to 1992

A. S. McLaren[1], R. H. Bourke[2], J. E. Walsh[3], R. L. Weaver[4]

Based upon 50 and 100-km-long segments of submarine upward looking echo-sounder measurements an analysis was performed on the data from 12 voyages centered over the North Pole conducted between 1958 and 1992 to establish the nature of the interannual variability in ice thickness and to discern if any thinning trend could be ascribed to anthropogenic warming of the atmosphere. Only at the North Pole do sufficient replicate submarine tracks exist to provide a temporal record of ice thickness. Over this 34-year period the overall mean draft was 3.6 m but demonstrated a large interannual variability, ranging from 2.8 m in 1986 to 4.4 m in 1970, sufficiently large to obscure any discernable thinning of the Arctic sea ice cover. Thick ice years were associated with years of increased amounts of ridged ice. During thin ice years, ~20% more young and first year ice was present along with reduced amounts of ridged ice. The percentage of open water/young ice (< 0. 3 m) in the vicinity of the North Pole was relatively small in the mean (~ 1%).

The interannual variability in ice thickness at the Pole is undoubtedly related to the interannual variability noted in the strength and direction of the Transpolar Drift Stream as reflected in the variability of the mean annual atmospheric pressure pattern over the central Arctic Basin.

A main probability density function (pdf) of ice draft was computed from the 12 data sets. Its variability within each bin width was substantial (~ 10%) reflecting the large year-to-year variability in ice thickness. Climate and ice prediction modelers will have to take this variability in ice draft (pdf) into account when incorporating such pdf's into their models.

INTRODUCTION

The Arctic Ocean is generally regarded as a potentially sensitive indicator of global change, especially since the predicted result of continued greenhouse-gas emissions is global warming amplified by changing conditions occurring at high latitudes (Intergovernmental Panel on Climate Change, 1990). Such warming could have important consequences for the thickness and extent of Arctic sea ice and for the volume of fresh water supplied to the Greenland Sea and its consequent impact on deep convection and global thermohaline circulation. The presumption is that ice melting could lead to a decrease in both thickness and areal extent, significantly affecting the heat flux at high latitudes.

In simulations using coupled ocean-atmospheric models the Arctic ice thickness has been shown to decrease considerably when the atmospheric CO_2 content is increased (Manabe et al., 1992). It is important to establish the natural variability inherent in Arctic sea-ice draft, however, before drawing conclusions about climate induced changes in ice thickness.

At present, most measurements of sea-ice thickness are based on submarine upward-looking sonar data which record the ice draft, that is, the portion submerged below sea level. True ice thickness includes the above-water freeboard which can be estimated theoretically or empirically. Bourke and Garrett (1987) used unclassified submarine cruise logs and sonar data to develop seasonal climatologies of ice draft in the Arctic Basin. The ice draft recorded by the submarines NAUTILUS (in 1958) and QUEENFISH (in 1970) along the

[1]*Science Service, Science News Magazine, Washington, D.C.*
[2]*Department of Oceanography Naval Post Graduate School, Monterey, CA*
[3]*Department of Atmospheric Science, University of Illinois*
[4]*Cryospheric and Polar Processes Division of CIRIES, University of Colorado, Boulder*

The Polar Oceans and Their Role in Shaping the Global Environment
Geophysical Monograph 85

same transect through the Canadian Basin and over the North Pole during early August was 0.7 - 0.8 m thinner in 1970 than in 1958, and the areal coverage of open water and thin ice was greater in 1970 than in 1958 (McLaren, 1989). Submarine sonar measurements indicated significantly thinner sea ice north of Greenland in 1987 than in 1976 (Wadhams, 1990).

The findings of McLaren and Wadhams raise intriguing questions about recent variations of sea ice thickness in the context of climate change. From the evidence so far available, McLaren et al. (1990) concluded that too little is known about the natural variability of Arctic sea-ice thickness to draw any conclusions about climate-induced changes in the thickness based on comparative analyses of only a few sea-ice draft datasets collected within the Arctic Basin. Recently, however, Gloersen and Campbell (1991) reported that satellite-borne scanning multispectral microwave radiometer (SMMR) data collected by Nimbus 7 between October 1978 and August 1987 reveal significant decreases of ice extent and open water areas within the Arctic. Unfortunately, Nimbus 7 provides no SMMR data on ice extent and open water north of 84°N (Parkinson and Cavalieri, 1989), precluding findings concerning the portion of the Arctic Basin around the North Pole.

In an effort to learn more about the natural variability of sea-ice draft and the extent of open water or thin ice in the interior of the Arctic Basin, we have used the ice draft measurements acquired from 12 submarines that have operated in the vicinity of the North Pole. Ice draft measurements are routinely obtained at approximately 1.0 m intervals along the submarine track from an upward looking sonar installed primarily for ice avoidance.

More than 200,000 km of sea-ice draft data have been collected within the Arctic Basin by submarines since 1958. Less than 15% of this data has so far been analyzed. The region surrounding the North Pole was selected for analysis of interannual changes in sea ice thickness because it is the only location that has been repeatedly surveyed by U.S. nuclear submarines since 1958. Only in the vicinity of the North Pole do nearcoincident submarine tracks exist.

To date, only contours of the mean ice draft, either annual (LeSchack, published in Hibler, 1980) or seasonal (Bourke and Garrett, 1987; Bourke and McLaren, 1992), have been constructed to illustrate the basin-wide geographic distribution of the ice thickness. Of potentially more value to climate and ice thickness prediction models is to represent the thickness of the ice cover as a distribution or probability density function (pdf) (Thorndike et al., 1975; Wadhams, 1983). For example, most of the heat flow between the ocean and atmosphere over the Arctic Ocean takes place through open water or thin ice, neither of which is adequately represented for a given area solely by the mean ice draft. Similar arguments can be offered concerning the need for information about the thickness distribution of multiyear or deformed (ridged/rafted) ice. In this paper we present data illustrating how the ice draft pdf at a given location can vary from year-to-year. Our findings point to a substantial interannual variability, thus limiting the utility of data derived from an individual cruise or experiment.

The 12 submarine cruises examined in this study span a period of 34 years (1958-1992). Each consists of a 50 or 100-km-long segment centered over the North Pole, with the exception of the older NAUTILUS and QUEENFISH data which were 192 and 338 km-long segments, respectively, and were thus omitted from Table 1. The early NAUTILUS wider beam (11.5°) sonar data have been corrected to provide an accurate equivalence to the data from the narrow beam sonars (<2.0°) used by the other 11 submarines (McLaren, 1986). Most of the cruises took place in late winter/early spring conditions (April or May). The NAUTILUS and QUEENFISH cruises took place in August while the PINTADO and SILVERSIDES cruises occurred in October. Since at the North Pole the seasonal variation in mean ice draft is expected to be small, these four cruises were included to augment an otherwise already sparse data set and to expand the length of the data record. We recognize that an approximate 0.3 m seasonal variation in ice draft may he present (Maykut and Untersteiner, 1971) for the central Arctic region and have thus applied a 0.3 m correction (increase) to the mean ice draft of the above four summer/fall data sets.

DATA PROCESSING

The underice draft data recorded on each set of analog recordings (NAUTILUS and QUEENFISH) were manually digitized using a follower and a digitizer tablet which produced one data point every 0.05 mm of the recording chart (McLaren, 1989). This process was not required for data from 1977 and later for which the underice drafts were digitally recorded on magnetic tapes and diskettes. The basic reference of underice draft measurements should be the actual sea level. No independent measurements of sea level were available, however. In order to maintain acuracy sea level had to be determined through frequent reference (every 3-4 km) to indications of open water on the analog and digital recordings and inserted during initial data processing. Since digitization and the ship's speed variations can produce irregular spacing between individual ice draft data points, all underice draft distribution data were initially subjected to an interpolation process. This ensured that individual data points were equally spaced at 1.5-m intervals, permitting consistent processing of raw data from all cruises in subsequent statistical analyses.

All digitized data were edited and analyzed statistically by McLaren using data processing and statistical software packages originally developed by Wadhams and Home (1980) and later modified by McLaren (1986). Since the ice draft data are recorded for operational purposes, the raw data can contain a variety of sensor and transmission system errors. The data used here were provided by the U. S. Navy's Arctic Submarine Laboratory, San Diego, after careful editing by their staff to remove these errors through frequent calibration to open water/new ice indications from coincidentally recorded analog records. The resultant accuracy of the edited data is estimated to be $\pm$0.3 m (T. Luallin, personal communication). When averaged over 50- and 100-km segments, the errors associated with mean statistics are estimated to be $\pm$0.15 m (McLaren et al., 1992).

The data, averaged over 50 or 100-km long segments, were screened and sorted to generate mean, standard deviation and other statistics shown in Tables 1 and 2. In Table 1, ice having drafts less than 3.0 m is assumed to be "first-year ice" and ice with drafts greater than 3.0 m is assumed to be "multiyear or deformed ice." Additionally, any ice draft less than 0.3 m is assigned to the "open water/new ice" category. The pressure ridge keel statistics listed in Table 2 were computed for keel drafts exceeding a 9-m threshold (Wadhams and Home, 1980). The "level" ice statistic provides a measure of the amount of underformed (non-ridged) ice present. (Level ice may be first-year or multiyear. It is defined as a segment of relatively smooth ice wherein the draft of a measurement point differs from that of a point 10 m to either side by less than 25 cm {Wadhams and Home, 1980}).

RESULTS

Mean and Comparative Ice Statistics

The mean ice draft recorded during each submarine cruise from 1977 to 1992, computed at 5 km intervals for 100-km-long segments centered over the North Pole, is shown in Figure 1. The mean ice draft is approximately 3.6 m (Table 2) but considerable year-to-year variability is quite evident. In general, one can expect portions of the ice cover near the Pole to vary in thickness from 2.5 m to 5 m. However, for any given year the range is generally less, of the order of 1 to 1.5 m. Over the 100 km path length the ice draft can sometimes be remarkably uniform (e. g., PARGO 1991, ARCHERFISH 1988) or quite variable, indicative of extensive areas of thin or ridged ice (e. g., PINTADO 1978, GURNARD 1990). The beginning and end of the GURNARD track shows the mean ice draft to be in excess of 5 m. Unlike all the other tracks, GURNARD's track upon leaving the North Pole was almost a reciprocal of her inbound track. Thus the higher mean draft at 90 km could represent a re-sampling of an extensive heavily ridged area encountered during the first 20 km of her approach to the North Pole.

The 50 km mean ice drafts presented in this paper may be regarded as instantaneous measurements, not annual or seasonal averages. Nor can the spatial representativeness be inferred from the mean draft of a single 50 km segment. Because this study focuses on the interannual variations of ice draft near the North Pole, a key issue is the magnitude of the sampling variability relative to the interannual variability. If the former were comparable to the latter, the data sets used here would provide little basis for conclusions about the interannual variability of drafts averaged over 50 km at the North Pole.

In order to address this issue, we have evaluated the differences between the mean drafts of adjacent 50-km segments of data from six late-April/early-May cruises: FLYING FISH 1977, ARCHERFISH 1979, ARCHERFISH 1986, BILLFISH 1987, ARCHERFISH 1988 and GURNARD 1990. McLaren et al. (1992) discuss the data from these cruises in more detail. The mean absolute difference in the draft of *adjacent* 50-km segments of these six cruises is 0.33 m (standard deviation = 0.20 m). By contrast, the mean absolute difference in the drafts of 50-km segments from *different* cruises (in different years) is 0.64 m. Thus the variability across cruises made at the same time of year exceeds the same-cruise variability by approximately two standard deviations of the latter. One may reason that the variability across cruises would be even larger, relative to the same-cruise variability, had cruises from other parts of the year been included. This comparison indicates that the 0.5-1.5 m differences discussed here are largely manifestations of interannual variations, while some of the differences smaller than approximately 0.5 m may be attributable to sampling variations.

Table 1 indicates that the two fall cruises (PINTADO 1978 and SILVERSIDES 1989) reported the highest incidence of leads for both 50- and 100-km-long segments, about 2 to 3 times greater than that noted during most of the spring cruises. This expected seasonal difference has been routinely observed during submarine cruises, in satellite images (Barry and Maslanik, 1989) and in ice thickness analyses (Wadhams, 1983; Bourke and Garrett, 1987). The seasonal trend is obscured in 1986 as the higher-than-normal percentages of open water/new ice conditions (Table 1) occurred in the year of thinnest mean ice draft among the 12 data sets examined. The 1986 cruise was conducted in May, prior to any significant melting activity. Hence, it is not

TABLE 1: Comparative Sea-Ice Statistics for 50-km and 100-km-Long Transects Centered Over the North Pole

	Flyingfish 1977	Pintado 1978	Archerfish 1979	Archerfish 1986	Billfish 1987	Archerfish 1988	Silversides 1989	Gurnard 1990	Pargo Grayling 1991	1992
					Ice Draft					
Mean Draft (m)										
50 km	4.2	2.9	4.0	2.8	4.1	3.3	3.5	3.6	3.5	3.1
100 km	4.0	2.9	5.5	3.1	4.3	3.4	3.7	-	3.5	3.3
Standard Deviation (m)										
50 km	2.2	2.1	2.8	1.9	2.2	1.6	2.3	2.2	2.4	2.4
100 km	2.2	1.8	2.8	2.0	2.3	1.7	2.5	-	2.3	2.4
					Ice Types					
First Year (%)										
<3 m										
50 km	40.6	68.6	49.8	72.9	27.2	57.4	51.2	54.7	57.2	67.2
100 km	46.0	66.9	48.4	60.2	25.3	54.1	47.1	-	57.8	62.4
Mulityear/Deformed (%)										
>3 m										
50 km	59.4	31.4	50.2	27.1	73.8	43.6	48.8	45.3	42.8	32.8
100 km	54.0	33.1	51.7	39.8	74.7	45.9	52.9	-	42.2	37.6
					Polynyas/Leads					
Open Water/New Ice										
<0.3 m										
50 km	0.0	3.0	0.8	2.5	0.1	0.0	2.8	0.1	0.7	1.8
100 km	0.0	1.9	1.6	1.4	0.1	0.0	2.4	-	0.6	1.4

TABLE 2: Percentages of Ice Cover Within various Draft Ranges for 50 km Segments Centered on the North Pole

	0-0.3 m	0.3-0.7 m	0.7-1.2 m	1.2-2.0 m	2-3 m	3-4 m	>4 m	Mean (m)	Keels (m)	#Keels km	Level %
Nautilus 58[1]	1.1	5.4	8.6	16.9	18.4	14.8	34.7	3.1	--	1.1	44.1
Queenfish 70[1]	1.3	3.0	1.2	2.4	36.7	22.0	33.4	4.1	--	2.2	39.3
Flyingfish 77	0.0	0.0	0.3	1.9	38.3	21.9	37.6	4.2	11.4	3.1	25.2
Pintado 78	3.0	4.2	2.7	27.3	31.5	14.0	17.4	2.9	11.3	1.3	40.1
Archerfish 79	0.8	0.3	1.3	14.2	33.2	17.8	32.5	4.0	12.5	1.5	45.5
Archerfish 86	2.5	1.5	5.6	22.6	40.7	11.2	15.9	2.8	11.5	1.3	41.6
Billfish 87	0.1	1.4	0.9	5.5	18.3	35.9	37.9	4.1	11.7	2.3	34.2
Archerfish 88	0.0	0.2	0.5	6.8	49.8	23.2	19.4	3.3	10.7	1.1	42.8
Silverside 89	2.8	1.3	2.7	6.7	37.8	23.1	25.7	3.5	11.7	1.6	36.9
Gurnard 90	0.1	0.7	2.3	7.9	43.7	19.5	25.8	3.6	11.3	2.2	41.4
Pargo 91	0.7	1.1	2.0	7.2	46.2	19.4	23.4	3.5	12.1	1.8	35.7
Grayling 92	1.8	1.0	4.5	36.6	23.4	10.2	22.7	3.1	11.9	2.1	42.9
Mean	1.2	1.7	2.7	13.0	34.8	19.4	27.2	3.5	11.6	1.8	39.1
Stn Dev	1.1	1.7	2.4	10.9	10.4	6.9	7.8	0.5	0.5	0.6	5.6

1. 192 km - long segment
2. 338 km - long segment

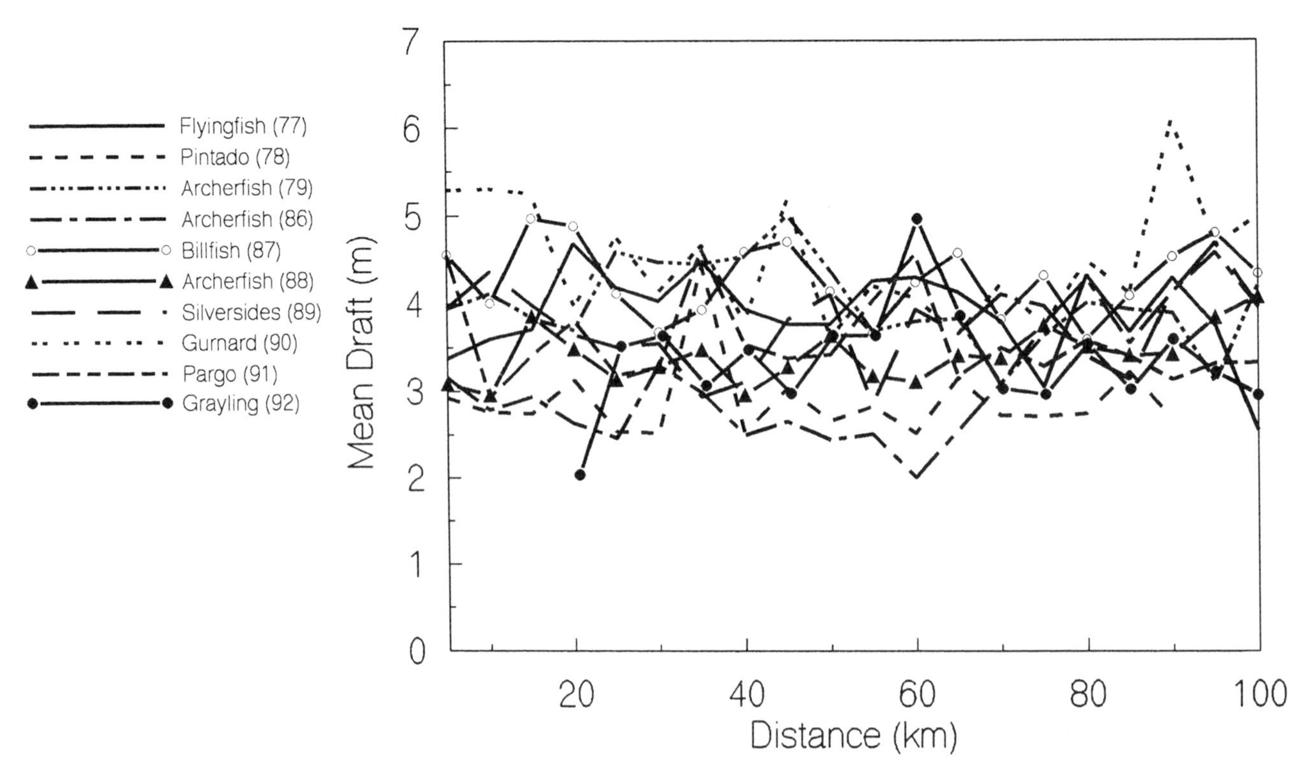

Figure 1. Mean ice draft at 5-km intervals for 100-km-long transects across the North Pole derived from submarine-mounted upward-looking sonars.

certain whether the excessive lead/polynya occurrences noted are responding to a seasonal or interannual forcing.

Analysis of lead/polynya occurrence in the immediate vicinity of the North Pole indicates that they appear to be fairly evenly spread out over the 50 or 100 km path length with no one single area of predominant open water or thin ice. Also, as can be seen in Table 2, the summer/fall increase in lead/polynya occurrence (i. e., ice draft <0. 3 m) does not appear to influence the overall mean ice draft. The mean appears to respond more to changes in occurrence of ice 1 to 2 m thick or greater than 4 m thick.

The mean ice draft for each 50 km segment centered over the Pole for each of the 12 cruises is listed in Table 2 and plotted as a function of time in Figure 2. Four years (1970, 1977, 1979 and 1987) stand out as periods of relatively thicker ice (>4 m) while 1978, 1986 and 1992 were years of relatively thinner ice (≤3.1 m). The remainder were within 0.4 m of the overall mean, 3.6 m.

No discernable trend in mean ice draft over this 34-year period (Figure 2) is evident which would suggest either a tendency towards thinning or thickening of the ice cover at the North Pole. A linear regression indicates that the uncertainty (standard error) of the slope of a least squares fit to the data is larger than the slope itself (slope = -0.0126 m yr^1; standard error = 0.0147). A Monte Carlo test performed to assess the significance of the slope indicates that a slope of at least this magnitude is likely to occur by chance with a probability of more than 70%. Thus the slope is clearly insignificant in a statistical sense.

Examination of Tables 1 and 2 indicates that years of relatively thicker ice are associated with years of excessive amounts (~10% more) of multiyear/deformed ice, while thin-ice years result from substantial increases (~20% more) in young and first-year ice along with reduced quantities of multiyear/deformed ice. These icetype percentages are plotted in Figure 2. The correlation of years having high quantities of multiyear/deformed ice and greater than average mean ice draft is exceptional. The inverse relation between thin ice years and excessive amounts of ice < 2 m thick is also good.

Table 2 indicates that there is little variation in the mean keel draft (> 9 m) from year-to-year. The standard deviation of 0.5 m is small relative to the mean draft of 11.6 m. Hence, those years (1970, 1977, 1979 and 1987) exhibiting large percentages of deformed ice (>4 m) are not solely the result of the presence of deep-draft ice keels but appear to be the result of increased deformation and rafting. These years are

Probability Density Function

Thorndike et al. (1975) defined the ice thickness distribution, g(h), over a range of ice thicknesses h_1 <h <h_2

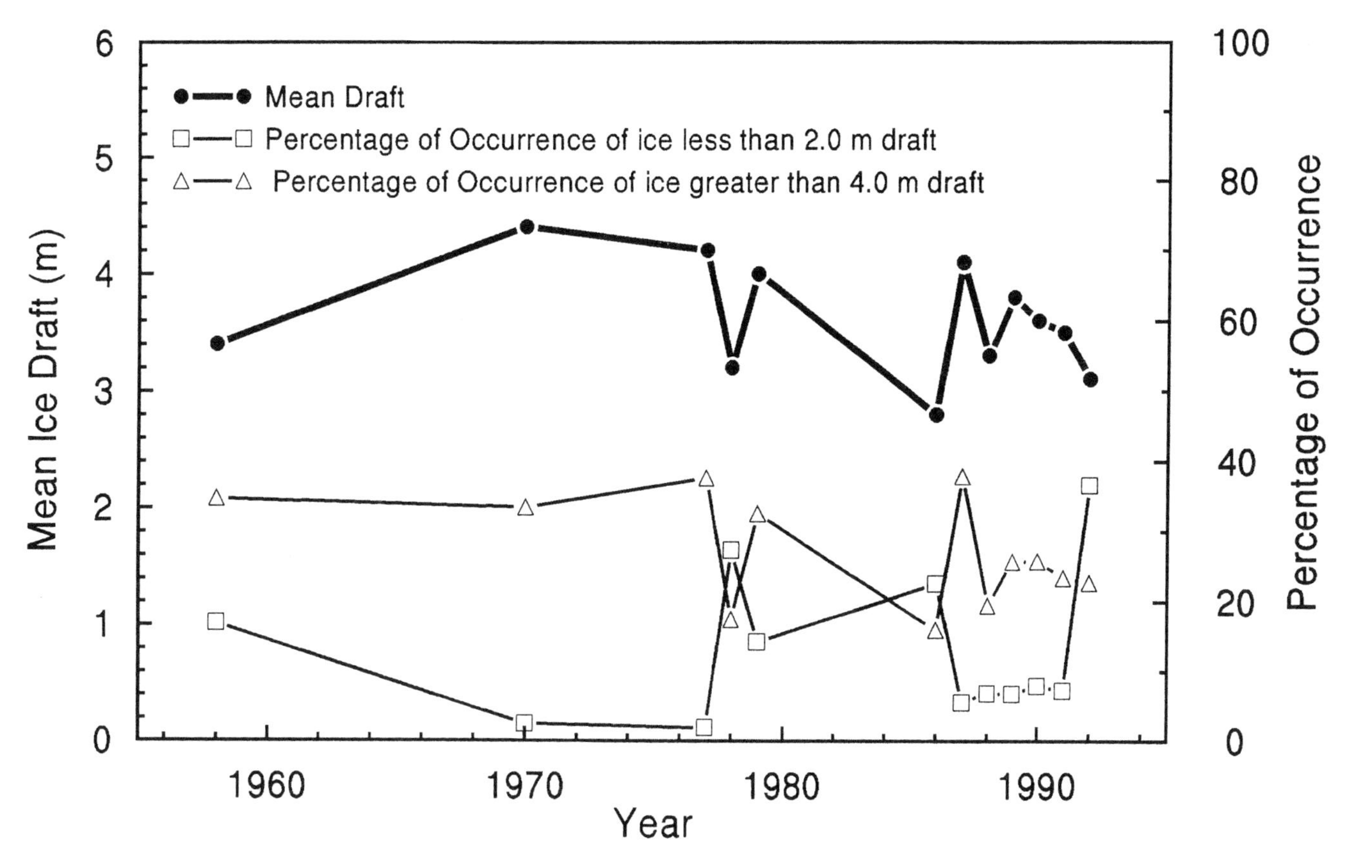

Figure 2. Time series of mean ice draft and frequency of ice draft less than 2.0 m and greater than 4.0 m for 50-kmlong segments centered over the North Pole.

to be $\int g(h)\,dh = 1$, valid for a given area of ice ranging in thickness from open water ($h = 0$) to an upper thickness limit (h_{max}). The distribution, $g(h)$, can also be interpreted as a probability density function which varies in time and space over a broad range of scales.

As a component of the statistical analysis typically performed on the submarine sonar data, ice thickness pdf's are routinely determined for selected track segments of each submarine under-ice sonar data package analyzed (Wadhams and Horne, 1980; McLaren, 1989). Wadhams (1983) shows that most pdf's initially rise rapidly to a peak thickness of 2 to 3 m and then tail off slowly towards thicker values with a shape that can be described by a negative exponential distribution. Subsequent work by Wadhams (1986, 1992) and by McLaren (1984, 1988, 1989) have confirmed this. Sea ice thickness pdf's demonstrate both temporal and spatial variability, exhibiting higher probabilities of thin ice in fall and early winter at the onset of freezing and of thicker ice off the north coasts of Greenland and the Canadian Archipelago due to convergence of the ice by winds and currents on the continental land masses (Bourke and Garrett, 1987). Basin-wide depictions of seasonal ice thickness distributions have not been previously presented due to an insufficient number of suitable coincidental submarine under-ice sonar data sets. The 12 North Pole data sets analyzed for this report provide, for the first time, an excellent example of the size and character of the variability in ice thickness pdf that may be encountered at a given location.

The percentage of ice contained in bins of varying ice draft is listed in Table 2 for each submarine voyage analyzed. The bin size is not constant but reflects the thickness values or ranges routinely tabulated in the literature, permitting intercomparison with older data sets. The mean pdf from Table 2 is plotted as a histogram with error bars illustrating the standard deviation within each bin (Figure 3). One sees that the mean ice draft of 3.6m is not associated with the modal value and that the percentage of deformed ice (>4 m) exceeds that of ice near the mean. The standard deviations listed in Table 2 represent a substantial fraction of the total in each bin width, indicative of the interannual variability described earlier. Because of these large standard deviations, mean sea ice thickness pdf's should be used cautiously when incorporated in climate models in place of more traditional long-term mean ice drafts.

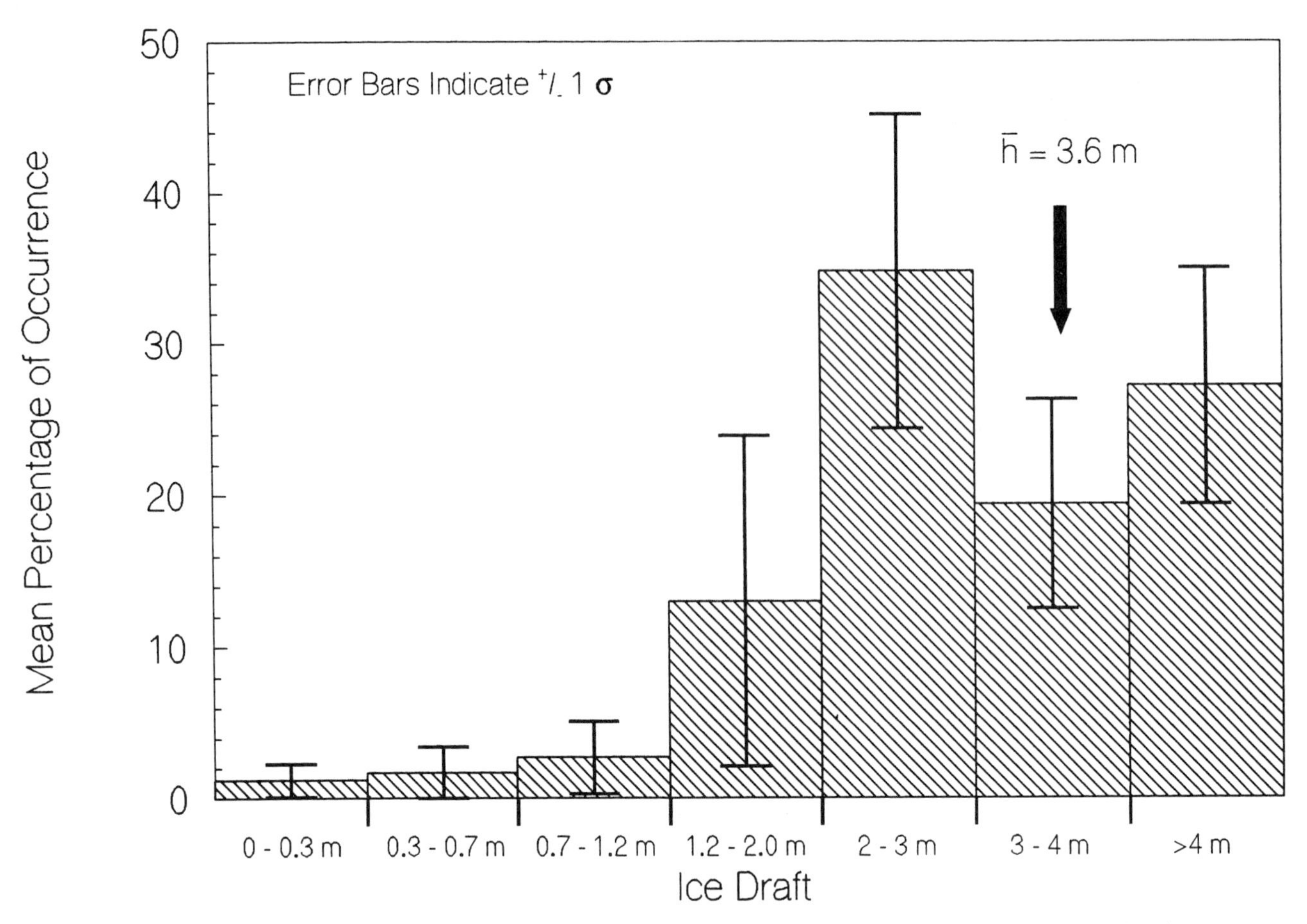

Figure 3. Histogram of ice draft averaged from twelve 50km-long segments centered over the North Pole taken between 1958 and 1992.

That the ice thickness pdf at the North Pole exhibits significant year-to-year variability is not surprising given its location relative to the Transpolar Drift Stream (TDS). Examination of the mean annual atmospheric pressure pattern over the Arctic Ocean (Colony and Rigor, 1989), reveals large variations in the strength and direction of the TDS (as influenced predominantly by atmospheric forcing) in the vicinity of the Pole. The pressure field near the Pole shows years when the cross-basin flow is weak and other years when it is well-organized and intense. Obviously, the amount of deformed ice observed in a given year will reflect the forcing experienced during the months of active ice formation.

It is only at the North Pole that replicate data exist to formulate mean thickness pdf's. In order to establish mean thickness pdf's for other regions within the Arctic basin, more submarine echo-sounder data must be obtained, edited and analyzed to identify regions of spatial coherency before any averaging can be performed. This should be possible in the next few years as more submarine-derived data are made available to the research community. In the meantime, we intend to re-examine the North Pole data sets to derive true thickness pdf's having a constant bin width. This is especially necessary for the thicker ice "tail" of the distribution.

SUMMARY

Twelve data sets of submarine upward-looking sonar measurements of sea ice draft in the vicinity of the North Pole, spanning a period of 34 years (1958-1992), have been examined in order to establish the nature of its interannual variability there and to determine if any trend can be discerned that might be ascribed to warming of the atmosphere. The North Pole is the only location in the Arctic Basin where repeated observations of the ice draft permit such an analysis. The mostly late winter (late April/early May) ice draft data for each submarine cruise, measured at approximately 1.5 m intervals along the submarine track, were averaged over 50 or 100 km-long segments in order to establish comprehensive sets of mean statistics for analysis and comparison.

The overall mean ice draft at the North Pole during this period was 3.6 m but a large year-to-year variability, ranging from 2.8 m in 1986 to 4.4 m in 1970, was observed. This is more than enough to obscure any evidence of ice cover thinning. The slope of a least squares fit to the 34-year record exhibited no statistical significance. Large changes in mean ice draft and open water/new ice are noted to occur on scales as short as one year. Years of anomalously high mean draft (1970, 1977, 1979 and 1987) are associated with years of excessive amounts of multiyear or ridged ice (~10% more). On the other hand, thin ice years (1978 and 1986) are characterized by increased percentages (~20% more) of young and first-year ice along with reduced quantities of ridged ice.

Because the heat flux through open or thin-ice areas is two orders of magnitude greater than that through thick ice (Badgley, 1966; Maykut 1978), the considerable year-to-year variability noted within the ice thickness categories of Table 2 obtained during 12 different years demonstrates the importance of using an ice thickness probability distribution function in climate models rather than mean ice thickness values. As more pdf data become available for other regions of the Arctic Ocean, climate and ice prediction modelers should consider incorporating spatially varying pdf's into their models in order to more accurately describe the thickness of the sea ice cover over the Arctic Ocean.

Acknowledgements. This work was supported by an NSF Arctic Programs SGER grant (DPP-9014884) and a NOAA grant (NA26GPO1O1-01).

REFERENCES

Badgley, F. I., Heat balance at the surface of the Arctic Ocean, in *Proceedings of the Symposium on the Arctic Heat Budget and Atmospheric Circulation,* J. 0. Fletcher, editor, RM-5233-NSF, Rand Corp., Santa Monica, 1966.

Barry, R. G. and J. A. Maslanik, Arctic sea ice characteristics and associated atmosphere-ice interactions in summer inferred from SMMR data and drifting buoys, 1979-1984, *Geojournal, 18,* 35-44, 1989.

Bourke, R. H. and R. P. Garrett, Sea ice thickness distribution in the Arctic Ocean, *Cold Reg. Sci. Technol.,* 13, 259-280, 1987.

Bourke, R. H. and A. S. McLaren, Contour mapping of Arctic Basin ice draft and roughness parameters, *J. Geophys. Res., 97(C* 1 1), 17,715-17,728, 1992.

Colony, R. L. and 1. Rigor, Arctic Ocean buoy program data report for 1 January 1986-31 December 1986, APL-UW TM 6-89, Applied Phys. Lab., Univ. of Washington, Seattle, 1989.

Gloersen, P. and W. J. Campbell, Recent variations in Arctic and Antarctic sea-ice covers, *Nature, 352,* 33-36, 1991.

Hibler, W. D. III, Modeling a variable thickness sea ice cover, *Mon. Weather Rev.,* 108(12), 1943-1973, 1980.

Hibler, W. D. Ill, S. J. Mock and W. B. Tucker Ill,Classification and variation of sea ice ridging in the western Arctic Basin, *J. Geophys. Res.,* 79(18), 2735-2743, 1974.

Intergovernmental Panel of Climate Change, Climate change: The IPCC assessment, Houghton, J. T., G. J. Jenkins and J. J. Ephraums (editors), Cambridge Univ. Press, Cambridge, 1990.

Manabe, S., M. J. Spelman and R. J. Stouffer, Transient responses of a coupled ocean-atmosphere model to gradual changes of atmospheric CO_2 Part II: Seasonal response, *J. Climate,* 5, 105-126, 1992.

Maykut, G. A., Energy exchange over young sea ice in the central Arctic, J. Geophys. Res., 83(C7), 36463658, 1978.

Maykut, G. A. and N. Untersteiner, Some results from a time dependent, thermodynamic model of sea ice, *J. Geophys. Res.,* 76, 1550-1575, 1971.

McLaren, A. S., Analysis of the under-ice topography of the Arctic Basin as recorded by USS NAUTILUS in 1958 and USS QUEENFISH in 1970, Ph. D. Thesis, 163 pp., Univ. of Colorado, Boulder, 1986.

McLaren, A. S., Analysis of the under-ice topography in the Arctic Basin as recorded by the USS NAUTILUS, August 1958, *Arctic,* 41, 117-126, 1988.

McLaren, A. S., The under-ice thickness distribution of the Arctic Basin as recorded in 1958 and 1970, *J. Geophys. Res.,* 94(C4), 4971-4983, 1989.

McLaren, A. S, P. Wadhams and R. Weintraub, The sea ice topography of M'Clure Strait in winter and summer of 1960 from submarine profiles, *Arc-tic,* 37(2), 110-120, 1984.

McLaren, A. S., R. G. Barry and R. H. Bourke, Could Arctic ice be thinning?, *Nature,* 345, 762, 1990.

McLaren, A. S., J. E. Walsh, R. H. Bourke, R. L. Weaver and W. Wittmann, Variability in sea-ice thickness over the North Pole from 1977 to 1990, *Nature,* 358, 224-226, 1992.

Moritz, R. E., Seasonal and regional variability of sea ice thickness distribution, in Proceedings of the Third *Conference on Polar Meteorology and* Oceanography, pp. 68-71, Portland, OR, Am. Meteor. Soc., Boston, 1992.

Parkinson, C. L. and D. L. *Cavalieri, Arctic sea ice 1973 1987: seasonal, regional and interannual variability, *J. Geophys. Res.,* 94(CIO), 14,49914,523, 1989.

Thorndike, A. S., D. A. Rothrock, G. A. Maykut and R. L. Colony, The thickness distribution of sea ice, *J. Geophys. Res.,* 80(33), 4501-4513, 1975.

Thorndike. A. S., A toy model linking atmospheric thermal radiation and sea ice growth, *J. Geophys. Res.,* 97(C6), 9401-9410, 1992.

Wadhams, P., Arctic sea ice morphology and its measurement, *J. Soc. Underw. Technol.,* 9(2) 112, 1983.

Wadhams, P., Evidence for thinning of the Arctic ice cover north of Greenland, *Nature,* 345, 795-797, 1990.

Wadhams, P., Sea ice thickness distribution in the Greenland Sea and Eurasian Basin, May 1987, *J. Geophys. Res.,* 97(C4), 5331-5348, 1992.

Wadhams, P. and R. J. Home, An analysis of ice profiles obtained by submarine sonar in the Beaufort Sea, *J. Glaciology,* 25(93), 401-424, 1980.

Wadhams, P. and T. Davy, On the spacing and draft distributions for pressure ridge keels, *J. Geophys.Res.,* 91(C9), 10697-10708, 1986.

On the Required Accuracy of Atmospheric Forcing Fields for Driving Dynamic-Thermodynamic Sea Ice Models

Holger Fischer

Alfred-Wegener-Institut für Polar- und Meeresforschung, Bremerhaven, Germany

Peter Lemke

Alfred-Wegener-Institut für Polar- und Meeresforschung, Bremerhaven, Germany and Institut für Umweltphysik, Universität Bremen, Germany

Presently several improved sea ice - ocean models are being developed in order to investigate the role of sea ice in determining the boundary conditions at the sea surface and to assess its influence on the global ocean circulation. For driving sea ice - ocean models atmospheric data sets are being compiled using analyses of numerical weather prediction models, in situ observations and remote sensing products. Since these forcing data (wind, temperature, humidity, precipitation and cloud cover) originate from a variety of sources with different accuracy, the question arises how accurate the atmospheric data has to be in order to be useful for a realistic simulation of sea ice variability on a seasonal and interannual time scale. This paper presents an error analysis on the basis of numerical experiments with a dynamic-thermodynamic sea ice - mixed layer model for the Weddell Sea. The standard model experiment was optimized to reproduce the observed ice concentrations derived from satellite passive microwave data for the years 1986 and 1987. In addition a long term comparison with six drifting buoys deployed on the sea ice during the same period shows a good agreement between model drift and buoy trajectories. Further model experiments show that a 10 % deviation from the standard seasonal cycle is produced by errors of the order of $0.6°C$ for surface air temperature, 0.2 for cloud cover, 16 % for relative humidity, 27 W/m^2 for incoming shortwave radiation, 18 W/m^2 for incoming longwave radiation and 24 % of the wind speed. This indicates that improved analyses are required especially for the surface air temperature field.

1 INTRODUCTION

In recent years sea ice models for polar regions have received increasing attention as a component of models simulating climate processes. Sophisticated sea ice models include thermodynamic and dynamic algorithms which are coupled to the ocean and the atmosphere. It has been shown by various authors [Hibler, 1979; Stössel et al.,1990] that climatologically forced dynamic-thermodynamic sea ice models are able to simulate the seasonal retreat and advance of the sea ice cover reasonably well. During the last years sea ice simulations have been carried out using daily forcing data from the analyses of numerical weather prediction models [Hibler and Ackley,1983; Lemke et al.,1990; Stössel, 1992]. The quality of these forcing data varies regionally, especially with the density of the observing stations. Since there are only a few stations in polar regions, the forcing data there largely represent model products. With the goal of seasonal and interannual simulations of sea ice, the accuracy of daily forcing data in polar regions plays an important role as input for sea ice models.

Sea ice thickness, concentration and drift are the main prognostic variables of sea ice models. It is known from various sensitivity studies with sea ice models [e.g. Hibler, 1984; Owens and Lemke, 1990; Chapman et al., 1992], that the sea ice variables are rather sensitive to changes of certain atmospheric forcing data or model parameters. A comparitive study which relates the influence of the various atmospheric forcing variables on sea ice thickness, concentration and drift has not been presented yet. Two of the main questions are: How accurate do we have to know the forcing data and how precise do we have to define the optimal model parameter set?

This study calculates the allowable error of the atmospheric forcing data and of some model parameters such that the deviation of the sea ice variables is less than 10% of the standard simulation. These calculations are carried out for the Weddell Sea region. After a short description of the sea ice model (section 2)

The Polar Oceans and Their Role in Shaping the Global Environment
Geophysical Monograph 85

the results of the standard run (section 3.1), which is compared to observations, and sensitivity experiments (section 3.2) are presented. These sensitivity results are discussed and summarized in section 4.

2 DESCRIPTION OF THE SEA ICE MODEL

2.1 *Sea ice model*

The present sea ice model is based on the dynamic-thermodynamic model presented by Hibler [1979] and is coupled to a 1-dimensional oceanic mixed-layer model which determines the oceanic heat fluxes prognostically (see Lemke et al. [1990] for further details). The thermodynamic formulation is primarily a version of the heat budget calculation from Parkinson and Washington [1979] which treats the incoming longwave and shortwave radiation fluxes with standard bulk formulations based on Idso and Jackson [1969] and Zillman [1972]. A snow layer is included based on Owens and Lemke [1990]. According to Semtner [1976] a linear temperature profile is assumed in the snow and ice layers.

A scheme for snow/ice conversion, similar to Leppäranta [1983], is added to the thermodynamic part of the model. This conversion considers the flooding effect of sea ice floes due to a negative freeboard which is often observed in Antarctic regions because of the heavy snow load. The snow/ice conversion embeded in the numerical scheme is required for a more realistic simulation of the observed snow thicknesses.

2.2 *Model domain*

The model domain with respect to the velocity grid points represents the South Atlantic region from 60° W to 60° E and from 80° S to 47.5° S. The spatial resolution is 2.5° × 2.5° degrees on a polar stereographic grid. The model is integrated with a daily timestep to obtain an update of the sea ice conditions. At continental boundaries a no-slip condition is applied, whereas the horizontal oceanic boundaries are represented by outflow cells where ice may enter or leave the model domain due to advection or diffusion.

2.3 *Forcing data*

The daily forcing data originate from the European Centre for Medium Range Weather Forecasts (ECMWF) and include the wind components, air temperature, relative humidity and air pressure for the period of the years 1986 and 1987. The surface forcing data are extra- or interpolated from 1000 hPa and 850 hPa pressure levels (Trenberth and Olson [1988]), with the exception of the wind components at 1000 hPa which are directly applied as surface winds [Stössel, 1992]. The twelve-hourly surface values are averaged to daily data.

Additionally, the annual mean of the zonally averaged total cloud amount, based on the analyses of van Loon [1972] is used. The precipitation rate is set to a constant value of 35 cm/year. From Olbers and Wübber [1991] the annually averaged geostrophic currents are interpolated on to the model grid. The geostrophic currents are coupled to the sea ice model through the oceanic drag coefficient c_w and a turning angle of 25° degrees. Precipitation, total cloud amount and ocean currents are used as climatological forcing because seasonal and daily variations are insufficiently known.

3 RESULTS

3.1 *Standard run*

In order to reach an equilibrium seasonal cycle the model was integrated for five years with the atmospheric forcing from 1986. Using the 31 December sea ice conditions as a restart a transient run was started for the years 1986 and 1987. The analysis of the 1987 integration was used to describe the results. The main model parameters and their standard values are listed in Table 1.

The values of some of the model parameters correspond well to those of previous large scale antarctic sea ice models [Hibler and Ackley,1983; Lemke et al., 1990; Stössel et al.,1990], while some dynamic parameters are slightly different. These modifications result from a comparison of the simulated velocities to the drift of ARGOS buoys deployed during the Winter Weddell Sea Project 1986 (WWSP'86) [Kottmeier and Hartig,1990] and a drift buoy (No. 534 in Figure 1) deployed by the Scott Polar Research Institute / British Antarctic Survey (SPRI/BAS) in 1986 [Rowe et al.,1989]. The observed daily drift tracks are analysed for the fall, winter and spring seasons. The long-term comparison of modelled and observed drift uses a Lagrangian technique similiar to Hibler and Ackley [1983] and Ip et al. [1991].

One of the main differences to previous investigations are the modified values of the atmospheric and oceanic drag coefficients,

Table 1. Model parameters of the standard run

No.	Parameter	Meaning	Standard value
1.	α_i	ice albedo	0.75
2.	α_s	snow albedo	0.85
3.	α_{im}	melting ice albedo	0.66
4.	α_{sm}	melting snow albedo	0.75
5.	α_w	water albedo	0.1
6.	h_o	lead closing parameter	0.5 $[m]$
7.	P^*	ice strength parameter	2.0×10^4 $[N/m^2]$
8.	C	dynamic ice parameter	20
9.	c_w	oceanic drag coefficient	3.0×10^{-3}
10.	c_a	atmospheric drag coefficient	1.5×10^{-3}

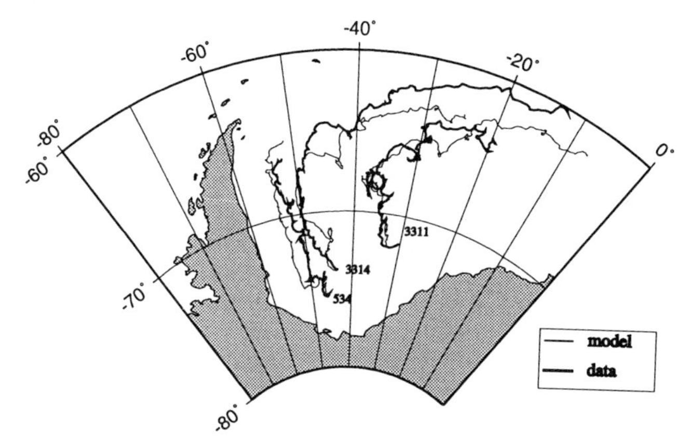

Figure 1: *Comparison of observed (thick lines) and simulated (thin lines) buoy drift for the standard run. Buoy No. 534 represents the period from 1 May to 31 December 1986. Buoys No. 3311 and 3314 are examples from a cluster of buoys drifting from 1 April to 31 December 1987.*

c_w and c_a, respectively. McPhee [1980] and Stössel [1992] pointed out, that the ratio of the drag coefficients of $c_w/c_a = 2$ (using surface winds) is more important than their actual values. An oceanic drag coefficient of $c_w = 5.5 \times 10^{-3}$ derived by McPhee [1980] from manned drift stations during the Arctic Ice Dynamics Joint Experiment (AIDJEX) is commonly used in Arctic sea ice models. This value of c_w was also applied to previous Antarctic sea ice simulations [Hibler and Ackley,1983; Lemke et al., 1990; Stössel et al.,1990]. In our simulations a reduction of the drag coefficients to $c_w = 3.0 \times 10^{-3}$ and $c_a = 1.5 \times 10^{-3}$ such that the ratio remains $c_w/c_a = 2$ yields the best correspondence between model and buoy drift for the periods of 1 May 86 to 31 December 86 and 1 April 87 to 31 December 87 (see Figure 1). These periods were chosen because the buoys then certainly drifted within the sea ice pack and not in open water.

Wamser and Martinson [1993] analysed ship drift data during WWSP'86 and obtained an atmospheric drag coefficient of $c_a = 1.72 \times 10^{-3}$. Overland [1985] summarized observed atmospheric drag coefficients (10 m) for different locations and surface roughnesses. His values range from $c_a = 1.2 - 3.7 \times 10^{-3}$ and agree with our buoy-fitted model value of $c_a = 1.5 \times 10^{-3}$.

Usually a value of $2.75 \times 10^4\ N/m^2$ for the ice strength parameter P^* is used in sea ice models with daily forcing [e.g. Hibler and Ackley, 1983]. Since variations of P^* have a stronger effect in regions of convergent or shearing motion like the western Weddell Sea, the buoy which drifted in 1986 in northerly direction along the Antarctic peninsula (buoy no.534 in Figure 1) was used to optimize the ice strength parameter. The best model and buoy track correspondence was obtained with a reduced value of $P^* = 2.0 \times 10^4\ N/m^2$. The buoys in areas with predominantly divergent conditions like the eastern Weddell Sea are less influenced by changes of P^*. They are more controlled by the ratio of the drag coefficients.

The simulated variations of the ice extent (area southward of the ice edge, Figure 2) and ice covered area (grid area times concentration, Figure 3) are compared to data from passive microwave satellite sensors. The Scanning Multichannel Microwave Radiometer (SMMR) data were used from 1 January 86 to 8 July 87, whereas the Special Sensor Microwave/Imager (SSMI) data represent the period from 9 July 87 to 2 December 87. The remote sensing data were calculated from a pixelsize of $50 \times 50\ km^2$.

For both years Figures 2 and 3 indicate higher minima for the ice model as compared to the data. The difference in 1987 is approximately $1.5 \times 10^6\ km^2$ for the ice extent and $1.3 \times 10^6\ km^2$ for the ice covered area. This is primarily due to an enhanced horizontal ice distribution in the south-eastern Weddell Sea (see Figure 4) which indicates that the temperature analyses of the ECMWF in summertime may have been too cold. Figure 5 shows time series of surface air temperature from buoys and ECMWF data during the period 1 January to 30 April 1987 . The buoys which drifted near the summer ice edge region [Kottmeier and Hartig,1990] show generally higher air temperatures during the melting season as compared to the ECMWF data. At the end of April, with the beginning of the winter season, buoy data and ECMWF analyses agree on average.

A comparison of the simulated and observed ice covered area indicates that the model seems to simulate larger areas of high ice concentration (Figure 6). The remote sensing data for the ice concentration are derived from National Air and Space Administration (NASA) Team algorithm and underestimate the total ice concentration at high concentration [Allison et al., 1993]. This effect is probably due to an underestimation of young ice in the ice concentration range of 85% to 100% [Steffen et al., 1992]. Figure 7 shows the total ice concentration from September 1987 with the same SSMI-data as Figure 6b, but derived with a competitive algorithm [Comiso et al.,1992]. This algorithm gives a better agreement with simulated ice concentrations (Figure 6a) which show larger areas of high ice concentrations together with a small marginal ice zone (see also dotted line in Figure 3). The differences between Figure 6b and 7 call for a greater effort to validate existing algorithms for sea ice concentration using passive microwave data.

The geographical distribution of sea ice thickness (not shown here) is similar to Fig. 1 in Harder and Lemke (this volume).

3.2 Sensitivity studies

3.2.1. Model experiments. Differences between simulations and observations are partly due to errors both in the model physics as well as in the atmospheric forcing conditions. In order to determine the response of the model to those errors sensitivity experiment with modified model parameters and surface forcing were performed. The degree of modification is shown in Table 2. Generally four experiments were run in addition to the standard experiment: two with plus/minus the maximum modification and two with half the change.

Table 2 lists the maximum range of variations of forcing or model parameters and terms which are strongly connected to forcing pa-

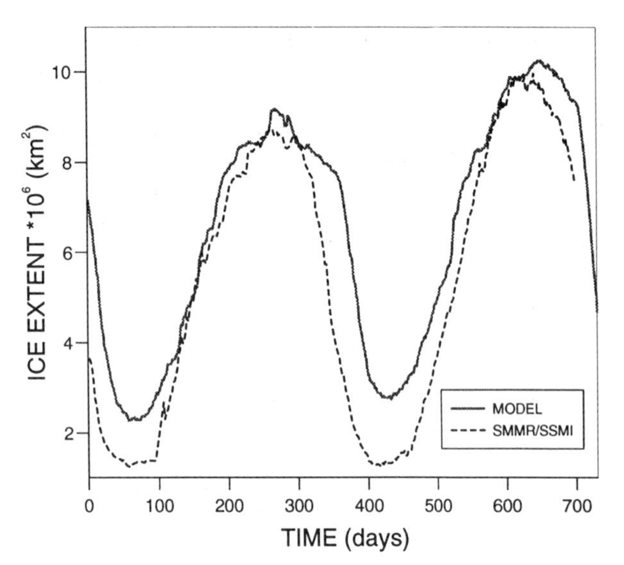

Figure 2: *Seasonal cycle for 1986 and 1987 of the integrated ice extent (> 15%) for the standard experiment (solid line) and the SMMR/SSMI data (dashed line).*

rameters. No. 1 to 6 in Table 2 are variables which depend on the daily forcing, whereas No. 7 to 10 are specified model parameters. The radiation terms $F_s\downarrow$ and $F_l\downarrow$ are not directly read into the model. Their bulk formulation [Parkinson and Washington, 1979] depends strongly on the forcing parameters (such as cloud cover, humidity, air temperature) and plays an important role for the thermodynamic characteristics of the sea ice, especially the local growth and melting. The radiation terms represent a good indicator for the interaction of the forcing parameters and provide the possibility for a comparison to variations of single forcing parameters. The model

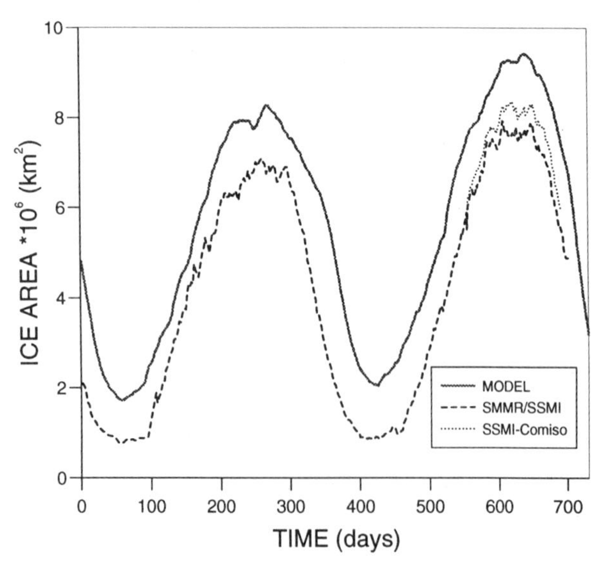

Figure 3: *Same as Fig.2 for the sea ice covered area using the NASA-Team-Algorithm (dashed line) and the Comiso-Algorithm (dotted line).*

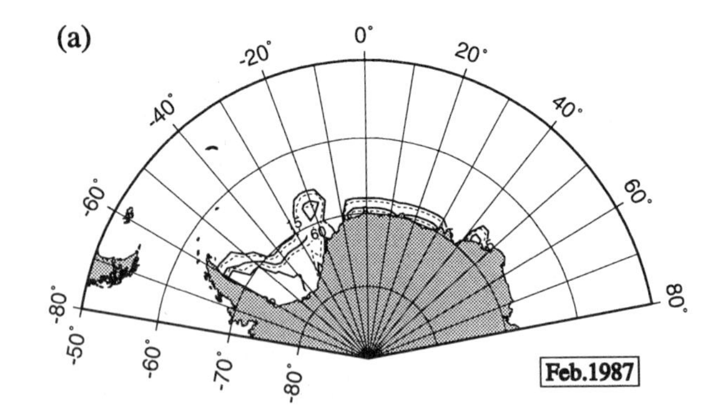

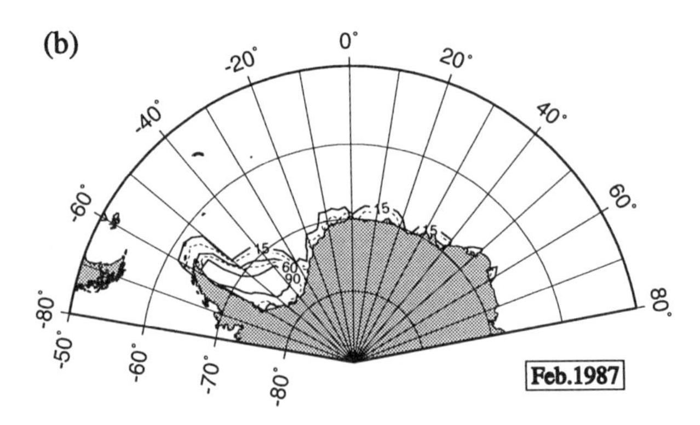

Figure 4: *Mean February total ice concentration in percent derived from the standard model (a) and from SMMR data according to the NASA-Team-Algorithm (b).*

parameters α and h_0 also influence the thermodynamic part of the sea ice model. The parameter variations take into account that the total cloud amount Cl is limited by $0 \leq Cl \leq 1$ and the shortwave radiation $F_s\downarrow \geq 0$. The surface wind (No.6) is only changed concerning its magnitude to maintain the general wind pattern for the model region. Albedo variations (No.7) are only considered for dry and melting snow and ice. The water albedo α_w is taken to be constant at 0.1.

The sensitivity experiments represent two-year integrations with the 1986/1987 atmospheric forcing and with modified parameters. The common starting point for all sensitivity runs is the fifth year equilibrium run for 1986. The final year 1987 is used in the analysis of the model results. The response of the model is analysed in terms of the maximum and minimum ice extent and ice volume. The modified maximum and minimum values are analysed with a linear regression. Furthermore the mean September drift pattern for 1987 is used to analyse the response of the ice drift. To exclude boundary effects in the drift pattern only velocities south of 60° S are considered. The velocities of all grid boxes inside this region are added to obtain an areal average of the mean September drift velocity $\overline{v}$.

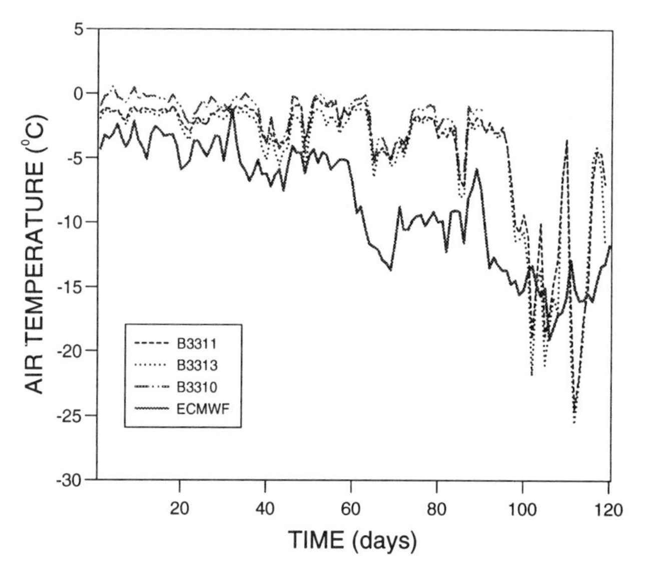

Figure 5: *Surface air temperatures from three drifting buoys No. 3310,3311 and 3313 (located near the summer ice edge) and corresponding ECMWF data for the period 1 January to 30 April 1987.*

As an example Figure 8 represents the response of the maximum and minimum ice volume and ice extent to variations of the surface air temperature. The slopes (b, solid lines) derived from a linear fit indicate the sensitivity of the model to parameter modifications.

$$b^V_{S,W} = \frac{\Delta V_{S,W}}{\Delta P} \quad (1)$$

$$b^E_{S,W} = \frac{\Delta E_{S,W}}{\Delta P} \quad (2)$$

Where V is the sea ice volume, E the sea ice extent, S and W denote the summer and winter extrema and P the forcing variable or the model parameter.

The annual cycle of the sea ice volume for 1987 with respect to variations of the surface air temperature is shown in Figure 9. The analog figure for the ice extent with the same variations is shown in Figure 10. The curve for air temperatures increased by $2°C$ ($T_{air} + 2$) shows the minimum ice extent at approximately $1.6 \times 10^6\ km^2$ which is in better correspondence with the observed summer ice extent derived from SSMI-data in 1987 (Figure 2).

3.2.2. Sensitivity results. For all variations of the forcing and model parameters the days of the maximum (wintertime) and minimum (summertime) ice volume and ice extent nearly coincide with those of the standard run. Nearly all experiments indicate a negligible phase shift of the seasonal cycle, and only in a few cases the largest deviation is about ±15 days for the ice volume and about ±10 days for the ice extent.

Table 3 and 4 show the maximum allowable parameter range $\Delta^* P$ for a 10% deviation from the standard run concerning ice volume as well as ice extent.

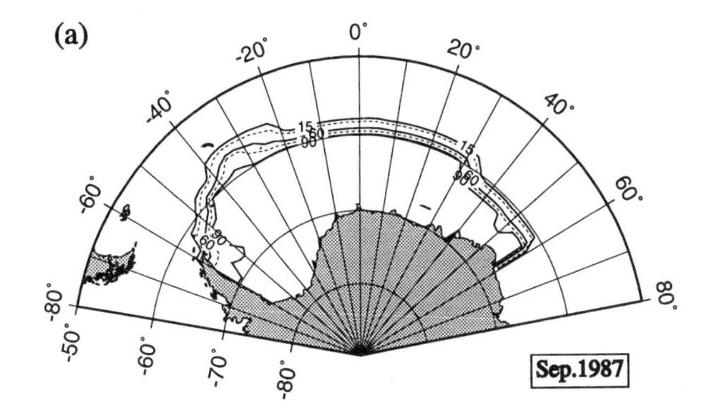

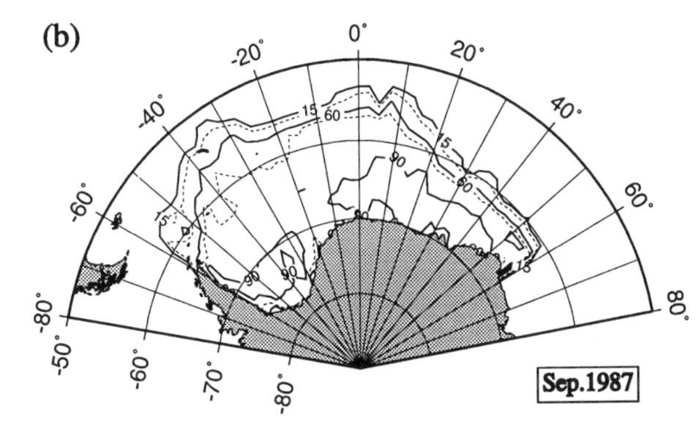

Figure 6: *Same as Fig.4 for September.*

$$\Delta^* P^V_{S,W} = \frac{\Delta^* V}{b^V_{S,W}} \quad (3)$$

$$\Delta^* P^E_{S,W} = \frac{\Delta^* E}{b^E_{S,W}} \quad (4)$$

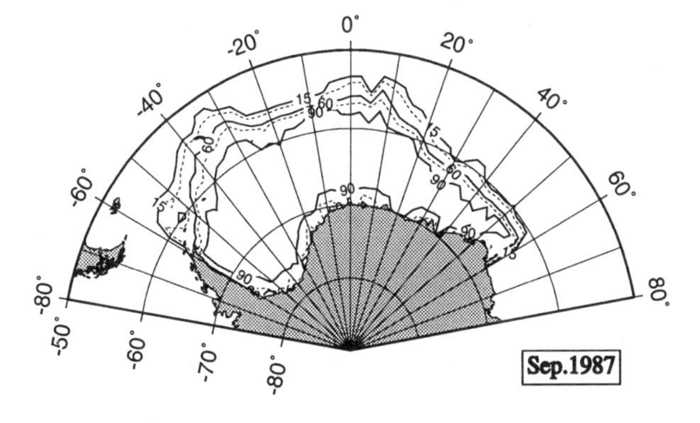

Figure 7: *Mean September total ice concentration in percent derived from same SSMI data as in Figure 6b using the Comiso-Algorithm.*

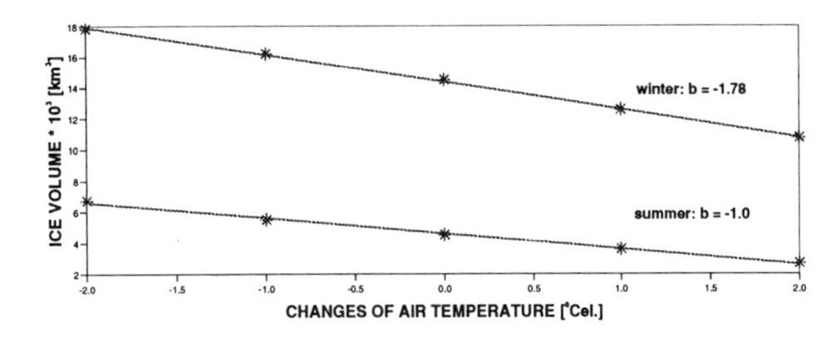

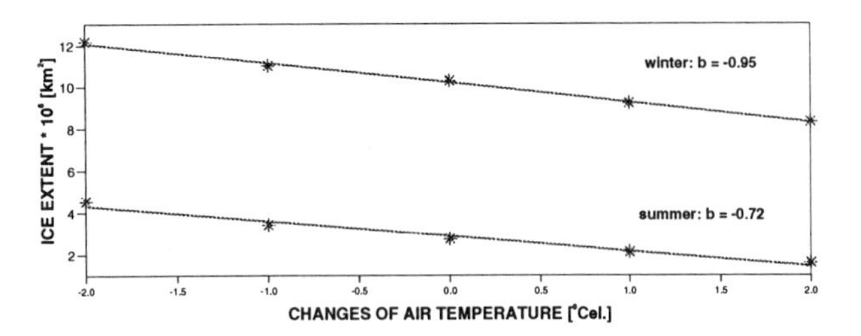

Figure 8: *Maximum and minimum ice volume (upper panel) and ice extent (lower panel) as a function of changes of surface air temperature (asterisks). The slopes b (solid lines) are derived from a linear fit and are given in terms of* $10^3 km^3/°C$ *and* $10^6 km^2/°C$, *respectively.*

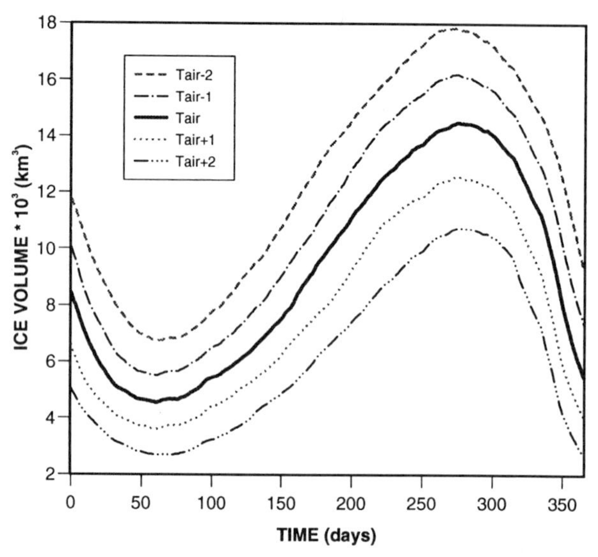

Figure 9: *Response of the seasonal cycle (1987) of the ice volume to variations of the surface air temperature* T_{air} *in the range of* $-2°$ *to* $+2°C$.

In Table 3 the reference value ($\Delta^* V$, $\Delta^* E$) for the 10% deviation is the annual mean value, which is represented by the mean of standard minimum and maximum values of the ice volume and ice extent (i.e. $\Delta^* V = 0.1(V_S + V_W)/2$ in Eq. 3). This normalisation suppresses the effect of the areal and volumetric difference of the summer and winter ice extent and volume. Table 4 shows the same parameter range for a 10% deviation from the actual standard ice situation in summer and in winter (i.e. $\Delta^* V = 0.1 V_{S,W}$ in Eq. 3). A $\pm$ sign indicates that ice volume or ice extent increase with increasing parameter values, while a $\mp$ sign indicates decreasing volume or extent with increasing parameter value.

Although dynamic-thermodynamic sea ice models are less sensitive to modifications of the boundary conditions than purely thermodynamic models [see e.g. Hibler, 1984; Lemke et al.,1990] there is still a strong dependency of the model results on the variation of the surface air temperature T_{air}. For a 10% deviation from the standard run the air temperature has to be known to an accuracy of ± 0.5 to $\pm 1.1°C$.

Although the total cloud amount is only crudely known, the experiments demonstrated the relative importance of a well predicted cloud cover as input data for sea ice models. The required accuracy of the cloud cover is ± 0.2. This result is in a good agreement with a sensitivity study of a one-dimensional thermodynamic sea ice model by Shine and Crane [1984]. Similarily the relative humidity rh has a rather strong influence on the sea ice (via latent heat flux (bulk formula) and shortwave radiation). It has to be known to about $\pm 10\%$ in winter and $\pm 20\%$ in summer for an acceptable prediction of ice volume and extent.

Due to the bulk formulations the radiation terms $F_s\downarrow$ and $F_l\downarrow$ are functions of air temperature, total cloud amount and humidity. As a consequence the radiation terms have also a strong influence on ice volume and extent. The sensitivity experiments indicate that errors of the incoming longwave radiation $F_l\downarrow$ are more significant than errors of the incoming shortwave radiation $F_s\downarrow$. The required accuracy of the incoming longwave radiation ranges between 15 to 25 W/m^2, whereas the incoming solar radiation has to be known to about 20 to 40 W/m^2.

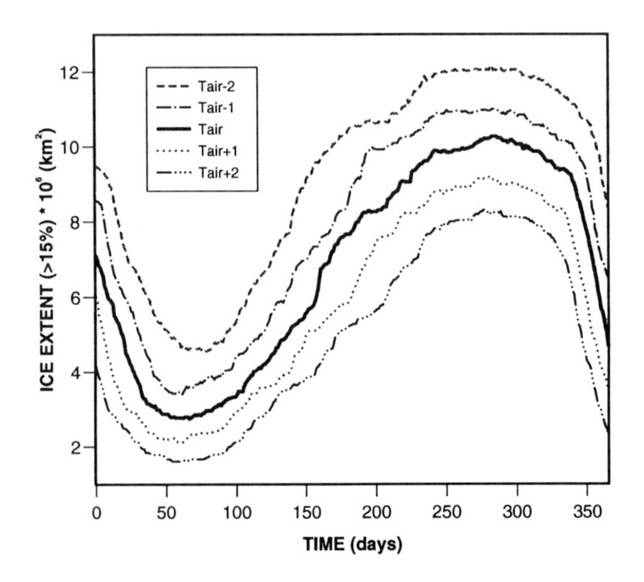

Figure 10: *Same as Fig.9 for the ice extent.*

Table 2. Maximum variations of forcing and model parameters

No.	Parameter	max. range of variation
1.	air temperature T_{air}	$\pm$ 2.0 $[°C]$
2.	total cloud amount Cl	$\pm$ 0.2
3.	relative humidity rh	$\pm$ 20 [%]
4.	incoming shortwave radiation $F_s\downarrow$	$\pm$ 60 $[W/m^2]$
5.	incoming longwave radiation $F_l\downarrow$	$\pm$ 60 $[W/m^2]$
6.	wind speed u_{wind}	$\pm$ 40 [%]
7.	albedo of snow and ice α	$\pm$ 0.1
8.	ice strength parameter P^*	$\pm$ 1.5 $\times$ 10^4 $[N/m^2]$
9.	dynamic ice parameter C	$\pm$ 10
10.	lead closing parameter h_o	$\pm$ 0.2 $[m]$

Changes of the surface wind speed u_{wind} strongly affect the ice volume and extent in wintertime due to modification of the ice thickness in convergent drift regions and an increased Weddell Gyre circulation with enhanced northern flow. Calculations of the mean ice velocity $\overline{v}$ in September 1987 indicates that the wind speed u_{wind} is the only parameter which noticably influences the mean ice velocity $\overline{v}$. All other sensitivity experiments yield only 0 - 5% modifications of $\overline{v}$ in comparison to the value of the standard run of $\overline{v}$ = 10.5 cm/s for September 1987. An increase of the wind speed of 40% increases $\overline{v}$ to 14.4 cm/s whereas a reduction of u_{wind} of 40% decreases $\overline{v}$ to a value of 6.3 cm/s. This result shows that the mean drift speed $\overline{v}$ varies linearly with the wind speed variation.

The ice extent seems to be rather insensitive to small variations of the model parameters, only the albedo α seems to be important for the summer ice extent. The absorbed shortwave solar energy is a function of the surface albedo α. As Tables 3 and 4 indicate the albedo for the summer situation has to be known for ± 0.1 to ± 0.2. Melt ponds, snow layer or flooding have an impact on the surface albedo which is in the range of the model derived deviation of 0.2. The lead closing parameter h_o is of importance only for the winter ice volume and has to be known to an accuracy of 0.1 to $0.3m$.

The summer values in Table 4 are generally smaller and the winter values are generally larger as compare to Table 3. This is the result of smaller errors applied in summer (10% of summer value) and larger errors allowed in winter (10% of winter value). General requirements for the accuracy of the forcing data are given in Table 5 as a summary of the results of Tables 3 and 4.

4 CONCLUSIONS

The winter extent of the sea ice cover in the Weddell Sea is generally well reproduced with a dynamic-thermodynamic sea ice model using modified model parameters and atmospheric boundary conditions from the ECMWF-analyses. The modelled summer extent is slightly too large, presumably due to too cold air temperatures. Comparison to ice concentrations indicate that the model seems to produce larger compactnesses. The different existing algorithms to derive sea ice concentrations from brightness temperatures, on the other hand, are still under debate.The lack of well known tie-points in Antarctic regions leads to uncertainties in determining the total ice concentration. A more complete model comparison and tuning is only justified when improved concentraion algorithms are available for the antarctic sea ice.

The general ice drift pattern is in good agreement with the buoy drift for the Weddell Sea region and depends strongly on the ratio of the drag coefficients c_w/c_a. The role of c_w/c_a is particularly important in regions of divergent drift, whereas the balance of c_w/c_a and P^* becomes more significant in regions of convergence. The analyses of the mean September drift velocity in the eastern Weddell Sea supports this argument, because of its low sensitivity to

Table 3. Maximum allowable parameter range for a deviation of the summer and winter ice volume and extent from the standard run for 1987 of the order of 10% of the annual mean values.

No.	Parameter	ice volume winter(max)	ice volume summer(min)	ice extent winter(max)	ice extent summer(min)
1.	T_{air} $[°C]$	$\mp$0.5	$\mp$1.0	$\mp$0.7	$\mp$0.9
2.	Cl	$\pm$0.2	$\pm$0.2	$\pm$0.2	$\pm$0.2
3.	rh [%]	$\mp$12	$\mp$20	$\mp$12	$\mp$24
4.	$F_s\downarrow$ $[W/m^2]$	$\mp$20	$\mp$35	$\mp$22	$\mp$40
5.	$F_l\downarrow$ $[W/m^2]$	$\mp$13	$\mp$21	$\mp$18	$\mp$26
6.	u_{wind} [%]	$\pm$17	–[a]	$\pm$17	–[a]
7.	α	–[a]	$\pm$0.2	–[a]	$\pm$0.2
8.	P^* $[N/m^2]$	$\mp$0.8 $\times$ 10^4	$\mp$1.3 $\times$ 10^4	–[a]	–[a]
9.	C	$\pm$19	$\pm$18	–[a]	–[a]
10.	h_o $[m]$	$\pm$0.1	–[a]	–[a]	–[a]

[a] less than 10% change for max. parameter variation

Table 4. Maximum allowable parameter range for a deviation of the summer and winter ice volume and extent from the standard run for 1987 of the order of 10% of the summer and winter values, respectively.

No.	Parameter	ice volume winter(max)	ice volume summer(min)	ice extent winter(max)	ice extent summer(min)
1.	T_{air} $[°C]$	∓0.8	∓0.5	∓1.1	∓0.4
2.	Cl	±0.3	±0.1	±0.3	±0.1
3.	rh [%]	∓18	∓9	∓20	∓9
4.	$F_s\downarrow$ $[W/m^2]$	∓31	∓16	∓36	∓16
5.	$F_l\downarrow$ $[W/m^2]$	∓20	∓9	∓29	∓10
6.	u_{wind} [%]	±26	–[a]	±27	∓33
7.	α	–[a]	±0.1	–[a]	±0.1
8.	P^* $[N/m^2]$	$\mp1.3 \times 10^4$	$\mp0.6 \times 10^4$	–[a]	$\pm1.4 \times 10^4$
9.	C	–[a]	±8	–[a]	–[a]
10.	h_o $[m]$	±0.2	±0.3	–[a]	–[a]

[a] less than 10% change for max. parameter variation

variations of P^* and strong sensitivity to changes of the wind forcing. If only convergent regions (for example west of $45°W$) are considered, the variations of P^* become more significant.

Sensitivity studies indicate that the summer ice extent and volume strongly depend on the thermodynamic forcing, most importantly on the surface air temperature. But also little-known forcing parameters as total cloud amount and humidity play an important role as atmospheric boundary conditions. The summer sea ice variables do not seem to be significantly influenced by the wind field. The available energy for the melting processes determines the reduction of the ice thickness and thus the position of the ice edge in summer. The winter extent and volume on the other hand are affected by the dynamics as well as by the thermodynamics.

The modelled ice extent is nearly independent of variations of the model parameters. Only the summer albedo seems to be important. This underlines the thermodynamic influence on the summer sea ice extent. The ice volume (particularly the winter value), on the other hand, is rather sensitive to variations of the model parameters. This shows that ice thickness observations would be most valuable for model optimization.

This study has demonstrated the generally good applicability of the ECMWF forcing fields in regions of sparse observational data. It has also shown the problems of the surface temperature field which is largely a model product (buoy temperatures are not assimilated). The air temperatures generated by the numerical weather prediction model are rather dependent on the lower boundary conditions applied (seasonal climatological ice edge, slab of ice with fixed thickness and 100 % coverage). The strong sensitivity of the sea ice model results on temperature errors require an improvement of the temperature forcing fields via a re-analysis with modified boundary conditions (observed ice edge, fractional ice cover). A re-analysis of the last decade of observations with the most recent version of the atmospheric circulation model using the observed sea ice extent will be performed in 1994/95 by the European Centre for Medium Range Weather Forecasts and the American Numerical Meteorological Center.

Table 5. The averaged required accuracy of forcing data for a 10% deviation from the sea ice model standard run for the year 1987

No.	Parameter	mean required accuracy
1.	ΔT_{air}	$0.6°C$
2.	Δ Cl	0.2
3.	Δrh	16 %
4.	$\Delta F_s\downarrow$	$27\ W/m^2$
5.	$\Delta F_l\downarrow$	$18\ W/m^2$
6.	Δu_{wind}	24 %

Acknowledgements. The authors would like to thank M. Harder for his effective programming support and valuable recommendations. Thanks are due to G. König-Langlo, C. Kottmeier, M. Scheduikat, A. Stössel and T. Viehoff for constructive discussions and suggestions, D. Crane and C. Kottmeier for the preparation of the buoy data and K. Saheicha for preparing the SMMR/SSMI data. The Deutsche Wetterdienst (DWD), provided the ECMWF data, which were preprocessed by R. Schnur. This is AWI publication no.732.

REFERENCES

Allison, I., Brandt R.E. and S.G. Warren. East Anarctic sea ice: albedo, thickness distribution, and snow cover. *Journal of Geophysical Research*, 98(C7):12417–12429, 1993.

Chapman, W.L., Welch, W., Bowman, K.P. Sacks, J. and J.E. Walsh. Multivariate sensitivities of a dynamic-thermodynamic sea ice model. In *fifth conference on climate variations*. American Meteorological Society, 1992.

Comiso, J.C., Grenfell, T.C., Lange, M., Lohanick A.W., Moore, R.K. and P. Wadhams. Microwave remote sensing of the Southern Ocean ice cover. In *Microwave remote sensing of sea ice, Chapter 12*, volume 68, pages 243–259. Geophysical Monograph, American Geophysical Union, Washington, USA, 1992.

Harder, M. and P. Lemke. Modelling the extent of sea ice ridging in the Weddell Sea. this volume.

Hibler, W.D., III. A dynamic thermodynamic sea ice model. *Journal of Physical Oceanography*, 9(C4):815–846, 1979.

Hibler, W.D., III. The role of sea ice dynamics in modeling CO_2 increases. In J. Hansen and T. Takahashi, editors, *Climate processes and climate sensitivity*, volume 29, pages 238–253. Geophysical Monograph, American Geophysical Union, Washington, D.C., USA, 1984.

Hibler, W.D., III. and S.F. Ackley. Numerical simulation of the Weddell Sea pack ice. *Journal of Geophysical Research*, 88(C5):2873–2887, 1983.

Idso, S.B. and R.D. Jackson. Thermal radiation from the atmosphere. *Journal of Geophysical Research*, 74:5379–5403, 1969.

Ip, C.F., Hibler W.D.,III and G.M. Flato. On the effect of rheology on seasonal sea ice simulations. *Annals of Glaciology*, 15:17–25, 1991.

Kottmeier, C. and R. Hartig. Wind observations of the atmosphere over Antarctic sea ice. *Journal of Geophysical Research*, 95(D10):16551–16560, 1990.

Lemke, P., Owens, W.B. and W.D. Hibler III. A coupled sea ice - mixed layer - pycnocline model for the Weddell Sea. *Journal of Geophysical Research*, 95(C6):9513–9525, 1990.

Leppäranta M. A growth model for black ice, snow ice and snow thickness in subarctic basins. *Nordic Hydrology*, 14:59–70, 1983.

McPhee, M.G. An analysis of pack ice drift in summer. In R. Pritchard, editor, *Sea ice processes and models*, pages 62–75. AIDJEX International commision on snow and ice symposium, University of Washington, Seattle, USA, 1980.

Olbers, D. and C. Wübber. The role of wind and buoyancy forcing of the Antarctic circumpolar current. Technical Report No. 22, Berichte aus dem Fachbereich Physik, Alfred Wegener-Institut für Polar- und Meeresforschung, Bremerhaven, Germany, 1991.

Overland, J.E. Atmospheric boundary layer structure and drag coefficient over sea ice. *Journal of Geophysical Research*, 90(C5):9029–9049, 1985.

Owens, W.B. and P. Lemke. Sensitivity studies with a sea ice - mixed layer - pycnocline model in the Weddell Sea. *Journal of Geophysical Research*, 95(C6):9527–9538, 1990.

Parkinson, C.L. and W.M. Washington. A large-scale numerical model of sea ice. *Journal of Geophysical Research*, 84(C1):311–337, 1979.

Rowe, M.A., Sear C.B., Morrison, S.J., Wadhams, P., Limbert, D.W.S. and D.R.J. Crane. Periodic motions in Weddell sea pack ice. *Annals of Glaciology*, 12:145–151, 1989.

Semtner, A.J. A model for the thermodynamic growth of sea ice in numerical investigations of climate. *Journal of Physical Oceanography*, 6(3):379–389, 1976.

Shine K.P. and R.G. Crane. The sensitivity of a one-dimensional thermodynamic sea ice model to changes in cloudiness. *Journal of Geophysical Research*, 89(C6):10615–10622, 1984.

Steffen, K., Key, J., Cavalieri, D.J., Comiso, J., Gloersen, P., Germain, K.St. and I. Rubinstein. The estimation of geophysical parameters using passive microwave algorithms. In *Microwave remote sensing of sea ice, Chapter 10*, volume 68, pages 201–231. Geophysical Monograph, American Geophysical Union, Washington, USA, 1992.

Stössel A. Sensitivity of Southern Ocean sea ice simulations to different atmospheric forcing algorithms. *Tellus*, 44A:395–413, 1992.

Stössel A., Lemke, P. and W.B. Owens. Coupled sea ice - mixed layer - pycnocline simulations for the Southern Ocean. *Journal of Geophysical Research*, 95(C6):9539–9555, 1990.

Trenberth, K.E. and J.G. Olson. ECMWF Global analyses 1979-1986: Circulation Statistics and Data Evaluation. Technical Report TN-300+STR, National Center for Atmospheric Research (NCAR), Boulder, USA, 1988.

van Loon, H. Cloudiness and precipitation in the southern hemisphere. In *Meteorological Monograph*, volume 13, pages 101–111, 1972.

Wamser, C. and D.G. Martinson. Drag coefficients for winter Antarctic pack ice. *Journal of Geophysical Research*, 98(C7):12431–12437, 1993.

Zillman, J.W. A study of some aspects of the radiation and heat budgets of the southern hemisphere oceans. In *Meteorological study*, volume 26, 562pp. Bureau of Meteorology, Dept. of the Interior, Canberra, Australia, 1972.

H. Fischer, P. Lemke, Alfred-Wegener-Institut für Polar- und Meeresforschung, Postfach 120161, D - 27515 Bremerhaven, Germany.

On the Effect of Ocean Circulation on Arctic Ice-Margin Variations

W. D. Hibler, III and Jinlun Zhang

Thayer School of Engineering, Dartmouth College, Hanover, NH 03755

A high resolution ice ocean circulation model with 40 km horizontal resolution and 21 vertical levels is used to simulate the seasonal and interannual characteristics of the ice cover and ocean circulation of the Arctic, Greenland, Norwegian and Barents seas over the time interval 1979-1985. The model yields a mean annual heat budget close to balance and a realistic ocean circulation. Results from a hierarchy of numerical simulations using this model are analyzed to help determine the dominant mechanisms responsible for Arctic ice margin variations. The correlation between simulated and observed seasonally averaged interannual variations of the ice edge is good in the Arctic Basin and Barents Sea with about 50% of the observed variance explained by the model. In the Greenland Sea, the model explains about 50% of the observed variance in the summer and fall but poorly predicts the winter and spring ice margin variations. Simulations with a motionless thermodynamic only ice model without interannually varying ocean heat fluxes yield very poor correlations with observed ice margin variations everywhere and reduce the ice extent about 200 km compared to the full ice-ocean model. A simulation without open ocean boundaries, on the other hand, results in an increase of ice extent about 100 km. Except for the Arctic Basin where ice transport effects are found to be dominant, the interannual results portray a complex system with the effects of ice transport and oceanic heat flux often competing so that neither one totally dominates. As a consequence both terms need to be included for realistic simulations of ice margin variations.

1. INTRODUCTION

The Arctic sea ice margin, especially in the Barents and Greenland Seas, is controlled by a complex interaction between the atmosphere, ocean and sea ice circulation. Seasonal and interannual variations of these ice margins are often correlated with salinity anomalies (*Marsden and Mysak*, 1991) in the high latitude North Atlantic and hence form an integral part of the global ocean climate system. As a consequence understanding the basic mechanisms responsible for ice margin variations becomes important for understanding and modeling the global oceans, especially as regards climate change.

Numerical modeling of the ice-ocean system in conjunction with observations of the sea ice margin provides a useful means to deduce the role of different components of this coupled system on variations of the ice margin. A framework for ice ocean circulation modeling was developed by Hibler and Bryan (1979). The Hibler and Bryan model consists of a dynamic thermodynamic sea ice model coupled to a baroclinic ocean model with all but the top one or two layers of the ocean model weakly relaxed to mean annual temperature and salinity data (*Levitus*, 1982). In the results described here this relaxation is performed with a 5 year time constant which allows interannual and seasonal variability of the ice-ocean system to be simulated while still constraining the long term climatic characteristics of the ocean model to climatology.

A major result of the Hibler and Bryan (1987) study was that inclusion of ocean circulation was critical for simulating seasonal ice margin variations in the Greenland and Barents seas. However, Hibler and Bryan (1987) did not examine the extent to which this model could be used to simulate interannual variations of the ice margin. Also the crude resolution of the model together with lack of inflow boundary conditions produced poorly balanced salt and heat budgets together with certain unsatisfactory aspects of the ocean circulation.

To address some of these shortcomings and to begin to sort out the dominant mechanisms responsible for the seasonal and interannual variations of the ice margin, we have constructed a high resolution ice ocean circulation model. This model is used here to examine the Arctic ice ocean system over the six year period 1979-1986. Efficient numerical integrations of this higher resolution model over this multi-year period were made possible by an improved

The Polar Oceans and Their Role in Shaping the Global Environment
Geophysical Monograph 85

numerical procedure (*Zhang and Hibler*, 1994) for solving the nonlinear viscous-plastic sea ice model equations of Hibler (1979). In practical terms this implicit numerical method results in an order of magnitude faster solution of the ice model so that the sea ice portion of the coupled model takes considerably less time than the optimized ocean circulation model.

In this paper we focus on determining the physical mechanisms responsible for interannual and seasonal ice edge variations of the Arctic ice margin by means of analysis of a series of numerical sensitivity studies. In addition, analysis of the ice mass budget characteristics in the Greenland Sea provides insight into the role of sea ice transport on high latitude salinity anomalies.

2. MODEL DESCRIPTION

The basic equations and numerical framework of the ice ocean model are essentially similar to that of Hibler and Bryan (1987) except for a different numerical method for solving the sea ice momentum equations. However, a number of changes to the model configuration have been made. Foremost, the vertical and horizontal resolution of the model has been increased from 180 km lateral resolution and 15 vertical levels to 40 km horizontal resolution and 21 vertical levels. This has allowed the horizontal kinematic eddy viscosity and diffusion constants to be lowered to 1.25×10^8 cm^2/s and 1.5×10^6 cm^2/s as compared to 10^9 cm^2/s and 10^7 cm^2/s as used by Hibler and Bryan (1987). In addition, lateral inflow boundary conditions through both the Faero Shetland passage and Bering Strait are implemented with a net inflow of one Sverdrup through the Bering Strait and 6 Sverdrups through the Faero Shetland passage. These inflows are balanced by a net outflow through the Denmark Strait of 7 Sverdrups. In addition the vertical profiles of velocity are calculated geostrophically from temperature and salinity values which are strongly relaxed to observed values in these passages.

As discussed by Hibler and Bryan, the sea ice model is a dynamic-thermodynamic model with a plastic sea ice rheology coupled to a two layer (ice and open water) sea ice thickness distribution. A full heat budget calculation (*Hibler*, 1980) is performed to determine the growth rate of the sea ice over both the ice covered and open water portions of the grid cell. One essential change is that for thermodynamic calculations (following *Walsh et. al.*, 1982), a uniform distribution of sea ice is assumed ranging from zero to twice the mean thickness. The growth rate of the ice covered portion of a grid cell is then calculated by averaging thermodynamic calculations over seven thickness levels spanning this distribution. This procedure supplies more realistic thickness and ice margin results closer to those obtained using a full thickness distribution model (*Flato and Hibler*, 1993)

The horizontal model grid is shown in Figure 1 and the thicknesses of the vertical levels are shown in Table 1. Also illustrated in Figure 1 is the demarcation of three geographical regions: the Arctic Basin, the Barents Sea, and the Greenland and Norwegian Seas. These three separate regions are used to examine in some detail the seasonal ice and ocean heat budget characteristics of the model.

3. ATMOSPHERIC FORCING FIELDS

A critical issue in forcing ice-ocean circulation models is the specification of atmospheric temperature fields. Mean monthly data sets useful for examining interannual variability have been compiled by Jones et. al. (1986) and Hansen and Lebedeff (1987). For this study we have used the Hansen data which were obtained from John Walsh (private communication). However these data have been modified to take into account the over amplification of warming temperatures over the Arctic ocean in summer due to the lack of ice stations used in the analysis. The procedure used for this modification together with comparisons of the effect of unmodified temperature fields on the simulations is given in Hibler and Zhang (1993). Briefly by making use of observed ice edge variations the Hansen mean temperature over the period 1979-85 was replaced by a seven year mean formed by averaging summer temperatures which were not allowed to go above freezing when there was an observed ice cover. This procedure is consistent with detailed analyses of drifting buoy temperatures (*R. Colony*, private communication) which show that the air temperature over the ice in summer rarely goes above freezing until the ice disappears. This procedure yields forcing fields with interannual variability but with a more realistic mean.

For wind forcing, daily geostrophic winds were calculated from atmospheric pressure fields incorporating Arctic data buoy pressure measurements (John Walsh, private communication). Other forcing fields were obtained in the same manner as discussed by Hibler and Walsh (1982) which follow very closely the treatment of Parkinson and Washington (1979).

4. SIMULATION RESULTS

To obtain results relatively independent of the initial conditions the coupled diagnostic model was initialized with Levitus (1982) mean annual temperature and salinity data and integrated for seven years using interannual varying atmospheric forcing over the time interval January 1979 to December 1985. In this standard model run, inflow boundary conditions were utilized together with an interactive ice model so that both seasonal and interannual variations in the ocean and atmosphere were simulated. In this and all other simulations a relaxation to mean annual

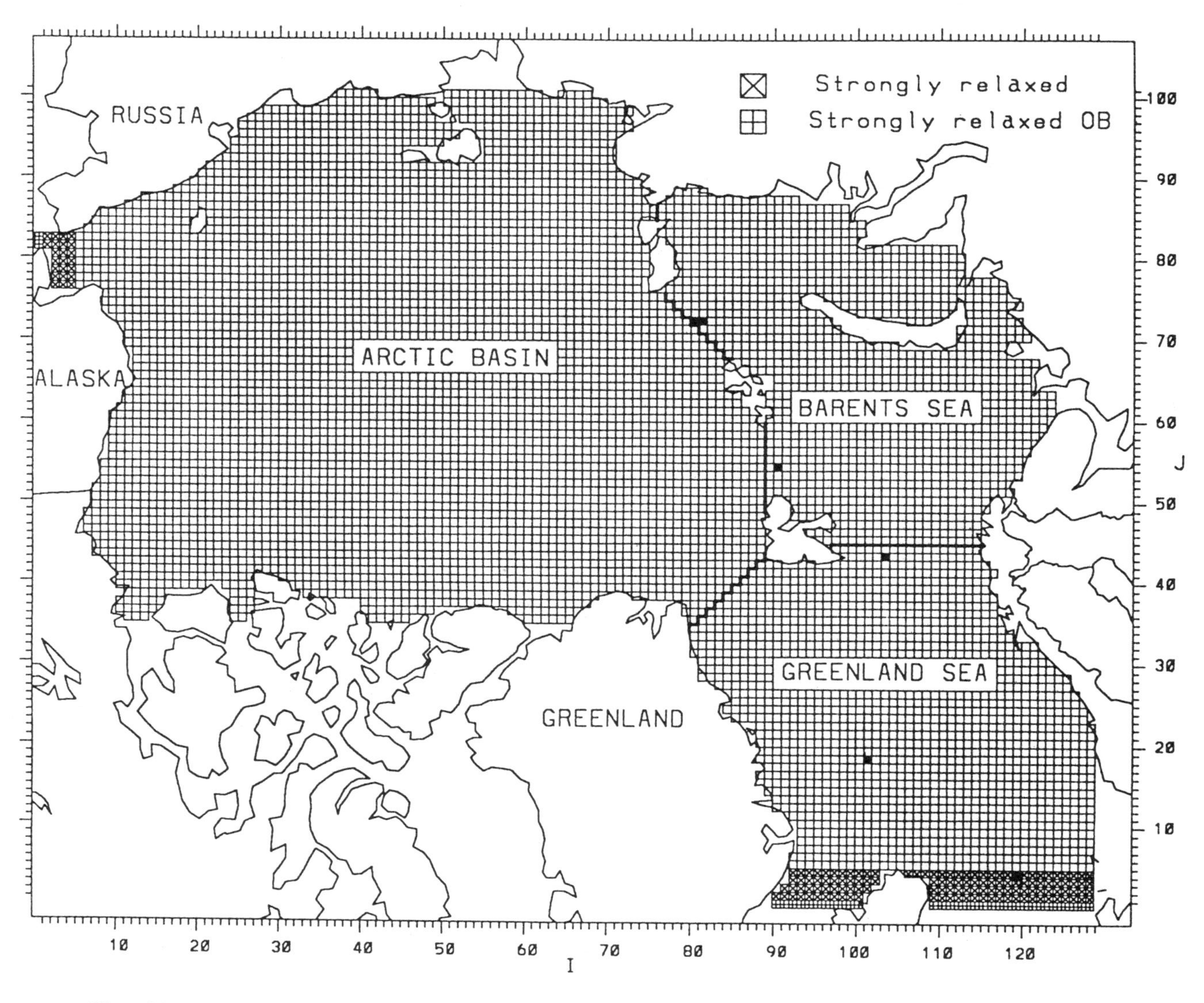

Fig. 1 Model grid used for ice ocean simulations. The dark lines denote boundaries of the geographical regions used in later analysis. A weak 5 year relaxation to observed temperature and salinity data was used below 30 meters except at the noted lateral boundary cells where a 30 day relaxation was used together with fixed specified vertically averaged inflows.

Table 1. The Vertical Thicknesses for the Ocean Model

10.00
15.43
22.50
31.47
42.83
56.83
74.00
94.50
118.7
146.8
178.9
214.9
254.3
296.9
345.8
405.8
469.2
537.5
613.3
694.2
792.5

temperature and salinity data with a five year time constant was employed below the top two layers of the ocean. The ice and ocean model results at the end of the seven year period were then utilized as initial conditions for a series of 7 year comparative simulations.

To help in examining the relative role of different processes, a hierarchy of three (counting the standard model run) interactive model simulations were run over this seven year period. To retain some degree of full seasonal variability in both the sea ice and ocean while removing some of the physical mechanisms, a simulation with no inflow at the boundaries and a simulation with motionless thermodynamic sea ice were performed. In this second thermodynamic only coupled simulation the sea ice growth and decay were modeled, but the variations in salt fluxes due to melting and growth were much different due to no ice advection. In particular, in the standard Hibler and Bryan (1987) model all the salt from the sea water is removed and put into the top layer of the ocean upon freezing. Thus ice formed in the Arctic Basin which is transported out of the Arctic and melted in the Greenland Sea will deliver a substantial net freshwater flux in that region. In contrast, over an annual cycle, the thermodynamic only model sensitivity run will deliver hardly any net freshwater to the Greenland Sea.

In addition to these simulations with interactive ocean, two "ice model only" sensitivity studies were carried out with fixed seasonally varying oceanic heat fluxes and currents so that no interannual variability in the ocean was allowed to affect the ice margin and thickness characteristics. These two runs consisted of a full ice model with dynamics, and hence transport and deformation; and a thermodynamic only ice model sensitivity run.

In the following sections we first examine the circulation and heat budget characteristics of the standard model. Following that the seasonal and interannual characteristics of these various model runs relative to ice margin variations are discussed.

4.1 *Ocean Circulation Characteristics*

Some of the general characteristics of the ocean circulation from the standard model simulation are illustrated in Figures 2, 3 and 4 which show the seven year

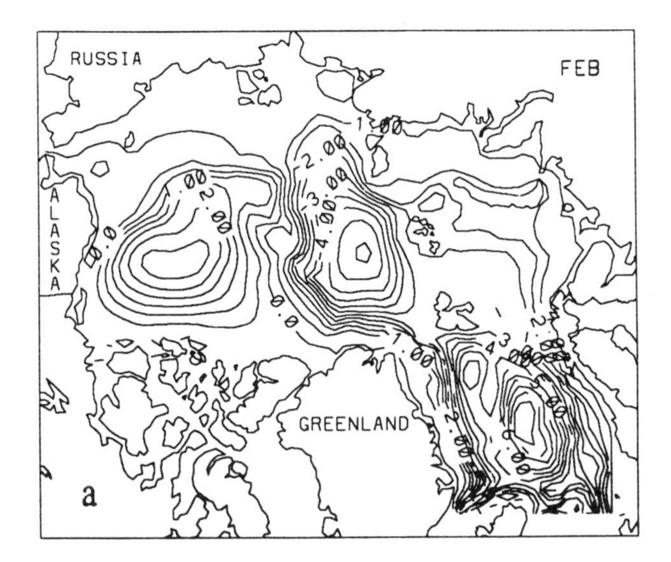

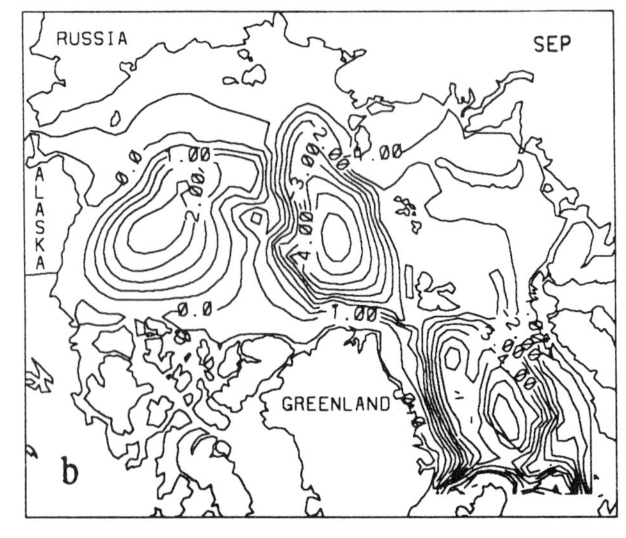

Fig. 2. Seven year mean February and September streamfunctions for the vertically integrated oceanic flow over the time period 1979-1985. Contour interval is 0.5 Sv.

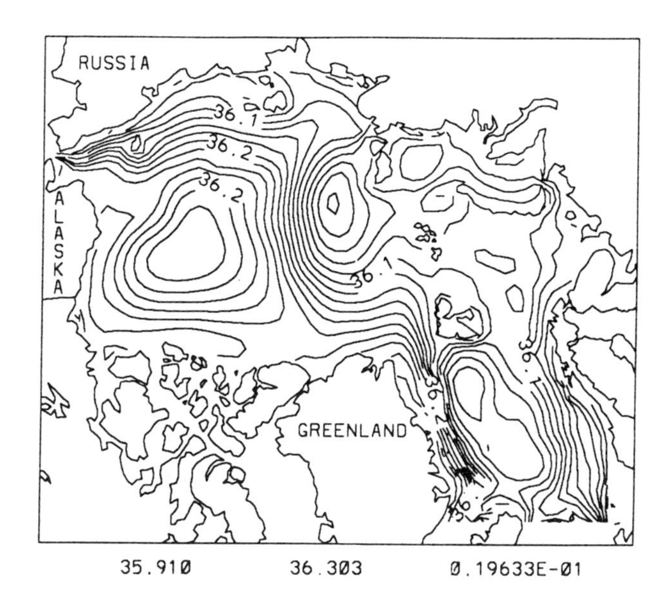

Fig. 3. Mean annual pressure contours at the third vertical level (37m) of the standard Ice-Ocean circulation model over the time interval 1979-1985. The numbers at the bottom reading from the left to the right are the minimum and maximum contour values, and the contour interval, respectively.

mean vertically averaged stream function for February and September, the annually averaged pressure at level 3 of the ocean model and the annually averaged vertical cross section of ocean velocities across the Fram Strait. Basically the ocean circulation consists of 3 gyres, a clockwise Beaufort Sea gyre, a counterclockwise flow in the Eurasian Basin and a counterclockwise flow in the Greenland Sea. While not apparent in the integrated flow,

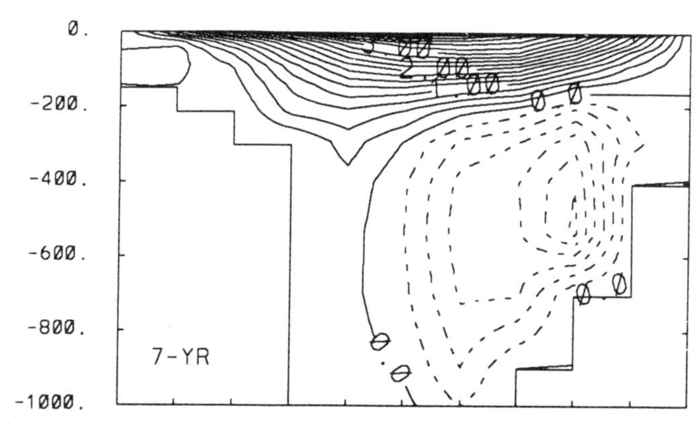

Fig. 4. Mean annual cross section of the ocean velocity (cm/s) across the Fram Strait over the time period 1979-1985. Solid contours denote outflow and dashed contours inflow.

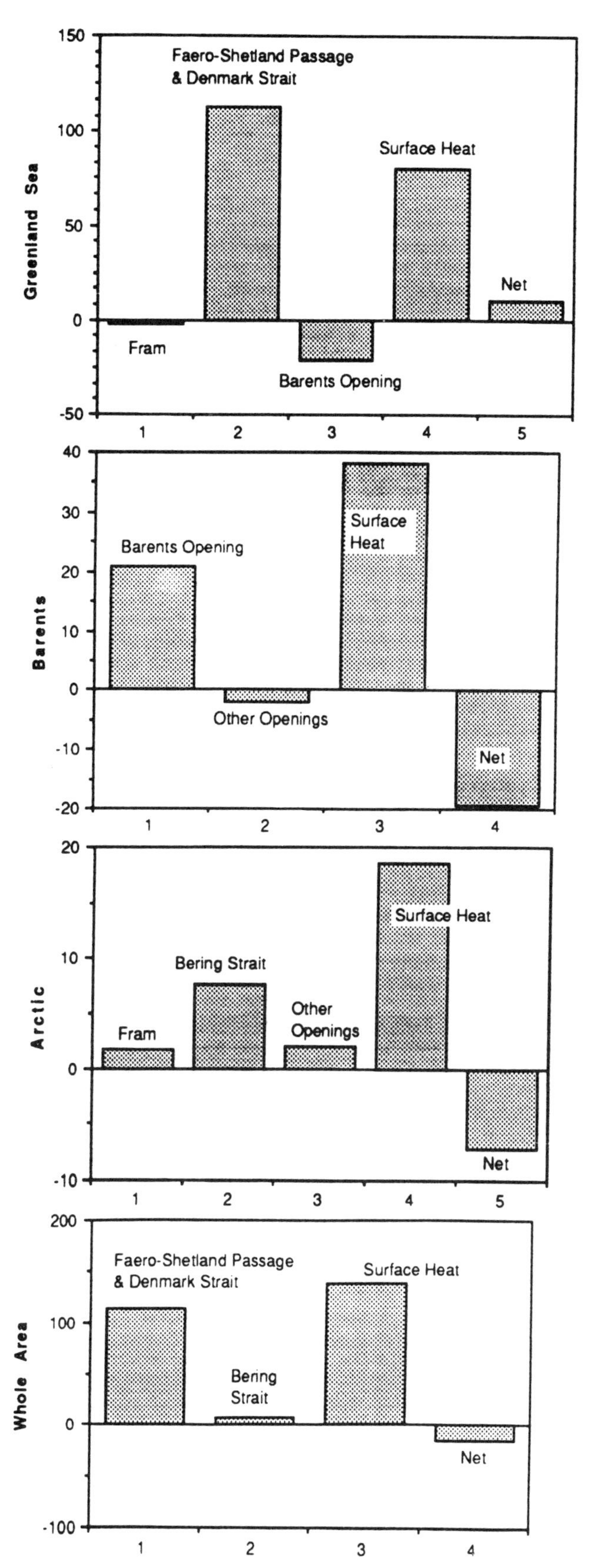

Fig. 5. Mean annual lateral heat transports and surface heat losses in different regions for the standard ice ocean model over the time period 1979-1985. Units: 10^{12} watts.

Figure 4 shows a substantial subsurface flow of Atlantic water into the Arctic Ocean on the Spitsbergen side of the Fram Strait. While somewhat smaller than observational estimates, this transport is realistic in character and depth.

The vertically integrated flow in Figure 2 is largely forced by the bottom torque term, that is the interaction of the density fields and the bottom topography and is concentrated in the upper 1000 meters of the ocean. Since the flow is largely geostrophic it is not greatly affected by wind, and the summer and winter vertically integrated flow are similar with some deviations. One notable exception is a coastal current along the Beaufort Sea coast that is weakly apparent in the contours of integrated flow. This current arises partially from the inflow through the Bering Strait where the inflow bifurcates into two currents following topography, one along the North Alaskan coast and one along the Siberian Shelf. Sensitivity studies show this Eastward current to fluctuate substantially and to be substantially enhanced by ice interaction which modifies the wind stress transmitted into the ocean.

4.2 *Ocean Heat Budget and Ice Mass Balance Characteristics*

The heat budget characteristics of the model are illustrated in Figure 5 which shows the 7 year averaged lateral heat transport and the heat flux into the upper two layers of the ocean. In this figure lateral heat transport was determined by using their temperature difference from the freezing point. Over a seasonal cycle the lateral heat flux is effectively balanced by atmospheric heat fluxes and the freezing or melting of sea ice. The dominant locations for the lateral transport are the Faero Shetland passage which is the main source of Oceanic heat for the Greenland and Barents Seas and the Spitzbergen to Norway passage which allows a substantial amount of heat to be transferred from the Greenland Sea to the Barents Sea. For the Arctic Basin the Fram Strait supplies some heat but not as much as the Bering Strait or the other passages in and out of the Greenland-Norwegian Sea.

The character of the Greenland-Norwegian Sea and the Barents Sea regions can be contrasted with the Arctic. In both the Greenland-Norwegian Sea and the Barents Sea regions there is a strong seasonal cycle of oceanic heat flux (Figure 6) which prevents sea ice from forming and is largely balanced by lateral heat transport. In the Arctic Basin, on the other hand there is little heat loss at the surface and hence much less heat is needed to be supplied by inflow.

The balance of the overall heat budgets is much closer than obtained by Hibler and Bryan (1987) where much of the heat loss was supplied by diagnostic terms due partially

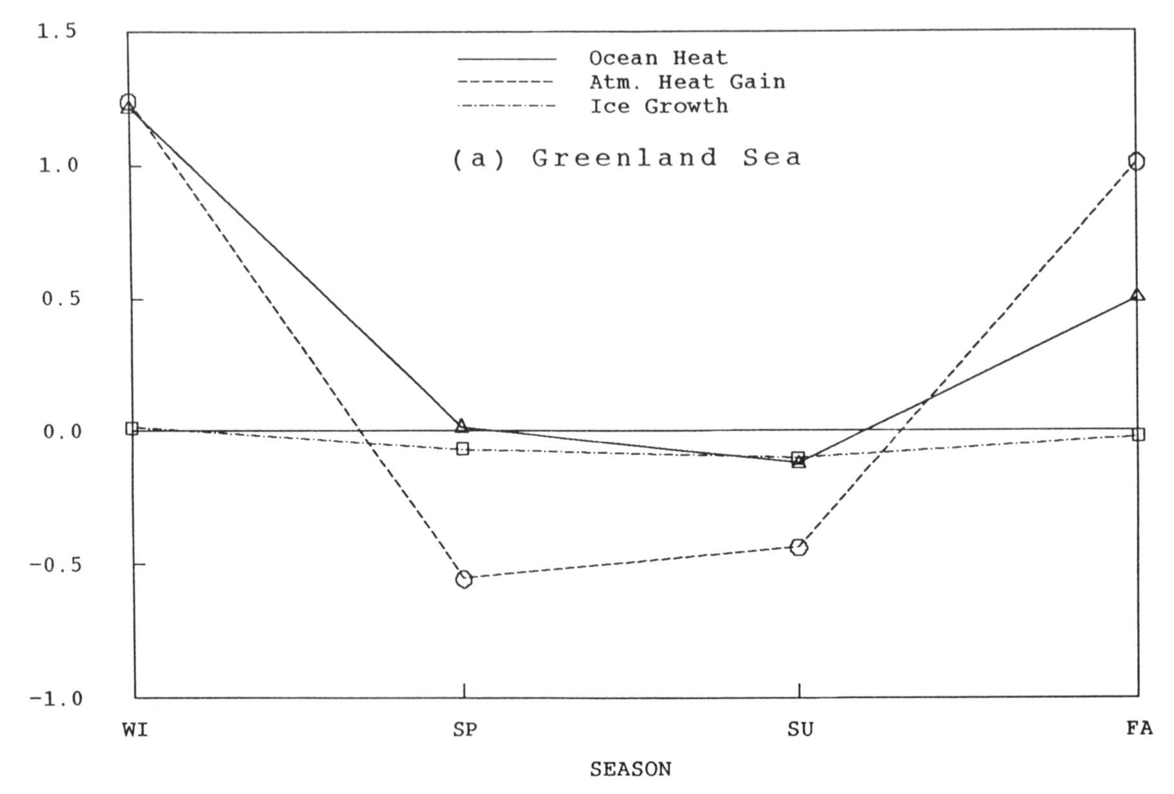

Fig. 6. Seven year mean season atmospheric heat gain, oceanic heat flux and ice growth in the three different regions. Units are in meters of ice per month averaged over the whole domain of each region.

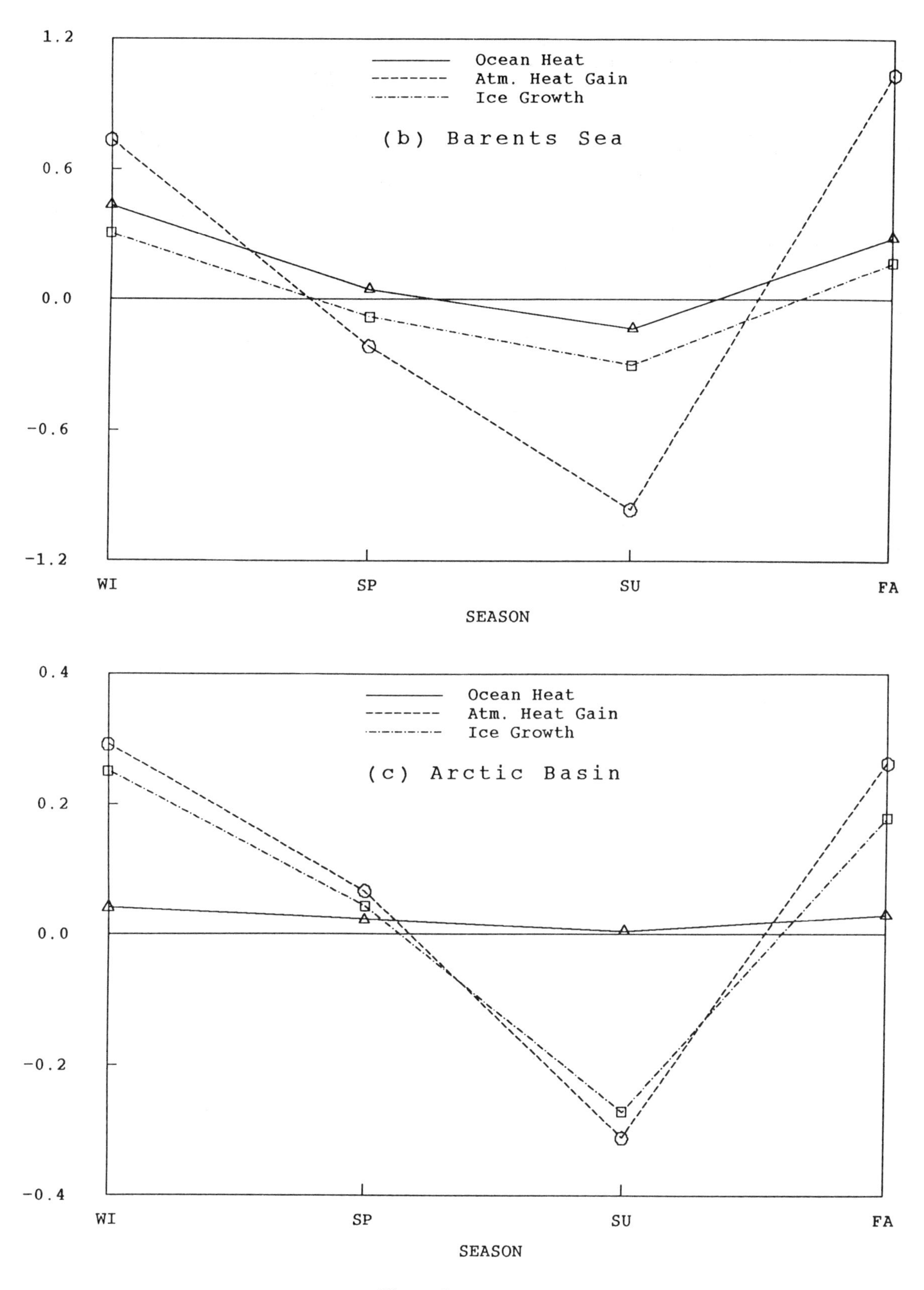

Figure 6. (continued)

to lack of inflow through the southern boundaries. In particular the heat transport through the southern boundary is about 2 and 1/2 times that obtained by Hibler and Bryan (1987).

To yield some insight into the physical mechanisms of ice growth and decay for the different regions, Figure 6 shows the average seasonal mean ice growth, atmospheric heat gain and oceanic heat flux for each of the three regions. In the Arctic Basin, there is a net ice growth which is compensated by a net atmospheric heat gain that is larger than the small oceanic heat flux. In the Greenland Sea, on the other hand, there is a net ice melt which arises because of the large oceanic heat flux. The Barents Sea is intermediate between these two extremes with a small amount of net ice growth. Note that, except for the Greenland Sea where there is hardly any net ice growth even in winter, there is a seasonal variation of the ice growth with ice growing in winter and melting in summer.

Overall these results depict a sea ice cover where ice mass is created in the Arctic Basin and is then transported into the Greenland Sea where it is largely melted. Indeed this transport through the Fram Strait, as discussed later, is so critical that fluctuations in simulated ice extent are largely controlled by ice transport variations. However it should be emphasized that these types of effects can certainly be deviated from locally with episodic events causing growth events near the East Greenland Sea ice margin.

4.3 *Seasonal and Interannual Variations of the Ice Margin*

The effects of different lateral boundary conditions and sea ice parameterizations on the sea ice margin in the fully coupled ice ocean models are shown in Figures 7 and 8 which illustrate the Greenland and Barents Sea ice margin for February and September and the average seasonal cycle of the observed and simulated ice margins for the three different regions. Basically the effects of removing ice transport and lateral heat transport in the ocean tend to be opposite in character with effect of removing lateral ice transport having a larger magnitude. This is especially true in the Greenland Sea where removing the ice transport (and hence indirectly the stabilizing fresh water flux into the ocean) causes a very much reduced ice edge expansion. While the ice transport does contribute to this, it is the absence of salt flux that dominates here. This was further verified by examining the results from a no salt flux sensitivity simulation (not shown) where there was also a substantially reduced expansion. In this sensitivity simulation ice advection was included but all surface salt fluxes were set equal to zero.

It is also notable that the thermodynamic only simulation yields a large difference in the Arctic Basin where large amounts of open water form in the Bering-Chuckchi seas when there is no ice advection to close up these regions in summer. Indeed as discussed below, ice

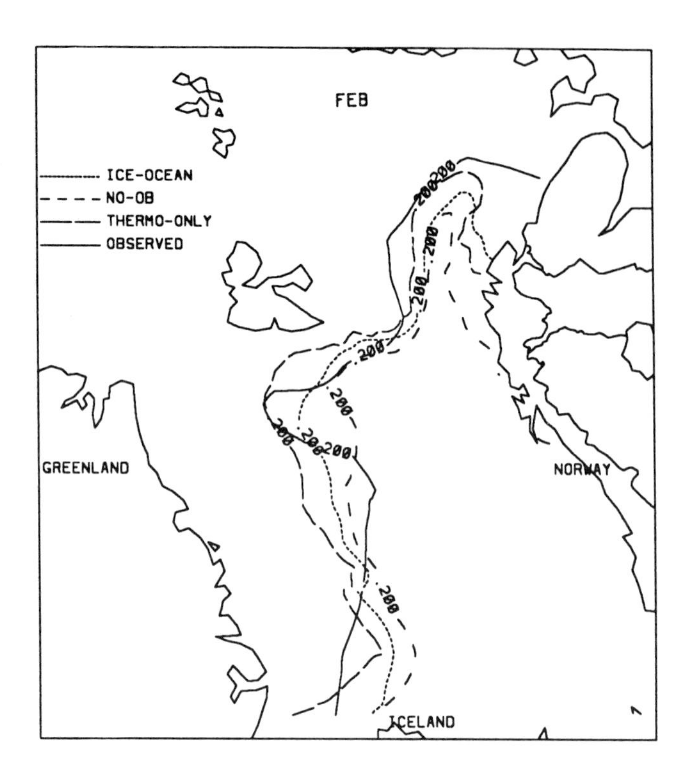

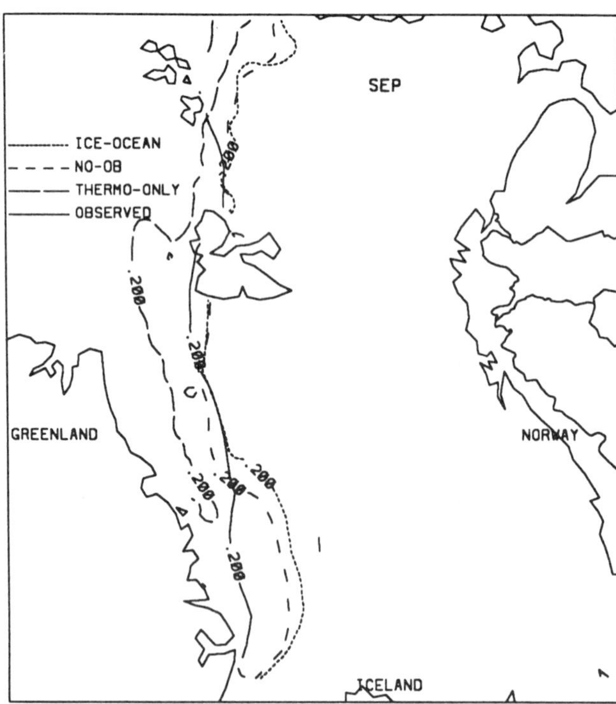

Fig. 7. Simulated and observed seven year mean ice margin for the Greenland-Norwegian and Barents seas for February and September. The three simulated curves shown are for the standard ice ocean model (Ice-Ocean), the ice ocean model without inflow (No-OB), and the ice ocean model with thermodynamic only sea ice (Thermo-Only).

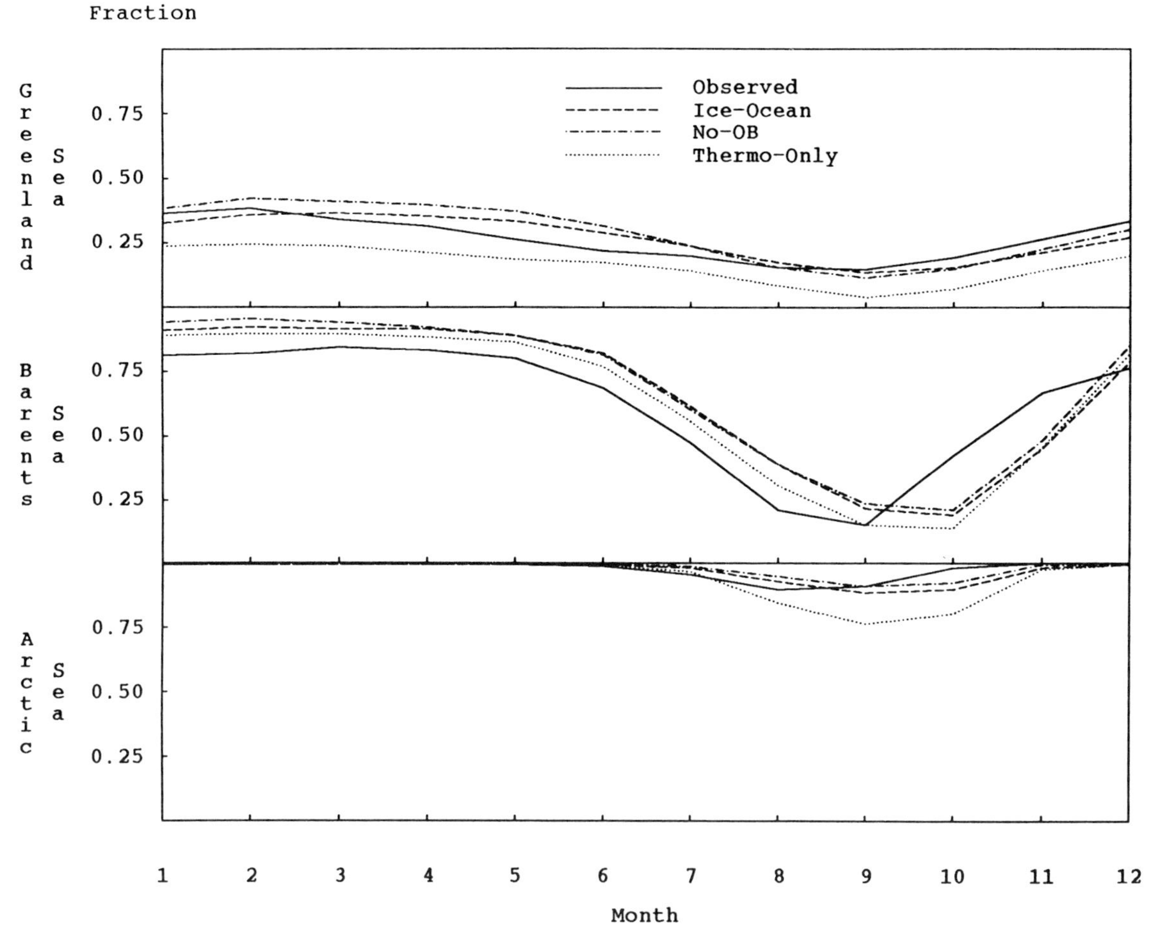

Fig. 8. Simulated and observed 7 year mean monthly fractions of are covered by ice with compactness greater than 0.2 for the Arctic Basin, Greenland Sea and Barents Sea. The three simulated curves shown area for the standard ice ocean model (Ice-Ocean), the ice ocean model without inflow (No-OB), and the ice ocean model with thermodynamic only sea ice (Thermo-Only).

advection is the most critical mechanism for simulating interannual variability of ice margin in the Arctic Basin. The lack of inflow through the Bering Strait on the other hand tends to increase the ice coverage time series only slightly. However interannual variations of this inflow (not included here) are likely not negligible for the interannual variability.

From Figures 7 and 8 it is also notable that the main effect of lateral inflow boundary conditions occurs in February in both the Barents and Greenland Sea. The thermodynamic effects on the other hand are not very pronounced at all in the Barents Sea in winter but do cause somewhat less ice to form in summer.

While many of the effects of ocean circulation can be ascertained from the seasonal cycle of the ice margin it is in the interannual variability that allows mechanisms to be better elucidated. Many of these mechanisms are apparent from examining the correlation between the observed and

Table 2. Correlation Coefficients Between Simulated and Observed Interannual Variations of the Seasonally Averaged Ice Margin Variations for Different Regions and Models

	Ice ocean	No OB	Thermo only	Ice Only	Thermo Only Ice only
Greenland	0.33	0.42	0.14	0.38	0.10
Bar. Sea	0.81	0.78	0.74	0.64	0.25
Arc. Basin	0.68	0.70	0.33	0.64	0.23
Whole	0.55	0.61	0.37	0.48	0.15

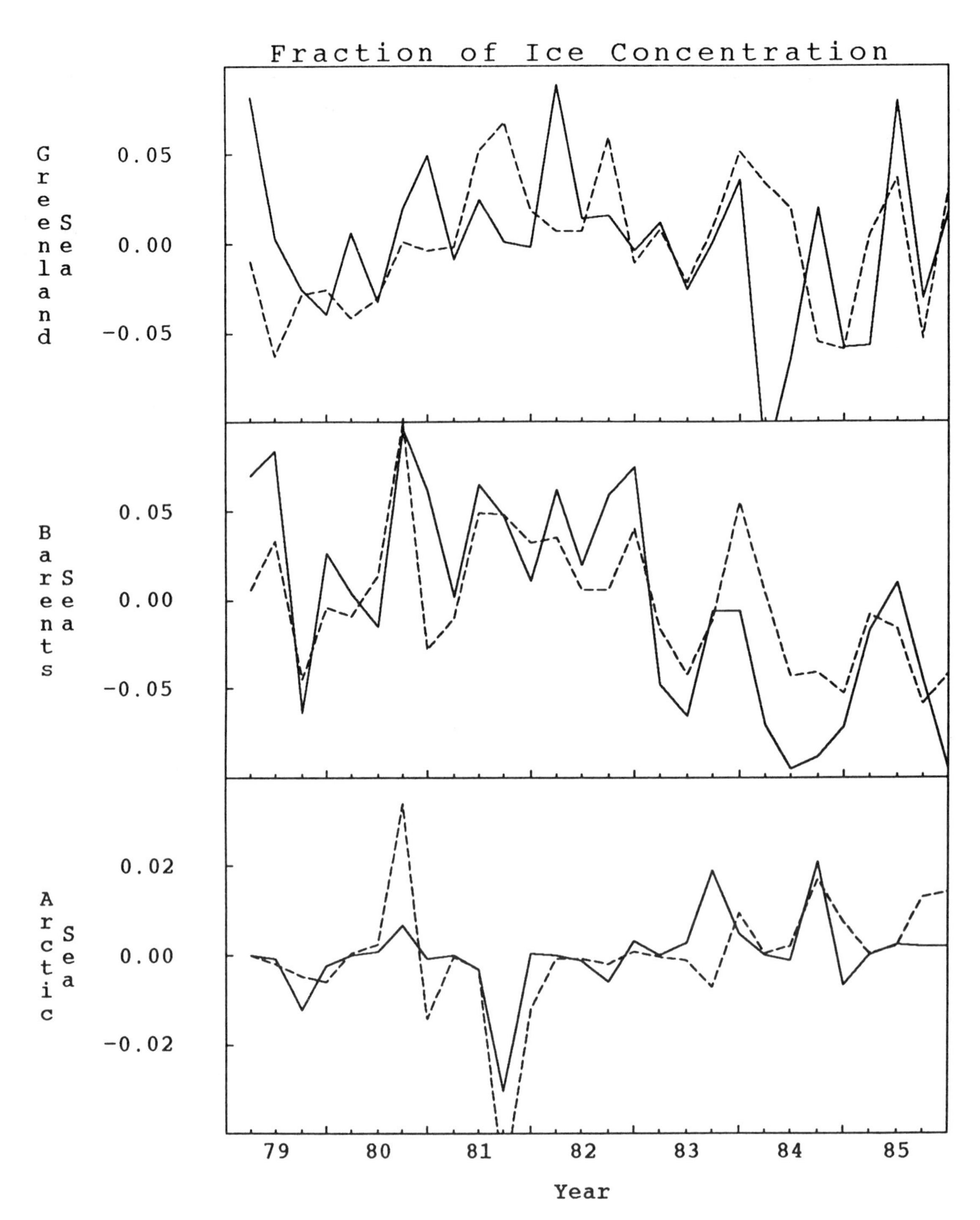

Fig. 9. Interannual deviations of simulated (dashed) and observed (solid) seasonally averaged ice margin variations for the three geographical regions.

modeled ice edges as shown in Figure 9 and Table 2. Table 2 shows correlations between the simulated and observed interannual variations of the ice margin while Figure 9 shows the visual time series comparison for the standard model.

Analysis of Table 2 tends to amplify the combined importance of ice advection and ocean circulation in affecting the ice margin variations. These two effects can interact in complex ways however. For example, if we look at the interannual variability of only the effect of sea

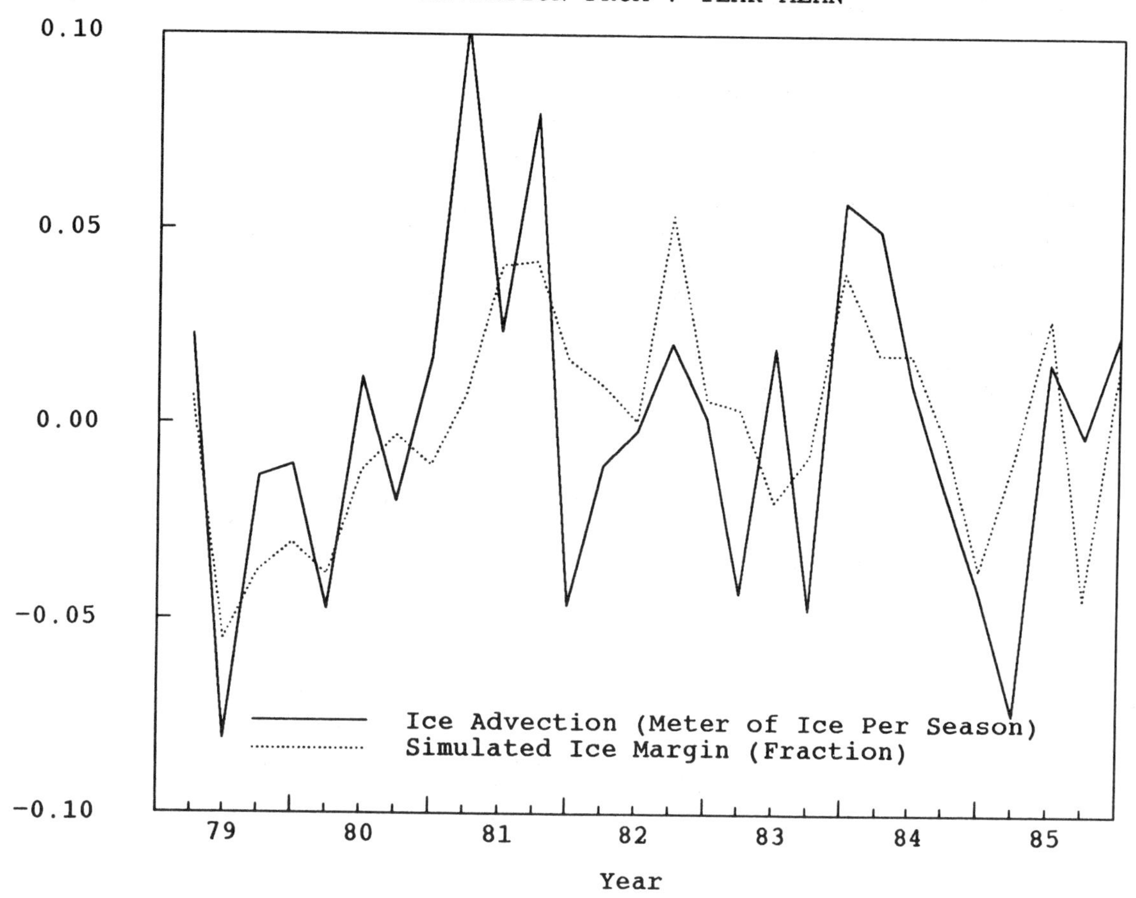

Fig. 10. Interannual deviations of seasonally averaged ice advection and simulated ice margin. The ice advection time series has been shifted by one season to the right (i.e., the winter ice advection is plotted with the spring ice margin)

ice thermodynamics, where there is no interannual variability in the ocean heat flux, the correlations with the observed ice edge are uniformly poor everywhere. Contrasting this with the thermo only model with interactive ocean circulation and the dynamic thermodynamic sea ice model with no interannually varying ocean circulation two critical needs can be identified: the need for ocean circulation effects in the Barents Sea region and the need for ice advection effects in all three regions but most notably the Arctic Basin and Greenland Sea.

The importance of ice advection in the Greenland Sea is perhaps even more graphically shown by the high correlation between the ice advection fluctuations with the simulated ice margin in the Greenland sea which is illustrated in Figure 10 and Table 3. These results show the dominance in the model between ice transport simulated out of the Fram Strait and ice margin variations in the Greenland Sea. However, while the correlations in the fall season are higher, this dominance does not clearly carry over to the observed ice margin. Consequently, there is also likely some control by variations in the ocean circulation not included here.

More insight into what is happening in the Greenland Sea may be gotten by examining the correlations for each individual season which are compiled in Table 4 and the actual seasonal scatter diagram of points which are shown in Figure 11. The seasonal scatter plots (Figure 11) show the simulated winter time ice margin in the Greenland Sea to be much smaller in winter than observed. The other seasons, on the other hand, have some reasonable variation in simulated ice edge. Since the effects of different boundary conditions (Figures 7 and 8) show lateral boundary conditions to significantly modify the ice margin, the results in Figure 11 suggest that the variations of the lateral transfer of heat through the Southern boundary is likely important to simulations of winter ice margins. This conclusion is also consistent with the higher correlations (Table 3) between ice advection and observed ice margin in the fall but not the winter.

Table 3. Correlation Coefficients Between Interannual Deviations of Ice Advection and Simulated and Observed Ice Margins in the Greenland Sea Region

	Seasonal shift *		No seasonal shift	
	Obse. Ice Edge	Simu. Ice Edge	Obse. Ice Edge	Simu. Ice Edge
Winter	-0.04	0.56	-0.56	-0.18
Spring	0.03	0.78	0.51	0.60
Summer	0.15	0.62	-0.20	-0.35
Fall	0.72	0.59	0.93	0.49
Year	0.22	0.61	0.17	0.19

* I.e., fall ice advection is correlated with winter ice margin.

Note also from Figures 8 and 11 that the order of magnitude modeled changes needed in winter are consistent with the effects of reducing the southern boundary heat fluxes into the Greenland Sea. Specifically, shutting off the lateral flow of heat at the southern boundary (Figure 8) results in about a decrease of .08 in the fraction of ice coverage in the Greenland Sea which is commensurate with deviations in the observed results in winter. Since it is known that the northward transport of heat through the Faero Shetland passage fluctuates, our results here suggest that fluctuations of northward heat transport may well be needed to obtain realistic variations of the ice margin.

While ice advection is found to be crucial to both the Greenland Sea region and the Arctic Basin the results for the Barents Sea are more complex. In particular Table 2 shows that inclusion of either ocean transport alone (thermo only model with interactive ocean) or ice advection alone (dynamic thermodynamic ice only model) yields reasonably good ice margin results in the Barents Sea. This complexity is also apparent in the correlations of the lateral heat flux with the ice edge variations (Table 5) where only a weak negative correlation with oceanic lateral heat transports occurs in the Barents Sea. (Note however that this correlation is higher in winter and spring when seasonal numbers are used). The likely explanation here is that Northward oceanic heat transports are associated with northward winds which will tend to impede ice advection southward. Including either one of the effects will cause the ice to move southward more slowly.

Some insight into the complexity of the processes accounting for the ice margin variations, and the very poor capability of air temperature thermodynamic considerations alone for predicting the ice margin variations, can be gotten by examining the correlations between ice growth, atmospheric heat gain and oceanic heat flux in Table 6. As can be seen, there is a very good correlation between oceanic heat flux and atmospheric heat gain in all regions. However, there is very poor correlation between oceanic heat flux and ice growth in the Greenland Sea and Barents Sea. These results reflect the fact that colder conditions yield greater convection and hence oceanic heat flux which in fact cancels out some of the effects of cooling. In the special case of the Arctic Basin, there is a high correlation. However, in this case the correlation is largely caused by higher growth rates and hence salt expulsion which in turn leads to greater oceanic heat fluxes. The main point here is that in the ice margin variations there is a competition between cooling effects, ice advection and northward transport of oceanic heat. As a consequence there is not a consistent simple correlation between either oceanic heat flux or northward oceanic heat transport and the ice margin.

5. CONCLUDING REMARKS

The main thrust of this paper has been to identify through numerical modeling experiments and analysis the main mechanisms responsible for variations of Arctic Sea ice margins with particular emphasis on interannual variations. The results show that the ice margin variations in the marginal ice zones in the Greenland and Barents seas can be viewed as a competition between ice advection, northward transport of heat and atmospheric cooling of the sea surface. In the case of the seasonal variation of the ice margin the role of ocean circulation and ice advection play relatively well defined roles in these marginal seas. Basically the northward transport of oceanic heat moves the ice edge much further back. However, since the advance of the ice margin is more advective in character, especially in the Greenland Sea, the melting of the southward moving ice delivers a substantial amount of freshwater flux to the ocean which stratifies the ocean and hence allows the ice to move further south by mitigating the heat flux. Hence if a motionless thermodynamic ice model is used the ice edge extent is reduced by several hundred kilometers.

On the interannual time scale there is a less clear picture of dominant terms because increases in ice advection and oceanic heat flux tend to occur together and often cancel each other. It does seem clear though, that within a background of mean ocean heat fluxes, the ice advection is particularly critical. If, for example, one employs a

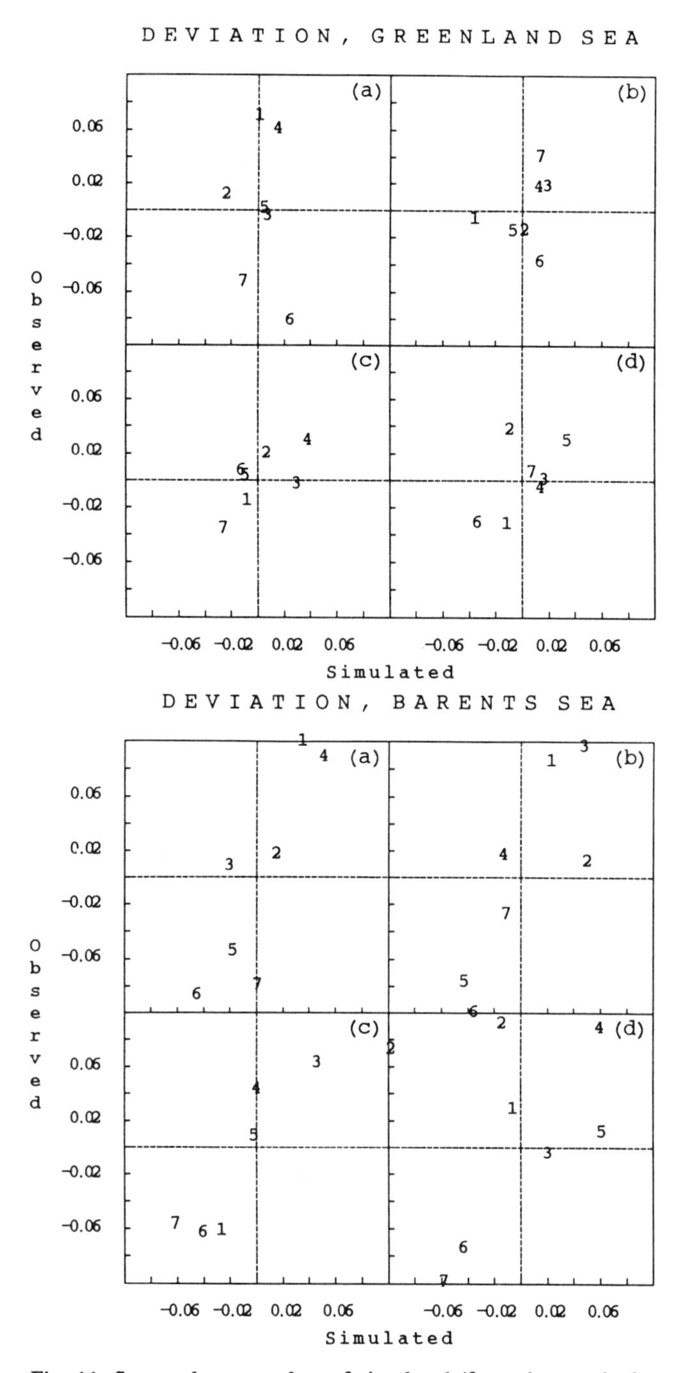

Fig. 11. Seasonal scatter plots of simulated (from the standard ice ocean model) and observed interannual deviations of seasonally averaged ice margin results for the three regions. (a) Winter; (b) Spring; (c) Summer; (d) Fall. The numbers denote the year: e.g. 1 is 1979, 2 is 1980, etc.

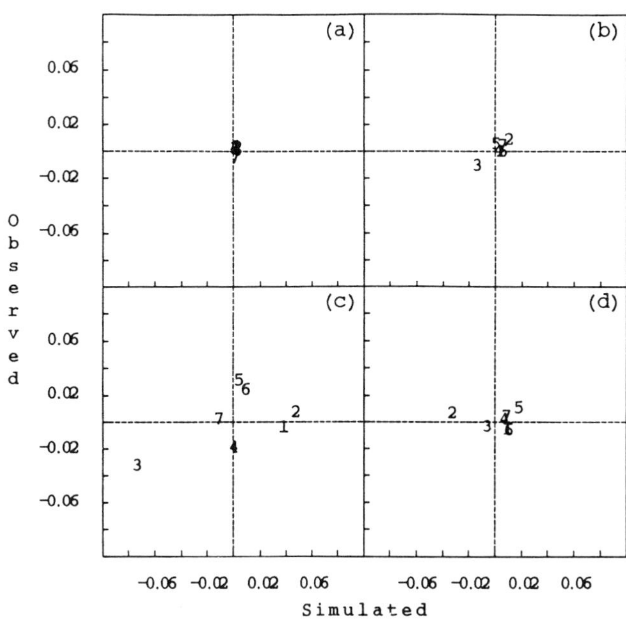

Figure 11. (continued)

thermodynamic only ice model with a background of seasonally varying oceanic heat fluxes, extremely poor correlations with observations occur in all regions. With ice advection included, however, correlations close to the full model results occur.

This importance of ice advection is amplified within the model based ice margin and transport results in the Greenland Sea. In particular the simulated variations of the ice margin follow very closely the advection of ice (with a seasonal lag) through the Fram Strait. This result in turn suggests that the extent of melt fluxes induced by sea ice outside the Arctic Basin may be largely a result of variations of ice transport southward. However, it should be emphasized that correlations of simulated and observed ice margin variations in the Greenland Sea are poor in winter and spring, during which time the effects of ice advection may be overshadowed by oceanic heat transport. In winter in particular very little modeled variation in the ice margin occurs. This can be contrasted with observed interannual variations of up to 15% of the average winter areal coverage. Interestingly this observed variation is about the same magnitude as the increase in average simulated areal coverage in winter if lateral oceanic heat transport at the model boundaries is set to zero. In particular in the winter and spring, there is a strong negative correlation between the lateral oceanic heat transport and ice margin variations. Variations in the lateral winter heat transport, for example, explain 60% of the variance of the spring ice margin variations. It is therefore our conclusion that variations of the heat flux through the southern boundary may be important

Table 4. Seasonal Correlation Coefficients Between Interannual Variations of the Seasonally Averaged Observed and Simulated Ice Margin Variations for Different Regions

	Winter	Spring	Summer	Fall
Greenland	-0.03	0.33	0.82	0.71
Bar. Sea	0.93	0.92	0.90	0.61
Arc. Basin	0.00	0.43	0.74	0.29
Whole	0.77	0.63	0.84	0.30

especially in the Greenland Sea ice margin. Inclusion of such variations is needed in future simulations.

In the Barents Sea, oceanic heat flux into the region takes on a significant role in this model. Here variations in lateral heat flux into the region play an important role on the ice margin similar to the role of ice advection in the Greenland Sea. Hence, in this region, as long as one has the interannually varying ocean circulation included, reasonable variations of the ice margin can be modeled.

Because of the relatively small amount of heat flowing into the Arctic Basin, the ice advection there almost totally dominates the ice margin variations which become significant in summer. Put quite simply an onshore wind pushes the ice into the coast and removes the open water. With an offshore wind the opposite effect occurs. Other mechanisms are largely second order.

Table 5. Correlation Between Lateral Heat Flux and Ice Margin Variations with One Season Shift (i.e. Fall Heat Transport with Winter Ice Margin)

	Year	Winter	Spring	Summer	Fall
Greenland	-0.32	-0.36	-0.29	0.23	0.41
Bar. Sea	-0.20	-0.47	-0.80	0.	0.38

Table 6. Correlations Between the Interannual Variability of Seasonally Averaged Oceanic Heat Flux, Atmospheric Heat Gain and Sea Ice Growth. Numbers Shown Here are, in Descending Order for the Greenland Sea, Barents Sea, Arctic Basin, and the Whole Region

	sea ice growth	oceanic heat flux
Atm. heat gain	.31 .65 .96 .77	.74 .68 .74 .72
Sea ice growth		.10 .27 .63 .37

Overall our model results portray ocean circulation as critical to the variations of the ice margin in the Greenland and, Norwegian and Barents Seas. The role of ocean circulation, somehow, is made more complex by the presence of southward ice advection. The ocean model used here has specified a fixed lateral oceanic heat flux at lateral boundaries. Consequently, in addition to penetrative convection and ocean eddy effects, what has not been included are the effects of increased ice transport and melting on salinity anomalies and perhaps on the larger scale ocean circulation. Numerical investigations of such interactions are currently in progress. What does seem clear, though, is that the use of thermodynamic only ice models in such ice ocean studies is not realistic.

Acknowledgments. We would like to thank Ola Johannessen for first interesting one of us (W. Hibler) in marginal ice zone sea ice problems and for encouragement to continue this work. Much of the model development work was made possible by earlier lower resolution ice-ocean model development by Peter Ranelli and John Ries. Insightful comments on this work by Greg Flato and Jim Overland, and computer graphics and software support by Jim Waugh were also most helpful. This work was supported by NSF Grant DPP9203470.

REFERENCES

Flato, G. M. and W. D. Hibler III, A multi-level sea-ice model with explicit inclusion of pressure ridging, submitted to *J. Geophys. Res.*, 1993.

Hansen, J. and S. Lebedeff. Global trends of mean surface air temperatures, *J. Geophys. Res.* 92, 13,345-13,372, 1987.

Hibler, W. D. III. A dynamic thermodynamic sea ice model, *J. Phys. Oceanogr.*, 9, 815-846, 1979.

Hibler, W. D. III Modelling a variable thickness sea ice cover, *Mon. Wea. Rev.*, 1943-1973, 1980.

Hibler, W. D. III and J. E. Walsh. On modeling seasonal and interannual fluctuations of Arctic sea ice, *J. Phys. Oceanogr.*, 12, 1514-1523, 1982.

Hibler, W. D. III and K. Bryan. A diagnostic ice-ocean model, *J. Phys. Oceanogr.*, 17, 987-1015, 1987.

Hibler, W. D. III and J. Zhang. Interannual and climatic characteristics of an ice ocean circulation model, in *Ice in the Climate System*, W.R. Peltier, Ed., Springer Verlag, 633-652, 1993.

Jones, P. D., S. C. B. Raper, R. S. Bradley, H. F. Diaz, P. M. Kelly and T. M. Wigley. North hemisphere surface air temperature variations. *J. Clim. Appl. Met.*, 25, 161-179, 1986.

Levitus, S. Climatological atlas of the world ocean, *NOAA Publ.* 13, U.S. Dept. of Commerce, Washington, DC, 1982.

Marsden, R. F. and L. A. Mysak. Evidence for stability enhancement of sea ice in the Greenland and Labrador sea *J. Geophys. Res.* 96, 4783-4789, 1991.

Parkinson, C. L. and W. M. Washington. A large scale numerical model of sea ice, *J. Geophys. Res.* 84, 311-337, 1979.

Walsh, J. E., W. D. Hibler III and B. Ross. Numerical simulation of northern hemisphere sea ice variability, 1951-1980, *J. Geophys. Res.* 90, 4847-4856, 1985.

Zhang, J. and Hibler, W. D. III. On an efficient numerical method for modeling sea ice dynamics, submitted to *J. Geophys. Res.*, 1994.

On the Surface Heat Fluxes in the Weddell Sea

Jouko Launiainen

Finnish Institute of Marine Research, Helsinki, Finland

Timo Vihma

Department of Geophysics, University of Helsinki, Finland

Turbulent surface fluxes of sensible and latent heat in the Weddell Sea were studied using drifting marine meteorological buoys with satellite telemetry. In 1990-1992 a total of 5 buoys were deployed on the sea ice, in the open ocean, and on the edge of a floating continental ice shelf. The buoys measured, among others, wind speed, air temperature and humidity with duplicate sensors and yielded year-round time series. The heat fluxes were calculated by the gradient and bulk methods based on the Monin-Obukhov similarity theory. Over the sea ice, a downward flux of 15 to 20 W/m^2 was observed in winter (with typical variations of 10 to 20 W/m^2 between successive days) and 5 W/m^2 in summer. For the latent heat flux, the results suggested a small evaporation of 0 to 5 W/m^2 in summer and weak condensation in winter. The highest diurnal values, up to 20 W/m^2, were connected with evaporation. Because of stable stratification, the transfer coefficients of heat and moisture were reduced to 80% of their neutral values, on the average. Over the leads and coastal polynyas, an upward sensible heat flux of 100 to 300 W/m^2 was typical, except in summer when the air temperature was close to the sea surface temperature. Over the continental shelf ice, the sensible heat flux was predominantly downwards (15 to 20 W/m^2), compensating the negative radiation balance of the snow surface. Over the snow and ice surfaces the magnitude of turbulent fluxes was smaller than that of radiative fluxes, while over the open water in winter sensible heat flux was the largest term. Modification of the continental air-mass flowing out from the shelf ice to the open sea was studied with aerological soundings made from a research vessel. Associated turbulent heat exchange was estimated on the basis of three methods: modification in the temperature profiles, surface observations, and diabatic resistance laws for the atmospheric boundary layer. If we estimate an area-averaged turbulent heat exchange between the surface and the atmosphere for the whole Weddell Sea on the basis of our data, the large upward fluxes from leads and coastal polynyas (with an areal coverage of 5 to 7% in wintertime) approximately balance the downward fluxes over the sea ice. A first-order estimate for the annual area-averaged total vertical heat loss from the water mass is 20 to 30 W/m^2.

1. INTRODUCTION

Polar oceans are areas of extreme differences in the temperature between the ocean and the atmosphere, but the heat exchange is frequently restricted by sea ice cover. Over ice-free areas, cracks, leads and polynyas, the heat flux from the sea to the air may reach several hundreds of watts per square meter [*Bromwich and Kurtz,* 1984; *Cavalieri and Martin,* 1985; *Schumacher et al.*, 1983]. On the other hand, the heat and moisture exchange affects the amount of sea ice, and controls the structure of the atmospheric and oceanic boundary layer. In certain areas of the polar oceans, the heat loss from the ocean also affects the deep water formation, especially when sea ice is produced increasing the surface salinity in the water.

It is well known that, despite the small areal coverage of leads and polynyas, they have a great effect on the total

The Polar Oceans and Their Role in Shaping the Global Environment
Geophysical Monograph 85

heat loss from the ocean and even a few percent of open water or thin young ice make a dominant contribution to the regional heat budget in winter [*Maykut*, 1978; *Ledley*, 1988]. The effect is also detectable in the output of global climate models [*Simmons and Budd*, 1991]. In Arctic regions, heat exchange over sea ice or over leads and polynyas has been studied in several field experiments, but observations on the meteorological variables required to compute the heat and moisture fluxes over Antarctic seas are still rare, especially from the winter period. A few analyses of heat exchange over Antarctic sea ice and coastal polynyas have been reported [*Bromwich and Kurz*, 1984; *Andreas and Makshtas*, 1985; *König-Langlo et al.*, 1990; *Kottmeier and Engelbart*, 1992].

The Weddell Sea is totally covered by ice during the austral winter, although the ice field is broken and leads are generated within the drifting divergent ice field; wider polynyas are frequent in the southeast and south near the ice shelves and the coast of the Antarctic continent [*Zwally and Comiso*, 1985]. The areal coverage of the polynyas and leads is estimated to be about 5% [*Schnack-Schiel*, 1987; *Augstein et al.*, 1991]. On the basis of Arctic data, the atmospheric surface layer in winter over old sea ice is typically stably stratified owing to large heat losses via longwave radiation. The turbulent heat flux is therefore generally directed downwards [*Vowinckel and Orvig*, 1973; *Untersteiner*, 1986; *Makshtas*, 1991; *Serreze et al.*, 1992]. The simple model results of *Makshtas* [1991] for Arctic sea ice in winter suggest that over sea ice less than 1 m thick the sensible heat flux is directed upwards, and over thicker ice downwards. In the Antarctic, the sea ice tends to be thinner and the stratification in the upper ocean weaker than in the Arctic, which should result in an increased heat flux from the ocean through the ice. Accordingly, it is not generally known whether the wintertime atmospheric surface layer over Antarctic sea ice is usually stably stratified or not. *Andreas and Makshtas* [1985] found the sensible heat flux to be as often to the atmosphere as to the ice in the northeastern Weddell Sea in the spring, whereas *König-Langlo et al.* [1990] found, in the late winter of 1989, the turbulent fluxes directed into the atmosphere. *Kottmeier and Engelbart* [1992] reported sensible heat fluxes in both directions over pack ice in late winter and spring, but with downward-directed fluxes prevailing. However, *Wamser and Martinson* [1993] report predominantly near-neutral or unstable atmospheric surface layer stratification in the northeastern Weddell Sea in winter.

In the Weddell Sea, the surface-layer air temperature over sea ice is typically -10 to -25°C in winter [*Hoeber*, 1989; *Launiainen et al.*, 1991]. When this cold air is advected over open leads, the temperature difference becomes extreme, resulting in intensive heat exchange. The heat loss from the leads is compensated by the enthalpy of freezing of the sea, and according to Arctic observations the leads typically remain open for a few hours only [*Lebedev*, 1968; *Bauer and Martin*, 1983; *Makshtas*, 1991]. After that new ice is formed, but it still permits considerable heat flux from the ocean.

In summer, the marginal ice zone retreats southward and westward and the coastal polynyas in the eastern Weddell Sea become part of the open ocean, but the sea ice remains in the central and western Weddell Sea. Summer conditions over the sea ice are very different from those in winter, because incoming solar radiation warms the snow and even small melt-water ponds may appear. Thus the wintry contrast between the temperature of the sea ice and the leads often vanishes and the air-sea partition of turbulent fluxes may vary in time. For example, according to the arctic data of *Leavitt et al.* [1978, also cf. *Andreas*, 1989] over the Beaufort sea ice, in summer the flux of sensible heat is as often directed upwards as downwards, and the latent heat flux usually has the same direction. In the Weddell Sea, the most intensive air-sea exchange still takes place near the coasts of the continent or continental ice shelves. The dominant wind direction on the eastern coast of the Weddell Sea is towards the ocean. The wind blowing from the continent brings rather cold and dry air to the open ocean, and even in summer the difference between the sea surface and air temperatures can reach 10 to 20°C. The winds in the more central Weddell Sea are more variable in direction and the summertime temperature difference between the sea and the atmosphere is smaller (some 5°C).

In this paper, data obtained by automatic marine meteorological buoys are analyzed to compute estimates for fluxes of heat and moisture over various surfaces: the sea ice, winter leads, summertime open ocean in the eastern Weddell Sea, and the continental ice-shelf edge. Although involving some apparent methodological weaknesses, the extensive data sets are presumably the first ones allowing year-round estimates for the area.

2. OBSERVATIONS

Three automatic marine meteorological buoy stations were deployed in the eastern Weddell Sea area from R/V *Aranda* during the first Finnish Antarctic Expedition in 1989-1990 (FINNARP-89). One of the buoys was deployed on a small sea-ice floe, one in the open ocean 100 km from the continental ice shelf, and one on the edge of a floating continental ice shelf, 250 m from the "shoreline". During the FINNARP-91 expedition on the Russian R/V *Academik*

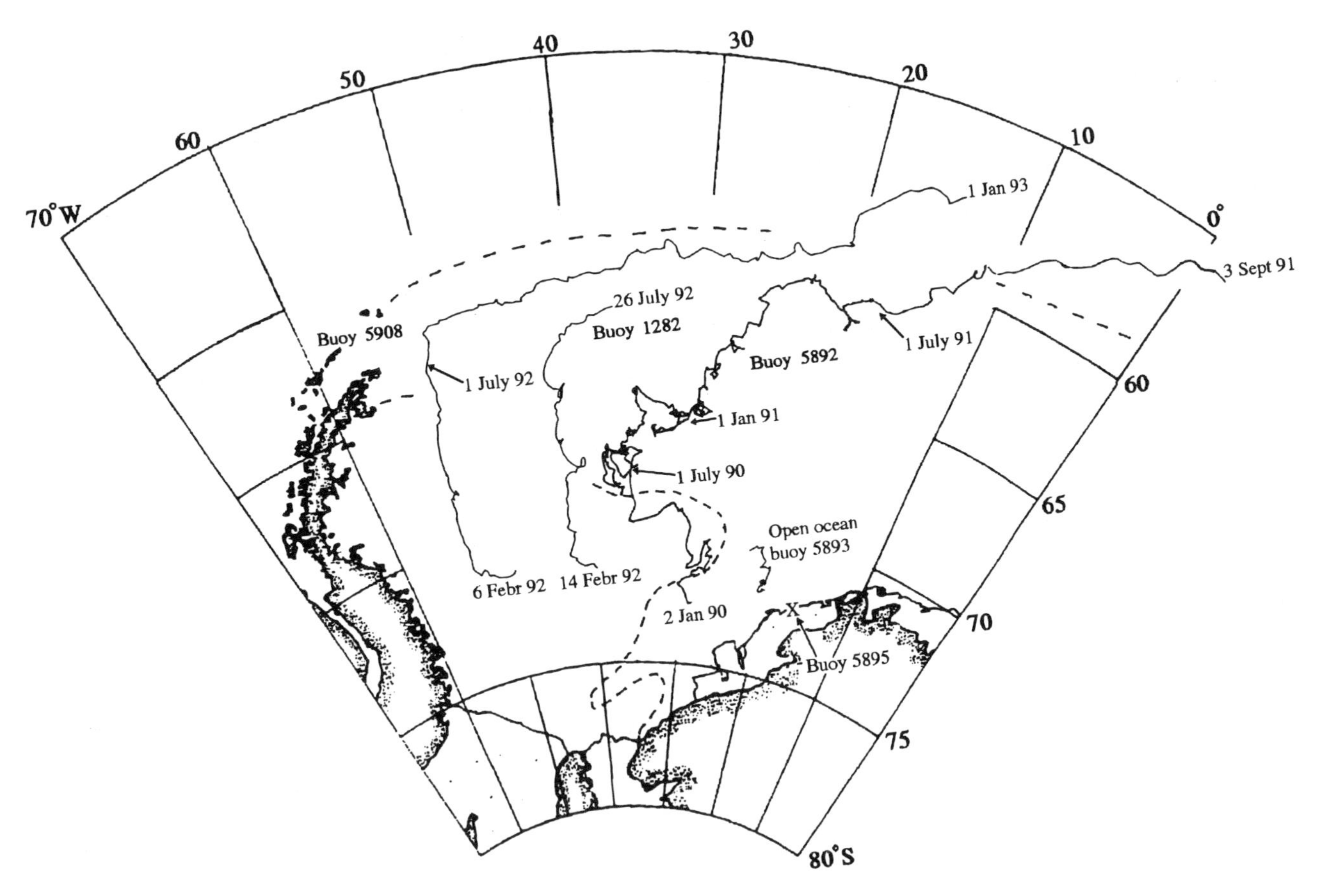

Fig. 1. Research area, deployment sites and the drift trajectories of the marine meteorological buoys in the Weddell Sea. Dashed lines indicate approximate sea ice margins in February 1990 and in September 1992.

Fedorov two more buoys were deployed in February 1992 on sea ice floes in more western parts of the Weddell Sea, in connection with the U.S.-Russian *Ice Station Weddell-1* experiment. Trajectories of the buoys are presented in Figure 1. The buoys made observations of the following quantities: atmospheric pressure, air temperature and humidity, wind speed and direction, surface temperature of snow, ice or ocean, and buoy orientation. In addition, the temperature profile in the ocean was measured. Multiple sensors were used to ensure data quality and to measure gradients of temperature and humidity in the atmospheric surface layer. The measurement heights, number and accuracy of sensors, and functioning periods of the buoys are presented in Table 1. The buoys were manufactured by Defense Systems Inc. (McLean, Virginia, USA). The choice and layout of the sensors were agreed in discussions between the research group and the manufacturer. A more detailed description of the configuration, types and calibration of the sensors used is given in the technical data reports by *Launiainen et al.* [1991] and *Vihma et al.* [1994].

The buoys were located and the data was transmitted by the Argos satellite survey. In polar regions there are 26 satellite passes a day, about 20 of which were orbitally suitable to receive data from our buoys, yielding an average observation interval of 1.2 h.

Additionally, surface meteorological observations and aerological balloon soundings (DigiCora Rawinsonde, Vaisala Co.) were made from the research vessel during the expeditions in the austral summers of 1989/1990 and 1991/1992.

3. COMPUTATION METHOD FOR THE FLUXES

The ice drift dynamics and physical aspects of momentum exchange in the Weddell Sea were discussed in *Vihma and Launiainen* [1993]. For this study, the fluxes of sensible and latent heat were computed by bulk methods on the basis of the *Monin-Obukhov* [1954] similarity theory using an algorithm described in detail by *Launiainen and Vihma* [1990]. Based on the universal profile gradients of velocity (V), temperature (θ) and specific humidity (q), the Monin-Obukhov similarity theory yields for the turbulent

TABLE 1. Marine meteorological buoy observations. Buoy identification number and lifetime, observation site, measurement heights for wind (V), temperature (T) and relative humidity (RH), observation quantities, number of sensors in each buoy (n) and accuracy are given.

Buoy ID	Lifetime	obs. site	height V	T & RH		obs. quantities	n	Accuracy
						Atmospheric pressure	1	1 hPa
5892	2 Jan 1990 - 28 Jan 1991	ice floe, drifting	3.5	1.9	3.2	air temperature	4	0.05°C
5893	11 Febr 1990 - 1 Apr 1990	open sea, drifting	3.4	1.7	3.0	relative humidity of air	2	2%
5895	25 Dec 1989 - still operating[a]	shelf edge, fixed	3.9	2.3	3.7	buoy hull temperature	1	0.2°C
1282	14 Febr 1992 - 26 July 1992	ice floe,drifting	4.0	2.0	3.6	water temperature	10[b]	0.1°C
5908	6 Febr 1992 - 5 Jan 1993	ice floe,drifting	3.6	2.0	3.4		20[c]	0.05°C
						wind speed	2[d]	0.3 m/s
						wind direction	1	5°
						snow depth	1[e]	2 cm

[a] 15 Dec. 1993, [b] for buoys 5892 and 5893, [c] for buoys 1282 and 5908, [d] only 1 for buoy 1282, [e] only for 1282.

fluxes of momentum (τ), sensible heat (H) and moisture (E) the familiar formulae of the gradient method (a) and the bulk-aerodynamic method (b) of

$$\tau = \rho u_*^2 = \rho K_M \frac{\partial V}{\partial z} = \rho C_D V_z^2 \tag{1}$$

$$H = -\rho c_p K_H \frac{\partial \theta}{\partial z} = \rho c_p C_H (\theta_s - \theta_z) V_z \tag{2}$$

$$E = -\rho K_E \frac{\partial q}{\partial z} = \rho C_E (q_s - q_z) V_z \tag{3}$$

(a) (b)

where V_z is the mean wind speed at a height of z, and u_* is the friction velocity. ρ is the air density, and c_p is the specific heat capacity of air. θ_s - θ_z and q_s - q_z are the differences in potential temperature and specific humidity between the atmosphere and the surface, respectively. (λE gives the flux of latent heat, λ being the enthalpy of vaporization.)

The transfer coefficients above depend on the measurement level, on the surface roughnesses for velocity and scalar quantities of θ and q, and on the surface layer stratification. Accordingly, the transfer coefficients are given as a set of functions:

$$\begin{aligned} K_M &= K_M(z/z_0, \Phi_M(z/L)), \\ K_H &= K_H(z/z_0, z/z_T, \Phi_M(z/L), \Phi_H(z/L)), \\ K_E &= K_E(z/z_0, z/z_q, \Phi_M(z/L), \Phi_E(z/L)) \end{aligned} \tag{4}$$

or

$$\begin{aligned} C_D &= C_D(z/z_0, \Psi_M(z/L)), \\ C_H &= C_H(z/z_0, z/z_T, \Psi_M(z/L), \Psi_H(z/L)), \\ C_E &= C_E(z/z_0, z/z_q, \Psi_M(z/L), \Psi_E(z/L)) \end{aligned} \tag{5}$$

where z_0, z_T and z_q are the roughness lengths for velocity, temperature and moisture. In neutral stratification, z_0, z_T and z_q define the transfer coefficients in question [e.g. *Launiainen and Vihma*, 1990]. Φ_M, Φ_H and Φ_E are the gradient forms and Ψ_M, Ψ_H and Ψ_E the integrated forms of the universal functions which give a stability correction for the profiles and transfer coefficients. In the argument z/L of the universal functions, L is the stratification parameter, the Monin-Obukhov length.

For our study, the roughness lengths z_0 defining the transfer coefficients K_M and the drag coefficients C_D were determined as follows (for further details see the Appendix):

1) For cases over the open sea the well-known slightly wind-dependent C_D values by *Smith* [1980, 1988] were used. For coastal polynyas, involving fetch-limited young waves, a somewhat more strongly wind-dependent drag form by *Wu* [1980] was used.

2) For leads in the sea-ice zone, for taking into account the form drag from the ice edges and the growth of waves in narrow leads, an overall C_D form by *Andreas and Murphy* [1986] was adopted.

3) For ice and snow, the roughness length z_0 and drag coefficient C_D were estimated using the model by *Banke et al.*, [1980], based on the mean geometric surface roughness (R), which was estimated to be slightly over 10 cm. Considering the buoy 5892, R for the specific ice floe was larger, but its average value in the region surrounding the floe was of the order of 10 cm.

TABLE 2. The roughness lengths and neutral bulk-transfer coefficients (referred to a height of 10 m) used in the flux calculations. Observed mean wind speed of 6 m/s used for the table.

obs. site	C_D and z_0	C_{HE} and z_{Tq}
open sea	$C_D = 1.09 \times 10^{-3}$	$C_{HE} = 1.01 \times 10^{-3}$
(typical case)	$z_0 = 0.05$ mm	$z_{Tq} = 0.02$ mm
coastal polynyas	$C_D = 1.22 \times 10^{-3}$	$C_{HE} = 1.09 \times 10^{-3}$
(typical case)	$z_0 = 0.09$ mm	$z_{Tq} = 0.02$ mm
leads	$C_D = 1.49 \times 10^{-3}$	$C_{HE} = 1.26 \times 10^{-3}$
	$z_0 = 0.3$ mm	$z_{Tq} = 0.04$ mm
ice and snow	$C_D = 1.90 \times 10^{-3}$	$C_{HE} = 1.44 \times 10^{-3}$
	$z_0 = 1$ mm	$z_{Tq} = 0.05$ mm
ice shelf	$C_D = 1.46 \times 10^{-3}$	$C_{HE} = 1.33 \times 10^{-3}$
	$z_0 = 0.3$ mm	$z_{Tq} = 0.09$ mm

As for the vapour and heat exchange coefficients, we feel the common assumption of $K_H \approx K_E$ and $C_H \approx C_E$ i.e $z_T \approx z_q$ to be reasonable, as suggested by *Monin and Yaglom* [1971], *Andreas and Murphy* [1986], and *Makshtas* [1991]. Accordingly, we compute the vapour and heat exchange coefficients as follows:

4) For open water surfaces a data compilation result by *Launiainen* [1983] based on a comparison of simultaneous C_D/C_E results, derived from direct eddy flux results in literature is used. For the neutral transfer coefficients of vapour and heat exchange this gives $C_{HE} = 0.63 \cdot C_D + 0.32 \cdot 10^{-3}$.

5) For ice and snow the model by *Andreas* [1987] is used, based on the aerodynamic roughness z_0 from above. The mean numerical values of the various roughness lengths and neutral bulk transfer coefficients are listed in Table 2. It may be anticipated that an accurate estimation of roughness lengths and transfer coefficients for leads presents the greatest difficulty.

For the universal functions, the so-called *Businger et al.* [1971] - *Dyer* [1974] type formulae were used for unstable stratification. For the stable region, a form by *Holtslag and de Bruin* [1988], not very much different from the well-known earlier form by *Webb* [1970], was used. Because the Monin-Obukhov length L, in the argument of the universal functions, contains the fluxes to be determined, the system of equations (1) to (5) leads to an iterative solution, described in detail in *Launiainen and Vihma* [1990]. Formulae for the transfer coefficients and universal functions are given in the Appendix.

One great difficulty in the estimation of heat and moisture fluxes over the sea ice was that the actual snow surface temperature was unknown. Thus, the bulk-method (i.e. "b"-equations of 1 to 3) could not be used for the sensible heat flux. Instead, the level-difference or gradient method ("a"-equations) was used. The level difference of temperature was obtained from two independent sensor pairs on each buoy. After the sensible heat flux and the snow surface temperature were estimated by the level difference method using the flux-profile relationships, the flux of latent heat was computed by the bulk method. The same approach has been applied by e.g. *Thorpe et al.* [1973] and *Leavitt et al.* [1978].

4. SOURCES OF INACCURACY IN FLUX ESTIMATES

Various factors may be the potential cause of errors in the flux estimates. A major difficulty in the computation over ice was that the actual snow surface temperature was unknown. Inaccuracies resulting from the sources (a) to (g) discussed below are summarized in Table 3. The inaccuracy may change with meteorological conditions, but typical situations were considered when calculating the values given.

a) The gradient or level difference method had to be applied to estimate the sensible heat flux over ice, and the result is very sensitive to the accuracy of the temperature difference between the observation levels. In our case, this may result in an inaccuracy of 15-20 W/m^2 in the sensible heat flux and 10-15 W/m^2 in the latent heat flux. The following discussion considers the problems in more detail.

The height difference between the two observation levels was small, of only 1.3 to 1.6 m centered at a height of 2.5 to 3 m, depending on the buoy. Especially above a relatively smooth surface the temperature difference within such a distance tends to be extremely small, except in strongly stably stratified conditions. To determine the fluxes to an accuracy of 20-30 W/m^2, the temperature difference should be measured with an accuracy of 0.05-0.1°C. This accuracy was easily achieved in calibrations in Finland before the expeditions, because the buoys measured in fact the temperature difference between the two heights (and the

TABLE 3. Inaccuracy in the flux estimates resulting from various potential error sources. Absolute or relative inaccuracy is given, whichever more relevant in the particular case.

factor (see text)	measurement inaccuracy	inaccuracy in sensible heat flux estimate	inaccuracy in latent heat flux estimate
a)	0.05°C	15-20 W/m²	10-15 W/m²
b)	1%		1 W/m²
c)	ice 1°C		5 W/m²
	water 0.1°C	1-2 W/m²	1 W/m²
d)	10%	10%	10%
e)	100%	4-8 W/m²	4-8 W/m²
f)	20 cm snow accum.	+ 5%	- 1%

temperature at the lower level). Still, a very slight electrical drift detected from the data during the measuring periods of two buoys decreased the accuracy. Using several indirect methods of checking and correcting, and using mean values for a day or longer, an accuracy significantly better than 0.1°C is believed to have been achieved. For example, one of the methods for finding the correction to the measured level difference in the air temperature was to study the behaviour of the level difference of temperature with respect to the bulk-difference between the temperature of air and the buoy hull inside the ice. Furthermore, comparison of the results based on the level difference method and bulk-aerodynamic method, both of which the open ocean buoy data with a measured surface temperature made possible, also gave a rather good reference for the accuracy of the level difference method (Figure 2). The bulk-method was also applicable for comparison for one of the ice floe buoys (5892, Figure 1) when it reached the open ocean after overwintering in the Weddell Sea.

The above refers to the principal air temperature sensors. In addition, the buoys had temperature probes in the humidity sensors. Since the resolution of these probes was only 0.7°C, they could not be used to calculate an instantaneous heat flux. However, median values for the heat fluxes based on these probes corresponded to those based on the principal air temperature sensors to an accuracy of 5 to 10 W/m² for the buoys 5892, 5893, 5895 and 1282. For the buoy 5908, results from the two sets of probes had a higher discrepancy. Finally, a filtering method for obvious errors was applied.

b) It was discovered that the measurements of relative humidity were not accurate enough to allow an estimate of the latent heat flux using the level difference method to be made. In calculations using the bulk-aerodynamic method, the observed mean humidity from the two sensors was used. It was measured initially to an accuracy of 1 to 2%, but the electrical sensors did not function properly when

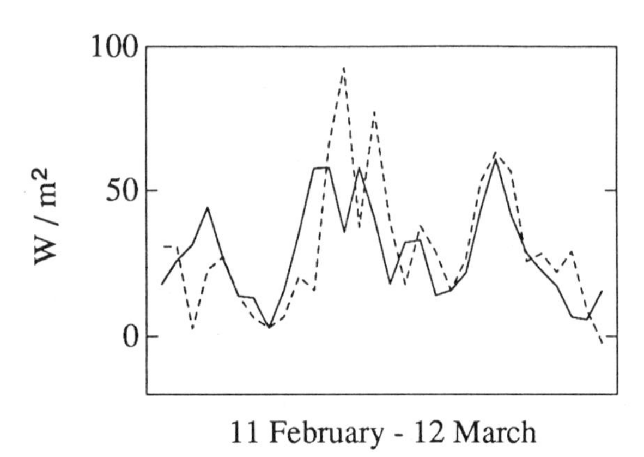

Fig. 2. Sensible heat flux as calculated using the bulk-aerodynamic (solid line) and the level difference method (dashed line) for the open sea buoy (5893). Diurnal means from February 11 to March 12, 1990.

used in temperatures lower than -20°C. In this context one should note that the possible range of moisture content in the air at low temperatures, say below -10°C, is so small that the latent heat flux is in practice controlled by the saturation moisture of the surface (i.e. by the surface temperature).

c) Accuracy of the surface temperature. The accuracy of the calculated snow surface temperature affects the latent heat flux calculated by the bulk method. However, because of rather large air-snow temperature differences the latent heat flux is not very sensitive to the inaccuracy of snow surface temperature, see Table 3. Over open water surfaces the accuracy of the difference between the air and surface temperature was of the order of 0.1°C. This gives an error of only 1 to 2 W/m² in the sensible heat flux and 1 W/m² in the latent heat flux.

d) Ice accretion and accumulation of snow sometimes prevented the collection of wind data. The data quality was improved using two wind sensors working on different principles (propeller and cup anemometers). The cup anemometer was found to be more liable to errors due to ice accretion. Periods when both anemometers continously showed wind speeds less than 1 m/s were removed from the analysis. A good correlation between the winds observed by the buoys and the geostrophic winds derived from ECMWF pressure analyses gave confidence in the wind data.

e) Estimation of the surface roughness affects the flux results. For the open ocean, more consensus in values is to be found, but for cases above narrow leads and ice, values are more poorly known. Estimation of the geometric roughness (R) of the ice and snow surface with an inaccuracy of 100% would, however, result in an inaccuracy of no more than 4-8% in the heat fluxes.

f) Changes in the observation heights due to snow accumulation or melt. Exact observation heights were known at the time of the buoy deployments. Later, only two buoys were visited, and the one on the ice shelf edge showed snow accumulation of 15 to 25 cm in two years. The snow accumulation was directly measured by one buoy only (1282), but the rate of change, 7 to 10 cm in half a year, is assumed to be comparable to that at the sites of the other ice floe buoys as well. The sensitivity of estimated fluxes to the measurement heights was found to be rather low. The sign of error in λE is opposite to that in H, because the former is calculated by the bulk-method and the latter by the gradient method.

g) Finally, the requirements of horizontal quasi-homogeneity and stationarity to satisfy the Monin-Obukhov similarity theory should have been reasonably well met over the sea ice and over the open ocean in the eastern Weddell Sea, but the narrow winter leads are far more problematic in this sense. In practice, however, no theory provides better methods for computing turbulent fluxes when direct eddy-covariance or dissipation measurements are not available. Moreover, the results of *Andreas et al.* [1979], *Smith et al.* [1983] and *Andreas and Murphy* [1986] suggest that the effects of horizontal non-homogeneity on the flux-gradient relations near the surface are perhaps not too severe in areas where changes in roughness remain small, as is the case for ice and open water.

In the light of the various potential error sources above, one may take the inaccuracy in various flux estimates to be around 20 W/m^2 or, in the case of large fluxes, even significantly larger. Accordingly, the absolute values of the results given by the study should be regarded as first-order estimates. On the other hand, relative values such as mean variations between successive days and periods are to be considered more characteristic and correct. The relative inaccuracy is largest over ice and snow-covered surfaces. The results over the open ocean should be the most reliable ones, while the estimates over polynyas and leads should be regarded as having rather good relative accuracy, although the absolute errors may be large, because the fluxes are large. However, the maximum values of over 400 to 600 W/m^2 found in our study may be erroneous, because they are met within extremely unstable conditions for which e.g. the validity of the universal functions has not been adequately proved.

5. RESULTS

The time series of air temperature and wind speed (at a height of 3 m) over the sea ice and at the edge of the floating ice shelf in 1990-1992 are given as diurnal means in Figure 3. The numerical data and further discussion are to be found in the technical reports [*Launiainen et al.*, 1991; *Vihma et al.*, 1994]. When considering the annual differences over the sea ice one should note that in 1990 the observations (of buoy 5892) were from the central Weddell Sea, whereas in 1992 these were from more western areas (Figure 1). Additionally, in July 1992 after the two drifting buoys (5908 and 1282) crossed latitude 63°S and started to drift east in the circumpolar current, they left the wintertime Weddell Sea, which may be seen e.g. as an increase in the air temperature (compare Figures 3 a and b).

5.1. Fluxes over the sea ice

Figure 4 gives estimates of the sensible and latent heat flux above the sea ice as time series of diurnal means. During summer, the sensible heat seems to be rather small and may be directed either upwards or downwards. During winter, the sensible heat flux tends to be downwards, from the air to the ice, as a consequence of radiatively-based surface cooling, reaching -50 W/m^2 or even more. On the other hand, cases of upward fluxes of 30 to 50 W/m^2 also exist during winter. Large variations in the air temperature, wind speed and cloudiness yield large variations in the heat flux, typically 10-20 W/m^2 between successive days. The results from the two buoys operating in 1992 are rather coherent, giving more reliability to an estimation method liable to potential inaccuracies. For the winter 1992 the mean sensible heat flux was -10 to -15 W/m^2 downwards. The estimates for latent heat flux suggest a small evaporation, with λE of 0 to 5 W/m^2, during the summer and very weak condensation during the winter. The highest diurnal values, up to 20 W/m^2, are connected with evaporation. The stability parameter 10/L was typically 0.05-0.5. Because of the stable stratification, the transfer

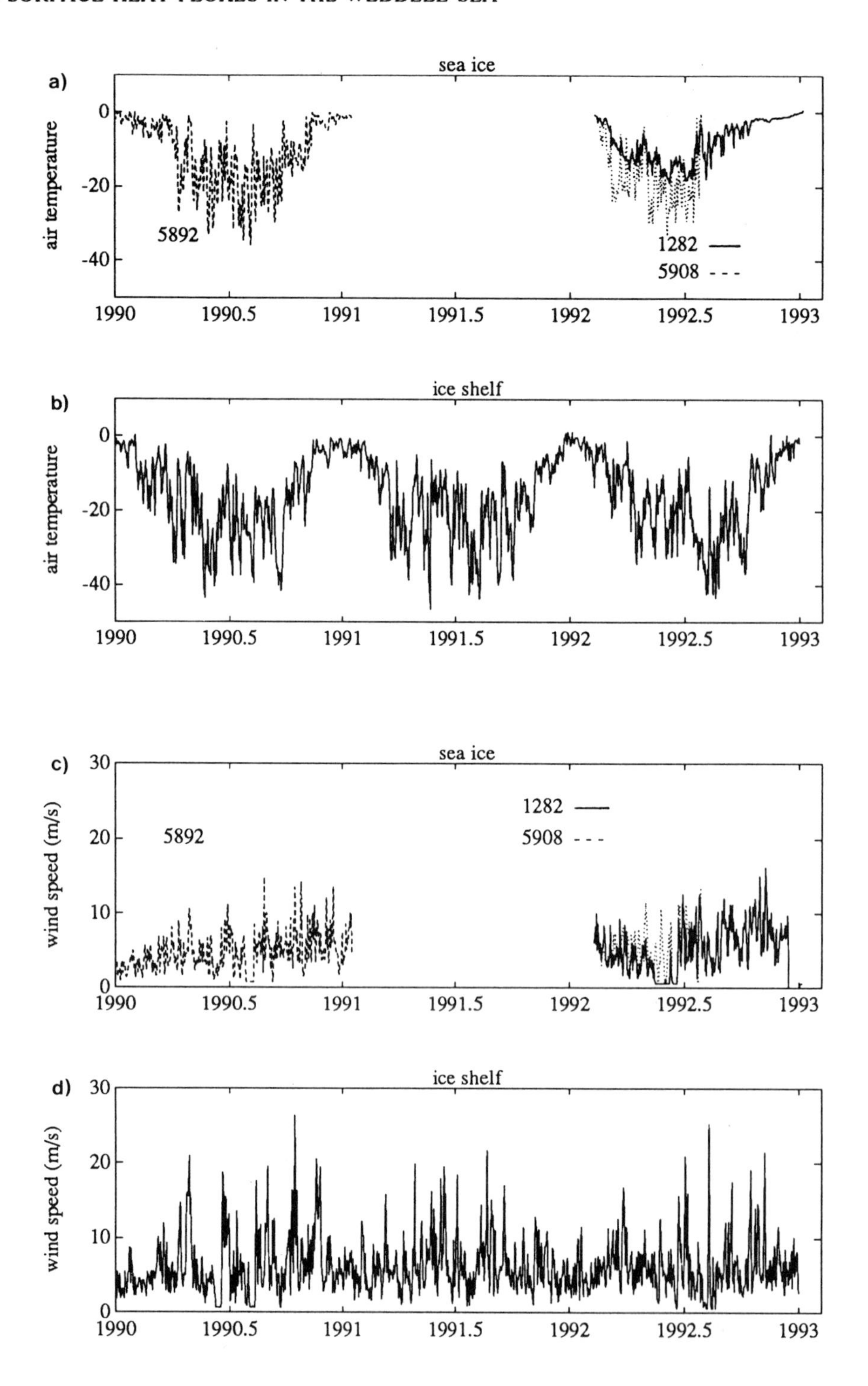

Fig. 3. Time series of the air temperature and wind speed as observed by four buoys during the three-year period 1990-1992 (diurnal means) *(a)* T_a over sea ice (buoys 5892, 1282 and 5908), *(b)* T_a over ice shelf edge (buoy 5895), *(c)* V over sea ice, and *(d)* V over ice shelf.

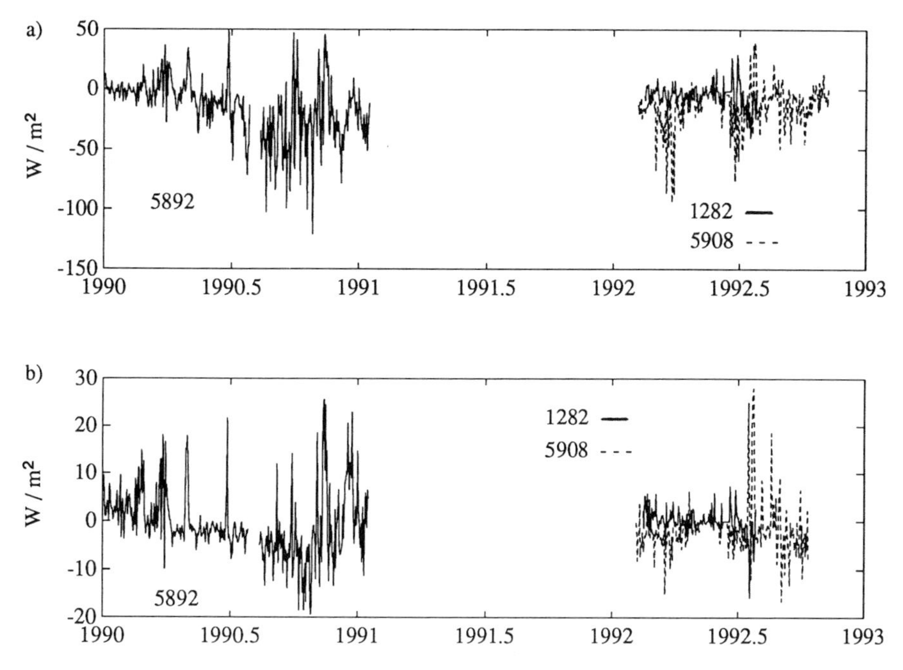

Fig. 4. Time series of heat fluxes over the sea ice (positive upwards). For geographic locations of the buoys, see Fig. 1. *(a)* sensible heat flux and *(b)* latent heat flux.

coefficients of heat and moisture were reduced, on the average, to 80% of their neutral values.

The results are in qualitative agreement with *Andreas and Makshtas* [1985] and *Kottmeier and Engelbart* [1992], although differences in geographical areas and observation seasons make the comparison somewhat difficult. The larger downward fluxes we obtained, especially in 1990, may partly result from the fact that the ice floe of buoy 5892 was much thicker than the typical pack ice in the Weddell Sea. Over this floe the sensible heat flux was -25 W/m^2 during the winter with a latent heat flux of -4 W/m^2, whereas over thin floes (1282 and 5908) the winter values were -12 W/m^2 and -2 W/m^2, respectively. The critical effect of ice thickness was pointed out by *Makshtas* [1991] and observed also by *König-Langlo and Zachek* [1991], whose data from the Weddell Sea in September 1989 indicated that the air temperature exceeded the surface temperature over old ice and was lower than this over new ice, when an area containing both old and new ice was transversed.

5.2. *Fluxes over leads and coastal polynyas*

Observations allowed us to roughly estimate the heat fluxes over leads, which are known to exist in the Weddell Sea ice field throughout the year. In the central Weddell Sea, leads are narrow and formed by the divergent drift of ice. The typical width of leads in the Weddell Sea is 10 to 10^3 m. Near the continental ice shelves in the eastern Weddell Sea, coastal polynyas are observed throughout the year [*Kottmeier and Engelbart*, 1992]. In summer when the ice edge retreats to the south they are often joined to the open ocean. *Zwally and Comiso* [1985] estimated 10% concentrations of open water in winter along the coast nearby the Halley station, while *Kottmeier and Engelbart* [1992] reported polynyas 4-15 km wide. *Pease* [1987] has developed a method of calculating the equilibrium width for a coastal polynya in winter, depending on the air temperature and wind speed away from the coast. Using the wintertime means for those quantities as measured by our buoy (-25°C and 4 m/s, buoy 5895), the method would give about 10 km for the equilibrium width.

On the eastern coast of the Weddell Sea, the wind is primarily directed away from the continent or along the shoreline direction [*van Loon et al.*, 1972; *Kottmeier*, 1988]. The ice is therefore advected away from the coast more rapidly than it is formed and the polynyas remain open. According to *Fahrbach and Rohardt* [1992], heat flux from deeper layers of the ocean may also be significant in the region. The air-sea temperature difference reaches extreme values over the coastal polynyas, because the wind nearly always has at least a moderate component off the continent [*Launiainen et al.*, 1991]. For this reason the air-mass measured at the buoy (5895) is advected to the open sea.

This allows us to use, as a first approach, the properties of the air mass as observed by the buoy to compute the fluxes near the shelf, for estimation of fluxes from coastal polynyas. The buoy is located at the floating shelf edge some 35 m above the sea surface, and the cold air mass flowing out from the continent drops more or less directly from here down to the sea surface. It is therefore difficult to accurately define a height corresponding to the air-sea temperature difference. To study the sensitivity of the flux estimates to this height, both the observing height of the buoy (Table 1) and that increased by 35 m were used. We found, however, that the errors are not very significant for these conditions.

The surface temperature in the leads and polynyas was estimated to be practically at the freezing point (-1.8°C), and for the leads the properties of the air mass as observed over the ice floes were used to compute the fluxes. The same method was used e.g. by *Maykut* [1986] to compute the fluxes over young ice. In summer, fluxes are more difficult to estimate, because the exact surface temperature is unknown, usually exceeding the freezing temperature and being closer to the air temperature. In this case, both the sensible and latent heat flux tend to be small, and a relative error may be very large - even the sign i.e. the direction may be unknown, if the actual lead surface temperature is not measured. Thus we do not even attempt to estimate fluxes over leads for summer.

The results for heat and moisture fluxes over winter leads and polynyas are presented in Figure 5. The fluxes from leads in the central Weddell Sea are generally smaller than those from coastal polynyas, although larger transfer coefficients were used for the former (Table 2). The reason is that the air over the Weddell Sea is generally warmer than the air over the coastal polynyas, the latter flowing out from the Antarctic continent. Wintertime sensible heat fluxes over 200 W/m^2 from leads and over 300 W/m^2 from coastal polynyas are frequent and values of latent heat up to 100 W/m^2 and 150 W/m^2 are to be found, respectively. Fluxes over leads were somewhat larger in winter 1992 than those in 1990 in more eastern areas of the Weddell Sea. In spite of this, the average heat flux from the ocean to the atmosphere is probably smaller in the western parts of the Weddell Sea, because the leads are propably rarer and more short-lived there, where the general ice drift is not so divergent.

It should be emphasized that the fluxes over leads and polynyas depend not only on fetch but also on time. The coastal polynyas may be semi-permanent [*Kottmeier and Engelbart*, 1992], but over leads fluxes computed with the assumption of the surface temperature at the freezing point typically continue for only a few hours, anyway for less than a day, until the formation of new ice [*Lebedev*, 1968; *Bauer and Martin*, 1983]. According to *Makshtas* [1991], however, fluxes through the new ice may be 2 to 3 times larger than those over multiyear ice, even two months after the freezing of a polynya. In the light of the above, the fluxes computed are to be seen as the potential contribution of leads, continously appearing and disappearing, to the ocean-air exchange. Fluxes somewhat larger than our estimates would be obtained for very narrow leads (less than 100 m wide), if fetch-dependent transfer coefficients were used. In our opinion, fetch-dependence is not crucial, as we do not know the distribution of the leads by width. *Makshtas* [1991] came to a similar conclusion for the Arctic Ocean. On the other hand, an air mass flowing over a wider polynya becomes warmer and moister during its traverse, and thus the temperature and moisture differences above the polynya decrease, resulting in reduced fluxes. Therefore the initial upwind properties of an air mass cannot be used for polynyas several kilometers wide (compare section 5.3). The problems have been studied in some more detail by *Andreas and Murphy* [1986], *Makshtas* [1991] and *Serreze et al.* [1992].

To compute the latent heat flux over polynyas and leads, we assumed that the relative humidity of the air mass flowing out from over the ice remains constant at a height of 3 m. In practise this means that the humidity used in the calculations is close to saturation. As pointed out by *Andreas et al.* [1979], the flux of latent heat over a lead can be estimated to within a (moisture-related) accuracy of 10% without any measurements of air humidity by assuming saturation, because the flux is controlled by the large difference in temperature between the air and the surface. We can therefore assume either the same relative humidity as measured over the sea ice or saturation humidity. According to our data, the difference in the resulting evaporation would not exceed 2% in conditions typical for winter leads. On the other hand, a crude assumption of completely dry air would typically lead to an error of 50%. Thus, for determining evaporation over a winter lead, it is much more important to get reasonable estimates for the transfer coefficient C_E and the wind speed, than to measure the air humidity. The situation is different when evaporation is measured over the sea ice or over the open ocean.

5.3. Fluxes over the open ocean in autumn

The data from buoy 5893 was used to compute the fluxes for the cooling period from 11 February until the freezing date of 31 March, 1990. During this period the buoy drifted at a distance of 100-150 km off the ice shelf (Figure 1). The fluxes were computed by the bulk-method using the air

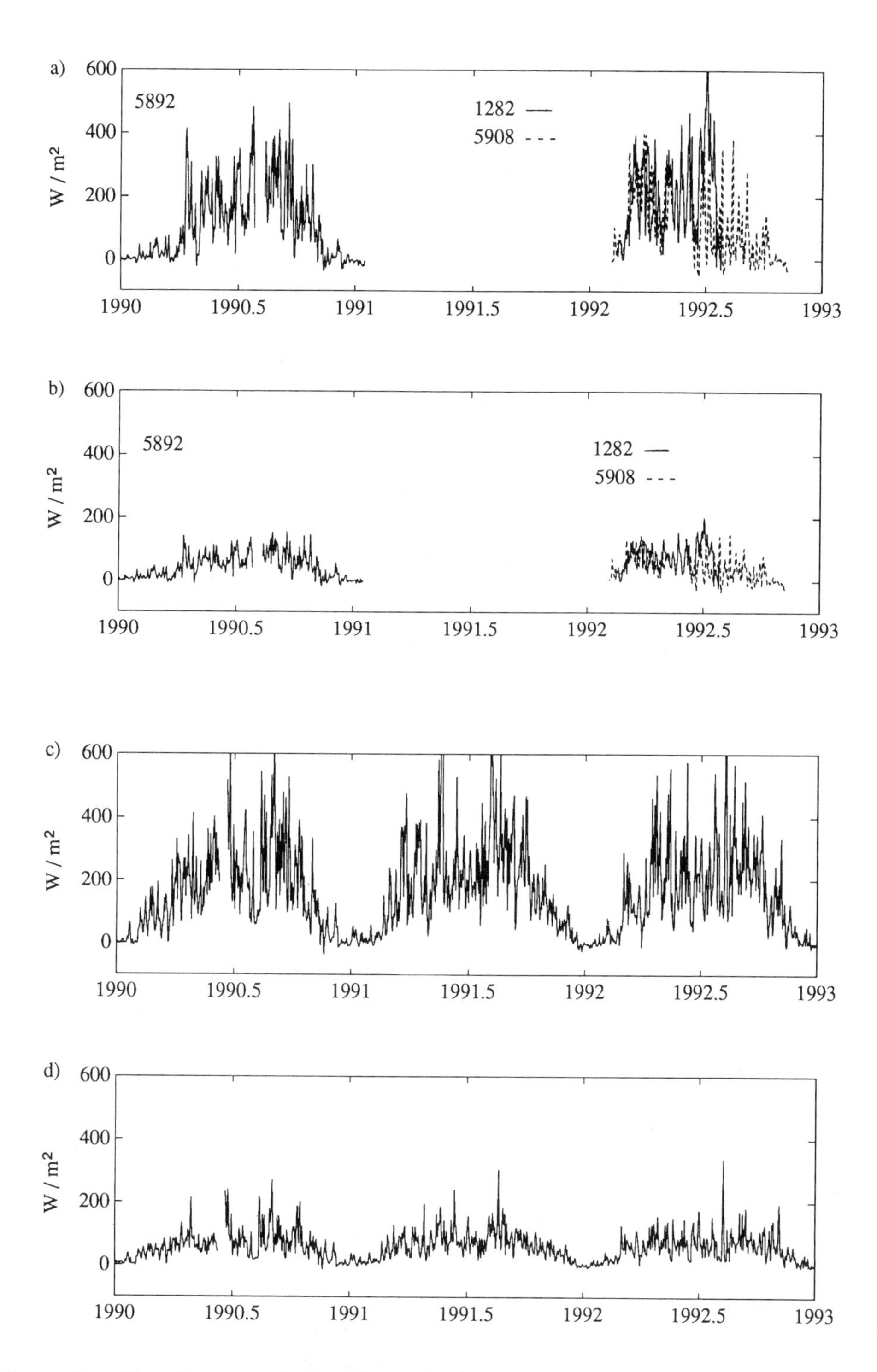

Fig. 5. Time series of heat fluxes over leads and coastal polynyas (positive upwards). For geographic locations, see Fig. 1 and the text. *(a)* sensible heat flux over leads, *(b)* latent heat flux over leads, *(c)* sensible heat flux over coastal polynyas, and *(d)* latent heat flux over coastal polynyas.

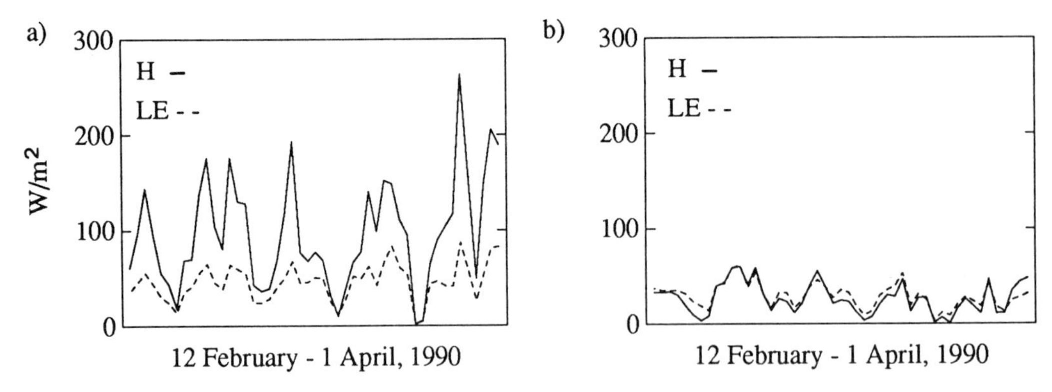

Fig. 6. Time series of heat fluxes over the open sea in late summer and autumn. *(a)* latent and sensible heat fluxes nearby the continental ice shelf and *(b)* fluxes at the open sea buoy 100 to 150 km off the ice shelf (see Fig. 1).

temperature, humidity, wind speed and sea surface temperature observed by the buoy (compare Figure 2). Additionally, fluxes over the sea close to the ice shelf were estimated on the basis of observations of buoy 5895 at the edge of a floating ice shelf (see section 5.2 above). There, the sea surface temperature was assumed to be practically at the freezing point. The resulting flux estimates are given in Figure 6.

It is distinctly seen from Figure 6 that the sensible heat flux is greatest close to the ice shelf and decreases further away towards the open ocean. The latent heat flux decreases slightly, but far less than the sensible heat flux. Thus the Bowen ratio, being about 2 in the vicinity of the ice shelf, diminishes to about 1 in the open sea. The total turbulent heat flux from the ocean was typically 100-150 W/m^2 close to the ice shelf and about 60 W/m^2 at a distance of 100 to 150 km from the shelf. The topic is further discussed in section 7.

5.4. *Fluxes over the floating continental ice shelf edge*

Stable stratification typically prevails over the continental ice shelves of Antarctica. The data of buoy 5895 was used to compute the fluxes at the edge of an ice shelf. The level-difference method was used to estimate the sensible heat flux and the snow surface temperature. Thereafter, the latent heat flux was obtained by the bulk method with the help of the calculated snow surface temperature. The results shown in Figure 7 suggest the sensible heat flux to be on average -17 W/m^2 (downwards), being strongest in the wintertime. During the summertime of 4 to 5 months the latent heat flux tends to be almost zero, but suggesting very small evaporation. The results remain rather comparable throughout the three-year period.

The heat balance of the snow surface was estimated (see section 6 for the radiation budget) to get a reference for the turbulent flux results. A good set of satellite images was available from January to the end of March 1990, and for the rest of the period the energy balance was estimated using a cloudiness of 2/8, giving a net radiation of 20 W/m^2 (effective outgoing longwave radiation 50 W/m^2, absorbed shortwave radiation -30 W/m^2, positive upwards). This is approximately balanced by the downwards-directed sensible heat flux. A heat flux from the shelf affects the surface energy budget as well, but its effect is small, as will be discussed below.

5.5. *Heat flux through ice and snow*

Measurements of the temperature at a depth of 0.6 to 0.7 m below the snow surface allowed us to roughly estimate the heat flux from the ocean through the ice and snow. In a stationary situation with homogeneous layers of snow and ice, and in the absence of phase transitions, the heat flux (*EH*) from ice and snow can be presented as [e.g., *Makshtas*, 1991]:

$$EH = -\lambda_i \frac{T_s - T_i}{h_i + (\lambda_i/\lambda_s)h_s} \tag{6}$$

where h_s is the snow thickness, and h_i is the ice thickness between the ice surface and the ice temperature measurement level. T_i is the ice temperature, T_s the snow surface temperature, and λ_i and λ_s are the heat conduction coefficients of ice and snow, for which we used values λ_i = 2.1 Wm^{-1}K^{-1} and λ_s = 0.3 Wm^{-1}K^{-1} [*Maykut*, 1986; *Makshtas*, 1991].

The temperature at the snow surface was calculated by the gradient method as described in section 3. Unfortunately, the snow thickness was measured only by buoy 1282. The inaccuracy estimate is based on sensitivity

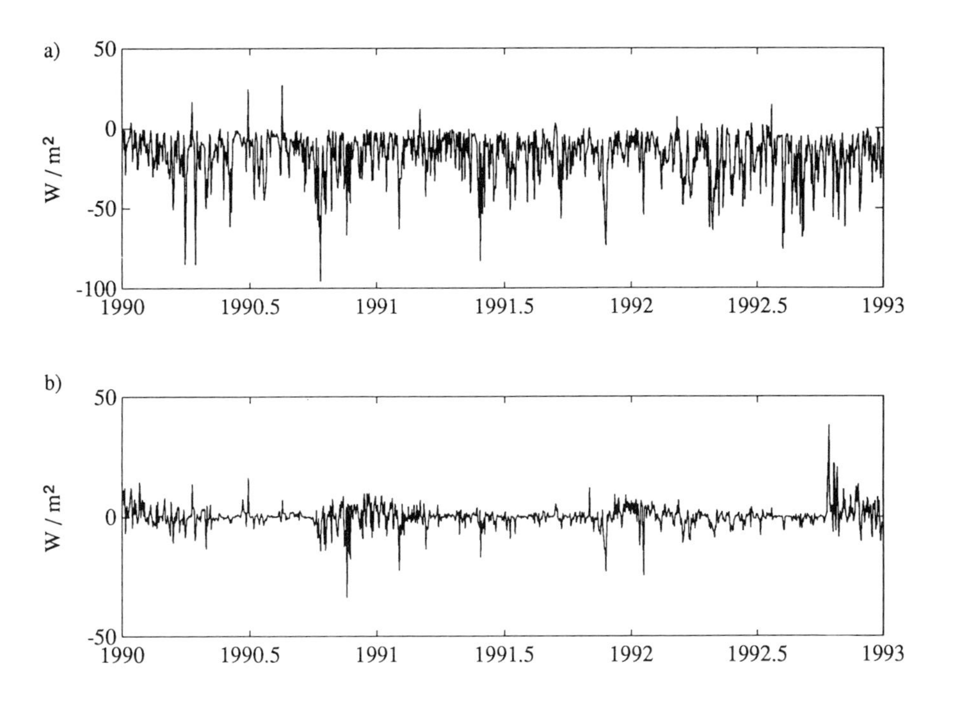

Fig. 7. Time series of heat fluxes over the edge of a floating continental ice shelf. *(a)* sensible heat flux and *(b)* latent heat flux.

tests for various sources of error (i.e. values of λ_i and λ_s, measurement of h_s, and T_s as calculated by the gradient method). An inaccuracy of ±30% in the heat conduction coefficients would cause an inaccuracy of ±2 W/m^2 in the heat flux. This added to the inaccuracy caused by T_s would result in a total inaccuracy of about ±4 W/m^2. Additionally, the assumption of a linear temperature profile in snow and ice (6) may itself cause some error. Based on the analyses of *Makshtas* [1991], we estimate that the error for ice about 1 m thick would be some 20-30%, or perhaps less when average (smoothed) meteorological parameters are used (as for Table 5).

On the basis of our calculations the conductive heat flux from the Weddell Sea through the ice and snow should be of the order of 10 W/m^2, with a standard deviation of diurnal means of 5 to 10 W/m^2 due to synoptic scale variations in the snow surface temperature. At buoy 1282 the heat flux remained fairly constant through the autumn to winter period from February 15 to July 26, because the snow accumulation in winter compensated the increase in T_i-T_s and prevented the flux from increasing.

For the continental ice shelf (buoy 5895), the equation for heat flux reduces to $EH = -\lambda_s(T_s-T_z)/h_s$, where h_s is now to be interpreted as the depth of the snow temperature sensor. Based on two visits to the buoy after its deployment, the average rate of snow accumulation is known, and calculations resulted in a weak mean heat flux of 1-2 W/m^2 upwards from the snow cover. Almost zero fluxes prevailed in summer and in winter the fluxes varied in direction, reflecting the rapid response of snow surface temperature to changes in air temperature, most pronounced in winter. The results are summarized in Table 5.

6. BOWEN RATIO AND SUMMARY OF THE SURFACE HEAT BALANCE

It is customary to analyze the ratio of fluxes of sensible and latent heat using the concept of the Bowen ratio, $Bo = H/\lambda E$, assumed to remain rather constant under certain meteorological conditions. Knowledge of the Bowen ratio would then allow us to estimate either H or λE on the basis of measurements of one or the other. Unfortunately, data regarding the Bowen ratio over sea ice and polynyas is very sparse. The Bowen ratios for wintertime (in this context defined as the period from April to the end of October) resulting from this study are given in Table 4.

The results suggest somewhat larger winter values for coastal polynyas (based on the data from the buoy 5895) than for leads in the central Weddell Sea. The reason is to be found in the outflow of the continental air-mass over the coastal polynyas, resulting in an intensive flux of sensible heat near the ice shelf. Over leads and polynyas in winter, values ranging from 2 to 6 have been reported by *Andreas et al.*, [1979], *den Hartog et al.* [1983] and *Serreze et al.*

TABLE 4. Bowen ratios ($Bo = H/\lambda E$)

surface type	*Bo*	
	mean	std
coastal polynyas	2.9	0.9
winter leads	2.5	0.7
open ocean, autumn	0.8	0.3

[1992].

As for conditions above the sea ice, *Andreas* [1989] computed Bowen ratios based on the measurements of *Leavitt et al.* [1978] over multiyear ice in the Beaufort Sea and found that Bo was variable in sign, but that in summer positive values were more than twice as common as negative values. We found that the Bowen ratio was positive over sea ice for 80-90% of the time. The mean values over ice are not listed above, because the flux of latent heat is frequently close to zero.

A summary of the estimated values of the various heat balance components is given in Table 5. The turbulent and conductive heat fluxes in the table are based on our data, but the radiative fluxes are only first-order estimates to get a reference for the magnitudes of the heat balance terms. Our data of air and surface temperatures are used when calculating the radiation terms, but cloudiness is only estimated, and is set to a constant value throughout the year (2/8 for the ice shelf, 4/8 for the coastal polynyas, and 6/8 for the sea ice and leads) except for those daily values derived from satellite images for the open ocean buoy. The shortwave radiation was estimated using the method of *van Ulden and Holtslag* [1985] with the albedo over water surfaces depending on the sun's altitude according to *Payne* [1972]. Several methods were tested to estimate the longwave radiation [*Berliand and Berliand*, 1952; *Budyko*, 1963; *Parkinson and Washington*, 1979; *van Ulden and Holtslag*, 1985; *Omstedt*, 1990]. However, an accurate estimate for the effective outgoing longwave radiation

TABLE 5. Estimates of surface heat balance components in the Weddell Sea (in W/m^2, positive upwards). Net shortwave radiation is denoted by SWR and the effective outgoing longwave radiation by LWR.

Surface type	Sensible heat flux	latent heat flux	heat flux through ice	SWR	LWR
Sea ice	-15	0	10	-20	20-40
MIZ	+1	4			
interior	-17	-2	10		
winter	-17	-3	10	0	20-40
summer	-5	3		-40	30-50
Ice shelf	-17	0	1	-30	40-60
winter	-16	0	1	0	30-50
summer	-12	2	1	-60	50-60
open ocean (autumn)	27	30		-100	60-70
coastal polynyas	160	60		-100	80-110
winter	240	80		-2	100-130
summer	(30)	(20)		-240	60-70
leads	140	60		-90	60-90
MIZ	70	30			
interior	150	60			
winter	190	70		-5	70-110
summer	(20)	(14)		-190	30-50

Note: Summer defined as a period from December to end of February, winter from June to end of August. MIZ = marginal ice zone; the buoy 5892 drifted in MIZ from January to the end of April, 1990, and the buoy 5908 in November, 1992. During other periods analyzed the buoys drifted in the interior of the ice field. Heat flux through ice calculated on the basis of the buoys 1282 (sea ice) and 5895 (ice shelf) only. Summer values over leads and coastal polynyas are uncertain due to unknown sea surface temperature.

(LWR) without knowledge of vertical profiles of temperature and humidity (reaching above the surface layer) is impossible. Therefore the methods based on surface conditions and cloudiness contain some discrepancies giving rise to the uncertainity ranges shown in Table 5. (We estimated the inversion strength to be 5°C when using the method of *van Ulden and Holtslag*, [1985].)

Keeping in mind the potential error sources discussed and the great variability in meteorological and ice conditions, the results for various energy balance components in Table 5 should be regarded as nothing more than a rough overall estimate. However, the importance of coastal polynyas and leads for the local heat balance can be seen. As to their effect on the overall heat exchange in the Weddell Sea, the results must be used in connection with the areal fraction of polynyas and leads (normally less than 10%). The turbulent fluxes are somewhat larger than the radiative fluxes over narrow open water areas, i.e. polynyas and leads, except in summer when shortwave radiation dominates. Over the open ocean the air temperature is adjusted to the sea surface temperature and radiative fluxes dominate turbulent fluxes. Radiative fluxes are larger over the sea ice and ice shelf, too. The effective outgoing longwave radiation, in particular, plays a role of essential importance throughout the year.

7. AIR-MASS MODIFICATION NEAR THE ICE SHELF

The location and alignment of two meteorological buoys and the research vessel offshore of the Riiser-Larsen ice shelf (at 72.5°S, 17°W) offered a situation for the study of air-mass modification over the sea near the ice shelf, under conditions of cold and dry air flowing out from over the shelf towards the sea, during the period February 11 to 12, 1990. The temperature and wind history of the buoy at the shelf edge show a period of katabatic wind to have been started then, during late summer in the area.

7.1. Calculation of the heat flux from the observed profiles

Conservation of enthalpy ($\rho c_p T$) in a turbulent field gives, under an assumption of two-dimensionality:

$$\rho c_p \frac{\partial T}{\partial t} = -\rho c_p u \frac{\partial T}{\partial x} - \rho c_p \frac{\partial (\overline{T'w'})}{\partial z} - \rho c_p \frac{\partial (\overline{T'u'})}{\partial x} + Q_0$$

(1) (2) (3a) (3b) (4)

where the term (1) denotes local change and (2) advection. Terms (3a) and (3b) describe turbulent exchange, in which T', w' and u' denote turbulent fluctuations of temperature and of vertical and horizontal velocity, respectively. (4) denotes diabatic sources and sinks of enthalpy, such as radiation and condensation.

Assuming semistationarity and (3a) >> (3b) and neglecting (4) we get

$$\rho c_p u \frac{\partial T}{\partial x} \approx -\rho c_p \frac{\partial (\overline{T'w'})}{\partial z} = -\frac{\partial H(x,z)}{\partial z}$$

i.e. the vertical local change in heat flux is mostly balanced by advection. In our case this holds especially well, because a strong inversion prevents the upward heat flux from penetrating through the inversion, i.e. through the height of z_h below which $H(z) \to 0$ (see Figure 8b). Integrating up to the layer we have

$$\rho c_p \int_0^{z_h} u \frac{\partial T}{\partial x} dz \approx H(x,0)$$

and

$$\rho c_p \int_0^X \int_0^{z_h} u \frac{\partial T}{\partial x} dz\, dx = \int_0^X H(x,0)\, dx$$

Considering the horizontal changes with distance e.g. over a polynya or, as in our case, from the shelf edge downwind (Figure 8a) to the research vessel and to the open sea buoy area, the above may be integrated to

$$\rho c_p \int_0^{z_h} u\, [T(X,z) - T(0,z)]\, dz = X\bar{H} \tag{7}$$

to give the "integral" method heat flux, used by various authors for arctic leads [*Badgley*, 1966; *Andreas et al.*, 1979; *Makstash*, 1991].

Accordingly, assuming the velocity not to be too dependent on x, the left-hand side of (7) may be easily calculated, from the temperature profiles at the shelf edge and over the sea, to give the mean heat flux. The integral heat flux provides a basis for comparing the bulk-aerodynamic based fluxes at various observation sites.

If we replace T and T' by the specific humidity and its fluctuations, and c_p by the enthalpy of vaporization, (7) gives the latent heat flux.

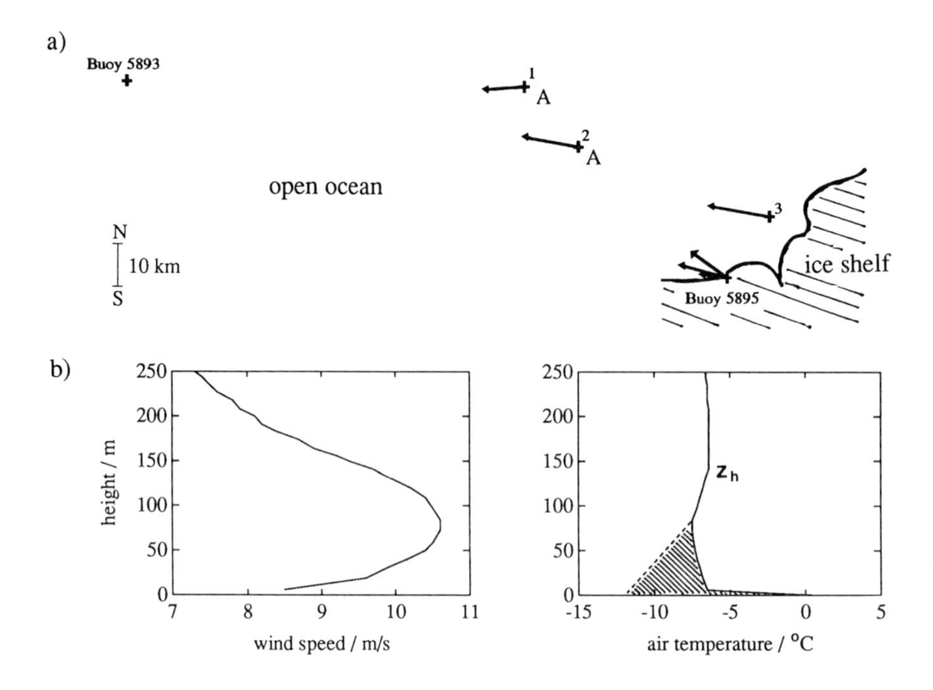

Fig. 8. *(a)* Location of meteorological buoys (5893, 5895) and the R/V *Aranda* (A) during an air-mass modification study on February 12, 1990. Arrows give wind vectors at the shelf buoy and at the research vessel in the various case studies 1 to 3. *(b)* Vertical profile of wind speed and temperature given by meteorological sounding for case 2. The hatched area represents the increase in the heat content of an air column after flowing out from the shelf to the sea.

TABLE 6. Aerological sounding cases on February 12, 1990. z_h = height of the inversion layer at the ice shelf, s_i = inversion strength = $T(h_i)$-$T(3m)$, d = distance from the shelf to R/V *Aranda*, h_m = height of the modified surface layer at the site of *Aranda*.

case no.	sounding time	V (m/s)	air temperature 5895 (3m)	air temperature *Aranda* (13m)	z_h(m)	s_i (K)	d (km)	h_m (m)
1	00^h	6.0	-7.8	-4.6	230	5	55	165
2	05^h	8.5	-11.9	-6.9	140	7.5	41	83
3	09^h	9.0	-10.7	-8.1	130	7	6	43

7.2 Case studies

Figure 8a shows the research area, giving the site of the meteorological buoys and the location of the research vessel during the cases (1 to 3) studied on February 12, 1990, and also shows the wind vectors observed at various places. The cases correspond to situations when aerological balloon soundings were made from the research vessel.

Figure 8b gives the wind and temperature profiles above the shelf edge and above the sea for case 2. The hatched area shows heating of the cool air flowing out over the sea. At sea, the temperature profile was obtained from ship observations and a balloon sounding, directly. For the upwind shelf edge, the profile was constructed by extrapolating the ship-based balloon-observed profile downwards from the inversion base, to correspond to the surface-layer temperature observed by the buoy at the shelf edge. A comparison in case 3, when the ship was nearest to the shelf, showed the method to be rather accurate. The wind profile (Figure 8b) indicates a katabatic-type wind with velocity increasing down to the surface friction layer. The observations during the case studies allow us a first-order comparison of estimates of heat exchange as calculated from (7). The mean characteristics of the modification cases studied are given in Table 6.

In addition to the bulk measurements and an analysis of

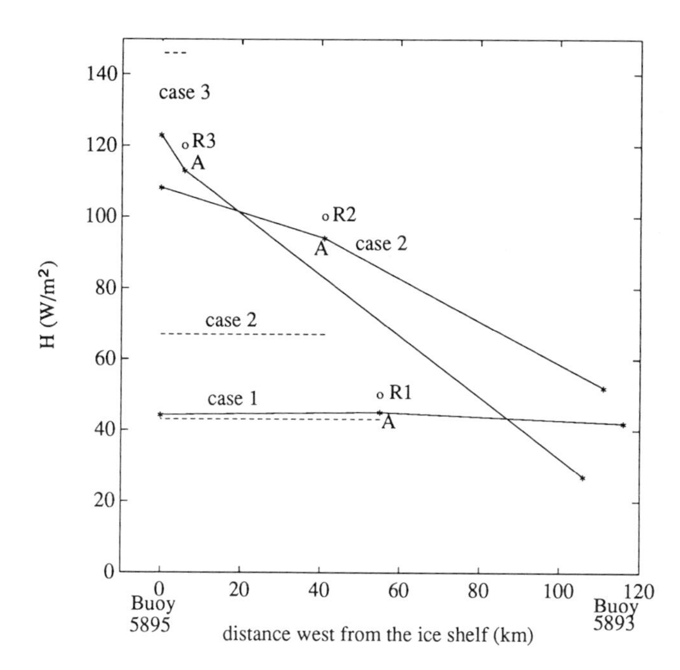

Fig. 9. Sensible heat flux near the ice shelf (site of the buoy 5895), at the research vessel (A) and at the open ocean buoy (5893) as estimated by three different methods: (1) the bulk-aerodynamic method (solid lines), (2) the modification rate of temperature profiles (dashed lines), the integral method giving the average heat flux between the ice shelf and the research vessel, (3) the diabatic resistance laws for the atmospheric boundary layer (symbols R1, R2, and R3 for the three sounding cases described in Table 6).

air-mass modification using eq. (7), the fluxes were also estimated on the basis of diabatic resistance laws for the atmospheric boundary layer. For such an estimation, one needs to know the sea surface temperature, the height of the boundary layer and the wind speed and potential temperature at that height. The height of the boundary layer is taken as the inversion base (h_m in Table 6) following *Heinemann and Rose* [1990] and *Garratt* [1992, p. 186]. For details of the iterative method, see e.g. *Kottmeier and Engelbart* [1992]. The resistance laws are applicable only for the site of R/V *Aranda*. The results based on all the three methods are presented in Figure 9.

The results of the three methods are reasonably in accordance with each other. The bulk-aerodynamic calculations in particular correspond surprisingly well to the results based on the resistance laws, as was also found by *Heinemann and Rose* [1990] in the southern Weddell Sea. The results from air-mass modification differ more from the others in cases 2 and 3. The accuracy of the results based on the modification of the temperature profiles is decreased e.g. by the fact that the temperature profile at the edge of the ice shelf was not measured. The temperature measured by buoy 5895 was assumed to represent the air temperature 3 m above the sea surface (which was assumed to be at ~ 0°C), but the ice shelf had a steep 35 m high edge and a certain amount of mixing must take place when the air flows out from the ice shelf. Additionally, the exact height of the inversion base was not clearly defined in cases 2 and 3, which affects the estimates of the heat surplus (i.e. the hatched area in Figure 8b). In case 1, the modified surface layer was thick, with a well-defined inversion base, and the modification analysis gave almost the same result as the bulk-aerodynamic calculations. Similarity of the estimates given by the two methods would also require that the air temperature was constant along the shoreline of the ice shelf, as the wind did not blow directly from buoy 5895 to the research vessel. The discrepancy found in the results is thus quite natural. The results give, however, three almost independent estimates of the heat flux close to the continental ice shelf and the following conclusion may be drawn: the sensible heat flux varied between 40 and 150 W/m^2 and decreased with increasing distance over the open water. The problematics and data will be the subject of a specific modelling study and, hopefully, of an experimental verification within the next few years.

8. CONCLUSIONS

The study was based essentially on data from 5 marine meteorological buoys. Two of the buoys yielded high-quality data from the Weddell Sea over a period of about a year, and one buoy standing on an ice shelf edge has yielded data for over 4 years so far. Among others, the data permit estimation of the components of heat and moisture exchange in the area. On the other hand, although being based upon very reliably-measured local quantities (air temperature, air pressure, surface wind, humidity), the calculated results suffer from various potential methodological error sources (section 4). This is especially true for the measurement and calculation of sensible heat flux by the gradient or level-difference method above the ice. For this reason, the results given in Table 5 should be regarded as first estimates, although two independent sensor pairs were used in each buoy and indirect checking methods showed the sensors to have remained very stable. Also e.g. the buoys in 1992 in the western Weddell Sea showed mutually comparable and coherent results.

Except in the marginal ice zone, the results suggest that the sensible heat flux above the sea ice is most frequently directed downwards. This is because the radiation balance tends to be negative and it cannot be balanced by the heat flux through the ice alone. In summertime, the average sensible heat flux is close to zero within the accuracy of the measurements. For the future, a more accurate measuring or modelling of the surface temperature and radiation balance (or at least observations on cloudiness) form the key problems for getting more reliable estimates of the heat balance over snow and ice.

The turbulent heat fluxes over leads and polynyas were large, even larger than the heat losses via longwave radiation. In summer, the lower atmosphere was almost in thermal equilibrium with the sea surface (except close to the continent) resulting in reduced fluxes. Thus the time series showed apparent annual cycles in the fluxes (Figure 5) and temporal variations over shorter periods were most pronounced in winter, when the air temperature varied far more than in summer. Perhaps the most important question in the estimation of the absolute magnitudes of the fluxes above leads is to get reasonable estimates for the transfer coefficients. The importance of leads and polynyas for the regional heat balance is evident, and we may anticipate that the heat flux, especially from coastal polynyas, should contribute convection and deep water formation near the ice shelves. The above, together with studies of air-modification near the ice shelf with encouraging bulk-flux comparisons, offers an interesting basis for further modelling.

If we wish to provide a rough sketch of the overall vertical heat exchange in the Weddell Sea, a few estimates can be derived from the data given in Table 5. Estimating the areal coverage of leads and polynyas to be 5 to 7%, except during a 3-month-long summer with an order of 60% ice coverage, annual area-averages can be calculated. First, as to the sensible heat, the significant heat flux from the leads seems to be balanced by the downward flux above the sea ice. In winter, the area averaged sensible heat flux would be about -5 W/m^2, which is approximately in accordance with the model results of *Stössel and Claussen* [1993]. Secondly, our data yields an estimate of 20 to 30 W/m^2 for the mean annual vertical heat loss from the Weddell Sea. Finally, an overall estimate for the latent heat flux, 3 to 5 W/m^2, accounts for a water vapour flux of $1.2 \cdot 10^{-6}$ to $2.0 \cdot 10^{-6}$ kgm^{-2}s^{-1}.

Finally, we believe that significant advances will be made in the near future, when our data are considered with other recent data sets, especially those obtained by the German buoy studies in the Weddell Sea and by the studies in the U.S.-Russian *Weddell Sea Ice Station 1*.

APPENDIX: EMPIRICAL FORMULAE FOR COMPUTATION OF TURBULENT SURFACE FLUXES

The computation algorithm is described in detail by *Launiainen and Vihma* [1990]. Here we give the formulae for transfer coefficients and universal functions used in the computations according to the equations (1) to (5) with notations as in the main text.

A. For practical calculations the transfer coefficients enter into eqs. (4) and (5) as the roughness lengths as

$$\ln z_0 = \ln z - kC_D^{-1/2} \qquad (A1)$$

$$\ln z_T \approx \ln z_q = \ln z - kC_D^{1/2}C_H^{-1} \qquad (A2)$$

where z is the height the transfer coefficients C_D and C_{HE} are referred to, and the von Karman constant $k \approx 0.40$. The coefficients used (referred to 10 m) for the above read:

a) over open sea:

$C_D \times 10^3 = 0.61 + 0.063V$ [*Smith*, 1980]

where V is the wind speed.

$C_{HE} \times 10^3 = 0.63 C_D \times 10^3 + 0.32$ [*Launiainen*, 1983]

b) over coastal polynyas:

$C_D \times 10^3 = 0.8 + 0.065V$ [*Wu*, 1980]

C_{HE} as above

c) over leads in the sea-ice zone:

$C_D \times 10^3 = 1.49$ [*Andreas and Murphy*, 1986]

C_{HE} as above yielding to $=1.26 \times 10^{-3}$

d) over ice and snow:

Drag coefficient from the on-site mean geometric roughness R as

$C_D \times 10^3 = 1.1 + 0.072R$ [*Banke et al.*, 1980]

$\ln z_{Tq} = \ln z_0 + b_0 + b_1 \ln(Re) + b_2(\ln(Re))^2$ [*Andreas*, 1987]

where Re is the roughness Reynolds number.

$$Re = z_0 C_D^{1/2} \frac{V}{\nu}$$

ν is the kinematic viscosity of air.

B. Transfer coefficients covering diabatic cases as

$$C_D = \frac{k^2}{[\ln(z/z_0) - \Psi_M(z/L) + \Psi_M(z_0/L)]^2} \qquad (A3)$$

$$C_{HE} = \frac{k^2}{(\ln(z/z_0) - \Psi_M(z/L) + \Psi_M(z_0/L))} \times \frac{1}{(\ln(z/z_{Tq}) - \Psi_{HE}(z/L) + \Psi_{HE}(z_0/L))} \tag{A4}$$

The Monin-Obukhov length L calculated as

$$L = \frac{u_*^3 T_0 c_p}{gkH(1+0.61 T_0 c_p E/H)} \tag{A5}$$

where $T_0 = (T_a+T_s)/2$.

The formulae we use for universal functions read:
a) for the stable region (ζ = z/L > 0) from *Holtslag and deBruin* [1988]:

$$\Psi_M = \Psi_{HE} = -a\zeta - b(\zeta - c/d)\exp(-d\zeta) - bc/d \tag{A6}$$

where a = 0.7, b = 0.75, c = 5 and d = 0.35.
b) for the unstable region (ζ < 0) from *Businger et al.* [1971] and *Dyer* [1974] type forms of:

$$\Psi_M = 2\log(\frac{1+\phi_M^{-1}}{2}) + \log(\frac{1+\phi_M^{-2}}{2}) - 2\arctan\phi_M^{-1} + \frac{\pi}{2} \tag{A7}$$

where $\phi_M = (1-19.3\zeta)^{-1/4}$ and

$$\Psi_{HE} = 2\log(\frac{1}{2}(1+\phi_M^{-1})) \tag{A8}$$

where $\phi_{HE} = (1-12\zeta)^{-1/2}$. In the above, the profile coefficients 19.3 and 12 are those obtained by *Högström*, [1988].
The system of equations (1) to (3) and (A3) to (A8) leads to an iterative solution, the algorithm being given in *Launiainen and Vihma* [1990].

Acknowledgments. We wish to thank Kalevi Rantanen (IVO Service Co.), Seppo Kivimaa (Technical Research Centre of Finland) and the crews of R/V *Aranda* and R/V *Academik Fedorov* (Russia) for field assistance. Juha Uotila is acknowledged for his contribution in processing the data and drafting figures. The field phases of the project were financially supported by FINNARP (Finnish Antarctic Research Program, Ministry of Trade and Industry), and the Academy of Finland has given some funding support for the scientific work. Pentti Mälkki, chair of FINNARP and director of the Finnish Institute of Marine Research, is acknowledged for help and encouragement.

REFERENCES

Andreas, E.L., A theory for the scalar roughness and the scalar transfer coefficients over snow and sea ice, *Boundary-Layer Meteorol., 38,* 159-184, 1987.

Andreas, E.L., A year of Bowen ratios over the frozen Beaufort Sea, *J. Geophys. Res., 94,* 12,721-12,724, 1989.

Andreas, E.L., C.A. Paulson, R.M. Williams, R.W. Lindsay, and J.A. Businger, The turbulent heat flux from Arctic leads, *Bound.-Layer Meteorol., 17,* 57-91, 1979.

Andreas, E.L., and A. P. Makshtas, Energy exchange over Antarctic sea ice in the spring, *J. Geophys. Res., 90,* 7119-7212, 1985.

Andreas, E.L., and B. Murphy, Bulk transfer coefficients for heat and momentum over leads and polynyas, *J. Phys. Oceanogr., 16,* 1875-1883, 1986.

Augstein, E., N. Bagriantsev, and H. W. Schenke, The expedition Antarktis VIII/1-2, 1989 with the Winter Weddell Gyre Study of the research vessels Polarstern and Akademik Fedorov, *Ber. Polarforch., 84,* 134 pp., 1991.

Badgley, F. J., Heat budget at the surface of the Arctic Ocean, *Proc. on the Symposium on the Arctic heat budget and atmospheric circulation,* edited by J. O. Fletcher, pp. 267-277, Rand Corporation, Santa Monica, Calif., 1966.

Banke, E. G., S. D. Smith, and R. J. Anderson, Drag coefficients at AIDJEX from sonic anemometer measurements, in *Sea Ice Processes and Models,* edited by R. S. Pritchard, pp. 430-442, Univ. of Washington Press, Seattle, 1980.

Bauer, J., and S. Martin, A model of grease ice growth in small leads, *J. Geophys. Res., 88,* 2917-2925, 1983.

Berliand, M.E., and T.G. Berliand, Determining the net long-wave radiation of the Earth with consideration of the effect of cloudiness, *Isv. Akad. Nauk. SSSR Ser. Geofis.*, No. 1, 1952.

Bromwich, D.H., and D.D. Kurtz, Katabatic wind forcing of the Terra Nova Bay polynya, *J. Geophys. Res., 89,* 3561-3572, 1984.

Budyko, M.I., *Atlas of the heat Balance of the World,* Glabnaja Geofiz. Observ., Moscow, 69 pp., Also: Guide to the Atlas of the Heat Balance of the Earth, Translated by I.A. Donehoo, U.S. Wea. Bur., WB/T-106, Washington D.C., 1963.

Businger, J.A., J.C. Wyngaard, Y. Izumi, and E.F. Bradley, Flux-profile relationships, *J.Atmos.Sci.,* 28, 181-189, 1971.

Cavalieri, D. J., and S. Martin, A passive microwave study of polynyas along the Antarctic Wilkes Land Coast, in *Oceanology of the Antarctic Continental Shelf, Antarctic Res. Ser.,* vol. 43, edited by S. S. Jacobs, pp. 227-252, AGU, Washington, D.C., 1985.

Dyer, A.J., A review of flux-profile relationships, *Boundary-Layer Meteorol., 7,* 363-372, 1974.

Fahrbach, E., and G. Rohardt, Supression of bottom water formation in the southeastern Weddell Sea Shelf due to melting of glacial ice, *Deep Sea Research,* 1992

Garratt, J.R., *The atmospheric boundary layer,* Cambridge University Press, 1992.

den Hartog, G., S.D. Smith, R.J. Anderson, D.R. Topham, and R.G. Perkin, An investigation of a polynya in the Canadian archipelago 3, surface heat flux, *J. Geophys. Res., 88,* 2911-2916, 1983.

Heinemann, G., and L. Rose, Surface energy balance, parameterizations of boundary-layer heights and the application of resistance laws near an Antarctic ice shelf

front, *Boundary-Layer Meteorol., 51,* 123-158, 1990.
Hoeber, H., One year temperature records in the atmospheric surface layer above sea ice and open water, *Bound.-Layer Meteorol., 48,* 293-297, 1989.
Holtslag, A.A.M, and H.A.R. de Bruin, Applied modeling of the nighttime surface energy balance over land, *J. Appl. Meteorol., 37,* 689-704, 1988.
Högström, U., Non-dimensional wind and temperature profiles in the atmospheric surface layer: a re-evaluation, *Bound.-Layer Meteorol.*, 42, 55-78, 1988.
Kottmeier, Ch., Atmosphärische Strömungsvorgänge am Rande der Antarktis, *Ber. des Inst. fur Met. und Klimat.* Hannover, 33, 153 pp., 1988.
Kottmeier, Ch., and D. Engelbart, Generation and atmospheric heat exchange of coastal polynyas in the Weddell Sea, *Bound.-Layer Meteorol., 60,* 207-234, 1992.
König-Langlo, G., and A. Zachek, Radiation budget measurements over Antarctic sea ice in late winter, World Climate Programme, Research, *WCRP-62,* (Appendix C), 41-44, 1991.
König-Langlo, G., B. Ivanov, and A. Zachek, Energy exchange over Antarctic sea ice in late winter, in *The Role of the Polar Regions in Global Change,* Proceedings from an International Conference, Fairbanks, Alaska, 11-15 June, 1990, edited by Weller, G., L. McCauley, and C. Wilson, 1990.
Launiainen, J., Parameterization of the water vapour flux over a water surface by the bulk aerodynamic method, *Annales Geophysicae, 1,* 481-492, 1983.
Launiainen, J., and T. Vihma, Derivation of turbulent surface fluxes - an iterative flux-profile method allowing arbitrary observing heights. *Environmental Software, 5,* 113-124, 1990.
Launiainen, J., T. Vihma, J. Aho, and K. Rantanen, Air-sea interaction experiment in the Weddell Sea, Argos-Buoy Report from FINNARP-5/89, 1990-1991. *Antarctic Reports of Finland no. 2,* Ministry of Trade and Industry, 1991.
Leavitt, E., M. Albright, and F. Carsey, Report on the AIDJEX meteorological experiment, *AIDJEX Bull., 39,* 121-147, 1978.
Lebedev, V.L., Maximum size of a wind-generated lead during sea freezing, *Oceanology,* Engl. Transl., 8, 313-318, 1968.
Ledley, T.S., A coupled energy balance climate-sea ice model: impact of sea ice and leads on climate, *J. Geophys. Res., 93,* 15,919-15,932, 1988.
Makshtas, A.P., *The heat budget of Arctic ice in the winter,* Int. Glaciol. Soc., Cambridge, 77 pp., 1991.
Maykut, G.A., Energy exchange over young sea ice in the central Arctic, *J. Geophys. Res., 83,* 3646-3658, 1978.
Maykut, G.A., The surface heat and mass balance, in *Geophysics of Sea Ice,* edited by N. Untersteiner, pp. 3xx-463, Plenum Press, New York, 1986.
Monin, A. S., and A. M. Obukhov, Dimensionless characteristics of turbulence in the surface layer of the atmosphere (in Russian), *Trudy Geofiz. Inst. Akad. Nauk SSSR, 24,* 163-187, 1954.
Monin, A.S., and A.M. Yaglom, *Statistical Fluid Mechanics,* Vol. 1. The MIT press, 769 p., 1971.
Omstedt, A., A coupled one-dimensional sea ice-ocean model applied to a semi-enclosed basin, *Tellus*, 42A, 568-582, 1990.
Parkinson, C.L., and W.M. Washington, A large-scale numerical model of sea ice, *J. Geophys. Res.*, 84, 311-337, 1979.
Payne, R.E., Albedo of the sea surface, *J. Atmos. Sci.*, 29, 959-970, 1972.
Pease, C.H., The size of wind-driven coastal polynyas, *J. Geophys. Res., 92,* 7049-7059, 1987.
Schnack-Schiel, S., The winter expedition of R. V. Polarstern to the Antarctic (ANT V/1-3), *Ber. Polarforsch., 39,* 259 p., Bremerhaven, Germany, 1987.
Schumacher, J. D., K. Aagard, C. H. Pease, and R. B. Tripp, Effects of a shelf polynya on flow and water properties in the northern Bering Sea, *J. Geophys. Res., 88,* 2723-2732, 1983.
Serreze, M.C., J.A. Maslanik, M.C. Rehder, R.C. Schnell, J.D. Kahl, and E.L. Andreas, Theoretical heights of buoyant convection above open leads in the winter Arctic pack ice cover, *J. Geophys. Res., 97,* 9411-9422, 1992.
Simmons, I., and W. F. Budd, Sensitivuty of the southern hemispehre circulation to leads in the Antarctic pack ice, *Quart. J. Roy. Meteorol. Soc., 117,* 1003-1024, 1991.
Smith, S.D., Wind stress and heat flux over the open ocean in gale force winds, *J. Phys. Oceanogr.*, 10, 709-726, 1980.
Smith, S.D, Coefficients for sea surface wind stress, heat flux, and wind profiles as a function of wind speed and temperature, *J. Geophys. Res.*, 93, 15467-15472, 1988.
Smith, S.D., R.J. Anderson, G. den Hartog, D.R. Topham, and R.G. Perkin, An investigation of a polynya in the Canadian archipelago 2, structure of turbulence and sensible heat flux, *J. Geophys. Res., 88,* 2900-2910, 1983.
Stössel, A. and M. Claussen, On the momentum forcing of a large-scale sea-ice model, *Climate Dynamics*, 9, 71-80, 1993.
Thorpe, M.R., E.G. Banke, and S.D. Smith, Eddy correlation measurements of evaporation and sensible heat flux over Arctic sea ice, *J. Geophys. Res., 78,* 3573-3584, 1973.
Untersteiner, N. The geophysics of sea ice: overwiev, in *Geophysics of Sea Ice,* edited by N. Untersteiner, pp. 1-8, Plenum Press, New York, 1986.
Wamser, C., and D. Martinson, Drag coefficients for winter Antarctic pack ice, *J. Geophys. Res.*, 98, 12431-12437, 1993.
van Loon, H., J. J. Taljaard, T. Sasamori, J. London, D. V. Hoyt, K. Labitzke, and C. W. Newton, Meteorology of the southern hemisphere, *Meteorological Monographs, 13,* No. 35, 1972.
van Ulden, A. P., and A. A. M. Holtslag, Estimation of atmospheric boundary layer parameters for diffusion applications, *J. Clim. Appl. Meteorol., 24,* 1196-1207, 1985.
Webb, E.K., Profile relationships: the log-linear range and extension to strong stability. *Quart. J. Roy. Meteorol. Soc., 96,* 67-90, 1970.
Vihma, T., and J. Launiainen, Ice drift in the Weddell Sea in 1990-1991 as tracked by a satellite buoy, *J. Geophys. Res., 98,* 14,471-14,485, 1993.
Vihma, T., J. Launiainen, J. Uotila, and K. Rantanen, Air-sea interaction experiment in the Weddell Sea, 2nd Meteorological Argos-Buoy Data Report from FINNARP, 1990-1993. *Antarctic Reports of Finland no. 4.* Ministry of Trade and Industry, 1994, in press.
Wu, J., Wind stress coefficients over sea surface in near-neutral

conditions, a revisit, *J. Phys. Oceanogr., 10,* 727-740, 1980.

Vowinckel, E., and S. Orvig, Synoptic energy budgets from the Beaufort Sea, in *Energy Fluxes over Polar Surfaces,* edited by S. Orvig, 299 pp. World Meteorological Organization, Geneva, 1973.

Zwally, H.J., and J.C. Comiso, Antarctic offshore leads and polynyas and oceanographic effects, in *Oceanology of the Antarctic Continental Shelf, Antarctic Research Series, vol. 43.* edited by S. S. Jacobs, AGU, Washington, D.C., 1985.

J. Launiainen, Finnish Institute of Marine Research, P.O. Box 33, FIN-00931 Helsinki, Finland.

T. Vihma, Department of Geophysics, P.O. Box 4 (Fabianinkatu 24A), FIN-00014 University of Helsinki, Finland.

Southern Ocean Wave Fields During the Austral Winters, 1985-1988, by Geosat Radar Altimeter

William J. Campbell, Edward G. Josberger

United States Geological Survey, University of Puget Sound, Tacoma, Washington

Nelly M. Mognard

CNES-GRGS, Toulouse, France

The Geosat radar altimeter is the first satellite sensor to have acquired a global sea state data set over a period of several years. The Geosat altimeter observations over the Southern Ocean have been processed to derive the austral winter seasonally averaged wave fields from 1985 through 1988. The seasonal data were obtained by binning the orbital swath data acquired during the month of July, August and September into 2° latitude by 2° longitude pixels. A comparison of the satellite derived significant wave height estimates for the 4-year data set, to the Navy climatological fields derived from a compilation of observations from 1854 to 1969 reveals a global underestimation of the climatological mean wave heights by more than 1 meter over large regions of the Southern Ocean. Mesoscale features of the order of 1000 km are observed in the austral winter satellite fields for all years that are not found in the climatological fields. The location and the intensity of these mesoscale features exhibit strong variations from one year to the next. With the satellite observations, the highest sea states, for each of the four austral winters, are found south of latitude 40°S and can reach the Antarctic ice edge. In general, the highest Geosat mean significant wave heights are of the order of 4 to 5 m and within each ocean, there is a significant annual variation in the extent of these regions with large wave heights. In the region south of 50°S, 30 to 50% of the measured wave heights are greater than 5 m. The Indian Ocean south of 40°S, is almost entirely covered with these large waves while the Atlantic Ocean is significantly calmer, with only the eastern portion containing such large waves, which do not occur every year. The observations from the Pacific Ocean show that there are regions of high waves whose position varies from year to year.

1. INTRODUCTION

The Southern Ocean is the most hostile marine environment in the world. It is a part of the world oceans where, due to the lack of ship traffic and observations, few in-situ weather and sea state observations are available. In these regions, as well as elsewhere in the world ocean, microwave sensors onboard satellites provide the only means of acquiring synoptic, repetitive, and uniformly accurate sea surface data. Chelton et al. [1990], used the 3-month data set acquired by the Seasat-A scatterometer (SASS) to estimate monthly wind stress fields and compared them to the climatological fields derived from individual ship observations by Hellerman and Rosenstein [1983]. In this comparison, the largest differences were obtained across the Southern Ocean in data deficient areas where individual ship observations had to be supplemented with wind-rose data from the Antarctic [*Hellerman and Rosenstein*, 1983]. The wind stress magnitudes across nearly all the Southern

The Polar Oceans and Their Role in Shaping
the Global Environment
Geophysical Monograph 85

Ocean were much greater with the SASS and the differences were as large as 2 $dyn.cm^{-2}$. With such differences in the wind fields between satellite versus classical climatology, one might expect to find large differences when comparing wave fields derived from satellite measurements versus classical climatology based on ship reports.

Mognard et al. [1986] used the data acquired by the Seasat radar altimeter to investigate wind and wave conditions in the Southern Ocean on a synoptic scale. These data, acquired during the austral winter of 1978, showed that 10-12 m significant wave height (SWH) were measured in regions of the Southern Ocean every few days and that background levels of 2-4-m-high waves were constantly present. The analysis of the Seasat altimeter data set also showed that large swell events associated with intense storms could be measured, and that these swells propagating across the oceans could be directly observed for the first time with the Seasat altimeter [*Mognard*, 1984]. With the 3-month data set acquired by Seasat during the austral winter 1978, the seasonal wind speed and SWH were obtained by Chelton et al. [1981]. The monthly mean wind speed and SWH that were analyzed from July to September 1978 revealed a clockwise rotation of the regions of highest sea state from the Eastern Atlantic in July to the Western Pacific in September [*Mognard et al.*, 1983].

Geosat is the first satellite to have acquired a long-term global data set, over more than 4 years, from April 1985 to January 1990. While the altimeter is the only satellite sensor to provide SWH measurements, ocean surface wind speed can also be estimated by different satellite sensors such as scatterometer, and multifrequency microwave radiometer which have larger footprint than the altimeter. In this paper we have thus chosen to specifically analyze the altimeter SWH measurements and not the wind speed, because for climatological analyses the wind speed measurements from sensors with larger footprints would be better suited. The Geosat altimeter data set has been processed to obtain the austral winter SWH fields in the Southern Ocean from the latitude 30°S to the Antarctic ice edge. The winter wave height fields are analyzed for the years 1985 to 1988 and compared to classical climatological seasonal fields derived from individual ship reports.

2. THE DATA SETS

2.1. *The Geosat Data Set*

The significant wave height (SWH) is computed onboard the Geosat satellite from the shape of the average returned radar pulse. This measurement is available every second along the satellite track, from a footprint of approximately 10 km in diameter. MacArthur et al. [1987] give a complete description of the Geosat radar altimeter system. A variety of studies have been undertaken to validate the Geosat significant wave height measurements. These include comparisons with the National Data Buoy Center (NDBC) network by Dobson et al. [1987] and with numerical wave models [*Guillaume and Mognard*, 1992]. According to these two comparisons, the Geosat altimeter measurements of SWH has a 0.4 m low bias and a standard deviation of 0.3 m when compared to buoys and the model. Carter et al. [1992] found an overall low bias of 13% between the SWH measured by the buoys when compared to the Geosat altimeter estimates. This 13% low bias was also found in comparisons with hindcasts from the METEO FRANCE VAG wave model by Guillaume and Mognard [1992], and this bias has been applied to correct the Geosat significant wave height measurements used in this analysis.

The Geosat altimeter functioned from April 1985 to early January 1990, and in 1989 the satellite measurements started to degrade, resulting in large regions with sparse or no measurements. Hence, the 1989 wave fields have spatial structures that are predominantly produced by the 10-km-wide orbital swaths. Therefore we used only the observations from 1985 through 1988 to derive the seasonal fields and a 4-year Geosat climatology of significant wave height.

Figures 1a-d show the seasonal significant wave height fields from 1985 through 1988 extending from the Antarctic ice edge north to 30°S. A multiple step procedure was employed to derive the seasonal fields. The procedure begins with the 13% bias correction which is applied to the once per second SWH measurements. For each austral winter season (July, August, and September), the individual measurements are binned into 2°x2° latitude-longitude bins for the entire world, from 72°N to 72°S. We used this binning procedure to be consistent with the analysis by Mognard and Campbell [1984] who compared the Seasat altimeter monthly wind speed fields to the SASS and the Scanning Multichannel Microwave Radiometer (SMMR) fields. For the 3-month periods considered, each 2°x2° bin contains from 400 to over 600 individual altimeter measurements. A spatial low-pass filter was then applied to the average wave fields to remove residual sampling errors resulting from the irregular distribution of the altimeter observations due to Geosat orbital characteristics. An average of the 4 years of Geosat observations produced a satellite climatology, that is shown in Figure 3. Finally, a coordinate transformation into a south polar projection extending northward to 30°S produced a data set that was then automatically contoured.

2.2. *The Austral Winter Ice Edge Location*

An important factor in analyzing Southern Ocean sea state variations is the position of the sea ice edge because

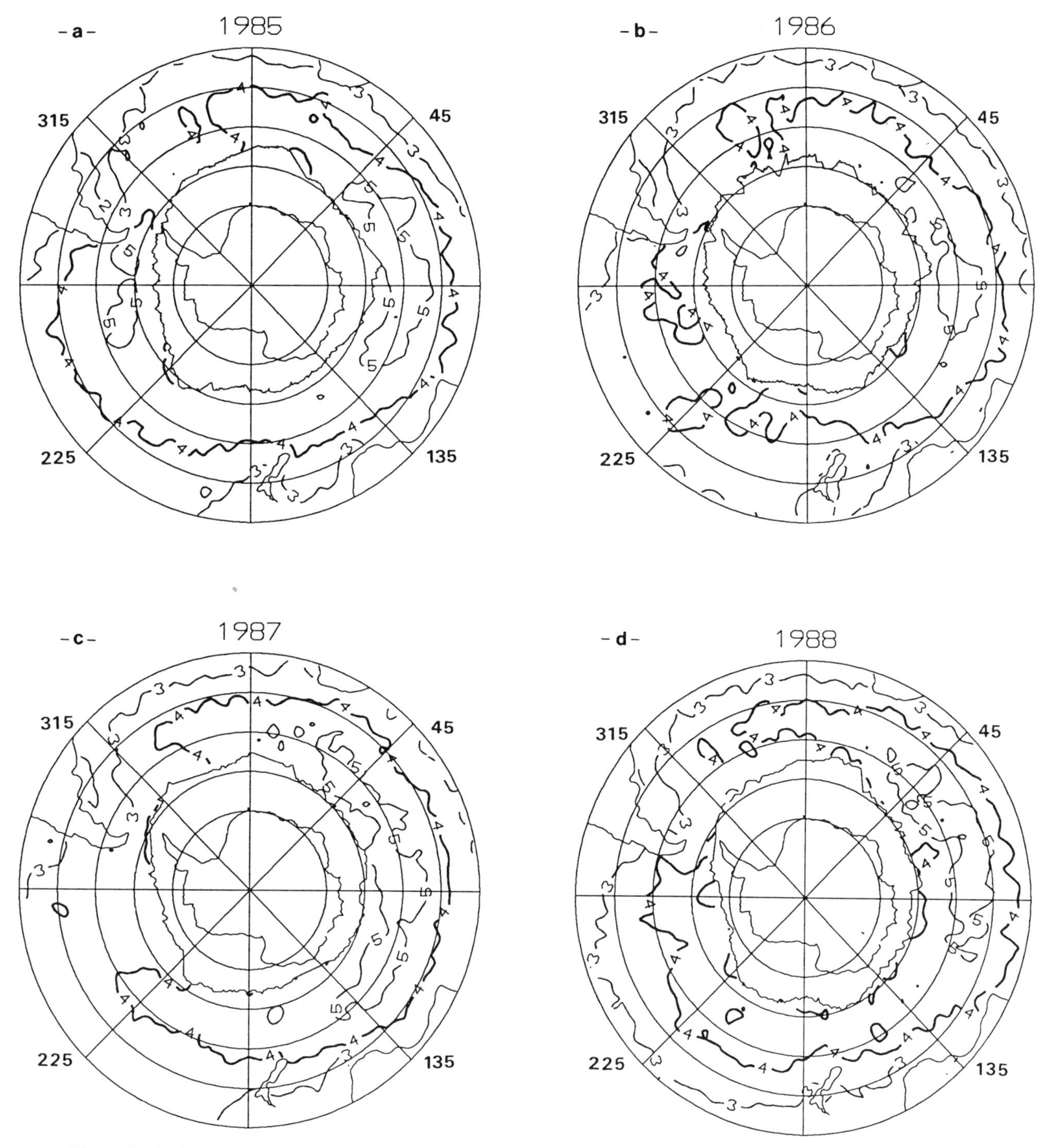

Figure 1a-d. Contour maps of the austral winter Geosat wave heights in meters for 1985-1988 (a-d) plotted south of latitude 30°S (the latitude circles are drawn every 10° up to 70°S). The heavy line is the 4 m contour, and the thin line surrounding the Antarctic continent is the extent of the sea ice derived from SMMR or SSM/I data sets.

altimeter observations are not valid over sea ice. Among the different parameters measured by the altimeter, the most sensitive to the presence of sea ice in the altimeter footprint is the standard deviation of the surface height. This measurement could be use to determine the ice edge; however, because the Geosat surface height data were classified before 1987 and were thus not available, we used the passive microwave measurements from the SMMR onboard the Nimbus-7 satellite and the SSM/I onboard a DMSP satellite to determine the Southern Ocean sea ice extent for the four Geosat years . Gloersen and Campbell [1988], among others, show that passive microwave observations accurately map sea ice extent, especially on the global and regional scales. Furthermore, Laxon [1990] shows that there is a close agreement between the determination of the Antarctic sea ice extent with the Geosat altimeter and the SMMR.

According to Antarctic satellite passive microwave observations by Zwally et al. [1983] and Gloersen et al. [1992], the maximum sea ice extent occurs in mid-September. To minimize the sea ice contamination of the Geosat measurements, we did not use any altimeter observations within the September 15 ice edge. Because we were primarily interested in wave fields for the entire Southern Ocean, rather than just at the ice edge, the ice edges from September 15 are sufficient for this study and a more thorough ice edge determination from the maximum ice extent at each longitude within each winter period has not been attempted.

For 1985 and 1986 we used the SMMR total ice concentration maps which are distributed on CD-ROM, SMMR Polar Data, Volume 7, by NASA Goddard Space Flight Center or the National Snow and Ice Data Center. The ice edge was determined by the location of the 30% total ice concentration line. For 1987 and 1988, we used the SSM/I observations which are also available on CD-ROM, DMSP SSM/I Brightness Temperature Grids, Polar Regions, Volumes 1, 5, and 9, from the National Snow and Ice Data Center. In the SSM/I case, because only brightness temperatures are available, the NASA team algorithm, Gloersen and Campbell [1988] was used to generate total ice concentrations. The ice edge was then determined by the location of the 30% total ice concentration line. On Figures 1 and 5 the ice edge is the thin line surrounding the Antarctic continent.

2.3. *Climatology*

To assess the differences between the Geosat altimeter seasonal fields, we compared these fields to the wave height field from the climatology. The climatological wave field was obtained from the U.S. Navy Climatic Atlas of the World [1992] produced by the Naval Oceanography Command Detachment, Asheville, NC. This digital atlas gives the monthly mean values of sea state-derived data covering the period from 1854 to 1969 on both a 5°x5° grid and a 1°x1° grid. To compare these values to the Geosat observations, we used the 1°x1° values, which were resampled to yield 2°x2° monthly maps of wave height. The months of July, August and September were then averaged to produce seasonal austral winter conditions to compare to the altimeter fields.

3. CLIMATOLOGY AND GEOSAT AUSTRAL WINTER FIELD

Figure 2 shows the climatological austral winter wave height field; the most notable feature is the nearly uniform zonal banding. The circumpolar region between 55°S and 60°S to 40°S, has waves greater than 3.5 m. This band of waves narrows dramatically in the Drake Passage yielding two regions with waves less than 3 m, one on each side of the South American continent. The region of waves greater than 3.5 m recedes poleward south of Australia and New Zealand where the northern boundary of the 3.5 m isopleth oscillates between 50° and 52°S. The highest waves are found in a small region of the central Pacific, between 50°S and 60°S, and reach values greater than 4 m but less than 4.5 m. Regions with waves less than 2 m are located along the American continent, mostly on the Atlantic side, and on the Pacific coast of Australia. To indicate an approximate ice edge position, the 1987 SSM/I ice edge has been superimposed to the climatology. A puzzling decrease in the wave height from the climatology is observed near the ice edge, especially in the Atlantic and Indian Oceans. This decrease in wave height begins 500 km or more from the ice edge, much further north than the ice edge could extend.

The 4-year Geosat average SWH field, Figure 3, shows the classical zonal banding observed in the climatological field, but with a superimposed spatial variability in the order of 500-1000 km. While the dominant feature in the climatology field is the circumpolar band of waves greater than 3.5 m, the dominant feature in the Geosat field is the band of waves greater than 4 m located across most of the Southern Ocean between 40°S and the ice edge with the exception of the western Atlantic region. In the Geosat field, the Central Indian Ocean is covered with waves greater than 5 m, in a region where the climatology only shows wave heights between 3.5 m and 4 m.

The difference in magnitude between the Geosat 4-year average and the climatology is shown in Figure 4. The largest differences range between 1 and 2 m and are found in large regions of the Central and Eastern Indian oceans. Smaller regions with differences between 1 and 1.5 m are found along the American continent on the Atlantic and Pacific sides and against the tip of Africa. Most of the Atlantic and Pacific have differences ranging between 1 and 0.5 m. The standard deviations of the individual Geosat wave heights in the 2°x2° bins range between 0.8 and 1.2 m, with the largest values located southward.

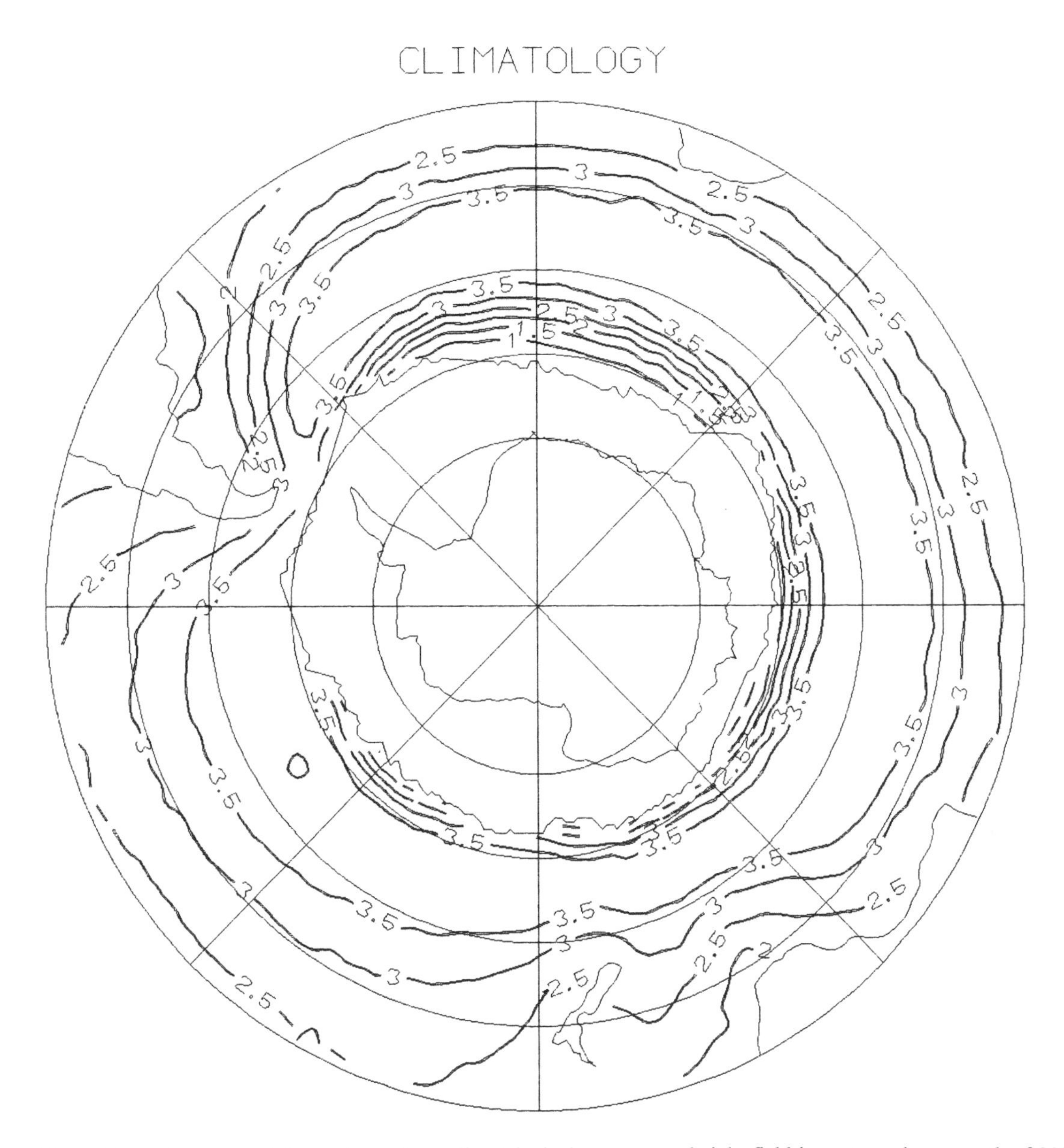

Figure 2. Contour map of the austral winter climatological mean wave height field in meters, shown south of 30°S, with latitude circles drawn every 10° up to 70°S.

To assess the interannual variability in the Geosat measurements and to compare to the climatology, Table 1 gives the climatological SWH, the Geosat 4-year average and the Geosat yearly mean wave height and standard deviation associated with each ocean in the latitude band 40°S to 60°S, for 1985 to 1988. The Western Pacific is bounded by 150°E to 140°W and the Eastern Pacific is bounded by 140°W to 70°W. For the Atlantic and the Eastern Pacific, Geosat estimates mean waves ranging between 0.5 and 1 m above the climatology estimates, while the difference increases to 0.75 to 1.25 m for the Western Pacific, and to 1 to 1.5 m for the Indian ocean. The standard deviations associated to the yearly Geosat mean values vary in each of the oceanic sectors between 1 and 1.5 m. These standard deviations, when combined with the large number of observations per pixel, yield a 95% confidence interval of ±0.15 m or less, about the mean wave height. Therefore, for the Geosat period, the climatological significant wave height field underestimates the observed wave heights throughout the Southern Ocean. The next section discusses the temporal and spatial variability of the satellite-derived wave height fields.

Figure 3. Contour map of the Geosat average wave height field in meters for the period 1985 to 1988, in the Southern Ocean south of 30°S, with latitude circles drawn every 10° up to 70°S.

4. GEOSAT SEASONAL FIELDS FROM 1985 TO 1988

The wave height fields, for each year show significant interannual variations (Figures 1a-d). For 1985 (Figure 1a), the Central Indian Ocean, between 45°S-55°S and 45°E-130°E, and a smaller region in the Eastern Pacific contain waves higher than 5 m. For 1986, Figure 1b shows a global decrease in SWH when compared to 1985. Only the Eastern Indian Ocean has a region with waves higher than 5 m, and in the Pacific Ocean, 3 and 4 m high waves dominate. The Atlantic differs little from 1985. For 1987 (Figure 1c), 4 m waves or greater are located in a continuous band extending clockwise from the Central Atlantic to the Central Pacific. 1987 is the year with the largest continuous region covered with waves larger than 5 m extending from the region south of the tip of Africa to a region south of the eastern portion of Australia. In 1988, Figure 1d, as was the case in 1985, the region with waves larger than 4 m extends continuously around the Southern Ocean with the exception of the Western Atlantic. A region with

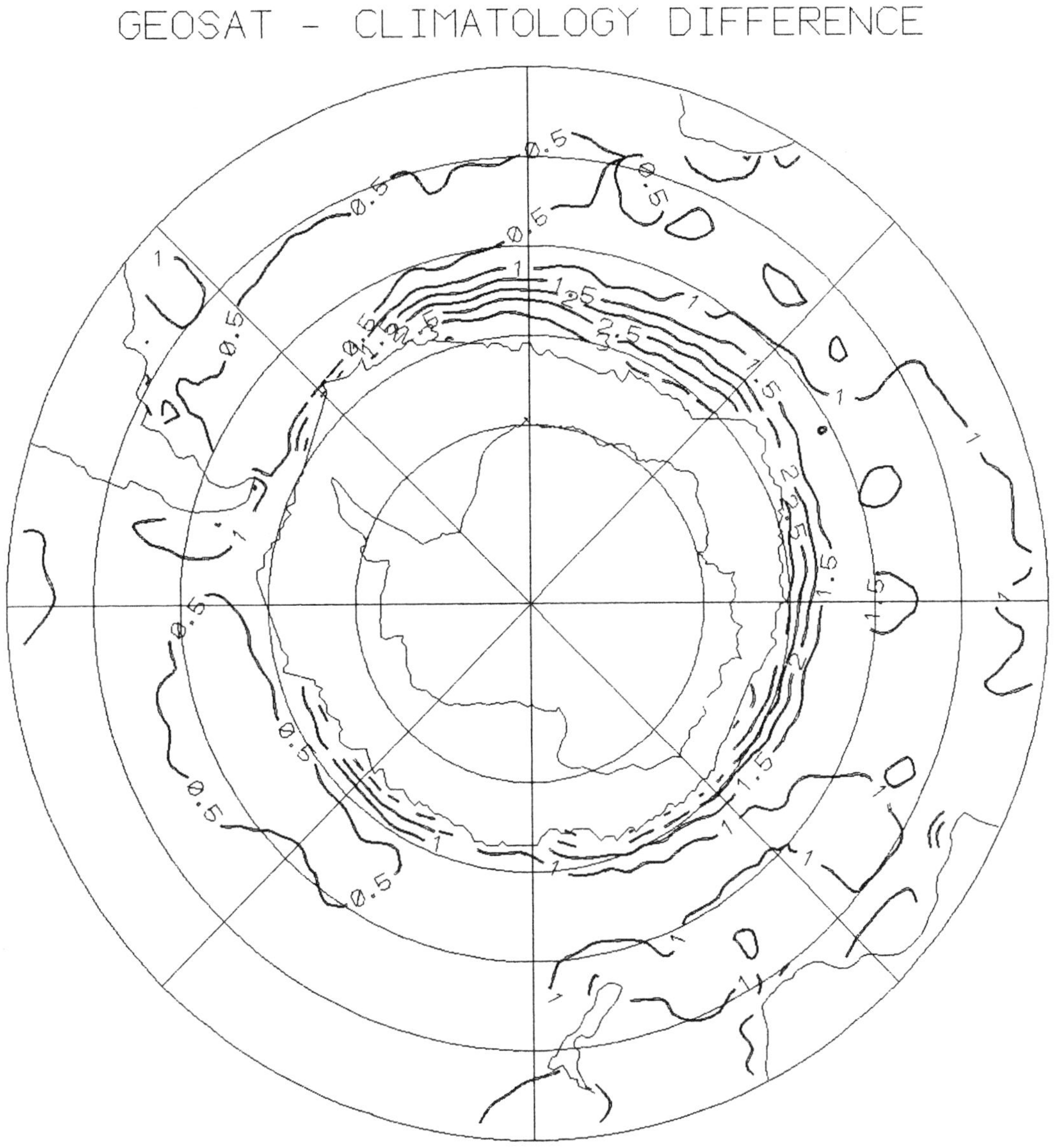

GEOSAT - CLIMATOLOGY DIFFERENCE

Figure 4. Contour map of the differences between Geosat and climatology austral winter wave heights in meters.

Table 1. For each ocean, the average significant wave height (SWH) from climatology, the 4-year Geosat average and the Geosat seasonal observations (in meters), have been computed in the region between 40°S and 60°S. For the seasonal observations, the average standard deviation of wave height is given.

Data Set	Atlantic		Indian		W. Pacific		E. Pacific	
	SWH	$\overline{\sigma}$	SWH	$\overline{\sigma}$	SWH	$\overline{\sigma}$	SWH	$\overline{\sigma}$
Climatology	3.09		3.55		3.13		3.48	
4-yr. Geosat	3.88		4.85		4.14		4.20	
1985	3.78	1.23	4.82	1.23	4.17	1.17	4.60	1.25
1986	3.92	1.22	4.70	1.20	3.88	0.99	4.03	1.25
1987	4.02	1.55	5.02	1.36	4.40	1.42	4.01	1.14
1988	3.79	1.30	4.84	1.21	4.06	1.15	4.14	1.06

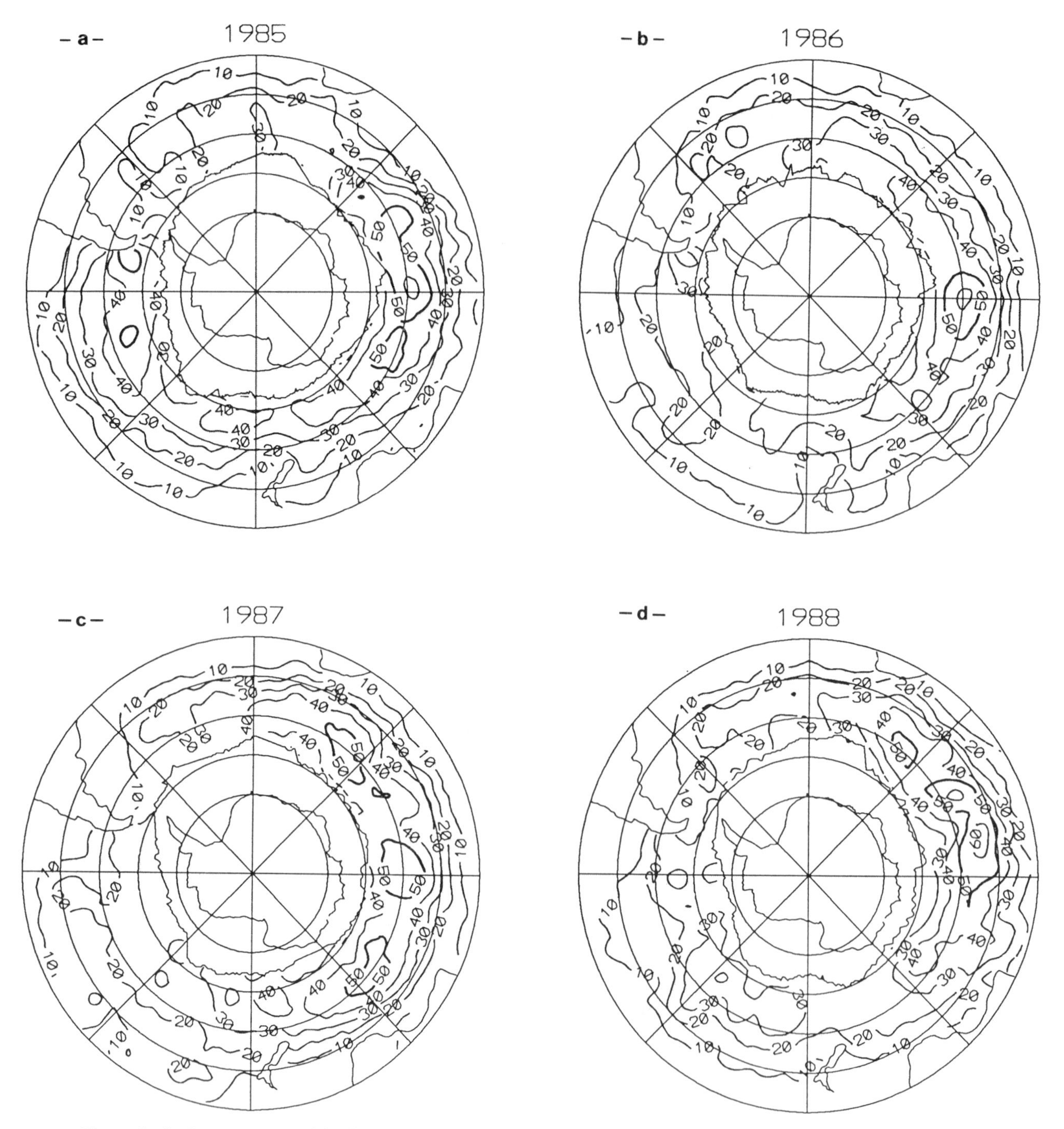

Figure 5a-d. Contour maps of the frequency of waves greater than 5 m in each bin for 1985-1988 (a-d). The contour interval is 10% and the latitude circles are drawn every 10° southward of 30°S.

waves higher than 5 m is found across the Central Indian Ocean (Figure 1d). The standard deviation computed (but not shown) for each of the fields exhibit values ranging from 0.8 to 1.6 m with the highest values mostly located southward, along the ice edge.

Geosat observations not only provide information on the mean sea state but also on the wave height distributions. Figures 5a-d show the frequency distribution of waves greater than 5 m for each year (contour interval 10%). For 1985, the 30% isopleth (Figure 5a) extends from the West-

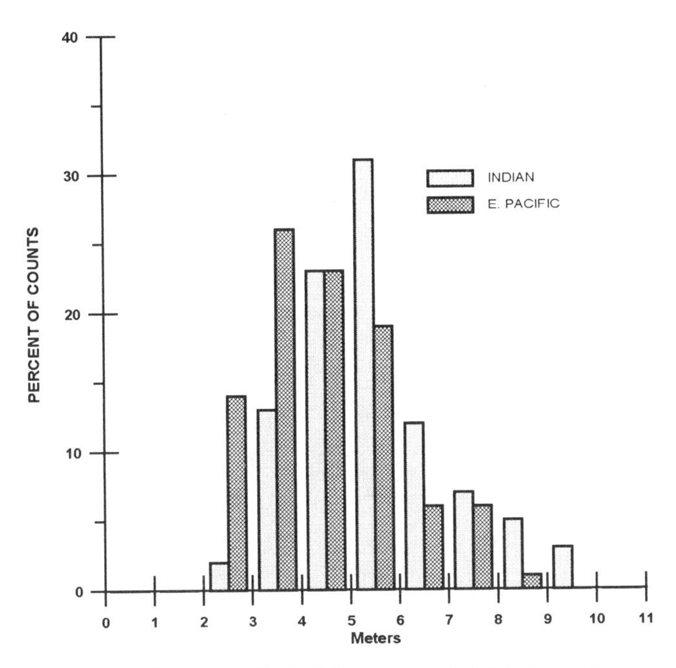

Figure 6. Histogram of significant wave heights in meters as measured by the Geosat altimeter from 1985-1988 in two 2°x2° pixels, one located in the Central Indian Ocean centered at 49°S and 91°E and the other one in the Eastern Pacific at 55°S and 255°E.

ern Indian Ocean eastward all the way to Drake Passage. Large regions where more than 50% of individual waves are greater than 5 m are measured in the Central Indian Ocean and in the Western Pacific. In 1986, the extent of regions delineated by the 30% isopleth decreases by about half and there is a westward shift towards the Atlantic Ocean. Only the Indian Ocean and the Eastern Atlantic have a frequency greater than 30% for waves higher than 5 m, while for the whole Pacific Ocean, the frequency for these waves is less than 30% (Figure 5b). In 1987, the Pacific Ocean has still only about 20% of waves greater than 5 m while in the Indian Ocean, three large regions have more than 50% of waves higher than 5 m (Figure 5c). In 1988, there are two regions where 30% of the waves are greater than 5 m, in the Central Pacific and over most of the Southern Indian Ocean. Like the case for the two preceding years, only the Eastern and Central Indian Ocean have a region where 50% of the waves are greater than 5 m (Figure 5d).

Over the 4-year period, the Central Indian Ocean has the largest seasonal sea states, while the Eastern Pacific Ocean shows the most variability. Figure 6 gives the 4-year average of the wave height distributions for a pixel at 49°S and 91°E, in the middle of the high sea state region found in the Indian Ocean, and for a pixel at 55°S and 255°E, in the region of high annual variability observed in the Eastern Pacific Ocean. The differences between the two regions are clear, the peak for the Indian Ocean occurs in the 5-6 m range while the peak in the Eastern Pacific Ocean occurs in the 3-4 m range. Also, for each bin greater than 5-6 m, the Indian Ocean always contains larger waves than the Eastern Pacific Ocean. Figure 7 compares the interannual variability at these two locations. For the Indian Ocean the peak always remains at 5-6 m, while in the Eastern Pacific, the peak shifts from 4-5 m in 1985, to 3-4 m in 1986 and 1988, and to 2-3 m in 1987. The distributions are most dissimilar in 1987, the year when a large continuous band of waves greater than 5 m extended from the Eastern Atlantic Ocean across the Indian Ocean to the Western Pacific Ocean (Figure 1c). The two regions were most similar in 1985, when the Eastern Pacific contained waves larger than 5 m (Figure 1a).

The regional and interannual variations of the wave distributions are shown in Figures 8, 9, and 10. These figures give the percentage of waves in each ocean that are greater than, 7 m, 5 m and less than 2 m, for three latitude bands, and for the 4 years. (Each ocean has been subdivided into three latitude bands: 30°S to 40°S, 40°S to 50°S, and 50°S to 60°S.) As expected, for all years the percentage of waves greater than 7 m and than 5 m increases with increasing latitude (Figures 8 and 9). In Figure 8, between 30°S and 40°S, the percentage of waves greater than 7 m is less than 1% except in the Indian Ocean where it reaches 2 and 3% in 1987 and 1988. The Indian Ocean has the greatest incidence (between 6 and 10%) of these high waves in the "Roaring Forties", while for the other oceans, the incidence is of the order of 5% or less. Between 50°S and 60°S, in the vicinity of the ice edge, the Indian Ocean and the Western Pacific show the highest interannual variability (between 7% and 14% for the Indian Ocean and between 3% and 12% for the Western Pacific). In this high-latitude region, the maximum frequency of these high waves occurs in 1985 for the Eastern Pacific, while in the other oceans the maximum occurs in 1987.

For waves greater than 5 m, (Figure 9), the number of occurrences of these waves in each ocean increases approximately by a factor of four, when compared to waves greater than 7 m (Figure 8). As expected, the trends are similar for the two wave heights, the percentage of these high waves increases towards the pole. Figures 8 and 9 also show that, for all years, at all latitude bands, the Indian Ocean is the roughest. In the 30-40°S latitude band, the percentage of waves greater than 5 m varies from 5% to 22%, while for the same region, the percentage of waves greater than 7 m is less than 3%. 1988 was an exceptional case, in this latitude band of the Indian Ocean, the wave

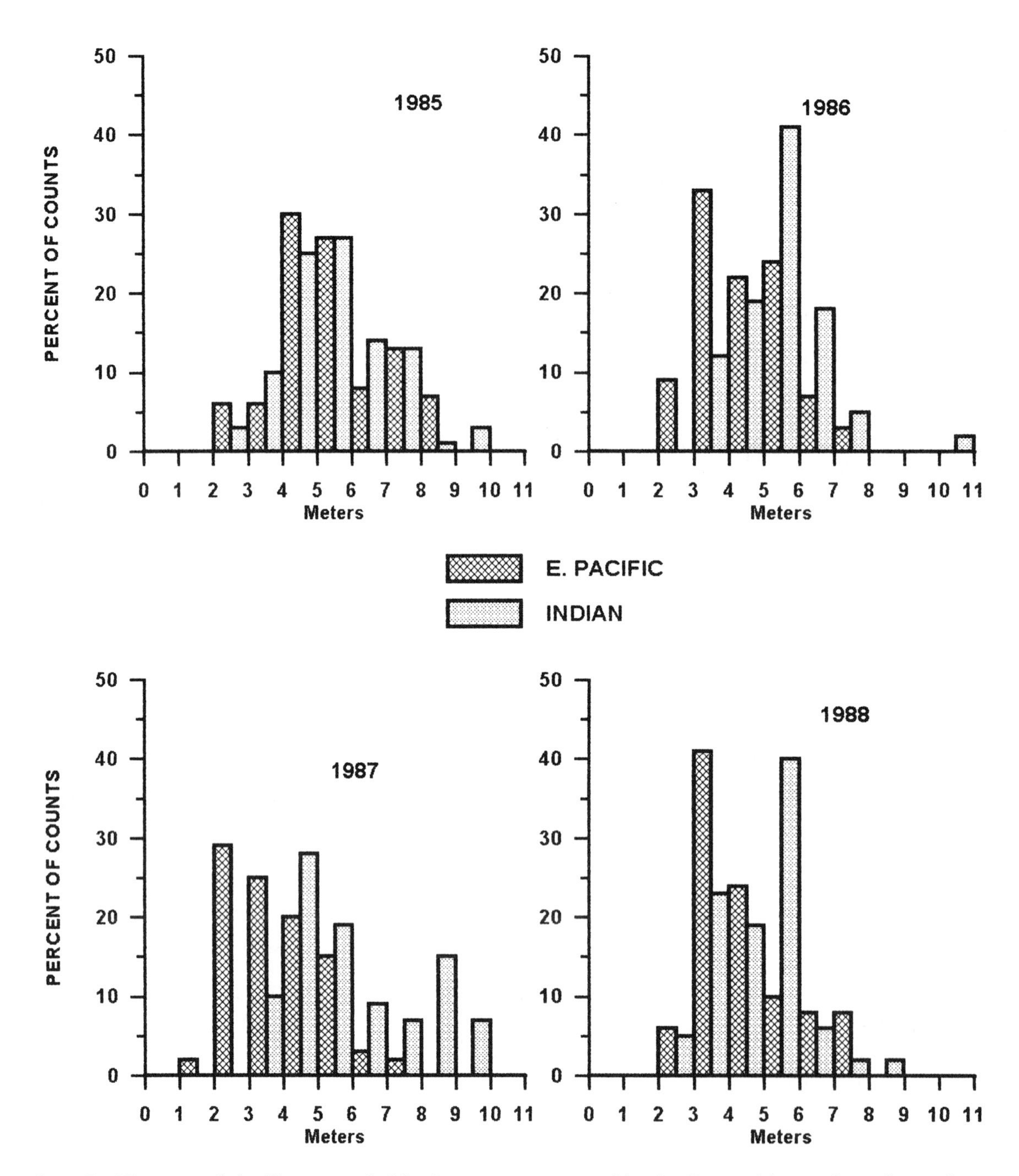

Figure 7. Histogram of significant wave heights in meters as measured by the Geosat altimeter for each year from 1985 to 1988 in two 2°x2° pixels, one located in the Central Indian Ocean centered at 49°S and 91°E and the other one in the Eastern Pacific at 55°S and 255°E.

percentages reached a maximum, this behavior was not observed elsewhere. This trend persists in the 40°-50°S latitude band. In the "Roaring Forties", the percentage of waves greater than 5 m in the Indian Ocean reaches between 35 and 44%, far more than any other ocean for all years. In the 50 to 60°S latitude band, the Indian Ocean is still the roughest, although the percentages in the West and East Pacific increase to near those observed in the Indian Ocean. The Atlantic is by far the calmest of the four regions when the 5 and 7 m waves are considered.

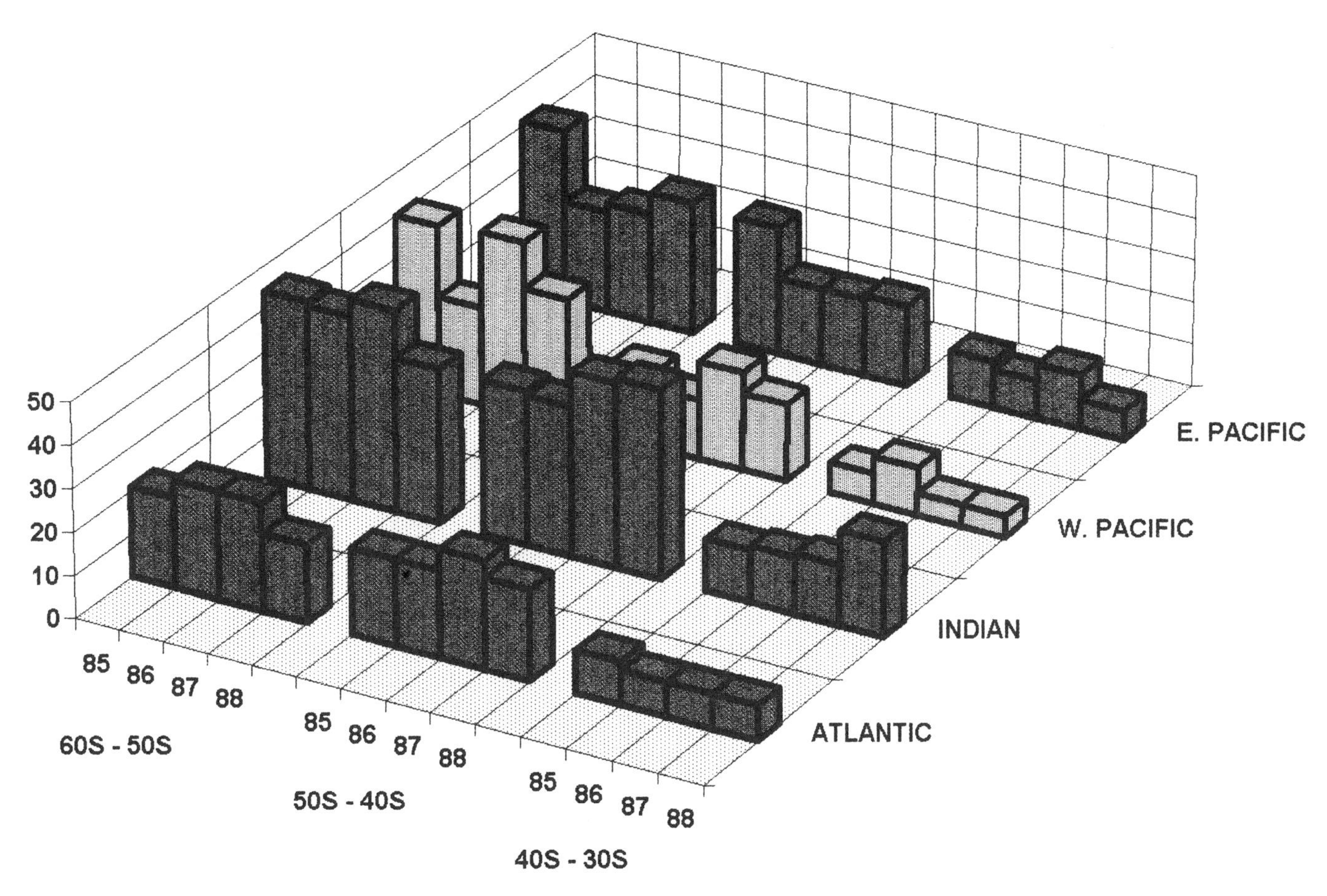

Figure 8. Fraction of waves greater than 7 m in the Atlantic Ocean, Indian Ocean, West Pacific Ocean, and the East Pacific Ocean, for 1985 through 1988 for three latitudinal bands from 30°S to 40°S; from 40°S to 50°S; and from 50°S to 60°S.

The Geosat observations also allow for the investigation of relatively calm conditions. Figure 10 shows the percentage of waves less than 2 m for the four oceans. The Atlantic Ocean has highest percentage of low waves in the two most southern latitude bands (40° to 60°S). For all years, the northern band of the Western Pacific contains the greatest percentage of waves less than 2 m, between 18% and 25%. Overall the Indian Ocean has the least amount of waves less than 2 m high.

5. CONCLUSIONS

Comparison with the climatology derived from individual ship observations shows that classical climatology of significant wave height greatly underestimates the magnitude of the average seasonal wave height in the Southern Ocean for 1985 through 1988. Similar results were obtained in the comparison between wind stress climatologies derived from Seasat scatterometer and the wind stress fields estimated from individual ship measurements [*Chelton et al.*, 1990]. It is in the Southern Indian Ocean that the differences are the largest (Geosat measures mean waves with values ranging between 1 and 2 m above the values in the climatology). The first satellite seasonal wave field obtained with the Seasat altimeter in 1978 already showed that the Indian Ocean had the roughest sea states with mean waves greater than 4.6 m [*Chelton et al.*, 1981]. The global zonal banding of the seasonal wave field with smaller mean waves located along the eastern coast of South America and Australia is in agreement in both fields.

Interannual variability is observed in the altimeter fields during the four austral winters analyzed (Figures 1a-d and

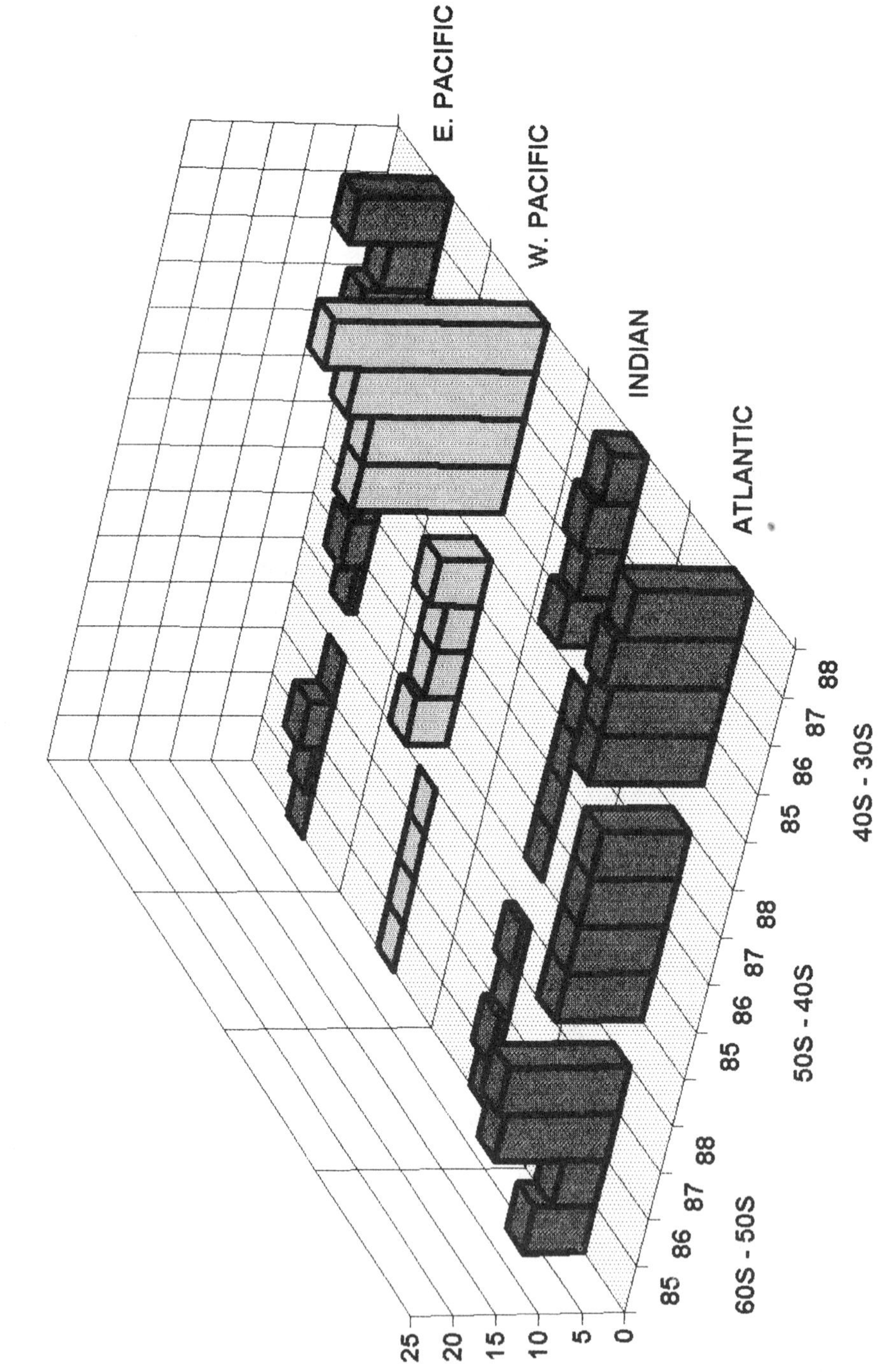

Figure 9. Fraction of waves greater than 5 m in the Atlantic Ocean, Indian Ocean, West Pacific Ocean, and the East Pacific Ocean, for 1985 through 1988 for three latitudinal bands from 30°S to 40°S; from 40°S to 50°S; and from 50°S to 60°S.

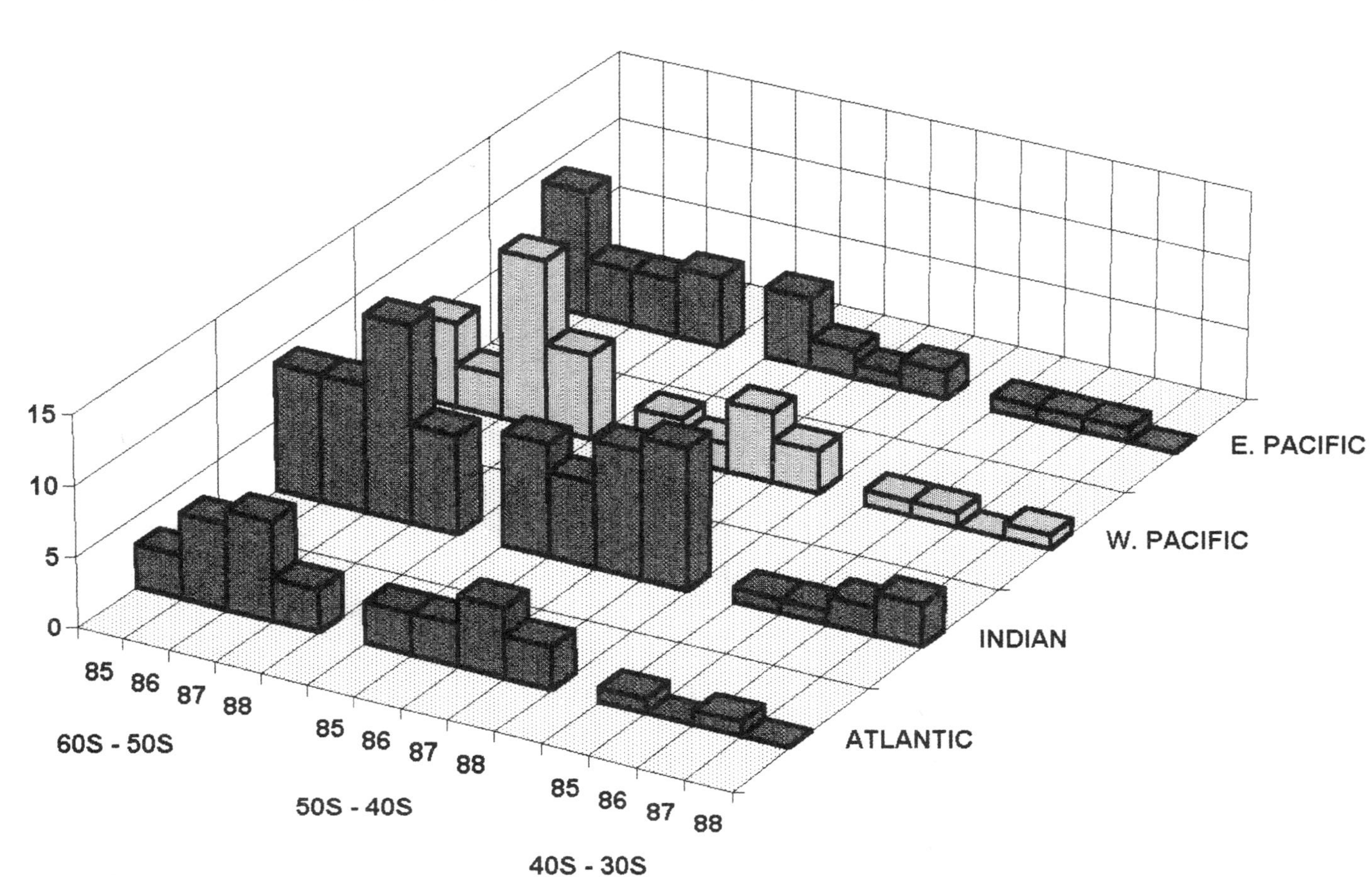

Figure 10. Fraction of waves less than 2 m in the Atlantic Ocean, Indian Ocean, West Pacific Ocean, and the East Pacific Ocean, for 1985 through 1988 for three latitudinal bands from 30°S to 40°S; from 40°S to 50°S; and from 50°S to 60°S.

5a-d). For instance 1985 is the only year when the Eastern Pacific has a large region covered with mean waves of the same order of magnitude as the Indian Ocean. 1986 is the year with the overall smallest coverage with high waves. In 1987, the largest continuous band of waves greater than 5 m extends over a band with a width of 110° in longitude across the Indian Ocean, while the Eastern Pacific is covered with the smallest mean waves (less than 4 m) for the 4-year period considered.

The spatial and temporal variations measured by the Geosat altimeter are important for our understanding of the austral winter climate patterns' variability. Satellite radar altimeters are the only instruments which provide synoptic, repetitive, and uniformly accurate significant wave heights measurements. With the data sets acquired with the next generation of radar altimeters onboard ERS-1 and Topex/Poseidon, a monitoring of sea state interannual variability over the oceans is becoming possible. The important differences with the classical climatological wave field show that, for shipping as well as for climate purposes, using satellite fields as input instead of climatological fields would lead to a better understanding of our environment and its variability.

Acknowledgments. We thank Robert Paradise for his efforts in data processing, and Jo Eggers for her help with plotting the Geosat wave fields and in the preparation of the manuscript.

REFERENCES

Carter, D. J. T., P. G. Challenor, and M. A. Srokosz, An assessment of Geosat Wave height and wind speed measurements, *J. Geophys. Res.*, *97*, 11383-11392, 1992.

Chelton, D. B., K. J. Hussey, and M. E. Parke, Global satellite measurements of water vapor, wind speed and wave height, *Nature*, *294*, 529-532, 1981.

Chelton, D. B., A. M. Mestas-Nunez, and M. H. Freilich, Global wind stress and Sverdrup circulation from the Seasat scatterometer, *J. Phys. Ocean.*, *20*, 1175-1205, 1990.

Dobson, E., F. Monaldo, J. Goldhirsh, and J. Wilkerson, Validation of Geosat altimeter-derived wind speeds and significant wave heights using buoy data, *J. Geophys. Res.*, *92*, 10,719-10,731, 1987.

Gloersen, P., W. J. Campbell, D. J. Cavalieri, J. C. Comiso, C. L. Parkinson, and H. J. Zwally, Arctic and Antarctic Sea Ice, 1978-1987: Satellite Passive Microwave Observations and Analysis, *Special Publication SP-511*, 290 pp., NASA, 1992.

Gloersen, P., and W. J. Campbell, Variations in the Arctic and Antarctic global sea ice covers during 1978-1987 as observed with the Nimbus 7 Scanning Multichannel Microwave Radiometer, *J. Geophys. Res.*, *93*, 10,666-10,674, 1988.

Guillaume, A., and N. M. Mognard, A new method for the validation of altimeter-derived sea state parameters with results from wind and wave models, *J. Geophys. Res.*, *97*, 9705-9717, 1992.

Hellerman, S., and M. Rosenstein, Normal monthly wind stress over the world ocean with error estimates, *J. Phys. Ocean.*, *13*, 1093-1104, 1983.

Laxon, S., Seasonal and inter-annual variations in Antarctic sea ice extent as mapped by radar altimetry, *Geophys. Res. Let.*, *17*, 1553-1556, 1990.

MacArthur, J. I., P. C. Marth, Jr., and J. G. Wall, The Geosat altimeter, *Tech. Dig.*, *8*, Johns Hopkins APL, pp. 176-181, 1987.

Mognard, N. M., W. J. Campbell, R. E. Cheney, and J. G. Marsh, Southern ocean mean monthly waves and surface winds for winter 1978 by Seasat radar altimeter, *J. Geophys. Res.*, *88*, 1736-1744, 1983.

Mognard, N. M., Swell in the Pacific Ocean observed by Seasat radar altimeter, *Marine Geodesy*, *8*, 183-210, 1984.

Mognard, N. M., and W. J. Campbell, Comparison of sea surface wind speed fields by Seasat radar altimeter, scatterometer, and scanning multichannel microwave radiometer, with an emphasis on the Southern Ocean, in *Proc. of IGARSS'84*, ESA SP-215, pp. 403-409, 1984.

Mognard, N. M., W. J. Campbell, R. E. Cheney, J. G. Marsh, and D. Ross, Southern ocean wind waves derived from Seasat radar altimeter measurements, in *Wave Dynamics and Radio Probing of the Ocean Surface*, edited by O. M. Phillips and K. Hasselmann, pp. 32,479-32,489, Plenum, New York, 1986.

U. S. Navy Marine Climatic Atlas of the World, version 1.0, Naval Oceanographic Command Detachment Asheville, Federal Building, Asheville, NC, 28801, 1992

Zwally, H. J., J. C. Comiso, C. L. Parkinson, W. J. Campbell, F. D. Carsey, and P. Gloersen, Antarctic sea ice 1973-1976: satellite passive microwave observations, *Special Publication SP-459*, NASA, 1983.

W. J. Campbell and E. G. Josberger, Ice and Climate Project, U.S. Geological Survey, University of Puget Sound, Tacoma, WA 98416.

N. M. Mognard, CNES-GRGS, 18 Avenue E. Belin, 31055 Toulouse Cedex, France.

Factors Affecting Variations of Snow Surface Temperature and Air Temperature Over Sea Ice In Winter

Peter S. Guest and Kenneth L. Davidson

Department of Meteorology, Naval Postgraduate School, Monterey, California

The temperature at the top of an ice floe depends on turbulent, radiative and conductive heat fluxes. Turbulent heat transfer closely links this temperature with the near-surface air temperature. Measurements during the CEAREX drift, northeast of Svalbard, show that the downward radiation during overcast periods corresponds to blackbody radiation at the near-surface air temperature. An analytical expression describes how conduction of heat through the ice is related to snow or ice thermal diffusivities and atmospheric forcing time scales. Temperature observations obtained during LEADEX detected time scales associated with conduction of heat through ice and snow.

1. INTRODUCTION

The most commonly used index of climate and climate change is the air temperature, T_{air}, in the surface layer of the atmosphere. Studies using general circulation models (GCMs) by Boer *et al.* [1992], Cao *et al.* [1992], Manabe *et al.* [1992], and others, predict that the polar regions are the most sensitive areas on Earth to surface warming induced by anthropogenic "greenhouse" gases. In this paper, we examine some of the factors which affect short-term variations in surface air temperature during the Arctic winter. The goal of the paper is to provide some tools, in the form of thermodynamic simplifications, which can be useful for determining surface energy balances and surface temperatures over sea ice regions. The issue of long-term temperature change is not addressed directly, but the results should be useful for designing and verifying numerical models of the Arctic ocean-ice-atmosphere system.

Specifically, we will show that T_{air} is usually closely linked to the temperature at the top of the snow or ice surface, T_{sfc}, during the Arctic winter, despite the presence of very stable atmospheric boundary layers (ABLs). Secondly, we will show an example of close linking between T_{air} and the surface downward longwave radiation, when the sky is overcast. Finally, we show how the contribution of conductive heat flux to the surface energy balance depends on the atmospheric time scale of interest.

The Polar Oceans and Their Role in Shaping the Global Environment
Geophysical Monograph 85

2. LINKING OF NEAR-SURFACE AIR AND AIR/SNOW INTERFACIAL TEMPERATURES

The AVHRR thermal channels are useful for determining surface temperatures (snow, ice or water) in Arctic regions, e.g. Key and Haefliger [1992], Lindsay and Rothrock [1994]. Directly measuring the near-surface air temperature, T_{air}, generally requires *in situ* instruments. However, in this section, we will show that the surface temperature, T_{sfc}, is usually closely linked to T_{air}; therefore a remotely-sensed T_{sfc} can be used to estimate air temperature fields.

Low-level temperature inversions (surface-based or elevated) are ubiquitous features of the Arctic atmosphere, particularly during dark seasons when they are present almost 100% of the time over the central Arctic pack ice [Belmont, 1957; Vowinckel and Orvig, 1970; Serreze *et al.*, 1992]. Wexler [1936] showed that temperature inversions can result from a radiation balance between the nearly blackbody emission of the snow surface compared to the graybody emission of the lower atmosphere. Wexler neglected a lot of important physics, but his concept that there is a strong radiational coupling between the snow surface and an isothermal temperature maximum layer in the lower atmosphere is the basic explanation for the presence of low-level temperature inversions in Arctic regions during clear weather [Curry, 1983; Overland and Guest, 1991]. During the central Arctic winter, the low-level inversions extend to the surface (surface-based) at least 60% of the time for the months November through April, according to studies based on data from the Russian drifting *North Pole* stations [Vowinckel and Orvig, 1970; Serreze *et al.*, 1992].

Based on the modeling and observational results discussed in the previous paragraph, one might conclude that extreme temperature gradients in the lowest few meters of the atmosphere are common during the central Arctic winter and, therefore, T_{air} and T_{sfc} are several degrees different and cannot be considered "closely coupled". However, the *North Pole* station data may over-predict the frequency of surface-based inversions because the camps were located on ice islands that may have had lower surface temperatures than the surrounding sea ice. The vertical resolution of the *North Pole* data may not have been sufficient to detect very shallow surface mixed layers. During the 1992 Leads Experiment in the Beaufort Sea, the authors detected surface-based inversions in only 33% of the high-resolution (20 m) soundings in April, in contrast to the 65% reported by the Serreze *et al.* [1992] *North Pole* climatology.

Even when a surface-based inversion is present, the temperature change is generally not concentrated in the lowest few meters, except during very light winds. During a 30-day period in October and November (solar radiation was negligible), the authors measured the temperature at the snow/air interface, T_{sfc}, and at five levels up to 15 m elevation, as part of the Coordinated Eastern Arctic Experiment drift phase (CEAREX drift), which occurred in the Arctic Ocean northeast of Svalbard in 1988. With the exception of a 3-day calm period, the difference between T_{sfc} and T_{air} (5.2 m above the surface) was less than 2 °C for the entire period [Overland and Guest, 1991, Fig. 18], even when rawinsondes detected inversion bases below 100 m, which was 22% of the soundings. A study being prepared by Yu, Rothrock and Lindsay (U. Washington) shows that T_{sfc} and T_{air} (2 m) were usually within 2 °C of each other, based on data from *North Pole* stations 6 and 7, 1957-1958. The largest monthly median difference was 1.5 °C; it occurred in January. These data included calm wind conditions and did not include periods with greater than 0.3 cloud coverage.

Maintaining the strong stability of the lower atmosphere during the Arctic winter does not require large turbulent cooling at the surface. The January average of the downward turbulent surface heat flux over thick, solid, pack ice in the Arctic is approximately 16 Wm^{-2}, with lower average values occurring in other months [Maykut, 1986]. Due to upward heat fluxes from leads and thin ice, the areal-average turbulent fluxes for the central Arctic are less than the above value. These low values for the turbulent heat fluxes mean that, either the areal averages of T_{sfc} and T_{air} are very close to each other, or surface layer conditions are so stable that turbulence is suppressed and the turbulent heat transfer coefficient becomes small. Profile and turbulence measurements in the surface layer of the Arctic winter atmosphere at the AIDJEX camps [e.g. Carsey, Leavitt, 1980], the *North Pole* Stations [e.g. Makshtas, 1984] and the CEAREX drift [Overland and Guest, 1991], and other locations, demonstrate that, unless surface winds are very light (approximately 2 ms^{-1} or less), continuous turbulence will extend at least a few meters above the surface, allowing T_{sfc} and T_{air} (at a typical instrument enclosure level of 2 meters) to closely track each other.

Overland and Guest [1991] demonstrated the coupling between T_{sfc} and T_{air} (at 10 m elevation in this case) using Overland's [1988] numerical boundary layer model with Semtner's [1976] thermodynamic sea ice model. Downward radiation was suddenly decreased by 80 Wm^{-2} to simulate a change from overcast to clear sky conditions. The surface temperature, T_{sfc}, rapidly decreased, but T_{air} also decreased as a shallow internal boundary layer was formed. After 12 hours, T_{air} minus T_{sfc} was less than 2 °C and the downward heat flux was less than 10 Wm^{-1}. The geostrophic wind for this case was 3 ms^{-1}. The coupling between T_{sfc} and T_{air} would have been even closer with a more typical (higher) geostrophic wind speed.

There are three basic reasons for the close coupling between T_{sfc} and T_{air} in the observations and models described above. First, turbulence is more effective at transferring heat than radiation, for most situations. For example, unless the wind speed at 2 m elevation is less than about 3 ms^{-1}, the sensible heat flux produced by a T_{sfc} and T_{air} difference of 1 °C will be larger than the net longwave radiational flux resulting from a difference between T_{sfc} and the longwave radiative equilibrium temperature, T_{rad}, of 1 °C. The definition of T_{rad} is $(F_L/\sigma)^{1/4}$ where F_L is the downwelling longwave radiation and σ is the Stefan-Boltzmann constant. In the above example, we used Monin-Obukhov surface layer theory using coefficients for pack ice conditions suggested by Andreas [1988] and assumed that the surface was a blackbody radiator. This shows that unless winds are light, T_{sfc} is more affected by changes in T_{air} than by changes in downward radiation. The second reason for the close coupling is that most of the surface area of the central Arctic is covered with snow in the winter. Snow has a relatively low thermal conductance and heat capacity which means that T_{sfc} will rapidly adjust to changes in forcing, such as a change in T_{air}. The third reason for the close coupling is that because the Arctic atmospheric boundary layer is so shallow during the winter, it has a low bulk heat capacity and ABL temperatures (including the surface layer T_{air}) will quickly adjust to forcing changes, such as a turbulent heat flux induced by a change in T_{sfc}. Thus, the stable Arctic atmosphere increases the coupling by limiting the depth to which heat can be stored or extracted. Only when winds are so light that continuous turbulence does not extend to the air temperature measurement height will T_{air} and T_{sfc} become effectively de-coupled.

For numerical modeling applications, T_{air} and T_{sfc} should not be considered independent of each other. For example, if an ice model uses a fixed T_{air}, the ice (or snow) surface temperature and conductive ice fluxes are strongly constrained by T_{air}. In reality, T_{air} changes in response to surface fluxes, so that a change in downward radiation, for example, will affect T_{sfc} directly by radiation and T_{air} indirectly by surface turbulent fluxes. Similarly, T_{sfc} adjusts to changes in T_{air}, so an atmospheric model which assumes a fixed T_{sfc} may produce incorrect turbulent heat fluxes when T_{air} is rapidly changing due to horizontal advection or diabatic heating.

This discussion also shows that AVHRR snow surface temperatures can be used to provide estimates of near-surface air temperature. The error in T_{sfc} derived from AVHRR, as estimated by Lindsay and Rothrock [1994] is 2 - 5 °C, which is larger than the typical differences between T_{sfc} and T_{air} during non-calm conditions (less than 2 °C).

3. LINKING LONGWAVE RADIATION AND NEAR-SURFACE AIR TEMPERATURE

During central Arctic dark seasons (fall and winter), the absolute (as opposed to relative) values of T_{sfc} and T_{air} for multi-year ice are primarily controlled by the longwave radiation balance between the surface and the atmosphere, with some contribution from conductive heat fluxes [Overland and Guest, 1991]. The downward longwave radiation at the surface, F_L, depends on the vertical structure of temperature, humidity and various particulates, such as ice crystals, water droplets and other components of the atmospheric aerosol. Downward radiation in the Arctic is particularly sensitive to clouds and ice crystals [e.g. Curry *et al.*, 1990; Curry and Ebert, 1992], which are often difficult or impossible to detect.

Unfortunately, in many situations, the only available information concerning atmospheric conditions is from surface observations. For this reason, large-scale ice modelers often use parameterizations of F_L that are based only on parameters observed at the surface, such as T_{air} and cloud cover. In this section, we compare some recent measurements of F_L in the Arctic Ocean near Svalbard with T_{air} and cloud cover values, in order to determine the validity of the simple F_L parameterizations for this region.

We plot F_L vs. T_{air} (5.2 m) from the CEAREX drift for totally clear (all stars visible) and totally overcast (no stars even dimly visible) sky conditions as determined by visual observations from the ground. We also plot dotted lines showing the Maykut and Church [1973] empirical fit

$$F_L = \sigma T_{air}^4 (0.7855(1 + 0.22232\, C^{2.75}) \quad (1)$$

and dashed lines showing the empirical fit suggested by Idso and Jackson [1969] using the Marshunova [1961] cloud cover correction as suggested by Parkinson and Washington [1979]

$$F_L = \sigma T_{air}^4 \, (1 - 0.261 \exp(-7.77 \times 10^{-4} (T_{air} - 273)^2)) \, (1 + 0.3C) \quad (2)$$

where C is the cloud cover. Two lines for each of the above fits are plotted, the upper line represents overcast (C = 1) conditions, while the lower line represents clear conditions (C = 0). Finally a solid line shows the relation

$$F_L = \sigma T_{air}^4 \quad (3)$$

which represents the condition T_{air} equals T_{rad}.

Note that during clear conditions, there is considerable scatter in the T_{air} vs. F_L relationship (Figure 1, lower part). Although all stars were sharply visible during reported clear conditions, ice crystals ("diamond dust") were virtually always observed at the surface during these periods. Curry et al. [1990] showed that ice crystals in the lower troposphere may increase F_L by as much as 80 Wm^{-2} under otherwise clear skies. There were no quantitative measurements of the ice crystals at the

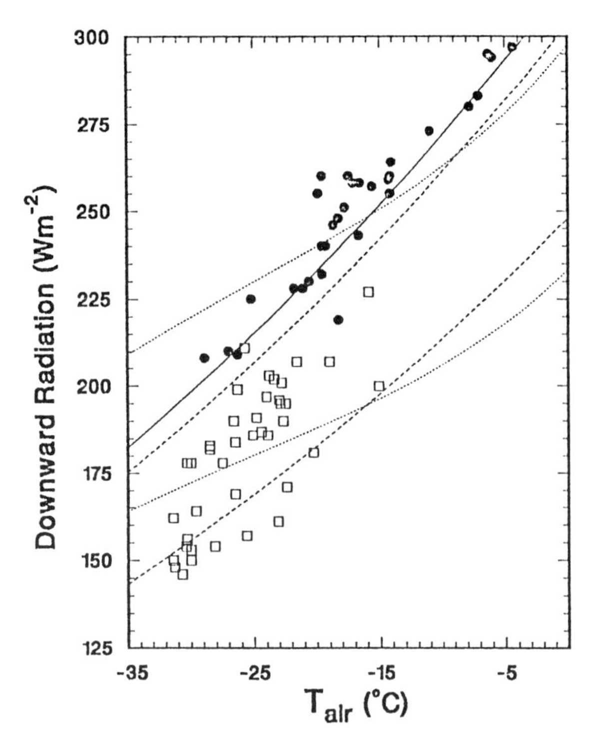

Fig. 1. The relationship between surface air temperature, T_{air}, (X-axis) and downward longwave radiation, F_L, (Y-axis) based on measurements obtained during the CEAREX drift (individual points) and empirical fits (lines). Two sky categories are shown: totally clear (open squares, lower two lines) and overcast (solid circles, upper three lines). The formulas for the empirical fits and other details are explained in the text.

surface or at upper levels, but it is possible that variations in ice crystal amounts contributed to variations in F_L, which were not reflected by T_{air} variations, thus causing scatter in the T_{air} vs. F_L relationship. Neither of the two empirical fits for clear skies (bottom two lines in Figure 1), nor any other fit based only on T_{air} could predict F_L with an accuracy of better than 25 Wm^{-2} during the CEAREX drift.

During overcast conditions, there was considerably less scatter in the T_{air} vs. F_L relationship (Figure 1, upper part). Almost 80% of the observed values were within 10 Wm^{-2} of the value predicted by (3) (solid line in Figure 1), which is close to the estimated accuracy of the measurements. The (1) and (2) predictions (dashed and dotted lines in Figure 1) do not fit the data as well as the simple T_{air} equals T_{rad} relationship expressed by (1). There may be less scatter during overcast skies because cloud emissions dominate F_L and the variable ice crystal effect is relatively less important. When overcast conditions are present, the sources for F_L are closer to the surface and turbulence is likely to extend higher than during clear weather. During the CEAREX drift, overcast conditions were invariably associated with low stratus or strato-cumulus conditions and cloud bases were usually below 500 m. If a turbulent connection exists between the surface layer where T_{air} is measured and the levels within the cloud where F_L sources exist, the coupling between the two will be especially close. Although summer clouds do not appear to be closely connected to the surface via turbulence [Curry, 1986], the winter situation may be different because the cloud layer cannot become de-coupled from the surface by shortwave radiational cloud warming effects.

These results show that, during the CEAREX drift overcast periods, using a radiative transfer model with detailed cloud effects, such as Curry and Ebert [1992], would be unlikely to improve greatly the prediction of F_L obtained by simply assuming it was equal to the T_{air} blackbody emission value. If T_{air} is not known, some type of radiative transfer model is required to determine F_L, which then becomes a good predictor for T_{air}. During clear skies, T_{air} alone is not a good predictor for F_L and radiative transfer models should provide better estimates of F_L, if upper-air information, including ice crystal quantities, can be obtained.

4. CONDUCTION AND ATMOSPHERIC TIME SCALES

Conduction of heat through ice or snow at the surface, F_c, is an important component of the surface energy balance in the central Arctic winter, particularly when the ice is thin or when atmospheric forcing is rapidly changing, as might occur when cloud conditions change. In this section, we present some simple expressions which demonstrate relationships between heat conduction at the surface, the thickness of an ice floe, and the time scales of atmospheric forcing. Although the results in this section can used to quantify F_c be for certain situations, the main purpose is to provide the reader with some conceptual ideas concerning heat conduction within ice floes which may be useful for developing numerical models or interpreting measurements.

One can assume that a steady-state situation exists, so that the conductive heat flux, F_c, is constant through an ice floe. The flux is driven by the temperature difference between the ocean, T_{sea} and the top of the floe, T_{sfc}.

$$F_c = (T_{sea} - T_{sfc})(k_{floe}/d_{floe}) \qquad (4)$$

where k_{floe} is the thermal conductivity of the floe and

d_{floe} is the thickness of the ice floe. If several snow or ice layers with different thermal properties are present, k_{floe}/d_{floe} represents the bulk conductivity for the entire floe, defined as the inverse of the sum of the resistances (d_n/k_n) of each layer n, where d_n and k_n are the thickness and conductivity for that layer.

The steady-state assumption is not always correct, even in a general qualitative sense. After a change in forcing conditions, the heat flux through the ice floe is not constant with depth. In this situation, the heat capacity of the snow/ice becomes an important parameter, in addition to the conductivity. A multi-layer time-dependent snow/ice model with inclusion of salinity effects can be used to determine the relationship between F_c and T_{sfc}. But, for purposes involving a single atmospheric forcing event or time scale, a much simpler expression for F_c is often adequate.

We define a depth scale, z_{ice}, which represents the depth within an ice flow which thermally interacts with the surface.

$$z_{ice} = 2\,(t\nu/\pi)^{1/2} \quad (5)$$

The scale depends on a time scale of atmospheric forcing, t, and the thermal diffusivity ν, of the ice floe.

When $t < t_{ice}$ where

$$t_{ice} = d_{floe}^2\,\pi/4\nu \quad (6)$$

then $z_{ice} < d_{floe}$ and the surface temperature is not greatly affected by ice thickness, d_{floe}. For this case, d_{floe} should be replaced with z_{ice}, and T_{sea} replaced with the temperature at that depth, T_{bot}, in (4). But when $t > t_{ice}$ the bottom temperature is "felt" by the surface and using d_{floe} is more accurate than using z_{ice} to determine F_c by (4).

If more than one time scale of atmospheric forcing is of interest, a multi-layer thermodynamic sea ice model such as Semtner [1976], Maykut [1982] or Ebert and Curry [1993] is required to determine T_{sfc}. The multi-layer models incorporate expressions for T_{sfc} based on a balance of heat fluxes at the ice/atmosphere interface. A time scale associated with the upper layer of a model, t_1, can be determined by substituting one-half of the model layer thickness for d_{floe} on the right side of (6). A multi-layer ice model will not correctly determine changes in surface temperature due to atmospheric forcing on time scales less than t_1. This shows how the z_{ice} vs time relationship can be used to determine the maximum upper layer thickness required for a particular application of a multi-layer sea ice model.

A related problem with multi-layer model predictions of T_{sfc} occurs after a rapid change in atmospheric forcing (e.g. a step increase in downward longwave radiation due to cloud formation). According to multi-layer ice models which assume that all the heat fluxes at the surface interface are in balance, such as Semtner [1976], the surface temperature can change by an unrealistically large amount in the time step immediately following the change in atmospheric forcing. The magnitude of the change predicted by the models depends on the thickness and conductivity specified for the upper layer in the ice model. But in nature, the heat capacity of the upper part of the top layer prevents an immediate jump in T_{sfc}, so that the time required to reach the temperature predicted by the model is different from the actual amount of time required. The time scale t_1 is a measure of the possible error in the time associated with changes in T_{sfc} due to rapid changes in atmospheric forcing.

We described above how z_{ice} can be used directly for certain modeling purposes. The expression for z_{ice} can also be used to provide a simple graphical representation of several concepts related to ice floe characteristics and time scales of atmospheric forcing phenomena (Figure 2). The thick solid lines in Figure 2 represent the value of z_{ice} for three surface types: ice, old snow and new snow.

Points above and to the left of the thick solid lines in Figure 2 represent "thick" ice or snow which has an internal heat flux which varies with depth. The surface bulk thermal conductivity is defined as k/h, where k is the effective conductivity of the floe above depth h, which equals either z_{ice} (thick ice) or d_{floe} (thin ice). The bulk conductivity for "thick" ice depends on the time scale of atmospheric forcing. The atmospheric forcing does not affect heat fluxes and ice growth at the ice/ocean interface. Points below and to the right of the thick solid lines represent "thin" ice which has a constant heat flux throughout. For thin ice, T_{bot} equals the ocean temperature and the effective thickness, h, equals the floe thickness, d_{floe}. The heat flux at the ice/ocean interface is the same as at the ice/atmosphere interface and therefore ice formation at the bottom of the ice is closely connected to atmospheric thermodynamic forcing.

In the regions to the right and above the dotted (sea ice) or dashed (old snow) lines, $k/h < 2\varepsilon\sigma(T_{rad}^3 + T_{sfc}^3)$, where the latter term approximates the change in upward radiation resulting from a change in T_{sfc}, which means that T_{sfc} is closer to T_{rad} than to T_{bot} (thick ice) or T_{sea} (thin ice), assuming no turbulent transport of heat. In this region, labelled "Radiation Dominates" in Figure 2, changes in radiation conditions are more important than

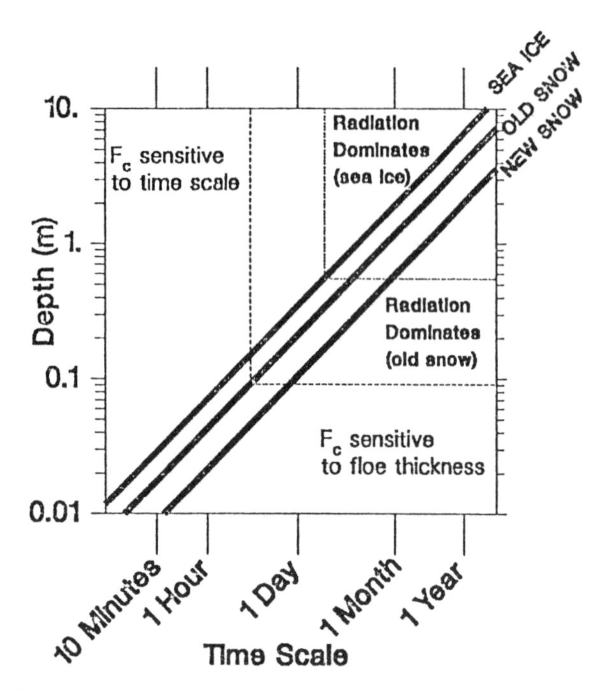

Fig. 2. The relation between depth within a floe (Y-axis) and time since a change in atmospheric forcing (X-axis) for solid sea ice (ν = 1.08 x 10^{-6}), old snow (ν = 0.4 x 10^{-6}), and light feathery new snow (ν = 0.1 x 10^{-6}) surfaces, represented by thick solid lines. Above and to the right of the dotted lines, radiation, F_L, dominates the surface temperature while below and to the left conduction, F_C, dominates, for a bare sea ice surface. The dashed line represents a similar boundary for an old snow surface. Radiation dominates everywhere for new snow. See text for further explanations.

changes in ice depth or atmospheric time scales for determining T_{sfc}. For new snow, the radiation dominant region is not shown because it covers virtually the entire plot area, i.e. practically any amount of new snow causes the surface temperature to be closer to T_{rad} than T_{bot}.

Time series of air temperature and ice/snow interfacial temperature, T_{ice}, (Figure 3) demonstrate the ice conduction concepts discussed above. Note that T_{ice} is measured below the surface and is separated from the surface by a well-insulated snow layer so that the coupling between T_{ice} and T_{air} is not as close as the T_{sfc} vs. T_{air} coupling discussed above. Following the cutoff of solar radiation in the fall, there was a general downward trend in T_{air} (and T_{ice}) for approximately six weeks before a constant average dark season temperature was reached in late December during the 1991-1992 Arctic Leads Experiment (LEADEX) in the Beaufort Sea. Six weeks is approximately the time scale for a thermal change to penetrate an ice floe with 2 m of ice and 0.20 m snow, typical LEADEX region values. (The combined snow and ice time scale is calculated by squaring the sum of the square roots of the respective time scales.) After six weeks, the ice changes from "thick" to "thin" and further cooling of the ice is limited by conduction of heat from the ocean. (The actual physical thickness of the ice hasn't changed; we are using the "thick" and "thin" concepts as discussed in this section.) This shows that one can ignore ice thickness when analyzing synoptic time scale temperatures and heat fluxes over Arctic multi-year ice. But for seasonal or climate time scales, variations in ice thickness must be considered. Factors other than conduction of heat through multi-year ice, such as cooling of the entire atmosphere [e.g. Overland and Guest, 1991], also affect the time scales associated with the fall cooling of the central Arctic. However, ice conduction is especially important early in the dark season when the floes are still relatively warm.

A time scale associated with a layer of snow or ice is determined by substituting the layer thickness for d_{floe} in (6). The time scale associated with 20 cm of old snow is about one day. The time lag between T_{air} and T_{ice}, as measured by the LEADEX buoys, was also about one day (Figures 3 and 4) indicating the relevance of this time scale to ice conduction and temperature changes within an ice floe in response to atmospheric forcing.

In a location such as the Weddell Sea, where d_{floe} values are around 0.5 m and little snow is present, the time scales associated with the bottom of the ice are in the same range as synoptic time scales (2 - 4 days); therefore,

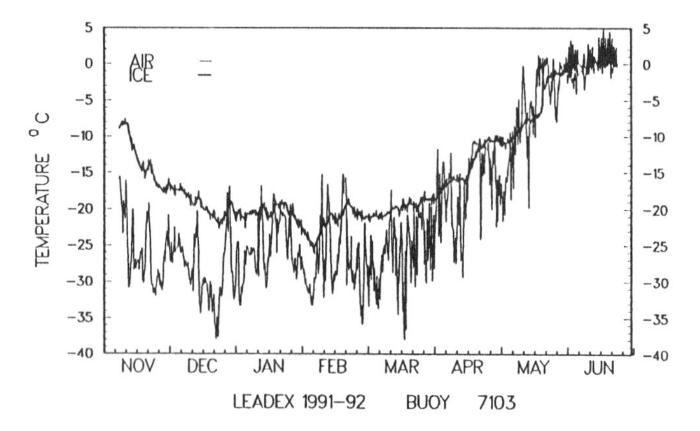

Fig 3. The air temperature, T_{air}, (thin) and ice/snow interfacial temperature, T_{ice}, (thick) from a buoy during the LEADEX project in the Beaufort Sea. The ice/snow interfacial temperatures were measured under 10 - 20 cm of snow and the air temperatures were 2.8 m above the surface.

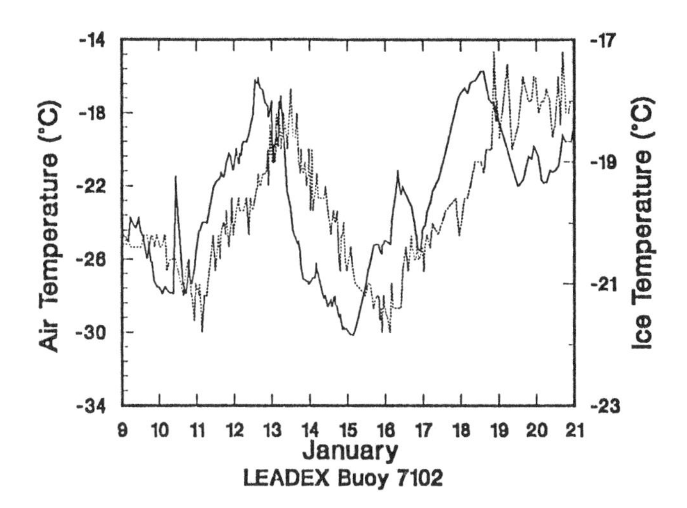

Fig. 4. The air temperature, T_{air}, (solid) and ice/snow interfacial temperature, T_{ice}, (dotted) from a buoy during the LEADEX project during a period when strong synoptic variations in air temperature were occurring. Note that the temperature scales are different. The variations in ice temperature on hourly time scales is measurement noise.

the ice is always "thin" and surface temperatures are strongly affected by variations in floe thickness. In addition, changes in ice/ocean interfacial heat fluxes and thermodynamic ice growth are affected by synoptic atmospheric forcing variations, unlike Arctic multi-year ice.

5. SUMMARY

We have shown some examples of how the ice/atmosphere interfacial temperature, T_{sfc}, and the near-surface air temperature, T_{air}, are affected by turbulent, radiative and conductive heat fluxes. Turbulence closely links T_{sfc} and T_{air}. The relation between T_{air} (and hence T_{sfc}) and downward longwave radiation, F_L, measured during the CEAREX drift, was so close during overcast conditions that T_{air} could have been used as a good (± 10 Wm^{-2}) predictor for F_L. An analytical expression describes how conduction of heat through the ice is related to snow/ice thermal diffusivities and atmospheric forcing time scales. Time scales associated with ice conduction are detectable from temperature observations obtained during LEADEX.

Acknowledgements. The support of the Naval Research Laboratory (Program Element 0601153N) and the Direct Funding Research Program at the Naval Postgraduate in Monterey California USA is gratefully acknowledged.

REFERENCES

Andreas, E. L., Estimating turbulent surface heat fluxes over polar, marine surfaces. *Preprints The Second Conference on Polar and Meteorology and Oceanography*, Am. Meteorol. Soc., Madison, WI, March 29-31, 65-68, 1988.

Belmont, A. D., Lower tropospheric inversions at ice island T-3, *J. Atmos. Terr. Phys.*, Special Supplement: Proceedings of the Polar Atmosphere Symposium, Part I, 215-284, 1957.

Boer, G. J., N. A. McFarlane, and M. Lazare, Greenhouse gas-induced climate change simulated with the CCC second-generation general circulation model, *J. Climate*, *5*, 1045-1077, 1992.

Carsey, F. D., The boundary layer height in air stress measurement, in *Sea Ice Processes and Models*, edited by R.S. Pritchard, pp. 443-451, University of Washington Press, Seattle, 1980.

Cao, H. X., J. F. B. Mitchell, and J. R. Lavery, Simulated diurnal range and variability of surface temperature in a global climate model for present and doubled CO_2 climates, *J. Climate*, *5*, 920-943, 1992.

Curry, J. A., 1983: On the formation of continental polar air, *J. Atmos. Sci.*, *40*, 2278-2292.

Curry, J. A., Interactions among turbulence, radiation and microphysics in Arctic stratus clouds. *J. Atmos. Sci.*, *43*, 90-106, 1986.

Curry, J. A., and E. E. Ebert, Annual cycle of radiation fluxes over the Arctic Ocean: sensitivity to cloud optical parameters, *J. Climate*, *5*, 1267-1280, 1992.

Curry, J. A., F. G. Meyer, L. F. Radke, C. A. Brock and E. E. Ebert, Occurrence and characteristics of lower tropospheric ice crystals in the Arctic, *International J. Clim.*, *10*, 749-764, 1990.

Ebert, E. E., and J. A. Curry, A intermediate one-dimensional thermodynamic sea ice model for investigating ice-atmosphere interactions, *J. Geophys. Res.*, *98*, 10085-10110, 1993.

Idso, S. B., and R. D. Jackson, Thermal radiation from the atmosphere, *J. Geophys. Res.*, *74*, 5397-5403, 1969.

Key, J., and M. Haefliger, Arctic ice surface temperature retrieval from AVHRR thermal channels. *J. Geophys. Res.*, *97*, 5885-5893, 1992.

Leavitt, E., Surface-based air stress measurements made during AIDJEX, in *Sea Ice Processes and Models*, edited by R.S. Pritchard, pp. 419-429, University of Washington Press, Seattle, 1980.

Lindsay, R. W., and D. A. Rothrock, Arctic sea ice surface temperature from AVHRR, *J. Climate*, *7*, 174-183, 1994.

Makshtas, A. P., The heat budget of Arctic ice in the winter, Arct. Antarc. Res. Inst., St. Petersburg, Russia, 1984. (English translation edited by E.L. Andreas, 1991. Available from Int. Glaciol. Soc., Cambridge, England)

Manabe, S., M. J. Spelman and R. J. Stouffer, Transient responses of a coupled ocean-atmosphere model to gradual changes of atmospheric CO_2, Part II: Seasonal response. *J. Climate*, *5*, 105-126, 1992.

Marshunova, M. S., Principal characteristics of the radiation balance of the underlying surface and of the atmosphere in the Arctic. *Trudy Arkt. Antarkt. Nauch. Issle. Inst., 229*, 1961. (Translated by the Rand Corp., Santa Monica, California, RM-5003-PR, 1966.)

Maykut, G. A., Large-scale heat exchange and ice production in the central Arctic, *J. Geophys. Res.*, *87*, 7971-7984, 1982.

Maykut, G. A., Surface heat and mass balance, in *The Geophysics of Sea Ice*, edited by N. Untersteiner, pp. 395-463, Plenum, New York, 1986.

Maykut, G. A., and P. E. Church, Radiation climate of Barrow, Alaska, 1962-66, *J. Appl. Met., 12*, 620-628, 1973.

Overland, J. E., A model of the atmospheric boundary layer over sea ice during winter, *Preprints The Second Conference on Polar and Meteorology and Oceanography*, Am. Meteorol. Soc., Madison, WI, March 29-31, 69-72, 1988.

Overland, J. E., and P. S. Guest, The Arctic snow and air temperature budget over sea ice during winter, *J. Geophys. Res.*, *96*, 4651-4662, 1991.

Parkinson, C. A., and W. M. Washington, A large-scale model of sea ice, *J. Geophys. Res., 84*, 311-337, 1979.

Semtner, A. J., Jr., A model for the thermodynamic growth of sea ice in numerical investigations of climate, *J. Phys. Ocean.*, *6*, 379-389, 1976.

Serreze, M. C., J. D. Kahl, and R. C. Schnell, Low-level temperature inversions of the Eurasian Arctic and comparisons with Soviet drifting station data, *J. Climate.*, *5*, 9411-9422, 1992.

Vowinckel, E., and S. Orvig, The climate of the north polar basin, *World Survey of Climatology, Vol. 14, Climates of Polar Regions,* edited by S. Orvig, Elsevier, 1970.

Wexler, H., Cooling in the lower atmosphere and the structure of polar continental air, *Mon. Weather Rev.*, *64*, 122-136, 1936.

P. S. Guest, K. L. Davidson, Meteorology Department, Naval Postgraduate School, Monterey, CA 93943-5114

Aircraft Measured Atmospheric Momentum, Heat and Radiation Fluxes over Arctic Sea Ice

Jörg Hartmann, Christoph Kottmeier*, Christian Wamser and Ernst Augstein*

Alfred-Wegener-Institut für Polar- und Meeresforschung, Bremerhaven, Germany
**Universität Bremen, Fachbereich Physik, Bremen, Germany*

The vertical turbulent momentum, sensible and latent heat fluxes and the surface radiation balance are derived from measurements of low level flights (<50 m height) with a highly instrumented aircraft over Fram Strait in September/October 1991. High resolution information on the sea ice cover is obtained with a digital line scan camera. It is found that the drag coefficient for neutral static stability at 10 m height can be composed of a skin drag ($c_{dns} = 1.1 \cdot 10^{-3}$), which coincides with the open water value, and a form drag which linearly increases with the mean ice area perpendicular to the surface wind vector per unit surface area. The ratio of the generally small sensible and latent heat fluxes (both $\leq 20\,\mathrm{Wm^{-2}}$) is close to unity for near neutral atmospheric stratification and no dependence of these fluxes on sea ice concentration can be detected, at least for the encountered ice concentrations larger than 50%. Measurements at about 40 m height are not sufficient to study cases with stable stratification since the flight level seems to be fully decoupled from the surface processes. In this autumn measurements 50% to 90% of the net energy flux at the surface is made up by the radiation balance. Therefore, radiative fluxes form important components in polar air-sea exchange processes. The long wave downward radiation can be parameterised using the $\epsilon\sigma T^4$ law with the near surface air temperature and the empirically determined values for the emissivity $\epsilon = 0.71$ and $\epsilon = 0.90$ for clear and cloudy skies, respectively. The standard deviations of our measurements from this parameterisation are $4.6\,\mathrm{Wm^{-2}}$ for clear and $8.6\,\mathrm{Wm^{-2}}$ for cloudy skies. These values fall into the range of the instrumental uncertainty.

1 INTRODUCTION

The dynamic, thermal and radiative properties of sea ice exert a major influence on the air-sea exchanges of momentum, sensible and latent heat as well as on the surface radiation balance in polar regions. Therefore, measured and computed vertical fluxes of radiation, heat and momentum across the ice covered ocean surface must be adequately related to the prevailing sea ice conditions. Although various authors have intensively investigated the air-sea interactions in the Antarctic and Arctic seas with the aid of different observational methods during the last two decades [*Andreas et al.,* 1984; *Walter, et al.,* 1984; *Anderson,* 1987; *Belitz et al.,* 1987; *Fairall and Markson,* 1987; *Hoeber and Gube-Lehnhardt,* 1987; *Kellner, et al.,* 1987; *Guest and Davidson,* 1987; *Martinson and Wamser,* 1990] considerable uncertainties still exist with respect to the roughness as well as to the atmospheric heat and moisture transfer coefficients of ice covered ocean surfaces.

The Polar Oceans and Their Role in Shaping the Global Environment
Geophysical Monograph 85

This deficiency may be primarily attributed to the limited validity of surface based single point measurements which provide the majority of the data and the lack of quantitative information on the relevant sea ice characteristics. Consequently the latter were only qualitatively described e.g. in the overview on stress measurements over sea ice by *Overland* [1985] and in the comprehensive study of Arctic sea ice roughness by *Guest and Davidson* [1991].

To overcome both the weakness of single point measurements and of poor sea ice information we have carried out a field programme with a highly instrumented research aircraft over Fram Strait (Figure 1) in autumn of 1991. Besides the basic scientific instruments the Dornier 228 turbo-prop aircraft (Figure 2) was equipped with a gust probe, short and long wave upward and downward looking radiation sensors, a radiation thermometer and — particularly important — a digital line scan camera for continuous high resolution recordings of the Earth's surface structure along the flight track. Flights were carried out at heights betweeen 30 and 50 m to measure the surface radiation balance and the vertical turbulent fluxes of heat, water vapour and momentum near the sea surface. On days with little cloudiness one leg was flown at 3000 m to get a broad view on the ice cover. The line scan camera data enabled

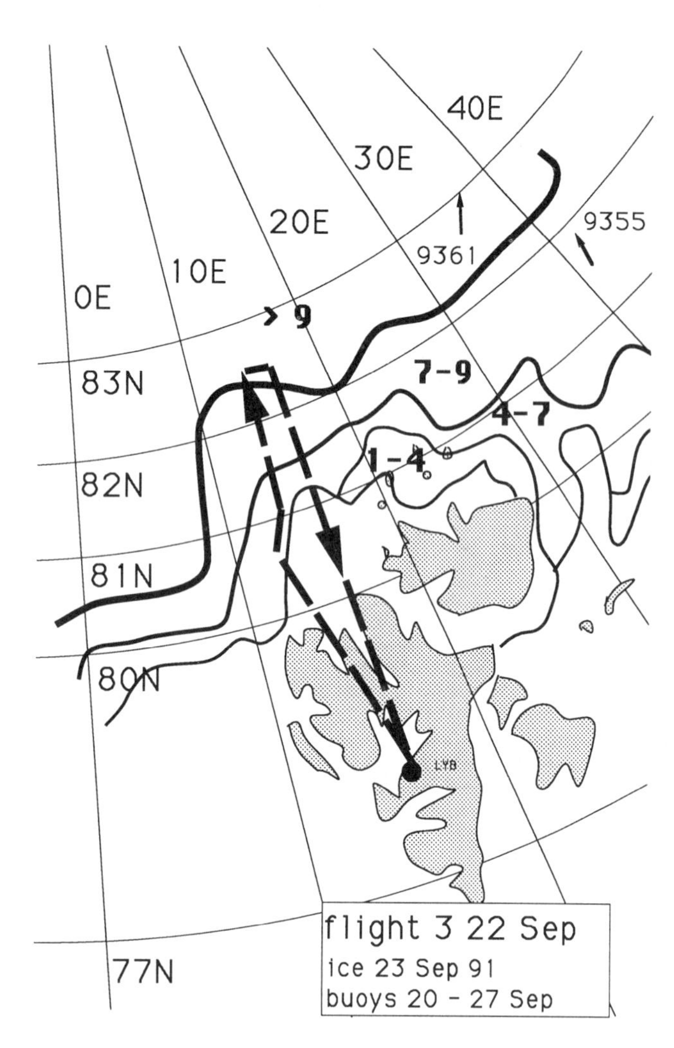

Figure 1: Area of the experiment and a typical flight path as thick broken line. Flights were carried out in the region between 0°E, 79°N and 20°E, 82°N. Solid lines mark the ice concentrations in tenth, denoted by the figures 1-4, 4-7, 7-9 and > 9, reproduced from the NOAA weekly ice charts. The arrows in the upper right corner indicate the initial and final positions of two automatic ARGOS buoys (No. 9361 and 9355) during the period from 20 to 27 Sept 1991. The northbound leg was flown at 3200 m, the southbound leg between 30 m and 150 m.

us to quantitatively determine different sea ice categories which could be assigned to specific flight sections. Examples from a northbound flight on 22 September are shown in Figure 3.

The mean vertical structure of the lower troposphere was measured during several ascents and descents of the aircraft during each mission. Representative examples for nearly neutral and for stable vertical density distributions in the lower troposphere are reproduced on Figure 4.

The Radiation and Eddy Flux Experiment 1991 (REFLEX I) took place from 16 September to 17 October 1991 in the marginal ice zone of Fram Strait. The following turbulent flux considerations are based on flight legs with a nearly neutral vertical density distribution within the lowest 40 m of the atmosphere. A broader data set will be used for the discussion of the surface radiation balance which does not depend directly on the static stability of the air flow.

The primary aim of this paper is to present the observed atmospheric momentum, heat, moisture and radiation fluxes for different sea ice conditions at the end of the Arctic summer. General aspects of the aircraft missions and instrumental details will be addressed only briefly in this framework since both are outlined comprehensively in a technical report by *Hartmann et al.* [1992].

2 THE REFLEX I CAMPAIGN

During REFLEX I the turbulent fluxes of momentum, heat and moisture were measured near the surface and the surface radiation balance was determined for a variety of surface and atmospheric conditions in a total of 17 aircraft missions.

The sea ice concentration, C_{ice}, and the floe size distribution were obtained with a downward looking digital line scan camera. The surface temperature distribution was quantitatively determined by a radiation thermometer. Simultaneously the turbulent fluxes were measured with a turbulence probe under the aircraft wing and the broadband up- and downwelling radiation fluxes were obtained from upward and downward looking pyranometers and pyrgeometers. The boundary layer vertical structure (mixed layer thickness z_i, wind vector $\vec{v}(z)$ and potential temperature $\theta(z)$) was derived from aircraft ascents and descents.

Of the 17 flights 9 cases had off-ice and 7 cases on-ice surface winds. One flight was carried out when the air flow was turning from an on-ice to an off-ice direction. Stratiform clouds with bases below 1000 m were present over the sea ice zone during 9 missions while larger cloud free areas were encountered by 5 flights with off-ice and 2 flights with on-ice wind directions.

From the aircraft observations the following quantities and parameters have been derived:

- surface radiation balances for short wave and long wave spectral ranges and short wave surface albedo
- number density distribution of floe sizes
- floe edge density
- concentrations of old and new ice
- vertical turbulent fluxes of momentum, sensible and latent heat
- drag coefficient, sensible and latent heat transfer coefficients
- bulk Richardson number.

Our subsequent discussion with regard to turbulence measurements will be restricted to observations over uni-

Figure 2: The research aircraft *Polar 2* of the Alfred Wegener Institute. The cylinder under the right wing houses the gust probe system.

form sea ice of at least 10 km length and neutral atmospheric stratification. These conditions are fulfilled by 16 flight legs on 4 days, which are identified in Table 1 together with 5 flight legs which fullfill the sea ice but not the static stability criterion.

3 DATA PROCESSING AND ANALYSIS

Aircraft measurements have generally to be processed by procedures which account for the movements of the platform besides the standard instrumental corrections and calibrations. Since the latter may be considered as routine methods we will subsequently refer only to some of the former ones.

3.1 Line Scan Camera

To obtain accurate information on the ice concentration and floe size structure we use a vertically downward looking line scan camera (LSC) to record digital images in the visible spectral range. The camera has an opening angle of 56° and records every 30 milliseconds one line of 1024 pixels orthogonal to the flight direction. Each pixel has an eight-bit value which indicates the intensity of the reflected sunlight in the range from 400 to 900 nm. The spatial resolution of the system depends on the flight velocity and on the height of the aircraft. At a height of 1000 m and a speed of 70 ms^{-1} ice floes of about 2 m^2 can be resolved.

Analysis of the LSC*–data.* The surface characteristics of ice, nilas and open water are identified by an interactive dynamic threshold method. The sea ice concentration C_{ice} is determined by the percentage of pixels recognised as ice. Characteristic floe sizes are found by calculating the averaged size distribution $\overline{g}(p)$ of chords with which the scan lines cut the ice floes. Here, p is the chord length in number of pixels, and $g(p)$ (dimension $pixel^{-1}$) the number of chords of size p in one scan line. The distribution $g(p)$ is averaged over 1024 consecutive scans. If we assume the floes are nearly circular the number density distribution

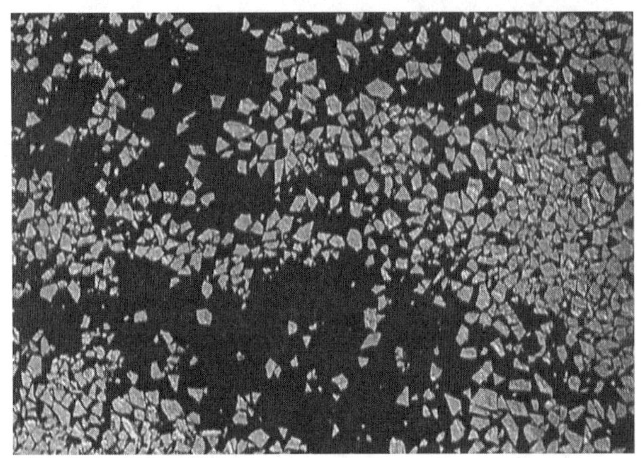

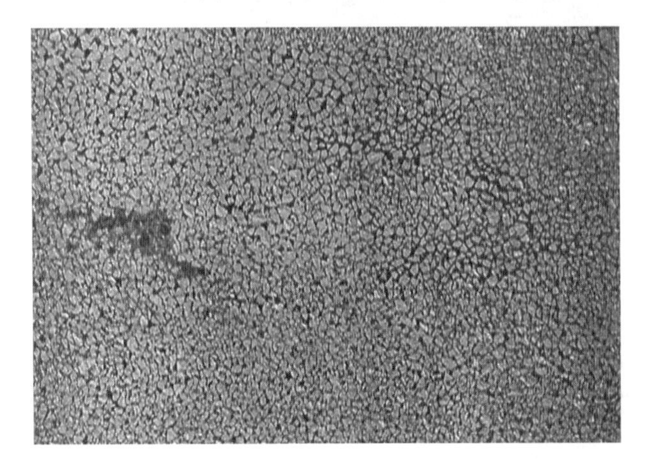

Figure 3: Line Scan Camera images of sea ice structure during an on-ice flight at 3000 m height. The frames show an area of 3 by 2 km, distances are counted from the ice edge.

- **a**: at 60 km: compact ice floes with diameters up to 250 m.
- **b**: at 25 km: scattered ice floes with diameters between 30 and 60 m.
- **c**: at 3 km: compressed ice floes with a typical diameter of 20 m with a patch of flooded and refrozen floes.

$f(d)$ (the number of floes of size d (in m) per unit area, $d = px$) is given by

$$f(d) = \frac{1}{p}\,\frac{y}{x}\,\bar{g}(p)\,\frac{1}{A_L}, \qquad (1)$$

where x and y are the respective lengths that one pixel represents in the direction parallel and orthogonal to the flight path, and A_L is the area in m^2 covered by one scan line. The characteristic floe size is then defined as the value of d where $d^2 f(d)$, the area density distribution, has its maximum. Thus, it corresponds to floes with the highest relative area coverage.

Floe edge density. The momentum exchange between air and sea ice depends on the surface roughness which may be expressed over sea ice as a sum of the skin and the form drag. The latter is assumed to be proportional to the length of the ice edges perpendicular to the wind direction. This quantity can be represented by the floe edge density r_e, i.e. the mean length of ice edges orthogonal to the mean wind vector per unit area. Values of r_e can been derived from line scan camera data by

$$r_e = \frac{n}{N}\,y\,\frac{1}{A_P} \qquad (2)$$

with n = the number of ice-pixels in each scan line with a water-pixel as a neighbour in flight direction, N is the total number of pixels in the scan line (1024) and A_P the area represented by one pixel.

The floe edge density and characteristic floe sizes as a function of distance from the ice edge are shown for a northbound flight on 22 September (cf. Figure 3) in Figure 5.

3.2 *Radiation Measurements*

The radiometer mounting supports are thermally insulated from the fuselage and positioned to keep aircraft shading effects to a minimum.

Before applying a correction for the aircraft movement, all radiation data R are corrected for the instrumental response time by

$$R_{\text{corr}} = R + \tau \frac{\partial R}{\partial t} \qquad (3)$$

with a response time $\tau = 1.2$s.

The upper pyranometer measures a radiation flux

$$R_S\!\downarrow = D + I\frac{\cos\beta}{\cos}, \qquad (4)$$

which is composed of the diffuse radiation D and the direct radiation I. In (4) is the angle between the local earth zenith and the sun, and β is the angle between the instrument zenith and the sun. Differentiation of (4) with respect to $\alpha = \frac{\cos\beta}{\cos}$ yields

$$I = \frac{\partial R_S\!\downarrow}{\partial \alpha}. \qquad (5)$$

For a section with horizontal flight conditions and when

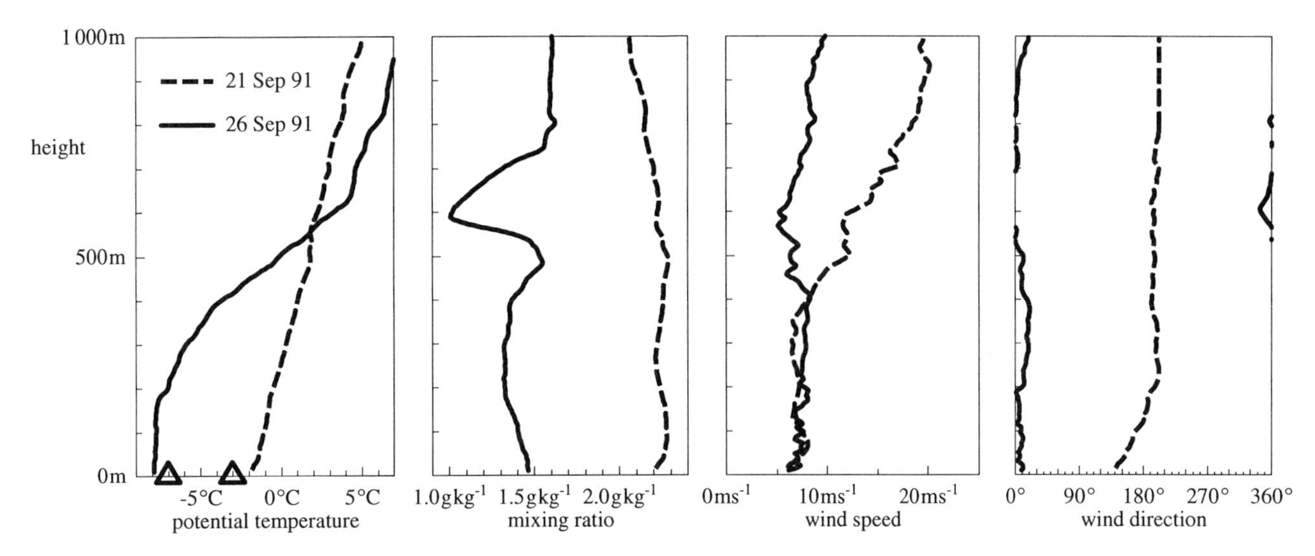

Figure 4: Vertical structure of the lowest 1000 m of the atmosphere representative of neutral or slightly unstable (full lines) and stable (broken lines) vertical density distributions obtained from aircraft ascents and descents. The triangles in the leftmost graph represent the surface (potential) temperature as measured by the radiation thermometer.

I and can be assumed constant, we exploit the effects of the small variations of the aircraft's pitch and roll angle on cos β, to derive the direct and diffuse radiation as the slope and ordinate intersection, respectively, of a plot of $R_S\downarrow$ versus α (Figure 6). The global radiation results as

$$G = D + I. \tag{6}$$

The average of $R_S\downarrow$ does not coincide with the sum of I and D if the aircraft is not parallel to the local earth surface.

A first order check of instrumental malfunctioning was carried out with the aid of the ratio of upward ($F_S\uparrow$) and downward ($F_S\downarrow$) shortwave fluxes, because the albedo $A = F_S\uparrow/F_S\downarrow$ derived from two independent sensors is rather sensitive to observational errors. The calibration uncertainty of the pyranometers was established through various recalibration procedures to be less than $\pm 5\mathrm{Wm}^{-2}$.

The downward looking pyrgeometer measurements can be compared to those of the radiation thermometer. Since the latter has a very small opening angle, area representative surface temperatures are determined through the following procedure. We first derive mean open water ($\overline{T_w}$) and mean ice surface ($\overline{T_i}$) temperatures from the radiation thermometer and then calculate total area averages

$$\overline{T_0} = C_{\text{ice}}\overline{T_i} + (1 - C_{\text{ice}})\overline{T_w} \tag{7}$$

taking the ice concentration C_{ice} from the line scan data. The mean black body surface radiation

$$F_0 = \sigma\overline{T_0}^4 \tag{8}$$

with σ = Stefan Boltzmann constant, and the directly measured surface radiation values differ by 2%. If we refer this difference totally to the pyrgeometer we find uncertainties of the longwave upward radiation flux of $\pm 5\,\mathrm{Wm}^{-2}$.

3.3 *Turbulence Measurements*

The turbulence sensors are mounted in a pod ("Meteopod") under the right wing of *Polar 2*. The flow sensor of the Meteopod is a de-icable five-hole-probe, mounted roughly 1.5 m in front of the leading edge of the wing. The 5-hole-probe measures angle-of-attack, angle-of-sideslip, airspeed and static pressure. Positioning and attitude information is provided by a fuselage-bound inertial navigation system (INS) and an additional *Attitude and Heading Reference System* (AHRS) inside the Meteopod. Humidity fluctuations are measured by the aircraft-version of the A.I.R. Lyman-alpha instrument and a dew-point mirror. A Vaisala-humicap is used for absolute reference. A very fast open wire Pt100 in a reverse-flow housing measures the temperature fluctuations. The Meteopod system is described in detail by *Vörsmann et al.* [1989].

4 OBSERVATIONAL RESULTS

4.1 *Turbulent Vertical Fluxes*

The main objective of this study is to detect the influence of different sea ice conditions on the air-sea momentum and energy exchanges. Therefore, we have to restrict our considerations to measurements which can be clearly related to reasonably defined sea ice characteristics. The requirements for this purpose are

- the flight level must be below 50 m height

TABLE 1. Low-level flights used for turbulence measurements. h is the mean altitude of the aircraft, l the length of the run, z_i the depth of the mixed layer, defined as the height range over which $\partial\theta/\partial z < 0.001\mathrm{Km}^{-1}$, U the wind speed at altitude h, $|\nabla_h \vec{U}|$ the absolute value of the horizontal gradient of the wind speed and Ri_b the bulk-Richardson number. The symbols • and ∘ denote near neutral and stable conditions, respectively.

date	h m	l km	z_i m	U m/s	$\lvert\nabla_h \vec{U}\rvert$ $10^4\mathrm{s}^{-1}$	Ri_b
∘ 22 Sep	38	13	–	4.9	0.28	0.349
∘ 23 Sep	48	29	–	4.5	0.50	0.572
∘ 23 Sep	48	10	–	3.3	0.45	0.790
∘ 23 Sep	47	11	–	3.2	0.15	0.964
• 26 Sep	46	10	220	6.8	0.08	-0.046
• 26 Sep	44	18	190	3.2	0.96	-0.272
• 26 Sep	32	15	180	7.7	0.51	-0.022
• 26 Sep	34	11	200	8.1	1.05	-0.031
• 27 Sep	42	19	600	6.2	0.46	-0.016
• 27 Sep	38	10	200	6.9	0.55	-0.034
• 27 Sep	39	29	200	6.7	0.15	-0.017
• 28 Sep	42	19	170	9.5	0.08	0.004
• 28 Sep	44	12	170	9.2	0.77	0.010
• 28 Sep	47	11	140	8.7	0.28	-0.000
• 28 Sep	40	15	150	9.9	0.54	-0.010
• 28 Sep	38	16	150	9.5	0.14	-0.011
• 28 Sep	40	33	150	9.6	0.15	-0.012
• 28 Sep	47	16	180	11.7	0.14	-0.111
• 29 Sep	39	14	280	13.1	0.62	-0.002
• 29 Sep	32	41	600	10.6	0.48	-0.059
∘ 5 Oct	45	36	–	4.2	0.23	0.336

- similar sea ice characteristics must exist for at least 10 km intervals along the track line
- near neutral stratification of the air column must exist below 50 m height

As we will demonstrate by some observations, measurements in situations with a distinctly stable vertical density distribution within the layer below flight level are not valid to investigate surface exchange processes.

The statistical representativeness of covariances derived from space or time series depends on the averaging length l, the measurement height z and the atmospheric stratification. *Wyngaard* [1973] showed that a for covariance statistics $(\overline{u'w'})$ under neutral stratification the accuracy a is

$$a = \sqrt{\frac{20z}{l}}. \tag{9}$$

For our conditions with l ≈10 km to 40 km and 30 m< a <50 m a ranges between 0.15 and 0.3. Thus, most of the scatter of the data in Figure 7 could be due to the statistical uncertainty of the flux measurements.

This consideration is only valid for horizontally homogeneous lower boundary and large scale flow conditions. The flight sections listed in Tabel 1 are carefully selected to cover only portions of the low-level flights where the ice concentration, the edge density and surface temperature did not show marked changes over the length of the run. The large scale flow conditions are verified from aircraft wind vector and air temperature measuremnts. The horizontal gradients of the wind speed are listed in Table 1 for each run which amount on average to 0.4 m/s per 100 km. The horizontal gradients of air and surface temperature were derived as 0.2 and 0.6 K per 100 km, respectively.

Momentum fluxes. Under the aforementioned restrictions there are 16 cases left with neutral or unstable and

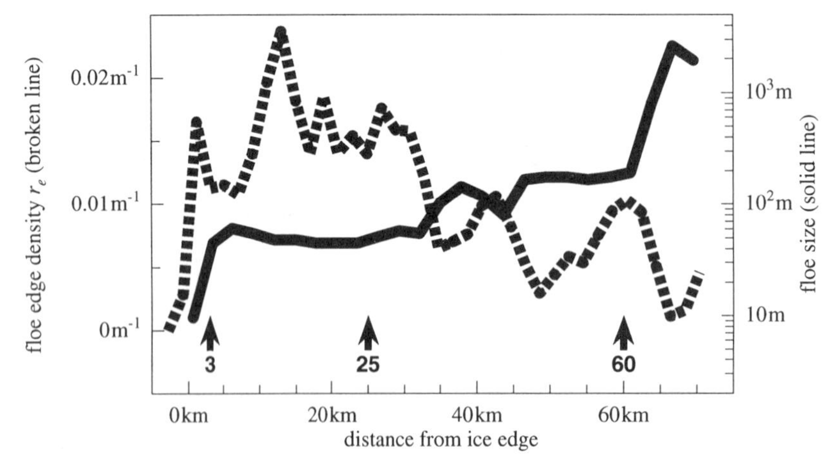

Figure 5: Characteristic floe sizes and mean length of floe edges as a function of distance from the ice edge for the northbound flight on 22 September 1991. Arrows indicate the positions of the images of Figure 3.

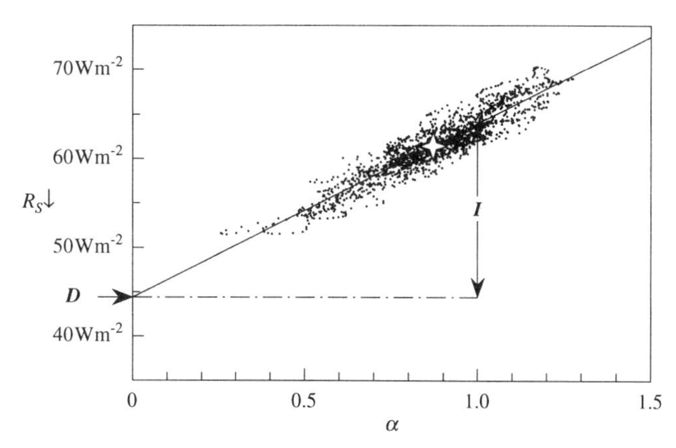

Figure 6: Short wave radiation flux measured $R_S\downarrow$ by the upward looking pyranometer versus $\alpha = \frac{\cos\beta}{\cos}$ for a run under cloud free conditions. The solid straight line is a least-squares fit through the data. The direct and diffuse radiations are indicated by the symbols I and D, respectively. The star marks the average of all data points.

4 with stable density distributions as shown for the drag coefficient at 10 m height, c_{d10}, in Figure 7 a by dots or open circles, respectively.

The static stability of the atmospheric surface layer is represented by the bulk Richardson number Ri_b defined as

$$\mathrm{Ri}_b = \frac{g}{\overline{\theta}} \frac{z(\overline{\theta_z} - \overline{\theta_0})}{\overline{U_z}^2} \tag{10}$$

with g the acceleration of gravity, $\overline{\theta}$ the potential air temperature, $\overline{U_z}$ the wind speed at height z, and $\overline{\theta_0}$ the potential surface temperature. The overbars indicate area averages along flight sections.

First the drag coefficient c_{dz} is determined for the level z with the aid of the equation

$$\overline{\tau_z} = \overline{\rho} c_{dz} \overline{U}_z^2 \tag{11}$$

with $\overline{\tau_z}$ = mean downward flux of horizontal momentum, derived from the gust probe measurements via

$$\overline{\tau_z} = \overline{\rho}\sqrt{\overline{u'w'}^2 + \overline{v'w'}^2} \tag{12}$$

$\overline{\rho}$ is the air density.

Then for general comparison c_{dz} is reduced to the 10 m level (c_{d10}) with aid of a one-dimensional numerical model. The model [*Raasch,* 1988] uses Monin-Obukhov similarity in the surface layer, and further upwards eddy diffusivities calculated by a Prandtl-Kolmogorov approach as a function of turbulent kinetic energy. After initialisation with steady state wind profiles and neutral stratification up to a height of 200 m and using the measured temperature profile above, the surface temperature is adapted

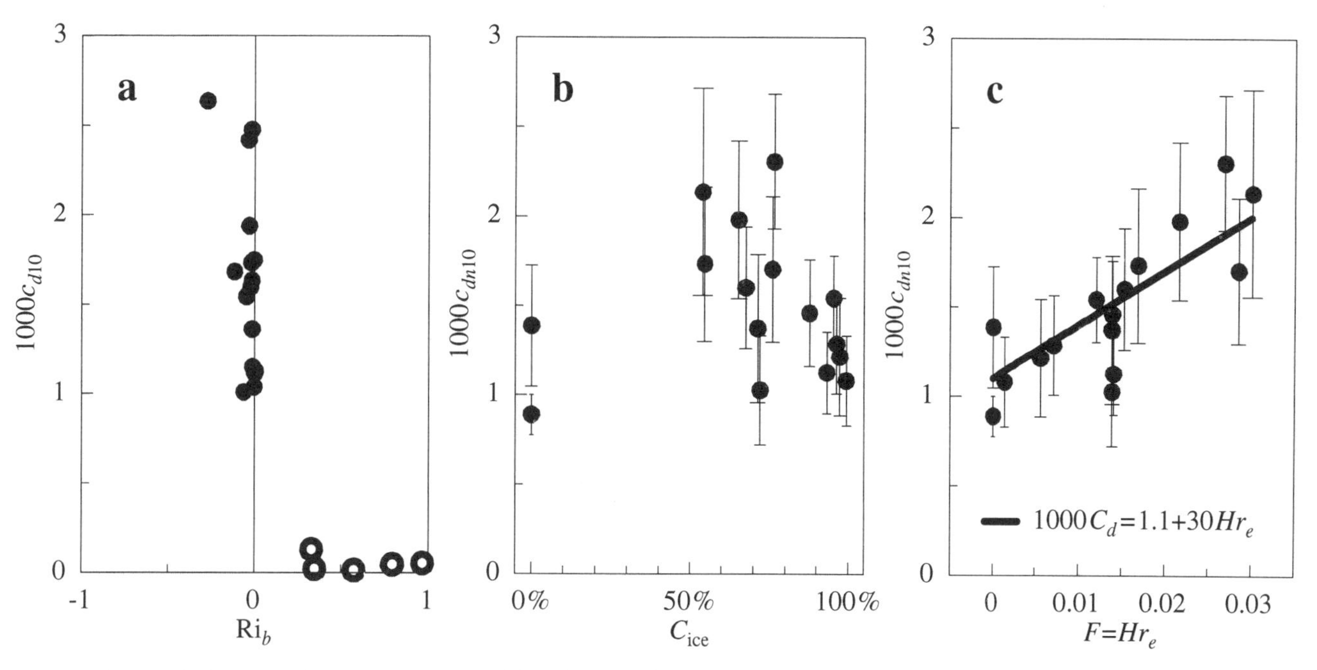

Figure 7: Drag coefficients referenced to 10 m height. The values in figures b and c are corrected for stability. The vertical bars in figures b and c represent the statistical uncertainty due to the limited sampling length as specified by equation (9) in the text.

a: c_{d10} versus bulk Richardson number Ri_b.

b: c_{dn10} as a function of ice concentration.

c: c_{dn10} as a function of the floe form parameter $F = Hr_e$.

to the measured value, and the model is run for roughly two hours to achieve nearly stationary conditions. Then the vertical profiles of temperature and wind are found to match the observed profiles reasonably well and the turbulent fluxes and turbulent kinetic energy at flight level agree with the measurements. We therefore use the model ratio of c_{dz} to c_{d10} to reduce the drag coefficient derived from observations at flight level to 10 m height.

For the slightly unstable density distributions the drag coefficients have been reduced to neutral conditions with the aid of the bulk Richardson number and the well known profile relationships suggested e.g. by *Brutsaert* [1982] to obtain c_{dn10}. These latter values are displayed in Figures 7 b and 7 c. The nearly zero c_{d10}-values (open circles in Figure 7 a) for statically stable conditions do not necessarily indicate zero momentum fluxes at the air-ice interface, but rather reflect the fact that under these circumstances the flight level (30 m to 50 m height) is more or less decoupled from the surface processes. This view is also supported by the sensible and latent heat fluxes in Figure 8.

Sea ice concentration is one of the routinely observed quantities in polar oceanic regions and it can also be satisfactorily estimated with the aid of actual sea ice models. Therefore, it is worthwhile to explore to which extent C_{ice} can be used for simplified descriptions of the air-sea exchange processes in the presence of sea ice. Earlier attempts [*Andreas et al.,* 1984; *Guest and Davidson,* 1987] provide hints that the drag coefficient c_{dn10} increases with growing ice concentration. When the latter exeeds about 80% c_{dn10} tends to decrease again. Our measurements (Figure 7 b) neither support nor contradict such a result. Although the lack of data for ice concentrations less than 50% reduces the validity of our conclusions, the large scatter of the data which generally governs the findings of nearly all other published observations support our doubts in the usefulness of C_{ice} as a parameter for sea ice roughness. Since we are not able to establish any useful relationship between the form parameter F defined below and the ice concentration even on the basis of the full REFLEX I data set, we are convinced that ice concentration alone is not an appropriate indicator of the hydrodynamic roughness of ice covered ocean surfaces.

A more physical approach has been proposed by *Hanssen-Bauer and Gjessing* [1988], who similar to *Arya* [1973] and *Brown* [1981] distinguish between the skin and the form drag of ice floes. But since the theoretical functions of the first mentioned authors do not fit our observations satisfactorily we modify their concept on the basis of the assumptions that the skin drag of open water and of sea ice surfaces coincides and that the form drag is proportional to the dynamic pressure which is exerted by the wind vector on the ice floes. We consequently propose for a drag coefficient under neutral static stability conditions

$$c_{dn} = c_{dns} + a\, r_e\, H \tag{13}$$

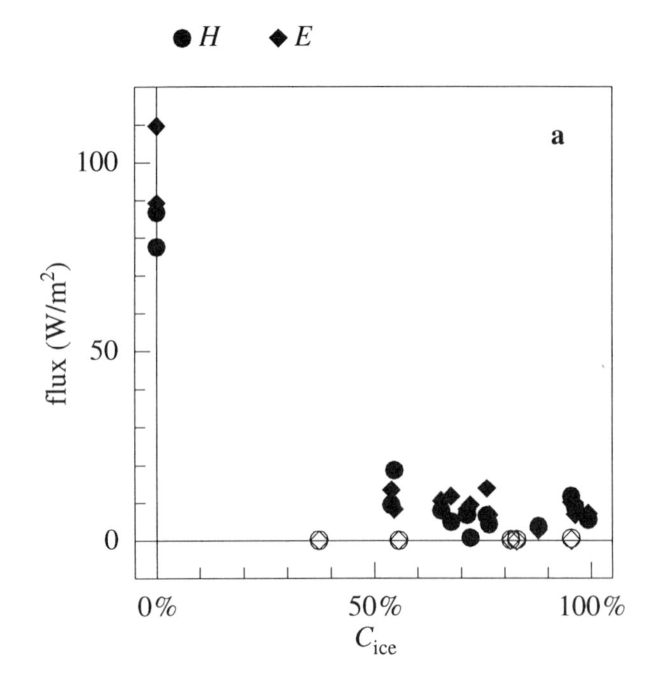

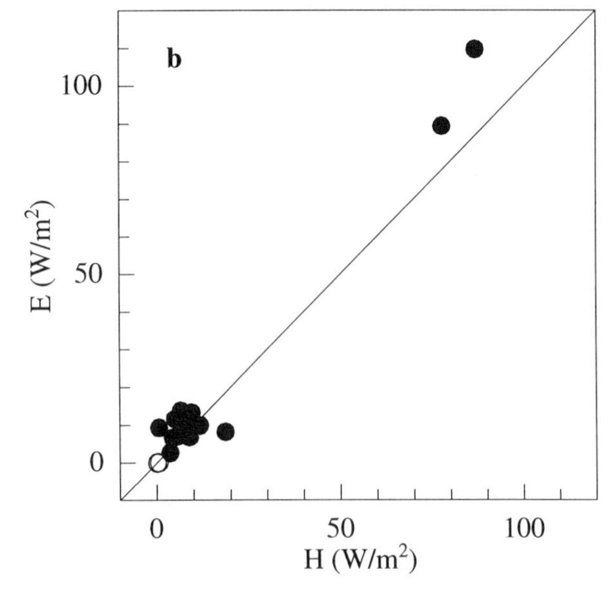

Figure 8:

a: Turbulent vertical sensible (H) and latent (E) heat fluxes versus ice concentration.

b: Turbulent latent versus sensible heat flux indicating a near unity Bowen ratio.

with the skin drag coefficient c_{dns} being the value for open water under neutral stability conditions. The floe edge density r_e represents the length of the ice floe edges perpendicular to the wind vector per unit horizontal area and H is the effective free board height of the floes which may also account for the effect of ridges. The dimensionless coefficient $F = r_e H$ parameterises the form resistance of the ice floes on the wind force. The coefficient $0 < a < 1$ contains the influence of all other aerodynamic effects of the ice cover.

The observed relation between the drag coefficient and the floe form parameter F is approximated in Figure 7 c by a straight line which is characterised by $c_{dns} = 1.1 \cdot 10^{-3}$ and $a = 30 \cdot 10^{-3}$. Since the mean freeboard height could not be directly measured during this campaign we have assumed H to increase linearly from 20 cm at the ice edge to 80 cm at 100 km inside the interior pack ice and to stay constant further into the ice belt. Similar freeboard heights have been reported by *Anderson* [1987] and *Guest and Davidson* [1987]. Even with this crude approximation of H and without accounting for specific ice deformations or for wind shading effects, the straight line in Figure 7 c fits the observations quite satisfactorily.

The drag coefficients derived from our measurements are somewhat lower than those reported e.g. by *Fairall and Markson* [1987] from aircraft measurements and by *Guest and Davidson* [1987] and *Anderson* [1987] from ship measurements for high ice concentrations. Since we found a good agreement of wind and flux measurements in comparison flights with other aircraft we have no reason to assume that systematic instrumental errors affect our results.

Sensible and latent heat fluxes. Atmospheric turbulent vertical sensible and latent heat fluxes over sea ice have been determined by several authors for Arctic [*Andreas,* 1980; *Maykut,* 1978; *Thorpe, et al.,* 1973; *Kellner, et al.,* 1987] and for Antarctic [*Andreas et al.,* 1984; *Andreas and Makshtas,* 1985; *Allison et al.,* 1982; *Gube-Lenhardt and Hoeber,* 1986] sea ice areas. Mainly due to lack of information it is frequently assumed that the transfer coefficients for sensible and latent heat c_h and c_e, respectively are identical and that they do not significantly depend on ice concentration. These general results are confirmed by our measurements during REFLEX I. In this campaign at the end of summer the sensible and the latent heat fluxes at about 40 m height over sea ice were too weak to study their dependency on ice concentration (at least for $C_{\text{ice}} > 50\%$), or on the flow form parameter F. The heat and moisture flux transfer coefficients at the mean flight level are found to be $c_h = 0.91 \cdot 10^{-3} \pm 0.31 \cdot 10^{-3}$ and $c_e = 0.73 \cdot 10^{-3} \pm 0.35 \cdot 10^{-3}$ with a considerable scatter. Similar to findings of *Andreas and Makhstas* [1985] over the Antarctic ice belt in early spring our measurements over Fram Strait in early autumn (Figure 8 b) are characterised by a mean Bowen ratio of about 1 over sea ice as well as over open water near the ice edge during off-ice flow conditions.

When the surface layer is stably stratified the measured fluxes are not significantly different from zero, and consequently the transfer coefficients at flight level are negligibly small. In these situations the fluxes at the flight level are obviously entirely decoupled from the surface processes and they cannot be reasonably related to the surface values.

4.2 *Surface Radiation Balance*

The REFLEX I measurements provide the short and long wave downward and upward radiative fluxes near the sea surface in the marginal sea ice zone of Fram Strait for different atmospheric and ice conditions. The radiative fluxes contribute significantly to the surface energy balance and also affect the turbulent exchange. The observations enable us to determine the short wave albedo and the long wave emission for different sea ice concentrations and to study cloud and clear sky effects on the downward fluxes. In the framework of the full data set, we will subsequently consider 4 cases in more detail which represent conditions with

- □ clear sky and cold air
- ○ clear sky and warm air
- ■ sky overcast with stratus clouds and cold air
- ● sky overcast with stratus clouds and warm air.

The same symbols apply to Figure 9. For all other cases only cloudy and cloud free conditions are distinguished.

Short wave radiation. The direct downward short wave radiation (Figure 9 a and b) naturally depends on the elevation angle of the sun. If this geometric effect is eliminated in regarding the extinction $\kappa = \cos() \ln(F_S{\downarrow}_0 / F_S{\downarrow})$, shown in Table 2, for cloudy and clear skies, significant differences become obvious: while cloudy situations are governed by a large scatter due to differences in the optical thickness of the clouds and varying amounts of cloud cover, the clear sky values are rather constant. The standard deviations correspond to variations of the normalised radiation fluxes of 50% for cloudy and 11% for clear sky conditions.

TABLE 2. Extinction coefficient $\kappa = \cos() \ln(F_S{\downarrow}_0 / F_S{\downarrow})$ for cloudy and clear skies. $F_S{\downarrow}_0$ and $F_S{\downarrow}$ are the downward shortwave radiative fluxes at the outer boundary of the atmosphere and at the surface, respectively. The average $\bar{\kappa}$ and standard deviations σ_κ are calculated for all low-level runs with either 8/8 or 0/8 cloud cover.

	$\bar{\kappa}$	σ_κ
cloudy(8/8)	0.161	0.060
clear (0/8)	0.066	0.013

The albedo values, shown in Figure 10, reasonably reflect the generally expected correlation of short wave reflection and ice concentration but both cases with a clear sky indicate extremely high albedo values between 25% and 35% also for open water. We attribute this behaviour to the total reflection of the direct solar radiation due to the low sun elevation angles ($< 5°$). *Payne* [1972] and *Katsaros et al.* [1985] have observed similar high values of ocean albedo for sun elevation angles lower than 15°.

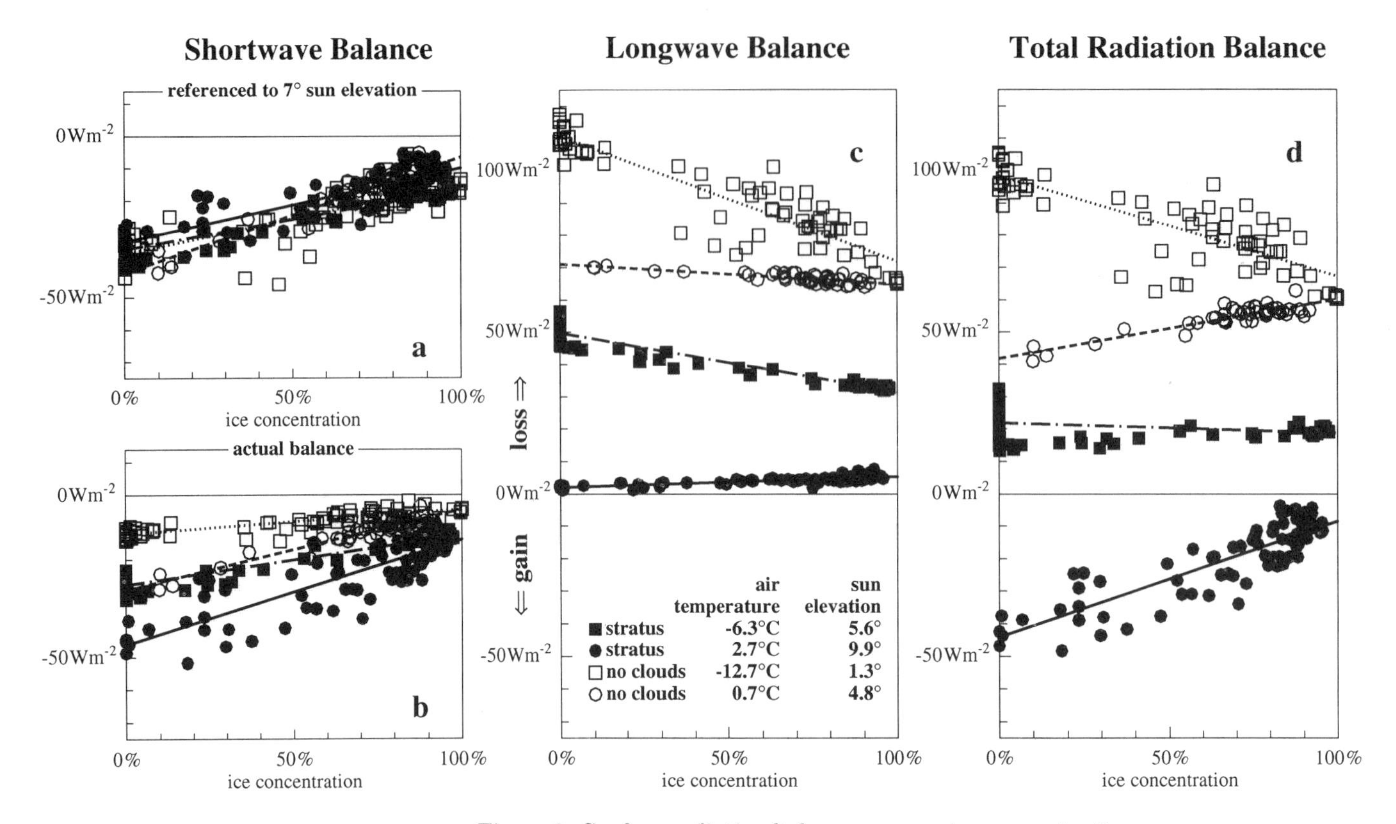

Figure 9: Surface radiation balances versus ice concentration.
a: shortwave balance referenced to a sun elevation angle of 7°
b: shortwave balance as observed
c: longwave balance
d: total radiation balance

In addition to the total reflection, systematic instrumental uncertainties may have contributed to the high albedo values as well. During both runs with overcast skies when no total reflection occurs the albedo values drop to less than 10% over open water. The albedo for clear skies is significantly larger (≈ 20%) than for the diffuse radiation in cloudy situations. This difference together with the increased extinction in the presence of clouds leads to rather uniform short wave surface balances for all four cases (Figure 9 a) when the effect of different sun elevation angles is

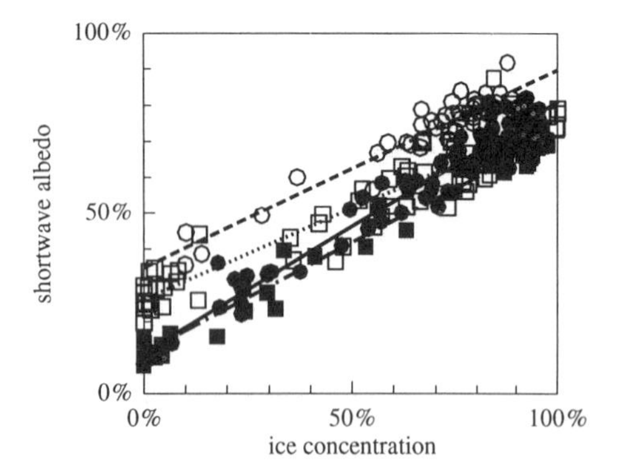

Figure 10: Shortwave albedo versus ice concentration for two clear sky (open symbols) and two overcast (full symbols) conditions. The symbol coding is the same as in Figure 9 c.

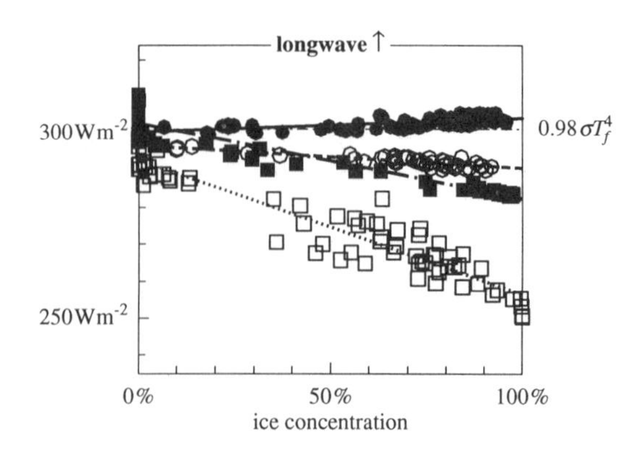

Figure 11: Longwave upwelling radiation versus ice concentration. The symbol coding is the same as in Figure 9 c.

removed. The scatter of the cloud free data reflects mainly variations of the optical surface conditions.

Long wave radiation. The long wave upward radiation is controlled by the water, ice or snow surface temperatures. The decreasing values of upward long wave fluxes with increasing ice concentration (Figure 11) are caused by the simultaneously decreasing mean surface temperatures.

The most pronounced variations of the long wave surface radiation balance, shown in Figure 9 c, result from the downward flux component which thus gains importance e.g. in model considerations. As known from various studies (e.g. *Marshunova* [1961]) the long wave downward radiation is primarily modified by changes of the temperature dependent water vapour concentration in clear air and by the various types of cloudiness as to be seen on Figure 12. Our measurements lie between $0.67\sigma T_{40}^4$ and $0.97\sigma T_{40}^4$, with T_{40} being the air temperature at 40 m height. The clear sky conditions are best approximated with $\epsilon = 0.71 \pm 0.016$ and the cloudy cases with $\epsilon = 0.90 \pm 0.030$. The corresponding standard deviations are equivalent to radiation fluxes of 4.6 Wm^{-2} and 8.6 Wm^{-2} for cloud free and cloudy conditions, respectively. In spite of the large scatter particularly for the cloudy cases it may nevertheless be appropriate to apply $\epsilon\sigma T_{40}^4$ with the above values for ϵ e.g. to sea ice and ocean models which are not explicitely coupled to the atmosphere.

Compared with the total surface radiation balances (Figure 9 d) the surface energy loss through sensible and latent heat fluxes (Figure 8) has rather small values. This features the predominance of the radiative fluxes in the energy balance of ice covered sea surfaces.

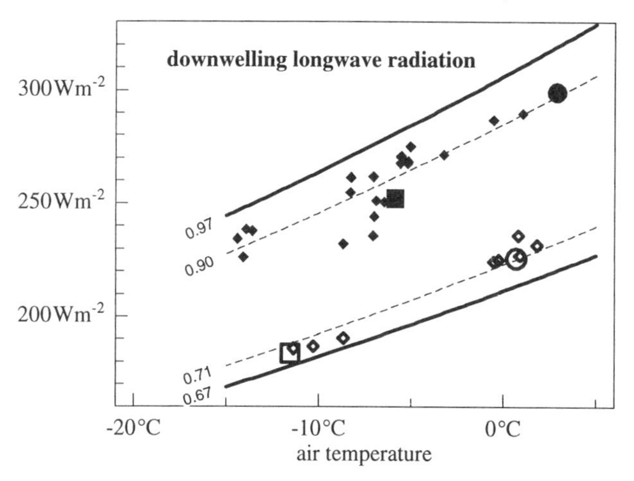

Figure 12: Downwelling longwave radiation versus air temperature at flight level. Open symbols indicate clear and full symbols cloudy skies. The four cases discussed above are marked by large sybmols. The curves represent the temperature dependence of the black body radiation with emissivities of $\epsilon = 0.97$ for the maximum, $\epsilon = 0.67$ for the minimum observed values, and with $\epsilon = 0.90$ and $\epsilon = 0.71$ for the best fit for cloudy and clear cases, respectively.

5 CONCLUSIONS

Low level flights (< 50 m height) with a suitably instrumented aircraft provide satisfactory quantitative estimates of the surface radiation balance and of the momentum, sensible and latent heat exchanges between the atmosphere and the polar ocean under non-stable atmospheric conditions. The vertical turbulent and radiative fluxes can be appropriately related to characteristics of a partly or fully ice covered sea surface with the aid of a digital line scan camera images. In stably stratified air the dynamic and thermal lower boundary effects on the air column obviously terminate distinctly below 40 m height. For flux measurements in this situation a new gust probe device is under development which can be attached by cable to a helicopter to be flown at or even below 10 m above the ice surface.

The main results of our field work during REFLEX I are:

- In the presence of sea ice the surface drag coefficient can be partitioned into a skin and a form drag component. The skin drag is assumed to be the same for sea ice and water surfaces. The form drag is then proportional to a dimensionless floe form parameter $F = Hr_e$ which results as the ice floe area perpendicular to the wind vector per unit horizontal surface area.
- The generally low sensible and latent heat fluxes (both $\leq$ 20 Wm^{-2}) over sea ice in autumn do not depend significantly on the observed sea ice concentration (> 50%) and the Bowen ratio is close to 1. The respective transfer coefficients $c_h = 0.91$ and $c_e = 0.73$ differ by 15% but since both show a large scatter this difference is not significant.
- The radiation balance forms the major component of the atmospheric part of the surface energy balance in the Arctic pack ice region during early autumn. For the downwelling long wave radiation emissivity coefficients ($\epsilon = 0.71$) for clear and ($\epsilon = 0.90$) cloudy skies are derived which can be applied in the Stefan-Boltzmann law together with the near surface air temperature for first order downward longwave radiation flux approximations.

Acknowledgments. The success of the REFLEX I campaign is to a large extent based on the continuous efforts of the technical crew. We particularly thank the four pilots of the Polar 2, Halu Meyer, Hans-Peter Joenk, Hans-Jürgen Berns and Tom Wede, who have carried out the low-level flights with a high degree of competence and responsibility. Thanks are also due to Christof Lüpkes for performing the calculation with the one-dimensional model. We also wish to extend our thanks to two anonymous reviewers who provided valuable hints for improvements of the manuscript. AWI publication number: 672.

References

Anderson, R.J., Wind Stress Measurements Over Rough Ice During the 1984 Marginal Ice Zone Experiment, *J. Geophys. Res., 92*, 6933–6941, 1987.

Andreas, E.L., Estimation of heat and mass fluxes over Arctic leads, *Mon. Wea. Rev., 108,* 2057–2063, 1980.

Andreas, E.L. and Makshtas, A.P., Energy exchange over Antarktic sea ice in the spring, *J. Geophys. Res., 90*, 7199–7212, 1985.

Andreas, E.L., Tucker III, W.B. and Ackley, S.F., Atmospheric Boundary-Layer Modification, Drag Coefficient, and Surface Heat Flux in the Antarktic Marginal Ice Zone, *J. Geophys. Res., 89*, 649–661, 1984.

Allison, I.F., Tivendale, C.M., Akerman, G.J., Tann, J.M., and Wills, R.H., Seasonal variations in the surface energy exchanges over Antarctic sea ice and coastal waters, *Ann. Glaciol., 3*, 12–16, 1982.

Arya, S.P.S, Contribution of Form Drag on Pressure Ridges to the Air Stress on Arctic Ice, *J. Geophys. Res., 78*, 7092–7099, 1973.

Belitz, H.J., Kottmeier, C., Hartig, R. and Stuckenberg, H.U., Zur aerodynamischen Rauhigkeit arktischer Meereisflächen, *Meteorol. Rdsch., 40*, 97–107, 1987.

Brown, R.A., Modeling the geostrophic drag coefficient for AIDJEX, *J. Geophys. Res., 86* 1989–1994, 1981.

Brutsaert, W., *Evaporation into the Atmosphere*, 299 pp., D. Reidel, Dordrecht, Holland, 1982.

Fairall, C.W. and Markson, R., Mesoscale Variations in Surface Stress, Heat Fluxes, and Drag Coefficient in the Marginal Ice Zone During the 1983 Marginal Ice Zone Experiment, *J. Geophys. Res., 92*, 6921–6932, 1987.

Gube-Lenhardt, M. and Hoeber, H., The development of the atmospheric boundary layer over the coastal region of the Weddell-Sea during offshore winds, *J. Rech. Atmos., 19*, 47–59, 1986.

Guest, P.S. and Davidson, K.L., The Effect of Observed Ice Conditions on the Drag Coefficient in the Summer East Greenland Sea Marginal Ice Zone, *J. Geophys. Res., 92*, 6943–6954, 1987.

Guest, P.S. and Davidson, K.L., The Aerodynamic Roughness of Different Types of Sea Ice, *J. Geophys. Res., 96*, 4709–4721, 1991.

Hanssen-Bauer, I. and Gjessing, Y.T., Observations and model calculations of aerodynamic drag on sea ice in the Fram Strait, *Tellus, 40A*, 151-161, 1988.

Hartmann, J., Kottmeier, Ch., and Wamser, C., Radiation and Eddy Flux Experiment 1991 (REFLEX I), *Reports on Polar Research, 105*, Alfred-Wegener-Institut für Polar- und Meeresforschung, Bremerhaven, 1992.

Hoeber, H. and Gube-Lehnhardt, M., The Eastern Weddell Sea Drifting Buoy Data Set of the Winter Weddell Sea Project (WWSP) 1986. *Reports on Polar Research, 37*, Alfred-Wegener-Institut für Polar- und Meeresforschung, Bremerhaven, 1987.

Katsaros, K.B., McMurdie, L.A., Lind, R.J. and DeVault, J.E., Albedo of a Water Surface, Spectral Variation, Effects of Atmopsheric Transmittance, Sun Angle and Wind Speed, *J. Geophys. Res., 90*, 7313–7321, 1985.

Kellner, G., Wamser, C. and Brown, R.A., An Observation of the Planetary Boundary Layer in the Marginal Ice Zone, *J. Geophys. Res., 92C7*, 6955–6965, 1987.

Marshunova, M.S., Principle Regularities of the Radiation Balance of the Underlying Surface and of the Atmosphere in the Arctic, (English translation) *Soviet Data on the Arctic Heat Budget and its Climatic Influence*, J.O. Fletcher, B. Keller, and S.M. Olenicoff, Eds., The Rand Corp. RM-5003-PR, 51–132, 1961.

Martinson, D. and Wamser, C., Ice drift and momentum exchange in winter Antarktik pack ice, *J. Geophys. Res., 95*, 1741–1755, 1990.

Maykut, G.A., Energy exchange over young sea ice, *J. Geophys. Res., 83*, 3646–3658, 1978.

Overland, J.E., Atmospheric Boundary Layer Structure and Drag Coefficients Over Sea Ice, *J. Geophys. Res., 90*, 9029–9049, 1985.

Payne, R.E., Albedo of the sea surface, *J. Atmos. Sci., 29*, 959–970, 1972.

Raasch, S., *Numerische Simulation zur Entwicklung von Wirbelrollen und konvektiver Grenzschicht bei Kaltluftausbrüchen über dem Meer*. Ph.D thesis, Universität Hannover, 1988.

Thorpe, M.R., Banke, E.G. and Smith, S.D., Eddy correlation measurements of evaporation and sensible heat flux over arctic sea ice, *J. Geophys. Res., 78*, 3573–3584, 1973.

Vörsmann et al.P., Friderici, B. and Hoff, A.M., METEOPOD – ein flugzeuggestütztes Trubulenzmeßsystem, *Promet, 1/2*, 57–64, 1989.

Walter, B.A., Overland, J.E. and Gilmer, R.O., Air-Ice Drag Coefficients for First Year Sea Ice Derived From Aircraft Measurements, *J. Geophys. Res., 89*, 3550–3560, 1984.

Wyngaard, J., On Surface-Layer-Turbulence, *Workshop on Micrometeorology*, edited by D. Haugen, American Meteorological Society, Science Press, PA, 101–149, 1973.

J. Hartmann, C. Kottmeier, C. Wamser and E. Augstein, Alfred-Wegener-Institut für Polar- und Meeresforschung, Postfach 12 01 61, D 27515 Bremerhaven, Germany.

Opening and Closing of the "Husky 1" Lead Complex

Robert W. Fett

Consultant Meteorologist, Watersmeet, Michigan

Kenneth L. Davidson

Naval Postgraduate School, Monterey, California

James E. Overland

Pacific Marine Environmental Laboratory/NOAA, Seattle, Washington

Synthetic Aperture Radar (SAR) data were collected over the central Beaufort Sea during the Leadex experiment which extended from early March to the end of April 1992. The data revealed no evidence of a lead on 21 March, followed by a significant lead opening in the same region on 24 March. The lead, again, was virtually closed by 27 March. Fracture patterns of identifiable multiyear floes on each day permitted precise location of the opening and closing. Opening of the lead coincided with the onset of moderate east-southeasterly winds in advance of a low pressure system moving through the area. The reason for the highly reflective portions of the lead region as they appeared in the SAR data cannot be precisely determined. Potential causes are: (1) Bragg scattering from wind-roughened water; (2) radar return from jagged edges of rafted ice; and (3) radar return from "frost flowers", small highly saline, frond-like protuberances which grow rapidly on thin ice in open lead (high humidity) conditions. Closing of the lead occurred as the low-pressure system moved northeastward past the area, bringing northerly flow on the western side. We use ice station surface observations, high resolution satellite data, and conventional analyses to document this case study.

1. INTRODUCTION

The Leads Experiment (LEADEX) was conducted at and in the vicinity of an ice camp established in the Beaufort Sea (near 73°N, 145°W) during the period 24 March to 25 April 1992. The purpose of the experiment was to document meteorological and oceanographic parameters associated with the opening and closing of leads. Details of the experiment are described in a paper by Morison, et al., (1993).

LEADEX was the first Arctic experiment in which the highest quality direct readout satellite data from both the NOAA and DMSP satellite systems were continuously archived for later research while also being utilized through immediate processing for operational use. This data set was further embellished by the availability of Synthetic Aperture Radar (SAR) data over the LEADEX

The Polar Oceans and Their Role in Shaping the Global Environment
Geophysical Monograph 85

area of operations. These data, acquired from the European Space Agency's satellite ERS-1, were especially useful in yielding information relating to ice type, age, and movement. This paper focuses on a SAR-observed lead opening and closing event, further documented by NOAA and DMSP data, automated ice station data from buoys implanted on the ice, and conventional analyses.

The comparison of meteorological data and high resolution SAR images provide information on regional and small scale processes which can form the basis of a new generation of ice models. It is important for that purpose to build up a set of case studies such as this and earlier examples (Marko and Thompson 1977, Overland et al., 1992).

2. REGIONAL METEOROLOGICAL AND SEA ICE ANALYSES 21-24 MARCH 1992

Time series of meteorological parameters obtained from the LEADEX ice station array, as well as ice station movement describe the atmospheric forcing during the SAR detected event. The positions of ice stations in the array on 23 March, are shown in Fig. 1. Time series of pressure and vector wind from 0000 UTC, 21 March through 28 March, obtained from the ice stations, are shown in Fig. 2. Panels for the ice station are arranged, top to bottom, in Fig. 2 according to the passage of time of the pressure minimum which also corresponds to times of most of the significant wind shifts.

Throughout the 21-22 March period the ice stations reported very light, 1-2.5 ms^{-1}, southwesterly winds, becoming northwesterly (Fig. 2b). The light wind prevailed until the afternoon of 23 March when the direction veered to northeast and then east in response to an approaching low-pressure system from the east. The pressure also began to decrease at this time (Fig. 2a). By 0000 UTC on 24 March wind speeds at the ice stations increased to above 5 m s^{-1} and were becoming east-southeasterly.

Figure 3 shows the Anchorage, Alaska, National Weather Service (NWS) analysis for 24 March at 1200

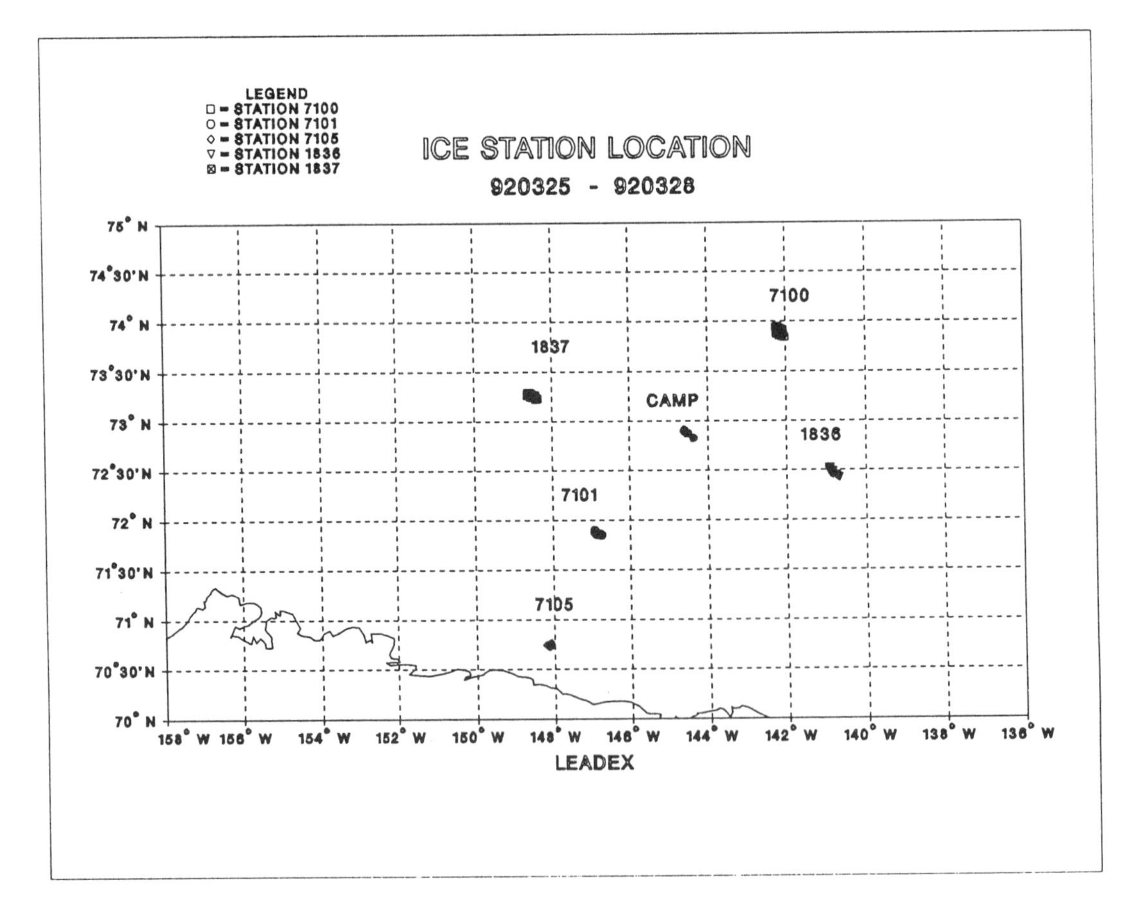

Fig. 1. Locations of LEADEX ice stations (7101, 1836, 1837, and 7100) from 23-28 March 1992.

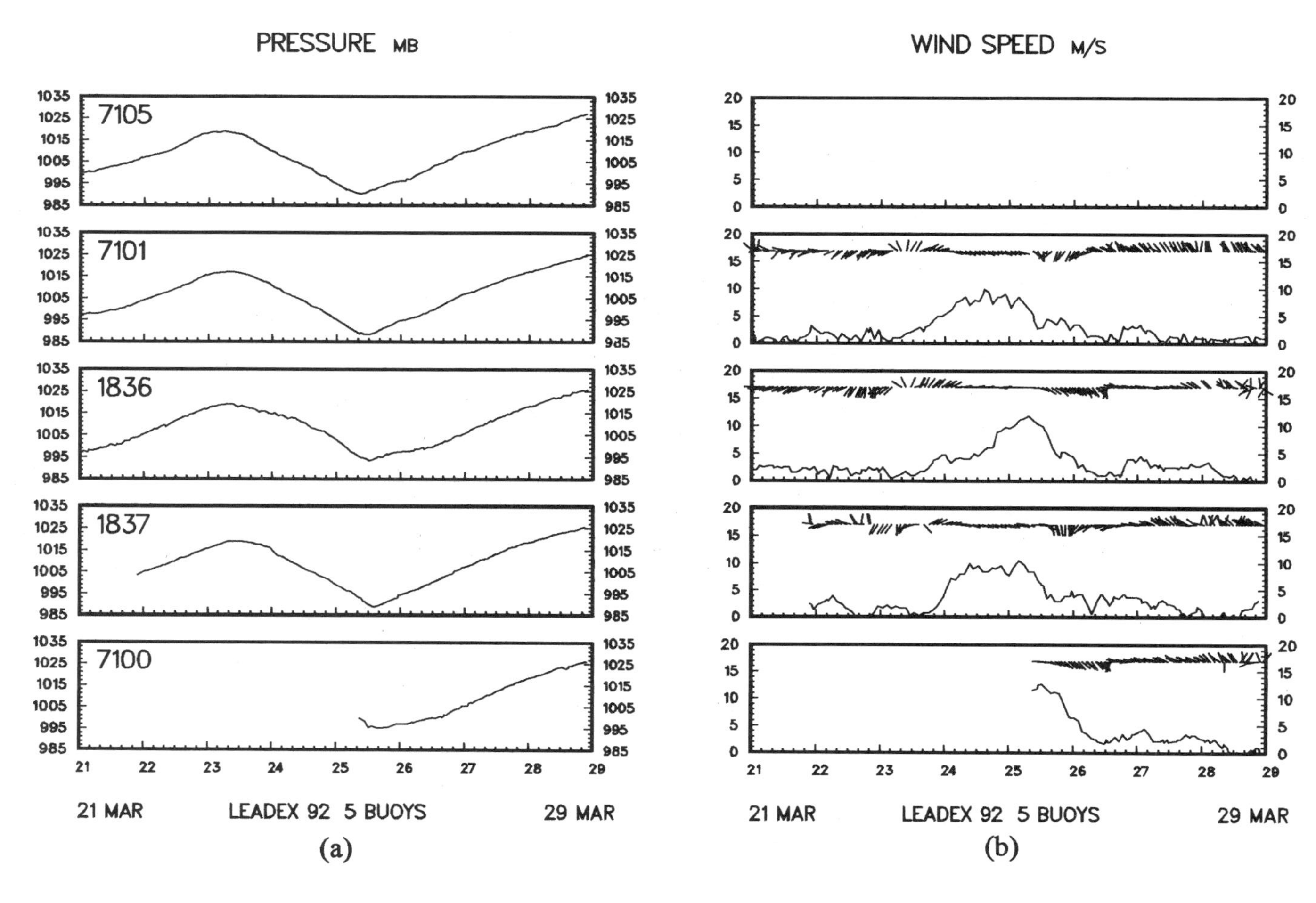

Fig. 2. Time series of (a) barometric pressure and (b) vector winds from LEADEX ice stations for 21-29 March 1993. See Figure 1 for relative locations.

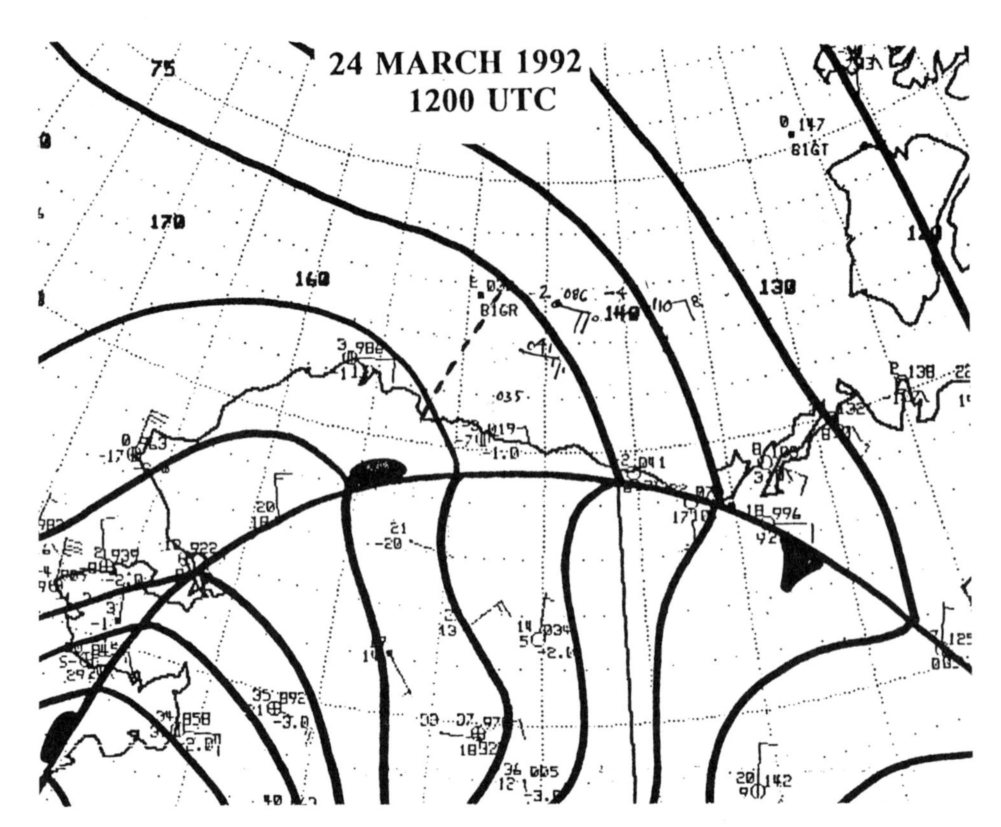

Fig. 3. NWS (Anchorage) surface analysis. 24 March 1992, 1200 UTC. A cylonic wind shift is shown associated with a trough east of Barrow.

UTC. A cyclonic wind shift between the ice stations and Barrow provides evidence of the approaching low pressure system.

A DMSP visible image acquired on at 0158 UTC on 24 March is shown in Fig. 4. There is no definite indication of a low center or cloud vortex in the vicinity of Barrow. Note that the region between Deadhorse and the ice camp shows no strong evidence of lead features. A pronounced lead is evident, however, west of the ice camp extending northeastward from Barrow.

Wind speeds over the area of concern continued to increase during the day, reaching 7 to 10 m s^{-1} at the western-most ice stations (7101 and 1837) near 1200 UTC, as direction shifted to the southeast (Fig. 2).

Figure 5 shows an additional DMSP visible image acquired on 24 March at 1723 UTC. Comparing this image with Fig. 4 shows that the data have captured a newly opened lead between 71° and 72°N, immediately north of Deadhorse. Although the Fig. 4 view is partially obscured in that region the thinness of the cloud cover is sufficient such that this lead would have been visible if it had been open. Note that the lead extending northeast of Barrow is visible through similarly obscuring thin cloudiness. The implication is that the lead opened dramatically within the 15-hr interval between the two views. Cyclonic circulation around a low pressure center is implied by the cloud band curvature in Fig. 5, and verified on the Navy Operational Regional Atmospheric Prediction System (NORAPS) 925 mb analysis for 25 March at 0000 UTC (Fig. 6). This analysis shows a low-pressure trough extending north of Barrow and eastward along the North Slope. Note the strong southeasterly winds east of the trough in this analysis.

3. SMALL SCALE ICE FEATURES (10 - 100 KM)

On 21 March 1992 at 2129 UTC a SAR image (Fig. 7) revealed a region of multiyear floes just north of the fast ice between Deadhorse and the ice camp. Center point of this image is 71.63°N, 147.52°W. Dashed lines drawn on this figure indicate the approximate positions where leads formed a few days later (24 March). Features A, B, C, D, and E are labeled to aid in comparison with subsequent SAR images. An ARGOS transmitting ice-

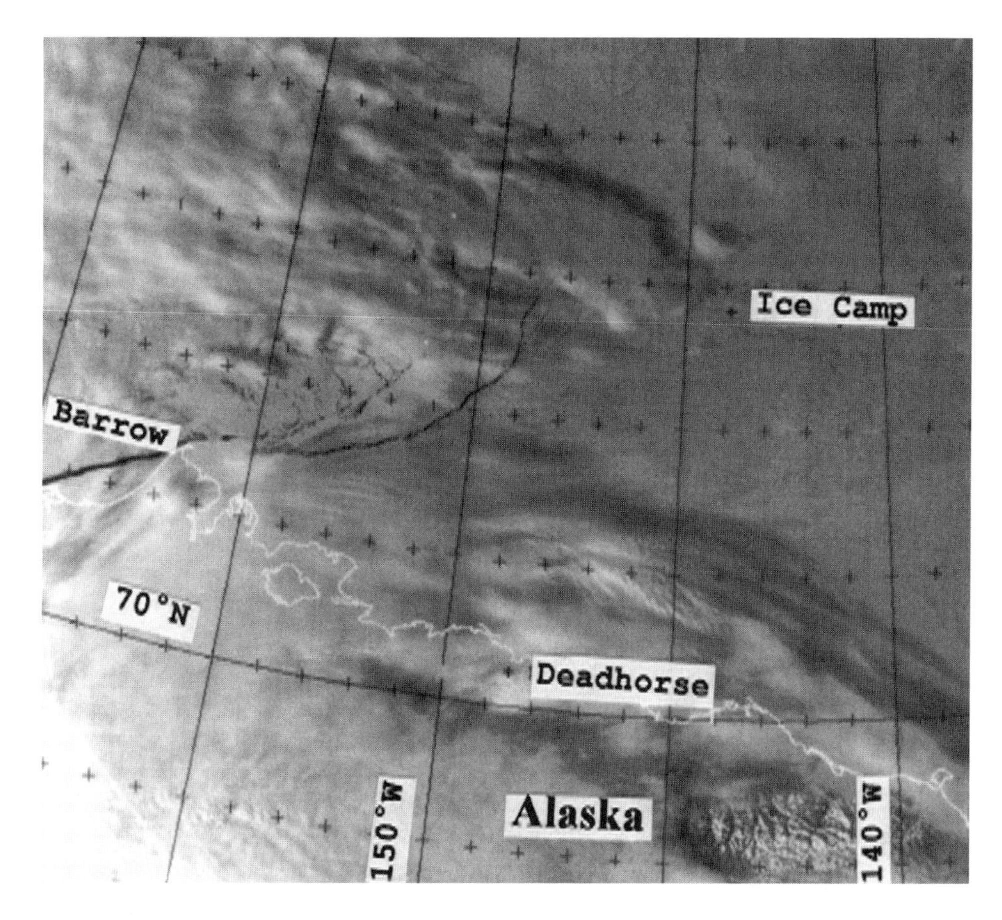

Fig. 4 A DMSP visible image. 24 March 1992, 0158 UTC. Note lead west of ice camp extending norteastward from Barrow.

station array with pressure, vector wind, temperature, and current sensors was deployed in the region. One of these stations, 7101, was placed on the ice just north of the northernmost lead. Its position is also shown on this figure.

Figure 8 shows SAR data received on 24 March at 2129 UTC. These data are centered near 71.6°N, 147°W, and show the major lead openings in that region. The three leads shown in the SAR image are visible in the DMSP image (Fig. 6) diverging from one another toward the east, in both images.

In comparing the SAR image of Fig. 8 with that of Fig. 7 you can see that the leads have formed roughly along the dashed lines as indicated in Fig. 7; however, the leads south of the larger lead appear to have been shear-induced rather than wind driven. In particular the southern portions of multiyear floes A, B, and D, were sheared off and moved eastward without moving southward to permit a larger lead opening. The ability of the leads to separate further was prevented by the strong wind stress from the southeast and the apparent inability of floes immediately to the north to move further northward.

In contrast to the shear leads, the major lead to the north opened significantly, and caused a fracture and northwestward movement of the northern portion of multiyear floe A of approximately 3 km. Further to the southwest the opening of the lead reaches a maximum width of about 5 km. This lead was later named "Husky 1".

4. REGIONAL METEOROLOGICAL AND SEA ICE ANALYSIS 25 - 28 MARCH 1992

A low pressure center was analyzed on the NWS surface analysis for 25 March at 1200 UTC (Fig. 9), approximately 15 hours after the time of the SAR image. NWS did not have access to the ice station data for this analysis without the ice station reports which have since been superimposed on the analysis. Ice station pressure

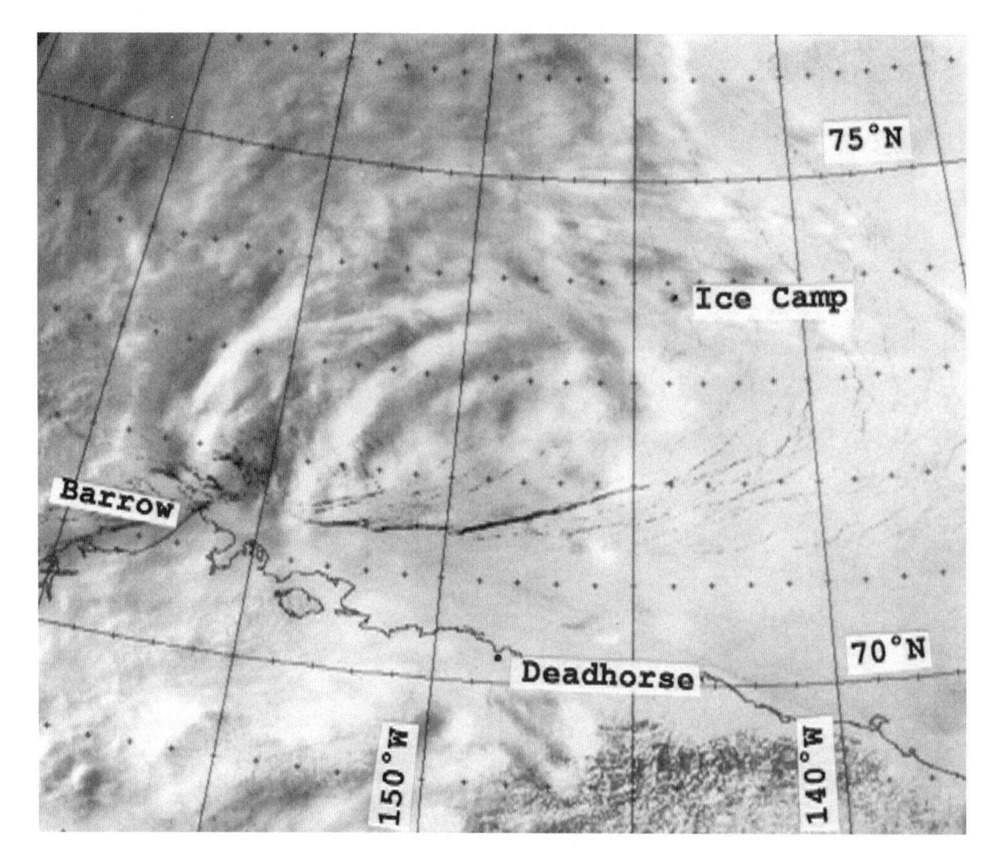

Fig. 5. A DMSP visible image. 24 March 1992, 1723 UTC. Note new lead formation south of ice camp.

time series show that pressure minima were observed near 1200 UTC at 7101 and 1836, 2 hours later (near 1400 UTC) at 1837, and 6 hours later (near 1800 UTC) at 7100. The western-most ice-stations, 7101 and 1837, reported lower pressure values than 1836 and 7100, 987 mbs versus 995 mbs. The ice-station observed pressure and vector wind trends and relative values agree with the 1200 UTC NWS analysis for a northward propagating closed low with a center located west of the array. Further, DMSP and NOAA satellite data gave clear evidence of a low center formation in the region much earlier (shortly after 25/0000 UTC).

Figure 10 is a NOAA infrared satellite image showing clear evidence of the low formation on 25 March at 0532 UTC. Development and movement of this low was not classical. It appeared to form in the northern lee of the Brooks Range and then move rapidly north-northwestward passing west of the ice camp location in the space of only a few hours. Further movement was to the northeast. A secondary low is also apparent in this image forming just north of Barrow, Alaska.

The low-pressure system continued its northward movement throughout the 25th of March. Winds became southerly and decreased as the low center passed to the west of the array (Figure 2). Winds at 7101 became southerly at 0800 UTC 25 March, and began shifting to westerly at 0000 UTC 26 March. Winds at the other western ice station, 1837, became southerly 4 hours later (1200 UTC) confirming the northward propagation. Winds at both ice stations remained southerly for 12 hours.

The track of 7101 is shown in Fig. 11 for 22 to 28 March and reveals the closeness between the ice response to atmospheric forcing from 22 through 25 March. ARGOS-determined ice station locations were interpolated to hourly points and plotted at 12-hour intervals for this track display. The minimal ice station movement during the very light wind period (22-23 March), and the sudden and rapid southeasterly wind maximum early on 24 March are evident in Fig. 11. The northwest movement of the ice in response to the east-southeasterly flow is consistent with the generally

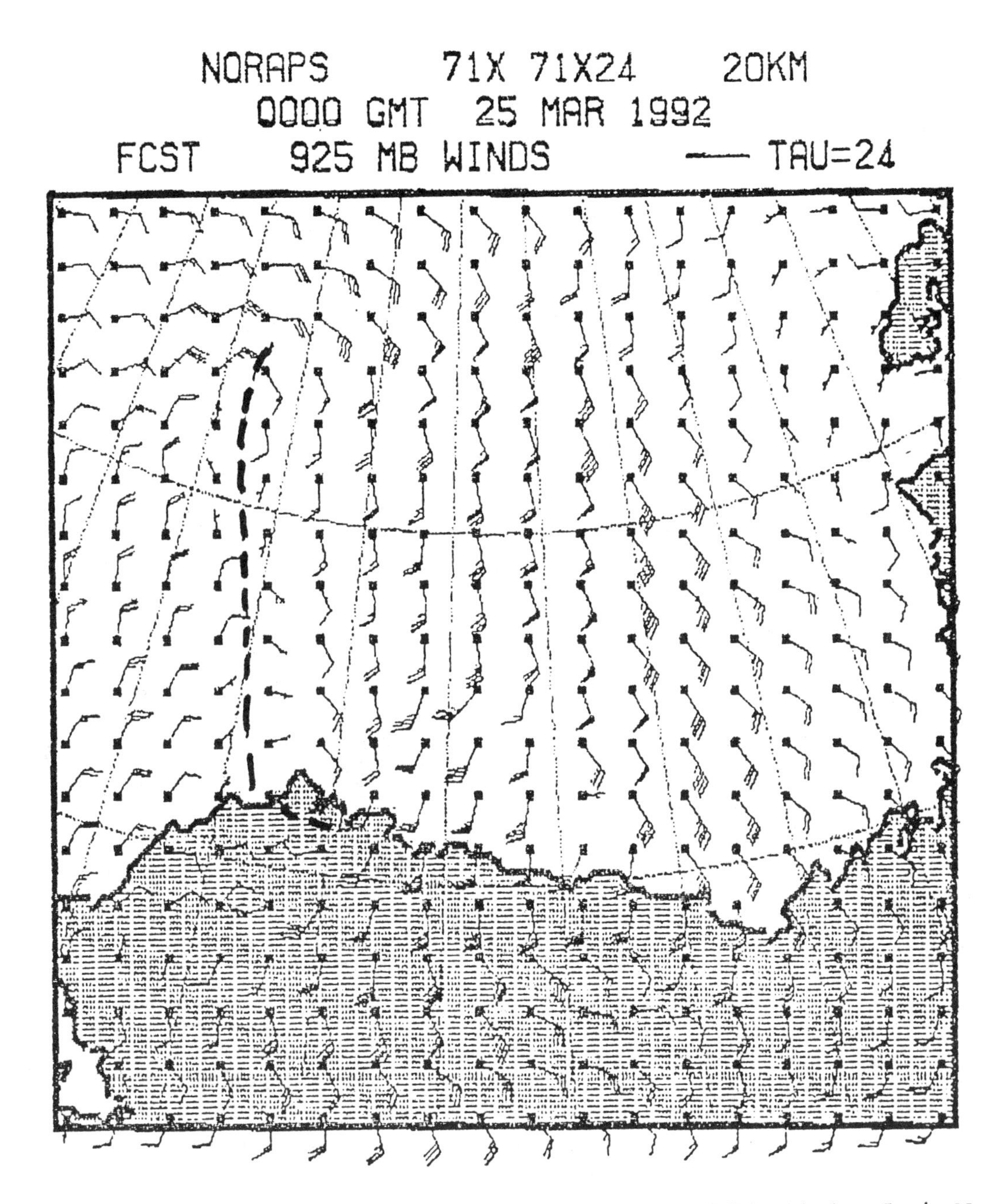

Fig. 6. Navy mesoscale model 925 mb analysis. 25 March 1992, 0000 UTC. 1 barb = 5ms^{-1}. Note the strong southeast winds east of the trough axis.

accepted observation that ice tends to move in a direction 30° - 40° to the right of the surface wind direction (Stringer, 1984).

A NOAA infrared image at 1606 UTC on 25 March (Fig. 12) showed many new lead openings in the region between Deadhorse and the ice camp. The 1606 UTC image was taken after winds began to decrease within the ice station array (Fig. 2). The low center in this image is implied near 74.5°N, 151.5°W.

Analyses were performed on ice station movement relative to their centroid. The relative motions from 25 March 0800 UTC to 27 March 0000 UTC follow the wind shear zone existing between 7101 and 7100, (Fig. 13). Ice station 7101 regressed, 7100 advanced, and

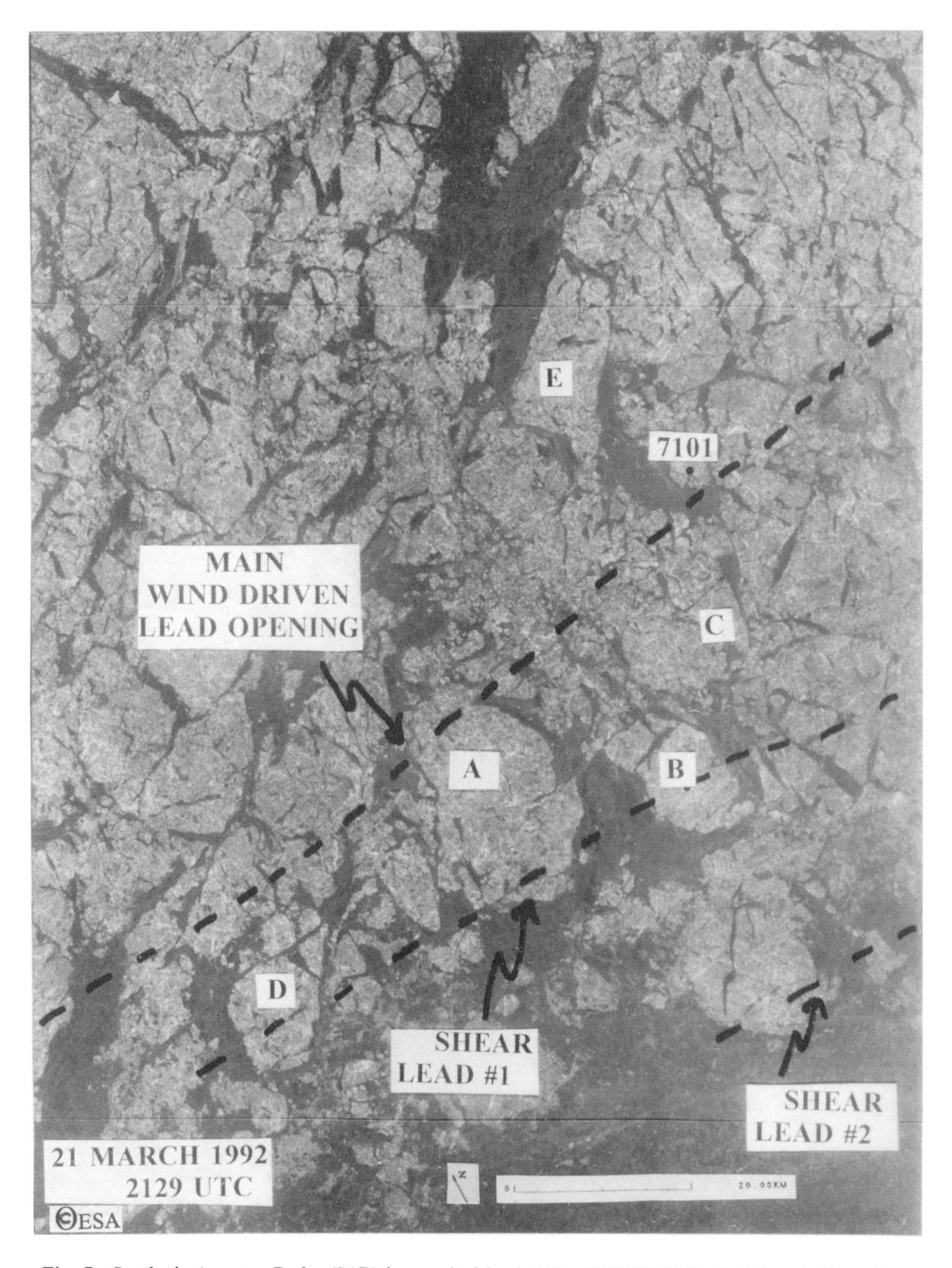

Fig. 7. Synthetic Aperture Radar (SAR) image. 21 March 1992, 2129 UTC. Dashed lines indicate future lead locations. Copyright 1992 ESA; used by permission.

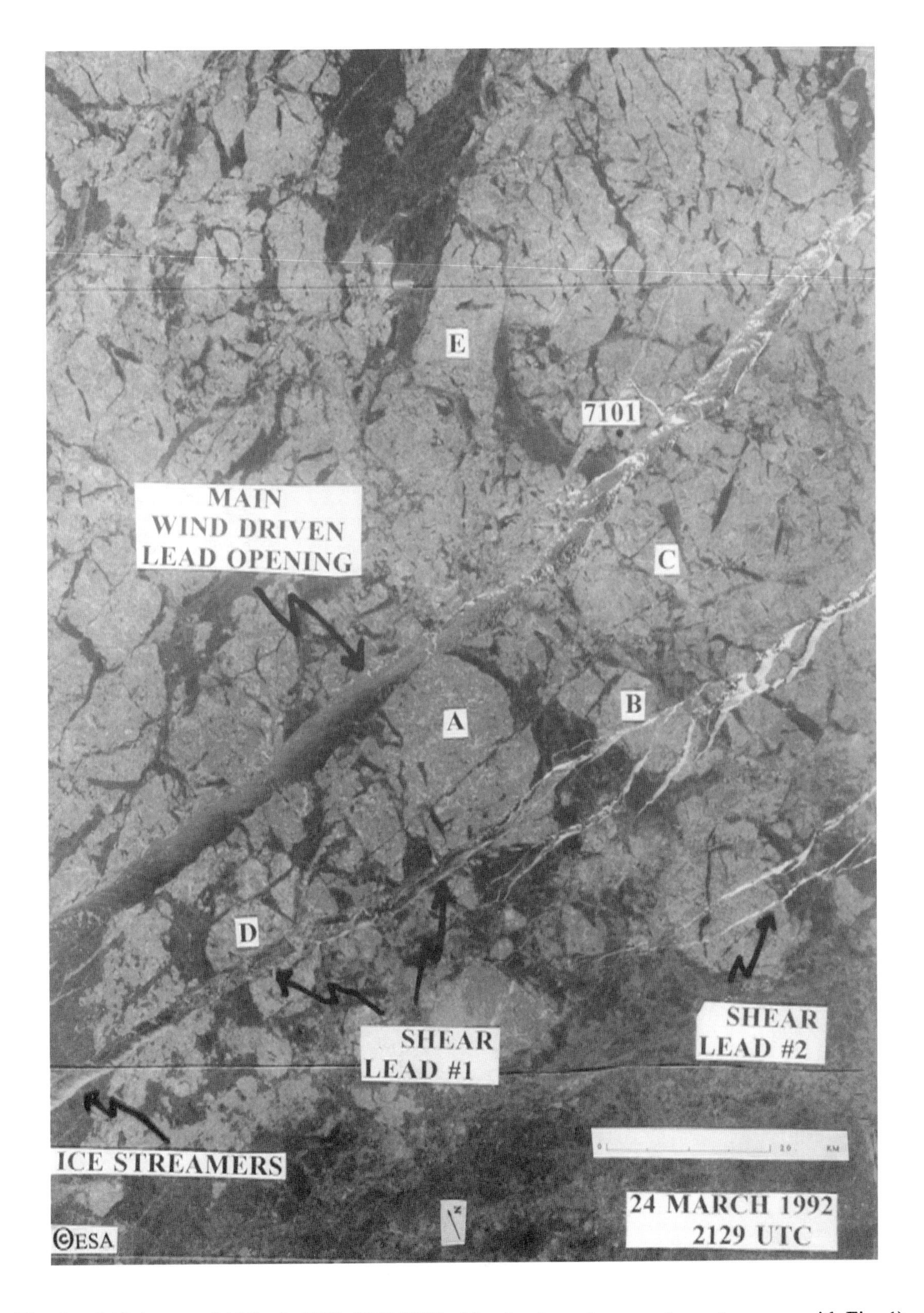

Fig. 8. SAR image. 24 March 1992, 2129 UTC. New lead openings are shown (compare with Fig. 1). Copyright 1992 ESA; used by permission.

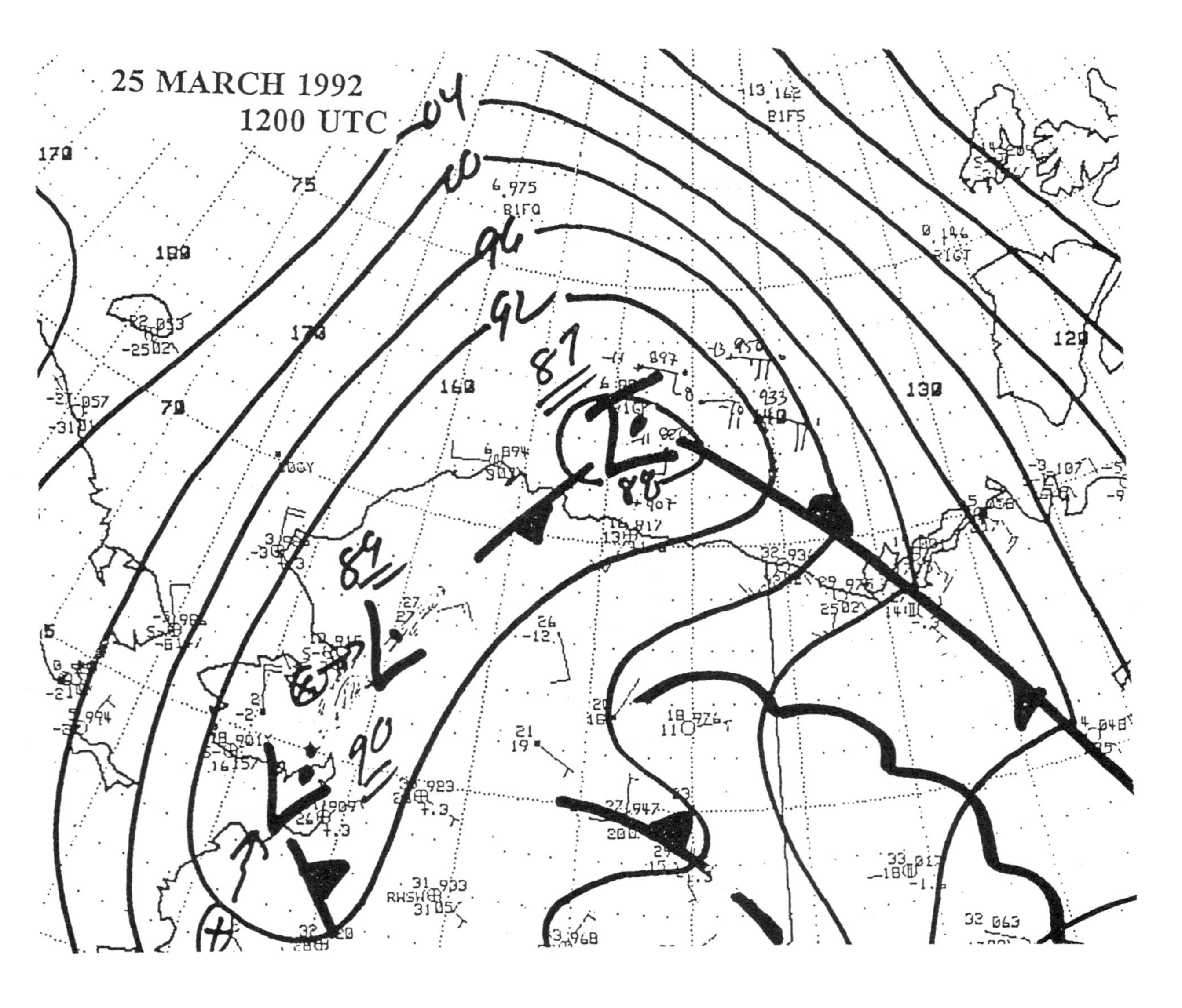

Fig. 9. NWS (Anchorage) surface analysis. 25 March 1992, 1200 UTC. A new low pressure center is analyzed east of Barrow.

Fig. 10. A NOAA infrared (Ch4) image. 25 March 1992, 0532 UTC. Cloud curvature indicates a low pressure system forming south ot the ice camp.

1836 and 1837 moved minimally relative to the centroid. The effect is consistent with wind-driven ice movement due to a cyclonic weather system moving through the region.

The NWS surface analysis on the following day, 26 March at 0600 UTC (Fig. 14), agrees with ice station winds (Fig. 3) that shift to westerly after the low center passed. The shift was accompanied by very low wind speeds and occurred first at the western-most stations (7101 and 1837; Fig. 3). By 26 March, 1800 UTC, winds shifted to west-northwest at all ice stations and remained below 5 ms^{-1}. Station 7101, nearest Husky 1, shifted first and had northwest winds after 0800 UTC. The low center at this time moved north-northeastward to near 77°N, 143°W. Comparing the ice station 7101's track and vector winds in Figs. 2 and 11 shows that the station changed direction almost immediately with the wind reversal even though wind speeds were quite light.

5. INTERPRETATION OF THE SAR IMAGERY

What is the cause for the heightened reflectivity of features of the leads in the northeast region? It has been suggested that such heightened reflectivity may be the result of Bragg scattering of wind roughened water in a relatively high wind speed area (Shuchman, et al., 1991). Since ice stations 7101 and 1837 reported southeast winds at near 10 m s^{-1}, the hypothesis of Bragg scattering

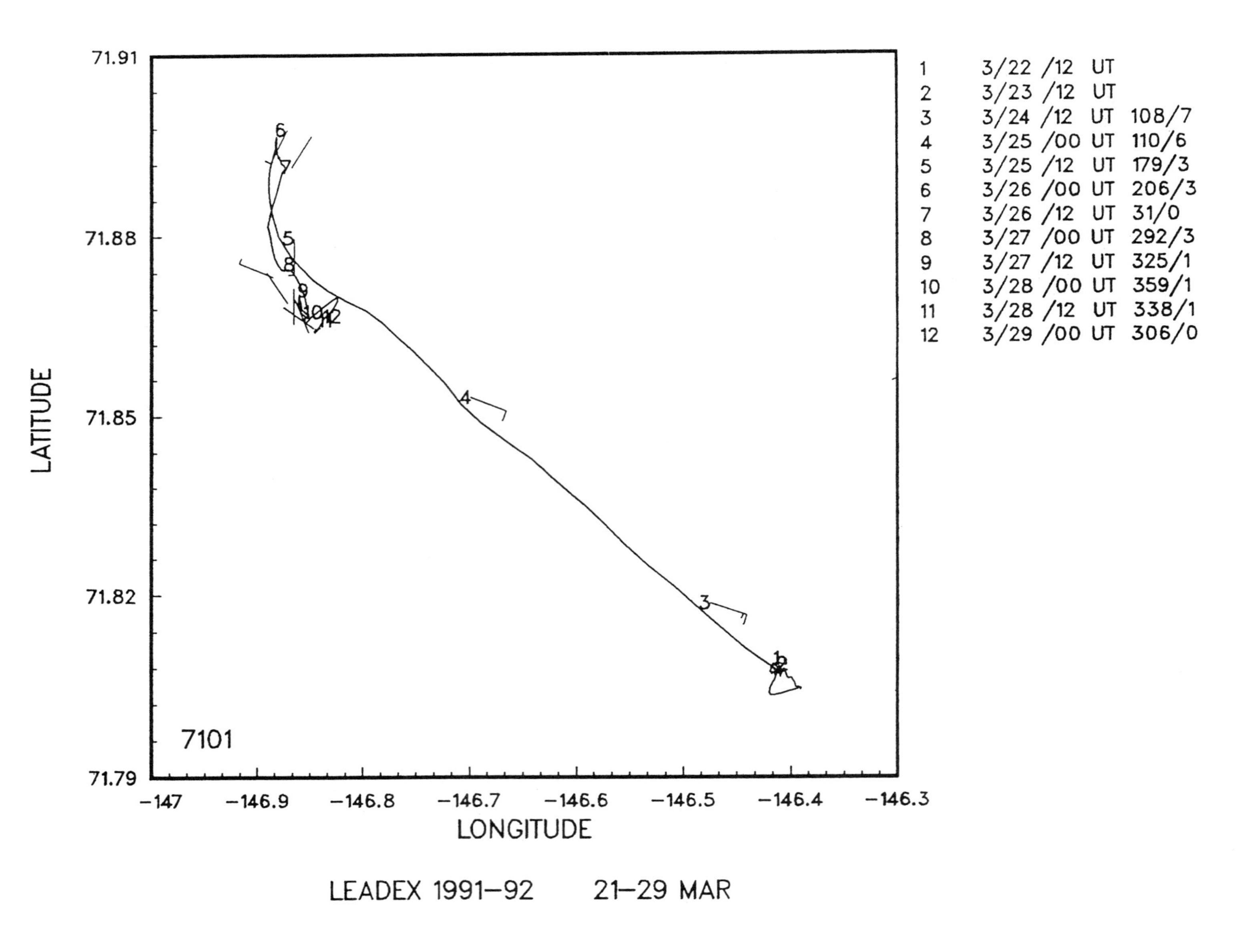

Fig. 11. Track of ice-station 7101 from 22-28 March 1992. The ice station moved rapidly northwestward with the onset of moderate southeasterly winds.

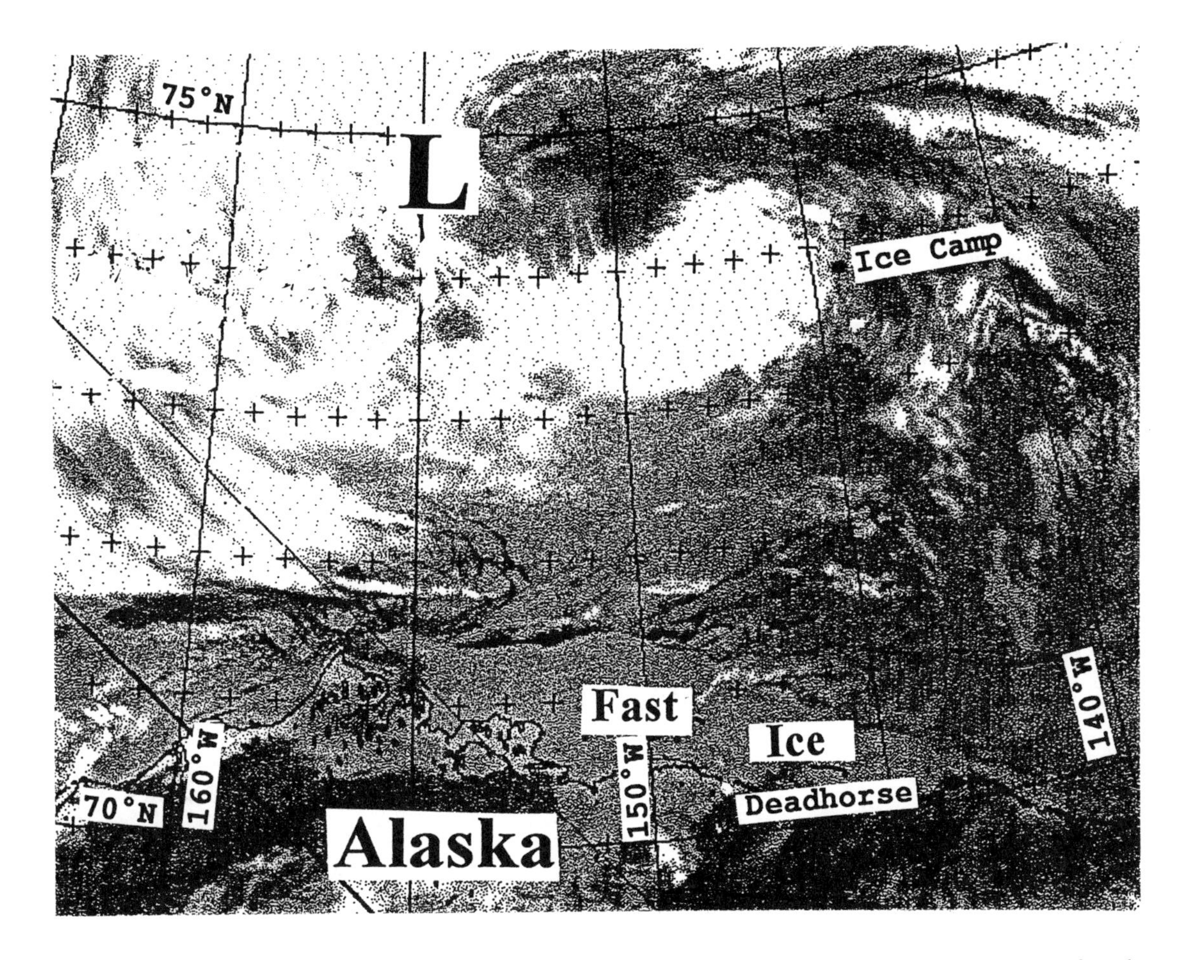

Fig. 12. NOAA infrared (Ch4) data. 25 March 1992, 1606 UTC. The low center has now moved to the northeast of the ice camp.

seems reasonable. Open water regions in SAR imagery which under calm conditions, reverse tonality at higher wind speeds and appear very bright. It is important to realize, that once leads are open they refreeze very rapidly. Refrozen areas in the -16°C air temperature which prevailed at the time, also would appear relatively brighter in the SAR image, though not of peak intensity due to the thinness of the newly-formed ice. The eastern portions of shear lead #1 and shear lead #2 show a consistent bright return which may be related to Bragg scattering or backscatter from areas of refrozen ice (Shuchman, et al., op cit.) The main lead near buoy 7101 shows a much more complicated pattern of brighter and darker tones. Additional factors may influence the radar return in that region. A rafting of ice floes within the lead and ridging of ice along the northern and southern boundaries are potential causes for heightened reflectivity. Radar return from such features is non-uniform and could account for much of the signal particularly evident in the main lead.

Figure 15 is an aerial photograph of a lead in the southern Beaufort during spring 1991, in which jagged edges of ice floes in a rafted condition are shown. Note also the pressure ridges and shadows apparent on either side of this lead indicating a ridge height of at least a meter or more, judging from the 2134 m altitude from which the photo was taken. Such features would cause a very bright SAR return of the type and intensity shown in the northernmost lead (Fig. 8).

Finally, "Frost flowers" have been identified as a cause of heightened SAR reflectivity (Drinkwater and Crocker, 1988). Such features develop over refrozen leads when the ice is generally greater than 10 cm thick in temperatures of -18°C or less. The "flowers" consist of frozen high salinity water that rises out of the ice in stems extending to as much as 3 cm. Supercooled cloud

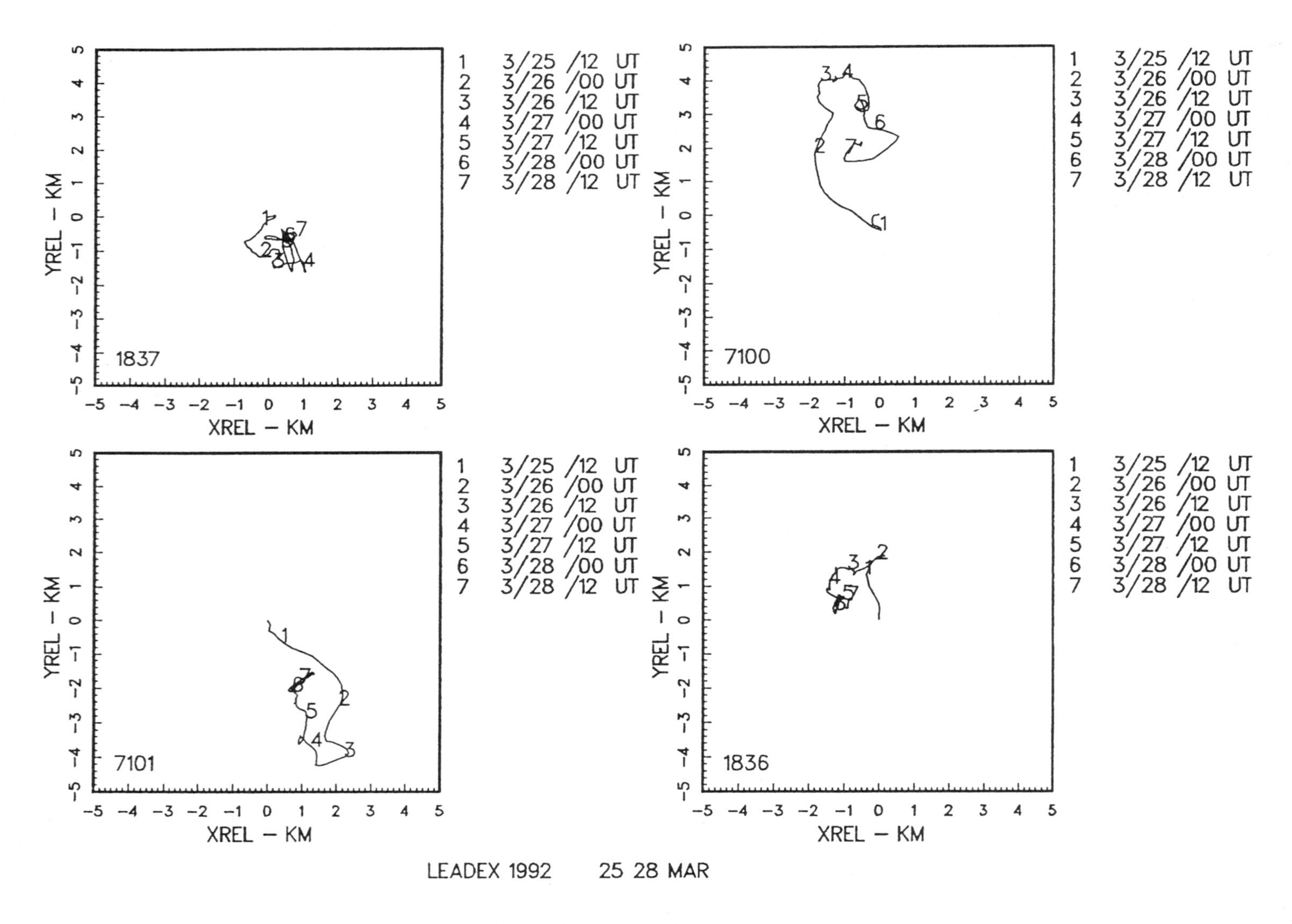

Fig. 13. Movement of ice stations relative to their centroid, 25-28 March 1992.

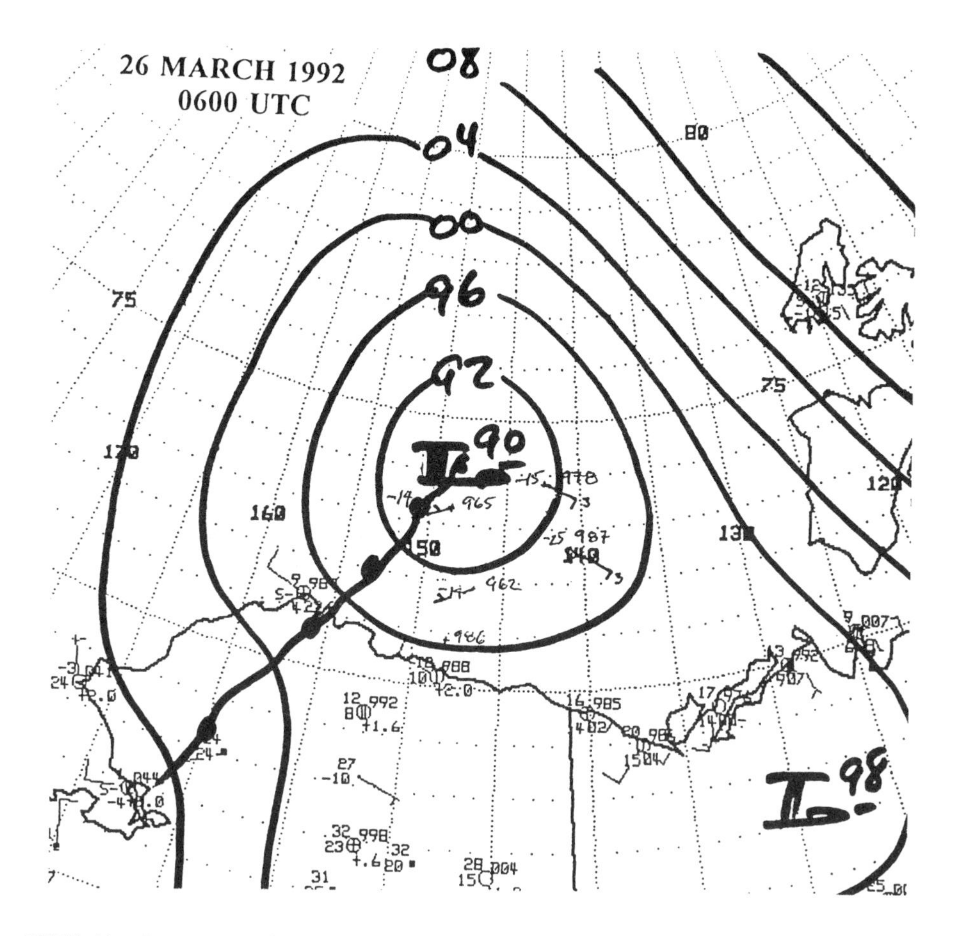

Fig. 14. NWS (Anchorage) surface analysis. 26 March 1992, 0600 UTC. Wind speeds diminished as the low center moved north of the ice camp.

plumes and increased humidity generated from an open lead may be a factor contributing to rapid frost flower development. The heightened SAR reflectivity is a combined effect of increased surface roughness caused by the "flowers" and the increased dielectric constant related to their salty composition (Drinkwater and Crocker, op cit).

It is likely that one or more of the above factors were involved in the heightened reflectivity in the leads of Fig. 8. More research is required to understand the predominant cause.

The tonality of the main wind driven lead in Fig. 8 changes from almost black on the south side (left of ice floe A) to a light gray shade on its northern border. This change in tonality suggests open (non-roughened) water on the south edge giving way to thin nilas and gray ice on the north side. Such an arrangement is typical of ice formation in a lead under persistent flow from the southern quadrant. Figure 16 is an aerial photograph of a similar type of ice formation over a southwest-northeast oriented lead, observed over the southern Beaufort during spring of 1991. The ice streamers (streets of frazil ice) in the center portion of the photograph are aligned with the wind flow coming from the left (south) side of the image.

Using the above "ice streamer" concept to infer wind direction from imagery data, and testing its application to SAR data, a feature of great interest appears in the lower left corner of Fig. 8. In this region what appears to be ice streamers, or rows of frazil ice, have formed as resolved by the SAR data. The SAR data are much coarser resolution (100 m) than the aerial photograph so that individual lines of ice are not resolved; however, the pattern and configuration of the SAR view is very suggestive of this effect. Ten-meter resolution SAR data have clearly revealed such features in other areas (e.g.

Fig. 15. Aerial photograph of a refrozen lead in the south-central Beaufort Sea. Spring 1991. Note ridges formed on either side of lead and jagged edges of rafted ice.

Fig. 16. An aerial photograph of a SW-NE oriented lead in the south-central Beaufort Sea. Spring 1991. Wind flow is from left (south) side of image. Lighter gray shade tonalities indicates progressively thicker new ice formation.

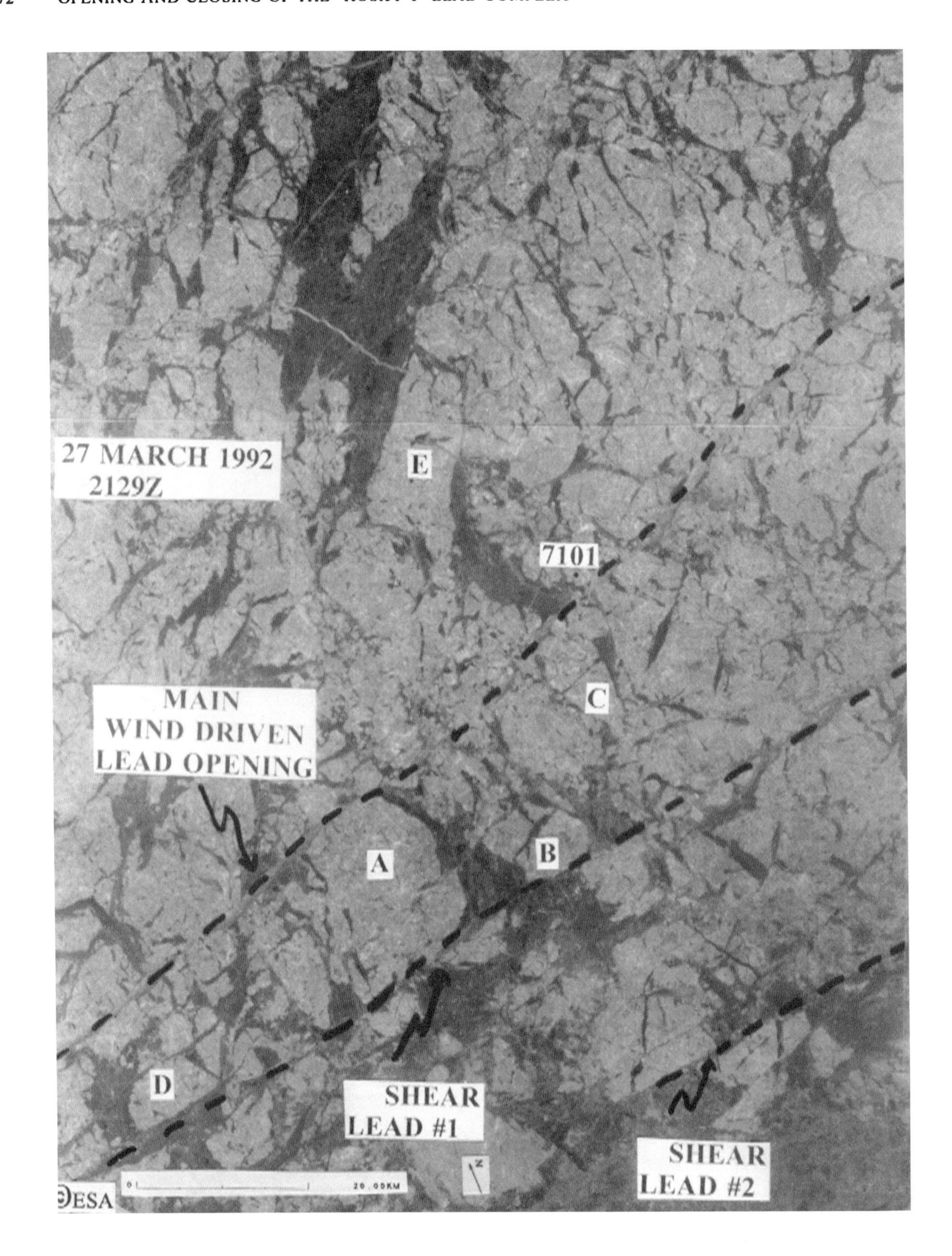

Fig. 17. SAR image. 27 March 1992, 2129 UTC. Leads have now largely closed (compare with Fig. 8).

SAR image ID 2086110, 17 April 92). The wind surface wind direction in all such instances was closely aligned with the major streamer axis. The streamers at this time according to this interpretation would indicate a north-northwesterly wind flow over the ice at that location, with open water on the north side of the lead.. At about the same time ice station 7101 was recording southeast winds at 8 m s^{-1} (Fig. 2). The separate indications imply a low pressure center located very near the left center portion of the SAR image.

The next available SAR image over the area was 27 March at 2129 UTC (Fig. 17). This image revealed that the main wind driven lead to the north closed to a maximum width of only 1 or 2 km and was largely refrozen. This conclusion follows since, 1.) winds were nearly calm over the area at this time; 2.) open water under such conditions should appear black; yet 3.) nevertheless, the lead is moderately reflective. Newly formed thin ice condition could explain the increase in reflectivity. The fact that even higher reflectivity, like Fig. 8 is not observed also suggests that frost flower development or rafting was not a prevalent condition within the lead at this time.

In comparing this image with that of Fig. 7, which shows the ice area before lead formation, you can be see that the ice floes to the north have not returned precisely to their positions before the break. They have shifted somewhat to the west relative to the position of floes A and D. Conversely, ice floes on the south side of shear lead 1 and shear lead 2 have shifted eastward in relative position by almost 10 km.

6. SUMMARY

An important objective of LEADEX was to gain insight into environmental conditions forcing leads to open and close. SAR images used in this study show an unambiguous example of leads opening and coming to a near close in the space of only a few days. Ice station reports in the immediate area show clear evidence that leads opened in response to moderate southeasterly flow in advance of an approaching low pressure system and closed as the system moved past while light northerly winds replaced the southerly flow.

Of considerable interest is that much lower resolution weather satellite data were also able to resolve the major lead opening, suggesting the potential for monitoring such events on a near real-time basis, using data routinely available at major weather centers.

The highly reflective SAR radar signals appearing in this example appear to have multiple possible causes which require further investigation to maximize understanding and further potential for operational use.

Acknowledgements. SAR data used in this study were obtained through specific authorization from the Alaska SAR Facility, Fairbanks, Alaska. The ice stations and the data collection were funded by the Naval Postgraduate School laboratory funding program and ONR. Support for analyses/interpretations was provided by NRL.

REFERENCES

Drinkwater, M. R., and G. B. Crocker, 1988: Modeling changes in the dielectric and scattering properties of young snow-covered sea ice at GHz frequencies, *J. Glaciol.*, 34(118), 274-282.

Marco, J. , and R. Thompson, 1977: Rectilinear leads and internal motions in the ice pack of the western Arctic Ocean, *J. Geophys. Res.*, 82, 979-987.

Morison, et al., 1993: The LeadEx Experiment, *EOS, Trans. Amer. Geophy. Union*, 35, 393, 396-397.

Overland, J. E. , B. A. Walter, and K. L. Davidson, 1992: Sea-ice deformation in the Beaufort Sea, *AMS proceedings of the Third Conference on Polar Meteorology and Oceanography*, Portland, OR., 64-67.

Shuchman, R. A., C.C. Wackerman, and L. L. Sutherland, 1991: The use of synthetic aperture to map the polar oceans, *ERIM*, P. O. Box 134001, Ann Arbor, MI, 48113-4001, 337 pp..

Stringer, W. J., D. J. Barnett, and R. H. Godin, Handbook for Sea Ice Analysis and Forecasting, *Rep. NAVENVPREDSCHFAC CR 84-03*, 324 PP., Naval Research Laboratory, Monterey, CA., 1984.

Polynyas as a Possible Source for Enigmatic Bennett Island Atmospheric Plumes

DIRK DETHLEFF

GEOMAR Research Center for Marine Geosciences, Kiel, Germany

During polar winter open waters are well defined energy source areas, which release significant amounts of oceanic heat and water vapor forming buoyant atmospheric plumes. In this study the coincidence of a large polynya and the presence of high altitude plumes in the vicinity of Bennett Island on February 18, 1983, is discussed. Heat flux data derived from a balance-model reveal that the released energy of approximately 650 W/m^2 from the open water surface during the initial phase of the events by far exceeded the amounts necessary for turbulent upward air movements. The net heat flux as high as $2.4x10^{11}W$ from the entire polynya of 300 to 375 km^2 extent is suggested to be the energy source for the origin of strong local thermal updrafts and subsequent upper tropospheric plume formation. Submarine methane outbursts from sediments or underlying coal beds were mainly suggested as a valid source for such features near Bennett Island. However, all plumes of that remote area documented in the literature were observed during winter or early spring, when temperatures are extremely low and the Polar Ocean is ice covered, whereas the release of natural gases due to thermally weakened permafrost should rather happen during summer.

INTRODUCTION

Atmospheric plumes are warm, moist and buoyant *or* extremely cold and horizontally extended clouds of vapor or even ice crystals [*Matson*, 1986; *Schnell et al.*, 1989; *Sechrist et al.*, 1989]. Some of these features rise from point sources of different diameter and reach altitudes as high as 7-13 kilometers [*Kienle et al.*, 1983; *Clarke et al.*, 1986; *Matson*, 1986]. They have been described since the early 70s as a significant meteorological feature in the high Arctic [*Kienle et al.*, 1983; *St. Amand et al.*, 1985; *Clarke et al.*, 1986; *Matson*, 1986; *Sechrist et al.*, 1989; *Kerr*, 1992]. Frequently occurring at the northern tip of Novaya Zemlya and off the eastern and southeastern coast of Bennett Island (Figure 1, Figures 2 and 2a), those phenomena were explained as volcanic eruptions, methane and gas hydrate releases from shelf sediments or underlying coal beds, as orographic clouds or even as anthropogenic impacts. However, neither seismic records revealing tectonic, eruptive or explosive events in those regions have ever been reported, nor atmospheric concentrations of methane and relevant trace metals or elements were ever elevated during the occurrence of a Bennett Island or any other atmospheric plume [*Kienle et al.*, 1983; *St. Amand et al.*, 1985; *Schnell et al.*, 1992; *Paull and Buelow*, 1993].

According to Hansen and Schnell [quoted after *Kerr*, 1992] and *Schnell et al.* [1992], the scientific mystery of, at least, Bennett Island plume formation was solved as an orographic and meteorological cloud phenomenon. However, concerning Hansen's and Schnell's observations [quoted after *Kerr*, 1992], *Paull and Buelow* [1992] admitted that the distinction between plumes and "normal" clouds is rather uncertain. Nevertheless, apart from the question "plume" or "cloud", at present there is still one link missing: What is the energy source causing plumes or "normal" clouds in the vicinity of Bennett Island?

SHORT NOTES ON TERMINOLOGY

The term "plume" in this scientific context is either used for (i) a large-volume, buoyant, turbulent, and continuing convective upward heat and moisture flux over leads [e. g. *Glendening and Burk*, 1992; *Schnell et al.*, 1989] or (ii) for

The Polar Oceans and Their Role in Shaping the Global Environment
Geophysical Monograph 85

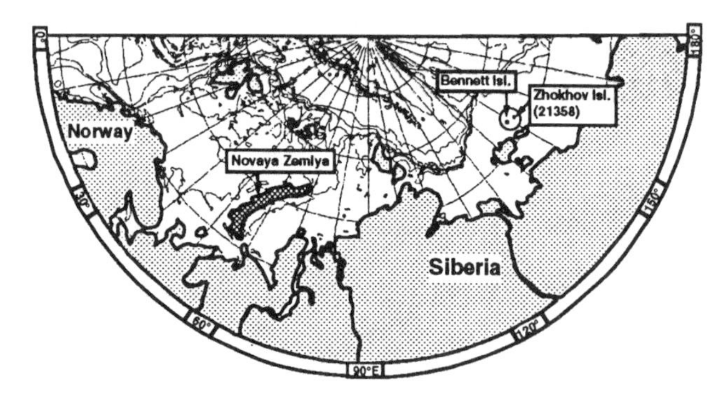

Fig. 1. Map of the eastern Arctic. Bennett and Zhokhov Islands are part of the De Long Islands. Zhokhov weather station is identified by the five digit code of the World Meteorological Organisation (WMO).

long extended ice clouds developing from high-altitude point sources [e. g. *Kienle et al.*, 1983; *Matson*, 1986]. The author assumes that both kinds of phenomena, at least near Bennett Island, are closely connected. Thus, both features are called "atmospheric plumes" and for simplification here will be named "plumes".

COINCIDENCE OF PLUMES AND POLYNYAS

Polynyas or flaw leads in the East Siberian Arctic can be closed or opened within few days or even less, induced by strong changing winds [*Dethleff et al.*, 1993; *Pease*, 1987]. Due to extreme sea/air temperature differences such open water areas in the Arctic ice cover are important sources for oceanic heat loss and release of water vapor [e. g. *Zakharov*, 1966; *Maykut*, 1982; *Pease*, 1987; *Martin and Cavalieri*, 1989; *Smith et al.*, 1990]. As modelled by *Schnell et al.* [1989] an energy flux of approximately 300 W/m^2 over a 10 km wide polynya could warm and moisten the air sufficiently for convective plume formation reaching altitudes of 4 kilometers, and thus, penetrate the Arctic boundary layer inversion.

On February 18, 1983, large plumes rose partly at the same time from different sources approximately 10 kilometers east and southeast of Bennett Island (Figure 3, Figure 4). The source regions were about 10 kilometer in diameter and the plumes reached altitudes of probably 7 kilometers [*Kienle et al.*, 1983]. During that day, a large polynya existed east of Bennett Island (Figure 3, Figure 4). It was situated directly below the punctiform plume sources as indicated on NOAA-6 and NOAA-7 satellite images. As will be presented in this study, the formation of plumes near Bennett Island on February 18, 1983 is suggested to be closely connected to the thermal energy and moisture released by the polynya.

MATERIAL AND METHODS

Meteorological and Satellite Data

Basic data were collected in order to balance and quantify the heat flux from ocean to atmosphere and to discuss a possible source and mechanism for the origin of the Bennett Island plumes. Synoptical 6-hour weather data (wind direction and speed, air temperature, humidity, cloudiness) from Zhokhov Island were obtained from the World Meteorological Organization (WMO, Figure 1). The Eastern Arctic pressure charts were adapted from the *European Meteorological Bulletin* [1983]. The satellite images for estimating the polynya extent east of Bennett Island on February 18, 1983, were taken from *Kienle et al.* [1983].

Heat Flux Calculations

The heat flux model used in this work was forced with the 6 hour meteorological data record. The calculations of the ocean-to-atmosphere energy fluxes through the polynya are based on the following heat balance equation:

$$F_{net} = F_S + F_L + R_A - R_L \qquad (1)$$

where F_{net} is the net heat flux, F_S and F_L are the sensible and latent heat flux, respectively, R_L is the outgoing longwave radiation, and R_A is the backscattered longwave radiation from the atmosphere. No incoming shortwave radiation occurs at high latitudes during winter and thus, this term was neglected in the calculations.

Sensible heat flux. The sensible heat flux can be written as:

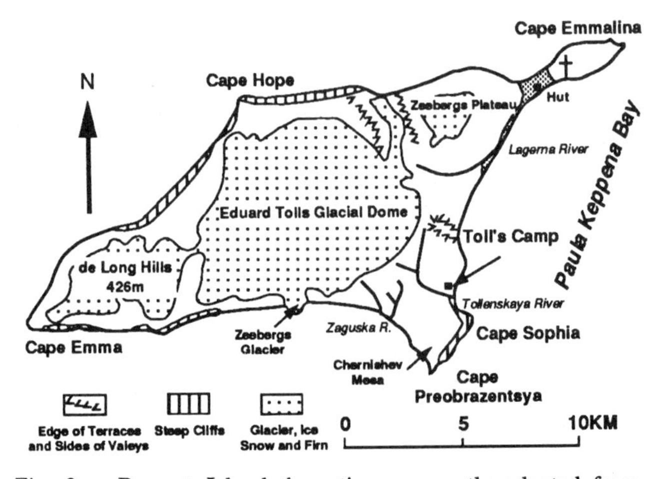

Fig. 2. Bennett Island thematic map, partly adapted from *Vol'nov et al.* [1970] and *Verkulich et al.* [1989].

Fig. 2a. The image, taken on April 23, 1992 from helicopter, shows the southwesterly tip of Bennett Island. Direction of view is to the northeast across Eduard Tolls Glacial Dome. Note the snow- and ice-capped, plateau-like surface of the island where no particular high rising peak or lateral extended mountain range is recognizable. Large areas of open water (lower image section) were generated due to a preceding 6 day period of northeasterly winds.

$$F_s = \rho_{air} c_p C_s V_w (T_{air} - T_{sea}) \quad (2)$$

where ρ_{air} is the air density (1.25 kg/m^3), c_p represents the specific heat of air at constant pressure (1004 J deg^{-1} kg^{-1}), and C_s is the sensible heat transfer coefficient (1.75x10^{-3}) according to *Parkinson and Washington* [1979]. The term V_w represents the wind velocity, and T_{air} is the synoptically recorded air temperature at Zhokov Island WMO-station. The sea surface temperature (T_{sea}) is set to a freezing point of -1.6° C.

Latent heat flux. The flux of latent heat (F_L) is mainly driven by the wind velocity and the difference of the specific humidity between the air at the water surface and the air at 10 m altitude (q_{10} - q_0). This relationship can be expressed as:

$$F_L = \rho_{air} C_L L_v V_w (q_{10} - q_0) \quad (3)$$

where C_L represents a latent heat transfer coefficient (1.75x10^{-3}) and L_v is the latent heat of vaporization (2.5x10^6 J/kg).

Longwave radiation. The longwave radiative energy loss (R_L) of the polynya may be approximated by the following equation:

$$R_L = \varepsilon_s \sigma T_{sea}^4 \quad (4)$$

where ε_s is the surface emissivity (0.97) and σ represents the Stefan-Boltzmann constant (5.67x10^{-8}Wm^{-2}deg^{-4}).

Atmospheric longwave radiation. The atmospheric longwave radiation (R_A) is obtained from the following term:

$$R_A = E_{air} \sigma T_{air}^4. \quad (5)$$

E_{air} is parameterized as a function of air temperature and fractional cloud cover (Cl, given in 1 to 8 eighth):

$$E_{air} = 0.99(1 - 0.261(1\text{-}0.75Cl^2) \exp(-7.7x10^{-4}(T_{air} - 273)^2)). \quad (6)$$

RESULTS

The meteorological data reveal a 6 day period from February 15 to 20, 1983, predominated by southwesterly,

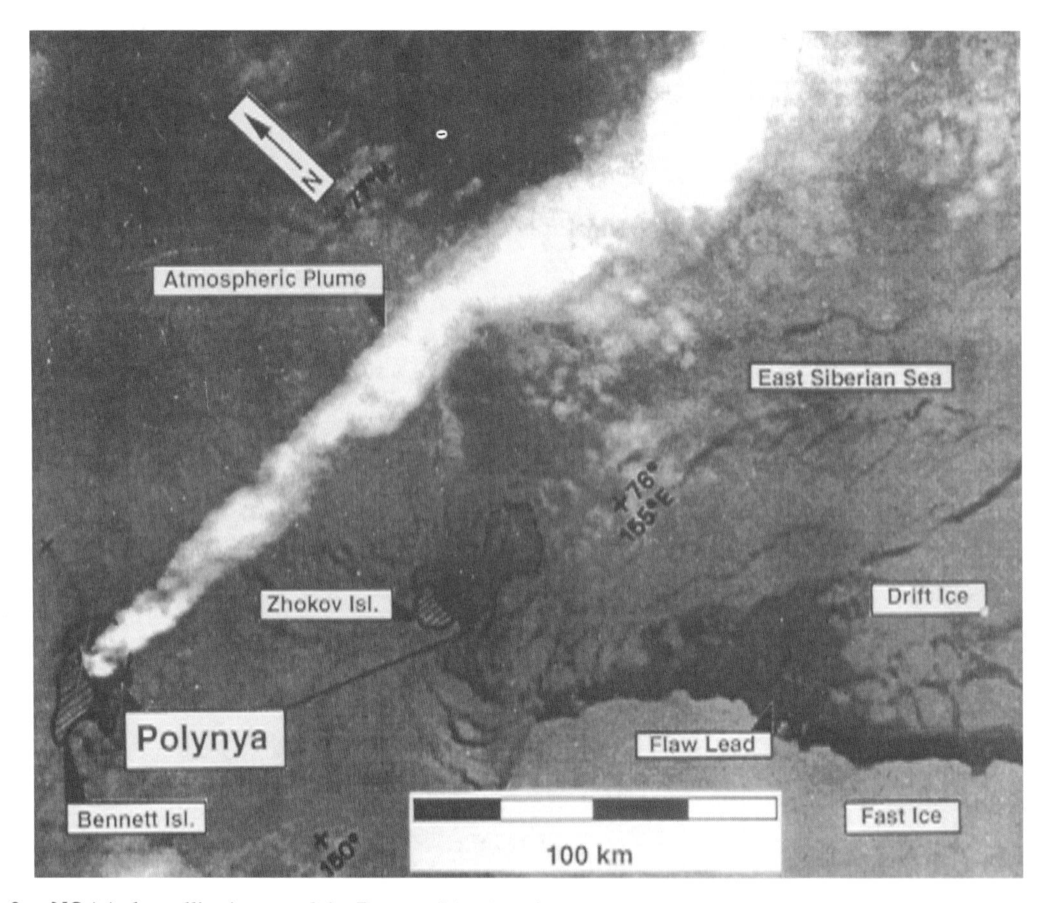

Fig. 3. NOAA-6 satellite image of the Bennett Island area at 06:16 UT on February 18, 1983 [adapted from *Kienle et al.*, 1983].

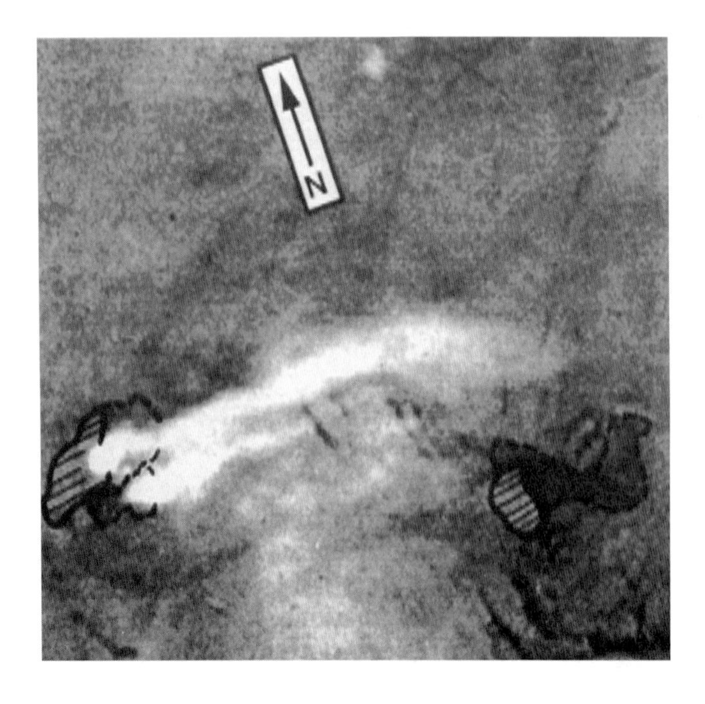

westerly and northwesterly winds with mean speeds of approximately 10 m/s and temporarily high net heat fluxes from the open water near Bennett Island (Table 1, Figure 6). The air temperatures range between -23° and -33° C (Figure 5). Strong westerly winds namely between February 16, 18.00 UT and February 17, 18.00 UT formed a large polynya east of Bennett Island as evident in Figures 3 and 4. According to the mean temperature (-27° C) and wind speed (10 m/s) during that 24-30 hrs period a maximum and stable polynya width of roughly 15 km can be estimated considering the studies of *Pease* [1987, see Figure 5., p. 7051]. As calculated by the same author, a 30 hrs offshore-

Fig. 4. NOAA-7 satellite image of the initial phase of plume development at 00:47 UT on February 18, 1983, reproduced from *Kienle et al.* [1983]. Note several source areas for the plumes. Overlapping of one of the circular plume sources and the shape of the island is assumably caused by the position of NOAA-satellites and subsequently, the angle of view.

TABLE 1. Basic meteorological data from Zhokhov Station and energy fluxes over the polynya as calculated from equation (1) to (6) in the Bennett Island area during February 15 to 20, 1983.

Date	Time (UT)	Air Temperature (°C)	Wind Speed (ms^{-1})	Wind Direction (°)	Heat Flux (Wm^{-2}) Sensible	Latent	Longwave	Backscatter	Total
15	0.00	-33.0	6	220	-292	-38	298	163	-466
	6.00	-31.4	7	210	-324	-44	298	171	-495
	12.00	-28.3	12	190	-497	-73	298	185	-683
	18.00	-24.4	12	210	-424	-70	298	195	-598
16	0.00	-25.9	13	200	-490	-78	298	192	-675
	6.00	-26.5	10	220	-386	-60	298	172	-572
	12.00	-24.8	12	200	-432	-71	298	200	-601
	18.00	-23.3	14	280	-471	-81	298	177	-673
17	0.00	-26.5	10	280	-386	-60	298	172	-572
	6.00	-28.6	8	300	-335	-49	298	169	-513
	12.00	-28.6	8	300	-335	-49	298	169	-513
	18.00	-28.0	10	250	-409	-61	298	173	-596
18	0.00	-24.0	13	210	-452	-76	298	178	-648
	6.00	-23.9	12	230	-415	-70	298	191	-592
	12.00	-23.1	11	230	-367	-63	298	205	-523
	18.00	-25.5	9	240	-334	-54	298	198	-487
19	0.00	-24.5	11	240	-391	-65	298	201	-553
	6.00	-24.1	1	270	-35	-6	298	202	-137
	12.00	-25.6	8	240	-298	-48	298	198	-446
	18.00	-25.7	8	240	-299	-48	298	198	-447
20	0.00	-24.4	12	240	-424	-70	298	201	-592
	6.00	-25.2	11	250	-403	-65	298	199	-567
	12.00	-24.7	9	250	-322	-53	298	195	-479
	18.00	-24.7	9	250	-322	-53	298	195	-479

wind period was necessary to open such a large area of ice free water. Both values, the polynya width and the opening time, derived from the paper mentioned above, are in convincing agreement with (i) the observed lateral extent of the polynya east of Bennett Island on February 18, 1983 (Figure 3, Figure 4) and (ii) the duration of the preceding period of westerly winds between February 16, 18.00 UT and February 17, 18.00 UT (Table 1, Figure 5). The longitudinal size of the open water area can be estimated to as much as 20-25 km (Figure 3). Consequently, a total polynya area of roughly 300-375 km^2 occurring during the plume events can be evaluated.

The results of the energy balance calculations are presented in Figure 6. The data reveal a net heat flux of 648 W/m^2 continuing from the water surface at 00.00 UT on February 18, solely 47 minutes before the initial phase of plume formation was noticed (Figure 4). An amount of 592 W/m^2 was released at 06.00 UT on the same day while the main plume was emanating from the open water area east of Bennett Island (Figure 3). Due to Bennett Island local meteorological and oceanographic conditions on February 18, 1983, an oceanic energy loss of approximately 1.8-2.4×10^{11} W can be evaluated for the entire polynya during the plume events.

According to the high pressure gradient in about 5 kilometers altitude (≈ 500 hPa level, Figure 7) and above, westerly winds reached velocities comparable to jet streams (> 200 km/h). These storms promoted the far eastward extension of the plumes discussed in this study. According to weather charts from the *European Meteorological Bulletin* [1983] the cloud cover at Zhokov Island was mapped as Cirrocumulus clouds, normally occurring at altitudes as high as 6 to 12 kilometers.

Main errors in the evaluation of the net heat flux from the polynya are based on the reliability of the balance calculations and on the open water estimate. Due to lower heat transfer coefficients used in the equations in respect to other authors [e. g. *Lindsay*, 1976], the sensible and latent energy flux in this study can be wrong by 30 %. Slight differences (lower values) of 5-10 % occur in the calculation of the longwave atmospheric radiation in respect to the term applied by *Parkinson and Washington* [1979]. Under consideration of the heat flux errors and a slightly variable open water extent the estimate of the net energy released

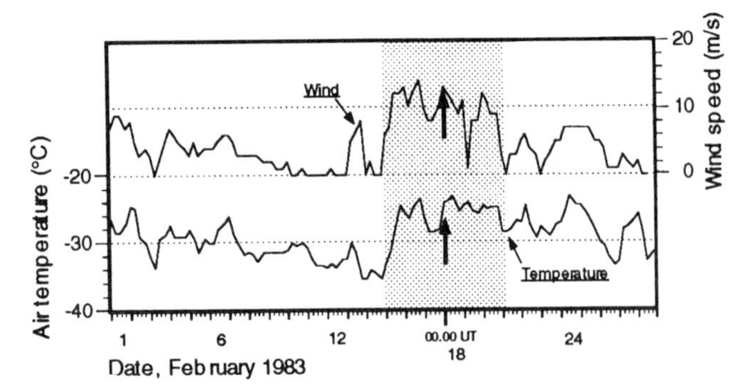

Fig. 5. Run of 6 hours wind speed and air temperature record during February 1983. Bold arrows mark the phase of plume occurrence. The six-day period of strong westerly winds is indicated by a stippled area.

from the entire polynya have an uncertainty of about 30-35 %. The values presented in this work are presumably to low, however, are probably not wrong by more than 50 %.

PHYSICAL CONDITIONS OF PLUME FORMATION

The phenomenon of plume origin on February 18, 1983 east of Bennett Island is suggested to be in causal connection with the significant ocean-to-atmosphere net energy flux from the local polynya. The feature is here proposed to be explained considering basic principles of atmospheric convection and thermal elevation of heated air masses (Figure 8). Due to friction by the surrounding colder atmosphere and additional lateral entrainment, the accelerated body of warm and moist air is slowed down at its margins, promoting the formation of a cylindrical thermal updraft (in short: thermal) of about 10 km in diameter.

The updraft enforces cold air masses from above and the surroundings to descend and substitute for the air at the base of the convective column. Sweeping over the open water area towards the center of thermal rise and withdrawing heat energy from the sea surface, the descended and formerly cold air now becomes part of the circulation. Due to very low surrounding temperatures at the condensation level and above, large volumes of small frozen water droplets are formed which generate ice clouds or plumes as observed on satellite images (Figure 3, Figure 4). According to the thermal energy released during the event, the condensation level of the warm updraft over the polynya is suggested to reach a height of approximately 1-3 km. Caused by its energy excess in respect to surrounding cold atmosphere, the rising warm air penetrated the Arctic boundary layer inversion. Thus, it can be assumed that, in analogy to convective clouds in cold air outbreaks over polar waters, the plume may have reached tropopause height, which was

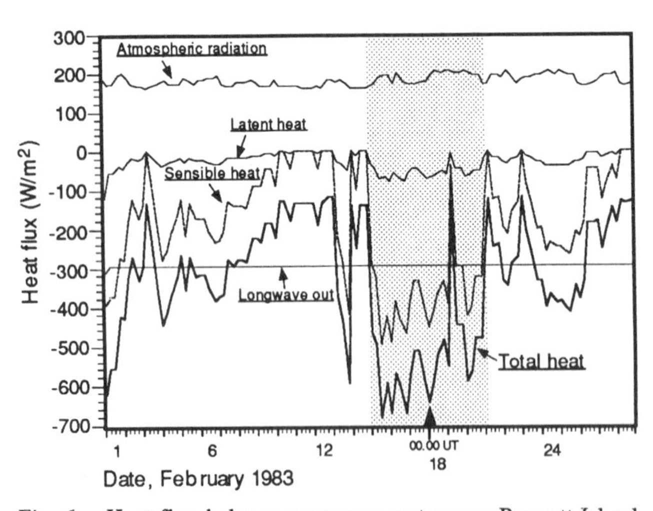

Fig. 6. Heat flux balance over open water near Bennett Island during February 1983. The arrow indicates the beginning of the plume event. The stippled area points out the six-day period of high net heat fluxes.

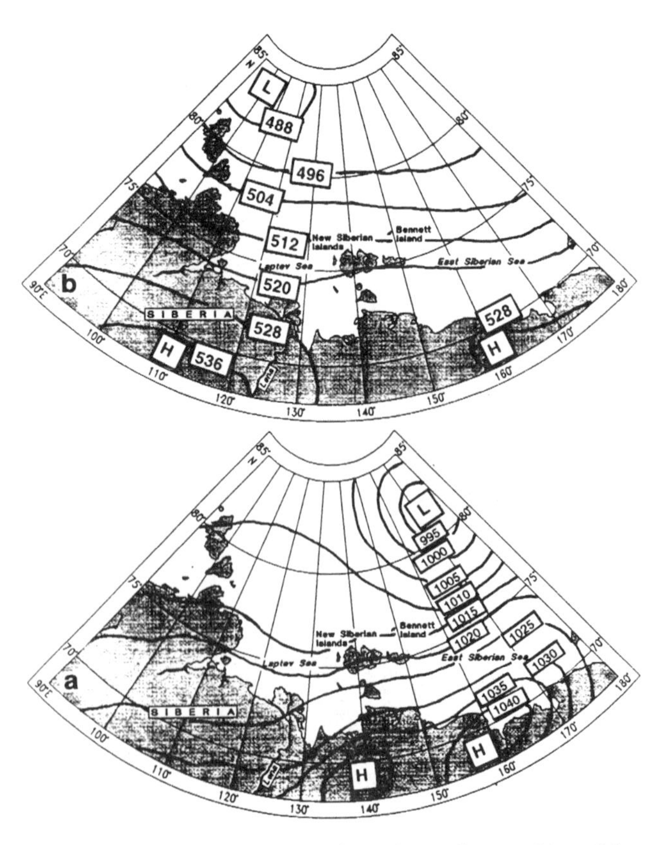

Fig. 7. Atmospheric pressure charts (a: surface, and b: ≈ 5 km altitude) of the east Siberian Arctic at 00:00 UT on February 18, 1983.

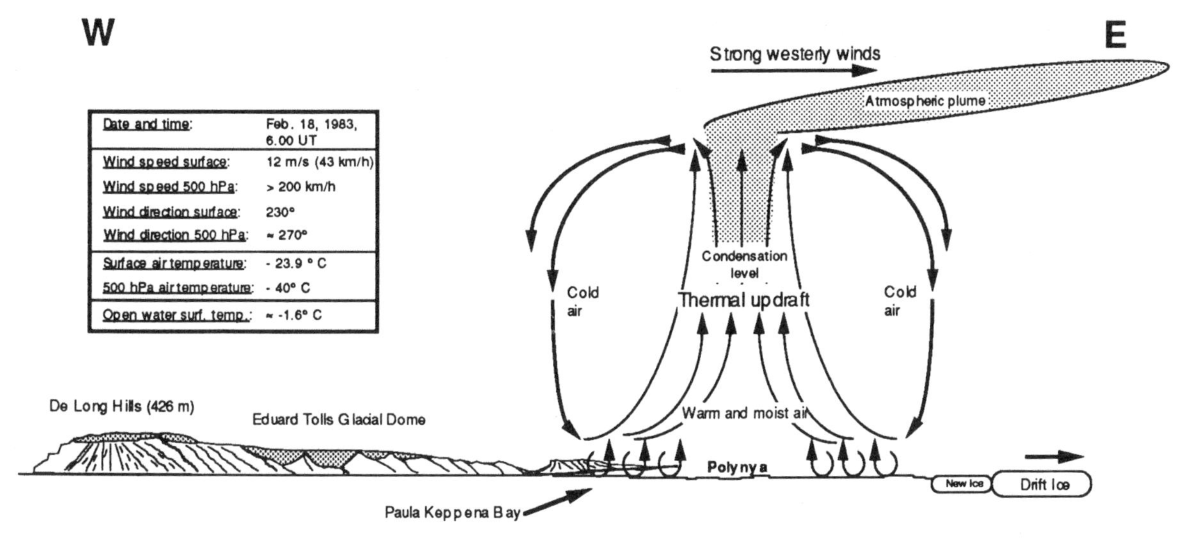

Fig. 8. Sketch of primary convection, thermal air mass circulation and replacement, and plume formation east of Bennett Island. The island profile is reworked from *Makeyev et al.* [1992]. Dotted areas represent snow, firn and glaciers. Note that the figure is not in scale.

at 8.5 km altitude on February 18, 1983 [*Kienle et al.*, 1983].

The appearance of two individual plumes over separated open water areas east of Bennett Island at 00:47 UT on February 18 implies the occurrence of distinct thermals during the initial phase of plume formation (Figure 4). Figure 9 shows, how several columns of rising air circulating in the same direction (a), will merge due to rotational expansion and, consequently, marginal collison and friction-induced turbulence (b). Finally, one single thermal updraft is formed typically tending to lie approximately in the center above the heating area (c).

DISCUSSION

As argued above, westerly winds are necessary for polynya opening and plume formation east of Bennett Island on February 18, 1983. According to long term observations [*Gorshkov*, 1983], however, prevailing winds in that area are from easterly and southerly directions with velocities as high as 16 m/s during the winter period (October to May), whereas westerly winds are infrequent. The relatively rare appearance of plumes east of Bennett Island in general may thus probably be due to a significant lack of strong westerly winds in the area of interest.

A faint low level plume was noticed at 18.04 UT on February 17, 1983 [*Kienle et al.*, 1983]. As shown in Figure 6 and Table 1, the net heat flux from the polynya was about 600 W/m^2 during that time and thus, sufficient energy was released for initiating strong upward movement of buoyant moist air. However, the frame conditions such as polynya width and wind direction were not favorable to maintain or even progress that feature and hence, the plume faded. Subsequently, the formation of plumes over open water east of Bennett Island, at least on the occasion of February 18, 1983, is suggested to depend on the succession and temporary coincidence of some special conditions: (i) the air temperature is significantly lower than -20° C and the mean wind velocity amounts to roughly or even exceeds 10 m/s during the event and before; (ii) a preceding period of strong westerly winds form large *stable-sized* areas of open water east and southeast of Bennett Island within typical synoptical time scales, which depend on air temperature and wind speed; (iii) the polynya opening period (ii) is succeeded by prevailing strong southeasterly winds which are not reduced by the shape of the island and thus, directly sweep over the open water enhancing significant heat loss from the ocean to atmosphere; (iv) extremely cold atmospheric temperatures and jet streams occur in the upper troposphere to promote the conversion of the convected moisture to long extended plume clouds consisting of supercooled small water droplets.

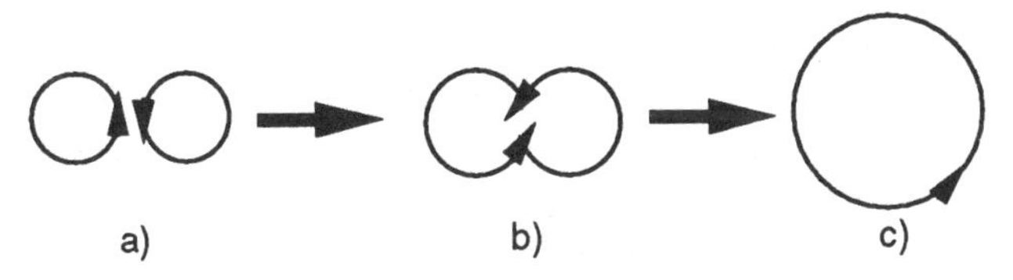

Fig. 9. The sketch shows a horizontal section of plume development.

The *perpendicular* and *undisturbed* thermal updraft over the heating polynya and the near surface peripheric inflow, once induced, is suggested to supply itself by the mechanism of thermal air mass circulation and replacement as presented in Figure 8. The jet streams at 5 kilometer altitude additionally may support the vertical circulation processes by drawing air masses upward inside the thermal comparable to a strong breeze sucking smoke from a chimney. Temporary oceano-graphic energy sources such as upwelling of relatively warm water masses or local currents could intensify the oceanic heat loss from the polynya and thus, may support con-trolling the mechanism of plume formation. Zhokov Island, located roughly 100 miles southeast of Bennett (Figure 3), faces a large east coastal polynya on February 18, 1983 as well, however, no plumes were detected in the vicinity. It can be speculated that Zhokov Island is characterized by local oceanographic conditions which do not promote the formation of visible plumes. The collapse of thermals, and thus, sudden fading of the Bennett Island plumes could be caused by the wrong frame conditions as mentioned above or by a reduced energy release from the water surface due to a cover of newly formed frazil and slush ice.

According to *Verkulich et al.* [1989] the maximum topographic height of Bennett Island is 426 m. This reference is in a good agreement with estimations of approximately 400 m height for Bennett Island, performed during GEOMAR expedition in April 1992 from about 700 m flight altitude (helicopter). Hence, cloud formation at an altitude of around 3000-4000 m (or even more) on the lee side of an "only" 400 m high island due to enforced orographic air movements, as proposed by Schnell [quoted after *Kerr*, 1992], seems to be not *that* likely. However, this unconvincing solution of the plume enigma is presumably attributed to the wrong assumption, that the Island has a height of 1000 m.

The strong surface wind blowing from southwesterly directions (210° and 230°, respectively, see Table 1) during the plume events on February 18, 1983 evidently could not sweep over mountain or glacier peaks on the island kicking up clouds off the eastern or even southeastern coast of Bennett. Orographic clouds, if kicked up at the lee side of the island during that day at all, must have developed *north* of Bennett in accordance to the surface wind directions.

Considering the *circular* base of the plume origin over open water at relatively high altitudes, a release of natural gases from *linear* faults on Bennett Island itself, as postulated by different authors, seems to be most unlikely. A simply forceless "bubbling" release of methane or gas hydrates from narrow, *extended* cracks in the submarine permafrost can be dismissed as a valid source for *punctiform* plumes, as well. Additionally, no fuel burn can generate such amounts of energy which were necessary for Bennett Island plume formation on February 18, 1983. A burning town, as presented by *Priestley* [1959] in a drastic model, would provide "solely" energy amounts of roughly 2.5×10^{10} W for smoke plume formation reaching altitudes of about 2,400 m.

CONCLUSIONS

1) On February 18, 1983, extended atmospheric plumes emanated from punctiform sources directly above a large polynya east of Bennett Island.

2) Polynyas are important sources of thermal energy during polar winter. Approximately $1.8\text{-}2.4 \times 10^{11}$ W were released from the open water area near Bennett Island during that day. The heat flux likely promoted a strong turbulent upward air movement (thermal) and was by far sufficient for high altitude plume formation.

3) The initial development, maintainance and subsequent progress of such thermals depend on the succession and temporary coincidence of distinct frame conditions, which were fulfilled on February 18, 1983: (i) low temperatures and high wind speeds during the event and before, (ii) preceding westerly wind for opening the polynya, (iii) southwesterly winds sweeping directly over the polynya during the event, and (iv) extremely cold temperatures and the occurrence of jet streams in the upper troposhere.

4) The *perpendicular* and *undisturbed* updraft of warm and moist air is suggested to supply itself by the mechanism of thermal air mass circulation and replacement.

5) Additionally, the plume formation may be supported by the chimney-effect attributed to jet-streams from westerly directions. These strong winds also control the far eastward extension of the plume at the 500 hPa level.

6) Faint development or break-down of thermals and thus, sudden fading of the plumes, could be due to unfavorable local meteorological and ice conditions: (i) lack of open water, (ii) reduced release of oceanic heat due to calming winds or due to a cover of newly produced frazil and slush ice, and (iii) changing wind direction.

7) A simply forceless bubbling release of methane or gas hydrates from shelf sediments, as well as volcanic activities or anthropogenic impacts (fuel burns) can be assumably excluded as a valid source for high altitude plume formation in the vicinity of Bennett Island.

Acknowledgments. This study was financially supported by research funds appropriated by the "Bundesministerium für Forschung und Technologie" of the Federal Republic of Germany to the GEOMAR Forschungszentrum in Kiel,

Germany. Furtheron, I am indebted to thank Prof. Dr. Lutz Hasse, Dr. Erk Reimnitz, Dr. Dirk Nürnberg, Dr. Robert Spielhagen, Niels Noergaard-Pedersen and Ortrud Runze for reviewing this paper and giving new ideas. Special thanks are given to Eckhard Kleine and Peter Löwe from the "Bundesamt für Seeschiffahrt und Hydrographie" at Hamburg, for performing the heat flux calculations. This paper is dedicated to my dear friend and colleague Erk Reimnitz.

REFERENCES

Clarke, J. W., P. St. Amand, and M. Matson, Possible Cause of Plumes from Bennett Island, Soviet Far Arctic, *AAPG Bull., 70*, 574, 1986.

Dethleff, D., D. Nürnberg, E. Reimnitz, M. Saarso, and Y. P. Savchenko, The Laptev Sea - Its significance for Arctic sea-ice formation and transpolar flux of sediments, *Rep. Polar Res., 120*, 44 pp, 1993.

European Meteorological Bulletin, *Amtsblatt des Deutschen Wetterdienstes*, Februar 1983, Deutscher Wetterdienst, Zentralamt Offenbach, Germany, 1983.

Glendening, J. W., and S. D. Burk, Turbulent Transport from an Arctic Lead: A Large-Eddy Simulation, *Boundary Layer Meteorology*, 59, 315-339, 1992.

Gorshkov, S. G., *World Ocean Atlas*, 3, Arctic Ocean, 80-103, 1983.

Kerr, R. A., U. S. - Russian Team Solves Arctic Mystery, *Science, 257*, 35, 1992.

Kienle, J., J. G. Roederer, and G. E. Shaw, Volcanic Event in Soviet Arctic? *EOS Trans. AGU, 64, 20*, 376-378, 1983.

Lindsay, R. W., Wind and temperature profiles taken during Arctic Lead Experiment, *M. S. thesis*, 89 pp, Dep. of Atm. Sc., Univ. of Wash., Seattle, 1976.

Makeyev, V., V. Pitul'ko, and A. Kasparov, Ostrova De-Longa: An analysis of palaeoenvironmental data, *Polar Record, 28, 167*, 301-306, 1992.

Martin, S., and D. J. Cavalieri, Contributions of the Siberian Shelf Polynyas to the Arctic Ocean Intermediate and Deep Water, *J. Geophys. Res., 94*, 12,725-12,738, 1989.

Matson, M., Large Plume Events in the Soviet Arctic, *EOS Trans. AGU, 67*, 1372-1373, 1986.

Maykut, G. A., Large-scale heat exchange and ice production in the Central Arctic, *J. Geophys. Res., 87*, 7971-7984, 1982.

Parkinson, C. L., and W. M. Washington, A large-scale numerical model of sea ice, *J. Geophys. Res., 84, C1*, 311-337, 1979.

Paull, C. K., and W. J. Buelow, Arctic Mystery: Plumes or Clouds, *Science, 258*, 725, 1992.

Paull, C. K., and W. J. Buelow, Enigmatic Arctic Cloud Plumes, *Science, 259*, 164, 1993.

Pease, C. H., The Size of Wind-Driven Coastal Polynyas, *J. Geophys. Res., 92*, 7049-7059, 1987.

Priestley, C. H. B., *Turbulent Transfer in the Lower Atmosphere*, The University of Chicago Press, 130 pp, Chicago, Illinois, U.S.A, 1959.

Schnell, R. C., R. G. Barry, M. W. Miles, E. L. Andreas, L. F. Radke, C. A. M. Brock, M. P. McCormick, and J. L. Moore, Lidar detection of leads in Arctic sea ice, *Nature, 339*, 530-532, 1989.

Schnell, R. C., A. D. A. Hansen, E. Dlugokencky, T. J. Conway, A. V. Polissar, and G. S. Golitsyn, Airborne Investigation of the Bennett Island Plume, in *Abstracts for the Fifth Symposium on Arctic Air Chemistry, September 8-10*, Copenhagen, 1992.

Sechrist, F. S., R. W. Fett, and D. C. Perryman, *Forecasters Handbook for the Arctic,* Naval Environmental Prediction Research Facility, Technical Report TR 89-12, Monterey, CA, 1989.

Smith, S. D., R. D. Muench, and C. H. Pease, Polynyas and leads: An overview of physical processes and environment, *J. Geophys. Res., 95*, 9461-9479, 1990.

St. Amand, P., P. J. Clarke, and M. Matson, Curious plumes from Bennett Island, in *Proceedings of the Arctic Oceanography Conference and Workshop, June 11-14, 1985*, Naval Ocean Research and Development Activity (NORDA), Bay St. Louis, 1985.

Verkulich, S. R., A. G. Krusanov, and M. A. Anisimov, Recent State and Changes in the Glaciation of Bennett Island During the Last 40 Years (in Russian), *Arctic and Antarctic Research Institute*, St. Petersburg, Russia, 1989.

Vol'nov, D. A., D. S. Sorokov, and O. A. Ivanov, *Geologicheskaya karta Novosibirskikh Ostrovov* (Geological map of the New Siberian Islands), Moscow, Ministry of Geology, 1970.

Zakharov, V. F., The role of flaw leads off the edge of fast ice in the hydrological and ice regime of the Laptev Sea, *Academy of Sciences of the USSR - Oceanology, 6*, 815-821, 1966.

Dirk Dethleff, GEOMAR Research Center for Marine Geosciences, Wischhofstraße 1-3, D-24148 Kiel, Germany.

Interannual Variability of the Thermohaline Structure in the Convective Gyre of the Greenland Sea

G.V.Alekseev, V.V.Ivanov and A.A.Korablev

Arctic and Antarctic Research Institute, St.Petersburg, Russia

The temporal variability of thermohaline conditions in the Greenland Sea Convective gyre is examined on the basis of the long term observational series. The existence of two stable types of winter thermohaline structure is discovered. The transition from one type to another occurs through the pre-convective state and consequent convection. The characteristic feature of the pre-convective state is an increased (about 0.07 PSU above normal) surface salinity, caused by the external salt water influx. Potential temperature and salinity time series joint analysis confirms the crucial role of the surface salinity in the convection realization. An explanation of the surface to bottom overturning events and of the low frequency variability of convection activity is suggested on this basis.

1. INTRODUCTION

In recent years, an increased attention has been given to the studies of the polar oceans effect on the global climate changes [*ACSYS*,1992]. Among the mechanisms of this influence is a deep winter convection by means of which deep and bottom waters are renewed and the global vertical circulation is maintained. In the Northern hemisphere the central part of the Greenland Sea is considered to be the most suitable region for the deep and bottom waters formation due to the convection process [*Nansen*,1906; *Nikiforov, Shpayher,* 1980; *Johanessen*,1986]. Interaction between different current systems upon the complicated bottom topography creates the cyclonic circulation field. Together with the intense surface cooling and the weak density stratification in the winter time it provides necessary preconditions for the deep convection development [*Killworth*,1983]. However, a remarkable irregularity of deep convection events in the past is a reason to presume the existence of additional, essentially variable forces that can contribute much in the convection realization. Evidently, the action of those forces should be reflected in the convective thermohaline structure features.

The Polar Oceans and Their Role in Shaping the Global Environment
Geophysical Monograph 85

Four water masses are traditionally distinguished in the Greenland Basin [*Swift*,1986]. The surface water extends down to 100-120 m and is extremely sensitive to the hydrometeorological conditions annual evolution. The temperature and salinity annual oscillation ranges are about 4°C and 1.2 PSU (practical salinity units, permille salt by weight) correspondingly. The intermediate layer is occupied by the warmer (in the winter time) and more saline modified Atlantic water. It provides a sort of "heat-screen", isolating deep and bottom waters from external influences. The warming of the bottom water, which is observed within the recent twenty years, is the evidence of the deep convection activity reduction - the only reasonable mechanism of bottom water ventilation [*Meincke et al.*,1992].

In the present paper, we are studying the hypothesis that the deep convection development is significantly controlled by the surface salinity. In particular, surface to bottom convection requires the surface salinity to be higher than 34.82 PSU [*Alekseev et al.*, submitted to *Oceanology*]. If thus is true, the convective transformation of the thermohaline structure in the intermediate and deep layers should be correlated to the increase of the surface salinity above the crucial value. The traces of such intercommunication are demonstrative only for the two cases of the deep convection confirmed by direct measurements (1984 and 1989). To obtain a more valid conclusion we examine the correctness of this conformity on the basis of the continuous observational series.

2. DATA COLLECTION

The data presented here were gathered from different sources (Figure 1). A historical data set including the deep oceanographic observations through the time period 1950 - 1984 was received from WODC-2 (Obninsk,Russia). Since 1984, AARI research vessels have explored the Central Greenland Sea more thoroughly on a regular station net. At present, the data set contains 21 specialized surveys with an increased spatial resolution. The high quality data obtained in the framework of the Greenland Sea Project (GSP) comprise the years 1987-1989 and 1993 (Table 1). The region under investigation is shown in Figure 2 with smoothed bottom topography as background information and the total number of hydrographical stations used for the analysis.

All deep sea CTD and Nansen bottle data were loaded into special hydrographical data base for convective region. A uniform data quality control procedure has been applied for the data checking and validation. Data were checked for random and systematic errors. For random errors correction, the standard deviations for different layers inside convection region were calculated. The data with values outside the defined limits for each layer were eliminated.

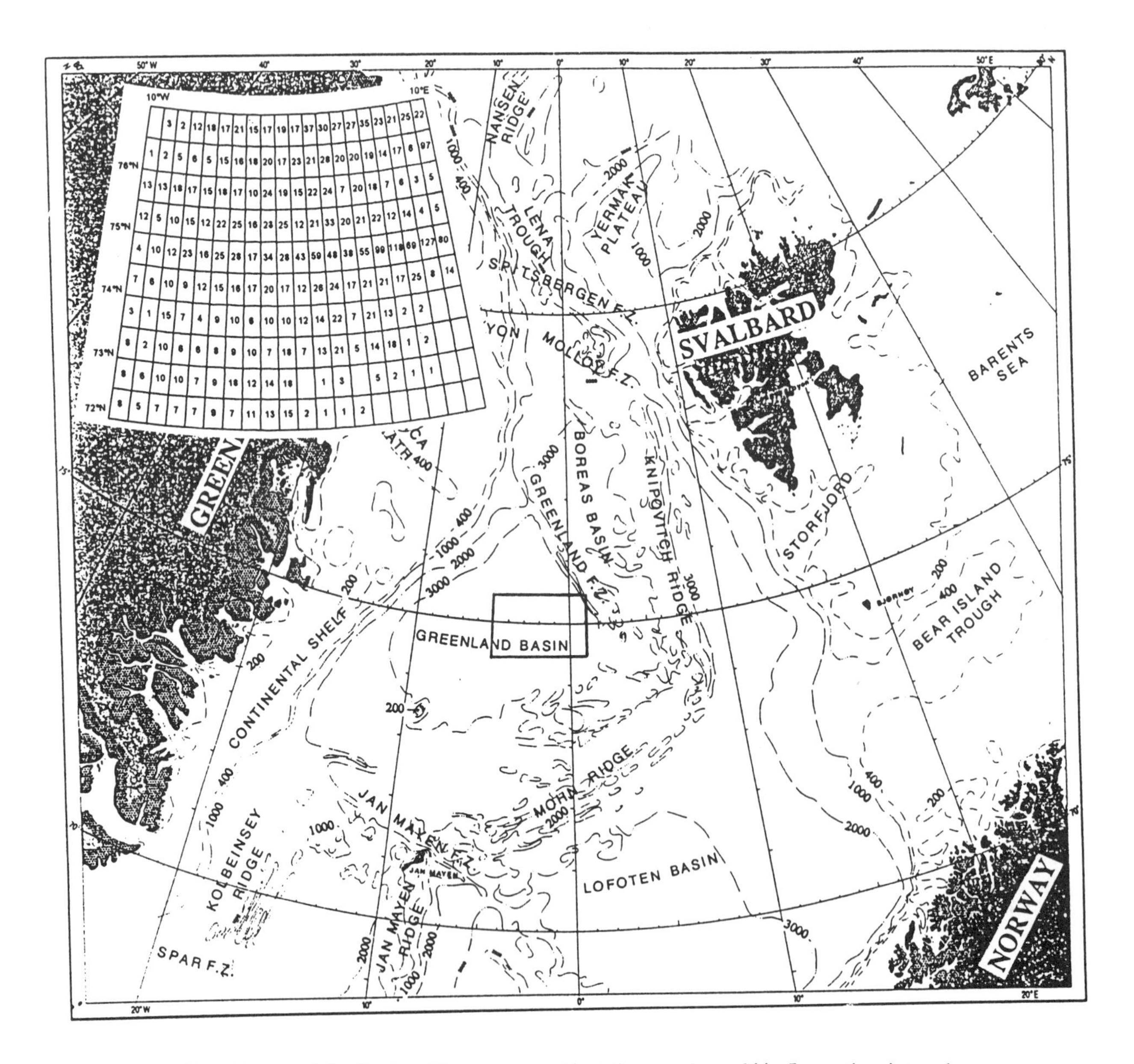

Fig.1. Temporal distribution of the oceanographic stations number within 5-year time intervals.

Table 1. Time table of the surveys included in data collection in the region limited by 72°-77°N and 10°W-10°E

N	R/V	num of st.	start	finish
1	Shuleykin	60	28.03.84	11.04.84
2	Shuleykin	59	20.01.86	4.02.86
3	Shuleykin	53	22.04.86	2.05.86
4	Multanovsky	41	28.11.86	8.12.86
5	Shuleykin	42	15.01.87	25.01.87
6	Multanovsky	80	16.04.87	1.05.87
7	Viese	9	7.08.87	12.08.87
8	Viese	40	14.08.87	21.08.87
9	Multanovsky	56	19.10.87	30.10.87
10	Shuleykin	53	17.01.88	29.01.88
11	Multanovsky	13	2.11.88	4.11.88
12	Multanovsky	40	7.03.89	16.03.89
13	Multanovsky	15	17.03.89	19.03.89
14	Viese	60	30.08.89	9.09.89
15	Multanovsky	35	7.12.89	13.12.89
16	Shuleykin	60	18.02.90	25.02.90
17	Multanovsky	20	6.03.90	9.03.90
18	Multanovsky	29	5.06.90	11.06.90
19	Viese	29	12.07.90	17.07.90
20	Viese	41	10.08.90	17.08.90
21	Shuleykin	67	15.03.91	31.03.91
22	Multanovsky	54	6.03.93	21.03.93
1	Hudson	29	3.03.82	29.03.82
2	Meteor	30	2.06.82	19.06.82
3	Polarstern	21	18.06.87	29.06.87
4	Valdivia	14	11.02.88	16.02.88
5	Valdivia	31	7.06.88	16.06.88
6	Polarstern	24	14.06.88	2.07.88
7	Meteor	29	26.11.88	12.12.88
8	Valdivia	23	26.02.89	12.03.89
9	Polarstern	17	18.05.89	30.05.89
10	Mosby	25	27.06.89	5.07.89
11	Polarstern	23	2.04.93	12.04.93

The coefficients for limits identification depend on the depth level. For surveys with insufficient information for standard deviation calculation, the corrections were made after the comparison of vertical profiles with the average one for a defined region. Where it was possible, intercalibration with high quality GSP data was performed. The systematic errors correction procedures for each survey were carried out for Greenland Sea Deep Water mass. This water mass exhibits extremely small variations for all oceanographic parameters. Hence, the surveys with inexplicable deep water mean values were excluded from the analysis.

3. THERMOHALINE STRUCTURE CONVECTIVE TRANSFORMATION

The potential temperature and salinity mean values and standard deviations for different water layers in the Central Greenland Sea are shown in Table 2. They were calculated by means of a standard averaging procedure using oceanographic data for the two years when the deep convection was observed (1984, 1989) and for the two years when it was absent (1987, 1990). The corresponding vertical profiles are presented in Figure 3. The averaging area was limited by the 3200 m isobath in order to eliminate the Arctic Frontal zone, which has an essentially different thermohaline structure. The upper and intermediate layer characteristics are mostly different for the convective and non-convective cases. The deviations in deep and bottom layers are hardly visible, but nevertheless they are still significant taking into account the number of the averaged values. The homogenization of the water column is the

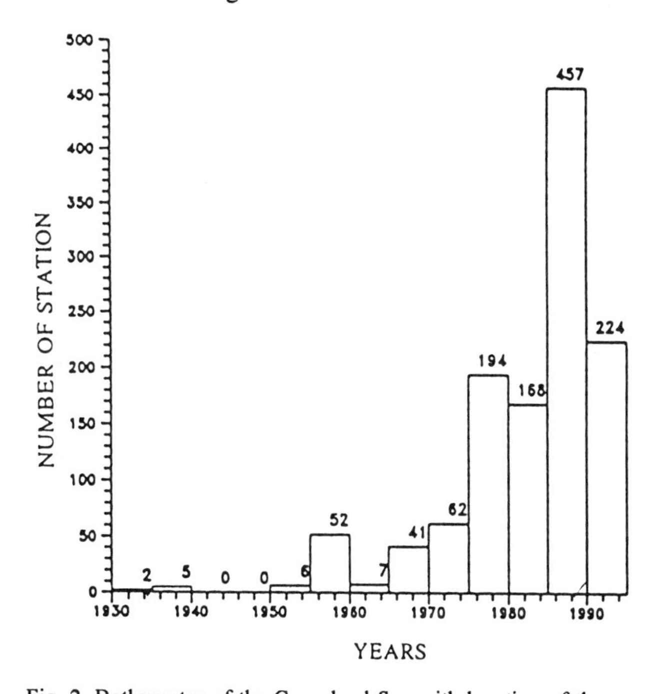

Fig. 2. Bathymetry of the Greenland Sea with location of the area under investigation. The figures in the 0.5° x 1° squares represent the total number of the deep oceanographic stations inside each square.

Table 2. Mean potential temperature and salinity averaged in different layers for surveys when deep convection was observed (1984, 1989) and for surveys when it was absent (1987,1990). The region for mean values and simple standard deviations (SSD) calculation is limited by 74°30-75°30 N and 5°W-1°E. Stations were chosen for bottom depth more then 3200 m and for winter time (January-April).

LAYER [m]	convection				no convection			
	θ [°C]/SSD		S [PSU]/SSD		θ [°C]/SSD		S [PSU]/SSD	
0 - 40	-1.485	0.179	34.832	0.051	-1.784	0.090	34.749	0.031
40 - 90	-1.414	0.286	34.842	0.040	-1.771	0.103	34.752	0.029
90 - 125	-1.381	0.285	34.848	0.032	-1.611	0.459	34.770	0.035
125 -175	-1.320	0.249	34.855	0.024	-1.034	0.644	34.826	0.042
175 - 225	-1.285	0.226	34.859	0.020	-0.587	0.281	34.870	0.018
225 - 275	-1.261	0.225	34.863	0.016	-0.608	0.205	34.880	0.013
275 - 350	-1.247	0.192	34.866	0.011	-0.674	0.180	34.883	0.011
350 - 450	-1.223	0.184	34.867	0.011	-0.834	0.144	34.881	0.012
450 - 550	-1.212	0.160	34.870	0.012	-0.947	0.119	34.881	0.010
550 - 700	-1.235	0.181	34.874	0.012	-1.020	0.098	34.881	0.010
700 - 900	-1.212	0.108	34.876	0.011	-1.084	0.044	34.883	0.008
900 - 1100	-1.221	0.107	34.880	0.009	-1.109	0.028	34.885	0.007
1100 - 1350	-1.193	0.076	34.882	0.008	-1.130	0.021	34.887	0.005
1350 - 1750	-1.214	0.069	34.885	0.007	-1.151	0.021	34.889	0.004
1750 - 2100	-1.208	0.053	34.888	0.005	-1.166	0.025	34.891	0.005
2350 - 2750	-1.239	0.042	34.887	0.005	-1.207	0.032	34.891	0.005
2750 - 3250	-1.261	0.035	34.887	0.005	-1.240	0.028	34.891	0.005
3250 - 3700	-1.281	0.024	34.887	0.004	-1.261	0.028	34.888	0.005

general tendency of the convective transformation. It is especially pronounced within the upper and intermediate layers. The first one is warmed and salinated while the second one is cooled. The absence of total vertical uniformity in the convective case may seem odd for the first glance. However, as the deep convection is not an instantaneous process, the scattered homogeneous water columns, created by the small convective events, are rapidly collapsed due to lateral inflow and mixing with the ambience [*Worcester et al.*,1993]. Besides, the data used for the Table 2 rather characterize the change of general vertical thermohaline structure caused by convection within the relatively large region exceeding the area of single downwelling cores, where the vertical uniformity is more pronounced [*Greenland Sea Project Group*, 1990].

The mean salinity values within the surface to bottom water columns for convective and non-convective cases are 34.881 PSU and 34.883 PSU respectively. The increase of the mean salinity is observed within the upper 150 m layer (ΔS=0.075 psu) while its decrease occurs within the rest of the water column (ΔS=-0.005 PSU). The convective evolution of the salinity field is supposed to be caused by external sources, lateral mixing and vertical mixing. Obviously, in the absence of the two first effects, the total

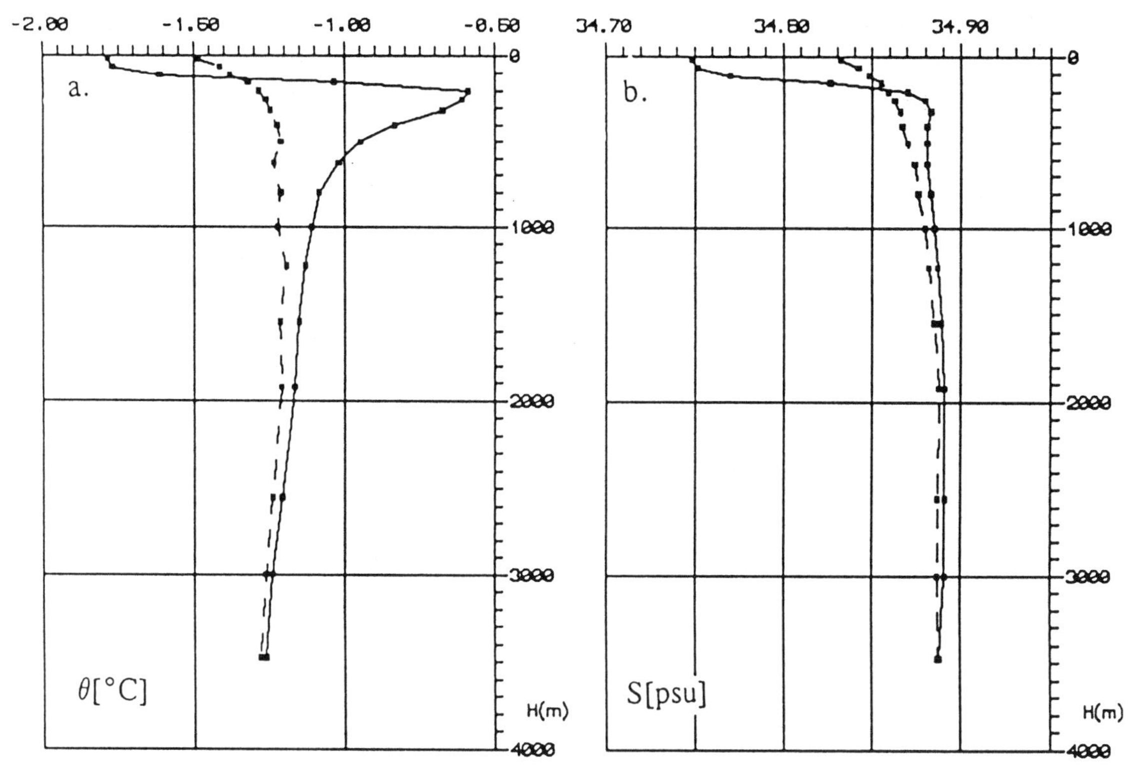

Fig.3. Vertical distribution of mean water properties in the Central Greenland Sea (74°30'-75°30'N, 5°W - 1°E) : (a) potential temperature (°C), (b) salinity (PSU) for convective years (1984, 1989) - dashed lines; for years without convection - solid lines (see also Table 2).

salt amount within the water column ought to be preserved. Therefore, the imbalance found is likely to be connected with the difference between the salt outflow due to the lateral mixing and the salt influx from external sources. The upper layer is the most suitable for an external salt injection, judging by the experimental data (Figure 3) and by the theoretical presumption concerning the necessity of an initial surplus salinization of the upper layer, needed to start the overturning [*Aagaard and Carmack,* 1989; *Rudels,*1990]. However, the salt redistribution due to the vertical mixing is also significant. Its effect is clearly seen in the surface distribution of the oceanographic characteristics in 1984 - a unique surface to bottom convection event confirmed by direct measurements [*Nagurny et.al.*,1985]. In Figure 4 one can see the local ellipse-shaped area (about 100 km in diameter) with extremal values in its centre. The signs of temperature, salinity, dissolved oxygen and silicate anomalies show that the source of these anomalies is the underlying water.

To estimate the relative contribution of external sources and convective mixing to the reformation of the salinity field, two sequential probes, carried out in spring 1989 in the climatic centre of the Greenland Convective gyre were examined (Figure 5). The first probe (A) corresponds to pre-convective conditions. In the upper 1000-m layer, there was a stable stratification, in the lower layer, a quasi-homogeneous, mean-like temperature and salinity distribution. The average salinity within the upper 0-100 m layer was 34.817 PSU, i.e. 0.06 PSU higher than the non-convective average value (see Table 2). In the next four days the upper 800 m layer was mixed but a number of thin inverse layers still remained (B). A noticeable temperature and salinity decrease in the intermediate and deep layers indicates that convection reached the 1500 m horizon. This conclusion is also confirmed by the mean salinity values calculated for typical layers before and after the overturning (Table 3). Salt redistribution due to the contra-directed vertical motions leads to the noticeable increase in the salinity in the upper 400 m layer. The salinity increase is maximum at the surface (about 0.05 PSU). Salt balance is not maintained within the convective layer due to lateral mixing. The latter accounts for the decrease of the average salinity of 0.004 PSU This effect is hardly seen at the surface, but it is quite significant in the deep layer. Similar conclusions can be made from the comparison of the two probes carried out a few days earlier, about 200 km to the

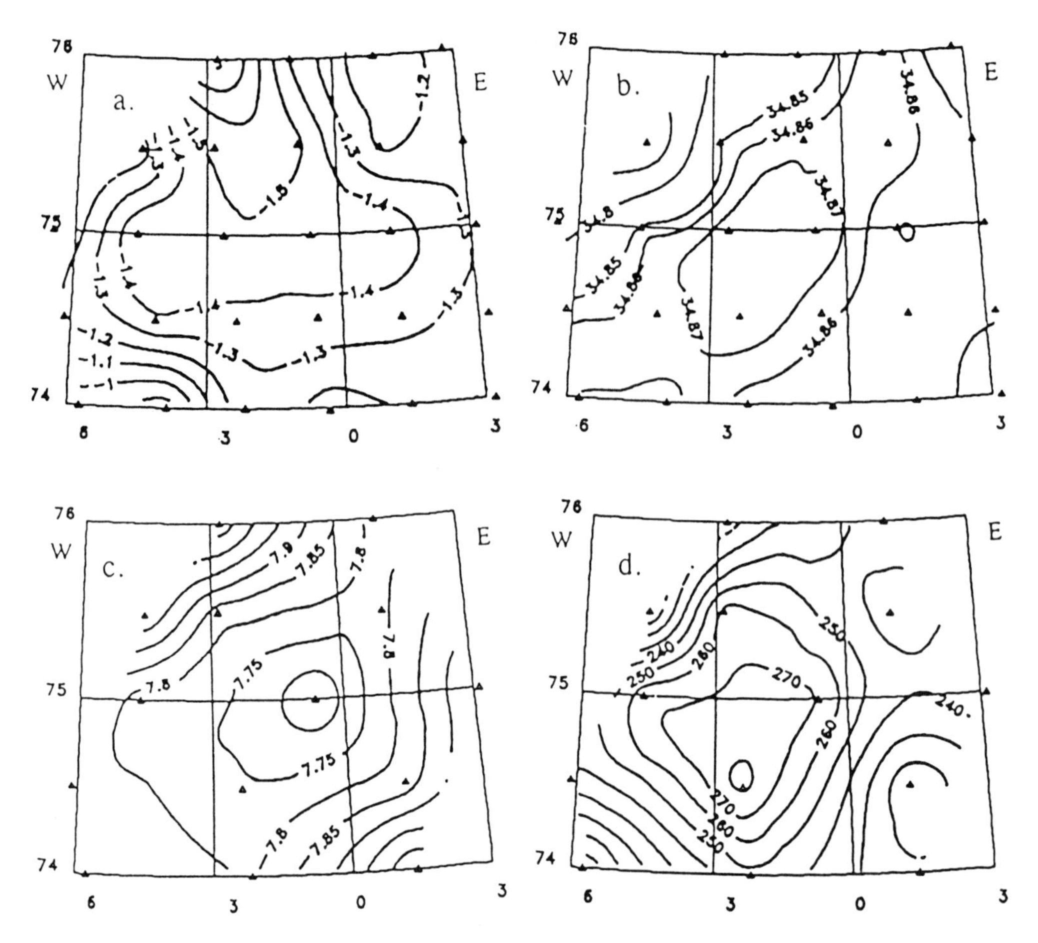

Fig.4. Distribution of the oceanographic characteristics at the surface in April,1984: (a) temperature (°C), (b) salinity (psu), (c) dissolved oxygen (ml/l), (d) silicates (mcg-at/l).

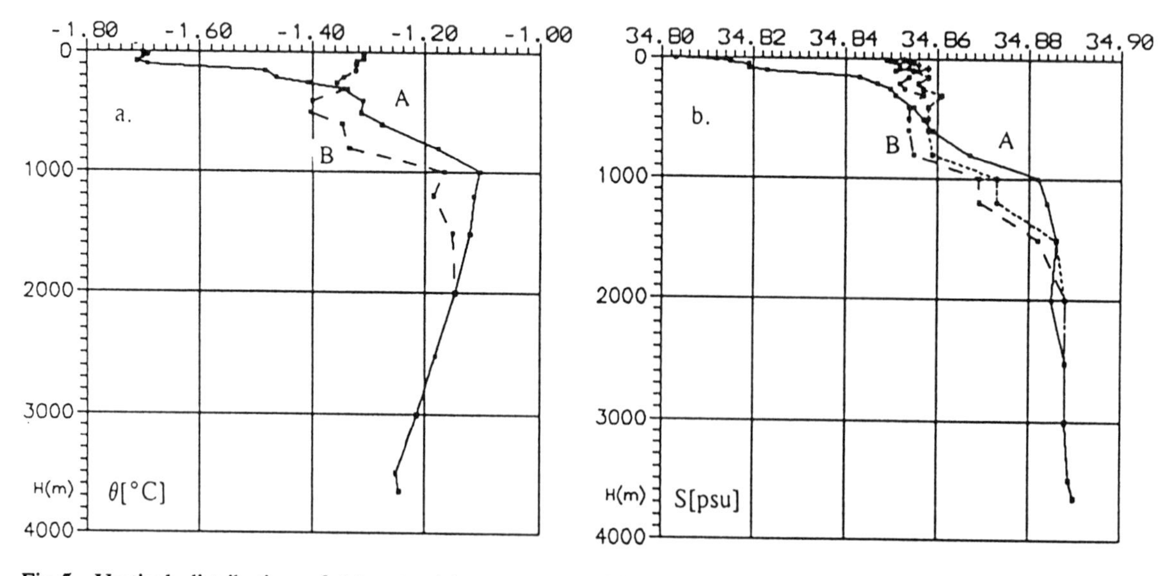

Fig.5. Vertical distribution of (a) potential temperature (°C) and (b) salinity (PSU) for the two consecutive oceanographic stations located at 75°00'N, 0°30'W (R/V "Professor Multanovsky"): A - March,13,1989 (solid line); B - March,17,1989 (dashed line); calculated profile with excluded lateral mixing effect (dotted line).

Table 3. Mean salinity within the typical layers for the sequantial pairs of stations in March, 1989 (see Figures 5,6)

LAYER [m]	St. A	St. B	LAYER [m]	St. C	St. D
0 - 400	34.840	34.854	50 - 500	34.867	34.873
400 -1500	34.874	34.864	500 -2000	34.892	34.879
0 -1500	34.865	34.861	50 -2000	34.886	34.878
1500 -bot.	34.887	34.887	2000 -bot.	-	-

west by R/V "Valdivia" (Figure 6). The distance between the stations C and D was about 27 km. Thus, one can assume that vertical distributions obtained at these stations also describe the different stages of the same process. The balance estimates for these probes are of the same order and tendency as for the previous ones (Table 3). The essential difference is the existence of 250 m homogeneous layer at station C. Consequently, the near surface salinity (the data above 50 m are absent) cannot be regarded as a pre-convective parameter. Using the balance estimates for stations A and B we can approximately determine the pre-convective surface salinity in the vicinity of C and D stations as 34.83 PSU

To summarise, the difference between the average convective and non-convective surface salinity values is about 0.08 PSU (Table 2), but the salinity value directly in the centre of the convective zone is about 0.04 PSU higher than the average one (see Figures 4-6). Therefore, we may assume the actual difference between the convective and non-convective values to be approximately 0.12 PSU On the other hand, the difference between the convective and pre-convective surface salinity values is about 0.05 PSU (according to the field data in 1989). Thus, the transference to the convective thermohaline structure is accompanied by an external salinization about 0.07 PSU, while the relative salinization due to the vertical mixing is about 40% of the total one. To obtain the balance, the salinity decrease within the entire water column due to lateral mixing ought to be about 0.0025 PSU (see Table 2 and Figure 5 -dotted line). Hence, the pre-convective surface salinity can be estimated as 34.819 PSU The surface salinity histogram, calculated using the entire winter data set, confirms this conclusion (Figure 7). Three frequency maximums evidently correspond to the most probable thermohaline conditions in the Central Greenland Sea. Two additional conclusions should be made according to the accomplished analysis. Firstly, the needed initial salinization is relatively small and can be simply achieved either by means of advection, or by freezing out fresh water. Secondly, the convective transition would disturb deep and bottom layers only in the case of

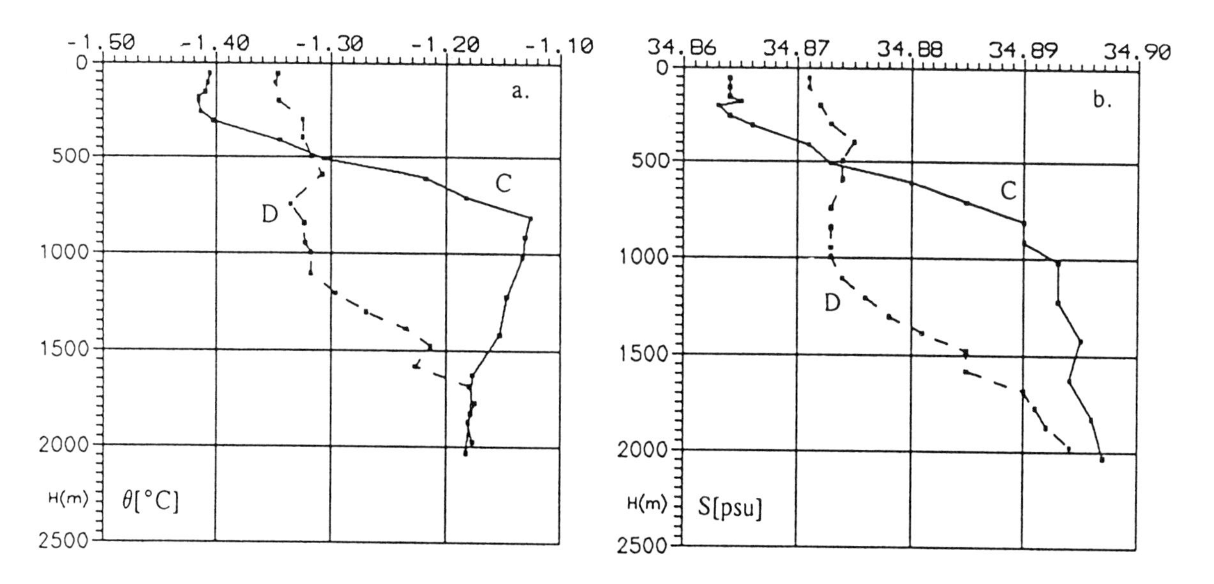

Fig.6. Vertical distribution of (a) potential temperature (°C) and (b) salinity (PSU) for the two oceanographic stations, carried out by R/V "Valdivia": C - 75°08'N, 3°28'W, March,9,1989 (solid line); D - 74°52'N, 3°24'W, March,9,1989 (dashed line).

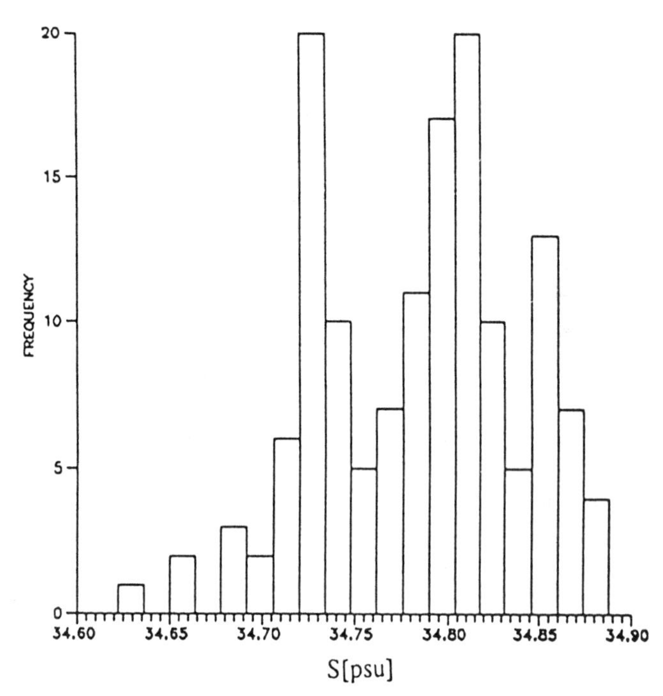

Fig.7. Histogram of the winter surface salinity (PSU) based on the entire data set.

deep (lower than 1500 m) convection, while the transformation of the thermohaline structure in the intermediate layer would occur even for a weak convection process.

4. THERMOHALINE PARAMETERS TIME SERIES ANALYSIS

On the basis of the obtained results, the 35-year observational series was examined in order to find out the appropriateness of temperature and salinity variability within the typical layers of the Greenland Sea Convective gyre. The area for spatial averaging was the same as that described in the previous section. The following vertical intervals were chosen for calculations: the surface - A (0-12 m), the intermediate layer with winter temperature maximum - B (240-420 m), the deep layer - C (1000 -2000 m) and the bottom water mass - D (2000 m - bottom).

In the time series presented in Figure 8-a, the mean values of surface salinity corresponding to the winter time (January-April) are used, while the averaged potential temperature in the intermediate and deep layers corresponds to the entire year. It is clearly seen that potential temperature curves in the layers B, C and D qualitatively coincide with each other. On these curves, one can notice a non-monotonic trend resembling the branch of a low-frequency oscillation with a period exceeding 80 years and a minimum value in 1975. The oscillation's range strongly decreases with depth. Its value in the D-layer is about one fifth of that in the intermediate one. Smaller interannual variations are present against this background. In the B-layer their range may exceed the total low-frequency variability within a single year (0.70°C in 1989-1990). The confirmed surface to bottom convection event (1984) is distinguished by an anomalously low potential temperature in all layers. Besides this year coordinated falls on the potential temperature curves in the B and C layers are observed in 1981, 1986, 1988 and 1989. In 1974, the absolute potential temperature minimum (-1.297°C) was obtained in the bottom layer.

The same curves, corresponding to a shorter time interval (1986 -1991) with an increased temporal resolution are presented in Figure 8-b. It is clearly seen that in 1986, 1988 and 1989 the potential temperature decrease in the intermediate and deep layers was caused by the winter convection. On the other hand, in 1987, 1990 and 1991 weak temperature falls were observed only within the B-layer while the deep water mass remained practically undisturbed. Hence, a threshold value of the potential temperature in the B-layer which characterizes the convection extending lower than the 1000 m horizon can be approximately estimated as -1.2° C (compare with Table 2). The potential temperature decreasing lower than this limit actually means that the intermediate temperature maximum is broken down and nothing, except external conditions, is able to prevent the further convection development. However, the quasi-monotonic increase of the bottom water potential temperature shows that there were no events of surface to bottom convection within the last decade (since 1984). The post-convective regeneration of the thermohaline structure contributes much to the future convection development. The intense intrusion of the modified Atlantic water into the intermediate layer totally restores the non-convective thermohaline structure in the B-layer as it happened in 1986 and 1989, while, in the case of weaker intrusion, only partial regeneration occurs (1988). The latter provides more favourable thermohaline conditions for convection development in the next year and promotes the formation of clusters of deep convection events [*Killworth*, 1983].

In the time series of the surface salinity (Figures 8-a,b) under the background of the general salinity decrease (0.04 psu from 1959 to 1993), the short-term fluctuations are also present. In several cases, the coincidence between the maxima and minima of the surface salinity and opposite anomalies of potential temperature in the layers B and C is observed. In particular, it is true for convective and non-convective years, presented in the previous section.

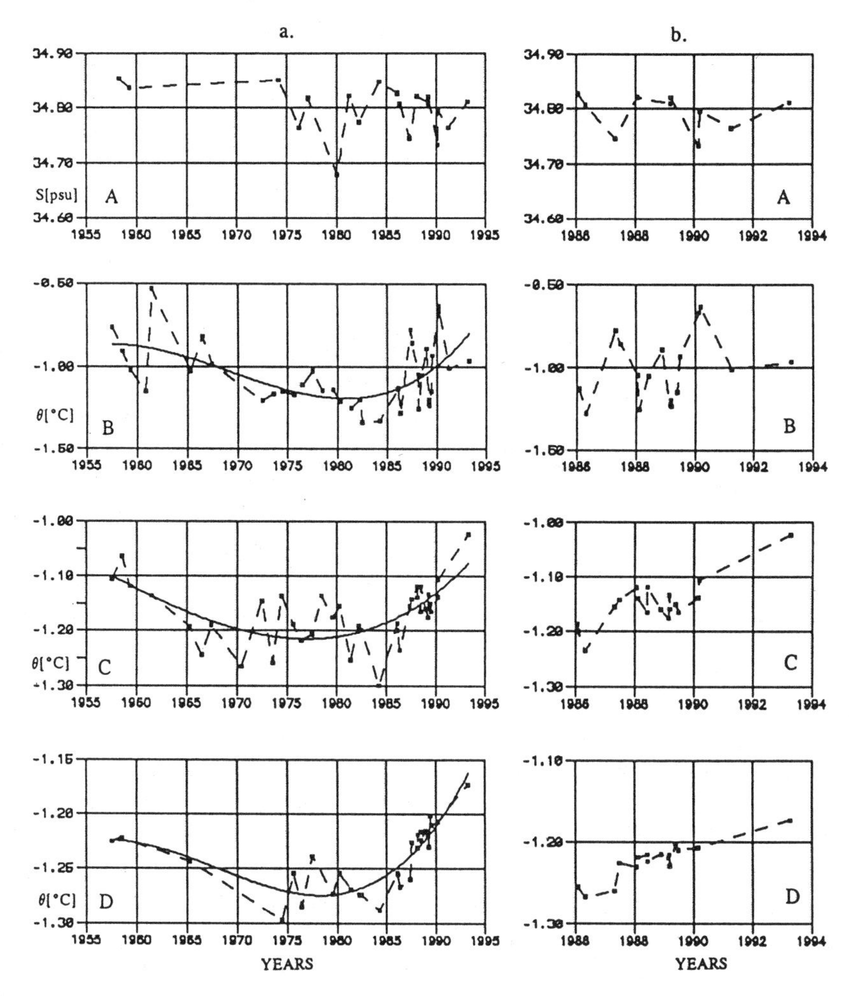

Fig.8. Time series of the mean water properties in different water layers of the Central Greenland Sea for the entire observational series (a) and for its portion with an increased temporal resolution (b): A - salinity (0-12 m); B - potential temperature (240 -420 m); C - potential temperature (1000 -2000 m); D - potential temperature (2000 m - bottom). Solid line: the third degree polynomial approximation.

The time series cross-correlation analysis results are presented in Table 4. The small amount of a reliable data within the suitable time intervals caused the large 95% confidence intervals. The relatively low cross-correlation coefficient between the surface salinity and the potential temperature in the C-layer, to our mind, reflects the fact that convection penetration into the deep layer is limited by the crucial value of the surface salinity (approximately 34.82

Table.4 Time series cross-correlation coefficients (ref. Fig.8). The bound of 95% confidence level is shown in brackets.

	Pot.temperature (240 - 420 m)	Pot.temperature (1000 -2000 m)
Salinity (0 -12 m)	-0.80 (0.55)	-0.66 (0.61)
Pot.temperature (240 -420 m)		0.61 (0.43)

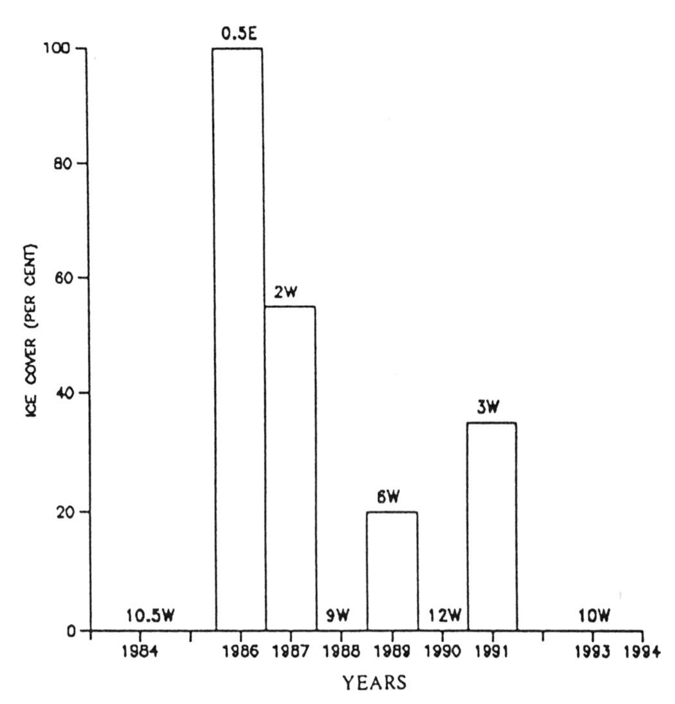

Fig.9. Interannual variability of the weekly averaged ice cover (per cent) and the ice edge longitude at 75°N in February- April in the Central Greenland Sea (74°30'- 75°30'N, 5°00'W - 1°00'E).
The ice edge location written above the columns.

PSU). Using for analysis only extremal values of surface salinity and corresponding potential temperatures in the C-layer promotes the increasing of cross-correlation coefficient up to -0.97 with 95% bound equal to 0.88. The connection between the surface salinity and the potential temperature in the intermediate layer is more direct. It does not depend upon the surface salinity range interval.

Thus, the crucial role of the surface salinity in the Greenland Sea convection occurrence is confirmed by the long term observations. It provides a good basis for the explanations of the continuous bottom water mass warming in the large scale surface salinity variability.

5. DISCUSSION

As it was already mentioned, the surplus salinization of the surface layer, which is a prerequisite condition for the overturning, can be achieved in different ways, each causing a particular type of convection. Theoretically, one can assume three possible scenarios of the convection development against the background of intense surface cooling:

- thermal convection - the compression of the surface layer is caused by the temperature decrease. The surface salinity level sufficient for the deep overturning is reached due to the external saline water influx;
- haline convection - the mixed layer thickness before freezing is extremely small. The salinization and convection are controlled by phase transitions [*Rudels*,1990];
- thermohaline convection - after the termination of the first stage of the haline convection, the ice is driven away from the convective zone. The underlying water raised towards the surface is sufficiently saline for the convection to proceed without further freezing.

Thus, the most reliable indicator of the convection type is the detailed data concerning the ice distribution. The available visual reports on the sea surface conditions in the 1984 and 1989 cruises include no information about ice in the convective zones. However, it is worth mentioning that detection of the ice thickness variations (and even its presence or absence) accompanying the haline convection [*Rudels*,1990], is on extremely difficult task from the technical point of view. Indirect information concerning this problem was received from the weekly-averaged Breknell's ice maps. From these maps, for the spring months, the ice cover of the Central Greenland Sea and the corresponding locations of the ice edge at 75°N were estimated (Figure 9). The influence of these parameters is not dominant. Among the four years with deep convection the ice edge was shifted westwards and the ice cover decreased twice (1984, 1988), while twice the situation was opposite (1986,1989).

Thermohaline conditions in the surface layer are controlled by two competitive processes: strengthening of the Jan Mayen Polar Current that causes the invasion of cool freshened water from the south-west [*Bourke et al.*,1992] and meandering of the West Spitsbergen current - the source of the warmer and saline water. The realization of the thermohaline conditions favourable for a certain convection type depends upon the result of this competition. The traces of saline intrusions into the convective gyre domain from the north-east are clearly seen in 1984 and 1989 (Figure 10). So, the advective source of pre-convective surface salinization, and, consequently, the thermal convection type in these years, is quite evident. The efficiency of the cross-frontal salt intrusions into the Central Greenland Sea is confirmed by the average surface salinity distribution (Figure 11) and by the data obtained in summer cruises [*Kotlerman*,1991].

Summing up, the following explanation can be proposed: the events of intense surface to bottom convection are the results of single, strong, salt water intrusions, while the low frequency oscillations of convection activity are caused by the influence of the large scale surface salinity anomalies. The long lifetime of such anomalies in the northern North Atlantic and the Nordic Seas is confirmed by direct

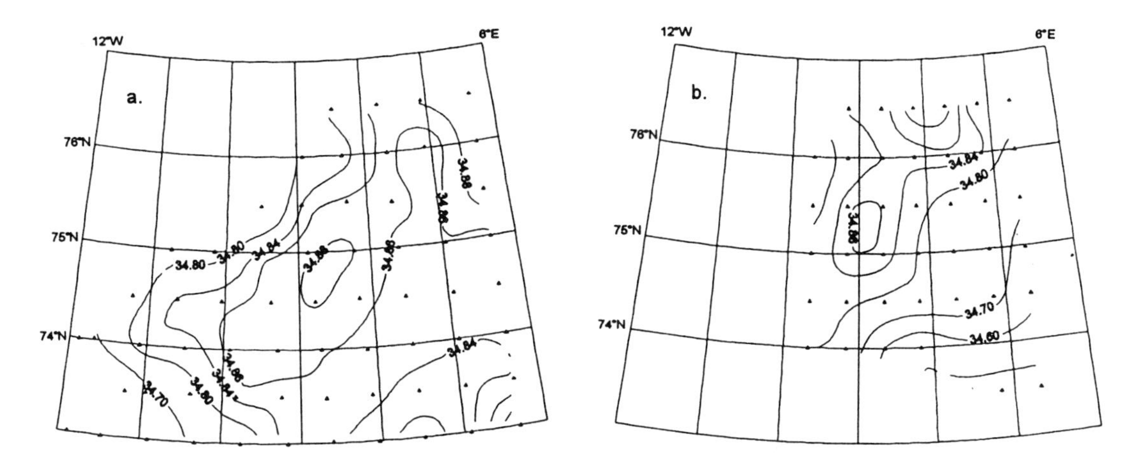

Fig.10. Distribution of the salinity (PSU) at the surface in April,1984 (a) and in March,1989 (b).

observations [*Dickson et al.*,1988]. Their origin is reasonably explained by the variability of fresh water discharge through the Fram strait [*Aagaard and Carmack*, 1989]. The frontal and current systems in the Nordic Seas prevent an immediate influence of the anomaly on thermohaline conditions in the Central Greenland Sea. Favourable conditions for an increase of the fresh water discharge from the Arctic Basin appear when the Arctic ice

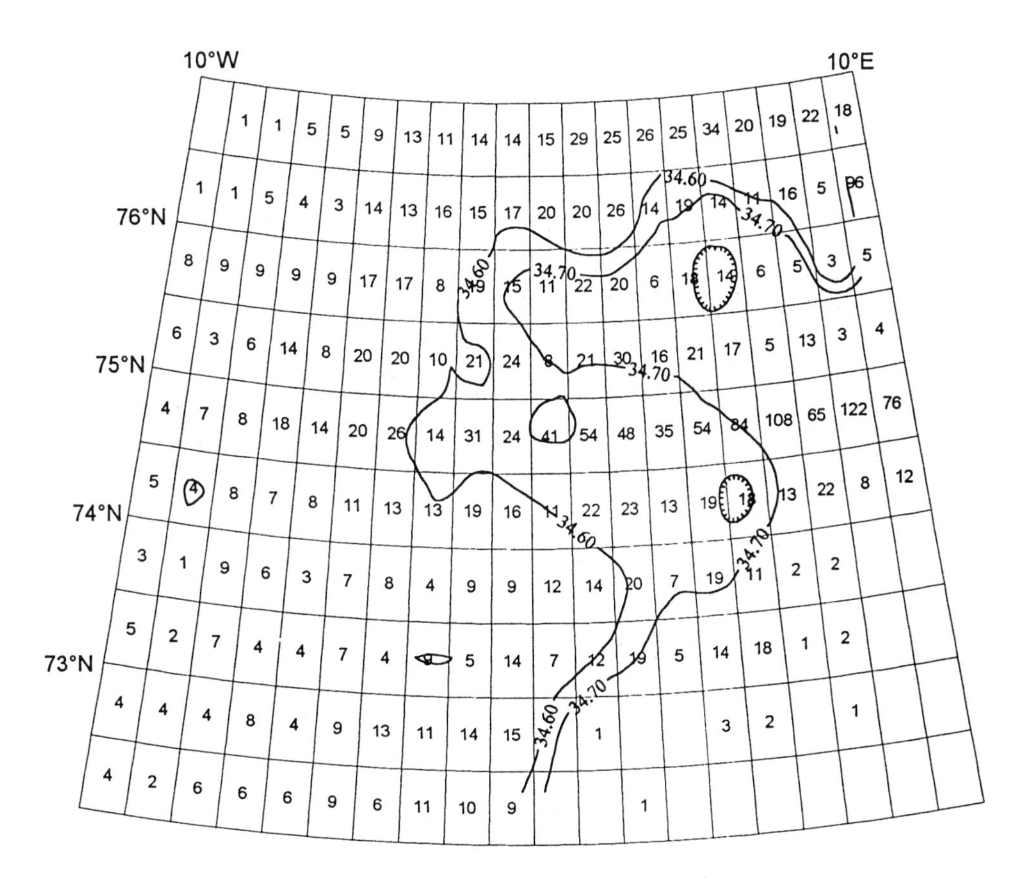

Fig.11. Mean annual distribution of the salinity (PSU) at the surface based on the entire data set. The figures in the 0.5° × 1° squares represent the number of salinity measurements inside each square.

cover diminishes. The latter my be linked with the global warming effect. In the present century the transition from cooling to warming occurred approximately in 1965. Therefore, the time interval between the beginning of the global warming and the beginning of the Greenland Sea deep water temperature increase, equal to ten years, coincides with the time interval needed for the large scale salinity anomaly to return back to the Nordic Seas after its initial creation in the East Greenland Current [*Aagaard and Carmack*, 1989].

6. CONCLUSIONS

Three possible types of thermohaline structure are discovered within the Greenland Sea Convective gyre domain in winter time: non-convective (normal), convective and pre-convective. The first and the second ones are stable while the third type is transient between them. The evolution to convective type is controlled by the winter convection. The thermohaline structure transformation degree is determined by the intensity of the preceding convection. Post-convective regeneration is caused by external forces: summer warming and freshening near the surface, advection of the modified Atlantic water into intermediate layer and lateral mixing in deep and bottom layers. The characteristic feature of pre-convective thermohaline conditions is an increased surface salinity (0.07 PSU above normal). Additional salt injection - the necessary pre-condition to start overturning- occurs to the advective intrusion of the West Spitsbergen Current water across the Arctic front. In this case, the convection is thermal and the additional increase of the surface salinity due to salt redistribution within the convective column is about 40 % of the total one. Pure haline convection, maintained only by the phase transitions in the Greenland Sea Convective gyre is hardly probable. In the time series of the spatially averaged surface salinity and potential temperature in the intermediate and deep layers, correlated variations are detected. A high correlation coefficient of -0.80 between the surface salinity and potential temperature in the intermediate layer confirms the crucial role of the surface salinity for the convection. Explains the single case of surface to bottom convection as a consequences of the intense salt water intrusions and the low frequency variability of convective activity by the effect of the large scale surface salinity anomalies.

Acknowledgments. This work has been supported by the Ministry of Science of Russia and by Russian Fund of Fundamental Research (Grant N 93-05-14106). We are grateful to our colleagues from the Alfred Wegener Institute (Bremerhaven) for the opportunity to use GSP data. We are grateful to Pr. Ola M. Johannessen for an attention to the manuscript and useful remarks. The publication of this paper has been sponsored by the organizers of the Nansen Centennial Symposium.

REFERENCES

Aagaard K. and Carmack E.C., The role of sea ice and other fresh water in the Arctic circulation, *J.Geophys.Res.*, 94, C10, 14485-14498, 1989.

(ACSYS) Scientific concept of the Arctic System Study, WCRP-72, 1992.

Bourke R.H., Paquette R.G., Blythe R.F., The Jan Mayen Current of The Greenland Sea, *J.Geophys.Res.*, 97, 7241-7250, 1992.

Dickson R.R., Meincke J., Malmberg S.-A. and Lee A.J.,The "great salinity anomaly" in the northern North Atlantic 1968-1982, *Progr. Oceanogr.*, 20, 103-151, 1988.

Greenland Sea Project Group, Greenland Sea Project - A venture towards improved understanding of the ocean role in climate, *EOS*,71, 750-751, 754-755, 1990.

Johannessen O.M., Breef overview of the Physical oceanography, In: *The Nordic Seas*. Ed.G.Hurdel., Springer-Verlag. N-4. Berlin, Heidelberg,Tokyo. Chapter 4, 103-124, 1986.

Kilworth P.D., Deep convection in the World ocean, *Reviews of Geophysics and Space physics*, 21, N1, 1-26, 1983.

Koltermann K.P., The circulation of the Greenland Sea as a consequence of the thermohaline system of the European Polar Seas, *Deuthsche Hydrograph. Zeitchrift*, 23, 181, pp., 1991.

Meincke J., Jonsson S., Swift H. Variability of convective conditions in the Greenland Sea, *ICES mar. Sci. Symp.*, 195, 32-39, 1992.

Nagurny A.P., Bogorodsky P.V., Popov A.V., Svyschenikov P.N., Intensive formation of cold bottom water at the surface of the Greenland Sea, *Dokl.AN SSSR*, 284, 478-480, 1985.

Nikiforov E.G., Shpayher A.O. *Features of the formation of hydrological regime large scale variations in the Arctic ocean* (in Russian), Gidrometeoizdat, Leningrad, 269, pp.,1980.

Nansen F.,Northern waters: Captain Roald Amundsen's oceanographic observations in the Arctic Seas 1901., *Videnskabs Selkabets*, 1, Math.-Natur.Klasse, Christiania,1906.

Rudels B. Haline convection in the Greenland Sea, *Deep Sea Res.*, 37, 9, 1491-1511, 1990.

Swift J.H. The arctic waters. In: *The Nordic Seas.*, Ed.G.Hurdel. Springer-Verlag. N-4, Berlin, Heidelberg, Tokyo. Chapter 5, 129-153, 1986.

Worcester P.F., Lynch J.F., Morawitz W.M.L., Pawlowicz R., Sutton P.J., Cornuelle B.D., Johannessen O.M., Munk W.H., Owens W.B., Shuchman R. and Spindel R.C., Evolution of the large-scale temperature field in the Greenland Sea during 1988-89 from tomographic measurements, *Geophys.Res.Lett.*,20, 1993, 2211-2214.

G.V.Alekseev, V.V.Ivanov, A.A.Korablev, Arctic and Antarctic Research Institute, Bering st., 38, St.Petersburg, 199397, Russia.

Microwave Remote Sensing of the Snow and Ice Cover: The Russian Experience

K. Ya. Kondratyev and V.V. Melentyev

St. Petersburg Center for Ecological Safety, Russian Academy of Sciences, and Nansen International Environmental and Remote Sensing Center, St. Petersburg, Russia

Microwave remote sensing techniques are useful for deriving properties of snow and ice. There has been substantial Russian research in developing such techniques, as well as their scientific application. The main centers of such activities are described, and results of fundamental research are summarized. Results from selected case studies are presented and compared with those from western research. Included are results on retrieving ice concentration, ice type, ice thickness, and ice state during the melt period. These airborne microwave remote sensing investigations provide information on the ice cover in several regions in the eastern Arctic.

1. INTRODUCTION

Remote sensing is a well-established means to acquire data on the components of the global environment, including processes occurring in the polar regions. Because these regions are under conditions of a permanent, dynamic sea ice cover, the presence of continental snow and ice, and extremely difficult conditions for standard hydrometeorological and oceanographic observations, it is especially important to develop remote sensing techniques to retrieve parameters in the polar regions. Such techniques can be developed to analyze some of the most important climate-related parameters such as: the size distribution of ice floes, the ice type and thickness, its concentration and dynamics, as well as the processes of ice formation and melting, transfer of heat fluxes through the snow and ice cover, and the state of the polar atmosphere.

Sensors operating in the visible and thermal spectral ranges have been used extensively to get information on the spatial distribution of the ice and snow cover since the early days of satellite meteorology, and are now used operationally. However, the usefulness within these spectral ranges is substantially limited due to: (i) insufficient (or absent) illumination of the polar regions of the Earth by solar radiation; (ii) frequent cloud cover; (iii) the dependence of the measurements on the viewing geometry and solar elevation; (iv) the need of a strict account of the atmospheric correction; and (v) inadequate knowledge of the reflectance and emittance properties of natural substances.

The main advantages of microwave remote sensing are that such sensors are not limited by darkness and are generally not cloud-affected. As such, microwave sensors provide solutions to most of the aforementioned problems, though uncertainties remain concerning some microwave properties. Microwave sensors have been used on aircraft for nearly 30 years, and have been satellite-borne since 1968. During this period, there has been extensive use of microwave remote sensing for sea ice applications. The technological and methodological progress has been well-summarized recently [*Carsey*, 1992], except that Russian research remains generally unreported.

This paper presents Russian progress in microwave remote sensing of snow and ice, and thus serves as a supplement to *Carsey* [1992]. First, we present an overview of the main centers of microwave remote sensing research in Russia. Then, we briefly describe the Russian research that helped establish the fundamentals of microwave remote sensing of ice. Then, we present results from selected Russian case studies using microwave measurements to derive snow and ice parameters; included are

The Polar Oceans and Their Role in Shaping the Global Environment
Geophysical Monograph 84

results using microwave frequencies lower than reported from other remote sensing studies.

2. MICROWAVE REMOTE SENSING RESEARCH CENTERS IN RUSSIA

Research on the microwave properties of natural surfaces began in Russia in the mid-1960s. The research has included theoretical and observational studies in meteorology, oceanography, and terrestrial studies. The National Interdisciplinary Program has supported the existing and recently-formed scientific groups, operating in the systems of Hydrometeorological Service, Academy of Sciences, Ministry of Education, and other Russian governmental departments.

Researchers at the Main Geophysical Observatory (MGO) in St. Petersburg (formerly Leningrad) were the first in Russia to develop techniques for microwave remote sensing of sea ice. The activities have been primarily focused on developing a complex of airborne sensors to map the state of the sea ice cover. Scientists there have studied the relationships between the airborne microwave brightness temperature (T_B) of the sea surface and the ice cover concentration, and mapped the distribution of ice and its properties along the aircraft transects [*Rabinovich et al.*, 1970, 1975; *Kondratyev et al.*, 1973, 1992].

The Institute of Radio Engineering and Electronics of the Academy of Sciences (Moscow) carried out a number of theoretical and empirical studies of the emittance properties of sea ice of different types and thicknesses, which suggested the capabilities of microwave remote sensing. Included are empirical studies of sea ice emissivity, and airborne microwave measurements of sea ice brightness temperature in different Arctic regions. Scientists at the institute have also analyzed multi-channel passive microwave data from *Kosmos-243*, the first satellite with microwave sensors, launched in 1968 [*Basharinov and Gurvich,* 1970].

Research on the electrophysical parameters of sea ice, continental ice, snow and frozen soils, as well as thematic processing of aircraft- and satellite- passive microwave data, has been carrried out at the Arctic and Antarctic Research Institute (AARI) in St. Petersburg. Scientists at this institute also have extensive experience with active microwave data [*Bogorodsky and Oganesyan*, 1987].

The development of techniques to study the characteristics of multi-year sea ice has been carried out at the National Center for Studies of Natural Resources of the USSR Committee on Hydrometeorology in Moscow. Theoretical and experimental studies have emphasized the retrieval of sea ice parameters throughout the different seasons [*Bukharov et al.*, 1990].

Theoretical and observational studies of snow and ice parameters in the inland water basins have been made at the Institute for Lake Studies of the Russian Academy of Sciences in St. Petersburg. Analysis of microwave data on snow, ice, and frozen soil led to the development and inclusion of the two-channel (6.0 and 37.5 GHz) scanning microwave radiometer on the *Almaz-1* satellite [*Kondratyev et al.*, 1985, 1989].

3. MICROWAVE REMOTE SENSING OF SEA ICE: FUNDAMENTAL RESEARCH

The development of microwave remote sensing techniques in Russia started with studies of the emittance properties of snow, ice, and water, and development of the respective instruments and techniques to retrieve the characteristics of such surfaces. The IL-18 research aircraft, with microwave radiometers at several frequencies, was used extensively. The initial research flights served to estimate the relationships between brightness temperatures (T_B) and the concentration of sea ice [*Rabinovich et al.*, 1970].

Model calculations of sea ice emissivity [*Kondratyev et al.*, 1992] using data on the dielectric constants of different structures of ice (Table 1) indicated the potential for microwave mapping of different ice types. Similar calculations and direct radiation measurements have been made elsewhere, as summarized in *Eppler et al.* [1992]. The values for multi-year ice are in general agreement; however, our values calculated for frazil and columnar ice are much higher than measured for "new ice" [*Eppler et al.*, 1992]. The largest part of the differences is presumably due to the use of terms that represent different ice conditions; indeed, new ice emissivities increase rapidly with growth [*Grenfell and Comiso*, 1986]. Remaining differences may be explained by differences in frequency (9.4 vs. 10 GHz) and the temperature-dependency of emissivity, as there are seasonal differences in the measurements presented in *Eppler et al.* [1992].

The data of Table 1 suggested that combining vertical and horizontal polarizations could be useful for ice type identification. For each ice type at any thermodynamic temperature, the change in the emittance (ε) as a function of angle (θ) is polarization-dependent such that: $|\Delta\varepsilon_V/\Delta\theta| < |\Delta\varepsilon_H/\Delta\theta|$.

The usefulness of an aircraft microwave system with high spatial resolution and high sensitivity, as well as an in-flight system for analysis and thematic processsing was demonstrated during the Bering Sea Experiment (BESEX), a joint Soviet - U.S. project [*Kondratyev et al.,* 1973; *Kondratyev,* 1982]. The key in the development of such a system was the development of an antenna with an

TABLE 1. 9.4 GHz emissivities (ε) of sea ice as a function of angle, polarization, ice temperature, and ice type. [From *Kondratyev et al.,* 1992]

Ice	nadir angle	0°			20°			30°			40°		
Form	T_s (°C)	-10°	-20°	-30°	-10°	-20°	-30°	-10°	-20°	-30°	-10°	-20°	-30°
Columnar	ε_V	0.80	0.85	0.91	0.82	0.87	0.92	0.84	0.90	0.94	0.88	0.92	0.96
	ε_H	0.80	0.85	0.91	0.78	0.84	0.90	0.75	0.81	0.88	0.71	0.78	0.85
Frazil	ε_V	0.80	0.86	0.92	0.82	0.87	0.93	0.84	0.89	0.95	0.88	0.92	0.96
	ε_H	0.80	0.86	0.92	0.78	0.84	0.91	0.75	0.81	0.89	0.71	0.78	0.86
Multiyear	ε_V	0.94	0.95	0.95	0.95	0.96	0.96	0.96	0.97	0.97	0.98	0.98	0.98
	ε_H	0.94	0.95	0.95	0.93	0.94	0.95	0.92	0.93	0.93	0.90	0.90	0.91

electrically-scanning beam (phase grid). The scanning radiometer made it possible to outline the water-land interface, map the spatial distribution of sea ice, estimate ice concentration, and estimate the size of large floes and leads. Further data analysis revealed the ability to identify four intermediate structures in the transition zone from open water to compact ice: (i) newly-forming ice, (ii) the ice-water interface, (iii) the adjacent small floes, and (iv) fields of large floes of different ice types. It was also shown that changes in the sea surface roughness result in variations of T_B and affect the degree of angular dependence. The results of these calculations and measurements, combined with similar investigations elsewhere, e.g., *Carsey* [1992] established the basis for all-weather mapping of the Arctic seas using microwave sensors.

4. RESULTS FROM SELECTED CASE STUDIES

The fundamentals of microwave remote sensing of sea ice having been established, further investigations were made to refine and expand this understanding, and to investigate the snow and ice cover in the Arctic. From 1983-1985, three large-scale IL-18 microwave missions were undertaken to survey the marginal seas and coastlines of the eastern Arctic. Investigations were also made outside the Arctic basin, in the Sea of Okhotsk, the Kuril-Hokkaido region, and the Sea of Japan.

In the initial period, February and March 1983, a series of flights was made systematically from the Chukchi Sea to the Barents Sea, with the aircraft based at several places along the coast. The research team included an expert on visual ice analysis from the Arctic and Antarctic Research Institute, which made it possible to map in detail situation and to monitor the ice dynamics. The airborne observations were complemented by multi-spectral satellite data. Each stage of the 3-year project was coordinated with field studies, providing surface validation.

4.1. *Sea Ice Parameter Retrieval in the Sea of Okhotsk*

Scanning radiometer measurements were used to estimate sea ice conditions in the southeast margin of the Sea of Okhotsk. Frequency distributions of 14.6 GHz (2.1 cm) brightness temperature were derived from two transects in the Freeze Strait (46°N, 149°E) and Katherine Strait (45°N, 147°E), near the Kuril Islands. The histograms for each region were bimodal; the lower mode corresponds to open water, whereas the upper mode represents first-year ice. The bimodality is due to the large differences in emissivities of water and first-year ice [*Eppler*, 1992]. The T_B distributions suggest that the proportion of open water and new ice is higher in the Freeze Strait region than in the Katherine Strait region.

The next step in parameter retrieval is the quantitative estimation of ice concentration. The 14.6 GHz measurements over the Sea of Okhotsk revealed that water surface T_B ranged from 110-125 K. These measurements agree well with T_Bs calculated using realistic values of sea roughness and atmospheric state. Measurements were then made over extended, homogeneous fields of first-year ice (from visual assessment); the mean T_B = 221 K. The T_Bs from open water and ice were then used to estimate ice concentration from the histograms from the entire transect. Based on the emissivity of first-year ice, a threshold level (T_{Bcrit}) was selected. Ice concentration is then retrieved simply by dividing the number of values > T_{Bcrit} by the number of values < T_{Bcrit}.

The uncertainties in T_B estimation due in errors in reducing the scanning radiometer data to nadir, result in a scatter in the ice concentration estimation not more than 5%. Because the antenna field-of-view can include water and sea ice of different type, the probabilities of receiving emission from the different surfaces largely determine the character of the T_B distributions. This can limit the effectiveness of ice concentration estimation and ice classification based on T_B frequency distributions.

In this investigation, the estimate of ice concentration along the Freeze Strait transect is 48%; along the Katherine Strait transect it is 71%. In this case, the microwave-derived estimates agree well with the 10-day ice map from the Russian Hydrometeorological Service, compiled from an aircraft visual survey.

4.2. Sea Ice Thickness Estimation from Low-Frequency Microwaves in the Sea of Okhotsk and Chukchi Sea

The potential retrieval of other sea ice parameters was investigated using low-frequency microwave radiometer data. Included on some of the IL-18 flights were radiometers operating at 0.9 and 1.7 GHz (35 and 18.1 cm, respectively). The acquisition of measurements at 0.9 GHz was limited due to the unwieldiness of the large antenna required. Radiometer data at 1.7 GHz were analyzed and found to show potential for the estimation of first-year ice thickness, which is related to the depth of the sensed layer, which increases at this wavelength.

Figure 1 shows the distribution of the Sea of Ohkotsk ice cover near Magadan (60°N, 151°E) on February 8, 1983. The ice concentration, estimated using the histogram technique described in section 4.1, is 100%. The following

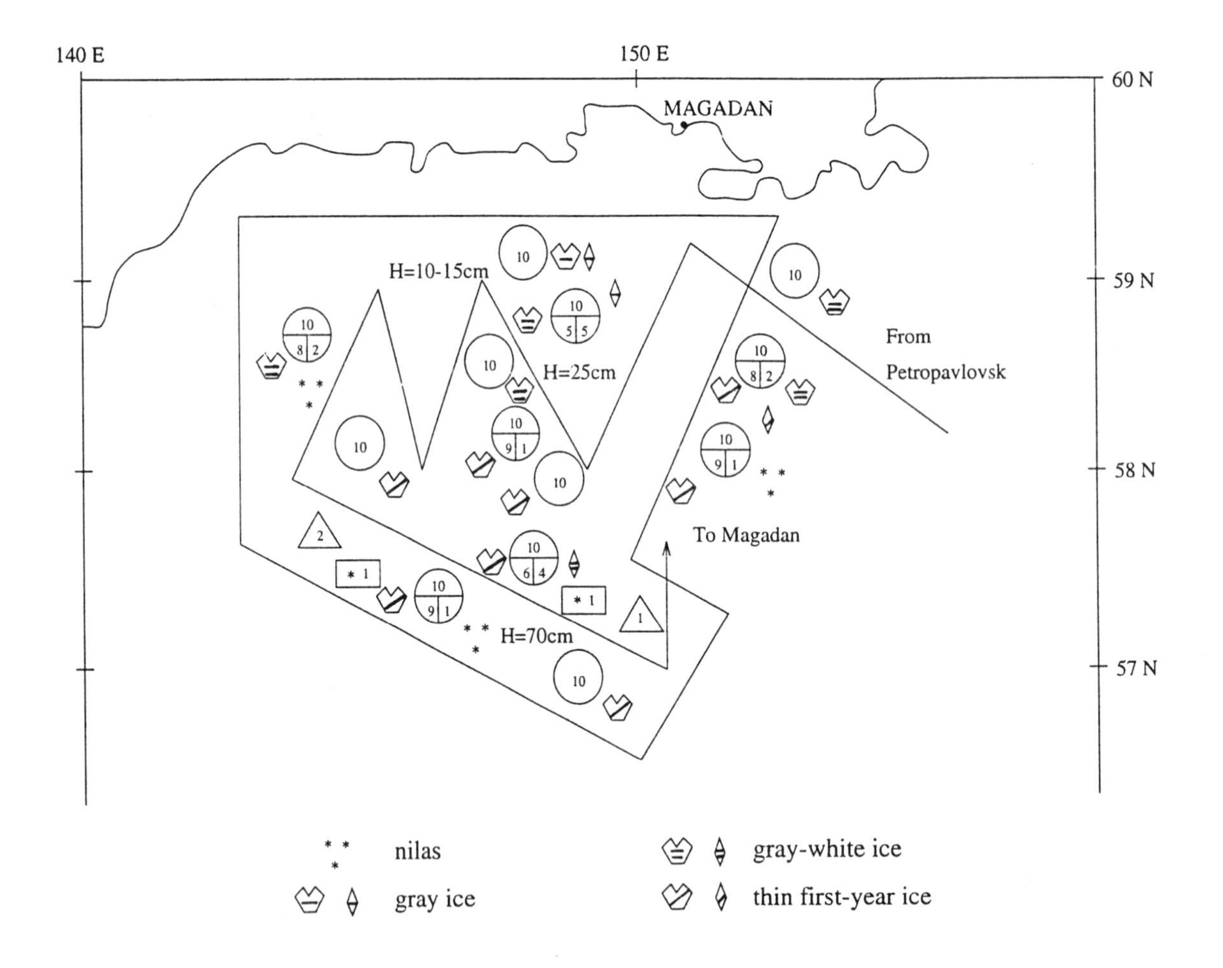

Fig. 1. Map of sea ice parameters during the 1.7 GHz microwave survey of the Sea of Okhotsk, February 8, 1983. The codes and symbols are: numbers in circles = ice concentration (in tenths) in the upper part, and the components of thick ice and thin ice in the lower left- and right-parts, respectively; numbers in triangles = degree of ridging (0 to 5); numbers in rectangles = degree of snow cover (0 to 3); H = ice thickness.

ice types (with associated thicknesses) were identified: dark nilas (5 cm), fields of gray ice (10-15 cm), fields of gray-white ice (15-30 cm), and thin first-year ice (30-70 cm). In the spring of 1985, a survey of the same region was repeated, revealing the presence of two other ice types in the northeastern part of the region: moderate first-year ice (70-120 cm) and thick first-year ice (> 120 cm, T_B = 215-220 K).

Analysis of these microwave measurements and visual observations shows that 1.7 GHz brightness temperatures enable the distinction between thick first-year ice (120-150 cm), prevailing in this region at this time, and white ice (≈ 70 cm). The white ice, connecting the first-year ice fields, forms where polynyas had earlier been; the white ice fields are on the order of several hundred km^2. The mean 1.7 GHz T_Bs were 242 K for thick first-year ice, and 234 K for white ice.

In a separate investigation, multi-frequency measurements were made in the Chukchi Sea on February 28, 1984. The 3-channel (1.7, 2.2, and 2.7 GHz) measurements in the decimeter range made it possible to identify 5 levels of first-year ice thickness. The T_Bs correspond to the ice classification summarized in Table 2.

In these investigations and others, e.g., *Comiso* [1986], ice thickness is inferred through ice classsification and associated ice thickness. There is presently no microwave system to directly retrieve this parameter [*Wadhams and Comiso*, 1992]. Because ice thickness is important climatologically and for understanding ice dynamics, the ability to reliably retrieve ice thickness from microwave data would be a major breakthrough. The relative lack of similar research using such low microwave frequencies is largely due to the practical restrictions of the antenna size required for such low-energy measurements. That is, even if low-frequency microwave measurements exhibit a strong dependence on ice thickness, the implementation of such a sensor system appears unfeasible.

4.3 Active and Passive Microwave Sensing of Fast Ice in the East Siberian Sea

The usefulness of combining active and passive microwave measurements was investigated in the fast ice zone in the East Siberian Sea in February 1983. The low-altitude (400 m) mission started over land (15-20 km from the coast) and ended at a distance of 60 km over the sea ice in Long Strait, connecting the East Siberian and Chuckchi seas. The subsequent flight transects, including those in the opposite direction, were made 10 to 20 km eastward The passive microwave measurements were accompanied by active microwave observations from the ground-based meteorological radar station (MRS-2) at Cape Schmidt (68°N, 179°W). The radar station is on an islet 60 m from the coastline. The height of the 9.4 GHz, circular-scanning antenna system was 7 m above sea level, with an antenna angle of 0.7°, permitting a range of 16 km under normal conditions of refraction.

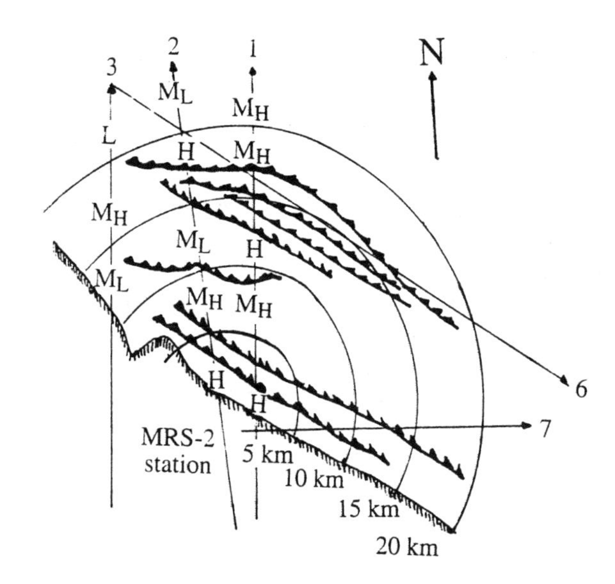

Fig. 2. Schematic of combined active and passive microwave sensing in the Cape Schmidt region. Flight transects are indicated with lines. Ice ridges are indicated with cerrated lines. Codes refer to microwave-estimated ice concentrations: L ≤ 20%, M_L = 20-40%, M_H = 40-60%, H ≤ 60%.

The presence of ice is detectable using radar because of differences in surface roughness, which influences radar return. The MRS-2 radar best identifies older, deformed ice, which has a high return. The MRS-2 data from February 4, 1983, revealed the presence of near-shore fast ice, with a series of ice ridges oriented mainly along the coastline (Figure 2). The ridges were formed in clearings

TABLE 2. Microwave brightness temperatures (1.7 GHz) of ice in the Sea of Okhotsk, February 28, 1984.

Ice Type	1st-year >150 cm	1st-year <150 cm	White	Grey-White	Nilas	Water 0°C, 30‰
T_B (K)	245	235-245	225-235	210-235	105-120	93-96

beyond the near-shore ice, with subsequent compression in the 2-month period from December 1982 to the time of survey. The first zone, resulting from the compression of the near-shore ice, about 2 km wide, is 2-2.5 km from the coast. The second zone is at a distance of 8 km. The third zone detected by the radar is 1 to 5 km wide, 13 km from the coast; it consists of 4-5 ridges formed at different times.

An attempt was made to find out the relationship between the passive microwave T_B variability (standard deviation of the brightness temperature, σT_B) of homogeneous, extended ice areas and variations in the first-year ice with ridges. Four levels of ice concentration are considered: (i) < 20% (L-low); (ii) 20-40% (M_L-moderate/low); (iii) 40-60% (M_H-moderate/high); and (iv) 60-90% (H-high), corresponding to the following values of σT_B at 1.7 GHz: ±1 K; ±2 K; ±3 K; and ±4 K. The ice concentration in the region of the IL-18 mission on February 4, 1983 varied from 20% to 90%, in agreement with the MRS-2 data (Figure 2).

The experience using combined active and passive microwave data has shown that these techniques supplement each other, and can provide information about the sea ice conditions needed to navigate in the Arctic coastal regions. Indeed, combined satellite passive- and active-microwave systems are beginning to be used for such purposes in the Siberian marginal seas [*Johannessen et al.*, 1992].

4.4. Microwave Studies of Freshwater Ice and Snow: High-Latitude and High-Altitude Case Studies

In winter, vast expanses of northern Siberia are largely covered with tundra, generally snow-covered, and frozen rivers and lakes. These high-latitude features can also be found in high-altitude settings in winter. The microwave properties of such features were investigated using the IL-18 aircraft. The February 4, 1983 flight was based at Anadyr (64°N, 178°E). Passive microwave measurements were made over the region of high-mountain tundra north from Anadyr to Cape Schmidt. The terrestrial part of the transect of the survey revealed the ability to identify the areas of fresh-water ice (vast, frozen lakes in the tundra), with mean T_B = 206 K, as well as lake ice fractures. The continental snow-covered tundra is characterized by mean T_B = 230 K, with standard deviation (σT_B) = ± 6 K, which is somewhat higher than the respective σT_B for the sea ice cover in the adjacent Chuckchi Sea (Long Strait region).

Remote sensing of ice during the melt period also has important applications. Examples of these are studies of the thermodynamic and radiative regimes of water basins, atmosphere-ice-water interaction processes, the ice regime of the inland water basins, and assessments of river runoff and flooding. A technique has been proposed [*Kondratyev et al.*, 1985, 1992] to retrieve the characteristics of melting ice using microwave radiometers. This approach uses a cross-track scanning antenna system, thus exploiting the angular dependence of the emittance. The angular emittance characteristics of ice and water are used to assess the state of the melting ice. This technique was developed using a series of 14.6 GHz passive microwave measurements made in April 1983, at altitude 2900 m above the high-mountain lake Sevan in the Caucausus. Flights were made about the longitudinal axis of the lake and around its margins.

The results for one transect are shown in Figure 3. As seen from these data, the distribution of the lake ice cover emission is heterogeneous. The highest T_Bs correspond to the ice cover in the extreme southeastern part of the lake; here the ice appears white, generally smooth with some ridging. Several dark lineaments and streaks 3-5 m wide were seen, and were also readily identified as inhomogeneities on the on-board real-time T_B plot; these fractures and streaks are marked with vertical dashes of different length, characterizing different widths of streaks (Figure 4). A polynya observed beyond the near-shore ice was also detected on the T_B plot.

Smaller ice fields of similar structure can be identified from the visual and microwave survey over the Great Sevan in the part adjacent to the strait separating it from the Small Sevan. This springtime fresh-water ice, conditionally attributed as white ice, has a mean $T_B \approx 220$

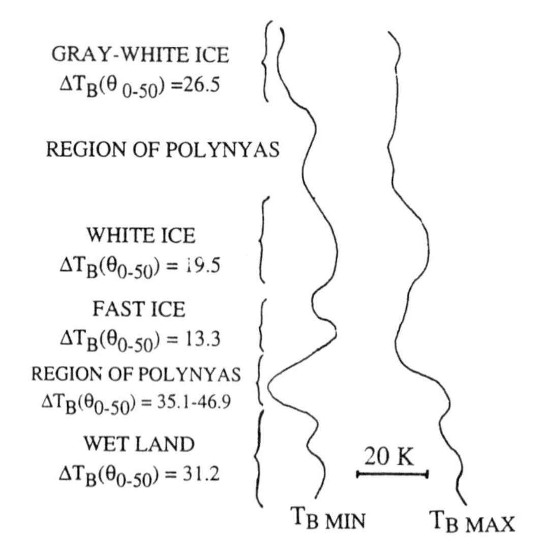

Fig. 3. Distribution of Lake Sevan ice types and brightness temperatures measured with an airborne radiometer. Surface codes are: L (land), F (fast ice), Wh (white ice), Gr-Wh (gray-white ice), and Gr (gray ice).

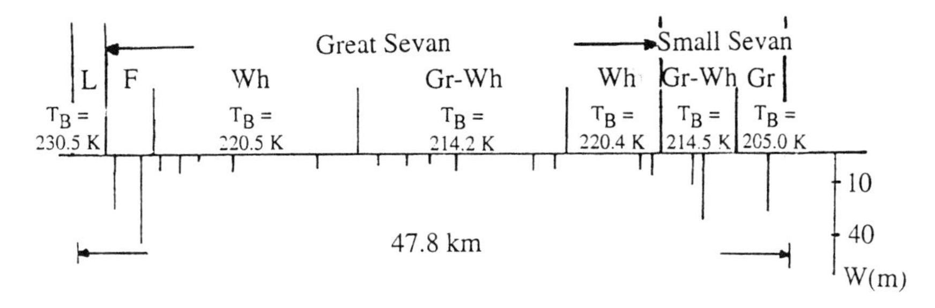

Fig. 4. Along-transect plot of Lake Sevan ice types and associated microwave parameters. $\Delta T_B(\theta_{0-50})$ is the mean difference in horizontally-polarized brightness temperature as measured between 0°-50°.

K. Most of the Great Sevan was covered with darker-appearing ice, termed gray-white ice. As the aircraft passed over the Small Sevan, a large band of disturbed ice was seen; it was attributed to gray-white and gray ice. The mean T_Bs for these ice types are 214 K and 205 K, respectively. The flight over this region also revealed several fractures with a corresponding T_B decrease.

Horizontally-polarized measurements at 14.6 GHz were made on the following three levels of melting, spring ice: (i) small; (ii) moderate; and (iii) large (the ice has nearly disintegrated). The following values of the parameter $\Delta T_B(\theta_{0-50})$ were ascribed to these levels: (i) 15 K; (ii) 15-25 K; (iii) 25-35 K. Values of $\Delta T_B(\theta_{0-50})$ >35 K correspond to the open-water areas. Figure 4 is an example of the Great Sevan T_Bs in the 0°-50° scanning regime. The minimum $\Delta T_B(\theta_{0-50})$ refers to the near-shore ice; white ice and gray-white ice differ in both the absolute value of emission ($T_{Bw} > T_{Bg/w}$) and $\Delta T_B(\theta_{0-50})$. Gray-white ice has a greater liquid water content, and a correspondingly higher value of $\Delta T_B(\theta_{0-50})$.

The results of these airborne investigations of land, snow, and freshwater ice demonstrated that various surface characteristics can be assessed using their microwave emission properties. These characteristics include the ice structure and the presence of liquid water in the ice, which makes it possible to assess indirectly such parameters as its density. It is also possible to identify fractures, streaks, polynyas, fresh-water, ice-land interfaces, snow-covered land areas, and snow-free land areas. Here, the angular dependence of microwave emission was shown to be useful, as suggested by laboratory measurements [*Grenfell and Comiso,* 1986] and models [*Winebrenner et al.*, 1992].

5. CONCLUSION

This paper has served to present the experience of the Russian research programs in the microwave remote sensing of snow and ice. The nearly 30 years of research has involved the development of microwave instruments and techniques, as well as investigations in an extensive range of geographic regions. The particular case studies presented represent just a small part of that research.

The principles of using microwave remote sensing techniques for snow and ice studies are presently well-established [*Carsey*, 1992; *Kondratyev et al.,* 1992]. The development of microwave sensor systems has progressed, with the latest generation of satellite-borne sensors including the Special Sensor Microwave Imager (SSM/I) and the ERS-1 Synthetic Aperture Radar (SAR). Despite these technical advances, there remain uncertainties in parameter retrieval. Further advances require contributions from the entire microwave remote sensing community, and will include both theoretical and observational research such as that presented here.

Acknowledgements. The authors acknowledge the useful suggestions from the reviewers, In particular, we thank Martin Miles, Nansen Environmental and Remote Sensing Center, Bergen, Norway, for editing assistance.

REFERENCES

Basharinov, A.E., and A.S. Gurvich, Studies of the earth's microwave emission and that of the atmosphere using the Kosmos-243 satellite, (in Russian), *Vestnik AN SSSR, N10,* 37-42, 1970.

Bogorodsky, V.V., and A.G. Oganesyan, *The penetrating radar sounding of sea ice with digitally processed signals,* (in Russian), 343 pp., Gidrometeoizdat, St. Petersburg, 1987.

Bukarov, M.V., P.A. Nikitin and V.A. Golovnya, Special assessment of the state of sea ice from radar and radiometric images from the oceanographic satellites, (in Russian), *Proceedings of the National Center on Studies of Natural Resources*, *37*, 175-185, 1990.

Carsey, F.D. (ed.), *Microwave Remote Sensing of Sea Ice. AGU Monograph 68.*, 462 pp., American Geophysical Union, Washington, D.C. , 1992.

Carsey, F.D., R.G. Barry, D.A. Rothrock, and W.F. Weeks, Status and future directions for sea ice remote sensing, in *Microwave Remote Sensing of Sea Ice., AGU Monograph 68,* edited by F. Carsey, pp. 443-446, American Geophysical Union, Washington, D.C., 1992.

Comiso, J.C., Characteristics of Arctic winter sea ice from satellite multispectral microwave observations, *Journal of Geophysical Research, 91(C8), pp. 975-994, 1986.*

Eppler, D.T., L.D. Farmer, A.W. Lohanick, M.R. Anderson, D.J. Cavalieri, J. Comiso, P. Gloersen, C. Garrity, T.C. Grenfell, M. Hallikainen, J.A. Maslanik, C. Mätzler, R.A. Melloh, I. Rubenstein, and C.T. Swift, Passive microwave signatures of sea ice, in *Microwave Remote Sensing of Sea Ice, AGU Monograph 68,* edited by F. Carsey, pp. 47-72, American Geophysical Union, Washington, D.C., 1992.

Grenfell, T.C., and J.C. Comiso, Multi-frequency passive microwave observations of the first-year sea ice grown in a tank. *IEEE Transactions on Geosciences and Remote Sensing GE-24(6)*, 826-831, 1986.

Johannessen, O.M., S. Sandven, Ø. Skagseth, K. Kloster, Z. Kovacs, P. Sauvader, L. Geli, W. Weeks, and J. Louet, ERS-1 SAR ice routing of "L'Astrolabe" through the Northeast Passage, in *Proceedings of the Central Symposium of the International Space Year Conference, Munich, Germany 30 March-4 April 1992,* ESA SP-341, pp. 997-1002, 1992.

Kondratyev, K.Ya. (ed.), *The USSR / USA Bering Sea Experiment,* 307 pp., Amering Publ., New Delhi, 1982.

Kondratyev, K. Ya., V. V. Melentyev, and Yu. I. Rabinovich, The Soviet-American Bering Sea Experiment, (in Russian), *Meteorology and Hydrology, N11*, 3-10, 1973.

Kondratyev, K.Ya., V.V. Melentyev, and V.Yu. Alexandrov, Microwave emittance properties of different types of the surface for negative temperatures, (in Russian), *Reports of the USSR Academy of Science, 306*, 67-70, 1989.

Kondratyev, K.Ya., V.V. Melentyev, and B.A. Nazarkin, *The Spaceborne Remote Sensing of Water Basins and Catchment Areas,* (in Russian), 248 pp., Gidrometeoizdat, St. Petersburg, 1992.

Kondratyev, K.Ya., V.P. Vlasov, and V.V. Melentyev, The radiothermal emission of the melting ice cover as an indicator of its state with lake Sevan as an example, (in Russian), *Reports of the USSR Academy of Science, 280*, 839-842, 1985.

Rabinovich, Yu. I., V.S. Loshchilov, and E.M. Shulgina, Analysis of the results of measuring the ice cover characteristics (Option C), (in Russian), in *The Soviet-American Bering Strait Experiment. Proc. Final Symp. on the results of the Soviet-American expedition*, pp. 284-313, Gidrometeoizdat, Leningrad, 1975.

Rabinovich, Yu.I., G.G. Shchukin, and A.I. Novoselov, Application of radar to the ice survey, (in Russian), *Proceedings of the Main Geophysical Observatory, 235*, 67-71, 1970.

Wadhams, P., and J.C. Comiso, The ice thickness distribution inferred using remote sensing techniques, in *Microwave Remote Sensing of Sea Ice, AGU Monograph 68,* edited by F. Carsey, pp. 375-383, American Geophysical Union, Washington, D.C., 1992.

Winnebrenner, D.P., J. Bredlow, A.K. Fung, M.R. Drinkwater, S. Nghiem, A.J. Gow, D.K. Perovich, T.C. Grenfell, H.C. Han, J.A. Kong, J.K. Lee, S. Mudaliar, R.G. Onstott, L. Tsang, and R.D. West, Microwave sea ice signature modeling, in *Microwave Remote Sensing of Sea Ice, AGU Monograph 68,* edited by F. Carsey, pp. 137-176, American Geophysical Union, Washing-ton, D.C., 1992.

K.Ya. Kondratyev and V.V. Melentyev, Nansen International Environmental and Remote Sensing Center, Korpusnaya str. 18, 197042 St. Petersburg, Russia.

Short- and Long-Term Temporal Behavior of Polar Sea Ice Covers From Satellite Passive-Microwave Observations

William J. Campbell

Ice and Climate Project, U.S. Geological Survey, University of Puget Sound, Tacoma, Washington

Per Gloersen and H. Jay Zwally

Laboratory for Hydrospheric Process, NASA Goddard Space Flight Center, Greenbelt, Maryland

All-weather and all-season observations of polar sea ice have been made with imaging passive-microwave radiometers (ESMR, SMMR, and SSMI) nearly continuously since 1973. In the Arctic Ocean, persistent areas of low ice concentration, the result of unresolved polynyas and leads, have been observed in passive-microwave data since 1978, as a result of the improved capabilities of the SMMR and SSMI. These areas appear during the latter part of the fall freeze-up and during the winter. The onset usually occurs in a few days or less, and the persistence varies from a week to over a month. Comparison of the temporal behavior of the SMMR sea ice concentrations, in four selected locations of these persistent areas, with concurrent geostrophic winds and ice divergence calculated from concurrent Arctic Ocean buoy data, gives mixed results. On a longer time scale, the combined SMMR-SSMI record shows several repetitions of the low concentration events in these locations. During a 7-week overlap period, the SMMR and SSMI sea ice concentrations agree to within a few percentage points. Overall, measurements by the ESMR, SMMR, and SSMI have initiated acquisition of a reliable long-term record needed for assessing the state of the polar sea ice, particularly with regard to detecting sea ice changes associated with warming or cooling trends.

INTRODUCTION

Observations of polar sea ice, unhindered by cloud cover and lack of illumination, have been made with imaging passive-microwave radiometers nearly continuously since 1973. Satellite passive-microwave imaging began with the Electrically Scanned Microwave Radiometer (ESMR) onboard the NASA Nimbus 5 satellite, which provided coverage during 1973-1976. ESMR was followed by the Scanning Multichannel Microwave Radiometer (SMMR) on the Nimbus 7, which provided coverage during 1978-1987, and the Special Sensor Microwave/Imagers (SSMIs) on various Department of Defense meteorological satellites from 1987 to the present. SSMI coverage overlapped that of SMMR for about 7 weeks. The combined SMMR/SSMI sea ice data set represents a reliable baseline for a longer-term record needed for assessing the state of the polar sea ice, particularly changes in sea ice coverage associated with warming or cooling trends.

In this paper, we describe and analyze four regions of low central Arctic pack ice concentrations of varying persistence in the time interval 1978-1992. This analysis includes comparison of the time histories of the ice concentrations from SMMR and weather systems in these regions to examine the relationship of reduced concentrations and the wind-forcing of ice motion and divergence. We scan the combined SMMR-SSMI record for multiple occurrences of these events in the four Arctic locations.

The Polar Oceans and Their Role in Shaping the Global Environment
Geophysical Monograph 85

Additionally, we review aspects of combining sea ice data sets from different spacecraft passive microwave sensors, their respective properties and estimated accuracies, and the benefits of overlap data between the sensors.

ESMR, SMMR, AND SSMI AND THEIR SEA ICE ALGORITHMS

The NASA Nimbus 5 ESMR [Gloersen et al., 1974] was a single-channel instrument operating at a wavelength of 1.55 cm. ESMR scanned cross-track and received horizontally polarized radiation from the off-nadir pixels. Its thermal noise was about 3 K. Sea ice concentrations were calculated by a linear interpolation between the average brightness temperature observed over open ocean and 100% sea ice. The open ocean value includes the average effects of atmospheric moisture and surface roughness caused by winds. The 100% sea ice value used is an average sea ice emissivity multiplied by a monthly climatological value of surface temperature. All of the foregoing factors result in estimated ice concentration errors of 15% in regions where the sea ice is all first-year [Zwally et al., 1983a; Parkinson et al., 1987]. In regions of Arctic multiyear ice, the concentration accuracy is less, depending upon the fraction of the ice that is multiyear, because of its lower emissivity. In areas of nearly 100% total ice concentration, typical of the central Arctic in winter, the multiyear ice concentration may be estimated. Another estimate of the accuracy was obtained by comparison with Landsat data, yielding a value of 12% after fitting the Landsat vs. ESMR data about a line with a slope of 0.67 and an intercept of 25% [Comiso and Zwally, 1982]. On single-day polar maps of the sea ice from ESMR, storm systems over the open seas give numerous instances of false indications of sea ice caused by the increase in atmospheric moisture during those events. A large contribution to this error results from the location of the 1.55-cm channel part way up on the wing of the water vapor line at 1.36 cm. In monthly averaged sea ice maps, the weather effects are greatly reduced because of the spatial motion of the weather systems and the use of the observed average brightness temperature over open ocean to represent 0% ice concentration [Zwally et al., 1983a; Parkinson et al., 1987].

The NASA Nimbus 7 SMMR [Gloersen and Barath, 1977; Gloersen et al., 1984] was a ten-channel radiometer that measured both horizontally and vertically polarized components at the wavelengths of 0.81, 1.43, 1.67, 2.80, and 4.55 cm. A conical scanning arrangement provided a constant incidence angle at the surface of the earth throughout the swath. The thermal noise of the various channels ranged from 0.4 to 1.1 K, the largest corresponding to the 0.81-cm channels. Three of these channels, 0.81V, 1.67V, and 1.67H, are combined to calculate sea ice concentrations using the SMMR Team Algorithm [Cavalieri et al., 1984; Gloersen and Cavalieri, 1986; Gloersen et al., 1992]. The SMMR algorithm interpolates among three tie points for open water, first-year sea ice, and, in the Arctic, multiyear sea ice. In the Antarctic, the multiyear ice does not have emissivity characteristics that allow it to be distinguished from first-year ice. However, another ice type appears, which may be ice with a heavy snow cover, and has emissivity characteristics distinct from most first-year ice. Therefore, the algorithm uses the third tie point to accommodate this ice type in the Antarctic instead of a multiyear ice tie point. An important feature of this algorithm is the incorporation of radiance ratios, which eliminates the effects of temperature variations to first order. The effects of weather-related phenomena, such as wet snow and its subsequent refreezing, on the microwave emissivity of sea ice [Mätzler et al., 1984; Onstadt et al., 1987] are also reduced in these ratios (e.g., see Figure 2.1.2 in Gloersen et al., 1992); however, the amount of reduction is not well established. The differing microwave spectral properties of first-year and multiyear sea ice permit their separate detection with this algorithm. Also, a distinct improvement in the accuracy of total sea ice concentrations (includes all ice types) is achieved, as compared to the ESMR. Comparisons between sea ice concentration obtained with SMMR, SSMI, and Landsat in selected areas yield estimates of SMMR algorithm accuracies ranging from about 3% in the Beaufort Sea and the Antarctic seas to an overall Arctic accuracy of about 7% under frozen conditions [Steffen and Schweiger, 1991]. Under melting conditions, the accuracy degrades to about 10%. The effects of storm systems observed on single-day maps of ice concentration are greatly reduced compared to the ESMR, because of the selection of 1.67 cm rather than 1.55 cm for the wavelength of the Ku-band channels and the application of the weather filter in the algorithm. As a result of these improvements, it is feasible to produce reliable time series of average sea ice concentrations and sea ice extents from single-day SMMR data [Gloersen and Campbell, 1988, 1991a, 1991b; Gloersen et al., 1992].

The DMSP SSMI is a seven-channel radiometer that measures both polarized components at the wavelengths 0.35, 0.81, and 1.55 cm and the vertically polarized component at 1.36 cm. The SSMI design provides improvements over the SMMR in the means of calibrating the radiometers and in the separation of the polarized components. A slight degradation in the spatial resolution, as a result of a smaller antenna dish, is partly compensated by the lower orbit of the satellite. Additionally, the reversion to 1.55 cm for the Ku-band channels, rather than the SMMR wavelength of 1.67 cm, results in much more pronounced weather effects. Nevertheless, the similarity of the SSMI and SMMR channel wavelengths permits the utilization of the SMMR Team sea ice algorithm, with modified coefficients, for the SSMI data [Cavalieri et al., 1991]. The modifications consist of SSMI-specific tie points for open water, first-year ice, and multiyear ice for the Arctic and a second ice type for the Antarctic. Additional modifications consist of an adjustment of the original weather filter parameter and the addition of an extra weather filter based on the 1.36-cm channel (D. J. Cavalieri, personal communication, 1993); reprocessing the SSMI data to accommodate these changes is in progress. Estimates of SSMI accuracies are comparable to those of SMMR [Steffen and Schweiger, 1991].

LONG-TERM BEHAVIOR OF THE POLAR SEA ICE COVERS

The extent of the polar sea ice covers has long been considered an indication of the state of the climate, dating back to the early Icelandic records of the proximity of ice to their shores. Determining climatic trends is generally considered to require records of at least 30 years in length, in order to assess long-term trends in light of significant interannual variability. Sea ice extent records from satellites date as far back as 1964, but the sporadic nature of the sea ice data obtained from the early visible and infrared sensors precludes obtaining any reliable trends from the data. The ESMR record (1973-1976) is much too short by itself to perceive any trends. Zwally et al. [1983b] noted a distinct decline in the Antarctic ice extent maxima in the mid-1970s, but a significant recovery by 1981. Gloersen and Campbell [1988] noted that there was a visually discernible decline in the global ice extent maxima over the combined operating lifetimes of ESMR and SMMR (Figure 1), with no obvious trends in either the separate Arctic or Antarctic ice extent maxima. Parkinson and Cavalieri [1989] reported that annually-averaged SMMR Arctic ice extents had a negative trend from 1979-1986, but also a large interannual variability.

In comparing sea ice extent data from different sensors, differences in the sensor characteristics, algorithms, and mapping techniques must be considered [Gloersen et al., 1992]. For instance, the ice extent minima in the ESMR Antarctic record (Figure 1) are about 2×10^6 km^2 greater than those for the SMMR, on the average. Correspondingly, the Antarctic ice extent maxima are different for ESMR and SMMR. However, the fourth ESMR ice extent maximum is similar to the SMMR maxima. On the other hand, the seasonal cycles observed with either instrument have about the same amplitude. These observations, by themselves, fail to provide sufficient evidence for separating real ice coverage changes from instrument differences. Some of the ESMR-SMMR differences may be attributed to the different map grid sizes used for ESMR and SMMR, and to a different means of assigning land values vs. ocean values to borderline map pixels [Gloersen et al., 1992]. The differing spatial resolutions and sea ice algorithms of the two instruments may also contribute to these discrepancies to a lesser extent; however, the sea ice extent calculation is relatively insensitive to the sea ice concentration algorithm used. An overlap period of operation of the ESMR and SMMR would have permitted unambiguous sorting out of these discrepancies. Remapping of the ESMR data with the gridding scheme used for the SMMR and SSMI may be necessary before serious comparisons can be made.

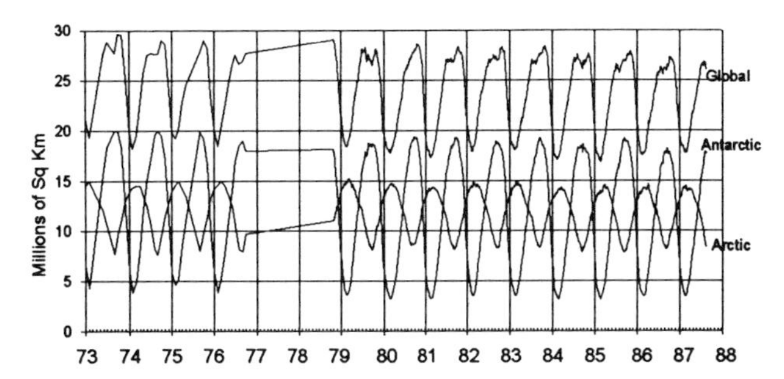

Figure 1. Combined ESMR-SMMR records of sea ice extents for the Arctic, Antarctic, and their sum, the global curve [from Gloersen and Campbell, 1988].

The 7-week overlap of the SMMR and SSMI data sets provides the opportunity of comparing the results of sea ice concentration calculations from both data sets. Despite the overlap periods, this sort of comparison requires taking into account SMMR-SSMI visit time differences of 8 to 16 hours (depending on the orbital node) and the large variations of sea ice concentration within these short times. To identify areas of high daily ice variability where these comparisons would be invalid, differences in the polar ice concentration maps obtained from the SMMR ascending and descending orbital nodes were calculated, and a mask prepared of all sea ice map pixels with large differences in concentration from one orbital node to another for a given day. The resulting mask covers from 10%-20% of the sea ice area on each of

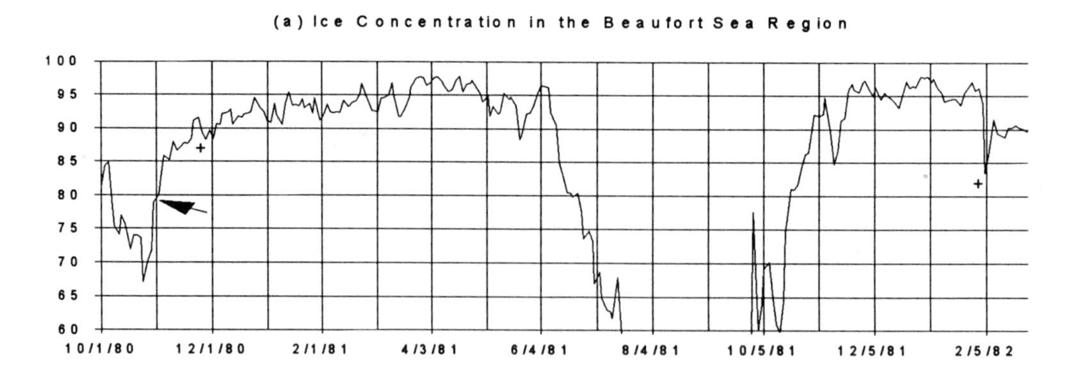

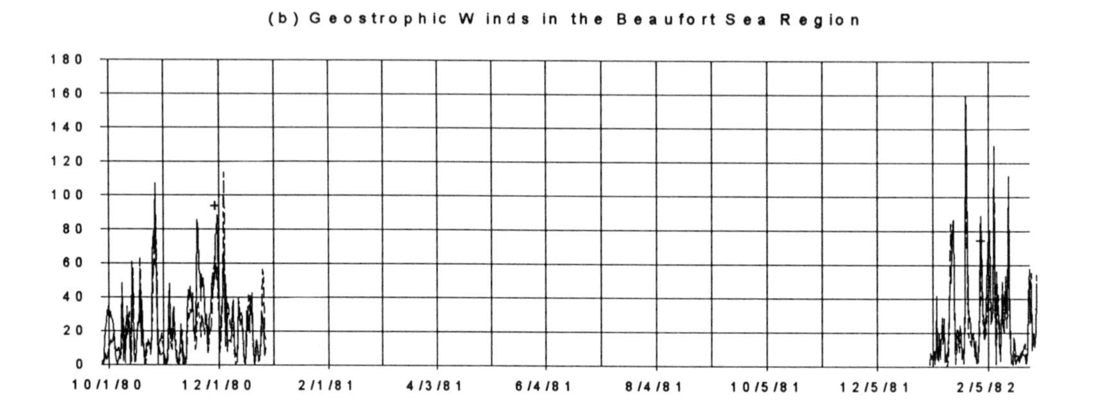

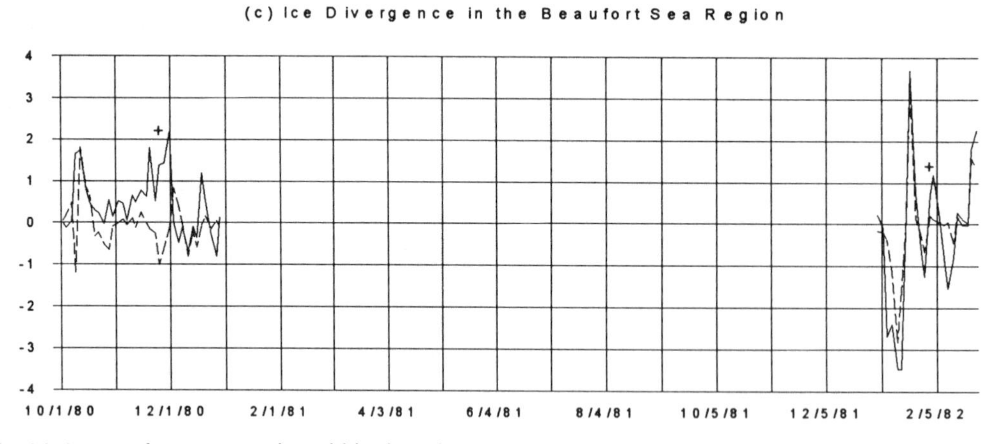

Figure 2. (a) Average ice concentration within the polygon shown in Plate 1 and for a similar time period, with time extensions on either end. (b) Corresponding geostrophic wind (in arbitrary units -- see text) computed from AOBP data. (c) Corresponding ice divergence (in arbitrary units -- see text) computed from AOBP data.

the days. These masks were then used to exclude the highly variable areas from both the SMMR and SSMI daily averaged single-day data before the comparison was made. (The SSMI data available at the time of the comparison were already averaged over a day.) The pixel-by-pixel differences (for those not masked) were accumulated for the 20 days of overlap during August 1987. The resulting Arctic SMMR and SSMI ice concentration differences have a bias of 0.5% and a standard deviation of 5% [Gloersen et al., 1992]. This is remarkably good, considering that boreal summer melting conditions and high ice-signature variability occurred in the Arctic during the overlap period. The ice-concentration differences for austral winter conditions in the Antarctic have a 0.2% bias and a 2.5% standard deviation. Application of the second weather filter to the SSMI data might reduce the observed SMMR-SSMI differences.

An earlier application of a band-limited regression technique [Lindberg, 1990] to the SMMR-only data set [Gloersen and Campbell, 1991b] revealed a 2% decline in sea ice extent in the Arctic over the 9-year SMMR lifetime, at a confidence level of 96%, and no significant trend in the Antarctic. A similar analysis is in progress on the existing 13-year combined SMMR and SSMI data set. However, in light of the large interannual variations in the Arctic and Antarctic ice extents and the small ice extent trends thus far observed, combining these data sets will still not result in a sufficiently long record for establishing a climate trend, but will provide a more solid baseline against which future data may be compared.

REGIONS OF PERSISTENT LOW ICE CONCENTRATIONS IN THE ARCTIC

Earlier analysis of maps of sea ice concentration obtained from the SMMR [Gloersen et al., 1992] had identified some regions of the Arctic Ocean where lower-than-average values of sea ice concentration persist over periods of weeks to several months. These low values were observed in ice concentration maps of single-day, monthly average, and mean monthly anomalies. We expand upon those observations here for four specific regions designated by their locations in the Beaufort Sea, Canadian Basin, and the Chukchi and East Siberian Seas, and indicated by the white polygons in Plates 1-4. These regions are not the storm-induced short-period coastal polynyas analyzed by Cavalieri and Martin [1993] for the purpose of estimating halocline salt fluxes, but extend further from the coastlines or are over deeper portions of the Arctic Ocean.

Beaufort Sea Region

There are two time sequences shown for the low ice concentration area in the Beaufort Sea (Plate 1), one starting on November 2, 1980, and the other on February 3, 1982. On November 2, 1980, there are still shore polynyas remaining from the previous summer's open water condition. A low concentration region of interest lies farther north, between the latitudes of 72° and 76° N, where the concentrations are as low as 72%. In the next three scenes, November 6, 10, and 24, the ice concentrations gradually increase to a maximum of 80%-84% on November 24 when the low ice concentration pattern shifts to the west (and partly out of the polygon region). The concentrations decrease again to 76%-80% on December 2, and then vary through the remaining scenes shown. As will be illustrated later, this oscillatory behavior persists for the remainder of the winter. In the following winter, a similar decrease occurs, most prominently later in the season and farther south in the polygon region. The decrease begins about February 3, 1982,

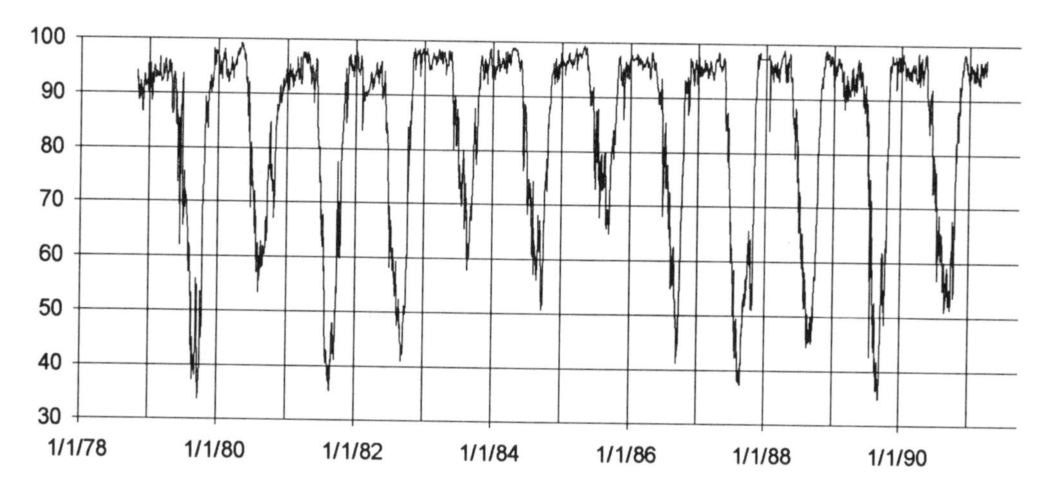

Figure 3. The combined SMMR-SSMI record of average ice concentration within the polygon shown in Plate 1.

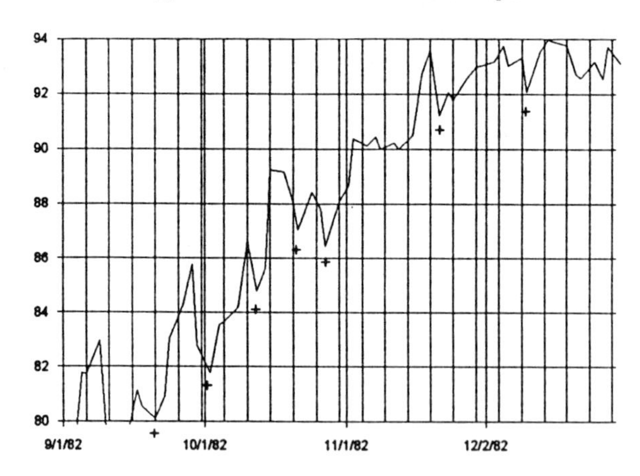

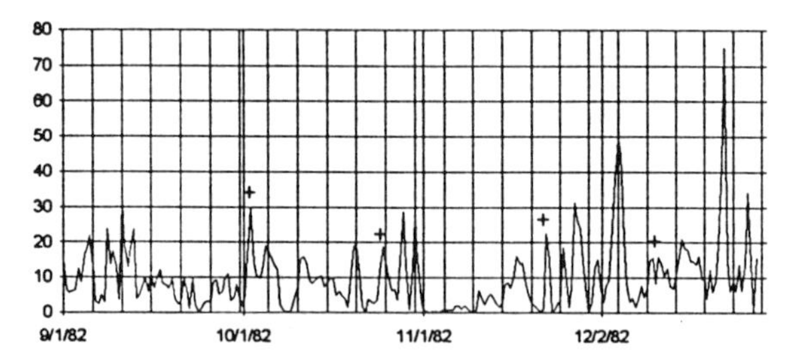

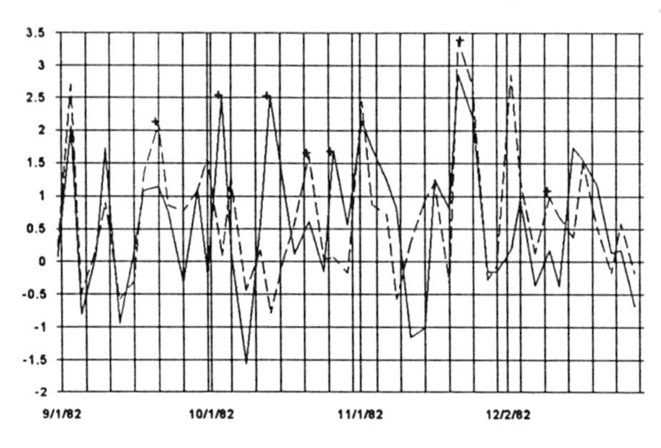

Figure 4 (a) Average ice concentration within the polygon shown in Plate 2 and for a similar time period in 1982, with time extensions on either end. (b) Corresponding geostrophic wind (in arbitrary units -- see text) computed from AOBP data. (c) Corresponding ice divergence (in arbitrary units -- see text) computed from AOBP data.

reaching a minimum on February 4, then concentration increases again as shown on February 7 and 9, but to a lower level than before. As seen in the monthly average and monthly mean anomaly maps of ice concentration [Gloersen et al., 1992], this indication of lower ice con-

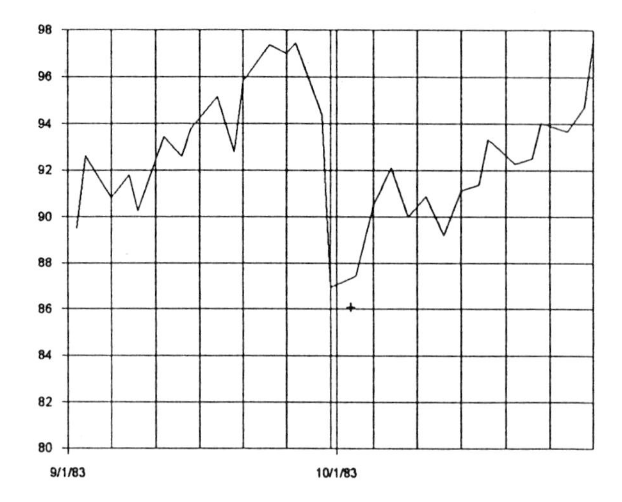

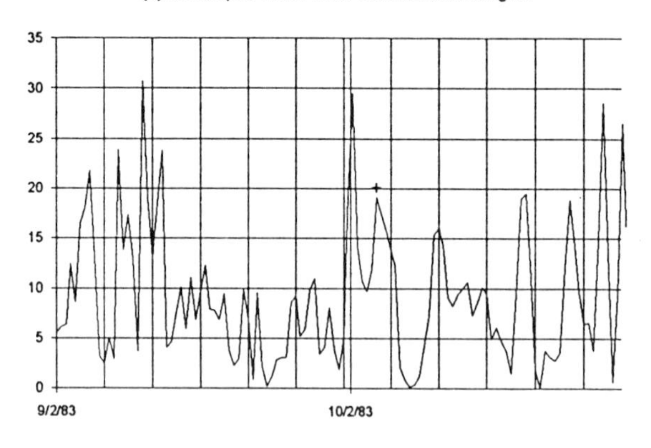

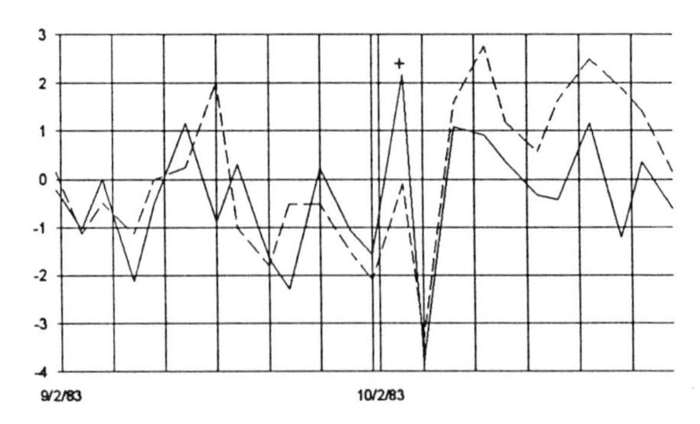

Figure 5. (a) Average ice concentration within the polygon shown in Plate 2 and for a similar time period in 1983, with time extensions on either end. (b) Corresponding geostrophic wind (in arbitrary units -- see text) computed from AOBP data. (c) Corresponding ice divergence (in arbitrary units -- see text) computed from AOBP data.

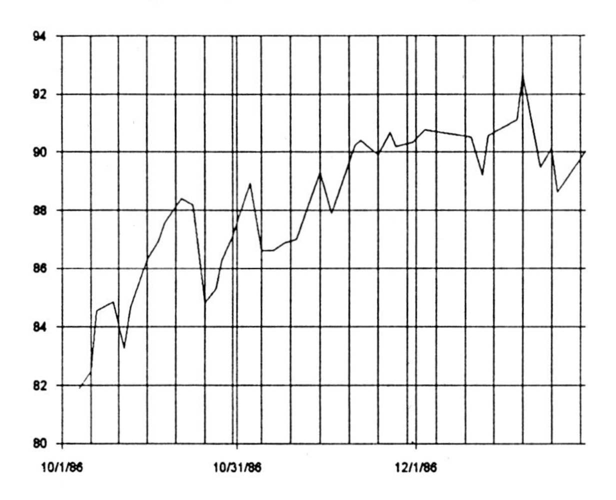

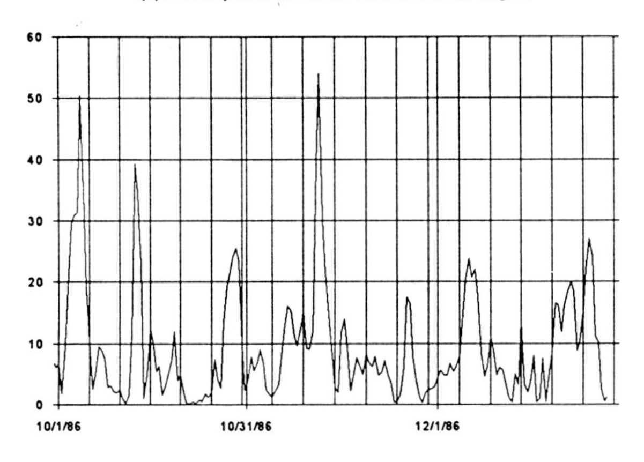

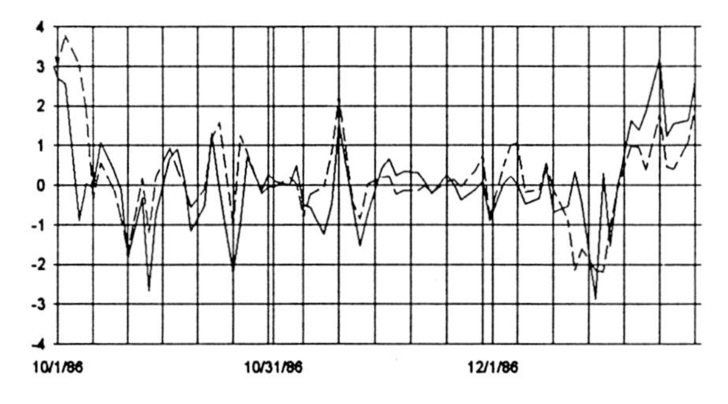

Figure 6. (a) Average ice concentration within the polygon shown in Plate 2 and for a similar time period in 1986, with time extensions on either end. (b) Corresponding geostrophic wind (in arbitrary units -- see text) computed from AOBP data. (c) Corresponding ice divergence (in arbitrary units -- see text) computed from AOBP data.

centration persists for unknown reasons until May 1992, at which time the concentration again increases.

Another way to examine the persistence of these low concentration areas is to plot the average ice concentration within the polygon as a function of time, as shown in Figure 2a. In Figure 2a, the time sequence starts before and ends after the events illustrated in Plate 1. At the beginning of this sequence, there is a minimum average ice concentration occurring late in October 1980 that results from a large area of open water along the shoreline. A similar opening occurs in early October 1981 as part of the winter freeze-up. The event on November 2, 1980, illustrated in Plate 1, shows up as a small bump on the curve, indicated by an arrow in Figure 2a, as the average concentration in the polygon is rapidly increasing during the freeze-up. A subsequent ice concentration minimum in Plate 1 (December 2, 1980) can also be seen in the average ice concentration curve in Figure 2, indicated by a "+," near the end of November 1980. While this drop is smaller than the estimated 5% accuracy of the algorithm [Gloersen et al., 1992], this curve is an average of many pixel areas that improves the accuracy of the average, and in any event, the drop is larger than the estimated precision of 1% or better. During the winter of 1980-81, the average ice concentration in the polygon (Figure 2) reaches at least nine minima. In 1982, the February minimum shown in Plate 1 also appears in the average ice concentration in Figure 2a, again indicated by a "+." A 12.5-year record (including SSMI data -- note the seamless transition in August 1987) of average ice concentrations is shown in Figure 3. The 1980-81 winter event is difficult to discern, but the one in February 1982 clearly stands out. Similar features are seen in late fall 1985 and 1986, and early in the year in 1988 and 1989.

Divergence of the ice pack and consequent reduced concentrations may be caused by storm events [Zwally and Walsh, 1987] and oceanic currents from tidal forcing or upwelling. To explore these mechanisms further, we have used some concurrent data from the Arctic Ocean Buoy Program (AOBP). Included in the AOBP data [e.g., Thorndike and Colony, 1980, 1981, 1982] are buoy surface pressures and buoy locations, analyzed fields of surface pressure in 2° latitude by 10° longitude bins, and fields of ice velocities separated into u,v components, as well as their derivatives along x,y, in 4° latitude by 20° longitude bins. Although these fields are based on sparse data and are on much coarser grids than the SMMR data, we have used them to obtain additional fields of geostrophic winds (in arbitrary units, obtained simply by differencing adjacent surface pressure bins in each direction and then using the rms of both pressure-difference com-

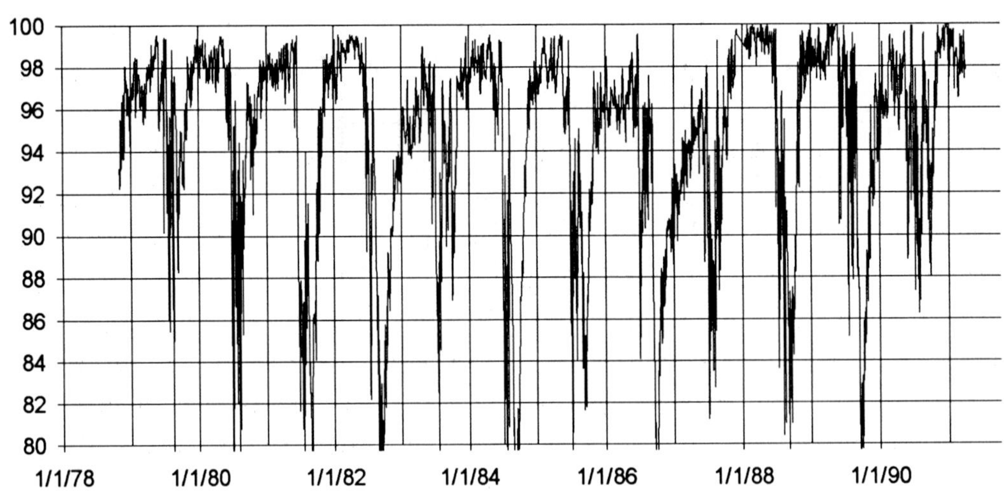

Figure 7. The combined SMMR-SSMI record of average ice concentration in the Canadian Basin within the polygon shown in Plate 2.

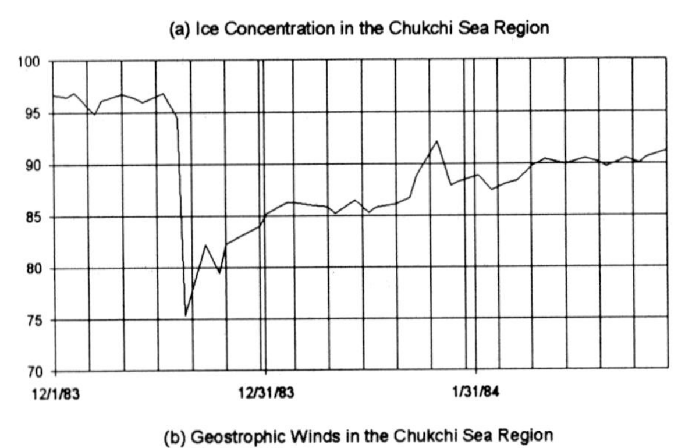

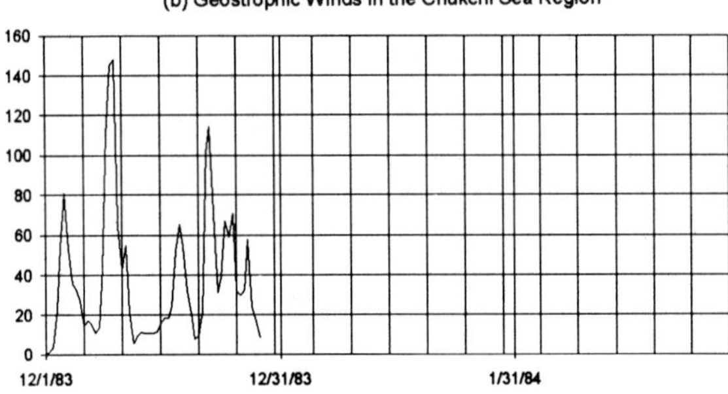

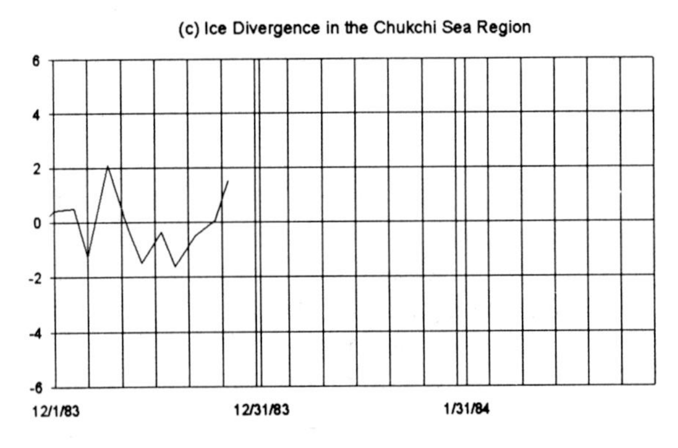

ponents as the scalar wind) and ice divergence, obtained directly from the ice velocity derivative fields. In Figure 2b, geostrophic winds from two of the bins are plotted on the same time scale as the ice concentrations in Figure 2a, but are limited to the time intervals near the two examples shown in Plate 1. The center of one bin is located at 72° N, 150° W (solid curve), and the other at 74° N, 150° W (dashed curve), between the two meridians passing through the polygon. (The two parallels passing through the polygon are 70° N and 75° N.) The lack of coincidence between the solid and dashed curves and their rapid fluctuations may indicate the deleterious effect of the scarcity of data points available for use in the AOBP analyzed fields. The minima shown in Plate 1 on December 2, 1980, and February 5, 1982, (marked with "+"s in Figure 2a) appear to be associated with a high wind event. The occurrence of an ice concentration minimum on February 4, 1982 is also bracketed by two stronger wind events occurring several days before and after. However, in this case, the direction of the wind and the proximity of the shoreline definitely play an important role. Referring to Colony et al. (1983), the winds were offshore from February 1 -- February 4, followed by a reversal to onshore winds on February 5, and variable

Figure 8. (a) Average ice concentration within the polygon shown in Plate 3 and for a similar time period, with time extensions on either end. (b) Corresponding geostrophic wind (in arbitrary units -- see text) computed from AOBP data. (c) Corresponding ice divergence (in arbitrary units -- see text) computed from AOBP data.

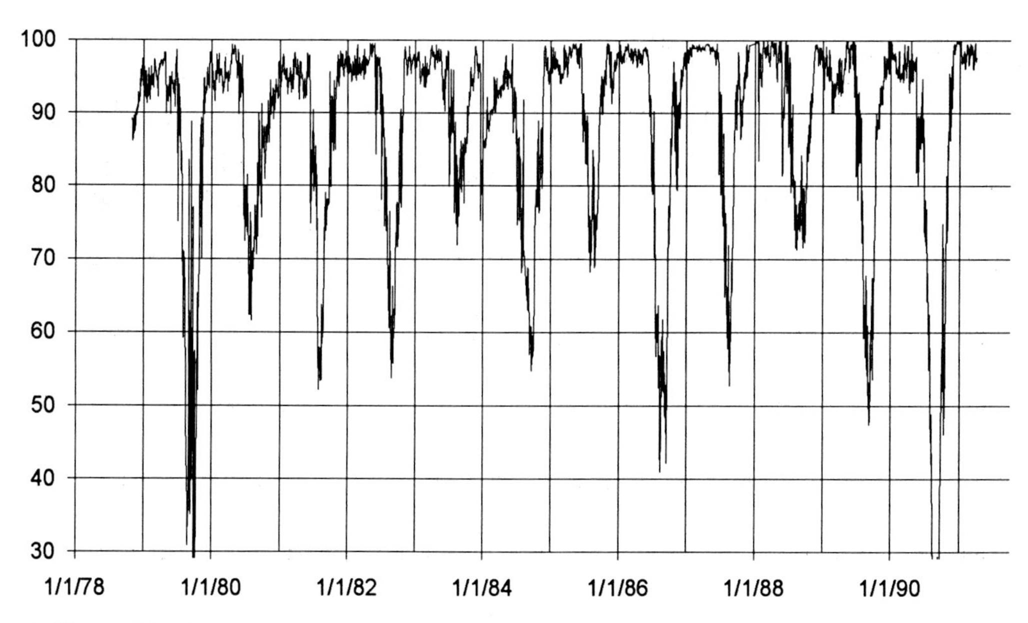

Figure 9. The combined SMMR-SSMI record of average ice concentration in the Chukchi Sea within the polygon shown in Plate 3.

directions thereafter. The average ice concentration minimum occurring near the end of October 1980 (before the arrow in Figure 2a) also appears to coincide with a major wind event. It should be kept in mind that a high wind event can lead to either compression or divergence, depending upon such things as wind direction, locations of boundaries, and the average ice concentration.

Comparison of AOBP ice divergences with SMMR or SSMI ice concentrations involves an implicit test of the validity of the correspondence between them, a reasonable assumption under conditions of low ice concentration gradients. Also implicit in this test is the neglect of any changes in ice concentration caused by concurrent freezing and melting of the ice. Generally, ice divergence causes a decrease in ice concentration, which is followed by an increasing ice concentration as new ice is formed when freezing conditions exist. In the fall and winter cases examined here, freezing conditions should exist. Therefore, observed reductions in ice concentration associated with divergence should be somewhat smaller than they would be if no concurrent ice production occurred. The AOBP ice divergences are shown in Figure 2c, also for two bins, one centered at 74° N, 160° W (solid curve) and the other at 74°N, 140° W (dashed curve), corresponding to the two meridians passing through the polygon. Again, the curves in Figure 2c apparently do not correlate highly with those in Figures 2a and 2b, but there appears to be a coincidence between the maxima in ice divergence and the minima in average ice concentration identified earlier in Plate 1 and Figure 2a on December 2, 1980, and February 4 - 5, 1982.

Although there is some correspondence between the reduced ice concentrations as observed with SMMR and the ice divergence calculated from the AOBP data, the low overall correlation may be the result of a sparse data set and the interpolation scheme used to fill in the grids of analyzed AOBP data, and of the neglect of ice freezing and melting. In addition, the timing of the ice divergence, as recorded by the buoys with respect to the next observation by SMMR, is critical, because freezing following the ice divergence can rapidly increase the ice concentration.

Canadian Basin Region

The Canadian Basin region of reduced sea ice concentrations is more or less due north of the Beaufort Sea region, as illustrated in Plate 2 and Figures 4a, 5a, and 6a. In this case, we show three separate series occurring in 1982, 1983, and 1986. For easier visualization, the AOBP analyses are shown separately for these three periods. This time, a single bin of the AOBP data was chosen for the geostrophic winds, at 82°-84° N, 120°-130° W. Two adjacent bins were chosen for the ice diver-

(a) Ice Concentration in the East Siberian Sea Region

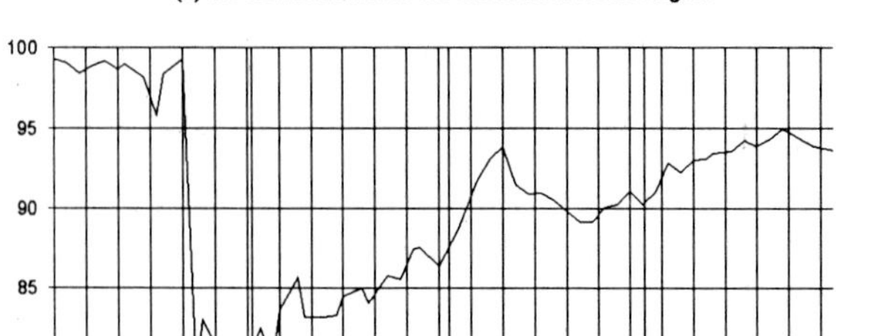

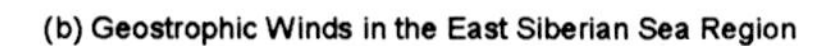

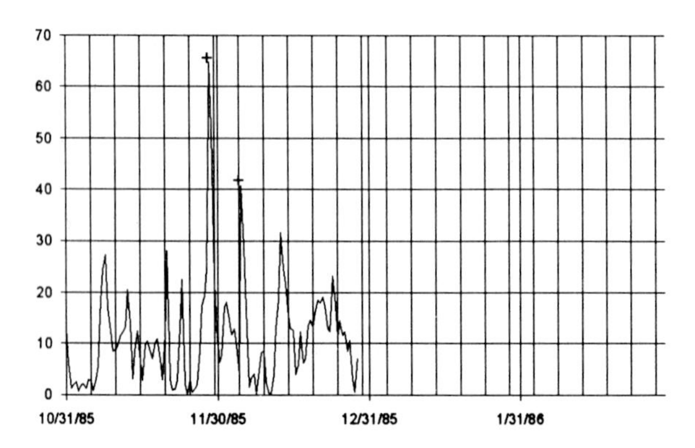

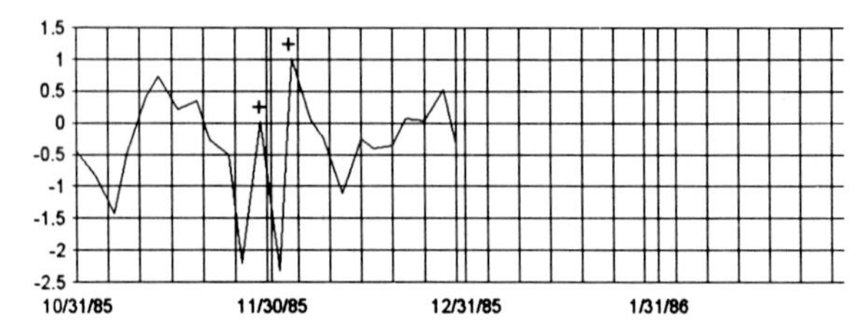

Figure 10. (a) Average ice concentration within the polygon shown in Plate 4 and for a similar time period, with time extensions on either end. (b) Corresponding geostrophic wind (in arbitrary units -- see text) computed from AOBP data. (c) Corresponding ice divergence (in arbitrary units -- see text) computed from AOBP data.

gence, centered at 86° N, 160° W (solid curve) and the other at 86° N, 120° W (dashed curve), near the eastern and western edges of the polygon. In 1982, the ice concentration maps in Plate 2 show the reduced concentrations around October 9 and October 13, 1982. Comparison of the average ice concentration, the geostrophic wind, and the ice divergence in Figure 4 shows seven coincidences between the maxima in the ice divergence and the minima in the average ice concentrations (indicated by "+"s on each curve), and four of those also corresponded to high wind events. However, several examples of a lack of coincidence are also noted in the figure. Perhaps of more significance is the predominance of divergent conditions in this region of reduced sea ice concentration throughout much of this period.

In 1983, the ice concentration minimum appears on October 2 in Plate 2, and appears to persist for several weeks in Figure 5a. The correlations with the geostrophic wind and divergence are very good in this case, taking into account the location and size of the AOBP grid cell. During the period prior to the October 2 event, the conditions are predominantly convergent while the concentration is increasing. During this time, the ice concentrations are generally increasing because the open leads and polynyas are freezing. After the event, the conditions are predominantly divergent, and the ice concentration increases at a slower rate than the increase prior to the event.

In 1986, the ice concentrations are lower throughout much of the winter season, as also shown in Figure 6a. This time, the AOBP ice divergences, on average, are close to zero and show little correlation with the changes in ice concentration.

The 12.5-year record of average ice concentration (Figure 7) shows that the oscillations in the average ice concentrations are present at least every winter. The oscillations in summer may include freeze/thaw cycles and the formation of melt ponds. The interannual variability in the mean winter ice concentrations is large, ranging from about 92% in 1986/87 to more than 98% in 1989/90.

Chukchi Sea Region

The wintertime low ice-concentration event described in Plate 3 and Figure 8 for the winter of 1983-84 in the Chukchi Sea seems to be unique during the 12.5-year record (Figure 9), except possibly for a less extreme event in the first quarter of 1989. This event persists for about 3 months, longer than the prior two cases. For this region, the cells for both the geostrophic winds and the ice divergence are centered at 74° N and 180° W. On December 20, 1983, the ice concentration (Figure 8a) drops markedly, but there is no evidence for a corresponding increase in the AOBP ice divergence (Figure 8c). However, in this location there were no actual AOBP buoys; the pressure and velocity fields were based on nearby buoys and Russian coastal weather stations.

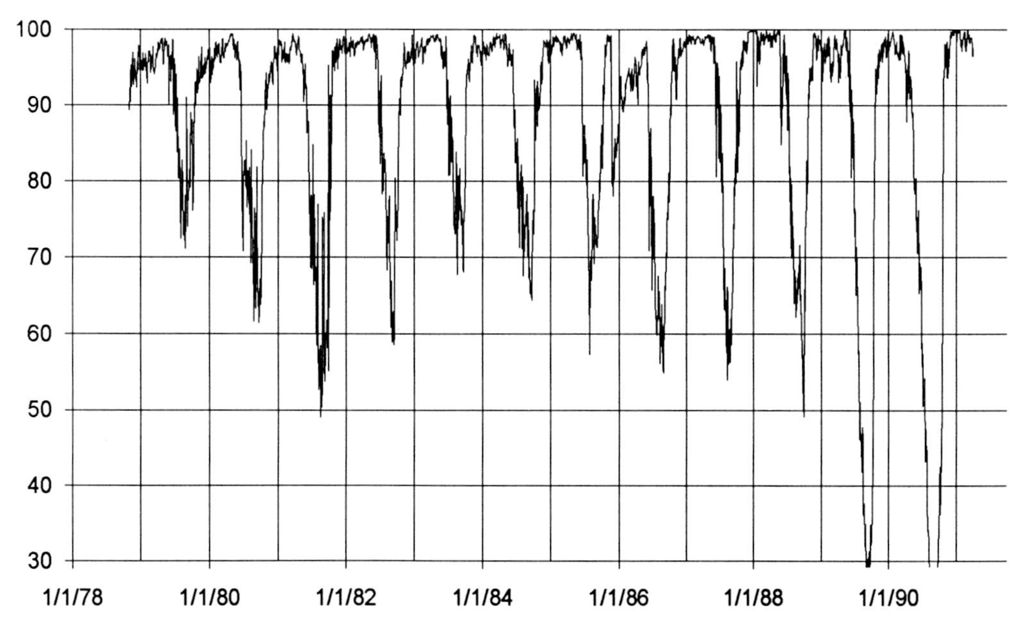

Figure 11. The combined SMMR-SSMI record of average ice concentration in the East Siberian Sea within the polygon shown in Plate 4.

The lack of actual buoy velocity data in this area seriously limits the accuracy of the calculated ice divergences.

East Siberian Sea Region

As in the previous case, the low ice-concentration event in the winter of 1985-1986 in the East Siberian Sea (Plate 4 and Figure 10) is apparently also unique, except possibly for two much smaller events in the first quarter of 1989 (Figure 11). It is also a very persistent phenomenon, lasting for several months. The initial onset of the low ice concentration-event (Figure 10a) on about November 19 seems to bear no relation to the geostrophic wind or ice divergence. However, on about November 28 and December 3-4, increases in the geostrophic wind (Figure 10b) and the ice divergence (Figure 10c) seem to be significantly correlated with a decrease in the ice concentration. In this case, the cell for the geostrophic wind is centered at 74° N, 170° E, and the cell for the ice divergence at 74° N, 160° E. As in the previous location, the AOBP buoy data were absent here, and the nearby Russian coastal stations provided the major inputs for the binned geostrophic winds (but not the ice divergences).

CONCLUDING REMARKS

In the four low ice-concentration events discussed, there are some instances when weather systems and associated ice divergences clearly affect the sea ice concentration. However, there is not a one-to-one correspondence between the average SMMR ice concentrations and the AOBP ice divergences. This lack of correspondence may result from the scarcity of the AOBP data for calculating the ice divergence and the coarseness of the gridded surface analyses, and the neglect of freezing and melting effects on ice-concentration changes. On the other hand, the regular oscillations noted in the averaged ice concentrations, of longer period than those in the geostrophic wind records, and the several-month persistence noted in the last two cases may not be weather-related. The possibilities of relating these phenomena to oceanic upwelling, tidal currents, basin resonances, or inertial motion in the ice remain as subjects for future study.

REFERENCES

Cavalieri, D. J., P. Gloersen, and W. J. Campbell, Determination of sea ice parameters with the Nimbus 7 SMMR, *J. Geophys. Res.* **89**, 5355-5369, 1984.

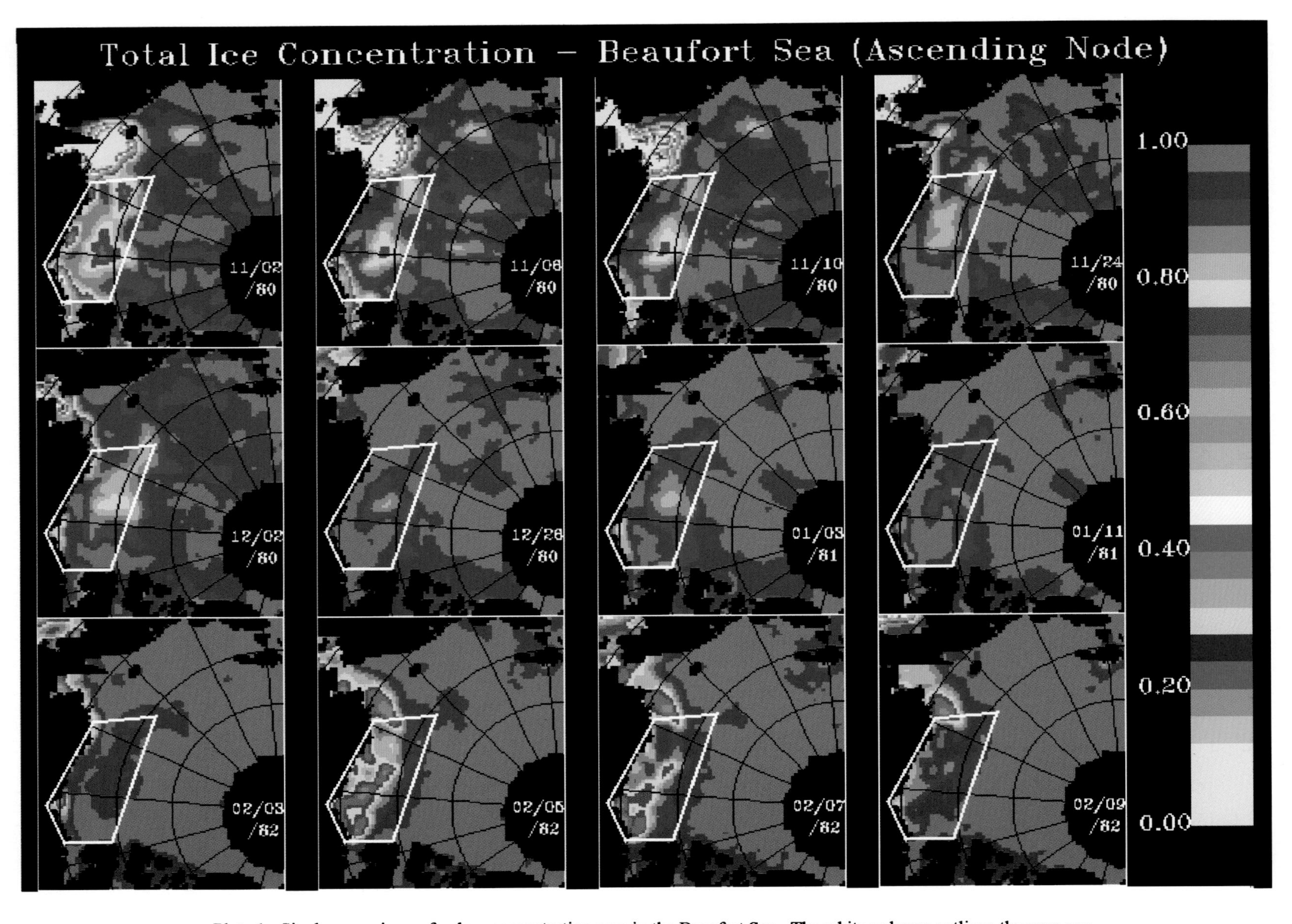

Plate 1. Single-pass views of a low-concentration area in the Beaufort Sea. The white polygon outlines the area analyzed in Figure 3a. The meridians passing through the polygon are 140° W and 160° W; the parallels are 70° N and 75° N.

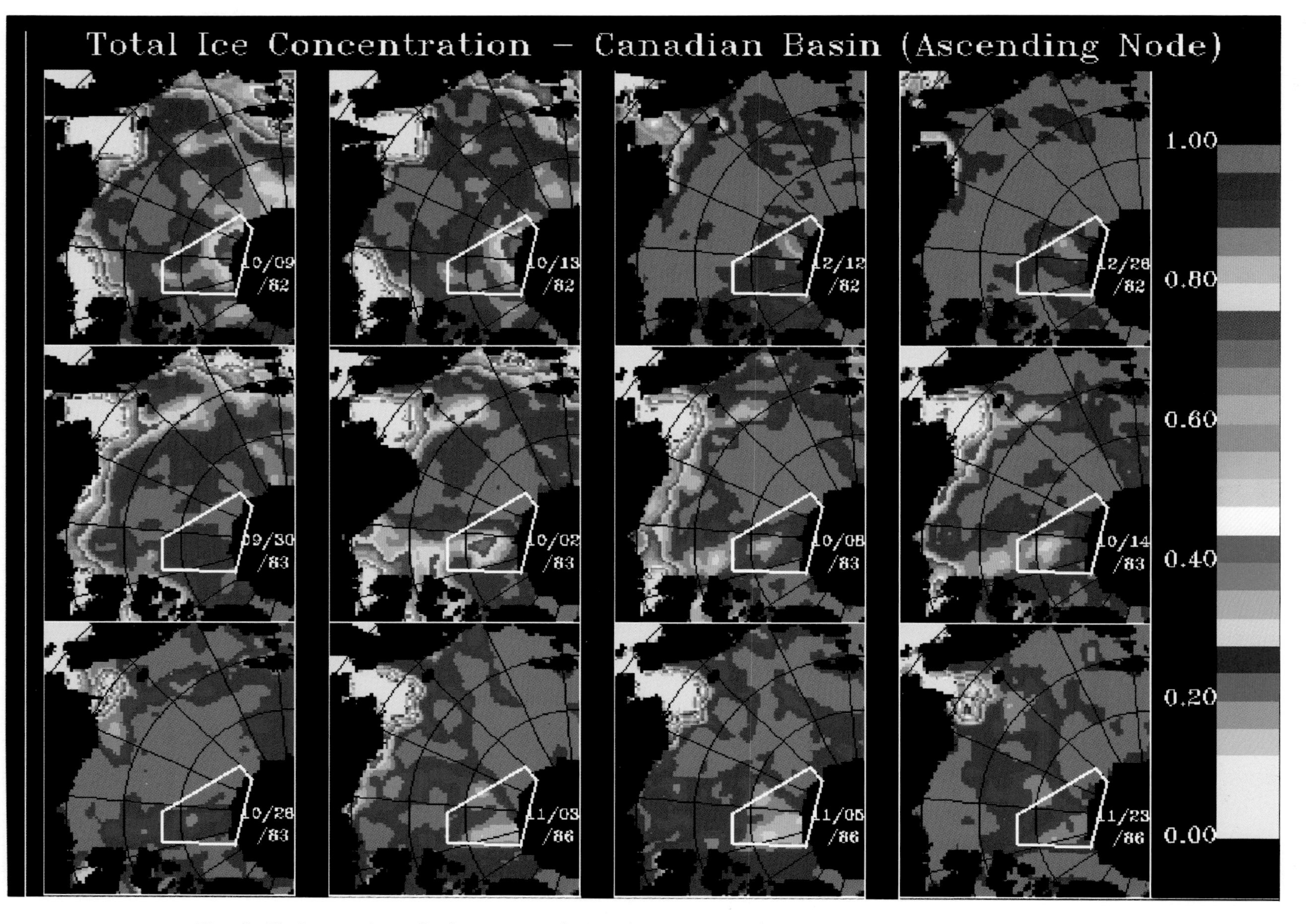

Plate 2. Single-pass views of a low-concentration area in the Canadian Basin. The white polygon outlines the area analyzed in Figures 4a, 5a, and 6a. The meridians passing through the polygon are 120° W, 140° W, and 160° W; the parallel is 75° N.

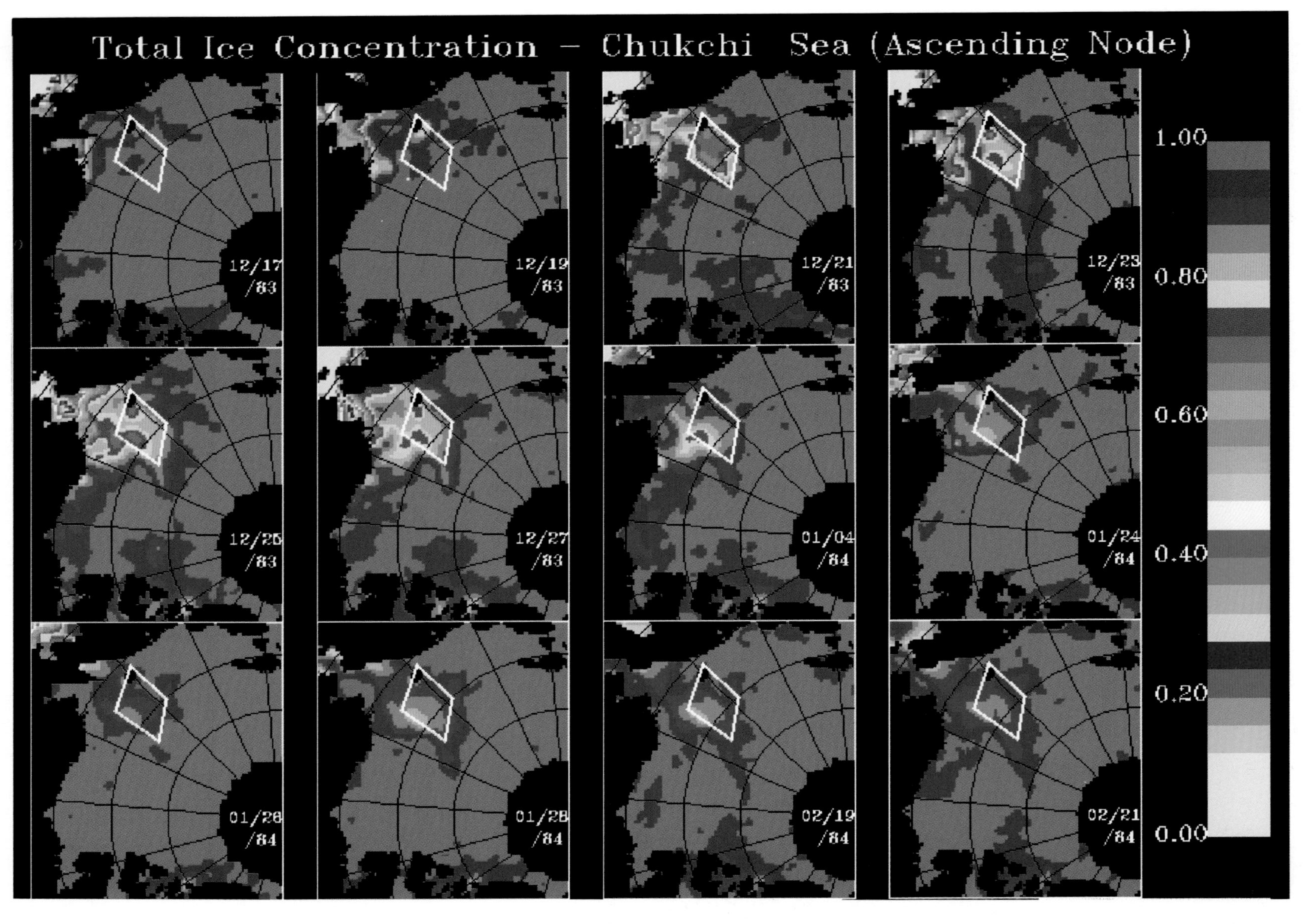

Plate 3. Single-pass views of a low-concentration area in the Chukchi Sea. The white polygon outlines the area analyzed in Figure 8a. The meridian passing through the polygon is 180° ; the parallel is 75° N.

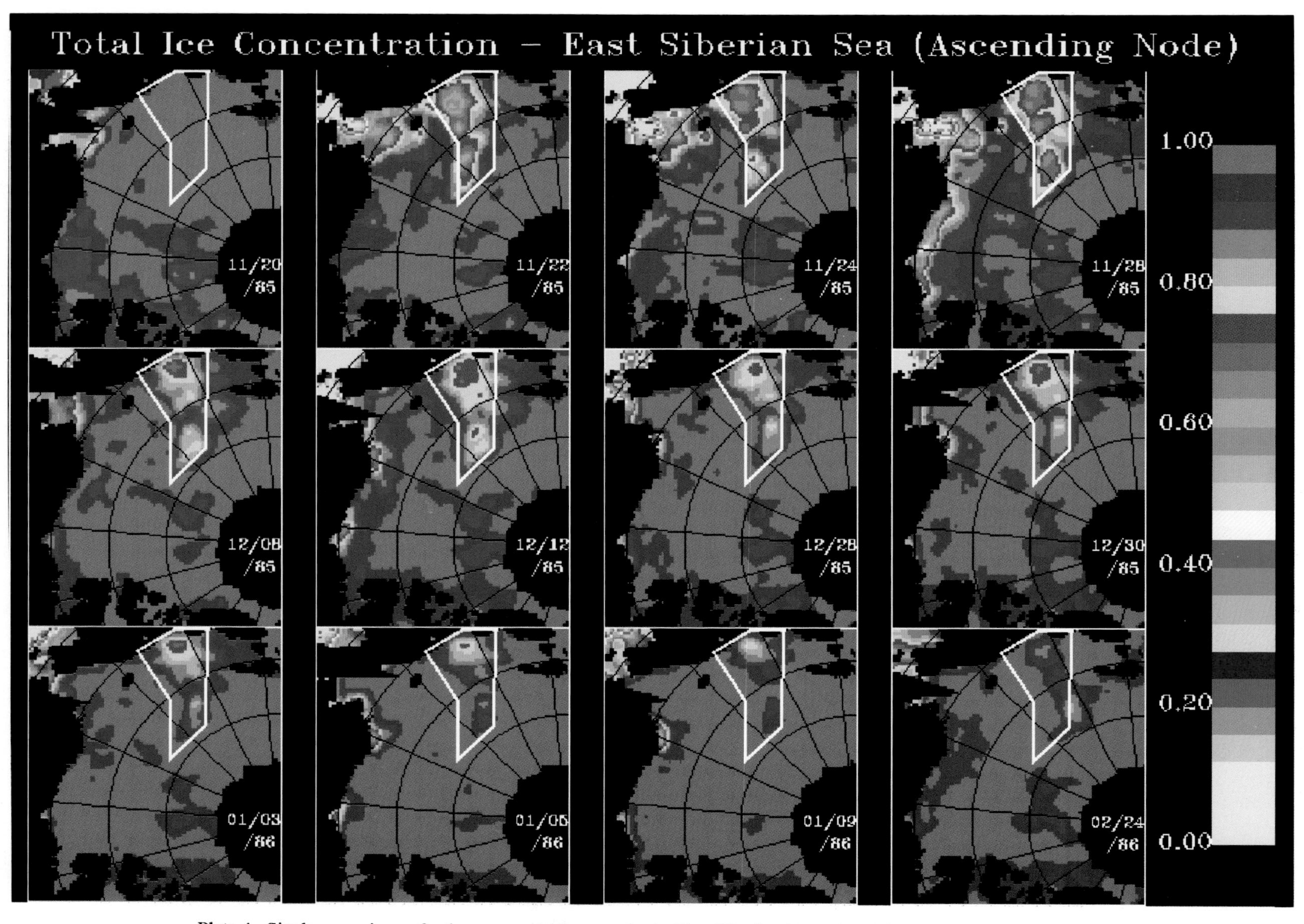

Plate 4. Single-pass views of a low-concentration area in the East Siberian Sea. The white polygon outlines the area analyzed in Figure 10a. The meridians passing through the polygon are 180° and 160° E; the parallel is 75° N.

Cavalieri, D. J., J. P. Crawford, M. R. Drinkwater, D. T. Eppler, L. D. Farmer, R. R. Jentz, and C. C. Wackerman, Aircraft active and passive microwave validation of sea ice concentration from the DMSP SSMI, *J. Geophys. Res.* **96**, 21,989-22,008, 1991.

Cavalieri, D. J., and S. Martin, The contribution of Alaskan, Siberian, and Canadian Coastal polynyas to the cold halocline layer of the Arctic Ocean, *J. Geophys. Res.*, in press, 1993.

Comiso, J. C., and H. J. Zwally, Antarctic sea ice concentrations inferred from Nimbus 5 ESMR and Landsat imagery, *J. Geophys. Res.* **87**, 5836-5844, 1982.

Gloersen, P., T. T. Wilheit, T. C. Chang, W. Nordberg, and W. J. Campbell, Microwave maps of the polar ice of the Earth, *Bull. Am. Met. Soc.* **55**, 1442-1448, 1974.

Gloersen, P., and F. T. Barath, A scanning multichannel microwave radiometer for Nimbus G and Seasat A, *IEEE J. of Oceanic Eng.* **OE-2**, 172-178, 1977.

Gloersen, P., D. J. Cavalieri, A. T. C. Chang, T. T. Wilheit, W. J. Campbell, O. M. Johannessen, K. B. Katsaros, K. F. Kunzi, D. B. Ross, D. Staelin, E. P. L. Windsor, F. T. Barath, P. Gudmandsen, E. Langham, and R. O. Ramseier, A summary of results from the first Nimbus 7 SMMR observations, *J. Geophys. Res.* **89**, 5335-5344, 1984.

Gloersen, P., and D. J. Cavalieri, Reduction of weather effects in the calculation of sea ice concentration from microwave radiances, *J. Geophys. Res.* **91**, 3913-3919, 1986.

Gloersen, P., and W. J. Campbell, Variations in the Arctic, Antarctic, and global sea ice covers during 1978-1987 as observed with the Nimbus 7 Scanning Multichannel Microwave Radiometer, *J. Geophys. Res.* **93**, 10,666-10,674, 1988.

Gloersen, P., and W. J. Campbell, Variations of extent, area, and open water of the polar sea ice covers: 1978-1987, *Proc. of the Int. Conf. on the Role of the Polar Regions in Global Change,* edited by G. Weller, C. L. Wilson, and B. A. B. Severin, pp. 28-34, Geophysical Institute, University of Fairbanks, Alaska, 1991a.

Gloersen, P., and W. J. Campbell, Recent variations in Arctic and Antarctic sea ice covers, *Nature* **352**, 33-36, 1991b.

Gloersen, P., W. J. Campbell, D. J. Cavalieri, J. C. Comiso, C. L. Parkinson, and H. J. Zwally, *Arctic and Antarctic sea ice, 1978-1987: Satellite Passive Microwave Observations and Analysis*, 293 pp., NASA SP-511, National Aeronautics and Space Administration, Washington, D.C., 1992.

Lindberg, C. R., Band limited regression. Part I: Simple linear models, AT&T Bell Labs Tech. Memo. 11217-901207-25TM, 1990.

Mätzler, C., R. O. Ramseier, and E. Svendsen, Polarization effects in sea ice signatures, *IEEE J. Oceanic Engr.* **OE-9**, 333-338. 1984.

Onstott, R. G., T. C. Grenfell, C. Mätzler, C. A. Luther, and E. A. Svendsen, Evolution of microwave sea ice signatures during early summer and midsummer in the marginal ice zone, *J. Geophys. Res.* **92**, 6825-6835, 1987.

Parkinson, C. L., J. C. Comiso, H. J. Zwally, D. J. Cavalieri, P. Gloersen, and W. J. Campbell, *Arctic Sea Ice, 1973-1976: Satellite Passive-Microwave Observations,* NASA SP-489, National Aeronautics and Space Administration, Washington, D.C., 296 pp., 1987.

Parkinson, C. L., and D. J. Cavalieri, Arctic sea ice, 1973-1987: Seasonal, regional, and interannual variability, *J. Geophys. Res.* **94**, 14,499-14,523, 1989.

Steffen, K., and A. Schweiger, DMSP-SSMI NASA team algorithm for sea ice concentration retrieval: Comparison with Landsat satellite imagery, *J. Geophys. Res.* **96**, 21,971-21,987, 1991.

Thorndike, A. S., and R. Colony, *Arctic Ocean Buoy Program Data Report, 19 January 1979-31 December 1979,* University of Washington Polar Science Center, Seattle, 1980.

Thorndike, A. S., and R. Colony, *Arctic Ocean Buoy Program Data Report, 1 January 1980-31 December 1980,* University of Washington Polar Science Center, Seattle, 1981.

Thorndike, A. S., and R. Colony, Sea ice motion in response to geostrophic winds, *J. Geophys. Res.* **87**, 5845-5852, 1982.

Thorndike, A. S., R. Colony, and E. A. Muñoz, *Arctic Ocean Buoy Program Data Report, 1 January 1981-31 December 1982,* University of Washington Polar Science Center, Seattle, 1983.

Zwally, H. J., J. C. Comiso, C. L. Parkinson, W. J. Campbell, F. D. Carsey, and P. Gloersen, *Antarctic Sea Ice, 1973-1976: Satellite Passive-Microwave Observations,* NASA SP-459, National Aeronautics and Space Administration, Washington, D.C., 206 pp., 1983a.

Zwally, H. J., C. L. Parkinson, and J. C. Comiso, Variability of Antarctic sea ice and changes in carbon dioxide, *Science* **220**, 1005-1012, 1983b.

Zwally, H. J., and J. E. Walsh, Comparison of observed and modeled ice motion in the Arctic Ocean, *Ann. Glaciol.* **9**, 136-144, 1987.

William J. Campbell (deceased), Ice and Climate Project, U.S. Geological Survey, University of Puget Sound, Tacoma, WA 98416

Per Gloersen (corresponding author)/Code 971, Laboratory for Hydrospheric Process, NASA Goddard Space Flight Center, Greenbelt, MD 20771

H. Jay Zwally/Code 971, Laboratory for Hydrospheric Process, NASA Goddard Space Flight Center, Greenbelt, MD 20771

Acoustic Remote Sensing

Walter H. Munk

Scripps Institution of Oceanography, University of California, San Diego, La Jolla, California

It is important to measure, rather than to speculate on climatic changes in the oceans. We discuss a program for measuring temperature changes in the interior ocean on a basin scale.

1. INTRODUCTION

When Roger Revelle in 1959 set into motion the measurements of atmospheric CO_2 that led to the now famed Keeling curves, [Keeling *et al.*, 1989] the proposal was greeted with skepticism: how could activities by puny mankind alter the planet?

The Keeling measurements indicated an atmospheric storage of CO_2 by 4 gigatons per year. How does this compare to the burning of fossil fuel? The earliest estimates yielded 6 gtons/y, a number larger than the measured atmospheric storage. The budget was closed by assuming 2 gtons/y of oceanic storage. Some ten years later the situation changed when it was estimated that the clearing of forests are an important consideration, and changed once again when the increased photosynthesis due to the enhanced atmospheric CO_2 was taken into account:

	Fossil Fuel		Clearing		Increased Photosyn.		Atmos. Storage		Ocean Storage
1960	6					=	4	+	2
1970	6	+	2			=	4	+	4
1985	6	+	2	-	2	=	4	+	2

These numbers (in gigatons per year) are rough and normalized. They are presented here just to make a point: the oceans are an important reservoir of CO_2. They are also an important reservoir of heat. Finally, they are a reservoir of ignorance: it is intolerable that ocean storage should be estimated by closing a highly uncertain budget. The ocean's role has to be documented by direct measurements. In 1985 a scientific officer of a funding agency declined a proposal for such measurements by noting that the role of the oceans had recently declined from 4 to 2 gtons/y.

2. DIRECT MEASUREMENTS

What does the direct evidence show? The only well documented oceanic section is in the subtropical gyre of the North Atlantic. The 1957 IGY section along 24° N was re-occupied in 1984 by Roemmich and Wunsch [1984] in 1981 and then again in 1992 [Parrilla *et al.*, in press]. It shows a increase in temperature between 500 and 2000 m across the entire ocean basin. At 1 km depth the increase is of order 0.1°C between 1957 and 1992, or 6 millidegrees per year (m°C/y). I am not claiming that this is a measure of oceanic greenhouse warming; it could well be an expression of ambient climate variability.

The indicated depth of penetration by several km is in agreement with geochemical measurements. For example, Freons have been released into the atmosphere only since 1950, yet the work of Raymond Weiss and others [Warner and Weiss, 1992]) clearly shows that by 1988 they can be detected to depths of 1 km and greater.

3. CLIMATE MODELING

What do the models suggest? Model estimates have been made by Manabe at Princeton [Manabe and Stouffer, in press] and by Mayer-Reimer and others at Hamburg [Mikolajewicz *et al.*, 1990]. Estimates are based on comparing coupled atmosphere-ocean runs without atmospheric greenhouse gases with runs including greenhouse gases. A discussion of the crucial underlying assumptions is here out of place. In all events, the models yield an ocean warming in the 1990's by something like 20 m°C/y at the surface, decreasing exponentially to 5 m°C/y at 1 km depth. The important point to be made is that the spatial variability is of order one. One must not visualize the greenhouse effect as a uniform diffusive downward flux of heat. We use the above numbers

The Polar Oceans and Their Role in Shaping the Global Environment
Geophysical Monograph 85

for orientation only. They are of the same order as the interior Atlantic warming discussed above, perhaps by accident. Two "sanity checks" can be made as follows the associated sea level rise by thermal expansion is 2 mm/y; the additional heat flux required is 2 W/m^2. Both numbers are acceptable.

4. ACOUSTIC THERMOMETRY

We have started a program for measuring the interior ocean temperature variability on a climatic scale. The procedure is to measure the travel time of acoustic pulses between remote acoustic sources and receivers. The speed of sound increases with temperature, so the travel time would diminish with calendar time in a warming ocean environment.

Local spot measurements at 1 km depth are subject to a month-to-month variability of order 1°C rms mostly associated with mesoscale eddies. This is a large number compared to our estimates of a yearly change by 0.005°C from climatic (ambient or greenhouse) variability. However, gyre-scale **averages** will reduce the mesoscale "noise" by a factor of ten to hundred and so provide acceptable signal to noise ratios. The existence of an ocean sound channel centered typically at 1 km depth makes it possible to perform gyre scale and basin scale transmissions. For the interior warming in the subtropical north Atlantic this translates to a decrease in travel time by 0.1 s/y. Previous work in "Ocean Acoustic Tomography" has achieved millisecond precision at 1000 km ranges.

5. THE HEARD ISLAND FEASIBILITY TEST

Our goal is then to achieve a precision of 20 to 50 milliseconds. Explosive sources do not permit measurements of arrival time to this accuracy. Phase-coherent processing of a coded signal by an electrically driven acoustic projector (as used in Ocean Acoustic Tomography) does permit such a precision. Accordingly a feasibility test was conducted in early 1991 to ascertain whether (i) existing sources could be detected at very large ranges, (ii) the coded signal could still be deciphered, and (iii) whether this could be done at an intensity that did not prove harmful to local marine life. It appears that all three questions can be answered in the affirmative.

The available sources were on loan from the U. S. Navy and could not be subjected to depths exceeding 300 m. Accordingly we had to work at high latitudes where the sound channel is shallow. We chose a high southern latitude site at Heard Island in the Indian Ocean, an uninhabited Australian Island. Sources were lowered from a vessel. The site provides for great-circle access into both the Atlantic and Pacific Oceans, and the signals were in fact detected on both the western and eastern coasts of the North American continent at a range of about 18,000 km (nearly half way around the globe) with a travel time of 3 1/2 hours. Altogether the signal was received at 14 sites in all major ocean basins, with receivers lowered from vessels of 9 nations.

Transmissions were scheduled for 10 days on a 1 hour on, 2 hours off schedule. After 5 days we encountered a storm with 10 m waves, and the equipment was essentially demolished. By then we had collected sufficient data to answer the posed questions.

It is obvious that Heard Island was an acoustic feasibility test, not a climate experiment. We encountered some difficulties. The Antarctic Circumpolar Front is sufficiently sharp so as to lead to acoustic mode-to-mode scattering which makes the interpretation of the record difficult. Shoals and seamounts lead to refracted multipaths. These are problems that will have to be solved in a future program now being prepared.

6. ACOUSTIC THERMOMETRY OF OCEAN CLIMATE (ATOC)

We are now engaged in a program of future measurements. The ultimate purpose is to provide information on changes in the temperature profile along many tens of sections, and with adequate temporal sampling. The section in the north Atlantic previously discussed was sampled at decade intervals, and may be the only such section giving evidence of decadal changes on a gyre scale.

The transmissions will be from fixed sources to fixed receivers (leading to great simplification in the analysis) and avoid crossing sharp fronts. We are taking great care to avoid interference from bathymetric features.

Sources will be cable connected to shore with the aim to transmit for many years. Some of the receivers are vertical arrays with many elements so as to permit the resolution of acoustic modes. Different acoustic modes subtend different depths within the water column, and accordingly the results should give information on the depth-dependent changes in the upper ocean, not just on what happens at the sound axis.

Early work is centered in the Pacific. There will be one transmission from the California coast to New Zealand. A second source will be installed in the Hawaiian Islands to be received at stations to the north, west and east. This will give information on the coherence of climatic changes on a gyre scale. We will make standard hydrographic sections along some of the acoustic paths in order to compare measured changes in acoustic travel time with those inferred from traditional measurements.

Much of this early work consists of development and testing of sources and receivers, under the guidance of R. Spindel, D. Hyde, A. Baggeroer and P. Worcester. Signal strategy is being prepared by T. Birdsall and K. Metzger. At the same time an Arctic transmission program is being planned by P. Mikhalevsky, O. Johannessen, M. Slavinsky, D. Farmer and others. This will involve a source and receiver based on the *Fram*. Plans are being formulated for an Atlantic program under the leadership of J. Gould, F. Schott and D. Palmer. Some plans for the Southern Oceans are being developed by A. Forbes and G. Brundrit. It is premature to go into any detail about these future plans.

7. THE CLIMATE PROBLEM

I need to end this account by a word of caution. Even if all the problems of acoustic generation, propagation and analysis are adequately handled, there remains the more formidable problem of the climatological interpretation of the data. The separation of the greenhouse changes from the ambient variability (such as ENSO) will be difficult; Mikolajewicz *et al.*. [1990] estimate that it will take two or three decades. The ocean response to greenhouse warming lags the atmosphere by several decades. At the same time the signal to noise situation might be more favorable in the ocean. These are unsolved problems.

From the very start we will work closely with the modeling community, for validation and assimilation. Further, the acoustic data set will not stand alone; in particular, a joint analysis with the satellite altimetry data should prove fruitful. Altimetry provides good horizontal resolution, fair time resolution and poor depth resolution. The acoustic program has poor horizontal resolution, good time resolution and fair depth resolution. In a few years we hope to be a long way from accounting for the oceans by the closure of an uncertain global heat budget.

REFERENCES

Keeling, C., R. Bacastow, A. Carter, S. Piper, T. Whorf, M. Heimann, W. Mook, and H. Roeloffzen, A three-dimensional model of atmospheric CO_2 transport based on observed winds: I Observational data and preliminary analysis, in *Aspects of Climate Variability in the Pacific and the Western Americans, Geophys. Monogr. Ser.*, vol. **55**, edited by D.H. Peterson, pp. 165-236, AGU, Washington, D.C., 1989.

Manabe, S., and R.J. Stouffer, Greenhouse gas-induced evolution of the coupled ocean-atmosphere system over several centuries, *Nature*, in press.

Mikolajewicz, U., B.D. Santer, and E. Maler-Reiner, Ocean response to greenhouse warming, *Nature, 345*, 589-593, 1990.

Parrilla, G., A. Lavin, H. Bryden, and M. Garcia, Rising temperatures in the subtropical North Atlantic Ocean, *Nature,* in press.

Roemmich, D., and C. Wunsch, Apparent changes in the climate state of the deep North Atlantic Ocean, *Nature*, *307*, 447-450, 1984.

Warner, M.J., and R.F. Weiss, Chlorofluoromethane in South Atlantic Intermediate Water, *Deep-Sea Research*, *39, 2053-2075,* 1992.

Closing Note: A Tribute to Fridtjof Nansen

Norbert Untersteiner

University of Washington, Seattle, Washington

The purpose of the Nansen symposium and volume has been to report the results of scientific research, to celebrate the memory of Fridtjof Nansen, and to remind ourselves of his exemplary life as a scientist and humanitarian.

In this context it seems appropriate for the international community to remember that the Norwegian Sea and the Arctic Ocean were the cradle of modern oceanography. Mohn, Knudsen, Helland-Hansen, Nansen, Ekman, Fjeldstad, Sverdrup, and many others from this part of the world did their work in the North and became the "founding fathers" of oceanography.

Looking at the work that is being done today it seems easy to be optimistic about the future. We have new techniques to observe, from space and by other remote and automatic methods, and we have unprecedented means of computing to analyze our data and to calculate our models. More important, we have gained a world view of science that includes the polar regions as integral and interactive parts rather than as remote places where men with ice and pemmican in their beards take readings with exotic instruments. International collaboration has become a hallmark of polar science, and in this regard, too, research in the Arctic played a pivotal role.

Today, the task of both developing arctic resources and protecting the arctic environment is unthinkable without the solid basis of a scientific community that transcends political boundaries.

What makes this homage to Nansen so particularly timely is that his life and career remind us, like that of few other people, how science and scientific research must not be allowed to become ends in themselves. Especially during the past decade, more probing questions are being asked about the definition and relative merits of basic versus applied science, and about where to draw the line between research that we *could* do, and research that we *should* do.

The realization that World War II would be, and was, won by science and technology was a shock whose effects are still with us today. The Manhattan Project and Hiroshima catapulted science onto center stage, and left us with decades of recriminations, moral dilemmas, and political discord, but it also introduced an era of unprecedented advance and societal support for science.

After the end of World War I, Nansen became a champion of the persecuted, the hungry, the homeless and hopeless, and all other victims of world events. His unrelenting, selfless dedication and magnetic personality succeeded in moving the conscience of the World. Fridtjof Nansen would be appalled at the prospect, 20 or 30 years from now, of ten billion people on Earth, most of whom will be unlikely to share in the resources enjoyed by the lucky minority. Our species is being threatened, and we ourselves are the threat.

It is time to gather up all our energies and follow the great pace-setters of the human spirit. Fridtjof Nansen was one of them. He condensed in a single lifetime that wonderful progression from adventurous youth, to methodical and focused research, to teacher, educator, and finally to diplomat and champion of international justice and humanism. Besides everything else, Nansen was a supremely fortunate man, born into an environment of sold ethical values and material security, gifted with a keen intelligence, a strong body, a healthy joy in self-discipline and competition, and in addition to all that, he had an uncommon share of luck in all his many undertakings. It was the perfect raw material from which heroes are made, and he did not fail to become one.

Fridtjof Nansen is a *Norwegian* national treasure. But men of that magnitude belong to the whole world. Let us remember his spirit and dedication, and rise to the challenge.

The Polar Oceans and Their Role in Shaping
the Global Environment
Geophysical Monograph 85

N. Untersteiner, University of Washington, Seattle, WA 98105.